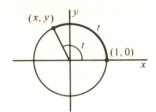

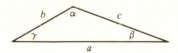

Formulas from Trigonometry

1. If t is directed arc length on the unit circle (also radian measure of the angle subtending the arc), then

$$\sin t = y \qquad \cot t = \frac{x}{y}$$

$$\cos t = x \qquad \sec t = \frac{1}{x}$$

$$\tan t = \frac{y}{x} \qquad \csc t = \frac{1}{y}$$

(x, y) $(1, 0)$

2. $\sin^2 t + \cos^2 t = 1 \qquad \sec^2 t - \tan^2 t = 1$

$$\csc^2 t - \cot^2 t = 1$$

3. $\sin t \csc t = 1 \qquad \cos t \sec t = 1$

$$\tan t \cot t = 1$$

4. $\tan t = \dfrac{\sin t}{\cos t} \qquad \cot t = \dfrac{\cos t}{\sin t}$

5. $\sin(-t) = -\sin t \qquad \cot(-t) = -\cot t$

$\cos(-t) = \cos t \qquad \sec(-t) = \sec t$

$\tan(-t) = -\tan t \qquad \csc(-t) = -\csc t$

6. $\sin\left(\dfrac{\pi}{2} - t\right) = \cos t \qquad \cos\left(\dfrac{\pi}{2} - t\right) = \sin t$

$\tan\left(\dfrac{\pi}{2} - t\right) = \cot t \qquad \cot\left(\dfrac{\pi}{2} - t\right) = \tan t$

$\sec\left(\dfrac{\pi}{2} - t\right) = \csc t \qquad \csc\left(\dfrac{\pi}{2} - t\right) = \sec t$

7. $\sin(u + v) = \sin u \cos v + \cos u \sin v$

$\sin(u - v) = \sin u \cos v - \cos u \sin v$

$\cos(u + v) = \cos u \cos v - \sin u \sin v$

$\cos(u - v) = \cos u \cos v + \sin u \sin v$

$$\tan(u + v) = \frac{\tan u + \tan v}{1 - \tan u \tan v}$$

$$\tan(u - v) = \frac{\tan u - \tan v}{1 + \tan u \tan v}$$

8. $\sin 2t = 2 \sin t \cos t$

$$\cos 2t = \cos^2 t - \sin^2 t = 1 - 2 \sin^2 t$$
$$= 2 \cos^2 t - 1$$

$$\tan 2t = \frac{2 \tan t}{1 - \tan^2 t}$$

9. $\sin^2 \dfrac{t}{2} = \dfrac{1}{2}(1 - \cos t) \qquad \cos^2 \dfrac{t}{2} = \dfrac{1}{2}(1 + \cos t)$

$$\tan \frac{t}{2} = \frac{1 - \cos t}{\sin t} = \frac{\sin t}{1 + \cos t}$$

10. Law of Sines: $\dfrac{\sin \alpha}{a} = \dfrac{\sin \beta}{b} = \dfrac{\sin \gamma}{c}$

Law of Cosines: $c^2 = a^2 + b^2 - 2ab \cos \gamma$

11. $y = \sin^{-1} x \Leftrightarrow x = \sin y, \ -\pi/2 \le y \le \pi/2$

$y = \cos^{-1} x \Leftrightarrow x = \cos y, \ 0 \le y \le \pi$

$y = \tan^{-1} x \Leftrightarrow x = \tan y, \ -\pi/2 < y < \pi/2$

$y = \cot^{-1} x \Leftrightarrow x = \cot y, \ 0 < y < \pi$

$y = \sec^{-1} x \Leftrightarrow x = \sec y,$

$$0 \le y < \pi/2 \text{ or } \pi \le y < 3\pi/2$$

$y = \csc^{-1} x \Leftrightarrow x = \csc y,$

$$0 < y \le \pi/2 \text{ or } \pi < y \le 3\pi/2$$

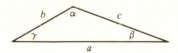

Calculus
and
Analytic
Geometry

Second Edition

Calculus and Analytic Geometry

Philip Gillett
University of Wisconsin

D. C. Heath and Company
Lexington, Massachusetts Toronto

For my wife, Dooie, and our children,
Robert, Mary, Bernard, John,
Joan, Maria, Anne, and Patricia

Cover photograph: © Michael Furman, 1980

Copyright © 1984 by D. C. Heath and Company.
Previous edition copyright © 1981

Published simultaneously in Canada.

Printed in the United States of America.

International Standard Book Number: 0-669-06059-3

International Standard Book Number: 0-669-07639-2 (Heath International Edition)

Library of Congress Catalog Card Number: 83-813439

Preface to the Second Edition

In the preface to the first edition of this book it is suggested that writing a new calculus book may be like climbing Everest. People do it because calculus is *there,* a huge, compelling, rich, rewarding mass of mathematics that exerts an enormous attraction on those who have struggled to master it, to teach it, or to contribute to it. As the book goes into its second edition, I still feel that way (about this and other matters mentioned in the preface to the first edition). For that reason the earlier preface is reproduced unchanged in this edition.

The aim, depth, and scope of the second edition are nearly the same as in the first. I say "nearly" because the addition of three optional sections has broadened the scope for those who choose to cover them. The depth of coverage is no different, however, and the aim (to communicate calculus to undergraduates as effectively as possible, particularly with regard to matters of style, exposition, and motivation) remains the same. I am pleased beyond the bounds of modesty by the number of students who have written to say how easy it is to read the book; that was the main reason I had for writing it.

The most visible change in this new edition is the large number of new problems. Many of these come at the beginning of problem sets, to provide more paired and graded drill exercises from which to choose. The sets of additional problems at the end of each chapter have also been enlarged, so that now there are more than 1000 such problems.

The most important changes in the text itself are these:

- Sections 3.1 and 3.2 (on the Power Rule and the algebra of derivatives) have been interchanged and to some extent combined.
- Section 6.2 (The Integral of a Real Function) is more clearly oriented toward Riemann sums than before.
- A new Section 10.2 (Integrals Involving Trigonometric Functions) is designed to make life easier when trigonometric substitutions are discussed in the following section.
- Chapter 12 (Infinite Series) now begins with a section on sequences.
- An optional section on series solutions of differential equations now ends Chapter 13 (Power Series).
- Chapter 14 in the first edition (Geometry in the Plane) proved to be too long. It is now confined to polar coordinates and conics; the vector analysis in the plane that previously ended it has been made the subject of a new Chapter 15 (with a new optional section on planetary motion at the end).
- The penultimate chapter on vector fields has been reorganized into shorter sections.
- The last chapter on differential equations now ends with a new optional section on vibrations.

As in the first edition, the first seven chapters of this book cover the standard topics of first-semester calculus in 38 sections (which in a typical course leaves room for two-day treatment of many sections). The next seven chapters constitute a solid second-semester course and have 37 sections. The rest of the book (on vectors, multivariable calculus, vector fields, and differential equations) has six chapters and 37 sections.

My thanks to the many users who responded to the publisher's survey, and to the reviewers who supplied more detailed suggestions. I adopted most of these, to the great benefit of the book. The reviewers were:

David Crystal, Rochester Institute of Technology

Clare Friedman, University of San Diego

Ronald B. Guenther, Oregon State University

Eleanor Killam, University of Massachusetts

Steven G. Krantz, The Pennsylvania State University

Eugene Krause, University of Michigan

John W. Lee, Montana State

Marcus M. McWaters, University of South Florida

Leonard J. Putnick, Siena College

M. S. Ramanujan, University of Michigan

Edward C. Turner, State University of New York at Albany

Once again I am indebted to the staff at D. C. Heath, especially former editorial director Robert Runck and mathematics editor Mary Lu Walsh (for reasons known to each) and production editor Cathy Cantin (for keeping the enterprise on schedule).

Philip Gillett

Preface to the First Edition

When George Mallory was asked why he wanted to climb Everest, he gave the now classic answer "Because it is there." Writing a new book on calculus may be a similar undertaking. To those who question the point of it, one wants to respond in the style of Mallory. Calculus is *there,* a huge, compelling, rich, rewarding mass of mathematics that exerts an enormous attraction on those who have struggled to master it, to teach it, or to contribute to it. Add the urge to write about it and the temptation is overwhelming. One produces the book first and explains it afterward.

Since this book is designed for the traditional three-semester calculus sequence, there is little that is new about its topics or novel in the order of their presentation. I wrote it because I believe that after twenty-eight years of teaching, I have learned how to communicate with undergraduates. Inability to communicate is the most common charge students bring against mathematicians and their textbooks, particularly in matters of style, exposition, and motivation. I am not concerned with the justice of such complaints but with the problem of responding to them. It is the main reason I had for writing this book.

There is nothing incompatible between good mathematics and a decent literary style. The reader need only look at the quotations at the beginning of each chapter of this book to see that. When G. H. Hardy said "there is no permanent place in the world for ugly mathematics," he might have added "nor for tedious explanations of it." Of course, one cannot discuss the Mean Value Theorem by coining epigrams, but there is no reason for a book to be altogether ponderous. I have made a serious effort in this book to keep the reader awake, stimulated, and even eager to continue.

More important than literary style is the critical matter of clarity. Calculus need not be obscure. Every sentence in this book has been constructed with care, to convey the intended message with a minimum of confusion. Although clear writing cannot make a tough subject easy, it can go a long way toward removing unnecessary obstacles to understanding.

Most important of all is what I must call, for want of a better term, motivation. It is devastating to encounter one major idea after another that has been simply announced in a dreary procession of definitions, theorems, proofs, examples, and exercises. If this book does not impress its users with the motivation offered for important ideas, I will be disappointed. These efforts are not spectacular; the seasoned teacher who does such things routinely in class will not be surprised by any of them. But they run through the whole book. For illustrations of what I mean, the prospective user might look at how the student is introduced to the following topics.

The derivative. Analytic trigonometry. Limits. The Chain Rule. Continuity. Differentials. The integral. The Fundamental Theorem of Calculus. Length of a curve. The natural logarithm. Inverse functions. The number *e*. Inverse trigonometric functions. Hyperbolic functions. Decomposition of rational functions. Improper integrals. Infinite series. Power series. Vectors. Curvature. The cross product. Chain rules in multivariable calculus. The gradient. Directional derivatives. Double and triple integrals. Line integrals. Green's Theorem, the Divergence Theorem, and Stokes' Theorem.

The treatment of some of these topics deserves more detailed comment.

- Limits are used informally long before they are defined. When they are presented with precision, the discussion is divided between an optional section and a section in which the essential theorems can be understood without proof. The results are sufficiently flexible for instructors of widely varying taste.

- Differentials are introduced as natural shorthand in connection with linear approximation, with no fuss about their ultimate meaning. They are put to the good uses for which they were intended; Leibniz was not as misguided as we have sometimes been led to believe.

- The notation and terminology required for a careful treatment of integration are replaced as soon as possible by Riemann sums that involve a minimum of symbolic apparatus. This simplified point of view is maintained throughout the rest of the book, particularly in applications. In every context where the utility of the integral is of paramount importance to the student, I have tried to avoid imposing meaningless rigor.

- I was especially concerned to give the student a feeling for the coherence of the "elementary functions of analysis." Transcendental functions are introduced as natural solutions of unsolved integration problems, and the analogies between them are exposed by means of simple differential equations and Taylor series. In this context the inverse trigonometric, hyperbolic, and inverse hyperbolic functions are less mysterious (and assume more of their true importance.)

- Chapter 10, on techniques of integration, is concise (much of the debris having been cleared away earlier in places where it is natural for it to collect). Even though numerical integration is included at the end, this chapter is sufficiently uncluttered to represent a fresh approach to a subject that can be dull and forbidding.

- The representation problem in Taylor series is largely avoided by solving appropriate differential equations (which is a natural way to arrive at the series for common functions). This removes a major obstacle to comprehension.

- Vectors are presented from the outset as elements of $\Re^2$ (or $\Re^3$), which can be *interpreted* as arrows but whose properties are more easily understood algebraically. This simplifies the student's life, particularly when vector functions are used in multivariable calculus to motivate the discussion of chain rules, tangent planes, directional derivatives, and Lagrange multipliers.

- I have tried to rescue the subject of vector fields from the obscurity in which it seems to languish. Green's Theorem is drawn almost painlessly from a discussion of independence of path (which is itself motivated by a search for conditions leading to conservative fields). Surface integrals and the Divergence Theorem are developed by analogy, and Stokes' Theorem by another analogy. The idea is to prevent the student from sinking without trace in a sea of jargon that obscures the true coherence of these important results.

All of this reflects a fundamental optimism about the teaching of calculus. I think a substantial three-semester sequence can be put between the covers of a book without mortally wounding the student or compromising the legitimate aspirations of the instructor. Calculus can be the watershed of the student's scientific education that we have always said it is.

None of the above affects students where they live, which is in the problem sets. I have put a great deal of effort into providing good problems (ranging from the routine to the challenging in every section). They are constructed to come at the subject in every way I could think of (short of a summary of the hard and soft sciences, which I believe is out of place in a calculus book). They are also designed to minimize numerical and algebraic manipulation, particularly in the earlier parts of the book. While it is true that problems concocted to be as painless as possible are liable to distort the reader's perception of mathematics in relation to the real world, the typical calculus student has a need to experience success. On the other hand, there is ample opportunity for the exercise of pocket calculators and many explicit references to their use.

I hesitate to offer concrete suggestions for a syllabus covering the book, because calculus courses vary so widely in their organization. Nevertheless, I had a definite scenario in mind. The first seven chapters cover the standard topics of first-semester calculus in 38 sections (which in a typical course leaves room for two-day treatment of many sections). The next seven chapters constitute a solid second-semester course and also have 38 sections. The rest of the book (on multivariable calculus, vector fields, and differential equations) has 29 sections. There is room for other arrangements, however, depending on the instructor and the time. It is worth mentioning that Chapters 11, 12, and 13 need not be left in place. The instructor who prefers to delay infinite series may easily do so.

Responsibility for this book is shared by many, as is always the case with so large an undertaking. I am particularly indebted to Professor David A. Smith of Duke University, who helped plan the book, read every word with a critical eye, suggested good problems, and kept reminding me of what we were trying to do. Professor Larry Olson (a colleague of many years) influenced the book more indirectly by showing me better ways to teach. He also graciously consented to my use of notes he and I wrote for a treatment of differential equations at the end of our calculus sequence.

One of the reviews I found most perceptive turned out to have been written by a former student of mine, Professor Donald Sherbert of the University of Illinois at Urbana–Champaign. His comments helped immensely to improve the book.

Other reviewers who criticized various stages of the manuscript, and encouraged its development by positive suggestions, were also helpful. I acknowledge their contributions with gratitude:

Paul Wm. Davis, Worcester Polytechnic Institute
Murray Eisenberg, University of Massachusetts, Amherst
Robert E. Fullerton, Palomar College
Larry Goldstein, University of Maryland
Mark Hale, University of Florida—Gainesville

Joseph Krebs, Boston College
Melvin Lax, California State University—Long Beach
Maurice L. Monahan, South Dakota State University
Willard A. Parker, Kansas State University
Dorothy Ryan, Bunker Hill Community College
Leonard Shapiro, North Dakota State University

I am grateful to several special people at D. C. Heath: acquisitions editor Irving Rockwood, whose encouragement and criticism started the book in the right direction; mathematics editor Robert Macek and editorial director Robert Runck, who understood the needs of an author; designer Libby Griffiths, who conceived the attractive form in which the book appears; and production editor Pamela Starr Nisetich, whose remarkable ability to pull things together transformed a pile of paper into a finished textbook.

I will be pleased to hear from instructors and students who use this book. Feedback is the lifeblood of a teacher and writer.

Philip Gillett

Contents

Note to the Reader • *xix*

1 Prologue to Calculus 1

1.1 Real Numbers and the Coordinate Line • 2
1.2 The Coordinate Plane and Straight Lines • 11
1.3 Some Important Curves in the Plane • 21
1.4 Functions • 28
1.5 Real Functions • 36
Additional Problems • 42

2 The Derivative 44

2.1 Two Equivalent Problems • 44
2.2 The Derivative of a Real Function; Limit Notation • 53
2.3 Trigonometric Functions • 61
2.4 Limits (Optional) • 74
2.5 Properties of Limits • 85
Additional Problems • 95

3 Techniques of Differentiation 97

3.1 The Power Rule and the Algebra of Derivatives (I) • 97
3.2 The Power Rule and the Algebra of Derivatives (II) • 104
3.3 The Chain Rule • 111
3.4 Derivatives of Implicitly Defined Functions • 118
Additional Problems • 126

4 Continuity and Differentiation 127

4.1 Continuity • 127
4.2 Properties of Continuous Functions • 136
4.3 The Mean Value Theorem • 142
4.4 Applications of the Mean Value Theorem • 148

4.5 Higher-Order Derivatives • 155

4.6 Linear Approximation • 162

4.7 The Differential of a Real Function • 168

Additional Problems • 177

5 Applications of Differentiation   180

5.1 The Derivative as Rate of Change • 180

5.2 Extreme Values • 187

5.3 Curve Sketching • 194

5.4 Solving Equations by Newton's Method (Optional) • 204

5.5 Maximum–Minimum Problems • 210

Additional Problems • 233

6 The Integral   226

6.1 Two Equivalent Problems • 226

6.2 The Integral of a Real Function • 238

6.3 Properties of the Integral • 246

6.4 The Fundamental Theorem of Calculus • 254

6.5 Integration by Substitution • 265

Additional Problems • 274

7 Applications of Integration   276

7.1 Area Between Curves • 276

7.2 Volume of a Solid of Revolution • 283

7.3 Length of a Curve • 290

7.4 Area of a Surface of Revolution • 299

7.5 Work • 304

7.6 Moments and Centroids • 311

7.7 Introduction to Differential Equations • 322

Additional Problems • 333

8 Exponential and Logarithmic Functions   335

8.1 The Natural Logarithm • 336

8.2 Inverse Functions • 343

8.3 The Natural Exponential Function • 353

8.4 General Exponential and Logarithmic Functions • 359

8.5 Integration Involving Exponential and Logarithmic Functions • 367

8.6 Applications of Exponential and Logarithmic Functions • 372

Additional Problems • 381

9 Inverse Trigonometric, Hyperbolic, and Inverse Hyperbolic Functions   385

9.1 Inverse Trigonometric Functions • 386

9.2 Integration Involving Inverse Trigonometric Functions • 396

9.3 Hyperbolic Functions • 402

9.4 Inverse Hyperbolic Functions • 408

9.5 The Catenary (Optional) • 416

Additional Problems • 419

10 Techniques of Integration   422

10.1 Integration by Parts • 423

10.2 Integrals Involving Trigonometric Functions • 429

10.3 Trigonometric Substitutions • 433

10.4 Decomposition of Rational Functions into Partial Fractions • 442

10.5 Miscellaneous Integration Problems • 451

10.6 Numerical Integration • 456

Additional Problems • 462

11 More about Limits   465

11.1 Limits Involving Infinity • 465

11.2 L'Hôpital's Rule and Indeterminate Forms • 473

11.3 Improper Integrals • 482

11.4 Other Types of Improper Integrals • 488

11.5 Taylor Polynomials • 494

Additional Problems • 504

12 Infinite Series   506

12.1 A Brief Look at Sequences • 507

12.2 Infinite Series • 511

12.3 The Integral Test • 520

12.4 Comparison Tests • 527

12.5 Alternating Series • 534

12.6 The Ratio Test and Root Test • 542

Additional Problems • 549

13 Power Series   551

13.1 Functions Defined by Power Series • 522

13.2 Taylor Series • 562

13.3 More About Power Series • 570

13.4 The Binomial Series • 575

13.5 Series Solutions of Differential Equations (Optional) • 581

Additional Problems • 588

14 Geometry in the Plane   591

14.1 Polar Coordinates • 591

14.2 Area and Length in Polar Coordinates • 597

14.3 Conics: The Parabola • 603

14.4 Conics: The Ellipse and Hyperbola • 608

Additional Problems • 618

15 Vector Analysis in the Plane   620

15.1 Vectors in the Plane • 620

15.2 Vector Functions • 629

15.3 Motion in the Plane • 636

15.4 More on Curvature and Motion • 644

15.5 Planetary Motion (Optional) • 651

Additional Problems • 657

16 Geometry in Space   659

16.1 Coordinates and Vectors in Space • 659

16.2 Lines and Planes • 664

16.3 The Cross Product • 671

16.4 Quadric Surfaces • 679

16.5 Cylindrical and Spherical Coordinates • 685

16.6 Curves in Space • 690

Additional Problems • 702

17 Functions of Several Variables   704

17.1 Partial Derivatives • 705

17.2 Chain Rules and the Gradient • 715

17.3 Tangent Planes and Tangent Lines • 723

17.4 The Directional Derivative • 730

17.5 Extreme Values • 739

17.6 Lagrange Multipliers (Optional) • 751

Additional Problems • 757

18 Multiple Integrals   760

18.1 Iterated Integrals in the Plane • 761

18.2 Double Integrals • 769

18.3 Double Integrals in Polar Coordinates • 780

18.4 Surface Area • 785

18.5 Triple Integrals in Rectangular and Cylindrical Coordinates • 790

18.6 Triple Integrals in Spherical Coordinates • 797

Additional Problems • 803

19 Vector Fields   806

19.1 Line Integrals in the Plane • 807

19.2 Gradient Fields • 816

19.3 Green's Theorem • 822

19.4 Green's Theorem Revisited • 830

19.5 Surface Integrals and the Divergence Theorem • 840

19.6 Line Integrals in Space and Stokes' Theorem • 850

Additional Problems • 859

20 Differential Equations   862

20.1 Ordinary Differential Equations • 862

20.2 Exact Differential Equations • 869

20.3 Linear First-Order Equations • 872

20.4 Second-Order Equations • 879

20.5 Linear Second-Order Equations • 886

20.6 Constant Coefficients • 894

20.7 The Nonhomogeneous Case • 901

20.8 Vibrations • 910

Additional Problems • 914

Answers to Selected Odd-Numbered Problems (and all True-False Quizzes)   A-1

Appendix Tables   T-1

Square Roots and Cube Roots • T-1

Trigonometric Functions of Numerical Input • T-3

Natural Logarithms • T-4

Exponential and Hyperbolic Functions • T-7

Index   I-1

Note to the Reader

One of the best teachers I ever had was in the habit of announcing on the first day of class that "the time to start studying for the final exam is now." Good advice. Since you are about to embark on a calculus course, I would add that the best way to stay afloat is to reserve a solid block of time for it on a regular basis.

That time ought not to be spent in only one way, however. This book is designed to be *read* (not merely consulted when difficulties in a problem assignment drive you to it). A quick first reading of each section (to give you the basic ideas) should be followed by a careful, line-by-line study (with pencil and paper at hand to fill in details that may not be clear to you). If you are not in the habit of this kind of preliminary study, you will be pleasantly surprised, I think, by the difference it makes.

Enough sermon. You would not be in a calculus course at all if you were not already accustomed to success in mathematics. Calculus was a great watershed in my education; I never saw the world in quite the same way again. I hope you enjoy the subject. There is nothing else like it (with the possible exception of a good climb of a lovely mountain). Use your ice axe and crampons when the going gets rough; when you get to the end of your rope, tie a knot in it and hang on. And don't forget to look at the scenery along the way.

1 Prologue To Calculus

God created the natural numbers; everything else is the work of man.
LEOPOLD KRONECKER (1823–1891)
(in a reference to the development of mathematics)

Whenever you can, count.
FRANCIS GALTON (1822–1911)

Calculus is the mathematics of change and motion. To appreciate its power, it is enough to realize that college students with ordinary mathematical preparation can use calculus to solve problems that once baffled the finest minds of Europe. To cite a famous example, suppose we assume that the sun exerts a force on the earth that varies inversely as the square of the distance between them. What can we infer about the orbit of the earth?

The answer, that it must be an ellipse, was given by Isaac Newton (1642–1727) in his celebrated *Principia Mathematica,* written at the urging of the astronomer Edmund Halley. Halley had visited Newton at Cambridge to put the question to him, and was astonished to learn that Newton had long ago worked out the answer. Having mislaid his notes, he did the work again; the publication of the finished masterpiece was to revolutionize mathematics and science.

Of course that was in the seventeenth century, which seems a long time ago. The power of calculus has not diminished since then, but has grown until today it is hardly possible to get along without it. Every scientific discipline uses it, and mathematics itself builds on calculus as a cornerstone of the magnificent modern edifice known as analysis.

As important as applications are, however, the central theme of this book is not celestial mechanics, or physics, or indeed any of the sciences. You need not be interested in science at all to understand and appreciate calculus. Like much else in mathematics it has a life of its own, with an aesthetic beauty and logical coherence that are independent of its connection with practical problems. The best way to learn it is to share the intellectual excitement of its creators and developers, who were fascinated by both the problems they solved and the means they used to solve them.

This chapter is a prologue to all that. It is designed to help get you in shape, on the premise that one needs some conditioning at the lower altitudes before expecting to enjoy a good climb. Even the well-prepared student should look through it (to become familiar with the terminology and notation if nothing else). Others, who may need a review of preliminary topics, should spend some time with the problem sets as well.

No reader, whether browsing or studying, should regard this chapter as typical of the book. It is not calculus any more than hiking through the foothills is mountain climbing. It may even be more difficult than the real thing if you are out of shape! Calculus itself is as comprehensible as any other part of mathematics, but you must be ready for it.

1.1 REAL NUMBERS AND THE COORDINATE LINE

Every statement in calculus can be defended by appealing to the properties of the set of **real numbers,** a set which we denote by the letter $\Re$. Some of these properties are axioms and others may be proved as theorems; together they serve as a logical foundation for all the arguments to come. On the whole we regard this foundation as already in place. The purpose of this section is to reinforce your understanding of real numbers as they are used in calculus.

The real numbers arose historically in response to the problem of measurement. If we want to count sheep, the **natural numbers** $1, 2, 3, 4, \ldots$ are adequate. Keeping track of profit and loss in the exchange of goods requires the ideas of "zero" and "negative," which lead to the larger system of **integers** $0, 1, -1, 2, -2, 3, -3, \ldots$. To measure anything continuous, however (like temperature), we need a scale that contains all values in between. **Rational numbers** (quotients of integers like $\frac{2}{3}$ or $-\frac{8}{5}$) appear to serve this need, but even they are inadequate for some measurements, as the early Greeks learned to their dismay. The ratio of circumference to diameter of a circle, for example, is a number that cannot be expressed as a quotient of integers. This number has acquired a special name (the Greek letter π) because no ordinary symbol of arithmetic can be found to express its value. (The rational number $\frac{22}{7}$ is a good approximation, but no such number actually equals π because π is irrational.)

The continuous number scale is called a **coordinate line.** It is a straight line marked like a thermometer, with a point 0 selected as the zero point (the **origin**) and a point 1 whose location determines the unit of measurement and the positive direction. (See Figure 1.) Other integers are located by marking off consecutive unit lengths in one direction or the other, whereas a fraction like $\frac{22}{7}$ is plotted by dividing the unit length into seven equal lengths and marking off twenty-two of these lengths in the positive direction. In this way we can imagine associating a point of the coordinate line with each rational number.

Figure 1
Coordinate line

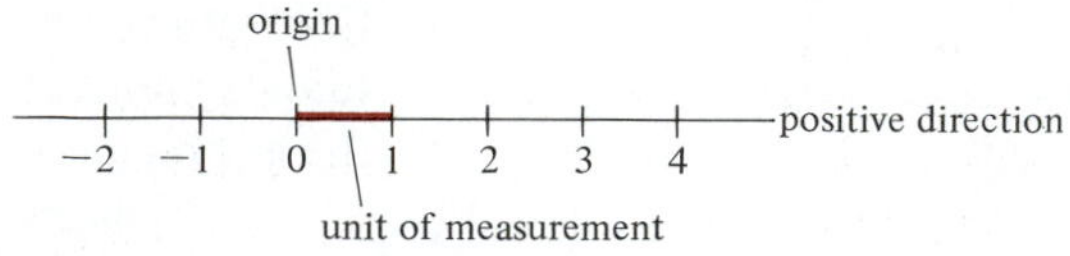

Irrational numbers (like π) are harder to discuss, as is the reverse procedure of associating a number with each point of the coordinate line. What is needed (we omit the details) is a process of indefinite subdivision of intervals corresponding to the decimal representation of the number (or generating it, if we start with the point). The result is important, and may be summarized in two statements:

> Every real number may be plotted as a point of the coordinate line.
>
> Every point of the coordinate line may be labeled by a real number.

For all practical purposes we then treat "real numbers" and "points of the coordinate line" as the same thing.

A closer analysis of the correspondence between $\Re$ (the set of real numbers) and the coordinate line reveals that *every real number is an infinite decimal.* Examples from the set of rational numbers are

$$3 = 3.000 \cdots = 3.\overline{0} \qquad \text{(the bar means repetition)}$$
$$-\tfrac{5}{2} = -2.5000 \cdots = -2.5\overline{0}$$
$$\tfrac{2}{3} = 0.666 \cdots = 0.\overline{6}$$
$$\tfrac{7}{11} = 0.636363 \cdots = 0.\overline{63}$$

Each of these decimals is **periodic** (meaning that a digit or block of digits is indefinitely repeated). Moreover, every periodic decimal represents a rational number, as the following example suggests.

◻ **Example 1**   Find the rational number represented by the periodic decimal $0.181818 \cdots$.

Solution: Let $x = 0.181818 \cdots$. Then

$$\begin{aligned} 100x &= 18.181818\cdots \\ x &= 0.181818\cdots \\ \hline 99x &= 18 \qquad \text{(by subtraction)} \\ x &= \tfrac{18}{99} = \tfrac{2}{11} \end{aligned}$$

Thus the periodic decimal $0.\overline{18}$ is the rational number $\tfrac{2}{11}$.   ◻

This identification of rational numbers with periodic decimals throws considerable light on the nature of real numbers. For it follows that *irrational numbers are nonperiodic decimals.* Moreover, since periodic decimals are the exception rather than the rule, we are obliged to regard the "typical" real number as irrational. Only a few of these numbers (relatively speaking) have names, for example,

- $\pi = 3.1415926535\cdots$ (ratio of circumference to diameter of a circle)
- $\sqrt{2} = 1.4142135623\cdots$ (positive number whose square is 2)
- $e = 2.7182818284\cdots$ (base of natural logarithms, to be defined in Chapter 8)

To see why periodic decimals are exceptional, imagine throwing a ten-sided die repeatedly, each time recording the digit on the top face. The resulting sequence of digits (like $9033278 \cdots$) may be regarded as the decimal representation of a real number. How likely is it that this decimal will be periodic?

For that reason (and because we ordinarily use only rational numbers in arithmetic) it may seem unnatural to think of irrational numbers as typical. Nevertheless the set of real numbers is incomplete without them. Geometrically speaking, the coordinate line would be "full of holes" if we plotted only rational points. To make sense of the intuitive idea of a continuous numerical scale, we must include irrational numbers.

We assume that addition and multiplication in $\Re$ (together with the laws of algebra governing these operations) are familiar. You should realize, however, that in practice we tend to overlook the difficulties. For example, what is $\pi + \sqrt{2}$? Geometrically we can see what is going on; the point $\pi + \sqrt{2}$ is obtained by juxtaposition of directed line segments, as shown in Figure 2. The problem of obtaining a "numerical answer," however, is that real numbers are infinite decimals (in this case nonperiodic, which makes matters worse). Nobody actually carries out the addition, as in a case like $5 + 3 = 8$. The sum is either approximated (by using only a finite number of decimal places) or simply left in the form $\pi + \sqrt{2}$ and treated in the context of the usual algebraic laws. (For example, it is the same as $\sqrt{2} + \pi$.)

We also assume some familiarity with the properties of $\Re$ based on **order,** which from a geometric point of view is merely the left-to-right arrangement of points on a horizontal coordinate line. If a and b are as shown in Figure 3, we write $a < b$ (equivalently, $b > a$) and say that *a is less than b* (or *b is greater than a*). The idea is the same as the notion of "colder" in connection with a thermometer, as when we say that a temperature of -10 is colder than a temperature of 5.

Because calculus requires facility in the writing and solving of inequalities, we list here some basic properties of order that will be useful later. First, however, we need some notation (to shorten the statement of the laws).

(i) If A and B are equivalent statements (meaning that each implies the other), we write $A \Leftrightarrow B$. This may be translated "*A if and only if B*" and is really two implications in one:

$A \Rightarrow B$ (A implies B, that is, A is true only if B is true.)
$B \Rightarrow A$ (B implies A, that is, A is true if B is true.)

(ii) If S is a set and x is an element of S, we write $x \in S$ ("*x belongs to S*").

Figure 2
Addition of directed line segments

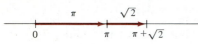

Figure 3
Order on the number line

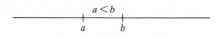

Properties of Order in $\Re$

1. $a < b \Leftrightarrow b - a$ is positive.
2. $a < b$ and $b < c \Rightarrow a < c$.   (Transitive Law)
3. $a < b \Rightarrow a + c < b + c$.   (Addition preserves order.)
4. $a < b$ and $c > 0 \Rightarrow ac < bc$.   (Multiplication by a positive number preserves order.)
5. $a < b$ and $c < 0 \Rightarrow ac > bc$.   (Multiplication by a negative number *reverses* order.)
6. $a < b$ and $c < d \Rightarrow a + c < b + d$.

7. $a < b \Leftrightarrow 1/a > 1/b$ (if a and b are positive).
8. $a < b \Leftrightarrow a^2 < b^2$ (if a and b are positive).
9. $a^2 \geq 0$ for every $a \in \mathcal{R}$.

□ **Example 2**   In calculus it is often important to determine when certain expressions are positive, zero, or negative. If $m = 5 - 2x$, for example, for what values of x is m positive? zero? negative?

Solution: To find when m is positive, we solve the inequality $5 - 2x > 0$, using the above properties of order:

$5 - 2x > 0$
$5 > 2x$ (Addition of $2x$ preserves order.)
$\frac{5}{2} > x$ (Division by the positive number 2, equivalent to multiplication by $\frac{1}{2}$, preserves order.)
$x < \frac{5}{2}$ (The statements $a < b$ and $b > a$ are equivalent.)

Thus m is positive for all values of x less than $\frac{5}{2}$. Note that we could have arrived at the same result by writing

$5 - 2x > 0$
$-2x > -5$ (Subtraction of 5, equivalent to addition of -5, preserves order.)
$x < \frac{5}{2}$ (Division by the negative number -2 *reverses* order.)

As you can see, it is important to be careful about multiplication, preserving or reversing the sense of the inequality depending on whether we multiply by a positive or negative number.

To find when m is zero, we solve the equation $5 - 2x = 0$, obtaining $x = \frac{5}{2}$. Finally (by solving the inequality $5 - 2x < 0$ in the same way as already shown), we find that m is negative when $x > \frac{5}{2}$. The behavior of m corresponding to various values of x is shown in Figure 4.   □

Figure 4
Behavior of $m = 5 - 2x$

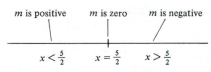

m is positive m is zero m is negative

$x < \frac{5}{2}$ $x = \frac{5}{2}$ $x > \frac{5}{2}$

Example 2 illustrates how **intervals** of the coordinate line may arise in the study of variables (like $m = 5 - 2x$). Another example is the equation $y = \sqrt{4 - x^2}$, which expresses the variable y in terms of the variable x. Since the square root of a negative number is meaningless in $\mathcal{R}$, we must impose the condition

$4 - x^2 \geq 0$ ("$4 - x^2$ is nonnegative")

in order for y to be a real number. This condition is equivalent to the statement

$-2 \leq x \leq 2$ ("x is between -2 and 2")

So the set of numbers represented by x is the interval with endpoints ± 2 on the coordinate line. (See Figure 5.)

Figure 5
Solution set of $4 - x^2 \geq 0$

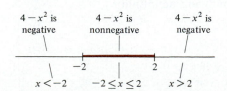

$4 - x^2$ is $4 - x^2$ is $4 - x^2$ is
negative   nonnegative   negative

-2 2

$x < -2$ $-2 \leq x \leq 2$ $x > 2$

Figure 6
Closed intervals

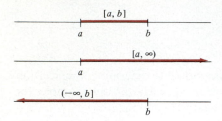

Figure 7
Open intervals

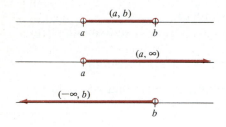

Figure 8
Half-open intervals

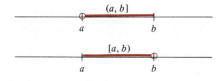

Figure 9
Distance from x to 0 when $x > 0$

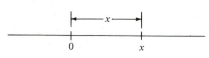

Figure 10
Distance from x to 0 when $x < 0$

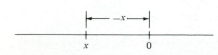

These examples are typical. More often than not, the set of numbers represented by a given variable is an interval (or union of intervals). There are several types, each with its own notation.

Closed intervals (see Figure 6):

- The set of all real numbers x satisfying $a \le x \le b$, denoted by $[a,b]$.
- The set of all $x \in \mathfrak{R}$ such that $x \ge a$, denoted by $[a,\infty)$.
- The set of all $x \in \mathfrak{R}$ such that $x \le b$, denoted by $(-\infty,b]$.

Open intervals (see Figure 7):

- $\{x \in \mathfrak{R}: a < x < b\}$, denoted by (a,b).
- $\{x \in \mathfrak{R}: x > a\}$, denoted by (a,∞).
- $\{x \in \mathfrak{R}: x < b\}$, denoted by $(-\infty,b)$.

Half-open intervals (see Figure 8):

- $\{x \in \mathfrak{R}: a < x \le b\}$, denoted by $(a,b]$.
- $\{x \in \mathfrak{R}: a \le x < b\}$, denoted by $[a,b)$.

Remark   The notation given for intervals may need some explanation. First, the symbols ∞ and $-\infty$ do not represent real numbers (nor endpoints of intervals). They are a kind of mathematical shorthand designed to indicate that the interval in question is unbounded. Second, the notation $\{x \in \mathfrak{R}: a < x < b\}$ may be read "the set of x in $\mathfrak{R}$ such that $a < x < b$," or simply "the set of real numbers between a and b." In general, braces $\{\ \}$ are a signal that we are talking about a set, either by listing its elements or (as in the present case) by stating a property that characterizes its elements. The colon (:) is a signal that such a property is about to be given (usually translated "such that"). Set notation is not used often in this book, but sometimes it is convenient (particularly when we are trying to save space).

With this notation we may express the set of solutions of the inequality $5 - 2x > 0$ as the open interval $(-\infty, \frac{5}{2})$ and the solution set of $4 - x^2 \ge 0$ as the closed interval $[-2,2]$. While such notation is not essential (we may simply write $x < \frac{5}{2}$ to represent the solutions of the first inequality and $-2 \le x \le 2$ for the second), it is sufficiently useful to be worth learning.

Given a point x of the coordinate line, it is often important to know how far it is from the origin. Our first thought might be that the answer is simply x, as shown in Figure 9. If $x < 0$, however, the distance is not x, but $-x$. (See Figure 10.) For example, if $x = -3$, the distance is $-x = -(-3) = 3$.

To cover all cases by a single formula, we may write $\sqrt{x^2}$, since

$$\sqrt{x^2} = \begin{cases} x & \text{if } x \ge 0 \\ -x & \text{if } x < 0 \end{cases}$$

An alternate symbol is $|x|$, as in the following definition.

If $x \in \mathfrak{R}$, the **absolute value** of x is

$$|x| = \sqrt{x^2} = \begin{cases} x & \text{if } x \ge 0 \\ -x & \text{if } x < 0 \end{cases}$$

□ **Example 3**   Use both parts of the definition to find the absolute value of -3.

Solution: The first part of the definition gives $|-3| = \sqrt{(-3)^2} = \sqrt{9} = 3$, while the second part (in which we use the bottom line because -3 is negative) gives $|-3| = -(-3) = 3$. Note particularly that the bottom line, $|x| = -x$ if $x < 0$, does not yield a negative answer, despite appearances. One of the persistent misconceptions of algebra students is that a symbol like $-x$ represents a negative number. It depends on x! If $x > 0$, then $-x$ is negative, but not otherwise.   □

To find the distance between any points of the number line, we must consider several cases. If $0 < x < y$ (Figure 11), it is just a matter of computing the difference $y - x$. The same idea works for any arrangement of the points, provided that the difference is computed in the right order. Absolute value is tailor-made for the situation; you should confirm that the following formula is correct in all cases.

Figure 11

Distance between x and y

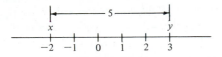

Distance on the Coordinate Line

If x and y are any points of the real number line, the distance between them is

$$d(x,y) = |x - y| = \sqrt{(x - y)^2}$$

□ **Example 4**   Find the distance between the points $x = -2$ and $y = 3$.

Solution: It is clear from Figure 12 that the distance between x and y is 5. The formula cranks out that answer without pictures:

$$d(x,y) = |x - y| = |-2 - 3| = |-5| = 5$$ □

Figure 12

Distance between -2 and 3

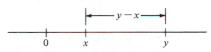

□ **Example 5**   Derive the *Midpoint Formula:*

The **midpoint** of any interval with endpoints a and b is

$$x = \tfrac{1}{2}(a + b)$$

(the *average* of the endpoints).

Solution: Assuming that $a < b$ (as in Figure 13), we find the midpoint by solving the equation $d(a,x) = d(x,b)$ for x:

$$\begin{aligned}
|a - x| &= |x - b| \\
x - a &= b - x \qquad \text{(because } a < x \text{ and } x < b\text{)} \\
2x &= a + b \\
x &= \tfrac{1}{2}(a + b)
\end{aligned}$$

Figure 13

Midpoint of an interval

While a single point may not seem to qualify as an interval, the definition says that $[a,a] = \{x : a \leq x \leq a\}$. The only value of x satisfying $a \leq x \leq a$ is $x = a$. This special case of a closed interval will be convenient later.

Figure 14
Solution set of $|x| < a$

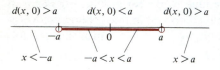

Figure 15
Solution set of $|x| > a$

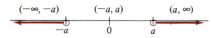

If $a > b$, the same argument with a and b interchanged yields $x = \frac{1}{2}(b + a)$, which is the same as before. We can even allow $a = b$, since in that case

$$x = \tfrac{1}{2}(a + b) = \tfrac{1}{2}(2a) = a$$

The "interval" is now $[a,a]$, which consists of the single point a.   □

□ **Example 6**   Assuming that $a > 0$, solve the inequalities $|x| < a$ and $|x| > a$.

Solution: Since $|x| = d(x,0)$, the solutions of the inequality $|x| < a$ are the points for which $d(x,0) < a$. These are the points satisfying $-a < x < a$, as you can see from Figure 14. While an algebraic argument can be given proving that

$$|x| < a \Leftrightarrow -a < x < a$$

the picture is entirely convincing.

The solution set of $|x| > a$ must be the rest of the number line (excluding the points $x = a$ and $x = -a$, which are the solutions of $|x| = a$). This set is the union of the intervals (a,∞) and $(-\infty,-a)$, as shown in Figure 15. Hence

$$|x| > a \Leftrightarrow x > a \text{ or } x < -a$$ □

These results are sufficiently important to be worth recording:

Suppose that $a > 0$. Then
$$|x| < a \Leftrightarrow -a < x < a$$
$$|x| = a \Leftrightarrow x = a \text{ or } x = -a$$
$$|x| > a \Leftrightarrow x > a \text{ or } x < -a$$

□ **Example 7**   Solve the inequality $|2x - 3| < 5$.

Solution: We appeal to the above statement, replacing x by $2x - 3$ and a by 5:

$$|2x - 3| < 5 \Leftrightarrow -5 < 2x - 3 < 5$$
$$\Leftrightarrow -2 < 2x < 8 \qquad \text{(Addition of 3 preserves order.)}$$
$$\Leftrightarrow -1 < x < 4 \qquad \text{(Division by 2 preserves order.)}$$

The solution set is the open interval $(-1,4)$.   □

□ **Example 8**   Solve the inequality $|1 - 2x| \geq 7$.

Solution: We write the inequality as follows:

$$|1 - 2x| \geq 7 \Leftrightarrow 1 - 2x \geq 7 \text{ or } 1 - 2x \leq -7$$
$$\Leftrightarrow -6 \geq 2x \text{ or } 8 \leq 2x$$
$$\Leftrightarrow x \leq -3 \text{ or } x \geq 4$$

The solution set is the union of the intervals $(-\infty, -3]$ and $[4, \infty)$. (See Figure 16.)

Figure 16
Solution set of $|1 - 2x| \geq 7$

We end this section by stating the *Triangle Inequality* in $\mathcal{R}$, namely

$$|x + y| \leq |x| + |y| \qquad \text{for all } x \text{ and } y$$

You can convince yourself that this inequality is true by considering such cases as x and y both positive, both negative, opposite in sign, and so on. A more elegant argument is outlined in the problem set. The Triangle Inequality is important in calculus.

Problem Set 1.1

Solve each of the following inequalities.

1. $3 - 5x > 1$ **2.** $7 - 6x < 3x + 1$

3. $8x - 7 < 3 - 2x$ **4.** $2x + 1 > 3x - 5$

5. $x^2 > 6x - 9$ **6.** $x^2 < 4x - 4$

7. $x^2 < x^3$ **8.** $x^3 + x > 0$

9. Draw a picture illustrating the *Transitive Law:* If $a < b$ and $b < c$, then $a < c$.

10. Convince yourself that addition preserves order, as follows.

     (a) Draw a coordinate line showing two points a and b satisfying $a < b$.

     (b) Assuming that $c > 0$, show the points $a + c$ and $b + c$ in your picture. Does it look like $a + c < b + c$?

     (c) Repeat part (b) assuming that $c < 0$.

     (d) We left out $c = 0$. What happens in that case?

11. Explain why $|3 - \pi|$ is equal to $3 - \pi$ or $\pi - 3$. Which is correct?

12. What is the value of $|\frac{2}{3} - 0.\overline{6}|$?

In each of the following, find the distance between the given points.

13. 2 and -7 **14.** -5 and 0

15. $\sqrt{2}$ and $\sqrt[3]{3}$ **16.** π and $\sqrt{10}$

17. $\frac{3}{11}$ and $0.\overline{27}$ **18.** e and π

Find the midpoint of each of the following intervals.

19. $[-2, 6]$ **20.** $[2, 2]$ **21.** $(0, \frac{2}{3}]$

22. $(-\frac{7}{2}, \frac{5}{6})$ **23.** $(1/\sqrt{2}, \sqrt{2})$ **24.** $[\pi/6, 5\pi/6]$

25. Suppose that a student writes $(1, -2)$ instead of $(-2, 1)$ for the set of real numbers x satisfying $-2 < x < 1$. According to our definition of the notation for intervals, $(1, -2)$ would have to mean $\{x \in \mathcal{R}: 1 < x < -2\}$, but the student argues that either way we are talking about the set of real numbers between -2 and 1. Do you agree? Why or why not?

26. Many students try to solve an inequality like $|x| > a$ (where $a > 0$) by writing $-a > x > a$ (imitating the procedure used in the solution of $|x| < a$, namely $-a < x < a$). Then they argue that $-a > x > a$ is equivalent to the correct answer, $x > a$ or $x < -a$. Why is this wrong?

Solve each of the following inequalities.

27. $|3x - 1| \leq 7$ **28.** $|2 - x| < 3$

29. $|5x - 2| > 3$ **30.** $|3x + 1| < 8$

31. $|5 - 2x| \leq 9$ **32.** $|2x - 1| > 5$

33. $|5x + 8| \geq 1$ **34.** $|x^2 - 1| < 3$

35. $|x + 4| = 4$ **36.** $|x^2 - 2| = 2$

37. Show that $x^2 < y^2 \Leftrightarrow |x| < |y|$. (This is a more general version of Order Property 8. When absolute value is used, we may assert equivalence of the inequalities without assuming that x and y are positive.)

38. Why is it incorrect to solve the inequality $x^2 < 25$ by writing $x < 5$? What is the correct solution? What is the solution set of $x^2 > 25$?

Use Problem 37 to solve each of the following inequalities.

39. $x^2 < 4$ **40.** $x^2 \leq 3$ **41.** $x^2 > 16$

42. $x^2 \geq 1$ **43.** $(x - 3)^2 \leq 25$ **44.** $(x + 5)^2 > 1$

Interpret each of the following inequalities in terms of the distance between x and another point. Then find the solution set. For example, $|x - 3| < 2$ means $d(x,3) < 2$ (that is, x is less than 2 units from 3). The solution set is the open interval (1,5).

45. $|x - 1| < 3$ **46.** $|x + 2| \leq 4$

47. $|x| > 1$ **48.** $|x - 5| \geq 1$

49. What is the solution set of $|x - 3| < 0$? of $|x - 3| \leq 0$?

50. What is the solution set of $|x + 1| > 0$? of $|x + 1| \geq 0$?

51. Divide 5 by 13 to generate the periodic decimal $\frac{5}{13} = 0.\overline{384615}$ and note the sequence of remainders in your division ($11, 6, 8, 2, 7, 5, \ldots$). When one of these repeats, the division starts over in the same pattern. Why is such repetition inevitable when division is used to find the decimal representation of a rational number?

52. Problem 51 indicates that every rational number can be represented by a periodic decimal. To illustrate the converse (that every periodic decimal represents a rational number), let $x = 0.\overline{45}$. Then $100x = 45.\overline{45}$. Subtract x from $100x$ and solve for x.

53. Let $x = 0.1010010001 \cdots$ (where the nth 1 is followed by n zeros). Is x rational or irrational? Explain.

54. Show that $a < b$ and $c < d \Rightarrow a + c < b + d$. *Hint:* Add c to each side of $a < b$ and b to each side of $c < d$. What law is needed to complete the argument?

55. Convince yourself that the distance between x and y on the number line is $|x - y|$ by considering the following cases.

(a) x and y are positive.

(b) x and y are negative. *Hint:* The distance between x and 0 is $-x$ and the distance between y and 0 is $-y$. The distance between x and y is therefore $-x - (-y)$ or $-y - (-x)$, depending on the arrangement of the points.

(c) x and y have opposite signs.

(d) x or y is 0.

56. Explain why $d(x,y) = d(y,x)$. (Hence the distance between two points does not depend on order.)

57. Use the definition $|x| = \sqrt{x^2}$ to explain why $|xy| = |x||y|$ and (if $y \neq 0$) $|x/y| = |x|/|y|$.

58. Derive the Triangle Inequality $|x + y| \leq |x| + |y|$ as follows.

(a) Explain why $-|x| \leq x \leq |x|$ for every $x \in \mathcal{R}$.

(b) Use Order Property 6 to show that

$$-|x| \leq x \leq |x| \quad \text{and} \quad -|y| \leq y \leq |y| \Rightarrow$$
$$-a \leq x + y \leq a$$

where $a = |x| + |y|$. Why does the Triangle Inequality follow?

59. Explain why the Triangle Inequality is an equality if x and y have the same sign and a strict inequality if x and y have opposite signs. What if x or y is zero?

60. Suppose that x and y are real numbers with the property that

$$|x - y| < \varepsilon \qquad \text{for every } \varepsilon > 0$$

(The symbol ε is the lowercase Greek letter *epsilon*.) Explain why $x = y$. (This is a fancy way of showing that two numbers are equal, when a direct approach might be hard. We'll use it in calculus.)

61. If $p \in \mathcal{R}$, any open interval containing p is called a **neighborhood** *of p*.

(a) Given $\delta > 0$ (the symbol δ is the lowercase Greek letter delta), explain why the solution set of the inequality $|x - p| < \delta$ is a neighborhood of p. What are its endpoints? What is its midpoint?

(b) Draw a picture of the solution set of the inequality $0 < |x - p| < \delta$. How does this differ from the neighborhood in part (a)?

62. Explain why every point of an open interval I has the property that some neighborhood of the point is contained in I. (Such a point is called an *interior* point of I; an open interval consists entirely of interior points.) What points of a closed interval are not interior points?

63. Given $p \in \mathcal{R}$ and $\delta > 0$, the solution sets of the inequalities $0 < x - p < \delta$ and $0 < p - x < \delta$ are called *right* and *left* neighborhoods of p, respectively. Draw pictures to explain why. What are their endpoints?

1.2
THE COORDINATE PLANE AND STRAIGHT LINES

One of the historical landmarks of science and mathematics was the introduction of coordinates into geometry (by the French philosopher and mathematician René Descartes in the seventeenth century). The numerical and algebraic flavor this brought to the subject was responsible for major advances after a long period of relatively little progress; without it there would be no calculus. The key idea is the characterization of a given set of points (straight line, curve, surface, or whatever) by equations or inequalities. When the geometric properties of the set are studied by an algebraic analysis of these statements, we say that we are doing **analytic geometry.**

An **ordered pair** of real numbers is exactly what the words suggest, namely two real numbers (not necessarily different) written in such a way that we can distinguish one as "first" and the other as "second." The usual notation is (x, y), where $x \in \mathcal{R}$ and $y \in \mathcal{R}$. Of course, this should not be confused with the notation for an open interval given in Section 1.1. Usually the context makes the meaning clear.

The set of all ordered pairs of real numbers is written

$$\mathcal{R}^2 = \{(x, y): x \in \mathcal{R}, y \in \mathcal{R}\}$$

and is called the **coordinate plane** because of its identification with the set of points of a plane. Just as real numbers are associated with points of a straight line, ordered pairs of real numbers are located by means of a coordinate system imposed on a plane. More precisely, suppose that the ordered pair (x, y) is given. The first number of the pair (see Figure 1) is located in the usual way on a horizontal line with positive direction to the right (the x axis). A vertical line L_1 is drawn through this point. The second number is located on a vertical line (the y axis) with positive direction upward and origin coinciding with the origin of the x axis. (Each axis has the same unit of length unless otherwise specified.) A horizontal line L_2 is drawn through this point. The point P where L_1 and L_2 intersect is labeled (x, y).

Conversely, any given point of the plane determines a pair (x, y) by reversal of the above procedure. Hence we regard points of the plane and ordered pairs of real numbers as essentially the same thing and we write $P = (x, y)$ without worrying about the distinction. The numbers x and y are called **rectangular coordinates** of the point P.

To find the distance between two points (x_1, y_1) and (x_2, y_2), we draw a triangle as shown in Figure 2. The Pythagorean Theorem yields

$$d^2 = |x_2 - x_1|^2 + |y_2 - y_1|^2 = (x_2 - x_1)^2 + (y_2 - y_1)^2$$

from which we obtain the following important formula.

Figure 1
Ordered pair as a point of the plane

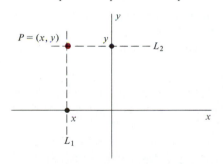

Figure 2
Distance in the coordinate plane

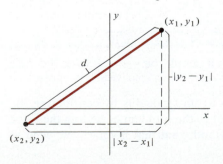

Distance in the Coordinate Plane

If $P = (x_1, y_1)$ and $Q = (x_2, y_2)$ are any points of $\Re^2$, the distance between them is

$$d(P,Q) = \sqrt{(x_2 - x_1)^2 + (y_2 - y_1)^2}$$

Our proof of the Distance Formula breaks down if P and Q lie on a line parallel to a coordinate axis (because no triangle can be drawn in that case). By using the formula for distance on a coordinate line, however, you can confirm that the conclusion is still correct.

Another generalization of Section 1.1 is the following.

The midpoint of the line segment with endpoints (x_1, y_1) and (x_2, y_2) has coordinates

$$x = \tfrac{1}{2}(x_1 + x_2), \ y = \tfrac{1}{2}(y_1 + y_2)$$

In other words (as in the Midpoint Formula of Section 1.1) we *average* the coordinates of the endpoints.

Figure 3
Midpoint of a line segment

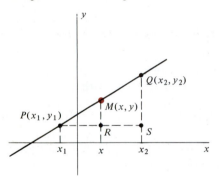

To derive these formulas, let $M = (x,y)$ be the midpoint of PQ (as shown in Figure 3). Then x is the midpoint of the interval on the x axis having endpoints x_1 and x_2. (Use similar triangles in Figure 3 to see why.) The Midpoint Formula in $\Re$ yields $x = \tfrac{1}{2}(x_1 + x_2)$. The formula for y is derived in the same way (by projections onto the y axis).

□ **Example 1**   Find the distance between the points $P = (1, -2)$ and $Q = (3,4)$ and find the midpoint of PQ.

Solution: The Distance Formula yields

$$d(P,Q) = \sqrt{(3-1)^2 + (4+2)^2} = \sqrt{4 + 36} = \sqrt{40} = 2\sqrt{10}$$

The Midpoint Formula gives $x = \tfrac{1}{2}(1+3) = 2$ and $y = \tfrac{1}{2}(-2+4) = 1$ as the coordinates of the midpoint $M = (x,y)$. As a check, note that

$$d(P,M) = \sqrt{(2-1)^2 + (1+2)^2} = \sqrt{10}$$

and

$$d(M,Q) = \sqrt{(3-2)^2 + (4-1)^2} = \sqrt{10}$$

so M is halfway between P and Q, as it should be.   □

In the next chapter we are going to start calculus by considering the **problem of tangents,** which is equivalent to the problem of finding the slope of a curve at a given point. To understand what that involves, we need to

Figure 4
Slope of a straight line

$$\text{slope} = \frac{\text{rise}}{\text{run}}$$

know something simpler, namely the *slope of a straight line*. Highway engineers measure this by computing the ratio of *rise* to *run* (as shown in Figure 4). For example, if an entrance ramp rises 2 meters in a horizontal run of 50 meters, the slope is $\frac{2}{50} = 0.04$ (meters of rise per meter of run). Mathematicians define slope in the same way if the line rises from left to right, but if it falls they call the slope negative. The formula that covers both cases is given in the following definition.

Suppose that L is a nonvertical straight line in the coordinate plane. The **slope** of L is

$$m = \frac{y_2 - y_1}{x_2 - x_1}$$

where (x_1, y_1) and (x_2, y_2) are any two (distinct) points of the line.

□ **Example 2**   Find the slope of the line containing the points $(-1, 3)$ and $(2, 1)$.

Solution: The slope is

$$m = \frac{1 - 3}{2 - (-1)} = -\frac{2}{3}$$

Note that it makes no difference which point is considered to be (x_1, y_1) and which is (x_2, y_2). We could just as well compute

$$m = \frac{3 - 1}{-1 - 2} = -\frac{2}{3}$$

This is true in general because

$$\frac{y_2 - y_1}{x_2 - x_1} = \frac{y_1 - y_2}{x_1 - x_2}$$

It is also immaterial which points of the line are used to compute slope, as you can see by looking at Figure 5. Here we show two points $P = (x_1, y_1)$ and $Q = (x_2, y_2)$ other than the given points $(-1, 3)$ and $(2, 1)$. Using similar triangles, we conclude that

$$\frac{y_2 - y_1}{x_2 - x_1} = \frac{3 - 1}{-1 - 2} = -\frac{2}{3}$$

Figure 5
Line containing $(-1, 3)$ and $(2, 1)$

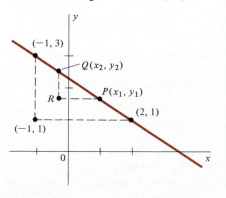

as before. This illustrates the fact that slope is a *global* property of a straight line, in the sense that the same slope is obtained no matter where we choose to compute it. In calculus we will define slope as a *local* property (of a given curve), in the sense that (in general) the slope depends on where we compute it. The slope of a straight line will then be seen as a special case. □

The following observations about slope are left for the problem set. (See Figure 6.)

A horizontal line has zero slope.

A line that rises from left to right has positive slope.

A vertical line has no slope.

A line that falls from left to right has negative slope.

Figure 6
Slope of lines

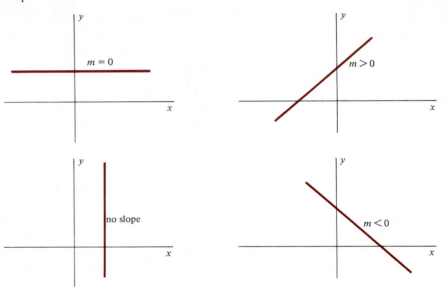

Note the distinction between *zero slope* and *no slope*. Every nonvertical line has a slope (zero in the case of horizontal lines). A vertical line, on the other hand, is literally "slopeless."

The concept of slope is useful in dealing with parallel and perpendicular lines. To be precise, we state the following.

Let L_1 and L_2 be nonvertical lines with slopes m_1 and m_2, respectively. Then

$$L_1 \text{ and } L_2 \text{ are parallel} \Leftrightarrow m_1 = m_2$$
$$L_1 \text{ and } L_2 \text{ are perpendicular} \Leftrightarrow m_1 m_2 = -1$$

Figure 7
Perpendicular lines

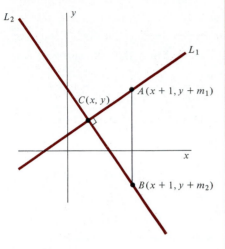

We leave the first of these for the problem set. To prove the second, look at Figure 7, in which L_1 and L_2 are shown as intersecting lines, (x, y) is their point of intersection, and two other points are shown with coordinates corresponding to the given slopes. Note that when we move along L_1 to the point with first coordinate $x + 1$, the second coordinate must become $y + m_1$. (Why?) Similarly, a change from x to $x + 1$ along L_2 is accompanied by a change from y to $y + m_2$.

Now we appeal to the Pythagorean Theorem (and its converse). In triangle ABC shown in Figure 7, we may assert that

$$\text{The angle at } C \text{ is } 90° \Leftrightarrow d(A,B)^2 = d(A,C)^2 + d(B,C)^2$$

Hence L_1 and L_2 are perpendicular if and only if

$$[(x + 1) - (x + 1)]^2 + [(y + m_1) - (y + m_2)]^2$$
$$= [(x + 1) - x]^2 + [(y + m_1) - y]^2 + [(x + 1) - x]^2 + [(y + m_2) - y]^2$$
$$(m_1 - m_2)^2 = (1)^2 + (m_1)^2 + (1)^2 + (m_2)^2$$
$$m_1{}^2 - 2m_1m_2 + m_2{}^2 = m_1{}^2 + m_2{}^2 + 2$$
$$-2m_1m_2 = 2$$
$$m_1m_2 = -1$$

□ **Example 3**   Find the slope of the line through $(2,1)$ perpendicular to the line containing $(-1,3)$ and $(2,1)$. (See Figure 8.)

Solution: The slope of L_1 is $m_1 = -\frac{2}{3}$ (as we found in Example 2). Hence the slope of L_2 is

$$m_2 = \frac{3}{2} \quad \text{(the } \textit{negative reciprocal} \text{ of } m_1)$$

Note that L_2 is drawn in Figure 8 by locating a second point, say $(4,4)$. Such a point may be found by starting at $(2,1)$ and changing x to some other value (in this case 4). The corresponding change in y must then be $2m_2 = 3$. (Why?) Hence y goes from 1 to 4. (Such primitive methods of locating points of a straight line are not needed if we know an equation of the line. That is the next subject to be discussed.)   □

Figure 8
Line through $(2,1)$ perpendicular to L_1

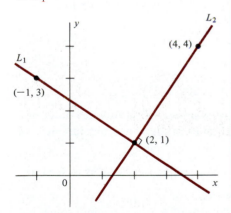

Now we are ready to develop equations of straight lines in the plane. The main things to learn are the *standard forms* in which these equations may be written, and the elegant relation between straight lines and first-degree equations (to be given at the end of the section).

□ **Example 4**   Find an equation of the straight line determined by the points $(1,5)$ and $(-1,1)$.

Solution: If L is the line, we seek a way of characterizing it algebraically, say

$$L = \{(x,y): \underline{\hspace{2cm}}\}$$

where the blank is to be filled in by a defining property that distinguishes between points of L and other points of the coordinate plane. In other words we want to be able to write

$$(x,y) \in L \Longleftrightarrow \underline{\hspace{3cm}}$$

The tool is at hand for doing this. For we know that the slope of L is

$$m = \frac{5 - 1}{1 - (-1)} = 2$$

If the slope is computed using, say, $(-1,1)$ and any other point $(x,y) \in L$, we obtain

$$\frac{y - 1}{x + 1} = 2$$

whereas if $(x,y) \notin L$, then

$$\frac{y - 1}{x + 1} \neq 2$$

(See Figure 9.) The equation

$$\frac{y - 1}{x + 1} = 2$$

appears to be the defining property we seek. One small problem is that it excludes the point $(x,y) = (-1,1)$. (Why?) But this is easily remedied by writing $y - 1 = 2(x + 1)$ instead; here is a condition that every point of L satisfies and no other point does. Hence

$$(x,y) \in L \Longleftrightarrow y - 1 = 2(x + 1)$$

that is,

$$L = \{(x,y): y - 1 = 2(x + 1)\}$$

We refer to $y - 1 = 2(x + 1)$ as an *equation of L*. It is worth observing that there are infinitely many others. For example, we could just as well write

$$L = \{(x,y): 2x - y + 3 = 0\} \qquad \text{or} \qquad L = \{(x,y): 4x - 2y + 6 = 0\}$$

It is therefore imprecise to refer to *the* equation of L as though it were unique. On the other hand, any two equations of L are equivalent (meaning that each implies the other), so the difficulty is not serious.

Note that the characterization of L by an equation enables us to find points of the graph with less trouble than in Example 3 (where we used slope alone). We simply insert convenient values of x or y into the equation and compute corresponding values of the other variable. Writing $y = 2x + 3$, for example, we may let x have values like $-3, -2, -1, 0, 1$ and find the corresponding values of y (namely $-3, -1, 1, 3, 5$). The graph is shown in Figure 10. □

Figure 9
Characterization of L

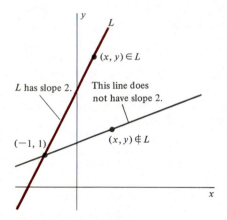

Figure 10
Graph of $2x - y + 3 = 0$

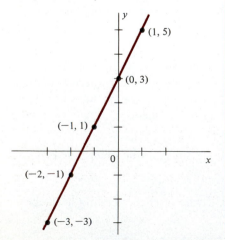

With Example 4 in mind, suppose that we are dealing with any line L whose slope m is known (or can be computed). If we also know a point of L, say (x_0, y_0), an equation of L can be found by observing that

$$(x, y) \in L \Leftrightarrow y - y_0 = m(x - x_0)$$

Note that we do not write

$$\frac{y - y_0}{x - x_0} = m$$

because this would exclude $(x, y) = (x_0, y_0)$.

Point-Slope Form

An equation of the line with slope m containing the point (x_0, y_0) is $y - y_0 = m(x - x_0)$.

□ **Example 5**   Find an equation of the line containing $(1,1)$ and $(-1,4)$.

Solution: The slope is

$$m = \frac{4 - 1}{-1 - 1} = -\frac{3}{2}$$

Using $(1,1)$ as the point (x_0, y_0) in the point-slope form, we find the equation

$$y - 1 = -\tfrac{3}{2}(x - 1) \Leftrightarrow 3x + 2y - 5 = 0$$

Note that if (x_0, y_0) is taken to be $(-1,4)$ instead, we obtain

$$y - 4 = -\tfrac{3}{2}(x + 1) \Leftrightarrow 3x + 2y - 5 = 0$$

as before.   □

The line in Figure 10, namely

$$L = \{(x,y): 2x - y + 3 = 0\}$$

has **y intercept** 3 (the y coordinate of the point where L intersects the y axis, obtained by setting $x = 0$). If we solve $2x - y + 3 = 0$ for $y = 2x + 3$, the y intercept appears as the second term on the right. It is interesting to observe that the coefficient of the first term, namely 2, is the slope. To see that this is true in general, let L be any line with slope m and y intercept b. Then L contains the point $(0,b)$; the point-slope form of its equation is

$$y - b = m(x - 0) \qquad \text{or} \qquad y = mx + b$$

Slope-Intercept Form

An equation of the line with slope m and y intercept b is $y = mx + b$.

An important application of the slope-intercept form is in finding the slope of a line whose equation is given. For example, suppose that L is known to have equation $x - 2y + 4 = 0$. To find the slope (if we are unaware of the slope-intercept form), we might use the equation to locate two points of L, say $(0,2)$ and $(2,3)$. Then

$$m = \frac{3 - 2}{2 - 0} = \frac{1}{2}$$

However, that's the hard way. It is easier to solve for y in the equation $x - 2y + 4 = 0$ to obtain the slope-intercept form $y = \frac{1}{2}x + 2$. The slope $m = \frac{1}{2}$ can be read off by inspection.

□ **Example 6**   Find an equation of the vertical line L containing the point $(2,0)$.

Solution: Vertical lines cannot be represented by any of the equations discussed so far. (Why?) However, each point of L is of the form $(2,y)$, as shown in Figure 11. Hence

$$L = \{(x,y): x = 2\}$$

More generally, the vertical line containing the point $(a,0)$ is

$$L = \{(x,y): x = a\}$$

that is, its defining equation is $x = a$.

In the same vein, note that the horizontal line containing the point $(0,b)$ is $L = \{(x,y): y = b\}$, as shown in Figure 12. However, this need not be considered separately as in the case of vertical lines. A horizontal line has slope $m = 0$; the slope-intercept form $y = mx + b$ reduces to $y = b$.   □

To pull things together, we assert that *every* line has an equation of the form $Ax + By + C = 0$. To confirm this, observe that if the line is nonvertical, it has a slope m and a y intercept b. An equation is

$$y = mx + b \quad \text{or} \quad mx + (-1)y + b = 0$$

which is of the required form with $A = m$, $B = -1$, and $C = b$. If the line is vertical, it has an x intercept a. An equation is

$$x = a \quad \text{or} \quad 1x + 0y + (-a) = 0$$

which is of the required form with $A = 1$, $B = 0$, and $C = -a$.

Conversely, every equation of the form $Ax + By + C = 0$ (where A and B are not both zero) represents a straight line. For if $B \neq 0$, we have

Figure 11
$L = \{(x,y): x = 2\}$

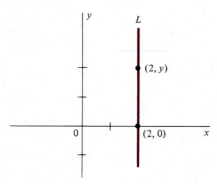

Figure 12
$L = \{(x,y): y = b\}$

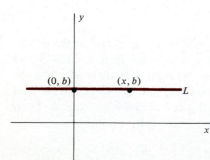

$$Ax + By + C = 0 \Leftrightarrow y = -\frac{A}{B}x - \frac{C}{B}$$

which is of the form $y = mx + b$. If $B = 0$, then $A \neq 0$ and we have

$$Ax + By + C = 0 \Leftrightarrow x = -\frac{C}{A}$$

which is of the form $x = a$. In both cases the set $L = \{(x,y): Ax + By + C = 0\}$ is a straight line.

The conclusion is an elegant statement that is typical of the power of analytic geometry to tie neat knots:

Linear Equations and Straight Lines

An equation of the form $Ax + By + C = 0$ (A and B not both zero) is called **linear in x and y.** Every straight line in the plane can be represented by a linear equation and every such equation represents a straight line.

Problem Set 1.2

Find the distance between each of the following pairs of points, the midpoint of the segment joining them, and the slope of the line containing them.

1. $(-1,3)$ and $(2,5)$
2. $(\frac{1}{2},\frac{2}{3})$ and $(\frac{3}{2},-\frac{4}{3})$
3. $(\sqrt{2},0)$ and $(0,\sqrt{3})$
4. $(3,1)$ and $(-1,1)$
5. $(2,-1)$ and $(2,3)$
6. Suppose that P and Q are two points of a line parallel to a coordinate axis. Confirm that the distance formulas in $\Re^2$ and $\Re$ give the same result for $d(P,Q)$.
7. Use the Distance Formula to show that the triangle with vertices $(-1,0)$, $(1,2)$, $(-2,5)$ is a right triangle.
8. Show that the triangle with vertices $(-3,1)$, $(2,6)$, $(1,-1)$ is isosceles. Then confirm that the straight line through $(2,6)$ and the midpoint of the opposite side is perpendicular to that side.
9. A line contains $(3,5)$ and the midpoint of the segment joining $(-1,2)$ and $(3,0)$. Is this line perpendicular to the segment?
10. Use the concept of slope to show that the points $(-1,3)$, $(0,1)$, $(1,-1)$ lie in a straight line.

Find the slope of each of the following lines (represented by the given equation).

11. $y = 3x - 2$
12. $y = -2x + 3$
13. $2x - 3y = 5$
14. $2x + 3y = 5$
15. $ax + by + c = 0$ $(b \neq 0)$
16. $ax - by = 1$ $(b \neq 0)$

Find an equation of each of the following straight lines.

17. Through the origin with slope 1
18. Through the origin with slope -1
19. Through the point $(3,-1)$ with slope $-\frac{1}{2}$
20. Through the point $(-2,5)$ with slope $\frac{2}{3}$
21. Slope 3 and y intercept 1
22. Slope -2 and y intercept 0
23. Slope 2 and x intercept 1
24. x intercept 2 and y intercept 3
25. x intercept -1 and y intercept 1
26. y intercept 2, parallel to the line $2x + y - 3 = 0$
27. x intercept 3, parallel to the line $2x - y + 5 = 0$
28. Through the points $(1,1)$ and $(2,3)$
29. Through the points $(4,-3)$ and $(0,5)$
30. Through the points $(1,3)$ and $(-1,2)$

31. Through the points $(2,1)$ and $(2,-3)$

32. Through the points $(-1,3)$ and $(1,3)$

33. Containing the points of intersection of the graphs of $y = x^2$ and $y = 2 - x^2$

34. Through the origin, perpendicular to the line $5x + 3y = 15$

35. Through $(0,4)$, perpendicular to the line $x - 3y = 0$

36. Perpendicular bisector of the segment with endpoints $(3,0)$ and $(0,2)$

37. Perpendicular bisector of the segment with endpoints $(-1,5)$ and $(3,3)$

38. Tangent to the circle with center $(0,0)$ and radius 1 at the point $(0,1)$. *Hint:* A theorem from plane geometry says that the tangent at a given point of a circle is perpendicular to the radius drawn to the point.

39. Tangent to the circle with center $(0,0)$ and radius 5 at the point $(3,-4)$.

40. Tangent to the circle with center $(1,-2)$ and radius 5 at the point $(5,-5)$.

41. Through $(0,-1)$ and tangent to the circle with center $(0,1)$ and radius 1. *Hint:* Let (a,b) be the unknown point of tangency and write two equations involving a and b. Note that there are two lines fitting the description. (Why?)

42. Through $(-2,0)$ and tangent to the circle with center $(0,0)$ and radius 1

43. Let L be a line with x intercept $a \neq 0$ and y intercept $b \neq 0$. Show that an equation of L is

$$\frac{x}{a} + \frac{y}{b} = 1$$

(This is called the **intercept form** of the equation of L.)

The **angle of inclination** of a straight line is the positive angle measured counterclockwise from the x axis to the line (unless the line is horizontal, in which case the angle is zero). Find the angle of inclination of each of the following lines.

44. The line through $(0,0)$ and $(-1,1)$

45. The line with slope 1 containing the origin

46. The line through $(2,0)$ and $(2,3)$

47. Explain why the slope of a nonvertical line is $m = \tan \theta$, where θ is the angle of inclination of the line.

Explain each of the following statements.

48. A horizontal line has zero slope.

49. A line that rises from left to right has positive slope.

50. A vertical line has no slope.

51. A line that falls from left to right has negative slope.

52. Let L_1 and L_2 be nonvertical lines with slopes m_1 and m_2, respectively. Explain why L_1 and L_2 are parallel if and only if $m_1 = m_2$.

53. Some people like to say that the slope of a vertical line is ∞. If L_1 and L_2 are horizontal and vertical, respectively, and we write $m_1 = 0$ and $m_2 = \infty$, what do we get when the condition $m_1 m_2 = -1$ is invoked?

54. (*Skip this problem if your trigonometry is shaky!*) The perpendicularity condition $m_1 m_2 = -1$ was derived in the text by appealing to the Pythagorean Theorem. Another proof may be based on Figure 13, beginning with the statement

$$L_1 \text{ and } L_2 \text{ are perpendicular} \Leftrightarrow \theta_2 = \theta_1 + 90°$$

Figure 13
Perpendicular lines

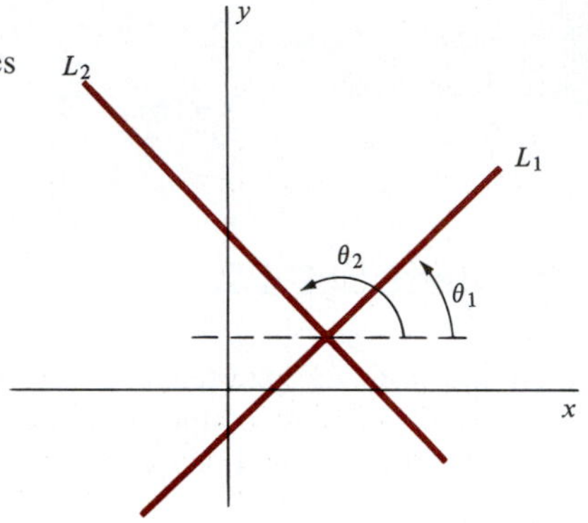

Finish the argument as follows.

(a) Explain why $\theta_2 = \theta_1 + 90° \Leftrightarrow \tan \theta_2 = \tan(\theta_1 + 90°)$.

(b) Use the identities $\tan(90° - \theta) = \cot \theta$ and $\cot(-\theta) = -\cot \theta$ from trigonometry to show that $\tan(\theta_1 + 90°) = -\cot \theta_1$.

(c) At this point we know that

$$L_1 \text{ and } L_2 \text{ are perpendicular} \Leftrightarrow \tan \theta_2 = -\cot \theta_1$$

Complete the proof.

55. Find the distance between the point $(1,-2)$ and the line $3x - 2y = 0$. *Hint:* First find the point of intersection of the given line and the perpendicular line through $(1,-2)$.

56. The method suggested in Problem 55 can be generalized to prove that the distance between the point (x_0, y_0) and the line L with equation $Ax + By + C = 0$ is

$$d = \frac{|Ax_0 + By_0 + C|}{\sqrt{A^2 + B^2}}$$

Derive this formula as follows.

(a) Confirm that an equation of L is

$$A(x - x_0) + B(y - y_0) = -(Ax_0 + By_0 + C)$$

(b) Explain why an equation of the line through (x_0, y_0) perpendicular to L is

$$B(x - x_0) - A(y - y_0) = 0$$

Note: Keep in mind that A or B may be zero. Your argument must cover those cases.

(c) Solve these equations simultaneously to show that if (x, y) is the point of intersection of the lines, then

$$x - x_0 = \frac{-A}{A^2 + B^2}(Ax_0 + By_0 + C)$$

$$y - y_0 = \frac{-B}{A^2 + B^2}(Ax_0 + By_0 + C)$$

(d) Now find the distance d between (x, y) and (x_0, y_0).

Use the formula in Problem 56 to find the following distances.

57. Between the origin and the line $x - y + 1 = 0$
58. Between the origin and the line $x + y = 1$
59. Between $(2,5)$ and the line $x - 2y + 3 = 0$
60. Between $(-1,2)$ and the line $y = x$

61. Between $(1,1)$ and the line $2x - 3y + 1 = 0$
62. Show that the line segment joining the midpoints of two sides of a triangle is parallel to the third side and is half its length. *Hint:* Choose the coordinate system so that the third side lies in the x axis with one endpoint at the origin.
63. A *median* of a triangle is the line segment joining a vertex and the midpoint of the opposite side. In the triangle with vertices $(4,1)$, $(1,3)$, $(-1,-1)$ find the length of the median containing $(4,1)$.
64. Show that the medians of the triangle with vertices $(0,0)$, $(6,0)$, $(2,4)$ intersect in a point. (That this is true in general is a well-known theorem of plane geometry.)
65. What if we allowed both A and B to be zero in the equation $Ax + By + C = 0$? What can you say about $\{(x,y): Ax + By + C = 0\}$ in that case?
66. Explain why the nonlinear equation $x^2 - 2xy + y^2 = 0$ represents a straight line. Does that contradict our statement in the text about linear equations and straight lines?

1.3
SOME IMPORTANT CURVES IN THE PLANE

In this section we discuss some common curves that are important in calculus, together with a few facts about symmetry in the plane. The graphing procedures discussed will appear again in Section 5.3, when we go into more detail.

□ **Example 1**   Let S be the circle with center $C = (5,1)$ and radius 4. (See Figure 1.) Find an equation of S.

Solution: Let $P = (x, y)$ be any point of the plane. Then

$$P \in S \Leftrightarrow d(C, P) = 4 \Leftrightarrow \sqrt{(x-5)^2 + (y-1)^2} = 4$$
$$\Leftrightarrow (x-5)^2 + (y-1)^2 = 16 \qquad □$$

The same argument with $(5,1)$ replaced by an arbitrary point (h,k) and with 4 replaced by a positive number r leads to the following general statement.

Figure 1
Graph of $(x - 5)^2 + (y - 1)^2 = 16$

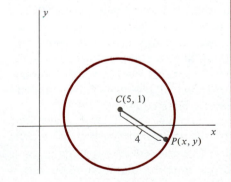

Every circle has an equation of the form $(x - h)^2 + (y - k)^2 = r^2$, where (h,k) is the center and $r > 0$ is the radius.

□ **Example 2**   Find the center and radius of the circle $x^2 + y^2 - 10x - 2y + 10 = 0$.

Solution: We "complete the square" in x and y, a method whose source is the formula $(x + a)^2 = x^2 + 2ax + a^2$. (Note that the leading coefficient is 1 and the constant term is the *square of half the coefficient of x.*) The given equation can be put in the standard form of the equation of a circle by writing

$$(x^2 - 10x + \underline{\quad}) + (y^2 - 2y + \underline{\quad}) = -10 + \underline{\quad}$$

where the blanks are to be filled in by completing the square. The complete statement is

$$x^2 + y^2 - 10x - 2y + 10 = 0$$
$$\Leftrightarrow (x^2 - 10x + 25) + (y^2 - 2y + 1) = -10 + 26$$
$$\Leftrightarrow (x - 5)^2 + (y - 1)^2 = 16$$

Hence the graph is the circle with center (5,1) and radius 4 shown in Figure 1.   □

Figure 2
Symmetry of a set S about a line L

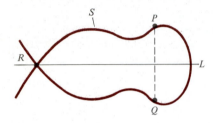

Before proceeding to other common types of curves, we pause for some remarks about symmetry. A set S in the plane is said to be *symmetric about the line L* if L acts as a mirror in which half of S is reflected into the other half. (See Figure 2.) More precisely, each point $P \in S$ has a "twin" $Q \in S$ with the property that L is the perpendicular bisector of the segment PQ (unless P is a point of L, like R in the figure, in which case $Q = P$). Three special cases of this definition are of particular interest.

Figure 3
Symmetry about the x axis

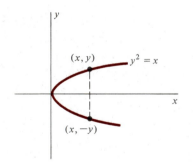

> S is symmetric about the x axis if $(x, -y) \in S$ whenever $(x, y) \in S$, that is, if the equation of S is equivalent to the equation obtained when y is replaced by $-y$.

As an example, consider the set $S = \{(x,y): y^2 = x\}$ shown in Figure 3. By plotting a few points satisfying the equation $y^2 = x$, such as (0,0), $(1,\pm 1)$, and $(4,\pm 2)$, you can see why its graph is the curve shown in the figure. Even without the figure, however, we can confirm that S is symmetric about the x axis. Any point (x,y) satisfying the equation has a twin $(x, -y)$ also satisfying the equation because y appears to an even power: $(\pm y)^2 = y^2$. Since the x axis is the perpendicular bisector of the segment joining these points, we have the symmetry claimed.

Figure 4
Symmetry about the y axis

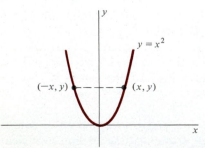

> S is symmetric about the y axis if $(-x, y) \in S$ whenever $(x, y) \in S$, that is, if the equation of S is equivalent to the equation obtained when x is replaced by $-x$.

An example of this kind of symmetry is the set $S = \{(x,y): y = x^2\}$ shown in Figure 4. This time the twin points are (x,y) and $(-x,y)$ because $(\pm x)^2 = x^2$.

The presence of an even power is not the only way such symmetry can occur. If you recall the graph of $y = \cos x$ from trigonometry (Figure 5), you can see the same kind of symmetry, this time due to the fact that $\cos(-x) = \cos x$ for all x. (We will review trigonometry in Chapter 2.)

Figure 5
Symmetry of the cosine curve

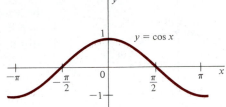

> S is symmetric about the line $y = x$ if $(y,x) \in S$ whenever $(x,y) \in S$, that is, if the equation of S is equivalent to the equation obtained when x and y are interchanged.

An example is the set $S = \{(x,y): xy = 1\}$ shown in Figure 6 (which we will discuss in more detail later in this section). It is clear from the equation $xy = 1$ that interchanging x and y does not affect the graph, since the equation simply becomes $yx = 1$. The reason this implies symmetry about the line L with equation $y = x$ is that the midpoint of the segment joining $P = (x,y)$ and $Q = (y,x)$ has equal coordinates $\frac{1}{2}(x + y)$ and $\frac{1}{2}(y + x)$. Hence it lies on L. Moreover, the slope of PQ is -1, the negative reciprocal of the slope of L. Thus L is the perpendicular bisector of PQ.

A set S is said to be *symmetric about the point* C if each point $P \in S$ has a twin $Q \in S$ with the property that C is the midpoint of PQ. The most important special case is the following.

Figure 6
Symmetry about $L = \{(x,y): y = x\}$

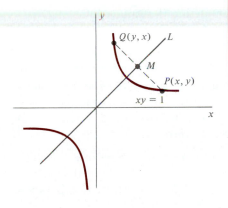

> S is symmetric about the origin if $(-x,-y) \in S$ whenever $(x,y) \in S$, that is, if the equation of S is equivalent to the equation obtained when x and y are replaced by $-x$ and $-y$, respectively.

An example is the set $S = \{(x,y): y = x^3\}$ shown in Figure 7. Since $(-x)^3 = -x^3$, any point (x,y) satisfying the equation $y = x^3$ has a twin $(-x,-y)$ also satisfying the equation; if $y = x^3$, then $-y = -x^3 = (-x)^3$. The origin is the midpoint of the segment joining these points (why?), so we have the symmetry claimed.

Now we return to our discussion of important curves in the plane. An equation of the circle with center at the origin and radius a is $x^2 + y^2 = a^2$, which can be written in the form $x^2/a^2 + y^2/a^2 = 1$. The equation

$$\frac{x^2}{a^2} + \frac{y^2}{b^2} = 1 \qquad (a > b > 0)$$

Figure 7
Symmetry about the origin

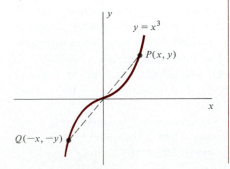

Figure 8

Graph of $\dfrac{x^2}{a^2} + \dfrac{y^2}{b^2} = 1$

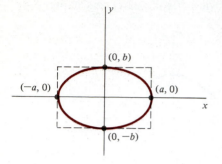

represents a similar curve called an **ellipse.** (See Figure 8.) Like the circle it is symmetric about each coordinate axis (and the origin) because of the squared terms, its intercepts being $(\pm a, 0)$ and $(0, \pm b)$. Solving for

$$y = \pm \frac{b}{a}\sqrt{a^2 - x^2} \quad \text{and} \quad x = \pm \frac{a}{b}\sqrt{b^2 - y^2}$$

observe that the radicals are real if and only if $-a \leq x \leq a$ and $-b \leq y \leq b$. Hence the ellipse is confined to the rectangle shown in Figure 8.

The equation $y^2/a^2 + x^2/b^2 = 1$ $(a > b > 0)$ also represents an ellipse, but this time the intercepts are $(\pm b, 0)$ and $(0, \pm a)$, the longer diameter (of length $2a$) being vertical instead of horizontal.

Another important curve in the plane is the **parabola,** a prototype of which is the graph of $y = x^2$. As noted in our remarks about symmetry, the vertical line $x = 0$ (the y axis) is a line of symmetry. Since $y \geq 0$ for all x (and $y = 0$ when $x = 0$), the curve opens upward from its lowest point $(0,0)$. This point is called the **vertex** and the line of symmetry is called the **axis** of the parabola. (See Figure 9.)

Figure 9

Graph of $y = x^2$

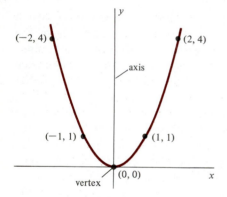

□ **Example 3**   Discuss and sketch the graph of $x^2 - 4x + y + 3 = 0$.

Solution: Although this equation is more complicated than $y = x^2$, its graph is similar, as you can see by completing the square in x:

$$x^2 - 4x + y + 3 = 0 \Leftrightarrow x^2 - 4x + 4 = -y - 3 + 4$$
$$\Leftrightarrow (x - 2)^2 = 1 - y$$

This time the squared term, $(x - 2)^2$, leads to symmetry about the vertical line $x - 2 = 0$, that is, $x = 2$. (See Figure 10.) Moreover, $1 - y = (x - 2)^2 \geq 0$ for all x, that is, $y \leq 1$ (with $y = 1$ when $x = 2$). Hence the curve opens downward from its vertex $(2,1)$.   □

The equation in Example 3 is quadratic in x and linear in y. A similar example (quadratic in y and linear in x) is $y^2 + 4x - 6y + 1 = 0$. Completing the square in y, we obtain $(y - 3)^2 = 4(2 - x)$, which is a parabola opening to the *left* from its vertex $(2,3)$. (See Figure 11.)

The last type of curve we discuss in this section is the **hyperbola,** the simplest example of which is the graph of $xy = 1$. This differs from our earlier examples because it has two branches with the coordinate axes as **asymptotes** (straight lines that approximate the graph ever more closely as we follow it into the remote parts of the plane). To see why, write the equation in the form $y = 1/x$ and note that y is undefined when $x = 0$. Values of x close to 0 result in numerically large values of y (positive or negative depending on whether $x > 0$ or $x < 0$). For example, $y = 1000$ when $x = 0.001$ and $y = -1000$ when $x = -0.001$.

In other words, *y increases without bound* when x approaches 0 through positive values and *decreases without bound* when x approaches 0 through negative values. We express these facts symbolically by writing

$$y \to \infty \quad \text{when } x \downarrow 0 \quad \text{and} \quad y \to -\infty \quad \text{when } x \uparrow 0$$

Figure 10

Graph of $(x - 2)^2 = -(y - 1)$

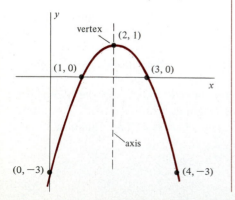

Figure 11
Graph of $(y - 3)^2 = -4(x - 2)$

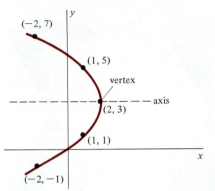

(Recall from Section 1.1 that ∞ and $-\infty$ are merely shorthand ways of indicating unbounded behavior, not real numbers or points of the coordinate line. The notation $x \downarrow 0$ and $x \uparrow 0$ means that x decreases toward 0 and increases toward 0, respectively. Sometimes the notation $x \to 0^+$ and $x \to 0^-$ is used instead.)

Similarly, values of x that are numerically large result in values of y close to 0. In symbols, $y \to 0$ when $x \to \infty$ or when $x \to -\infty$. As noted in our remarks about symmetry, the equation $xy = 1$ is unaffected when x and y are interchanged, so the graph is symmetric about the line $y = x$. It is also symmetric about the origin. (Why?) See Figure 12.

An example of a hyperbola with lines of symmetry parallel to the coordinate axes is the graph of $16x^2 - 9y^2 = 144$. (See Figure 13). Since

$$y = \pm \tfrac{4}{3} \sqrt{x^2 - 9} \qquad \text{and} \qquad x = \pm \tfrac{3}{4} \sqrt{y^2 + 16}$$

we must restrict $x \geq 3$ or $x \leq -3$, whereas y is unrestricted. The graph is symmetric about each coordinate axis (and about the origin).

More generally, equations of the type $x^2/a^2 - y^2/b^2 = 1$ or $y^2/a^2 - x^2/b^2 = 1$ represent hyperbolas, the first with branches opening horizontally (as in Figure 13), the second opening vertically.

Much more can be said about circles, ellipses, parabolas, and hyperbolas, but we prefer to postpone further discussion until their special properties are needed in calculus. The Greek geometers perceived these curves as **conic sections** obtained by slicing an infinite cone at various angles by a plane. (See Figure 14.) One of the achievements of Descartes (1596–1650) was to show that they are **second-degree curves,** in the sense that each can be characterized by a quadratic equation $Ax^2 + Bxy + Cy^2 + Dx + Ey + F = 0$. This brought a new coherence and powerful analytic techniques to an ancient subject.

For our present purposes it is sufficient for you to be familiar with the names of these curves and to be able to sketch the graphs of simple second-degree equations. Most calculus students have encountered this material before; if you need additional review, the problem set should be helpful.

Figure 12
Graph of $xy = 1$

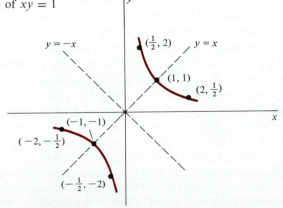

Figure 13
Graph of $16x^2 - 9y^2 = 144$

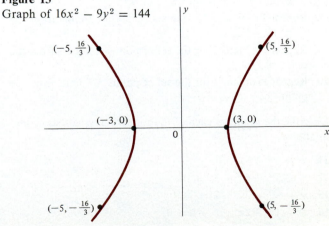

Figure 14
Conic sections

Ellipse   Parabola   Hyperbola

Problem Set 1.3

In each of the following, complete the square to fill in ☐ and △ so that an identity results.

1. $x^2 - 10x + \square = (x - \triangle)^2$
2. $x^2 + 6x + \square = (x + \triangle)^2$
3. $x^2 + 5x + \square = (x + \triangle)^2$
4. $x^2 - 5x + \square = (x - \triangle)^2$
5. $x^2 - 12x + \square = (x - \triangle)^2$
6. $x^2 + 20x + \square = (x + \triangle)^2$

7. Let S be the set of points that are 2 units from the point $(1, -2)$. Describe S geometrically and find an equation satisfied by each point (x, y) in S.

8. Repeat Problem 7 for the set of points 3 units from the point $(2, -1)$.

9. Repeat Problem 7 for the set of points $\sqrt{2}$ units from the point $(0, 3)$.

10. Repeat Problem 7 for the set of points $\sqrt{5}$ units from the point $(1, 0)$.

Describe the graph of each of the following equations or inequalities.

11. $x^2 + y^2 = 9$
12. $x^2 + y^2 - 6x + 8y = 0$
13. $x^2 + y^2 + 3x + 2 = 0$
14. $x^2 + y^2 - 2x + y + 2 = 0$

15. $3x^2 + 3y^2 + 12x - 6y + 14 = 0$
16. $x^2 + y^2 - 10y + 25 = 0$
17. $x^2 + y^2 - 2xy = 0$
18. $y = \sqrt{4 - x^2}$
19. $x = -\sqrt{1 - y^2}$
20. $x^2 + y^2 < 1$
21. $x^2 + y^2 \geq 1$

Find an equation of each of the following circles.

22. Center $(2, 0)$, tangent to the y axis
23. Center $(1, -1)$, containing the point $(3, 1)$
24. Diameter with endpoints $(1, 3)$ and $(-3, 1)$
25. Radius 5, containing the points $(4, 1)$ and $(-3, 0)$. *Hint:* Let (h, k) be the center and write two equations involving h and k. How many circles fit the given description?
26. Center $(2, -1)$, tangent to the line $3x + 4y - 12 = 0$. *Hint:* Use Problem 56, Section 1.2.
27. To find the circle containing $(2, 1)$, $(-2, 3)$, $(4, 5)$, assume that its equation has the form $x^2 + y^2 + Dx + Ey + F = 0$. Determine D, E, F by using the fact that each of the given points must satisfy this equation.
28. Another way to do Problem 27 is to observe that the perpendicular bisectors of the line segments joining the given points intersect at the center of the circle. (Draw a pic-

ture!) Find the center and radius by making use of that fact.

29. Find the points of intersection of the line $x - 2y = 0$ and the circle $x^2 + y^2 - 4x - 2y = 0$.

30. Use the equation of a circle to prove that a triangle inscribed in a semicircle is a right triangle.

31. Figure 15 shows an ellipse with equation

$$\frac{y^2}{a^2} + \frac{x^2}{b^2} = 1 \qquad (a > b > 0)$$

(a) Explain why the intercepts of the ellipse are as shown, namely $(\pm b,0)$ and $(0,\pm a)$.

Figure 15

Graph of $\dfrac{y^2}{a^2} + \dfrac{x^2}{b^2} = 1$

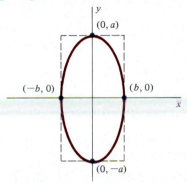

(b) Why is the ellipse symmetric about each coordinate axis and the origin?

(c) Explain why the ellipse is boxed in by the rectangle shown in the figure.

Each of the following equations represents an ellipse. Name its center (the intersection of its lines of symmetry) and sketch the ellipse after locating the endpoints of its horizontal and vertical diameters.

32. $16x^2 + 25y^2 = 400$ 33. $9x^2 + 4y^2 = 36$
34. $x^2 + 4y^2 = 1$
35. $x^2 + 4y^2 + 2x - 16y + 1 = 0$
36. $9x^2 + 4y^2 - 24y = 0$
37. $9x^2 + y^2 - 54x - 2y + 73 = 0$
38. $9x^2 + 16y^2 - 96y + 108 = 0$

39. Show that the area of the square inscribed in the ellipse

$$\frac{x^2}{a^2} + \frac{y^2}{b^2} = 1 \qquad \text{is} \qquad \frac{4a^2b^2}{a^2 + b^2}$$

40. Let $F = (1,0)$ and suppose that L is the vertical line con-

taining the point $(4,0)$. If S is the set of points satisfying $d(P,F) = \frac{1}{2}d(P,L)$, show that

$$(x,y) \in S \Leftrightarrow \frac{x^2}{4} + \frac{y^2}{3} = 1$$

Each of the following equations represents a parabola. Name the vertex and axis and sketch the parabola by plotting the vertex and a few other strategically located points.

41. $y^2 = x$ 42. $y = -x^2$
43. $y = 1 - x^2$ 44. $y^2 - x - 4 = 0$
45. $y^2 - 4x - 4y = 0$ 46. $4x^2 - 4x - 24y - 23 = 0$
47. $y^2 + 2x = 0$

Find an equation of each of the following.

48. The horizontal parabola with vertex $(1,0)$ containing the point $(2,2)$

49. The vertical parabola with vertex $(2,-1)$ containing the origin

50. The horizontal parabola containing the points $(0,0)$, $(0,2)$, $(1,3)$

Each of the following equations represents part of one or more parabolas. What parabolas? Sketch the graph of each equation.

51. $y = \sqrt{x}$ 52. $x = -\sqrt{y}$ 53. $y = x|x|$

54. Let $F = (0,2)$ and suppose that L is the horizontal line containing the point $(0,-2)$. Sketch the set S of points that are equidistant from F and L and show that

$$(x,y) \in S \Leftrightarrow x^2 = 8y$$

55. An artillery piece placed at the origin fires a shell in the coordinate plane. (The x axis is at ground level; the positive direction of the y axis is upward.)

(a) Assuming that the coordinates of the shell t seconds after firing are $x = 50t$ and $y = 50t - 5t^2$ (in meters), explain why the path of the shell is a parabola. What is its equation in x and y?

(b) What is the horizontal range, that is, how far from the origin does the shell land?

(c) When does the shell reach its highest point and how high does it go?

56. In calculus we will show that if the angle of elevation of the artillery piece in Problem 55 is α and the muzzle velocity is v_0 meters per second, then (neglecting air resistance)

$$x = (v_0 \cos \alpha)t \qquad \text{and} \qquad y = (v_0 \sin \alpha)t - \frac{1}{2}gt^2$$

where g is the acceleration due to gravity.

(a) Show that the horizontal range is $(2v_0{}^2/g)\sin\alpha\cos\alpha$.

(b) What angle of elevation gives maximum horizontal range? *Hint:* Use the identity $\sin 2\alpha = 2\sin\alpha\cos\alpha$ from trigonometry.

57. Figure 16 shows a hyperbola with equation

$$\frac{x^2}{a^2} - \frac{y^2}{b^2} = 1$$

(a) Explain why the hyperbola has x intercepts $(\pm a,0)$ and lies outside the vertical strip bounded by the lines

Figure 16

Graph of $\dfrac{x^2}{a^2} - \dfrac{y^2}{b^2} = 1$

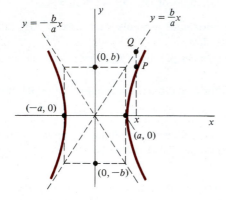

$x = \pm a$. Why are there no y intercepts, and no restrictions on y?

(b) Why is the hyperbola symmetric about each coordinate axis and about the origin?

(c) If x is the first coordinate of the points P and Q shown in the figure, show that

$$d(P,Q) = \frac{b}{a}(x - \sqrt{x^2 - a^2})$$

Why is it reasonable to conclude that $d(P,Q)$ is close to zero when x is large? (This shows that the lines $y = \pm(b/a)x$ in Figure 16 are asymptotes of the hyperbola.)

(d) A device for remembering the asymptotes is to replace 1 by 0 on the right side of the equation of the hyperbola. Show that this yields the equations $y = \pm(b/a)x$.

Each of the following equations represents a hyperbola. Sketch the graph after drawing its asymptotes.

58. $x^2 - y^2 = 1$ 59. $y^2 - 4x^2 = 4$

60. $9x^2 - 16y^2 - 144 = 0$ 61. $16x^2 - 9y^2 + 144 = 0$

62. $xy = 2$

63. $xy - y = 1$ *Hint:* Solve for y in terms of x.

64. $xy - x = 1$ 65. $xy = -1$

Each of the following equations represents a "degenerate case" of the second-degree curves discussed in this section. Describe the graph.

66. $x^2 - 2x = 0$ 67. $3x^2 + y^2 - 6x + 6y + 12 = 0$

68. $x^2 - y^2 = 0$ 69. $xy = 0$

1.4 FUNCTIONS

The concept of **function** is central to calculus (which in its abstract setting might even be called the *Theory of Functions*). Although you have probably encountered the idea in earlier courses, we are going to start from scratch. We begin with a simple example.

□ **Example 1** Suppose that we have constructed a computer that squares each number we feed into it. Such a machine would have certain built-in limitations. For example, its keyboard might permit the typing of only ten inputs, the integers $0, 1, 2, \ldots, 9$. The corresponding outputs would be $0^2 = 0, 1^2 = 1, 2^2 = 4, \ldots, 9^2 = 81$, and we could display the entire operation of the computer in a table:

x	0	1	2	3	4	5	6	7	8	9
y	0	1	4	9	16	25	36	49	64	81

Note that in this table we have indicated the typical input by the letter x and the corresponding output by y. Each of these letters is a **variable,** in the sense that x may be any element of the set

$$A = \{0, 1, 2, \ldots, 9\}$$

and y is an element of the set

$$B = \{0, 1, 4, \ldots, 81\}$$

Because x may be selected arbitrarily, while y is determined by the choice of x, we use different names for the variables and for the sets whose elements they represent. We call x the **independent variable** and A the **domain,** while y is the **dependent variable** and B is the **range.** What the computer does is to convert each element of the domain into a definite element of the range.

$\square$

There are several ways of describing the situation in Example 1. One is to give the *table of values* already shown. Equivalently, we could write down the *set of ordered pairs* (x,y) displayed in the table:

$$\{(0,0), (1,1), (2,4), \ldots, (9,81)\}$$

Or we might plot these pairs as points in the coordinate plane; the resulting *graph* (consisting of just ten points in this example) would serve as a picture of the set of ordered pairs (and hence as a description of the operation of the computer). Or we might simply observe that $y = x^2$, where $x \in A$. This formula could be pasted on the computer and would characterize it completely (together with the keyboard displaying the elements of A).

However we choose to describe it (by a table, a set of ordered pairs, a graph, a formula, or even a purely verbal description), we are dealing with a *rule by which an arbitrary input* (selected from the set A) *is converted to a definite output* (in the set B). Such a rule is called a **function.**

In a room full of computers it would be irresistible to give them names. Big Bertha might be the one with the greatest capacity for typing inputs, or the most complicated mode of operation. Our squaring machine might be called Little John. Mathematicians are more prosaic and simply use letters. Suppose, for example, that Little John is called f. If x is the input, the output (instead of being called Little John's) is written $f(x)$, read "f of x." In other words $f(x) = x^2$, which means that $f(x)$ is another name for the dependent variable y. Note the distinction between f, the name of the function, and $f(x)$, the label for its output. See Figure 1, which shows Little John as a machine labeled f, the output of the machine being $y = f(x) = x^2$.

Figure 1
Squaring machine

$x \in A = \{0, 1, 2, \ldots, 9\}$

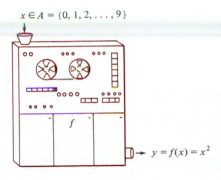

$\rightarrow y = f(x) = x^2$

Figure 2
Graph of $f(x) = x^2$, $x \in \Re$

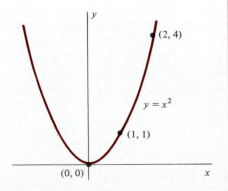

$\square$ **Example 2**   Suppose that the domain of the function f in Example 1 is enlarged to $\Re$. Sketch its graph.

Solution: The graph is the parabola $y = x^2$ shown in Figure 2. It contains the graph of the function in Example 1 as a subset, namely the set of isolated points $(0,0), (1,1), (2,4), \ldots, (9,81)$. A bug negotiating the latter graph would have some heroic jumping to do, whereas the complete parabola presents no problem. □

Examples 1 and 2 illustrate the fact that no function is properly defined by a formula alone. A domain is always lurking in the background (either explicitly stated or understood). To display this information concisely, it is common practice to write $f: A \to B$. This is supposed to name a function f (with domain A) that converts each element of A into an element of B. We do not require that B must be the range (as it was in Example 1) but simply any set containing the range. An abstract diagram like Figure 3 helps us visualize what the function does. We say that *f sends A into B*.

These preliminary remarks help explain the following definition.

Suppose that A and B are (nonempty) sets. A **function** from A to B is a rule by which each $x \in A$ is converted into a definite $y \in B$. We call x the *input* (or **independent variable**) and y the corresponding *output* (or **dependent variable**). If f is the name of the function, the output corresponding to an input x is written $y = f(x)$. The set of inputs (namely A) is called the **domain** of f, while the set of outputs (a subset of B, perhaps B itself) is called the **range.** To indicate that f sends A into B we write $f: A \to B$.

Figure 3
Diagram of a function

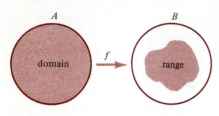

□ **Example 3**   A notation describing the function in Example 2 is

$$f: \Re \to \Re \qquad \text{defined by } f(x) = x^2$$

The formula tells us how the function works; the first $\Re$ tells us that any real number is an acceptable input; the second $\Re$ says that the outputs are real numbers. □

The outputs in Example 3 are actually *nonnegative* real numbers ($x^2 \geq 0$ for all $x \in \Re$). Moreover, every nonnegative real number is an output (since every $y \geq 0$ has a square root, that is, $x^2 = y$ for some $x \in \Re$). The range is therefore $[0,\infty)$ rather than all of $\Re$. (This is geometrically apparent from Figure 2, in which we look at the vertical extent of the curve, namely all $y \geq 0$.) Thus we could write $f: \Re \to [0,\infty)$ instead of $f: \Re \to \Re$. In many cases, however, the range is hard to find. That is why we allow B in the notation $f: A \to B$ to be vague; any set containing the outputs is allowed.

□ **Example 4**   The function

$$f: [0,\infty) \to \Re \qquad \text{defined by } f(x) = x^2 - 4\sqrt{x}$$

Figure 4
Graph of $f(x) = x^2 - 4\sqrt{x}$

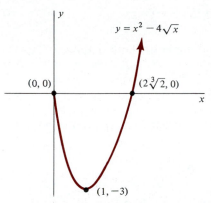

converts each nonnegative real number into a real number. If we had to write $f: [0, \infty) \to B$, where B is the range, there would be a long pause while we figured out what the range is. By methods we have not yet discussed, it can be shown to be the interval $[-3, \infty)$. (See Figure 4.) More precisely, the values of f start at 0 (when $x = 0$), decrease to -3 as x goes from 0 to 1, then increase without bound as x increases. But this is not easy to confirm without calculus. □

□ **Example 5**   Why is it incorrect to denote the square root function $f(x) = \sqrt{x}$ by $f: \mathfrak{R} \to \mathfrak{R}$?

Solution: According to our definition, every element of the domain must be a legitimate input. Negative numbers, however, do not have real square roots; if the outputs are to be in $\mathfrak{R}$, the domain must be restricted to $[0, \infty)$. The notation $f: [0, \infty) \to \mathfrak{R}$ is therefore acceptable. So is the notation $f: \mathfrak{R} \to C$ (where C is the set of complex numbers), for this would allow the outputs to include square roots of negative numbers and we could relax the restrictions on the domain. □

Despite the preceding remarks, it is customary in a calculus book to allow a certain amount of imprecision about functions. We often lapse into language like, "Consider the function $y = x^2$," without specifying the inputs. To get away with this, we make the following agreement.

> A function $f: A \to B$ is said to be **real** if both A and B are subsets of $\mathfrak{R}$. When its mode of operation is given by the formula $y = f(x)$ without specifying A, its domain is understood to be the set of legitimate inputs. That is, the domain consists of all real values of x for which $f(x)$ is a real number.

□ **Example 6**   Find the understood domain of the real function Q defined by

$$Q(x) = \frac{x^2 - 1}{x - 1}$$

Solution: The number 1 is not a legitimate input, since $Q(1) = 0/0$ is meaningless. Hence the understood domain is the set of real numbers $x \neq 1$. Of course it might occur to you to write

$$Q(x) = \frac{(x - 1)(x + 1)}{x - 1} = x + 1$$

and then claim that $Q(1) = 2$. The functions defined by

$$Q(x) = \frac{x^2 - 1}{x - 1} \qquad \text{and} \qquad f(x) = x + 1$$

Figure 5

Graph of $y = \dfrac{x^2 - 1}{x - 1}$

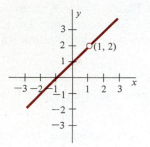

are not the same, however, as you can see by looking at their graphs. The first is a straight line with a hole in it, namely (1,2), while the second is the same line with the hole filled in. (See Figure 5.)   □

□ **Example 7**   Find the understood domain of the function

$$f(x) = \sqrt{\frac{4 - x^2}{x}}$$

Solution: To guarantee that $f(x)$ is a real number, we must impose the restriction

$$\frac{4 - x^2}{x} \geq 0$$

(Why?) Since the solutions of this inequality are the values of x satisfying $0 < x \leq 2$ or $x \leq -2$ (confirm!), the understood domain of f is the union of the half-open interval (0,2] and the closed interval $(-\infty, -2]$.   □

Remark   Many students need help in solving inequalities like

$$\frac{4 - x^2}{x} \geq 0$$

What is needed (in view of the fact that the right side is zero) is an analysis of the signs of the expression

$$r(x) = \frac{4 - x^2}{x} = \frac{(2 - x)(2 + x)}{x}$$

Figure 6

Sign of $r(x) = \dfrac{(2 - x)(2 + x)}{x}$

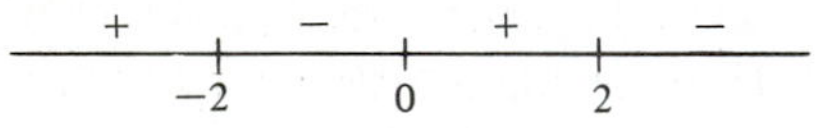

```
    +        −        +        −
  ──┼────────┼────────┼────────
   −2        0        2
```

The key to this is the fact that the linear expressions $2 - x$, x, and $2 + x$ change sign at the points 2, 0, and -2, respectively. Hence we look at the intervals shown in Figure 6. The signs marked in this figure correspond to the behavior of $r(x)$ in the indicated intervals. To discover, for example, that $r(x) < 0$ when $x > 2$, we observe that $2 - x$ is negative in this interval while x and $2 + x$ are positive. When x moves into the interval (0,2) (crossing the point 2), the factor $2 - x$ becomes positive (while x and $2 + x$ remain positive). Hence $r(x)$ is now positive. Similarly, we find that $r(x) < 0$ in $(-2,0)$ and $r(x) > 0$ in $(-\infty, -2)$. Since we are interested in the inequality $r(x) \geq 0$, we simply read off from our figure where the plus signs occur, and observe that $r(x)$ is zero at $x = \pm 2$ and meaningless at $x = 0$. Thus the solution set is $(-\infty, -2] \cup (0,2]$.

Let $f: A \to B$ be a real function. The **graph** of f is the subset of $\mathcal{R}^2$ defined by

$$\{(x,y): x \in A \text{ and } y = f(x)\}$$

A little thought should convince you that the graph of a real function is by itself an adequate definition of the function. For example, the set

$$\{(0,0),\ (1,1),\ (2,4),\ \ldots,\ (9,81)\}$$

contains all the information needed to describe the function in Example 1. The domain is the set of first members of the pairs, namely

$$A = \{0, 1, 2, \ldots, 9\}$$

The range is the set of second members,

$$B = \{0, 1, 4, \ldots, 81\}$$

The mode of operation need not be specified by a formula (although in this case it is easy to see that it is $y = x^2$). The formula is superfluous when each input is already paired with the corresponding output.

In view of this observation we often make no distinction between a real function and its graph. One need not say that a function is a "rule" or "computer"; it is sufficient to name the pairs (x, y). Ordinarily one thinks of the function $y = x^2$ and the parabola shown in Figure 2 as distinct concepts. But if the function is really

$$\{(x, y): x \in \mathfrak{R} \text{ and } y = x^2\}$$

then it *is* the parabola. It is often helpful to think of real functions that way, as nothing more than subsets of $\mathfrak{R}^2$.

It is important to note, however, that a real function is not just any subset of $\mathfrak{R}^2$. Consider the following example.

□ **Example 8**   The *unit circle* in the coordinate plane is the set

$$S = \{(x, y): x^2 + y^2 = 1\}$$

This is a subset of $\mathfrak{R}^2$, but it is not a function. For if we treat it as a computer that prints out y when we feed in x, we blow a fuse. For example, what is y when $x = 0$? The equation $x^2 + y^2 = 1$ tells us nothing more than $y = \pm 1$, which violates the definition of function. Each input is supposed to yield a definite output, that is, $f(x)$ is supposed to be unambiguous. The graphical interpretation of the problem is that vertical lines sometimes intersect the circle more than once, as shown in Figure 7. Here we have marked an input x and the corresponding outputs y_1 and y_2 obtained from x by moving along the vertical line until the graph is encountered.

Such a situation is intolerable from the standpoint of "function as computer." For if x is given, what is $f(x)$? We cannot say $f(x) = y_1$ on Tuesday and $f(x) = y_2$ on Thursday.   □

Figure 7

Is the unit circle a function?

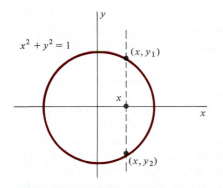

A real function, therefore, may be described as *a subset of $\mathfrak{R}^2$ intersected at most once by any vertical line.* When this condition is not imposed, we say that we are dealing with a **relation.** The intuitive idea is that x and y are somehow connected (as in the equation $x^2 + y^2 = 1$), but not necessarily in such a way that y is uniquely determined by x. In other words, a relation is any subset of $\mathfrak{R}^2$, while a (real) function is a special kind of relation in which each x leads to exactly one corresponding y.

□ **Example 9**   Erase the bottom half of the circle in Figure 7 (that is, impose the restriction $y \geq 0$) and the equation $x^2 + y^2 = 1$ does define a function. For now we can solve for y without any ambiguity, namely $y = \sqrt{1 - x^2}$. Hence the upper semicircle is a function

$$f_1 \colon [-1,1] \to \mathcal{R} \qquad \text{defined by } f_1(x) = \sqrt{1 - x^2}$$

Similarly, the lower semicircle is a function

$$f_2 \colon [-1,1] \to \mathcal{R} \qquad \text{defined by } f_2(x) = -\sqrt{1 - x^2}$$

This illustrates the fact that a relation may give rise to several functions, depending on which part of it we want to keep.   □

Problem Set 1.4

1. What is the (understood) domain of the real function $f(x) = x^3$? Find $f(-2)$, $f(-1)$, $f(0)$, $f(1)$, $f(2)$, and sketch the graph of f.

2. What is the (understood) domain of the real function $f(x) = 1 - x$? Find $f(-1)$, $f(0)$, $f(1)$, $f(2)$, and sketch the graph of f.

3. What is the (understood) domain of the real function $f(x) = 1/x$? Find $f(1)$, $f(2)$, $f(\frac{1}{2})$, $f(-1)$, $f(-2)$, $f(-\frac{1}{2})$, and sketch the graph of f.

4. What is the (understood) domain of the real function

$$f(x) = \frac{x^2 - 4}{x - 2}?$$

Find $f(-1)$, $f(0)$, $f(1)$, $f(3)$, $f(4)$, and sketch the graph of f.

5. If $f(x) = x^2$, find
   (a) $f(x + h)$
   (b) $\dfrac{f(x + h) - f(x)}{h}$

6. Repeat Problem 5 with $f(x) = 1 - x$.
7. Repeat Problem 5 with $f(x) = x^2 - 1$.
8. Repeat Problem 5 with $f(x) = x^3$.
9. Define the function f by the table

x	0	1	2	3	4
$f(x)$	0	2	4	6	8

   (a) What is the domain of f? What is the range?

(b) Find a formula for $f(x)$ in terms of x. (This question is mainly for entertainment. Since the table describes the function completely, no formula is needed.)

10. Define the function f by the table

x	1	2	3	4	5
$f(x)$	-1	1	-1	1	-1

    (a) What is the domain of f? What is the range?
    (b) Find a formula for $f(x)$ in terms of x.

11. Let f be the function $\{(-2,\frac{1}{4}), (-1,\frac{1}{2}), (0,1), (1,2), (2,4)\}$.
    (a) What is the domain of f? What is the range?
    (b) Find a formula for $f(x)$ in terms of x. *Hint:* Look at the pattern in the list of outputs $\frac{1}{4}$, $\frac{1}{2}$, 1, 2, 4.
    (c) Draw a smooth curve through the points of f in the coordinate plane.
    (d) Assuming that this curve extends f to a function whose domain is $\mathcal{R}$, use it to estimate $f(\frac{1}{2})$.
    (e) Assuming that the formula in part (b) applies to this extended function, what is $f(\frac{1}{2})$? Are parts (d) and (e) in approximate agreement?
    (f) The formula serves as a check on the graph in finding $f(\frac{1}{2})$. It is no help in finding $f(\pi)$, however. Why? Use the graph to estimate $f(\pi)$.

    In part (f) we are assuming that you don't know what $f(\pi)$ is because we have not discussed such inputs in the formula in part (b). However, you can bypass the question with a calculator by just punching the right keys. The result is $f(\pi) = 8.824 \cdots$. As a preliminary definition, therefore, we may take f to be the function the calculator says it is! Later we will define it by the methods of calculus.

12. Define f by the rule that $f(n)$ is the nth digit (after the decimal point) in the decimal representation of $\frac{3}{4}$.

  (a) Find $f(1), f(2), f(3), f(4)$.

  (b) What is the domain of f? What is the range?

13. Define f by the rule that $f(n)$ is the nth digit (after the decimal point) in the decimal representation of π.

  (a) Find $f(1), f(2), f(3), f(4)$. (For the decimal representation of π, see Section 1.1.)

  (b) What is the domain of f? Do you know the range? (If you do, you must have π worked out to at least 32 decimal places!)

  (c) You may not know $f(10)$; perhaps nobody knows $f(10^6)$. Does ignorance of a function's output affect the question of whether it is properly defined?

  (Mathematicians, not to mention philosophers, differ on the answer to Problem 13(c). Does π really exist? If so, we are gradually discovering its decimal representation. Or does our inability to construct it mean that it isn't there?)

14. A *prime number* is a positive integer greater than 1 that cannot be factored except as a product of itself and 1. The first few primes are 2, 3, 5, 7, 11, 13, 17, ... For each positive integer n, define $f(n)$ to be the number of primes less than or equal to n.

  (a) Find $f(1), f(2), f(4), f(8), f(16), f(32)$.

  (b) What is the range of f? *Hint:* It can be proved that there are infinitely many primes.

15. Define the function f in the domain $N = \{1, 2, 3, 4 \ldots\}$ by

$$f(n) = \begin{cases} 1 \text{ if } n = 1 \text{ or } n = 2 \\ f(n-1) + f(n-2) \text{ if } n > 2 \end{cases}$$

Find $f(1), f(2), f(3), f(4), f(5), f(6), f(7)$.

The functional values in Problem 15 constitute the famous *Fibonacci sequence,* which is so interesting (and has so many applications) that there is even a journal (the *Fibonacci Quarterly*) devoted to publication of articles about it.

16. For each $x \geq 0$ define $f(x) = n$, where n is the greatest integer not exceeding x, that is, $n \leq x < n + 1$.

  (a) Find $f(0), f(\frac{1}{2}), f(1), f(\sqrt{2}), f(3), f(\pi)$.

  (b) What is the range of f?

  (c) Draw the graph of f for $0 \leq x < 4$.

The function in Problem 16 is called the *greatest integer function* and is denoted by $f(x) = [x]$.

17. A taxi costs \$1.25 for the first mile and 75¢ for each additional mile (or fraction thereof). It will not go farther than 5 miles. Let $f(x)$ be the cost of riding the taxi for x miles.

  (a) Find $f(\frac{1}{2}), f(1), f(\frac{3}{2}), f(2), f(\frac{5}{2}), f(3)$.

  (b) What is the domain of f? the range?

  (c) Draw the graph of f.

Find the (understood) domain of each of the following real functions.

18. $f(x) = 1 - x^2$ What is the range?

19. $f(x) = x^2 - 2x$ What is the range?

20. $f(x) = 1 - |x|$ What is the range?

21. $f(x) = |x|$ What is the range?

22. $f(x) = \sqrt{x - 1}$ What is the range?

23. $f(x) = \sqrt{4 - x^2}$ What is the range?

24. $f(x) = \dfrac{1}{x - 1}$

25. $f(x) = \dfrac{x^2 - 9}{x - 3}$

26. $f(x) = \dfrac{\sqrt{x}}{x^2 - x - 6}$

27. $f(x) = \dfrac{x}{9 - x^2}$

28. $f(x) = \sqrt{\dfrac{x}{1 - x}}$

29. $f(x) = \sqrt{\dfrac{2 - x}{x}}$

30. $f(x) = \dfrac{1}{x^2 - x + 1}$ *Hint:* What is the solution set (in $\mathcal{R}$) of the equation $x^2 - x + 1 = 0$?

31. $f(x) = \dfrac{2}{1 + x^2}$

Sketch the graph of each of the following relations and determine whether it is a function. If it is (say f), give a formula for $f(x)$.

32. $\{(x, y): x = 2\}$

33. $\{(x, y): y = 1\}$

34. $\{(x, y): y^2 = x\}$

35. $\{(x,y): y^2 = x \text{ and } y \geq 0\}$
36. $\{(x,y): y^2 = x \text{ and } y \leq 0\}$
37. $\{(x,y): y^3 = x\}$
38. $\{(x,y): 2x + y = 1\}$
39. $\{(x,y): x^2 + y^2 = 4\}$
40. $\{(x,y): x^2 + y^2 = 4 \text{ and } y \geq 0\}$
41. $\{(x,y): x^2 + y^2 = 4 \text{ and } y \leq 0\}$
42. $\{(x,y): x^2 + y^2 < 4\}$
43. $\{(x,y): xy = 4\}$
44. $\{(x,y): |x| + |y| = 1\}$
45. $\{(x,y): |x| + |y| = 1 \text{ and } y \geq 0\}$
46. $\{(x,y): |x| + |y| = 1 \text{ and } y \leq 0\}$
47. $\{(x,y): y > x\}$ *Hint:* The graph is a region having the line $y = x$ as boundary.

In each of the following, a bug's position in the coordinate plane at time t is given by the function $P: \Re \to \Re^2$. Describe the bug's path.

48. $P(t) = (t, t^2)$ *Hint:* Eliminate t from the equations $x = t$ and $y = t^2$.
49. $P(t) = (2t, t - 1)$

50. $P(t) = (t^2, t^3)$
51. $P(t) = (\cos t, \sin t)$ *Hint:* $\cos^2 t + \sin^2 t = 1$.
52. The temperature at each point of the circular disk

$$D = \{(x,y): x^2 + y^2 \leq 25\}$$

is given by $T: D \to \Re$, where $T(x,y) = \sqrt{x^2 + y^2}$.
   (a) Which point of the disk is coolest?
   (b) A bug starts at $(1,0)$ and (disliking change) follows a path along which the temperature is constant. What is the temperature and what is the path? In general, what are the isothermal ("equal temperature") paths?

53. In Problem 52 a bug starts at $(3,4)$ and moves across the disk along a diameter. What real function describes the temperature variation along this path? What are the domain and range of this function?

54. In Problem 52 a bug lands on the disk D and then jumps straight upward a distance equal to T^2, where T is the temperature at its landing point (x,y). Using triples (x,y,z) to locate points in three-dimensional space (the z axis is perpendicular to the coordinate plane and has its zero point at the origin), define the function $P: D \to \Re^3$ by $P(x,y) = $ position of bug after jumping. What is the formula for $P(x,y)$ in terms of x and y?

1.5
REAL FUNCTIONS

In the last section we said that a function $f: A \to B$ is *real* if both A and B are subsets of $\Re$. In other words, both the input and output are real numbers. In this section we describe several common types of real functions that are encountered repeatedly in calculus.

□ **Example 1**   If $c \in \Re$, the function $f: \Re \to \Re$ defined by $f(x) = c$ is called a **constant function.** Its graph is

$$\{(x,y): x \in \Re \text{ and } y = c\}$$

which is a horizontal straight line. (See Figure 1.)

Figure 1
Constant function

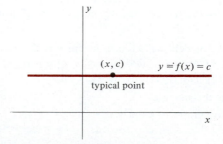

□

Figure 2
Identity function

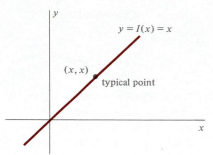

$y = I(x) = x$

(x, x)
typical point

□ **Example 2**   The function $I(x) = x$ is called the **identity function.** Its graph is

$$\{(x, y): y = x\}$$

the straight line shown in Figure 2.   □

□ **Example 3**   The function $f(x) = mx + b$ (where m and b are real numbers) is called a **linear function** because its graph (Figure 3) is a straight line (as shown in Section 1.2). Note that constant functions and the identity function are special cases.

Figure 3
Linear function

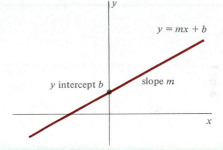

$y = mx + b$

y intercept b slope m

□

□ **Example 4**   If n is a positive integer greater than 1, then $f(x) = x^n$ is called a **power function.** Its graph has one of the shapes shown in Figures 4 and 5.

Figure 4
Even-power function

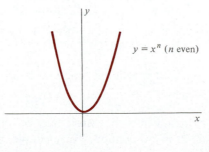

$y = x^n$ (n even)

Figure 5
Odd-power function

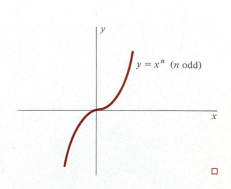

$y = x^n$ (n odd)

□

□ **Example 5**   If n is a positive integer greater than 1, the function $f(x) = x^{1/n} = \sqrt[n]{x}$ is called a **root function.** Its domain is $\Re$ if n is odd (Figure 6) and $[0, \infty)$ if n is even (Figure 7). (Why?)

Figure 6
Odd-root function

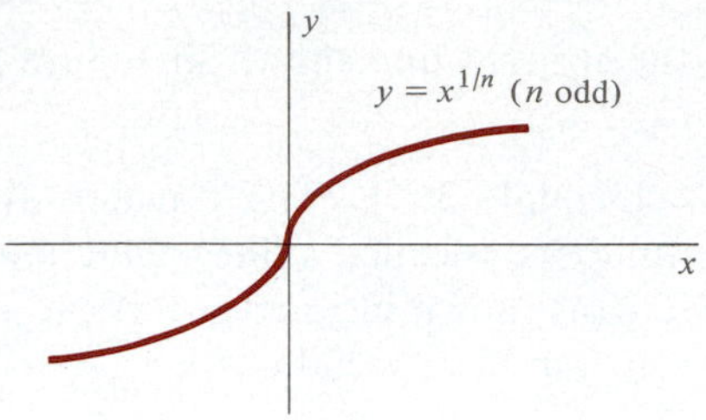

$y = x^{1/n}$ (*n* odd)

Figure 7
Even-root function

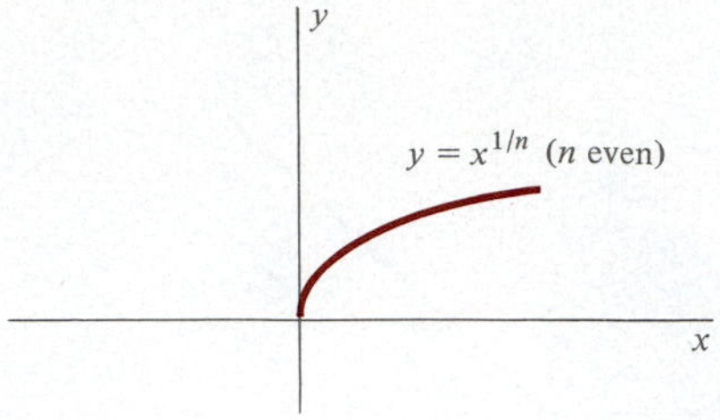

$y = x^{1/n}$ (*n* even)

□

□ **Example 6**   If *n* is a nonnegative integer and $a_0, a_1, \ldots, a_n$ are real numbers, then

$$P(x) = a_0 + a_1 x + a_2 x^2 + \cdots + a_n x^n$$

is called a **polynomial.** Special cases are $y = c$, $y = x$, $y = mx + b$, and $y = x^n$, already discussed. In general the domain is ℜ and the graph is a continuous smooth curve (definitions later!). An example is $P(x) = 3x^4 + 4x^3 - 12x^2$, whose graph is shown in Figure 8. While anybody can locate points of the graph (simply by computing values of P corresponding to various values of x), it is not an easy matter to do it intelligently. It is a great help to know where the **turning points** are, namely $(-2,-32)$, $(0,0)$, $(1,-5)$. For it is in the neighborhood of these points that the graph changes from rising to falling or vice versa. Moreover, a glance at the figure tells us that the range of the function is $[-32,\infty)$, a fact not easily discovered without knowing that $P(-2) = -32$ is the minimum value of the function. Such details of graphing are among the important applications of calculus. (See Chapter 5.)   □

Figure 8
Graph of $y = 3x^4 + 4x^3 - 12x^2$

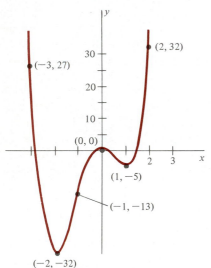

□ **Example 7**   If $P(x)$ and $Q(x)$ are polynomials,

$$f(x) = \frac{P(x)}{Q(x)}$$

is called a **rational function.** Its domain is

$$\{x: Q(x) \neq 0\}$$

which means that there may be values of x in the neighborhood of which $f(x)$ is unbounded. For example, the domain of

$$f(x) = \frac{x^2}{(x-2)^2}$$

is $\{x: x \neq 2\}$. When x is close to 2, the numerator of $f(x)$ is nearly 4 while the denominator is positive and close to 0. Hence $f(x)$ increases without bound as x approaches 2. The graphical interpretation of this is shown in Figure 9, where we indicate how the curve rises on either side of the vertical line $x = 2$. (Such a guide line is called an **asymptote.**)

Figure 9

Graph of $y = \dfrac{x^2}{(x-2)^2}$

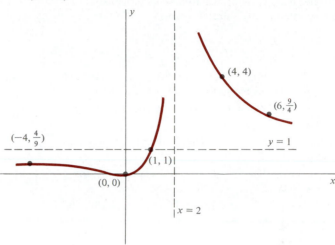

A similar observation is that when x is numerically large (positive or negative), $f(x)$ is nearly equal to 1. To see why, write

$$f(x) = \left(\frac{x}{x-2}\right)^2 = \left(\frac{1}{1-2/x}\right)^2$$

(We divided numerator and denominator by x.) The term $2/x$ is negligible when x is numerically large, which means that $f(x)$ approaches 1. Hence the curve follows the horizontal line $y = 1$ as we travel far to the right or left.

Finally we observe that $(0,0)$ is the lowest point of the curve because

$$f(x) = \left(\frac{x}{x-2}\right)^2 \geq 0 \qquad \text{for all } x \neq 2$$

and $f(0) = 0$. Thus the main features of the graph are clear; it only remains to plot a few conveniently located points to make a good sketch.   □

Figure 10

Graph of $y^2 = x^2(3-x)$

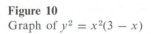

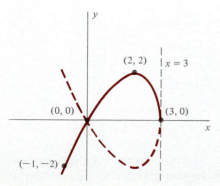

□ **Example 8**   The domain of the function $y = x\sqrt{3-x}$ is $(-\infty, 3]$. (Why?) Hence there is no graph to the right of the vertical line $x = 3$. (See Figure 10.) The endpoint $(3,0)$ is of some interest, as is the turning point $(2,2)$ (discovered by methods not yet discussed). One can see by inspection that large negative values of x yield large negative values of y, so we plot a couple of points like $(0,0)$ and $(-1,-2)$ and sketch the graph. We have also drawn its reflection in the x axis (the dashed curve), because it is worthwhile to observe that the whole loop is the relation

$$\{(x,y): y^2 = x^2(3-x)\}$$

(Square both sides of $y = x\sqrt{3-x}$ to obtain the relation.) The fact that the function is part of this loop is a help to our intuition. For example, it makes plausible the guess that there is a vertical tangent at $(3,0)$.   □

Problem Set 1.5

Sketch the graph of each of the following polynomials by plotting a few intelligently selected points.

1. $y = x^2 - 1$ **2.** $y = 1 - x^2$ **3.** $y = x^3$

4. $y = x^3 - 1$ **5.** $y = x^4$ **6.** $y = 16 - x^4$

7. $y = x^3 - 3x$ *Hint:* There are "turning points" at $x = \pm 1$ (discovered by methods not yet discussed).

8. $y = 2x^3 - 3x^2 - 12x + 6$ *Hint:* There are turning points at $x = -1$ and $x = 2$.

9. $y = x^4 - 8x^2 + 16$ *Hint:* There are turning points at $x = 0$ and $x = \pm 2$.

10. $y = 3x^4 + 4x^3 - 12x^2 + 10$ *Hint:* There are turning points at $x = 0, 1, -2$.

11. Sketch the graph of $y = 1/(x^2 + 1)$ after answering the following questions.

(a) What is the domain?

(b) What symmetry does the graph have?

(c) Why is $(0,1)$ the highest point?

(d) How does y behave when x is numerically large?

12. The graph of $y = x + (1/x)$ has turning points at $x = \pm 1$. Use that fact and your answers to the following questions to sketch the graph.

(a) What is the domain?

(b) Why is the graph symmetric about the origin?

(c) How does y behave when x is near 0 and positive? near 0 and negative?

(d) Explain why the line $y = x$ is an asymptote.

13. The graph of

$$y = \frac{4(x - 1)}{x^2}$$

has a turning point at $x = 2$. Use that fact, together with your answers to the following questions, to sketch the graph.

(a) What is the domain?

(b) How does y behave when x approaches 0?

(c) Where does the curve intersect the x axis?

(d) How does y behave when x is numerically large? *Hint:* Divide numerator and denominator by x to obtain

$$y = \frac{4(1 - 1/x)}{x}$$

14. The graph of $y = \sqrt{x(2 - x)}$ is part of a circle. What circle? Sketch the graph of this function.

15. Sketch the graph of

$$y = \sqrt{\frac{x}{x - 1}}$$

after answering the following questions.

(a) What is the domain?

(b) Explain why the line $x = 1$ is an asymptote.

(c) Explain why the line $y = 1$ is an asymptote.

(d) Where does the graph intersect the x axis?

(e) What part of the relation $y^2 = x/(x - 1)$ is this graph? In view of your answer, what kind of tangent would you expect at the origin?

16. Sketch the graph of the relation

$$y^2 = \frac{4x}{2 - x}$$

by answering the kind of questions we have been asking in the preceding problems.

17. A superficial glance at the function

$$y = \frac{x^2 - 4}{x - 2}$$

might lead one to expect the vertical asymptote $x = 2$. That is not the case, however. What is the graph?

18. The graph of

$$y = \frac{x - 1}{\sqrt{x} - 1}$$

can be described as a "half-parabola with a hole in it." What parabola are we talking about and where is the hole? Sketch the graph. *Hint:* $x - 1 = (\sqrt{x} - 1)(\sqrt{x} + 1)$.

19. Sketch the graph of $y = |x|/x$. (Don't just plot points! Analyze the function by making use of the definition of $|x|$.)

20. The graph of $y = |1 - x^2|$ is shaped something like a W; it is made up of parts of two parabolas. What parabolas? Sketch the graph.

21. Sketch the graph of the function defined by the rule

$$y = \begin{cases} x & \text{if } x \le 1 \\ 2 - x^2 & \text{if } x > 1 \end{cases}$$

22. Sketch the graph of the function

$$y = \begin{cases} -1 & \text{if } x < -1 \\ x & \text{if } -1 \le x \le 1 \\ 1 & \text{if } x > 1 \end{cases}$$

23. Assume that postal regulations in The Shire are as uncomplicated as the hobbits who live there, say 1¢ per ounce or fraction thereof. Sketch the graph of the postage function, $f(x)$ = cost of mailing a letter weighing x oz. Confine the domain to $(0,5]$.

24. Sketch the graph of the "age function," $f(x)$ = age at most recent birthday of person born x years ago. Confine the domain to $[0,4)$.

25. A real function is said to be *even* if $f(-x) = f(x)$ for all x in the domain. (An example is $f(x) = x^2$.) Explain why the graph of an even function is symmetric about the y axis.

26. A real function is said to be *odd* if $f(-x) = -f(x)$ for all x in the domain. (An example is $f(x) = x^3$.) Explain why the graph of an odd function is symmetric about the origin.

Which of the following functions are even? odd? neither?

27. $f(x) = x^2 - 1$ 28. $f(x) = x^3 - 1$
29. $f(x) = x^3 - 3x$ 30. $f(x) = |x|$

31. $f(x) = \sqrt{9 - x^2}$ 32. $f(x) = \dfrac{1}{1 + x^2}$

33. $f(x) = \dfrac{x}{1 + x^2}$ 34. $f(x) = \dfrac{x}{x - 1}$

35. $f(x) = 2 \cos x - 1$ 36. $f(x) = 5 \sin 2x$

37. $f(x) = \sin x - \cos x$ 38. $f(x) = \tan x + 2 \sin x$

39. There is one real function that is both even and odd. What is it?

In the following problems we assume that f and g are real functions with domains that have a nonempty intersection. (Hence there is a domain in which f and g are both defined.)

40. Suppose that f and g are odd. Which of the following functions are even? odd? not necessarily either one?
    (a) $f + g$, defined by $(f + g)(x) = f(x) + g(x)$
    (b) $f - g$, defined by $(f - g)(x) = f(x) - g(x)$
    (c) fg, defined by $(fg)(x) = f(x)g(x)$
    (d) f/g, defined by $(f/g)(x) = f(x)/g(x)$ (Assume that the common domain of f and g contains points at which g is not zero, so that f/g has a domain.)

41. Repeat Problem 40 when f and g are even.
42. Repeat Problem 40 when f is odd and g is even.

True–False Quiz

1. Each point of the coordinate line may be represented by a periodic decimal.
2. $\pi = 22/7$.
3. $x \geq 0$ and $x \leq 0 \Rightarrow x = 0$.
4. If z is an integer and $x < y$, then $xz < yz$.
5. If x and y are real numbers, then $(x - y)(y - x) \leq 0$.
6. The inequality $x^2 + 1 < 2x$ has no solutions in $\mathcal{R}$.
7. The solution set of the inequality $6/x < 1$ is the interval $(6, \infty)$.
8. $\sqrt{x^2} = x$ for all $x \in \mathcal{R}$.
9. If $x > y$, then $d(x,y) = x - y$.
10. If $\delta > 0$ and $p \in \mathcal{R}$, the solution set of $|x - p| < \delta$ is an open interval having p as midpoint.
11. $|x - 2| > 1 \Rightarrow 1 > x > 3$.
12. $|x + y| = |x| + |y|$ for all x and y in $\mathcal{R}$.
13. The distance between $(0,3)$ and $(2,-3)$ in $\mathcal{R}^2$ is $2\sqrt{10}$.
14. If (a,b) is any point of a line with slope $\frac{3}{2}$, so is $(a + 2, b + 3)$.
15. If two lines have slopes m_1 and m_2, the equations $m_1 = m_2$ and $m_1 m_2 = -1$ cannot both be true.

16. The slope of a horizontal line times the slope of a vertical line is -1.
17. An equation of the line through $(-1,1)$ with slope $\frac{1}{2}$ is $x - 2y + 3 = 0$.
18. The line containing $(1,-1)$ and $(2,2)$ is parallel to the line $x - 3y = 7$.
19. The lines $x - 2y = 5$ and $2x + y = 1$ are perpendicular.
20. The graph of $y = x^2 + 1$ is symmetric about the x axis.
21. The graph of $x^2 - y^2 = 1$ is symmetric about the line $y = x$.
22. The graph of $x^2 + y^2 - 2x + 2 = 0$ is a circle.
23. $\{(x,y): x^2 + y^2 < 1\}$ is the interior of a circle.
24. The longest diameter of the ellipse $2x^2 + y^2 = 2$ is horizontal.
25. The vertex of the parabola $y = 2 - x^2$ is $(0,2)$.
26. The graph of $y = \sqrt{x}$ is part of a parabola.
27. The coordinate axes are asymptotes of the hyperbola $xy = 2$.
28. The domain of
$$f(x) = \sqrt{\frac{x}{2 - x}}$$
is the interval $[0,2)$.

29. The range of $f(x) = 4 - x^2$ is the interval $(-\infty, 4]$.

30. If

$$f(x) = \frac{x^2 - 4}{x - 2}$$

then $f(2) = 4$.

31. If $f(x) = x^2 - x$, then $f(-x) = f(x)$ for all x.

32. The parabola $y^2 = x$ is the graph of a function.

33. $\{(x,y): x^2 + y^2 = 1,\ y \geq 0\}$ is the graph of a function.

34. The function

$$f(x) = \frac{x^2}{x^2 + 1}$$

is rational.

Additional Problems

Solve each of the following inequalities.

1. $|1 - 3x| < 5$ **2.** $x^2 < 9$

3. $|x^2 - 2| > 3$ **4.** $x^2 > -1$

5. Suppose that the weight of an object x miles from the center of the earth is given by the formula

$$F = \frac{24 \times 10^8}{x^2}$$

where F is in pounds. Assuming that the radius of the earth is 4000 miles, find the range of altitudes in which the object weighs at least 96 pounds.

6. One way to solve the inequality $x^2 < 25$ is to write the equivalent inequality $|x| < 5$. Solve it instead by factoring the left side of $x^2 - 25 < 0$ and analyzing signs as suggested in the *Remark* in Section 1.4.

7. It is tempting to solve the inequality

$$\frac{x - 2}{x} < 0$$

by multiplying each side by x, obtaining $x - 2 < 0$. Why is this incorrect? What is the correct solution?

Solve each of the following inequalities by analyzing signs as in the *Remark* in Section 1.4.

8. $x^2 < 6x - 5$

9. $x^4 > x^2$

10. $x^3 > 27$ *Hint:* $x^3 - 27 = (x - 3)(x^2 + 3x + 9)$.

11. $\dfrac{x - 8}{x + 2} \geq 0$ **12.** $\dfrac{x(2 - x)}{x + 1} \leq 0$

13. $\dfrac{(x + 2)(x - 3)}{x(x^2 + 1)} < 0$ **14.** $\dfrac{x - x^2}{1 + x} > 0$

15. $x + \dfrac{1}{x} < 2$ **16.** $\dfrac{8}{x} < x - 2$

17. $\dfrac{3}{x - 2} < \dfrac{1}{2x + 1}$ **18.** $\dfrac{2 - x}{x} \geq 1$

19. $\dfrac{3}{x^2 - 4} > \dfrac{x}{x - 2}$ **20.** $\dfrac{x^2}{x - 2} - 1 \geq \dfrac{x^2 + 3}{x^2 - 4}$

21. Find the distance between the points $(\sqrt{2}, 2)$ and $(2, -\sqrt{2})$.

Find an equation of each of the following lines.

22. With slope -2 containing the point $(3, -1)$

23. Through $(-2, 3)$ perpendicular to the line $5x - 2y = 2$

24. Perpendicular bisector of the line segment joining the points $(2, -5)$ and $(0, 1)$

25. Tangent to the circle $x^2 + y^2 = 25$ at the point $(-3, 4)$

26. Suppose that the coordinates of a moving point (x, y) are given in terms of the time t by the equations $x = 2t + 1$ and $y = -t$.

   (a) Where is the point when $t = 0$? when $t = 1$?

   (b) Explain why the "track" of the moving point is the line $x + 2y - 1 = 0$. In what direction is the point moving on this line?

27. Explain why the lines $Ax + By + C_1 = 0$ and $Ax + By + C_2 = 0$ are parallel. Then use Problem 56, Section 1.2, to show that the distance between them is

$$d = \frac{|C_1 - C_2|}{\sqrt{A^2 + B^2}}$$

Hint: First dispose of the case where the lines are vertical. Then, assuming they are not, use the y intercept of one of them as the point (x_0, y_0) in Problem 56, Section 1.2.

28. Use Problem 27 to find the distance between the lines $2x - y = 4$ and $y = 2x + 1$.

Test each of the following sets for symmetry about the x axis, the y axis, the line $y = x$, and the origin. In each case draw a picture of S.

29. $S = \{(x, y): x^2 + y^2 = 1\}$ **30.** $S = \{(x, y): x = 4\}$

31. $S = \{(x, y): y = 1\}$ **32.** $S = \{(x, y): x^2 = 8y\}$

33. $S = \left\{(x,y): \dfrac{x^2}{4} + \dfrac{y^2}{3} = 1\right\}$ **34.** $S = \{(x,y): y^2 = 4x\}$

35. $S = \{(x,y): xy = 2\}$ **36.** $S = \{(x,y): x = y^3\}$

37. $S = \{(x,y): y = \sin x\}$

38. Find an equation of the circle having $(5,-2)$ and $(1,4)$ as endpoints of a diameter.

39. Find the center and radius of the circle $x^2 + y^2 - 4x + 2 = 0$.

40. Sketch the graph of $2x^2 + y^2 = 8$.

41. Find the vertex and axis of the parabola $y^2 + 8x + 2y - 15 = 0$, and sketch.

42. Sketch the curve $4x^2 - y^2 = 4$ together with its asymptotes.

43. Let $t \in \mathcal{R}$.

   (a) Find the distance between the points $(t^2,2t)$ and $(1,0)$.

   (b) If $S = \{(x,y): x = t^2 \text{ and } y = 2t, \text{ where } t \in \mathcal{R}\}$, what point of S is closest to $(1,0)$?

   (c) Draw a picture of S. *Hint:* Eliminate t from the equations $x = t^2$ and $y = 2t$ to find an equation relating x and y.

Name the domain and range of each of the following functions, and sketch its graph.

44. $f(x) = \sqrt{1 - x^2}$ **45.** $f(x) = x^2 - 2x$

46. $f(x) = \dfrac{1}{x^2}$ **47.** $f(x) = x + |x|$

48. $f(x) = \dfrac{|x|}{x}$

49. If $f(x) = x^2$, reduce the expression

$$\frac{f(x + h) - f(x)}{h}$$

to simplest form.

50. Which of the relations

$$\{(x,y): y^2 = x\} \quad \text{and} \quad \{(x,y): y^3 = x\}$$

is a function? Explain.

2 | The Derivative

I do not know what I may appear to the world, but to myself I seem to have been only a boy playing on the sea-shore, and diverting myself in now and then finding a smoother pebble or a prettier shell than ordinary, whilst the great ocean of truth lay all undiscovered before me.

SIR ISAAC NEWTON (1642–1727)

Taking mathematics from the beginning of the world to the time of Newton, what he has done is much the better half.

GOTTFRIED WILHELM VON LEIBNIZ (1646–1716)
(generally credited with having created calculus independently of Newton)

Nature and Nature's laws lay hid in night:
God said, Let Newton be! and all was light.

ALEXANDER POPE (1688–1744)

2.1
TWO EQUIVALENT PROBLEMS

Calculus has its origin in several related problems. Two of these are the *problem of tangents* and the determination of the *velocity of a moving object*. We begin this chapter with a discussion of each of these questions (which turn out to be equivalent). The results lead directly to one of the fundamental ideas of calculus, the *derivative*. This involves the concept of *limit*, which we develop intuitively; at the end of the chapter we discuss some of the details. The chapter also includes a review of trigonometry.

Tangent to a Curve

We know from plane geometry that the tangent to a circle at a given point is the straight line through the point perpendicular to the radius drawn to the point. See Figure 1, in which we show the circle $x^2 + y^2 = 25$ and the tangent at $(-3, 4)$. Since the slope of the radius is $-\frac{4}{3}$, the slope of the tangent is $\frac{3}{4}$ and hence its equation is $y - 4 = \frac{3}{4}(x + 3)$, or $3x - 4y + 25 = 0$.

Figure 1
Tangent to a circle

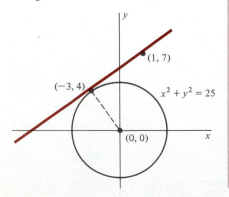

44

Figure 2
Tangent making contact with a curve
more than once

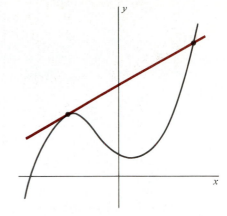

Figure 3
Tangent to $y = x^2$ at (1,1)

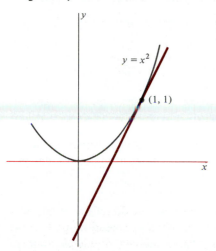

Things are not so simple when the given curve is not a circle. Sometimes (by analogy with the circle) it is suggested that a tangent is a "straight line making contact with the curve at exactly one point," but you can see from Figure 2 that such a definition is inadequate. Nevertheless most people have an intuitive idea of what a tangent is; the problem is to be precise about it.

□ **Example 1**   Discuss the problem of finding the tangent to the curve $y = x^2$ at the point (1,1).

Solution: The graph, together with the tangent at (1,1), is shown in Figure 3. Of course that begs the question; it is hardly fair to draw it before we know what it is! Nevertheless we are going to begin by assuming that a definite tangent exists, our plan being to sneak up on it.

There is no difficulty in drawing a line through (1,1) and a neighboring point (x,y) on the parabola. (See Figure 4.) This line is not the tangent, but it is a good substitute if (x,y) is near (1,1). Its slope is

$$Q(x) = \frac{y - 1}{x - 1} = \frac{x^2 - 1}{x - 1}$$

where we use the functional notation $Q(x)$ to indicate that the slope is a quotient whose value depends on x. Note that $Q(x)$ is defined only for values of $x \neq 1$, since (x,y) must be distinct from (1,1) if we are to draw the line shown in Figure 4.

Now imagine the point (x,y) brought closer to the point (1,1). The nearer it gets, the closer x is to 1. The question is, how does the slope $Q(x)$ behave during this process? Observe that for all $x \neq 1$ we can write

Figure 4
Approximation to tangent at (1,1)

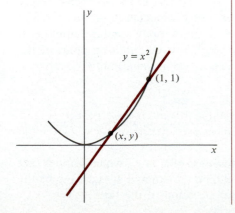

$$Q(x) = \frac{(x - 1)(x + 1)}{x - 1} = x + 1$$

and from this it is evident that as x gets closer to 1, the slope approaches 2. While x cannot be allowed to *equal* 1 (and hence the slope never equals 2), there is nevertheless no doubt about the number $Q(x)$ is approaching. It is not, for example, approaching 3, or 2.01, or 1.9997; it is approaching 2. Why not agree that the tangent is the line through (1,1) with slope 2? No other agreement would be sensible, so we adopt this as the *definition* of the tangent.

Thus the problem of finding the tangent at (1,1) is solved. The line through (1,1) with slope 2 is represented by the equation $y - 1 = 2(x - 1)$ or $2x - y - 1 = 0$. In Figure 3 it may appear that we had to guess the direction of the line, but now we can plot two of its points, say (1,1) and (0, -1), and sketch it accurately.   □

Remark   In Example 1 the reader has a right to object that we began by evading the question of what a tangent is. We said that we would sneak up on it, but what line did we actually have in mind? One answer is the line through (1,1) with the same direction as the curve at that point, but what does "direction" mean? Is it our line of sight as we move along the curve looking straight ahead? That is simply the tangent!

The difficulty is that we began by talking about a concept that is intuitively clear but lacks precise definition. We used our intuition to lead us to the definition, but once the definition is adopted we don't need intuition. We can appeal to the definition instead and say, "That is the line we had in mind." Study Example 1 carefully to appreciate this point. What looks like circular reasoning is really motivation for a definition. If we were merely trying to be logical, we would give the definition and be done with it. (Definitions require no introduction or defense.) We would say, "The tangent to the curve $y = x^2$ at the point (1,1) is the line through (1,1) with slope 2." But then you might justly accuse us of being arbitrary. In the end we *are;* every definition is arbitrary! But mathematicians are no different from other people in their desire to be understood.

Note that in the end there is no mystery about direction. When you reach (1,1) in your travel along the curve, simply look ahead along the line $2x - y - 1 = 0$ we found in Example 1. More precisely, define the **slope of the curve** (previously an ambiguous idea) to be the slope of the tangent, namely 2.

□ **Example 2**   The slope of the curve $y = x^2$ at (1,1) is 2, as we agreed in Example 1. There is no reason why we cannot adopt a similar definition at any point of the curve, say (x_0, y_0). If (x, y) is a neighboring point of the curve, the line through (x_0, y_0) and (x, y) is an approximation to the tangent. Its slope is

$$Q(x) = \frac{y - y_0}{x - x_0} = \frac{x^2 - x_0^2}{x - x_0} = \frac{(x - x_0)(x + x_0)}{x - x_0} = x + x_0 \qquad (x \neq x_0)$$

Closeness of (x, y) to (x_0, y_0) implies closeness of x to x_0, which implies that $Q(x)$ is nearly equal to $2x_0$. Hence we define the tangent at (x_0, y_0) to be the line through (x_0, y_0) with slope $2x_0$. An equation of the tangent is

Figure 5
Graph of $y = |x|$

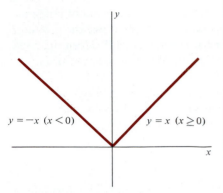

$y = -x \ (x < 0)$ $y = x \ (x \geq 0)$

Figure 6
What is the direction at (0,0)?

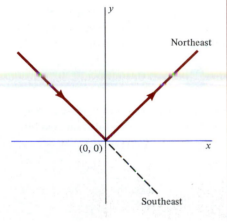

Northeast

(0, 0)

Southeast

Figure 7
Vertical tangent at (0,0)

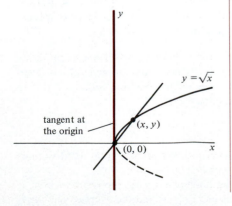

$y = \sqrt{x}$

tangent at
the origin

(x, y)

(0, 0)

$$y - y_0 = 2x_0(x - x_0)$$

Thus the slope of the curve $y = x^2$ at any point (x_0, y_0) is $2x_0$. To say the same thing without subscripts, the slope of the curve $y = x^2$ at any point (x, y) is $m(x) = 2x$. The slope depends on x. At $(1,1)$ it is 2; at $(0,0)$ it is 0; at $(2,4)$ it is 4; and so on.   □

□ **Example 3**   The graph of

$$y = |x| = \begin{cases} x \text{ if } x \geq 0 \\ -x \text{ if } x < 0 \end{cases}$$

consists of two rays meeting at the origin, as shown in Figure 5. Discuss its slope at the origin.

Solution: There is a "corner" at $(0,0)$, which suggests that we will have difficulty defining a tangent at that point. To find out what happens, let (x, y) be a neighboring point of the graph. The slope of the line through $(0,0)$ and (x, y) is

$$Q(x) = \frac{y - 0}{x - 0} = \frac{y}{x} = \frac{|x|}{x} = \begin{cases} 1 \text{ if } x > 0 \\ -1 \text{ if } x < 0 \end{cases}$$

(Note that x cannot be 0. Why?) When (x, y) is close to $(0,0)$, what is $Q(x)$ close to? It is 1 or -1 depending on whether $x > 0$ or $x < 0$. There is no number m we can name that $Q(x)$ approaches as x approaches 0.

We describe this situation by saying that there is no tangent at $(0,0)$. Nor does the graph have a slope at that point. A traveler moving from left to right on this path would be puzzled if you asked about the direction of the path at $(0,0)$. (See Figure 6.) The traveler might say, "Well, I was going southeast. But this is ridiculous. I had to come to a screeching halt and now apparently I am about to go northeast. The least they could do is put up some signs."   □

□ **Example 4**   The graph of $y = \sqrt{x}$ is the upper half of the parabola $y^2 = x$ (Figure 7). It is geometrically apparent that its tangent at the origin is the y axis, which leads us to believe that the slope at $(0,0)$ is undefined. We confirm this by looking at

$$Q(x) = \frac{y - 0}{x - 0} = \frac{\sqrt{x}}{x} = \frac{1}{\sqrt{x}}   (x > 0)$$

When x approaches 0, $Q(x)$ increases without bound, which is what we should expect. For this means that the line through $(0,0)$ and (x, y) gets steeper as (x, y) approaches the origin. The line it approaches is the vertical line $x = 0$. Thus in this case the curve has a tangent, but its slope is undefined.   □

Velocity of Motion in a Line

Now we turn to a question that is closely related to the problem of tangents. Suppose that a ball is thrown straight upward, its height above the ground t seconds later being $h = 64t - 16t^2$ (in feet). The source of this formula need not concern us now; we simply take it as given. What we want to discuss is the *velocity* of the ball.

Presumably you already know how to compute an *average rate,* as in the case of a car making a trip of 150 miles in 3 hours. When we say that its average rate is 50 mph we are using the formula

$$\text{Rate} = \frac{\text{Distance}}{\text{Time}}$$

(sometimes written in the form Distance = Rate × Time). This formula is not much help in the case of a ball thrown upward from the ground, because the rate is not constant. Moreover, the ball reverses direction at its highest point; our discussion of velocity should include a way of distinguishing between upward and downward motion.

What we need is a mathematical refinement of the operation of a radar unit, which measures the *instantaneous rate* at which an object is moving. The reason it works so well is that the radar pulse (from the unit to the moving object and back) travels at the speed of light. Hence the object moves a very short distance in a very small interval of time while its rate is being measured.

Let's apply that idea to the ball. When $t = t_0$, the height of the ball is $h_0 = 64t_0 - 16t_0{}^2$. If t is a later clock reading ($t > t_0$), the height has become $h = 64t - 16t^2$. The *change in position* (called "displacement") is

$$h - h_0 = (64t - 16t^2) - (64t_0 - 16t_0{}^2) = 64(t - t_0) - 16(t^2 - t_0{}^2)$$

The corresponding change in time is $t - t_0$; we define the *average velocity* during the time interval $[t_0, t]$ to be

$$\frac{h - h_0}{t - t_0} = \frac{64(t - t_0) - 16(t^2 - t_0{}^2)}{t - t_0} = 64 - 16(t + t_0)$$

This is the same idea as our computation of average rate in the case of a car going 150 miles in 3 hours, but since it allows for negative (or zero) displacement we call it *velocity*. (*Speed* is the absolute value of velocity.)

Remark   If the ball is rising during the time interval $[t_0, t]$, displacement is the same as distance traveled, and the average velocity is positive. In general, however, it may be positive, negative, or even zero. If $t_0 = 1.9$, for example, and $t = 2.1$, then $h_0 = 63.84$ and $h = 63.84$, so the average velocity is

$$\frac{h - h_0}{t - t_0} = \frac{0}{0.2} = 0$$

This does not mean that the ball is motionless; instead it rises from $h_0 = 63.84$ to its highest point and then falls back to $h = 63.84$.

It is also worth noting that there is no mathematical reason to restrict $t > t_0$. If we allow $t < t_0$, the time interval is $[t, t_0]$ instead of $[t_0, t]$ and the average velocity is

$$\frac{(\text{terminal position}) - (\text{initial position})}{(\text{later time}) - (\text{earlier time})} = \frac{h_0 - h}{t_0 - t} = \frac{h - h_0}{t - t_0}$$

The ratio is the same either way.

The next step should be apparent. To make the average velocity a good approximation to the *instantaneous velocity* at time t_0, we choose t close to t_0 (as is done in a radar unit). More precisely, we evaluate the *limit* of average velocity as t approaches t_0. As you can see from our formula for average velocity, the limit is

$$v_0 = 64 - 16(t_0 + t_0) = 64 - 32t_0$$

Dropping the subscript (the only purpose of which was to distinguish the fixed instant t_0 from the variable time t), we have a formula for instantaneous velocity v at time t, namely

$$v = 64 - 32t = 32(2 - t)$$

This formula is a precise instrument for discussing the motion of the ball. For it tells us not only *how fast* the ball is moving at any time t, but also *in what direction*. The following table of heights and velocities illustrates what we mean.

t	0	1	2	3	4
h	0	48	64	48	0
v	64	32	0	-32	-64

When $t = 0$ the ball is leaving the ground ($h = 0$) with speed 64 ft/sec, moving upward. One second later it has reached a height of 48 ft and is still moving upward, with speed 32 ft/sec. At $t = 2$ it has reached its highest point (64 ft) and has come to a momentary stop. When $t = 3$ the ball is back to height $h = 48$ and is *falling* with speed 32 ft/sec (because the velocity is negative). It strikes the ground when $t = 4$ with speed 64 ft/sec (the same speed with which it was thrown, but the motion is opposite in direction).

A graph of h as a function of t is shown in Figure 8 (not to be confused with the path of the ball, which is in a vertical straight line). The formula $v = 64 - 32t$ is nothing more than the slope of this graph, as we can check by the methods described in the first part of this section. For if (t_0, h_0) is an arbitrary point of the graph and (t, h) is a nearby point, the slope of the line joining them is

$$Q(t) = \frac{h - h_0}{t - t_0} = 64 - 16(t + t_0)$$

(the average velocity computed earlier). Its limit as t approaches t_0 is the slope of the graph at (t_0, h_0), namely $m(t_0) = 64 - 32t_0$. The slope at any point (t, h) is therefore $m(t) = 64 - 32t$, which is the formula for velocity at time t.

Figure 8
Graph of $h = 64t - 16t^2$, $0 \le t \le 4$

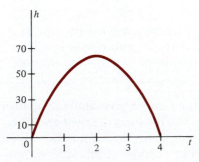

□ **Example 5**   Suppose that an object is moving along a coordinate line (the s axis) in such a way that its position at time t is $s = p(t) = t^2$. Find a formula for velocity.

Solution: Let t_0 be the instant at which we are going to compute veloc-ity and suppose that $t \neq t_0$. The average velocity during the time interval with endpoints t_0 and t is

$$Q(t) = \frac{\text{change in position}}{\text{change in time}} = \frac{p(t) - p(t_0)}{t - t_0}$$

$$= \frac{t^2 - t_0{}^2}{t - t_0} = \frac{(t - t_0)(t + t_0)}{t - t_0} = t + t_0$$

When t approaches t_0, $Q(t)$ approaches $2t_0$, so the velocity at t_0 is $v(t_0) = 2t_0$. The same statement without subscripts is that the velocity at time t is $v(t) = 2t$.   □

This derivation of $v(t) = 2t$ from the law of motion $s = t^2$ is mathemat-ically indistinguishable from Example 2, where we derived the slope $m(x) = 2x$ from the equation $y = x^2$. We may use that fact to interpret the meaning of positive and negative velocity. When $t = -1$, for example, the velocity is $v = -2$. The negative clock reading is no problem; it is like the year 1 B.C. (simply an earlier time than the one we choose to call 0). But what are we to make of negative velocity?

Recall from Section 1.2 that negative *slope* (of a straight line) means that the line is falling from left to right. If the equation of the line is $y = mx + b$, this implies that y is decreasing as x increases. It is geometrically apparent (somewhat harder to prove!) that negative slope of a *curve* means the same thing. Thus the parabola $y = x^2$ is falling from left to right as we pass through the point $(-1,1)$ because the slope is negative ($m = -2$ at $x = -1$). (See Figure 9.) The same statement about the law of motion $s = t^2$ is that s is decreasing as the clock ticks off $t = -1$ because the veloc-ity is negative ($v = -2$ at $t = -1$). Since s is the coordinate of an object moving along the s axis, the object must be going in the negative direction of that axis. Similarly, positive velocity means that the object is moving in the positive direction of the s axis.

We may summarize as follows.

Suppose that an object is moving along a coordinate line with velocity v at time t. Its **speed** is the absolute value of v. Its **direction** is positive if $v > 0$ and negative if $v < 0$. (If $v = 0$, it has come to a momentary stop.)

When we get to the subject of motion in a curve, you should remember what we have said about velocity here. Its two qualities of speed and direc-tion (which in linear motion require nothing more than a number with a sign) will be described by using *vectors*. The purpose of this section, how-ever, is merely to exhibit the equivalence of the problem of tangents and the problem of velocity, in preparation for the definition of derivative in the next section.

Figure 9
Interpretation of slope

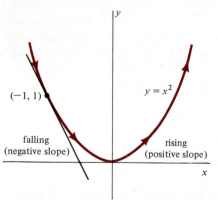

$(-1, 1)$

$y = x^2$

falling
(negative slope)

rising
(positive slope)

Problem Set 2.1

1. Let $Q(x)$ be the slope of the line through (2,4) and a neighboring point (x,y) of the parabola $y = x^2$.
   (a) What is the formula defining Q? For what values of x is $Q(x)$ defined?
   (b) What number does $Q(x)$ approach as (x,y) approaches (2,4)?
   (c) Find an equation of the tangent at (2,4).

2. Let $Q(x)$ be the slope of the line through $(1,-1)$ and a neighboring point (x,y) of the parabola $y = -x^2$.
   (a) What is the formula defining Q? For what values of x is $Q(x)$ defined?
   (b) What number does $Q(x)$ approach as (x,y) approaches $(1,-1)$?
   (c) Find an equation of the tangent at $(1,-1)$.

In each of the following, find the slope of the graph at the given point and an equation of the tangent at that point.

3. $y = 2x^2$ at (1,2)
4. $y = 3x^2$ at (2,12)
5. $y = 1 - x^2$ at (0,1)   How could the result have been predicted graphically?
6. $y = \frac{1}{2}x^2$ at (0,0)   How could the result have been predicted graphically?
7. $y = x^3$ at (1,1)
   Hint: $a^3 - b^3 = (a - b)(a^2 + ab + b^2)$
8. $y = x^4$ at (1,1)
9. $y = \sqrt{x}$ at (1,1)   Hint: $x - 1 = (\sqrt{x} - 1)(\sqrt{x} + 1)$
10. $y = x|x|$ at (0,0)   How could the result have been predicted graphically?

In each of the following, s is the position of a moving object at time t. Find the velocity at the given instant.

11. $s = t^2 + 1$ at 1
12. $s = t^2 - 1$ at 1
13. $s = t^3 - 4$ at 2
14. $s = \frac{1}{2}t^3$ at 2
15. $s = 1/t$ at 2
16. $s = 2/t$ at 1
17. $s = \sqrt{t}$ at 4
18. $s = 1 - \sqrt{t}$ at 4

19. If (x_0, y_0) is a point of the curve $y = 2x^2$, let $Q(x)$ be the slope of the line through (x_0, y_0) and a neighboring point (x,y) of the curve.
    (a) What is the formula defining Q? For what values of x is $Q(x)$ defined?
    (b) Find the slope of the curve at (x_0, y_0) by determining what $Q(x)$ approaches as (x,y) approaches (x_0, y_0).
    (c) Drop the subscripts in the answer to part (b) to obtain the slope at any point (x,y). Does the result check with Problem 3 when $x = 1$?

20. Repeat Problem 19 for the curve $y = 3x^2$. (Compare part (c) with Problem 4.)

In each of the following, find the slope of the graph at (x_0, y_0). Then drop the subscripts to obtain the slope at any point (x,y).

21. $y = 1 - x^2$ (Compare with Problem 5.)
22. $y = \frac{1}{2}x^2$ (Compare with Problem 6.)
23. $y = x^3$ (Compare with Problem 7.)
24. $y = x^4$ (Compare with Problem 8.)
25. $y = \sqrt{x}$ For what values of x is the slope at (x,y) defined?
26. $y = x|x|$ Hint: Consider the cases $x_0 = 0$, $x_0 > 0$, and $x_0 < 0$ separately.

In each of the following, a law of motion is given. Find the velocity at time t_0, then drop the subscript to obtain the velocity at time t. (The result should check with the corresponding special case in Problems 11 through 18.)

27. $s = t^2 + 1$
28. $s = t^2 - 1$
29. $s = t^3 - 4$
30. $s = \frac{1}{2}t^3$
31. $s = 1/t$
32. $s = 2/t$
33. $s = \sqrt{t}$
34. $s = 1 - \sqrt{t}$

In each of the following, use formulas derived in Problems 19 through 26 to determine where the graph is rising (positive slope), where it is falling (negative slope), and where it flattens out (zero slope). Sketch the graph using this information.

35. $y = 2x^2$
36. $y = 3x^2$
37. $y = 1 - x^2$
38. $y = \frac{1}{2}x^2$
39. $y = x^3$
40. $y = x^4$
41. $y = \sqrt{x}$
42. $y = x|x|$

In each of the following, use formulas derived in Problems 27 through 34 to determine when the object is moving in the positive direction, when it is moving in the negative direction, and when it comes to a momentary stop.

43. $s = t^2 + 1$
44. $s = t^2 - 1$
45. $s = t^3 - 4$
46. $s = \frac{1}{2}t^3$
47. $s = 1/t$ (Assume that $t > 0$.)

48. $s = 2/t$ (Assume that $t > 0$.)

49. $s = \sqrt{t}$ **50.** $s = 1 - \sqrt{t}$

51. An object dropped near the surface of the earth (and encountering no air resistance) falls a distance $s = \frac{1}{2}gt^2$ in t seconds (where g is a constant). Show that the velocity of the object t seconds after it is dropped is $v(t) = gt$.

52. A stone is thrown straight upward. After t seconds its height above the ground (in feet) is $s = 32t - 16t^2$.

(a) Show that the velocity of the stone at time t is $v(t) = 32 - 32t$. What is its initial velocity?

(b) When does the stone reach its highest point and how high does it rise?

(c) When does the stone return to the ground and what is its velocity when it hits the ground? What is its speed at that instant?

53. A ball is thrown straight upward from the top of a building. After t seconds its height above the ground (in feet) is $s = 96 + 16t - 16t^2$.

(a) How tall is the building?

(b) Show that the velocity of the ball at time t is $v(t) = 16 - 32t$. What is its initial velocity?

(c) When does the ball reach its highest point and how high (above the ground) does it rise?

(d) Assuming that the ball returns to the roof of the building, find when it lands. What is its velocity at that instant? its speed?

(e) The ball could have been thrown by a person whose arm was extended beyond the edge of the roof. In that case it would land on the ground. When would it land and with what velocity? with what speed?

54. Show that the slope of the graph of $y = x^3 - 3x$ at (x, y) is $m(x) = 3x^2 - 3$. Then confirm that there are turning points at $x = \pm 1$. (See Problem 7, Section 1.5.)

55. Show that the slope of the graph of $y = x + (1/x)$ at (x, y) is $m(x) = 1 - (1/x^2)$. Use the result to confirm that there are turning points at $x = \pm 1$. (See Problem 12, Section 1.5.)

56. Show that the slope of the graph of $y = x + |x|$ is undefined at the origin. Sketch the graph.

57. Explain why the tangent to the curve $y = x^{1/3}$ at the origin is vertical. Sketch the graph. *Hint:* The curve is symmetric about the origin. (Why?)

58. Explain why the graph of $y = 1 - x^{2/3}$ has a vertical tangent at $(0,1)$. Sketch the graph. *Hint:* The curve is symmetric about the y axis and $(0,1)$ is its highest point. (Why?)

59. Show that the tangent to the curve $y = x^{3/2}$ at the origin is horizontal.

60. Show that if $x_0 > 0$, the slope of the curve $y = x^{3/2}$ at (x_0, y_0) is $\frac{3}{2}\sqrt{x_0}$. *Hint:* To simplify the formula for slope of the line through (x_0, y_0) and (x, y), let $a = x_0^{1/2}$ and $b = x^{1/2}$. The formula becomes

$$Q(x) = (b^3 - a^3)/(b^2 - a^2)$$

61. From Problem 60 we may conclude that the slope of the graph of $y = x^{3/2}$ at (x, y) is $m(x) = \frac{3}{2}\sqrt{x}$ $(x > 0)$. Why does this formula also apply when $x = 0$? Sketch the graph of $y = x^{3/2}$ by noting where it rises, falls, and flattens out.

62. Show that if the law of motion of a moving object is linear $(s = at + b)$, the velocity is constant. What is the average velocity during any given interval in this case?

63. Show that if the law of motion of a moving object is quadratic $(s = at^2 + bt + c)$, the velocity is linear, $v(t) = 2at + b$.

64. In Problem 63 let t_1 and t_2 be endpoints of a time interval and let t be its midpoint. Show that the velocity at time t is the average of the velocities at t_1 and t_2.

65. In Problem 64 show that the velocity at the midpoint is also the average velocity during the interval. (Thus in a law of motion $s = at^2 + bt + c$ the "average velocity" as defined in the text can be computed by averaging the instantaneous velocities at the endpoints of the interval.)

66. Give an example of a law of motion for which neither of the statements in Problems 64 and 65 is true.

67. Suppose that the periodic motion of an object bobbing up and down at the end of a spring is represented by $s = \sin t$.

(a) Where is the object when $t = 0$, $\pi/2$, π, $3\pi/2$, 2π?

(b) Show that the average velocity during the time interval with endpoints t and $t + h$ $(h \neq 0)$ is

$$\sin t \left(\frac{\cos h - 1}{h}\right) + \cos t \left(\frac{\sin h}{h}\right)$$

Hint: Use the addition formula

$$\sin(u + v) = \sin u \cos v + \cos u \sin v$$

from trigonometry.

(c) It can be shown that as h approaches 0 the expressions $(\cos h - 1)/h$ and $(\sin h)/h$ approach 0 and 1, respectively. Use that fact to find the velocity at time t.

(d) How fast, and in what direction, is the object going when $t = 0$, $\pi/2$, π, $3\pi/2$, 2π?

2.2
THE DERIVATIVE OF A REAL FUNCTION; LIMIT NOTATION

In the last section we discussed the problem of tangents and the problem of velocity. Let us summarize those discussions and make some general observations.

1. *Slope of a Curve.* If (x_0, y_0) is a point of the graph of the real function $y = f(x)$, we find the slope at (x_0, y_0) as follows.

   (a) Let (x, y) be a point of the graph "near" (x_0, y_0). The slope of the line containing (x_0, y_0) and (x, y) is given by the *difference quotient*

$$Q(x) = \frac{y - y_0}{x - x_0} = \frac{f(x) - f(x_0)}{x - x_0} \qquad (x \neq x_0)$$

   (b) Bring (x, y) closer to (x_0, y_0), thereby making x approach x_0. If this causes $Q(x)$ to converge to a definite value m, we write

$$m = \lim_{x \to x_0} Q(x) = \lim_{x \to x_0} \frac{f(x) - f(x_0)}{x - x_0}$$

   and call m the *slope of the curve* at (x_0, y_0). The notation

$$\lim_{x \to x_0} Q(x)$$

   may be read "limit of $Q(x)$ as x approaches x_0." We are assuming, as we did in the last section, that intuition is enough to support our present aims in connection with this concept. For a more substantial discussion, see Sections 2.4 and 2.5.

   Since the value of the above limit depends on x_0, the slope is a new function, $m(x_0) = $ slope at (x_0, y_0), or simply $m(x) = $ slope at (x, y) if we drop the subscript.

2. *Velocity of a Moving Object.* Let $s = p(t)$ be the position at time t of an object moving in a straight line (where p is a real function). We find the velocity at time t_0 as follows.

   (a) Let t be a slightly different time. The average velocity during the time interval with endpoints t_0 and t is

$$Q(t) = \frac{\text{change in position}}{\text{change in time}} = \frac{p(t) - p(t_0)}{t - t_0} \qquad (t \neq t_0)$$

   (b) Let t approach t_0. If this causes $Q(t)$ to converge to a definite value v, we write

$$v = \lim_{t \to t_0} Q(t) = \lim_{t \to t_0} \frac{p(t) - p(t_0)}{t - t_0}$$

   and call v the *velocity* at time t_0. Since the answer depends on t_0, the velocity is a new function, $v(t_0) = $ velocity at time t_0, or simply $v(t) = $ velocity at time t if we drop the subscript.

It is a principle of human thought and language that when we encounter distinct problems with the same answer, we ought to ignore the inessential details of the problems and invent a word for the idea they have in common. The forms of the answers in (1) and (2) are identical. So we forget about geometry (slope of a curve) and physics (velocity of a moving object) and concentrate on the mathematical substance of the process illustrated. In each case a new function is derived from a given one by evaluating the limit of a difference quotient. The new function is called the **derivative** of the given function, as in the following definition.

Let f be a real function with domain D and suppose that x_0 is an interior point of D. The **difference quotient** associated with f at x_0 is

$$Q(x) = \frac{f(x) - f(x_0)}{x - x_0} \qquad (x \neq x_0 \text{ and } x \in D)$$

The **derivative of f at x_0** is the number

$$f'(x_0) = \lim_{x \to x_0} Q(x)$$

(provided this limit exists, in which case we call f **differentiable** at x_0).

If S is a subset of the domain and f is differentiable at each point of S, we say that f is *differentiable in S*.

The function f' whose domain is the set of points at which f is differentiable and whose value at x is $f'(x)$ is called the *derivative* of f.

Remark   The above definition requires x_0 to be an "interior point" of D, that is, a point with the property that neighboring points are also in D. More precisely, there is an open interval containing x_0 (a *neighborhood* of x_0) that lies in D. This guarantees that $f(x)$ is defined for all x near x_0, which is essential for evaluation of the limit of $Q(x)$ as x approaches x_0. The requirement can be relaxed in some circumstances, however. If x_0 is an endpoint of an interval in D, the limit is *one-sided;* sometimes we can still make sense of it. (See Example 3.)

□ **Example 1**   Find the derivative of $f(x) = x^3$.

Solution: The difference quotient associated with f at x_0 is

$$Q(x) = \frac{f(x) - f(x_0)}{x - x_0} = \frac{x^3 - x_0^3}{x - x_0} = \frac{(x - x_0)(x^2 + x_0 x + x_0^2)}{x - x_0}$$

$$= x^2 + x_0 x + x_0^2 \qquad (x \neq x_0)$$

Since

$$\lim_{x \to x_0} Q(x) = \lim_{x \to x_0} (x^2 + x_0 x + x_0^2) = 3x_0^2$$

the derivative at x_0 is $f'(x_0) = 3x_0^2$. Having arrived at this, we may observe that in the beginning x_0 was any real number. The subscript is superfluous; we might as well say that f' is defined by $f'(x) = 3x^2$.   □

The observation at the end of Example 1 sometimes confuses people. What you should note is that the letter used is immaterial. We may write $f'(t) = 3t^2$ or $f'(\alpha) = 3\alpha^2$ or $f'(\square) = 3\square^2$. The *form* is what counts in the definition of a function, not the symbol used to identify the independent variable. The reason for the subscript in the first place is that in the process of examining the difference quotient we needed a label for *its* independent variable. We chose x, which means that we needed a different label for the point at which the derivative is to be evaluated.

To clarify this matter further, suppose we feel like calling the original point x. Then we need another letter for the independent variable in Q; suppose we adopt z. The difference quotient associated with f at x is then

$$Q(z) = \frac{f(z) - f(x)}{z - x} = \frac{z^3 - x^3}{z - x} = \frac{(z - x)(z^2 + xz + x^2)}{z - x}$$

$$= z^2 + xz + x^2 \qquad (z \neq x)$$

Since

$$\lim_{z \to x} Q(z) = \lim_{z \to x} (z^2 + xz + x^2) = 3x^2$$

the derivative at x is $f'(x) = 3x^2$.

□ **Example 2**   Find the derivative of $f(x) = \sqrt{x}$.

Solution: The difference quotient associated with f at x (where $x > 0$) is

$$Q(z) = \frac{f(z) - f(x)}{z - x} = \frac{\sqrt{z} - \sqrt{x}}{z - x} = \frac{\sqrt{z} - \sqrt{x}}{(\sqrt{z} - \sqrt{x})(\sqrt{z} + \sqrt{x})}$$

$$= \frac{1}{\sqrt{z} + \sqrt{x}} \qquad (z \neq x \text{ and } z \geq 0)$$

The derivative at x is

$$f'(x) = \lim_{z \to x} Q(z) = \lim_{z \to x} \frac{1}{\sqrt{z} + \sqrt{x}} = \frac{1}{2\sqrt{x}}$$

Since x was any positive number, we conclude that f is differentiable in the interval $(0, \infty)$.   □

Note in Example 2 that the domain of f is $[0, \infty)$. We did not consider the derivative at $x = 0$ because 0 is not an interior point of the domain. Suppose we stretch the definition, however, and try it. The difference quotient associated with f at 0 is

$$Q(z) = \frac{f(z) - f(0)}{z - 0} = \frac{\sqrt{z}}{z} = \frac{1}{\sqrt{z}} \qquad (z > 0)$$

Since $Q(z)$ increases without bound as $z \to 0$,

$$\lim_{z \to 0} Q(z) \text{ does not exist}$$

and we conclude that f is not differentiable at 0. (See Example 4, Section 2.1, where we observed that the tangent at the origin is vertical and the slope is undefined.)

Despite the failure of this attempt to broaden the definition, there are times when it makes sense. See Example 3.

□ **Example 3**   The domain of $f(x) = x^{3/2}$ is $[0, \infty)$. Even though 0 is not an interior point, let's include it in our analysis and see what happens. The difference quotient associated with f at x (where $x \geq 0$) is

$$Q(z) = \frac{f(z) - f(x)}{z - x} = \frac{z^{3/2} - x^{3/2}}{z - x} \qquad (z \neq x \text{ and } z \geq 0)$$

To simplify this, let $a = x^{1/2}$ and $b = z^{1/2}$. Then

$$Q(z) = \frac{b^3 - a^3}{b^2 - a^2}$$

$$= \frac{(b - a)(b^2 + ba + a^2)}{(b - a)(b + a)}$$

$$= \frac{b^2 + ba + a^2}{b + a} = \frac{z + \sqrt{zx} + x}{\sqrt{z} + \sqrt{x}}$$

Now we have to distinguish cases. If $x > 0$, we can write

$$f'(x) = \lim_{z \to x} Q(z) = \lim_{z \to x} \frac{z + \sqrt{zx} + x}{\sqrt{z} + \sqrt{x}}$$

$$= \frac{x + \sqrt{x^2} + x}{\sqrt{x} + \sqrt{x}}$$

$$= \frac{3x}{2\sqrt{x}} \qquad (\sqrt{x^2} = x \text{ because } x > 0)$$

$$= \frac{3}{2}\sqrt{x}$$

This argument breaks down if $x = 0$ (because $3x/(2\sqrt{x})$ is meaningless). At $x = 0$, however, the difference quotient reduces to

$$Q(z) = \frac{z^{3/2} - 0}{z - 0} = \sqrt{z} \qquad (z > 0)$$

and hence $\qquad\qquad f'(0) = \lim_{z \to 0} Q(z) = 0$

According to our definition of derivative, the endpoint $x = 0$ should not have been included in this analysis. Let's agree to examine endpoints, however, and to accept the results when the limit of the difference quotient exists. Then we may summarize Example 3 by saying that $f(x) = x^{3/2}$ is differentiable in $[0, \infty)$ and

$$f'(x) = \begin{cases} \frac{3}{2}\sqrt{x} \text{ if } x > 0 \\ 0 \text{ if } x = 0 \end{cases}$$

As you can see, it is unnecessary to separate the cases $x > 0$ and $x = 0$ in the end. The formula $f'(x) = \frac{3}{2}\sqrt{x}$ (if $x \geq 0$) covers both possibilities. The graph of $y = x^{3/2}$ (unlike that of $y = x^{1/2}$, which has a vertical tangent at the origin) should be drawn with slope 0 at the origin. (See Figure 1.) □

Figure 1

Horizontal and vertical tangents at the origin

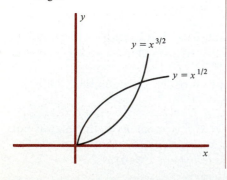

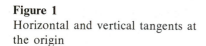

The effect of our agreement in Example 3 is that we are extending the definition of derivative to include endpoints. The limit of the difference quotient in such a case is "one-sided," in the sense that

$$Q(z) = \frac{f(z) - f(x)}{z - x} \qquad \text{(where } x \text{ is an endpoint)}$$

is examined only for values of z on one side of x. In Example 3, for instance, where we wrote

$$f'(0) = \lim_{z \to 0} \frac{z^{3/2} - 0}{z - 0} = \lim_{z \to 0} \sqrt{z} = 0$$

the limit is evaluated by making z approach 0 "through positive values." (Otherwise $\sqrt{z}$ would be meaningless.)

Sometimes this sort of thing complicates the theory. (We will point it out when it does.) But on the whole it is helpful to allow it. Would it not offend your intuition to say that $f(x) = x^{3/2}$ is differentiable only for $x > 0$? That would prevent the natural extension of the formula $f'(x) = \frac{3}{2}\sqrt{x}$ to the point $x = 0$, when it is plain from the graph that there is a horizontal tangent at the origin.

We have defined the difference quotient associated with f at x to be

$$Q(z) = \frac{f(z) - f(x)}{z - x} \qquad (z \neq x)$$

Figure 2

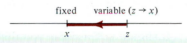

fixed   variable $(z \to x)$

x z

Figure 3

fixed   variable $(h \to 0)$

x $x + h$

(See Figure 2.) Sometimes it is more convenient to denote the point z by the label $x + h$ ($h \neq 0$), the idea being that when h is small, $x + h$ is near x. (See Figure 3.) Then the difference quotient is a function of h, namely

$$q(h) = \frac{f(x + h) - f(x)}{(x + h) - x} = \frac{f(x + h) - f(x)}{h} \qquad (h \neq 0)$$

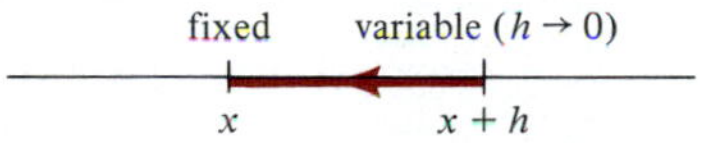

and the derivative of f at x is

$$f'(x) = \lim_{h \to 0} q(h) = \lim_{h \to 0} \frac{f(x + h) - f(x)}{h}$$

□ **Example 4**   Discuss the problem of finding the derivative of $f(x) = \sin x$.

Solution: The difference quotient associated with f at x is

$$Q(z) = \frac{\sin z - \sin x}{z - x} \qquad (z \neq x)$$

which does not simplify in any obvious way. If we substitute $x + h$ for z, however, the difference quotient becomes

$$q(h) = \frac{\sin(x + h) - \sin x}{h} \qquad (h \neq 0)$$

and a possible simplification suggests itself. Using the addition formula

$$\sin(u + v) = \sin u \cos v + \cos u \sin v$$

we have

$$q(h) = \frac{\sin x \cos h + \cos x \sin h - \sin x}{h} = \frac{\sin x (\cos h - 1) + \cos x \sin h}{h}$$

$$= \sin x \left(\frac{\cos h - 1}{h}\right) + \cos x \left(\frac{\sin h}{h}\right)$$

It is not obvious how the expressions $(\cos h - 1)/h$ and $(\sin h)/h$ behave when h approaches zero, but for the sake of argument suppose that their limits are a and b, respectively. Then

$$f'(x) = \lim_{h \to 0} q(h) = (\sin x)(a) + (\cos x)(b)$$

and the problem is solved. If you have an electronic calculator, you can probably guess what a and b are by computing the above expressions for values of h near 0. For the present, however, we will leave the problem at this stage. It is a good example of the kind of trigonometry needed in calculus, and may whet your appetite for the review given in the next section.

□

According to the terminology we have now adopted, the two equivalent problems of Section 2.1 are solved by finding a derivative:

1. Given a curve with equation $y = f(x)$, the slope at (x, y) is the derivative $m(x) = f'(x)$.
2. Given a position function $s = p(t)$, the velocity at time t is the derivative $v(t) = p'(t)$.

These are *two* interpretations of the derivative. There are many others, as we shall see.

Calculus is a subject with a rich history. It should not surprise you that there are many alternate symbols for the derivative; you should learn to live with the ones in common use. Chief among these, besides the notation $f'(x)$ already introduced for the derivative of $y = f(x)$, are

- y' (read "y prime")

- $\dfrac{dy}{dx}$ (read "dee y, dee x"—not "dy over dx")

- $D_x y$ (read "derivative of y as a function of x")

The notation dy/dx is due to Leibniz, who regarded dy and dx as separate entities and dy/dx as a true fraction. Eventually we'll do the same; the symbol is very useful. For the present, however, you should not think of dy/dx as a fraction; it is just another notation for the derivative.

Depending on the context, the above symbols are of varying precision. They must be used with due attention to the message intended.

□ **Example 5**   If $y = f(x) = x^4$, then

$$f'(1) = \lim_{x \to 1} \frac{f(x) - f(1)}{x - 1} = \lim_{x \to 1} \frac{x^4 - 1}{x - 1} = \lim_{x \to 1} \frac{(x - 1)(x + 1)(x^2 + 1)}{x - 1}$$
$$= \lim_{x \to 1} (x + 1)(x^2 + 1) = 4$$

The notation $f'(1) = 4$ describes the situation precisely: We have evaluated the function f' at the point 1 to obtain the number 4. The other symbols mentioned above are not as useful in this context. For example, if we were to write $y' = 4$ or $D_x y = 4$, the reader might infer that the derivative is the constant function $f'(x) = 4$, whereas the correct formula is $f'(x) = 4x^3$. (Confirm!) Some writers introduce notation like

$$y'|_{x=1} = 4 \qquad \text{or} \qquad D_x y|_{x=1} = 4$$

to represent the situation, but the functional notation $f'(1) = 4$ is simpler.

□

□ **Example 6**   The derivative of $y = f(x) = (2x - 1)^2$ is

$$f'(x) = 4(2x - 1)$$

(as you can check). The same message is conveyed by

$$\frac{dy}{dx} = 4(2x - 1) \qquad \text{or} \qquad D_x y = 4(2x - 1)$$

but the notation $y' = 4(2x - 1)$ is risky in some situations (because the symbol y' makes no reference to the independent variable). Suppose, for example, that we substitute $u = 2x - 1$ in $y = (2x - 1)^2$ to obtain $y = u^2$. The derivative of $y = u^2$ *as a function of u* is $dy/du = 2u$ or $D_u y = 2u$ (Example 2, Section 2.1). If we were to write $y' = 2u$, the reader would have a right to complain. For there is a distinction between $y' = 4(2x - 1)$ (the derivative of y as a function of x) and $y' = 2u$ (the derivative of y as a function of u). It is for this reason (among others) that mathematicians have developed the following terminology (which sometimes strikes the beginner as jargon).

The symbols dy/dx and $D_x y$ are translated *derivative of y with respect to x*. To *differentiate y with respect to x* means to find the derivative of $y = f(x)$, and the process of finding it is called *differentiation*.

It is also common practice to regard the symbols d/dx and D_x as *operators* (the operation being differentiation). Whatever is written immediately following them is the function being differentiated. For example,

$$\frac{d}{dx}(x^2) = 2x \quad \text{(Example 2, Section 2.1)}$$

and
$$D_x(x^{3/2}) = \frac{3}{2}\sqrt{x} \quad \text{(Example 3)}$$

One advantage of this notation is that the expression in parentheses need not be identified by some other symbol, say y or $f(x)$. That saves writing. Do not, however, mix the notation! Sometimes students write

$$\frac{dy}{dx}(x^2) = 2x$$

which is confused. We should either let $y = x^2$ and say that $dy/dx = 2x$, or forget about y and simply write

$$\frac{d}{dx}(x^2) = 2x$$

□

Problem Set 2.2

In each of the following, find the indicated derivative by set-ting up the difference quotient associated with the function at the specified point and then evaluating the appropriate limit.

1. $f'(0)$ if $f(x) = x^2 - 4$ How could the result have been predicted from the graph?

2. $f'(1)$ if $f(x) = 9 - x^2$

3. $g'(2)$ if $g(x) = x^2 - 2x$

4. $\phi'(-1)$ if $\phi(t) = 3t^2 - t$

5. $H'(2)$ if $H(x) = 8 - x^3$

6. $f'(0)$ if $f(x) = x^3 + 5$

7. $f'(2)$ if $f(u) = u^5$

8. $g'(1)$ if $g(v) = v^6$

9. $p'(0)$ if $p(t) = t\sqrt{t + 2}$

10. $F'(1)$ if $F(x) = (x - 1)\sqrt{x}$ **11.** $p'(1)$ if $p(t) = \dfrac{1}{t - 2}$

12. $g'(0)$ if $g(x) = \dfrac{1}{x^2 + 1}$ In view of the result, how would you draw the graph through $(0,1)$?

13. $f'(0)$ if $f(x) = x^{5/2}$ In view of the result, how would you draw the graph near $(0,0)$?

14. $f'(0)$ if $f(x) = x^{2/3}$ In view of the result, how would you draw the graph through $(0,0)$?

15. $F'(0)$ if $F(x) = \sqrt[3]{x}$ In view of the result, how would you draw the graph through $(0,0)$?

16. $G'(1)$ if $G(x) = 1/\sqrt{x}$

17. $f'(1)$ if $f(t) = \sqrt{1 - t}$ How could the result have been predicted from the graph?

18. $g'(0)$ if $g(x) = x + |x|$ How could the result have been predicted from the graph?

19. The difference quotient associated with $f(x) = \sqrt{25 - x^2}$ at 3 is
$$Q(x) = \frac{\sqrt{25 - x^2} - 4}{x - 3}$$
and its limit as $x \to 3$ is $f'(3)$. Use a calculator to evaluate $Q(3.1)$, $Q(3.01)$, and $Q(3.001)$, thus obtaining successive approximations to $f'(3)$.

20. Repeat Problem 19 by evaluating $Q(2.9)$, $Q(2.99)$, and $Q(2.999)$.

21. Let $Q(x)$ be the difference quotient associated with $f(x) = (x^2 + 1)^4$ at 1. Approximate $f'(1)$ by evaluating $Q(0.9)$, $Q(0.99)$, and $Q(0.999)$.

22. Repeat Problem 21 by evaluating $Q(1.1)$, $Q(1.01)$, and $Q(1.001)$.

23. Let $Q(x)$ be the difference quotient associated with $f(x) = \sin x$ at 0. Approximate $f'(0)$ by evaluating $Q(\pm 0.1)$, $Q(\pm 0.01)$, and $Q(\pm 0.001)$. *Note:* The angle se-lector on your calculator must be set on radians (for rea-sons to be explained in the next section).

24. Repeat Problem 23 in the case of $f(x) = \cos x$.

25. Repeat Problem 23 in the case of $f(x) = 2^x$. *Note:* You may not know how 2^x is defined for irrational values of x. But a calculator will supply approximations (if it has a y^x key).

26. Repeat Problem 23 in the case of $f(x) = e^x$. *Note:* e is the base of natural logarithms mentioned in Section 1.1. Use the e^x key on your calculator.

In each of the following, find the indicated derivative as the limit of an appropriate difference quotient. (Several forms are mentioned in the text. Use the one that seems most conven-ient.)

27. $f'(x)$ if $f(x) = x^2 + x$

28. $g'(x)$ if $g(x) = 5x^2 - x$

29. dy/dx if $y = 2x^3 - 7$

30. $D_x y$ if $y = 4 - x^4$

31. $D_t s$ if $s = t^4 - 5t$

32. ds/dt if $s = t^3 - t + 1$

33. y' if $y = x^4 + x^2 + 2$

34. y' if $y = x^4 - 3x + 1$

35. $D_t u$ if $u = 1/t^2$

36. dv/dt if $v = -6/t^2$

37. $f'(x)$ if $f(x) = \dfrac{x}{x + 5}$

38. $F'(x)$ if $F(x) = \dfrac{x - 3}{x}$

39. dv/dt if $v = t^{5/2}$

40. $g'(x)$ if $g(x) = \sqrt[3]{x}$

41. $f'(x)$ if $f(x) = x^{2/3}$

42. $p'(t)$ if $p(t) = \sqrt{1 - t}$

43. Use the definition to show that the derivative of a constant function is the zero function. How could this have been predicted from a graph?

44. Use the definition to show that the derivative of a linear function, $f(x) = mx + b$, is $f'(x) = m$. How could this have been predicted from the graph?

45. Derive the formula $D_x(ax^2 + bx + c) = 2ax + b$.

46. The graph of $y = |x|$ is shown in Figure 5, Section 2.1. What is its slope when $x > 0$? when $x < 0$? Use your an-swers to explain the formula
$$D_x|x| = \frac{x}{|x|}$$
What is the domain of this derivative?

47. A painless way to differentiate $f(x) = \sqrt{1 - x^2}$ is to in-terpret the derivative as slope.
   (a) What is the graph of f?
   (b) Reasoning from the graph, what would you expect $f'(x)$ to be at $x = 0$? at $x = \pm 1$?
   (c) If (x,y) is any interior point of the graph, explain why the slope of the tangent at (x,y) is $-x/y$. Why does it follow that
$$f'(x) = \frac{-x}{\sqrt{1 - x^2}} \ ?$$
What is the domain of f'?

2.3
TRIGONOMETRIC FUNCTIONS

Figure 1

Motion of an object attached to a spring

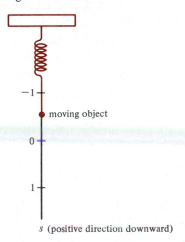

−1

moving object

0

1

s (positive direction downward)

A student beginning a course in calculus is often expected to know something about trigonometry. It is not unusual, however, for this knowledge to be confined largely to "right-triangle trigonometry," in which sines, cosines, and tangents are defined in terms of acute angles of a right triangle, angles are measured in degrees, and the problems are mostly of the kind that surveyors or navigators are interested in solving. There is nothing wrong with this. But in calculus the uses of trigonometry are somewhat broader and deeper. This section provides an introduction to "analytic trigonometry," in which the emphasis is more numerical than geometrical, and the definitions are more general.

An example of what we mean by analytic trigonometry is the motion of an object bobbing up and down at the end of a spring. (See Figure 1.) If this motion is represented by the position function $s = p(t) = \sin t$ (where t is time), observe that it makes no sense to give values of t in degrees. That raises the question of what $\sin t$ means when t is not an angle but a *number*. Moreover, if we are interested in the velocity $ds/dt = p'(t)$, we must examine the difference quotient

$$Q(z) = \frac{p(z) - p(t)}{z - t} = \frac{\sin z - \sin t}{z - t}$$

or (substituting $t + h$ for z to obtain a more manageable expression)

$$q(h) = \frac{\sin (t + h) - \sin t}{h}$$

This will not go anywhere unless we know the addition formula

$$\sin (u + v) = \sin u \cos v + \cos u \sin v$$

and even then we can get no farther than

$$q(h) = \frac{\sin t \cos h + \cos t \sin h - \sin t}{h}$$

$$= \frac{\sin t (\cos h - 1) + \cos t \sin h}{h}$$

$$= \sin t \left(\frac{\cos h - 1}{h}\right) + \cos t \left(\frac{\sin h}{h}\right)$$

To finish the problem, we must investigate

$$\lim_{h \to 0} \frac{\cos h - 1}{h} \quad \text{and} \quad \lim_{h \to 0} \frac{\sin h}{h}$$

(whose values are not obvious).

Let us therefore start at the beginning, defining the trigonometric functions in such a way that they can be discussed in the context of calculus. In Figure 2 we have drawn the *unit circle* in the xy plane, that is, the circle of radius 1 and center at the origin. Through the point $A = (1,0)$ we have drawn a vertical line labeled as a coordinate axis with origin at A and posi-

Figure 2
Wrapping real numbers onto the
unit circle

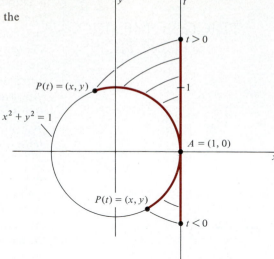

tive direction upward. We will think of this as a number line on which we
locate real numbers t.

Let $t \in \mathfrak{R}$ be given, that is, suppose we are presented with a point of the
t axis. Imagining this axis as a line of thread that can be wrapped around the
circle, let $P(t) = (x,y)$ be the point where t ends up when this wrapping
takes place. In other words, the line segment from A to t becomes the arc
from A to P, positive or negative (that is, counterclockwise or clockwise)
depending on whether $t > 0$ or $t < 0$. (Of course when $t = 0$, $P = A$.)

The **trigonometric functions** sine, cosine, tangent, cotangent, secant,
cosecant (abbreviated sin, cos, tan, cot, sec, csc) are defined in terms of t by
writing

$$\sin t = y \qquad\qquad \cot t = \frac{x}{y} \qquad (y \neq 0)$$

$$\cos t = x \qquad\qquad \sec t = \frac{1}{x} \qquad (x \neq 0)$$

$$\tan t = \frac{y}{x} \qquad (x \neq 0) \qquad \csc t = \frac{1}{y} \qquad (y \neq 0)$$

Observe that since sin, cos, tan, . . . are names of functions, proper notation
would be sin (t), cos (t), tan (t), . . . , just as we write $f(t)$ when the function
is f. The parentheses are commonly omitted, however.

Although the input t in each of these functions is a number (not an
angle), we can associate an angle θ with t by the scheme illustrated in Figure
3. The **initial side** of θ is the positive x axis; the **terminal side** is the ray from
$(0,0)$ through (x,y). In plane geometry the angle is the union of these rays,
but in trigonometry the term "angle" is understood to mean "rotation" (of

Figure 3
Angle associated with a real number

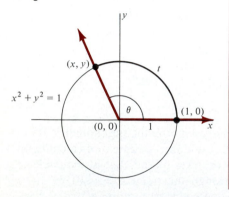

the initial side into the terminal side, positive or negative depending on whether the rotation is counterclockwise or clockwise, and possibly involving more than one revolution depending on the size of t).

How should an angle (or rotation) be measured? The Babylonian system (with which we are all familiar) is to divide one revolution into 360 equal parts and call each part a **degree**. In calculus, however, it turns out to be more natural to use the number t, that is, to agree that t *is the measure of* θ. This enables us to regard θ and t as virtually interchangeable.

Remark   The unit of measure in this system is called a **radian**, but actually it is unnecessary to give it a name. When we say that "θ is an angle of measure 2" (and are asked "2 what?"), the simplest answer is "2 units," where a *unit* is the radius of the unit circle, namely 1. There is no more need to invent a name for it than for the unit on the coordinate line. When we say that the distance between 3 and 5 on this line is 2, we do not ask "2 what?" The distance is just a real number.

We may convert from one system of angle measurement to the other by observing that a straight angle (180°) cuts off an arc of length π on the unit circle (because the circumference is 2π). Hence

$$\boxed{\pi \text{ radians} = 180 \text{ degrees}}$$

This means that

$$1 \text{ radian} = \frac{180}{\pi} \text{ (or approximately 57.3) degrees}$$

$$1 \text{ degree} = \frac{\pi}{180} \text{ (or approximately 0.0175) radians}$$

To see the connection between our definition of the trigonometric functions (which may be new to you) and the usual formulas of right-triangle trigonometry, suppose that $0 < t < \pi/2$. Then t is the radian measure of an acute angle θ (Figure 4) and we define

Figure 4
Trigonometric functions of an acute angle

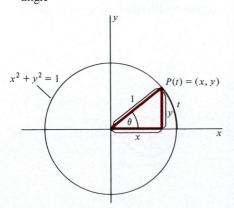

$$\sin\theta = \sin t = \frac{y}{1} = \frac{\text{opp}}{\text{hyp}} \qquad \cot\theta = \cot t = \frac{x}{y} = \frac{\text{adj}}{\text{opp}}$$

$$\cos\theta = \cos t = \frac{x}{1} = \frac{\text{adj}}{\text{hyp}} \qquad \sec\theta = \sec t = \frac{1}{x} = \frac{\text{hyp}}{\text{adj}}$$

$$\tan\theta = \tan t = \frac{y}{x} = \frac{\text{opp}}{\text{adj}} \qquad \csc\theta = \csc t = \frac{1}{y} = \frac{\text{hyp}}{\text{opp}}$$

(where we have used the abbreviations *opp, adj,* and *hyp* for *side opposite* θ, *side adjacent to* θ, and *hypotenuse,* respectively). More generally, if t is any real number, we may take θ to be the "angle" (rotation) whose radian measure is t. If f is any one of the six trigonometric functions, we define $f(\theta) = f(t)$.

A symbol like sin 2 therefore has two meanings. The simplest (from the standpoint of calculus) is that it is the output of the sine function corresponding to the input 2. (No need to mention angles.) When it is convenient, however, we may interpret 2 as the radian measure of an angle θ and regard sin 2 as the sine of this angle as understood in high school trigonometry.

Note that degrees are virtually banished from this scheme of things! Since most people are more comfortable with them, however, it is common to allow notation like sin 30°. The understanding is that if this is to be incorporated into our definition of sine as a real function, the numerical input is not 30 but $\pi/6$. It is important to remember these remarks when we use the trigonometric functions in calculus. The inputs are *numbers*. If angles are to be associated with these numbers, they must be measured in radians.

Except for special values of t (like 0, $\pi/6$, $\pi/4$, $\pi/3$, $\pi/2$, and their integral multiples), the trigonometric functions of t are hard to compute. We assume that you have encountered these special values before (perhaps in terms of degree measure) and that you have learned, or can quickly figure out, the following table. (See the problem set for details.)

t	$\sin t$	$\cos t$	$\tan t$
0	0	1	0
$\pi/6$	$1/2$	$\sqrt{3}/2$	$1/\sqrt{3}$
$\pi/4$	$1/\sqrt{2}$	$1/\sqrt{2}$	1
$\pi/3$	$\sqrt{3}/2$	$1/2$	$\sqrt{3}$
$\pi/2$	1	0	—

Other values of the trigonometric functions can be found from tables, or (more easily) from an electronic calculator. How these values are computed is another matter; we will discuss it in Chapter 13.

A great deal of information about the trigonometric functions may be remembered by learning what their graphs look like. By plotting intelligently selected points, you can verify Figures 5–10 (the main features of

Figure 5
Graph of $u = \sin t$

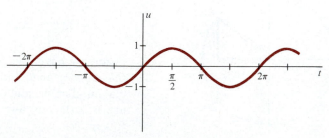

Figure 6
Graph of $u = \cos t$

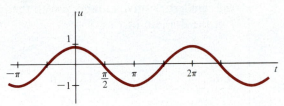

Figure 7
Graph of $u = \tan t$

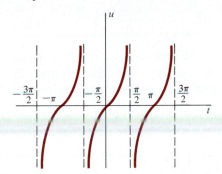

Figure 8
Graph of $u = \cot t$

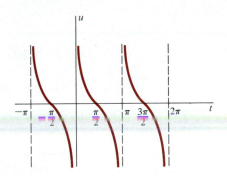

Figure 9
Graph of $u = \sec t$
(cosine curve is dashed)

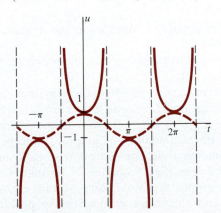

Figure 10
Graph of $u = \csc t$
(sine curve is dashed)

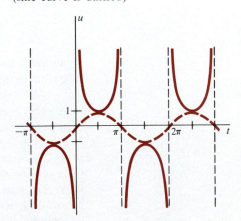

which should be memorized). The following table of domains, ranges, and periods is also worth checking out. (The *period* of a function f, if it exists, is the smallest positive number p with the property that $f(t + p) = f(t)$ for all t in the domain of f.)

Function	Domain	Range	Period
sine	$\Re$	$[-1,1]$	2π
cosine	$\Re$	$[-1,1]$	2π
tangent	$\left\{t \in \Re : t \neq \pm\dfrac{\pi}{2}, \pm\dfrac{3\pi}{2}, \pm\dfrac{5\pi}{2}, \dots\right\}$	$\Re$	π
cotangent	$\{t \in \Re : t \neq 0, \pm\pi, \pm 2\pi, \dots\}$	$\Re$	π
secant	$\left\{t \in \Re : t \neq \pm\dfrac{\pi}{2}, \pm\dfrac{3\pi}{2}, \pm\dfrac{5\pi}{2}, \dots\right\}$	$(-\infty, -1] \cup [1, \infty)$	2π
cosecant	$\{t \in \Re : t \neq 0, \pm\pi, \pm 2\pi, \dots\}$	$(-\infty, -1] \cup [1, \infty)$	2π

While many of the statements in this table are apparent from the definitions, some are not. How do we know, for example, that the range of tangent is $\Re$? This is equivalent to asking whether, given any real number u, there is a real number t such that $\tan t = u$. In other words $\tan t$ "takes on all real values." Of course the graph indicates that this is the case (every horizontal line intersects it). An analytic argument is needed, however, to back up the evidence of the graph (which after all is sketched merely on the basis of a few "typical" points).

Similarly, how do we know that the period of tangent is π? It is easy to see that if f is any trigonometric function, then

$$f(t + 2\pi) = f(t) \qquad \text{for all } t \text{ in the domain}$$

Hence each trigonometric function is periodic, with a period no larger than 2π. To show that π is the period of tangent, we have to prove that

$$\tan(t + \pi) = \tan t \qquad \text{for all } t \text{ in the domain}$$

and also that no smaller positive number has this property. We leave such questions for the problem set.

Another important property of the trigonometric functions is that each is either *even* or *odd* (unlike most functions, which are neither). A real function f is said to be *even* if $f(-t) = f(t)$ for all t in the domain, and *odd* if $f(-t) = -f(t)$ for all t in the domain. The graphical interpretation is that an even function is symmetric about the vertical axis, while an odd function is symmetric about the origin. (See Problems 25 through 42, Section 1.5.) You can see from their graphs that sine, cosecant, tangent, and cotangent are odd, while cosine and secant are even. (See the problem set for an analytic argument.)

Even–Odd Identities

$$\sin(-t) = -\sin t \qquad \cot(-t) = -\cot t$$
$$\cos(-t) = \cos t \qquad \sec(-t) = \sec t$$
$$\tan(-t) = -\tan t \qquad \csc(-t) = -\csc t$$

The following *fundamental identities* occur so often in applications that you should memorize them. Their proofs are based directly on the definitions (together with the equation of the unit circle, $x^2 + y^2 = 1$).

Fundamental Identities

$$\sin t \csc t = 1 \qquad \sin^2 t + \cos^2 t = 1 \qquad \tan t = \frac{\sin t}{\cos t}$$

$$\cos t \sec t = 1 \qquad \sec^2 t - \tan^2 t = 1$$

$$\tan t \cot t = 1 \qquad \csc^2 t - \cot^2 t = 1 \qquad \cot t = \frac{\cos t}{\sin t}$$

Remark   The notation $\sin^2 t$ is shorthand for $(\sin t)^2$, the idea being to avoid parentheses. It also arises from the symbol fg for the product of two functions:

$$(fg)(t) = f(t)g(t) \qquad \text{(Problem 40, Section 1.5)}$$

If f and g are the same function, it is natural to write f^2 for fg; then $f^2(t) = [f(t)]^2$. Unfortunately mathematicians are inconsistent about notation of this kind, since we do not write $\sin^{-1} t$ in place of $(\sin t)^{-1}$. The reason is that $\sin^{-1}$ is the *inverse sine* function (to be defined in Chapter 9).

Most of the remaining trigonometric identities that are important in calculus may be derived from one crucial result, the formula

$$\cos(u - v) = \cos u \cos v + \sin u \sin v$$

We list them here and give suggestions for their proof in the problem set.

Cofunction Identities

$$\sin\left(\frac{\pi}{2} - t\right) = \cos t \quad \text{and} \quad \cos\left(\frac{\pi}{2} - t\right) = \sin t$$

$$\tan\left(\frac{\pi}{2} - t\right) = \cot t \quad \text{and} \quad \cot\left(\frac{\pi}{2} - t\right) = \tan t$$

$$\sec\left(\frac{\pi}{2} - t\right) = \csc t \quad \text{and} \quad \csc\left(\frac{\pi}{2} - t\right) = \sec t$$

Addition Formulas

$$\sin(u + v) = \sin u \cos v + \cos u \sin v \qquad \sin(u - v) = \sin u \cos v - \cos u \sin v$$
$$\cos(u + v) = \cos u \cos v - \sin u \sin v \qquad \cos(u - v) = \cos u \cos v + \sin u \sin v$$

$$\tan(u + v) = \frac{\tan u + \tan v}{1 - \tan u \tan v} \qquad \tan(u - v) = \frac{\tan u - \tan v}{1 + \tan u \tan v}$$

Multiplication Formulas

$$\sin 2t = 2 \sin t \cos t \qquad \cos 2t = \cos^2 t - \sin^2 t \qquad \tan 2t = \frac{2 \tan t}{1 - \tan^2 t}$$

$$\sin \tfrac{1}{2}t = \pm\sqrt{\frac{1 - \cos t}{2}} \qquad \cos \tfrac{1}{2}t = \pm\sqrt{\frac{1 + \cos t}{2}} \qquad \tan \tfrac{1}{2}t = \frac{1 - \cos t}{\sin t}$$

Now we turn to one of the most important limits of calculus, namely

$$\lim_{t \to 0} \frac{\sin t}{t}$$

We have already seen (in the example at the beginning of this section) why this limit is important; the derivative of sine depends on it. All the other trigonometric functions can be differentiated in terms of sine (as we will see in the next chapter), so the whole calculus of these functions flows from this limit.

To investigate the limit, look at the unit circle. Assuming that $0 < t < \pi/2$ (Figure 11), we claim that

$$BP < \text{arc } AP < AQ \tag{1}$$

The first of these inequalities is apparent from Figure 11. (The segment BP is a shorter route from P to the x axis than the arc AP.) To prove the second inequality, observe that

$$\text{area sector } AOP < \text{area } \triangle OAQ \tag{2}$$

The area of a circular sector of radius r and central angle θ (in radians) is $A = \frac{1}{2}r^2\theta$. (See Problem 16.) Hence

$$\text{area sector } AOP = \tfrac{1}{2}t = \tfrac{1}{2}\text{arc } AP$$

Since

$$\text{area } \triangle OAQ = \tfrac{1}{2}(OA)(AQ) = \tfrac{1}{2}AQ$$

inequality (2) reads $\frac{1}{2}\text{arc } AP < \frac{1}{2}AQ$, from which arc $AP < AQ$ as claimed in (1).

Now observe that $BP = y = \sin t$ and (by similar triangles)

$$AQ = \frac{AQ}{OA} = \frac{BP}{OB} = \frac{y}{x} = \frac{\sin t}{\cos t}$$

Thus the inequalities in (1) become

$$\sin t < t < \frac{\sin t}{\cos t}$$

Dividing by $\sin t$ (which is positive because $0 < t < \pi/2$), we obtain

Figure 11

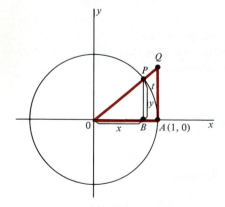

$$1 < \frac{t}{\sin t} < \frac{1}{\cos t}$$

or (taking reciprocals)

$$1 > \frac{\sin t}{t} > \cos t \qquad \text{(Order Property 7, Section 1.1)} \tag{3}$$

Now suppose that in the beginning we had assumed $-\pi/2 < t < 0$ instead of $0 < t < \pi/2$. The preceding argument with t replaced by $-t$ yields

$$1 > \frac{\sin(-t)}{-t} > \cos(-t)$$

which reduces to (3) again because $\sin(-t) = -\sin t$ and $\cos(-t) = \cos t$. Hence (3) is true for all t close to 0 (whether positive or negative).

Evidently the heart of the matter is (3). For these inequalities show that $(\sin t)/t$ is boxed in between $\cos t$ and 1. When t approaches 0, $\cos t$ approaches 1 (see Problem 66). Hence

$$\boxed{\lim_{t \to 0} \frac{\sin t}{t} = 1}$$

Note that in the preceding argument t is simply a real number (or the radian measure of angle AOP). It is important to realize that if angle AOP were measured in degrees, the limit would no longer be 1, but $\pi/180$. (Why?) Since we use the limit to find the derivative of sine (see Example 2), everything in calculus that involves this derivative is predicated on the assumption that the independent variable is either a real number or an angle measured in radians.

If you have an electronic calculator, you might check the entries in the table on this page, which indicates how $(\sin t)/t$ approaches 1 as $t \to 0$. Be sure to set the angle selector on radians! If you set it on degrees, you will find that

$$\frac{\sin t}{t} \to \frac{\pi}{180} = 0.0174532925199\cdots$$

□ **Example 1**   In the example at the beginning of this section, another important limit occurred, namely

$$\lim_{h \to 0} \frac{\cos h - 1}{h} = 0$$

This may be confirmed by making use of the limit just discussed:

$$\frac{\cos h - 1}{h} = \frac{(\cos h - 1)(\cos h + 1)}{h(\cos h + 1)} = \frac{\cos^2 h - 1}{h(\cos h + 1)}$$

$$= \frac{-\sin^2 h}{h(\cos h + 1)} \qquad \text{(because } \sin^2 h + \cos^2 h = 1\text{)}$$

$$= -\left(\frac{\sin h}{h}\right)\left(\frac{\sin h}{\cos h + 1}\right)$$

Convergence of $(\sin t)/t$ to 1 as $t \to 0$

t	$\dfrac{\sin t}{t}$
± 0.5	0.959
± 0.4	0.974
± 0.3	0.985
± 0.2	0.993
± 0.1	0.998
± 0.01	0.99998
± 0.001	0.9999998
$\vdots$	$\vdots$

When $h \to 0$ the first of these fractions approaches 1 (as we have just shown). The second approaches 0 because $\sin h \to 0$ and $\cos h \to 1$. Hence

$$\lim_{h \to 0} \frac{\cos h - 1}{h} = 0 \qquad \square$$

□ **Example 2**   Show that the derivative of sine is cosine.

Solution: We need only put together what we know. If $p(t) = \sin t$, the difference quotient associated with p at t is

$$q(h) = \sin t \left(\frac{\cos h - 1}{h} \right) + \cos t \left(\frac{\sin h}{h} \right)$$

as we showed at the beginning of this section. (Also see Example 4, Section 2.2, which you should now read if you skipped it earlier.) Since

$$\lim_{h \to 0} \frac{\cos h - 1}{h} = 0 \qquad \text{and} \qquad \lim_{h \to 0} \frac{\sin h}{h} = 1$$

we have

$$p'(t) = \lim_{h \to 0} q(h) = (\sin t)(0) + (\cos t)(1) = \cos t \qquad \square$$

In the problem set we ask you to use a similar argument to find the derivative of cosine. The results are fundamental in calculus involving the trigonometric functions:

$$D_x(\sin x) = \cos x \qquad \text{and} \qquad D_x(\cos x) = -\sin x$$

Problem Set 2.3

1. Fill in the following table.

Degrees	0°	30°		60°		120°	135°	150°
Radians			$\pi/4$		$\pi/2$			

180°			240°			315°	330°	
	$7\pi/6$	$5\pi/4$		$3\pi/2$	$5\pi/3$			2π

2. Find the radian measure of each of the following angles.
   (a) 50° **(b)** 36° **(c)** 20° **(d)** 75° **(e)** 54° **(f)** 15°

3. Find the degree measure of each of the following angles (given in radians).
   (a) $5\pi/9$ **(b)** $13\pi/4$ **(c)** 6π **(d)** $7\pi/18$ **(e)** $\pi/36$ **(f)** $8\pi/9$

4. Use an isosceles right triangle to verify the values given in the text for sine, cosine, and tangent of $\pi/4$.

5. Repeat Problem 4 for cotangent, secant, and cosecant of $\pi/4$.

6. In a right triangle with acute angles 30° and 60° and hypotenuse 2, why are the sides opposite these angles 1 and $\sqrt{3}$, respectively?

7. Use Problem 6 to verify the values given in the text for sine, cosine, and tangent of $\pi/6$.

8. Repeat Problem 7 for cotangent, secant, and cosecant of $\pi/6$.

9. Use Problem 6 to verify the values given in the text for sine, cosine, and tangent of $\pi/3$.

10. Repeat Problem 9 for cotangent, secant, and cosecant of $\pi/3$.

11. Suppose that t is a multiple of $\pi/2$. Explain why the point $P(t)$ obtained by wrapping t onto the unit circle lies on a coordinate axis. Then fill in the following table.

t	$\sin t$	$\cos t$	$\tan t$	$\cot t$	$\sec t$	$\csc t$
0						
$\pm\pi/2$						
$\pm\pi$						
$\pm 3\pi/2$						
$\pm 2\pi$						

12. Label the quadrants of the coordinate plane 1, 2, 3, 4 (starting with the quadrant in which $x > 0$ and $y > 0$ and going counterclockwise). Then fill in the following table with the proper signs ($+$ or $-$).

$P(t)$ in Quadrant				
$\sin t$ and $\csc t$				
$\cos t$ and $\sec t$				
$\tan t$ and $\cot t$				

13. Explain why the length of arc subtended by a central angle of θ radians in a circle of radius r is $s = r\theta$. (See Figure 12.) *Hint:* In a circle of fixed radius, s is proportional to θ. What is its value when $\theta = 2\pi$?

Figure 12
Arc length on a circle

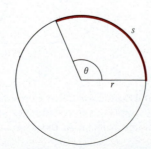

14. The village of Last Chance is 36° north of the equator. Assuming that the earth is spherical (with a radius of 6357 kilometers), find the distance from Last Chance to the North Pole (measured along the earth's surface).

15. A television camera at the center of a circular track of radius 100 meters follows a woman as she runs 50 meters along the track. Through how many degrees does the camera turn?

16. Show that the area of a circular sector of radius r and central angle θ radians is $A = \frac{1}{2}r^2\theta$. *Hint:* In a circle of fixed radius, A is proportional to θ. What is its value when $\theta = 2\pi$? (See Figure 13.)

Figure 13
Area of a circular sector

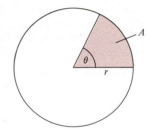

17. Find the area of the circular sector of radius 3 and central angle 120°.

18. Find the area of the circular sector of radius 5 and central angle 20°.

The graph of each of the following functions is shown in the text. Verify it by sketching your own (plotting a few convenient points and using the period).

19. $u = \sin t$ **20.** $u = \cos t$

21. $u = \tan t$ **22.** $u = \cot t$

23. $u = \sec t$ **24.** $u = \csc t$

Verify each of the following fundamental identities by using the unit circle definition of the trigonometric functions.

25. $\sin t \csc t = 1$ **26.** $\cos t \sec t = 1$

27. $\tan t \cot t = 1$ **28.** $\tan t = \dfrac{\sin t}{\cos t}$

29. $\cot t = \dfrac{\cos t}{\sin t}$ **30.** $\sin^2 t + \cos^2 t = 1$

31. $\sec^2 t - \tan^2 t = 1$ **32.** $\csc^2 t - \cot^2 t = 1$

33. Draw pictures to show that if $P(t) = (x, y)$, then $P(-t) = (x, -y)$. Why does it follow that $\sin(-t) = -\sin t$ and $\cos(-t) = \cos t$? (Thus sine is an odd function and cosine is even.)

34. Use Problem 33 to show that tangent, cotangent, and cosecant are odd, while secant is even.

In Problems 35 through 48 assume that the formula $\cos(u - v) = \cos u \cos v + \sin u \sin v$ is already proved. (See Problem 50 for an outline of its proof.)

35. Explain why $\cos(\pi/2 - t) = \sin t$. Then replace $\pi/2 - t$ by t to obtain $\sin(\pi/2 - t) = \cos t$. What is a geometric interpretation of these formulas when $0 < t < \pi/2$?

36. Use the formulas in Problem 35 to derive the identities $\tan(\pi/2 - t) = \cot t$ and $\cot(\pi/2 - t) = \tan t$.

37. Use the formulas in Problem 35 to derive the identities $\sec(\pi/2 - t) = \csc t$ and $\csc(\pi/2 - t) = \sec t$.

38. Derive the formula for $\cos(u + v)$ by writing $u + v = u - (-v)$ and using the formula for $\cos(u - v)$.

39. Derive the formula for $\sin(u + v)$ by writing

$$\sin(u + v) = \cos\left[\frac{\pi}{2} - (u + v)\right] = \cos\left[\left(\frac{\pi}{2} - u\right) - v\right]$$

and using the formula for $\cos(u - v)$.

40. Use Problem 39 to derive the formula for $\sin(u - v)$.

41. Use the formulas for $\sin(u + v)$ and $\cos(u + v)$ to derive the formula for $\tan(u + v)$.

42. Use Problem 41 to derive the formula for $\tan(u - v)$.

43. Derive the formula $\sin 2t = 2 \sin t \cos t$ by writing $2t = t + t$ and using an addition formula.

44. Derive the formula $\cos 2t = \cos^2 t - \sin^2 t$ by proceeding as in Problem 43.

45. Derive the formula for $\tan 2t$ by proceeding as in Problem 43.

46. Explain why $\cos 2t = 1 - 2 \sin^2 t$ and use the result to derive the formula $\sin^2 t = \frac{1}{2}(1 - \cos 2t)$. Then replace t by $t/2$ to obtain the formula for $\sin \frac{1}{2}t$ listed in the text.

47. Explain why $\cos 2t = 2 \cos^2 t - 1$ and use the result to derive the formula $\cos^2 t = \frac{1}{2}(1 + \cos 2t)$. Then replace t by $t/2$ to obtain the formula for $\cos \frac{1}{2}t$ listed in the text.

48. Multiply numerator and denominator of $\tan t = \sin t/\cos t$ by $2 \sin t$ to show that

$$\tan t = \frac{1 - \cos 2t}{\sin 2t}$$

Then obtain the formula for $\tan \frac{1}{2}t$ given in the text.

49. In Problem 48 multiply by $2 \cos t$ instead to obtain

$$\tan \tfrac{1}{2}t = \frac{\sin t}{1 + \cos t}$$

50. Let u and v be any real numbers. Derive the formula for $\cos(u - v)$ as follows:

(a) The point $P(u - v)$ on the unit circle is obtained by wrapping $u - v$ from the point $P(0) = (1,0)$ as origin. Why is this equivalent to wrapping u from the point $P(v)$ as origin?

(b) Explain why it follows from (a) that the chord with endpoints $P(0)$ and $P(u - v)$ has the same length as the chord with endpoints $P(u)$ and $P(v)$. What are the coordinates of $P(u)$, $P(v)$, and $P(u - v)$?

(c) Use (b) and the Distance Formula (together with the identity $\sin^2 t + \cos^2 t = 1$) to finish the proof.

51. Let θ be an angle in standard position (initial side along the positive x axis) and suppose that (x, y) is a point on its terminal side r units from the origin ($r > 0$). Explain why

$$\sin \theta = \frac{y}{r} \qquad\qquad \cot \theta = \frac{x}{y} \quad (y \neq 0)$$

$$\cos \theta = \frac{x}{r} \qquad\qquad \sec \theta = \frac{r}{x} \quad (x \neq 0)$$

$$\tan \theta = \frac{y}{x} \quad (x \neq 0) \qquad \csc \theta = \frac{r}{y} \quad (y \neq 0)$$

(The trigonometric functions are sometimes *defined* this way. The definitions in the text correspond to the choice $r = 1$.)

52. The *Law of Cosines* says that if a, b, c are the sides of a triangle and a is the side opposite α, then $a^2 = b^2 + c^2 - 2bc \cos \alpha$. Derive it as follows.

(a) Place the triangle in a coordinate system as shown in Figure 14. Then use Problem 51 to explain why $C = (b \cos \alpha, b \sin \alpha)$.

Figure 14
Triangle with α
in standard position

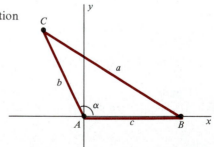

(b) Use the Distance Formula to express $a = d(B, C)$ in terms of the coordinates of B and C and simplify the result to finish the proof.

53. Explain why the range of sine and cosine is the interval $[-1,1]$. Why does it follow that the range of secant and cosecant is $(-\infty,-1] \cup [1,\infty)$?

54. To show that the range of tangent is $\mathcal{R}$, let u be any real number. The problem is to name $t \in \mathcal{R}$ such that $\tan t = u$.

   (a) Why is this problem equivalent to finding a point (x,y) of the unit circle satisfying $y/x = u$?

   (b) Why is part (a) equivalent to finding a real number x satisfying

$$\frac{1-x^2}{x^2} = u^2?$$

   (c) Show that such a real number exists by solving for x in terms of u.

55. Why does it follow from Problem 54 that the range of cotangent is $\mathcal{R}$?

56. Verify that the period of sine is 2π as follows.

   (a) Explain why $\sin(t + 2\pi) = \sin t$ for all t. (Hence the period, if one exists, is no larger than 2π.)

   (b) Suppose there were a positive number $p < 2\pi$ such that $\sin(t + p) = \sin t$ for all t. Replace t by 0 and $\pi/2$ to arrive at a contradiction. (Hence no number less than 2π serves as the period; 2π is the period! Similar reasoning shows that cosine, secant, and cosecant also have period 2π.)

57. Verify that the period of tangent is π as follows.

   (a) Show that if $P(t) = (x,y)$, then $P(t + \pi) = (-x,-y)$.

   (b) Why does it follow that $\tan(t + \pi) = \tan t$ for all t in the domain of tangent?

   (c) Suppose that p is a positive number with the property that $\tan(t + p) = \tan t$ for all t in the domain of tangent. Let $t = 0$ to show that p cannot be less than π. (A similar argument shows that cotangent has period π.)

Use the fact that $\lim\limits_{t \to 0} \dfrac{\sin t}{t} = 1$ to evaluate each of the following limits.

58. $\lim\limits_{t \to 0} \dfrac{t}{\sin t}$

59. $\lim\limits_{x \to 0} \dfrac{\sin 3x}{x}$ *Hint:* Multiply numerator and denominator by 3.

60. $\lim\limits_{x \to 0} \dfrac{x}{\sin 2x}$ **61.** $\lim\limits_{t \to 0} \dfrac{t^2}{\sin t}$ **62.** $\lim\limits_{t \to 0} t \cot t$ **63.** $\lim\limits_{t \to 0} \dfrac{\tan t}{t}$

64. Show that the difference quotient associated with $f(x) = \cos x$ at x is

$$q(h) = \cos x \left(\frac{\cos h - 1}{h}\right) - \sin x \left(\frac{\sin h}{h}\right)$$

and use the result to conclude that $D_x(\cos x) = -\sin x$.

65. Show that the difference quotient associated with $f(x) = \tan x$ at x is

$$q(h) = \frac{\sec^2 x}{1 - \tan x \tan h} \left(\frac{\tan h}{h}\right)$$

and use the result to conclude that $D_x(\tan x) = \sec^2 x$. *Hint:* Refer to Problem 63 after using the addition formula for $\tan(u + v)$.

We have been regarding it as obvious (from the graphs) that

$$\lim\limits_{t \to 0} \sin t = 0 \quad \text{and} \quad \lim\limits_{t \to 0} \cos t = 1$$

The next problem outlines a more analytic argument.

66. The *contraction property* of sine and cosine is

$$|\sin u - \sin v| \le |u - v|$$

and

$$|\cos u - \cos v| \le |u - v|$$

for all u and v. (That is, when two real numbers are fed into sine or cosine, the outputs are no farther apart than the inputs.)

   (a) Use Figure 15 to explain this property. *Hint:* Each side of the triangle shown in the figure is shorter than the arc joining $P(u)$ and $P(v)$.

Figure 15
The contraction property

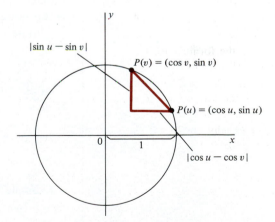

   (b) Why does it follow from the contraction property that

$$|\sin t| \le |t| \quad \text{and} \quad |\cos t - 1| \le |t|$$

   for all t?

   (c) Why does it follow from part (b) that when $t \to 0$, $\sin t$ gets close to 0 and $\cos t$ gets close to 1?

2.4
LIMITS
(Optional)

This section (and the next) are designed to lend precision to the concept of limit. The results (in the form of theorems about limits) will be used as the book progresses. The reader may simply learn the theorems, however (they are stated in the next section), and still understand what happens in succeeding chapters. Hence this section is optional.

In Section 2.1 (Example 1) we agreed to find the slope of the curve $y = x^2$ at (1,1) by examining the difference quotient

$$Q(x) = \frac{y - 1}{x - 1} = \frac{x^2 - 1}{x - 1} = \frac{(x - 1)(x + 1)}{x - 1}$$

$$= x + 1 \qquad (x \neq 1)$$

We said that $Q(x)$, while undefined at $x = 1$, approaches 2 as x approaches 1, and we adopted the number $m = 2$ as the answer to the question.

The statement "$Q(x)$ approaches 2 as x approaches 1" may be interpreted geometrically by looking at the graph of Q. Since $Q(1)$ is undefined, while $Q(x) = x + 1$ for all $x \neq 1$, the graph is a straight line with a hole in it. (See Figure 1.) The coordinates of the hole are (1,2), so the y coordinate of a bug traveling on the graph gets closer to 2 as its x coordinate approaches 1. We express this in somewhat different language by saying that "the limit of $Q(x)$ as x approaches 1 is 2" or (in symbolic form)

$$\lim_{x \to 1} Q(x) = 2$$

This statement cannot be interpreted as an evaluation of $Q(x)$ at $x = 1$. While it is true that $Q(x) = x + 1$, and this formula yields 2 when x is replaced by 1, it is nevertheless meaningless to say that $Q(1) = 2$. For $Q(x)$ is the slope of the line through (1,1) and (x,y), which is not defined unless (x,y) and (1,1) are distinct. Thus the statement

$$\lim_{x \to 1} Q(x) = 2$$

does not refer to what happens at $x = 1$ but to the behavior of $Q(x)$ when x is near 1.

Perhaps this is clear enough, at least on intuitive grounds. However, there are difficulties.

Figure 1

Graph of $Q(x) = \dfrac{x^2 - 1}{x - 1}$

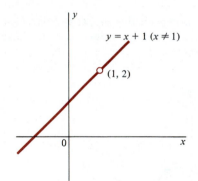

$y = x + 1 \ (x \neq 1)$

$(1, 2)$

□ **Example 1(a)**   We have argued that if

$$Q(x) = \frac{x^2 - 1}{x - 1}$$

then

$$\lim_{x \to 1} Q(x) = 2$$

our reasoning being that when x is close to 1, $Q(x) = x + 1$ is close to 2. But suppose a critic suggests that

$$\lim_{x \to 1} Q(x) = 2.001$$

arguing that when x is close to 1, $Q(x)$ is near 2.001. On what grounds do we say that the critic is wrong?

While this may seem to be a perverse question, it is a difficulty we must meet and overcome. No mathematical concept is useful if it is so imprecise as to allow two people to come up with different answers. Perhaps we do not think of 2 and 2.001 as very far apart, but if we are dealing in light years, the difference is enormous (more than nine billion kilometers!). We cannot afford to disagree at all; the limit is either 2 or it isn't, and we must settle on some definitive way to reach a decision.

Mathematicians struggled for a long time to develop the definition we are going to give, but it is simple enough as it applies to this example. The idea is that if 2 is the answer, we should be able to force $Q(x)$ *as close to 2 as our critic desires.* (Closer, for example, than 2.001, if required.) So we let our critic define "close." And we remember that $Q(x)$ depends on x; we control its behavior by placing restrictions on x.

It is like a contest. Our critic goes first, naming a neighborhood of 2 in which $Q(x)$ is to lie. We go second, responding to the challenge by naming a neighborhood of 1 in which x should lie in order to keep $Q(x)$ where our critic wants it. *If we can respond to every challenge,* our critic must agree that the answer is 2. But *if there is even one neighborhood of 2 to which we are unable to confine $Q(x)$ by keeping x close to 1,* we must admit that the answer is not 2.

To understand the definition we draw from this, you should recall that a *neighborhood* of a point p is an open interval containing p. If p is deleted, the neighborhood is said to be *punctured.* Occasionally we will also refer to a *right neighborhood* of p and a *left neighborhood* of p, which are open intervals of the form (p,b) and (a,p), respectively. (See Problems 61 and 63, Section 1.1.)

Our definition says that

$$\lim_{x \to 1} Q(x) = 2$$

provided that the following condition is met:

> Given any neighborhood of 2 (say N), there is a punctured neighborhood of 1 (say M) with the property that $x \in M \Rightarrow Q(x) \in N$.

The reason we puncture the neighborhood of 1 is that the domain of Q excludes 1. The implication

$$x \in M \Rightarrow Q(x) \in N$$

would fail if we allowed $x = 1$.   □

□ **Example 1(b)**   Suppose that our critic wants $Q(x)$ to lie in the interval $N = (1.98, 2.05)$, that is $1.98 < Q(x) < 2.05$. To discover how we should restrict x in response to this challenge, we solve the inequalities

$$1.98 < \frac{x^2 - 1}{x - 1} < 2.05$$

that is, $1.98 < x + 1 < 2.05$ $(x \neq 1)$. Subtracting 1 from each side, we find

$$0.98 < x < 1.05 \qquad (x \neq 1)$$

which means that we may choose M to be the interval $(0.98,1.05)$ with 1 deleted. Since

$$x \in M \Rightarrow 0.98 < x < 1.05 \qquad (x \neq 1)$$
$$\Rightarrow 1.98 < x + 1 < 2.05 \qquad (x \neq 1)$$
$$\Rightarrow Q(x) \in N$$

our critic should be satisfied. □

□ **Example 1(c)**   We must convince ourselves that our critic cannot baffle us by *any* challenge. Instead of $N = (1.98,2.05)$ suppose that an arbitrary neighborhood of 2 is named. It ought to be apparent that we can assume this neighborhood is symmetric about 2, simply by using the closer endpoint to compute the radius. For example, there should be no complaint if we replace $(1.98,2.05)$ by $(1.98,2.02)$, which has midpoint 2 and radius 0.02. (See Figure 2.) If we can keep $Q(x)$ in the second of these, we are automatically keeping it in the first, and that is what our critic demands.

Hence let us assume that the challenge is of the form

$$N = (2 - \varepsilon, 2 + \varepsilon)$$

where ε is any positive number (the radius of N). Our problem is to name a punctured neighborhood of 1 (say M) such that

$$x \in M \Rightarrow Q(x) \in N$$

As in Example 1(b) it is just a matter of solving the appropriate inequalities. Our critic is asking us to satisfy

$$2 - \varepsilon < Q(x) < 2 + \varepsilon$$

so we write

$$2 - \varepsilon < \frac{x^2 - 1}{x - 1} < 2 + \varepsilon \Leftrightarrow 2 - \varepsilon < x + 1 < 2 + \varepsilon \qquad (x \neq 1)$$
$$\Leftrightarrow 1 - \varepsilon < x < 1 + \varepsilon \qquad (x \neq 1)$$

Our response should now be clear. We choose M to be the neighborhood $(1 - \varepsilon, 1 + \varepsilon)$ with 1 deleted. Then (following the above implications backwards) we have

$$x \in M \Rightarrow Q(x) \in N$$

which is what our critic must believe to be satisfied.

A graph helps clarify the procedure in Example 1(c). See Figure 3, in which we show M and N on the x and y axes, respectively. Our critic names a

Figure 2
Replacing an arbitrary neighborhood by a symmetric neighborhood

Figure 3
Sending M into N

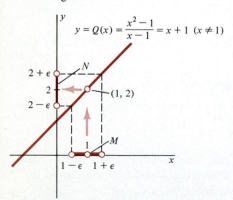

symmetric neighborhood of 2 with radius $\varepsilon > 0$. We respond by naming a (punctured) neighborhood of 1. The arrows indicate that the function Q sends the points of M into N, which is what our critic demands.

Note that our choice of M is not unique. If we were to take M to be the neighborhood $(1 - \varepsilon/2, 1 + \varepsilon/2)$ with 1 deleted, the implication

$$x \in M \Rightarrow Q(x) \in N$$

would still be correct. The punctured neighborhood shown in Figure 3 is the largest (and simplest) we can select, but any smaller neighborhood serves as well.   □

□ **Example 1(d)**   In Example 1(a) we asked how we could argue with a critic who suggests that

$$\lim_{x \to 1} Q(x) = 2.001$$

Figure 4

Boxing $Q(x)$ away from 2.001

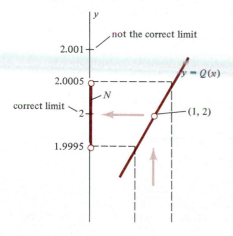

Now we are in a position to say that the critic is definitely wrong. For we know from Example 1(c) that $Q(x)$ can be made arbitrarily close to 2 by taking x sufficiently close to 1. In particular we can box $Q(x)$ away from 2.001 by forcing it into the neighborhood $N = (1.9995, 2.0005)$. (See Figure 4.) In other words the statement

$$\lim_{x \to 1} Q(x) = 2.001$$

is false. By a similar argument we can show that if L is any number except 2, the statement

$$\lim_{x \to 1} Q(x) = L$$

is false. For we can box $Q(x)$ away from L by keeping x close to 1.   □

□ **Example 2**   Sometimes the neighborhoods involved in evaluating a limit are *one-sided*. Consider, for example, the statement

$$\lim_{x \to 1} \sqrt{x - 1} = 0$$

Figure 5

$\lim_{x \to 1} \sqrt{x - 1} = 0$

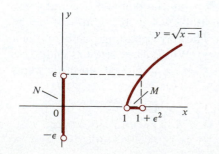

The graph of $f(x) = \sqrt{x - 1}$ is shown in Figure 5; it is the upper half of the parabola $y^2 = x - 1$. (Why?) A critic who doubts that the limit of $f(x)$ as $x \to 1$ is 0 would challenge us to confine $f(x)$ to a neighborhood of 0, say $N = (-\varepsilon, \varepsilon)$, where $\varepsilon > 0$. Since the domain of f is $\{x : x \geq 1\}$, we cannot work with an ordinary neighborhood of 1 in response. Values of x to the left of 1 cannot be used at all. What we do instead is to name a *right neighborhood* of 1, as shown in Figure 5. To figure out what M should be, we follow the horizontal line from ε on the y axis until we hit the graph, then proceed down to the x axis. The point we hit is the right-hand endpoint of M, say b. To find b, we observe that it must satisfy $f(b) = \varepsilon$, that is,

$$\sqrt{b - 1} = \varepsilon \Leftrightarrow b - 1 = \varepsilon^2$$
$$\Leftrightarrow b = 1 + \varepsilon^2$$

Hence M is the open interval $(1, 1 + \varepsilon^2)$. To confirm that it works (without depending on the picture), we write

$$
\begin{aligned}
x \in M &\Rightarrow 1 < x < 1 + \varepsilon^2 \\
&\Rightarrow 0 < x - 1 < \varepsilon^2 \\
&\Rightarrow 0 < \sqrt{x - 1} < \varepsilon \qquad \text{(Order Property 8, Section 1.1)} \\
&\Rightarrow f(x) \in N
\end{aligned}
$$

☐

These examples should help explain the following definition of limit.

Limit of a Real Function

Let f be a real function with domain D and suppose that a is a real number having a punctured neighborhood in D. (This insures that all points near a, with the possible exception of a itself, are points of D. In other words, $f(x)$ is defined for all x near a, but $f(a)$ may be undefined.) The statement

$$
\lim_{x \to a} f(x) = L
$$

where L is a real number, is defined to mean the following:

Given any neighborhood of L (say N), there is a punctured neighborhood of a (say M) such that $x \in M \Rightarrow f(x) \in N$.

If a has only a right [left] neighborhood in D, we replace "punctured neighborhood" by "right [left] neighborhood."

☐ **Example 3**   Prove that $\lim\limits_{x \to 2} x^2 = 4$.

Figure 6
Naming M when N is given

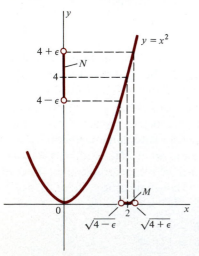

Solution: We start by assuming that a critic has named a neighborhood of 4, say

$$
N = (4 - \varepsilon, 4 + \varepsilon) \qquad \text{where } \varepsilon > 0
$$

Our problem is to name a neighborhood of 2 (say M) such that

$$
x \in M \Rightarrow x^2 \in N
$$

(No need to puncture it this time. Why?) From the points $4 - \varepsilon$ and $4 + \varepsilon$ on the y axis (Figure 6), follow horizontal lines to the graph of $y = x^2$ ($x \geq 0$), then vertical lines to the x axis. The corresponding points on the x axis are $\sqrt{4 - \varepsilon}$ and $\sqrt{4 + \varepsilon}$, respectively. (Why?) They serve as endpoints of M. It is geometrically apparent that $x \in M \Rightarrow x^2 \in N$; if our critic wants algebraic confirmation, we write

$$
\begin{aligned}
x \in M &\Rightarrow \sqrt{4 - \varepsilon} < x < \sqrt{4 + \varepsilon} \\
&\Rightarrow 4 - \varepsilon < x^2 < 4 + \varepsilon \\
&\Rightarrow x^2 \in N
\end{aligned}
$$

☐

Many calculus books give a definition of limit which does not mention neighborhoods. To see how this is done, assume that the neighborhood N is symmetric about L with radius $\varepsilon > 0$. Then

$$f(x) \in N \Leftrightarrow L - \varepsilon < f(x) < L + \varepsilon$$
$$\Leftrightarrow -\varepsilon < f(x) - L < \varepsilon$$
$$\Leftrightarrow |f(x) - L| < \varepsilon$$

Similarly, assume that M is symmetric about a with radius $\delta > 0$. Then

$$x \in M \Leftrightarrow a - \delta < x < a + \delta \qquad (x \neq a)$$
$$\Leftrightarrow -\delta < x - a < \delta \qquad (x \neq a)$$
$$\Leftrightarrow 0 < |x - a| < \delta$$

The statement $\lim_{x \to a} f(x) = L$ is therefore equivalent to the following:

Given $\varepsilon > 0$, there is a $\delta > 0$ such that

$$0 < |x - a| < \delta \Rightarrow |f(x) - L| < \varepsilon$$

If a has only a right neighborhood in the domain of f, we replace

$$0 < |x - a| < \delta \qquad \text{by} \qquad 0 < x - a < \delta$$

(to keep $x > a$). If there is only a left neighborhood of a in the domain, we replace

$$0 < |x - a| < \delta \qquad \text{by} \qquad 0 < a - x < \delta$$

(to keep $x < a$).

To see how this "ε-δ definition" works in practice, refer to our earlier examples. If $\varepsilon > 0$ is named in Example 1(c), where we proved that

$$\lim_{x \to 1} \frac{x^2 - 1}{x - 1} = 2$$

the corresponding $\delta > 0$ is $\delta = \varepsilon$. That is,

$$0 < |x - 1| < \varepsilon \Rightarrow \left| \frac{x^2 - 1}{x - 1} - 2 \right| < \varepsilon$$

In Example 2, $\lim_{x \to 1} \sqrt{x - 1} = 0$, we name $\delta = \varepsilon^2$ (and use a right neighborhood). That is,

$$0 < x - 1 < \varepsilon^2 \Rightarrow |\sqrt{x - 1} - 0| < \varepsilon$$

In Example 3, on the other hand, the ε-δ definition is not as convenient. The neighborhood $M = (\sqrt{4 - \varepsilon}, \sqrt{4 + \varepsilon})$ is not symmetric about the point 2. To name $\delta > 0$ such that

$$|x - 2| < \delta \Rightarrow |x^2 - 4| < \varepsilon$$

we would have to figure out which of the endpoints $\sqrt{4 - \varepsilon}$ and $\sqrt{4 + \varepsilon}$ is closer to 2 and cut down M to a symmetric neighborhood with the smaller distance as radius. There is not much point in taking the trouble! This illustrates the fact that we may choose between the ε-δ definition and the neighborhood definition as the situation demands.

□ **Example 4**   To prove that $\lim_{x \to 2} (3x - 2) = 4$, we suppose that a critic has named $\varepsilon > 0$. Our problem is to name $\delta > 0$ such that

$$|x - 2| < \delta \Rightarrow |(3x - 2) - 4| < \varepsilon$$

(Note that we do not insist on $0 < |x - 2|$ in this case, because it is unnecessary to keep $x \neq 2$.) Since

$$|(3x - 2) - 4| < \varepsilon \Leftrightarrow |3(x - 2)| < \varepsilon$$

$$\Leftrightarrow |x - 2| < \frac{\varepsilon}{3}$$

we name $\delta = \varepsilon/3$. Then (following the implications backwards) we have

$$|x - 2| < \delta \Rightarrow |x - 2| < \frac{\varepsilon}{3} \Rightarrow |3(x - 2)| < \varepsilon \Rightarrow |(3x - 2) - 4| < \varepsilon \quad □$$

□ **Example 5**   To show that $\lim_{x \to 2} (1/x) = \frac{1}{2}$, we suppose as usual that we are confronting a skeptic, who gives us a neighborhood of $\frac{1}{2}$ with radius $\varepsilon > 0$. Our problem is to force $1/x$ to lie in this neighborhood by restricting x to be near 2.

The easiest procedure is to look at the graph (Figure 7). Here we have shown the challenger's neighborhood (cut down, if necessary, to exclude 0) as an interval on the y axis with $\frac{1}{2}$ as midpoint. To keep $y = 1/x$ in this neighborhood, we restrict x to the neighborhood of 2 shown on the x axis. This neighborhood is not symmetric about 2, but our first definition of limit does not require it to be.   □

Figure 7

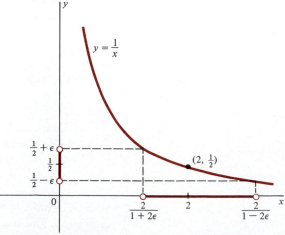

Now we look at some limits that fail to exist.

□ **Example 6**   Let $f(x) = |x|/x$ and consider

$$\lim_{x \to 0} f(x)$$

As you can see from Figure 8, this limit does not exist, since $f(x)$ is 1 or -1 depending on whether $x > 0$ or $x < 0$. To convince a skeptic by appealing to the definition, we would assume the contrary, namely

$$\lim_{x \to 0} f(x) = L$$

where L is a real number. No matter what L is, there is no way to confine $f(x)$ to a small neighborhood of L by keeping x close to 0 (because every neighborhood of 0 contains both positive and negative values of x, the corresponding values of f being 1 and -1).   □

Figure 8

What does $\dfrac{|x|}{x}$ approach when $x \to 0$?

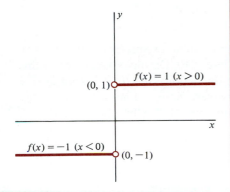

In Example 6 we may confine x to a right neighborhood of 0 and evaluate

$$\lim_{x \downarrow 0} f(x) = 1$$

The notation $x \downarrow 0$ means that x decreases toward 0; such a limit is called a **right-handed limit.** Similarly, the **left-handed limit** in this example is

$$\lim_{x \uparrow 0} f(x) = -1$$

where $x \uparrow 0$ means that x increases toward 0. One way of deciding that

$$\lim_{x \to 0} f(x) \text{ does not exist}$$

is to compute the one-sided limits and observe that they are different. (It can be proved that when the domain permits approach from both sides, the ordinary limit exists if and only if the one-sided limits are equal.)

In view of these remarks, you may want to go back to Example 2, where we evaluated

$$\lim_{x \to 1} \sqrt{x - 1} = 0$$

by confining x to a right neighborhood of 1. Is this an ordinary limit or a right-handed limit? Our answer is that it is both! The domain of $f(x) = \sqrt{x - 1}$ does not permit x to be less than 1, so it is a matter of indifference whether we write

$$\lim_{x \to 1} \sqrt{x - 1} \quad \text{or} \quad \lim_{x \downarrow 1} \sqrt{x - 1}$$

In Example 6, on the other hand, the domain allows $x \to 0$, $x \downarrow 0$, or $x \uparrow 0$, and hence it is necessary to distinguish between them.

Figure 9

Infinite jump at the origin

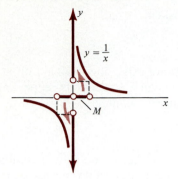

□ **Example 7**   The failure of the limit to exist in Example 6 is due to a finite jump in the graph. However, there are several ways a limit may fail to exist. Consider

$$\lim_{x \to 0} f(x) \qquad \text{where } f(x) = \frac{1}{x}$$

(See Figure 9.) It is clear from the picture that if M is any (punctured) neighborhood of 0, the statement $x \in M$ does not imply any statement of the type $|f(x) - L| < \varepsilon$. The most we can say is that $f(x)$ is unbounded when $x \to 0$ (increasing or decreasing depending on whether $x \downarrow 0$ or $x \uparrow 0$). There is no number L such that $f(x)$ is near L for all $x \in M$.   □

□ **Example 8**   Another way a limit may fail to exist is by oscillation. The graph of $f(x) = \sin(1/x)$ is shown in Figure 10. On any given piece of this curve there is no problem in describing the action; we are on a kind of sine wave between $y = 1$ and $y = -1$. However, there is a compression of the wave near the y axis, an increase in frequency that is boundless as $x \to 0$.

Figure 10

Graph of $y = \sin \dfrac{1}{x}$

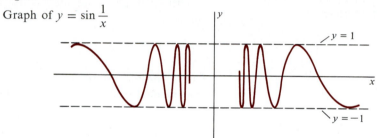

Between any two points of the curve on opposite sides of the y axis there are infinitely many vibrations, a situation that is hard to visualize and impossible to draw. We conclude that

$$\lim_{x \to 0} \sin \frac{1}{x} \qquad \text{does not exist}$$ □

Figure 11

Graph of $y = x \sin \dfrac{1}{x}$

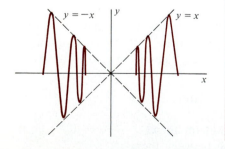

□ **Example 9**   The difficulty in Example 8 may be removed by "damping" the sine wave. Let $f(x) = x \sin(1/x)$, the damping factor being x. This has the same frequency of vibration as the function $y = \sin(1/x)$, so we are not curing the infinity of oscillations. But now the graph lies between the lines $y = x$ and $y = -x$, as shown in Figure 11. To see why, observe first that

$$-1 \leq \sin \frac{1}{x} \leq 1 \qquad \text{for all } x \neq 0 \quad \text{(Why?)}$$

Now multiply each side by x (preserving the inequalities if $x > 0$ and reversing them if $x < 0$). This yields

$$-x \le f(x) \le x \qquad \text{if } x > 0$$
$$-x \ge f(x) \ge x \qquad \text{if } x < 0$$

Hence in any case $f(x)$ is between x and $-x$. It follows that

$$\lim_{x \to 0} f(x) = 0$$

because $f(x)$ is boxed in between two functions ($y = x$ and $y = -x$) whose common limit as $x \to 0$ is 0.   □

Problem Set 2.4

1. To verify that $\lim_{x \to 2} (3x + 1) = 7$, assume that a critic has named a neighborhood of 7 of the form $N = (7 - \varepsilon, 7 + \varepsilon)$, where $\varepsilon > 0$. Name a neighborhood of 2 (say M) such that $x \in M \Rightarrow 3x + 1 \in N$. Why is it unnecessary to puncture M?

2. To verify that $\lim_{x \to 3} (6x - 13) = 5$, assume that a critic has named a neighborhood of 5 of the form $N = (5 - \varepsilon, 5 + \varepsilon)$, where $\varepsilon > 0$. Name a neighborhood of 3 (say M) such that $x \in M \Rightarrow 6x - 13 \in N$. Why is it unnecessary to puncture M?

3. To verify that

$$\lim_{x \to 3} \frac{x^2 - 9}{x - 3} = 6$$

let $N = (6 - \varepsilon, 6 + \varepsilon)$ be a given neighborhood of 6. Name a punctured neighborhood of 3 (say M) such that

$$x \in M \Rightarrow \frac{x^2 - 9}{x - 3} \in N$$

Why must M be punctured?

4. To verify that

$$\lim_{x \to 5} \frac{x^2 - 25}{x - 5} = 10$$

let $N = (10 - \varepsilon, 10 + \varepsilon)$ be a given neighborhood of 10. Name a punctured neighborhood of 5 (say M) such that

$$x \in M \Rightarrow \frac{x^2 - 25}{x - 5} \in N$$

Why must M be punctured?

Each of the following limits is of the form $\lim_{x \to a} f(x) = L$. As in Problems 1 through 4, verify the limit by assuming as given a neighborhood of L of the form $N = (L - \varepsilon, L + \varepsilon)$ and then naming a neighborhood of a (say M) such that $x \in M \Rightarrow f(x) \in N$. Puncture M when necessary.

5. $\lim_{x \to 1} (5x - 1) = 4$

6. $\lim_{x \to 3} (2x + 4) = 10$

7. $\lim_{x \to 3/2} \dfrac{4x^2 - 9}{2x - 3} = 6$

8. $\lim_{x \to 4/3} \dfrac{9x^2 - 16}{3x - 4} = 8$

9. $\lim_{x \to 4} \sqrt{x} = 2$

10. $\lim_{x \to 3} x^2 = 9$

11. $\lim_{x \to 1} \dfrac{1}{x} = 1$

12. $\lim_{x \to 2} \dfrac{6}{x} = 3$

13. $\lim_{x \downarrow 5} \sqrt{x - 5} = 0$ (M must be a right neighborhood of 5.)

14. $\lim_{x \uparrow 4} \sqrt{4 - x} = 0$ (M must be a left neighborhood of 4.)

15. $\lim_{x \uparrow 1} \sqrt{1 - x^2} = 0$ (M must be a left neighborhood of 1.)

16. $\lim_{x \downarrow 2} \sqrt{x^2 - 4} = 0$ (M must be a right neighborhood of 2.)

17. To verify that $\lim_{x \to 1} 3x = 3$, let $\varepsilon > 0$ be given. Name $\delta > 0$ such that

$$|x - 1| < \delta \Rightarrow |3x - 3| < \varepsilon.$$

18. To verify that

$$\lim_{x \to 2} \frac{x^2 - 4}{x - 2} = 4$$

let $\varepsilon > 0$ be given. Name $\delta > 0$ such that

$$0 < |x - 2| < \delta \Rightarrow \left| \frac{x^2 - 4}{x - 2} - 4 \right| < \varepsilon$$

19. To verify that $\lim_{x \to 0} (1 - x^2) = 1$, let $\varepsilon > 0$ be given. Name $\delta > 0$ such that $|x| < \delta \Rightarrow |(1 - x^2) - 1| < \varepsilon$.

20. To verify that $\lim_{x \to 0} \sin x = 0$, let $\varepsilon > 0$ be given. Name $\delta > 0$ such that $|x| < \delta \Rightarrow |\sin x| < \varepsilon$.   *Hint:* Use the contraction property of sine (Problem 66, Section 2.3).

21. To verify that $\lim_{x \downarrow 3} (1 + \sqrt{x - 3}) = 1$, let $\varepsilon > 0$ be given. Name $\delta > 0$ such that

$$0 < x - 3 < \delta \Rightarrow |(1 + \sqrt{x - 3}) - 1| < \varepsilon$$

Why do we write $0 < x - 3 < \delta$ instead of $0 < |x - 3| < \delta$?

22. To verify that $\lim_{x \uparrow 2} (1 + \sqrt{4 - x^2}) = 1$, let $\varepsilon > 0$ be given. Name $\delta > 0$ such that

$$0 < 2 - x < \delta \Rightarrow |(1 + \sqrt{4 - x^2}) - 1| < \varepsilon$$

Why do we write $0 < 2 - x < \delta$ instead of $0 < |x - 2| < \delta$?

In each of the following, evaluate the given limit, then use an appropriate definition to prove that you are correct.

23. $\lim_{x \to -2} \dfrac{x^2 - 4}{x + 2}$

24. $\lim_{x \to -1} \dfrac{x^2 - 1}{x + 1}$

25. $\lim_{x \to 2} x^3$

26. $\lim_{x \to 0} (4 - x^2)$

27. $\lim_{x \to 0} \dfrac{(1 + x)^2 - 1}{x}$

28. $\lim_{x \to 1} \dfrac{2}{x}$

29. $\lim_{x \to 0} \cos x$ *Hint:* Use the fact that $|\cos t - 1| \le |t|$ for all t (Problem 66, Section 2.3).

Note: The results of Problems 20 and 29,

$$\lim_{x \to 0} \sin x = 0 \quad \text{and} \quad \lim_{x \to 0} \cos x = 1$$

have appeared before (Problem 66, Section 2.3). They were used in Section 2.3 to establish the important fact that

$$\lim_{t \to 0} \frac{\sin t}{t} = 1$$

and they will be used in the next section to derive a general property of the six trigonometric functions.

30. $\lim_{x \to 0} x^2 \sin \dfrac{1}{x}$

31. $\lim_{x \to 0} (1 - \sqrt{x})$

32. $\lim_{x \to 3} \sqrt{9 - x^2}$

33. Show that if $f(x) = c$ is a constant function, then

$$\lim_{x \to a} f(x) = c$$

In each of the following, evaluate the limit (or decide that it does not exist). You need not prove that your answer is correct.

34. $\lim_{x \to 0} \dfrac{(3 + x)^2 - 9}{x}$

35. $\lim_{x \to 1} \left(\dfrac{x^2}{x - 1} - \dfrac{x}{x - 1} \right)$

36. $\lim_{x \to 2} \dfrac{x^3 - 8}{x - 2}$

37. $\lim_{x \to 4} \dfrac{x - 4}{\sqrt{x} - 2}$

38. $\lim_{x \to 2} \dfrac{x^2}{x - 2}$

39. $\lim_{x \to 2} \sqrt{1 - x}$

40. $\lim_{x \to 0} \tan x$

41. $\lim_{x \to 0} \tan x \cot x$

42. $\lim_{x \to 0} x^0$

43. $\lim_{x \to 3} \dfrac{|x|}{x}$

44. $\lim_{x \downarrow 2} \dfrac{|x - 2|}{x - 2}$

45. $\lim_{x \uparrow 2} \dfrac{|x - 2|}{x - 2}$

46. Is it correct to say that

$$\lim_{x \to 2} \sqrt{x} = 1.414?$$

Explain.

47. Explain why

$$\lim_{x \to 2} x^2 \ne 3.99$$

by naming a neighborhood of 3.99 to which $f(x) = x^2$ cannot be confined by keeping x near 2.

48. Suppose that

$$\lim_{x \to a} f(x) = L > 0$$

Explain why there is a punctured neighborhood of a (say M) such that $f(x) > 0$ for all $x \in M$.

Note: Problem 48 shows that a function with a positive limit must have positive values for all x near the point of approach. The same statement is true with "positive" replaced by "negative." (Why?)

49. Suppose that the domain of $f(x)$ permits x to approach a from either side.

(a) Show that if the one-sided limits $\lim_{x \downarrow a} f(x)$ and $\lim_{x \uparrow a} f(x)$ exist and have a common value L, then

$$\lim_{x \to a} f(x) = L$$

(b) Conversely, if $\lim_{x \to a} f(x) = L$, why are the one-sided limits both equal to L?

50. In view of the results of Problems 44 and 45, what can you say about

$$\lim_{x \to 2} \frac{|x - 2|}{x - 2}?$$

51. If f is defined by the rule

$$f(x) = \begin{cases} x - 1 \text{ when } x \geq 0 \\ x + 1 \text{ when } x < 0 \end{cases}$$

find $\lim_{x \downarrow 0} f(x)$ and $\lim_{x \uparrow 0} f(x)$. What can you say about

$$\lim_{x \to 0} f(x)?$$

52. If f is defined by the rule

$$f(x) = \begin{cases} x \text{ when } x < 1 \\ 2 - x \text{ when } x > 1 \end{cases}$$

find $\lim_{x \downarrow 1} f(x)$ and $\lim_{x \uparrow 1} f(x)$. What can you say about

$$\lim_{x \to 1} f(x)?$$

53. If $f(x) = \sqrt{x - 3} + \sqrt{3 - x}$, what can you say about

$$\lim_{x \to 3} f(x), \lim_{x \downarrow 3} f(x), \lim_{x \uparrow 3} f(x)?$$

2.5
PROPERTIES OF LIMITS

The last section (an optional discussion of the formal definition of limit) is not essential for the evaluation of most limits encountered in the early parts of calculus. It is unnecessary to be technical when the value of a limit is apparent. Moreover, a difficult limit may often be reduced to easy ones by an appropriate application of the properties of limits. Suppose, for example, that you feel like balking at the statement $\lim_{x \to 0} \tan x = 0$. Of course it is clear from the graph of tangent (Figure 7, Section 2.3) that the answer is 0. But it is not trivial to prove it. One way to avoid a confrontation with the formal definition of the last section is to observe that we *have* proved

$$\lim_{x \to 0} \sin x = 0 \qquad \text{and} \qquad \lim_{x \to 0} \cos x = 1$$

(See Problems 20 and 29 in the last section, or Problem 66, Section 2.3.) It seems reasonable to conclude that

$$\tan x = \frac{\sin x}{\cos x} \to \frac{0}{1} = 0 \qquad \text{as } x \to 0$$

or (more formally)

$$\lim \tan x = \lim \frac{\sin x}{\cos x} = \frac{\lim \sin x}{\lim \cos x} = \frac{0}{1} = 0$$

(We suppress $x \to 0$ in each limit to simplify the notation.) From this it is clear that we are making an assumption: How do we know that the limit of a quotient is the quotient of the limits?

Another example is

$$\lim_{x \to 4} \frac{x - 4}{\sqrt{x} - 2} = \lim_{x \to 4} (\sqrt{x} + 2) = 4$$

To defend the last step, we may use the definition directly. But if we have already done that in the case of

$$\lim_{x \to 4} \sqrt{x} = 2 \qquad \text{(Problem 9, Section 2.4)}$$

it hardly seems worthwhile to suffer through it again. A better approach is to write

$$\lim_{x \to 4} (\sqrt{x} + 2) = \lim_{x \to 4} \sqrt{x} + \lim_{x \to 4} 2 = 2 + 2 = 4$$

Again, however, note the assumptions: The limit of a sum is the sum of the limits; the limit of a constant is the constant.

Perhaps these assumptions strike you as obvious. But some of them are not easy to establish in general. In this section we state theorems about limits for future reference, offering only incomplete proofs (with more details in the problem set).

□ **Theorem 1** (*Algebra of Limits*)   Let f and g be real functions whose sum, difference, product, and quotient are defined, and suppose that

$$\lim_{x \to a} f(x) \qquad \text{and} \qquad \lim_{x \to a} g(x)$$

exist. Then

1. $\lim (f + g)(x) = \lim f(x) + \lim g(x)$
2. $\lim (f - g)(x) = \lim f(x) - \lim g(x)$
3. $\lim (fg)(x) = [\lim f(x)][\lim g(x)]$
4. $\lim (f/g)(x) = \dfrac{\lim f(x)}{\lim g(x)}, \qquad$ provided that $\lim g(x) \neq 0$

That is, the limit of a sum (difference, product, quotient) of two functions is the sum (difference, product, quotient) of their limits.   □

In Theorem 1 we are assuming that the domains of f and g overlap, so that there is a common domain where they are both defined. Their sum, difference, product, and quotient are the functions $f + g, f - g, fg, f/g$ defined by

$$(f + g)(x) = f(x) + g(x) \qquad (f - g)(x) = f(x) - g(x)$$
$$(fg)(x) = f(x)g(x) \qquad (f/g)(x) = f(x)/g(x)$$

(The quotient requires the additional assumption that the common domain of f and g contains points at which g is not zero, so that f/g has a domain.)

It is doubtful whether a straightforward proof of Theorem 1 is very enlightening, particularly since the technical details are gruesome in places. In the problem set we will outline some ingenious ways that mathematicians have devised to avoid the difficulties; for those students who did not skip the last section, we offer a proof of (1) in an optional note at the end of this section.

In many applications of Theorem 1 two special limits are needed, namely

$$\lim_{x \to a} x = a \quad \text{and} \quad \lim_{x \to a} c = c$$

Each of these sounds obvious when put into words:

• The function $f(x) = x$ approaches a when x approaches a.
• The constant function $f(x) = c$ approaches c when x approaches a.

We will take them to be obvious, with the observation that proofs based on the technical definition of limit in the last section can also be given.

A special case of (3) in Theorem 1 is worth noting, namely

$$\lim c\, g(x) = c \lim g(x)$$

In other words a constant factor may be "moved across the limit sign." This follows from (3) by taking $f(x) = c$ and using the fact that $\lim c = c$.

▫ **Example 1**   Use the algebra of limits to evaluate $\lim_{x \to 1} (5x - 2)$.

Solution: You should supply a reason for each step in the following:

$$\lim_{x \to 1} (5x - 2) = \lim_{x \to 1} 5x - \lim_{x \to 1} 2 = 5 \lim_{x \to 1} x - 2 = 5(1) - 2 = 3 \qquad ▫$$

More generally, if $f(x) = ax + b$ is any linear function and $x_0 \in \mathfrak{R}$, then

$$\lim_{x \to x_0} f(x) = \lim_{x \to x_0} (ax + b) = \lim_{x \to x_0} ax + \lim_{x \to x_0} b = a \lim_{x \to x_0} x + b = ax_0 + b$$

Having gone through this tedium once, you should not repeat it! For the answer is simply $f(x_0)$. In other words, if f is a linear function and $x_0 \in \mathfrak{R}$,

$$\lim_{x \to x_0} f(x) = f(x_0)$$

The limit may be found by evaluating the function at the point approached.
A function with this property is called *continuous at* x_0, an idea of such importance that we shall return to it repeatedly throughout calculus.

Continuity of a Real Function

Suppose that x_0 is a point of the domain of the real function f. We say that f is *continuous at* x_0 if

$$\lim_{x \to x_0} f(x) = f(x_0)$$

We will return to this idea in Chapter 4. For the present we remark (and you may assume) that virtually every (real) function we have named in this book is continuous at each point of its domain. This includes the functions classified in Section 1.5 (constant, linear, power, root, polynomial, and rational) and the six trigonometric functions (sine, cosine, tangent, cotangent, secant, and cosecant).

On the other hand, you should not use the formula

$$\lim_{x \to x_0} f(x) = f(x_0)$$

indiscriminately. If

$$f(x) = \frac{x^2 - 1}{x - 1}$$

for example, we cannot find $\lim_{x \to 1} f(x)$ by evaluating $f(1)$, because 1 is not in the domain. Even when the point of approach is in the domain, the formula is not necessarily correct, as the next example shows.

□ **Example 2**   Define the function f by

$$f(x) = \begin{cases} \dfrac{x^2 - 4}{x - 2} & \text{if } x \neq 2 \\ 1 & \text{if } x = 2 \end{cases}$$

Since $f(x) = x + 2$ when $x \neq 2$, while $f(2) = 1$, the graph of f is a straight line with a point displaced (Figure 1). As you can see,

$$\lim_{x \to 2} f(x) = 4 \neq f(2)$$

so f is not continuous at 2.   □

Figure 1

Straight line with a point displaced

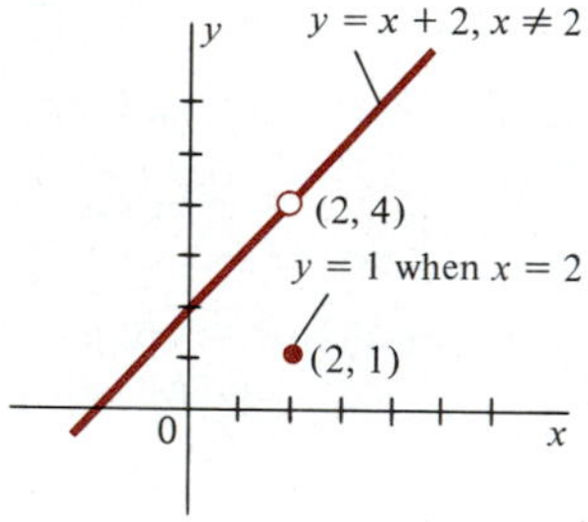

□ **Example 3**   Confirm that the function

$$f(x) = \frac{3x^2 + 1}{x - 2}$$

is continuous at each point $x_0 \neq 2$.

Solution: The problem is to show that

$$\lim_{x \to x_0} f(x) = f(x_0)$$

Suppressing $x \to x_0$ in each limit to save writing, we have

$$\lim f(x) = \lim \frac{3x^2 + 1}{x - 2} = \frac{\lim (3x^2 + 1)}{\lim (x - 2)} = \frac{\lim 3x^2 + \lim 1}{\lim x - \lim 2} = \frac{3 \lim x^2 + 1}{x_0 - 2}$$

$$= \frac{3(\lim x)(\lim x) + 1}{x_0 - 2} = \frac{3(x_0)(x_0) + 1}{x_0 - 2} = \frac{3x_0^2 + 1}{x_0 - 2} = f(x_0)$$

As you can see, establishing continuity of a rational function is just a matter of applying the algebra of limits repeatedly, until we reduce the problem to the evaluation of "obvious" limits.   □

Figure 2
Graph of $y = \tan x \cot x$

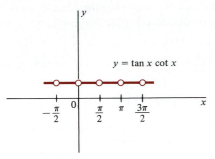

□ **Example 4**   The function $\tan x \cot x$ has the constant value 1 wherever it is defined. (See Figure 2.) Its graph is the horizontal line $y = 1$ with holes punched out at

$$(0,1), \left(\pm\frac{\pi}{2},1\right), (\pm\pi,1), \ldots$$

from which it is apparent that

$$\lim_{x \to 0} \tan x \cot x = 1$$

On the other hand, suppose we apply (3) of Theorem 1, writing

$$\lim_{x \to 0} \tan x \cot x = \left(\lim_{x \to 0} \tan x\right)\left(\lim_{x \to 0} \cot x\right) = (0)\left(\lim_{x \to 0} \cot x\right) = 0$$

If this is correct, we have proved that $1 = 0$. Before reading on, can you find the fallacy?

The point is that $\lim f(x)g(x) = [\lim f(x)][\lim g(x)]$ only when the limits on the right side exist. It is tempting to argue that $(0)(\lim \cot x) = 0$ on the premise that "zero times anything is zero." The correct statement, of course, is that zero times any real number is zero. Since

$$\lim_{x \to 0} \cot x \text{ does not exist} \qquad \text{(Figure 8, Section 2.3)}$$

our "equation" reads $1 = (0)(\text{horseradish})$.   □

The next theorem is easier to illustrate than it is to state. Hence we present some examples first.

□ **Example 5**   To find $\lim_{x \to 0} \sqrt{\cos x}$, we reason intuitively as follows. As $x \to 0$, $\cos x$ approaches 1. The square root of a number close to 1 is itself close to 1. Hence

$$\lim_{x \to 0} \sqrt{\cos x} = 1$$

An analysis of this reasoning reveals that we are treating $\sqrt{\cos x}$ as a *composite function* (cosine followed by square root). The "inside function" is $g(x) = \cos x$, the "outside function" is $f(x) = \sqrt{x}$; the composition is

$$f[g(x)] = f(\cos x) = \sqrt{\cos x}$$

What we did to evaluate the limit was to work from the inside out. That is, we ignored the outside function temporarily while we determined that $\cos x \to 1$ as $x \to 0$. Then we applied the outside function to obtain $\sqrt{\cos x} \to \sqrt{1}$ as $x \to 0$. In symbols,

$$\lim \sqrt{\cos x} = \sqrt{\lim \cos x} = \sqrt{1} = 1$$

or $\lim f[g(x)] = f[\lim g(x)]$.

The heart of the matter is the formula

$$\lim f[g(x)] = f[\lim g(x)]$$

which indicates how the limit of a composition is found by working from the inside out. An informal way of putting it is that the symbols "lim" and "f" are interchanged. Instead of finding the limit of f we evaluate f of the limit. (Sufficient conditions for this to work are discussed after Example 6.) □

□ **Example 6**   To find $\lim_{x \to 1} (3x - 1)^3$, we could apply (3) of Theorem 1 repeatedly, writing

$$\lim (3x - 1)^3 = [\lim (3x - 1)][\lim (3x - 1)][\lim (3x - 1)] = 2^3 = 8$$

It is easier, however, to think of $(3x - 1)^3$ as a composition, namely the linear function $g(x) = 3x - 1$ followed by the power function $f(x) = x^3$. Then

$$\lim (3x - 1)^3 = \lim f[g(x)] = f[\lim g(x)] = f(2) = 8$$

When you get used to this idea, you will not need to introduce the functional symbols f and g. Just interchange "lim" and "outside function," writing

$$\lim (3x - 1)^3 = [\lim (3x - 1)]^3 = 2^3 = 8$$

In other words, work from the inside out. □

The obvious question to raise at this point is under what conditions does the formula

$$\lim_{x \to a} f[g(x)] = f\left[\lim_{x \to a} g(x)\right]$$

apply? The obvious answer is that $\lim g(x)$ must exist and lie in the domain of f, since otherwise the right side would not make sense. It is not quite that simple, however. The interchange of "lim" and "f" requires f to have the property

$$\lim_{u \to u_0} f(u) = f(u_0) \qquad \text{where } u_0 = \lim_{x \to a} g(x)$$

(That is, f should be continuous at u_0.) For this condition is equivalent to

$$\lim_{u \to u_0} f(u) = f\left(\lim_{u \to u_0} u\right) \qquad \text{(why?)}$$

which is precisely the interchange property needed.

□ **Theorem 2** (*Composite Function Theorem*)   Let f and g be real functions whose composition $f[g(x)]$ is defined. Then

$$\lim_{x \to a} f[g(x)] = f\left[\lim_{x \to a} g(x)\right]$$

provided that $u_0 = \lim_{x \to a} g(x)$ exists and f is continuous at u_0. □

For those students who read the last section, we offer an informal proof of Theorem 2 in an optional note at the end of this section. Also see Problems 50 and 51 in the problem set.

□ **Example 7**   Explain why the Composite Function Theorem does not apply to

$$\lim_{x \to \pi} \tan \frac{x}{2}$$

Solution: It is incorrect to write

$$\lim_{x \to \pi} \tan \frac{x}{2} = \tan \left(\lim_{x \to \pi} \frac{x}{2} \right) = \tan \frac{\pi}{2}$$

For although the limit of the inside function exists, namely

$$u_0 = \lim_{x \to \pi} \frac{x}{2} = \frac{\pi}{2}$$

the outside function (tangent) is not continuous at u_0. [In fact, $\tan (\pi/2)$ is not even defined.]

Despite this remark, it is still possible to use the idea of the Composite Function Theorem. When x approaches π, the inside function $x/2$ approaches $\pi/2$. Since tangent is unbounded when its input is allowed to approach $\pi/2$ (Figure 7, Section 2.3), we conclude that

$$\lim_{x \to \pi} \tan \frac{x}{2} \text{ does not exist} \qquad \qquad \square$$

Our next theorem is so geometrically apparent (Figure 3) that we omit its proof altogether. (See the problem set, however, for hints on how to construct one.)

□ **Theorem 3** (*Squeeze Play Theorem*)   Suppose that $f(x)$ is between $g(x)$ and $h(x)$ for all x near a. If $g(x)$ and $h(x)$ have a common limit as $x \to a$ (say L), then

$$\boxed{\lim_{x \to a} f(x) = L} \qquad \qquad \square$$

We have already used the Squeeze Play Theorem in our argument that

$$\lim_{t \to 0} \frac{\sin t}{t} = 1 \qquad \text{(Section 2.3)}$$

Recall that we wrote

$$1 > \frac{\sin t}{t} > \cos t \qquad \text{for all } t \text{ near } 0$$

Since $\lim_{t \to 0} 1 = 1$ and $\lim_{t \to 0} \cos t = 1$

we have $(\sin t)/t$ squeezed between two functions with a common limit. Hence its limit is the same.

Another application of the Squeeze Play Theorem may be seen in Example 9, Section 2.4, where we evaluated

$$\lim_{x \to 0} x \sin \frac{1}{x}$$

Figure 3
Squeeze Play Theorem

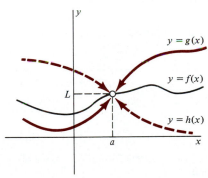

$y = g(x)$

$y = f(x)$

L

$y = h(x)$

a

Since $f(x) = x \sin(1/x)$ is between $g(x) = x$ and $h(x) = -x$ for all $x \neq 0$ (Figure 11, Section 2.4) and since

$$\lim_{x \to 0} g(x) = \lim_{x \to 0} h(x) = 0$$

the limit of $f(x)$ must be 0.

Optional Note (*on the proof of* (1) *in Theorem* 1)

Let
$$A = \lim_{x \to a} f(x) \qquad \text{and} \qquad B = \lim_{x \to a} g(x)$$

The problem is to prove that

$$\lim_{x \to a} (f + g)(x) = A + B$$

In other words, given $\varepsilon > 0$, we must name $\delta > 0$ such that

$$0 < |x - a| < \delta \Rightarrow |(f + g)(x) - (A + B)| < \varepsilon$$

But

$$|(f + g)(x) - (A + B)| = |[f(x) - A] + [g(x) - B]| \leq |f(x) - A| + |g(x) - B|$$

(See the Triangle Inequality in Section 1.1.) To force this to be less than ε, we need only force $f(x)$ and $g(x)$ to be within $\varepsilon/2$ units of A and B, respectively. Since we know from the definitions of A and B that this can be done by keeping x sufficiently close to a, the proof should be clear in outline.

　　To be precise in detail, we know there are positive numbers δ_1 and δ_2 such that

$$0 < |x - a| < \delta_1 \Rightarrow |f(x) - A| < \frac{\varepsilon}{2}$$

and
$$0 < |x - a| < \delta_2 \Rightarrow |g(x) - B| < \frac{\varepsilon}{2}$$

Let δ be the smaller of δ_1 and δ_2. Then

$$0 < |x - a| < \delta \Rightarrow |f(x) - A| < \frac{\varepsilon}{2} \text{ and } |g(x) - B| < \frac{\varepsilon}{2}$$

$$\Rightarrow |(f + g)(x) - (A + B)| < \frac{\varepsilon}{2} + \frac{\varepsilon}{2} = \varepsilon$$

Optional Note (*on the proof of Theorem* 2)　　Instead of a formal proof, we offer the following intuitive argument. To show that

$$\lim_{x \to a} f[g(x)] = f(u_0)$$

we suppose that a skeptic has named a neighborhood of $f(u_0)$. (See Figure 4.)

Figure 4
Intuitive argument for Composite
Function Theorem

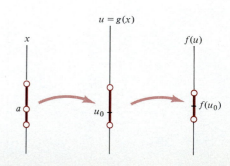

Our problem is to confine $f[g(x)]$ to this neighborhood by keeping x sufficiently close to a. We do this in a chain of steps:

1. Confine $f[g(x)]$ to the given neighborhood of $f(u_0)$ by controlling $u = g(x)$. We know this can be done because

$$\lim_{u \to u_0} f(u) = f(u_0)$$

2. Force $g(x)$ as close to u_0 as required in (1) by keeping x near a. We know this can be done because

$$\lim_{x \to a} g(x) = u_0$$

A formal version of this argument is outlined in Problems 50 and 51 in the problem set.

Problem Set 2.5

Evaluate each of the following by using properties of limits.

1. $\lim\limits_{x \to 1} (7x + 3)$

2. $\lim\limits_{x \to 2} (9x - 5)$

3. $\lim\limits_{x \to 3} \dfrac{x^2 - 9}{x - 3}$

4. $\lim\limits_{x \to 1} \dfrac{x^2 - 1}{x - 1}$

5. $\lim\limits_{x \to 2} (5 - x^2)$

6. $\lim\limits_{x \to 2} (x^3 + x)$

7. $\lim\limits_{x \to 5} \sqrt{x - 5}$ *Hint:* Use Theorem 2 with continuity of root functions.

8. $\lim\limits_{x \to 1} \sqrt{9 - x^2}$

9. $\lim\limits_{x \to 4} x^{3/2}$

10. $\lim\limits_{x \to 8} x^{2/3}$

11. $\lim\limits_{x \to 1} \dfrac{1}{x}$

12. $\lim\limits_{x \to 2} \dfrac{4}{x}$

13. $\lim\limits_{x \to 0} \dfrac{x - 3}{2x + 1}$

14. $\lim\limits_{x \to 1} \dfrac{3x + 1}{3x - 1}$

15. $\lim\limits_{x \to 1} \dfrac{x^3 - 1}{x - 1}$

16. $\lim\limits_{x \to 3} \dfrac{x - 3}{3 - x}$

17. $\lim\limits_{x \to 2} \sqrt{2 - x}$

18. $\lim\limits_{x \to 4} \sqrt{x(x - 3)}$

19. $\lim\limits_{x \to 0} \sin 2x$ *Hint:* Use Theorem 2 with continuity of trigonometric functions.

20. $\lim\limits_{x \to 0} \cos 2x$

21. $\lim\limits_{x \to \pi} \cot \dfrac{x}{2}$

22. $\lim\limits_{x \to 0} \sec \sqrt{x}$

23. $\lim\limits_{x \to 0} x \cos \dfrac{1}{x}$ *Hint:* Use Theorem 3.

24. $\lim\limits_{x \to 0} x^2 \sin \dfrac{1}{x}$

25. $\lim\limits_{x \to \pi} \sqrt{1 - \sin^2 x}$

26. $\lim\limits_{x \to 3\pi/2} \sqrt{1 - \cos^2 x}$

27. $\lim\limits_{x \to 0} |\sin x - \cos x|$

28. $\lim\limits_{x \to \pi/4} |\sec x + \tan x|$

29. $\lim\limits_{x \to 0} \dfrac{\sin x + x}{x}$

30. $\lim\limits_{x \to 0} \dfrac{2x - 3 \sin x}{x}$

31. In Section 2.3 (Example 1) we argued that

$$\lim_{h \to 0} \frac{\cos h - 1}{h} = 0$$

Look at the argument again, this time noting what theorems about limits were used.

In Problem Set 2.3 you were asked to evaluate the following limits. Look at them again, this time indicating what theorems about limits are used.

32. $\lim\limits_{t \to 0} \dfrac{t}{\sin t}$

33. $\lim\limits_{x \to 0} \dfrac{\sin 3x}{x}$

34. $\lim\limits_{x \to 0} \dfrac{x}{\sin 2x}$

35. $\lim\limits_{t \to 0} \dfrac{t^2}{\sin t}$

36. $\lim\limits_{t \to 0} t \cot t$

37. $\lim\limits_{t \to 0} \dfrac{\tan t}{t}$

In the text we said that virtually every function named in this book is continuous at every point of its domain. The following problems are designed to justify that statement (in part).

38. A polynomial is a function of the form

$$P(x) = a_0 + a_1 x + a_2 x^2 + \cdots + a_n x^n$$

where n is a nonnegative integer and each a_k is a real number. Show that if $x_0 \in \mathcal{R}$, then

$$\lim_{x \to x_0} P(x) = P(x_0)$$

(Thus a polynomial is continuous at every point $x_0 \in \mathcal{R}$.)

Hint: We already know that

$$\lim_{x \to x_0} c = c \quad \text{and} \quad \lim_{x \to x_0} x = x_0$$

39. Let x_0 be any point of the domain of the rational function

$$f(x) = \frac{P(x)}{Q(x)} \quad (P \text{ and } Q \text{ being polynomials})$$

Why is f continuous at x_0?

40. We know that sine and cosine are continuous at 0, that is,

$$\lim_{x \to 0} \sin x = \sin 0 \quad \text{and} \quad \lim_{x \to 0} \cos x = \cos 0$$

(See Problems 20 and 29, Section 2.4, or Problem 66, Section 2.3.) To prove that sine is continuous at any $x_0 \in \mathcal{R}$, we lift ourselves by our bootstraps:

(a) Confirm that

$$\sin x = \sin [(x - x_0) + x_0]$$
$$= \sin (x - x_0) \cos x_0 + \cos (x - x_0) \sin x_0$$

(b) Why does it follow that

$$\lim_{x \to x_0} \sin x = \cos x_0 \cdot \lim_{x \to x_0} \sin (x - x_0)$$
$$+ \sin x_0 \cdot \lim_{x \to x_0} \cos (x - x_0)?$$

(c) Now explain why $\lim_{x \to x_0} \sin x = \sin x_0$.

41. Prove that cosine is continuous at every $x_0 \in \mathcal{R}$ by imitating Problem 40.

42. Why does it follow from Problems 40 and 41 that tangent, cotangent, secant, and cosecant are continuous wherever they are defined?

In the text we said that the technical details of the proof of Theorem 1 are gruesome in places. There are clever ways to avoid them, however; if you are interested (and have read Section 2.4) do Problems 43–49.

43. Use the definition of limit to show that if

$$\lim_{x \to a} g(x)$$

exists, then (for each $c \in \mathcal{R}$)

$$\lim_{x \to a} cg(x) = c \lim_{x \to a} g(x)$$

44. Now prove (2) of Theorem 1.

45. Use the definition of limit to show that if

$$\lim_{x \to a} f(x) = 0 \quad \text{and} \quad \lim_{x \to a} g(x) = 0$$

then $\lim_{x \to a} (fg)(x) = 0$. (This proves (3) of Theorem 1 in the special case where the limits of f and g are 0.)

46. Now prove (3) of Theorem 1 by letting

$$A = \lim_{x \to a} f(x) \quad \text{and} \quad B = \lim_{x \to a} g(x)$$

and by writing

$$(fg)(x) = [f(x) - A][g(x) - B] + Bf(x) + Ag(x) - AB$$

47. Use the definition of limit to show that if $u_0 \neq 0$, then

$$\lim_{u \to u_0} \frac{1}{u} = \frac{1}{u_0}$$

Hint: Imitate Example 5, Section 2.4.

48. Suppose that $\lim_{x \to a} g(x)$ exists and is not 0. Why does it follow from Problem 47 that

$$\lim_{x \to a} \frac{1}{g(x)} = \frac{1}{\lim_{x \to a} g(x)}?$$

Hint: Use the Composite Function Theorem. (*Objection:* That comes after Theorem 1. *Answer:* Its proof is independent of Theorem 1. See Problems 50 and 51.)

49. Now prove (4) of Theorem 1.

50. Explain why the Composite Function Theorem in the text is equivalent to the statement that if

$$\lim_{x \to a} g(x) = u_0 \quad \text{and} \quad \lim_{u \to u_0} f(u) = f(u_0)$$

then $\lim_{x \to a} f[g(x)] = f(u_0)$.

51. To prove the version of the Composite Function Theorem stated in Problem 50, let $L = f(u_0)$ and assume that $\varepsilon > 0$ has been given. The problem is to name $\delta > 0$ such that

$$0 < |x - a| < \delta \Rightarrow |f[g(x)] - L| < \varepsilon$$

(a) Explain why there is a $\rho > 0$ such that

$$|u - u_0| < \rho \Rightarrow |f(u) - L| < \varepsilon$$

(The symbol ρ is the lowercase Greek letter *rho*.)

(b) Explain why there is a $\delta > 0$ such that

$$0 < |x - a| < \delta \Rightarrow |g(x) - u_0| < \rho$$

(c) Why does it follow from parts (a) and (b) that

$$0 < |x - a| < \delta \Rightarrow |f[g(x)] - L| < \varepsilon?$$

52. To prove the Squeeze Play Theorem, we suppose that a skeptic has named a neighborhood of L (say N). The problem is to force $f(x) \in N$ by keeping x near a.

(a) Explain why $g(x)$ and $h(x)$ can be forced into N by keeping x near a.

(b) Why does it follow that (for these values of x) $f(x)$ is in N?

(c) Can you make this argument more formal, so that a confirmed skeptic would believe it?

True–False Quiz

1. The difference quotient associated with $f(x) = x^2$ at $x = 3$ is $Q(x) = x + 3$ $(x \neq 3)$.

2. The tangent to the curve $y = 1 - x^2$ at $(1,0)$ is the line $2x + y - 2 = 0$.

3. The slope of the graph of $y = x - |x|$ at the origin is 0.

4. The slope of the curve $y = \sqrt{4 - x^2}$ at $(0,2)$ is 0.

5. An object whose position at time t is $s = t^3$ has speed 3 at $t = 1$.

6. An object whose position at time t is $s = \sqrt{t}$ has velocity 0 at $t = 0$.

7. If $f(x) = 2x - 5$, then

$$\lim_{x \to 3} \frac{f(x) - f(3)}{x - 3} = 2$$

8. If $y = ax + b$, then $dy/dx = a$.

9. If $u = t^2 - 1$, then $D_t u = 2t$.

10. $D_x(x - x^2) = 1 - 2x$.

11. If $f(x) = x^3$, then $f'(-x) = f'(x)$ for all x.

12. The function $f(x) = 1/x$ is differentiable in its domain.

13. The function $f(x) = x|x|$ is differentiable at $x = 0$.

14. $\cos 2 > \cos 2°$.

15. The function $f(t) = \tan 2t$ is even.

16. The range of the function $f(t) = \sin t / \cos t$ is $\mathcal{R}$.

17. The distance between the points $(\tan 2, 3)$ and $(0,4)$ is greater than 1.

18. $(\sin t + \cos t)^2 - 1 = \sin 2t$ for all t.

19. If $f(x) = \sin x$, then $f(u + v) = f(u) + f(v)$ for all u and v.

20. $\displaystyle \lim_{t \to 0} \frac{t}{\sin t} = 0$.

21. If $f(t) = \sin t$, then $f'(0) = 0$.

22. If $f(x) = \sin x$, then

$$\lim_{h \to 0} \frac{f(x + h) - f(x)}{h} = \cos x$$

23. $\dfrac{d}{dt}(\sin^2 t + \cos^2 t) = 0$.

24. $\displaystyle \lim_{x \to \pi} \tan \frac{3x}{4} = 1$.

25. $\displaystyle \lim_{t \to \pi} \frac{\sqrt{1 - \sin^2 t}}{\cos t} = 1$.

26. $\displaystyle \lim_{t \to 0} \cos t = 0$.

27. The coordinates of the hole in the graph of $y = \dfrac{x^2 - 16}{x - 4}$ are $(4, 8)$.

28. $\displaystyle \lim_{x \downarrow 3} \frac{|3 - x|}{3 - x} = 1$.

29. If $P(x)$ is a polynomial, then $\displaystyle \lim_{x \to c} P(x) = P(c)$.

30. $\displaystyle \lim_{x \to 1} \left(\frac{x}{x - 1} - \frac{1}{x - 1} \right) = \lim_{x \to 1} \frac{x}{x - 1} - \lim_{x \to 1} \frac{1}{x - 1}$.

31. If $f(x) = 1/x$ and $g(x) = \sin x$, then

$$\lim_{x \to 0} f[g(x)] = f\left[\lim_{x \to 0} g(x) \right]$$

Additional Problems

In each of the following, find an equation of the tangent to the graph at the given point.

1. $y = x^2 - 3x$ at $(1, -2)$

2. $y = 4 - x^2$ at $(2, 0)$

3. $y = \dfrac{1}{x - 1}$ at $(2, 1)$

4. $y = \dfrac{x}{x - 1}$ at $(2, 2)$

5. $y = \sqrt{x - 2}$ at $(2, 0)$

6. $y = \sqrt[3]{x}$ at $(0, 0)$

7. $y = x^{5/3}$ at $(0, 0)$

8. $y = (x - 1)\sqrt[4]{x}$ at $(1, 0)$

9. $y = \cos x$ at $(\pi/2, 0)$

10. $y = 3 \sin x$ at $(0, 0)$

In each of the following, s is the position of a moving object at time t. Find the velocity at the given instant.

11. $s = t^3 - t$ at 1

12. $s = t^4 - t$ at 1

13. $s = 2/\sqrt{t}$ at 1

14. $s = 4/t^2$ at 2

15. $s = \sin 2t$ at 0

16. $s = \tan t$ at 0

In each of the following, find $f'(x)$ by evaluating the limit of an appropriate difference quotient.

17. $f(x) = x^2 + x$

18. $f(x) = x^2 - 3x$

19. $f(x) = x^3 + x^2$

20. $f(x) = x^4 - 3x$

21. $f(x) = \dfrac{x}{x - 5}$

22. $f(x) = \dfrac{x - 2}{x}$

23. $f(x) = x^{4/3}$

24. $f(x) = x^{3/4}$

25. Use the definition of derivative to show that if $f(x) = 3x - x^3$, then $f'(x) = 3(1 - x^2)$. Find where the graph of f is rising and where it is falling and sketch it.

26. Use the definition of derivative to show that if $f(x) = x^3 - 12x$, then $f'(x) = 3(x^2 - 4)$. Find where the graph of f is rising and falling and sketch it.

27. Use the definition of derivative to show that if $f(x) = 1 - |x|$, then

$$f'(x) = \begin{cases} -1 \text{ when } x > 0 \\ 1 \text{ when } x < 0 \end{cases}$$

while $f'(0)$ is undefined. Sketch the graph of f.

28. Use the interpretation of derivative as slope to explain why the derivative of $y = \sqrt{a^2 - x^2}$ is

$$\frac{dy}{dx} = \frac{-x}{\sqrt{a^2 - x^2}}$$

Where is the function $f(x) = \sqrt{a^2 - x^2}$ differentiable?

29. A ball is thrown straight upward. After t seconds its height above the ground (in feet) is $h = 96t - 16t^2$.

   (a) Show that the velocity of the ball at time t is $v(t) = 96 - 32t$.

   (b) When does the ball reach its highest point and how high does it rise?

   (c) When does the ball return to the ground and what is its velocity then?

30. A ball is thrown straight upward from the top of a tower. After t seconds its height above the ground (in feet) is $h = 48 + 32t - 16t^2$.

   (a) Show that the velocity of the ball at time t is $v(t) = 32 - 32t$.

   (b) When does the ball reach its highest point and how high (above the ground) does it rise?

   (c) When does the ball return to the top of the tower? with what speed?

31. If the position of a moving object at time t is $s = \sin t$, how fast is the object moving when $t = 2\pi/3$? in what direction?

32. Repeat Problem 31 with $s = \cos t$.

33. If $f(x) = \cos x$, evaluate

$$\lim_{x \to 0} \frac{f(x) - f(0)}{x}$$

by making use of the definition of derivative.

34. If $f(x) = \sin x$, evaluate

$$\lim_{x \to \pi} \frac{f(x) - f(\pi)}{x - \pi}$$

by making use of the definition of derivative.

In each of the following, find $f'(t)$.

35. $f(t) = \sec^2 t - \tan^2 t$
36. $f(t) = \cot^2 t - \csc^2 t$
37. $f(t) = \sin t \csc t$
38. $f(t) = \cos t \sec t$
39. $f(t) = \sin(-t)$
40. $f(t) = \cos(-t)$
41. $f(t) = \sin(\pi/2 - t)$
42. $f(t) = \cos(\pi/2 - t)$
43. $f(t) = \sin(t + \pi)$
44. $f(t) = \cos(\pi - t)$

45. If $f(t) = \sin 2t$, show that $f'(t) = 2 \cos 2t$.

46. If $f(t) = \cos 2t$, show that $f'(t) = -2 \sin 2t$.

47. What are the coordinates of the hole in the graph of

$$y = \frac{x^3 - 8}{x - 2}?$$

48. Evaluate

$$\lim_{t \to 0} \frac{t}{\cot t}$$

Is it legitimate to write

$$\lim \frac{t}{\cot t} = \frac{\lim t}{\lim \cot t}$$

in this case? Explain.

49. Evaluate

$$\lim_{x \downarrow 0} \frac{|x|}{x} \cos x \quad \text{and} \quad \lim_{x \uparrow 0} \frac{|x|}{x} \cos x$$

Does the limit as $x \to 0$ exist?

50. Replace $\cos x$ by $\sin x$ in the preceding problem and answer the same questions.

51. Evaluate

$$\lim_{x \to 0} \frac{\sin x + 2x}{x}$$

52. Assuming that the function $f(x) = 2^x$ is continuous, evaluate

$$\lim_{x \to 0} \cos x$$

3 Techniques of Differentiation

The emphasis of the last chapter was on the definition of the derivative and the origins of that definition in problems of geometry and mechanics. In this chapter we are interested in the more mundane question of how the derivatives of common functions may be found quickly and painlessly. The idea is to become proficient in the art of differentiation, so that later applications of the derivative will be undistracted by technical problems.

By the end of this chapter you should be able to differentiate powers, roots, and algebraic combinations of functions, compositions of functions, "implicitly defined" functions, and the functions of trigonometry. In short, you should have developed a repertoire.

3.1
THE POWER RULE AND THE ALGEBRA OF DERIVATIVES (I)

One of the remarkable formulas of calculus is

$$D_x(x^r) = rx^{r-1}$$

which is known as the *Power Rule*. When you have used the definition of derivative to work out a few examples like

$$D_x(x^2) = 2x, \qquad D_x(x^3) = 3x^2, \qquad D_x(x^4) = 4x^3$$

the formula may not seem especially noteworthy; what is remarkable is that it holds for any real number r. For example:

$$D_x(x^{-2}) = -2x^{-3} \qquad \text{(Problem 35, Section 2.2)}$$
$$D_x(x^{2/3}) = \tfrac{2}{3}x^{-1/3} \qquad \text{(Problem 41, Section 2.2)}$$
$$D_x(x^{\pi}) = \pi x^{\pi-1} \qquad \text{(to be proved when we discuss}$$
$$\text{irrational exponents in Chapter 8)}$$

Naturally the formula has certain built-in limitations, as in

$$D_x(x^{1/2}) = \tfrac{1}{2}x^{-1/2}$$

The domain in this example is understood to consist of those values of x for which both $x^{1/2} = \sqrt{x}$ and $\tfrac{1}{2}x^{-1/2} = 1/(2\sqrt{x})$ are defined. Hence the domain is $(0, \infty)$.

We will prove the Power Rule in several stages by considering cases corresponding to the various types of exponent r may represent:

- *Case 1:* $r = n$, where n is a nonnegative integer ($n = 0, 1, 2, 3, \ldots$).
- *Case 2:* $r = 1/n$, where n is a positive integer greater than 1.
- *Case 3:* $r = p/q$, where p and q are positive integers ($q > 1$).
- *Case 4:* $r = -s$, where s is a positive rational number.
- *Case 5:* r is irrational.

As you can see, these cases cover every possible real exponent. In this section, however, we will be able to prove only the first two. We begin with Case 1, in which $r = n$ ($n = 0, 1, 2, 3, \ldots$).

When $n = 0$ and $n = 1$ the Power Rule reads $D_x(x^0) = 0x^{-1}$ and $D_x(x^1) = 1x^0$, or simply $D_x(1) = 0$ and $D_x(x) = 1$ when $x \neq 0$. Hence it is correct in the agreed domain. Of course in practice nobody uses it in these cases, since the formulas $D_x(1) = 0$ and $D_x(x) = 1$ are obvious without reference to the Power Rule (and they hold in $\Re$). Hence we assume that $n \geq 2$. The difference quotient associated with $f(x) = x^n$ at x is

$$Q(z) = \frac{f(z) - f(x)}{z - x} = \frac{z^n - x^n}{z - x} \qquad (z \neq x)$$

This reduces to

$$\frac{z^2 - x^2}{z - x} = \frac{(z - x)(z + x)}{z - x} = z + x \qquad \text{if } n = 2,$$

$$\frac{z^3 - x^3}{z - x} = \frac{(z - x)(z^2 + zx + x^2)}{z - x} = z^2 + zx + x^2 \qquad \text{if } n = 3,$$

$$\frac{z^4 - x^4}{z - x} = \frac{(z - x)(z^3 + z^2x + zx^2 + x^3)}{z - x}$$
$$= z^3 + z^2x + zx^2 + x^3 \qquad \text{if } n = 4,$$

and so on. The derivative is $f'(x) = \lim_{z \to x} Q(z)$, which turns out to be

$$\lim_{z \to x} (z + x) = 2x \qquad \text{if } n = 2,$$

$$\lim_{z \to x} (z^2 + zx + x^2) = 3x^2 \qquad \text{if } n = 3,$$

$$\lim_{z \to x} (z^3 + z^2x + zx^2 + x^3) = 4x^3 \qquad \text{if } n = 4,$$

and so on. In general we use the factoring formula

$$z^n - x^n = (z - x)(z^{n-1} + z^{n-2}x + z^{n-3}x^2 + \cdots + z^2x^{n-3} + zx^{n-2} + x^{n-1})$$

(which can be formally established by mathematical induction) to write

$$Q(z) = \frac{z^n - x^n}{z - x}$$

$$= z^{n-1} + z^{n-2}x + z^{n-3}x^2 + \cdots + z^2x^{n-3} + zx^{n-2} + x^{n-1}$$

When $z \to x$ this expression approaches $x^{n-1} + x^{n-1} + \cdots + x^{n-1}$ (n terms), so we conclude that

$$f'(x) = \lim_{z \to x} Q(z) = nx^{n-1}$$

Having proved Case 1, we are ready for Case 2. We defer the proof until the end of this section, however, so that we may present some examples and do some algebra of derivatives. For that purpose (and also in the problem set) we assume that all cases of the Power Rule are proved.

□ **Example 1**   Find the derivative of $y = 1/x^2$.

Solution: In the next section we will develop a Quotient Rule for the differentiation of fractions. In this example, however, it is not needed; the constant numerator enables us to write the fraction as a power, $y = x^{-2}$. The Power Rule then yields

$$\frac{dy}{dx} = -2x^{-3} = -\frac{2}{x^3} \qquad \qquad □$$

□ **Example 2**   The graph of $y = x^{3/2}$ (domain $x \geq 0$) contains the origin. To determine its slope there, we write

$$y' = \tfrac{3}{2}x^{1/2} = \tfrac{3}{2}\sqrt{x} = 0 \qquad \text{at } x = 0$$

Hence the curve should be drawn with a horizontal tangent at (0,0), as shown in Figure 1.   □

Figure 1
Graph of $y = x^{3/2}$

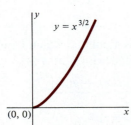

□ **Example 3**   As a contrast to Example 2, consider the behavior of $y = x^{2/3}$ (domain $\mathcal{R}$) at the origin. Since

$$y' = \tfrac{2}{3}x^{-1/3} = \frac{2}{3\sqrt[3]{x}}$$

Figure 2
Graph of $y = x^{2/3}$

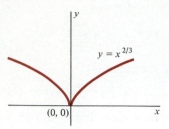

the slope is numerically large when x is near 0 (undefined at $x = 0$). The graph has a cusp at (0,0), as shown in Figure 2.   □

Now we turn to the algebra of derivatives, reminding you that if f and g are real functions with overlapping domains, their *sum, difference, product,* and *quotient* are the functions $f + g, f - g, fg, f/g$ defined by

$$(f + g)(x) = f(x) + g(x) \qquad (fg)(x) = f(x)g(x)$$
$$(f - g)(x) = f(x) - g(x) \qquad (f/g)(x) = f(x)/g(x) \qquad \text{if } g(x) \neq 0$$

(See Section 2.5.) The "algebra of derivatives" refers to formulas for differentiating such combinations of functions. Suppose, for example, that $f(x) = x^2$ and $g(x) = \sin x$. Then

$$(f + g)(x) = x^2 + \sin x$$

a sum whose derivative is tedious to find by a direct application of the definition. Since we already know that

$$D_x(x^2) = 2x \qquad \text{(Power Rule)}$$
$$D_x(\sin x) = \cos x \qquad \text{(Example 2, Section 2.3)}$$

it is easier to write

$$D_x(x^2 + \sin x) = D_x(x^2) + D_x(\sin x) = 2x + \cos x$$

Of course this raises a question! How do we know that the derivative of a sum is the sum of the derivatives? It sounds reasonable. On the other hand, the derivative of a product is not the product of the derivatives, as we will see in the next section. Thus, for example,

$$D_x(x^2 \sin x) \neq D_x(x^2) \cdot D_x(\sin x)$$

Evidently what we need are some theorems telling us what to expect. (To appreciate the need, consider the fact that the great Leibniz himself did not at first know how to differentiate a product!) We begin with the derivative of a sum.

In the domain where f and g are both differentiable,

$$\frac{d}{dx}[f(x) + g(x)] = \frac{d}{dx}f(x) + \frac{d}{dx}g(x)$$

Proof: Let $h = f + g$ and suppose that x is a point at which f and g are differentiable. The difference quotient associated with h at x is

$$\frac{h(z) - h(x)}{z - x} = \frac{[f(z) + g(z)] - [f(x) + g(x)]}{z - x}$$

$$= \frac{[f(z) - f(x)] + [g(z) - g(x)]}{z - x}$$

$$= \frac{f(z) - f(x)}{z - x} + \frac{g(z) - g(x)}{z - x}$$

Hence

$$h'(x) = \lim_{z \to x} \frac{h(z) - h(x)}{z - x}$$

$$= \lim_{z \to x} \left[\frac{f(z) - f(x)}{z - x} + \frac{g(z) - g(x)}{z - x} \right]$$

$$= \lim_{z \to x} \frac{f(z) - f(x)}{z - x} + \lim_{z \to x} \frac{g(z) - g(x)}{z - x} \qquad \text{(why?)}$$

$$= f'(x) + g'(x) \qquad \qquad \square$$

The derivative of a sum of functions is the sum of their derivatives.

This justifies the procedure suggested earlier, namely

$$D_x(x^2 + \sin x) = D_x(x^2) + D_x(\sin x)$$

The same idea applies to the sum of any (finite) number of functions, as in

$$\frac{d}{dx}[f(x) + g(x) + h(x)] = f'(x) + g'(x) + h'(x)$$

(To see why, group the first two functions and apply our sum rule to the sum of $f + g$ and h.)

In the domain where f is differentiable,

$$\frac{d}{dx}[cf(x)] = c \frac{d}{dx} f(x) \qquad \text{for every } c \in \mathcal{R}$$

In other words, a constant may be moved outside the differentiation sign.

Proof: See the problem set. $\square$

In the domain where f and g are both differentiable,

$$\frac{d}{dx}[f(x) - g(x)] = \frac{d}{dx} f(x) - \frac{d}{dx} g(x)$$

Proof: See the problem set. $\square$

These theorems may be used together to obtain the derivative of any *linear combination* of functions:

$$\frac{d}{dx}[af(x) + bg(x)] = a \frac{d}{dx} f(x) + b \frac{d}{dx} g(x)$$

where a and b are numbers and f and g are (differentiable) functions. Sometimes that fact is expressed by saying that the operation of differentiation is *linear*.

□ **Example 4**   Find the derivative of $f(x) = 5x^2 + 2 \sin x$.

Solution: Since $D_x(x^2) = 2x$ and $D_x(\sin x) = \cos x$, the linearity of the operator D_x allows us to write

$$f'(x) = D_x(5x^2 + 2 \sin x) = D_x(5x^2) + D_x(2 \sin x)$$
$$= 5D_x(x^2) + 2D_x(\sin x) = 5(2x) + 2(\cos x)$$
$$= 10x + 2 \cos x \qquad\qquad □$$

Optional Note (*on the proof of Case 2 of the Power Rule*)   Suppose that $r = 1/n$, where n is a positive integer greater than 1. The problem is to prove that

$$D_x(x^{1/n}) = \frac{1}{n} x^{(1/n)-1}$$

The difference quotient associated with $f(x) = x^{1/n}$ at x is

$$Q(z) = \frac{f(z) - f(x)}{z - x} = \frac{z^{1/n} - x^{1/n}}{z - x}$$

$$= \frac{b - a}{b^n - a^n} \qquad \text{(where } a = x^{1/n} \text{ and } b = z^{1/n})$$

$$= \left(\frac{b^n - a^n}{b - a}\right)^{-1}$$

We assume that root functions are continuous in their domain, that is,

$$\lim_{z \to x} z^{1/n} = x^{1/n}$$

In other words, $b \to a$ as $z \to x$. Hence

$$f'(x) = \lim_{z \to x} Q(z) = \lim_{b \to a} \left(\frac{b^n - a^n}{b - a}\right)^{-1}$$

Now apply the Composite Function Theorem (Section 2.5) to write

$$f'(x) = \left(\lim_{b \to a} \frac{b^n - a^n}{b - a}\right)^{-1}$$

The expression in parentheses is the derivative of x^n at the point a. (Why?) Hence by Case 1 it is na^{n-1} and we find

$$f'(x) = (na^{n-1})^{-1} = \frac{1}{n} a^{1-n} = \frac{1}{n} (x^{1/n})^{1-n} = \frac{1}{n} x^{(1/n)-1}$$

The continuity of root functions (assumed in Case 2) is not easy to establish by a direct application of the definition of limit. Can you prove, for example, that

$$\lim_{z \to x} \sqrt[3]{z} = \sqrt[3]{x} \;?$$

Power functions, on the other hand, are clearly continuous (the algebra of limits!).

Figure 3
Power and root functions

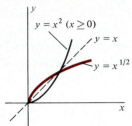

$y = x^2 \ (x \geq 0)$
$y = x$
$y = x^{1/2}$

For example,

$$\lim_{z \to x} z^3 = \lim_{z \to x} (z \cdot z \cdot z) = \left(\lim_{z \to x} z\right)\left(\lim_{z \to x} z\right)\left(\lim_{z \to x} z\right)$$

$$= x \cdot x \cdot x = x^3$$

Graphically, a root function $y = x^{1/n}$ is just the reflection in the line $y = x$ of the power function $y = x^n$ (restricted to $x \geq 0$ if n is even). If the one is continuous, surely the other is, too. (See Figure 3, which shows the case $n = 2$.) This argument can be made more rigorous, but it is hardly worthwhile when the geometry is so convincing.

Problem Set 3.1

Find the derivative of each of the following functions.

1. $y = x^2 - 2x$

2. $y = 1 - x^2$

3. $y = 2x^3 + x - 5$

4. $y = x^3 - 5x^2 + 2$

5. $y = x^7 - 2x^5 + 4x^3 - 5x$

6. $y = \frac{1}{2}x^6 + \frac{3}{4}x^4 - 3x^2 + 10$

7. $y = 2 \sin x + 3 \cos x$

8. $y = 5 \sin x - 6 \cos x$

9. $y = \dfrac{3}{x^2}$

10. $y = x^2 + \dfrac{1}{x^2}$

11. $y = x^3 + \dfrac{1}{x^3}$

12. $y = 3x^{-4}$

13. $y = \sqrt[3]{x}$

14. $y = 2\sqrt[4]{x}$

15. $y = \dfrac{3}{\sqrt{x}} + \sqrt{x}$

16. $y = x^{5/3} - x^{-1/3}$

In each of the following, find the indicated derivative.

17. $f'(1)$ if $f(x) = x^4 - 2x^3 + 2x - 8$

18. $F'(4)$ if $F(x) = x^{5/2}$

19. $g'(16)$ if $g(t) = 16\sqrt[4]{t}$

20. $\dfrac{du}{dx}$ if $u = \dfrac{1}{x^5}$

21. $D_x(5\sqrt{x})$

22. $\dfrac{d}{dt}\left(6 \cos t - \dfrac{5}{\sqrt{t}}\right)$

In each of the following, find an equation of the tangent to the curve at the indicated point.

23. $y = x^3 - x + 1$ at $x = 1$

24. $y = x^6 - 2x^4 + 2x^2 - 1$ at $x = 1$

25. $y = 4 - \sqrt{x}$ at $x = 1$

26. $y = 12/\sqrt[3]{x}$ at $x = 8$

27. Show that the graphs of $y = x^3/3$ and $y = 1/x$ are perpendicular at each point of intersection. (First decide what it must mean to say that two curves are perpendicular.)

28. Explain why the curve $y = x^7 + x^5 + x^3 + x + 1$ is always rising.

29. Show that the slope of the curve $y = x^5 - (1/x)$ is always positive. Where does the curve intersect the x axis? Sketch the curve, noting that the y axis is an asymptote. (Why?)

30. Use the derivative to help you sketch the graph of $y = 1 - x^{2/3}$.

31. Let $f(x) = x^3 + 3x - 2$.

(a) In view of the fact that $f(0) = -2$ and $f(1) = 2$, why would you expect the equation $x^3 + 3x - 2 = 0$ to have at least one real root?

(b) Explain why the graph of f has no horizontal tangent.

(c) Why do you suppose it follows from part (b) that the equation $x^3 + 3x - 2 = 0$ has no more than one real root?

32. Suppose that $f(x) = x^n$, where n is an even positive integer. Explain why $f'(x) < 0$ if $x < 0$, $f'(x) = 0$ if $x = 0$, and $f'(x) > 0$ if $x > 0$. What do these statements imply about the graph of f?

33. Suppose that $f(x) = x^n$, where n is an odd positive integer (greater than 1). Explain why $f'(x) > 0$ if $x \neq 0$ and $f'(x) = 0$ if $x = 0$. What do these statements imply about the graph of f?

34. Name a function whose derivative is $3x^2 - 2x + 5$. How many such functions are there?

35. A certain curve has slope $4x^3$ at each point (x, y) and contains the point $(0,1)$. What is its equation?

36. The coordinate of a moving object at time t is

$$s = \frac{t^3}{48} + \frac{48}{\sqrt[3]{t}}$$

What is the velocity of the object at $t = 8$?

37. The coordinate of a moving object at time t is $s = t^{3/2} - 3t + 5$.

(a) Where does the object start? with what speed?

(b) In what direction does the object move for a while after it starts?

(c) When and where does it come to a momentary stop?

(d) When does it return to its starting point and in what direction is it going when it does?

(e) What happens after that?

38. The Power Rule says that $D_x(x^2) = 2x$. Is it therefore correct to write, say, $D_x(2x - 3)^2 = 2(2x - 3)$? *Hint:* Compare with $D_x(4x^2 - 12x + 9)$.

39. Show that no tangent to the curve $y = 1/x$ intersects the curve twice. (This is apparent from the graph. However, prove it analytically.) *Hint:* Label the point of tangency (a,b), not (x,y), to avoid confusion with x and y in the equation of the tangent.

40. There are two lines through $(0, -1)$ tangent to the parabola $y = x^2$. (Draw a picture!)

(a) Find the points of tangency. *Hint:* If (a,b) is such a point, the slope of the tangent is $2a = (b + 1)/a$.

(b) Find equations of the tangents.

41. Use addition formulas for sine and cosine (Section 2.3) to show that $D_x \sin(x + c) = \cos(x + c)$ for every $c \in \mathcal{R}$.

42. Derive the formula

$$\frac{d}{dx}[cf(x)] = c\frac{d}{dx}f(x)$$

by using the definition of derivative.

43. Derive the formula

$$\frac{d}{dx}[f(x) - g(x)] = \frac{d}{dx}f(x) - \frac{d}{dx}g(x)$$

by using differentiation formulas already proved.

44. Use the definition of derivative to do Problem 43.

45. Derive the linearity property

$$\frac{d}{dx}[af(x) + bg(x)] = a\frac{d}{dx}f(x) + b\frac{d}{dx}g(x)$$

where a and b are numbers and f and g are differentiable functions.

46. Let $f(x) = (x - 1)^{3/2}$ and $g(x) = (1 - x)^{3/2}$.

(a) Use the definition of derivative to show that $f'(1) = 0$ and $g'(1) = 0$.

(b) If $h = f + g$ and we attempt to evaluate

$$h'(1) = \lim_{z \to 1} \frac{h(z) - h(1)}{z - 1}$$

what goes wrong? Does $h'(1) = f'(1) + g'(1) = 0$?

(This problem illustrates how one-sided derivatives can complicate the theory, as mentioned in Section 2.2. Rather than stating hypotheses that exclude such cases, we assume that algebraic combinations of functions are legitimate only when the domains overlap in nontrivial ways.)

3.2
THE POWER RULE AND THE ALGEBRA OF DERIVATIVES (II)

In this section we develop formulas for the differentiation of products and quotients and apply the results to prove another case of the Power Rule.

Product Rule

In the domain where f and g are both differentiable,

$$\frac{d}{dx}[f(x)g(x)] = f(x)g'(x) + g(x)f'(x)$$

Proof: Let $h = fg$ and suppose that x is a point at which f and g are both differentiable. The difference quotient associated with h at x is

$$\frac{h(z) - h(x)}{z - x} = \frac{f(z)g(z) - f(x)g(x)}{z - x} \tag{1}$$

To change this into a form involving the difference quotients associated with f and g at x, namely

$$\frac{f(z) - f(x)}{z - x} \quad \text{and} \quad \frac{g(z) - g(x)}{z - x}$$

we employ a devious algebraic device. Subtract and add $f(z)g(x)$ in the numerator of Equation (1):

$$\frac{h(z) - h(x)}{z - x} = \frac{f(z)g(z) - f(z)g(x) + f(z)g(x) - f(x)g(x)}{z - x}$$

$$= \frac{f(z)[g(z) - g(x)] + g(x)[f(z) - f(x)]}{z - x}$$

$$= f(z) \cdot \frac{g(z) - g(x)}{z - x} + g(x) \cdot \frac{f(z) - f(x)}{z - x}$$

Now use the algebra of limits to write

$$h'(x) = \left[\lim_{z \to x} f(z)\right]\left[\lim_{z \to x} \frac{g(z) - g(x)}{z - x}\right] + \left[\lim_{z \to x} g(x)\right]\left[\lim_{z \to x} \frac{f(z) - f(x)}{z - x}\right]$$

The last three of these limits are $g'(x)$, $g(x)$, and $f'(x)$, respectively. (Why?) To evaluate $\lim_{z \to x} f(z)$, however, we need the important theorem that *f is continuous at x because it is differentiable there.* That is,

$$\lim_{z \to x} f(z) = f(x) \qquad \text{because } f'(x) \text{ exists}$$

Assuming that fact, we find $h'(x) = f(x)g'(x) + g(x)f'(x)$.   □

As you can see, there is a serious gap in the proof of the Product Rule. Recall from Section 2.5 that a function f is *continuous at x_0* if

$$\lim_{x \to x_0} f(x) = f(x_0)$$

that is, if the limit can be found by substitution of x_0 in the function. We assert that this is a necessary consequence of the existence of $f'(x_0)$, that is,

Differentiability $\Rightarrow$ Continuity

To see why, observe that

$$\lim_{x \to x_0} [f(x) - f(x_0)] = \lim_{x \to x_0} \left[\frac{f(x) - f(x_0)}{x - x_0} \cdot (x - x_0)\right]$$

$$= \left[\lim_{x \to x_0} \frac{f(x) - f(x_0)}{x - x_0}\right]\left[\lim_{x \to x_0} (x - x_0)\right]$$

$$= f'(x_0) \cdot 0 = 0$$

Hence $\lim_{x \to x_0} f(x) = f(x_0)$.

The converse of this implication is false, that is,

$$\text{Continuity} \nRightarrow \text{Differentiability}$$

For example, $f(x) = |x|$ is continuous at 0 but not differentiable there. The same is true of $f(x) = x^{2/3}$. (See Figures 1 and 2, which show a "corner" and

Figure 1
Graph of $y = |x|$

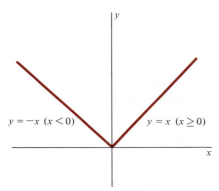

$y = -x \ (x < 0)$ $y = x \ (x \geq 0)$

Figure 2
Graph of $y = x^{2/3}$

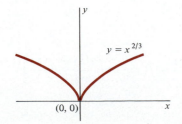

$y = x^{2/3}$

$(0, 0)$

a "cusp," respectively, at the origin. Such points do not occur on the graph of a function that is differentiable in its domain.)

Product Rule in Words

The derivative of a product is the first function times the derivative of the second plus the second function times the derivative of the first.

□ **Example 1**   Since $D_x(x^2) = 2x$ and $D_x(\sin x) = \cos x$, we can write

$$D_x(x^2 \sin x) = (x^2)D_x(\sin x) + (\sin x)D_x(x^2)$$
$$= (x^2)(\cos x) + (\sin x)(2x)$$
$$= x^2 \cos x + 2x \sin x \qquad\qquad □$$

Quotient Rule

In the domain where f and g are both differentiable and $g(x) \neq 0$,

$$\frac{d}{dx}\frac{f(x)}{g(x)} = \frac{g(x)f'(x) - f(x)g'(x)}{g(x)^2}$$

Proof: Let $h = f/g$ and suppose that x is a point at which f and g are differentiable and $g(x) \neq 0$. The difference quotient associated with h at x is

$$\frac{h(z) - h(x)}{z - x} = \frac{1}{z - x}\left[\frac{f(z)}{g(z)} - \frac{f(x)}{g(x)}\right] = \frac{g(x)f(z) - f(x)g(z)}{(z - x)g(x)g(z)}$$

Now subtract and add $f(x)g(x)$ in the numerator to obtain

$$\frac{h(z) - h(x)}{z - x} = \frac{g(x)f(z) - f(x)g(x) + f(x)g(x) - f(x)g(z)}{(z - x)g(x)g(z)}$$

$$= \frac{g(x)[f(z) - f(x)] - f(x)[g(z) - g(x)]}{(z - x)g(x)g(z)}$$

$$= \frac{1}{g(x)g(z)}\left[g(x) \cdot \frac{f(z) - f(x)}{z - x} - f(x) \cdot \frac{g(z) - g(x)}{z - x}\right]$$

Since g is differentiable (hence continous) at x, we know that

$$\lim_{z \to x} g(z) = g(x)$$

Hence the algebra of limits applied to

$$\lim_{z \to x} \frac{h(z) - h(x)}{z - x}$$

yields

$$h'(x) = \frac{1}{g(x)^2}\left[g(x) \cdot \lim_{z \to x}\frac{f(z) - f(x)}{z - x} - f(x) \cdot \lim_{z \to x}\frac{g(z) - g(x)}{z - x}\right]$$

$$= \frac{g(x)f'(x) - f(x)g'(x)}{g(x)^2} \qquad\qquad □$$

Remark   In the preceding proof the difference quotient associated with h at x is defined only when $g(z) \neq 0$. We are given that $g(x) \neq 0$ and we know that g is differentiable (hence continuous) at x. Therefore $g(z)$ is close to $g(x)$ when z is close to x. More precisely, we may confine z to a neighborhood of x in which $g(z)$ is so close to $g(x)$ that it is also not zero, thus guaranteeing that the difference quotient is defined during the process of finding its limit. (See Problem 48, Section 2.4.)

Quotient Rule in Words

The derivative of a quotient is the denominator times the derivative of the numerator minus the numerator times the derivative of the denominator, all divided by the denominator squared.

□ **Example 2**   Use the Quotient Rule to find the derivative of $\tan x$.

Solution: Since $D_x(\sin x) = \cos x$ and $D_x(\cos x) = -\sin x$, we have

$$D_x(\tan x) = D_x\left(\frac{\sin x}{\cos x}\right) = \frac{(\cos x)D_x(\sin x) - (\sin x)D_x(\cos x)}{\cos^2 x}$$

$$= \frac{(\cos x)(\cos x) - (\sin x)(-\sin x)}{\cos^2 x} = \frac{\cos^2 x + \sin^2 x}{\cos^2 x}$$

$$= \frac{1}{\cos^2 x} = \sec^2 x \qquad \qquad □$$

The derivatives of the other trigonometric functions are left for you to do in the problem set. The results are important and should be learned.

$D_x(\sin x) = \cos x$	$D_x(\cos x) = -\sin x$
$D_x(\tan x) = \sec^2 x$	$D_x(\cot x) = -\csc^2 x$
$D_x(\sec x) = \sec x \tan x$	$D_x(\csc x) = -\csc x \cot x$

As an aid to your memory, note that the right-hand column of this table can be obtained from the left-hand column by replacing each function by its cofunction and introducing a minus sign.

□ **Example 3**   Find the derivative of

$$f(x) = \frac{4\sqrt{x}}{x^2 + 3}$$

Solution: It will simplify the work if we factor out the 4 before applying the Quotient Rule:

$$f'(x) = 4D_x\left(\frac{\sqrt{x}}{x^2+3}\right) = 4 \cdot \frac{(x^2+3)D_x(\sqrt{x}) - \sqrt{x}\,D_x(x^2+3)}{(x^2+3)^2}$$

Since
$$D_x(\sqrt{x}) = \frac{1}{2\sqrt{x}} \qquad \text{(Power Rule)}$$

and
$$D_x(x^2+3) = 2x \qquad \text{(why?)}$$

the numerator in the above expression for $f'(x)$ becomes

$$\frac{x^2+3}{2\sqrt{x}} - \sqrt{x}\,(2x) = \frac{(x^2+3) - 4x^2}{2\sqrt{x}} = \frac{3(1-x^2)}{2\sqrt{x}}$$

Hence
$$D_x\left(\frac{4\sqrt{x}}{x^2+3}\right) = \frac{6(1-x)(1+x)}{\sqrt{x}\,(x^2+3)^2} \qquad \square$$

A worry that students often have in problems like Example 3 is that the calculus is easy but the algebra is hard. The only cure for that is to work through the details of the examples carefully. It is essential to become proficient in this kind of thing, as the next example illustrates.

□ **Example 4**   Suppose that we are interested in graphing

$$y = \frac{4\sqrt{x}}{x^2+3} \qquad \text{(defined for } x \geq 0\text{)}$$

The derivative will tell us a great deal about the graph, but not until it is worked into the form given in Example 3, namely

$$y' = \frac{6(1-x)(1+x)}{\sqrt{x}\,(x^2+3)^2}$$

Then it is not hard to figure out that

$$\begin{aligned}
y' &> 0 &&\text{when } 0 < x < 1 \\
y' &= 0 &&\text{when } x = 1 \qquad \text{(why not also } x = -1\text{?)} \\
y' &< 0 &&\text{when } x > 1
\end{aligned}$$

(See the table of signs in Figure 3. For a review of this kind of analysis, see the Remark following Example 7 in Section 1.4.) Hence the graph rises, flattens out, and falls as shown in Figure 4. Moreover, there is a vertical tangent at $(0,0)$ because y' increases without bound as $x \to 0$. To see why, note that the numerator of y' approaches 6 when $x \to 0$, while the denominator approaches 0. The tangent to the curve is therefore steep when x is close to 0. At the origin it is vertical (and the slope is undefined).   □

Figure 3
Sign of y'

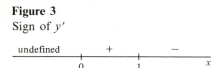

Figure 4
Graph of $y = \dfrac{4\sqrt{x}}{x^2+3}$

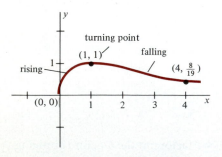

□ **Example 5**   Prove Case 4 of the Power Rule ($r = -s$, where s is a positive rational number).

Solution: Assuming that Case 3 is done, we know that $D_x(x^s) = sx^{s-1}$. Hence

$$D_x(x^r) = D_x\left(\frac{1}{x^s}\right) = \frac{(x^s)(0) - (1)(sx^{s-1})}{(x^s)^2} \quad \text{(Quotient Rule)}$$

$$= \frac{-sx^{s-1}}{x^{2s}} = -sx^{-s-1} = rx^{r-1} \qquad \square$$

Problem Set 3.2

Find the derivative of each of the following functions.

1. $y = \tan x - 3 \sec x$ **2.** $y = \cot x + 2 \csc x$

3. $y = 3 \cos x + 2 \cot x$ **4.** $y = x \tan x$

5. $y = x^3 \cos x$ **6.** $y = \sin x \cos x$

7. $y = \sec x \tan x$ **8.** $y = \sqrt{x}(x^2 + 1)$

9. $y = x^2(\sqrt{x} - 1)$ **10.** $y = (x^3 + 1)(x^2 - 2x + 1)$

11. $y = (x^2 - 1)(x^4 + 1)$ **12.** $y = \sqrt[3]{x}(1 - x^2)$

13. $y = \sqrt[4]{x}(x^2 + 1)$ **14.** $y = \dfrac{x^2}{2x + 1}$

15. $y = \dfrac{x - 1}{x^2 + x + 1}$ **16.** $y = \dfrac{x^2 - 4}{\sqrt{x}}$

17. $y = \dfrac{\sqrt{x}}{3x + 1}$ **18.** $y = \dfrac{\sqrt[3]{x} - 1}{x^3 + 1}$

19. $y = \dfrac{\sqrt[4]{x} - 1}{x^4 + 1}$ **20.** $y = x^{2/3} \sin x$

21. $y = x^{3/2} \cos x$ **22.** $y = \dfrac{x}{\cos x}$

23. $y = \dfrac{\sec x}{x^2}$ **24.** $y = \dfrac{\sin x - \cos x}{\tan x}$

25. $y = \dfrac{\sin x}{\tan x - \sec x}$ **26.** $y = \dfrac{x^{5/2}}{\sin x}$

27. $y = \dfrac{x^{2/5}}{\cos x}$

28. $y = x^{-1/2} \tan x + x^{1/2} \cot x$

29. $y = x^{1/3} \cos x - x^{-1/3} \sin x$

Find each of the following derivatives.

30. $\dfrac{d}{dt}(2 \sin t + \tan t)$ **31.** $\dfrac{d}{d\theta}(\cot \theta - \csc \theta)$

32. $D_u(u \sin u)$ **33.** $D_x(x^2 \sec x)$

34. $D_t(\sqrt{t} \cos t)$ **35.** $\dfrac{d}{dx}\left(\dfrac{\sin x}{x}\right)$

36. $\dfrac{d}{dt}\left(\dfrac{\tan t}{t}\right)$ **37.** $D_t\left(\dfrac{t}{t - 2}\right)$

38. $\dfrac{d}{du}\left(\dfrac{5u^2}{1 - u}\right)$ **39.** $\dfrac{d}{dx}\left(\dfrac{x^2 - 1}{x^2 + 1}\right)$

40. $D_x \sqrt{x}(x^2 - 1)$ **41.** $D_x\left(\dfrac{\sqrt{x}}{x^2 - 1}\right)$

42. $D_x(x^{3/2} \sin x)$ **43.** $\dfrac{d}{dx}\left(\dfrac{x^2 - 1}{\sqrt[3]{x}}\right)$

44. $p'(0)$ if $p(t) = t^{7/3} \tan t$

45. $\phi'(\pi)$ if $\phi(u) = \dfrac{\sin u}{u^{3/2}}$

Find each of the following derivatives by using the formula

$$D_x \sqrt{1 - x^2} = \frac{-x}{\sqrt{1 - x^2}}$$

(derived in Problem 47, Section 2.2, by geometrical reasoning).

46. $D_x(x \sqrt{1 - x^2})$ **47.** $D_x(x^2 \sqrt{1 - x^2})$

48. $D_x\left(\dfrac{x}{\sqrt{1 - x^2}}\right)$ **49.** $D_x\left(\dfrac{x^2}{\sqrt{1 - x^2}}\right)$

50. Use the answer to Problem 48 to explain why the graph of

$$y = \frac{x}{\sqrt{1 - x^2}}$$

is always rising. What is the domain? Why are the lines $x = \pm 1$ asymptotes of the graph? Sketch the graph.

51. Find the derivative of $f(x) = (x^2 - 1)/x^2$ by using **(a)** the Quotient Rule; **(b)** the Product Rule, writing $f(x) = x^{-2}(x^2 - 1)$; and **(c)** the Power Rule, writing $f(x) = 1 - x^{-2}$. Which is easiest?

52. Sketch the graph of $y = x^{2/3}(x^2 - 4)$ after answering the following questions.

(a) What is the domain?

(b) For what values of x is $y' > 0$? $y' < 0$?

(c) Why is there a vertical tangent (and a cusp) at the origin?

(d) What are the lowest points of the graph?

(e) Where does the graph intersect the x axis?

53. Find the x coordinate of the point at which the graph of

$$y = \frac{\sqrt[3]{x}}{1 + x^3}$$

has a horizontal tangent.

54. Show that $D_x(\tan x \cot x) = 0$ in two ways. Sketch the graph of $y = \tan x \cot x$.

Use the Quotient Rule to confirm each of the following formulas.

55. $D_x(\cot x) = -\csc^2 x$

56. $D_x(\sec x) = \sec x \tan x$

57. $D_x(\csc x) = -\csc x \cot x$

58. Use the Product Rule on $\cot x = \cos x \csc x$ to confirm the formula

$$D_x(\cot x) = -\csc^2 x$$

Hint: $\csc^2 x - \cot^2 x = 1$.

59. Use the Quotient Rule on $\sec x = \tan x / \sin x$ to confirm the formula

$$D_x(\sec x) = \sec x \tan x$$

60. Use multiplication formulas for $\sin 2x$ and $\cos 2x$ (Section 2.3) to show that

$$D_x(\sin 2x) = 2 \cos 2x$$

(Note that the answer is not $\cos 2x$!)

61. Use the identity

$$\tan \tfrac{1}{2}x = \frac{1 - \cos x}{\sin x} \qquad \text{(Section 2.3)}$$

to show that $D_x(\tan \tfrac{1}{2}x) = \tfrac{1}{2} \sec^2 \tfrac{1}{2}x$. *Hint:* You will also need the identity

$$\cos^2 \tfrac{1}{2}x = \tfrac{1}{2}(1 + \cos x) \qquad \text{(Section 2.3)}$$

62. Use the Product Rule to derive the formula

$$\frac{d}{dx}[cf(x)] = c\frac{d}{dx}f(x)$$

(See Problem 42, Section 3.1.)

63. Let $f(x) = x|x|$.

(a) Explain why the Product Rule cannot be used to find $f'(0)$.

(b) Show directly from the definition of derivative that $f'(0) = 0$.

(c) If $x \neq 0$, then $D_x|x| = |x|/x$ (Problem 46, Section 2.2). Use this formula and the Product Rule to show that if $x \neq 0$, then $f'(x) = 2|x|$.

(d) Explain why the formula in part (c) (derived for $x \neq 0$) actually holds for all x.

(e) Sketch the graph of f.

64. Prove Case 1 of the Power Rule (Section 3.1) by mathematical induction. *Hint:* We already know that the rule is valid when $n = 1$. To complete the induction, show that for every positive integer k,

$$D_x(x^k) = kx^{k-1} \Rightarrow D_x(x^{k+1}) = (k + 1)x^k$$

(Use the Product Rule on $x^k \cdot x$.) It follows from the Principle of Mathematical Induction that the formula holds for every positive integer n.

65. Suppose that $y = [f(x)]^n$, where f is differentiable and n is a positive integer. If $f(x) = x$, the Power Rule says that $y' = nx^{n-1}$, but in general it is incorrect to write

$$y' = n[f(x)]^{n-1}$$

(See Problem 38, Section 3.1.)

(a) Use mathematical induction to show that

$$y' = n[f(x)]^{n-1}f'(x)$$

(This is a special case of the *Chain Rule*, to be discussed in the next section.)

(b) In view of part (a), what is $D_x(\sin^4 x)$?

66. Suppose that f is differentiable at x_0. Prove that f is continuous at x_0, that is, $\lim_{x \to x_0} f(x) = f(x_0)$, as follows.

(a) We know that

$$f'(x_0) = \lim_{x \to x_0} Q(x)$$

where $\quad Q(x) = \dfrac{f(x) - f(x_0)}{x - x_0} \qquad (x \neq x_0)$

Confirm that if $x \neq x_0, f(x) = f(x_0) + Q(x)(x - x_0)$.

(b) Apply the algebra of limits to the expression in part (a) to obtain

$$\lim_{x \to x_0} f(x) = f(x_0)$$

(c) Compare this argument to the one given in the text.

In this section we are going to pursue the question of differentiating *composite functions* (first defined in Section 2.5). An example of such a function is $h(x) = x^{2/3}$ (arising in Case 3 of the Power Rule, which we have not yet proved). For we can write

$$h(x) = (x^{1/3})^2 = f[g(x)]$$

where $f(x) = x^2$ and $g(x) = x^{1/3}$. Knowing that

$$f'(x) = 2x \qquad \text{and} \qquad g'(x) = \tfrac{1}{3}x^{-2/3}$$

how do we find $h'(x)$?

Another example is

$$h(x) = \sin(x^2 + 1) = f[g(x)]$$

where $f(x) = \sin x$ and $g(x) = x^2 + 1$. The constituent parts of this composition have derivatives

$$f'(x) = \cos x \qquad \text{and} \qquad g'(x) = 2x$$

But what is the derivative of the composition?

Notation for the composition of f and g is $h = f \circ g$, defined by

$$h(x) = (f \circ g)(x) = f[g(x)]$$

Evidently what is needed in the preceding examples is a "composite function rule" (usually called the *Chain Rule*) for finding the derivative of h in terms of the derivatives of f and g. We develop it as follows.

The difference quotient associated with h at x is

$$\frac{h(z) - h(x)}{z - x} = \frac{f[g(z)] - f[g(x)]}{z - x} = \frac{f(v) - f(u)}{z - x}$$

where $u = g(x)$ and $v = g(z)$. If $v \neq u$, we can write

$$\frac{h(z) - h(x)}{z - x} = \frac{f(v) - f(u)}{v - u} \cdot \frac{v - u}{z - x} = \frac{f(v) - f(u)}{v - u} \cdot \frac{g(z) - g(x)}{z - x}$$

from which

$$h'(x) = \lim_{z \to x} \frac{h(z) - h(x)}{z - x} = \left[\lim_{z \to x} \frac{f(v) - f(u)}{v - u} \right]\left[\lim_{z \to x} \frac{g(z) - g(x)}{z - x} \right]$$

The second of these limits is $g'(x)$. (We are assuming that g is differentiable at x.) To evaluate the first, observe that g is also continuous at x, that is,

$$\lim_{z \to x} g(z) = g(x)$$

Hence $v \to u$ as $z \to x$ and we can write

$$\lim_{z \to x} \frac{f(v) - f(u)}{v - u} = \lim_{v \to u} \frac{f(v) - f(u)}{v - u} = f'(u) = f'[g(x)]$$

(provided that f is differentiable at u). Thus we have arrived at the *Chain Rule*

$$h'(x) = f'[g(x)]g'(x) \qquad \text{where } h = f \circ g$$

Remark   In the above argument it is essential that $v \neq u$, since we multiplied numerator and denominator of

$$\frac{f(v) - f(u)}{z - x}$$

by $v - u$. For most functions encountered in elementary calculus we can safely assume that $v = g(z)$ is different from $u = g(x)$ for z in some (punctured) neighborhood of x. Except for constant functions it is unusual for a function to have the same value at more than a finite number of points in a finite interval; we need only keep z sufficiently close to x to avoid $g(z) = g(x)$ altogether. (See Figure 1.) Hence for these functions the argument given is valid. (For an argument that is valid in general, see the problem set.)

The Chain Rule says that the derivative of $h = f \circ g$ is the product of the derivatives of f and g (properly evaluated). Note that f' is not evaluated at x, but at $u = g(x)$, a fact we must take into account when using the formula.

□ **Example 1**   Find the derivative of $x^{2/3}$.

Solution: Let
$$h(x) = x^{2/3} = (x^{1/3})^2 = f[g(x)]$$
where $f(x) = x^2$ and $g(x) = x^{1/3}$. Since
$$f'(x) = 2x \qquad \text{and} \qquad g'(x) = \tfrac{1}{3}x^{-2/3}$$
the Chain Rule formula yields
$$h'(x) = f'(x^{1/3})g'(x) = (2x^{1/3})(\tfrac{1}{3}x^{-2/3}) = \tfrac{2}{3}x^{-1/3} \qquad\qquad □$$

By imitating Example 1 we can now finish off Case 3 of the Power Rule (Section 3.1). Let
$$h(x) = x^{p/q} = (x^{1/q})^p = f[g(x)]$$
where $f(x) = x^p$ and $g(x) = x^{1/q}$ (p and q positive integers, $q > 1$). Since
$$f'(x) = px^{p-1} \qquad \text{(Case 1 of the Power Rule)}$$
$$g'(x) = \frac{1}{q}x^{(1/q)-1} \qquad \text{(Case 2 of the Power Rule)}$$
we find
$$h'(x) = f'[g(x)]g'(x)$$
$$= p(x^{1/q})^{p-1} \cdot \frac{1}{q}x^{(1/q)-1} = \frac{p}{q}x^{(p/q)-1}$$

Figure 1

Punctured neighborhood of x in which $g(z) \neq g(x)$

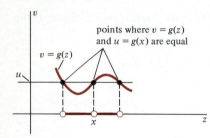

points where $v = g(z)$ and $u = g(x)$ are equal

$v = g(z)$

We have now proved the first four cases of the Power Rule, corresponding to *rational* exponents. The remaining case is when the exponent is *irrational*, as in $D_x(x^\pi) = \pi x^{\pi-1}$. There is no way we can handle this case now, for we have not defined what a symbol like x^π means. We will not return to the problem until Chapter 8.

□ **Example 2**   Find the derivative of $\sin(x^2 + 1)$.

Solution: Let

$$h(x) = \sin(x^2 + 1) = f[g(x)]$$

where $f(x) = \sin x$ and $g(x) = x^2 + 1$. Since

$$f'(x) = \cos x \qquad \text{and} \qquad g'(x) = 2x$$

we find

$$h'(x) = f'(x^2 + 1)g'(x) = \cos(x^2 + 1) \cdot 2x = 2x \cos(x^2 + 1) \qquad □$$

□ **Example 3**   In Problem 47, Section 2.2, we found

$$D_x \sqrt{1 - x^2} = \frac{-x}{\sqrt{1 - x^2}}$$

by geometrical reasoning. With the Power Rule and Chain Rule at our disposal, we may write

$$D_x \sqrt{1 - x^2} = D_x(1 - x^2)^{1/2} = \tfrac{1}{2}(1 - x^2)^{-1/2}(-2x) = \frac{-x}{\sqrt{1 - x^2}} \qquad □$$

It is worthwhile to render the formula $h'(x) = f'[g(x)]g'(x)$ in words.

To differentiate $h(x) = f[g(x)]$, find the derivative of the outside function and evaluate it at the inside function. Then multiply by the derivative of the inside function.

□ **Example 4**   Find $D_x(\sin^3 x)$.

Solution: The notation $\sin^3 x$ means $(\sin x)^3$. The derivative of $f(x) = x^3$ (the outside function) is $3x^2$; evaluated at $g(x) = \sin x$ (the inside function), it is $3 \sin^2 x$. Multiply by the derivative of the inside function to get

$$D_x(\sin^3 x) = 3 \sin^2 x \cos x \qquad □$$

Another way of thinking of the Chain Rule is to write

$$y = h(x) = f[g(x)] = f(u) \qquad \text{where } u = g(x)$$

Then the formula $h'(x) = f'[g(x)]g'(x)$ becomes

$$\frac{dy}{dx} = f'(u)g'(x)$$

or, since $y = f(u)$ and $u = g(x)$,

$$\frac{dy}{dx} = \frac{dy}{du}\frac{du}{dx}$$

In this version (due to Leibniz) it looks like the du cancels, leaving us with a trivial identity. Of course the Leibniz symbol dy/dx is not really a fraction (unless we give separate meanings to dx and dy, which we have not done yet). The above equation should therefore be regarded as a helpful memory device, not a relation between fractions. In words it says

To differentiate y with respect to x, first differentiate with respect to u, then multiply by the derivative of u with respect to x.

□ **Example 5**   Find the derivative of $y = (x^2 - 1)^{5/2}$.

Solution: Let $u = x^2 - 1$. The derivative of $y = u^{5/2}$ with respect to u is

$$\frac{dy}{du} = \tfrac{5}{2}u^{3/2}$$

Multiply by the derivative of u with respect to x $(du/dx = 2x)$ to get

$$\frac{dy}{dx} = \frac{dy}{du}\frac{du}{dx} = \tfrac{5}{2}u^{3/2} \cdot 2x$$

$$= \tfrac{5}{2}(x^2 - 1)^{3/2} \cdot 2x = 5x(x^2 - 1)^{3/2}$$

Note that u does not appear in the final result. It is a mental crutch that is used and then discarded.   □

□ **Example 6**   To differentiate $y = \tan^3 x = (\tan x)^3$, we can let $u = \tan x$ and proceed as in Example 5. It is easier, however, to avoid explicit substitution. Think in the following pattern:

1. Differentiate $(\ \ )^3$ to get $3(\ \ )^2$, inserting $\tan x$.
2. Differentiate $\tan x$ to get $\sec^2 x$.

The product of these two derivatives is the answer,

$$\frac{dy}{dx} = 3(\tan x)^2 \sec^2 x = 3\tan^2 x \sec^2 x \qquad \square$$

□ **Example 7**   The "chain" in the Chain Rule may involve more than two steps, as in the following computation of

$$D_x \sqrt{\sin 2x} = D_x(\sin 2x)^{1/2}$$

1. Differentiate $(\ \ )^{1/2}$ to get $\frac{1}{2}(\ \ )^{-1/2}$, inserting $\sin 2x$.
2. Differentiate $\sin(\ \ )$ to get $\cos(\ \ )$, inserting $2x$.
3. Differentiate $2x$ to get 2.

The product of these derivatives is the answer,

$$D_x \sqrt{\sin 2x} = \frac{1}{2}(\sin 2x)^{-1/2} \cdot \cos 2x \cdot 2 = \frac{\cos 2x}{\sqrt{\sin 2x}} \qquad \square$$

□ **Example 8**   The Quotient Rule, Power Rule, and Chain Rule enable us to write

$$D_x\left(\frac{x^2}{\sqrt{1-x^2}}\right) = D_x\left[\frac{x^2}{(1-x^2)^{1/2}}\right]$$

$$= \frac{(1-x^2)^{1/2}(2x) - x^2 \cdot \frac{1}{2}(1-x^2)^{-1/2}(-2x)}{1-x^2}$$

Multiplying numerator and denominator by $(1-x^2)^{1/2}$, we find

$$D_x\left(\frac{x^2}{\sqrt{1-x^2}}\right) = \frac{2x(1-x^2) + x^3}{(1-x^2)^{3/2}} = \frac{2x - x^3}{(1-x^2)^{3/2}} = \frac{x(2-x^2)}{(1-x^2)^{3/2}} \qquad \square$$

□ **Example 9**   In Problem 46, Section 2.2, we used the interpretation of the derivative as slope to find

$$D_x|x| = \frac{|x|}{x} = \frac{x}{|x|}$$

An interesting way to confirm this is to write $|x| = \sqrt{x^2}$ and apply the Chain Rule:

$$D_x|x| = D_x(x^2)^{1/2} = \frac{1}{2}(x^2)^{-1/2}(2x) = \frac{x}{\sqrt{x^2}} = \frac{x}{|x|} \qquad \square$$

Generally speaking, whenever a differentiation formula is discovered, we should "build it into the Chain Rule," as indicated by the following summary of previously derived results.

Formula	Built into the Chain Rule
$D_x\|x\| = \dfrac{x}{\|x\|}$	$\dfrac{d}{dx}\|u\| = \dfrac{u}{\|u\|}\dfrac{du}{dx}$
$D_x(x^r) = rx^{r-1}$	$\dfrac{d}{dx}(u^r) = ru^{r-1}\dfrac{du}{dx}$
$D_x(\sin x) = \cos x$	$\dfrac{d}{dx}(\sin u) = (\cos u)\dfrac{du}{dx}$
$D_x(\cos x) = -\sin x$	$\dfrac{d}{dx}(\cos u) = (-\sin u)\dfrac{du}{dx}$
$D_x(\tan x) = \sec^2 x$	$\dfrac{d}{dx}(\tan u) = (\sec^2 u)\dfrac{du}{dx}$
$D_x(\cot x) = -\csc^2 x$	$\dfrac{d}{dx}(\cot u) = (-\csc^2 u)\dfrac{du}{dx}$
$D_x(\sec x) = \sec x \tan x$	$\dfrac{d}{dx}(\sec u) = (\sec u \tan u)\dfrac{du}{dx}$
$D_x(\csc x) = -\csc x \cot x$	$\dfrac{d}{dx}(\csc u) = (-\csc u \cot u)\dfrac{du}{dx}$

Only the left-hand column should be learned; the right-hand column embodies a general principle that applies to any differentiation formula.

Problem Set 3.3

Find the derivative of each of the following functions.

1. $y = (1 - x)^3$ **2.** $y = (x^2 - 4)^6$

3. $y = \tan^2 x$ **4.** $y = \cos^4 x$

5. $y = \sqrt{x^4 + 2}$ **6.** $y = (3x + 4)^{3/2}$

7. Differentiate $y = 1/(x^2 - 2)$ as it stands (by the Quotient Rule). Then write $y = (x^2 - 2)^{-1}$ and differentiate by the Chain Rule. Which is easier?

8. Repeat Problem 7 for $y = 5/(1 - 2x)^3$.

Differentiate each of the following functions.

9. $y = \sin(2\pi x)$ **10.** $y = \cos(1 - x)$

11. $y = \cos^3(2x)$ **12.** $y = \sin^2(x/2)$

13. $y = |2x - 1|$ **14.** $y = |\sec x|$

15. $y = |\sin x|$ **16.** $y = |1 - x|$

17. $y = \sin^2 x \cos x$ **18.** $y = \dfrac{\sin^2 x}{\cos x}$

19. $y = \dfrac{x}{(x^2 + 3)^2}$ **20.** $y = \dfrac{x^2}{(x^2 + 2)^3}$

21. $y = \dfrac{x^2}{\tan^2 x}$ **22.** $y = x \sin(1/x)$

23. $y = x^3(2x - 1)^4$ **24.** $y = x\sqrt{4 - x^2}$

25. $y = \dfrac{x}{\sqrt{9 - x^2}}$ **26.** $y = \dfrac{\sqrt{x^2 + 1}}{x}$

27. $y = \left(x + \dfrac{2}{x}\right)^3$ **28.** $y = \left(1 - \dfrac{1}{x^2}\right)^4$

29. $y = \left(\dfrac{2x - 1}{x + 3}\right)^4$ **30.** $y = \sqrt{\dfrac{1 + x}{1 - x}}$

31. Find the velocity and speed at $t = 1$ of an object whose position at time t is $p(t) = \sin(\pi t)$.

32. Find an equation of the tangent to the curve

$$y = (x^3 - 2)^4 \quad \text{at } x = 1$$

33. Sketch the graph of $y = x\sqrt{3 - x}$ after answering the following questions.

(a) What is the domain?

(b) Where does the graph intersect the x axis?

(c) For what values of x is $y' > 0$? $y' = 0$? $y' < 0$?

(d) What is the highest point on the graph?

(e) Why is there a vertical tangent at $(3,0)$?

34. Suppose that $f(x) = x^2/\sqrt{2x - 1}$.

(a) Confirm that $f'(x) = x(3x - 2)/(2x - 1)^{3/2}$.

(b) Why is it incorrect to say that $f'(0) = 0$?

(c) For what values of x is $f'(x) > 0$? $f'(x) = 0$? $f'(x) < 0$?

(d) Explain why the range of f is $[4\sqrt{3}/9, \infty)$.

35. Suppose that $f(x) = |1 - x^2|$.

(a) Use the formula $D_x|x| = x/|x|$ (and the Chain Rule) to confirm that

$$f'(x) = \frac{-2x(1 - x^2)}{|1 - x^2|}$$

(b) Why does the graph of f have "sharp turning points" at $x = \pm 1$?

(c) Why is there a "smooth turning point" at $x = 0$?

(d) Sketch the graph.

(e) The graph could have been sketched without calculus by observing that

$$f(x) = \begin{cases} 1 - x^2 & \text{if } -1 \le x \le 1 \\ x^2 - 1 & \text{if } x > 1 \text{ or } x < -1 \end{cases}$$

Which is easier?

36. Sketch the curve $y = |\sin x|$, noting the points at which the graph appears not to be smooth. Confirm by examining the derivative.

37. Show in two ways that $D_x(\sin^2 x + \cos^2 x) = 0$.

38. Differentiate each side of the identity

$$\sin 2x = 2 \sin x \cos x$$

to obtain an identity involving $\cos 2x$.

39. One has to be careful in problems like the preceding one. For example, suppose we differentiate each side of $x^2 - x - 6 = 0$ to obtain $2x - 1 = 0$. Then $x = \frac{1}{2}$. But

$$x^2 - x - 6 = 0 \Rightarrow x = 3 \text{ or } x = -2$$

Hence $3 = \frac{1}{2}$ or $-2 = \frac{1}{2}$. Where is the fallacy?

40. Use the cofunction identities

$$\cos x = \sin\left(\frac{\pi}{2} - x\right) \quad \text{and} \quad \cos\left(\frac{\pi}{2} - x\right) = \sin x$$

together with the Chain Rule to derive the formula for $D_x(\cos x)$ from the derivative of sine. Compare with the argument in Problem 64, Section 2.3, based on the definition of derivative!

41. Use the idea of Problem 40 to derive the formula for $D_x(\cot x)$ from the derivative of tangent.

42. Use the idea of Problem 40 to derive the formula for $D_x(\csc x)$ from the derivative of secant.

43. A point $P = (x, y)$ is moving along the curve $y = 1/x$. If $dx/dt = -4$ at $(2, \frac{1}{2})$, what is dy/dt at that point? Which way is P moving as it passes through $(2, \frac{1}{2})$?

44. The volume of a spherical balloon is $V = \frac{4}{3}\pi r^3$, where r is the radius. If $r = 5$ and $dr/dt = 2$ at time $t = 1$, what is dV/dt at that instant? Is the balloon expanding or contracting?

45. Figure 2 shows a TV camera (located 5 meters from a straight track) following an Olympic runner P. If P is moving 10 meters per second, how fast is the camera turning when $x = 5$? *Hint:* Differentiate each side of $\tan\theta = x/5$ with respect to time t (using the Chain Rule) and solve for $d\theta/dt$. What are the units of the answer?

Figure 2
How fast is θ changing?

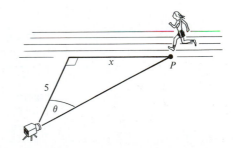

46. In Chapter 8 we will prove that $D_x(e^x) = e^x$ (where e is the base of natural logarithms, approximately 2.718). Explain why

$$D_x(e^{x^2}) = 2xe^{x^2}$$

47. In Section 3.2 we proved the Quotient Rule by a direct application of the definition of derivative. Give a different proof by applying the Product Rule and Chain Rule to $f \cdot g^{-1}$.

48. Our first argument for the Chain Rule, in which we assumed $g(z) \neq g(x)$ for z close to x, breaks down if g is a constant function. What can you say about the Chain Rule formula

$$h'(x) = f'[g(x)]g'(x) \qquad \text{(where } h = f \circ g\text{)}$$

if $g(x) = c$?

49. (General argument for the Chain Rule) Suppose that g is differentiable at x and f is differentiable at $u = g(x)$. Prove that $(f \circ g)'(x) = f'(u)g'(x)$ as follows.

(a) The difference quotient associated with $h = f \circ g$ at x is

$$\frac{h(z) - h(x)}{z - x} = \frac{f(v) - f(u)}{z - x} \qquad \text{where } v = g(z)$$

Letting

$$Q^*(v) = \frac{f(v) - f(u)}{v - u} \qquad \text{if } v \neq u$$

and $Q^*(v) = f'(u)$ if $v = u$ explain why $f(v) - f(u) = Q^*(v)(v - u)$ regardless of whether $v = u$ or $v \neq u$.

(b) Show that $\lim_{z \to x} Q^*(v) = f'(u)$.

(c) Use (a) and (b) to conclude that $h'(x) = f'(u)g'(x)$.

3.4
DERIVATIVES OF IMPLICITLY DEFINED FUNCTIONS

In Section 1.4 we pointed out that $\{(x, y): x^2 + y^2 = 1\}$, the unit circle in the coordinate plane, is a relation but not a function. Geometrically speaking, the reason for this is that vertical lines sometimes intersect the graph more than once, whereas a (real) function is a subset of $\mathcal{R}^2$ intersected at most once by any vertical line. Algebraically speaking, we do not find a unique y corresponding to each x, but rather the ambiguous solution $y = \pm\sqrt{1 - x^2}$.

We went on to point out that a relation may give rise to several functions, depending on which part of it we want to keep. The upper half of the unit circle, for example, is the function

$$f_1: [-1, 1] \to \mathcal{R} \text{ defined by } f_1(x) = \sqrt{1 - x^2}$$

whereas the lower semicircle is the function

$$f_2: [-1, 1] \to \mathcal{R} \text{ defined by } f_2(x) = -\sqrt{1 - x^2}$$

(See Figure 1.) Such functions are said to be defined *implicitly* by the relation, the idea being that $x^2 + y^2 = 1$, unlike an equation of the form $y = f(x)$, does not give y explicitly in terms of x.

Figure 1
Two functions defined by the relation $x^2 + y^2 = 1$

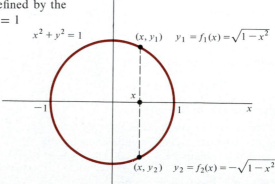

What we want to investigate in this section is the question of differentiating implicitly defined functions. Of course in the example $x^2 + y^2 = 1$ there is no problem. Simply write y explicitly in terms of x and differentiate as usual:

$$f'_1(x) = D_x(\sqrt{1 - x^2}) = \frac{-x}{\sqrt{1 - x^2}}$$

or

$$f'_2(x) = D_x(-\sqrt{1 - x^2}) = \frac{x}{\sqrt{1 - x^2}}$$

Suppose, however, that we are either unwilling or unable to solve for y in terms of x. What then? The answer is deceptively simple. The equation $x^2 + y^2 = 1$ becomes an identity when $f_1(x) = \sqrt{1 - x^2}$ (for example) is substituted for y:

$$x^2 + (\sqrt{1 - x^2})^2 = 1$$

Why not regard this as an equality between two identical functions and differentiate each side? The left side is the function

$$x^2 + (\sqrt{1 - x^2})^2 = x^2 + (1 - x^2)$$

in the domain $[-1,1]$, while the right side is the constant function 1. The derivatives are, respectively,

$$D_x[x^2 + (1 - x^2)] = 2x + (-2x) = 0 \qquad \text{and} \qquad D_x(1) = 0$$

which of course are equal in the domain.

We seem not to have said anything. If we did not know what to substitute for y, however, the above program would have to be carried out in abstract terms. More precisely, we assume that $y = f(x)$ is implicitly defined by $x^2 + y^2 = 1$ and write

$$x^2 + [f(x)]^2 = 1 \qquad (-1 \leq x \leq 1)$$

Differentiating each side with respect to x (using the Chain Rule in the second term on the left), we find

$$2x + 2f(x)f'(x) = 0$$

from which

$$f'(x) = -\frac{x}{f(x)} \qquad (-1 < x < 1)$$

Note that this yields

$$f'_1(x) = \frac{-x}{f_1(x)} = \frac{-x}{\sqrt{1 - x^2}} \qquad \text{or} \qquad f'_2(x) = \frac{-x}{f_2(x)} = \frac{x}{\sqrt{1 - x^2}}$$

depending on whether $f = f_1$ or $f = f_2$ (which checks with our previous results when the differentiation was explicit).

It is not really necessary to substitute $f(x)$ for y. Just differentiate each side of $x^2 + y^2 = 1$ as is, keeping in mind that y is supposed to be a function of x:

$$\frac{d}{dx}(x^2) + \frac{d}{dx}(y^2) = \frac{d}{dx}(1)$$

$$2x + 2y\frac{dy}{dx} = 0 \qquad \text{(Chain Rule in the second term!)}$$

from which $dy/dx = -x/y$.

This sort of thing is called **implicit differentiation** (a misnomer because it suggests a new kind of differentiation). The only thing new about it is that y is not explicitly given in terms of x. The rules of differentiation, however, are the same as ever.

□ **Example 1**   Assuming that the equation

$$xy - y - x = 0 \qquad (1)$$

defines one or more differentiable functions (y in terms of x), we differentiate each side with respect to x (using the Product Rule in the first term on the left):

$$\frac{d}{dx}(xy) - \frac{d}{dx}(y) - \frac{d}{dx}(x) = \frac{d}{dx}(0)$$

$$x\frac{dy}{dx} + y - \frac{dy}{dx} - 1 = 0$$

$$(x - 1)\frac{dy}{dx} = 1 - y$$

$$\frac{dy}{dx} = \frac{1 - y}{x - 1} \qquad (x \neq 1) \qquad (2)$$

While Equation (2) may not seem to be a complete answer (because y is not known in terms of x), it is nevertheless useful. For example, to find the slope of the graph of $xy - y - x = 0$ at $x = 3$, we start by using Equation (1):

$$x = 3 \Rightarrow 3y - y - 3 = 0 \Rightarrow y = \tfrac{3}{2}$$

Hence we are talking about the point $(3,\tfrac{3}{2})$ on the graph. According to Equation (2), the slope at that point is

$$\frac{dy}{dx} = \frac{1 - y}{x - 1} = \frac{1 - \tfrac{3}{2}}{3 - 1} = -\tfrac{1}{4} \qquad \qquad \square$$

Example 1 raises a question. How do we know that the equation $xy - y - x = 0$ defines a function at all? If it does not, the process of implicit differentiation is an empty exercise, since the symbol dy/dx is meaningless if y is not a well-defined (and differentiable) function of x. Of course in Example 1 we can answer the question by solving for y:

$$xy - y = x$$
$$(x - 1)y = x$$
$$y = \frac{x}{x - 1} \tag{3}$$

We can check Equation (2) by explicit differentiation of Equation (3):

$$\frac{dy}{dx} = \frac{(x - 1)(1) - x(1)}{(x - 1)^2} \quad \text{(Quotient Rule)}$$

$$= \frac{-1}{(x - 1)^2}$$

To see that Equation (2) is equivalent to this, substitute Equation (3) in Equation (2):

$$\frac{dy}{dx} = \frac{1 - y}{x - 1} = \frac{1 - \dfrac{x}{x - 1}}{x - 1} = \frac{(x - 1) - x}{(x - 1)^2} = \frac{-1}{(x - 1)^2}$$

The trouble with checking by solving for y first is that it defeats the purpose of implicit differentiation (which is designed to find the derivative without explicitly determining the function). Moreover, it is difficult or impossible to carry out in all but the simplest cases.

□ **Example 2**   Find y' from

$$2x^2 - 2xy + y^2 = 5 \tag{4}$$

Solution: We assume that Equation (4) defines one or more differentiable functions of x. Differentiate each side with respect to x:

$$\frac{d}{dx}(2x^2) - 2\frac{d}{dx}(xy) + \frac{d}{dx}(y^2) = \frac{d}{dx}(5) \tag{5}$$

$$4x - 2(xy' + y) + 2yy' = 0 \quad \text{(Product Rule and Chain Rule)}$$
$$2x - (xy' + y) + yy' = 0 \quad \text{(divide by 2)}$$
$$(y - x)y' = y - 2x \quad \text{(rearrange terms)}$$
$$y' = \frac{y - 2x}{y - x} \quad \text{(provided that } y - x \neq 0\text{)}$$

If the slope is wanted at (say) $x = 1$, we observe from Equation (4) that

$$x = 1 \Rightarrow y^2 - 2y - 3 = 0 \Rightarrow (y + 1)(y - 3) = 0$$
$$\Rightarrow y = -1 \text{ or } y = 3$$

The slope at the points $(1, -1)$ or $(1, 3)$ is

$$\frac{dy}{dx} = \frac{y - 2x}{y - x} = \tfrac{3}{2} \text{ or } \tfrac{1}{2}$$

depending on which point we use.   □

□ **Example 3**   How do we solve for y in the equation

$$y - x \sin y + 1 = 0?$$

There is no algebraic procedure that will serve! Assuming, however, that a differentiable function of x is lurking somewhere, it is not hard to find the derivative:

$$\frac{dy}{dx} - \left(x \cos y \frac{dy}{dx} + \sin y \right) = 0$$

$$(1 - x \cos y) \frac{dy}{dx} = \sin y$$

$$\frac{dy}{dx} = \frac{\sin y}{1 - x \cos y}$$

□

These examples illustrate the fact that while implicit differentiation itself is not very complicated, we do not always know what function we are differentiating, nor even whether such a function exists. Evidently what is needed is an *Implicit Function Theorem* to tell us when to expect an equation in x and y to define a differentiable function of x. You can see that such a theorem is essential by examining the equation

$$(x - y)^2 + 2xy = y^2 + 1$$

Since this reduces to $x^2 = 1$ (the graph of which consists of the two vertical lines $x = \pm 1$), it certainly does not define y as a function of x. Implicit differentiation would be nonsense in this case, yet the nonsense is not altogether obvious. (You should try differentiating and see what happens!)

Unfortunately the theorem we have in mind cannot be proved with the tools we have at our disposal. (See the following optional note for a descriptive version of it.) We ask you to wait for multivariable calculus; meanwhile you will have to trust us not to present any foolish problems.

Optional Note (*on the Implicit Function Theorem*)   The general form of an equation in x and y is $F(x, y) = 0$, where F is a function defined in some domain in R^2 (and having real values). For example, the equation $x^2 + y^2 = 1$ can be written $F(x, y) = 0$, where $F(x, y) = x^2 + y^2 - 1$. The **partial derivative** of such a function (with respect to x) is the derivative of F regarded as a function of x alone (thinking of y as fixed). For example, the partial derivative of

$$F(x, y) = y - x \sin y + 1 \qquad \text{(Example 3)}$$

with respect to x is $D_x F(x, y) = -\sin y$. Similarly, the partial derivative with respect to y is $D_y F(x, y) = 1 - x \cos y$ (obtained by keeping x fixed and regarding F as a function of y alone).

In nontechnical terms the **Implicit Function Theorem** says that if (x_0, y_0) is a point of the graph of $F(x, y) = 0$ in a neighborhood of which F and its partial derivatives are continuous, and if

$$D_y F(x_0, y_0) \neq 0$$

then the equation $F(x, y) = 0$ defines y as a differentiable function of x in a neighborhood of x_0. (See Figure 2.) The idea is that a smooth portion of the graph of $F(x,y) = 0$ can be picked out (avoiding points where the tangent is vertical) and that this subset of $\mathfrak{R}^2$ is a differentiable function $y = f(x)$.

Figure 2
Implicit Function Theorem

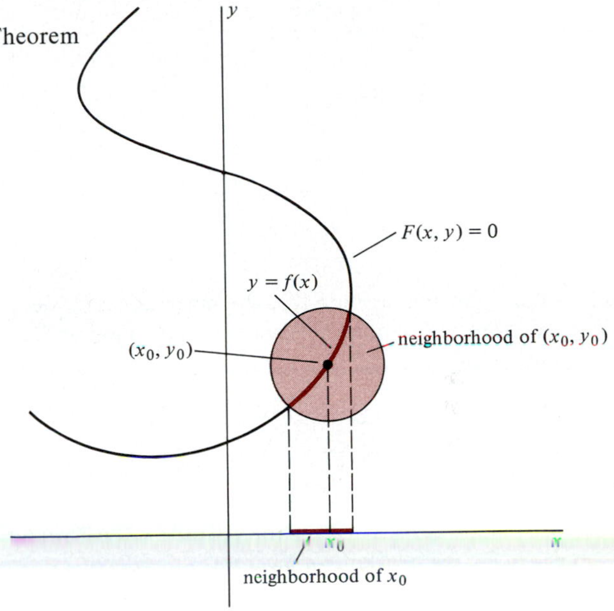

The equation

$$2x^2 - 2xy + y^2 = 5$$

for example, can be written in the form $F(x,y) = 0$, where

$$F(x,y) = 2x^2 - 2xy + y^2 - 5$$

Since $D_y F(x,y) = -2x + 2y = 2(y - x)$, we can expect implicit differentiation to succeed wherever $y - x \neq 0$. Note that this is the condition we gave in Example 2, where we found

$$\frac{dy}{dx} = \frac{y - 2x}{y - x} \qquad (y - x \neq 0)$$

On the other hand, the equation

$$(x - y)^2 + 2xy = y^2 + 1$$

(which does not define y as a function of x) can be written in the form $F(x,y) = 0$, where

$$F(x,y) = (x - y)^2 + 2xy - y^2 - 1$$

Since $D_y F(x,y) = 2(x - y)(-1) + 2x - 2y = 0$, there is no point at which we can apply the Implicit Function Theorem. The graph of $F(x,y) = 0$ is a pair of vertical lines ($x = \pm 1$), so it is not surprising that no subset of it is a real function.

Problem Set 3.4

In each of the following, use implicit differentiation to find y' from the equation as it stands. Then check by solving for y before differentiating.

1. $x + 2y = 5$ **2.** $5x - 3y = 1$

3. $xy = 1$ **4.** $\dfrac{x}{y} = 2$

5. $y^2 = x$ **6.** $y^5 = x^2$

7. $x^3 + y^3 = 1$ **8.** $x^2 - y^3 = x$

9. $x - \sqrt{y} = 2$ **10.** $xy - 2x + 3y = 0$

In each of the following, assume that the given equation defines one or more differentiable functions of x. Find y' by implicit differentiation.

11. $x^2 + y^2 = 3$ **12.** $(x - 2)^2 + y^2 = 4$

13. $x^2 - xy + y^2 = 1$ **14.** $2x^2 - 5xy - y^2 = 2$

15. $\dfrac{1}{x} - \dfrac{1}{y} = 1$ **16.** $\sqrt[3]{x} + \sqrt[3]{y} = 1$

17. $y = (x + y)^2$ **18.** $x = \dfrac{y - 1}{y + 1}$

19. $\sin(x + y) = \sin x + \sin y$

20. $y^5 - x \sin y - 3y^2 + 1 = 0$

In each of the following, find an equation of the tangent at the indicated point.

21. $y^2 + 3xy = 4$ at $(1, -4)$

22. $x^2y - y^2 + 6 = 0$ at $(1, 3)$

23. $xy + y^2 = 6$ at $(1, 2)$

24. $x^2 - xy + y^2 = 3$ at $(2, 1)$ *Hint:* What is the slope near $(2, 1)$?

25. The curve $x^2 + xy + y^2 = 3$ has two tangents at $x = 1$. What are their equations?

26. Use implicit differentiation to confirm that the hyperbola $x^2 - y^2 = 1$ has vertical tangents at the points where it intersects the x axis.

27. Use implicit differentiation to confirm that the ellipse

$$\frac{x^2}{a^2} + \frac{y^2}{b^2} = 1$$

has vertical tangents at the points where it intersects the x axis and horizontal tangents at the points where it intersects the y axis.

28. Find where the curves $xy = 2$ and $x^2 - y^2 = 3$ intersect and show that they are perpendicular at these points.

29. There are two lines through $(-2, 0)$ tangent to the parabola $y^2 = 4x$. Find their equations.

The next three problems are theorems from analytic geometry. Note that they can be done without any detailed knowledge of parabolas, ellipses, and hyperbolas! When you do study such curves, these theorems will be useful.

30. Show that the tangent to the parabola $y^2 = 4cx$ at (x_0, y_0) is

$$y_0 y = 2c(x + x_0)$$

Hint: Evaluate y' at (x_0, y_0) before using it as slope in the point-slope form of the equation of a straight line. You will also need the fact that (x_0, y_0) satisfies the equation of the parabola, that is, $y_0^2 = 4cx_0$.

31. Show that the tangent to the ellipse

$$\frac{x^2}{a^2} + \frac{y^2}{b^2} = 1$$

at (x_0, y_0) is

$$\frac{x_0 x}{a^2} + \frac{y_0 y}{b^2} = 1$$

32. Show that the tangent to the hyperbola

$$\frac{x^2}{a^2} - \frac{y^2}{b^2} = 1$$

at (x_0, y_0) is

$$\frac{x_0 x}{a^2} - \frac{y_0 y}{b^2} = 1$$

33. Use Problem 30 to find the tangent to the parabola $y^2 = 8x$ at $(2, -4)$. Check by doing the problem directly.

34. Use Problem 31 to find the tangent to the ellipse $9x^2 + 16y^2 = 144$ at $(4, 0)$. Check by referring to Problem 27.

35. Use Problem 32 to find the tangent to the hyperbola $x^2 - 9y^2 = 9$ at $(5, \frac{4}{3})$.

36. The graph of $x^{2/3} + y^{2/3} = a^{2/3}$ (where a is a positive constant) is known as a "hypocycloid of four cusps."

  (a) Explain why the curve is boxed in by the lines $x = \pm a$ and $y = \pm a$, and find its points of intersection with the coordinate axes. What kind of symmetry does it have?

  (b) Find y' by implicit differentiation and determine where the horizontal and vertical tangents occur.

  (c) Sketch the curve.

37. Consider the equation $\tan(y/x) = 1$.

   (a) Explain why the graph of this equation consists of straight lines of the form $y = mx$, where $m = \pi/4 + n\pi$, n an integer (except that the origin is missing from each line).

   (b) The slope of $y = mx$ is m. Confirm that implicit differentiation in the original equation yields $y' = m$.

38. Suppose that $y = x^{p/q}$ (p and q positive integers, $q > 1$). To prove that

$$y' = \frac{p}{q}x^{(p/q)-1}$$

(Case 3 of the Power Rule, Section 3.1), we might observe that

$$y = x^{p/q} \Rightarrow y^q = x^p$$

and use implicit differentiation to find y'.

   (a) Carry this out to obtain the formula for y'.

   (b) Do you see any hidden assumption that needs justifying before this argument can be accepted as valid?

39. The straight line $y = 2x$ is also the graph of $(y - 2x)^2 = 0$. What happens when we try to find its slope by implicit differentiation? What does the Implicit Function Theorem say about the situation?

40. Show that implicit differentiation in

$$(x - y)^2 + 2xy = y^2 + 1$$

yields $x = 0$, while the given equation implies $x = \pm 1$. It would seem to follow that $0 = \pm 1$. Explain.

True–False Quiz

1. If f and g are differentiable, then

$$\frac{d}{dx}[f(x)g(x)] = f'(x)g'(x)$$

2. If f and g are differentiable, then

$$\frac{d}{dx}\frac{f(x)}{g(x)} = \frac{f'(x)}{g'(x)}$$

3. If $f'(c)$ exists, then $\lim_{x \to c} f(x) = f(c)$.

4. If $f(x) = x^{3/2}$, then $f'(4) = 3$.

5. If $f(x) = \dfrac{x}{x-2}$, then $f'(3) = -2$.

6. $D_x\left(\dfrac{x}{\tan x}\right) = \cot x - x\csc^2 x$.

7. $D_x x|x| = 2|x|$ for all x.

8. $D_x\left(\dfrac{1}{x^2+1}\right) = \dfrac{-1}{(x^2+1)^2}$.

9. If $f(x) = x^7$, then

$$\lim_{x \to 1}\frac{f(x) - f(1)}{x - 1} = 7$$

10. If $f(x) = \sec x$, then

$$\lim_{h \to 0}\frac{f(x+h) - f(x)}{h} = \sec x \tan x$$

11. If $f(x) = (x^2 + 1)^4$, then

$$\lim_{x \to 0}\frac{f(x) - f(0)}{x} = 4$$

12. The graph of $y = x\sqrt{2 - x^2}$ has horizontal tangents at $(1,1)$ and $(-1,-1)$.

13. If f and g are differentiable and $h(x) = f[g(x)]$, then $h'(x) = f'(x)g'(x)$.

14. $D_x \cos 2x = -\sin 2x$.

15. $D_x|\sin x| = |\cos x|$.

16. If $f(t) = \cot\dfrac{t}{2}$, then $f'(\pi) = -1$.

17. The slope of the curve $x^2 - y^2 = 1$ at (x,y) is x/y $(y \neq 0)$.

18. The slope of the curve $xy^2 - x^2 + 6 = 0$ at $(3,1)$ is $\frac{5}{2}$.

Additional Problems

Find the derivative of each of the following functions.

1. $y = x + \dfrac{1}{x}$

2. $y = 2\sqrt{x} + \dfrac{2}{\sqrt{x}}$

3. $y = \frac{2}{3}x^{3/2} - \frac{3}{2}x^{2/3}$

4. $y = \frac{4}{5}x^{5/4} + \frac{5}{4}x^{4/5}$

5. $y = x^3 \sin x$

6. $y = \sqrt{x}\cos x$

7. $y = \sin^3 2x$

8. $y = \tan^2 \frac{1}{2}x$

9. $y = (x^2 + 1)^2(x^3 - 1)$

10. $y = x(x + 3)^4$

11. $y = \sqrt{x}(x - 3)$

12. $y = x^2\sqrt{x^2 + 1}$

13. $y = \sqrt{x^2 + 1} + \dfrac{1}{\sqrt{x^2 + 1}}$

14. $y = \dfrac{x}{x^2 + 4}$

15. $y = \dfrac{x(x - 3)}{x - 2}$

16. $y = \dfrac{1}{x^2 - 1}$

17. $y = \sqrt{\dfrac{x}{x^2 + 1}}$

18. $y = (x^3 - 1)^2$

19. $y = x\cos(1/x)$

20. $y = \sin 2x - \csc 2x$

21. $y = |\cos x|$

22. $y = |\tan 2x|$

23. $y = \dfrac{x^2}{\sqrt{9 - x^2}}$

24. $y = x^2\sqrt{4 - x^2}$

25. Use the Quotient Rule on $\csc x = \cot x / \cos x$ to confirm the formula

$$D_x(\csc x) = -\csc x \cot x$$

26. Show that $D_x(\sin x \cos x) = \cos 2x$ in two ways.

27. If $y = \tan x$, show that $y' = 1 + y^2$.

28. Find the speed at $t = \frac{1}{2}$ of a particle whose position at time t is $s = \cos \pi t$. In what direction is the particle traveling at that instant?

29. Find the y intercept of the tangent to the curve $y = x \sin x$ at $(\pi, 0)$.

30. Find the lines through $(0,2)$ tangent to the parabola $y = 1 - x^2$.

31. Show that the graph of $y = 4 - x^{2/3}$ has a vertical tangent at $(0,4)$ and sketch the graph.

32. Explain why the graph of $y = x/(x - 1)$ is always falling. Sketch it together with its asymptotes.

33. Sketch the graph of $y = x^2/\sqrt{1 - x^2}$ after answering the following questions.

(a) What is the domain?

(b) What are the asymptotes of the graph?

(c) Where is the graph rising? falling?

34. Sketch the graph of $y = \sec x - \tan x$, $-\pi/2 < x < \pi/2$, after answering the following questions.

(a) Why is the graph always falling? *Hint:* First show that

$$y' = \frac{\sin x - 1}{\cos^2 x}$$

(b) Why does the graph lie above the x axis?

(c) What is the y intercept?

(d) How does y behave when $x \downarrow -\pi/2$?

(e) Explain why $\lim_{x \uparrow \pi/2} y = 0$. *Hint:* First show that $y = (\cos x)/(1 + \sin x)$.

35. Show that no tangent to the curve $y = x^2$ intersects the curve twice.

36. A certain curve has slope $3x^2$ at each point (x,y) and passes through the point $(2,1)$. What is its equation?

37. Differentiate each side of the identity $\cos 2x = \cos^2 x - \sin^2 x$ to obtain an identity involving $\sin 2x$.

In each of the following, find dy/dx by implicit differentiation.

38. $x^2 - y^2 = 4$

39. $\sqrt{x} + \sqrt{y} = 1$

40. $|x| + |y| = 1$

41. $x^2 + 3xy + y^2 = 4$

42. $y = (x - y)^2$

43. $xy - y^2 = 2$

44. $x \sin y - y = 5$

45. $x = \dfrac{y + 4}{y - 4}$

46. $x + \dfrac{x}{y} + y = 2$

47. $x\sqrt{1 + y^2} = y$

48. Find an equation of the tangent to the curve $y/x + xy = 3x - y$ at $(1,1)$.

49. Find an equation of the tangent to the curve $x^2 + xy + y^2 = 3$ at $(-2,1)$.

50. Find y' from the equation $(y/x) - (x/y) = 1$ in two ways, as follows.

(a) Differentiate implicitly in the equation as it stands, obtaining $y' = y/x$.

(b) Rewrite the equation in the form $y^2 - x^2 = xy$ and differentiate to obtain $y' = (2x + y)/(2y - x)$.

(c) Reconcile the results.

51. Find y' from the equation $xy + 1/(xy) = 2$ in two ways, as follows.

(a) Differentiate implicitly in the equation as it stands.

(b) Rewrite the equation in the form $x^2y^2 - 2xy + 1 = 0$ before differentiating.

(c) Reconcile the results.

4 Continuity and Differentiation

4.1 CONTINUITY

In Section 2.5 we agreed to call a function f *continuous* at a point x_0 of its domain if $\lim_{x \to x_0} f(x) = f(x_0)$, that is, if the limit of the function can be found by evaluating the function at the point of approach. Now we want to return to primitive ideas and inquire into the reasons why this definition makes sense.

□ **Example 1**   Define the function f by

$$f(x) = \begin{cases} \dfrac{x^2 - 4}{x - 2} & \text{if } x \neq 2 \\ 1 & \text{if } x = 2 \end{cases}$$

Since $f(x) = x + 2$ when $x \neq 2$, while $f(2) = 1$, the graph of f is a straight line with a point displaced. (See Figure 1.) Most people would not need technical definitions of continuity and discontinuity to agree that this graph is continuous everywhere except at $x = 2$, where it is discontinuous. If asked why they agree, they might say, "You can draw the graph on the left of the

Figure 1
Straight line with a point displaced

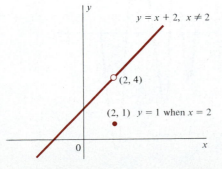

127

hole (2,4) without lifting your pencil from the paper, as well as the graph on the right of the hole. But the displaced point requires a separate action."

There are at least two implications of this kind of statement. First, even an intuitive notion of continuity involves the fact that it is a *local* property. One refers to the graph as continuous at one point and discontinuous at another. Second (and this sounds paradoxical), it is not so local that one can determine continuity or discontinuity at a point by ignoring the behavior of the graph at nearby points. There is a cohesion of the graph in the neighborhood of a point where it is continuous, in the sense that the graph is "all in one piece" near the point.

Mathematicians learned a long time ago, however, not to trust such intuitive language indiscriminately. Consider the next example.   □

□ **Example 2**   Define the function f by

$$f(x) = \begin{cases} x \sin \dfrac{1}{x} \text{ if } x \neq 0 \\ 0 \text{ if } x = 0 \end{cases}$$

The graph of f lies between the lines $y = x$ and $y = -x$ (why?) and has infinitely many "wiggles" in any neighborhood of the origin. (See Figure 2.) Note that the origin itself is a point of the graph because $f(0) = 0$. (We added this point to make the domain complete.)

Does your intuitive feeling for continuity suggest that this graph is continuous at the origin? If not, we can sympathize with you. For it is impossible to trace the curve from one side of the origin to the other with a pencil. Sooner or later (as Figure 2 indicates) you must give up trying to follow the

Figure 2
Damped sine wave with origin included

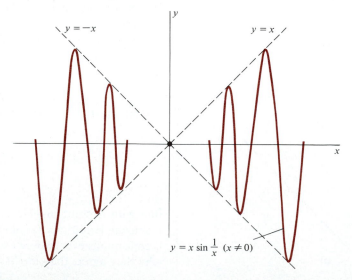

wiggles of the curve, lift your pencil from the paper, put in the point (0,0), then start in again on the other side.

On the other hand, perhaps you feel that the curve is continuous. We can defend that, too, by observing that there is no identifiable break in the graph. There are no holes, displaced points, or other evidences of discontinuity.

What are we to do in such a situation? It seems plain that we have to abandon pictorial thinking, at least for purposes of a definition, and try to say something in analytic language that will reflect as nearly as possible our intuitive grasp of the subject.   □

Suppose, therefore, that f is a real function and we want to explain what we mean by the statement f *is continuous at* x_0, where $x_0 \in \mathcal{R}$. There are three requirements that seem reasonable to impose.

1. The graph of $y = f(x)$ must contain a point corresponding to $x = x_0$, that is, x_0 should be in the domain of f. This means, for example, that continuity of $f(x) = \sqrt{x}$ at $x = -1$ is not even a possibility, since $f(-1)$ is undefined.

2. A traveler on the graph must be able to identify a definite point toward which he is heading as x approaches x_0, that is,

$$\lim_{x \to x_0} f(x) \text{ should exist}$$

Note that this is independent of requirement (1), since the definition of limit does not involve x_0 itself, but only points near x_0. A traveler on the graph of

$$y = \frac{x^2 - 4}{x - 2} \qquad \text{(straight line with a hole in it)}$$

would identify (2,4) as the point he is approaching when $x \to 2$, even though there is no point on the graph corresponding to $x = 2$.

3. Here is the crucial condition! The point required by (1) and the point of approach identified in (2) must coincide. That is,

$$\lim_{x \to x_0} f(x) = f(x_0)$$

Another way of saying this is that the point that "should" be there (as anticipated by a traveler on the graph) actually *is* there. Neither a hole nor a displaced point greets him as he arrives, but the point he expected.

It is this third condition that fails in Example 1. For although the graph contains a point corresponding to $x = 2$, namely (2,1), and although a traveler would have no doubt he is approaching (2,4) as $x \to 2$, the points do not coincide. That is,

$$\lim_{x \to 2} f(x) = \lim_{x \to 2} (x + 2) = 4 \neq f(2)$$

Therefore f is discontinuous at $x = 2$.

On the other hand, the function in Example 2 is continuous at $x = 0$:

1. There is a point on the graph corresponding to $x = 0$, namely $(0,0)$.
2. A traveler on the graph (despite a likely motion sickness induced by the increasing frequency of oscillations) would perceive that he is approaching a limit as $x \to 0$.
3. The point he expects to be there *is* there, that is,

$$\lim_{x \to 0} f(x) = \lim_{x \to 0} x \sin \frac{1}{x} = 0 = f(0)$$

Figure 3

Continuity at boundary points

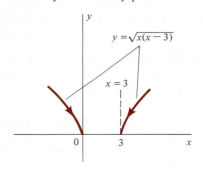

□ **Example 3**   The function $f(x) = \sqrt{x(x - 3)}$ is shown in Figure 3. Since

$$\lim_{x \to 0} f(x) = f(0) \qquad \text{and} \qquad \lim_{x \to 3} f(x) = f(3)$$

the function is continuous at $x = 0$ and at $x = 3$. A traveler on the left branch of the graph would identify $(0,0)$ as the point he is approaching as $x \to 0$. Since $f(0) = 0$, he would experience no discontinuity at the endpoint of this branch. Similarly, a traveler on the right branch is heading for $(3,0)$ as $x \to 3$. Since $f(3) = 0$, he would judge the right branch to be continuous at its endpoint. It is true that there is a break in the graph, but that is not due to any failure of the function to behave in its domain. It is the domain itself that is the culprit: there are no functional values corresponding to $x \in (0,3)$.   □

These examples suggest that our definition of continuity does two things a good definition should do. It fits our intuition in those cases where we are in no doubt about the facts (and therefore need no formal definition). More importantly, it settles ambiguous cases where our perception is not so clear (and where arguments may arise).

Let f be a real function and suppose that x_0 is a point of its domain. We say that f is *continuous at x_0* if

$$\lim_{x \to x_0} f(x) = f(x_0)$$

If S is a subset of the domain and f is continuous at each point of S, we say that f is *continuous in S*.

Virtually every (real) function important enough to be given a name is continuous in its domain. The following examples indicate what we mean.

□ **Example 4**   Power functions are continuous in $\mathcal{R}$. For if $f(x) = x^n$ (where n is a positive integer) and $x_0 \in \mathcal{R}$, we may use the multiplication rule for limits repeatedly to write

$$\lim_{x \to x_0} f(x) = \lim (x)(x) \cdots (x) \qquad (n \text{ factors in the product})$$
$$= (\lim x)(\lim x) \cdots (\lim x) = (x_0)(x_0) \cdots (x_0) = f(x_0)$$

(We have suppressed $x \to x_0$ in the limits to save writing.)   □

□ **Example 5**   Polynomials are continuous in $\mathfrak{R}$. To see why, let

$$P(x) = a_0 + a_1 x + a_2 x^2 + \cdots + a_n x^n$$

where each a_k is a real number and n is a nonnegative integer. If $x_0 \in \mathfrak{R}$, then the algebra of limits yields

$$\lim_{x \to x_0} P(x) = \lim a_0 + a_1 \lim x + a_2 \lim x^2 + \cdots + a_n \lim x^n$$

$$= a_0 + a_1 x_0 + a_2 x_0^2 + \cdots + a_n x_0^n$$

(power functions are continuous)

$$= P(x_0)$$ □

□ **Example 6**   The reciprocal function $f(x) = 1/x$ is continuous in its domain. For if $x_0 \neq 0$, we may use the division rule for limits to write

$$\lim_{x \to x_0} f(x) = \frac{\lim 1}{\lim x} = \frac{1}{x_0} = f(x_0)$$ □

Figure 4

Graph of $y = \dfrac{1}{x}$

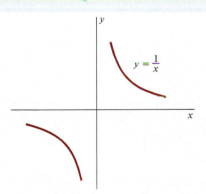

Remark   When a function is continuous in its domain, many writers simply call it **continuous,** using the global term (unqualified) to mean that there is no point of the domain where the function is discontinuous. This is not unreasonable, since a traveler on the graph would never encounter a point where there is any difficulty. But it can be misleading. To say that $f(x) = 1/x$ is continuous (without any qualification) is to risk overlooking the infinite jump at $x = 0$. (See Figure 4.) It also sounds paradoxical to say "$f(x)$ is continuous" in one breath and "$f(x)$ is discontinuous at $x = 0$" in the next. That is why we prefer the language "continuous in its domain." It leaves the door open for further remarks.

□ **Example 7**   Every rational function is continuous in its domain. For if

$$f(x) = \frac{P(x)}{Q(x)}   \text{where } P \text{ and } Q \text{ are polynomials}$$

and if x_0 is a point at which $Q(x_0) \neq 0$, then

$$\lim_{x \to x_0} f(x) = \frac{\lim P(x)}{\lim Q(x)} = \frac{P(x_0)}{Q(x_0)} = f(x_0)$$ □

Figure 5

Power and root functions

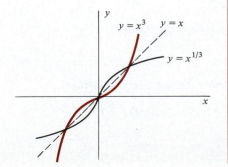

□ **Example 8**   Every root function is continuous in its domain. For if

$$f(x) = x^{1/n}   \text{(where } n \text{ is a positive integer greater than 1)}$$

the graph of f is the reflection in the line $y = x$ of the graph of

$$y = x^n \text{ (restricted to the domain } x \geq 0 \text{ if } n \text{ is even)}$$

It is geometrically apparent that a continuous curve remains continuous if it is reflected in a line, and we know power functions are continuous. Hence so are root functions. (See Figure 5, which shows the case $n = 3$.)   □

□ **Example 9**   Each of the six trigonometric functions is continuous in its domain. For we already know from Problem 66, Section 2.3, that sine and cosine are continuous at 0, that is,

$$\lim_{x \to 0} \sin x = 0 = \sin 0 \quad \text{and} \quad \lim_{x \to 0} \cos x = 1 = \cos 0$$

It is a remarkable fact that this is all we need! If $x_0 \in \mathfrak{R}$, we can write

$$\sin x = \sin [(x - x_0) + x_0]$$
$$= \sin (x - x_0) \cos x_0 + \cos (x - x_0) \sin x_0$$

and hence

$$\lim_{x \to x_0} \sin x = (\cos x_0) \cdot \lim_{x \to x_0} \sin (x - x_0) + (\sin x_0) \cdot \lim_{x \to x_0} \cos (x - x_0)$$

Letting $u = x - x_0$ and observing that $u \to 0$ when $x \to x_0$, we obtain

$$\lim_{x \to x_0} \sin x = (\cos x_0) \cdot \lim_{u \to 0} \sin u + (\sin x_0) \cdot \lim_{u \to 0} \cos u$$
$$= (\cos x_0)(0) + (\sin x_0)(1) \quad \text{(continuity of sine and cosine at 0)}$$
$$= \sin x_0$$

In the problem set we will ask you to show (similarly) that

$$\lim_{x \to x_0} \cos x = \cos x_0$$

so both sine and cosine are continuous in $\mathfrak{R}$. It follows by the algebra of limits that tangent, cotangent, secant, and cosecant are continuous wherever they are defined. (See the problem set.)   □

Once it is known that all these "elementary" functions are continuous in their domains, it is an easy matter to discuss other functions that are built up from them. For example, the function

$$f(x) = x^3 \sqrt{\cos x}$$

is continuous in its domain because x^3, $\sqrt{x}$, and $\cos x$ are continuous in theirs. The argument is based on the algebra of limits and the Composite Function Theorem (Section 2.5). If x_0 is a point of the domain of f, then

$$\lim_{x \to x_0} f(x) = (\lim x^3)(\lim \sqrt{\cos x}) = (\lim x^3)(\sqrt{\lim \cos x})$$
$$= x_0^3 \sqrt{\cos x_0} = f(x_0)$$

More generally, we may state the following principles.

Preservation of Continuity Under Arithmetic Operations

If f and g are continuous at x_0, so are $f + g, f - g, fg, f/g$ (provided they are defined at x_0).

Preservation of Continuity Under Composition

If g is continuous at x_0 and f is continuous at $u_0 = g(x_0)$, then $f \circ g$ is continuous at x_0.

□ **Example 10**   Show that the function

$$f(x) = \tan\left(\sqrt[4]{\frac{x-1}{x^3+2}}\right)$$

is continuous wherever it is defined.

Solution:

1. $(x-1)/(x^3+2)$ is continuous in its domain because it is a rational function.
2. $\sqrt[4]{(x-1)/(x^3+2)}$ is continuous in its domain because it is a composition of a root function and a rational function.
3. $f(x)$ is continuous in its domain because it is the composition of tangent and the function in (2).

In other words the only question (given $x_0 \in \mathcal{R}$) is whether x_0 is in the domain of f. If it is, f is continuous at x_0.   □

□ **Example 11**   Not all functions are defined by simple formulas. Consider the function

$$f(x) = \begin{cases} \dfrac{2}{x^2+1} & \text{if } x < 0 \text{ or } x > 1 \\ x & \text{if } 0 \leq x \leq 1 \end{cases}$$

which is defined differently in different parts of its domain. Where is f continuous?

Solution: It often helps in questions of this kind to sketch a graph. (See Figure 6.) The arrows in the figure indicate the progress of an imaginary bug moving from left to right along the graph. The bug is on the curve

$$y = \frac{2}{x^2+1}$$

when $x < 0$, approaching the point $(0,2)$. That point is missing from the graph, however; when $0 \leq x \leq 1$ the bug is on the line $y = x$. The jump from $(0,2)$ to $(0,0)$ represents a discontinuity in the function. At $(1,1)$ there is a sudden change in direction as the bug goes from the

Figure 6
Graph of f in Example 11

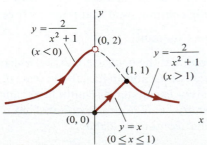

line $y = x$ back to the curve $y = 2/(x^2 + 1)$, but no discontinuity. (The *derivative* of f is discontinuous at $x = 1$, but that is a different matter.) Thus we conclude from the graph that f is continuous everywhere except at $x = 0$.

A check by the definition is hardly worthwhile, but of course it should give the same results. Discontinuity at $x = 0$, for example, is due to the fact that

$$\lim_{x \uparrow 0} f(x) = \lim_{x \uparrow 0} \frac{2}{x^2 + 1} = 2 \quad \text{and} \quad \lim_{x \downarrow 0} f(x) = \lim_{x \downarrow 0} x = 0$$

Hence $\lim_{x \to 0} f(x)$ does not exist, and f is not continuous at 0. Continuity at $x = 1$, on the other hand, is checked by noting that

$$\lim_{x \uparrow 1} f(x) = \lim_{x \uparrow 1} x = 1 \quad \text{and} \quad \lim_{x \downarrow 1} f(x) = \lim_{x \downarrow 1} \frac{2}{x^2 + 1} = 1$$

Hence $\lim_{x \to 1} f(x) = 1$. Since $f(1) = 1$, the requirements for continuity at $x = 1$ are satisfied. Continuity at points other than $x = 0$ and $x = 1$ is due to the fact that rational functions, in this case $y = 2/(x^2 + 1)$, and polynomials (in this case $y = x$) are known to be continuous in their domains.

□

Remark   The trouble with the discussion in this section is that it gives a false impression of the class of real functions. It looks like most of them are continuous almost everywhere! A real function selected at random, however, is apt to be wild, like

$$f(x) = \begin{cases} 0 \text{ if } x \text{ is rational} \\ 1 \text{ if } x \text{ is irrational} \end{cases}$$

This function is discontinuous at every point. (Why?) Moreover, even continuous functions are usually uncivilized, consisting perhaps entirely of "pathological" points like the origin on the damped sine wave of Example 2. Students often find this statement hard to believe, because they are accustomed to drawing curves which are not only continuous but "smooth." (A *smooth* curve has a tangent at every point.) Mathematicians of a century ago were in pretty much the same state of mind. But Karl Weierstrass (1815–1897), who defined continuity in terms of limits, revolutionized their thinking by coming up with a continuous curve which is *nowhere smooth*. (Loosely speaking, every point of the Weierstrass curve is a center of infinitely many wiggles.) It was that sort of thing that led mathematicians away from pictorial thinking into the modern era of rigorous definitions and proofs.

It is worthwhile to recall as we end this section that a function with a derivative at a given point is automatically continuous at that point, that is,

Differentiability $\Rightarrow$ Continuity

(See Section 3.2, where we also pointed out that the converse is false.) Since the typical function studied in calculus is differentiable in its domain (except possibly at a few isolated points), we rarely encounter discontinuities. On the other hand, the ones we do encounter are likely to be important. That is why we need to have a clear idea of what continuity means.

Problem Set 4.1

In each of the following, determine where f is continuous. (In some cases a sketch of the graph will help.)

1. $f(x) = x^2 - x + 1$ **2.** $f(x) = x^4 + 3x^2 - 2$

3. $f(x) = \dfrac{2}{x}$ **4.** $f(x) = \dfrac{1}{x^2}$

5. $f(x) = \dfrac{1}{x^2 + 1}$ **6.** $f(x) = \dfrac{x}{x^2 + 4}$

7. $f(x) = \dfrac{x}{x^2 - 1}$ **8.** $f(x) = \dfrac{1}{9 - x^2}$

9. $f(x) = \dfrac{x}{x^2 - 2x}$ **10.** $f(x) = \dfrac{x^2}{x^2 - x - 6}$

11. $f(x) = \sqrt{x - 1}$ **12.** $f(x) = \sqrt{4 - x}$

13. $f(x) = \sqrt{x^2 - x}$ **14.** $f(x) = \sqrt{\dfrac{x}{2 - x}}$

15. $f(x) = x^{-1/2}$ **16.** $f(x) = x^{-1/3}$

17. $f(x) = (8 - x)^{2/3}$ **18.** $f(x) = (1 - x)^{9/2}$

19. $f(x) = \dfrac{\cos x}{1 - \sin x},\ 0 \le x \le 2\pi\ (x \neq \pi/2)$

20. $f(x) = \dfrac{\sin x}{1 + \cos x},\ 0 \le x \le 2\pi\ (x \neq \pi)$

21. $f(x) = \tan 2x$ **22.** $f(x) = \sec \tfrac{1}{2}x$

23. $f(x) = \dfrac{x^2 - 9}{x - 3}$ **24.** $f(x) = \dfrac{x^2 - 4}{x + 2}$

25. $f(x) = \dfrac{x - 1}{1 - x}$ **26.** $f(x) = \tan x \cot x$

27. $f(x) = x - |x|$ **28.** $f(x) = \sin |x|$

29. $f(x) = \begin{cases} 1/x & \text{if } x \neq 0 \\ 0 & \text{if } x = 0 \end{cases}$ **30.** $f(x) = \begin{cases} x/4 & \text{if } x \le 2 \\ 1/x & \text{if } x > 2 \end{cases}$

31. $f(x) = \begin{cases} 1 & \text{if } x < 0 \\ 1 - x^2 & \text{if } x \ge 0 \end{cases}$

32. $f(x) = \begin{cases} \sqrt{1 - x^2} & \text{if } |x| \le 1 \\ 0 & \text{if } |x| > 1 \end{cases}$

33. $f(x) = \begin{cases} \dfrac{x^2 + 2x - 3}{x^2 - 1} & \text{if } x \neq \pm 1 \\ 2 & \text{if } x = 1 \end{cases}$

34. $f(x) = \begin{cases} \dfrac{|x|}{x} & \text{if } x \neq 0 \\ 0 & \text{if } x = 0 \end{cases}$

35. $f(x) = \begin{cases} \dfrac{|x|}{x}\cos x & \text{if } x \neq 0 \\ 0 & \text{if } x = 0 \end{cases}$

36. $f(x) = \begin{cases} \dfrac{\sin x}{x} & \text{if } x \neq 0 \\ 1 & \text{if } x = 0 \end{cases}$

37. $f(x) = \begin{cases} x & \text{if } x \text{ is an integer} \\ 0 & \text{if } x \text{ is not an integer} \end{cases}$

38. $f(x) = [x]$ [For each $x \ge 0$, $f(x) = n$, where n is the greatest integer not exceeding x, that is, $n \le x < n + 1$. See Problem 16, Section 1.4.]

39. According to Hooke's Law, the force needed to stretch a spring x meters from its natural length is proportional to x, say $F = kx$. Eventually, however, the "elastic limit" is reached and the spring is permanently distorted or even broken. If this happens when $x = a$, in what domain is the function $F = kx$ continuous?

40. The "state" of H_2O (solid, liquid, gas) depends on the temperature. Assigning the numbers 1, 2, 3 to solid, liquid, gas, respectively, we may define the "state function"

$$S(t) = \begin{cases} 1 & \text{if } -273 < t \le 0 \\ 2 & \text{if } 0 < t \le 100 \\ 3 & \text{if } t > 100 \end{cases}$$

where t is (Celsius) temperature. Sketch the graph of S and determine where it is continuous.

41. A company producing x kilograms of a certain substance per day realizes a profit (in thousands of dollars per day) of

$$P(x) = 2x - 1 \qquad (0 \le x \le 2)$$

To produce more than 2 kilograms per day, the company must start up the night shift, which increases its fixed costs; then its profit is

$$P(x) = 2x - 2 \qquad (2 < x \le 4)$$

(a) Sketch the graph of P and determine where it is continuous.

(b) On a day when the company's plant is shut down, what is its "profit"?

(c) What daily production will enable the company to break even?

(d) What production from the night shift makes it worthwhile to start up?

42. In Example 9 we showed that sine is continuous in $\Re$. Use a similar argument to show that cosine is continuous in $\Re$.

43. Do Problem 42 by using the fact that

$$\cos x = \sin\left(\frac{\pi}{2} - x\right)$$

44. Use the algebra of limits to show that tangent, cotangent, secant, and cosecant are continuous in their domains.

45. Suppose that f is continuous in the interval I and let a and m be real numbers. Why is the function

$$F(x) = f(x) - f(a) - m(x - a)$$

continuous in I?

46. Suppose that f is continuous at x_0 and $f(x_0) > 0$. Why is $f(x)$ positive for all x (in the domain of f) "near" x_0?

47. Give arguments supporting "preservation of continuity under arithmetic operations."

48. Give an argument supporting "preservation of continuity under composition."

4.2
PROPERTIES OF CONTINUOUS FUNCTIONS

The theorems of this section are all "obvious." They are surprisingly difficult to prove, however. We are going to state one of them (on which the others can be made to depend) as a fundamental principle that we ask you to accept without proof. The only way to prove it is to quarry down to the deepest axioms that govern $\Re$, and that is not ordinarily done in an elementary calculus course. As a matter of fact, some of the theorems of this section are *equivalent* to axioms, in the sense that it is a matter of indifference whether we start with them or with other fundamental assumptions when we study $\Re$.

Suppose that f is a real function continuous in the finite closed interval $[a,b]$. An intuitive description of its graph (Figure 1) might go as follows:

The graph is in one piece, is tied to definite endpoints, and is contained in a finite region of the plane. It has a highest point and a lowest point, and intersects all horizontal lines in between.

Figure 1
Graph of a function continuous in a closed interval

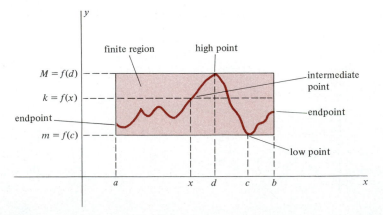

Figure 2
Damped sine wave

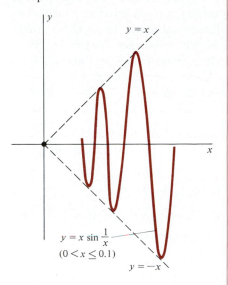

Figure 3
Range of $f(x) = \sin x$, $0 \leq x \leq \pi$

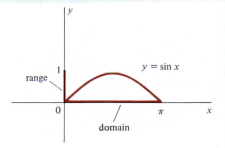

Figure 4
Function with unbounded range

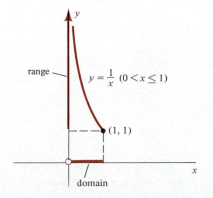

Of course in view of the bizarre types of continuity mentioned in the last section (like the Weierstrass curve), one should not take this description, or the picture, too seriously. The graph of

$$f(x) = \begin{cases} x \sin \dfrac{1}{x} & \text{if } 0 < x \leq 0.1 \\ 0 & \text{if } x = 0 \end{cases}$$

is a case in point. (See Figure 2, which shows part of the damped sine wave in Example 2 of the last section, from $x = 0$ to $x = 0.1$.) Is it really in one piece? Is it tied to the endpoint $(0,0)$? It would be hard to deny that it is, since there is no identifiable break in it. On the other hand, one cannot *draw* it in one piece, nor visualize *how* it is tied to the left endpoint. As we have suggested before, the only way to be precise is to give analytic definitions and theorems that are based on intuition, and then rely on these when intuition is hazy.

□ **Theorem 1** (*Fundamental Principle of Continuity*)   The range of a function continuous in a finite closed interval is itself a finite closed interval.

□

This theorem (which we do not prove) says more than immediately meets the eye. The following examples will help explain it.

□ **Example 1**   The function $f(x) = \sin x$ is continuous in the finite closed interval $[0,\pi]$. Its range, as you can see from Figure 3, is the finite closed interval $[0,1]$. Note that Theorem 1 does not tell us how to find it (we must do that by examining the function), but it does guarantee that the range is a finite closed interval.

□

□ **Example 2**   The function $f(x) = 1/x$ is continuous in the half-open interval $(0,1]$. Its range corresponding to that domain is $[1,\infty)$, which is not a finite closed interval. (See Figure 4.) Of course the reason that Theorem 1 does not apply is that the domain $(0,1]$ is not a closed interval.

□

□ **Example 3**   Define the function f by

$$f(x) = \begin{cases} \dfrac{|x|}{x} & \text{if } x \neq 0 \\ 0 & \text{if } x = 0 \end{cases}$$

and consider f in the domain $[-2,2]$. (See Figure 5.) The range of f is not an interval at all, but the finite set $\{-1, 0, 1\}$. The domain is a finite closed interval, but Theorem 1 does not apply because f is not continuous at 0.

□

Figure 5

Function whose range is not an interval

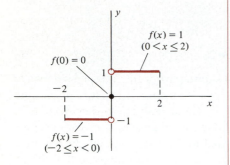

A closer look at Figure 1 (in which the range of f is shown as an interval $[m,M]$ on the y axis) reveals an important aspect of Theorem 1. The range is not merely contained in the interval $[m,M]$, *it is the interval*. This means that every number of the interval is a functional value. In particular there are points c and d in $[a,b]$ such that $f(c) = m$ and $f(d) = M$. Since $m \leq f(x) \leq M$ for every x in $[a,b]$, the functional values $f(c)$ and $f(d)$ are the smallest and largest, respectively, of all functional values. We call them "extreme values" of f, as in the following definition.

An *extreme value* of a function f with domain D is a functional value that is either the largest of all functional values or the smallest. More precisely, if $c \in D$ and

$$f(x) \geq f(c) \qquad \text{for all } x \in D$$

the number $m = f(c)$ is called the *minimum value* of f. If $d \in D$ and

$$f(x) \leq f(d) \qquad \text{for all } x \in D$$

the number $M = f(d)$ is called the *maximum value* of f.

The numbers m and M in this definition are often called *global* extreme values because all other functional values are $\geq m$ and $\leq M$. Sometimes we are not interested in all functional values, but only those corresponding to values of x "near" c (or "near" d). If there is a neighborhood of c in which $f(x) \geq f(c)$ (not necessarily for all $x \in D$ but only for x in this neighborhood), we call $m = f(c)$ a *local minimum* of f. If $f(x) \leq f(d)$ for all x in a neighborhood of d, then $M = f(d)$ is a *local maximum* of f. (See Figure 6.) Later on we will discuss local extreme values in some detail. Our next theorem, however, is a statement about global maximum and minimum values.

Figure 6

Local extreme values

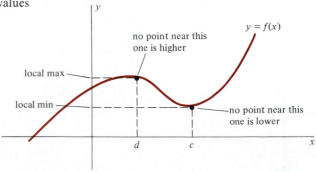

□ **Theorem 2** (*Extreme Value Theorem*)   A function continuous in a finite closed interval has a maximum value and a minimum value.

Proof: Already given. The extreme values are the endpoints m and M of the range (guaranteed by Theorem 1 to be a finite closed interval).   ☐

☐ **Example 4**   The extreme values of the function $f(x) = \sin x$, $0 \leq x \leq \pi$, are the numbers $m = f(0) = 0$ and $M = f(\pi/2) = 1$. (See Example 1.) It is also correct to say that $m = f(\pi)$, that is, the minimum occurs at $x = \pi$ as well as at $x = 0$.   ☐

☐ **Example 5**   The function $f(x) = 1/x$, $0 < x \leq 1$, has a minimum value $m = f(1) = 1$, but no maximum. (See Example 2.) Theorem 2 does not apply because the domain $(0,1]$ is not a closed interval.   ☐

☐ **Example 6**   As in Example 3, restrict the function

$$f(x) = \begin{cases} |x|/x \text{ if } x \neq 0 \\ 0 \text{ if } x = 0 \end{cases}$$

to the domain $[-2,2]$. Since f is not continuous at 0, Theorem 2 does not apply. Nevertheless f has extreme values, namely $m = -1$ and $M = 1$. (Note that Theorem 2 is not thereby contradicted! A function that does not satisfy the hypothesis may or may not obey the conclusion.)   ☐

☐ **Example 7**   Discuss extreme values of the function $f(x) = x^2$ in the domain $-1 \leq x < 2$.

Solution: The graph of f is shown in Figure 7. As you can see, f has the property that

$$0 \leq f(x) < 4 \qquad \text{for all } x \text{ in } [-1,2)$$

Its minimum value is $m = f(0) = 0$, but there is no maximum. For although the number $M = 4$ exceeds every functional value, it is not in the range; there is no point x in $[-1,2)$ such that $f(x) = M$. The most we can say is that M is the *least upper bound* of the range (sort of a substitute for the missing maximum value). More precisely, M is an *upper bound* (meaning that no functional value exceeds it) and among all the upper bounds there are, it is the *least*. (The number 5, for example, is an upper bound, but it is not the smallest.)   ☐

☐ **Theorem 3** (*Intermediate Value Theorem*)   Let m and M be the extreme values of a function f continuous in $[a,b]$. If k is any intermediate value ($m \leq k \leq M$), there is at least one number x in $[a,b]$ such that $f(x) = k$.

Proof: See the discussion preceding the definition of extreme value, where we observed that every number in the interval $[m,M]$ is a functional value.   ☐

Figure 7
Bounded function with no maximum

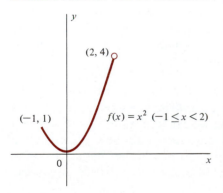

In plain English, Theorem 3 says that "f takes on all values" from its minimum to its maximum. In other words, its graph intersects every horizontal line between the low point and the high point. (See Figure 1.) For example, the number $k = \frac{1}{2}$ is an intermediate value of the function $f(x) = \sin x$, $0 \leq x \leq \pi$. (See Example 1.) A number x in $[0,\pi]$ with the property that $f(x) = k$ is $x = \pi/6$; another possibility is $x = 5\pi/6$. In any case $f(x) = \frac{1}{2}$.

An example of a function that does not have the intermediate value property is

$$f(x) = \begin{cases} \dfrac{|x|}{x} & \text{if } x \neq 0 \\ 0 & \text{if } x = 0 \end{cases}$$

(See Example 3.) The extreme values of f are $m = -1$ and $M = 1$, but f does not hit all values in between. It does not, for example, ever equal $\frac{1}{2}$.

Our next theorem is a corollary of Theorem 3; in fact it is equivalent to it, as we ask you to show in the problem set. The reason we state the theorem separately is that it is important in its own right (and has a distinctive name).

□ **Theorem 4** (*Zero Theorem*)   If f is continuous in $[a,b]$ and $f(a)$ and $f(b)$ have opposite signs, there is at least one $x \in (a,b)$ such that $f(x) = 0$.

□

What Theorem 4 says geometrically is that the graph of a function continuous in $[a,b]$ must hit the x axis somewhere between a and b if its endpoints are on opposite sides of the x axis. (See Figure 8.) This is significant in the location of (real) roots of an equation, as the next example suggests.

□ **Example 8**   The equation $x^3 + x - 1 = 0$ cannot be solved by the elementary methods of high school algebra. (The left side has no obvious factors; there are no rational roots.) But we can let $f(x) = x^3 + x - 1$ and observe that $f(0) = -1$ and $f(1) = 1$. Hence there must be a root r between 0 and 1. By repeating the procedure, we can box in the root as closely as we please. For example, $f(0.6) = -0.184$ and $f(0.7) = 0.043$, as you can check with a calculator. Hence $0.6 < r < 0.7$.

The Zero Theorem points up a fundamental distinction between the set Q of rational numbers and the set $\mathcal{R}$ of real numbers. For it is only in $\mathcal{R}$ that we can assert the theorem. If our "universe" of numbers were confined to Q, the theorem would be false. Example 8 shows this; a simpler example (and a classic one) is the equation $x^2 - 2 = 0$. No rational number satisfies this equation. (It is often proved in high school algebra that $\sqrt{2}$ is irrational.) The Zero Theorem, however, guarantees the existence of a real solution (because $x^2 - 2$ is continuous and has opposite signs for, say, $x = 1$ and $x = 2$).

Figure 8

Zeros of a continuous function

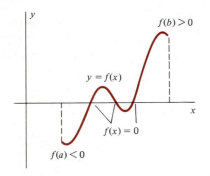

Problem Set 4.2

According to the Fundamental Principle of Continuity, the range of each of the following functions is a finite closed interval. Name the interval. (In some cases a graph will help.)

1. $f(x) = 2x - 1$ in the domain [0,2]
2. $f(x) = 1 - x$ in the domain [0,2]
3. $f(x) = x^2$, $-2 \leq x \leq 1$
4. $f(x) = 1 - x^2$, $-1 \leq x \leq 2$
5. $f(x) = 3/x$, $1 \leq x \leq 3$
6. $f(x) = \sqrt{x}$, $0 \leq x \leq 4$
7. $f(x) = |x|$, $-1 \leq x \leq 1$

8. $f(x) = \dfrac{2}{1 + x^2}$, $-1 \leq x \leq 1$

9. $f(x) = \dfrac{x - 2}{2 - x}$, $0 \leq x \leq 1$

10. $f(x) = \dfrac{2x - 6}{x - 3}$, $0 \leq x \leq 2$

11. $f(x) = |x^2 - 4|$, $-3 \leq x \leq 3$
12. $f(x) = \sin x$, $-\pi \leq x \leq \pi$
13. $f(x) = 2 \cos x$, $-\pi/2 \leq x \leq \pi/2$
14. $f(x) = \sin x \cos x$, $-\pi/4 \leq x \leq \pi/12$ *Hint:* Use the identity $\sin 2x = 2 \sin x \cos x$.
15. $f(x) = \sec^2 x - \tan^2 x$, $-\pi/4 \leq x \leq \pi/4$
16. $f(x) = |\tan x|$, $-\pi/3 \leq x \leq \pi/3$
17. Name the range of $f(x) = 4 - x^2$, $-1 < x \leq 2$, and note that it is a finite closed interval. The domain, however, is not closed. Why doesn't this contradict Theorem 1?
18. Name the maximum and minimum values of $f(x) = |x - 1|$, $-1 \leq x < 2$, noting that they exist even though the domain is not closed. Why doesn't this contradict Theorem 2?
19. Although you may be unable to name it, how can you be sure that the function $f(x) = \sqrt{x(1 - x)}$ has a maximum value? What is its minimum value and where does it occur?
20. Repeat Problem 19 for the function $f(x) = \sqrt{x(3 - x)}$.

In each of the following, find the extreme values (if any) of the function, and determine whether it is bounded.

21. $f(x) = x^2 - 1$, $-1 \leq x < 2$
22. $f(x) = \tan x$, $0 \leq x < \pi/4$
23. $f(x) = \cot x$, $0 < x \leq \pi/2$
24. $f(x) = 4/x^2$, $0 < x \leq 2$
25. Name the extreme values of the function

$$f(x) = \frac{x^2 - 1}{x - 1}, \quad 0 \leq x \leq 2 \qquad (x \neq 1)$$

Does the function take on every value in between? Explain.

26. Repeat Problem 25 for the function

$$f(x) = \frac{4 - x^2}{2 + x}, \quad -3 \leq x \leq 0 \qquad (x \neq -2)$$

27. When we use tables or a calculator to approximate the number t between 0 and $\pi/2$ such that $\sin t = 0.7$, we assume that such a number exists. What justifies this assumption?
28. What guarantees the existence of a number t between $\pi/2$ and π such that $\cos t = -3/4$?
29. The cube root of 2 is a number x satisfying the equation $x^3 = 2$. How do we know that such a number exists?
30. The symbol $\sqrt{3}$ represents a positive number x such that $x^2 = 3$. How do we know that such a number exists?
31. Explain how you can be sure that the equation $x^4 + x - 1 = 0$ has a real solution. Does every fourth-degree equation have a real solution?
32. Explain how you can be sure that the equation $x^3 - x - 1 = 0$ has a real solution. Does every cubic equation have a real solution?
33. One of the solutions of the equation in Problem 31 is between 0 and 1.
   (a) Use the Zero Theorem to locate this solution between consecutive tenths.
   (b) (For a calculator) Locate the solution between consecutive hundredths.
34. The equation in Problem 32 has only one real solution. Use the Zero Theorem to locate this solution between consecutive tenths.
35. Why do the graphs of $y = x$ and $y = \cos x$ intersect somewhere between $x = 0$ and $x = \pi/2$? *Hint:* The intersection occurs when $x = \cos x$.
36. Generalize Problem 35 to show that if f and g are continuous in [a,b], $f(a) < g(a)$, and $f(b) > g(b)$, their graphs intersect somewhere between a and b.
37. Show how the Zero Theorem follows from the Intermediate Value Theorem.
38. Had we proved the Zero Theorem first, the Intermediate Value Theorem would be a corollary. Show why. *Hint:* Given k satisfying $m \leq k \leq M$, where m and M are the extreme values of a function f continuous in [a,b], the problem is to name $x \in [a,b]$ such that $f(x) = k$. Apply the Zero Theorem to the function $g(x) = f(x) - k$.

4.3
THE MEAN VALUE THEOREM

In this section our main objective is to prove that between any two points of a smooth curve (Figure 1) there is a point at which the tangent is parallel to the chord joining the points. This innocent-sounding statement is one of the most important theorems of calculus. (Recall from the last Remark in Section 4.1 that a smooth curve must have a tangent at every point. If the curve is the graph of a function f, then f must be differentiable at all points where the tangent is nonvertical.)

Figure 1
Tangent parallel to a chord

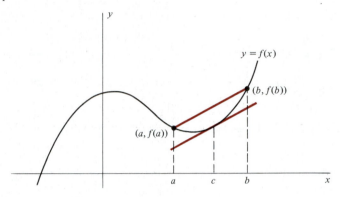

Since the slope of the chord joining $(a, f(a))$ and $(b, f(b))$ is

$$m = \frac{f(b) - f(a)}{b - a}$$

and the slope of the tangent at c is $f'(c)$, the analytic statement of the theorem is that there is a point c between a and b such that $f'(c) = m$, that is,

$$f'(c) = \frac{f(b) - f(a)}{b - a}$$

Note that the labels used for the endpoints are immaterial. For if a and b are interchanged, the above equation is unaffected:

$$f'(c) = \frac{f(a) - f(b)}{a - b} = \frac{f(b) - f(a)}{b - a}$$

In either case we may write the result in the form

$$f(b) - f(a) = f'(c)(b - a) \qquad \text{for some } c \text{ between } a \text{ and } b$$

which is the conclusion of the *Mean Value Theorem*. Its hypotheses and proof will be given after some preliminary discussion.

◻ **Example 1**   Find c in the case of the function $f(x) = x^2$ in the interval [1,4].

Solution: We write

$$f(4) - f(1) = f'(c)(4 - 1)$$

Since $f'(x) = 2x$, this reduces to $16 - 1 = 2c(3)$, from which $c = \frac{5}{2}$. Thus in this case the point at which the tangent is parallel to the chord corresponds to the midpoint of the given interval. In general we cannot expect things to be that simple; in fact we do not usually expect to find c at all. It is the *existence* (not the precise location) of c that the Mean Value Theorem asserts. That turns out to be the important thing, as we will see.   ◻

A special case of the Mean Value Theorem occurs when the endpoints of the curve $y = f(x)$, $a \le x \le b$, are on the same horizontal level, that is, when $f(a) = f(b)$. Then the slope of the chord joining $(a, f(a))$ and $(b, f(b))$ is zero and we seek a point c between a and b for which $f'(c) = 0$. (See Figure 2.) This special case is known as **Rolle's Theorem;** it says that a smooth curve cannot intersect a horizontal line twice without having a horizontal tangent somewhere in between. What is surprising about it is that even though it is a special case, it is in fact equivalent to the Mean Value Theorem. One way to see why is that a rotation of Figure 2 will change the slope of the chord from 0 to m; then we are back to the Mean Value Theorem.

Figure 2
Rolle's Theorem

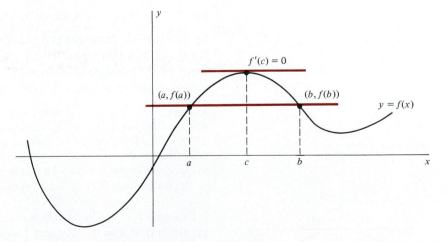

To prove Rolle's Theorem, we need a statement that is important in its own right, namely that a smooth curve must have a horizontal tangent at every high or low point that is not an endpoint. (See Figure 3.)

Figure 3
Illustration of
Theorem 1

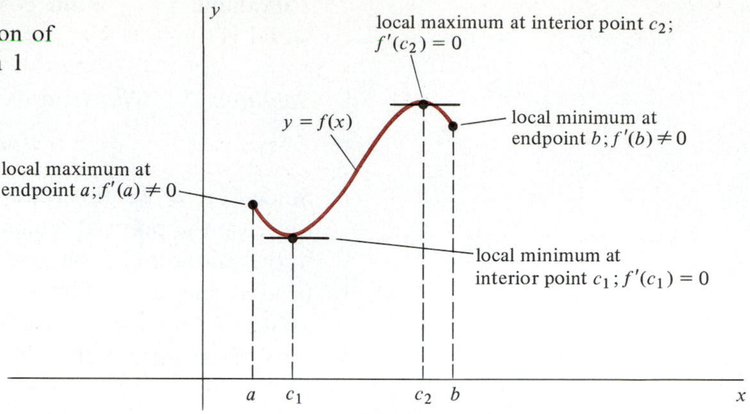

local maximum at interior point c_2;
$f'(c_2) = 0$

$y = f(x)$

local minimum at
endpoint b; $f'(b) \neq 0$

local maximum at
endpoint a; $f'(a) \neq 0$

local minimum at
interior point c_1; $f'(c_1) = 0$

a c_1 c_2 b x

□ **Theorem 1**   Suppose that f has an extreme value at c, where c is an interior point of the domain of f. If $f'(c)$ exists, it must be zero.

Proof: Suppose that $f'(c) > 0$. Since

$$f'(c) = \lim_{x \to c} Q(x) \qquad \text{where } Q(x) = \frac{f(x) - f(c)}{x - c}$$

$Q(x)$ must be positive for all x near c. (Why?) For these values of x, the numerator and denominator of $Q(x)$ must have the same sign, which means that

$$x < c \Rightarrow f(x) < f(c) \qquad \text{and} \qquad x > c \Rightarrow f(x) > f(c)$$

This conflicts, however, with the fact that $f(c)$ is an extreme value: either

$$f(x) \leq f(c) \text{ for all } x \text{ near } c \text{ [if } f(c) \text{ is a maximum]}$$

or $$f(x) \geq f(c) \text{ for all } x \text{ near } c \text{ [if } f(c) \text{ is a minimum]}$$

In view of these inequalities we cannot have $f(x) < f(c)$ on one side of c and $f(x) > f(c)$ on the other! Hence $f'(c)$ cannot be positive. The same argument (with a couple of inequalities reversed) shows that $f'(c)$ cannot be negative, either. Hence if it exists at all, it must be 0.   □

Remark   In Theorem 1 note the hypothesis that c is an *interior* point of the domain of f (that is, not a boundary point). This means that not only c itself but a neighborhood of c lies in the domain, which is crucial to the proof. Note, too, that the extreme value at c need not be "global" but may be "local." (See Section 4.2 for the meaning of these terms.) All that is required in the proof is that one or the other of the inequalities $f(x) \leq f(c)$ or $f(x) \geq f(c)$ should hold for all x near c.

Figure 4
Graph of $y = \dfrac{1}{x^2 + 1}$

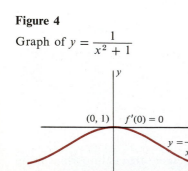

$(0, 1)$ $f'(0) = 0$

$y = \dfrac{1}{x^2 + 1}$

□ **Example 2**   The function $f(x) = 1/(x^2 + 1)$ has an extreme value (a maximum) at $x = 0$. Theorem 1 says that $f'(0)$ must be 0 (if it exists). You can check that

$$f'(x) = \frac{-2x}{(x^2 + 1)^2}$$

and hence $f'(0) = 0$. (See Figure 4.)   □

Figure 5
Graph of $y = |x|$

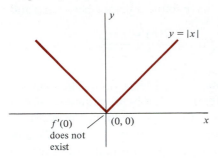

□ **Example 3**   The function $f(x) = |x|$ has an extreme value (a minimum) at $x = 0$. This time $f'(0)$ does not exist, and Theorem 1 does not apply. (See Figure 5.)   □

□ **Example 4**   The function $f(x) = x^2$, $0 \leq x \leq 1$, has a minimum at $x = 0$ and a maximum at $x = 1$. (See Figure 6.) Since these are endpoints of the domain, Theorem 1 does not apply. Anything can happen at endpoints! In this case $f'(0) = 0$ and $f'(1) \neq 0$, as you can check.   □

□ **Theorem 2** (*Rolle's Theorem*)   If f is differentiable in the open interval (a,b) and continuous in the closed interval $[a,b]$, and if $f(a) = f(b)$, there is a point c between a and b such that $f'(c) = 0$.

Figure 6
Graph of $y = x^2$, $0 \leq x \leq 1$

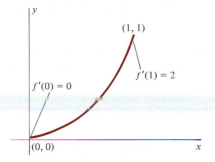

Proof: The Extreme Value Theorem (Section 4.2) says that f has a maximum M and a minimum m. If $M = m$, the function is constant and $f'(x) = 0$ for all x in the interval. If $M \neq m$, then at least one of M or m occurs at a point c strictly *between* a and b, since $f(a) = f(b)$. Apply Theorem 1 at this interior point to conclude that $f'(c) = 0$.   □

It is worth noting that the proof of Rolle's Theorem requires continuity in a closed interval (so that the Extreme Value Theorem of Section 4.2 may be used), but differentiability is needed only at interior points. That does not mean that f' is undefined at the endpoints, but that it is a matter of indifference whether it exists or not. The same remarks apply to the Mean Value Theorem, which we are now ready to state and prove.

□ **Theorem 3** (*Mean Value Theorem*)   If f is differentiable in the open interval (a,b) and continuous in the closed interval $[a,b]$, there is a point c between a and b such that

$$f(b) - f(a) = f'(c)(b - a)$$

Figure 7
Reducing the Mean Value Theorem
to Rolle's Theorem

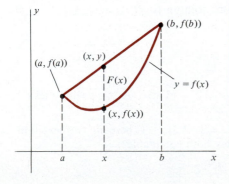

Proof: The idea is to replace f by a function with equal values at a and b, so that we may apply Rolle's Theorem. In Figure 7 we show what this function is, namely

$F(x) = (y$ coordinate on the chord$) - (y$ coordinate on the curve$)$

Since the chord meets the curve at the endpoints, we know that $F(a) = 0$ and $F(b) = 0$, so F has equal values at a and b. To find $F(x)$ in general, note that an equation of the chord is

$$y - f(a) = m(x - a) \qquad \text{where } m = \frac{f(b) - f(a)}{b - a}$$

Hence the y coordinate on the chord is $y = f(a) + m(x - a)$. The y coordinate on the curve is $f(x)$, so we have

$$F(x) = f(a) + m(x - a) - f(x) \qquad a \leq x \leq b$$

Since F is differentiable and continuous wherever f is (why?), we may apply Rolle's Theorem to F, obtaining a point c between a and b such that $F'(c) = 0$. But $F'(x) = m - f'(x)$, so we have found a point c between a and b such that

$$m - f'(c) = 0, \text{ that is, } f'(c) = \frac{f(b) - f(a)}{b - a} \qquad \square$$

We noted earlier that the hypotheses of the Mean Value Theorem are stated as economically as possible (differentiability being required only at interior points of the interval, but continuity at the endpoints as well). In view of the fact that differentiability implies continuity (Section 3.2), students sometimes wonder why we don't simply assume that $f'(x)$ exists in $[a,b]$ and be done with it. Our next example shows why the economical version (while it takes longer to state) is useful.

□ **Example 5**   Does the Mean Value Theorem apply to $f(x) = \sqrt{x}$ in the interval $[0,4]$? If so, find c; if not, explain why not.

Solution: Since $f(x) = x^{1/2}$ is a root function (Example 8, Section 4.1), we know that it is continuous in its domain; in particular it is continuous in the closed interval $[0,4]$. The derivative, however, is

$$f'(x) = \frac{1}{2}x^{-1/2} = \frac{1}{2\sqrt{x}}$$

which is undefined at $x = 0$. The hypotheses of the Mean Value Theorem are designed to accommodate such a case, for they do not require existence of $f'(x)$ at the endpoints of the interval. Thus the theorem applies.
To find c, we write

$$f(4) - f(0) = f'(c)(4 - 0)$$

$$2 = \frac{1}{2\sqrt{c}} \cdot 4$$

$$\sqrt{c} = 1$$

$$c = 1$$

Thus despite the fact that the graph of f has a vertical tangent at the origin (Figure 8), there is a point $c = 1$ between $a = 0$ and $b = 4$ at which the tangent has the same slope as the chord joining $(0, f(0))$ and $(4, f(4))$.   □

Figure 8
Graph of $f(x) = \sqrt{x}$, $0 \le x \le 4$

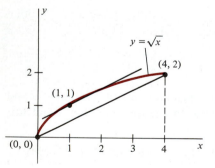

Problem Set 4.3

In each of the following, illustrate the Mean Value Theorem by finding an interior point c of the interval $I = [a,b]$ such that $f(b) - f(a) = f'(c)(b - a)$.

1. $f(x) = x^2$, $I = [0,2]$

2. $f(x) = x^2 - 2x$, $I = [1,3]$

3. $f(x) = x^3$, $I = [-2,1]$ (There are two possibilities for c. Are they both acceptable?)

4. $f(x) = x^3 - 2x$, $I = [-1,2]$ (There are two possibilities for c. Are they both acceptable?)

5. $f(x) = \sqrt{x}$, $I = [4,9]$

6. $f(x) = 1/x$, $I = [1,2]$
7. $f(x) = x - \sin x$, $I = [0,\pi]$
8. $f(x) = 2x + \cos x$, $I = [-\pi/2, \pi/2]$
9. Why are the instructions for Problems 1 through 8 impossible to carry out if $f(x) = |x|$ and $I = [-1,2]$?
10. If $f(x) = x^3 - 3x^2 + 5$, then $f(0) = f(3)$ and Rolle's Theorem guarantees a point c between 0 and 3 such that $f'(c) = 0$. Find c. What does the Mean Value Theorem say in this situation?
11. If $f(x) = \sec x$, then $f(-\pi/4) = f(\pi/4)$ and Rolle's Theorem guarantees a point c between $-\pi/4$ and $\pi/4$ such that $f'(c) = 0$. Find c. How could this have been predicted from the graph?
12. What happens when we try to apply Rolle's Theorem to $f(x) = x^{2/3}$ in the interval $[-1,1]$?
13. At 6:00 a policeman sees a car go by at 55 mph. He radios ahead to another policeman, five miles down the road, who sees the same car go by at 6:04 (doing 55 again).
    (a) What is the average speed of the car between 6:00 and 6:04?
    (b) The motorist is arrested for speeding, but argues that he was never clocked over 55. He is convicted by a judge who points out that he must have been doing 75 at least once. Is the judge correct? Explain.
14. The position at time t of an object moving along a coordinate line is $s = t - 2\sqrt{t}$.
    (a) What is the average velocity of the object during the time interval $[1,4]$?
    (b) At what time is the instantaneous velocity equal to the average velocity?
    (c) What do these questions have to do with the Mean Value Theorem?
15. Repeat Problem 14 for an object with position $s = \sin(\frac{1}{2}\pi t)$ during the time interval $[0,1]$. [A calculator is helpful in part (b).]
16. The position at time t of an object moving along a coordinate line is $s = t^2 - \sin t$.
    (a) What is the average velocity during the time interval $[0,1]$?
    (b) Show that the time t between 0 and 1 at which the instantaneous velocity is equal to the average velocity satisfies the equation
    $$2t - \cos t = 1 - \sin 1$$
    (This equation is hard to solve! The problem illustrates the fact that the "mean value" guaranteed by the Mean Value Theorem is not in general known.)
17. Let $f(x) = (x - 1)(x - 2)(x^2 + 1)$.

(a) Why must $f'(x)$ be 0 for some value of x between 1 and 2?
(b) Find $f'(x)$ and confirm by the Zero Theorem (Section 4.2) that the equation $f'(x) = 0$ has a root between 1 and 2.

18. Explain why the equation $x^3 + x - 1 = 0$ cannot have more than one real root. *Hint:* Suppose there were two. What would Rolle's Theorem say about $f(x) = x^3 + x - 1$?
19. Use Rolle's Theorem to show that a cubic equation
$$ax^3 + bx^2 + cx + d = 0 \qquad (a, b, c, d \in \mathfrak{R}, a \neq 0)$$
cannot have more than one real root if $b^2 < 3ac$. *Hint:* What does the Quadratic Formula tell you about the equation $f'(x) = 0$?
20. The graphs of $y = x$, $y = \sin x$, and $y = \tan x$ suggest that
$$\sin x < x < \tan x \qquad \text{if } 0 < x < 1$$
Confirm this statement by using the Mean Value Theorem. *Hint:* Apply the theorem to the functional differences $\sin x - \sin 0$ and $\tan x - \tan 0$.
21. Suppose that f is differentiable in an interval containing a and b. Why does the Mean Value Theorem apply to f in the interval having a and b as endpoints?
22. In the text (Theorem 3) we proved that Rolle's Theorem implies the Mean Value Theorem. Explain why the converse is true.
23. Show that if the Mean Value Theorem is applied to the quadratic function $f(x) = ax^2 + bx + c$ in any interval I, the value of x for which the tangent is parallel to the chord is the midpoint of I.
24. What is the maximum value of $f(x) = 1 - x^2$ in the domain $[-1,1]$? the minimum value? Does $f'(x) = 0$ at the points where these extreme values occur? What does Theorem 1 say?
25. What is the maximum value of $f(x) = 1 - x^{2/3}$ in the domain $[-1,1]$? What is the minimum value? Does $f'(x) = 0$ at the points where these extreme values occur? What does Theorem 1 say about this?
26. Suppose that f has a continuous derivative in a neighborhood of the point a and $f(a)$ and $f'(a)$ are known. Use the Mean Value Theorem to explain why the approximation formula
$$f(x) \approx f(a) + f'(a)(x - a)$$
is reasonable for x near a. Why is it important that f should have a *continuous* derivative?

Each of the following values was taken from a table (and can be checked by an electronic calculator). If the approximation formula in Problem 26 is used in lieu of the table or calculator, what is the result? (It is up to you to select the function f and the point a. Intelligent choices will yield good approximations.)

27. $\sin 0.2 = 0.1987$

28. $\sqrt{99} = 9.950$

29. $\sqrt[3]{8.5} = 2.041$

30. $\tan 0.8 = 1.0296$ *Hint:* 0.8 is close to $\pi/4$.

31. *Cauchy's Mean Value Theorem* says that if f and g are differentiable in (a,b) and continuous in $[a,b]$, there is a point c between a and b such that

$$g'(c)[f(b) - f(a)] = f'(c)[g(b) - g(a)]$$

(a) Confirm that Cauchy's Theorem reduces to the Mean Value Theorem of this section if $g(x) = x$.

(b) To prove Cauchy's Theorem, let
$$h(x) = g(x)[f(b) - f(a)] - f(x)[g(b) - g(a)]$$
Explain why h is a function to which Rolle's Theorem may be applied in the interval $[a,b]$. That is, explain why h is differentiable in (a,b) and continuous in $[a,b]$, and why $h(a) = h(b)$.

(c) Apply Rolle's Theorem to h and derive Cauchy's formula.

(d) Explain why Cauchy's Theorem may be written in the form

$$\frac{f(b) - f(a)}{g(b) - g(a)} = \frac{f'(c)}{g'(c)}$$

if $g'(x)$ is never 0 in the given interval.   *Hint:* The restriction on g' certainly means that $g'(c) \neq 0$. But why is $g(b) - g(a) \neq 0$?

In each of the following, find an interior point c of the interval $I = [a,b]$ such that $g'(c)[f(b) - f(a)] = f'(c)[g(b) - g(a)]$.

32. $f(x) = x^2$, $g(x) = \sqrt{x}$, $I = [0,4]$

33. $f(x) = \sin x$, $g(x) = \cos x$, $I = \left[0, \frac{\pi}{2}\right]$

34. $f(x) = x^{2/3}$, $g(x) = x^{3/2}$, $I = [0,1]$

4.4
APPLICATIONS OF THE MEAN VALUE THEOREM

When the Mean Value Theorem is applied to a function f in an interval with endpoints a and b, it guarantees the *existence* (without revealing the precise location) of a point c between a and b such that

$$f(b) - f(a) = f'(c)(b - a)$$

This formula is powerful in applications because it replaces a *difference of functional values*, $f(b) - f(a)$, by a derivative times the corresponding *difference of independent variable values*, $b - a$. The latter is easier to work with. For example, the difference $\sin 1.4 - \sin 1.2$ is hard to compute, but the difference $1.4 - 1.2$ is trivial. We shall exploit this aspect of the Mean Value Theorem in several ways, not only in this section but in things to come.

Our first application is a connection between the derivative of a function and the way in which the function changes. We have stated it before in geometric terms by saying that the graph of $y = f(x)$ is rising when $f'(x) > 0$ and falling when $f'(x) < 0$. Now we are in a position to explain what we mean in analytic terms.

Figure 1
Increasing function

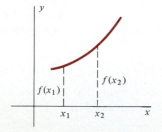

Let I be an interval in the domain of a function f. We say that f is *increasing* in I if

$$x_1 < x_2 \Rightarrow f(x_1) < f(x_2) \qquad \text{for all } x_1 \text{ and } x_2 \text{ in } I$$

and *decreasing* in I if

$$x_1 < x_2 \Rightarrow f(x_1) > f(x_2) \qquad \text{for all } x_1 \text{ and } x_2 \text{ in } I$$

Figure 2
Decreasing function

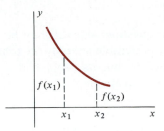

(See Figures 1 and 2.) If the inequalities involving f are not strict, that is, if

$$x_1 < x_2 \Rightarrow f(x_1) \leq f(x_2) \qquad \text{or} \qquad x_1 < x_2 \Rightarrow f(x_1) \geq f(x_2)$$

we use the terms *nondecreasing* or *nonincreasing*, respectively. In any case f is said to be *monotonic* in I.

□ **Example 1**   Each of the following statements can be quickly confirmed by looking at a graph. (See Figures 3 through 7.)

Figure 3
Graph of $f(x) = x^2$

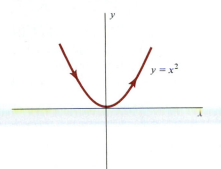

Figure 4
Graph of $f(x) = x^3$

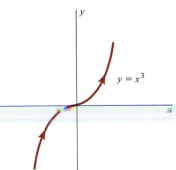

Figure 5
Graph of $f(x) = \tan x$,

$$-\frac{\pi}{2} < x < \frac{\pi}{2}$$

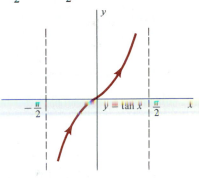

Figure 6
Graph of $f(x) = \dfrac{1}{x}$

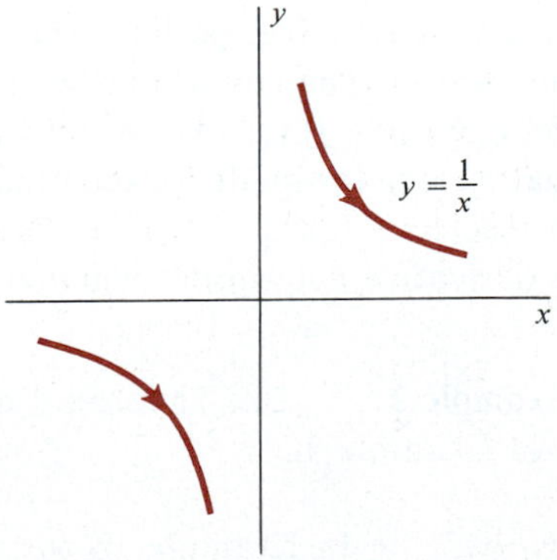

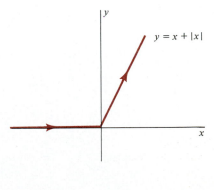

Figure 7
Graph of $f(x) = x + |x|$

(a) $f(x) = x^2$ is decreasing in $(-\infty, 0]$ and increasing in $[0, \infty)$.
(b) $f(x) = x^3$ is monotonic (in fact increasing) in its domain.
(c) $f(x) = \tan x$ is increasing in $(-\pi/2, \pi/2)$.
(d) $f(x) = 1/x$ is decreasing in $(-\infty, 0)$ and also in $(0, \infty)$. It is not, however, monotonic in its domain. (Why?)

(e) $f(x) = x + |x|$ is nondecreasing in $(-\infty,0]$ (also nonincreasing!) and increasing in $[0,\infty)$. It is monotonic in its domain.   □

Of course looking at a graph begs the question, since it may be the graph we are trying to figure out. It is usually preferable to apply the following analytic test. We have used this test before, but its proof had to be postponed until the Mean Value Theorem was available.

□ **Theorem 1**   If f is differentiable in the interval I, then

$$f'(x) > 0 \,[\geq 0] \text{ in } I \Rightarrow f \text{ is increasing [nondecreasing] in } I$$
$$f'(x) < 0 \,[\leq 0] \text{ in } I \Rightarrow f \text{ is decreasing [nonincreasing] in } I$$

(If I is not closed, the conclusion may be extended to include any endpoints of I at which f is continuous.)

Proof: Suppose that $f'(x) > 0$ in I. To prove that f is increasing in I, let x_1 and x_2 be any points of I such that $x_1 < x_2$. The desired inequality $f(x_1) < f(x_2)$ is equivalent to $f(x_2) - f(x_1) > 0$, so it seems reasonable to apply the Mean Value Theorem to f in the interval $[x_1,x_2]$. This yields a point c between x_1 and x_2 such that

$$f(x_2) - f(x_1) = f'(c)(x_2 - x_1)$$

Since $f'(c) > 0$ and $x_2 - x_1 > 0$, it follows that

$$f(x_2) - f(x_1) > 0$$

The remaining parts of the proof are left for the problem set.   □

In the proof of Theorem 1 we said that it seems reasonable to apply the Mean Value Theorem to f in the interval $[x_1,x_2]$. The reader may inquire *why* it is reasonable, perhaps thinking that the idea came from out of the blue. Not so! The clue is that we had to deal with a *difference of functional values,* $f(x_2) - f(x_1)$. As we said at the beginning of this section, such a situation is tailor-made for the Mean Value Theorem. When $f(x_2) - f(x_1)$ is replaced by $f'(c)(x_2 - x_1)$ we discover why Theorem 1 works; the sign of the derivative determines whether f is increasing or decreasing.

□ **Example 2**   Use Theorem 1 to confirm the statements in parts (b) and (d) of Example 1.

Solution: In Example 1b we said that $f(x) = x^3$ is monotonic (in fact increasing) in its domain. Since

$$f'(x) = 3x^2 \geq 0 \qquad \text{for all } x$$

Theorem 1 tells us that $f(x)$ is nondecreasing (hence monotonic) in its domain. To make that read *increasing,* we must be a little picky. For $f'(x)$ is positive only in the open intervals $(-\infty,0)$ and $(0,\infty)$. It follows from Theorem 1 that $f(x)$ is increasing in each of these intervals, but is it increasing in its domain? The graphical evidence (Figure 4) is unequivocal; the last sentence of Theorem 1 may be used to confirm it (by observing that f is continuous at the endpoint $x = 0$ of the above open intervals).

In Example 1d we said that $f(x) = 1/x$ is decreasing in $(-\infty,0)$ and also in $(0,\infty)$. We may confirm that statement by observing that

$$f'(x) = -\frac{1}{x^2} < 0 \qquad \text{for all } x \neq 0$$

Note that the endpoint $x = 0$ cannot be included this time because $f(x)$ is not continuous there. It would be unfortunate if it could be, for then we would conclude that f is monotonic in its domain, which is clearly not the case. (See Figure 6; points of the graph in the first quadrant are higher than the graph in the third quadrant, even though f is decreasing in every interval in its domain. Theorem 1 must be used with some care!) □

Another powerful application of the Mean Value Theorem occurs in the following context. Suppose that the slope of a graph is known at every point, that is, $f'(x)$ is given. Can we figure out what $f(x)$ is? For example, if $f'(x) = 2x$, what is $f(x)$? One answer that ought to come to mind is $f(x) = x^2$; others are $x^2 + 1$, $x^2 - 5$, or in general $x^2 + C$, where C is any real number. An equivalent problem is finding the position of a moving object when its velocity is known. For example, what is the coordinate of an object whose velocity at time t is $v = 2t$? Again, one obvious answer (or class of answers) is $s = t^2 + C$, where C is arbitrary.

The process of finding $f(x)$ when $f'(x)$ is given is called **antidifferentiation;** any function whose derivative is the given function is called an **antiderivative** of that function. A substantial part of every calculus course is devoted to the problem of finding antiderivatives. We are not going to discuss the methods now, but we will settle a basic question, namely, how many (essentially different) answers are there?

For example, when we put down $x^2 + C$ as a class of antiderivatives of $f'(x) = 2x$, we feel intuitively that there are no others. For the antiderivatives $y = x^2 + C$ form a class of "parallel curves," each with the same slope $2x$ at the typical point (x,y). Since these curves fill up the plane, it seems reasonable to believe that no other curve could have slope $2x$ and not be one of these. We may be mistaken, however. Just because we cannot think of other answers is no guarantee that none exists. What if there were some complicated function like

$$g(x) = \log(\sin\sqrt{x^2 + 1})$$

with the property that $g'(x) = 2x$?

What it boils down to is whether the condition $f'(x) = g'(x)$ implies anything about f and g. It is false to say that

$$f' = g' \Rightarrow f = g$$

since $f(x) = x^2$ and $g(x) = x^2 + 1$ (for example) are different functions with the same derivative. But $f' = g'$ does imply that f and g are identical "to within a constant." More precisely, we may state the following theorem.

□ **Theorem 2**   Two functions with the same derivative (in an interval I) differ by at most a constant. That is,

$$f'(x) = g'(x) \text{ for all } x \in I \Rightarrow f(x) - g(x) = C \text{ for all } x \in I$$

where C is some real number.

Proof: Let $h = f - g$. The problem is to prove that $h(x)$ has the same value for all $x \in I$. Select any two points of I, say x_1 and x_2, and apply the Mean Value Theorem to h in the interval with x_1 and x_2 as endpoints. This yields a point c between x_1 and x_2 such that

$$h(x_2) - h(x_1) = h'(c)(x_2 - x_1)$$

Since $h'(c) = f'(c) - g'(c) = 0$, it follows that

$$h(x_1) = h(x_2)$$

Let C be this constant value of h; then $f(x) - g(x) = C$ for all $x \in I$.   □

This theorem settles the question. If you can think of one antiderivative (by luck, skill, or inspiration), you have them all. Just add an arbitrary constant to get a symbolic representation of the whole family of answers.

□ **Example 3**   What are the antiderivatives of $\sin 2x$? To answer that, we need only guess one function f such that $f'(x) = \sin 2x$. A little trial and error suggests

$$f(x) = -\tfrac{1}{2}\cos 2x$$

(Check by differentiation!) Hence the entire class of antiderivatives is represented by the expression $-\tfrac{1}{2}\cos 2x + C$.   □

□ **Example 4**   If $dy/dx = 3x^2 - 5$ for all x, and $y = 3$ when $x = 1$, find y as a function of x.

Solution: The antiderivatives of $3x^2 - 5$ are of the form

$$y = x^3 - 5x + C$$

(Check by differentiation!) Since $y = 3$ when $x = 1$, we have $3 = 1 - 5 + C$, from which $C = 7$. Hence the function we seek is $y = x^3 - 5x + 7$.

   □

Remark   It is worth mentioning that an antiderivative of x^n (where $n \neq -1$) is $x^{n+1}/(n + 1)$. This helps in the antidifferentiation of polynomials (as in Example 4).

The velocity of an object moving in a straight line is the derivative of position, $v = ds/dt$, where s is the coordinate of the object at time t. **Acceleration** is the derivative of velocity, $a = dv/dt$, and is measured in velocity units per unit of time. If the units on the s axis are feet and time is measured in seconds, then v is in ft/sec and a is in ft/sec per second, commonly written ft/sec².

Acceleration may be complicated (depending on the motion), but one case is particularly simple. An object moving in a vertical straight line near the surface of the earth (with negligible air resistance) experiences an acceleration due to gravity which is virtually constant. It is positive or negative depending on the orientation of the coordinate line along which the object moves; its absolute value is denoted by g (approximately 32 ft/sec² in the English system of units).

□ **Example 5**   Suppose that a ball is thrown straight upward from the ground with a speed of 64 ft/sec. Find its height t seconds later.

Solution: Let h be the height of the ball t seconds after it was thrown. Then the coordinate line along which the ball moves is the h axis shown in Figure 8 (origin at ground level, positive direction upward). With this orientation of the axis, velocity is positive when the ball rises and negative when it falls, so v is always decreasing. Hence the acceleration is negative,

$$a = \frac{dv}{dt} = -32 \qquad \text{for all } t$$

Antidifferentiation yields $v = -32t + C_1$, where C_1 is a constant whose value may be determined by noting that $v = 64$ when $t = 0$. Thus $64 = -32(0) + C_1$ (from which $C_1 = 64$) and the formula for velocity is

$$v = -32t + 64$$

But $v = dh/dt$, so we antidifferentiate again to obtain

$$h = -16t^2 + 64t + C_2$$

where C_2 is a second constant. Since $h = 0$ when $t = 0$, we find $C_2 = 0$ and hence the formula for height is

$$h = -16t^2 + 64t$$

See Section 2.1, where we began our discussion of velocity with this formula. Its source should now be apparent; it is obtained from the acceleration due to gravity by repeated antidifferentiation.   □

□ **Example 6**   A car traveling 90 km/hr experiences a constant deceleration of 5 meters/sec² when the brakes are applied. Does it hit a barrier 65 meters down the road?

Solution: Choose the coordinate line (the s axis) with origin at the point

Figure 8
Ball traveling in a vertical line

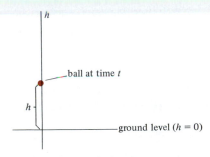

where the brakes are applied and positive direction in the direction of travel of the car. Then the acceleration is negative,

$$a = \frac{dv}{dt} = -5 \text{ meters/sec}^2 \qquad \text{for all } t$$

Start the clock ($t = 0$) when the brakes are applied. Then antidifferentiation yields

$$v = -5t + C_1 = -5t + 25 \text{ meters/sec}$$

(because $v = 90$ km/hr $= 25$ meters/sec when $t = 0$). Since $v = ds/dt$, we antidifferentiate again to obtain

$$s = -\tfrac{5}{2}t^2 + 25t + C_2 = -\tfrac{5}{2}t^2 + 25t$$

(because $s = 0$ when $t = 0$). The car stops when $v = 0$, that is, when $t = 5$. The distance traveled by the car from the point where the brakes were applied is the value of s when $t = 5$, namely $s = 62.5$ meters. Hence the car stops 2.5 meters short of the barrier.   □

Equations involving derivatives of an unknown function (like $dy/dx = 3x^2 - 5$ in Example 4 and $dv/dt = -5$ or $ds/dt = -5t + 25$ in Example 6) are called *differential equations*. We will encounter them repeatedly as we go on; antidifferentiation is the key to their solution.

Problem Set 4.4

In each of the following, use the derivative to find where f is increasing and where it is decreasing. Sketch the graph of f.

1. $f(x) = 1 - x$ **2.** $f(x) = 2x - 4$
3. $f(x) = 1 - x^2$ **4.** $f(x) = x^2 - 4$
5. $f(x) = x^3 + x$ **6.** $f(x) = x^3 - 1$
7. $f(x) = x - |x|$ **8.** $f(x) = |x - 1|$
9. $f(x) = x^{3/2}$ **10.** $f(x) = x^{2/3}$
11. $f(x) = 1/x^2$ **12.** $f(x) = 3/x$
13. $f(x) = \sin x, \ 0 \le x \le 2\pi$
14. $f(x) = \cos x, \ 0 \le x \le 2\pi$
15. $f(x) = \cot x, \ 0 < x < \pi$
16. $f(x) = \sec x, \ -\pi/2 < x < \pi/2$

Find all antiderivatives of each of the following functions.

17. $f(x) = 6x$ **18.** $f(x) = 5$
19. $f(x) = 3x + 2$ **20.** $f(x) = x^3 - 2x$
21. $f(x) = \sqrt{x}$ **22.** $f(x) = 5x^{3/2}$
23. $f(x) = \sin x$ **24.** $f(x) = \sec^2 x$

25. $f(x) = \cos 2x$ **26.** $f(x) = \sec 2x \tan 2x$
27. $f(x) = 3\cos x - \sin x$ **28.** $f(x) = 1 - \cos(x/2)$
29. $f(x) = 3x^2 + \dfrac{1}{x^2}$ **30.** $f(x) = \dfrac{x - 1}{x^3}$

31. If $dy/dx = 4x^3 - 1$ and $y = 5$ when $x = 2$, find y as a function of x.

32. If $f'(x) = 1/x^2$ and $f(1) = 0$, find $f(x)$.

33. Find an equation of the curve through the origin whose slope at each point (x, y) is $2(x - 1)$.

34. If $dy/dx = \sin x$ and $y = 1$ when $x = 0$, find y in terms of x.

35. If $f'(x) = \csc^2 x$ and $f(\pi/4) = 0$, find $f(x)$.

36. If $dy/dx = 6x^2 - 4x + 1$ and $y = 2$ when $x = 0$, find y in terms of x.

37. A ball is thrown upward from a window 80 ft above the ground with a speed of 64 ft/sec.

(a) Find the velocity and height of the ball t seconds after it was thrown.

(b) At what time does the ball reach its highest point and how high (above the ground) does it go?

(c) When does the ball hit the ground and with what speed?

38. A ball is thrown downward from a window 96 ft above the ground with a speed of 16 ft/sec. With what speed does it hit the ground?

39. A ball is dropped near the surface of the earth. Assuming that air resistance is negligible, derive the formula $s = \frac{1}{2}gt^2$ for the distance fallen in t seconds. (See Problem 51, Section 2.1, where we stated this formula without proof.)

40. A ball dropped from the top of a building takes 2 sec to hit the ground. How high is the building?

41. A ball dropped from the top of a building hits the ground with a speed of 80 ft/sec. How high is the building?

42. A balloon is rising at 8 ft/sec. When it is 80 ft off the ground a passenger drops her camera. How long does it take to hit the ground?

43. A car going 72 km/hr decelerates at the constant rate of 10 meters/sec^2 when the brakes are applied. Does it hit a barrier 15 meters down the road? If so, when? If not, how far is it from the barrier when it stops?

44. A car with constant acceleration goes from 0 to 60 mph in 11 sec. How far does it go in that time? *Hint:* 60 mph = 88 ft/sec.

45. Confirm that one antiderivative of $2 \sin x \cos x$ is $f(x) = \sin^2 x$. By writing $2 \sin x \cos x = \sin 2x$, you can see that another antiderivative is $g(x) = -\frac{1}{2} \cos 2x$. According to Theorem 2 the functions f and g differ by a constant. What is the constant?

46. Suppose that f has a constant derivative in $\mathfrak{R}$, say $f'(x) = m$.

(a) Why does it follow that $f(x) = mx + b$ for some real number b?

(b) If f also has the property that $f(u + v) = f(u) + f(v)$ for all u and v, why is $b = 0$?

47. Explain why a function whose derivative is 0 in an interval I must be constant in I.

48. The functions $f(x) = |x|/x$ and $g(x) = 2$ have the same derivative for all $x \neq 0$, yet they do not differ by a constant. Why doesn't this contradict Theorem 2?

49. The function $f(x) = \ln x$ (to be defined in Chapter 8) has domain $(0, \infty)$ and derivative $f'(x) = 1/x$. Moreover, $\ln 1 = 0$. Assuming these facts for the sake of argument, derive the laws of logarithms as follows.

(a) Confirm that

$$D_x(\ln ax) = \frac{1}{x} \quad \text{and} \quad D_x(\ln a + \ln x) = \frac{1}{x}$$

where a is any positive real number. Why does it follow that

$$\ln ax = \ln a + \ln x$$

for all $a > 0$ and $x > 0$?

(b) Confirm that

$$D_x(\ln x^r) = \frac{r}{x} \quad \text{and} \quad D_x(r \ln x) = \frac{r}{x}$$

where r is any rational number. Conclude that

$$\ln x^r = r \ln x$$

(c) Use parts (a) and (b) to show that

$$\ln \frac{a}{b} = \ln a - \ln b$$

for all $a > 0$ and $b > 0$.

50. Complete the proof of Theorem 1 as follows.

(a) Suppose that $f'(x) < 0$ in I. Use the Mean Value Theorem to show that f is decreasing in I.

(b) How should the arguments already given be modified if $f'(x) > 0$ is replaced by $f'(x) \geq 0$, or $f'(x) < 0$ by $f'(x) \leq 0$?

(c) What modifications (if any) are needed to extend the conclusion to include any endpoints of I at which f is continuous?

4.5

HIGHER-ORDER DERIVATIVES

In Section 4.4 we defined the *acceleration* of an object moving in a straight line to be $a = dv/dt$, where v is the velocity at time t. Since velocity is itself a derivative, $v = ds/dt$ (where s is the coordinate of the object at time t), acceleration is the *derivative of a derivative*, called a **second derivative**. This fact may be expressed in a variety of notations:

$$a = \frac{dv}{dt} = \frac{d}{dt}\left(\frac{ds}{dt}\right) = D_t(D_t s) = (s')'$$

These symbols suggest their own shorthand, namely

$$\frac{d}{dt}\left(\frac{ds}{dt}\right) = \frac{d^2s}{dt^2}, \qquad D_t(D_t s) = D_t{}^2 s, \qquad \text{and} \qquad (s')' = s''$$

Moreover, if $s = p(t)$, we write $s'' = p''(t)$.

There is no reason we cannot continue the process, obtaining third derivatives, fourth derivatives, and (in general) *nth-order* derivatives. When $n = 1$ we are talking about the derivative as originally defined, now called the **first derivative.** For later convenience we also allow $n = 0$, meaning the function we started with.

The notation $f', f'', f''', \ldots$ for the successive derivatives of a function f soon becomes unwieldy. Hence in general we write $f^{(n)}$, the parentheses being used to distinguish the nth derivative from f^n (the nth power of f). In this notation $f^{(0)}$ means f, $f^{(1)}$ is f', and so on. The other symbols need no modification:

$$\frac{d^n y}{dx^n}, \; D_x{}^n y \qquad \text{where } y = f(x)$$

□ **Example 1**   Find the successive derivatives of $y = f(x) = 2x^4 - 5x^2 + x + 1$.

Solution: You may confirm that

$$y' = f'(x) = 8x^3 - 10x + 1$$
$$y'' = f''(x) = 24x^2 - 10$$
$$y''' = f'''(x) = 48x$$
$$y^{(4)} = f^{(4)}(x) = 48$$

and all higher-order derivatives are zero. Functional notation is employed in the usual way, for example, $f''(1) = 14$.   □

□ **Example 2**   If $f(x) = \sin x$, then

$$f^{(0)}(x) = \sin x$$
$$f^{(1)}(x) = \cos x$$
$$f^{(2)}(x) = -\sin x$$
$$f^{(3)}(x) = -\cos x$$
$$f^{(4)}(x) = \sin x$$
$$\vdots$$

In general,

$$f^{(n)}(x) = \sin\left(x + \frac{n\pi}{2}\right)$$

a formula that is not obvious, but that may be checked by looking at the cases $n = 0$, 1, 2, 3 and appealing to the periodicity of sine.   □

□ **Example 3**   If $f(x) = 1/(1 - x)$, then

$$f^{(0)}(x) = (1 - x)^{-1}$$
$$f^{(1)}(x) = (1 - x)^{-2} \quad \text{(Chain Rule!)}$$
$$f^{(2)}(x) = 2(1 - x)^{-3}$$
$$f^{(3)}(x) = 2 \cdot 3(1 - x)^{-4}$$
$$f^{(4)}(x) = 2 \cdot 3 \cdot 4(1 - x)^{-5}$$

and in general

$$f^{(n)}(x) = n!(1 - x)^{-(n+1)}$$

where $n!$ ("n factorial") means $1 \cdot 2 \cdot 3 \cdots n$. Note that the general formula applies even when $n = 0$ if we adopt the convention $0! = 1$ as a special definition. Another reason for this convention is that $n! = n(n - 1)!$ for $n = 2, 3, 4, \ldots$ (Why?) The formula applies when $n = 1$ if we agree that $0! = 1$.   □

The second derivative of a function $y = f(x)$ has a geometric interpretation that is significant in graphing. Since it is the derivative of the first derivative, we may apply Theorem 1 of the last section (with f replaced by f') to conclude that

$$f''(x) > 0 \text{ in } I \Rightarrow f' \text{ is increasing in } I$$
$$f''(x) < 0 \text{ in } I \Rightarrow f' \text{ is decreasing in } I$$

But $f'(x)$ is the slope of the graph of f at (x, y). If it is increasing, the tangent is turning counterclockwise as we move from left to right along the curve, and we say that the graph is **concave up.** If $f'(x)$ is decreasing, the tangent is turning clockwise and we call the graph **concave down.** (See Figure 1.) A point of the graph where the concavity changes (and where f is continuous) is called a **point of inflection.**

Let f be a function whose domain includes the interval I. Then

$$f''(x) > 0 \text{ in } I \Rightarrow \text{The graph of } f \text{ is concave up in } I$$
$$f''(x) < 0 \text{ in } I \Rightarrow \text{The graph of } f \text{ is concave down in } I$$

If x_0 is a point at which f'' changes sign (and where f is continuous), then $(x_0, f(x_0))$ is a point of inflection.

□ **Example 4**   If $f(x) = x^3$, then $f''(x) = 6x$. Hence the graph of f is concave down in $(-\infty, 0)$, concave up in $(0, \infty)$, and $(0,0)$ is a point of inflection. (See Figure 2.)   □

□ **Example 5**   If $f(x) = x^{1/3}$, then

$$f''(x) = -\frac{2}{9\sqrt[3]{x^5}}$$

Figure 1
Concavity of a graph

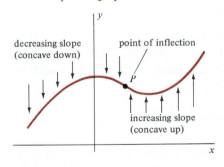

decreasing slope (concave down)

point of inflection

P

increasing slope (concave up)

Figure 2
Graph of $y = x^3$

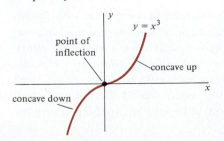

point of inflection

$y = x^3$

concave up

concave down

Figure 3
Graph of $y = x^{1/3}$

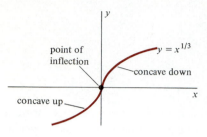

Figure 4
Graph of $y = x^4$

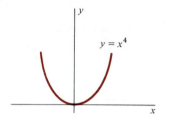

Figure 5
Graph of $y = \dfrac{1}{x}$

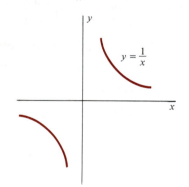

(Confirm!) Since this is positive when $x < 0$ and negative when $x > 0$, the graph of f is concave up in $(-\infty,0)$, concave down in $(0,\infty)$, and $(0,0)$ is a point of inflection. (See Figure 3.) □

Note that the point of inflection occurs when $f''(x) = 0$ in Example 4 and when f'' is undefined in Example 5. Generally speaking, a change in sign of f'' cannot be expected at points other than these two types.

□ **Example 6**   If $f(x) = x^4$, then $f''(x) = 12x^2$. Since this is positive for all $x \neq 0$, we conclude that the graph of f is concave up in $(-\infty,0)$ and in $(0,\infty)$. In fact it is geometrically apparent (Figure 4) that the graph is concave up in the whole domain. Note that although $f''(x) = 0$ at $x = 0$, there is no point of inflection because the concavity does not change. □

□ **Example 7**   If $f(x) = 1/x$, then $f''(x) = 2/x^3$. Hence the graph of f is concave down in $(-\infty,0)$ and concave up in $(0,\infty)$. (See Figure 5.) In this case f'' is undefined at $x = 0$ and there is a corresponding change of concavity. Yet there is no point of inflection, since there is no point on the graph corresponding to $x = 0$. (Contrast this with Example 5.) □

□ **Example 8**   The graph of

$$y = \frac{x^2}{(x-2)^2}$$

(which we discussed in Example 7, Section 1.5) is shown in Figure 6. It is clear that the graph is concave up in $(0,2)$ and in $(2,\infty)$. The concavity in $(-\infty,0)$ is not so apparent, but it looks like there is a point of inflection somewhere to the left of the origin. Since

$$y'' = \frac{8(x+1)}{(x-2)^4}$$

(a formula that requires considerable algebra to confirm), the graph is concave down in $(-\infty,-1)$, concave up in $(-1,2)$ and in $(2,\infty)$, and $(-1,\tfrac{1}{9})$ is a point of inflection. □

A legitimate question for the student to raise in connection with Example 8 (particularly if you did the algebra to find y'') is whether the information derived from the behavior of the second derivative is worth the effort. We drew the graph in Section 1.5 without mentioning concavity or points of inflection. Would this new information persuade us to draw it again? Hardly! The geometric interpretation of y'' is sufficiently useful to be worth learning, but you should exercise good judgment in applying it to the sketching of curves.

Figure 6

Graph of $y = \dfrac{x^2}{(x-2)^2}$

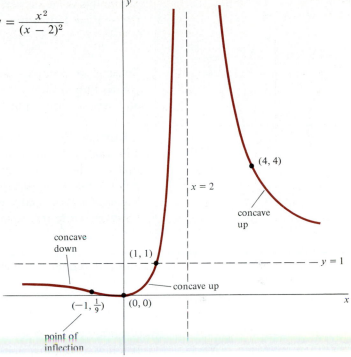

Higher-order derivatives of implicitly defined functions are found by the methods of Section 3.4, as in the following examples.

▫ **Example 9**   The graph of $x^2 - y^2 = 1$ is a hyperbola with asymptotes $y = \pm x$. (See Problem 57, Section 1.3.) Discuss its features by using calculus.

Solution: Differentiating each side of $x^2 - y^2 = 1$ with respect to x, we find

$$2x - 2yy' = 0$$

$$y' = \frac{x}{y}$$

Since y' increases without bound as we approach the points $(\pm 1, 0)$ of the graph, there are vertical tangents at these points. (See Problem 26, Section 3.4.) There are no horizontal tangents, for although y' appears to be zero at $x = 0$, the domain ($x \geq 1$ or $x \leq -1$) excludes that point.

Differentiating again (by the Quotient Rule), we obtain

$$y'' = \frac{y - xy'}{y^2} = \frac{1}{y^2}\left[y - x\left(\frac{x}{y}\right)\right] \qquad \left(\text{because } y' = \frac{x}{y}\right)$$

$$= \frac{y^2 - x^2}{y^3}$$

Figure 7

Graph of $x^2 - y^2 = 1$

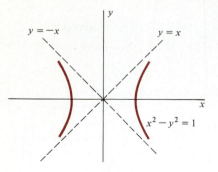

Figure 8

Graph of $\dfrac{x^2}{a^2} + \dfrac{y^2}{b^2} = 1$

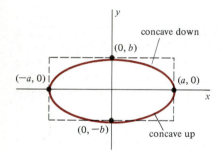

While this is an answer, it can be simplified by using the original equation $(x^2 - y^2 = 1)$ to write

$$y'' = -\frac{1}{y^3}$$

Since y'' is positive when $y < 0$ and negative when $y > 0$, the graph is concave up in the third and fourth quadrants and concave down in the first and second. See Figure 7.

□

□ **Example 10**   The graph of $x^2/a^2 + y^2/b^2 = 1$ is an ellipse. It is usually analyzed and sketched without calculus, by observing (for example) that it is boxed in between the lines $x = \pm a$ and $y = \pm b$, contains the points $(\pm a, 0)$ and $(0, \pm b)$, and is symmetric about the coordinate axes. (See Figure 8.) We usually guess the rest! To analyze the graph by calculus, however, write the equation in the form

$$b^2 x^2 + a^2 y^2 = a^2 b^2$$

and differentiate implicitly:

$$2b^2 x + 2a^2 y y' = 0$$

$$y' = -\frac{b^2 x}{a^2 y}$$

It follows from this that the tangent is horizontal at $(0, \pm b)$ and vertical at $(\pm a, 0)$. (See Problem 27, Section 3.4.) Moreover, the curve is falling from $(0, b)$ to $(a, 0)$ in the first quadrant because $y' < 0$ when x and y are positive.

Now differentiate again:

$$y'' = -\frac{b^2}{a^2}\left(\frac{y - xy'}{y^2}\right)$$

$$= -\frac{b^2}{a^2 y^2}\left[y - x\left(-\frac{b^2 x}{a^2 y}\right)\right] \quad \left(\text{from } y' = -\frac{b^2 x}{a^2 y}\right)$$

$$= -\frac{b^2}{a^4 y^3}(b^2 x^2 + a^2 y^2)$$

$$= -\frac{b^2}{a^4 y^3}(a^2 b^2) \quad \text{(why?)}$$

$$= -\frac{b^4}{a^2 y^3}$$

This shows that the upper half of the curve $(y > 0)$ is concave down (because $y'' < 0$) and the lower half $(y < 0)$ is concave up (because $y'' > 0$).

□

Problem Set 4.5

Find the second derivative of each of the following functions.

1. $y = 3x - 7$ **2.** $y = 5x^2 + x$

3. $y = 2x^3 - 4x$ **4.** $y = x^3 - x^2 + x - 1$

5. $y = \sqrt{x}$ **6.** $y = 4x^{3/2} - 3x$

7. $y = \dfrac{x - 1}{x}$ **8.** $y = \dfrac{x}{x - 1}$

9. $y = \tan x$ **10.** $y = \cot x$

11. $y = x \sin x$ **12.** $y = x \cos x$

In each of the following, a law of motion is given (position in terms of time). Find the acceleration and use it to determine when the velocity is increasing and when it is decreasing.

13. $s = 2t^2 + 20$ 14. $s = -16t^2 + 64t$

15. $s = t^3 - 3t^2$ 16. $s = t^4 - 4t^3 + 5$

17. $s = 3 \cos t,\ 0 \le t \le 2\pi$

18. $s = 2 \sin t,\ 0 \le t \le 2\pi$

19. The position of a moving object at time t is $s = \sqrt{t}$. Find the velocity and acceleration at $t = 1$ and describe the motion as the clock ticks off $t = 1$.

20. Repeat Problem 19 for the motion $s = \cos(\pi t)$.

21. Suppose that the coordinate of a moving object at time t (its "displacement" from the origin) is $x = A \cos \omega t + B \sin \omega t$, where A, B, and ω (lowercase Greek letter *omega*) are constants. Show that $d^2x/dt^2 = -\omega^2 x$.

22. The differential equation $d^2x/dt^2 = -\omega^2 x$ (Problem 21) is characteristic of *harmonic motion*. Why do physicists describe such motion by saying that "the force is proportional to the displacement and is in the opposite direction?" *Hint:* According to Newton's Second Law of Motion, force is mass times acceleration.

In each of the following, find a general formula for $f^{(n)}(x)$.

23. $f(x) = \dfrac{1}{2 - x}$

24. $f(x) = \dfrac{1}{1 + x}$ *Hint:* The alternating signs occurring in successive derivatives may be accounted for in the general formula by the factor $(-1)^n$, which takes on the values $1, -1, 1, -1, \ldots$ when $n = 0, 1, 2, 3, \ldots$.

25. $f(x) = \dfrac{1}{\sqrt{1 - x}}$ 26. $f(x) = \dfrac{1}{x}$

27. $f(x) = \sin 2x$ *Hint:* See Example 2.

28. Show that if $f(x) = \cos x$, then

$$f^{(n)}(x) = \cos\left(x + \frac{n\pi}{2}\right)$$

29. Suppose that you are driving along the graph of $y = f(x)$ from left to right.

(a) On portions of the road that are concave up, which way do you have to keep turning your car? What about portions that are concave down?

(b) What happens at a point of inflection?

30. The president's economic advisers report that although the general price level P is still increasing, it is rising more slowly. The president announces that "the inflation rate is going down" and predicts that "the high-water mark will soon be reached."

(a) There is still some bad news for consumers in the president's announcement. Describe it in terms of a derivative.

(b) State the good news in terms of a derivative.

(c) State the president's prediction in terms of the future behavior of a derivative.

(d) If the president is correct, what does the graph of P (as a function of the time t) look like?

31. Suppose that a certain population P cannot grow beyond the limiting value M. If the rate of growth is proportional to both P and $M - P$ (that is, to the number present and the number still to go), then

$$\frac{dP}{dt} = kP(M - P) \qquad (k = \text{constant})$$

where t is time. Show that the graph of P (as a function of t) has a point of inflection when the population is halfway to its upper limit.

32. The graphs of $y = 1 - x^2$ and $y = 1 - x^{2/3}$ both have highest point $(0,1)$ and both cross the x axis at $(\pm 1, 0)$. (Confirm!) Sketch the graphs after discussing the concavity, noting how it fits in with the fact that the first curve is smooth at $(0,1)$ while the second has a cusp there.

33. Why are there no points of inflection on the graph of $y = \sec x$?

In each of the following, use the first derivative to find where the graph is rising and falling, and the second derivative to find where it is concave up and concave down. Locate all turning points and points of inflection and sketch the graph.

34. $y = x^3 - 1$ 35. $y = x^3 - 3x + 2$

36. $y = 2x^3 - 3x^2 - 12x + 6$ 37. $y = x^4 - 8x^2 + 16$

38. $y = \dfrac{1}{x^2 + 1}$ 39. $y = x + \dfrac{1}{x}$

40. $y = \dfrac{4(x - 1)}{x^2}$ 41. $y = \sqrt{x}$

42. $y = x^{2/3}$ 43. $y = \sqrt{4 - x^2}$

44. $y = \sin x$ $(0 \le x \le 2\pi)$

45. $y = \tan x$ $\left(-\dfrac{\pi}{2} < x < \dfrac{\pi}{2}\right)$

46. $y = \cot x$ $(0 < x < \pi)$

47. If $y = x\sqrt{3 - x}$, then

$$y' = \frac{3(2 - x)}{2\sqrt{3 - x}}$$

(a) Use this result to show that

$$y'' = \frac{3(x - 4)}{4(3 - x)^{3/2}}$$

(b) Criticize the statement that there is a point of inflection at $x = 4$ because y'' changes sign there. What is the correct statement to make about the sign of y''?

(c) With the aid of y' and y'', sketch the graph. (You have seen this already in Example 8, Section 1.5. Has y'' added anything substantial to the discussion in that problem?)

48. The equation $y^2 = 4x$ represents a parabola. Use implicit differentiation to discuss its concavity and sketch the graph.

49. The equation $x^2/a^2 - y^2/b^2 = 1$ represents a hyperbola. Use implicit differentiation to show that

$$y'' = -\frac{b^4}{a^2 y^3}$$

and discuss concavity.

50. The graph of $x^{2/3} + y^{2/3} = a^{2/3}\, (a > 0)$ is a "hypocycloid of four cusps."

(a) Confirm that

$$y' = -\left(\frac{y}{x}\right)^{1/3} \quad \text{and} \quad y'' = \frac{a^{2/3}}{3x^{4/3}y^{1/3}}$$

(b) Use y' and y'' (together with remarks about domain, intercepts, and symmetry) to sketch the curve. (You may have done this already in Problem 36, Section 3.4. Is it worthwhile to find y''?)

51. A formula that will occur later in the book is

$$\kappa = \frac{|y''|}{(1 + y'^2)^{3/2}}$$

where κ (lowercase Greek *kappa*) is the "curvature" of the graph of $y = f(x)$. Use implicit differentiation to show that the circle $x^2 + y^2 = r^2\, (r > 0)$ has constant curvature $\kappa = 1/r$.

52. Referring to Problem 51, find the curvature of a straight line.

53. If $y = ax + b$ (a and b constant), then $y'' = 0$ for all x. Is the converse true? That is, if $f(x)$ is an unknown function whose second derivative is always 0, does it follow that $f(x) = ax + b$ for some constants a and b?

54. If f has a second derivative that is constant in $\Re$, show that $f(x)$ is a polynomial of degree ≤ 2.

4.6

LINEAR APPROXIMATION

Let $y = f(x)$ be a function that is differentiable at the point x_0. As Figure 1 suggests, the tangent line at $(x_0, f(x_0))$ serves as a reasonable approximation to the curve for values of x near x_0. Here we have written the equation of the tangent in the form $y = L(x)$. The idea is that if x is close to x_0, the point $(x, L(x))$ is not a bad substitute for the point $(x, f(x))$. By following the tangent (instead of the curve) as we move from the base point x_0 to the neighboring point x, we commit an error

$$|f(x) - L(x)|$$

which is presumably small when x is close to x_0.

Since the tangent has slope $f'(x_0)$ and contains the point $(x_0, f(x_0))$, its equation is

$$y - f(x_0) = f'(x_0)(x - x_0)$$

Solving for $y = L(x)$, we find

$$L(x) = f(x_0) + f'(x_0)(x - x_0)$$

Approximation by the tangent amounts to saying that $f(x) \approx L(x)$, that is,

$$\boxed{f(x) \approx f(x_0) + f'(x_0)(x - x_0)}$$

Figure 1
Approximation by the tangent

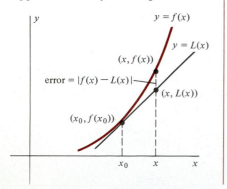

This is called **linear approximation** because $f(x)$ is replaced by a linear expression in x (that is, the original curve is replaced by a straight line).

□ **Example 1**   Engineers often use the approximation $\sin x \approx x$ for small values of x. Explain why.

Solution: Linear approximation of $f(x) = \sin x$ in a neighborhood of $x = 0$ is

$$f(x) \approx f(0) + f'(0)(x - 0)$$

Since $f(0) = 0$ and $f'(0) = \cos 0 = 1$, the approximation reduces to $\sin x \approx x$. Of course this is not surprising in view of the fact that

$$\lim_{x \to 0} \frac{\sin x}{x} = 1$$

(proved in Section 2.3).   □

□ **Example 2**   What linear function is a good approximation to $\sqrt{x}$ in a neighborhood of 1?

Solution: Let $f(x) = \sqrt{x}$ and compute $f(1) = 1$ and $f'(1) = \frac{1}{2}$. Our approximation formula then reads

$$f(x) \approx f(1) + f'(1)(x - 1) = 1 + \tfrac{1}{2}(x - 1)$$

so the linear function $L(x) = 1 + \frac{1}{2}(x - 1)$ will serve the purpose. We may use it, for example, to compute

$$\sqrt{0.98} \approx 1 + \tfrac{1}{2}(0.98 - 1) = 0.99$$ □

□ **Example 3**   Use linear approximation to compute $(5.103)^2$ and compare with the true value 26.040609.

Solution: Let $f(x) = x^2$ and select the "base point" $x_0 = 5$. Our linear approximation formula then reads

$$f(x) \approx f(5) + f'(5)(x - 5)$$

from which

$$(5.103)^2 = f(5.103) \approx 25 + 10(5.103 - 5) = 26.03$$

The error (compared to the true value) is $|26.040609 - 26.03| = 0.010609$. By itself this is not particularly meaningful. But its size relative to what we are trying to approximate (the **relative error**) is

$$\left| \frac{\text{error}}{\text{true value}} \right| = \frac{0.010609}{26.040609} \approx 0.0004$$

or about $\frac{1}{25}$ of one per cent.   □

□ **Example 4**   Use linear approximation to compute 1/9.78 and compare with the value 0.10224949 obtained from a calculator. What is the relative error?

Solution: Let $f(x) = 1/x$ and select the base point $x_0 = 10$. From the formula

$$f(x) \approx f(10) + f'(10)(x - 10)$$

we find

$$\frac{1}{9.78} = f(9.78) \approx \frac{1}{10} - \frac{1}{100}(9.78 - 10) = 0.1022$$

The relative error is

$$\left| \frac{0.10224949 - 0.1022}{0.10224949} \right| \approx 0.0005$$

or about $\frac{1}{20}$ of one per cent.   □

The student has a right to object that the error cannot be computed unless the true value is known. And of course if that is known, what is the point of approximating it? What is needed is an estimate of the error that is independent of the answer. In an optional note at the end of this section we prove that if f is twice-differentiable in a neighborhood of x_0, and x is a point of this neighborhood, there is a point c between x_0 and x such that

$$f(x) = f(x_0) + f'(x_0)(x - x_0) + \tfrac{1}{2}f''(c)(x - x_0)^2$$

When we use the approximation formula

$$f(x) \approx L(x) = f(x_0) + f'(x_0)(x - x_0)$$

we are neglecting the term involving $(x - x_0)^2$, which is not unreasonable when x is close to x_0. (How close must it be for $(x - x_0)^2$ to be smaller than $x - x_0$?)

More precisely, the error is

$$|f(x) - L(x)| = \tfrac{1}{2}|f''(c)|(x - x_0)^2$$

If we can find an upper bound for the numerical value of the second derivative, say $|f''(c)| \leq M$ for all c between x_0 and x, then an upper bound for the error is given by

$$|f(x) - L(x)| \leq \tfrac{1}{2}M(x - x_0)^2$$

To see how this works in practice, look at Example 1, where $f(x) = \sin x$ and $x_0 = 0$, and linear approximation gives $\sin x \approx x$. Since

$f''(x) = -\sin x$, we have

$$|f''(c)| = |-\sin c| \leq 1 \qquad \text{for all } c$$

so we may take $M = 1$ as an upper bound for the second derivative. The error is

$$|f(x) - L(x)| = |\sin x - x| \leq \tfrac{1}{2} \cdot 1(x - 0)^2 = \tfrac{1}{2}x^2$$

Thus (for example) without knowing the true value of $\sin 0.2$ we may conclude that $\sin 0.2 \approx 0.2$, with an error no larger than $\tfrac{1}{2}(0.2)^2 = 0.02$. This guarantees that $\sin 0.2 = 0.2$ correct to one decimal place (because the error would have to be at least 0.05 to change the first place).

□ **Example 5**   Use linear approximation to compute $\tan 0.8$ and estimate the error.

Solution: Let $f(x) = \tan x$ and choose the base point $x_0 = \pi/4$ (because 0.8 is close to $\pi/4$ and $\tan x$ and its derivatives are easy to compute at $\pi/4$). Since

$$f\left(\frac{\pi}{4}\right) = \tan\frac{\pi}{4} = 1 \qquad \text{and} \qquad f'\left(\frac{\pi}{4}\right) = \sec^2\frac{\pi}{4} = 2$$

our approximation formula becomes

$$\tan x \approx 1 + 2\left(x - \frac{\pi}{4}\right)$$

for x near $\pi/4$. Taking $x = 0.8$ (and replacing $\pi/4$ by 0.7854), we find

$$\tan 0.8 \approx 1 + 2(0.8 - 0.7854) = 1.0292$$

To estimate the error, we compute $f''(x) = 2\sec^2 x \tan x$ and try to name an upper bound M for

$$|f''(c)| = |2\sec^2 c \tan c| \qquad \text{(where } c \text{ is between } \pi/4 \text{ and } 0.8)$$

Both tangent and secant are positive and increasing in the interval $(0, \pi/2)$, as you can check from Figures 7 and 9 in Section 2.3. So we will choose some convenient value larger than c and use it to name M. How about $\pi/3$? Since $c < \pi/3$, we have

$$\tan c < \tan\frac{\pi}{3} = \sqrt{3} \qquad \text{and} \qquad \sec c < \sec\frac{\pi}{3} = 2$$

from which

$$|f''(c)| < 2(4)(\sqrt{3}) = 8\sqrt{3}$$

Thus we may take $M = 8\sqrt{3}$ and the error is

$$|f(x) - L(x)| = |\tan 0.8 - 1.0292| \leq \tfrac{1}{2}M(x - x_0)^2 = \tfrac{1}{2}(8\sqrt{3})\left(0.8 - \frac{\pi}{4}\right)^2$$

$$< 4(1.74)(0.8 - 0.78)^2 = 0.002784 < 0.003$$

In other words $\tan 0.8 \approx 1.0292$ with an error smaller than 0.003. This guarantees that $\tan 0.8 = 1.03$ correct to two decimal places. (Compare with the value $\tan 0.8 = 1.0296$ from a table.)   □

Remark   In estimating the error in Example 5 we kept trying to simplify the arithmetic while making sure that our inequalities were correct. Our choice of $\pi/3$ was for convenience in naming M, but of course other choices are possible. Once M was named, the exact value of our upper bound for the error was

$$\tfrac{1}{2}M(x - x_0)^2 = \tfrac{1}{2}(8\sqrt{3})\left(0.8 - \frac{\pi}{4}\right)^2$$

but that was hard to compute. So we replaced $\sqrt{3}$ by the *larger* value 1.74 and $\pi/4$ by the *smaller* value 0.78 to guarantee that

$$\tfrac{1}{2}M(x - x_0)^2 < 4(1.74)(0.8 - 0.78)^2$$

This was easy arithmetic; its value 0.002784 was replaced by the *larger* value 0.003 to yield the final estimate of the error.

In view of the availability of calculators, the student may be impatient with these details. What we are trying to illustrate in this section, however, is not efficiency but the *idea of approximation* (including the estimation of errors). It is an important subject.

Optional Note (*on the error in linear approximation*)   Suppose that f is twice-differentiable in a neighborhood of x_0 and x is a point of this neighborhood. If

$$L(x) = f(x_0) + f'(x_0)(x - x_0)$$

the error in using $L(x)$ as an approximation to $f(x)$ is the absolute value of $E(x) = f(x) - L(x)$. What we propose to show is that there is a point c between x_0 and x such that

$$E(x) = \tfrac{1}{2}f''(c)(x - x_0)^2$$

For then it will follow that

$$f(x) = f(x_0) + f'(x_0)(x - x_0) + \tfrac{1}{2}f''(c)(x - x_0)^2$$

which is what we want to prove.

Regarding x as fixed ($x \neq x_0$), define the function of t given by

$$F(t) = f(t) + f'(t)(x - t) + \frac{E(x)}{(x - x_0)^2}(x - t)^2$$

We assert that Rolle's Theorem (Section 4.3) applies to $F(t)$ in the interval with endpoints x_0 and x. To confirm this, observe that

$$F(x_0) = f(x_0) + f'(x_0)(x - x_0) + \frac{E(x)}{(x - x_0)^2}(x - x_0)^2 = L(x) + E(x) = f(x)$$

and

$$F(x) = f(x) + f'(x)(x - x) + \frac{E(x)}{(x - x_0)^2}(x - x)^2 = f(x)$$

Hence F has the same value at the endpoints of the interval (as required by Rolle's Theorem). Moreover, F is differentiable in this interval (including the endpoints) because f and f' are. Its derivative, in fact, is

$$F'(t) = f'(t) + f'(t)(-1) + f''(t)(x - t) + \frac{2E(x)}{(x - x_0)^2}(x - t)(-1)$$

$$= f''(t)(x - t) - \frac{2E(x)}{(x - x_0)^2}(x - t)$$

(Remember that x is fixed! The variable is t. Also note the use of the Product Rule and Chain Rule in the differentiation.)

Rolle's Theorem says that $F'(c) = 0$ for some point c between x_0 and x. Replacing t by c in the above formula for $F'(t)$ and setting the result equal to zero, we have

$$\frac{2E(x)}{(x - x_0)^2}(x - c) = f''(c)(x - c)$$

from which $E(x) = \frac{1}{2}f''(c)(x - x_0)^2$.

You should be able to follow the above argument. Thinking it up is another matter! The function $F(t)$ is the key to it, just as our proof of the Mean Value Theorem in Section 4.3 involved a tricky function to which we could apply Rolle's Theorem. Such devices do not come off the top of anybody's head; they are figured out after considerable thought about what is needed in the proof.

Problem Set 4.6

1. Let $f(x) = x^2$ and select the base point $x_0 = 3$.

   (a) What linear function $L(x)$ should be used to approximate $f(x)$ when x is close to x_0?

   (b) Use $L(x)$ to approximate $(3.12)^2$ and compare with the true value 9.7344.

2. Let $f(x) = \sqrt{x}$ and select the base point $x_0 = 4$.

   (a) What linear function $L(x)$ should be used to approximate $f(x)$ when x is close to x_0?

   (b) Use $L(x)$ to approximate $\sqrt{4.08}$ and compare with the calculator value 2.0199.

In each of the following, a function f is given, together with a base point x_0 and a value of $f(x)$ that is to be approximated. As in Problems 1 and 2, find $L(x)$ and use it to compute the indicated value. Compare with the value found from a calculator or table.

3. $f(x) = x^3$, $x_0 = 1$; compute $(1.06)^3$

4. $f(x) = 2/x$, $x_0 = 4$; compute $2/4.07$

5. $f(x) = \sqrt{x}$, $x_0 = 9$; compute $\sqrt{9.2}$

6. $f(x) = x^2$, $x_0 = 2$; compute $(1.91)^2$

7. $f(x) = 1/x$, $x_0 = 5$; compute $1/5.11$

8. $f(x) = x^4$, $x_0 = 1$; compute $(0.94)^4$

9. $f(x) = \cos x$, $x_0 = \pi/3$; compute $\cos 1$

10. $f(x) = \sin x$, $x_0 = \pi/6$; compute $\sin 0.5$

In each of the following, use linear approximation to compute the given number. Compare with the value obtained from a calculator or table. (It is up to you to select the function and the base point. Good choices will yield good approximations.)

11. $(3.026)^2$ **12.** $1/(0.97)^2$ **13.** $(4.1)^{3/2}$

14. $\sqrt[4]{15}$ **15.** $\tan 0.092$ **16.** $\sin 0.045$

17. $\cos 0.5$ **18.** $\tan 3$ **19.** $\sqrt[3]{7.88}$

20. $(8.12)^{4/3}$

Using the same notation as in Problems 1 through 10, let M be an upper bound for $|f''(c)|$, where c is between x_0 and x. Apply the formula

$$|f(x) - L(x)| \le \tfrac{1}{2}M(x - x_0)^2$$

to obtain an upper bound for the error in each of the following. Use the result to report the approximation to the appropriate number of places, then check by rounding off the calculator value to the same number of places.

21. Problem 1(b)

22. Problem 2(b)

23. Problem 3

24. Problem 4

25. Problem 5

26. Problem 6

27. Problem 7

28. Problem 8

29. Problem 9

30. Problem 10

In each of the following, determine how many decimal places can be guaranteed if the given linear approximation is used for x in the given interval.

31. $x^2 \approx 25 + 10(x - 5)$ in the interval $(4.8, 5.2)$

32. $x^3 \approx 8 + 12(x - 2)$ in the interval $(1.99, 2.01)$

33. $\sqrt{x} \approx 4 + \frac{1}{8}(x - 16)$ in the interval $(16, 17)$

34. $1/x \approx 0.1 - 0.01(x - 10)$ in the interval $(10, 10.3)$

35. $\sin x \approx x$ in the interval $(-0.02, 0.02)$ *Hint:* $\sin x < x$ if $x > 0$

36. Suppose that f is differentiable at x_0 and let

$$L(x) = f(x_0) + f'(x_0)(x - x_0)$$

be the linear function we have used in this section to approximate $f(x)$ for x near x_0. Show that the error in this approximation approaches 0 as $x \to x_0$, that is,

$$\lim_{x \to x_0} [f(x) - L(x)] = 0$$

37. While the statement in Problem 36 may sound reassuring, it does not amount to much. Show that *any* linear function whose graph contains the base point $(x_0, f(x_0))$ approximates $f(x)$ with an error that approaches 0 as $x \to x_0$. That is, if

$$y = f(x_0) + m(x - x_0)$$

where m is arbitrary, then

$$\lim_{x \to x_0} [f(x) - y] = 0$$

38. It is geometrically apparent that the tangent line $y = L(x)$ is a "better" approximation than any other line through the base point. The analytic reason for this is that the error in $f(x) \approx L(x)$ is small even when compared to the difference $x - x_0$. That is,

$$\lim_{x \to x_0} \frac{f(x) - L(x)}{x - x_0} = 0$$

Explain why this is true.

39. Show that no other line through the base point has the property derived in Problem 38. That is, if $y = f(x_0) + m(x - x_0)$ and

$$\lim_{x \to x_0} \frac{f(x) - y}{x - x_0} = 0$$

then $y = L(x)$. (It is in this sense that $L(x)$ is the "best" of all possible linear approximations.)

4.7
THE DIFFERENTIAL OF A REAL FUNCTION

In Section 4.6 we discussed the linear approximation formula

$$f(x) \approx f(x_0) + f'(x_0)(x - x_0)$$

where f is differentiable in a neighborhood of x_0 and x is any point of this neighborhood. Writing the approximation in the form

$$f(x) - f(x_0) \approx f'(x_0)(x - x_0)$$

we introduce the classical symbols

$$\Delta x = x - x_0 \quad \text{(increment of } x\text{)}$$
$$\Delta y = y - y_0 = f(x) - f(x_0) \quad \text{(increment of } y\text{)}$$

Our approximation formula may then be given in the abbreviated version

$$\Delta y \approx f'(x_0) \, \Delta x$$

(See Figure 1.) If $\Delta x \neq 0$ (that is, $x \neq x_0$), this can be written

$$\frac{\Delta y}{\Delta x} \approx f'(x_0)$$

Figure 1
The approximation $\Delta y \approx f'(x_0)\,\Delta x$

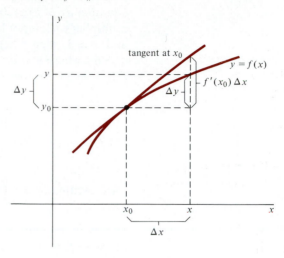

a formula that motivates (perhaps for the first time in this book) the Leibniz symbol dy/dx for the derivative.

Remark   Leibniz himself thought of the derivative as a quotient of infinitely small (but nonzero) quantities, a notion that was hard for his contemporaries to accept. George Berkeley (1685–1753) attacked the concepts and methods of the new analysis in a treatise that was notable for both its mathematical cogency and its skewering of religious skeptics. The subtitle read as follows:

> Wherein it is examined whether the object, principles, and inferences of the modern analysis are more distinctly conceived, or more evidently deduced, than religious mysteries and points of faith.

He was devastating, and went largely unanswered; it was another hundred years before mathematicians had much success with the foundations of analysis. More recently, the theory of infinitesimals (the evanescent "ghosts of departed quantities" as Berkeley gleefully called them) has been resurrected in a modern form known as *nonstandard analysis*. The success of the theory is a vindication of Leibniz, whose ideas now seem less naïve than formerly supposed.

We have never given separate meanings to the symbols dx and dy, and have mentioned once or twice that dy/dx is not to be regarded as a quotient. This has not been altogether honest, however. The fact is that there is nothing preventing us from defining dx and dy. We will do it in such a way that the equations

$$dy = f'(x)\,dx \qquad \text{and} \qquad \frac{dy}{dx} = f'(x)$$

are equivalent when $dx \neq 0$.

Note that if x is regarded as fixed, the equation $dy = f'(x)\,dx$ is nothing more than a linear relation between the variables dx and dy. It is not customary to use symbols like dx and dy for variables, but we cannot use x and y because they are preempted in the equation $y = f(x)$. Nevertheless the equation $dy = f'(x)\,dx$ defines an ordinary function, namely the one that multiplies each input by the (fixed) number $f'(x)$. For example, if $f(x) = x^2$ and $x = 3$, then $f'(x) = 6$ and we are talking about the linear function $v = 6u$, where $u = dx$ and $v = dy$. This function is called the *differential* of f at x, as in the following definition.

Let x be a point at which the function f is differentiable. The *differential* of f at x is the linear function

$$dy = f'(x)\,dx$$

also written $df = f'(x)\,dx$.

With this definition in hand, it is now legitimate to multiply each side of $dy/dx = f'(x)$ by dx as though the left side were a genuine fraction. The result,

$$dy = f'(x)\,dx$$

is meaningful because we have made a definition that gives it meaning! It may not seem to amount to much, but it is useful, as we will see. If it worries you because it is so arbitrary (designed to make something work that was previously meaningless), you should realize that every definition has something of the same arbitrary quality. The first time you encountered the symbol $\sqrt{a}$, for example, you were told that it is the nonnegative number whose square is a. When this is used to write $\sqrt{9} = 3$ and somebody asks why, the answer is that $3^2 = 9$ and we *defined* $\sqrt{9}$ to be the nonnegative number whose square is 9. In other words the statement $\sqrt{9} = 3$ is true by definition. In the same way we are now saying that dx is an independent variable in terms of which the dependent variable dy is given by the formula $dy = f'(x)\,dx$ (where x is fixed). If $f(x) = x^2$ and $x = 3$, then $dy = 6\,dx$ by definition.

□ **Example 1**   In each of the following, the differential is obtained by finding the corresponding derivative formula first, then multiplying each side by dx.

(a) If $y = x^n$, then $dy/dx = nx^{n-1}$ and hence $dy = nx^{n-1}\,dx$.

(b) If $y = \sin x$, then $dy/dx = \cos x$ and hence $dy = \cos x\,dx$.

(c) If $y = f(x) = \sqrt{x}$, then $f'(1) = \frac{1}{2}$ and hence the differential of f at 1 is $dy = \frac{1}{2}\,dx$.

(d) Since $\dfrac{d}{dx}(\tan x) = \sec^2 x$, we find $d(\tan x) = \sec^2 x\,dx$.

(e) Since $\dfrac{d}{dx}(u + v) = \dfrac{du}{dx} + \dfrac{dv}{dx}$, we find $d(u + v) = du + dv$.   □

The definition of the differential raises a theoretical question. Suppose that $y = f(u)$ and $u = g(x)$. If x is a point at which g is differentiable and f is differentiable at the point $u = g(x)$, we have

$$dy = f'(u)\, du \qquad \text{(definition of the differential of } f) \qquad (1)$$
$$du = g'(x)\, dx \qquad \text{(definition of the differential of } g) \qquad (2)$$

The symbol du plays two roles here. In Equation (1) it is supposed to be an independent variable in terms of which dy is given. But in Equation (2) it is a dependent variable (a function of the independent variable dx). Is there a conflict in these roles?

The answer is supplied by the Chain Rule, which says that the derivative of the composite function $h = f \circ g$ at x is

$$h'(x) = f'[g(x)]g'(x) = f'(u)g'(x) \qquad \text{(Section 3.3)}$$

Hence the differential of $y = h(x)$ is

$$dy = h'(x)\, dx = f'(u)g'(x)\, dx$$

Since this is precisely what one gets when du in Equation (2) is substituted for du in Equation (1), there is no conflict.

A different way of saying this is that the looser version of the Chain Rule given in Section 3.3, namely

$$\frac{dy}{dx} = \frac{dy}{du}\frac{du}{dx}$$

does after all work as a relation between fractions. The symbol du (independent in dy/du, dependent in du/dx) really does cancel, as the Leibniz notation so elegantly suggests.

Warning to the Reader

The definition of the differential, $dy = f'(x)\, dx$, calls for a Leibniz symbol *on each side of the equation*. While it is good form to write

$$D_x(\sin x) = \cos x$$

the similar-looking formula $d(\sin x) = \cos x$ is incorrect (and causes no end of confusion if indulged). The correct version is

$$d(\sin x) = \cos x\, dx$$

which of course is equivalent (for $dx \neq 0$) to

$$\frac{d}{dx}(\sin x) = \cos x \qquad \text{or} \qquad D_x(\sin x) = \cos x$$

Returning to the approximation formula

$$\Delta y = f(x) - f(x_0) \approx f'(x_0)(x - x_0) = f'(x_0)\, \Delta x$$

we need only take $dx = \Delta x$ to arrive at the statement

$$\boxed{\Delta y \approx dy}$$

where $dy = f'(x_0)\, dx$ is the differential of f at x_0. This says that the *increment* ("change") of a function whose independent variable changes from x_0 to x can be approximated by its *differential* at x_0. Note the distinction between the statements $dx = \Delta x$ (legitimate because dx is an independent variable whose value may be specified in any way we like) and $dy \approx \Delta y$ (an approximation rather than an equation).

□ **Example 2**   A spherical balloon has radius 2, which changes to 2.1 when the temperature rises. Approximate the corresponding change in volume.

Solution: The formula for volume of a sphere is $V = \frac{4}{3}\pi r^3$. Letting

$$f(r) = \tfrac{4}{3}\pi r^3$$

we see that we are being asked to approximate $\Delta V = f(2.1) - f(2)$. Since

$$\frac{dV}{dr} = 4\pi r^2 = 16\pi \qquad \text{at } r = 2$$

we can write

$$\Delta V \approx dV = 16\pi\, dr \qquad \text{where } dr = \Delta r = 2.1 - 2 = 0.1$$

Hence the change in volume is approximately 1.6π (or about 5).   □

□ **Example 3**   Suppose that the radius of the balloon in Example 2 is measured to be 2, with a possible error of 0.1. What is the (relative) error in the volume?

Solution: Take $r = 2$ and $dr = \Delta r = \pm 0.1$. The relative error in the volume is

$$\left|\frac{\text{error}}{\text{value}}\right| = \left|\frac{\Delta V}{V}\right| \approx \left|\frac{dV}{V}\right| = \left|\frac{4\pi r^2\, dr}{\frac{4}{3}\pi r^3}\right| = 3\left|\frac{dr}{r}\right| = 3\left(\frac{0.1}{2}\right) = 0.15$$

or about 15%. Note that the relative error in the radius is

$$\left|\frac{\Delta r}{r}\right| = \frac{0.1}{2} = 0.05$$

or 5%. Hence the relative error is approximately tripled. This is true in general, since

$$\left|\frac{\Delta r}{r}\right| = p\% \implies \left|\frac{dV}{V}\right| = 3\left|\frac{dr}{r}\right| = 3\left|\frac{\Delta r}{r}\right| = 3p\% \qquad \qquad □$$

□ **Example 4**   An angle of a triangle is measured to be 45°, with a possible relative error of 3%. How accurate is the sine of the angle?

Solution: Let $y = \sin x$. The relative error in the angle $x = \pi/4$ is

$$\left|\frac{\Delta x}{x}\right| = 0.03$$

from which $|dx| = |\Delta x| = 0.03|x|$. Hence the relative error in the sine of the angle is

$$\left|\frac{\Delta y}{y}\right| \approx \left|\frac{dy}{y}\right| = \left|\frac{\cos x \, dx}{\sin x}\right| = |\cot x| \, |dx| = 0.03|x| \, |\cot x| = 0.03\left(\frac{\pi}{4}\right)\left(\cot\frac{\pi}{4}\right)$$

$$\approx 0.0236$$

or about 2.4%.   □

All the derivative formulas previously developed can be restated in terms of differentials, sometimes with telling effect. For example, the Product Rule says that

$$\frac{d}{dx}[f(x)g(x)] = f(x)g'(x) + g(x)f'(x)$$

where the prime means differentiation with respect to x. If $u = f(x)$ and $v = g(x)$, the Product Rule takes the form

$$\frac{d}{dx}(uv) = u\frac{dv}{dx} + v\frac{du}{dx}$$

Having defined the Leibniz symbols as separate entities, we can multiply by dx to obtain $d(uv) = u\,dv + v\,du$, a formula which no longer refers to a specific independent variable. If u and v were functions of t instead of x, the formula would still be correct.

□ **Example 5**   Suppose that $x^3 + y^2 \sin x - y = 5$. "Implicit differentiation" (Section 3.4) is based on the assumption that this equation defines y as a function of x, so that we can differentiate each side with respect to x:

$$3x^2 + y^2 \cos x + \sin x \cdot 2y\frac{dy}{dx} - \frac{dy}{dx} = 0$$

This procedure discriminates between the variables (treating x as independent and y as dependent). The original equation, however, contains no such bias; we can just as well regard x as a function of y. Better yet, why not give x and y equal rights and regard them both as variables whose *differentials* are related?

$$d(x^3) + d(y^2 \sin x) - d(y) = d(5)$$
$$3x^2 \, dx + y^2 \, d(\sin x) + \sin x \cdot d(y^2) - dy = 0$$
$$3x^2 \, dx + y^2 \cos x \, dx + 2y \sin x \, dy - dy = 0$$
$$(3x^2 + y^2 \cos x) \, dx = (1 - 2y \sin x) \, dy$$

At this point we have two choices:

$$\frac{dy}{dx} = \frac{3x^2 + y^2 \cos x}{1 - 2y \sin x} \qquad \text{(derivative of } y \text{ as a function of } x\text{)}$$

$$\frac{dx}{dy} = \frac{1 - 2y \sin x}{3x^2 + y^2 \cos x} \qquad \text{(derivative of } x \text{ as a function of } y\text{)}$$

Another alternative is

$$(3x^2 + y^2 \cos x)\frac{dx}{dt} = (1 - 2y \sin x)\frac{dy}{dt}$$

which would be of interest if both x and y were dependent on some third variable t. Such a situation would occur, for example, if a particle were to move along the curve

$$x^3 + y^2 \sin x - y = 5$$

its coordinates (x, y) being functions of the time t. The moral is that differentials provide a certain freedom of interpretation. We are no longer so tied to the notion of one variable as a function of the other. This will be exploited in several ways as we proceed.   □

Problem Set 4.7

Find each of the following differentials.

1. dy if $y = 3x - 8$ **2.** dy if $y = 1 - 5x$

3. du if $u = \cos t$ **4.** dv if $v = \sin t \cos t$

5. dp if $p(t) = \dfrac{t}{1 - t}$ **6.** df if $f(x) = \sqrt{2x - 1}$

7. $d(x^2)$ **8.** $d(\sqrt{1 - u^2})$

9. $d(\sec t)$ **10.** $d(x)$

11. $d(1/x)$ **12.** $d(\sin^2 x + \cos^2 x)$

If $y = f(x)$ and x is fixed, dy is supposed to be a linear function of dx. In each of the following, find what this function is.

13. $y = x^3$ at $x = 2$ **14.** $y = \tan x$ at $x = \pi/4$

15. $y = \sin^2 x$ at $x = \pi/4$ **16.** $y = x^{3/2}$ at $x = 4$

17. $y = 4/x$ at $x = 2$ **18.** $y = |x|$ at $x = -3$

19. A square has sides of length x. If x is increased by an amount Δx, the situation is as shown in Figure 2.

   (a) Use the figure to show that the area of the enlarged square is

$$x^2 + 2x \Delta x + (\Delta x)^2$$

Figure 2
Increment in the area of a square

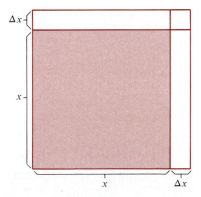

What is the increment of the area? (That is, how much has the area changed?)

(b) If the approximation $\Delta A \approx dA$ is used (where A is the original area), what is the result? What is the error? Interpret geometrically.

20. The increment in $y = x^3$ corresponding to the increment Δx in x is

$$\Delta y = (x + \Delta x)^3 - x^3$$

If the approximation $\Delta y \approx dy$ is used, what terms are being neglected? Why is this reasonable if Δx is small?

21. A box in the form of a cube has edges of length x. Its outside faces are coated with a layer of paint of thickness t.

(a) Explain geometrically why the volume of paint used is about $6x^2t$.

(b) Obtain the same approximation by using the differential of volume of a cube ($V = x^3$).

22. One side of a square is measured and its area is then computed. If the relative error in the measurement is 3%, approximately what relative error may be expected in the computed area?

23. A gravel path 1 foot wide goes around the edge of a circular pool 50 feet in diameter. Show that the exact area of the path is 51π, while its approximation by the differential of area is 50π. What is the relative error in the approximation?

24. If a ship could sail around the world on the equator and a plane could fly the same route at an altitude of 1 mile, show that the plane would travel about 6 miles farther than the ship. What would the answer be if the earth were the size of Jupiter? What if it were the size of a billiard ball?

25. The radius of a spherical ball contracts 2% due to the cold. Approximately what percentage decrease occurs in its surface area? (If r is the radius, surface area is $4\pi r^2$.)

26. A (hollow) spherical ball of radius r has surface area $4\pi r^2$. It is made of thin plastic of thickness t.

(a) Explain geometrically why the volume of plastic is about $4\pi r^2 t$.

(b) Obtain the same result by using the differential of volume of a sphere ($V = \frac{4}{3}\pi r^3$).

27. A rectangular sheet of metal is rolled into a cylindrical can of radius r and altitude h. (Ignore the lids; they are made separately.)

(a) If the thickness of the sheet is t, use geometry to approximate the volume of metal in the can.

(b) Obtain the same result by using the differential of volume of a cylinder ($V = \pi r^2 h$). Hint: h is constant; treat V as a function of r.

28. If an angle is measured to be $45°$ with a relative error of $p\%$, show that the relative error in its tangent is about $(\pi/2)p\%$.

29. Boyle's Law says that the pressure and volume of a confined gas at constant temperature are related by the equation $pv = k$, where k is constant. If $p = 30$ when $v = 1000$, approximate the change in p when v decreases to 990.

30. A classical experiment in physics is to measure the period T of a pendulum of length l and then compute the acceleration due to gravity from the formula $g = 4\pi^2 l/T^2$. Assuming that the length is accurately known, but that there is a relative error of 1% in measuring the period, how accurate is the value of g?

31. Some of the preceding problems could be shortened by observing that if x is measured with a certain relative error, the relative error in $y = cx^r$ is about $|r|$ times as much. Confirm this observation.

32. Draw a picture like Figure 1 in Section 4.6, letting $P = (x_0, f(x_0))$, $Q = (x, f(x_0))$, $R = (x, f(x))$, and $S = (x, L(x))$.

(a) What increments are proper labels for PQ and QR?

(b) What differentials are proper labels for PQ and QS? Hint: Consider the slope of the tangent line.

(c) Give a geometric interpretation of the approximation $\Delta y \approx dy$. What is the error in this approximation? Why may we expect the error to approach zero as $\Delta x \to 0$?

33. Suppose that y is a linear function of x, say $y = ax + b$.

(a) By referring to the geometric interpretation of Δy and dy in Problem 32, explain why the approximation $\Delta y \approx dy$ is exact.

(b) Confirm part (a) by computing Δy and dy.

Let x be a point at which $u = f(x)$ and $v = g(x)$ are differentiable. Verify the following properties of the differentials of f and g at x.

34. $d(u \pm v) = du \pm dv$

35. $d(cu) = c\,du \qquad (c \in \mathbb{R})$

36. $d(uv) = u\,dv + v\,du$

37. $d\left(\dfrac{u}{v}\right) = \dfrac{v\,du - u\,dv}{v^2} \qquad (v \neq 0)$

In each of the following, equate the differentials of each side of the given equation (as in Example 5). Use the result to find dy/dx and dx/dy.

38. $x^2 - y^2 = 4$ **39.** $x^3 - y^3 = x$

40. $x^2 - xy + y^2 = 1$ **41.** $y^2 + 2xy = 3y - x$

42. $y\cos x - x + y^2 = 2$ **43.** $\sin(xy) - y - 2x = 0$

44. Suppose that $y = f(x)$ is differentiable at x_0 and $f'(x_0) \neq 0$. If $dx = \Delta x \neq 0$, the approximation $\Delta y \approx dy$

may be written in the form $\Delta y/dy \approx 1$. Show that this approximation improves as $\Delta x \to 0$ by proving that

$$\lim_{\Delta x \to 0} \frac{\Delta y}{dy} = 1$$

45. Show that the increment in $y = \sqrt{x}$ corresponding to the increment Δx is

$$\Delta y = \frac{\Delta x}{\sqrt{x + \Delta x} + \sqrt{x}}$$

Then form the ratio $\Delta y/dy$ and confirm that its limit as $\Delta x \to 0$ is 1. (See Problem 44.)

46. The increment in $y = \sin x$ corresponding to the increment Δx is

$$\Delta y = \sin(x + \Delta x) - \sin x$$

Use an addition formula to work this out, then form the ratio $\Delta y/dy$ and confirm that its limit as $\Delta x \to 0$ is 1. (See Problem 44.)

47. Suppose that $x = 2t - 1$ and $y = t^2$.

(a) Find dx and dy as functions of dt and use the results to find dy/dx.

(b) Find y in terms of x by eliminating t and then find dy/dx. Is the result consistent with part (a)?

48. Problem 47 suggests that if x and y are given in terms of a parameter t, say $x = f(t)$ and $y = g(t)$, then

$$\frac{dy}{dx} = \frac{g'(t)}{f'(t)} = \frac{dy/dt}{dx/dt}$$

Explain why you might expect this to be true. Would you attach any conditions?

49. Suppose that $x = \cos t$ and $y = \sin t$.

(a) Explain why the path of the point (x, y) when t varies from 0 to 2π is the unit circle.

(b) Use Problem 48 to find dy/dx from the parametric equations and confirm that the same result is obtained by implicit differentiation in $x^2 + y^2 = 1$.

(c) The result in part (b) is $dy/dx = y' = -\cot t$. Find

$$y'' = \frac{d(y')}{dx}$$

from the parametric equations $x = \cos t$, $y' = -\cot t$ in the same way that you found y' in part (b). (Note that $y'' \neq \csc^2 t$! The primes mean differentiation with respect to x, not t.)

(d) Implicit differentiation in $x^2 + y^2 = 1$ yields $y' = -x/y$. Find y'' from this and confirm the result in part (c).

50. Suppose that $x = a \cos^3 t$ and $y = a \sin^3 t$, where a is a positive constant.

(a) Explain why the path of the point (x, y) corresponding to $0 \leq t < 2\pi$ is bounded by the lines $x = \pm a$ and $y = \pm a$.

(b) Obtain $y' = dy/dx = -\tan t$ from the parametric equations. What does this tell you about the appearance of the graph at the points corresponding to $t = 0$, $\pi/2$, π, $3\pi/2$?

(c) Derive

$$y'' = \frac{d^2 y}{dx^2} = \frac{1}{3a} \sec^4 t \csc t$$

from $x = a \cos^3 t$ and $y' = -\tan t$. What does this tell you about the concavity of the graph?

(d) Sketch the graph. Why is it the same as the graph of $x^{2/3} + y^{2/3} = a^{2/3}$? (This is the "hypocycloid of four cusps" mentioned in Problem 50, Section 4.5.)

51. Assuming that the theorem in Problem 48 is true, show that the parametric equations $x = f(t)$, $y = g(t)$ lead to

$$y'' = \frac{\dot{x}\ddot{y} - \dot{y}\ddot{x}}{\dot{x}^3}$$

where the primes mean differentiation with respect to x and the dots means differentiation with respect to t. (The dot notation is due to Newton.)

52. Use the formula in Problem 51 to find y'' from the parametric equations $x = t^3$, $y = t^2$. Check by expressing y as a function of x and then finding y''.

True–False Quiz

1. If $\lim_{x \to a} f(x)$ exists, then f is continuous at a.

2. To make $f(x) = (\sin x)/x$ continuous at $x = 0$, we should define $f(0) = 0$.

3. A function that is continuous at a certain point is automatically differentiable at that point.

4. If f is discontinuous at $x = a$, then $f'(a)$ does not exist.

5. If $T(x)$ is any one of the six trigonometric functions and a is a point of its domain, then $\lim_{x \to a} T(x) = T(a)$.

6. The range of $f(x) = 4 - x^2$ corresponding to the domain $[-1, 2]$ is $[0, 3]$.

7. The function $f(x) = \sqrt{x(2 - x)}$ has a maximum value.

8. The function $f(x) = \sqrt{x(x - 2)}$ has a maximum value.

9. The function $f(x) = 2/(x^2 + 1)$ is unbounded.

10. There is a value of x between 0 and π for which $(\sin x)/x = \frac{1}{2}$.

11. The equation $x^3 + 3x - 1 = 0$ has a root between 0 and 1.

12. The equation $x^3 + 3x - 1 = 0$ has three real roots.

13. The derivative of $f(x) = x(x - 1)(x^2 + 1)$ is zero for some value of x between 0 and 1.

14. If f is continuous in the interval $[a,b]$ and has an extreme value at a point c between a and b, then $f'(c) = 0$.

15. There is a tangent to the curve $y = x^{2/3}$ that is parallel to the chord joining $(0,0)$ and $(1,1)$.

16. The Mean Value Theorem applies to $f(x) = |x|$ in the interval $[-1,1]$.

17. If $f'(x) < 0$ in the interval I, then
$$x_1 > x_2 \Rightarrow f(x_1) > f(x_2) \qquad \text{for all } x_1 \text{ and } x_2 \text{ in } I$$

18. Two functions with the same derivative must be identical.

19. If $f'(x) = 2x$ for all x, then $f(x) - x^2$ has the same value for all x.

20. $D_x(\sec^2 x) = D_x(\tan^2 x)$ in the interval $(-\pi/2, \pi/2)$.

21. If $dy/dx = 2 \sin x$ and $y = 0$ when $x = 0$, then $y = -2 \cos x$.

22. If the position of a moving object at time t is $s = \cos \pi t$, the velocity is increasing when $t = 0$.

23. If the position of a moving object at time t is $s = t^3 + 4t$, the object is always moving in the positive direction.

24. A ball thrown upward with a speed of 32 ft/sec from a window 20 ft above the ground reaches a height of 36 ft above the ground.

25. If $y = \sec x$, then $y'' = \sec x \, (2 \sec^2 x - 1)$.

26. The graph of $y = x\sqrt{1 - x}$ is concave down.

27. Every cubic curve $y = ax^3 + bx^2 + cx + d$ has a point of inflection.

28. The graph of $y = 4 - x^{2/3}$ has a point of inflection at $(0,4)$.

29. If $y^2 - x^2 = 1$, then $y'' = 1/y^3$.

30. If $y^2 = 2x$, then $y''' = 3/y^4$.

31. Linear approximation of $\sqrt[3]{x}$ for x near 1 is given by $\sqrt[3]{x} \approx 1 + \frac{1}{3}(x - 1)$.

32. If $y = 3x + 2$, the approximation $\Delta y \approx dy$ is exact.

33. The increment of $y = x^4$ at $x = 1$ corresponding to $\Delta x = 0.02$ is approximately 0.08.

34. $d(\sin^2 x + \cos^2 x) = 0$.

35. $d(\sin x) = \cos x$.

Additional Problems

In each of the following, determine where f is continuous.

1. $f(x) = 1 - x^3$

2. $f(x) = \dfrac{x^2}{x^2 - 4}$

3. $f(x) = \dfrac{x}{x^2 - x + 2}$

4. $f(x) = \dfrac{x^2}{x^2 - 4x}$

5. $f(x) = (x - 2)^{3/2}$

6. $f(x) = \sqrt{\dfrac{x}{1 - x}}$

7. $f(x) = \dfrac{\tan x}{x}$

8. $f(x) = \dfrac{\cos x}{2 - \sin x}$

9. $f(x) = \left| \dfrac{5 - x}{x} \right|$

10. $f(x) = \begin{cases} |x| & \text{if } |x| \le 1 \\ 1 & \text{if } |x| > 1 \end{cases}$

11. $f(x) = \begin{cases} \sqrt{4 - x^2} & \text{if } |x| \le 2 \\ 0 & \text{if } |x| > 2 \end{cases}$

12. $f(x) = \begin{cases} x & \text{if } x < 0 \text{ or } x > 3 \\ x^2 - 2x & \text{if } 0 \le x \le 3 \end{cases}$

13. $f(x) = \begin{cases} 0 & \text{if } x \text{ is an integer} \\ x & \text{if } x \text{ is not an integer} \end{cases}$

14. How should $f(3)$ be defined to make $f(x) = (3 - x)/(x - 3)$ continuous everywhere?

15. How should $f(1)$ and $f(2)$ be defined to make $f(x) = (x - 2)/(x^2 - 3x + 2)$ continuous everywhere?

16. Assuming that sine and cosine are continuous, explain why $f(x) = \sec x + \tan x$ is continuous in its domain.

17. Assuming that tangent is continuous in its domain, explain why $f(x) = \sin 2x/(1 - \cos 2x)$ is continuous in its domain.

According to the Fundamental Principle of Continuity, the range of each of the following functions is a finite closed interval. Name the interval.

18. $f(x) = \dfrac{1}{x^2 + 1}$, $-1 \le x \le 1$

19. $f(x) = |x - 1|$, $0 \le x \le 2$

20. $f(x) = 9 - x^2$, $-1 \le x \le 3$

21. $f(x) = \cos^2 x - \sin^2 x$, $0 \le x \le \pi/4$

22. $f(x) = \csc^2 x - \cot^2 x$, $\pi/4 \le x \le 3\pi/4$

23. Suppose that the domain of f is the interval $[a,b]$. Under what conditions can you be sure that the range is the interval $[f(a), f(b)]$?

24. Name the extreme values of

$$f(x) = \cot x, \quad \pi/4 \le x \le \pi/2.$$

What is the least upper bound of the range?

25. Sketch the graph of

$$f(x) = \begin{cases} 3 & \text{if } x < 1 \\ x^2 - 6x + 8 & \text{if } x \ge 1 \end{cases}$$

Is f continuous in its domain? Does it have a global maximum? a global minimum?

26. If the domain of f is an open interval I and $f'(x) > 0$ for all x in I, explain why f cannot have an extreme value.

27. Without trying to find it, explain how you can be sure that the function $f(x) = \sqrt{(x - 1)(2 - x)}$ has a maximum value. What is its minimum value and where does it occur?

28. Name the extreme values of the function

$$f(x) = \frac{x^2 - 16}{x + 4}, \quad 0 \le x \le 4$$

Does the function take on every value in between? Explain.

29. What guarantees the existence of a number x between 0 and $\pi/2$ such that $\tan x = 100$?

30. The symbol $\sqrt{5}$ represents a positive number x such that $x^2 = 5$. How do we know that such a number exists?

31. Use Rolle's Theorem to prove that the equation $x^3 + 2x - 1 = 0$ cannot have more than one real solution. Then use the Zero Theorem to show that it does have a real solution and locate the solution between consecutive tenths.

32. If $f(x) = x(x - 3)(x^2 + 4)$, explain why $f'(x)$ must be 0 for some value of x between 0 and 3. Then find $f'(x)$ and confirm that $f'(2) = 0$.

33. If $f(x) = 2x - \sin x$, find a number c between 0 and π such that

$$f(\pi) - f(0) = f'(c)(\pi - 0)$$

What theorem does this illustrate?

34. The position at time t of a moving object is $s = 3t - t^3$.
  (a) What is the average velocity of the object during the time interval $[1,4]$?
  (b) At what time is the instantaneous velocity equal to the average velocity?
  (c) What theorem does this illustrate?

35. What is the maximum value of $f(x) = 1 - x^{3/2}$ in the domain $[0,1]$? What is the minimum value? Does $f'(x) = 0$ at the points where these extreme values occur? What does Theorem 1 in Section 4.3 say about this situation?

36. If $y' = 3x^2 + 2x + 1$ and $y = 4$ when $x = 1$, find y as a function of x.

37. If $y' = 2 - \cos x$ and $y = 1$ when $x = 0$, find y as a function of x.

38. Show that a particle with position $s = t^3 - 5t$ at time t is slowing down when $t = 1$.

39. An object moves along a coordinate line according to the law

$$s = t^3 - 3t^2 + 5 \qquad (t \ge 0)$$

How close does it come to the origin?

40. Let v be the velocity, r the speed, and a the acceleration of an object moving in a straight line. Show that $dr/dt = av/r$ and use the result to explain why the object is speeding up if v and a have the same sign, slowing down if they have opposite signs.

41. An object moves along a coordinate line according to the law

$$s = t^3 - 3t^2 + 4t + 5$$

  (a) Explain why the object is always moving in the positive direction.
  (b) Show that the object is slowing down until it reaches the point $s = 7$, after which it is speeding up.

42. An object is moving along a coordinate line (positive direction east) according to the law $s = t^3 - 3t$. By discussing the velocity and acceleration, confirm the following description of the motion.
  (a) Until it reaches the point $s = 2$ the object moves east with decreasing speed.
  (b) It stops when it reaches the point $s = 2$ and reverses direction, heading back west with increasing speed.

(c) At $s = 0$ it begins to slow down again, stopping at $s = -2$ and reversing direction.

(d) Thereafter it moves east with increasing speed.

43. An object moves along a coordinate line (positive direction east) according to the law $s = t^4 - 4t^3$. Use the velocity and acceleration to describe the motion in detail.

44. Show that the graph of $y = \csc x$ has no points of inflection.

45. Find a general formula for $f^{(n)}(x)$ if

$$f(x) = \frac{1}{3 - x}$$

46. Sketch the graph of $y = x^3 - 3x^2 + 1$ after answering the following questions.

(a) Where is the graph rising? falling?

(b) Locate all points of inflection.

(c) Use the Zero Theorem to locate x intercepts of the graph between consecutive integers.

47. Sketch the graph of $y = x^2/(x - 2)$ after answering the following questions.

(a) What are the asymptotes (if any)?

(b) Where is the graph rising? falling?

(c) Where is the graph concave up? concave down?

48. Use implicit differentiation to prove that the upper half of the parabola $y^2 = kx$ (where k is a nonzero constant) is concave down.

49. Use linear approximation to find $\sqrt{99.5}$ correct to three decimal places. Include a defense of the accuracy.

50. Repeat Problem 49 in the case of $(1.005)^4$.

51. Find dy as a linear function of dx if $y = 3 \sin x - x$ and $x = 0$.

52. Find dy in terms of x and dx if $y = 1/x^2$.

53. The increment in $y = x^4$ corresponding to the increment Δx in x is

$$\Delta y = (x + \Delta x)^4 - x^4$$

If the approximation $\Delta y \approx dy$ is used, what terms are being neglected? Why is this reasonable if Δx is small?

54. The increment in $y = \cos x$ corresponding to the increment Δx in x is

$$\Delta y = \cos (x + \Delta x) - \cos x$$

Use an addition formula to work this out, then form the ratio $\Delta y/dy$ and confirm that its limit as $\Delta x \to 0$ is 1. What can you conclude about the approximation $\Delta y \approx dy$?

55. If T is the period of a pendulum of length l, the acceleration due to gravity is $g = 4\pi^2 l/T^2$. Assuming that the period is accurately known, but that there is a relative error of 1% in measuring the length, how accurate is the value of g? (Compare with Problem 30, Section 4.7.)

56. If an angle is measured to be $60°$ with a relative error of 2%, find the (approximate) relative error in its cosine.

5 | Applications of Differentiation

All the effects of nature are only mathematical consequences of a small number of immutable laws.
P. S. LAPLACE

I never come across one of Laplace's 'Thus it plainly appears' without feeling that I have hours of hard work before me . . .
NATHANIEL BOWDITCH (1773–1838)
(American astronomer)

Since the mathematicians have invaded the theory of relativity, I do not understand it myself any more.
ALBERT EINSTEIN (1879–1955)

In this chapter we are going to put the derivative to work. You have already seen its interpretation as slope of a curve and velocity of a moving object; these are prototypes of a great many interesting and important ways in which the derivative turns out to be useful. Hardly any branch of human knowledge has remained untouched by mathematical analysis of the kind we are about to study.

5.1
THE DERIVATIVE AS RATE OF CHANGE

In Section 2.1 we discussed the problem of motion in a straight line. If $s = p(t)$ is the position of a moving object at time t, then

$$v = \frac{ds}{dt} = p'(t)$$

is the *velocity* at time t and

$$a = \frac{dv}{dt} = p''(t)$$

is the *acceleration* at time t. Each of these derivatives is a *rate*, the velocity being the "rate of change of position with respect to time" and the acceleration being the "rate of change of velocity with respect to time."

This terminology is useful in more general circumstances (not necessarily involving motion, or even time). If $y = f(x)$ is any (differentiable) function, we call $dy/dx = f'(x)$ the *rate of change of y with respect to x.* Many derivatives (interpreted as rates) are sufficiently important to have acquired special names. The following list will give you an idea of how broad the range of applications is.

- The momentum of a moving object is mv, where m is the object's mass and v is its velocity. Newton's Second Law of Motion says that the rate of change of momentum with respect to time is **force,** that is,

$$F = \frac{d}{dt}(mv)$$

If the mass is constant, this reduces to the famous formula $F = ma$, where a is acceleration.
- If s is distance measured along a curve and ϕ is the angle of inclination of the tangent line (Figure 1), then

$$\kappa = \left| \frac{d\phi}{ds} \right|$$

is called **curvature.** Appropriate units of measurement are (for example) radians per meter; what κ measures is the rate at which the curve turns as one travels along it.
- If m milligrams of a drug are given to a patient and the strength of the reaction is measured to be S, then dS/dm is called the **sensitivity** to the drug. It is an index of how much the patient's reaction changes per unit change in dosage.
- If q is the electrical charge in a circuit at time t, then $i = dq/dt$ is the **current,** measured in coulombs per second (amperes). It is the rate at which electric charge flows.
- In physics **work** is defined in terms of a force acting through a distance. The rate at which work is done, dW/dt, is called **power.** It is measured in joules per second (watts).
- If x units of merchandise are produced at a cost $C(x)$, then $C'(x)$ is called the **marginal cost** by economists. If the revenue resulting from the sale of x units is $R(x)$ and the profit is $P(x)$, then $R'(x)$ and $P'(x)$ are called **marginal revenue** and **marginal profit,** respectively. The word "marginal" in economics is a signal for a derivative.

Since this is not a book on physics or economics, we are not going to burden you with any more terminology from those sciences (or others) than we can help. Nevertheless one of the skills this chapter is designed to develop is the ability to read a scientific description of relations between changing quantities and translate the description into mathematical language.

In this section we are interested in problems where related quantities change with *time.* These are often called *related rate problems,* the idea being

Figure 1
Rate of change of inclination

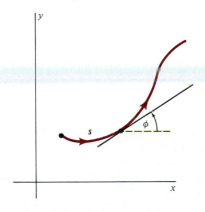

that if we know how the quantities are related, differentiation with respect to time will yield a relation between their rates.

□ **Example 1**   A man 6 ft tall is walking at the rate of 3 ft/sec toward a street light 15 ft above the ground. How fast is the length of his shadow changing? How fast is the tip of his shadow moving?

Solution: Referring to Figure 2, observe that the rate of change of x with respect to time is known, namely $dx/dt = -3$. (Why is the rate negative?) The first problem is to find ds/dt, the time rate of change of the length of the shadow. Hence we seek a relation between x and s (differentiation of which with respect to time will yield a relation between the known and unknown rates).

Using similar triangles in Figure 2, we obtain

$$\frac{15}{x + s} = \frac{6}{s}$$

from which $15s = 6(x + s)$, or $s = \frac{2}{3}x$. Both x and s are functions of t; differentiating each side of $s = \frac{2}{3}x$ *with respect to t,* we find

$$\frac{ds}{dt} = \frac{d}{dt}\left(\frac{2}{3}x\right) = \frac{2}{3}\frac{dx}{dt} = \frac{2}{3}(-3) = -2$$

Hence the length of the shadow is decreasing at 2 ft/sec.

The tip of the shadow has coordinate $x + s$ (measured from the base of the light). To find its time rate of change, we write

$$\frac{d}{dt}(x + s) = \frac{dx}{dt} + \frac{ds}{dt} = (-3) + (-2) = -5$$

Hence the tip of the shadow is moving toward the base of the light at 5 ft/sec.   □

Several aspects of Example 1 are worth selecting for general emphasis. (These are not so much "steps" to be followed in the order given, but just things worth mentioning.)

1. The question "how fast?" is usually a clue to the fact that *time* is involved. More precisely, a *time rate* (derivative with respect to time) is wanted. How fast is a snowball melting? Find dV/dt, where V is its volume. How fast is the water level in a reservoir rising? Find dh/dt, where h is the depth. How fast are prices rising? Find dP/dt, where P is the index that measures their level.
2. When the situation calls for it, *draw a picture!* Very often a picture is worth a thousand words. Avoid the common error of labeling any changing part of the picture as though it were constant; variables should always be shown as such.

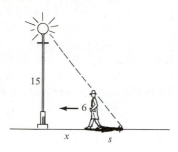

Figure 2
Man followed by shadow

3. Identify those variables whose rates are known or wanted. The crucial step in the problem is to figure out how these variables are related.
4. Once an equation involving the related variables is found, the rest is usually easy. Differentiate each side of the equation *with respect to time* (using whatever rules of differentiation are called for, notably the Chain Rule).
5. Solve for the desired rate, using whatever information is given (such as known rates and particular values of the variables).

Figure 3
How fast is the sandpile rising?

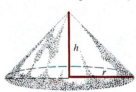

□ **Example 2**   Sand being dumped from a funnel forms a conical pile whose height is always one-third the diameter of the base. (See Figure 3.) If the sand is dumped at the rate of 2 cubic meters per minute, how fast is the pile rising when it is 1 meter deep?

Solution: Note that in Figure 3 we have not labeled the altitude 1. The altitude changes with time; the value $h = 1$ will not enter the problem until we are through analyzing the variables.

We are told that the pile is conical. Its volume is therefore

$$V = \tfrac{1}{3}\pi r^2 h \qquad \text{(for all } t\text{)}$$

We are also told that the volume is increasing at the rate of 2 cubic meters per minute; hence

$$dV/dt = 2 \qquad \text{(for all } t\text{)}$$

The problem is to evaluate the time rate of change of h at the instant when $h = 1$. Hence we seek a relation between V and h that can be differentiated with respect to time. The trouble is that V is a function of two variables, r and h, which suggests that we try to eliminate r. The means for doing that is the information

$$h = \tfrac{1}{3}(\text{diameter of base}) = \tfrac{1}{3}(2r) = \tfrac{2}{3}r \qquad \text{(for all } t\text{)}$$

Solving for $r = 3h/2$ and substituting in the formula for volume, we find

$$V = \frac{1}{3}\pi \left(\frac{3h}{2}\right)^2 \cdot h = \frac{3\pi}{4}h^3$$

$$\frac{dV}{dt} = \frac{dV}{dh}\frac{dh}{dt} \qquad \text{(Chain Rule!)}$$

$$2 = \frac{9\pi}{4}h^2 \frac{dh}{dt}$$

$$\frac{dh}{dt} = \frac{8}{9\pi h^2} \qquad \text{(for all } t\text{)}$$

At the instant when $h = 1$ the sandpile is rising at the rate

$$\frac{dh}{dt} = \frac{8}{9\pi} \approx 0.3 \text{ meters/min} \qquad \qquad \square$$

Figure 4
How fast is the camera turning?

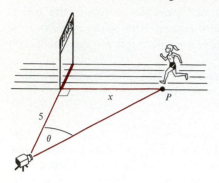

□ **Example 3**   Figure 4 shows a TV camera (located 5 meters from the finish line on a straight track) following an Olympic runner P. When the runner is 5 meters from the finish line she is moving 10 meters per second. How fast is the camera turning at that instant?

Solution: The problem is to find $d\theta/dt$ when $x = 5$. We are told that $dx/dt = -10$ when $x = 5$, so we need a relation between θ and x. Write

$$\tan \theta = \frac{x}{5}$$

$$\frac{d}{dt}(\tan \theta) = \frac{d}{dt}\left(\frac{1}{5}x\right)$$

$$\sec^2 \theta \, \frac{d\theta}{dt} = \frac{1}{5}\frac{dx}{dt}$$

$$\frac{d\theta}{dt} = \frac{1}{5}\cos^2 \theta \, \frac{dx}{dt}$$

At the instant when $x = 5$ we have $\theta = \pi/4$ and $dx/dt = -10$, from which

$$\frac{d\theta}{dt} = \frac{1}{5}\left(\frac{1}{2}\right)(-10) = -1$$

The units are radians per second. (Why?) The negative sign corresponds to the fact that θ is decreasing.   □

Figure 5
How fast is the boy falling?

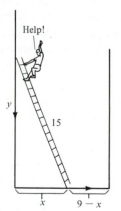

□ **Example 4**   A boy is at the top of a 15-ft ladder leaning against one wall of an alley 9 ft wide. (See Figure 5.) The ladder slips, its top sliding down the wall (with the boy hanging on), its foot sliding across the alley and striking the opposite wall at a speed of 4 ft/sec. How fast is the boy falling at that instant?

Solution: The "related rates" in this case are

$$\frac{dx}{dt} = 4 \quad \text{and} \quad \frac{dy}{dt} = ? \quad \text{at } x = 9$$

Since $x^2 + y^2 = 225$ for all t, we can write

$$\frac{d}{dt}(x^2 + y^2) = \frac{d}{dt}(225)$$

$$2x\frac{dx}{dt} + 2y\frac{dy}{dt} = 0$$

$$x\frac{dx}{dt} + y\frac{dy}{dt} = 0$$

At the instant when $x = 9$ this reads

$$9(4) + 12\frac{dy}{dt} = 0$$

from which $dy/dt = -3$ at $x = 9$. Hence the boy is falling at 3 ft/sec when the ladder hits the opposite side of the alley.   □

□ **Example 5**   A spherical drop of water loses moisture by evaporation at a rate proportional to its surface area. What can you say about its radius?

Solution: The loss of moisture is a change in the volume V; the rate at which it occurs is dV/dt. We are told that this rate is proportional to the surface area A, which means that it is some constant multiple of A. Since the volume is decreasing, dV/dt is negative, so we write $dV/dt = -kA$, where k is a positive constant. The volume of a sphere of radius r is $V = \frac{4}{3}\pi r^3$, from which

$$\frac{dV}{dt} = 4\pi r^2 \frac{dr}{dt}$$

Substituting $-kA$ for dV/dt and using the formula $A = 4\pi r^2$ for surface area, we find

$$-kA = A\frac{dr}{dt} \quad \text{or} \quad \frac{dr}{dt} = -k$$

Hence the radius decreases at a constant rate.   □

Problem Set 5.1

1. Pancake batter is poured into a pan to form a circular pancake whose area increases at the rate of 3 cm²/sec. How fast is the radius increasing when the diameter of the pancake is 10 cm?

2. A stone dropped into a pond causes a circular ripple to move outwards, the radius of the ripple increasing at the rate of 2 m/sec. How fast is the area inside the ripple changing when the radius is 5 meters?

3. The area of an isosceles right triangle decreases at the rate of 5 cm²/sec. At what rate is the hypotenuse changing when each of the other sides has length 10 cm?

4. The diagonal of a square increases at the rate of 3 m/sec. How fast is the area changing when the side of the square is 6 meters?

5. At t minutes after noon the radius of a melting (spherical) snowball is $7 - 2\sqrt{t}$ cm. How fast is the volume changing at 12:04 P.M.?

6. The volume of a sphere is changing at a constant rate. Show that the surface area changes at a rate that varies inversely as the radius.

7. A rectangular swimming pool (with a horizontal bottom) is being drained. If its length and width are 25 ft and 20 ft and the water level is falling at the rate of $\frac{1}{2}$ ft/min, how fast is the water draining?

8. Rain is falling into a cylindrical barrel at the rate of 20 cm³/min. If the radius of the base is 18 cm, how fast is the water rising?

9. The water level in a conical reservoir 50 ft deep and 200 ft across the top is falling at the rate of 0.002 ft/hr. How fast (in cubic feet per hour) is the reservoir losing water when the water is 30 ft deep?

10. A conical cup 6 cm across the top and 9 cm deep is filled with ginger ale at the rate of 3 cm³/sec. A bug originally at the bottom of the cup floats on the surface of the liquid as it rises.

    (a) Find the bug's velocity v as a function of the depth h of the ginger ale.

    (b) (Question for discussion) According to Einstein's theory of relativity, the velocity of light ($c \approx 3 \times 10^{10}$ cm/sec) is a universal speed limit. In view of part (a), however, it appears that $v > c$ when h is sufficiently small. Is Einstein discredited by a bug in a cup of ginger ale? *Hint:* How small must h be to guarantee $v > c$?

11. An icicle is in the shape of a right circular cone. At a certain point in time the height is 15 cm and is increasing at the rate of 1 cm/hr, while the radius of the base is 2 cm and is decreasing at $\frac{1}{10}$ cm/hr. Is the volume of ice increasing or decreasing at that instant? at what rate?

12. A block of ice has a square top and bottom and rectangular sides. At a certain point in time each dimension of the square is 30 cm and is decreasing at the rate of 2 cm/hr, while the depth is 20 cm and is decreasing at 3 cm/hr. How fast is the ice melting?

13. A boat floating several feet away from a dock is pulled in by a rope that is being wound up by a windlass at the rate of 3 ft/sec. If the windlass is 4 ft above the level of the boat, how fast is the boat moving through the water when it is 12 ft from the dock?

14. A man 6 ft tall walks away from a light 10 ft above the ground. If his shadow lengthens at the rate of 2 ft/sec, how fast is he walking?

15. A bug moves along the curve $y = 4 - \frac{1}{16}x^2$. Distance is measured in feet.

   (a) When the bug passes through the point (4,3) its y coordinate is increasing at 20 ft/sec. At what rate is its x coordinate changing?

   (b) How fast is the bug's y coordinate changing when it passes through the point (0, 4)?

   (c) The bug's x coordinate is decreasing at 25 ft/sec when it reaches the point $(-8,0)$. How fast is its y coordinate changing?

16. A portion of a roller coaster track has the shape of the curve

$$y = \frac{4}{1 + x^2} \qquad (-2 \le x \le 2)$$

As a car moves along the track, its shadow is cast on the ground (the x axis) by the sun's rays (parallel to the y axis). Distance is measured in meters.

   (a) When the car reaches the point $(-1,2)$ its shadow is moving to the right at 10 km/hr. How fast is the car gaining altitude at that instant?

   (b) What is the rate of change of altitude as the car passes through the point (0,4)? Does the answer depend on how fast the car is moving?

   (c) The car is losing altitude at the rate of 80 km/hr when it passes through the point (1,2). How fast is its shadow moving then?

17. A TV camera at the center of a circular track of radius 30 meters follows a runner around the track. At a certain instant the camera is turning at 10° per second. How fast is the runner moving at that instant? *Hint:* The length of arc subtended by a central angle of θ radians in a circle of radius r is $s = r\theta$.

18. Two cars leave the intersection of perpendicular highways at the same time, one going north at 30 mph, the other going east at 40 mph. How fast is the distance between them increasing?

19. At noon a ship sails due north from a certain point at 10 knots (nautical mph). Another ship leaves the same point at 1:00 P.M. on a course 60° east of north, sailing at 15 knots. How fast is the distance between the ships increasing at 3:00 P.M.? *Hint:* Use the Law of Cosines (Problem 52, Section 2.3).

20. Boyle's Law says that the pressure and volume of a confined gas at constant temperature are related by the equation $pv = k$, where k is a constant. At a certain instant the pressure is 25 and is increasing at 5 units per second. If the volume is 60 at that instant, what is its rate of change?

21. When a gas expands adiabatically (no energy change), its pressure and volume are related by the equation $pv^{1.4} = k$, where k is a constant. Find the rate of change of volume at the instant when $p = 10$ and $v = 20$, assuming that the pressure is decreasing 2 units per second at that instant. Is the volume increasing or decreasing?

22. A poker chip rolls away from the center of a circular table (radius 1 meter) at the rate of 2 meters per second. A light 1 meter above the center of the table shines on the chip with intensity

$$I = \frac{50 \cos \theta}{z^2}$$

(See Figure 6.) How fast is the intensity changing when the chip reaches the edge of the table?

Figure 6
Light shining on poker chip

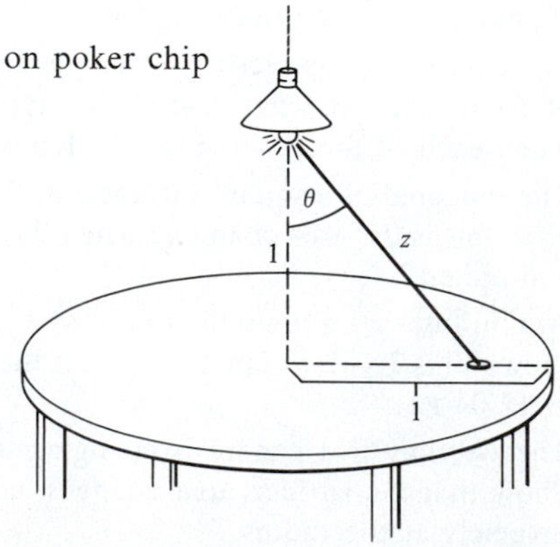

23. The beacon of a lighthouse 1 km from a straight shore revolves 5 times per minute. (See Figure 7.)

  (a) Find the speed of its beam along the shore in terms of θ.

  (b) What is the minimum speed of the beam along the shore?

  (c) Explain why the speed increases without limit as the beacon turns.

Figure 7
Lighthouse beam moving along a shore

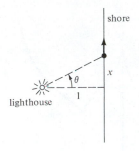

24. A dropped ball will fall about $16t^2$ ft in t seconds. If it is dropped from the same height as a street light 10 ft away and 16 ft off the ground, show that the speed of the ball's shadow along the ground is $20/t^3$ ft/sec. (It is interesting to think about where the shadow is at the start, and how fast it is moving!) What is the speed of the shadow when the ball hits the ground?

25. The cost (in dollars) of producing x units of a certain item is $C(x) = 15 + 80x - 0.3x^2$.

  (a) What is the change in cost corresponding to an increase of production from 100 to 101 units?

  (b) What is the marginal cost at $x = 100$?

  (c) Use the differential (Section 4.7) to explain why the answers to (a) and (b) are approximately the same. *Hint:* Take $dx = \Delta x = 1$ in the formula $\Delta C \approx dC$. (Thus we may interpret marginal cost, at a given production level x, as the *change in cost required to produce one more unit.*)

26. The revenue (in dollars) from the sale of x units of the commodity in Problem 25 is $R(x) = 90x$.

  (a) What is the marginal revenue? How does it compare to the increase in revenue due to the sale of one more item?

  (b) Find the marginal profit at $x = 100$. *Hint:* The profit is $P(x) = R(x) - C(x)$. (Why?)

  (c) About how much additional profit is made by increasing sales from 100 to 101 units? Why is the answer the same as the difference between marginal revenue and marginal cost?

5.2
EXTREME VALUES

One of the important applications of calculus is the determination of *extreme values* of a function. If the graph of the function is to be drawn without merely plotting points indiscriminately, we need to know its high and low points. If the range of the function is wanted, we have to find its largest and smallest values. If the function models the behavior of some physical quantity whose maximum or minimum value is sought, we seek its extreme values in order to answer the question.

In this section we do not have any particular applications in mind; our concern is with the *methods* of locating extreme values. Much of what we have to say is already familiar to you from earlier remarks. This section is designed to pull things together in a systematic way.

Recall from Section 4.2 that if

$$f(x) \leq f(x_0) \qquad \text{for all } x \text{ in a neighborhood of } x_0$$

then $f(x_0)$ is a *local maximum* of f (*global* if the inequality holds for all x in the domain). If the inequality is reversed, then $f(x_0)$ is a *minimum* (local or global as the case may be). To visualize some of the ways such extreme values may occur, look at Figure 1, which illustrates several things worth keeping in mind:

1. The simplest extreme values occur at interior points of the domain where the derivative exists, such as c_1 and c_2 in Figure 1. The derivative at such points must be zero (Theorem 1, Section 4.3).
2. On the other hand, a zero derivative does not guarantee an extreme value. It may indicate only a point where the curve flattens out, as at c_3 in Figure 1, where there is neither a maximum nor a minimum.

Figure 1
Extreme values

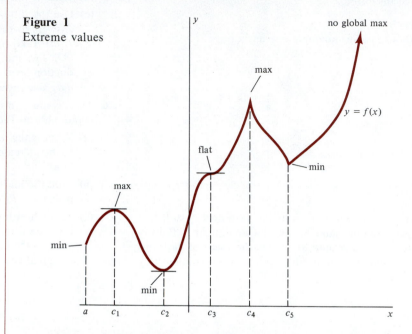

3. Extreme values can occur where the derivative is not zero, either because the point in question is not an interior point (like the endpoint a in Figure 1) or because the derivative does not exist (as at c_4 where there is a "cusp" or at c_5 where there is a "corner").
4. A minimum can be greater than a maximum. For example, $f(c_5) > f(c_1)$, the reason being that these are *local* extreme values.
5. Global extreme values need not exist unless the function is continuous in a finite closed interval (Theorem 2, Section 4.2). In Figure 1 there is no global maximum because the function is unbounded in the positive direction.

To reduce our search for extreme values to a systematic procedure, we adopt the following definition.

A point c in the domain of a function f is called a **critical point** of f if

(i) $f'(c) = 0$, or
(ii) $f'(c)$ is undefined, or
(iii) c is not an interior point of the domain.

The reason we single out such points and call them "critical" is that *extreme values cannot occur anywhere else.* To see why, suppose that c is a point at which f has an extreme value. If c is *not* a critical point, it must be an interior point of the domain, since otherwise it would be in Category (iii). Moreover, $f'(c)$ must exist, since otherwise c would be in Category (ii). But then Theorem 1, Section 4.3, implies that $f'(c) = 0$, which puts c in Category (i)!

These observations enable us to confine the search for extreme values to what (in this book) is a finite list of points. Of course if a function is constant in some interval, or repeats its behavior periodically (like $\sin x$), there are infinitely many critical points. Aside from these special cases, however, we will not consider any "wild" functions, such as $\sin(1/x)$, having more than a finite number of critical points. (See Figure 2.)

Figure 2

Graph of $y = \sin \dfrac{1}{x}$

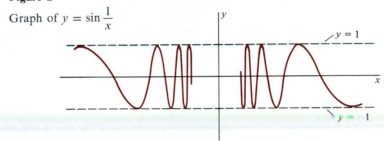

□ **Example 1**   Find the critical points of the function $f(x) = \sqrt{1 - x^2}$.

Solution: The domain of f is the closed interval $[-1,1]$, so $x = \pm 1$ are critical points because they are endpoints of the domain. Since

$$f'(x) = \frac{1}{2}(1 - x^2)^{-1/2}(-2x) = \frac{-x}{\sqrt{1 - x^2}}$$

$x = 0$ and $x = \pm 1$ are critical points because $f'(0) = 0$ while $f'(1)$ and $f'(-1)$ are undefined. (Observe that $x = \pm 1$ qualify as critical points in more than one way; the three categories in our definition are not mutually exclusive.)

Note the distinction between *critical points* (values of the independent variable) and *extreme values* (functional values corresponding to certain critical points). The number $x = 0$ is a critical point of $f(x) = \sqrt{1 - x^2}$; the corresponding functional value (a maximum) is $f(0) = 1$. Students sometimes make the mistake of calling 0 the maximum. The correct statement is that there is a maximum *at* 0, namely $f(0) = 1$. Similarly, there is a minimum at $x = \pm 1$, namely $f(\pm 1) = 0$. (See Figure 3.)   □

Figure 3

Critical points and extreme values of $f(x) = \sqrt{1 - x^2}$

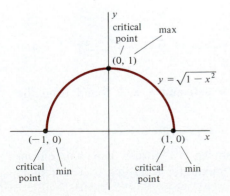

Once the critical points of a function f are located, the question that arises is whether they yield extreme values and if so, what kind. One widely used test is based on the fact that the sign of $f'(x)$ determines whether the graph of f is rising or falling. Look at the points c_1 and c_4 in Figure 1, for example. As we pass through these critical points from left to right, the graph changes from rising to falling, which indicates a high point of the graph. The corresponding analytic idea is that if the derivative changes sign from plus to minus, there is a local maximum. Similarly, a change from minus to plus (as at c_2 or c_5 in Figure 1) indicates a local minimum. On the other hand, if $f'(x)$ has the same sign on either side of a critical point (as at c_3 in Figure 1), no extreme value occurs.

We make these observations precise in the following theorem.

□ **Theorem 1** (*First Derivative Test*)   Let c be an interior critical point of the function f (that is, an interior point at which f' is zero or undefined) and suppose that f is continuous at c and differentiable near c.

1. If (for x near c) $f'(x)$ has one sign when $x < c$ and the opposite sign when $x > c$, then $f(c)$ is a local extreme value (maximum if the change is from plus to minus, minimum if it is from minus to plus).
2. If (for x near c) $f'(x)$ has the same sign when $x < c$ and when $x > c$, then $f(c)$ is not a local extreme value.

Proof: We will do the first part of (1) and leave the rest for the problems. Suppose that N is a neighborhood of c in which

$$f'(x) > 0 \quad \text{when } x < c \quad \text{and} \quad f'(x) < 0 \quad \text{when } x > c$$

We claim that $f(x) \le f(c)$ for all $x \in N$, which (if true) makes $f(c)$ a local maximum. To see that our claim is true, let x be any point of N (different from c) and let I be the closed interval having endpoints c and x. Since f is continuous at c and differentiable elsewhere in I, the Mean Value Theorem applies. Hence there is a point m between c and x such that

$$f(x) - f(c) = f'(m)(x - c)$$

But $f'(m)$ and $x - c$ have opposite signs whether $x < c$ or $x > c$. (See Figure 4.) Hence $f'(m)(x - c)$ is negative and we conclude that $f(x) - f(c) < 0$, that is, $f(x) < f(c)$.   □

Figure 4
Proof of Theorem 1

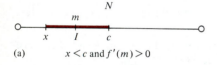

(a) $x < c$ and $f'(m) > 0$

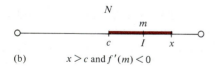

(b) $x > c$ and $f'(m) < 0$

□ **Example 2**   Find the extreme values of $f(x) = x^3 - 3x^2 + 1$ and sketch its graph.

Solution: The critical points are $x = 0$ and $x = 2$, found from the equation

$$f'(x) = 3x^2 - 6x = 3x(x - 2) = 0$$

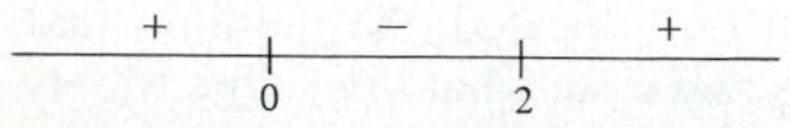

A "sign change diagram" for $f'(x)$ is shown in Figure 5. The change from plus to minus at $x = 0$ yields a local maximum according to Theorem 1;

Figure 5
Sign of $f'(x) = 3x(x - 2)$

Figure 6
Graph of $y = x^3 - 3x^2 + 1$

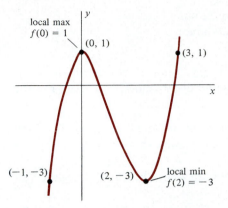

local max
$f(0) = 1$

$(0, 1)$

$(3, 1)$

y

x

$(-1, -3)$

$(2, -3)$

local min
$f(2) = -3$

Figure 7

Sign of $f'(x) = \dfrac{3(2 - x)}{2\sqrt{3 - x}}$

$+$ $-$ outside domain

2 3

there is a local minimum at $x = 2$ because of the change from minus to plus. The graph of f is shown in Figure 6. □

□ **Example 3**   Find the extreme values of $f(x) = x\sqrt{3 - x}$ and sketch its graph.

Solution: The domain of f is $(-\infty, 3]$, so $x = 3$ is one critical point. Since

$$f'(x) = x \cdot \frac{1}{2}(3 - x)^{-1/2}(-1) + \sqrt{3 - x} = \frac{-x}{2\sqrt{3 - x}} + \sqrt{3 - x}$$

$$= \frac{-x + 2(3 - x)}{2\sqrt{3 - x}} = \frac{3(2 - x)}{2\sqrt{3 - x}}$$

the only other critical point is $x = 2$. (The derivative is undefined at $x = 3$, but we already listed that one.) To apply the First Derivative Test at $x = 2$, we make the sign change diagram shown in Figure 7. Note that the signs are determined by the factor $2 - x$ in $f'(x)$; the other factors are all positive. Since there is a change from plus to minus at $x = 2$, the functional value $f(2) = 2$ is a local maximum.

Although there is no sign change at $x = 3$, $f(3) = 0$ is a local minimum. The graph is falling from $x = 2$ to $x = 3$; the endpoint $(3,0)$ is the lowest point in its vicinity. (See Figure 8.) There are no other extreme values. (Why?)

Figure 8
Graph of $y = x\sqrt{3 - x}$

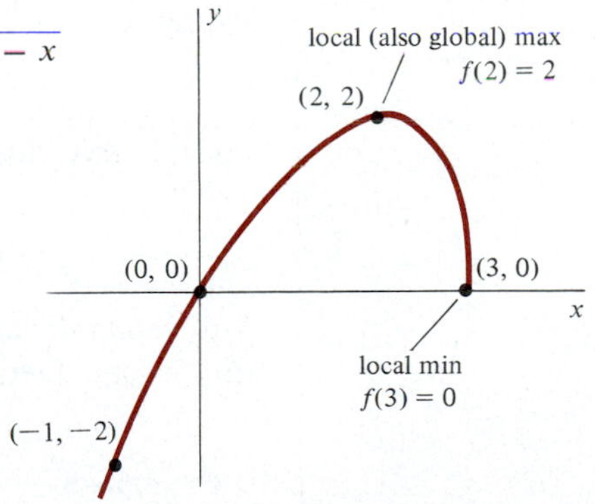

y

local (also global) max
$f(2) = 2$

$(2, 2)$

$(0, 0)$

$(3, 0)$

x

local min
$f(3) = 0$

$(-1, -2)$

□

There is another test for extreme values, this one involving the second derivative. Suppose we are dealing with a critical point c at which $f'(c) = 0$. If f'' is continuous at c, the graph of f is concave up or down near c depending on whether $f''(c)$ is positive or negative. It is geometrically apparent that the first of these implies a local minimum, while the second yields a maximum. (See Figure 9.) For an analytic proof (which we include not because the geometry is unconvincing but because it is an interesting application of linear approximation), see the optional note at the end of this section.

Figure 9
The Second Derivative Test

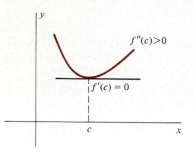

(a) Concave up, local minimum

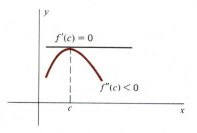

(b) Concave down, local maximum

□ **Theorem 2** (*Second Derivative Test*)   Let c be a critical point of f at which f'' is continuous and suppose that $f'(c) = 0$.

1. If $f''(c) > 0$, then $f(c)$ is a local minimum.
2. If $f''(c) < 0$, then $f(c)$ is a local maximum.
3. If $f''(c) = 0$, the test does not apply.   □

The Second Derivative Test has some disadvantages compared to the first:

(a) It applies only to "smooth" critical points (where f' is zero). The First Derivative Test can be used at critical points where f' is undefined as well.
(b) It fails at points where f' and f'' are both zero. The First Derivative Test leads to a definite conclusion.
(c) It requires the computation of two derivatives. In simple cases this is no problem, but often the work involved in finding f'' is more trouble than it's worth.

On the other hand, the Second Derivative Test does not require an analysis of sign changes in a neighborhood of the critical point. One need only compute f'' *at the point c.*

□ **Example 4**   Use the Second Derivative Test to find the extreme values of $f(x) = x^4 - 8x^2 + 16$.

Solution: Since $f'(x) = 4x^3 - 16x = 4x(x - 2)(x + 2)$, the critical values are $x = 0$ and $x = \pm 2$. To apply Theorem 2, write

$$f''(x) = 12x^2 - 16 \quad [\text{from } f'(x) = 4x^3 - 16x]$$

and observe that

$$f''(0) = -16 < 0 \quad \text{(local maximum at } x = 0)$$
$$f''(\pm 2) = 32 > 0 \quad \text{(local minimum at } x = \pm 2)$$

You can confirm these results by the First Derivative Test, but in this case the Second Derivative Test is less work).   □

□ **Example 5**   In Example 3 the derivative of $f(x) = x\sqrt{3 - x}$ required some nontrivial algebra before we got it in the form

$$f'(x) = \frac{3(2 - x)}{2\sqrt{3 - x}}$$

Once this is obtained, however, it is a matter of inspection to apply the First Derivative Test to the critical point $x = 2$. To achieve the same result from the Second Derivative Test, we must compute

$$f''(x) = \frac{3(x - 4)}{4(3 - x)^{3/2}}$$

(at the cost of considerably more work). Since $f''(2) < 0$, there is a local maximum at $x = 2$, but this is not the way to discover it! The Second Derivative Test is not recommended except in simple cases. ☐

Optional Note (*on the proof of Theorem 2*)   To prove part (1) of Theorem 2, assume that $f''(c) > 0$. The problem is to show that $f(c)$ is a local minimum, that is, $f(x) \geq f(c)$ for all x in some neighborhood of c. Since f'' is continuous at c, we know there is a neighborhood of c (call it N) in which f'' is positive. For each $x \in N$, use the linear approximation formula in Section 4.6 to write

$$f(x) = f(c) + f'(c)(x - c) + \tfrac{1}{2}f''(m)(x - c)^2 \qquad (m \text{ between } c \text{ and } x)$$

Since $f'(c) = 0$, this reduces to

$$f(x) - f(c) = \tfrac{1}{2}f''(m)(x - c)^2$$

But $f''(m) > 0$ (because m is in N) and $(x - c)^2 \geq 0$. Hence

$$f(x) \geq f(c) \text{ for all } x \in N$$

and $f(c)$ is a local minimum.

The proof of (2) is similar. To prove (3), we need examples showing that anything can happen when $f'(c)$ and $f''(c)$ are both zero. (See the problems.)

Problem Set 5.2

In each of the following, find the critical points and extreme values, and sketch the graph. Which of the extreme values are local and which are global?

1. $f(x) = 1 - x^2$
2. $f(x) = x^2 - 2x$
3. $f(x) = 5x - 1$
4. $f(x) = 3$
5. $f(x) = 2x^3 - 3x^2$
6. $f(x) = 3x - x^3$
7. $f(x) = (x - 1)^3$
8. $f(x) = x(x - 3)^2$
9. $f(x) = x^4 - 2x^2 + 1$
10. $f(x) = 3x^4 - 8x^3 + 9$
11. $f(x) = \dfrac{2}{1 + x^2}$
12. $f(x) = \dfrac{x}{1 + x^2}$
13. $f(x) = x - \dfrac{1}{x}$
14. $f(x) = x + \dfrac{1}{x}$
15. $f(x) = |x|$
16. $f(x) = 1 - |x|$
17. $f(x) = x - 2\sqrt{x}$
18. $f(x) = 4x^{3/2} - 6x$
19. $f(x) = 1 - x^{2/3}$
20. $f(x) = 3x^{2/3} - 2x$
21. $f(x) = \sin x + \cos x, \ 0 \leq x < 2\pi$
22. $f(x) = \sin x - \cos x, \ 0 \leq x < 2\pi$
23. The first derivative of $f(x) = 1/x^2$ is positive if $x < 0$ and negative if $x > 0$, yet f has no extreme values. Why doesn't this contradict Theorem 1? Is $x = 0$ a critical point?

24. Sketch the graph of

$$f(x) = \begin{cases} x & \text{if } x < 0 \\ 2x & \text{if } x \geq 0 \end{cases}$$

and explain why $x = 0$ is a critical point of f. What does the First Derivative Test say about $f(0)$?

25. Sketch the graph of $f(x) = x + |x|$ and explain why $x = 0$ is a critical point of f. What does the First Derivative Test say about $f(0)$?

26. One critical point of $f(x) = 3x^4 - 8x^3 + 9$ (Problem 10) is $x = 0$.

   (a) What does the First Derivative Test say about this point?

   (b) What does the Second Derivative Test say about it?

27. The function $f(x) = 1 - x^{2/3}$ (Problem 19) has critical point $x = 0$.

   (a) What does the First Derivative Test say about this point?

   (b) What does the Second Derivative Test say about it?

28. The First Derivative Test is not stated so as to apply to critical points that are boundary points of the domain of f.

Suppose, however, that c is such a point, and that f is continuous at c.

(a) Show that if $f'(x) > 0$ for all x in a left [right] neighborhood of c, then $f(c)$ is a local maximum [minimum].

(b) Show that if $f'(x) < 0$ for all x in a left [right] neighborhood of c, then $f(c)$ is a local minimum [maximum].

29. Apply Problem 28 to the critical point $x = 3$ of $f(x) = x\sqrt{3 - x}$ and confirm our observation in Example 3 that $f(3)$ is a local minimum.

30. Find the critical points of $f(x) = 2\sqrt{x} - x$ and use the First Derivative Test and Problem 28 to discuss them. Sketch the graph.

31. Define the function

$$f(x) = \frac{ax + b}{cx + d} \qquad (a, b, c, d \text{ real, } c \text{ and } d \text{ not both zero})$$

(a) Show that f has no critical points (and hence no extreme values) unless $ad - bc = 0$.

(b) What can you say about f if $ad - bc = 0$? *Hint:* If $ad - bc = 0$, there is a number k satisfying $a = kc$ and $b = kd$. (Why?)

32. Finish the proof of Theorem 1 as follows.

(a) Suppose that N is a neighborhood of c in which

$$f'(x) < 0 \qquad \text{when } x < c$$
$$\text{and} \qquad f'(x) > 0 \qquad \text{when } x > c$$

Show that $f(c)$ is a local minimum by proving that $f(x) \geq f(c)$ for all $x \in N$.

(b) Suppose that N is a punctured neighborhood of c in which $f'(x) > 0$ [or $f'(x) < 0$]. Show that $f(c)$ is not a local extreme value.

33. Finish the proof of Theorem 2 as follows.

(a) Suppose that $f'(c) = 0$ and $f''(c) < 0$ (where c is a point near which f is twice-differentiable and at which f'' is continuous). Show that $f(c)$ is a local maximum.

(b) Give examples showing that if $f'(c)$ and $f''(c)$ are both zero, $f(c)$ may be a local maximum, a local minimum, or neither.

5.3
CURVE SKETCHING

By now it should be plain that much of what we have to say about calculus and its applications is most easily understood in graphical terms. Pictures are not always as precise or complete as we would like, but they are an enormous aid to comprehension. In this section we propose to discuss some general graphing procedures. Most of them have come up before in one form or another; we will not be doing much that is new.

Suppose that an equation in x and y is given, not necessarily in the functional form $y = f(x)$, but simply a relation between x and y. To analyze its graph, we investigate the following questions (in whatever order seems natural, and with differing emphasis depending on the problem).

- Extent (What restrictions, if any, are there on x and y?)
- Symmetry (Is the graph symmetric about any lines or points?)
- Asymptotes (Does the graph look like a straight line in remote parts of the plane?)
- Monotonicity and Extreme Values (Where is the graph rising and where is it falling? Where are its high and low points?)
- Concavity (Where is the graph concave up and where is it concave down? Does it have any points of inflection?)

As you study these procedures, try to avoid the impression that we figure out a graph by going through a cookbook recipe! The idea is to do what needs doing; one learns what is important by experience with graphs. That is why remarks about graphing have been scattered through previous sections. By now you should know much of what we are talking about here.

Extent

This may refer to either x or y, depending on how the given equation is written. If the equation defines y as one or more functions of x, say $y = f(x)$, we discuss admissible values of x by finding the domain of f. Solving for x in terms of y, say $x = g(y)$, we determine admissible values of y by looking at the domain of g.

□ **Example 1**   Discuss the extent of the graph of $x^2 - y^2 = 1$.

Solution: Solving for y in terms of x, we have

$$y = \pm\sqrt{x^2 - 1}$$

from which we learn that $x \geq 1$ or $x \leq -1$. On the other hand,

$$x = \pm\sqrt{y^2 + 1}$$

which imposes no restrictions on y. The graph therefore lies outside the vertical strip $-1 < x < 1$ in the plane, as shown in Figure 1. The points $(\pm 1, 0)$ are obvious points to plot; note that we already have a good start on the graph! You can sketch the rest of it by recalling from earlier remarks that it is a hyperbola with asymptotes $y = \pm x$. □

Figure 1
Graph of $x^2 - y^2 = 1$

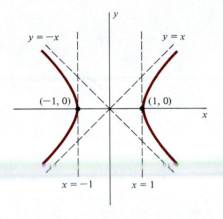

Symmetry

This has been discussed before, in Section 1.3. The main facts to recall are as follows.

1. If the given equation is unaffected when y is replaced by $-y$, the graph is symmetric about the x axis. An example is $y^2 = x$, shown in Figure 2.
2. If the equation is unaffected when x is replaced by $-x$, the graph is symmetric about the y axis. An example is $y = \cos x$, shown in Figure 3.

Figure 2
Graph of $y^2 = x$

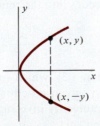

Figure 3
Graph of $y = \cos x$

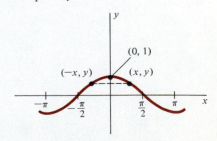

3. If the equation is unaffected when x and y are interchanged, the graph is symmetric about the line $y = x$. An example is $xy = 1$, shown in Figure 4.
4. If the equation is unaffected when x is replaced by $-x$ and y is replaced by $-y$, the graph is symmetric about the origin. An example is $y = x^3$, shown in Figure 5.

Figure 4
Graph of $xy = 1$

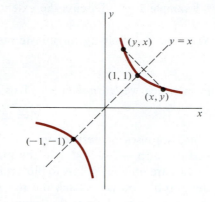

Figure 5
Graph of $y = x^3$

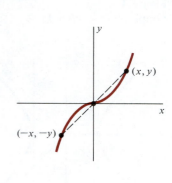

Figure 6
Graph of $y^2 = x^2(3 - x)$

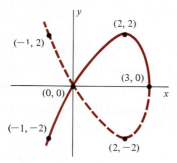

□ **Example 2**   The graph of $y^2 = x^2(3 - x)$ is symmetric about the x axis (because y may be replaced by $-y$ without affecting the equation). Solving for

$$y = \pm x \sqrt{3 - x}$$

we see that $x \leq 3$; the branch $y = x\sqrt{3 - x}$ is shown as the solid curve in Figure 6. (The maximum at $x = 2$ is found by the methods of Section 5.2.) The branch $y = -x\sqrt{3 - x}$ (the dashed curve in Figure 6) is obtained by reflection about the x axis. The complete graph is a loop.   □

Asymptotes

The most familiar example of a curve with asymptotes is the hyperbola $xy = 1$ (Figure 4), which becomes virtually indistinguishable from the coordinate axes as it recedes into the remote parts of the plane. Another example is the hyperbola

$$\frac{x^2}{a^2} - \frac{y^2}{b^2} = 1$$

(Problem 57, Section 1.3), which has asymptotes $y = \pm(b/a)x$.

More generally, given an equation in x and y, we discover horizontal and vertical asymptotes by investigating the behavior of the graph as one variable or the other increases or decreases without bound (assuming that the domain permits it).

□ **Example 3**   The graph of $y = x/(x^2 + 1)$ has the x axis as an asymptote. (See Figure 7.) To see why, divide numerator and denominator by x to obtain

$$y = \frac{1}{x + (1/x)} \approx \frac{1}{x} \qquad \text{for large } |x|$$

This shows that y approaches 0 as $x \to \infty$ or $x \to -\infty$. (Limits involving infinity will be discussed in Chapter 11. For the present we rely on your intuition, reminding you that the notation $x \to \infty$ does not mean that x is approaching a point called infinity, but is simply increasing without bound.) The high and low points $(1, \frac{1}{2})$ and $(-1, -\frac{1}{2})$ are found by the methods of Section 5.2. Also note that the graph is symmetric about the origin (because the equation is unaffected when x is replaced by $-x$ and y by $-y$).   □

□ **Example 4**   Consider the behavior of $y = x/\sqrt{x - 1}$ when $x \to \infty$. (Note that we are not interested in what happens when $x \to -\infty$ because the domain requires $x > 1$.) If x is large (say $x = 10,000$), we may safely ignore the 1 in the denominator, obtaining

$$y \approx \frac{x}{\sqrt{x}} = \sqrt{x}$$

Since $\sqrt{x}$ increases without bound as x does, y does not approach any limit as $x \to \infty$. Hence there is no horizontal asymptote. On the other hand, y is large when x is nearly 1, because the denominator of

$$y = \frac{x}{\sqrt{x - 1}}$$

is close to 0 (while the numerator is close to 1). Hence the vertical line $x = 1$ is an asymptote of the graph. (See Figure 8.) The minimum at $x = 2$ is found by the methods of Section 5.2.   □

□ **Example 5**   Find the asymptotes of the graph of

$$y = \frac{x + 1}{x - 1}$$

Solution: When x is numerically large (positive or negative) y is close to 1. Hence the line $y = 1$ is a horizontal asymptote. When x is close to 1, y is numerically large. More precisely, $y \to \infty$ when $x \downarrow 1$ and $y \to -\infty$ when $x \uparrow 1$. In the first case, when x decreases toward 1, y is large positively because the denominator is close to 0 (and positive) while the numerator is close to 2. The same thing happens when x increases toward 1, except that

Figure 7

Graph of $y = \dfrac{x}{x^2 + 1}$

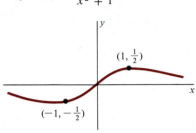

Figure 8

Graph of $y = \dfrac{x}{\sqrt{x - 1}}$

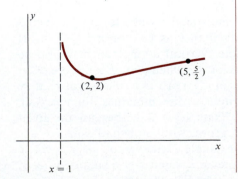

Figure 9

Graph of $y = \dfrac{x+1}{x-1}$

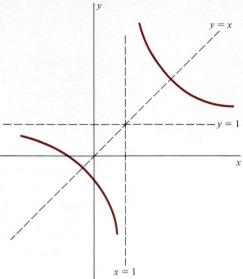

the denominator is negative. Hence the line $x = 1$ is a vertical asymptote. (See Figure 9.) Although it is not obvious, the graph is also symmetric about the line $y = x$, because interchanging x and y does not affect its equation:

$$y = \frac{x+1}{x-1} \Leftrightarrow x = \frac{y+1}{y-1}$$

To confirm this statement, observe that each of these equations is equivalent to $xy - x - y - 1 = 0$.

It is worth remarking that the latter equation is quadratic in x and y, so it must represent a familiar second-degree curve (Section 1.3). The graph is a hyperbola, but a rotation of axes through $45°$ is needed to reduce its equation to a recognizable standard form. We do not expect you to know about that; the problem is a good example of how we can get along without such technical matters from analytic geometry. The graph can be figured out by nothing more complicated than a discussion of asymptotes.   □

Figure 10

Graph of $y = \cot x$

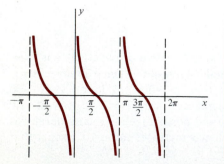

Examples 3 through 5 reveal that horizontal asymptotes are discovered by determining whether y approaches a limit L as $x \to \infty$ or $x \to -\infty$. If it does, the line $y = L$ is an asymptote. Vertical asymptotes may also be found this way (by looking at the behavior of x when $y \to \pm\infty$). More often than not, however (because of our bias in favor of y as a function of x), they are discovered in connection with "infinite discontinuities" in $y = f(x)$. Thus $x = 1$ is a vertical asymptote in Examples 4 and 5 because the given function is undefined at $x = 1$ and grows without bound when x approaches 1. Another (less obvious) example is the graph of $y = \cot x$ $(0 < x < \pi)$. This has vertical asymptotes $x = 0$ and $x = \pi$ because $y \to \infty$ when $x \downarrow 0$ and $y \to -\infty$ when $x \uparrow \pi$. (See Figure 10.)

□ **Example 6**   Sometimes an asymptote not parallel to a coordinate axis can be found by inspection. In the equation

$$y = x + \frac{1}{x}$$

it is clear that $y \approx x$ when x is numerically large. Hence $y = x$ is an asymptote. To sketch the curve, note that the y axis is also an asymptote (because $y \to \infty$ when $x \downarrow 0$ and $y \to -\infty$ when $x \uparrow 0$). Moreover, there are local extreme values at $x = \pm 1$ (discovered by the methods of Section 5.2) and the graph is symmetric about the origin (because the equation is unaffected when x is replaced by $-x$ and y by $-y$). See Figure 11.   □

Figure 11

Graph of $y = x + \dfrac{1}{x}$

Monotonicity and Extreme Values

As we said in Section 4.4, a function f is *monotonic* in an interval where it is increasing (or where it is decreasing). The analytic test is that

$$f'(x) > 0 \text{ in } I \Longrightarrow f \text{ is increasing in } I$$
$$f'(x) < 0 \text{ in } I \Longrightarrow f \text{ is decreasing in } I$$

Such behavior is usually discussed in connection with the identification of extreme values (particularly if the First Derivative Test is used to distinguish maxima from minima).

□ **Example 7**   Discuss monotonicity and extreme values of the function $y = 3x^4 + 4x^3 - 12x^2$ and sketch the graph.

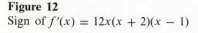

Figure 12
Sign of $f'(x) = 12x(x + 2)(x - 1)$

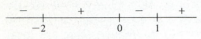

Figure 13
Graph of $y = 3x^4 + 4x^3 - 12x^2$

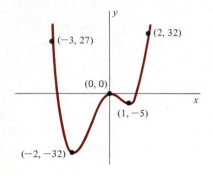

Solution: The derivative is

$$y' = 12x^3 + 12x^2 - 24x = 12x(x^2 + x - 2) = 12x(x + 2)(x - 1)$$

An analysis of sign changes (Figure 12) shows that the graph is falling when $x < -2$, rising when $-2 < x < 0$, falling when $0 < x < 1$, and rising when $x > 1$. The extreme values are

$$f(-2) = -32 \quad \text{(minimum)}$$
$$f(0) = 0 \quad \text{(maximum)}$$
$$f(1) = -5 \quad \text{(minimum)}$$

The graph is shown in Figure 13. Note that it may be sketched by plotting only a few points. The intervals of monotonicity guide us to an intelligent selection. (This discussion fills in the details of Example 6, Section 1.5.)   □

Concavity

In Section 4.5 we said that the graph of $y = f(x)$ is concave up in intervals where $f''(x) > 0$ and concave down when $f''(x) < 0$. Points of inflection occur where f is continuous and the concavity changes. Whether such information is useful depends on the problem. In Example 7 the second derivative is

$$y'' = 36x^2 + 24x - 24 = 12(3x^2 + 2x - 2)$$

To find the points of inflection we need the quadratic formula to learn where $y'' = 0$. In view of the adequate sketch already made in Figure 13, it hardly seems worthwhile.

In Example 4, on the other hand, the first two derivatives of $y = x/\sqrt{x - 1}$ are

$$y' = \frac{x - 2}{2(x - 1)^{3/2}} \quad \text{and} \quad y'' = \frac{4 - x}{4(x - 1)^{5/2}}$$

While it requires considerable work to find them, the results are interesting. Since y' changes sign from minus to plus at $x = 2$, the graph has a low point at $(2,2)$. (See Figure 8.) With no additional information we might be tempted to sketch the curve concave up everywhere. The second derivative, however, shows that the concavity changes from up to down at $x = 4$. Thus $(4, 4/\sqrt{3})$ is a point of inflection.

It is only honest to add that we could have guessed a point of inflection somewhere to the right of the minimum point. For as we pointed out in Example 4, $y \approx \sqrt{x}$ for large x. Hence the graph is virtually indistinguishable from the upper half of the parabola $y^2 = x$ (Figure 2) as we move to the right. It must therefore change from concave up near $x = 2$ to concave down for larger values of x.

It is worthwhile to expand on the preceding remark. Sometimes students become too dependent on calculus, perhaps because its power is seductive. It is possible to say a great deal without it. Consider the equation

$$y^2 = x^2(3 - x)$$

(which we have analyzed before). It is of course perfectly reasonable to apply calculus to $f(x) = x\sqrt{3 - x}$ (the branch of the graph shown as a solid curve in Figure 6). This provides a gold mine of information:

- The graph is rising when $x < 2$, falling when $2 < x < 3$, and is always concave down.
- $f(2) = 2$ is a local (also global) maximum.
- $f(3) = 0$ is a local minimum, and the graph has a vertical tangent at (3,0).

Suppose, however, that we knew none of these facts. We could discover the main features of the graph of the original equation, $y^2 = x^2(3 - x)$, as follows.

1. The domain is $(-\infty, 3]$, since $3 - x \geq 0$ if y is to be real.
2. Since the equation is unchanged when y is replaced by $-y$, the graph is symmetric about the x axis.
3. When x is near 3 (but less than 3), the significant factor (the one close to 0) is $3 - x$; the factor x^2 may be replaced by 9 without much error. Hence the graph is like

$$y^2 = 9(3 - x) \qquad \text{when } x \approx 3$$

Since this is a parabola opening to the left from its vertex (3,0), we have a good idea of how to draw the graph near this point. (For example, we expect a vertical tangent at $x = 3$.)
4. When $x \approx 0$ the significant factor is x^2; we can replace $3 - x$ by 3. Hence the graph is like $y^2 = 3x^2$ near the origin (two intersecting straight lines, $y = \pm\sqrt{3}x$, with slopes $\pm\sqrt{3}$).
5. It is now reasonable to guess that the graph flattens out somewhere between $x = 0$ and $x = 3$, and that the part in the first quadrant (at least) is concave down.
6. When x decreases without limit, $f(x)$ does, too.

Putting these primitive observations together, and plotting a few points like $(-1, \pm 2)$, $(0,0)$, $(2, \pm 2)$, $(3,0)$, we can sketch a pretty good graph. Not a derivative in sight!

We end this section by repeating that our suggestions for graphing should not be treated like a cookbook recipe. The idea is to discuss the graph intelligently, so that its general features are known before the point-plotting process begins. In most cases only a few points of the graph are needed, not a poorly conceived "table of values" with little relation to the significant aspects of the graph. A discriminating selection of points is the mark of a good curve-sketcher.

To understand the third observation, consider what happens when x changes from (say) 2.8 to 2.9. The "significant factor" $3 - x$ changes from 0.2 to 0.1, a decrease of 50%. The factor x^2 changes from 7.84 to 8.41, an increase of only 7.3%. The fourth observation may be similarly justified by letting x change from (say) 0.2 to 0.1. The factors x^2 and $3 - x$ change by 75% and 3.6%, respectively.

Problem Set 5.3

Sketch the graph of each of the following equations by discussing its significant features and then plotting a judicious selection of points. In some cases we have supplied questions for you to answer (as a guide to the discussion).

1. $y = x^3 - 3x + 2$
2. $y = 3 - 2x + 4x^2 - 2x^3$
3. $y = 3x^4 - 4x^3 - 12x^2 + 5$
4. $y = x^4 - 6x^2 + 8x + 10$ *Hint:* $x = 1$ is a critical point.

5. $y = \dfrac{2x}{1 + x^2}$

   (a) What symmetry does the graph have?
   (b) Why does $y \to 0$ when $x \to \pm\infty$?
   (c) Show that

   $$y' = \frac{2(1 - x)(1 + x)}{(1 + x^2)^2}$$

   and use the result to discuss monotonicity and extreme values.

6. $y = \dfrac{2x^2}{1 + x^2}$

7. $y = x - \dfrac{4}{x}$

   (a) What symmetry does the graph have?
   (b) Explain why the lines $x = 0$ and $y = x$ are asymptotes.
   (c) Show that the graph is always rising.
   (d) Where does the graph intersect the x axis?

8. $y = x + \dfrac{4}{x}$

9. $y = x^2 + \dfrac{2}{x}$

   (a) What symmetry does the graph have?
   (b) Explain why the y axis is an asymptote and why the graph follows the parabola $y = x^2$ for large $|x|$.
   (c) Where is the graph falling? rising? What are the extreme values?
   (d) Explain why there is a point of inflection where the graph crosses the x axis. (Use a calculator, or tables, to approximate this point.)

10. $y = x^2 + \dfrac{1}{x^2}$ 11. $y = x^3 + \dfrac{3}{x}$

12. $y = 1 - \dfrac{4}{x^2}$ 13. $y = \dfrac{2x}{x - 1}$

14. $y = \dfrac{x^2}{x - 1}$

   (a) Explain why $x = 1$ is a vertical asymptote and why there is no horizontal asymptote. (See Problem 34 for discussion of another asymptote.)
   (b) Show that

   $$y' = \frac{x(x - 2)}{(x - 1)^2}$$

   and use the result to discuss monotonicity and extreme values.

15. $y = \dfrac{x + 2}{x - 2}$

16. $y = \dfrac{1}{x(x - 2)}$ 17. $y = \dfrac{3x}{x^2 - 4}$

18. $y = \dfrac{x^2}{x^2 - 1}$

19. $y = x^{1/3}(4 - x)$

   (a) Where does the curve intersect the x axis?
   (b) Show that

   $$y' = \frac{4(1 - x)}{3x^{2/3}}$$

   and use the result to discuss monotonicity and extreme values. Why is there a vertical tangent at the origin?

20. $y = x^{2/3}(5 - x)$ *Hint:* There is a "cusp" at the origin.
21. $y = x^{2/3}(x^2 - 4)$
22. $y^2 = x^2(4 - x^2)$

   (a) Why does the graph lie in the vertical strip $-2 \le x \le 2$?
   (b) What symmetry does the graph have?
   (c) Where does it intersect the x axis?
   (d) Show that the slope of the branch $y = x\sqrt{4 - x^2}$ is

   $$y' = \frac{2(2 - x^2)}{\sqrt{4 - x^2}}$$

   and use the result to discuss monotonicity and extreme values. Why are there vertical tangents at $x = \pm 2$?
   (e) What mathematical symbol does the graph remind you of?

23. $y^2 = x^2(x^2 - 4)$ **24.** $y^2 = x(x - 3)^2$

25. $y^2 = \dfrac{x(x - 3)}{x - 2}$

    (a) Explain the restrictions $0 \le x < 2$ and $x \ge 3$.

    (b) Discuss symmetry and asymptotes.

    (c) Why is the graph like the parabola $y^2 = \frac{3}{2}x$ near the origin and like the parabola $y^2 = 3(x - 3)$ near the point $(3,0)$?

    (d) Sketch it while you're still ahead! (No derivatives needed.)

26. $y^2 = \dfrac{x^2(2 + x)}{2 - x}$

27. $y = \sqrt{x} - \sqrt{4 - x}$

28. $y = \sqrt{5 - x} - \sqrt{x - 1}$

29. $y = 2 \sin x + \cos 2x,\ 0 \le x < 2\pi$

    Hint: Use the identity $\sin 2x = 2 \sin x \cos x$ to show that $y' = 2 \cos x(1 - 2 \sin x)$.

30. $y = 2 \sin x + \sin 2x,\ 0 \le x < 2\pi$

31. $y = \tan x + \cot x,\ 0 < x < \dfrac{\pi}{2}$

32. $y = \sec x - \tan x,\ -\dfrac{\pi}{2} < x < \dfrac{\pi}{2}$

    (a) Show that the curve is always falling.

    (b) Explain why $\lim_{x \downarrow -\pi/2} (\sec x - \tan x) = \infty$ and hence the line $x = -\pi/2$ is a vertical asymptote.

    (c) Show that $\lim_{x \uparrow \pi/2} (\sec x - \tan x) = 0$.   *Hint:* Write $\sec x - \tan x$ in the form $\cos x/(1 + \sin x)$.

33. $y = x - \sin x$

    (a) Explain why the curve is never falling.

    (b) What happens at the critical points $x = 0, \pm 2\pi, \pm 4\pi,$ . . . ?

    (c) Show that there are points of inflection at $x = 0, \pm \pi, \pm 2\pi,$ . . . .

    (d) Rotate your paper through $45°$. What does the graph look like? How might this have been predicted?

34. One might argue that an asymptote of the graph of

$$y = \frac{x^2}{x - 1} \qquad \text{(Problem 14)}$$

is $y = x$, on the ground that when $|x|$ is large the 1 in the denominator may be ignored. On the other hand, division of x^2 by $x - 1$ yields

$$y = (x + 1) + \frac{1}{x - 1}$$

from which it appears that $y = x + 1$ is an asymptote.

    (a) Show that

$$\lim_{x \to \pm\infty} \left(\frac{x^2}{x - 1} - x\right) \ne 0$$

and hence $y = x$ is not an asymptote.

    (b) Show that

$$\lim_{x \to \pm\infty} \left[\frac{x^2}{x - 1} - (x + 1)\right] = 0$$

and hence $y = x + 1$ is an asymptote. (You might also check that the slope of the curve approaches the slope of the line $y = x + 1$ as $|x|$ increases.)

35. In Figure 14 we show the first-quadrant portions of the hyperbola

$$\frac{x^2}{a^2} - \frac{y^2}{b^2} = 1$$

and the line $y = (b/a)\, x$.

    (a) Show that

$$y_2 - y_1 = \frac{ab}{x + \sqrt{x^2 - a^2}}$$

and explain why it follows that

$$\lim_{x \to \infty} (y_2 - y_1) = 0$$

(Thus the hyperbola and the line are close together when x is large.)

    (b) Show that the slope of the hyperbola at P is

$$\frac{b}{a} \cdot \frac{x}{\sqrt{x^2 - a^2}}$$

and that this approaches the slope of the line as $x \to \infty$.

Figure 14
Hyberbola and its asymptote

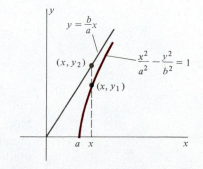

5.4
SOLVING
EQUATIONS
BY NEWTON'S
METHOD
(Optional)

To find the extreme values of $f(x) = x^4 - 8x^2 + 16$, we must solve the equation

$$f'(x) = 4x^3 - 16x = 0$$

Cubic equations, as you probably know from experience, are not easy to solve. Yet every time you have encountered one in this book, the difficulties have magically evaporated, as in the above example:

$$4x^3 - 16x = 0 \Rightarrow 4x(x^2 - 4) = 0 \Rightarrow x = 0 \text{ or } x = \pm 2$$

You will search in vain through the previous sections of this book for an example or exercise in which an equation that needs solving cannot be solved by elementary methods. It may not have occurred to you that this has been a conspiracy, but of course it has! Problems that lead to easy equations must be carefully selected.

It is time to discuss how to solve hard equations. In this section we present a method of successive approximations that enables us to compute the real roots of a given equation to as many decimal places as desired. The only assumption we make about the equation is that it can be written in the form $f(x) = 0$, where f is sufficiently well-behaved to have the derivatives needed in the discussion to follow.

Note first that the real roots of the equation $f(x) = 0$ are points where the graph of f intersects the x axis. Let r be such a root, that is, a real number such that $f(r) = 0$. While it may be impossible to find r exactly, we can usually guess a reasonable approximation; let x_0 be such a guess. (See Figure 1.) *Newton's Method* of improving the guess is to move from x_0 to the point $(x_0, f(x_0))$, then back along the tangent to its point of intersection with the x axis (say x_1). Of course if we move *along the curve* to the x axis, we will hit r. The idea of Newton's Method is to use the tangent to get close to r. Unless something goes wrong (you might think about how that could happen), repeating the procedure generates a sequence of improving approximations $x_0, x_1, x_2, \ldots$ that converge to r.

To make this geometric idea analytically useful, we observe that the equation of the tangent at $(x_0, f(x_0))$ is

$$y - f(x_0) = f'(x_0)(x - x_0)$$

The point $(x_1, 0)$ where the tangent intersects the x axis is found by substituting $(x_1, 0)$ for (x, y) in the equation and solving for x_1:

$$-f(x_0) = f'(x_0)(x_1 - x_0)$$
$$-f(x_0) = x_1 f'(x_0) - x_0 f'(x_0)$$
$$x_1 f'(x_0) = x_0 f'(x_0) - f(x_0)$$
$$x_1 = x_0 - \frac{f(x_0)}{f'(x_0)} \qquad \text{if } f'(x_0) \neq 0$$

Note that the formula is not applicable if $f'(x_0) = 0$. You should draw a picture of a situation where the slope is 0 at $(x_0, f(x_0))$ and observe how the method blows up!

Figure 1
Newton's Method of approximating a real root

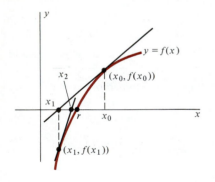

Repeating the process with x_1 in place of x_0 yields a new approximation

$$x_2 = x_1 - \frac{f(x_1)}{f'(x_1)} \quad \text{if } f'(x_1) \neq 0$$

In general, if x_n is a given approximation, the next one is given by:

Newton's Recursion Formula

$$x_{n+1} = x_n - \frac{f(x_n)}{f'(x_n)} \quad \text{if } f'(x_n) \neq 0, \, n = 0, 1, 2, \ldots$$

The reason this is called a recursion formula is that each output is a function of the preceding one. Such formulas are well-suited to a computer, since we need only write one program, then tell the computer to keep feeding back the output.

□ **Example 1**   Suppose that we want to compute $\sqrt{19}$. Use Newton's formula to derive the approximation

$$\sqrt{19} \approx x_{n+1} = \frac{1}{2}\left(x_n + \frac{19}{x_n}\right), \, n = 0, 1, 2, \ldots$$

where x_0 is a first guess.

Solution: Since $\sqrt{19}$ is a solution of the equation $x^2 - 19 = 0$, we let $f(x) = x^2 - 19$. Then $f'(x) = 2x$ and Newton's formula becomes

$$x_{n+1} = x_n - \frac{f(x_n)}{f'(x_n)} = x_n - \frac{x_n^2 - 19}{2x_n} = x_n - \frac{1}{2}x_n + \frac{1}{2}\cdot\frac{19}{x_n}$$

$$= \frac{1}{2}\left(x_n + \frac{19}{x_n}\right)$$

As an illustration, suppose that we use $x_0 = 4.5$ as a first approximation to $\sqrt{19}$. Then

$$x_1 = \frac{1}{2}\left(x_0 + \frac{19}{x_0}\right) = \frac{1}{2}\left(4.5 + \frac{19}{4.5}\right) = \frac{1}{2}(4.5 + 4.\overline{2}) = 4.36\overline{1}$$

A calculator gives $\sqrt{19} = 4.358 \cdots$, so we already have two-place accuracy.   □

Replacing 19 by $a > 0$ in the above argument, we obtain Newton's formula for approximating $\sqrt{a}$:

$$\sqrt{a} \approx x_{n+1} = \frac{1}{2}\left(x_n + \frac{a}{x_n}\right), \, n = 0, 1, 2, \ldots$$

Remark   The above formula is sometimes called the *Mechanic's Rule*. It is originally due to the ancient Babylonians. An intuitive argument for it (independent of calculus) is that if x_n is an approximation to $\sqrt{a}$, then $x_n^2 \approx a$ and hence $x_n \approx a/x_n$. If x_n is too small, a/x_n is too large (and vice versa), so $\sqrt{a}$ is boxed in between x_n and a/x_n. Hence a good choice for the next approximation is the average value

$$x_{n+1} = \frac{1}{2}\left(x_n + \frac{a}{x_n}\right)$$

□ **Example 2**   The equation $f(x) = x^3 + x - 1 = 0$ has a root r between 0 and 1 because $f(0) = -1 < 0$ and $f(1) = 1 > 0$. (See the Zero Theorem in Section 4.2.) Use Newton's Method to approximate r.

Solution: The graph of f (Figure 2) suggests that $r \approx 0.6$, so we take $x_0 = 0.6$ as our first guess. Since $f(x_0) = -0.184$ and $f'(x_0) = 2.08$, Newton's recursion formula yields

$$x_1 = x_0 - \frac{f(x_0)}{f'(x_0)} = 0.6 + \frac{0.184}{2.08} = 0.68846\cdots \qquad \text{(from a calculator)}$$

It is unlikely that this estimate is very accurate (for reasons that we will explain later). To save work, let's round it off to $x_1 = 0.7$ and try again:

$$x_2 = x_1 - \frac{f(x_1)}{f'(x_1)} = 0.7 - \frac{0.043}{2.47} = 0.68259\cdots$$

Comparing this with the previous approximation, we may reasonably guess that $x_2 = 0.68$ is correct to two places. The next approximation is then

$$x_3 = x_2 - \frac{f(x_2)}{f'(x_2)} = 0.68 + \frac{0.005568}{2.3872} = 0.682332\cdots$$

and it now appears that two-place accuracy (at least) is assured. If that is all we are trying to achieve, we can report the root to be $r \approx 0.68$.   □

□ **Example 3**   It is impossible to solve the equation $\sin x = x^2$ by algebraic means. Its roots, however, are x coordinates of the points of intersection of the curves $y = \sin x$ and $y = x^2$. (See Figure 3.) One root is clearly 0; there is also a positive root $r \approx 1$. Let $f(x) = x^2 - \sin x$ and choose $x_0 = 1$. Then Newton's formula gives

$$x_1 = 1 - \frac{(1)^2 - \sin 1}{2(1) - \cos 1} = 0.89139\cdots$$

Rounding this off to $x_1 = 0.9$, we feed it back in:

$$x_2 = 0.9 - \frac{(0.9)^2 - \sin 0.9}{2(0.9) - \cos 0.9} = 0.87736\cdots$$

It is doubtful that this is accurate beyond the second place, so let's take $x_2 = 0.88$ and run the program again:

$$x_3 = 0.88 - \frac{(0.88)^2 - \sin 0.88}{2(0.88) - \cos 0.88} = 0.8767394\cdots$$

Figure 2
Root of $x^3 + x - 1 = 0$

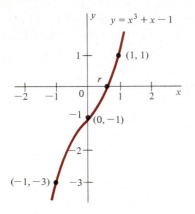

Figure 3
Intersections of $y = \sin x$ and $y = x^2$

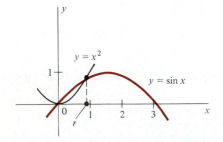

This is in fact correct to four places (see Example 5), but without an estimate of accuracy it is hard to be sure. We'll take $x_3 = 0.8767$ and use the formula one more time:

$$x_4 = 0.8767 - \frac{(0.8767)^2 - \sin 0.8767}{2(0.8767) - \cos 0.8767} = 0.8767262162 \cdots$$

It now seems safe to say that the root is $r \approx 0.8767$.   □

Remark   For most practical purposes the procedures illustrated in Examples 2 and 3 are adequate. We simply feed each approximation back into Newton's formula until we are satisfied that the decimal places we are interested in reporting have stabilized. The rest of this section is devoted to an estimate of accuracy that makes this remark less vague. You may omit it without any great loss, but it is worthwhile to note the results. (See *Convergence of Newton's Method* following this discussion.)

Let r be a real root of $f(x) = 0$ and suppose there is a neighborhood of r (say N) in which f is twice-differentiable and $f'(x) \neq 0$. Choose $x_0 \in N$ and assume that the approximations $x_1, x_2, x_3, \ldots$ generated by Newton's Method remain in N. For each $n = 0, 1, 2, \ldots$, the error in using x_{n+1} as an approximation to r is

$$|r - x_{n+1}| = \left| r - x_n + \frac{f(x_n)}{f'(x_n)} \right| \tag{1}$$

We now appeal to the error formula developed for linear approximation in Section 4.6, namely

$$f(x) = f(x_0) + f'(x_0)(x - x_0) + \tfrac{1}{2}f''(c)(x - x_0)^2 \qquad (c \text{ between } x_0 \text{ and } x)$$

where x_0 is the base point and x is a nearby point of application. In the present context we replace x_0 by x_n and x by r:

$$f(r) = f(x_n) + f'(x_n)(r - x_n) + \tfrac{1}{2}f''(c_n)(r - x_n)^2$$

where c_n is between x_n and r. But $f(r) = 0$. Dividing by $f'(x_n) \neq 0$, we obtain

$$0 = \frac{f(x_n)}{f'(x_n)} + (r - x_n) + \frac{1}{2}\frac{f''(c_n)}{f'(x_n)}(r - x_n)^2$$

from which

$$\left| r - x_n + \frac{f(x_n)}{f'(x_n)} \right| = \frac{1}{2}\left| \frac{f''(c_n)}{f'(x_n)} \right| |r - x_n|^2$$

Equation (1) now becomes

$$|r - x_{n+1}| = \frac{1}{2}\left| \frac{f''(c_n)}{f'(x_n)} \right| |r - x_n|^2, \, n = 0, 1, 2, \ldots \tag{2}$$

where c_n is between x_n and r. Now suppose that $|f''(x)|$ has an upper bound M and $|f'(x)|$ has a lower bound $m > 0$ for $x \in N$. Then Equation (2) yields

$$|r - x_{n+1}| \leq \frac{M}{2m}|r - x_n|^2, \, n = 0, 1, 2, \ldots \tag{3}$$

If x_n is correct to k decimal places, that is, if

$$|r - x_n| < \frac{1}{2} \times 10^{-k}$$

then

$$|r - x_{n+1}| < \frac{M}{2m}\left(\frac{1}{2} \times 10^{-k}\right)^2 = \frac{M}{8m} \times 10^{-2k}$$

Assuming that $M \leq 4m$, we find

$$|r - x_{n+1}| < \frac{1}{2} \times 10^{-2k}$$

which means that x_{n+1} is correct to $2k$ places.

Convergence of Newton's Method

Let r be a real root of $f(x) = 0$ and suppose that N is a neighborhood of r in which $|f''(x)| \leq M$ and $|f'(x)| \geq m > 0$, where M and m are numbers satisfying $M \leq 4m$. If $x_0 \in N$ and Newton's Method generates a sequence $x_1, x_2, x_3, \ldots$ in N, then each approximation to r is correct to twice as many places as the preceding one.

□ **Example 4**   Let r be the root of $f(x) = x^3 + x - 1 = 0$ between 0 and 1. In Example 2 our first approximation to r was $x_0 = 0.6$ and Newton's formula gave $x_1 = 0.68846 \cdots$. We said then that this estimate is unlikely to be very accurate; now we have a way of testing it. Taking $N = (0.5, 0.7)$ (a neighborhood of r containing $x_0 = 0.6$), we observe that for all $x \in N$,

$$|f''(x)| = |6x| < 6(0.7) = 4.2$$

Hence Equation (2) with $n = 0$ yields

$$|r - x_1| < \frac{1}{2}\left(\frac{4.2}{2.08}\right)|r - x_0|^2 < 0.02$$

because $|r - x_0| < 0.1$. This estimate tells us that $x_1 = 0.68846 \cdots$ is correct to one place (but not necessarily two). Thus our decision in Example 2 to round it off to $x_1 = 0.7$ was reasonable.

Now Inequality (3) comes into play. We already know that $M = 4.2$ is an upper bound of $|f''(x)|$ in N. A lower bound of $|f'(x)|$ is obtained by observing that

$$|f'(x)| = |3x^2 + 1| > 3(0.5)^2 + 1 = 1.75$$

Hence we take $m = 1.75$. Since $M \leq 4m$, we can guarantee from our theoretical discussion that the number of correct decimal places *doubles* as we proceed. It has taken us a while to get the process off the ground, but now watch it go!

We know that $x_1 = 0.7$ is correct to one place. Hence

$$x_2 = x_1 - \frac{f(x_1)}{f'(x_1)} = 0.7 - \frac{0.043}{2.47} = 0.68$$

is correct to two places and

$$x_3 = x_2 - \frac{f(x_2)}{f'(x_2)} = 0.68 + \frac{0.005568}{2.3872} = 0.6823$$

is accurate to four places. If that is not good enough for us, then

$$x_4 = x_3 - \frac{f(x_3)}{f'(x_3)}$$

will be correct to eight places!   □

□ **Example 5**   To estimate the accuracy of our approximations in Example 3, we may choose N to be the interval $(0.8, 1.1)$, which is a neighborhood of r containing $x_0 = 1$. For all $x \in N$ we observe that

$$|f''(x)| = |2 + \sin x| < 2 + \sin 1.1 < 2.9$$

so Equation (2) with $n = 0$ yields

$$|r - x_1| = \frac{1}{2} \left| \frac{f''(c_0)}{f'(x_0)} \right| |r - x_0|^2 \qquad (c_0 \text{ between } x_0 \text{ and } r)$$

$$< \frac{1}{2} \left(\frac{2.9}{1.45} \right) |r - x_0|^2 = |r - x_0|^2$$

Since x_0 and r are both in N (and $x_0 = 1$), they differ by less than 0.2, that is, $|r - x_0| < 0.2$. Hence

$$|r - x_1| < |r - x_0|^2 < 0.04$$

and we can be sure that $x_1 = 0.9$ is correct to one place.
For all $x \in N$ we have

$$|f'(x)| = |2x - \cos x| > 2(0.8) - \cos 0.8 > 0.9$$

so we may take $M = 2.9$ and $m = 0.9$ in Inequality (3). Since $M \leq 4m$, the correct number of decimal places will double at each stage. Hence $x_2 = 0.88$, $x_3 = 0.8767$, and $x_4 = 0.87672622$ are correct to two, four, and eight places, respectively.   □

Problem Set 5.4

1. What happens if Newton's Method is applied with a lucky first guess? That is, suppose $x_0 = r$, where r is the root.

2. In applying Newton's formula

$$x_{n+1} = x_n - \frac{f(x_n)}{f'(x_n)}$$

you can tell that something is wrong if $f(x_n)$ does not approach 0 as n increases. Why?

Use the formula in Example 1 to compute each of the following square roots correct to four places.

3. $\sqrt{10}$, starting with $x_0 = 3$
4. $\sqrt{89}$, starting with $x_0 = 9.5$
5. $\sqrt{17}$
6. $\sqrt{29}$

Use Newton's Method to compute each of the following roots correct to four places.

7. $\sqrt[3]{237}$ **8.** $\sqrt[4]{510}$ **9.** $\sqrt[4]{87}$ **10.** $\sqrt[3]{120}$

Each of the following equations has one positive root. Use Newton's Method to approximate it. (Also see Problems 21 through 30, which you may want to do along with these.)

11. $x^2 - 2x - 2 = 0$ (What is the answer in terms of radicals?)

12. $x^2 + 3x - 5 = 0$ (What is the answer in terms of radicals?)

13. $x^3 - x - 1 = 0$

14. $x^3 - x - 2 = 0$

15. $x^3 + 3x^2 - 2x - 5 = 0$

16. $2x^3 + 5x^2 + x - 4 = 0$

17. $\cos x = x$

18. $2 \sin x = x^2$

19. $\sin x = 1 - x^2$

20. $\sin x = 1 - x$

21.–30. In Problems 11 through 20, write the given equation in the form $f(x) = 0$ and let r be its positive root. Select a convenient neighborhood N of r and name M and m such that $|f''(x)| \leq M$ and $|f'(x)| \geq m > 0$ for $x \in N$, where $M \leq 4m$. Use Equation (2) in the text to determine when you have found an approximation correct to one place, and discuss the accuracy of subsequent approximations.

31. Computers divide more slowly than they add and multiply. Derive the recursion formula

$$x_{n+1} = x_n(2 - ax_n) \qquad n = 0, 1, 2, \ldots$$

for computing the reciprocal $1/a$ of a number $a \neq 0$. (Note that the formula does not involve any division.)

32. Use the formula in Problem 31 to compute $\frac{1}{13}$.

33. Suppose that $x_0 \neq 0$ is proposed as a first approximation to the root $r = 0$ of the equation $f(x) = x^{1/3} = 0$.

(a) Show that Newton's recursion formula yields

$$x_n = (-1)^n 2^n x_0 \qquad n = 0, 1, 2, \ldots$$

(b) What happens to x_n as n increases? Sketch the graph of f and explain what went wrong.

5.5

MAXIMUM–MINIMUM PROBLEMS

We now have all the artillery in place for an attack on some truly remarkable problems. In this section we are going to answer such questions as:

- What are the most economical dimensions of a tin can?
- If light rays travel from one medium to another in the shortest possible time, how are they refracted at the interface of the media?
- How many fruit trees should be planted in an orchard for maximum yield?
- How should a computer disk pack be manufactured to contain the largest possible amount of information?
- To minimize the cost per unit in a manufacturing process, what is the optimum level of production?
- What are the dimensions of the strongest beam that can be cut from a log?
- How high should a light be hung above a table to provide the most illumination?
- Where should you stand in front of a painting to get the best view?
- What is the best value to report when several different measurements of a quantity have been made in an experiment?
- When should a growing farm animal be sold in a falling market in order to get the highest price?

These questions have one thing in common. They all refer to something *extreme,* like best, shortest, cheapest, strongest, and so on. To answer them, we construct a *mathematical model* in which the thing we are trying to make extreme is expressed as a function of some independent variable. The problem is then reduced to finding the maximum or minimum functional value that answers the question.

□ **Example 1**   A rancher with 100 meters of fence intends to enclose a rectangular region along a river (which serves as a natural boundary requiring no fence). How much area can be enclosed?

Solution: To construct a mathematical model of the situation (Figure 1), we make some natural assumptions. Although the land may not be flat and the river may not be straight, we will act as though they are. We will also assume that x in Figure 1 may take on all real values between 0 and 50 (although even a mathematical rancher might find it difficult to cut the fence in such a way as to make $x = 0.0001$ or $x = 10\pi$!). This domain leaves out of account any complications like the space taken up by fence posts and corners. In other words it is unrealistic. But it is well suited for the mathematical model we wish to consider in order to answer the question in a reasonably simple way.

The area enclosed by the fence and the river is $A = xy$; the problem is to choose x and y so as to make A as large as possible. At first glance this appears to be a question involving a function of two variables (beyond our present capability to discuss). But we have not yet used the fact that there are 100 meters of fence. This yields the additional condition (called a *constraint*) that $2x + y = 100$. Solving for $y = 100 - 2x$ and substituting in $A = xy$, we find

$$A = x(100 - 2x) = 100x - 2x^2$$

This is a function of one variable, with domain $(0, 50)$. Hence we have reduced the problem to the purely mathematical one of finding the (global) maximum of

$$A = 100x - 2x^2, \; 0 < x < 50$$

Since the graph of A as a function of x is a parabola (Figure 2), we need not even use calculus; the vertex of the parabola will provide the answer. It is easier, however, to write

$$\frac{dA}{dx} = 100 - 4x = 4(25 - x) = 0 \qquad \text{when } x = 25 \text{ (critical point)}$$

$$\frac{d^2A}{dx^2} = -4 < 0 \qquad \text{(critical point yields a local maximum)}$$

The maximum is in fact global (why?), so the dimensions of the "critical rectangle" (the one with maximum area) are $x = 25$ and $y = 100 - 2x = 50$. The answer to the original question (what is the largest area that can be enclosed?) is

$$A = xy = (25)(50) = 1250 \text{ (square meters)} \qquad □$$

Figure 1
Rectangular field next to straight river

Figure 2
Graph of $A = 100x - 2x^2$

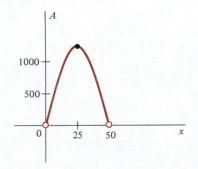

Example 1 illustrates some general principles that are worth noting.

1. To solve a maximum-minimum problem, begin by clearly determining what the question is. Some numerical measure is to be "maximized" or "minimized." Give it a name, label it by a letter, and think of it as a variable whose extreme values are involved in the solution of the problem.

2. Use whatever information is at your disposal (for example, geometrical formulas, scientific laws or relationships stated or implied in the problem) to express this variable in terms of one or more other variables. This defines a *function;* the problem is to investigate its extreme values.

3. At this stage of our treatment of calculus it is essential that our function should involve only one independent variable. Its dependence on that variable need not be explicit. (As we have seen earlier, it is possible to use "implicit differentiation" when the functional relationship is not explicit.) But conceptually at least, we must reduce the problem to the determination of the extreme values of a function of one variable. If several independent variables appear to be involved, use whatever "constraints" are available to eliminate all but one of them.

4. Be aware of the domain of the function. This is not always the set of real numbers for which the function is (mathematically) defined. It may be smaller due to physical limitations inherent in the problem itself.

5. When you have the function and its domain figured out, the problem is well on its way to solution. Discuss extreme values in the usual way by finding the critical points and investigating them. A graph of the function is often helpful; in any case the original question is usually answered when the appropriate extreme value is located.

□ **Example 2**   A manufacturer receives an order for oil cans that are to have a capacity of k cubic centimeters. Each can is made from a rectangular sheet of metal by rolling the sheet into a cylinder; the lids are stamped out from another rectangular sheet. What are the most economical proportions of the can?

Solution: We assume that the cost of a metal sheet is proportional to its area; hence we need only minimize the area of the two sheets to achieve the most economical can. In Figure 3 we have used the radius r and height h of the can to label the dimensions of the sheets. The first sheet is rolled into a cylinder, so its length is the circumference $2\pi r$. The second sheet must accommodate two lids of total diameter $4r$; its shorter dimension is one diameter, $2r$. The area of the sheets is therefore

$$A = 2\pi rh + 8r^2$$

where r and h are related by the constraint that the volume of the can is to be k cubic centimeters. The constraint is therefore

$$\pi r^2 h = k \qquad (r \text{ and } h \text{ in centimeters})$$

Figure 3
Sheets made into can and lids

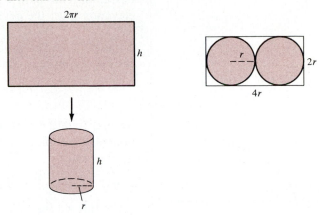

To minimize the area, we must express it in terms of one variable. That presents us with a choice. Shall we solve the constraint for h in terms of r, or vice versa? It is clearly easier to solve for h, so we write $h = k/\pi r^2$, from which

$$A = 2\pi r\left(\frac{k}{\pi r^2}\right) + 8r^2 = \frac{2k}{r} + 8r^2$$

The problem is to find the (global) minimum of A as a function of r in the domain $(0, \infty)$. (Realistically, of course, r cannot be allowed this much latitude. But we are studying a mathematical model, not the real thing.)

To minimize A, we seek critical values of r:

$$\frac{dA}{dr} = -\frac{2k}{r^2} + 16r = \frac{2(8r^3 - k)}{r^2}$$

$$= 0 \qquad \text{when } r = \frac{\sqrt[3]{k}}{2} \text{ (critical point)}$$

Since dA/dr changes sign from minus to plus at the critical point (why?), there is a local minimum at this point. It is in fact global in the domain $(0, \infty)$, because $A \to \infty$ when $r \to 0$ or $r \to \infty$ and there are no other critical points. (See Figure 4.)

The critical value of h is found (implicitly) by recalling that $h = k/\pi r^2$ and (at the critical point) $k = 8r^3$. Hence

$$h = \frac{8r^3}{\pi r^2} = \frac{8r}{\pi} = \frac{4}{\pi}(2r) \approx 1.27(2r)$$

that is, the height of the most economical can is about 1.27 times its diameter. Note that this is independent of k, the specified capacity. Hence it applies to any size can.   □

Figure 4
Graph of area in terms of radius

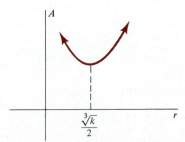
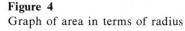

□ **Example 3**   A tracking device at $(-1,0)$ is kept trained on a missile descending the path $y^2 = x$. (Ground level is the x axis.) What angles of elevation must the device be capable of?

Solution: The problem is to find the extreme values of the angle θ shown in Figure 5. Since

$$\tan \theta = \frac{y}{1 + x} = \frac{y}{1 + y^2}$$

we may regard θ as a function of y (implicitly defined by the above equation). Differentiate with respect to y to obtain

$$(\sec^2 \theta)\frac{d\theta}{dy} = \frac{(1 + y^2) - y(2y)}{(1 + y^2)^2} = \frac{1 - y^2}{(1 + y^2)^2}$$

From this we see that

$$\frac{d\theta}{dy} = 0 \text{ when } y = 1 \qquad (\text{why not also } y = -1?)$$

a critical point that evidently yields maximum θ. Since $\tan \theta = \frac{1}{2}$ at $y = 1$, the maximum angle of elevation is $\theta \approx 27°$. The minimum angle is $\theta = 0°$ (occurring when $y = 0$).   □

□ **Example 4**   What is the maximum volume of a right circular cone inscribed in a sphere of radius a?

Solution: The volume of the cone (Figure 6) is $V = \frac{1}{3}\pi r^2 h$. To maximize this, we need a relation between r and h. From right triangle PQR in the figure, we find

$$a^2 = r^2 + (h - a)^2$$

and hence

$$V = \frac{1}{3}\pi[a^2 - (h - a)^2]h = \frac{\pi}{3}(2ah^2 - h^3)$$

$$\frac{dV}{dh} = \frac{\pi}{3}(4ah - 3h^2) = \frac{\pi h}{3}(4a - 3h)$$

$$= 0 \qquad \text{at } h = \frac{4}{3}a \text{ (critical point)}$$

(Why isn't $h = 0$ a critical point?) Since

$$\frac{d^2V}{dh^2} = \frac{\pi}{3}(4a - 6h) = -\frac{4\pi a}{3} < 0 \qquad (\text{at the critical point})$$

we have discovered a local (in fact global) maximum. Its value is

$$V = \frac{\pi}{3}h^2(2a - h) = \frac{16\pi a^2}{27}\left(2a - \frac{4}{3}a\right) = \frac{32\pi a^3}{81}$$

or $\frac{8}{27}$ of the volume of the sphere. (Confirm!)   □

Figure 5
Elevation of tracking device

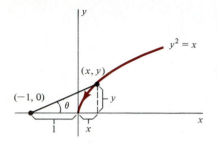

Figure 6
Cone inscribed in a sphere

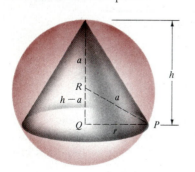

□ **Example 5**   An orchard owner has statistical records showing that if 25 apple trees are planted, each tree yields 500 apples (on the average), while the yield decreases about 10 apples per tree for each additional tree planted. How many trees should be planted for maximum total yield?

Solution: Let x be the number of trees planted in excess of 25. The yield per tree is then $500 - 10x$ and the total yield is

$$\begin{aligned}
y &= (25 + x)(500 - 10x) \\
&= 10(25 + x)(50 - x) \\
&= 10(1250 + 25x - x^2)
\end{aligned}$$

Since the owner can plant only a whole number of trees (and the yield is zero if 50 additional trees are planted), the domain of this function is $\{0, 1, 2, \ldots, 50\}$. Such a function cannot be differentiated (why?), but since it is a subset of the function defined by the same formula in the interval $[0,50]$, we can extend the domain and proceed as follows:

$$\frac{dy}{dx} = 10(25 - 2x) = 0 \qquad \text{at } x = \frac{25}{2} \text{ (critical point)}$$

$$\frac{dy}{dx} > 0 \text{ in } [0,\tfrac{25}{2}) \qquad \text{and} \qquad \frac{dy}{dx} < 0 \text{ in } (\tfrac{25}{2},50]$$

Hence the yield is increasing for $x = 0, 1, \ldots, 12$ and decreasing for $x = 13, 14, \ldots, 50$. The owner should plant 12 additional trees (37 altogether). He might also plant 13 additional trees; the same maximum is obtained.   □

□ **Example 6**   Mutt and Jeff are out of gas at A in Figure 7. Each thinks about how to reach B in the shortest possible time and then they set out simultaneously. Each is capable of hiking 3 mph through the woods and jogging 5 mph on the road.

Figure 7
What point D should Mutt head for?

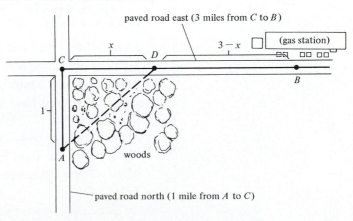

paved road east (3 miles from C to B)

(gas station)

paved road north (1 mile from A to C)

1. Mutt has decided to hike through the woods to some point D on the road and then jog on the road east to B. Find D to minimize his time.

At constant velocity the formula

$$\text{Distance} = \text{Rate} \times \text{Time} \qquad \left(\text{or Time} = \frac{\text{Distance}}{\text{Rate}}\right)$$

applies. So the time required for Mutt to go from A to D and then to B is

$$t = \frac{\sqrt{1 + x^2}}{3} + \frac{3 - x}{5} = \frac{1}{3}(1 + x^2)^{1/2} + \frac{1}{5}(3 - x)$$

The problem is to minimize t as a function of x in the domain $(0,3]$. (Why is the left endpoint omitted?) Hence we write

$$\frac{dt}{dx} = \frac{1}{3} \cdot \frac{1}{2}(1 + x^2)^{-1/2}(2x) + \frac{1}{5}(-1) = \frac{x}{3\sqrt{1 + x^2}} - \frac{1}{5} = 0$$

when
$$\frac{x}{3\sqrt{1 + x^2}} = \frac{1}{5}$$
$$5x = 3\sqrt{1 + x^2}$$
$$25x^2 = 9(1 + x^2)$$
$$16x^2 = 9$$
$$x = \frac{3}{4} \qquad \text{(critical point)}$$

Since the derivative

$$\frac{dt}{dx} = \frac{5x - 3\sqrt{1 + x^2}}{15\sqrt{1 + x^2}}$$

changes sign from minus to plus at $x = \frac{3}{4}$, we have found a local minimum, which is in fact global in $(0,3]$. (Why?) Mutt should head for a point $\frac{3}{4}$ of a mile east of C.

2. Jeff agrees that if Mutt insists on the woods, Mutt's answer is right. Nevertheless Jeff jogs north to C and east to B and gets there first.

Mutt: How long have you been here?

Jeff: Long enough to find out they're closed.

Mutt: You did 5 mph all the way?

Jeff: Yep. You have to look at time as a function of x in the right domain.

Mutt: Yeah, I know. There's a minimum at D. It took me 52 minutes.

Jeff: Yep. But I got here in 48.

Mutt: There's gotta be a reason.

Jeff: Yep.

Challenge to the Reader

Without looking ahead, explain how Mutt can be correct about his minimum time, yet Jeff got there sooner.

The explanation is to be found in the graph of t as a function of x in the *closed* domain $[0,3]$. (See Figure 8.) There is a discontinuity at $x = 0$, due to the fact that

$$\lim_{x \downarrow 0} \left(\frac{\sqrt{1 + x^2}}{3} + \frac{3 - x}{5} \right) = \frac{1}{3} + \frac{3}{5} = \frac{14}{15} \text{ hr} = 56 \text{ min}$$

while $t = 48$ min at $x = 0$. (When x is nearly 0, Mutt's route is virtually identical to Jeff's, but he is tramping through the woods at 3 mph while Jeff is doing 5 mph on the road.) It is true that $t = 52$ min is the global minimum of the time function in the domain $(0,3]$. But the global minimum in the domain $[0,3]$ is $t = 48$ min.   □

Figure 8
Discontinuity in time function

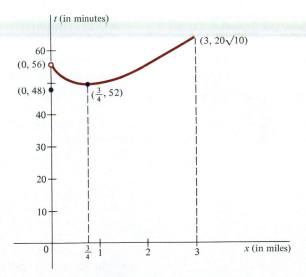

□ **Example 7**   Computer disk packs (Figure 9) contain information in units called *bytes* that are arranged in concentric circular tracks on each disk in the pack. (A typical disk is shown in Figure 10.) Manufacturing constraints limit the density to k bytes per centimeter along a given track, and p tracks per centimeter measured radially across the disk (where k and p are constants). If the number of bytes on each track must be the same (to achieve uniformity in reading the information), where should the innermost track be located to get the maximum number of bytes on the disk?

Figure 9
Structure of a disk pack

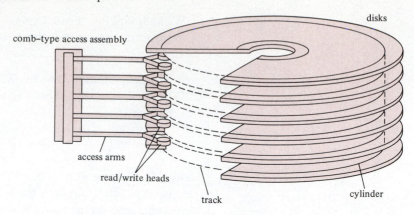

comb–type access assembly

disks

access arms

read/write heads

track

cylinder

Figure 10
Bytes on a computer disk

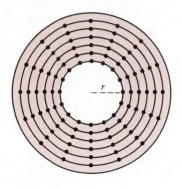

Solution: Since the number of bytes on each track is the same, the total number of bytes on the disk is

$$N = (\text{Number of bytes per track}) \times (\text{Number of tracks})$$

Let r be the radius of the innermost track; the problem is to maximize N by properly choosing r.

The length of the innermost track is $2\pi r$, so we can get $k(2\pi r)$ bytes on that track. (Other tracks have greater length, but since each track is supposed to have the same number of bytes, the innermost track determines the number of bytes per track.)

If R is the radius of the disk, the radial distance available for tracks is $R - r$. Hence the number of tracks we can put on the disk is $p(R - r)$. It follows that

$$N = k(2\pi r) \times p(R - r) = 2\pi k p(Rr - r^2)$$

Since

$$\frac{dN}{dr} = 2\pi k p(R - 2r) = 0 \qquad \text{when } r = \tfrac{1}{2}R$$

and since this choice of r yields a maximum (why?), the innermost track should be placed halfway out from the center of the disk.   □

Problem Set 5.5

1. A ball is thrown straight upward from the ground; t seconds later its height is $h = 48t - 16t^2$. What is the maximum height of the ball?

2. Find two numbers whose sum is 100 and whose product is maximum.

3. A manufacturer estimates that the profit from production of x units of a certain item is $P(x) = 40x - 0.02x^2 - 600$. How many units should be produced for maximum profit?

4. If x units of a certain item are manufactured, each unit can be sold for $200 - 0.01x$ dollars. How many units should be manufactured for maximum revenue? What is the corresponding unit price?

5. In Problem 4 suppose that the cost of manufacturing x units is $65x + 12,000$ dollars. How many units should be manufactured for maximum profit? What is the corresponding unit price?

6. In Problem 5 suppose that government taxes add $12 per unit to the cost of manufacturing x units. How many units should be manufactured for maximum profit? What is the corresponding unit price? Comparing this to the unit price in Problem 5, determine how much of the $12 unit tax is passed on to the consumer.

7. A rancher intends to fence off a rectangular region along a river (which serves as a natural boundary requiring no fence). If the enclosed area is to be 1800 square yards, what is the least amount of fence needed?

8. A rancher with 300 ft of fence intends to enclose a rectangular corral, dividing it in half by a fence parallel to the short sides of the corral. How much area can be enclosed?

9. A rectangular garden of area 75 ft² is bounded on three sides by a wall costing $8 per ft and on the fourth by a fence costing $4 per ft. What are the most economical dimensions of the garden?

10. An open box is made from a 16×16 cm piece of cardboard by cutting equal squares from each corner and folding up the sides. For maximum volume, what size squares should be cut out?

11. A cardboard box with a square base, rectangular sides, and no top is made to hold 4 ft³. What is the least amount of cardboard needed to make such a box?

12. A box with a square base, rectangular sides, and no top is to be made from 27 ft² of cardboard. What is the largest volume of such a box?

13. The Post Office will not accept a package whose combined length and girth (perimeter of a cross section perpendicular to the length) exceeds 84 in. What are the length and girth of the cylindrical tube of largest volume that can be mailed?

14. Repeat Problem 13 for a rectangular box with square cross sections perpendicular to its length.

15. At 12:00 noon a submarine 60 km north of a harbor and a frigate 50 km east of the harbor are both heading for the harbor, at 20 km/hr and 10 km/hr, respectively. When will they be closest to each other?

16. A rectangle with base on the x axis has its upper vertices on the curve $y = 3 - x^2$.
   (a) Find the maximum area of such a rectangle.
   (b) Does a rectangle of minimum area exist? What is the range of possible areas?

17. Repeat Problem 16 with "perimeter" in place of "area."

18. Show that the rectangle of largest area having a prescribed perimeter is a square.

19. Show that the rectangle of largest area that can be inscribed in a given circle is a square.

20. A rectangle is inscribed in a semicircular region of radius a (base on the diameter, upper vertices on the semicircle). Show that the maximum area of such a rectangle is a^2. Is this maximal rectangle a square?

21. In Problem 20 show that the maximum perimeter of the rectangle occurs when the ratio of base to altitude is 4.

22. A rectangular playground is to be placed inside an oval track in the shape of the ellipse $x^2/a^2 + y^2/b^2 = 1$. Show that the maximum area of the playground is $2ab$, and find its dimensions in terms of a and b.

23. A triangle is formed by the positive coordinate axes and a straight line through $(1,2)$. How large an area can such a triangle have? What is the minimum area?

24. A line segment passing through $(1,8)$ has endpoints $(a,0)$ and $(0,b)$, where a and b are positive. Find the shortest such segment.

25. A rectangular sheet of metal of fixed perimeter is rolled into a cylinder. What dimensions give maximum volume?

26. A right circular cylinder is inscribed in a sphere of radius a. Show that its maximum volume is

$$\frac{1}{\sqrt{3}}\left(\frac{4}{3}\pi a^3\right) = \frac{1}{\sqrt{3}} \times \text{(volume of the sphere)}$$

27. A right circular cylinder is inscribed in a right circular cone of fixed dimensions. Show that its maximum volume is $\frac{4}{9}$ the volume of the cone.

28. A metal can is made to hold k cubic centimeters. If "most economical" is understood to mean "minimum surface area" (including lids), show that the ratio of altitude to diameter of the most economical can is 1. (In Example 2, where some metal is wasted in making the lids, the ratio is $4/\pi \approx 1.27$.)

29. A cylindrical cup (no top) is made to hold k cubic centimeters. Show that the ratio of altitude to diameter of the most economical cup (the one with minimum surface area) is $\frac{1}{2}$.

30. Find the point on the unit circle $x^2 + y^2 = 1$ that is nearest the point $(2,1)$. Can you do this without calculus?

31. An observer at $(1,0)$ is watching an object descend the path $y^2 = x$. (Ground level is the x axis.) At what point of its path is the object closest to the observer?

32. What if the path of the object in Problem 31 is $y^2 = 2x$? $y^2 = 3x$?

33. A birdwatcher driving along the road $y = x^2$ spots a bird at $(1,0)$.

(a) To find the point of the road closest to the bird, what equation should be solved?

(b) Use Newton's Method to approximate the coordinates of the nearest point.

34. The strength of a rectangular wooden beam (of given length) varies directly as the product of its width and the square of its depth. ($S = kwh^2$, where k is a constant.) Show that the ratio of depth to width of the strongest beam that can be cut from a given circular log is $\sqrt{2}$. (See Figure 11.)

Figure 11
Beam cut from a log

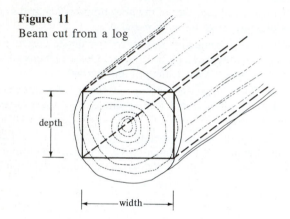

depth

width

35. The *stiffness* of the beam in Problem 34 varies directly as the product of its width and the *cube* of its depth. Show that the ratio of depth to width of the stiffest beam that can be cut from a given circular log is $\sqrt{3}$.

36. A wire of given length can be used to make a circle, or a square, or can be cut into two pieces to make both a circle and a square.

(a) How much of the wire should be used for the circle if the total enclosed area is to be a minimum?

(b) What if the enclosed area is to be a maximum?

37. A powerhouse is on one edge of a straight river and a factory is on the other edge, 100 meters downstream. The river is 50 meters wide. It costs $10 per meter to run electric cable across the river and $7 per meter on land.

(a) How should the cable be installed to minimize its cost?

(b) A local ordinance prohibits any cable on land near the factory and permits only 50 meters on land near the powerhouse. How does this affect the answer to part (a)?

38. Agent 00π is waiting at the edge of a straight canal 1 mile wide, in a motorboat capable of going 40 mph. There is a straight road along the opposite edge of the canal; her partner will have a 50 mph motorcycle waiting for her wherever she lands. At midnight she will receive a package to be delivered to a man in a Mercedes-Benz 5 miles down the road. (See Figure 12.)

Mission: Deliver the package in the shortest possible time.

Problem: Where should her partner park the motorcycle?

Figure 12
The spy who came across the canal

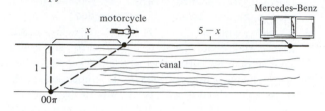

39. Thelma is driving north on an old road, trying to find a telephone. (See Figure 13.) The road becomes impassable at A, where a sign says that it is 2 miles to Highway Q (which Thelma knows goes east for 3 or 4 miles to Sleepy Hollow). She figures that she can hike to B at 4 mph and jog to Sleepy Hollow at 5 mph. Find x to minimize her time. (*Note:* k is the unknown distance from C to Sleepy Hollow. Besides being a cross-country champion, Thelma is good at calculus and doesn't panic merely because k is unknown.)

Figure 13
The road to Sleepy Hollow

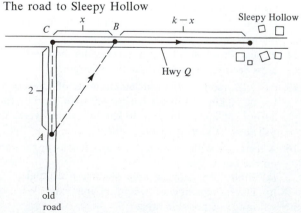

40. Suppose that the distance from C to B in the Mutt and Jeff problem (Example 6) is unknown, say k miles.

    (a) If $k > \frac{3}{4}$, explain why the answer to Mutt's minimum time problem is still $x = \frac{3}{4}$. Is it still true that Jeff gets there first?

    (b) What is Mutt's minimum time if $0 < k \leq \frac{3}{4}$ and how is it achieved? Is Jeff still the winner?

41. Fermat's Principle of Optics states that light travels from one point to another in such a way as to minimize the time.

    (a) In Figure 14 we show a source of light, a reflecting surface, and the path of a ray of light from source to observer. Assuming that the velocity of light is constant, show that $\theta_1 = \theta_2$. (The angle of incidence equals the angle of reflection.)

    (b) In Figure 15 we show a source of light, a refracting surface marking the interface between two media in which the velocity of light is v_1 and v_2, respectively, and the path of a ray of light from source to observer. Prove *Snell's Law of Refraction,*

$$\frac{\sin \theta_1}{\sin \theta_2} = \frac{v_1}{v_2}$$

Figure 14
Reflection of light

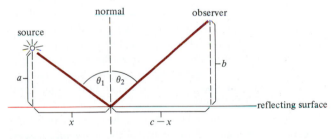

Figure 15
Refraction of light

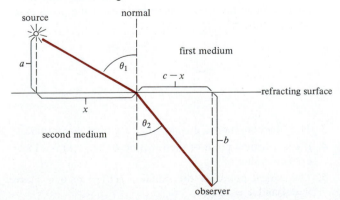

42. A light is to be hung above the center of a circular table (of radius 1) so as to give maximum illumination at the edge (Figure 16). If the illumination varies directly as $\sin \theta$ and inversely as d^2 ($I = k \sin \theta / d^2$), how high above the table should the light be placed?

Figure 16
"Fiat lux"

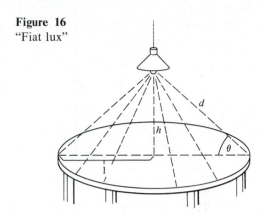

43. The illumination at a point due to a source of light varies directly as the strength of the source and inversely as the square of the distance from the source. Two sources of strengths a and b are c units apart. At what point on the line segment joining them is the illumination the least?

44. A picture 4 ft high is hung on a wall with its lower edge 2 ft above eye level. How far back from the wall should an observer stand to get the best view? ("Best view" means maximum θ as shown in Figure 17.) *Hint:* Maximize $\tan \theta$ instead of θ. (Why are the critical points the same?)

Figure 17
Picture at an exhibition

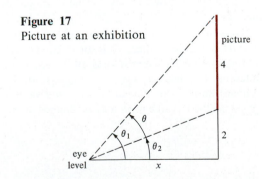

45. Two hallways in the physical education building meet at right angles as shown in Figure 18. A pole vaulter is carrying a pole (horizontally) around the corner.

    (a) Show that the critical length of the pole (for successful negotiation of the corner) is $(a^2 + b^2)^{3/2}$. *Hint:* Express things in terms of θ.

(b) (Easy arithmetic) If the hallways have widths 27″ and 64″, will a 10-ft pole get stuck?

(c) (For a calculator) If the hallways have widths 3 ft and 4 ft, will a 10-ft pole get stuck?

Figure 18
Will the pole get stuck?

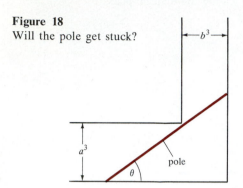

46. Suppose that n measurements $x_1, \ldots, x_n$ are made of a quantity whose true measure x is not known. The errors are then $|x - x_k|, k = 1, \ldots, n$. Statisticians often define the "best" value we can assign x to be that value for which the sum of the squares of the errors is a minimum. Show that this best value is simply the average of the measurements,

$$x = \frac{1}{n}(x_1 + \cdots + x_n)$$

(Note the metaphysical assumptions in the statement of this problem! We are assuming that some "true measure" exists. One might argue that it does not, that all we can ever do in reality is construct a "best" value based on the actual measurements.)

47. Suppose that x and y are known to be related by an equation of the form $y = mx$, where m is an unknown constant. To determine m, a scientist runs an experiment n times, obtaining the data $(x_1, y_1), \ldots, (x_n, y_n)$. Ideally, each y_k should be mx_k; its deviation from that value is $|mx_k - y_k|$. The "best" value of m may be defined as that

value for which the sum of the squares of the deviations is a minimum.

(a) Show that this best value is

$$m = \frac{x_1 y_1 + \cdots + x_n y_n}{x_1^2 + \cdots + x_n^2}$$

(b) If the experimentally determined data are $(0, 0.02)$, $(1, 1.71)$, $(2, 3.47)$, $(3, 5.11)$, $(4, 6.87)$, what is m? What would you predict y to be if $x = 5$?

48. It costs a bus company \$125 to run a bus on a certain tour, plus \$15 per passenger. The capacity of the bus is 20 persons and the company charges \$35 per ticket if the bus is full. For each empty seat, however, the company increases the ticket price by \$2. For maximum profit, how many empty seats would the company like to see?

49. It costs $25 + v^3 \times 10^{-4}$ dollars per hour to run a bus at v mph. Find the speed for the trip from Minneapolis to Chicago that will minimize the cost.

50. A cow weighing 300 pounds gains 8 pounds per day and costs 60 cents a day to keep. The market price for cows is 75 cents per pound, but is falling 1 cent per day. When should the cow be sold?

51. A manufacturer has fixed costs of \$100 per day. Production costs are 5 cents for each item manufactured, while miscellaneous additional costs come to $2x^2$ cents per day (where x is the number of items produced per day). What production level will minimize the cost per unit?

52. Let $C(x)$ be the (differentiable) cost of producing x units of a certain commodity. To minimize the average cost per unit, we set its derivative equal to zero. Show that this occurs when the marginal cost $C'(x)$ and the average cost per unit are the same.

53. If $R(x)$ is the revenue derived from the sale of x units of a certain commodity, $R'(x)$ is called the *marginal revenue*. The profit is $P(x) = R(x) - C(x)$, where $C(x)$ is the cost of producing x units. Why do economists say that marginal revenue must equal marginal cost if the profit is to be a maximum?

True–False Quiz

1. If the pressure of a gas is inversely proportional to its volume and is changing at a constant rate, the volume changes at a rate that is inversely proportional to the pressure.

2. Global extreme values are also local.

3. The function $f(x) = 2 - \sqrt{x}$ has a global maximum.

4. If c is a critical point of the function f, then $f(c)$ is a local extreme value of f.

5. The critical points of the function $f(x) = x\sqrt{6 - x}$ are $x = 4$ and $x = 6$.

6. The First Derivative Test shows that the critical point $x = 0$ of the function $f(x) = 1 - x^3$ yields neither a maximum nor a minimum.

7. The Second Derivative Test shows that the critical point $x = 0$ of the function $f(x) = 1 - x^3$ yields neither a maximum nor a minimum.

8. If $f'(c) = f''(c) = 0$, then $f(c)$ cannot be an extreme value of f.

9. The graph of $y = x^2/(x^2 + 1)$ has no asymptotes.

10. The y axis is an asymptote of the graph of $y = \csc x$.

11. The graph of $y = x/(x - 1)$ is symmetric about the line $y = x$.

12. The graph of $y = x - \sin x$ has infinitely many points of inflection.

13. The sum of any positive number and its reciprocal is at least 2.

14. The rectangle of maximum perimeter having a prescribed area is a square.

Additional Problems

1. A rectangular pool (with a horizontal bottom) is being filled with water at the rate of 10 ft³/min. If the length and width of the pool are 50 ft and 30 ft, how fast is the water level rising?

2. On a circular radar screen of diameter 8 inches a radial pointer turns at the rate of 2 revolutions per second. How fast does it sweep out area on the screen? *Hint:* The area of a circular sector of radius r and central angle θ is $\frac{1}{2}r^2\theta$.

3. A ship leaves port at noon, sailing north at 20 knots. A second ship leaves an hour later, sailing east at 30 knots. How fast is the distance between them changing at 2:00 P.M.?

4. Each side of an equilateral triangle is increasing at the rate of 6 cm/sec. How fast is the area changing when the sides are each 10 cm long?

5. One side of a triangle is increasing at 10 ft/sec and another is decreasing at 6 ft/sec. Their included angle is decreasing at 2 radians per second. At the instant when the sides have lengths 9 and 12, respectively, and the angle is 60°, how fast is the area changing? Is it increasing or decreasing?

6. Show that if the surface area of a sphere changes at a constant rate, the volume changes at a rate proportional to the radius.

7. At a certain instant an icicle in the shape of a right circular cone is 12 cm long and its length is increasing at the rate of 0.5 cm/hr, while the radius of its base is 1 cm and is decreasing at the rate of 0.05 cm/hr. Is the volume of the icicle increasing or decreasing at that instant? at what rate?

8. Water is entering a conical reservoir 10 meters deep and 20 meters across the top at 9 cubic meters per minute. How fast is the water rising when it is 3 meters deep?

9. A conical water tank 10 ft deep and 6 ft across the top is leaking. If the water level is falling at the rate of 2 ft/hr when the water is 3 ft deep, how fast is the water leaking at that instant?

10. A ball is moving along the curve $y^2 = 6x$. (Ground level is the x axis, the y axis points upward, and the sun is directly overhead.) At the instant when $x = 2$ the distance between the ball and the origin is decreasing at the rate of 5 units per second. How fast is the ball's shadow on the ground moving at that instant? in what direction? How fast is the altitude of the ball changing at that instant?

11. A rider on a Ferris wheel that is turning at 5 revolutions per minute casts a shadow on the ground directly beneath him. The radius of the wheel is 30 ft.

   (a) How fast is the shadow moving when the rider is at the top of the wheel? when he is halfway down?

   (b) How fast is the rider's altitude changing when he is at the top of the wheel? when he is halfway down?

12. A weight hangs from one end of a 25-ft rope which runs over a pulley 12 ft above the ground. The other end of the rope is pulled (at ground level) away from the pulley at the rate of 5 ft/sec. How fast is the weight rising when it is 7 ft off the ground?

13. A 15-ft ladder leans against a vertical wall. Its lower end slides away from the wall at the rate of 5 ft/sec. How fast is the (acute) angle between the ladder and the wall changing at the instant when it is 60°?

14. Show that a particle whose position at time t is $s = 3t^4 - 4t^3 + 3$ is always at least 2 units from the origin.

Sketch the graph of each of the following functions by discussing its significant features and then plotting a judicious selection of points.

15. $y = 4x^3 - 3x^2 + 1$

16. $y = x^3 - 3x^2 - 9x + 12$

17. $y = 3x^4 + 4x^3 - 12x^2 + 15$

18. $y = x^2(x - 2)^2$ **19.** $y = \dfrac{4x}{4 + x^2}$ **20.** $y = \dfrac{x^2 - 1}{x^2 + 1}$

21. $y = \dfrac{x}{\sqrt{x^2 + 1}}$ **22.** $y = x^2 - \dfrac{2}{x}$ **23.** $y = \dfrac{x}{1 - x^2}$

24. $y = \dfrac{x^2 - 1}{x^2 - 2x}$ **25.** $y = x^2 - 3x^{2/3} + 2$

26. $y = 2x^{3/2}(2 - x)$

27. $y = \tan x - \cot x,\ 0 < x < \pi/2$

28. $y = \csc x - \cot x,\ 0 < x < \pi$

29. Sketch the graph of the relation $y^2 = x^2(1 - x^2)$.

30. Sketch the graph of the relation $y^2 = \dfrac{x(1 + x)}{1 - x}$.

31. Sketch the graph of $y = x^4 - 4x^3 + 15$, including all turning points and points of inflection.

32. The graph in Problem 31 shows that the equation $x^4 - 4x^3 + 15 = 0$ has two positive solutions.

    (a) (Easy arithmetic) Use the graph to approximate the smaller solution and apply Newton's Method to improve the approximation.

    (b) (For a calculator) Do part (a) for the larger solution.

33. Two differentiable functions are positive for all x and have a local maximum at $x = c$. Prove that their product also has a local maximum at $x = c$.

34. Use calculus to prove that a quadratic function $y = ax^2 + bx + c\ (a > 0)$ is nonnegative for all x if and only if $b^2 - 4ac \leq 0$.

35. Do the preceding problem without calculus.

36. Find two positive numbers whose product is 100 and whose sum is minimum.

37. Find the most economical dimensions of a box of prescribed capacity with square top and bottom and rectangular sides.

38. A rectangular box with square base and no top is made from 12 square feet of cardboard. What are the dimensions of the box of largest volume?

39. An open box is made from a rectangular piece of cardboard (of dimensions 8 cm and 15 cm) by cutting square

pieces from each corner and bending up the sides. What is the maximum volume of such a box?

40. A practice field is to be laid out in the form of a rectangle with semicircular areas at each end. The track coach wants four laps around the field to be a mile and the physical education department wants the rectangular area to be as large as possible. Will the track team be able to run the 100-yd dash along a straight part of the track?

41. The rate at which a disease spreads in a certain town is proportional to the product of the number of infected people and the number of people not yet infected. Show that the disease is spreading most rapidly when half the town is infected.

42. An open cylindrical storage tank has volume 500 ft³. Its inside surface (walls and bottom) require annual coating with an expensive material, one gallon of which covers 100 ft². How many gallons are needed each year if the tank has been constructed to minimize the cost of coating?

43. A castle with vertical sides is surrounded by a moat and wall. The distance from the castle across the moat to the outside of the wall is a and the height of the wall is b. How long a ladder is needed to reach from the ground outside the wall to the side of the castle?

44. A farmer's house and barn are a meters and b meters, respectively, from a straight stream, and are c meters apart. (See Figure 1.) Show that the shortest path from the house to the stream and then to the barn (to fetch water for the animals) is to a point of the stream satisfying $\theta_1 = \theta_2$.

Figure 1

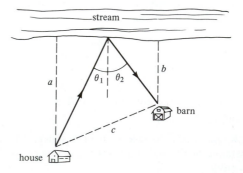

45. If the farmer in Problem 44 walks at speed v_1 to the stream and at speed v_2 away from it, show that the quickest route is to a point of the stream satisfying $v_1/v_2 = \sin \theta_1/\sin \theta_2$.

46. An observer at $(2,0)$ is watching an object descend the path $y^2 = x$. (Ground level is the x axis.) At what point of its path is the object closest to the observer?

47. A company producing x units of a certain item can sell them all at a price of $50 - 0.02x$ dollars per unit; it costs $1000 + 10x$ dollars to produce x units. How many units should be produced for maximum profit and what is the corresponding price per unit?

48. Answer the same questions as in Problem 47 assuming that the government imposes a tax of \$4 per unit. For maximum profit, how much of the tax does the company pass on to the consumer?

49. In Problem 48 suppose that the government not only imposes a tax of \$4 per unit, but also insists on increased production. At what level of production does the company start to lose money?

50. A company producing x units of a certain item can sell them all at a price of $a - bx$ dollars per unit; it costs $c + dx$ dollars to produce x units. Show that the price per unit should be $\frac{1}{2}(a + d)$ dollars for maximum profit.

51. In Problem 50 assume that the government imposes a tax of T dollars per unit and the company adjusts the price to continue to maximize its profit. How much of the tax is passed on to the consumer?

6 | The Integral

The derivative was introduced in Chapter 2 as a concept unifying the solutions of two equivalent problems. One (the problem of tangents) was purely geometrical; the other (velocity of a moving object) was a question from physics. As you have seen in the last three chapters, the theoretical and practical significance of the derivative goes far beyond these problems. It is a powerful tool with numerous applications.

Now we turn to the second great idea of calculus, the *integral*. We will introduce it in much the same way, as a mathematical concept that arises from the solution of apparently unrelated problems. Its remarkable connection with the derivative will be developed in the *Fundamental Theorem of Calculus,* a seventeenth-century discovery that is the keystone of the arch we are building. In the chapters to follow you will see the integral used for an even wider variety of applications (if possible) than the derivative. Taken together, the derivative and integral are what calculus is about.

6.1
TWO EQUIVALENT PROBLEMS

Area Under a Curve

Most people have an intuitive idea of what is meant by *area* (of a bounded region in the plane). If the boundary of the region is *polygonal* (consisting of straight pieces as shown in Figure 1), the area is readily calculated by cutting the region into triangular pieces. The problem is harder when the boundary is not polygonal; in that case we normally resort to approximations. We might, for example, cover the region with a grid of perpendicular lines, add up the areas we can compute, and ignore the rest. (See Figure 2.) It is

Figure 1
Polygonal region divided into triangles

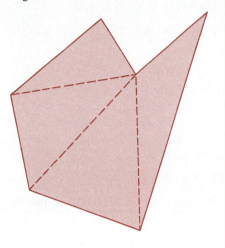

Figure 2
Approximation of area

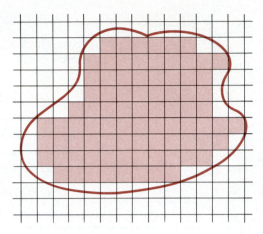

Figure 3
Region under the graph of a non-negative continuous function

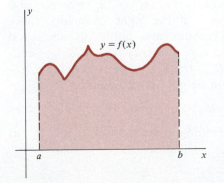

reasonable to expect that a limiting process of some kind (involving a refinement of the grid) is needed to find the exact answer. You might recall from geometry that such a process is used to derive the formula $A = \pi r^2$ for the area of a circle.

Most people would have no doubt that there *is* an exact answer! It usually comes as a shock to learn that some regions have such complicated boundaries as to be "unmeasurable," and that even a definition of area (in those cases where it exists) requires a fairly sophisticated mathematical apparatus.

We are going to simplify the problem by allowing only part of the boundary of the region to be curved. More precisely, let f be a (real) function continuous in the (finite) closed interval $I = [a,b]$, and suppose that $f(x) \geq 0$ for all $x \in I$. Then the region bounded by the graph of $y = f(x)$, the x axis, and the vertical lines $x = a$ and $x = b$ is like that shown in Figure 3. We refer to this as the "region under the curve," assuming that the straight parts of the boundary are clear from the context.

What we want to discuss is the area of this region, the "area under the curve." Even if you think you know what that means, try explaining it to somebody who does not! Intuition is good enough as far as it goes, but how do we come up with a *number?* The need for a definition is apparent.

□ **Example 1**   The region under the graph of

$$f(x) = x^2 + 1 \qquad (0 \leq x \leq 2)$$

is shown in Figure 4. One way to approximate its area is to "partition" the interval [0,2] and use rectangles based on the subintervals of the partition. We have shown four such rectangles in the figure; their heights are the

Figure 4
Overestimation of area

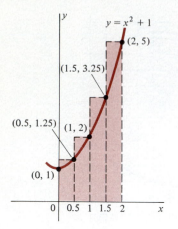

functional values $f(0.5) = 1.25$, $f(1) = 2$, $f(1.5) = 3.25$, $f(2) = 5$. Their widths are all the same, namely 0.5. The sum of their areas is

$$(1.25)(0.5) + (2)(0.5) + (3.25)(0.5) + (5)(0.5) = 5.75$$

a number which is clearly too large (because the region is contained in the rectangles).   □

□ **Example 2**   The approximation in Example 1 is too large because we chose the height of each rectangle to be the maximum value of f in the corresponding subinterval. In Figure 5 we show rectangles whose heights are *minimum* values of f, namely $f(0) = 1$, $f(0.5) = 1.25$, $f(1) = 2$, $f(1.5) = 3.25$. The sum of the rectangular areas is

$$(1)(0.5) + (1.25)(0.5) + (2)(0.5) + (3.25)(0.5) = 3.75$$

This number is too small, but at least we have the area boxed in, that is, $3.75 \leq A \leq 5.75$.   □

Figure 5
Underestimation of area

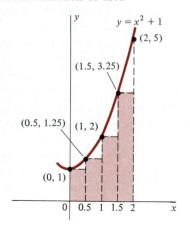

In Examples 1 and 2 we used extreme values of f on each subinterval to form the *upper sum* $U = 5.75$ and the *lower sum* $L = 3.75$. The same procedure with a finer partition (eight subintervals instead of four, for example) yields $U = 5.1875$ and $L = 4.1875$. It ought to be clear that any refinement of the partition will cause the upper sum to decrease (because it is too large, and gets closer to the true area as the partition is refined). Similarly, the lower sum will increase. No matter what partition is used, however, the area is always boxed in between the lower sum and upper sum:

$$L \leq A \leq U \qquad \text{for every partition}$$

To find the exact value of A, why not partition the interval into n subintervals (where n is arbitrary) and investigate the limit of these sums as n increases? (Equivalently, we may let $\Delta x = 2/n$ be the length of each subinterval and investigate the limit as $\Delta x \to 0$.) It is a simple idea, but the details are complicated; we have put them in an optional note at the end of the section. All we need here are the results, which are interesting.

Let $I = [0,2]$ be divided into n subintervals, each of length $\Delta x = 2/n$. (See Figure 6.) Label the points of subdivision $x_0, x_1, \ldots, x_n$ (where $x_0 = 0$ is the left endpoint of I and $x_n = 2$ is the right endpoint). Since $f(x) = x^2 + 1$ is increasing in I, its maximum value in the kth subinterval (where k is any one of the integers $1, 2, \ldots, n$) occurs at x_k, the right endpoint of the subinterval. Hence it is $M_k = f(x_k)$. The area of the taller rectangle shown in Figure 6 (the one with height M_k) is therefore

$$M_k \, \Delta x = f(x_k) \, \Delta x$$

The sum of all such rectangles (corresponding to the values $k = 1, 2, \ldots, n$) is the upper sum

$$U = M_1 \, \Delta x + M_2 \, \Delta x + \cdots + M_n \, \Delta x$$
$$= f(x_1) \, \Delta x + f(x_2) \, \Delta x + \cdots + f(x_n) \, \Delta x$$

Figure 6
Approximation of area by n rectangles

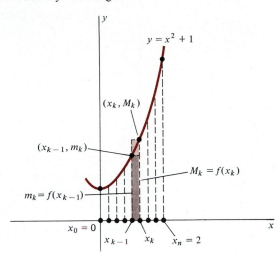

It turns out (see the note at the end of this section) that

$$U = \frac{4}{3}\left(1 + \frac{1}{n}\right)\left(2 + \frac{1}{n}\right) + 2$$

Similarly, the minimum value of f in the kth subinterval is $m_k = f(x_{k-1})$, the value of f at the left endpoint. Hence the lower sum is

$$L = m_1\,\Delta x + m_2\,\Delta x + \cdots + m_n\,\Delta x$$
$$= f(x_0)\,\Delta x + f(x_1)\,\Delta x + \cdots + f(x_{n-1})\,\Delta x$$

which turns out to be

$$L = \frac{4}{3}\left(1 - \frac{1}{n}\right)\left(2 - \frac{1}{n}\right) + 2$$

The area A is boxed in between L and U, that is,

$$\frac{4}{3}\left(1 - \frac{1}{n}\right)\left(2 - \frac{1}{n}\right) + 2 \le A \le \frac{4}{3}\left(1 + \frac{1}{n}\right)\left(2 + \frac{1}{n}\right) + 2$$

The intriguing thing about this is that it is true for every n ($n = 1, 2, 3, \ldots$). Since $1/n$ becomes negligible as n increases, it is apparent that A is the common limit of the upper and lower sums:

$$A = \frac{4}{3}(1)(2) + 2 = \frac{14}{3}$$

While this result is impressive (particularly if we plow through the omitted details!), a theoretical point needs clearing up. We said earlier that the problem is to *define A*, yet in the above discussion we took the position that it was already defined. Not that we knew its value, but we kept referring to it as something definite that was being overestimated, underestimated, boxed in, and so on. A skeptic would doubt the very existence of A! To satisfy such an objection, we purge our language of any reference to A until it is obvious that it exists (and is unique).

There is no doubt that the upper sums

$$U = \frac{4}{3}\left(1 + \frac{1}{n}\right)\left(2 + \frac{1}{n}\right) + 2$$

decrease as n gets larger, and the lower sums

$$L = \frac{4}{3}\left(1 - \frac{1}{n}\right)\left(2 - \frac{1}{n}\right) + 2$$

increase. Moreover, they decrease and increase to a common limit. The construction of these sums did not require any reference to the area under the curve, but only to the areas of rectangles based on the subintervals of our partition. Once the sums are constructed and simplified, however, it is clear that there *is* a number A such that

$$\frac{4}{3}\left(1 - \frac{1}{n}\right)\left(2 - \frac{1}{n}\right) + 2 \leq A \leq \frac{4}{3}\left(1 + \frac{1}{n}\right)\left(2 + \frac{1}{n}\right) + 2$$

for every n. Moreover, there is only one such number, namely $A = \frac{14}{3}$. Our skeptic would have to admit all that; we would then offer our definition. The *area under the curve* (hitherto vague, possibly even nonexistent) is the number $A = \frac{14}{3}$.

You should recall that we did much the same thing in Section 2.1, when we discussed the problem of tangents. The idea of slope of a curve seemed reasonable enough on intuitive grounds. But it was not actually defined until we had evaluated a limit.

To carry out this kind of analysis in general, we need some terminology and notation.

Let $I = [a,b]$ be a (finite) closed interval ($a < b$). A *partition* of I is a set of points

$$P = \{x_0, x_1, \ldots, x_n\}$$

such that $a = x_0 < x_1 < \cdots < x_{n-1} < x_n = b$, where n is a positive integer. These points divide I into a union of n *subintervals*

$$[x_{k-1}, x_k] \qquad k = 1, 2, \ldots, n$$

the lengths of which are denoted by

$$\Delta x_k = x_k - x_{k-1} \qquad k = 1, 2, \ldots, n$$

(or simply Δx if all the subintervals have the same length). The *norm* of the partition is the largest of these lengths, denoted by $\|P\|$.

▫ **Example 3**   A partition of $I = [-1,3]$ is

$$P = \{-1, 1, 1.5, 2, 3\}$$

Figure 7
Partition of an interval

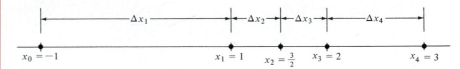

with norm $\|P\| = \Delta x_1 = 2$. (See Figure 7.) Note that the norm is a measure of the "fineness" of the partition (like the mesh of a screen, the largest aperture of which determines what size bug can get in). If $\|P\| < \delta$, every subinterval has length $\Delta x_k < \delta$; when δ is "small," the partition is "fine." In other words a partition with a small norm has a large number of points that are close together.   □

If $a_1, a_2, \ldots, a_n$ are real numbers, their sum is often abbreviated by writing

$$\sum_{k=1}^{n} a_k = a_1 + a_2 + \cdots + a_n$$

(The symbol Σ is the uppercase Greek letter **sigma**.) The letter k is called the **index of summation.**

In the problem set we will ask you to explain the following properties of the summation notation. They are based on ordinary algebraic laws, and may well strike you as obvious:

1. $\displaystyle\sum_{k=1}^{n} (a_k + b_k) = \sum_{k=1}^{n} a_k + \sum_{k=1}^{n} b_k.$

2. If $c \in \mathcal{R}$, then $\displaystyle\sum_{k=1}^{n} ca_k = c \sum_{k=1}^{n} a_k.$

3. If $c \in \mathcal{R}$, then $\displaystyle\sum_{k=1}^{n} c = nc.$

□ **Example 4**   Confirm that $\sum_{k=1}^{4} (2k^2 + 3) = 2 \sum_{k=1}^{4} k^2 + 4 \cdot 3$.

Solution: By definition,

$$\sum_{k=1}^{4} (2k^2 + 3) = (2 \cdot 1^2 + 3) + (2 \cdot 2^2 + 3) + (2 \cdot 3^2 + 3) + (2 \cdot 4^2 + 3)$$

$$= 2(1^2 + 2^2 + 3^2 + 4^2) + (3 + 3 + 3 + 3)$$

$$= 2 \sum_{k=1}^{4} k^2 + 4 \cdot 3$$

The same result is obtained from the above properties of summation notation:

$$\sum_{k=1}^{4} (2k^2 + 3) = \sum_{k=1}^{4} 2k^2 + \sum_{k=1}^{4} 3 = 2\sum_{k=1}^{4} k^2 + 4 \cdot 3 \qquad \square$$

The reason for introducing summation notation at this time is to simplify the writing of upper and lower sums. The following discussion will show what we mean. Suppose that f is continuous and nonnegative in $I = [a,b]$ and let

$$P = \{x_0, x_1, \ldots, x_n\}$$

be a partition of I. For each $k = 1, 2, \ldots, n$ let M_k and m_k be the maximum and minimum, respectively, of $f(x)$ in the kth subinterval $[x_{k-1}, x_k]$. (These extreme values are guaranteed to exist by Theorem 2, Section 4.2.)

It is geometrically apparent that if ΔA_k is the area of the shaded region in Figure 8, then

$$m_k \, \Delta x_k \le \Delta A_k \le M_k \, \Delta x_k$$

(For the moment, overlook the fact that we have no right to talk about this area!) Adding the n rectangular areas $M_k \, \Delta x_k$ ($k = 1, 2, \ldots, n$), we obtain the *upper sum* associated with P, namely

$$U(P) = M_1 \, \Delta x_1 + M_2 \, \Delta x_2 + \cdots + M_n \, \Delta x_n = \sum_{k=1}^{n} M_k \, \Delta x_k$$

Similarly, the *lower sum* associated with P is

$$L(P) = m_1 \, \Delta x_1 + m_2 \, \Delta x_2 + \cdots + m_n \, \Delta x_n = \sum_{k=1}^{n} m_k \, \Delta x_k$$

The sum

$$A = \Delta A_1 + \Delta A_2 + \cdots + \Delta A_n = \sum_{k=1}^{n} \Delta A_k$$

is the area under the curve and it is clear from Figure 9 that

$$L(P) \le A \le U(P)$$

Note that the partition P is arbitrary; hence these inequalities are true *for every P*.

Of course this begs the question, since it is A that we are trying to define. Apart from pictures, how do we know that such a number exists? It is not hard to prove that $L(P) \le U(P)$. It is also true (as we ask you to show in the problem set) that as the partition is refined the lower sums increase and the upper sums decrease. (More precisely, since in some circumstances they may not change, the lower sums are nondecreasing and the upper sums are nonincreasing.) The question is whether they have limits and if so, whether the limits are the same. The answer is yes, but the proof is not easy; we leave it for a more advanced course. Meanwhile we state for the record that a number A *does* exist such that

$$L(P) \le A \le U(P) \qquad \text{for every } P$$

Figure 8
Approximation of area on typical subinterval

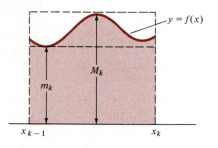

Figure 9

$L(P) \leq A \leq U(P)$

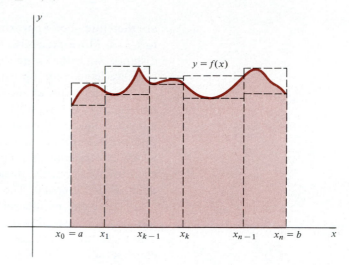

and that this number is unique. In a broader context than our present discussion of area, this statement is called the *Existence and Uniqueness Theorem*. Assuming its truth, we may *define* the area under the curve to be the number A.

Thus the problem of area is solved. If you want to explain to somebody what the area under a curve is, it is sufficient to describe what a partition is, what is meant by the upper and lower sums associated with a given partition, and then quote the theorem guaranteeing the existence and uniqueness of a number A between $L(P)$ and $U(P)$ for every P.

Distance Traveled in One Direction

Suppose that $v(t)$ is the velocity at time t of an object traveling in a straight line. If v is continuous and nonnegative in the time interval $[a,b]$, we can find the distance traveled during the interval as follows. Let

$$P = \{t_0, t_1, \ldots, t_n\}$$

be a partition of $[a,b]$ and consider the motion during the typical subinterval $[t_{k-1}, t_k]$. Since v is continuous, it has a maximum M_k and a minimum m_k in this interval. If $\Delta t_k = t_k - t_{k-1}$ is small, the numbers $M_k \Delta t_k$ and $m_k \Delta t_k$ are good approximations to the distance traveled from time t_{k-1} to time t_k. (Why?) Note that each of these numbers is a rate multiplied by a time. It is as if the velocity were constant (either M_k or m_k) during the typical subinterval. The sums

$$U(P) = M_1 \Delta t_1 + M_2 \Delta t_2 + \cdots + M_n \Delta t_n = \sum_{k=1}^{n} M_k \Delta t_k$$

$$L(P) = m_1 \, \Delta t_1 + m_2 \, \Delta t_2 + \cdots + m_n \, \Delta t_n = \sum_{k=1}^{n} m_k \, \Delta t_k$$

are therefore good approximations to the total distance traveled from $t = a$ to $t = b$. More precisely, $L(P) \leq D \leq U(P)$ for every P, where D is the distance. Since the existence and uniqueness of such a number D is guaranteed, we have come up with a theoretical description of distance.

The similarity of this discussion to our earlier description of area is apparent. As before, we made premature reference to something that was not yet defined (in this case distance). Most people have an intuitive grasp of distance in terms of velocity and time. But *unless the velocity is constant* (in which case distance is simply the rate multiplied by the time), we have no right to talk about the distance traveled from $t = a$ to $t = b$ until we have defined what it means. The definition parallels that of area under a curve in every detail except notation. The independent variable is t instead of x; the function is v instead of f; the solution of the problem is D instead of A. These are mere changes of letters. It is clear that we are dealing with a concept broader than the geometric or physical interpretation we have given to it. That concept is the subject of this chapter.

□ **Example 5**   Suppose that the velocity of a moving object at time t is $v(t) = \cos t$. To approximate the distance traveled from $t = 0$ to $t = \pi/2$, we may use the partition

$$P = \left\{ 0, \frac{\pi}{3}, \frac{\pi}{2} \right\}$$

This does not divide the interval $I = [0, \pi/2]$ into subintervals of equal length, but it is convenient because the functional values are simple at the points of the partition. Since $\cos t$ is decreasing in I (Figure 10), the lower and upper sums associated with P are

$$L(P) = \left(\cos \frac{\pi}{3} \right)\left(\frac{\pi}{3} - 0 \right) + \left(\cos \frac{\pi}{2} \right)\left(\frac{\pi}{2} - \frac{\pi}{3} \right) = \left(\frac{1}{2} \right)\left(\frac{\pi}{3} \right) + (0)\left(\frac{\pi}{6} \right) = \frac{\pi}{6}$$

$$U(P) = (\cos 0)\left(\frac{\pi}{3} \right) + \left(\cos \frac{\pi}{3} \right)\left(\frac{\pi}{6} \right) = \frac{\pi}{3} + \frac{\pi}{12} = \frac{5\pi}{12}$$

Hence $\pi/6 \leq D \leq 5\pi/12$ (or, approximately, $0.5 \leq D \leq 1.3$). The actual value (as we will show later) is $D = 1$, so these bounds are not very accurate. Their average, however, is about 0.9. This is not far off the mark (even though we used a partition with only two subintervals).   □

□ **Example 6**   Suppose that the velocity of a moving object at time t is $v(t) = 2t$. How far does the object travel during the time interval $I = [0, 1]$?

Solution: The question is equivalent to asking for the area under the graph of $v(t) = 2t$, $0 \leq t \leq 1$. (Why?) By elementary geometry (Figure 11) the answer is $D = \frac{1}{2}(1)(2) = 1$.   □

Figure 10

Graph of $v(t) = \cos t$, $0 \leq t \leq \pi/2$

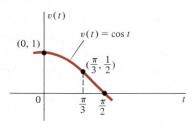

Figure 11

Area under the graph of $v(t) = 2t$, $0 \leq t \leq 1$

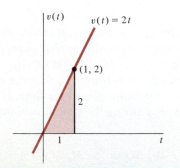

Optional Note (*on the upper and lower sums associated with* $f(x) = x^2 + 1$, $0 \leq x \leq 2$)   In the discussion following Example 2 we found

$$U = \sum_{k=1}^{n} M_k \, \Delta x = \sum_{k=1}^{n} f(x_k) \, \Delta x$$

where x_k is the right endpoint of the kth subinterval in the partition of $[0,2]$ into n subintervals of length Δx. Since $x_k = k \, (\Delta x)$ (why?), we have

$$f(x_k) = x_k^2 + 1 = k^2 \, (\Delta x)^2 + 1$$

and hence

$$U = \sum_{k=1}^{n} [k^2 \, (\Delta x)^2 + 1] \, \Delta x = \sum_{k=1}^{n} [k^2 \, (\Delta x)^3 + \Delta x] = (\Delta x)^3 \sum_{k=1}^{n} k^2 + n \, (\Delta x)$$

Now we use the formula

$$\sum_{k=1}^{n} k^2 = 1^2 + 2^2 + \cdots + n^2 = \frac{1}{6} n(n + 1)(2n + 1)$$

(often derived in algebra courses by mathematical induction). Since $\Delta x = 2/n$, we find

$$U = \frac{8}{n^3} \cdot \frac{n}{6}(n + 1)(2n + 1) + 2$$

$$= \frac{4}{3n^2}(n + 1)(2n + 1) + 2$$

$$= \frac{4}{3}\left(\frac{n + 1}{n}\right)\left(\frac{2n + 1}{n}\right) + 2$$

$$= \frac{4}{3}\left(1 + \frac{1}{n}\right)\left(2 + \frac{1}{n}\right) + 2$$

To find the lower sum

$$L = \sum_{k=1}^{n} m_k \, \Delta x = \sum_{k=1}^{n} f(x_{k-1}) \, \Delta x$$

we replace k by $k - 1$ in the upper sum to obtain

$$L = (\Delta x)^3 \sum_{k=1}^{n} (k - 1)^2 + n \, (\Delta x)$$

Since

$$\sum_{k=1}^{n} (k - 1)^2 = 0^2 + 1^2 + \cdots + (n - 1)^2$$

we may drop the term 0^2 and replace n by $n - 1$ in the above formula from algebra:

$$L = \frac{8}{n^3} \cdot \frac{1}{6}(n - 1)(n)(2n - 1) + 2$$

$$= \frac{4}{3n^2}(n - 1)(2n - 1) + 2$$

$$= \frac{4}{3}\left(1 - \frac{1}{n}\right)\left(2 - \frac{1}{n}\right) + 2$$

Problem Set 6.1

Compute each of the following sums.

1. $\displaystyle\sum_{k=1}^{5} (2k - 1)$ **2.** $\displaystyle\sum_{k=1}^{5} 3k$

3. $\displaystyle\sum_{k=1}^{4} 2^k$ **4.** $\displaystyle\sum_{k=1}^{4} 2^{-k}$

Express each of the following in summation notation.

5. $1^2 + 2^2 + 3^2 + \cdots + n^2$

6. $\sin 1 + \sin 2 + \sin 3 + \cdots + \sin n$

7. $f(x_1)\,\Delta x + f(x_2)\,\Delta x + f(x_3)\,\Delta x + \cdots + f(x_n)\,\Delta x$

8. $\cos x_0(\Delta x) + \cos x_1(\Delta x) + \cos x_2(\Delta x) + \cdots + \cos x_{n-1}(\Delta x)$

9. Let $f(x) = 4 - x$ and suppose that the interval [0,4] is subdivided by the partition $P = \{0, 1, 2, 3, 4\}$.

  (a) What is $\|P\|$?

  (b) Compute $L(P)$.

  (c) Compute $U(P)$.

  (d) Use elementary geometry to show that the area under the graph of $y = 4 - x$, $0 \le x \le 4$, is $A = 8$. Confirm that this is between $L(P)$ and $U(P)$. How accurate is the average of $L(P)$ and $U(P)$?

  (e) Another way to obtain an approximation to A is to use the functional value at the midpoint of each subinterval as the height of the corresponding rectangle. Compute this "midpoint approximation."

10. Let $f(x) = x + 1$ and suppose that the interval [0,2] is subdivided by the partition $P = \{0, \frac{1}{2}, 1, \frac{3}{2}, 2\}$. Answer the same questions as in Problem 9. (In part (d) show that $A = 4$.)

11. Let $f(x) = x^3$ and suppose that the interval [0,2] is subdivided by the partition $P = \{0, 1, 2\}$.

  (a) What is $\|P\|$?

  (b) Compute $L(P)$.

  (c) Compute $U(P)$.

  (d) Eventually we'll show that the area under the curve $y = x^3$, $0 \le x \le 2$, is $A = 4$. Confirm that this is between $L(P)$ and $U(P)$. How accurate is the average of $L(P)$ and $U(P)$?

  (e) Use midpoints of the subintervals of P to compute an approximation to A.

12. Let $f(x) = x^2$ and suppose that the interval $[-1, 2]$ is subdivided by the partition $P = \{-1, 1, 2\}$.

  (a) What is $\|P\|$?

  (b) Compute $L(P)$.

  (c) Compute $U(P)$.

  (d) Eventually we'll show that the area under the curve $y = x^2$, $-1 \le x \le 2$, is $A = 3$. Confirm that this is between $L(P)$ and $U(P)$. How accurate is the average of $L(P)$ and $U(P)$?

  (e) Use midpoints of the subintervals of P to compute an approximation to A.

13. The partition $Q = \{0, \frac{1}{2}, 1, 2\}$ is a "refinement" of the partition P in Problem 11, that is, $P \subset Q$. Compute $L(Q)$ and $U(Q)$ and confirm that $L(P) \le L(Q)$ and $U(Q) \le U(P)$. (This illustrates the fact that lower sums increase and upper sums decrease when the partition is refined. See Problem 34 for a more precise statement.)

14. The partition $Q = \{-1, 0, 1, 2\}$ is a refinement of the partition P in Problem 12. Compute $L(Q)$ and $U(Q)$ and confirm that $L(P) \le L(Q)$ and $U(Q) \le U(P)$. Did the lower sum actually increase or the upper sum decrease?

15. Let $f(x) = \sin x$ and suppose that the interval $[0,\pi]$ is subdivided by the partition

$$P = \left\{0, \frac{\pi}{6}, \frac{\pi}{2}, \frac{5\pi}{6}, \pi\right\}$$

  (a) What is $\|P\|$?

  (b) Compute $L(P)$.

  (c) Compute $U(P)$.

  (d) Eventually we'll show that the area under the curve $y = \sin x$, $0 \le x \le \pi$, is $A = 2$. Confirm that this is between $L(P)$ and $U(P)$. How accurate is the average of $L(P)$ and $U(P)$?

16. Let $v(t) = \cos t$ and suppose that the interval $[0,\pi/2]$ is subdivided by the partition

$$Q = \left\{0, \frac{\pi}{6}, \frac{\pi}{3}, \frac{\pi}{2}\right\}$$

  (a) What is $\|Q\|$?

  (b) Compute $L(Q)$.

  (c) Compute $U(Q)$.

  (d) Eventually we'll show that the distance traveled by an object with velocity $v(t) = \cos t$, $0 \le t \le \pi/2$, is $D = 1$. Confirm that this is between $L(Q)$ and $U(Q)$. How accurate is the average of $L(Q)$ and $U(Q)$?

  (e) Q is a refinement of the partition P in Example 5. Confirm that $L(P) \le L(Q)$ and $U(Q) \le U(P)$.

17. The partition

$$Q = \left\{0, \frac{\pi}{6}, \frac{\pi}{3}, \frac{\pi}{2}, \frac{2\pi}{3}, \frac{5\pi}{6}, \pi\right\}$$

is a refinement of the partition P in Problem 15. Compute $L(Q)$ and $U(Q)$ and confirm that $L(P) \leq L(Q)$ and $U(Q) \leq U(P)$.

18. We defined the area of a rectangle to be the product of its length and width, then used that idea to come up with the more general definition of area under a curve. What if the "curve" is the horizontal line $f(x) = c$ $(c > 0)$? Then the general definition of the area from a to b should reduce to $c(b - a)$. Show that it does.

19. When the velocity of a moving object is constant, the distance traveled is the rate multiplied by the time. We used that idea to define distance in the case of any continuous and nonnegative velocity function. What if the function is constant, say $v(t) = R > 0$? Then the general definition of the distance traveled from $t = 0$ to $t = T$ should reduce to $D = RT$. Show that it does.

20. Let A be the area under the curve $y = 2x$, $0 \leq x \leq 3$.
  (a) Use elementary geometry to show that $A = 9$.
  (b) Let P be a partition of $[0,3]$ into n subintervals of equal length. Show that

$$U(P) = 9\left(1 + \frac{1}{n}\right)$$

  Hint: Use the formula $1 + 2 + \cdots + n = \frac{1}{2}n(n + 1)$.
  (c) Show that $L(P) = 9(1 - 1/n)$.
  (d) What is the common limit of $U(P)$ and $L(P)$ as $n \to \infty$?

21. Let A be the area under the curve $y = 3 - x$, $0 \leq x \leq 2$.
  (a) Use elementary geometry to show that $A = 4$.
  (b) Let P be a partition of $[0,2]$ into n subintervals of equal length. Show that

$$L(P) = 6 - 2\left(1 + \frac{1}{n}\right)$$

  (c) Show that $U(P) = 6 - 2(1 - 1/n)$.
  (d) What is the common limit of $L(P)$ and $U(P)$ as $n \to \infty$?

22. Let $f(x) = 1 - x^2$ and suppose that P is a partition of $[0,1]$ into n subintervals of equal length.
  (a) Show that $L(P) = 1 - \frac{1}{6}(1 + 1/n)(2 + 1/n)$. *Hint:* Use the formula $1^2 + 2^2 + \cdots + n^2 = \frac{1}{6}n(n + 1) \cdot (2n + 1)$.
  (b) Show that $U(P) = 1 - \frac{1}{6}(1 - 1/n)(2 - 1/n)$.

(c) What is the area under the curve $y = 1 - x^2$, $0 \leq x \leq 1$?

23. Let $f(x) = 3x^2 + 1$ and suppose that P is a partition of $[0,4]$ into n subintervals of equal length.
  (a) Show that $U(P) = 32\left(1 + \frac{1}{n}\right)\left(2 + \frac{1}{n}\right) + 4$.
  (b) Show that $L(P) = 32\left(1 - \frac{1}{n}\right)\left(2 - \frac{1}{n}\right) + 4$.
  (c) What is the area under the curve $y = 3x^2 + 1$, $0 \leq x \leq 4$?

24. In Figure 12 we show a circle of radius r and two approximations to its area based on n congruent triangles.
  (a) Show that the areas of the shaded triangular regions are

$$\frac{1}{2}r^2 \sin \theta \qquad \text{and} \qquad r^2 \tan \frac{\theta}{2}$$

  respectively.
  (b) Show that the lower sum and upper sum are

$$\frac{1}{2}nr^2\left(\frac{\sin \theta}{\theta}\right) \qquad \text{and} \qquad \frac{2\pi r^2}{\theta}\left(\tan \frac{\theta}{2}\right)$$

  respectively.
  (c) What is the common limit of these sums as $n \to \infty$?

Figure 12
Lower and upper sums approximating the area of a circle

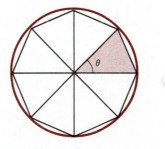

In Problems 25 through 31, use the definition that if a_1, $a_2, \ldots, a_n$ are real numbers, then

$$\sum_{k=1}^{n} a_k = a_1 + a_2 + \cdots + a_n$$

25. Show that $\sum_{k=1}^{n} (a_k + b_k) = \sum_{k=1}^{n} a_k + \sum_{k=1}^{n} b_k$.

26. Show that if $c \in \mathbb{R}$, then $\sum_{k=1}^{n} ca_k = c \sum_{k=1}^{n} a_k$.

27. Explain why it follows from Problems 25 and 26 that

$$\sum_{k=1}^{n} (ca_k + db_k) = c \sum_{k=1}^{n} a_k + d \sum_{k=1}^{n} b_k$$

28. Show that if $c \in \mathfrak{R}$, then $\sum_{k=1}^{n} c = nc$.

29. Explain why $\sum_{j=1}^{n} a_j = \sum_{k=1}^{n} a_k$.

(In other words the letter used for the index of summation is immaterial; it is a "dummy variable.")

30. Show that $\sum_{k=1}^{n} (x_k - x_{k-1}) = x_n - x_0$.

(This is a "telescoping sum," a type that occurs frequently in mathematics.)

31. If P is a partition of $[a,b]$, use Problem 30 to show that

$$\sum_{k=1}^{n} \Delta x_k = b - a$$

Interpret this result geometrically.

32. Suppose that P is a partition of the interval $[a,b]$ into n subintervals of equal length.

(a) What is $\|P\|$?

(b) Explain why $\|P\| \to 0 \Leftrightarrow n \to \infty$.

(c) If the subintervals are not necessarily of equal length, show that $\|P\| \to 0 \Rightarrow n \to \infty$, but that the reverse implication is false. (Thus in general we use the stronger condition $\|P\| \to 0$.)

33. If f is continuous in $I = [a,b]$ and P is a partition of I, argue without reference to pictures that $L(P) \leq U(P)$.

34. If P and Q are partitions of the interval $[a,b]$ such that $P \subset Q$, we call Q a *refinement* of P.

(a) Suppose that Q is obtained from P by inserting *one* additional point of subdivision. If f is continuous in $[a,b]$, explain why $L(P) \leq L(Q)$ and $U(Q) \leq U(P)$. *Hint:* Let $P = \{x_0, x_1, \ldots, x_n\}$ and suppose that the new point is inserted in the kth subinterval. Then $Q = \{X_0, X_1, \ldots, X_{n+1}\}$, where

$$X_i = x_i, \; i = 1, \ldots, k - 1$$
$$X_k = \text{new point } (x_{k-1} < X_k < x_k)$$
$$X_i = x_{i-1}, \; i = k + 1, \ldots, n + 1$$

To establish the required inequalities, you need only look at what happens when the kth terms $m_k \Delta x_k$ and $M_k \Delta x_k$ in $L(P)$ and $U(P)$ are replaced by the kth and $(k + 1)$th terms in $L(Q)$ and $U(Q)$. (Why?)

(b) Why does it follow that these inequalities are true if Q is any refinement of P?

6.2
THE INTEGRAL OF A REAL FUNCTION

In the last section we discussed two problems with the same solution, suggesting that their common characteristics would lead to a broader concept that includes them as special cases. To develop this concept, we begin with a more general notion of *approximating sum*.

In Problems 9 through 12 of Section 6.1 you were asked to compute not only $L(P)$ and $U(P)$, but also a "midpoint approximation" based on evaluation of f at the midpoint of each subinterval of P. The idea is that since $L(P)$ and $U(P)$ box in the area under the graph of f, it is reasonable to approximate the area by intermediate sums as well.

▫ **Example 1**   The region under the graph of

$$f(x) = x^2 + 1 \quad (0 \leq x \leq 2)$$

is shown in Figure 1. Compute the midpoint approximation to its area based on a partition of $[0,2]$ into four subintervals of equal length.

Solution: The partition called for is $P = \{0, 0.5, 1, 1.5, 2\}$. Midpoints of the subintervals are $c_1 = 0.25$, $c_2 = 0.75$, $c_3 = 1.25$, $c_4 = 1.75$. The corre-

Figure 1
Midpoint approximation of area

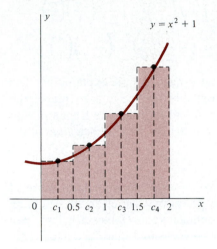

sponding functional values are

$$f(c_1) = 1.0625, \quad f(c_2) = 1.5625, \quad f(c_3) = 2.5625, \quad f(c_4) = 4.0625$$

and hence the midpoint approximation we seek is

$$\sum_{k=1}^{4} f(c_k)\,\Delta x_k = (1.0625)(0.5) + (1.5625)(0.5) + (2.5625)(0.5) + (4.0625)(0.5)$$

$$= 4.625$$

In Examples 1 and 2 of Section 6.1 we found the lower and upper sums associated with the partition P to be $L(P) = 3.75$ and $U(P) = 5.75$. The midpoint approximation is between them, as we should expect. Moreover, it seems reasonable to expect that as the partition is refined (which causes the lower and upper sums to approach a common limit), *the midpoint approximation will be squeezed toward the same limit.* □

There is no reason to confine the idea of intermediate sum to midpoint approximations. Why not allow c_1, c_2, c_3, c_4 in Example 1 to be *any* points of the respective subintervals? More generally, given any partition of the original interval, why not evaluate f at arbitrary points $c_1, c_2, \ldots, c_n$ of the n subintervals? If $f(c_k)$ happens to be the maximum value of f in the kth subinterval ($k = 1, 2, \ldots, n$), we are back to the upper sum $U(P)$, while if $f(c_k)$ is minimum, we obtain the lower sum $L(P)$. Ordinarily, however, the intermediate sum

$$\sum_{k=1}^{n} f(c_k)\,\Delta x_k = f(c_1)\,\Delta x_1 + f(c_2)\,\Delta x_2 + \cdots + f(c_n)\,\Delta x_n$$

is between $L(P)$ and $U(P)$, which makes it a good approximation to the area. We give it a special name, as in the following definition.

Suppose that f is defined in the (finite) closed interval $I = [a,b]$, where $a < b$. Given a partition

$$P = \{x_0, x_1, \ldots, x_n\}$$

of I, let c_k be any point of the kth subinterval $[x_{k-1}, x_k]$, $k = 1, 2, \ldots, n$. The expression

$$\sum_{k=1}^{n} f(c_k)\,\Delta x_k$$

is called a **Riemann sum** of f associated with P.

□ **Example 2**   Let $f(x) = \cos x$ and $I = [0,\pi]$. If $P = \left\{0, \dfrac{\pi}{3}, \dfrac{\pi}{2}, \dfrac{2\pi}{3}, \pi\right\}$

and we choose $c_1 = \pi/3$, $c_2 = \pi/3$, $c_3 = 2\pi/3$, $c_4 = 2\pi/3$, compute the corresponding Riemann sum of f associated with P.

Figure 2
Riemann sum with negative terms

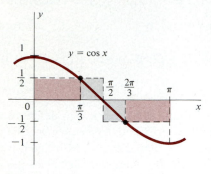

Solution: The desired sum is

$$\sum_{k=1}^{4} f(c_k)\,\Delta x_k = f\left(\frac{\pi}{3}\right)\left(\frac{\pi}{3}\right) + f\left(\frac{\pi}{3}\right)\left(\frac{\pi}{6}\right) + f\left(\frac{2\pi}{3}\right)\left(\frac{\pi}{6}\right) + f\left(\frac{2\pi}{3}\right)\left(\frac{\pi}{3}\right)$$

$$= \left(\frac{1}{2}\right)\left(\frac{\pi}{3}\right) + \left(\frac{1}{2}\right)\left(\frac{\pi}{6}\right) + \left(-\frac{1}{2}\right)\left(\frac{\pi}{6}\right) + \left(-\frac{1}{2}\right)\left(\frac{\pi}{3}\right) = 0$$

This may seem to be a strange result, but it has a simple explanation. The graph of cosine is partly above and partly below the x axis (unlike the graphs considered in Section 6.1). The first two terms of the Riemann sum represent areas of rectangles (Figure 2), but the last two are *negatives* of the areas of the rectangles below the x axis.   □

It is probably better to say that in computing Riemann sums we need not think about area at all; in fact the preceding definition says nothing about area (nor does it require f to be nonnegative in I). Nevertheless these sums have a limit as the partition is refined (provided that f is a reasonably well-behaved function). It is that limit to which we now give a name.

Let f be a real function with domain $I = [a,b]$, where $a < b$. If the Riemann sums of f associated with a partition P approach a limit S as $\|P\| \to 0$, we call S the *integral of f over I* (also the *integral of f from a to b*). This limit is written

$$S = \int_a^b f(x)\,dx = \lim_{\|P\|\to 0} \sum_{k=1}^{n} f(c_k)\,\Delta x_k$$

It is also denoted by either of the symbols

$$S = \int_a^b f \qquad \text{or} \qquad S = \int_I f$$

When the integral exists we call f *integrable over I*; to *integrate f* from a to b means to find S.

The symbol $\int$ is called the *integral sign*. It evolved from Σ and S and was regarded by the pioneers of calculus as the symbol for a sum. The idea was that

$$\int f(x)\,dx = \sum f(x)\,\Delta x$$

where the expression on the right is a sum of "infinitely many infinitely small" terms, each one having the form $f(x)\,\Delta x$. This notion was the object of considerable derision in the years following its formulation; mathematicians no longer talk that way. Nevertheless, you can see from the last section that the idea is sound. It is only a step removed from the more formal concept of the limit of a finite sum. Moreover (as we will explain in detail later), the symbolic formulation of "approximating sums" leading to an integral, namely

$$\sum f(x)\,\Delta x \to \int_a^b f(x)\,dx$$

is an enormous aid to comprehension, both theoretically and in practical applications.

The notation $\int_a^b f(x)\,dx$ has therefore survived, although for the present you should regard dx as an appendage with no meaning (other than as a reminder of Δx in the approximating sums). Later we will see that it is harmless to confuse it with the differential we discussed in Section 4.7.

The symbols $\int_I f$ and $\int_a^b f$ have the advantage of not referring to the independent variable, but they are normally used only in the abstract. When a specific function is named, say $f(x) = x^2$, we will always attach the dx, as in

$$\int_a^b x^2\,dx, \qquad not \quad \int_a^b x^2$$

If the independent variable is given another name, as in $f(t) = t^2$, the dx follows suit:

$$\int_a^b t^2\,dt, \qquad not \quad \int_a^b t^2\,dx$$

The importance of this appendage will become clear as we proceed.

The function f appearing in the integral notation is called the **integrand.** The numbers a and b are called the **limits of integration** (lower and upper, respectively). This terminology has nothing to do with "limit" in the usual sense, but merely refers to the endpoints of the interval I. They should always be included in the notation; the unadorned symbol $\int f(x)\,dx$ has come to have another meaning (to be introduced in Section 6.4).

Several things should be observed in connection with our definition of integral.

1. $f(x)$ is not required to be nonnegative (as in the area and distance problems of Section 6.1). Hence the integral of f need not represent area or distance; it is a concept independent of those ideas.
2. f is not required to be continuous in I. In Section 6.1 we assumed continuity to guarantee the existence of extreme values m_k and M_k in the typical subinterval, so that $L(P)$ and $U(P)$ could be computed. Riemann sums involve functional values $f(c_k)$ that are not necessarily extremes. The limit of these sums (as $\|P\| \to 0$) may exist even when f is not continuous. (See Example 7.)
3. The definition of integral is empty of content unless there are circumstances in which the existence of the limit S can be guaranteed.
4. The definition is ambiguous unless only one such limit exists.

The last two difficulties are removed by the theorem we mentioned in Section 6.1. (We are omitting the proof.)

Figure 3
Area under the graph of
$f(x) = c\,(c > 0),\ a \le x \le b$

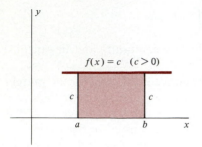

Figure 4
Area under the graph of
$f(x) = 3 - x,\ 0 \le x \le 2$

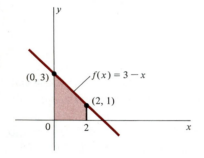

Figure 5
Area under the graph of
$f(x) = \sqrt{1 - x^2},\ -1 \le x \le 1$

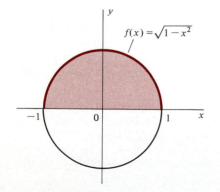

□ **Theorem 1** (*Existence and Uniqueness of the Integral*)   If f is continuous in $I = [a,b]$, the integral of f from a to b exists and is unique.   □

Remark   In Section 6.1 we referred to the existence and uniqueness of a number A such that $L(P) \le A \le U(P)$ for every partition P of I. Theorem 1, on the other hand, asserts the existence and uniqueness of the integral defined as a limit of Riemann sums. In more advanced courses it is proved that when f is continuous in I, the numbers A and $\int_a^b f$ are the same.

Concrete results based on the definition of integral are hard to come by, as you know from Section 6.1. The area interpretation of integral, however, is helpful in simple cases.

□ **Example 3**   The integral of a constant function $f(x) = c$ over any interval $[a,b]$ is

$$\int_a^b c\,dx = c(b - a)$$

This is geometrically apparent if $c \ge 0$, for then the integral is simply the area under $f(x) = c,\ a \le x \le b$. (See Figure 3.) If $c < 0$, the definition in this section must be used, but it is not hard. (See the problem set.)   □

□ **Example 4**   The area under the graph of $v(t) = 2t,\ 0 \le t \le 1$, is

$$A = \tfrac{1}{2}(1)(2) = 1$$

(See Figure 11, Section 6.1.) Hence $\int_0^1 2t\,dt = 1$.   □

□ **Example 5**   The area under the graph of $f(x) = 3 - x,\ 0 \le x \le 2$, is $A = 4$. (See Figure 4, and Problem 21 in Section 6.1.) Hence

$$\int_0^2 (3 - x)\,dx = 4$$ □

□ **Example 6**   Since the graph of $f(x) = \sqrt{1 - x^2}$ is the upper half of the unit circle $x^2 + y^2 = 1$ (Figure 5), we have

$$\int_{-1}^1 \sqrt{1 - x^2}\,dx = \frac{1}{2}(\text{area of unit circle}) = \frac{\pi}{2}$$

Note that this result (unlike those in the previous examples) would be difficult to confirm by a direct evaluation based on the definition of integral.   □

According to Theorem 1, continuity of a function f in $I = [a,b]$ is sufficient to guarantee that f is integrable over I. It is not a necessary condition,

Figure 6

Graph of $y = \dfrac{|x|}{x}$,

$-1 \leq x \leq 1\,(x \neq 0)$

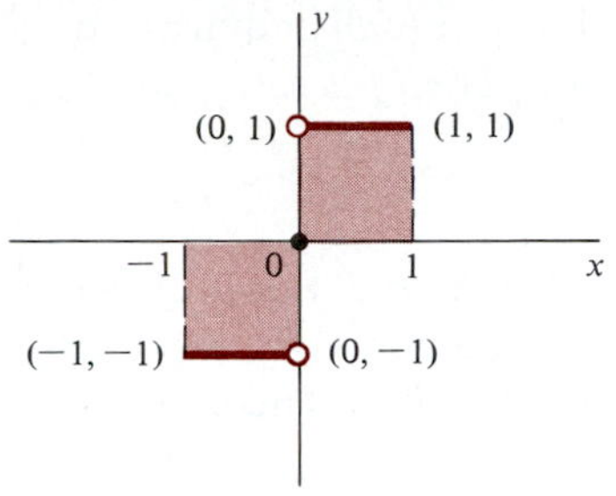

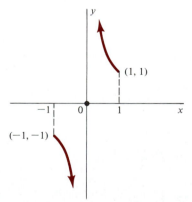

however; there are functions that can be integrated from a to b but are not continuous in I.

□ **Example 7**   The function f defined by

$$f(x) = \begin{cases} |x|/x & \text{if } x \neq 0 \\ 0 & \text{if } x = 0 \end{cases}$$

is discontinuous at 0. (See Figure 6.) Nevertheless its integral from -1 to 1 exists; in fact, since the negative of the area below the x axis cancels the area above the x axis, the integral is 0.   □

Remark   The area interpretation of integral does not apply when $f(x)$ is allowed to be negative. (There is no such thing as negative area!) We are leaning on your intuition when we say that the negative of the area below the x axis cancels the area above it. Note, too, that the discontinuity of f at 0 does not affect the integral interpreted in terms of area. The presence (or absence) of the point $(0,0)$ is immaterial; we might as well omit $f(0) = 0$ from the definition of f and consider $f(x) = |x|/x$ alone.

As a contrast to Example 7, consider the function f defined by

$$f(x) = \begin{cases} 1/x & \text{if } x \neq 0 \\ 0 & \text{if } x = 0 \end{cases}$$

Figure 7

Graph of $y = \dfrac{1}{x}$,

$-1 \leq x \leq 1\,(x \neq 0)$

Like $y = |x|/x$ in Example 7, this function is discontinuous at 0. (See Figure 7.) But it is radically different in that $|x|/x$ is bounded in $[-1,1]$ and $1/x$ is not.

It can be shown that in order for a function to be integrable (over an interval I), it must be bounded in I, that is, a number M must exist such that

$$|f(x)| \leq M \qquad \text{for all } x \in I \text{ (in the domain of } f)$$

This is a necessary condition for integrability:

> If f is not bounded in the interval I, it is not integrable over I.

Hence

$$\int_{-1}^{1} \frac{1}{x}\,dx \text{ does not exist}$$

(In Chapter 11 we will introduce the notion of *improper integral,* giving meaning to the symbol $\int_a^b f(x)\,dx$ in certain cases where f is unbounded. That will not affect our definition of *integrable* as given here, however.)

Where do we stand? Theorem 1 says that continuity in I is *sufficient* for integrability over I; we have now added the statement that boundedness is *necessary*. That leaves out of account functions that are bounded but discontinuous (like $|x|/x$ in Example 7). Are they integrable?

The answer is "sometimes." It is difficult to be more precise without a considerable digression. We can state the following theorem, however, which covers most cases of interest in elementary calculus.

□ **Theorem 2**   Suppose that f is bounded in $I = [a,b]$ and discontinuous at no more than a finite number of points of I. Then f is integrable over I.

□

Note that Theorem 1 is a special case of Theorem 2. For if f is continuous in I, it is automatically bounded (by the Fundamental Principle of Continuity in Section 4.2). Hence it is integrable over I by Theorem 2.

Problem Set 6.2

Use the area interpretation of integral to evaluate each of the following.

1. $\displaystyle\int_{-1}^{1} 3\,dx$

2. $\displaystyle\int_{0}^{\sqrt{2}} 5\,dt$

3. $\displaystyle\int_{2}^{3} (x - 2)\,dx$

4. $\displaystyle\int_{0}^{2} (2 - x)\,dx$

5. $\displaystyle\int_{-1}^{1} (1 - |x|)\,dx$

6. $\displaystyle\int_{0}^{3} (x + 2)\,dx$

7. $\displaystyle\int_{1}^{3} (4 - u)\,du$

8. $\displaystyle\int_{-3}^{3} \sqrt{9 - x^2}\,dx$

9. $\displaystyle\int_{0}^{5} \sqrt{25 - x^2}\,dx$

10. $\displaystyle\int_{0}^{1} (2 + \sqrt{1 - x^2})\,dx$

11. $\displaystyle\int_{0}^{1} (2 - \sqrt{1 - x^2})\,dx$

12. $\displaystyle\int_{1}^{4} [x]\,dx$, where $[x]$ is the "greatest integer in x" (Problem 16, Section 1.4).

13. Explain why the area of the ellipse $x^2/a^2 + y^2/b^2 = 1$ is

$$A = \frac{4b}{a} \int_{0}^{a} \sqrt{a^2 - x^2}\,dx$$

Then evaluate the integral to arrive at the formula $A = \pi ab$ for the area.

Each of the following is a Riemann sum associated with an arbitrary partition P of the given interval I. Evaluate the limit as $\|P\| \to 0$.

14. $I = [0,2]$, $\displaystyle\sum_{k=1}^{n} 3c_k\,\Delta x_k$

15. $I = [0,4]$, $\displaystyle\sum_{k=1}^{n} (4 - c_k)\,\Delta x_k$

16. $I = [-1,1]$, $\displaystyle\sum_{k=1}^{n} |c_k|\,\Delta x_k$

17. $I = [0,3]$, $\displaystyle\sum_{k=1}^{n} \sqrt{9 - c_k^2}\,\Delta x_k$

Which of the following integrals exist?

18. $\displaystyle\int_{0}^{2} \frac{|x - 1|}{x - 1}\,dx$

19. $\displaystyle\int_{0}^{2} \frac{x^2 - 1}{x - 1}\,dx$

20. $\displaystyle\int_{0}^{2} \frac{dx}{x - 1}$

21. $\displaystyle\int_{0}^{1} \frac{dx}{\sqrt{x}}$

22. $\displaystyle\int_{0}^{2\pi} \tan t \cot t\,dt$

23. $\displaystyle\int_{0}^{\pi/2} \sec t\,dt$

24. $\displaystyle\int_{0}^{\pi} \cot t\,dt$

25. $\displaystyle\int_{0}^{1} t \sin \frac{1}{t}\,dt$

26. $\int_0^1 \dfrac{\sin x}{x}\, dx$

27. Define the function f in the interval $I = [0,1]$ by the rule

$$f(x) = \begin{cases} 1 \text{ if } 0 < x \le 1 \\ c \text{ if } x = 0 \end{cases}$$

Then f is continuous if $c = 1$, discontinuous otherwise. By examining Riemann sums of f associated with an arbitrary partition of I, explain why

$$\int_0^1 f(x)\, dx = 1$$

in any case. (This problem illustrates the fact that if one point is displaced from the graph of a continuous function, the integral is unaffected. In fact any finite number of points can be displaced, or even removed altogether.)

28. Use Riemann sums to prove that if $c \in \mathcal{R}$, then

$$\int_a^b c\, dx = c(b - a)$$

as follows.

(a) If P is an arbitrary partition of $[a,b]$, explain why every Riemann sum of $f(x) = c$ associated with P reduces to $c(b - a)$.

(b) Why does it follow that the value of the integral of f is $c(b - a)$?

The remaining problems in this set are designed to show how Riemann sums can be used to derive important properties of the integral. They are good practice in setting up and interpreting Riemann sums. The properties themselves, however, will be developed somewhat differently (and more easily) in the next two sections.

29. Use Riemann sums to prove that if $c \in \mathcal{R}$ and f is continuous in $[a,b]$, then

$$\int_a^b cf(x)\, dx = c \int_a^b f(x)\, dx$$

("A constant may be moved across the integral sign.")

30. Use Riemann sums to prove that if f and g are continuous in $[a,b]$, then

$$\int_a^b [f(x) + g(x)]\, dx = \int_a^b f(x)\, dx + \int_a^b g(x)\, dx$$

("The integral of a sum is the sum of the integrals.")

31. Use Riemann sums to prove that if $f(x)$ is nonnegative and continuous in $[a,b]$, then

$$\int_a^b f(x)\, dx \ge 0$$

Interpret this result geometrically.

32. Suppose that f and g are continuous in $I = [a,b]$. Show that if $f(x) \le g(x)$ for all $x \in I$, then

$$\int_a^b f(x)\, dx \le \int_a^b g(x)\, dx$$

Hint: Apply Problem 31 to the function $g - f$ (and use Problems 29 and 30).

33. Suppose that f is continuous in $I = [a,b]$. Show that

$$\left| \int_a^b f(x)\, dx \right| \le \int_a^b |f(x)|\, dx$$

Hint: Apply Problem 32 to the inequalities $-|f(x)| \le f(x) \le |f(x)|$.

34. Let f be continuous in $I = [a,b]$ and suppose that F is an antiderivative of f in I, that is, $F'(x) = f(x)$ for all $x \in I$. Let $P = \{x_0, x_1, \ldots, x_n\}$ be any partition of I.

(a) Show that for each $k = 1, 2, \ldots, n$, there is a point $c_k \in [x_{k-1}, x_k]$ such that

$$F(x_k) - F(x_{k-1}) = f(c_k)\, \Delta x_k$$

Hint: Apply the Mean Value Theorem (Section 4.3) to F in the interval $[x_{k-1}, x_k]$.

(b) Explain why $\displaystyle\sum_{k=1}^n [F(x_k) - F(x_{k-1})] = F(b) - F(a)$.

Hint: This is a "telescoping sum" (as in Problem 30, Section 6.1).

(c) You have proved in parts (a) and (b) that for each partition P of I, there is a choice of points c_k such that the corresponding Riemann sum is constant, namely

$$\sum_{k=1}^n f(c_k)\, \Delta x_k = F(b) - F(a)$$

Why does it follow that the limit (as $\|P\| \to 0$) of Riemann sums of f associated with P is the number $F(b) - F(a)$?

(d) What is the value of $\int_a^b f$?

(This result is part of the *Fundamental Theorem of Calculus,* to be derived in Section 6.4. It reduces the evaluation of many integrals to a triviality, as the next problem illustrates.)

35. Confirm that $F(x) = \frac{1}{3}x^3 + x$ is an antiderivative of $f(x) = x^2 + 1$ and use Problem 34 to show that

$$\int_0^2 (x^2 + 1)\, dx = \frac{14}{3}$$

(Compare with our direct evaluation of this integral in Section 6.1!)

36. Prove that $f(x) = 1/x$ is not integrable over $I = [0,1]$ by showing that the Riemann sums of f associated with any partition of I are unbounded.

6.3
PROPERTIES OF THE INTEGRAL

Figure 1
Interval additivity

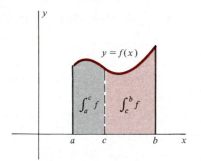

Figure 2
Graph of $y = |x|,\ -1 \le x \le 2$

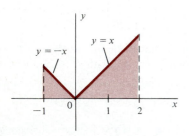

There are several situations in which the evaluation of an integral is made easier by breaking up the interval of integration. A glance at Figure 1 should convince you that the "interval additivity" formula

$$\int_a^b f(x)\, dx = \int_a^c f(x)\, dx + \int_c^b f(x)\, dx$$

makes sense when $f(x)$ is nonnegative and $a < c < b$, for then it is nothing more than addition of areas. It is not so obvious when the graph of f is allowed to go below the x axis, but in that case an argument based on Riemann sums can be given. (See Problem 38.)

□ **Example 1**   Use the interval additivity formula to evaluate

$$\int_{-1}^2 |x|\, dx$$

Solution: Since $|x|$ is x or $-x$ depending on whether $x \ge 0$ or $x < 0$,

$$\int_{-1}^2 |x|\, dx = \int_{-1}^0 |x|\, dx + \int_0^2 |x|\, dx = \int_{-1}^0 (-x)\, dx + \int_0^2 x\, dx$$

The area interpretation of integral (Figure 2) yields

$$\int_{-1}^0 (-x)\, dx = \tfrac{1}{2}(1)(1) = \tfrac{1}{2} \qquad \text{and} \qquad \int_0^2 x\, dx = \tfrac{1}{2}(2)(2) = 2$$

so

$$\int_{-1}^2 |x|\, dx = \tfrac{1}{2} + 2 = \tfrac{5}{2} \qquad\qquad □$$

In Example 1 it is easier to use the area interpretation directly:

$$\int_{-1}^2 |x|\, dx = \text{area under the graph} = \tfrac{1}{2} + 2 = \tfrac{5}{2}$$

Where the interval additivity formula is useful is in those cases where area interpretation is no help, as in

$$\int_0^{2\pi/3} |\cos t|\, dt = \int_0^{\pi/2} \cos t\, dt + \int_{\pi/2}^{2\pi/3} (-\cos t)\, dt$$

(Recall that $\cos t$ is positive in the first quadrant and negative in the second.) The integrals on the right side of this equation are easy to evaluate (as we will see in the next section).

Interval additivity is also important in the development of the Fundamental Theorem of Calculus (Section 6.4). For that purpose we need to extend it to arbitrary points a, b, c (not necessarily satisfying $a < c < b$). To do this, we must first define the meaning of $\int_a^b f$ when $a \geq b$.

The Integral with Arbitrary Limits

Suppose that f is integrable over the interval I and a and b are any points of I.

1. If $a < b$, then $\int_a^b f$ has already been defined.

2. If $a = b$, we define $\int_a^b f = \int_a^a f = 0$.

3. If $a > b$, we define $\int_a^b f = -\int_b^a f$.

Now consider the formula

$$\int_a^b f = \int_a^c f + \int_c^b f$$

when the points are in an order other than $a < c < b$. For example, suppose that $b < a < c$. Integrating in the normal order, we have

$$\int_b^c f = \int_b^a f + \int_a^c f = -\int_a^b f + \int_a^c f \qquad \text{(by the above definition)}$$

Hence

$$\int_a^b f = \int_a^c f - \int_b^c f = \int_a^c f + \int_c^b f$$

Other arrangements of the points (including cases in which they are not all distinct) are treated similarly, leading to the following theorem.

□ **Theorem 1**   Suppose that f is integrable over the interval I and a, b, c are any points of I (not necessarily distinct, and in no special order). Then

$$\int_a^b f = \int_a^c f + \int_c^b f \qquad\qquad\qquad □$$

Another important property of integration is that (like differentiation) it is *linear*. (See Section 3.1.) We express that fact in the following theorem.

Figure 3

$$\int_0^3 x \, dx = \text{area of triangle} = \tfrac{1}{2}(3)(3)$$

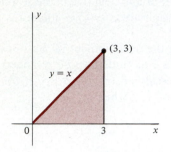

Figure 4

$$\int_0^3 \sqrt{9 - x^2} \, dx = \text{area of quarter-circle} = \tfrac{1}{4}\pi(3^2)$$

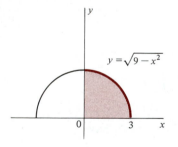

Figure 5
Theorem of the Mean for Integrals

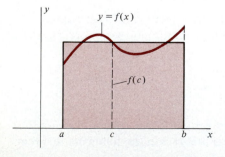

□ **Theorem 2**   Suppose that f and g are integrable in $[a,b]$ and $c \in \mathcal{R}$. Then

$$\int_a^b [f(x) + g(x)] \, dx = \int_a^b f(x) \, dx + \int_a^b g(x) \, dx$$

(the integral of a sum is the sum of the integrals) and

$$\int_a^b cf(x) \, dx = c \int_a^b f(x) \, dx$$

(a constant may be moved across the integral sign).   □

The most direct argument for Theorem 2 is to use the definition of integral in terms of Riemann sums. (See Problems 29 and 30, Section 6.2.) It is easier, however, to defer the proof until we have derived the Fundamental Theorem of Calculus in the next section.

□ **Example 2**   Use Theorem 2 and the area interpretation of integral to compute

$$\int_0^3 (x + 2\sqrt{9 - x^2}) \, dx$$

Solution: Since integration is a linear operation, we can write

$$\int_0^3 (x + 2\sqrt{9 - x^2}) \, dx = \int_0^3 x \, dx + 2 \int_0^3 \sqrt{9 - x^2} \, dx$$

$$= \tfrac{1}{2}(3)(3) + 2 \cdot \tfrac{1}{4}\pi(3^2) = \tfrac{9}{2}(1 + \pi)$$

(See Figures 3 and 4, in which we use the area interpretation to evaluate the integrals.)   □

Our next result is reminiscent of the Mean Value Theorem for derivatives (Section 4.3). As it applies to area under a curve, the theorem is easy to believe; all it says is that the right choice of altitude yields a rectangle whose area is the same as the area under the curve. (See Figure 5.) More precisely, there is a point c between a and b such that

$$\int_a^b f(x) \, dx = f(c)(b - a)$$

The left side is the area under the curve; the right side is the area of a rectangle with altitude $f(c)$ and base $b - a$.

□ **Theorem 3** (*Theorem of the Mean for Integrals*)   If f is continuous in the closed interval I with endpoints a and b, there exists a point $c \in I$ such that

$$\int_a^b f = f(c)(b - a)$$

Proof: Assume first that $a < b$. Since f is continuous in I, it has a maximum value M and a minimum value m (Theorem 2, Section 4.2). If $P = \{x_0, x_1, \ldots, x_n\}$ is any partition of I and c_k is any point in the kth subinterval ($k = 1, 2, \ldots, n$), then

$$m \leq f(c_k) \leq M \qquad \text{for each } k$$

Multiplying each side by Δx_k yields

$$m \, \Delta x_k \leq f(c_k) \, \Delta x_k \leq M \, \Delta x_k \qquad (k = 1, 2, \ldots, n)$$

Then, adding respective sides of these n inequalities (Order Property 6 in Section 1.1), we have

$$\sum_{k=1}^{n} m \, \Delta x_k \leq \sum_{k=1}^{n} f(c_k) \, \Delta x_k \leq \sum_{k=1}^{n} M \, \Delta x_k$$

that is,

$$m \sum_{k=1}^{n} \Delta x_k \leq \sum_{k=1}^{n} f(c_k) \, \Delta x_k \leq M \sum_{k=1}^{n} \Delta x_k$$

(See the second property of summation notation in Section 6.1.) Since the sum of $\Delta x_1, \Delta x_2, \ldots, \Delta x_n$ is the length of I (Problem 31, Section 6.1), we find

$$m(b - a) \leq \sum_{k=1}^{n} f(c_k) \, \Delta x_k \leq M(b - a)$$

The expression between $m(b - a)$ and $M(b - a)$ (which are both constants) is a Riemann sum of f associated with P. The limit of such sums as $\|P\| \to 0$ (which by definition is the integral of f over I) must also lie between $m(b - a)$ and $M(b - a)$, that is,

$$m(b - a) \leq \int_a^b f \leq M(b - a)$$

Dividing each side by $b - a$ (which is positive because $a < b$), we obtain

$$m \leq \frac{1}{b - a} \int_a^b f \leq M$$

The Intermediate Value Theorem (Section 4.2) guarantees a point $c \in I$ such that

$$f(c) = \frac{1}{b - a} \int_a^b f$$

and the conclusion of the theorem follows.

The cases $a > b$ and $a = b$ are left for the problem set.   □

The "right choice of altitude" mentioned before Theorem 3 is in some sense a *mean value* of $f(x)$ in the interval with endpoints a and b. To see

Figure 6
Water in an aquarium

(a)
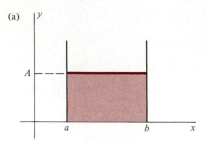

(b)
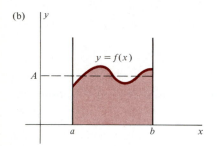

more clearly what we mean, imagine a rectangular aquarium filled with water to a depth A, as shown in Figure 6(a). If the water is disturbed, its surface takes the shape of a curve, say $y = f(x), a \leq x \leq b$, as in Figure 6(b). The "average depth" of the water is A in both cases (because the water settles back to depth A after being disturbed). In the second case it is impossible to compute the average in the usual way, because there are infinitely many depths to consider (one for each value of x between a and b). The average score on a calculus test, for example, is simply the sum of the scores divided by the number of scores. But we cannot use that idea if there are infinitely many scores!

The aquarium analogy suggests what to do. Since the amount of water showing in Figures 6(a) and 6(b) is the same, the shaded areas are equal. Hence

$$A(b - a) = \int_a^b f(x)\, dx$$

from which

$$A = \frac{1}{b - a} \int_a^b f(x)\, dx$$

This leads to the following definition.

> Suppose that f is continuous in the interval $I = [a,b]$, where $a < b$. The *average value of f in I* is defined to be the number
>
> $$A = \frac{1}{b - a} \int_a^b f$$

□ **Example 3**   Find the average value of the function $f(x) = x^2 + 1$ in the interval $I = [0,2]$.

Solution: The average value is

$$A = \frac{1}{2 - 0} \int_0^2 (x^2 + 1)\, dx = \frac{1}{2} \cdot \frac{14}{3} = \frac{7}{3}$$

(The integral was found in Section 6.1.) Note that this value occurs when

$$x^2 + 1 = \frac{7}{3}$$

$$x^2 = \frac{4}{3}$$

$$x = \frac{2}{\sqrt{3}} \approx 1.2$$

(The solution $x = -2/\sqrt{3}$ is not in I.) Thus the point $c \in I$ guaranteed to exist by Theorem 3 is $c = 2/\sqrt{3}$.   □

Example 3 illustrates the fact that the average value A is the same as $f(c)$ in Theorem 3. The Theorem of the Mean for Integrals therefore says that the average value of a function is actually achieved by the function at some point of the interval. In other words

$$f(c) = A = \frac{1}{b-a} \int_a^b f(x)\,dx \qquad \text{for some } c \in I$$

This is similar to the Mean Value Theorem for derivatives in Section 4.3, which in its interpretation involving motion says that the average velocity of an object with coordinate $s = p(t)$, $a \le t \le b$, is actually achieved by the object at some instant in the time interval $I = [a,b]$:

$$p'(c) = \frac{p(b) - p(a)}{b-a} \qquad \text{for some } c \in I$$

Differentiation and integration are evidently connected in some way! We will exploit that connection in the next section.

We end this section by stating three more properties of the integral that are occasionally useful.

□ **Theorem 4**   If f is nonnegative and continuous in $I = [a,b]$, then

$$\int_a^b f(x)\,dx \ge 0$$

Proof: Geometrically obvious! To illustrate the usefulness of the Theorem of the Mean for Integrals, however, we will prove it analytically. According to that theorem, there is a point $c \in I$ such that

$$\int_a^b f(x)\,dx = f(c)(b-a)$$

Since $f(c) \ge 0$ by hypothesis, and since $b - a$ is positive, the conclusion follows. (See Problem 31, Section 6.2, for an argument based on Riemann sums.)   □

□ **Theorem 5**   If f and g are continuous in $I = [a,b]$ and $f(x) \le g(x)$ for all $x \in I$, then

$$\int_a^b f(x)\,dx \le \int_a^b g(x)\,dx$$

Proof: The function $h(x) = g(x) - f(x)$ is nonnegative and continuous in I, so Theorem 4 says that

$$\int_a^b h(x)\,dx \ge 0$$

Since integration is a linear operation (Theorem 2), we have

$$\int_a^b h(x)\, dx = \int_a^b [g(x) - f(x)]\, dx = \int_a^b g(x)\, dx - \int_a^b f(x)\, dx \geq 0$$

from which the conclusion follows. (You may have done this already in Problem 32, Section 6.2.)   □

□ **Theorem 6**   If f is continuous in $[a,b]$, then

$$\left| \int_a^b f(x)\, dx \right| \leq \int_a^b |f(x)|\, dx$$

Proof: See Figure 7 for a geometric argument. The graph of f may go below the x axis, but the graph of $|f|$ does not. Therefore the integral of $|f|$ cannot be less than the integral of f. Similarly it can't be less than the integral of $-f$. Hence the conclusion follows.

For an analytic argument, observe that $-|f(x)| \leq f(x) \leq |f(x)|$ for all $x \in [a,b]$. Apply Theorem 5 to obtain

$$-\int_a^b |f(x)|\, dx \leq \int_a^b f(x)\, dx \leq \int_a^b |f(x)|\, dx$$

from which the conclusion follows. (Recall from Section 1.1 that $|u| \leq p$ if and only if $-p \leq u \leq p$.)   □

Figure 7
Graph of $|f|$ is graph of f with negative portions reflected in the x axis

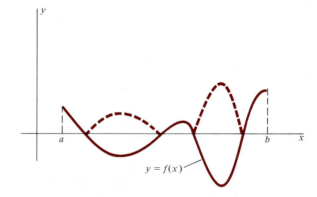

$y = f(x)$

Problem Set 6.3

1. Evaluate $\int_{-1}^2 (x + |x|)\, dx$ in two ways, as follows.

(a) Use the area interpretation of integral.

(b) Explain why

$$\int_{-1}^2 (x + |x|)\, dx = \int_{-1}^0 0\, dx + \int_0^2 2x\, dx$$

and evaluate each integral on the right.

2. Evaluate $\int_0^3 |x - 2|\, dx$ in two ways, as follows.

(a) Use the area interpretation of integral.

(b) Explain why

$$\int_0^3 |x - 2|\, dx = \int_0^2 (2 - x)\, dx + \int_2^3 (x - 2)\, dx$$

and evaluate each integral on the right. (Or use the results of Problems 3 and 4, Section 6.2.)

3. Evaluate $\int_0^2 (x - 1)\, dx$ in two ways, as follows.
   (a) Explain why

$$\int_0^2 (x - 1)\, dx = -\int_0^1 (1 - x)\, dx + \int_1^2 (x - 1)\, dx$$

   and use the area interpretation to evaluate each integral on the right.
   (b) Explain why

$$\int_0^2 (x - 1)\, dx = \int_0^2 x\, dx - \int_0^2 1\, dx$$

   and evaluate each integral on the right.

4. Evaluate $\int_0^3 (5 + 2x)\, dx$ in two ways, as follows.
   (a) Use the area interpretation of integral.
   (b) Explain why

$$\int_0^3 (5 + 2x)\, dx = \int_0^3 5\, dx + 2\int_0^3 x\, dx$$

   and evaluate each integral on the right.

Use the theorems of this section (and the area interpretation of integral) to evaluate each of the following.

5. $\int_0^2 (3x - 4)\, dx$

6. $\int_0^6 (4 - x)\, dx$

7. $\int_{-1}^1 (4x - 3)\, dx$

8. $\int_{-2}^2 (2 - 5x)\, dx$

9. $\int_{-1}^3 (2 - |x|)\, dx$

10. $\int_{-2}^2 (3x - |x|)\, dx$

11. $\int_0^3 \dfrac{|x - 2|}{x - 2}\, dx$

12. $\int_{-2}^2 \dfrac{|1 - x^2|}{1 - x^2}\, dx$

13. $\int_1^4 \dfrac{x^2 - 9}{x - 3}\, dx$

14. $\int_0^\pi \tan t \cot t\, dt$

15. $\int_0^2 (2x - 3\sqrt{4 - x^2})\, dx$ 16. $\int_0^1 (3 - 5x + 2\sqrt{1 - x^2})\, dx$

17. $\int_{-3}^3 (\sqrt{9 - x^2} + |x|)\, dx$ 18. $\int_{-4}^4 \left(\sqrt{16 - x^2} + \dfrac{|x|}{x}\right) dx$

In each of the following, find the average value of the given function in the given interval, and find a point in the interval at which the average value occurs.

19. $f(x) = 5,\ I = [0,2]$
20. $f(x) = 2 - x,\ I = [0,2]$
21. $f(x) = 4 - 2x,\ I = [0,2]$
22. $f(x) = 2x - 7,\ I = [-3,3]$

23. $f(x) = |x|,\ I = [-1,2]$
24. $f(x) = 4 - |x|,\ I = [-2,2]$
25. $f(x) = \sqrt{1 - x^2},\ I = [-1,1]$
26. $f(x) = \sqrt{16 - x^2},\ I = [0,4]$

Assuming that the formula

$$\int_a^b f = \int_a^c f + \int_c^b f$$

has been established for the case $a < c < b$, extend it to the following cases.

27. $b < c < a$ 28. $c < a < b$
29. $c < b < a$ 30. $a < c = b$
31. $b < c = a$

32. The Theorem of the Mean for Integrals is derived in the text for the case $a < b$. Why does the same formula hold in the cases $a > b$ and $a = b$?

33. Suppose that f is continuous in $I = [a,b]$ and $P = \{x_0, x_1, \ldots, x_n\}$ is a partition of I into n subintervals of equal length.
   (a) Letting c_k be any point of the kth subinterval $[x_{k-1}, x_k]$, we may find the ordinary arithmetic mean of the n functional values $f(c_1), f(c_2), \ldots, f(c_n)$ by computing

$$m = \frac{1}{n} \sum_{k=1}^n f(c_k)$$

   Show that

$$m = \frac{1}{b - a} \sum_{k=1}^n f(c_k)\, \Delta x_k$$

   (b) What is the limit of m as $n \to \infty$?
   This explains why the average value of f as defined in the text is a reasonable generalization of the ordinary arithmetic mean of a finite set of functional values.

34. Suppose that f is continuous in $I = [a,b]$. Use the Theorem of the Mean for Integrals to show that if

$$\int_a^b f(x)\, dx = 0$$

   there is at least one point $c \in I$ for which $f(c) = 0$.

35. Give an example showing that the statement in Problem 34 is false if f is not required to be continuous.

36. Draw a picture illustrating Theorem 5 when f and g are nonnegative in I.

37. The Triangle Inequality in Section 1.1 says that $|u + v| \leq |u| + |v|$.

(a) Explain why it follows that if $a_1, a_2, \ldots, a_n$ are real numbers, then

$$\left| \sum_{k=1}^{n} a_k \right| \leq \sum_{k=1}^{n} |a_k|$$

(b) Suppose that f is continuous in $I = [a,b]$ and partition I in the usual way. Explain why

$$\left| \sum_{k=1}^{n} f(c_k)\, \Delta x_k \right| \leq \sum_{k=1}^{n} |f(c_k)|\, \Delta x_k$$

where c_k is any point of the kth subinterval.

(c) Why does it follow that

$$\left| \int_{a}^{b} f(x)\, dx \right| \leq \int_{a}^{b} |f(x)|\, dx?$$

(This is a proof of Theorem 6 based on Riemann sums.)

38. Theorem 1 was not proved in the text in the case $a < c < b$. Fill in the details of the following argument.

(a) Partition the intervals $[a,c]$ and $[c,b]$ by

$$P = \{x_0, x_1, \ldots, x_m\} \quad \text{and} \quad Q = \{x_m, x_{m+1}, \ldots, x_n\}$$

respectively. Why is $R = P \cup Q$ a partition of $[a,b]$? Why do $\|P\| \to 0$ and $\|Q\| \to 0$ when $\|R\| \to 0$?

(b) If c_k is an arbitrary point of the kth subinterval $[x_{k-1}, x_k]$, explain why

$$\sum_{k=1}^{n} f(c_k)\, \Delta x_k = \sum_{k=1}^{m} f(c_k)\, \Delta x_k + \sum_{k=m+1}^{n} f(c_k)\, \Delta x_k$$

(c) Let $\|R\| \to 0$ in part (b) to obtain Theorem 1.

6.4
THE FUNDAMENTAL THEOREM OF CALCULUS

Evaluating the integral of a function over an interval is a formidable undertaking if we rely on a direct application of the definition (unless we program a computer to do approximations by Riemann sums). In this section we develop an alternative procedure that reduces the problem to manageable proportions in many cases.

To motivate the procedure, we return to the problem of motion in a straight line. Suppose that $v(t)$ is the velocity of a moving object at time t and assume that v is continuous and nonnegative in the time interval $I = [a,b]$. Then the distance traveled by the object from $t = a$ to $t = b$ is the number D satisfying $L(P) \leq D \leq U(P)$ for every partition P of I. (See Section 6.1.) In other words,

$$D = \int_{a}^{b} v(t)\, dt \qquad \text{(see the Remark following Theorem 1, Section 6.2)}$$

This number is hard to compute, as we have seen. A simpler way to find the distance is to use the position function $p(t)$. Since v is nonnegative, the object never reverses direction; in these circumstances it is clear that the distance it travels is $D = p(b) - p(a)$, the difference between its terminal position and its initial position. We therefore conclude that

$$\int_{a}^{b} v(t)\, dt = p(b) - p(a)$$

The fascinating thing about this is that p is an antiderivative of v, that is, $p'(t) = v(t)$ for all $t \in I$. Moreover, if q is any antiderivative of v, we know from Section 4.4 (Theorem 2) that $q(t) = p(t) + C$, where C is a constant. Hence

$$q(b) - q(a) = [p(b) + C] - [p(a) + C] = p(b) - p(a)$$

which means that the choice of antiderivative is immaterial:

$$\int_a^b v(t)\, dt = q(b) - q(a)$$

where q is *any* antiderivative of v.

This is too good not to be true in more general circumstances! It seems plausible to assert that if f is any function continuous in $I = [a,b]$, and if F is any antiderivative of f in I, then

$$\int_a^b f = F(b) - F(a)$$

Simply "antidifferentiate" f to get F and evaluate the difference between F at the upper limit and F at the lower limit. This expression occurs so often that mathematicians have devised a special notation for it:

$$\int_a^b f(x)\, dx = F(x)\Big|_a^b = F(b) - F(a)$$

Assuming this assertion to be true (we will prove it shortly), we are in a position to evaluate a good many integrals that were previously painful. (If you have done Problem 34 in Section 6.2, you have proved it already. The argument suggested there, however, is not easy to construct unaided.)

□ **Example 1**   In Section 6.1 we labored to show that the area under the curve $y = x^2 + 1, 0 \le x \le 2$, is $A = \frac{14}{3}$. Now we need only observe that an antiderivative of $f(x) = x^2 + 1$ is $F(x) = \frac{1}{3}x^3 + x$. Hence

$$A = \int_0^2 (x^2 + 1)\, dx = \left(\frac{1}{3}x^3 + x\right)\Big|_0^2 = \left(\frac{8}{3} + 2\right) - 0 = \frac{14}{3} \qquad \square$$

□ **Example 2**   In Example 5 of Section 6.1 we approximated the distance traveled from $t = 0$ to $t = \pi/2$ by an object whose velocity at time t is $v(t) = \cos t$. Now we can find it exactly:

$$D = \int_0^{\pi/2} \cos t\, dt = \sin t\Big|_0^{\pi/2} = \sin\frac{\pi}{2} - \sin 0 = 1 \qquad \square$$

□ **Example 3**   Compute

$$\int_1^3 \frac{dx}{\sqrt{x}}$$

Solution: It is not easy to come up with an antiderivative of $f(x) = 1/\sqrt{x}$ (unless you have had some practice with antidifferentiation).

We will simply tell you that $F(x) = 2\sqrt{x}$ differentiates back into $f(x)$. Hence

$$\int_1^3 \frac{dx}{\sqrt{x}} = 2\sqrt{x}\,\Big|_1^3 = 2\sqrt{3} - 2$$ □

□ **Example 4**   The graph of $y = \sqrt{r^2 - x^2}$ is the upper half of the circle $x^2 + y^2 = r^2$. Hence by the area interpretation of integral we find

$$\int_0^r \sqrt{r^2 - x^2}\, dx = \frac{1}{4}(\text{area of the circle}) = \frac{\pi r^2}{4}$$

Figure 1
Area under $y = \sqrt{r^2 - x^2}$,
$0 \le x \le r$

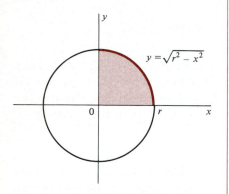

$$y = \sqrt{r^2 - x^2}$$

(See Figure 1.) To confirm this result by our new method, we come up against the problem of finding an antiderivative of $f(x) = \sqrt{r^2 - x^2}$. If you can name one, you are way ahead of us! Of course you may consult the "table of integrals" on the inside back cover, which will inform you that

$$\int \sqrt{r^2 - x^2}\, dx = \frac{1}{2}x\sqrt{r^2 - x^2} + \frac{1}{2}r^2 \sin^{-1}\frac{x}{r} + C$$

(The integral sign without limits means "antiderivative.") However, we have not introduced the function $\sin^{-1}$ ("inverse sine"). Even if we had, it is not easy to explain how the above formula is discovered. It can be checked by differentiation (if the derivative of the inverse sine is known), but that is not very enlightening.   □

Example 4 should bring us back down to earth. Not only are antiderivatives hard to find (in general), they may not even exist (unless we prove otherwise). Hence we confine the present discussion to simple cases.

One aspect of Example 4 is worth noting. We used the phrase "table of *integrals*" and then proceeded to write an integral sign without limits. In Section 6.2 we said that the unadorned symbol $\int f(x)\, dx$ has a different meaning than

$$\int_a^b f(x)\, dx$$

Mathematicians use it to represent the *class of all antiderivatives* of f. If F is one antiderivative, we know from Section 4.4 that the expression $F(x) + C$ represents them all. Hence we write

$$\int f(x)\, dx = F(x) + C$$

Now we are ready to state the *Fundamental Theorem of Calculus* (so named because the recognition of its importance by Newton and Leibniz marked the point in history when calculus is considered to have become a coherent subject).

Fundamental Theorem of Calculus

Suppose that f is continuous in the interval $I = [a,b]$.

1. f has an antiderivative in I. In fact the function $G(x) = \int_a^x f(t)\, dt$ qualifies as one:

$$G'(x) = \frac{d}{dx}\left[\int_a^x f(t)\, dt\right] = f(x) \qquad \text{for all } x \in I$$

2. If F is any antiderivative of f in I,

$$\int_a^b f = F(b) - F(a)$$

It is important to be clear about the function

$$G(x) = \int_a^x f(t)\, dt$$

The independent variable is x, which is why we use the letter t in the integrand (to avoid confusion). An example is

$$G(x) = \int_0^x 2t\, dt$$

in which the integrand is $f(t) = 2t$. The area interpretation of integral (see Figure 2) shows that for $x > 0$

$$G(x) = \int_0^x 2t\, dt = \text{area of the shaded region} = \frac{1}{2}(x)(2x) = x^2$$

As you can see, the derivative of this function is

$$G'(x) = \frac{d}{dx}\left[\int_0^x 2t\, dt\right] = \frac{d}{dx}(x^2) = 2x = f(x)$$

which is what Part 1 of the Fundamental Theorem says.

To see how to prove Part 1, return to the problem of distance. If $v(t)$ is the velocity of a moving object at time t, the distance traveled from $t = a$ to $t = x$ is

$$\int_a^x v(t)\, dt = p(x) - p(a)$$

where $p(t)$ is the position at time t. Assuming that a (the initial point) is fixed, the distance depends on x, that is, it is a function of x. Noting that $p(a)$

Figure 2
Function defined by an integral

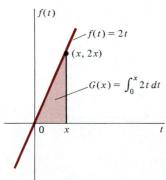

is a constant, we may differentiate each side of the above equation with respect to x to obtain

$$\frac{d}{dx}\left[\int_a^x v(t)\,dt\right] = p'(x) = v(x)$$

This says that the function

$$\int_a^x v(t)\,dt$$

is an antiderivative of $v(x)$. Of course that is no great news; we already know that p is an antiderivative of v. But why not use the same idea in the more general context of a function not known to have an antiderivative?

For simplicity of presentation, we state and prove the two parts of the Fundamental Theorem as separate theorems.

□ **Theorem 1** (*First Part of the Fundamental Theorem*)   Suppose that f is continuous in the interval $I = [a,b]$. The function

$$G(x) = \int_a^x f(t)\,dt \qquad (x \in I)$$

is an antiderivative of f in I, that is,

$$G'(x) = \frac{d}{dx}\left[\int_a^x f(t)\,dt\right] = f(x) \qquad \text{for all } x \in I$$

Proof: Let x be any point of I. We use the definition of derivative,

$$G'(x) = \lim_{z \to x}\frac{G(z) - G(x)}{z - x}$$

Since $G(x) = \int_a^x f \quad$ and $\quad G(z) = \int_a^z f$

Theorem 1 of the last section yields

$$G(z) - G(x) = \int_a^z f - \int_a^x f = \int_x^z f$$

The Theorem of the Mean for Integrals (Section 6.3) says that

$$\int_x^z f = f(c)(z - x) \qquad \text{where } c \text{ is between } x \text{ and } z$$

Hence $$\frac{G(z) - G(x)}{z - x} = f(c)$$

Letting $z \to x$ (which forces $c \to x$), we find

$$G'(x) = \lim_{z \to x}\frac{G(z) - G(x)}{z - x} = \lim_{c \to x} f(c) = f(x)$$

(because f is continuous at x).   □

□ **Theorem 2** (*Second Part of the Fundamental Theorem*)   Suppose that f is continuous in the interval $I = [a,b]$ and F is any antiderivative of f in I. Then

$$\int_a^b f = F(b) - F(a)$$

Proof: We know from Theorem 1 that the function

$$G(x) = \int_a^x f(t)\, dt$$

is an antiderivative of f in I. By Theorem 2, Section 4.4, we also know that $F(x) = G(x) + C$, where C is a constant. Hence

$$F(b) - F(a) = [G(b) + C] - [G(a) + C] = G(b) - G(a)$$

$$= \int_a^b f - \int_a^a f = \int_a^b f$$

Although this proof of Theorem 2 depends on Theorem 1, it is worth noting that the second part of the Fundamental Theorem is in fact independent of the first part. See Problem 34, Section 6.2, in which Theorem 2 is proved by using Riemann sums, and no reference is made to the function

$$G(x) = \int_a^x f(t)\, dt \qquad\qquad □$$

□ **Example 5**   Let $f(x) = \cos x$. The first part of the Fundamental Theorem says that

$$\frac{d}{dx}\left[\int_0^x f(t)\, dt \right] = f(x)$$

that is,

$$\frac{d}{dx}\left(\int_0^x \cos t\, dt \right) = \cos x$$

Since the second part of the Fundamental Theorem is independent of the first part (see the above remark), we can check this result by writing

$$\int_0^x \cos t\, dt = \sin t \Big|_0^x = \sin x$$

Hence

$$\frac{d}{dx}\left(\int_0^x \cos t\, dt \right) = \frac{d}{dx}(\sin x) = \cos x$$

as before.   □

Example 5 vividly illustrates what the Fundamental Theorem really says. *Differentiation and integration are inverse processes.* If one starts with the cosine function, integrating it and then differentiating the result merely returns the cosine function. That is why the integral sign without limits is used for antidifferentiation:

$$\int \cos x \, dx = \sin x$$

from which
$$\frac{d}{dx}\left[\int \cos x \, dx\right] = \frac{d}{dx}(\sin x) = \cos x$$

Of course we are guilty of some imprecision in the above. The correct antidifferentiation formula involves an arbitrary constant:

$$\int \cos x \, dx = \sin x + C$$

While this makes no difference in the final result,

$$\frac{d}{dx}\left[\int \cos x \, dx\right] = \frac{d}{dx}(\sin x + C) = \cos x$$

it does mean that the other way around is not quite so nice. If one starts with the cosine function, differentiating it and then integrating the result does not necessarily return the cosine function:

$$\int \left[\frac{d}{dx}(\cos x)\right] dx = \int (-\sin x) \, dx = \cos x + C$$

This technical failure of differentiation and integration to be inverse processes need not concern us. The crucial relationship between them (as stated in the Fundamental Theorem) is of enormous significance in calculus.

Remark 1   Although the first part of the Fundamental Theorem guarantees that a continuous function has an antiderivative, it does not tell us how to find a simple one. Consider, for example, the function $f(x) = 1/x$. If you did not know Theorem 1, could you name an antiderivative of f? It is in circumstances like these that the usefulness of Theorem 1 becomes clear. For it says that the function

$$F(x) = \int_1^x \frac{dt}{t}$$

is an antiderivative of $f(x)$, that is,

$$\frac{d}{dx}\left(\int_1^x \frac{dt}{t}\right) = \frac{1}{x}$$

Because this antiderivative is important, mathematicians have a special name for it (the *natural logarithm*) and they have worked out an extensive table of values for it. (More recently, they have programmed computers to evaluate it.) We will study this antiderivative in Chapter 8; you need not think about it now. The point we are making is that Theorem 1 provides an antiderivative of every (continuous) function. *Integration is a means of carrying out antidifferentiation.*

Remark 2   The second part of the Fundamental Theorem, on the other hand, says that *antidifferentiation is a way of carrying out integration.* Since it is so impressive to write, say,

$$\int_0^{\pi/2} \cos t\, dt = \sin t \Big|_0^{\pi/2} = \sin\frac{\pi}{2} - \sin 0 = 1$$

beginners often feel that this part of the Fundamental Theorem is the primary revelation. Indeed, a substantial chunk of every elementary calculus course is devoted to discovering antiderivatives that can be used to evaluate integrals. (See Chapter 10.) This can be misleading, however. In practice many integrals are evaluated by numerical methods that do not appeal to Theorem 2 at all (because an antiderivative other than the one supplied by Theorem 1 is difficult or impossible to name).

□ **Example 6**   Find $\dfrac{d}{dx} \displaystyle\int_0^{\sin x} (1 - t^2)\, dt$.

Solution: Theorem 1 is not directly applicable here, because the upper limit is not x but a function of x. We use the Chain Rule (Section 3.3):

$$\frac{d}{dx}\int_0^{\sin x}(1 - t^2)\,dt = \frac{d}{dx}\int_0^{u}(1 - t^2)\,dt \qquad (u = \sin x)$$

$$= \left[\frac{d}{du}\int_0^{u}(1 - t^2)\,dt\right]\frac{du}{dx} \qquad \text{(Chain Rule)}$$

$$= (1 - u^2)\frac{du}{dx} \qquad \text{(Theorem 1, with } u \text{ in place of } x)$$

$$= (1 - \sin^2 x)(\cos x) = \cos^3 x$$

Although it is gilding the lily, we can check this result by Theorem 2:

$$\int_0^{\sin x}(1 - t^2)\,dt = \left(t - \frac{1}{3}t^3\right)\Big|_0^{\sin x} = \sin x - \frac{1}{3}\sin^3 x$$

from which

$$\frac{d}{dx}\int_0^{\sin x}(1 - t^2)\,dt = \frac{d}{dx}\left(\sin x - \frac{1}{3}\sin^3 x\right)$$

$$= \cos x - \frac{1}{3}\cdot 3\sin^2 x \cdot \cos x$$

$$= \cos x(1 - \sin^2 x) = \cos^3 x \qquad\qquad □$$

As a final illustration of the power of the Fundamental Theorem, we derive the first part of Theorem 2, Section 6.3, namely

$$\int_a^b (f + g) = \int_a^b f + \int_a^b g$$

where f and g are continuous in $I = [a,b]$. (See Problem 30, Section 6.2, where a harder argument based on the definition of integral is suggested.)

According to Theorem 1, f and g both have antiderivatives in I, say F and G, respectively. Since $H = F + G$ is an antiderivative of $f + g$ (why?), Theorem 2 says that

$$\int_a^b (f + g) = H(b) - H(a) = [F(b) + G(b)] - [F(a) + G(a)]$$

$$= [F(b) - F(a)] + [G(b) - G(a)] = \int_a^b f + \int_a^b g$$

Trivial!

Since the second part of the Fundamental Theorem depends for its efficacy on our ability to find antiderivatives, it is important to develop a list of known formulas (a "table of integrals"). A good way to start is to check each of the following.

$$\int x^n \, dx = \frac{x^{n+1}}{n + 1} + C \qquad (n \neq -1)$$

$$\int \sin x \, dx = -\cos x + C$$

$$\int \cos x \, dx = \sin x + C$$

$$\int \sec^2 x \, dx = \tan x + C$$

$$\int \csc^2 x \, dx = -\cot x + C$$

$$\int \sec x \tan x \, dx = \sec x + C$$

$$\int \csc x \cot x \, dx = -\csc x + C$$

In each case the check consists in differentiating the right side to get the integrand on the left. For example,

$$D_x(-\csc x) = \csc x \cot x \qquad \text{(Section 3.2)}$$

which confirms the last formula.

The first formula in the preceding list is particularly important. It says that an antiderivative of x^n is found by increasing the exponent by 1 and dividing by the new exponent $(n + 1)$.

□ **Example 7**   Find $\int x^4 \, dx$.

Solution: Increase the exponent by 1 (to obtain x^5) and divide by the new exponent:

$$\int x^4 \, dx = \frac{x^5}{5} + C$$

While this is an easy operation, don't confuse it with differentiation! A common mistake is to think that $\int x^4 \, dx$ is $4x^3$. (!!)   □

□ **Example 8**   Compute

$$\int_1^8 \frac{dx}{\sqrt[3]{x}}$$

Solution: The thing to recognize is that $1/\sqrt[3]{x}$ is a power, namely $x^{-1/3}$. Then we can write

$$\int_1^8 \frac{dx}{\sqrt[3]{x}} = \int_1^8 x^{-1/3}\, dx = \frac{x^{2/3}}{2/3}\Big|_1^8 = \frac{3}{2}\sqrt[3]{x^2}\,\Big|_1^8 = \frac{3}{2}(4-1) = \frac{9}{2}$$

Now go back to Example 3. You should be able to see how we came up with the antiderivative $F(x) = 2\sqrt{x}$ of the integrand $f(x) = 1/\sqrt{x}$.   □

□ **Example 9**   Our brief table of integrals enables us to find

$$\int_0^{\pi/4} \sec^2 t\, dt = \tan t \,\Big|_0^{\pi/4} = \tan\frac{\pi}{4} - \tan 0 = 1 - 0 = 1$$ □

Problem Set 6.4

1. Use integration to find the area under the graph of $y = 4 - x$, $0 \le x \le 4$. (See Problem 9, Section 6.1.)

2. Use integration to find the area under the graph of $y = x + 1$, $0 \le x \le 2$. (See Problem 10, Section 6.1.)

3. Find the area under the curve $y = x^3$, $0 \le x \le 2$. (See Problem 11, Section 6.1.)

4. Find the area under the curve $y = x^2$, $-1 \le x \le 2$. (See Problem 12, Section 6.1.)

5. Find the area under the curve $y = \sin x$, $0 \le x \le \pi$. (See Problem 15, Section 6.1.)

6. Use integration to find the area under the graph of $y = 2x$, $0 \le x \le 3$. (See Problem 20, Section 6.1.)

7. Use integration to find the area under the graph of $y = 3 - x$, $0 \le x \le 2$. (See Problem 21, Section 6.1.)

8. Find the area under the curve $y = 1 - x^2$, $0 \le x \le 1$. (See Problem 22, Section 6.1.)

9. Find the area under the curve $y = 3x^2 + 1$, $0 \le x \le 4$. (See Problem 23, Section 6.1.)

Compute each of the following integrals.

10. $\displaystyle\int_1^4 \sqrt{x}\, dx$

11. $\displaystyle\int_4^9 \frac{dx}{\sqrt{x}}$

12. $\displaystyle\int_0^1 x^{3/2}\, dx$

13. $\displaystyle\int_0^1 x^{2/3}\, dx$

14. $\displaystyle\int_2^5 \frac{dt}{t^2}$

15. $\displaystyle\int_1^2 \frac{dt}{t^3}$

16. $\displaystyle\int_{-1}^1 (x - 2)(x + 2)\, dx$

17. $\displaystyle\int_0^1 (3u - 1)^2\, du$

18. $\displaystyle\int_{-\pi/2}^{\pi/2} \cos t\, dt$

19. $\displaystyle\int_{\pi/3}^{\pi/2} \sin t\, dt$

20. $\displaystyle\int_{\pi/6}^{3\pi/4} (4\cos t - 2\sin t)\, dt$

21. $\displaystyle\int_0^{\pi/3} (2\sin t + \cos t)\, dt$

22. $\displaystyle\int_0^{\pi/3} \sec t \tan t\, dt$

23. $\displaystyle\int_{\pi/6}^{\pi/2} \csc t \cot t\, dt$

24. $\displaystyle\int_{\pi/4}^{\pi/2} \csc^2 t\, dt$

25. $\displaystyle\int_0^{\pi/4} \tan^2 t\, dt$

    Hint: Use the identity $\sec^2 t - \tan^2 t = 1$.

26. $\displaystyle\int_{\pi/3}^{\pi/2} \cot^2 t\, dt$

    Hint: Use the identity $\csc^2 t - \cot^2 t = 1$.

Use integration to find the area of each of the following regions.

27. The region bounded by the curve $y = 9 - x^2$ and the x axis.

28. The region bounded by the curve $y = 16 - x^4$ and the x axis.

29. The region bounded by the curve $y = 1 - x^{2/3}$ and the x axis.

30. The region bounded by the curve $y = 1 - x^{3/2}$ and the coordinate axes.

31. The region bounded by the curve $y = 2 - |x|$ and the x axis. Check by using geometry.

32. The region bounded by the graph of

$$f(x) = \begin{cases} x^3, \ 0 \le x \le 1 \\ (x - 2)^2, \ 1 \le x \le 2 \end{cases}$$

and the x axis.

Find the average value of each of the following functions in the given interval.

33. x^2 in $[1,4]$ **34.** $4 - x^2$ in $[-2,2]$

35. $\sqrt[3]{x}$ in $[0,8]$ **36.** $\sqrt{x}$ in $[0,4]$

37. $\sin t$ in $[0,\pi/2]$

38. $\cos t$ in $[0,\pi]$ Interpret the result geometrically.

39. Let $p(t)$ be the position and $v(t)$ the velocity of a moving object at time t. "Average velocity" during the time interval $I = [a,b]$ is defined in Section 2.1 to be

$$\frac{p(b) - p(a)}{b - a}$$

Explain why this is the same as the "average value of v in I" as defined in Section 6.3.

40. Suppose that f has a continuous derivative in $[a,b]$. Find the average slope of the graph of $y = f(x)$, $a \le x \le b$, and interpret the result geometrically. Do you see any connection with the Mean Value Theorem of Section 4.3?

41. Show that the average value of a linear function $f(x) = ax + b$ in the interval $I = [c,d]$ occurs at the midpoint of I.

Use the first part of the Fundamental Theorem of Calculus to find each of the following derivatives.

42. $\dfrac{d}{dx} \displaystyle\int_0^x \dfrac{dt}{1 + t^2}$ **43.** $D_x \displaystyle\int_0^x \sin(t^2)\, dt$

44. $\dfrac{d}{du} \displaystyle\int_1^u \sqrt{4 - x^3}\, dx$ **45.** $D_t \displaystyle\int_2^t \sqrt{x^2 + 1}\, dx$

46. $D_x \displaystyle\int_x^1 \dfrac{dt}{\sqrt{4 - t^2}}$ **47.** $\dfrac{d}{dx} \displaystyle\int_x^0 \tan t\, dt$

48. $D_x \displaystyle\int_0^{x^2} 2t\, dt$ Check by another method.

49. $\dfrac{d}{dx} \displaystyle\int_0^{\tan x} (1 + t^2)\, dt$ Check by another method.

50. $\dfrac{d}{dx} \displaystyle\int_x^{x^2} |t|\, dt$ **51.** $D_x \displaystyle\int_x^{\sin x} \sqrt{1 - t^2}\, dt$

52. Prove that the function

$$f(x) = \int_x^{2x} \frac{dt}{t}$$

is constant in the interval $(0,\infty)$. *Hint:* What is $f'(x)$?

53. Suppose that f is continuous and nonnegative in $I = [a,b]$. For each $x \in I$ let $A(x)$ be the area under the graph of f from a to x. What is $A'(x)$?

Confirm each of the following formulas by differentiating the right side.

54. $\displaystyle\int x^n\, dx = \dfrac{x^{n+1}}{n + 1} + C$ $(n \ne -1)$

55. $\displaystyle\int \sin ax\, dx = -\dfrac{1}{a} \cos ax + C$ $(a \ne 0)$

56. $\displaystyle\int \cos ax\, dx = \dfrac{1}{a} \sin ax + C$ $(a \ne 0)$

57. $\displaystyle\int \sin^n x \cos x\, dx = \dfrac{\sin^{n+1} x}{n + 1} + C$ $(n \ne -1)$

58. $\displaystyle\int \cos^n x \sin x\, dx = -\dfrac{\cos^{n+1} x}{n + 1} + C$ $(n \ne -1)$

59. The formula $\int \sin 2x\, dx = -\frac{1}{2}\cos 2x + C$ can be checked by differentiation. (See Problem 55.) One can also write

$$\int \sin 2x\, dx = \int 2 \sin x \cos x\, dx$$

and use Problem 57. The answers are apparently different; explain how they can be reconciled.

60. Let $f(x) = |x|$.

(a) Show that $F(x) = \frac{1}{2}x|x|$ is an antiderivative of f in $(-\infty,\infty)$. *Hint:* The point $x = 0$ requires special consideration.

(b) Use the result of part (a) to evaluate $\int_{-1}^2 |x|\, dx$.

(c) Check by using the area interpretation of the integral.

61. The first part of the Fundamental Theorem says that

$$\frac{d}{dx} \int_a^x f(t)\, dt = f(x)$$

where a is the left-hand endpoint of an interval I in which f is continuous. It is also true that if c is any point of I,

$$\frac{d}{dx} \int_c^x f(t)\, dt = f(x)$$

Prove this in three ways, as follows.

(a) Imitate the proof of Theorem 1.

(b) Use Theorem 1, Section 6.3, to write

$$\int_c^x f(t)\,dt = \int_c^a f(t)\,dt + \int_a^x f(t)\,dt$$

and differentiate with respect to x.

(c) Use the second part of the Fundamental Theorem.

62. Suppose that f is continuous in $[a,b]$ and $c \in \mathcal{R}$. Use the Fundamental Theorem to prove that

$$\int_a^b cf = c\int_a^b f$$

(Compare this argument with the one suggested in Problem 29, Section 6.2.)

63. Use the Fundamental Theorem to show that if $c \in \mathcal{R}$, then

$$\int_a^b c\,dx = c(b-a)$$

64. Suppose that f is continuous in $[a,b]$ and $a < c < b$.

(a) Use the Fundamental Theorem to show that

$$\int_a^b f = \int_a^c f + \int_c^b f$$

(b) The argument in part (a) is not really legitimate, but is an example of circular reasoning. Why?

6.5

INTEGRATION BY SUBSTITUTION

The second part of the Fundamental Theorem (Theorem 2 in the last section) says that antidifferentiation can be used to carry out integration: If f is continuous in $[a,b]$ and F is an antiderivative of f, then

$$\int_a^b f(x)\,dx = F(b) - F(a)$$

Except in simple cases, we have not discussed how antiderivatives are found. Sometimes this is a difficult problem. Given the function $f(x) = \sqrt{x^3 + 1}$, for example, can we name a function $F(x)$ whose derivative is $f(x)$? Theoretically the answer is yes; the first part of the Fundamental Theorem guarantees that an antiderivative exists. But there is no "simple" antiderivative, that is, no familiar function that differentiates back into $f(x)$.

In those cases where a simple antiderivative can be found, it is certainly to our advantage to find one, as we saw in the last section. There are many techniques of antidifferentiation designed to help us in this task. (In Chapter 10 we will discuss them in detail.) Our purpose in this section is to show you how a *change of variable* can often be used with great effect.

▫ **Example 1**   Suppose that we want to find

$$\int \sin^2 x \cos x \, dx$$

that is, an antiderivative of

$$f(x) = \sin^2 x \cos x$$

You can probably see by inspection that $F(x) = \frac{1}{3}\sin^3 x$ will serve. (Also see Problem 57, Section 6.4.) Suppose, however, that for some reason this does not occur to you. It might occur to you that a substitution will help, namely $u = \sin x$. For the derivative of this "change of variable" function is $du/dx = \cos x$, or (using the differential notation of Section 4.7)

$$du = \cos x \, dx$$

Formal substitution of u and du in the original integral gives

$$\int \sin^2 x \cos x \, dx = \int u^2 \, du = \frac{u^3}{3} + C = \frac{1}{3} \sin^3 x + C$$

a result that can be checked by differentiation:

$$\frac{d}{dx}\left(\frac{1}{3} \sin^3 x\right) = \frac{1}{3} \cdot 3 \sin^2 x \cdot \cos x = \sin^2 x \cos x \qquad \square$$

This "formal substitution" of u for $\sin x$ and du for $\cos x \, dx$ should go a long way toward explaining the purpose of the "appendage" dx in the notation

$$\int f(x) \, dx$$

In Section 6.2 we said that you should regard this part of the notation as temporarily meaningless, but that eventually it would be harmless to confuse it with the differential. "Harmless" is too weak a word! It is evidently a great help to identify it with the differential.

□ **Example 2**   To find $\displaystyle\int \frac{x \, dx}{\sqrt{4 - x^2}}$

suppose we change the variable to $u = 4 - x^2$. Then $du = -2x \, dx$, which suggests that we should insert -2 in the integrand to go with the $x \, dx$ already present. Compensating for this by putting $-\frac{1}{2}$ in front of the integral, we write

$$\int \frac{x \, dx}{\sqrt{4 - x^2}} = -\frac{1}{2} \int \frac{-2x \, dx}{\sqrt{4 - x^2}} = -\frac{1}{2} \int \frac{du}{\sqrt{u}} = -\frac{1}{2} \int u^{-1/2} \, du$$

$$= -\frac{1}{2} \frac{u^{1/2}}{1/2} + C \qquad \text{(Power Rule)}$$

$$= -\sqrt{u} + C = -\sqrt{4 - x^2} + C$$

As you can check, this is an antiderivative of the original integrand.   □

Evidently what is happening here is that a substitution $u = h(x)$, together with $du = h'(x) \, dx$, changes the original integrand $f(x) \, dx$ to a new one, $g(u) \, du$. In other words,

$$\int f(x) \, dx \quad \text{becomes} \quad \int g(u) \, du$$

If an antiderivative of g can be found, say G, the problem is solved. For if $G'(u) = g(u)$, then (by the Chain Rule)

$$\frac{d}{dx} G(u) = \frac{d}{du} G(u) \frac{du}{dx} = G'(u) \frac{du}{dx}$$

$$= g(u) \frac{du}{dx} = f(x) \qquad \text{[because } g(u) \, du = f(x) \, dx\text{]}$$

But this means that we have found an antiderivative of f, namely

$$F(x) = G(u) = G[h(x)] \qquad [\text{because } F'(x) = \frac{d}{dx}G(u) = f(x)]$$

We state these observations in the form of a theorem.

□ **Theorem 1**   To find $\int f(x)\,dx$, suppose that we change the variable by substituting a differentiable function $u = h(x)$. If this changes $f(x)\,dx$ to $g(u)\,du$, where g is continuous, and if G is an antiderivative of g, then

$$\int f(x)\,dx = \int g(u)\,du = G(u) + C = G[h(x)] + C$$

that is, $G[h(x)]$ is an antiderivative of $f(x)$.   □

Of course the main point of an antiderivative formula is its use in evaluating integrals. Consider the following examples.

□ **Example 3**   To find $\int_0^{\pi/2} \sin^2 x \cos x\,dx$, we proceed as in Example 1 by letting $u = \sin x$, $du = \cos x\,dx$:

$$\int \sin^2 x \cos x\,dx = \int u^2\,du$$

$$= \frac{u^3}{3} = \frac{1}{3}\sin^3 x \qquad \text{(arbitrary constant omitted)}$$

Hence
$$\int_0^{\pi/2} \sin^2 x \cos x\,dx = \frac{1}{3}\sin^3 x \Big|_0^{\pi/2} = \frac{1}{3}$$

There is a shortcut, however. Observe that the new variable, $u = \sin x$, takes on the values $u = 0$ when $x = 0$ and $u = 1$ when $x = \pi/2$. It seems reasonable (we will justify it shortly) to write

$$\int_0^{\pi/2} \sin^2 x \cos x\,dx = \int_0^1 u^2\,du = \frac{u^3}{3}\Big|_0^1 = \frac{1}{3}$$

thereby avoiding the return to the original variable in our first method.
                                                                              □

□ **Example 4**   To find

$$\int_0^1 \frac{x\,dx}{\sqrt{4 - x^2}}$$

we may proceed as in Example 2 by substituting $u = 4 - x^2$, $du = -2x\,dx$.

This time, however, insert new limits as soon as the variable is changed:

$$\int_0^1 \frac{x\,dx}{\sqrt{4-x^2}} = -\frac{1}{2}\int_0^1 \frac{-2x\,dx}{\sqrt{4-x^2}}$$

$$= -\frac{1}{2}\int_4^3 \frac{du}{\sqrt{u}} \qquad [u=4 \text{ when } x=0,\ u=3 \text{ when } x=1]$$

$$= \frac{1}{2}\int_3^4 u^{-1/2}\,du \qquad \left(\int_b^a f = -\int_a^b f\right)$$

$$= \frac{1}{2}\cdot 2u^{1/2}\Big|_3^4 = \sqrt{u}\,\Big|_3^4 = 2-\sqrt{3} \qquad\qquad \square$$

We justify this shortcut by the following theorem.

□ **Theorem 2**   To find $\int_a^b f(x)\,dx$, suppose that we change the variable by substituting a differentiable function $u = h(x)$. If this changes $f(x)\,dx$ to $g(u)\,du$, where g is continuous, then

$$\int_a^b f(x)\,dx = \int_{h(a)}^{h(b)} g(u)\,du$$

Proof: Let G be an antiderivative of g (guaranteed to exist by the first part of the Fundamental Theorem). According to Theorem 1, the function $F(x) = G[h(x)]$ is an antiderivative of $f(x)$. Hence

$$\int_a^b f(x)\,dx = F(b) - F(a) = G[h(b)] - G[h(a)]$$

But this is the same as

$$\int_{h(a)}^{h(b)} g(u)\,du$$

because G is an antiderivative of g. $\qquad\qquad\qquad\qquad\qquad\qquad\qquad$ □

The reader may be troubled by one aspect of Theorems 1 and 2. Each of them proposes *as part of the hypothesis* that the substitution $u = h(x)$ changes $f(x)\,dx$ to $g(u)\,du$, where g seems to pop up from nowhere. We are free to choose the change of variable function $[u = h(x)]$ as we please, but how do we know what g is?

The answer is that when the substitution $u = h(x)$, $du = h'(x)\,dx$ is made in $\int f(x)\,dx$, the function g will automatically appear if the substitution is a "good" one. For example, it may not be apparent what g is when we put $u = 4 - x^2$, $du = -2x\,dx$ into

$$\int \frac{x\,dx}{\sqrt{4-x^2}}$$

But go ahead and make the substitution anyway:

$$\int \frac{x\,dx}{\sqrt{4-x^2}} = -\frac{1}{2}\int \frac{-2x\,dx}{\sqrt{4-x^2}} = -\frac{1}{2}\int \frac{du}{\sqrt{u}} = -\frac{1}{2}\int u^{-1/2}\,du$$

Now it is clear that $g(u) = -\frac{1}{2}u^{-1/2}$!

It is hard to tell ahead of time which substitutions are "good." The point is that the hypotheses of Theorems 1 and 2 take care of themselves when the substitution works; you need not worry about them while you are testing various possibilities. Choosing good substitutions is more of an art than a science, anyway. One learns it by practice.

□ **Example 5**   To evaluate

$$\int_0^\pi \sin x\,(1 + \cos^3 x)\,dx$$

we break the integral into a sum,

$$\int_0^\pi \sin x\,dx + \int_0^\pi \cos^3 x \sin x\,dx$$

The first of these is easy:

$$\int_0^\pi \sin x\,dx = -\cos x\,\Big|_0^\pi = \cos x\,\Big|_\pi^0 = 1 - (-1) = 2$$

To evaluate the second, let $u = \cos x$, $du = -\sin x\,dx$:

$$\int_0^\pi \cos^3 x \sin x\,dx = -\int_0^\pi \cos^3 x(-\sin x\,dx)$$

$$= -\int_1^{-1} u^3\,du \qquad [u = 1 \text{ when } x = 0,$$
$$u = -1 \text{ when } x = \pi]$$

$$= \int_{-1}^1 u^3\,du = \frac{u^4}{4}\Big|_{-1}^1 = \frac{1}{4} - \frac{1}{4} = 0$$

Hence

$$\int_0^\pi \sin x\,(1 + \cos^3 x)\,dx = 2 + 0 = 2 \qquad\qquad □$$

Remark   In Example 5 note the change from

$$-\cos x\,\Big|_0^\pi \qquad \text{to} \qquad \cos x\,\Big|_\pi^0$$

This is justified by the formula $\int_b^a f = -\int_a^b f$ and is often a way of avoiding too many minus signs. It is easier to write

$$\cos x\,\Big|_\pi^0 = \cos 0 - \cos \pi = 1 - (-1) = 2$$

than $-\cos x \Big|_0^\pi = -\cos \pi - (-\cos 0) = -(-1) - (-1) = 1 + 1 = 2$

It is also worth noting in Example 5 that

$$\int_{-1}^{1} u^3 \, du = 0$$

because substitution of 1 and -1 in $u^4/4$ gives the same number. More generally,

$$\boxed{\int_{-a}^{a} f(x) \, dx = 0 \qquad \text{if } f \text{ is odd}}$$

Recall that an odd function has the property $f(-x) = -f(x)$ for all x in its domain. An *even* function, on the other hand, satisfies the identity $f(-x) = f(x)$. In that case (as we ask you to prove in the problem set)

$$\boxed{\int_{-a}^{a} f(x) \, dx = 2 \int_{0}^{a} f(x) \, dx \qquad \text{if } f \text{ is even}}$$

These two formulas are worth remembering; they save time and trouble in some of the problems to come.

□ **Example 6**   To find $\int x \sqrt{x + 2} \, dx$, let $u = \sqrt{x + 2}$. The procedure used in previous examples calls for differentiation of this function to find

$$du = \frac{dx}{2\sqrt{x + 2}}$$

It is easier, however, to observe that $u^2 = x + 2$, from which $x = u^2 - 2$ and $dx = 2u \, du$. Hence

$$\int x \sqrt{x + 2} \, dx = \int (u^2 - 2) \cdot u \cdot 2u \, du = \int (2u^4 - 4u^2) \, du$$

$$= \frac{2}{5} u^5 - \frac{4}{3} u^3 + C$$

$$= \frac{2}{5} (x + 2)^{5/2} - \frac{4}{3} (x + 2)^{3/2} + C$$

Another substitution that may be used in this example is $u = x + 2$. You may find it shorter.   □

□ **Example 7**   Evaluate $\int_{-1}^{1} \sqrt{1 - x^2} \, dx$ by making use of the substitution $x = \sin u$.

Solution: This substitution is not of the type called for in Theorem 2 [which refers to a change of variable function $u = h(x)$ that gives the new variable in terms of the old]. The equation $x = \sin u$, however, defines u implicitly in terms of x. Figure 1 shows that this definition is unambiguous if $-\pi/2 \leq u \leq \pi/2$, that is, x determines a unique value of u in this interval.

Figure 1
Function $u = h(x)$ implicitly defined
by $x = \sin u$, $-\pi/2 \le u \le \pi/2$

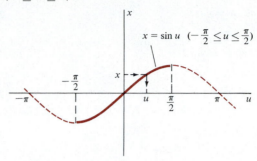

In other words, there is a function $u = h(x)$ lurking behind the equation $x = \sin u$ (if we agree to the restriction $-\pi/2 \le u \le \pi/2$). We need not name it explicitly, but may differentiate $x = \sin u$ as it stands to obtain $dx = \cos u \, du$. Since

$$\sqrt{1 - x^2} = \sqrt{1 - \sin^2 u} = \sqrt{\cos^2 u} = |\cos u| = \cos u$$

(because $\cos u \ge 0$ when $-\pi/2 \le u \le \pi/2$), the original integral is

$$\int_{-1}^{1} \sqrt{1 - x^2} \, dx = \int_{-\pi/2}^{\pi/2} \cos u \cdot \cos u \, du = \int_{-\pi/2}^{\pi/2} \cos^2 u \, du$$

The identity $\cos^2 u = \frac{1}{2}(1 + \cos 2u)$ (Section 2.3) reduces this to

$$\frac{1}{2} \int_{-\pi/2}^{\pi/2} (1 + \cos 2u) \, du = \frac{1}{2} \left(u + \frac{1}{2} \sin 2u \right) \Big|_{-\pi/2}^{\pi/2} = \frac{\pi}{2}$$

It is worth mentioning that in this example it is much easier to use the area interpretation of integral:

$$\int_{-1}^{1} \sqrt{1 - x^2} \, dx = \text{area of upper half of unit circle} = \frac{\pi}{2}$$

The point of the example is not to make a simple problem hard, but to illustrate the flexibility of our substitution theorems.   □

Problem Set 6.5

Find each of the following antiderivatives by making use of the suggested substitution.

1. $\int (2x - 1)^4 \, dx$, $u = 2x - 1$

2. $\int x^2(x^3 + 1)^2 \, dx$, $u = x^3 + 1$

3. $\int x\sqrt{1 + x^2} \, dx$, $u = 1 + x^2$

4. $\int \dfrac{dx}{\sqrt{3x - 1}}$, $u = 3x - 1$

5. $\int \dfrac{dx}{\sqrt{3x - 1}}$, $u = \sqrt{3x - 1}$

6. $\int \tan t \sec^2 t \, dt, \ u = \tan t$

7. $\int \cos^4 \theta \sin \theta \, d\theta, \ u = \cos \theta$

8. $\int \sin^3 2t \cos 2t \, dt, \ u = \sin 2t$

9. $\int \sec^2 \frac{1}{2} t \, dt, \ u = \frac{1}{2} t$

10. $\int \sec^3 t \tan t \, dt, \ u = \sec t$

Evaluate each of the following integrals by making use of the suggested substitution.

11. $\int_{\pi/6}^{\pi/4} \cos 2t \, dt, \ u = 2t$

12. $\int_0^1 \frac{x^2 \, dx}{(x^3 + 1)^{3/2}}, \ u = x^3 + 1$

13. $\int_0^2 \frac{x \, dx}{\sqrt{9 - x^2}}, \ u = 9 - x^2$

14. $\int_{-1}^1 x \sqrt{1 + x^2} \, dx, \ u = 1 + x^2$

15. $\int_{-1}^4 \frac{x \, dx}{\sqrt{x + 5}}, \ u = \sqrt{x + 5}$

16. $\int_0^3 \sqrt{9 - x^2} \, dx, \ x = 3 \sin u \quad (0 \leq u \leq \pi/2)$
Check by geometry.

17. $\int_0^1 \frac{dx}{\sqrt{4 - x^2}}, \ x = 2 \sin u \quad (0 \leq u \leq \pi/6)$

18. $\int_0^1 \frac{dx}{1 + x^2}, \ x = \tan u \quad (0 \leq u \leq \pi/4)$

19. $\int_0^1 \frac{x^3 \, dx}{\sqrt{4 - x^2}}, \ u = \sqrt{4 - x^2}$

20. $\int_0^1 \frac{x^3 \, dx}{\sqrt{x^2 + 1}}, \ u = \sqrt{x^2 + 1}$

Find each of the following by using an appropriate substitution.

21. $\int x^2 (1 + x^3)^4 \, dx$ **22.** $\int \frac{x^2 \, dx}{\sqrt{1 + x^3}}$

23. $\int t \sin t^2 \, dt$

24. $\int \frac{\sin \sqrt{t}}{\sqrt{t}} \, dt$

25. $\int \frac{dt}{\cos^2 2t}$

26. $\int \frac{(x - 1) \, dx}{(x^2 - 2x + 3)^2}$

27. $\int \frac{(2x - 1) \, dx}{\sqrt{x^2 - x}}$

28. $\int_3^4 x \sqrt{25 - x^2} \, dx$

29. $\int_0^3 \frac{x \, dx}{\sqrt{x^2 + 9}}$

30. $\int_0^3 \sqrt{3 - x} \, dx$

31. $\int_0^1 x \sqrt[3]{1 - x} \, dx$

32. $\int_{\pi/6}^{\pi/2} \frac{\cos x \, dx}{1 - \cos^2 x}$

33. $\int_0^\pi \cos x \sqrt{\sin x} \, dx$

34. $\int_{-\pi/4}^{\pi/4} \sec^2 \theta \tan^2 \theta \, d\theta$

35. $\int \cot t \csc^2 t \, dt$ can be found by either of the substitutions $u = \cot t$ or $u = \csc t$. Do it both ways and reconcile the answers.

36. Discuss the following evaluations of

$$\int_0^{\pi/4} \sin^2 x \, dx$$

(at least one of which must be wrong!).

(a) Since $\sin^2 x = \frac{1}{2}(1 - \cos 2x)$,

$$\int_0^{\pi/4} \sin^2 x \, dx = \frac{1}{2} \int_0^{\pi/4} (1 - \cos 2x) \, dx$$

$$= \frac{1}{2} \left(x - \frac{1}{2} \sin 2x \right) \Big|_0^{\pi/4} = \frac{\pi}{8} - \frac{1}{4}$$

(b) Let $u = \sin x, \ du = \cos x \, dx$, and write

$$\int \sin^2 x \, dx = \frac{1}{\cos x} \int \sin^2 x \cos x \, dx = \frac{1}{\cos x} \int u^2 \, du$$

$$= \frac{1}{\cos x} \cdot \frac{u^3}{3} = \frac{1}{3} \cdot \frac{\sin^3 x}{\cos x}$$

$$= \frac{1}{3} \sin^2 x \tan x \quad \text{(arbitrary constant omitted)}$$

Then

$$\int_0^{\pi/4} \sin^2 x \, dx = \frac{1}{3} \sin^2 x \tan x \Big|_0^{\pi/4} = \frac{1}{6}$$

37. Eventually we will define the function $\ln x$ (natural logarithm of x) in such a way that the formula

$$\int \frac{dx}{x} = \ln |x| + C$$

will hold. Assuming this to have been done, derive the following formulas by means of the suggested substitution.

(a) $\int \tan x \, dx = -\ln|\cos x| + C$ (Let $u = \cos x$.)

(b) $\int \cot x \, dx = \ln|\sin x| + C$ (Let $u = \sin x$.)

38. Find the area under the curve $y = x/\sqrt{x^2 + 4}$ between $x = 0$ and $x = 1$.

39. Find the area under one arch of the curve $y = \sin 2x$.

40. The velocity of a moving object at time t is $v(t) = \sin \pi t$. How far does the object travel from $t = 0$ to $t = 1$?

41. Due to an increasing wind, the speed of a sailboat t hours after the wind starts is $\sqrt{100 + 21t}$ km/hr. After an hour of sailing with the wind, how far has the sailboat gone?

42. Find the average value of $f(x) = x \cos x^2$ in the interval $-1 \le x \le 1$.

43. Find the average value of $f(x) = \sin^2 x$ in the interval $[0, \pi]$.

44. Suppose that f is continuous in $I = [-a, a]$. Explain why

$$\int_{-a}^{a} f(x) \, dx = 0 \qquad \text{if } f \text{ is odd}$$

Hint: Break the interval of integration into $[-a, 0]$ and $[0, a]$ and use the substitution $u = -x$ in the first integral.

45. In Problem 44 suppose that f is even. Show that

$$\int_{-a}^{a} f(x) \, dx = 2 \int_{0}^{a} f(x) \, dx$$

True–False Quiz

1. If $f(x) = 3$, each Riemann sum of f associated with a partition of $[0,2]$ is 6.

2. If $P = \{x_0, x_1, \dots, x_n\}$ is a partition of $[0, \pi/2]$, then

$$\lim_{\|P\| \to 0} \sum_{k=1}^{n} (\cos x_k) \, \Delta x_k = 1$$

3. If f is differentiable in $[a,b]$, it is integrable over $[a,b]$.

4. If $\int_I f$ exists, f must be bounded in the interval I.

5. $\int_0^1 \frac{\tan x}{x} \, dx$ exists.

6. $\int_0^1 (x - 1)^2 \, dx = \int_0^1 (t - 1)^2 \, dt$.

7. $\int_1^2 x^2 \, dx = \int_1^3 x^2 \, dx + \int_3^2 x^2 \, dx$.

8. The average value of $\sqrt{x}$ in the interval $[0,1]$ is $\frac{2}{3}$.

9. If c is a constant, then $\int_a^b c \, dx = c(b - a)$.

10. $\int_{-4}^{4} \sqrt{16 - x^2} \, dx = 8\pi$.

11. The area of the region bounded by the curve $y = 1 - x^2$ and the x axis is $\frac{4}{3}$.

12. The distance traveled during the time interval $[1,3]$ by a particle whose velocity at time t is $3t^2$ is 7.

13. The function $|x|$ has an antiderivative in the interval $[-1,1]$.

14. If $f(x) = \int_0^x \sin^2 t \, dt$, then $f''(x) = \sin 2x$.

15. If the graph of f passes through $(0,0)$ and $(1,1)$, and f' is continuous, then

$$\int_0^1 f'(x) \, dx = 1$$

16. $\dfrac{d}{dx} \int_x^1 t^3 \, dt = -x^3$.

17. If $F(x) = x^2 + 3$, then $\int_2^5 2x \, dx = F(5) - F(2)$.

18. If $F(x) = \int_0^x u \, du$ and $G(x) = \int_1^x v \, dv$, then $F(x) - G(x) = \frac{1}{2}$.

19. If $0 < x < 1$, then $\int_1^x \frac{dt}{t} < 0$.

20. $\int_0^{\pi/4} \sec^2 t \, dt = 1$.   21. $\int_1^4 x^{-1/2} \, dx = 2$.

22. The substitution $u = \sqrt{1 - x}$ changes $\int_0^1 \sqrt{1 - x} \, dx$ to $\int_0^1 2u^2 \, du$.

Additional Problems

1. If the interval $[1,3]$ is partitioned into four subintervals of equal length, what is the smallest Riemann sum of $f(x) = 1/x$ associated with the partition? the largest? Compare their average with the true value

$$\int_1^3 \frac{dx}{x} = 1.0986 \cdots$$

(to be discussed in Chapter 8).

2. If the interval $[0,2]$ is partitioned into four subintervals of equal length, what is the smallest Riemann sum of $f(x) = \sqrt{x}$ associated with the partition? the largest? Compare their average with the value of $\int_0^2 \sqrt{x}\, dx$.

3. In Problem 1 compute the midpoint approximation based on evaluation of f at the midpoint of each subinterval of the partition.

4. In Problem 2 compute the midpoint approximation.

5. Let $f(x) = 2x$ and partition the interval $[0,4]$ into four subintervals of equal length.

  (a) Compute the midpoint approximation.

  (b) Confirm that the answer in part (a) is not merely an approximation of $\int_0^4 2x\, dx$, but equals it.

6. Let $f(x) = 4x - 5$ and partition the interval $[0,3]$ into six subintervals of equal length.

  (a) Compute the midpoint approximation.

  (b) Confirm that the answer in part (a) is not merely an approximation of $\int_0^3 (4x - 5)\, dx$, but equals it.

7. Repeat Problem 5 for a partition of $[0,4]$ into n subintervals of equal length. *Hint:* Use the formula $1 + 3 + 5 + \cdots + (2n - 1) = n^2$ (proved in algebra by mathematical induction).

8. Repeat Problem 6 for a partition of $[0,3]$ into n subintervals of equal length.

9. Let D be the distance traveled during the time interval $[0,1]$ by an object whose velocity at time t is $v(t) = 2t$.

  (a) If P is a partition of $[0,1]$ into n subintervals of equal length, show that $U(P) = 1 + 1/n$. *Hint:* Use the formula $1 + 2 + 3 + \cdots + n = \frac{1}{2}n(n + 1)$.

  (b) Show that $L(P) = 1 - 1/n$.

  (c) Use parts (a) and (b) to find D.

  (d) Find D by integration via the Fundamental Theorem.

10. Let A be the area under the graph of $f(x) = 7 - 2x$, $0 \le x \le 3$.

  (a) If P is a partition of $[0,3]$ into n subintervals of equal length, show that $L(P) = 12 - 9/n$.

  (b) Show that $U(P) = 12 + 9/n$.

  (c) Use parts (a) and (b) to find A.

  (d) Find A by elementary geometry.

  (e) Use integration via the Fundamental Theorem to find A.

11. The interval $[0,2]$ is divided into n subintervals of equal length Δx and a point c_k is chosen in the kth subinterval, $k = 1, 2, \ldots, n$. What number does the sum $\sum_{k=1}^n (c_k^2 - 1)\, \Delta x$ approach as n increases?

12. The interval $[1,4]$ is divided into n subintervals of equal length Δx and a point c_k is chosen in the kth subinterval, $k = 1, 2, \ldots, n$. Find $\lim_{\Delta x \to 0} \sum_{k=1}^n (1/c_k^2)\, \Delta x$.

Evaluate each of the following integrals.

13. $\displaystyle\int_2^7 8\, dx$

14. $\displaystyle\int_0^6 (5 - 2x)\, dx$

15. $\displaystyle\int_{-2}^3 (x - |x|)\, dx$.

16. $\displaystyle\int_0^{10} [x]\, dx$. (See Problem 16, Section 1.4.)

17. $\displaystyle\int_{-5}^0 \sqrt{25 - x^2}\, dx$

18. $\displaystyle\int_0^1 (1 - x)^2\, dx$

19. $\displaystyle\int_0^\pi \sin^2 t\, dt + \int_0^\pi \cos^2 t\, dt$

20. $\displaystyle\int_0^{\pi/3} \tan^2 t\, dt + \int_0^{\pi/3} \sec^2 t\, dt$

21. $\displaystyle\int_0^8 \sqrt[3]{x}\, dx$

22. $\displaystyle\int_{-3}^3 (3 - x)(3 + x)\, dx$

23. $\displaystyle\int_0^{\pi/2} (\cos t - \sin t)\, dt$

24. $\displaystyle\int_0^{\pi/4} \sec t \tan t\, dt$

25. $\displaystyle\int_{\pi/4}^{\pi/2} \cot^2 t\, dt$

26. $\displaystyle\int_0^{\pi/2} \sin^2 t \cos t\, dt$

27. $\displaystyle\int_0^{\pi/4} \sin t \cos^3 t\, dt$

28. $\displaystyle\int_0^4 (5 - x^{3/2})\, dx$

29. $\displaystyle\int_0^3 \frac{x\, dx}{\sqrt{16 - x^2}}$

30. $\displaystyle\int_0^1 \frac{x\, dx}{\sqrt{1 + x^2}}$

31. $\displaystyle\int_{\pi/3}^{\pi/2} \csc^2 \tfrac{1}{2}t\, dt$

32. $\displaystyle\int_0^1 x^3 (2 - x^4)^3\, dx$

33. $\int \dfrac{x^3\,dx}{\sqrt{x^2-1}}$ *Hint:* Let $u = \sqrt{x^2-1}$

34. $\int \dfrac{dx}{\sqrt{2x+1}}$

35. Suppose you are asked to evaluate $\int_0^{\pi/2} \sec t \tan t\,dt$. What is your answer?

36. Repeat Problem 35 in the case of $\int_0^3 \dfrac{x^2-9}{x-3}\,dx$.

37. Suppose you are told that f is continuous and that

$$\int_0^2 f(x)\,dx = 6 \quad \text{and} \quad \int_2^5 f(x)\,dx = 4$$

What is the value of

$$\int_0^5 f(x)\,dx?$$

38. Let A be the area of the region under the graph of $y = \cos x$, $0 \le x \le \pi/2$. If one wants to draw a rectangle with base on the x axis from 0 to $\pi/2$, and area A, what should be its height? What is this number called?

39. Find the average value of $\sin x$ in the interval $[0,\pi]$. At how many points in the interval is this value attained?

40. Find the average height of a castle door bounded by the parabola $y = 4 - x^2$ and the x axis.

41. Show that the average value of $f'(x)$ in the interval $[x, x+h]$ is the same as the difference quotient

$$\dfrac{f(x+h) - f(x)}{h}$$

42. Use the definition of "average value of a function" to find the average speed of a ball dropped from the top of a 400-ft building.

43. Use the definition of "average value of a function" to show that the average speed of a car with constant acceleration from 0 mph to V mph is $V/2$.

44. If $A(x)$ is the area under the graph of $y = f(t)$, $a \le t \le x$, what is $A'(x)$?

45. Find $\dfrac{d}{dx} \int_0^x \sec^2 t\,dt$ in two ways.

46. Find $\dfrac{d}{dx} \int_1^x \cot t\,dt$. What is the domain of this function?

47. The velocity of a moving object at time t is $v(t) = \cos \pi t$. How far does the object travel from $t = 0$ to $t = \frac{1}{2}$?

48. Repeat Problem 47 for $v(t) = 16 - 32t$.

49. Find the area of the region in the first quadrant bounded by the graphs of $y = x^2$, $y = 1/x^2$, $y = 0$, and $x = 2$.

50. Find the area under the curve $y = \sqrt{1+x}$, $0 \le x \le 3$.

51. Find the area under the curve $y = 1/x^2$ from $x = 1$ to $x = b$ ($b > 1$). What does this area approach as the boundary $x = b$ is moved to the right?

52. $\int \tan t \sec^2 t\,dt$ can be found by either of the substitutions $u = \tan t$ or $u = \sec t$. Do it both ways and reconcile the answers.

53. Prove that the function

$$f(x) = \int_{2x}^{3x} \dfrac{dt}{t}$$

is constant in the interval $(0,\infty)$.

54. Use the substitution $u = x - a$ to show that

$$\int_a^b (x-a)(b-x)\,dx = \dfrac{1}{6}(b-a)^3$$

7 | Applications of Integration

Mathematics is the gate and key of the sciences.
ROGER BACON (c. 1214–c. 1294)

The essential fact is that all the pictures which science now draws of nature, and which alone seem capable of according with observational facts, are mathematical pictures.
JAMES JEANS (1877–1946)

To see what is general in what is particular, and what is permanent in what is transitory, is the aim of scientific thought.
ALFRED NORTH WHITEHEAD (1861–1947)

This chapter is devoted to a few of the numerous applications of integration. To avoid a digression into the terminology needed for sciences in which integration is important, we have not gone very far outside mathematics. The examples and problems are mostly confined to relatively straightforward geometrical and physical questions, such as area, volume, length, work, and center of mass. These are traditional questions to take up in a calculus course, both because they lend themselves to easy analysis and because the results are useful in later study. The student who is interested in more recent applications may not see the relevance of some of these topics. The purpose of the chapter, however, is to discuss a method of *measurement by integration* that is profoundly relevant to practical matters.

At the end of the chapter we also present an introduction to differential equations, a subject in which "integration by antidifferentiation" is the main tool. We will return to it frequently in the rest of the book.

7.1 AREA BETWEEN CURVES

We began our study of integration (Section 6.1) with a discussion of "area under a curve." If f is nonnegative and continuous in $[a,b]$, the area of the region under the graph of $y = f(x)$, $a \le x \le b$, is the number

$$A = \int_a^b f(x)\,dx$$

Of course the region in this discussion is of a special type, with three straight boundaries (the x axis and the vertical lines $x = a$ and $x = b$). Only its upper boundary is allowed to be curved. In this section we use Riemann sums to compute the area of regions with curved upper and lower boundaries.

Suppose that f and g are continuous in $I = [a,b]$ and $f(x) \geq g(x)$ for all $x \in I$. (See Figure 1.) To find the area of the shaded region (more precisely, to *define* it), let P be a partition of I into n subintervals. Choosing a point x in the typical subinterval, we draw a rectangle of base Δx and height $f(x) - g(x)$ as shown in the figure. Its area is

$$\Delta A = [f(x) - g(x)]\,\Delta x$$

where the symbol ΔA is intended to suggest a "piece" of the total area. Thinking of the region as a union of thin vertical strips, we are replacing the typical strip (which in general has curved upper and lower boundaries) by the rectangle shown in Figure 1. Our intuition suggests that the total area is approximately

$$A \approx \sum \Delta A = \sum [f(x) - g(x)]\,\Delta x$$

and that this approximation improves as the strips get thinner (that is, as the norm of the partition approaches zero). Letting $h(x) = f(x) - g(x)$, we therefore expect the area to be

$$A = \lim_{\|P\| \to 0} \sum h(x)\,\Delta x$$

where the sum is understood to consist of n terms of the form $\Delta A = h(x)\,\Delta x$.

Strictly speaking, only the right side of this equation has meaning; the area A is *defined* by the equation. (See our remarks about this point in

Figure 1
Region between two curves

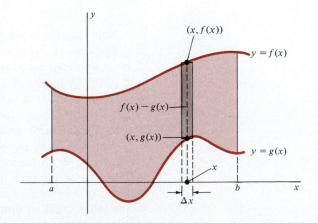

Section 6.1.) Since $\sum h(x)\,\Delta x$ is a Riemann sum of the function h associated with the partition P, we know that

$$\lim_{\|P\|\to 0} \sum h(x)\,\Delta x = \int_a^b h(x)\,dx = \int_a^b [f(x) - g(x)]\,dx$$

Hence the "area between the curves" is

$$A = \int_a^b [f(x) - g(x)]\,dx$$

It is worth observing that when $g(x) = 0$ in I (in which case the lower boundary of the region is the x axis and f is nonnegative) this definition reduces to our earlier one, namely

$$A = \int_a^b f(x)\,dx$$

Remark   The definition of "Riemann sum of h associated with P" in Section 6.2 is

$$\sum_{k=1}^n h(c_k)\,\Delta x_k$$

where c_k is an arbitrary point of the kth subinterval and Δx_k is the length of the subinterval. In practice it is easier to abbreviate the notation as we did in the above discussion, using x for the point c_k and Δx for Δx_k (and leaving out reference to k in the symbol for summation). In future applications we will even drop reference to the partition P, simply writing

$$\int_a^b h(x)\,dx = \lim \sum h(x)\,\Delta x$$

as a reminder that the integral of h is the limit of Riemann sums as the norm of P approaches zero.

Note the role of Riemann sums in this discussion; they arise naturally in the process of slicing up our region. It is this kind of thing that is so often done in applications of integration. A geometrical or physical entity is cut into pieces by means of a partition of an interval. The contribution of each piece is approximated in terms of a real function. The sum of these approximations then takes the form of a Riemann sum whose limit as the partition is refined is the measure of the original entity. By the definitions in Section 6.2, *this limit is the integral of the function involved in the Riemann sum.* Hence the measure we are seeking (area, volume, or any one of many other geometrical or physical quantities) is expressed as an integral. An understanding of Riemann sums and their relation to integrals is an essential part of the intellectual equipment of anyone who uses calculus.

Figure 2
Region bounded by $y = x$ and $y = x^2$

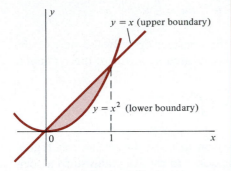

Figure 3
Region between sine and cosine curves

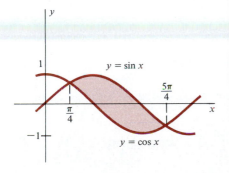

Figure 4
Region bounded by $x = 4 - y^2$ and the y axis

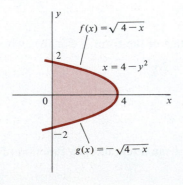

□ **Example 1**   Find the area of the region bounded by the curves $y = x$ and $y = x^2$.

Solution: The given curves intersect at points (x, y) satisfying the equations $y = x$ and $y = x^2$ simultaneously. We solve the system

$$\begin{cases} y = x \\ y = x^2 \end{cases}$$

by writing

$$x^2 = x$$
$$x^2 - x = 0$$
$$x(x - 1) = 0$$
$$x = 0 \quad \text{or} \quad x = 1$$

Since $y = 0$ when $x = 0$ and $y = 1$ when $x = 1$, the points of intersection are $(0,0)$ and $(1,1)$. (See Figure 2.) The area of the shaded region is

$$A = \int_0^1 (x - x^2)\, dx = \left(\frac{x^2}{2} - \frac{x^3}{3} \right)\Big|_0^1 = \frac{1}{6} \qquad \square$$

□ **Example 2**   The curves $y = \sin x$ and $y = \cos x$ intersect periodically. Find the area of the region bounded by these curves between consecutive points of intersection.

Solution: Since $\sin x = \cos x$ when $\tan x = 1$, the points of intersection of the given curves occur at

$x = \pi/4 + n\pi$, where n is an integer (because the period of tangent is π)

Two consecutive points of intersection are at $x = \pi/4$ and $x = 5\pi/4$, as shown in Figure 3. (The area does not depend on which pair of points we choose. Why?) The area of the shaded region is

$$\int_{\pi/4}^{5\pi/4} (\sin x - \cos x)\, dx = (-\cos x - \sin x)\Big|_{\pi/4}^{5\pi/4} = -(\cos x + \sin x)\Big|_{\pi/4}^{5\pi/4}$$

$$= -\left(-\frac{\sqrt{2}}{2} - \frac{\sqrt{2}}{2} \right) + \left(\frac{\sqrt{2}}{2} + \frac{\sqrt{2}}{2} \right) = 2\sqrt{2} \quad \square$$

□ **Example 3**   Find the area of the region bounded by $x = 4 - y^2$ and the y axis. (See Figure 4.)

Solution: We may think of the region as having upper boundary $f(x) = \sqrt{4 - x}$ and lower boundary $g(x) = -\sqrt{4 - x}$ (obtained by solving the equation $x = 4 - y^2$ for y in terms of x). Its vertical boundaries are the lines $x = 0$ and $x = 4$. Hence its area is

$$A = \int_0^4 [f(x) - g(x)]\, dx = \int_0^4 2\sqrt{4 - x}\, dx$$

Making the substitution $u = 4 - x$, $du = -dx$ (with $u = 4$ when $x = 0$ and $u = 0$ when $x = 4$), we have

$$A = -2 \int_4^0 u^{1/2} \, du = 2 \int_0^4 u^{1/2} \, du = \frac{4}{3} u^{3/2} \Big|_0^4 = \frac{32}{3}$$

This result can also be obtained by considering only the part of the shaded region above the x axis, doubling its area by virtue of the symmetry of the figure:

$$A = 2 \int_0^4 f(x) \, dx = 2 \int_0^4 \sqrt{4 - x} \, dx$$

Figure 5
Area by means of horizontal strips

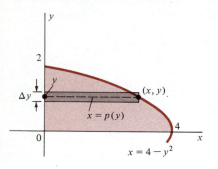

Alternate Solution: There is no reason to insist on cutting our region into vertical strips. In this example it is easier to use Riemann sums associated with a partition of the interval [0,2] on the y axis. (See Figure 5.) The region shown (the upper half of the original region) is cut into *horizontal* strips by means of the partition; the typical strip is then replaced by the rectangle in the figure. Selecting a point y in the typical subinterval, and letting $p(y) = 4 - y^2$, we see that the rectangle has base Δy and length $x = p(y)$, so its area is

$$\Delta A = x \, \Delta y = p(y) \, \Delta y$$

The total area is approximately

$$A \approx \sum \Delta A = \sum p(y) \, \Delta y$$

which is a Riemann sum of the function p associated with the given partition. Hence the exact area is

$$A = \lim \sum p(y) \, \Delta y = \int_0^2 p(y) \, dy = \int_0^2 (4 - y^2) \, dy = \left(4y - \frac{y^3}{3} \right) \Big|_0^2 = \frac{16}{3}$$

The area of the original region (shaded in Figure 4) is $2A = \frac{32}{3}$, as before.

$\square$

The two solutions of Example 3 illustrate one of the roles of the "appendage" dx in the notation

$$\int_a^b f(x) \, dx$$

Our first solution is based on Riemann sums of the form $\sum h(x) \, \Delta x$, where

$$h(x) = f(x) - g(x) = 2\sqrt{4 - x}$$

The limit of these sums is

$$\int_0^4 2\sqrt{4 - x} \, dx$$

in which we use dx to indicate the fact that the integrand is a function of x (and to match Δx in the Riemann sums).

The notation is different in our second solution (because the independent variable is y instead of x). Riemann sums of the form $\sum p(y)\,\Delta y$ lead to

$$\int_0^2 p(y)\,dy = \int_0^2 (4 - y^2)\,dy$$

where dy is used to match y in the integrand (and Δy in the Riemann sums). See Section 6.2, where we agreed that in the symbol

$$\int_a^b f(\ \ )\,d(\ \ )$$

the letter used in $d(\ \ )$ should match the independent variable in $f(\ \ )$.

Students who overlook this point (or who leave out the Leibniz symbol altogether) are prone to a common error. To see what we mean, return to the alternate solution of Example 3 and replace $4 - y^2$ by x. Then the area of the region shown in Figure 5 is

$$\int_0^2 x\,dy$$

a legitimate formula if it is understood that the integrand is the function $x = p(y) = 4 - y^2$. Suppose, however, that you were not in the habit of attaching the Leibniz symbol. Then it would be easy to write

$$\int_0^2 x = \frac{x^2}{2}\Big|_0^2 = 2$$

a result that is clearly wrong in view of our earlier evaluation of

$$\int_0^2 x\,dy = \int_0^2 (4 - y^2)\,dy = \frac{16}{3}$$

Figure 6
Region bounded by
$y = x(x - 1)(x - 3)$ and the x axis

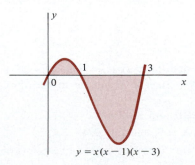

$y = x(x-1)(x-3)$

□ **Example 4**   Find the area of the region bounded by $y = x(x - 1)(x - 3)$ and the x axis. (See Figure 6.)

Solution: The given region is the union of two parts, in the first of which $y = x(x - 1)(x - 3)$ is the upper boundary, whereas in the second it is the lower boundary. (The x axis is the other boundary in each part.) Hence the area is

$$A = \int_0^1 [x(x - 1)(x - 3) - 0]\,dx + \int_1^3 [0 - x(x - 1)(x - 3)]\,dx$$

$$= \int_0^1 (x^3 - 4x^2 + 3x)\,dx + \int_1^3 (-x^3 + 4x^2 - 3x)\,dx$$

$$= \left(\frac{1}{4}x^4 - \frac{4}{3}x^3 + \frac{3}{2}x^2\right)\Big|_0^1 + \left(-\frac{1}{4}x^4 + \frac{4}{3}x^3 - \frac{3}{2}x^2\right)\Big|_1^3$$

It is not hard to finish the problem by evaluating these expressions as they stand. As problems of this kind get more complicated, however, it is worthwhile to cut down the arithmetic by using some preliminary algebra:

$$A = \frac{x^2}{12}(3x^2 - 16x + 18)\Big|_0^1 - \frac{x^2}{12}(3x^2 - 16x + 18)\Big|_1^3$$

$$= \frac{1}{12}(5) - \frac{9}{12}(-3) + \frac{1}{12}(5) = \frac{1}{12}(5 + 27 + 5) = \frac{37}{12} \qquad \square$$

Problem Set 7.1

In each of the following, a region is described by giving its boundaries. Use integration based on a cutting of the region into vertical strips to find the area. (Draw a picture!)

1. $y = x^2$ and $y = 2x$
2. $y = x$ and $y = x^3$
3. $y = 4 - x^2$ and $2x + y = 4$
4. $y = x^2$ and $x = y^2$
5. $y = 2 - x^2$ and $x + y = 0$
6. $x = y^2$, $y = 2$, and the y axis
7. $xy = 2$ and $x + y = 3$
8. $y^3 = x^2$ and $y = 1$
9. $y = \sqrt{x - 1}$, $x = 5$, and the x axis
10. $y = x^3$, $y = -x$, and $x = 1$

Use integration based on horizontal strips to find the area of each of the following regions (bounded by the given curves).

11. $y = x^2$ and $y = 2x$ (Compare with Problem 1.)
12. $y = x$ and $y = x^3$ (Compare with Problem 2.)
13. $y = 4 - x^2$ and $2x + y = 4$ (Compare with Problem 3.)
14. $y = x^2$ and $x = y^2$ (Compare with Problem 4.)
15. $y = 2 - x^2$ and $x + y = 0$ (Compare with Problem 5.)
16. $x = y^2$, $y = 2$, and the y axis   (Compare with Problem 6.)
17. $xy = 2$ and $x + y = 3$ (Compare with Problem 7.)
18. $y^3 = x^2$ and $y = 1$ (Compare with Problem 8.)
19. $y = \sqrt{x - 1}$, $x = 5$, and the x axis   (Compare with Problem 9.)
20. $y = x^3$, $y = -x$, and $x = 1$ (Compare with Problem 10.)

In each of the following, use whatever method seems convenient to find the area of the region bounded by the given curves.

21. $y = x^2 - 2x$ and the x axis
22. $y = x^3 - 2x$ and the x axis
23. $x = y^2 - 4y$ and the y axis
24. $y^2 = 4x$ and $y = 2x - 4$
25. $y = x^3 - 3x$ and $y = 2x^2$
26. $y = 1/x^2$, $y = x$, $y = 0$, and $x = 2$
27. $y = \sqrt{x}$, $y = 1/\sqrt{x}$, $y = 0$, and $x = 2$
28. $y^2 = x$ and $x - y = 2$
29. $y = x^3$ and its tangent line at $(1,1)$
30. $y^2 = x$, $(y - 2)^2 = x$, and the y axis
31. $y = 1/\sqrt{x + 1}$, $y = x + 1$, and $y = 2$
32. $y = \sqrt{x}$ and $y = x^3$
33. $y = x/\sqrt{9 - x^2}$, $x = \pm 2$, and $y = 0$
34. $y = x^2 - 2x$ and $y = 2x - x^2$
35. $\sqrt{x} + \sqrt{y} = 1$ and the coordinate axes
36. $|x| + |y| = 1$
37. $y = |x|$ and $x = 2 - y^2$
38. $y = x/\sqrt{4 - x^2}$ and $y = x$
39. $x^4 - 9x^2 + y^2 = 0$
40. $y^2 = x^2(3 - x)$
41. $y^2 = x + 2$, $y^3 = x$, and the x axis   *Hint:* $y^3 - y^2 + 2 = (y + 1)(y^2 - 2y + 2)$. The region is in the third quadrant.
42. $y = \cos x$, $y = 1$, $x = 0$, and $x = 2\pi$
43. $y = \sin x$, $y = \cos x$, $x = 0$, and $x = \pi/4$
44. $y = \sin 2x$ and $y = \cos x$ (between consecutive points of intersection on the x axis)
45. $y = \cos 2x$ and $y = \sin x$ (between consecutive points of intersection on opposite sides of the x axis)

7.2

VOLUME OF A SOLID OF REVOLUTION

Figure 1
Solid of revolution

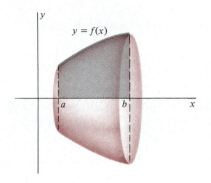

Let f be continuous and nonnegative in $[a,b]$. The region under the curve $y = f(x)$, $a \le x \le b$, is shaded in Figure 1. When this region is rotated about the x axis it generates a "solid of revolution" as shown in the figure. What is the volume of this solid?

The first thing to realize about this question is that a definition is needed. While we may know the volume of rectangular boxes, cylinders, cones, and spheres, the general concept of volume (like that of area) is beyond the reach of elementary geometry. Hence we shall use the methods of calculus to define it.

Partition $[a,b]$ in the usual way. Choosing a point x in the typical subinterval, we draw a rectangle with base Δx and height $f(x)$ as shown in Figure 2. Rotation of this rectangle about the x axis produces a cylindrical disk of thickness Δx and radius $f(x)$. The volume of the disk is

$$\Delta V = \pi f(x)^2 \, \Delta x$$

Thinking of the solid as cut into slices (like a stack of pancakes of various sizes), we approximate the total volume by writing

$$V \approx \sum \Delta V = \sum \pi f(x)^2 \, \Delta x$$

Since this is a Riemann sum of the function πf^2 associated with the given partition, the total volume is

Figure 2
Cylindrical disk produced by rotation of typical rectangle

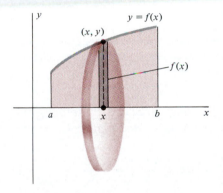

$$V = \lim \sum \pi f(x)^2 \, \Delta x = \int_a^b \pi f(x)^2 \, dx$$

Volume by Cylindrical Disks

Suppose that f is continuous and nonnegative in $[a,b]$. When the region under the graph of $y = f(x)$, $a \le x \le b$, is rotated about the x axis, a solid of revolution is generated whose volume is

$$V = \pi \int_a^b f(x)^2 \, dx$$

It sounds as though we have proved something. Actually, however, we have *defined* something, in this case volume of a solid of revolution (taking the volume of a cylinder to be already known). Even though we have boxed in the definition, you should not treat it as a general formula applicable to all solids of revolution. See Example 3, in which the formula is modified to apply to a region rotated about the y axis. Moreover, we will develop another method for finding volumes of solids of revolution, using an altogether different formula. The best approach to volume problems is to set up your own formula using Riemann sums as a guide.

Figure 3
Spherical ball as a solid of revolution

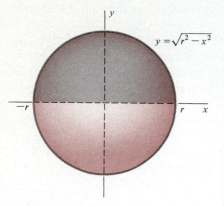

Figure 4
Parabolic loudspeaker horn

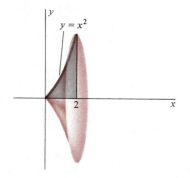

Figure 5
Rotation about the y axis

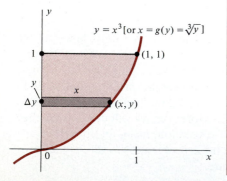

◻ **Example 1**   A spherical ball of radius r may be regarded as the solid generated by rotating the region under the semicircle $y = \sqrt{r^2 - x^2}$, $-r \le x \le r$, about the x axis. (See Figure 3.) According to the preceding discussion, its volume is

$$\int_{-r}^{r} \pi(\sqrt{r^2 - x^2})^2 \, dx = \pi \int_{-r}^{r} (r^2 - x^2) \, dx$$

$$= 2\pi \int_{0}^{r} (r^2 - x^2) \, dx \qquad \text{(Problem 45, Section 6.5)}$$

$$= 2\pi \left(r^2 x - \frac{x^3}{3} \right)\Big|_{0}^{r} = \frac{2\pi x}{3}(3r^2 - x^2)\Big|_{0}^{r} = \frac{4}{3}\pi r^3$$

This is the usual formula for the volume of a sphere; perhaps you have never seen it derived before. Of course from our present point of view that is not important. The volume is *defined* to be

$$V = \int_{a}^{b} \pi f^2$$

where (in this example) $[a,b] = [-r,r]$ and $f(x) = \sqrt{r^2 - x^2}$.   ◻

Remark   In Example 1 note that

$$\int r^2 \, dx = r^2 x \qquad \text{(arbitrary constant omitted)}$$

A common error is to write $r^3/3$ for this antiderivative, which overlooks the fact that r is constant. The Leibniz symbol dx helps avoid such errors; its role (as pointed out before) is to identify the variable (which is x, not r).

◻ **Example 2**   Find the volume of the solid generated by rotating the region under $y = x^2$, $0 \le x \le 2$, about the x axis. (See Figure 4.)

Solution: $V = \int_{0}^{2} \pi(x^2)^2 \, dx = \pi \int_{0}^{2} x^4 \, dx = \frac{\pi x^5}{5}\Big|_{0}^{2} = \frac{32\pi}{5}$ ◻

◻ **Example 3**   There is no reason to confine our discussion to solids of revolution about the x axis. Imagine rotating about the y axis the region bounded by the graph of $y = x^3$, the horizontal line $y = 1$, and the y axis. (See Figure 5.) The resulting solid of revolution can be cut into slices by means of a partition of the interval $[0,1]$ on the y axis. Choosing a point y in the typical subinterval, let $g(y)$ be the corresponding value of x, obtained from the equation $y = x^3$ by solving for $x = g(y) = \sqrt[3]{y}$. The rectangle of base Δy and length x produces a cylindrical disk when rotated about the y axis, of volume

$$\Delta V = \pi x^2 \, \Delta y = \pi g(y)^2 \, \Delta y$$

The total volume is approximately

$$V \approx \sum \Delta V = \sum \pi g(y)^2 \, \Delta y$$

which is a Riemann sum of the function πg^2 associated with the given partition. The exact volume is therefore

$$V = \lim \sum \pi g(y)^2 \, \Delta y = \int_0^1 \pi g(y)^2 \, dy = \pi \int_0^1 (\sqrt[3]{y})^2 \, dy = \pi \int_0^1 y^{2/3} \, dy$$

$$= \frac{3}{5} \pi y^{5/3} \Big|_0^1 = \frac{3}{5} \pi \qquad \square$$

□ **Example 4**   Find the volume of the solid generated by rotating the shaded region in Figure 6 about the y axis.

Solution: The horizontal strip shown in Figure 6 generates a disk with a hole in it when it is rotated about the y axis. The radius of the hole is x and the radius of the disk is 1. Hence the volume generated is

$$\Delta V = [\pi(1)^2 - \pi x^2] \Delta y = \pi(1 - x^2) \Delta y = \pi(1 - y^{2/3}) \Delta y$$

The total volume is

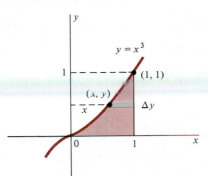

Figure 6
Rotation about the y axis

$$V = \pi \int_0^1 (1 - y^{2/3}) \, dy = \pi \left(y - \frac{3}{5} y^{5/3} \right) \Big|_0^1 = \pi \left(1 - \frac{3}{5} \right) = \frac{2}{5} \pi$$

As a check, note that the union of the regions shaded in Figures 5 and 6 is a square of side length 1. When this is rotated about the y axis, it generates a cylinder of radius 1 and altitude 1. The volume of the cylinder is $\pi(1)^2(1) = \pi$, which is the sum of the volumes found in Examples 3 and 4.

□

The preceding examples illustrate the *method of cylindrical disks* for finding the volume of a solid of revolution. There is another way (sometimes more convenient) called the *method of cylindrical shells*. Let f be continuous and nonnegative in $[a,b]$, where $a \geq 0$, and suppose that the region under the graph of $y = f(x)$, $a \leq x \leq b$, is rotated about the y axis. (See Figure 7.) Partitioning $[a,b]$ as usual, and choosing x in the typical subinterval, imagine rotating the rectangle of base Δx and altitude $f(x)$ about the y axis. This generates a "cylindrical shell," a region between two concentric cylinders of the same altitude. (See Figure 8.) Its volume (see the problem set) is

$$\Delta V = 2\pi(\text{mean radius})(\text{altitude})(\text{thickness}) = 2\pi r f(x) \, \Delta x$$

where r is the midpoint of the typical interval. (An intuitive argument for this formula is that the shell is like a tin can. Slit the can along an altitude and unroll it into a rectangular sheet. The volume of this sheet is its area $2\pi r h$ times its thickness.) The total volume is approximately

$$V \approx \sum \Delta V = \sum 2\pi r f(x) \, \Delta x$$

Figure 7
Volume by cylindrical shells

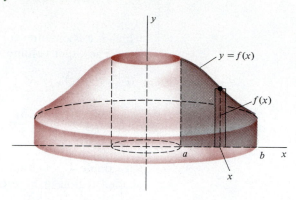

Figure 8
Cylindrical shell

which is not, as it stands, a Riemann sum. The reason is that r and x may not be the same point. However, the choice of x in the typical subinterval is up to us; let's take it to be r. Then

$$V \approx \sum 2\pi x f(x)\, \Delta x$$

which is a Riemann sum of the function $2\pi x f(x)$ associated with the given partition. The volume is therefore

$$V = \lim \sum 2\pi x f(x)\, \Delta x = \int_a^b 2\pi x f(x)\, dx$$

Volume by Cylindrical Shells

Suppose that f is continuous and nonnegative in $[a,b]$, where $a \geq 0$. When the region under the graph of $y = f(x)$, $a \leq x \leq b$, is rotated about the y axis, a solid of revolution is generated whose volume is

$$V = 2\pi \int_a^b x f(x)\, dx$$

Remark   It can be shown that the choice of $x = r$ in the above argument is unnecessary, that a sum of the form

$$\sum_{k=1}^{n} f(c_k) g(d_k)\, \Delta x_k$$

[which is a Riemann sum of $f(x)g(x)$ if $c_k = d_k$] has limit

$$\int_a^b f(x) g(x)\, dx$$

even if c_k and d_k are allowed to be different points of the kth subinterval. [In the above discussion of cylindrical shells, $g(x) = 2\pi x$.] This seems intuitively reasonable, since the length of the kth subinterval approaches 0 as $\|P\| \to 0$. Hence even if c_k and d_k are different, they become arbitrarily close as we refine the partition.

□ **Example 5**   Use the method of cylindrical shells to confirm that the volume of a sphere of radius r is $\frac{4}{3}\pi r^3$.

Solution: Referring to Figure 3, imagine rotating the right-hand half of the shaded region about the y axis (which generates the upper half of the spherical ball). According to the above discussion, the volume generated is

$$\int_0^r 2\pi x \sqrt{r^2 - x^2}\, dx = -\pi \int_0^r (r^2 - x^2)^{1/2}(-2x\, dx)$$

$$= -\pi \int_{r^2}^0 u^{1/2}\, du \qquad [u = r^2 - x^2,\ du = -2x\, dx]$$

$$= \pi \int_0^{r^2} u^{1/2}\, du = \frac{2}{3}\pi u^{3/2}\Big|_0^{r^2} = \frac{2}{3}\pi r^3$$

The volume of the sphere is therefore $\frac{4}{3}\pi r^3$.   ⬜

□ **Example 6**   Use the method of cylindrical shells to confirm the volume in Example 3.

Solution: Instead of rotating horizontal strips about the y axis (as in Figure 5), use vertical strips. (See Figure 9.) The typical strip generates a cylindrical shell of radius x, altitude $1 - y$, and thickness Δx; its volume is therefore

$$\Delta V = 2\pi x(1 - y)\, \Delta x = 2\pi x(1 - x^3)\, \Delta x$$

For the total volume we may write

$$V \approx \sum \Delta V = \sum 2\pi x(1 - x^3)\, \Delta x$$

and hence (taking the limit as the partition is refined)

$$V = \int_0^1 2\pi x(1 - x^3)\, dx = 2\pi \int_0^1 (x - x^4)\, dx = 2\pi \left(\frac{x^2}{2} - \frac{x^5}{5}\right)\Big|_0^1 = \frac{3}{5}\pi \quad □$$

When we say that the cylindrical shell in Example 6 has radius x, the reader has a right to complain. Do we mean the inner radius, the outer radius, the mean radius, or what? The answer is that it doesn't matter. For although our derivation of the method of shells depends on the formula

$$\Delta V = 2\pi(\text{mean radius})(\text{altitude})(\text{thickness})$$

Figure 9
Vertical strip to be rotated about y axis

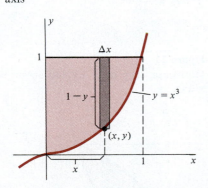

it can be shown (as mentioned in the preceding remark) that any choice of radius (that is, any value of x in the typical subinterval) leads to the same limit as the partition is refined. Hence we can afford to draw imprecise pictures (like Figure 9) in which the dimensions are labeled without clearly specifying whether x is the left-hand endpoint, the midpoint, the right-hand endpoint (or any other point) of the typical subinterval.

□ **Example 7**   The region bounded by $y = x^2$ and $x = y^2$ is rotated about the x axis. Use the method of disks and the method of shells to find the volume of the solid generated.

Solution (Method of Disks):   The typical "disk" is a washer (which has a hole in it). It is generated by rotating a vertical strip about the x axis. (See Figure 10.) Its volume is

$$\Delta V = \pi(\sqrt{x})^2 \, \Delta x - \pi(x^2)^2 \, \Delta x = \pi(x - x^4) \, \Delta x \qquad \text{(why?)}$$

and hence the total volume is

$$V = \pi \int_0^1 (x - x^4) \, dx = \frac{3}{10}\pi$$

Solution (Method of Shells):   This time we rotate a horizontal strip about the x axis. (See Figure 11.) The volume of the resulting cylindrical shell is

$$\Delta V = 2\pi y(\sqrt{y} - y^2) \, \Delta y = 2\pi(y^{3/2} - y^3) \, \Delta y$$

and hence the total volume is

$$V = 2\pi \int_0^1 (y^{3/2} - y^3) \, dy = \frac{3}{10}\pi \qquad\qquad\qquad □$$

Figure 10
Vertical strip to be rotated about x axis

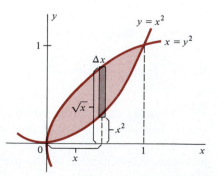

Figure 11
Horizontal strip to be rotated about x axis

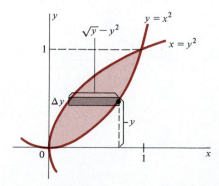

Problem Set 7.2

1. When the region bounded by $y = 2x$, $y = 4$, and the y axis is rotated about the y axis, a cone is generated. Find the volume of the cone by using **(a)** a formula from geometry, **(b)** the method of disks, and **(c)** the method of shells.

2. When the region bounded by $2x + y = 2$ and the coordinate axes is rotated about the x axis, a cone is generated. Find the volume of the cone by using **(a)** a formula from geometry, **(b)** the method of disks, and **(c)** the method of shells.

3. When the region bounded by $y = \sqrt{4x - x^2}$ and the x axis is rotated about the x axis, a sphere is generated. Find the volume of the sphere by using **(a)** a formula from geometry and **(b)** the method of disks.

4. The region in the first quadrant bounded by $y = 4 - x^2$ and the coordinate axes is rotated about the x axis. Find the volume of the resulting solid of revolution by using **(a)** the method of disks and **(b)** the method of shells.

5. The region of Problem 4 is rotated about the y axis. Find the volume of the resulting solid of revolution by using **(a)** the method of disks and **(b)** the method of shells.

6. The region bounded by $y = \sqrt{x}$, $x = 4$, and the x axis is rotated about the x axis. Find the volume of the resulting solid of revolution by using **(a)** the method of disks and **(b)** the method of shells.

7. The region of Problem 6 is rotated about the y axis. Find the volume of the resulting solid of revolution by using **(a)** the method of disks and **(b)** the method of shells.

8. The region of Problem 6 is rotated about the line $x = 4$. Find the volume of the resulting solid of revolution by using **(a)** the method of disks and **(b)** the method of shells.

9. The region of Problem 6 is rotated about the line $y = 2$. Find the volume of the resulting solid of revolution by using **(a)** the method of disks and **(b)** the method of shells.

10. The region in the first quadrant bounded by $y = \sin x$, $y = \cos x$, and the y axis is rotated about the x axis. Find the volume of the resulting solid of revolution. *Hint:* $\cos^2 x - \sin^2 x = \cos 2x$.

In each of the following, the region bounded by the given curves is rotated about the given line. Find the volume of the resulting solid of revolution by using whatever method seems most convenient.

11. $y = \sqrt{x - 1}$, $x = 5$, and the x axis (about the x axis)

12. $x^2 - y^2 = 1$ and $x = 2$ (about the x axis)

13. $y = 2x - x^2$ and the x axis (about the x axis)

14. $y = \sec x$, $x = \pi/3$, and the coordinate axes (about the x axis)

15. $y = \csc x$, $x = \pi/4$, $x = \pi/2$, and the x axis (about the x axis)

16. $y = \tan x$, $x = \pi/4$, and the x axis (about the x axis)

17. $y = \sin x$, $x = \pi$, and the coordinate axes (about the x axis)

18. $y = \cos x$, $x = \pi$, and the coordinate axes (about the x axis)

19. $xy = 1$, $x = 1$, $x = 2$, and the x axis (about the x axis)

20. $xy = 1$, $x = 1$, $x = 2$, and the x axis (about the y axis)

21. $y = 1/\sqrt{x}$, $x = 1$, $x = 4$, and the x axis (about the y axis)

22. $y = 5/\sqrt{9 + x^2}$, $x = 4$, and the coordinate axes (about the y axis)

23. $y = \sqrt{4 - x}$ and the coordinate axes (about the y axis)

24. $y = (\sin x)/x$, $x = \pi$, and the coordinate axes (about the y axis)

25. $x^2 - y^2 = 4$ and $y = \pm 1$ (about the y axis)

26. $x^2/a^2 + y^2/b^2 = 1$ (about the x axis)

27. $x^2/a^2 + y^2/b^2 = 1$ (about the y axis)

28. $y = x^2$ and $y = x$ (about the x axis)

29. $x = \sqrt{y}$, $x + y = 2$, and the y axis (about the y axis)

30. $y^2 = x^2(3 - x)$ (about the x axis)

31. $x^4 - 9x^2 + y^2 = 0$ (about the x axis)

32. $xy = 2$, $x = 1$, $x = 2$, and the x axis (about the line $x = 2$)

33. $y = x^2 + 1$, $x = 2$, and $y = 1$ (about the line $y = 1$)

34. $y = 4 - x^2$ and the x axis (about the line $y = 4$)

35. $\sqrt{x} + \sqrt{y} = 1$ and the coordinate axes (about the line $x = 1$)

36. Use integration to show that the volume of a right circular cone of base radius r and altitude h is $V = \frac{1}{3}\pi r^2 h$.

37. When the region inside the circle $x^2 + y^2 = a^2$ is rotated about the line $x = b$ $(a < b)$, a *torus* ("doughnut") is generated.

    (a) Find the volume of the torus.

    (b) Show that the volume of the torus is equal to the area of the circle multiplied by the distance traveled by its center when it is rotated. (This is an application of a famous theorem of the Greek geometer Pappus, who flourished around 330 A.D. Does it seem intuitively reasonable?)

38. A hole of radius 5 cm is drilled through the center of a bowling ball of radius 13 cm. How much volume is cut out?

39. A hemispherical bowl of radius 15 cm contains jello to a depth of 10 cm. How much jello is in the bowl? *Hint:* Turn the bowl upside down.

40. The method of cylindrical shells depends on the formula

$$2\pi(\text{mean radius})(\text{altitude})(\text{thickness})$$

for the volume of a shell. Derive this formula by regarding the shell as the region between two concentric cylinders of radii r_1 and r_2 $(r_1 < r_2)$ and altitude h (using the known formula for the volume of a cylinder). In other words show that the volume is

$$2\pi \cdot \tfrac{1}{2}(r_1 + r_2) \cdot h(r_2 - r_1)$$

The remaining problems do not refer to solids of revolution, but to solids with certain uniform cross sections. By slicing the solid in an appropriate way, you can set up a Riemann sum leading to an integral for its volume.

41. The base of a certain solid is circular, with radius 1. Cross sections of the solid perpendicular to a certain diameter of the base are square. (See Figure 12.) Find the volume of the solid. *Hint:* The volume of the typical slice is

$$\Delta V \approx (2x)(2x)\,\Delta y = 4(1 - y^2)\,\Delta y$$

Figure 12
Solid with circular base and square cross sections

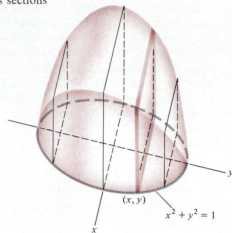

42. The base of a certain solid is circular, with radius 1. Cross sections of the solid perpendicular to a certain diameter of the base are equilateral triangles. Find the volume of the solid.

43. Use integration to show that the volume of a pyramid with square base of side a, and altitude h, is $V = \tfrac{1}{3}a^2h$.

7.3
LENGTH OF A CURVE

If you wanted to know the length of a garden hose lying in the shape of a curve, you would undoubtedly straighten it out and measure it on a linear scale. Mathematicians call this process *rectifying* the curve; in this section we shall discuss the details.

The first thing to settle is what we mean by a *curve*. We have used the term loosely until now (to refer to certain types of graphs), but in fact it has a more technical (and somewhat different) meaning which we shall have to define in order to talk intelligibly about length. First, however, a couple of examples to suggest what the difficulties are.

□ **Example 1**   Suppose that the position of a bug in the plane at time t is given by the *parametric equations*

$$x = t^3, \, y = t^2 \qquad 0 \le t \le 1$$

The graph of these equations is the set of points (x, y) corresponding to $t \in [0,1]$, that is, the set of positions of the bug. For example, the bug is at

Figure 1
Graph of $x = t^3$, $y = t^2$, $0 \le t \le 1$

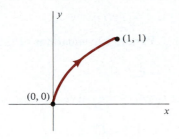

(0,0) when $t = 0$ and at (1,1) when $t = 1$; these are the *initial* and *terminal* points of the bug's path. See Figure 1, in which we indicate by an arrow the direction of the bug as t increases from 0 to 1.

Because t is an independent variable in terms of which the coordinates (x,y) of the bug are given, we refer to it as a *parameter;* hence the term *parametric equations.* (Note that values of t do not appear in Figure 1, but only the usual coordinates x and y.) Sometimes it is possible to eliminate the parameter and represent the graph in what is (at this stage) a more familiar way. Solving for t in the equation $x = t^3$, we find $t = x^{1/3}$ and hence $y = x^{2/3}$. While it is true that in this functional form we are more likely to feel at home with such things as slope, we have lost something by eliminating the parameter. For the equation $y = x^{2/3}$ merely describes the graph along which the bug moves. The equations $x = t^3$ and $y = t^2$ include not only that information but also the position of the bug at time t.   □

□ **Example 2**   The parametric equations $x = |t|^3$, $y = t^2$, $-1 \le t \le 1$, have the same graph as in Example 1, as you can see by eliminating the parameter:

$$|t| = x^{1/3} \quad \text{from which} \quad y = t^2 = |t|^2 = x^{2/3}$$

They describe a different motion, however. As t increases from -1 to 1, the bug goes from (1,1) to (0,0) and back again to (1,1), thus traversing the graph twice. If we are interested in how far the bug has traveled, the equation $y = x^{2/3}$ is of little help, because it does not distinguish between Examples 1 and 2. It appears that (at least in problems involving motion) parametric equations are the thing to study.   □

Examples 1 and 2 will help explain the terminology mathematicians have developed to cope with the problem of curves and their length. To a mathematician, a "curve" is not a graph (at least not in the present context), but rather its parametric representation. More precisely, the function

$$F: [0,1] \to \Re^2 \quad \text{defined by } F(t) = (t^3, t^2)$$

is the curve of Example 1. The curve in Example 2 is different, namely the function

$$G: [-1,1] \to \Re^2 \quad \text{defined by } G(t) = (|t|^3, t^2)$$

The *track* (or *trace*) of each curve is the same, namely the graph shown in Figure 1. But the *length* of a curve (not yet formally defined) is not necessarily the same as the distance along its track. In Example 1 it is, because the bug traverses the track only once. In Example 2, however, the length of the curve is twice the distance along the track.

Let $I = [a,b]$. A **curve** in $\Re^2$ is a function

$$F: I \to \Re^2 \qquad \text{defined by } F(t) = (f(t), g(t)), \, t \in I$$

where f and g are real functions defined in I. **Parametric equations** of the curve are

$$x = f(t), \, y = g(t) \qquad t \in I$$

t being called the **parameter.** The **track** (or *trace*) of the curve is the set of points

$$\{(x,y) \in \Re^2 : x = f(t), \, y = g(t), \, t \in I\}$$

(also called the *graph* of the parametric equations). The curve is said to be **continuous** if f and g are continuous in I, and **smooth** if dx/dt and dy/dt exist and are continuous in I, and if

$$\left(\frac{dx}{dt}\right)^2 + \left(\frac{dy}{dt}\right)^2 > 0 \qquad \text{for all interior points of } I$$

The term *smooth* is intended to suggest that a bug moving along the track of the curve suffers no abrupt changes of direction. The condition

$$\left(\frac{dx}{dt}\right)^2 + \left(\frac{dy}{dt}\right)^2 > 0 \qquad \text{for all interior points of } I$$

implies that dx/dt and dy/dt are never simultaneously zero (except possibly at an endpoint of I). Hence either $x = f(t)$ or $y = g(t)$ (or both) are always changing as t goes through the interval I. A bug moving along the track of the curve never stops (and in particular never reverses direction), because one or the other of its coordinates (or both) always has a nonzero rate of change.

Note that according to this terminology the curve in Example 1 is smooth, whereas the curve in Example 2 is not. In Example 1 we have $dx/dt = 3t^2$ and $dy/dt = 2t$, which are continuous in $I = [0,1]$ and are simultaneously zero only at the endpoint $t = 0$. In Example 2 we have

$$\frac{dx}{dt} = \begin{cases} 3t^2 & \text{if } t \geq 0 \\ -3t^2 & \text{if } t < 0 \end{cases} \qquad \text{and} \qquad \frac{dy}{dt} = 2t$$

While these derivatives exist and are continuous in $I = [-1,1]$, they are both zero at the interior point $t = 0$. This corresponds to the point at which the bug, having gone from $(1,1)$ to $(0,0)$, stops and reverses direction.

□ **Example 3**   The curves defined by

(a) $x = \cos t, \, y = \sin t, \, 0 \leq t \leq 2\pi$
(b) $x = \sin t, \, y = \cos t, \, 0 \leq t \leq 2\pi$
(c) $x = \cos 2t, \, y = \sin 2t, \, 0 \leq t \leq 2\pi$

all have the same track, namely the circle $x^2 + y^2 = 1$. (Why?) The first two have the same length (the circumference of the circle) because as t increases from 0 to 2π, the point (x, y) moves around the circle once. They have different initial and terminal points, however. In (a) the point (x, y) begins and ends at $(1, 0)$, whereas in (b) it begins and ends at $(0, 1)$. The length of the third curve is twice the circumference of the circle, because (x, y) traverses the track twice, beginning and ending at $(1, 0)$. All three curves are smooth. (Why?)   □

The above remarks about length are to be understood intuitively, because we have not yet defined what the length of a curve is. Returning to our opening idea of measuring a garden hose, we now want to discuss how to "rectify" (or "straighten out") a track, so as to come up with a reasonable number for length.

Suppose that a smooth curve is given, say $x = f(t), y = g(t), a \leq t \leq b$. It seems natural to straighten out its track by partitioning it as shown in Figure 2, approximating the length of each piece by means of a line segment whose length we already know how to compute. If the sum of these approximations has a limit as the segments are made shorter, we can take that limit as the definition of length.

To be more precise, we may accomplish the partitioning shown in Figure 2 by a partition of the parameter interval $I = [a, b]$. If t is the typical point of subdivision, the corresponding point on the track is

$$(x, y) = (f(t), g(t))$$

The next point of subdivision on the track is obtained by changing t to $t + \Delta t$, where Δt is the length of the typical subinterval of I. If Δx and Δy are the corresponding changes in x and y, this point is

$$(x + \Delta x, y + \Delta y) = (f(t + \Delta t), g(t + \Delta t))$$

Hence the typical arc in Figure 2 has length

$$\Delta s \approx \sqrt{(\Delta x)^2 + (\Delta y)^2}$$

Figure 2
(Approximate) rectification of a curve

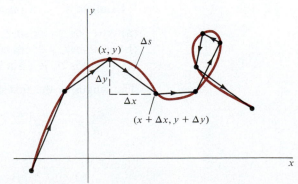

Anticipating the definition of *length of the curve,* we may say that it is given approximately by

$$s = \sum \Delta s \approx \sum \sqrt{(\Delta x)^2 + (\Delta y)^2}$$

It is not obvious what to do next. Note, however, that

$$\Delta x = f(t + \Delta t) - f(t) \qquad \text{and} \qquad \Delta y = g(t + \Delta t) - g(t)$$

are *functional differences* and recall from Section 4.4 that the Mean Value Theorem is tailor-made for such differences. Since f is differentiable in I, we can apply the theorem to f in the typical subinterval of I to obtain

$$\Delta x = f(t + \Delta t) - f(t) = f'(c)\,\Delta t$$

where c is somewhere between t and $t + \Delta t$. Similarly,

$$\Delta y = g(t + \Delta t) - g(t) = g'(d)\,\Delta t$$

where d is also between t and $t + \Delta t$. Hence

$$s \approx \sum \sqrt{[f'(c)\,\Delta t]^2 + [g'(d)\,\Delta t]^2} = \sum \sqrt{f'(c)^2 + g'(d)^2}\,\Delta t$$

This would be a Riemann sum of the function $\sqrt{f'(t)^2 + g'(t)^2}$ if c and d were the same point. As explained in the Remark in Section 7.2, it does not matter ultimately whether they are or not. Since f' and g' are continuous in I (because we are dealing with a smooth curve), we can be sure that the limit of the above sum exists and is equal to the integral of the above function. Hence we define the length of the curve as follows.

Suppose that the curve defined by

$$x = f(t),\, y = g(t) \qquad a \le t \le b$$

is smooth. Its **length** is the number

$$s = \int_a^b \sqrt{f'(t)^2 + g'(t)^2}\,dt = \int_a^b \sqrt{(dx/dt)^2 + (dy/dt)^2}\,dt$$

□ **Example 4**   Consider the three curves of Example 3,

(a) $x = \cos t,\, y = \sin t,\, 0 \le t \le 2\pi$
(b) $x = \sin t,\, y = \cos t,\, 0 \le t \le 2\pi$
(c) $x = \cos 2t,\, y = \sin 2t,\, 0 \le t \le 2\pi$

The length of the first one is found by computing

$$\frac{dx}{dt} = -\sin t, \frac{dy}{dt} = \cos t, \left(\frac{dx}{dt}\right)^2 + \left(\frac{dy}{dt}\right)^2 = \sin^2 t + \cos^2 t = 1$$

Hence

$$s = \int_0^{2\pi} 1\,dt = 2\pi$$

(the circumference of the track $x^2 + y^2 = 1$). The length of the second curve is clearly the same. To find the length of the third one, we compute

$$\frac{dx}{dt} = -2 \sin 2t, \quad \frac{dy}{dt} = 2 \cos 2t, \quad \left(\frac{dx}{dt}\right)^2 + \left(\frac{dy}{dt}\right)^2 = 4 \sin^2 2t + 4 \cos^2 2t = 4$$

Hence
$$s = \int_0^{2\pi} 2 \, dt = 4\pi$$

(twice the circumference of the track). □

□ **Example 5**   Find the length of the curve $x = t^3$, $y = t^2$, $0 \leq t \leq 1$. (See Example 1.)

Solution: We compute

$$\frac{dx}{dt} = 3t^2, \quad \frac{dy}{dt} = 2t, \quad \left(\frac{dx}{dt}\right)^2 + \left(\frac{dy}{dt}\right)^2 = 9t^4 + 4t^2 = t^2(9t^2 + 4)$$

Hence
$$s = \int_0^1 t \sqrt{9t^2 + 4} \, dt = \frac{1}{18} \int_0^1 (9t^2 + 4)^{1/2}(18t \, dt)$$

$$= \frac{1}{18} \int_4^{13} u^{1/2} \, du \qquad [u = 9t^2 + 4, \; du = 18t \, dt]$$

$$= \frac{1}{18} \cdot \frac{2}{3} u^{3/2} \Big|_4^{13} = \frac{1}{27}(13\sqrt{13} - 8) \qquad □$$

□ **Example 6**   Find the length of the curve $x = |t|^3$, $y = t^2$, $-1 \leq t \leq 1$. (See Example 2.)

Solution: Our definition of length does not directly apply to this curve, because it is not smooth. We can break it up, however, into the smooth curves

$$C_1: x = t^3, \; y = t^2, \; 0 \leq t \leq 1$$
$$C_2: x = -t^3, \; y = t^2, \; -1 \leq t \leq 0$$

These curves both have length $\frac{1}{27}(13\sqrt{13} - 8)$. (Why?) If we make the reasonable agreement that length is additive in such a case, we can say that the length of the original curve is $s = \frac{2}{27}(13\sqrt{13} - 8)$ (twice the length of the curve in Example 1). □

Although our discussion of length has been based on parametric representation, there is no reason why we cannot apply it to a graph defined by the more familiar functional expression $y = f(x)$, $a \leq x \leq b$. Simply "parametrize" by letting $x = t$! Parametric equations of the graph are then

$$x = t, \; y = f(t) \qquad a \leq t \leq b$$

Since $dx/dt = 1$ and $dy/dt = f'(t)$, we arrive at the following formula for length.

Suppose that $y = f(x)$ is a real function with a continuous derivative in the interval $[a,b]$. The length of its graph is

$$s = \int_a^b \sqrt{1 + f'(x)^2}\, dx = \int_a^b \sqrt{1 + y'^2}\, dx$$

□ **Example 7**   Find the length of the graph of $y = x^{3/2}$, $0 \le x \le 4$.

Solution: Since $y' = \frac{3}{2}x^{1/2}$ and $1 + y'^2 = 1 + \frac{9}{4}x = \frac{1}{4}(4 + 9x)$, we find

$$s = \frac{1}{2}\int_0^4 \sqrt{4 + 9x}\, dx = \frac{1}{18}\int_4^{40} u^{1/2}\, du \qquad [u = 4 + 9x,\ du = 9\, dx]$$

$$= \frac{1}{27} u^{3/2} \Big|_4^{40} = \frac{1}{27}(40\sqrt{40} - 8) = \frac{8}{27}(10\sqrt{10} - 1) \qquad\qquad □$$

□ **Example 8**   Find the length of the graph of $y = \sin x$, $0 \le x \le \pi$.

Solution: Since $y' = \cos x$ and $1 + y'^2 = 1 + \cos^2 x$, we find

$$s = \int_0^\pi \sqrt{1 + \cos^2 x}\, dx$$

No simple antiderivative of $\sqrt{1 + \cos^2 x}$ is available to us, so we leave the answer as an integral. Numerical methods of approximating integrals will be discussed in a later chapter.   □

We end this section by introducing the idea of "derivative of arc length." Suppose that we are interested in the length of that part of the curve

$$x = f(t),\ y = g(t) \qquad a \le t \le b$$

corresponding to the parameter interval $[a,t]$, where t is allowed to vary. (See Figure 3.) Evidently the length depends on t, that is, it is a function $s(t)$. According to our definition,

$$s(t) = \int_a^t \sqrt{f'(u)^2 + g'(u)^2}\, du$$

(Note the change of letter in the integrand, to avoid confusion with the variable t.)

By the first part of the Fundamental Theorem of Calculus (Section 6.4) we have

$$s'(t) = \sqrt{f'(t)^2 + g'(t)^2}$$

Figure 3
Length as a function of terminal point

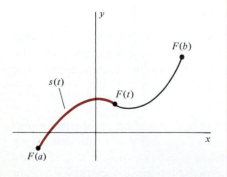

or simply
$$\frac{ds}{dt} = \sqrt{(dx/dt)^2 + (dy/dt)^2}$$

If t is time, this is a formula for the *rate of change of arc length with respect to time*. It is a measure of *how fast* the point (x, y) is moving along the track. In other words it is *speed* (as defined in Section 2.1, when we were discussing motion in a straight line).

In Section 2.1 speed was a creature of *velocity* (its absolute value). This raises the fascinating question of what velocity is in the case of motion along a curve (track). We are not going to get into it now, except to remark that velocity turns out to be a *vector* (with both magnitude and direction, as in the case of linear motion). We ask you to file that idea in the back of your head, as a line of inquiry that is going to be interesting to pursue.

It is also worth noting that $(ds/dt)^2 = (dx/dt)^2 + (dy/dt)^2$, from which [multiplying by $(dt)^2$] we have

$$(ds)^2 = (dx)^2 + (dy)^2$$

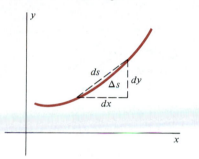

Figure 4
Differential of arc length

This Pythagorean relation between the differentials dx, dy, and ds may easily be remembered by referring to Figure 4. The actual change in s corresponding to the increments $dx = \Delta x$ and $dy = \Delta y$ is the arc length labeled Δs in the figure. When dx and dy are small, however, this length is closely approximated by the hypotenuse ds of the right triangle with legs dx and dy. In other words $\Delta s \approx ds$, which is what we should expect from the meaning of these symbols as explained in Section 4.7.

Problem Set 7.3

In each of the following, the position of a bug at time t is given by parametric equations. Find how far the bug travels by (a) ordinary geometry and (b) the formula for length of a curve.

1. $x = 2$, $y = t$, $0 \le t \le 3$
2. $x = t^2$, $y = 1$, $-2 \le t \le 2$
3. $x = 2t + 1$, $y = -3t + 2$, $0 \le t \le 1$
4. $x = t - 5$, $y = 2 - t$, $2 \le t \le 5$
5. $x = t^2 - 1$, $y = 1 - t^2$, $0 \le t \le 1$
6. $x = t^3$, $y = |t|^3$, $-1 \le t \le 1$
7. $x = 2 \sin^2 t$, $y = \cos 2t$, $0 \le t \le \pi$
8. $x = 2 \cos t$, $y = 2 \sin t$, $0 \le t \le \pi$
9. $x = \sin \pi t$, $y = \cos \pi t$, $0 \le t \le 3$
10. $x = \sec^2 t - 1$, $y = \tan^2 t$, $0 \le t \le \pi/4$

Find the length of each of the following curves.

11. $x = \frac{1}{3}t^3$, $y = \frac{1}{2}t^2$, $0 \le t \le 1$
12. $x = t^2$, $y = t^3$, $0 \le t \le 2$
13. $x = t - 1$, $y = t^{3/2}$, $0 \le t \le 1$
14. $x = t^{2/3}$, $y = t + 1$, $1 \le t \le 8$
15. $x = \frac{1}{2}t^2$, $y = \frac{1}{3}(2t + 1)^{3/2}$, $0 \le t \le 4$
16. $x = t^3 + 3t^2$, $y = t^3 - 3t^2$, $0 \le t \le 3$
17. $x = t^2 \cos t$, $y = t^2 \sin t$, $0 \le t \le 2$
18. $x = \cos^3 t$, $y = \sin^3 t$, $0 \le t \le \pi/2$
19. $x = \cos t + t \sin t$, $y = \sin t - t \cos t$, $0 \le t \le 2\pi$
20. $x = 1 - \cos t$, $y = t - \sin t$, $0 \le t \le 2\pi$

    Hint: $\sin^2 \dfrac{t}{2} = \dfrac{1}{2}(1 - \cos t)$

Find the length of each of the following graphs.

21. $y = 2x^{3/2}, 0 \leq x \leq 1$

22. $y = \frac{1}{3}(x^2 + 2)^{3/2}, 0 \leq x \leq 1$

23. $y = \frac{1}{3}x^{3/2} - \sqrt{x}, 0 \leq x \leq 4$

24. $y = \frac{1}{6}x^3 + \frac{1}{2}x^{-1}, 1 \leq x \leq 2$

25. $y = \frac{1}{8}x^4 + \frac{1}{4}x^{-2}, 1 \leq x \leq 2$

26. $y^2 = x^3$ between $(1, -1)$ and $(4, 8)$ *Hint:* Draw a picture!

27. Show that the length of the graph of $x = g(y), c \leq y \leq d$ (where g has a continuous derivative) is

$$s = \int_c^d \sqrt{1 + g'(y)^2} \, dy = \int_c^d \sqrt{1 + x'^2} \, dy$$

Use the formula in Problem 27 to find the length of each of the following graphs.

28. $x = y^{3/2}, 0 \leq y \leq 5$

29. $x = y^{2/3}, 1 \leq y \leq 8$

30. $x = (2y - 1)^{3/2}, 1 \leq y \leq 5$

31. $x = \frac{3}{5}y^{5/3} - \frac{3}{4}y^{1/3}, 1 \leq y \leq 8$

32. Explain why the graph of the parametric equations

$$x = at + x_0, \quad y = bt + y_0 \qquad (a \text{ and } b \text{ not both } 0)$$

is a straight line through the point (x_0, y_0). If $a \neq 0$, what is the slope of the line? What if $a = 0$?

33. Explain why the parametric equations

$$x = (x_2 - x_1)t + x_1, \quad y = (y_2 - y_1)t + y_1, \quad 0 \leq t \leq 1$$

represent the line segment with endpoints (x_1, y_1) and (x_2, y_2). (See Problem 32.) Then use the formula for length of a curve to derive the usual distance formula of analytic geometry.

34. If f has a continuous derivative in $[a, b]$, the curve $y = f(x), a \leq x \leq b$, is smooth. Why?

35. The position of a particle at time t is given by

$$x = a \cos^3 t, \quad y = a \sin^3 t, \quad 0 \leq t \leq 2\pi \quad (a > 0)$$

(a) Explain why the track of the particle is given by the equation $x^{2/3} + y^{2/3} = a^{2/3}$.

(b) Sketch the track, indicating the direction of travel of the particle. (See Problem 36, Section 3.4.)

(c) Is the curve defined by the given parametric equations smooth? Explain.

(d) Find the distance traveled by the particle. *Hint:* If you integrate from 0 to 2π, you may get 0. What should be done instead?

36. Find the length of the loop of the graph of

$$x = 3t^2, \quad y = t^3 - 3t$$

Hint: The loop is traversed during the parameter interval $[-\sqrt{3}, \sqrt{3}]$. Why?

37. A point P on the rim of a wheel of radius a is in contact with the road (the x axis) at the origin. As the wheel rolls to the right, P traces a path called a *cycloid*. Figure 5 shows the position of P when the wheel has turned through θ radians. Derive parametric equations of the cycloid as follows.

(a) Explain why $OT = a\theta$. *Hint:* See Problem 13, Section 2.3, for the length of arc PT.

(b) Confirm that $PQ = a \sin \theta$ and $QC = a \cos \theta$.

(c) Conclude that $x = a(\theta - \sin \theta), y = a(1 - \cos \theta)$.

Figure 5
Cycloid

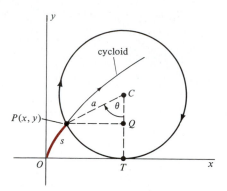

38. When the wheel of Problem 37 turns through one revolution, P traces out one arch of the cycloid

$$x = a(\theta - \sin \theta), \quad y = a(1 - \cos \theta)$$

How far does it travel? *Hint:* The parameter is θ, not t. Use appropriate trigonometric identities to obtain

$$\left(\frac{dx}{d\theta}\right)^2 + \left(\frac{dy}{d\theta}\right)^2 = 4a^2 \sin^2 \frac{\theta}{2}$$

39. Suppose that the wheel in Problem 37 turns at a constant rate, say $\omega = d\theta/dt$, where t is time. (Assume that $t = 0$ when P is at the origin.)

(a) Explain why $\theta = \omega t$.

(b) Show that the speed of P in its motion along the cycloid is

$$\frac{ds}{dt} = 2a\omega \left| \sin \frac{\omega t}{2} \right| \qquad (s \text{ is shown in Figure 5.})$$

(c) Explain why the speed of P varies from 0 to $2a\omega$. Where is P when these extreme values occur?

(d) If the wheel is on a car, why is the speed of the car equal to $a\omega$? *Hint:* The speed of the car is the rate of change of OT in Figure 5.

(e) One conclusion (perhaps surprising) is that a bug on the tire of a car going (say) 50 mph is sometimes traveling at 100 mph! Why?

40. The position of a particle at time t is given by $x = t^2$, $y = t^4$, $t \geq 0$.

(a) What is an equation in x and y of the track of the particle? Sketch the track and indicate the direction of travel.

(b) How fast is the particle moving at $t = 0$? at $t = 1$?

41. The position of a particle at time t is given by $x = \sqrt{t}$, $y = t$, $t \geq 0$.

(a) What is an equation in x and y of the track of the particle? Sketch the track and indicate the direction of travel.

(b) What can you say about the speed of the particle near the origin? at $t = 1$? (Note that the particles in Problems 40 and 41 travel the same path, but at radically different rates.)

42. The position of a particle at time t is given by

$$x = \sin t, \ y = \cos 2t, \ -\pi/2 \leq t \leq \pi/2$$

(a) What is an equation in x and y of the track of the particle? Sketch the track and indicate the direction of travel.

(b) How fast is the particle moving at $t = -\pi/2$? at $t = 0$? at $t = \pi/2$?

43. The position of a particle at time t is given by

$$x = a \cos t, \ y = b \sin t \qquad (a > b > 0)$$

(a) What is an equation in x and y of the track of the particle? Sketch the track and indicate the direction of travel.

(b) Find an expression for the speed of the particle at time t.

(c) What are the maximum and minimum speeds of the particle and where on the track do they occur?

(d) Write down an integral for the distance traveled by the particle during the time interval $[0, 2\pi]$.

7.4
AREA OF A SURFACE OF REVOLUTION

Suppose that the graph of the smooth curve $x = f(t)$, $y = g(t)$, $a \leq t \leq b$, lies above the x axis, as shown in Figure 1. If the graph is rotated about the x axis, it generates the surface of a solid of revolution. Our objective in this section is to find the area of such a surface. For example, the semicircle defined by

$$x = a \cos t, \ y = a \sin t \qquad 0 \leq t \leq \pi$$

generates a sphere of radius a when it is rotated about the x axis. (See Figure 2.) In this case we know the surface area $(A = 4\pi a^2)$, but in general the problem requires calculus.

We begin in a way that the reader should find familiar by now, using a partition that will enable us to compute the area as the limit of an approximating Riemann sum. Imagine the graph in Figure 1 cut into pieces by a partition of the parameter interval $[a,b]$. Let Δs be the length of the typical piece and let (x, y) be a point of this piece (as shown in Figure 1). When the graph is rotated about the x axis its typical piece generates a strip (or "ribbon") of the surface of revolution. The area of the strip is approximately the circumference of the circle traversed by (x, y) multiplied by the width Δs of the strip. (See Figure 3. If the ribbon were cut along a width and laid out flat on a table, its length would be approximately $2\pi y$.)

Figure 1
Graph to be rotated about the x axis

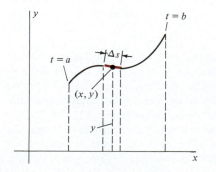

Figure 2
Sphere generated by semicircle
rotating about the x axis

Figure 3
Strip generated by rotation of typical
piece of curve

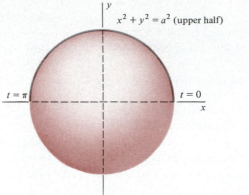

$x^2 + y^2 = a^2$ (upper half)

$t = \pi$ $t = 0$

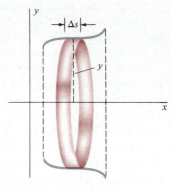

Δs

y

    In other words, the typical increment of surface area is $\Delta A \approx 2\pi y \, \Delta s$.
Add up the strips corresponding to the partition to obtain

$$A = \sum \Delta A \approx \sum 2\pi y \, \Delta s$$

an approximating sum which has a limit as the partition is refined (because
the given curve is smooth). This limit is defined to be the surface area,
that is,

$$A = \int_{t=a}^{t=b} 2\pi y \, ds$$

Since $x = f(t)$, $y = g(t)$, and $ds = \sqrt{(dx/dt)^2 + (dy/dt)^2} \, dt$ (Section 7.3),
the actual formula for A is

$$A = 2\pi \int_a^b g(t) \sqrt{[f'(t)]^2 + [g'(t)]^2} \, dt$$

There is no point in learning such a complicated result; it is better to think of
the formula $2\pi y \, ds$ for the area of the typical strip of surface.

Remark   Sometimes students wonder why the width of the typical strip is labeled
Δs, when it is Δx in volume problems. (See the method of cylindrical disks in
Section 7.2.) A quick answer is that it is a matter of definition and the one adopted
here gives familiar answers in those cases where we feel that we already know the
answers. (See Example 1 and similar questions in the problem set.)
    Another answer is that the typical ribbon is really the frustum of a cone, Δs
being the slant height (approximately). Figure 3 is misleading in that respect, for it
shows the typical piece of the curve as nearly flat, whereas in general it is slanted.

The surface area of the frustum is $\Delta A \approx 2\pi y \,\Delta s$, as above (a fact that is readily confirmed by geometry).

The best answer (in the long run) is that area of a surface of revolution is a special case of surface area in general. The formulas in this section can be derived from a more universal result to be discussed in Section 18.4.

□ **Example 1**   Find the surface area of a sphere of radius a.

Solution: The sphere may be generated by rotating the semicircle

$$x = a \cos t, \quad y = a \sin t \qquad 0 \le t \le \pi$$

about the x axis. Since $dx/dt = -a \sin t$ and $dy/dt = a \cos t$, we find

$$\left(\frac{ds}{dt}\right)^2 = \left(\frac{dx}{dt}\right)^2 + \left(\frac{dy}{dt}\right)^2 = a^2 \sin^2 t + a^2 \cos^2 t = a^2(\sin^2 t + \cos^2 t) = a^2$$

and hence $ds/dt = a$, that is, $ds = a \, dt$. Therefore

$$2\pi y \, ds = 2\pi(a \sin t) \cdot a \, dt = 2\pi a^2 \sin t \, dt$$

and the surface area is

$$A = \int_{t=0}^{t=\pi} 2\pi y \, ds = 2\pi a^2 \int_0^\pi \sin t \, dt = 4\pi a^2 \qquad\qquad □$$

□ **Example 2**   One arch of the cycloid

$$x = a(\theta - \sin\theta), \quad y = a(1 - \cos\theta)$$

is rotated about the x axis. Find the surface area generated.

Solution: This curve is discussed in Problem 37, Section 7.3. You need not have worked that problem, however, to figure out the present example. It is clear from the period of sine and cosine that one arch of the graph corresponds to the parameter interval $[0, 2\pi]$. (See Figure 4.) Since

$$dx = a(1 - \cos\theta) \, d\theta \qquad \text{and} \qquad dy = a \sin\theta \, d\theta$$

we find

$$
\begin{aligned}
(ds)^2 &= (dx)^2 + (dy)^2 \qquad \text{(Section 7.3)} \\
&= a^2(1 - 2\cos\theta + \cos^2\theta)(d\theta)^2 + a^2 \sin^2\theta \,(d\theta)^2 \\
&= a^2(1 - 2\cos\theta + \cos^2\theta + \sin^2\theta)(d\theta)^2 \\
&= a^2(2 - 2\cos\theta)(d\theta)^2 \\
&= 2a^2(1 - \cos\theta)(d\theta)^2 \\
&= 4a^2 \sin^2\frac{\theta}{2}(d\theta)^2 \qquad \left[\text{from the identity } \sin^2\frac{\theta}{2} = \frac{1}{2}(1 - \cos\theta), \right. \\
&\qquad\qquad\qquad\qquad\qquad \left. \text{Section 2.3}\right]
\end{aligned}
$$

Figure 4
Graph of the cycloid $x = a(\theta - \sin\theta), \ y = a(1 - \cos\theta)$

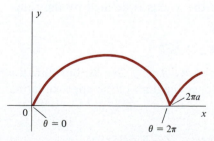

Since $0 \le \theta/2 \le \pi$, $\sin(\theta/2)$ is nonnegative and we have

$$ds = 2a \sin \frac{\theta}{2} \, d\theta$$

Thus

$$2\pi y \, ds = 2\pi \cdot a(1 - \cos\theta) \cdot 2a \sin \frac{\theta}{2} \, d\theta$$

$$= 8\pi a^2 \sin^3 \frac{\theta}{2} \, d\theta \qquad \left[1 - \cos\theta = 2\sin^2 \frac{\theta}{2}\right]$$

and the surface area is

$$A = 8\pi a^2 \int_0^{2\pi} \sin^3 \frac{\theta}{2} \, d\theta$$

$$= 16\pi a^2 \int_0^{\pi} \sin^3 u \, du \qquad \left[u = \frac{\theta}{2}, \, du = \frac{1}{2} d\theta\right]$$

$$= 16\pi a^2 \int_0^{\pi} (1 - \cos^2 u) \sin u \, du$$

$$= 16\pi a^2 \int_0^{\pi} (\sin u - \cos^2 u \sin u) \, du$$

$$= 16\pi a^2 \left(-\cos u + \frac{1}{3} \cos^3 u\right)\Big|_0^{\pi} = \frac{64\pi a^2}{3} \qquad \square$$

Remark   In Example 2 we encountered the antiderivative of $\sin^3 u$. The device for finding it is to break off $\sin u$, writing

$$\sin^3 u = \sin^2 u \cdot \sin u = (1 - \cos^2 u) \sin u = \sin u - \cos^2 u \sin u$$

This is easy to antidifferentiate because the differential of $\cos u$ is $-\sin u \, du$. Such devices are not obvious! When they occur you should file them away for future reference; we will discuss techniques of integration in Chapter 10.

As in the last section, we do not always use parametric representation. Sometimes a graph to be rotated about the x axis is defined by the functional expression

$$y = f(x) \qquad a \le x \le b$$

The surface area it generates is still given by $A \approx \sum 2\pi y \, \Delta s$, but now the independent variable is x instead of t. Hence the limit of the approximating sum (as the partition is refined) is

$$A = \int_{x=a}^{x=b} 2\pi y \, ds = 2\pi \int_a^b f(x) \sqrt{1 + [f'(x)]^2} \, dx$$

(See Section 7.3.) Again the thing to remember is not the final formula but the area $2\pi y \, ds$ of the typical strip of surface. The only difference in the analysis is that everything is expressed in terms of x.

□ **Example 3**   Find the surface area of the sphere generated by rotating the semicircle $y = \sqrt{a^2 - x^2}$ about the x axis.

Solution: Since $dy = \dfrac{-x\,dx}{\sqrt{a^2 - x^2}} = -\dfrac{x}{y}\,dx$

we find

$$(ds)^2 = (dx)^2 + (dy)^2 = (dx)^2 + \frac{x^2}{y^2}(dx)^2 = \frac{x^2 + y^2}{y^2}(dx)^2 = \frac{a^2}{y^2}(dx)^2$$

$$ds = \frac{a}{y}\,dx$$

$$2\pi y\,ds = 2\pi a\,dx$$

$$A = \int_{x=-a}^{x=a} 2\pi y\,ds = 2\pi a \int_{-a}^{a} dx = 4\pi a^2$$

Note that this is the same as the result in Example 1.

An alternate method of solution in this example is to differentiate implicitly in $x^2 + y^2 = a^2$ (the circle whose upper half is rotated to generate the sphere). This gives

$$2x\,dx + 2y\,dy = 0$$

and hence $dy = -\dfrac{x}{y}dx$

from which we proceed as before.   □

□ **Example 4**   Find the surface area generated by rotating the parabolic arc $y = x^2$, $0 \le x \le 1$, about the y axis.

Solution: When the typical piece of the graph (of length Δs) is rotated about the y axis, it generates a strip of area $\Delta A \approx 2\pi x\,\Delta s$. (See Figure 5.) Our formula for surface area this time is

$$A = \int_{x=0}^{x=1} 2\pi x\,ds$$

Since $dy = 2x\,dx$, we find

$$(ds)^2 = (dx)^2 + 4x^2\,(dx)^2 = (1 + 4x^2)(dx)^2$$

Hence $2\pi x\,ds = 2\pi x \sqrt{1 + 4x^2}\,dx$ and

$$A = 2\pi \int_0^1 x\sqrt{1 + 4x^2}\,dx = \frac{\pi}{4} \int_1^5 u^{1/2}\,du \quad [u = 1 + 4x^2,\ du = 8x\,dx]$$

$$= \frac{\pi}{4} \cdot \frac{2}{3} u^{3/2} \Big|_1^5 = \frac{\pi}{6}(5\sqrt{5} - 1)$$ □

Figure 5
Rotation of typical piece of a graph about the y axis

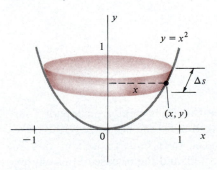

Problem Set 7.4

In each of the following, find the area of the surface generated by rotating the given curve about the x axis.

1. $x = t$, $y = 1$, $0 \leq t \leq 2$ (Check by geometry.)
2. $x = t^2$, $y = 2$, $0 \leq t \leq 2$ (Check by geometry.)
3. $x = t$, $y = 2t$, $0 \leq t \leq 1$ (Check by geometry.)
4. $x = 3t$, $y = 2t$, $0 \leq t \leq 1$ (Check by geometry.)
5. $x = t^2$, $y = t$, $0 \leq t \leq 1$
6. $x = 1 - t^2$, $y = 2t$, $0 \leq t \leq 1$
7. $x = t$, $y = t^3$, $0 \leq t \leq 1$
8. $x = 2t$, $y = \frac{1}{3}t^3$, $0 \leq t \leq 1$
9. $x = 3 \sin t$, $y = 3 \cos t$, $-\pi/2 \leq t \leq \pi/2$
10. $x = 1 - \cos t$, $y = \sin t$, $0 \leq t \leq \pi/2$
11. $x = a \cos^3 t$, $y = a \sin^3 t$ *Hint:* The upper half of the graph corresponds to $0 \leq t \leq \pi$. If you integrate from 0 to π, however, you may get zero unless you are careful about signs when $\pi/2 \leq t \leq \pi$. You can also integrate from 0 to $\pi/2$ and double the result. (Why?)
12. $x = t - \sin t$, $y = 1 - \cos t$, $0 \leq t \leq \pi$
13. $y = 2 - x$, $0 \leq x \leq 2$ (Check by geometry.)
14. $y = x$, $0 \leq x \leq 3$ (Check by geometry.)
15. $y = \frac{1}{3}x^3$, $0 \leq x \leq 1$
16. $y = x^3$, $0 \leq x \leq 2$
17. $y = 2\sqrt{x}$, $0 \leq x \leq 3$
18. $y^2 = 1 - x$, $0 \leq x \leq 1$
19. $y = x^{2/3}$, $0 \leq x \leq 8$ *Hint:* Let $u = (9x^{2/3} + 4)^{1/2}$
20. $y = x^3/12 + 1/x$, $1 \leq x \leq 2$
21. $y = x^4/8 + 1/(4x^2)$, $1 \leq x \leq 2$
22. $8y^2 = x^2(1 - x^2)$ *Hint:* The graph is a figure eight.

In each of the following, find the area of the surface generated by rotating the given curve about the y axis.

23. $x = 2t$, $y = t^2$, $0 \leq t \leq 1$
24. $x = t + 1$, $y = \frac{1}{2}t^2 + t$, $0 \leq t \leq 1$

25. $x = t^3$, $y = 2t$, $0 \leq t \leq 1$
26. $x = \frac{1}{2}t^2$, $y = \frac{1}{3}t^3$, $0 \leq t \leq 2$ *Hint:* Let $u = \sqrt{1 + t^2}$
27. $x = \cos t$, $y = \sin t$, $0 \leq t \leq \pi/2$
28. $x = \sin t$, $y = 1 - \cos t$, $0 \leq t \leq \pi$
29. $x = 1 - \cos t$, $y = t - \sin t$, $0 \leq t \leq 2\pi$
30. $y = 2x$, $0 \leq x \leq 1$ (Check by geometry.)
31. $y = \frac{1}{2}x^2$, $0 \leq x \leq 2$
32. $x = y^3/6 + 1/(2y)$, $1 \leq y \leq 2$
33. $y = \frac{1}{3}(x^2 - 2)^{3/2}$, $\sqrt{2} \leq x \leq 2$
34. The lateral surface area of a right circular cylinder of radius r and altitude h is given in geometry by $A = 2\pi rh$. Derive this formula by integration.
35. The lateral surface area of a right circular cone of base radius r and altitude h is given in geometry by $A = \pi r \sqrt{r^2 + h^2}$. Derive this formula by integration.
36. A *torus* ("doughnut") is generated by rotating a circle of radius a about a line b units from its center $(a < b)$. Find its area.
37. A *zone* of a sphere of radius a is the part of its surface between two parallel planes h units apart $(h \leq 2a)$.
    (a) Show that the area of the zone is $A = 2\pi ah$.
    (b) Assuming that the earth is spherical, if follows from part (a) that a polar zone and an equatorial zone have the same area if they lie between planes an equal distance apart. Why? (Thus a zone between planes one mile apart has the same area whether it contains the equator or is near the North Pole.)
    (c) Confirm that the formula in part (a) gives the correct result when the "zone" is the entire sphere.

7.5	You would probably be willing to place this book gently on top of your watch lying on a table. But if you were asked to lift the book above the table
WORK	and drop it on your watch, you might object. Everybody realizes intuitively that the book acquires energy when it is lifted, and the release of this energy when it is dropped will likely smash the watch.

Similarly, when a coiled spring is compressed by pushing against one end, it appears to store the energy imparted to it by the push. Like the book lifted above a table, it might well smash a watch when released to its original position.

Stored energy of this kind is hard to measure directly. Instead we calculate the *work* done by the force that moves an object from one position to another. In the case of a book lifted above a table, the force is exerted against the downward pull of gravity (the weight of the book) and is virtually constant. In such circumstances the work is defined to be the product of the force and the distance through which it acts. In other words (assuming the force to be constant)

$$\text{Work} = \text{Force} \times \text{Distance}$$

If force is measured in pounds and distance in feet, the units of work are *foot-pounds*. Physicists are more likely to use the metric system, in which force is in dynes (or newtons) and distance is in centimeters (or meters). The units of work are then *dyne-centimeters* (or *newton-meters*). We will not be particular about the units. The essential point is that work is defined as the product of force and distance when the force is constant.

□ **Example 1**   A book weighing 3 pounds is resting on a table. How much work is done in lifting the book 2 feet above the table?

Solution: The work is $W = (3)(2) = 6$ (ft-lb). This is not very interesting, but when it is converted to equivalent units of energy it serves as a measure of energy stored by the book when it is lifted (or released when it is dropped). Our focus in this section is not on energy as such (that is more a subject for a physics course), but on the application of integration to the case of work done by a variable force. For in that case the elementary definition of work as force times distance is not directly applicable. The key tool turns out to be the integral as a limit of Riemann sums.   □

□ **Example 2**   When a certain spring is stretched x units from its natural length, the force required to hold it stretched is $F(x) = 6x$. How much work is done in stretching the spring 10 units from its natural length?

Solution: The force varies from 0 to 60 as the free end of the spring is pulled from $x = 0$ to $x = 10$. (See Figure 1.) Hence we cannot simply compute the work as force times distance. Suppose, however, that we partition the interval [0,10] in the usual way. If the typical subinterval (of length Δx) is small, the force is nearly constant from one end of it to the other. Evaluating the force at some point x in the subinterval (and assuming that it has that value throughout the subinterval), we may approximate the work done in stretching the spring through the distance Δx by writing

$$\Delta W \approx \text{force} \times \text{distance} = F(x)\,\Delta x$$

The total work done in stretching the spring from 0 to 10 is

$$W = \sum \Delta W \approx \sum F(x)\,\Delta x$$

Figure 1
Coiled spring in three positions

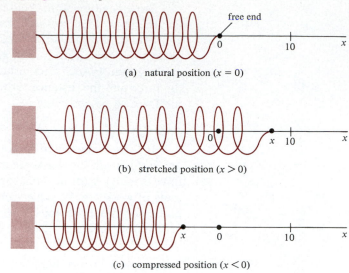

(a) natural position ($x = 0$)

(b) stretched position ($x > 0$)

(c) compressed position ($x < 0$)

which is a Riemann sum of $F(x)$ associated with the given partition. Hence

$$W = \lim \sum F(x)\,\Delta x = \int_0^{10} F(x)\,dx = \int_0^{10} 6x\,dx$$

$$= 3x^2 \Big|_0^{10} = 300 \text{ (units of work)} \qquad \square$$

Example 2 is based on *Hooke's Law,* which says that when a spring is stretched or compressed (within certain limits of elasticity), it exerts a restoring force proportional to the displacement. If its natural position (before any force is applied) is as shown in Figure 1, we may suppose that its free end is at point x after displacement ($x > 0$ or $x < 0$ depending on whether the spring is stretched or compressed). Since the restoring force is directed toward the origin, Hooke's Law may be expressed by writing

$$\text{Restoring Force} = -kx$$

where k is a positive constant. Note that when the spring is stretched ($x > 0$) the restoring force is in the negative direction, whereas it is in the positive direction when the spring is compressed ($x < 0$). The expression $-kx$ therefore gives both the magnitude and direction of the force.

The force that stretches or compresses the spring (and therefore does the work) is opposed to the restoring force. Hence it is given by $F(x) = kx$. In Example 2 we took $k = 6$; in general the value of k depends on the material and dimensions of the spring.

The analysis in Example 2 applies whenever an object is moved along a straight line by a force (in the direction of motion) that varies with position.

In general, if the force at point x is given by $F(x)$, and the object moves from $x = a$ to $x = b$, we define the work done by the force to be

$$W = \int_a^b F(x)\,dx$$

□ **Example 3**   A certain spring exerts a restoring force of 5 dynes when it is compressed 2 centimeters. How much work is done in compressing the spring 3 more centimeters?

Solution: According to Hooke's Law, the force of compression (opposite in direction to the restoring force) is

$$F(x) = kx \qquad (k > 0)$$

Since $x = -2$ when the spring is compressed 2 cm (see Figure 1), and since the restoring force at that point is 5 dynes, we have $F(-2) = -5$. Hence $-5 = k(-2)$, from which $k = \frac{5}{2}$. The force of compression is therefore $F(x) = \frac{5}{2}x$. The work done in compressing the spring 3 more centimeters (moving the free end from $x = -2$ to $x = -5$) is

$$W = \int_{-2}^{-5} F(x)\,dx = \int_{-2}^{-5} \frac{5}{2}x\,dx = \frac{5}{4}x^2 \Big|_{-2}^{-5} = \frac{105}{4} \text{ (dyne-centimeters)} \; \square$$

□ **Example 4**   A space capsule weighing 1000 lb on the surface of the earth is propelled to an altitude of 100 miles above the surface of the earth. How much work is done against the force of gravity?

Solution: We need a fact from physics, namely Newton's famous *Inverse-Square Law*. This says that the force exerted by the earth on the capsule (the *force of gravity*) is $f(x) = -k/x^2$, where x is the distance of the capsule from the center of the earth and k is a positive constant. (See Figure 2.) The negative sign is used because the x axis is directed upward and the force of gravity is downward. The lifting force (which opposes gravity) is therefore

$$F(x) = -f(x) = \frac{k}{x^2}$$

It is this force that does the work on the capsule.

To evaluate k, we assume that the radius of the earth is 4000 miles. Then $F(4000) = 1000$ because the capsule weighs 1000 lb on the surface of the earth. Hence

$$1000 = \frac{k}{(4000)^2}$$

from which $k = 16 \times 10^9$. Thus the lifting force is

$$F(x) = \frac{16 \times 10^9}{x^2}$$

Figure 2
Force of gravity on a space capsule

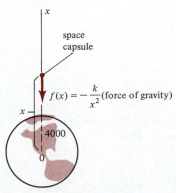

where x is in miles and $F(x)$ is in pounds. The work done in lifting the capsule from $x = 4000$ (the surface of the earth) to $x = 4100$ (100 miles higher) is

$$W = \int_{4000}^{4100} \frac{16 \times 10^9}{x^2}\, dx = -\frac{16 \times 10^9}{x}\Big|_{4000}^{4100} = 16 \times 10^9 \left(\frac{1}{4000} - \frac{1}{4100}\right)$$

$$= \frac{4}{41} \times 10^6 \ \text{(mile-pounds)} \approx 5 \times 10^8 \ \text{(ft-lb)} \qquad \square$$

Not all problems involving the calculation of work are solved by means of the formula

$$W = \int_a^b F(x)\, dx$$

The next example shows how we can return to first principles (using Riemann sums) to set up an integral for the work.

$\square$ **Example 5**   A tank in the shape of an inverted cone is full of water. The tank has diameter 20 ft at the top and is 15 ft deep. (See Figure 3.) If it is emptied by pumping the water over the rim, how much work is done?

Solution: The problem is that water at different depths must be lifted through different distances. To get around this, imagine the water in layers, the typical layer being so thin that we may regard the water in that layer as being all at the same depth (say h ft from the bottom of the tank). This layer is lifted through a distance $15 - h$ by working against the force of gravity (the weight of the layer). The volume of the layer is

$$\Delta V \approx \pi r^2 \, \Delta h$$

Since water weighs about 62.5 lb/ft³, the weight of the layer is

$$62.5 \, \Delta V \approx 62.5\pi r^2 \, \Delta h$$

Hence the work done in lifting this layer to the top of the tank is

$$\Delta W \approx \text{force} \times \text{distance} \approx 62.5\pi r^2 \, \Delta h(15 - h)$$

The total work done in lifting all the layers to the top of the tank is

$$W = \sum \Delta W \approx \sum 62.5\pi r^2(15 - h)\, \Delta h$$

Taking the limit as $\Delta h \to 0$, we find

$$W = \int_0^{15} 62.5\pi r^2(15 - h)\, dh$$

where it is understood that r is a function of h. Use similar triangles in Figure 3 to write $r/h = \frac{10}{15}$, from which $r = \frac{2}{3}h$. Hence

$$W = \frac{250\pi}{9} \int_0^{15} h^2(15 - h)\, dh = \frac{250\pi}{9} \int_0^{15} (15h^2 - h^3)\, dh$$

Figure 3
Typical layer of water in conical tank

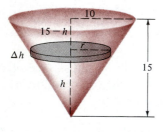

$$= \frac{250\pi}{9}\left(5h^3 - \frac{h^4}{4}\right)\Big|_0^{15} = \frac{125\pi h^3}{18}(20 - h)\Big|_0^{15}$$

$$= \frac{234{,}375\pi}{2} \approx 368{,}155 \text{ (ft-lb)} \qquad \square$$

□ **Example 6**   In certain circumstances the pressure and volume of a gas confined in a cylinder are related by the formula $PV^{1.4} = 50$, where P is in pounds per square inch and V is in cubic inches. How much work is done by a piston that compresses the gas from 10 cubic inches to 1 cubic inch?

Figure 4
Piston compressing gas in cylinder

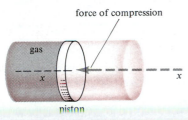

Solution: Let x be the distance of the piston from the fixed end of the cylinder, as shown in Figure 4. Let A be the area of the face of the piston. Since pressure is force per unit area, the force exerted by the gas on the piston is PA. The force that does the work of compression is therefore $F(x) = -PA$ (negative because it is opposite in direction to the positive x axis). The trouble with this expression is that $F(x)$ is not explicit in terms of x (nor do we know the interval through which the force acts). To get around this, let $x = a$ when $V = 10$ and $x = b$ when $V = 1$. The work done in compressing the gas from 10 in³ to 1 in³ is

$$W = \int_a^b F(x)\, dx = -\int_a^b PA\, dx$$

Since $V = Ax$, we have $dV = A\, dx$, and hence (changing the variable and limits)

$$W = -\int_{10}^1 P\, dV = -\int_{10}^1 50V^{-1.4}\, dV \qquad \text{(because } PV^{1.4} = 50\text{)}$$

$$= \frac{50V^{-0.4}}{0.4}\Big|_{10}^1 = 125(1 - 10^{-0.4}) \approx 75.2 \text{ (inch-pounds)} \qquad \square$$

Optional Note (*on kinetic energy*)   We end this section by deriving an expression for work in terms of *kinetic energy*, an interpretation that may be of interest if you are studying physics. Kinetic energy is the energy of an object due to its motion, and is defined to be

$$K = \tfrac{1}{2}mv^2$$

where m is the mass and v is the velocity of the object. We know that $v = dx/dt$, where x is the position of the object at time t. If $F(x)$ is the force on the object at point x, the work done in moving the object from $x = a$ to $x = b$ is

$$W = \int_a^b F(x)\, dx$$

Newton's Second Law of Motion says that force is mass times acceleration, that is, $F = m\,(dv/dt)$. Using the Chain Rule to write

$$\frac{dv}{dt} = \frac{dv}{dx}\frac{dx}{dt} = v\frac{dv}{dx}$$

we have

$$W = \int_a^b m \frac{dv}{dt} \, dx = m \int_a^b v \frac{dv}{dx} \, dx = m \int_{v(a)}^{v(b)} v \, dv$$

where $v(a)$ and $v(b)$ are the values of v corresponding to $x = a$ and $x = b$, respectively. Hence

$$W = \frac{1}{2} mv^2 \Big|_{v(a)}^{v(b)} = K(b) - K(a)$$

where $K(a)$ and $K(b)$ are the values of K at $x = a$ and $x = b$, respectively.

Thus the work done is equal to the *change in kinetic energy,* a result which may give some substance to our remarks about energy at the beginning of this section.

Problem Set 7.5

1. Suppose that an object is moved from $x = a$ to $x = b$ by a constant force $F(x) = c$. Show that the formula

$$W = \int_a^b F(x) \, dx$$

reduces to the elementary definition of work as force times distance.

2. A certain spring exerts a restoring force of 0.2 newtons when it is stretched 0.3 meters.

   (a) According to Hooke's Law, what is the formula for the restoring force? for the force that stretches the spring?

   (b) How much work is done in stretching the spring 0.5 meters?

3. In Problem 2 find the work done in compressing the spring 0.3 meters.

4. A force of 8 dynes is required to stretch a spring from its natural length of 10 cm to a length of 15 cm. How much work is done in stretching the spring to a length of 25 cm?

5. In Problem 4 find how much work is done in stretching the spring from a length of 12 cm to a length of 20 cm.

6. In Problem 4 find the work done in compressing the spring to a length of 6 cm.

7. In Problem 4 find how much work is done in compressing the spring from a length of 8 cm to a length of 5 cm.

8. A force of 10 lb is required to compress a spring from its natural length of 8 in to a length of 6 in. How much work is done in stretching the spring to a length of 12 in?

9. In Problem 8 find the work done in stretching the spring from a length of 9 in to a length of 14 in.

10. In Problem 8 how much work is done in compressing the spring to a length of 5 in?

11. In Problem 8 find the work done in compressing the spring from a length of 6 in to a length of 4 in.

12. Water weighing 40 lb is hauled up in a bucket weighing 3 lb from the bottom of a well 20 ft deep. The rope weighs 0.2 lb/ft.

    (a) Explain why the force working against gravity when the bucket is x ft from the bottom of the well is $F(x) = 47 - 0.2x$.

    (b) Find the work done against gravity.

13. A cable weighing 2 lb/ft is used to haul 100 lb of ore up a mine shaft 1000 ft deep. How much work is done?

14. A 120-lb mountaineer is hanging from a 30-ft rope weighing 0.1 lb/ft. How much work is done in pulling her up?

15. Find the work done against the force of gravity in firing a rocket weighing 5 tons on the surface of the earth to an altitude of 50 miles. (Assume that the radius of the earth is 4000 miles.)

16. A lunar module weighs 4 tons on the surface of the moon. Assuming the radius of the moon to be 2160 miles, find the work done against lunar gravity when the module is propelled to a height of 300 miles above the surface of the moon.

17. A lunar module weighs 2 tons when it is in orbit 100 miles above the surface of the moon. How much work is done by lunar gravity in pulling it down to the surface of the moon? (Assume that the radius of the moon is 2160 miles.)

18. Two oppositely charged particles attract each other with a force inversely proportional to the square of the distance between them. If the force is 10 dynes when they are 5 cm

apart, and the distance between them is increased to 25 cm, how much work is done?

19. Two similarly charged particles repel each other with a force inversely proportional to the square of the distance between them. If the force is 3 dynes when they are 10 cm apart, how much work is done in moving one of them from a distance of 20 cm to a distance of 15 cm of the other?

20. In Problem 18 suppose that the distance between the particles is increased from 5 cm to r cm, where $r > 5$.
    (a) Find the work done as a function of r.
    (b) What limiting value does the work approach as r increases without bound? (The result represents the work done in moving one particle from 5 cm "to infinity.")

21. In Problem 19 suppose that the distance between the particles is decreased from r cm to 10 cm, where $r > 10$.
    (a) Find the work done as a function of r.
    (b) What limiting value does the work approach as r increases without bound? (The result represents the work done in moving one particle "from infinity" to 10 cm.)

22. A cylindrical tank of radius 3 ft and height 10 ft is full of water weighing 62.5 lb/ft³. How much work is done in emptying the tank by pumping the water over the top?

23. In Problem 22 suppose that the tank is emptied by pumping the water to an outlet pipe 2 ft above the top. How much work is done?

24. A rectangular swimming pool is 50 ft long, 30 ft wide, and 8 ft deep. If the water is 7 ft deep, how much work is done in emptying the pool by pumping the water over the edge?

25. In Problem 24 suppose that the water is drained through an outlet at the bottom of the pool. How much work is done by the force of gravity?

26. A hemispherical tank of radius 5 ft is full of water. When the water is drained through an outlet at the bottom of the tank, how much work is done by the force of gravity?

27. In Problem 26 how much work is done in emptying the tank by pumping the water over the edge?

28. A cylindrical tank of radius 2 ft and length 5 ft is lying on its side. How much work is done in filling it with water through a hole in the bottom?

29. In Problem 28 find the work done in emptying the (full) tank by pumping the water to an outlet pipe 5 ft above the ground.

30. Gas expanding in a cylinder causes a piston to move so that the volume of the gas changes from 16 in³ to 25 in³. If the relation between pressure and volume is $PV^{3/2} = 60$, how much work is done by the expanding gas?

31. Gas expanding in a cylinder causes a piston to move so that the volume increases from V_1 to V_2. Show that no matter how the pressure and volume are related, the work done by the expanding gas is given by

$$W = \int_{V_1}^{V_2} P\, dV$$

32. An object of mass m (kilograms) is at rest. After 100 newton-meters of work is done on it, its velocity is 20 meters per second. What is m?

33. A car of mass 600 kilograms goes from 0 to 60 km/hr. How much work is done (in newton-meters)?

7.6
MOMENTS AND CENTROIDS

Suppose that a girl weighing 60 pounds and a boy weighing 90 pounds are at opposite ends of a teeter board 10 feet long, as shown in Figure 1. Where should the point of support be placed so that the board will balance?

To answer this question, we need the concept of *moment of force* from physics. Loosely speaking, this is a measure of the tendency of the force to produce rotation (in this case about the point of support). It depends on both the force and the point at which it is applied. More precisely, it is the product of the magnitude of the force and the distance of its line of action from the point at which rotation tends to occur. The principle of rotational equilibrium says that opposing moments must be equal if the board is to remain in balance.

Thus the weight of the girl (a downward force acting x feet to the right of the point of support) produces a moment of $60x$ tending to rotate the

Figure 1
Balancing a teeter board

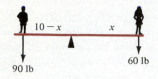

Figure 2
Moments in terms of coordinates

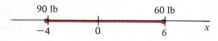

board clockwise. The boy's weight causes an opposing (counterclockwise) moment of $90(10 - x)$. To achieve equilibrium, we must have

$$60x = 90(10 - x)$$

from which $x = 6$. Hence the point of support should be 6 feet from the girl (and $10 - x = 4$ feet from the boy).

Another way to say the same thing is to introduce a coordinate line (with origin at the point of support) and to use coordinates (directed distances) instead of distances in the definition of moment. (See Figure 2.) The respective moments are then

$$(60)(6) = 360 \qquad \text{(from the girl's weight)}$$

$$(90)(-4) = -360 \qquad \text{(from the boy's weight)}$$

The principle of equilibrium in this context says that *the sum of the moments must be zero* for the board to balance. This has the effect of attaching opposite signs to opposing moments and is easier to deal with mathematically.

More generally, suppose that n particles of masses $m_1, m_2, \ldots, m_n$ are located at points $x_1, x_2, \ldots, x_n$ on the x axis. The moment of m_k relative to the origin is defined to be $m_k x_k$, while the total moment of the system is

$$M = m_1 x_1 + m_2 x_2 + \cdots + m_n x_n = \sum_{k=1}^{n} m_k x_k$$

The total mass of the system is

$$m = m_1 + m_2 + \cdots + m_n = \sum_{k=1}^{n} m_k$$

If this were concentrated at one point (with coordinate $\bar{x}$), its moment relative to the origin would be $m\bar{x}$. The point $\bar{x}$ for which this is equal to the total moment of the system, $m\bar{x} = M$, is called the **center of mass** of the system. Thus the formula for $\bar{x}$ is

$$\bar{x} = \frac{M}{m} = \frac{\sum m_k x_k}{\sum m_k}$$

Remark   Note that in this definition we have used mass instead of weight for the computation of moments. The former is a measure of inertia (resistance to change in the speed or direction of motion) and is independent of gravity. (Thus the mass of an object is the same on the moon as on the earth.) Weight, on the other hand, is the force of gravity on an object (and therefore depends on the gravitational field in which the object is placed). According to Newton's Second Law, the weight of an object of mass m is mg, where g is the acceleration due to gravity. At a given point in a gravitational field g is fixed, and hence (at that location) weight is just a constant multiple of mass. Had we used weight instead of mass in our definition of moment, the equation $m\bar{x} = M$ would have read $mg\bar{x} = Mg$. Since g cancels, it makes no difference in the computation of $\bar{x}$ whether we think of moments in terms of force (weight) or mass. Sometimes the center of mass is called the *center of gravity;* whatever its name, its location is determined solely by the distribution of mass in the system and does not depend on the gravitational field. In other words we may think of it as a purely mathematical concept and not worry about physics.

□ **Example 1**   Find the center of mass of the system consisting of a 60-lb girl and a 90-lb boy at opposite ends of a 10-ft teeter board.

Solution: Strictly speaking, we should convert weight to mass before proceeding. As we have explained, however, the final result is the same either way. We also neglect the contribution of the teeter board to the formulation of center of mass, assuming that its mass is uniformly distributed along its length (and is therefore irrelevant to the question).

With these agreements, we place the teeter board along a coordinate line with the origin at an arbitrary point (say midway between the girl and boy, as in Figure 3). The moment of the system relative to the origin is

$$M = m_1x_1 + m_2x_2 = (60)(5) + (90)(-5) = -150$$

while the mass is

$$m = m_1 + m_2 = 60 + 90 = 150$$

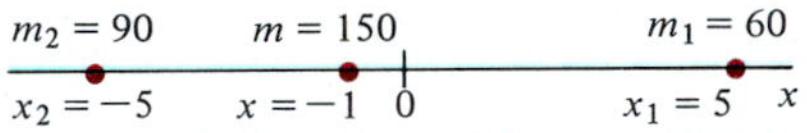

Hence the center of mass is

$$\bar{x} = \frac{M}{m} = \frac{-150}{150} = -1 \qquad \square$$

Figure 3
Center of mass of a two-point system

$m_2 = 90$ $m = 150$ $m_1 = 60$
$x_2 = -5$ $x = -1$ 0 $x_1 = 5$ x

Interestingly enough (perhaps you are way ahead of us, and predicted it!) the point $\bar{x} = -1$ is six feet from the girl (and four feet from the boy). It thus coincides with the point of support we found earlier (to balance the board). This is no accident; the concept of center of mass is designed to come out that way. In the general case we may imagine the x axis as an essentially weightless (but rigid) wire, and the masses $m_1, m_2, \ldots, m_n$ as beads strung on the wire. Then we may think of $\bar{x}$ as the point at which the system would balance on a knife edge. Putting it differently, we may replace the individual beads with a single bead at $\bar{x}$ (with mass $m = m_1 + m_2 + \cdots + m_n$) and the moment relative to the origin will be unchanged.

Remark   What if we had chosen a different origin in Example 1? That would be equivalent to translating the x axis (to a new origin at, say, the point h). A point with coordinate x relative to the first origin would have coordinate $x' = x - h$ relative to the new origin. The moment of the system would now be

$$M' = m_1x'_1 + m_2x'_2 = m_1(x_1 - h) + m_2(x_2 - h)$$
$$= m_1x_1 + m_2x_2 - (m_1 + m_2)h = M - mh$$

and the center of mass would be

$$\bar{x}' = \frac{M'}{m} = \frac{M - mh}{m} = \frac{M}{m} - h = \bar{x} - h$$

This is the new coordinate of the same point as before; you can generalize the argument to show that the center of mass does not depend on the point relative to which we compute moments. In other words it is a property of the mass distribution which is independent of the coordinate system we impose on it.

Continuous Mass Distribution Along a Straight Line

As long as we are dealing with a finite number of particles, ordinary arithmetic is adequate for the calculation of center of mass. Suppose, however, that mass is distributed *continuously* along some interval of the x axis, say $I = [a,b]$, with density $\delta(x)$ at each point $x \in I$. (Density is mass per unit length, the limit of $\Delta m/\Delta x$ as $\Delta x \to 0$, where Δm is the mass in a subinterval of length Δx.) Partitioning the interval in the usual way, let Δm be the mass in the typical subinterval, and choose any point x in this subinterval. If the partition is fine, the approximation

$$\delta(x) \approx \frac{\Delta m}{\Delta x}$$

is a good one. (Why?) The mass in the subinterval is then $\Delta m \approx \delta(x)\, \Delta x$ and the total mass is

$$m = \sum \Delta m \approx \sum \delta(x)\, \Delta x$$

This is a Riemann sum suggesting the formula

$$m = \int_{x=a}^{x=b} dm = \int_a^b \delta(x)\, dx$$

Hence we adopt this formula as a *definition* of total mass (distributed according to a continuous density function δ along an interval with endpoints a and b).

The moment of such a mass distribution (relative to the origin) is similarly defined. Thinking of the mass Δm as concentrated at the point x in the typical subinterval, we compute its moment to be

$$\Delta M = x\, \Delta m \approx x\delta(x)\, \Delta x$$

The total moment is

$$M = \sum \Delta M = \sum x\, \Delta m \approx \sum x\delta(x)\, \Delta x$$

which suggests the definition

$$M = \int_{x=a}^{x=b} x\, dm = \int_a^b x\delta(x)\, dx$$

The center of mass is defined as in the discrete case, but now it is a ratio of integrals instead of sums:

$$\bar{x} = \frac{M}{m} = \frac{\int x \, dm}{\int dm}$$

Remark   The missing limits of integration in the above formula are to be supplied when the integrands are expressed in terms of x. For example,

$$\int x \, dm = \int_a^b x \, \delta(x) \, dx$$

The reason we leave out the limits is that similar formulas for the center of mass of a region in the plane (to be discussed shortly) are more conveniently written this way.

□ **Example 2**   The density of a straight wire of length 5 cm is proportional to the distance from one end. Find its center of mass.

Solution: Place the wire so that the end referred to is at the origin ($a = 0$) and the other end is at $b = 5$. We are told that the density at point x ($0 \le x \le 5$) is $\delta(x) = kx$ (where k is a constant). The mass is therefore

$$m = \int_0^5 kx \, dx = \frac{25k}{2}$$

The moment relative to the origin is

$$M = \int_0^5 x \cdot kx \, dx = k \int_0^5 x^2 \, dx = \frac{125k}{3}$$

Hence the center of mass is

$$\bar{x} = \frac{M}{m} = \frac{125k}{3} \cdot \frac{2}{25k} = \frac{10}{3} \qquad \text{(cm from the given end of the wire)} \qquad □$$

Mass Distributions in the Plane

The terminology developed in connection with a linear mass distribution may be applied as well to a distribution of mass in a region of the plane. Density at a point is now defined as the limit of the ratio $\Delta m / \Delta A$ (where ΔA is the area of a subregion containing the point) and is expressed in units of mass per unit area (such as gm/cm^2). Then $\Delta m \approx \delta \, \Delta A$ and the total mass is

$$m = \lim \sum \Delta m = \int dm$$

as before. To define the center of mass, we use moments relative to the coordinate axes (instead of the origin) because the location of a point in the plane requires two coordinates. To see how this works, consider the typical element of mass Δm. Imagining the mass to be concentrated at a single point (x, y), observe that its (directed) distance from the y axis is x. By analogy to the linear case, its moment relative to that axis is $\Delta M_y = x \, \Delta m$. The total moment is defined to be

$$M_y = \lim \sum x \, \Delta m = \int x \, dm$$

Similarly, the moment relative to the x axis is

$$M_x = \lim \sum y \, \Delta m = \int y \, dm$$

The center of mass is defined to be the point $(\bar{x}, \bar{y})$ where the total mass m could be concentrated without changing these moments:

$$m\bar{x} = M_y \qquad \text{and} \qquad m\bar{y} = M_x$$

Hence the formulas for $\bar{x}$ and $\bar{y}$ are

$$\boxed{\bar{x} = \frac{M_y}{m} = \frac{\int x \, dm}{\int dm}, \quad \bar{y} = \frac{M_x}{m} = \frac{\int y \, dm}{\int dm}}$$

The terminology and notation in this discussion are deliberately vague compared to what we did in the linear case. The reason for this is that we are not prepared to fill in the details until multivariable calculus is developed in later chapters. As a case in point, consider the notion of "typical element of mass." You would have a right to expect (by analogy to the linear case) that this refers to a chunk of the mass distribution that is small in all dimensions (say a square whose diagonal approaches zero as our partitioning of the mass is refined). In this section, however, we are constrained by the limitations of single-variable calculus to work with elements of mass that are small in only one dimension (like a thin horizontal or vertical strip). The integrals obtained in this way involve functions of only one variable, and hence can be handled. It is worth mentioning (for future reference) that it is more natural to use double integrals (involving functions of two variables) in setting up moments of two-dimensional mass distributions.

Figure 4
Calculation of M_y by means of vertical strips

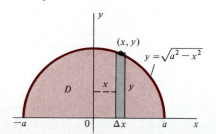

□ **Example 3**   Find the center of mass of a piece of tinfoil occupying the semicircular region

$$D = \{(x, y): x^2 + y^2 \leq a^2, y \geq 0\}$$

Solution: We assume that the tinfoil has density $\delta(x, y) = k$ at each point $(x, y) \in D$, where k is constant. Dividing the region into vertical strips (see Figure 4), observe that the area of the typical strip is $\Delta A \approx y \, \Delta x$ and hence its mass is

$$\Delta m = k \, \Delta A \approx ky \, \Delta x$$

Assuming the strip to be thin, we can speak of its (directed) distance from the y axis as approximately x. Hence its moment relative to the y axis is

$$\Delta M_y \approx x\, \Delta m \approx kxy\, \Delta x$$

Adding the contributions of all the strips (and taking the limit as they become thinner), we find

$$M_y = \int_{x=-a}^{x=a} x\, dm = \int_{-a}^{a} kxy\, dx$$

Since $y = \sqrt{a^2 - x^2}$ (because $y \geq 0$), we have

$$M_y = k \int_{-a}^{a} x\sqrt{a^2 - x^2}\, dx = 0 \qquad \text{(because the integrand is odd)}$$

Figure 5
Calculation of M_x by means of horizontal strips

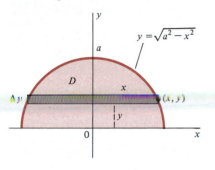

To find M_x, divide the region into horizontal strips (Figure 5). The mass of the typical strip is

$$\Delta m = k\, \Delta A \approx k(2x)\, \Delta y = 2kx\, \Delta y$$

and its moment relative to the x axis is

$$\Delta M_x \approx y\, \Delta m \approx 2kxy\, \Delta y$$

Hence

$$M_x = \int_{y=0}^{y=a} y\, dm = \int_{0}^{a} 2kxy\, dy$$

Since $x = \sqrt{a^2 - y^2}$ (the positive square root is chosen because $x > 0$), we find

$$M_x = 2k \int_{0}^{a} y\sqrt{a^2 - y^2}\, dy$$

$$= -k \int_{a^2}^{0} u^{1/2}\, du \qquad (u = a^2 - y^2,\ du = -2y\, dy)$$

$$= \frac{2}{3} ku^{3/2} \Big|_{0}^{a^2} = \frac{2}{3} ka^3$$

To finish the problem, we need the mass m of the tinfoil. While this may be found by integration,

$$m = \int_{x=-a}^{x=a} k\, dm = k \int_{-a}^{a} y\, dx = k \int_{-a}^{a} \sqrt{a^2 - x^2}\, dx$$

it is easier to observe that the density is constant and hence the mass is simply k times the area of the semicircle:

$$m = kA = k \cdot \frac{\pi a^2}{2} = \frac{k\pi a^2}{2}$$

Thus

$$\bar{x} = \frac{M_y}{m} = \frac{0}{m} = 0 \qquad \text{and} \qquad \bar{y} = \frac{M_x}{m} = \frac{2}{3} ka^3 \cdot \frac{2}{k\pi a^2} = \frac{4a}{3\pi}$$

and the center of mass is $(\bar{x}, \bar{y}) = (0, 4a/3\pi) \approx (0, 0.42a)$. □

Remarks (*about Example 3*)

1. Vertical strips are useful in finding M_y because it makes sense to refer to their distance from the y axis. The distance of a vertical strip from the x axis, on the other hand, is not a meaningful idea. Hence we must either use horizontal strips to find M_x or replace each vertical strip with an equivalent mass element whose distance from the x axis is unambiguous. (See Remark 3.)

2. You may have guessed ahead of time that $M_y = 0$. For the mass is distributed symmetrically about the y axis. In such circumstances it is geometrically apparent that the moment of each strip on one side of the y axis is cancelled by the moment of the symmetrically located strip on the other side of the axis. Hence

$$\bar{x} = \frac{M_y}{m} = 0$$

that is, the center of mass lies on the y axis. The computation of M_y as an integral is wasted effort; it is more economical to make use of symmetry whenever we can.

3. An alternate procedure for finding M_x is to stick with the vertical strips of Figure 4, observing that the center of mass of the typical strip is approximately $(x, y/2)$. (Why?) Imagining the mass of the strip to be concentrated at this point, we find its moment relative to the x axis to be

$$\Delta M_x \approx \frac{y}{2} \Delta m \approx \frac{y}{2}(ky \, \Delta x) = \frac{k}{2} y^2 \, \Delta x$$

Hence $\quad M_x = \int_{-a}^{a} \frac{k}{2} y^2 \, dx = \frac{k}{2} \int_{-a}^{a} (a^2 - x^2) \, dx$

$$= k \int_{0}^{a} (a^2 - x^2) \, dx \qquad \text{(because the integrand is even)}$$

$$= k\left(a^2 x - \frac{x^3}{3}\right)\Big|_{0}^{a} = \frac{2}{3} k a^3$$

4. The density k cancels in the fractions $\bar{x} = M_y/m$ and $\bar{y} = M_x/m$. This always happens in the case of constant density; the center of mass is independent of δ and is therefore completely determined by the geometry of the figure. For that reason the center of mass of a "homogeneous" mass distribution is frequently called its **centroid** instead. This term is intended to refer to the geometrical center of the distribution (assuming the density to be constant and hence irrelevant).

□ **Example 4**   Find the centroid of the region under the curve $y = \sin x$, $0 \le x \le \pi$.

Solution: Observe first that since the region is symmetric about the line $x = \pi/2$ (Figure 6), we have $\bar{x} = \pi/2$. To find $\bar{y} = M_x/m$, we shall consider the (constant) density to be $\delta(x, y) = 1$, since its value is irrelevant anyway. Then the mass is

$$m = 1 \cdot A = \int_{0}^{\pi} \sin x \, dx = 2$$

Figure 6
Calculation of M_x by means of vertical strips

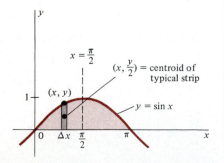

The moment relative to the x axis is

$$M_x = \int_{x=0}^{x=\pi} \frac{y}{2} \, dm = \int_0^\pi \frac{y}{2} \cdot 1 \cdot y \, dx$$

$$= \frac{1}{2} \int_0^\pi \sin^2 x \, dx = \frac{1}{4} \int_0^\pi (1 - \cos 2x) \, dx$$

$$= \frac{1}{4}\left(x - \frac{1}{2}\sin 2x\right)\Big|_0^\pi = \frac{\pi}{4}$$

Hence

$$\bar{y} = \frac{M_x}{m} = \frac{\pi/4}{2} = \frac{\pi}{8}$$

Thus the centroid is $(x, y) = (\pi/2, \pi/8) \approx (1.57, 0.39)$. (Note that we used vertical strips in our computation of M_x, which is the alternate procedure suggested in Remark 3. Horizontal strips are inconvenient, as you will see if you try them.)   □

Mass Distributions Along a Curve

The ideas of this section can be extended to a mass distribution along a curve in the plane (like a wire in the linear case, except that the context is two-dimensional because the wire is no longer required to be straight). The next example indicates what we mean.

□ **Example 5**   Find the centroid of the curve

$$x = a \cos^3 t, \quad y = a \sin^3 t \qquad 0 \le t \le \pi/2$$

Solution: The curve is shown in Figure 7 (the first-quadrant portion of the hypocycloid mentioned in Problem 36, Section 3.4, and Problem 35, Section 7.3). To find its centroid, let Δs be the length of a typical piece, and choose a point (x, y) in this piece. If the density at (x, y) is $\delta(x, y) = k$, the mass of the piece is $\Delta m = k \, \Delta s$ and its moments relative to the coordinate axes are

$$\Delta M_y \approx x \, \Delta m = kx \, \Delta s \qquad \text{and} \qquad \Delta M_x \approx y \, \Delta m = ky \, \Delta s$$

Hence

$$m = \int dm = \int k \, ds, \quad M_y = \int x \, dm = \int kx \, ds, \quad M_x = \int y \, dm = \int ky \, ds$$

and

$$\bar{x} = \frac{M_y}{m} = \frac{\int x \, ds}{\int ds}, \quad \bar{y} = \frac{M_x}{m} = \frac{\int y \, ds}{\int ds} \qquad \text{(because } k \text{ cancels)}$$

Figure 7
First-quadrant portion of a hypocycloid

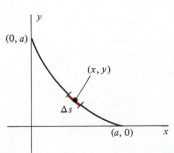

Since $(ds)^2 = (dx)^2 + (dy)^2$, we compute

$$dx = -3a \sin t \cos^2 t \, dt \qquad \text{and} \qquad dy = 3a \sin^2 t \cos t \, dt$$

Hence

$$(ds)^2 = (9a^2 \sin^2 t \cos^4 t + 9a^2 \sin^4 t \cos^2 t)(dt)^2 = 9a^2 \sin^2 t \cos^2 t \, (dt)^2$$

$$ds = 3a \sin t \cos t \, dt$$

$$\int ds = \int_0^{\pi/2} 3a \sin t \cos t \, dt = \frac{3a}{2} \sin^2 t \Big|_0^{\pi/2} = \frac{3a}{2}$$

$$\int x \, ds = \int_0^{\pi/2} a \cos^3 t \cdot 3a \sin t \cos t \, dt = 3a^2 \int_0^{\pi/2} \cos^4 t \sin t \, dt$$

$$= -\frac{3a^2}{5} \cos^5 t \Big|_0^{\pi/2} = \frac{3a^2}{5}$$

$$\bar{x} = \frac{\int x \, ds}{\int ds} = \frac{3a^2}{5} \cdot \frac{2}{3a} = \frac{2}{5} a$$

Since the curve is symmetric about the line $y = x$, we need not compute $\bar{y}$ separately; it is the same as $\bar{x}$. Hence the centroid is $(\bar{x}, \bar{y}) = (\frac{2}{5}a, \frac{2}{5}a)$.

A physical interpretation of this result may be obtained by imagining the curve to be a wire of constant density. Placing the wire on a membrane (essentially weightless), we may say that $(\bar{x}, \bar{y})$ is the point of the membrane at which the system would balance on a pin. $\quad\square$

Problem Set 7.6

1. Two people weighing 120 lb and 180 lb are at opposite ends of a teeter board 12 ft long. Where should the point of support be placed so that the board will balance?

2. Repeat Problem 1 in the case of two people weighing 150 lb and 200 lb, the teeter board being 14 ft long.

3. Particles of masses $m_1 = 3$, $m_2 = 2$, $m_3 = 7$ are located at $x_1 = 1$, $x_2 = -4$, $x_3 = 5$, respectively. Where is the center of mass of the system?

4. Repeat Problem 3 in the case of four masses, $m_1 = 5$, $m_2 = 6$, $m_3 = 11$, $m_4 = 10$, located at $x_1 = 2$, $x_2 = -5$, $x_3 = 4$, $x_4 = 2$, respectively.

5. The road from City A (population 35,000) to City B (population 80,000) is 40 miles long, and has two towns along the way that are 12 miles from A and 10 miles from B, respectively. The populations of the towns are 10,000 and 15,000. Where is the best place to locate an airport serving these communities?

6. The road from City A (population 40,000) to City B (population 60,000) is 50 miles long. There is one town (population 5000) along the way, 20 miles from A. Where is the best place to locate an airport serving these communities?

7. Show that the center of mass of a straight wire of constant density is at its midpoint.

8. A straight wire extending from $x = -1$ to $x = 1$ has density $\delta(x) = x^2$ at point x. Find its center of mass. How could the result have been predicted?

9. A straight wire extending from $x = 0$ to $x = 4$ has density $\delta(x) = \sqrt{x}$ at point x. Find its center of mass.

10. Repeat Problem 9 if $\delta(x) = 2 - \sqrt{x}$.

Find the centroid of each of the following regions.

11. The region bounded by $x + y = 1$ and the coordinate axes.

12. The region bounded by $y - x = 1$ and the coordinate axes.

13. The first-quadrant portion of the circular disk $x^2 + y^2 \leq a^2$.

14. The region bounded by $y = x^2$ and $y = 1$.

15. The region under the curve $y = x^3$, $0 \le x \le 1$.

16. The region bounded by $y = 4 - x^2$ and the x axis.

17. The region bounded by $y = x$ and $y = x^2$.

18. The region bounded by $y = x^2 - 2x$ and the x axis.

19. The region under the curve $y = \cos x$, $-\pi/2 \le x \le \pi/2$.

20. The region under the curve $y = \sin x$, $\pi/6 \le x \le 5\pi/6$.

21. The region bounded by $y = 1 - |x|$ and the x axis.

22. The region bounded by $y = 1 - |x - 1|$ and the x axis.

Each of the following regions is covered by a thin material whose density at (x, y) is $\delta(x, y)$.

23. The region in Problem 13, with $\delta(x, y) = x$. Find $\bar{y}$.

24. The region in Problem 14, with $\delta(x, y) = y^2$. Find the center of mass.

25. The region in Problem 15, with $\delta(x, y) = x$. Find the center of mass.

26. The region in Problem 16, with $\delta(x, y) = y$. Find the center of mass.

27. The region in Problem 17, with $\delta(x, y) = x$. Find $\bar{x}$.

28. The region in Problem 18, with $\delta(x, y) = |y|$. Find $\bar{y}$.

29. Show that the centroid of a rectangle is at its geometrical center.

30. Show that the centroid of a triangle is the point of intersection of its medians, as follows.

Figure 8
Finding $\bar{y}$ by means of horizontal strips

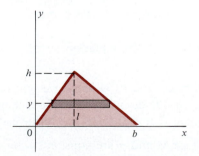

(a) Place the triangle in the coordinate plane as shown in Figure 8, and use horizontal strips to find $\bar{y} = h/3$. *Hint:* By similar triangles,

$$\frac{h - y}{l} = \frac{h}{b}$$

(b) Noting that in part (a) any side may be regarded as the base, use the theorem from plane geometry which says that the medians intersect in a point whose distance from any side is one-third of the altitude drawn to that side.

31. Suppose that a plane region is rotated about a line in its plane not intersecting the region. A theorem of the Greek geometer Pappus says that the volume generated is the area of the region times the distance traveled by its centroid. Prove this theorem as follows.

(a) The line about which the region is rotated may be taken to be the y axis. Referring to Figure 9, explain why the volume generated is

$$V = 2\pi \int_a^b x\, l(x)\, dx$$

Figure 9
Region to be rotated about the y axis

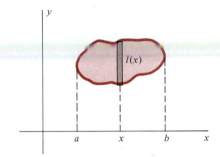

(b) Show that

$$A\bar{x} = \int_a^b x\, l(x)\, dx$$

where A is the area of the region and $(\bar{x}, \bar{y})$ is its centroid.

(c) Combine parts (a) and (b) to finish the proof.

32. Use the Theorem of Pappus to find the volume of the torus generated by rotating the region inside the circle $x^2 + y^2 = a^2$ about the line $x = b\,(a < b)$. (See Problem 37, Section 7.2.)

33. Use the Theorem of Pappus together with the known volume of a sphere to find the centroid of a semicircular region of radius a. (The result should confirm Example 3.)

34. Use the Theorem of Pappus together with the known volume of a cone to find the centroid of a right triangular region of base r and altitude h.

35. Suppose that the x and y axes are replaced by the lines $y = k$ and $x = h$, respectively, and suppose that moments of a plane region R are calculated relative to these new axes to find the centroid of R.

    (a) Show that the new moments are $M_h = M_y - mh$ and $M_k = M_x - mk$, where m is the mass of the region.

    (b) Explain why the centroid calculated in terms of these moments is $(\bar{x} - h, \bar{y} - k)$, where $(\bar{x}, \bar{y})$ is the centroid calculated as usual.

    (c) Why does it follow that the centroid of R is unaffected by a translation of axes?

36. Find the centroid of the semicircular arc of radius a defined by

$$x = a \cos t, \, y = a \sin t \qquad 0 \leq t \leq \pi$$

37. One arch of a cycloid (Problem 37, Section 7.3) is defined by

$$x = a(\theta - \sin \theta), \, y = a(1 - \cos \theta) \qquad 0 \leq \theta \leq 2\pi$$

Find its centroid. *Hint:* Use symmetry to avoid a complicated integration! The identity $\sin^2 \frac{1}{2}\theta = \frac{1}{2}(1 - \cos \theta)$ will also be helpful.

38. Find the first coordinate of the centroid of the curve $y = \frac{2}{3}x^{3/2}$, $0 \leq x \leq 1$.

39. A wire in the shape of the quarter circle

$$x = a \cos t, \, y = a \sin t \qquad 0 \leq t \leq \pi/2$$

has density $\delta(x, y) = kx$ at each point (x, y). Find its center of mass.

7.7
INTRODUCTION TO DIFFERENTIAL EQUATIONS

Galileo once conjectured that the velocity of a body falling from rest (in a vacuum) is proportional to the distance fallen. In other words,

$$\frac{ds}{dt} = ks$$

where s is the distance fallen in t seconds and k is a constant. This is an example of a *differential equation,* that is, an equation involving an unknown function and one or more of its derivatives.

To see what Galileo's conjecture implies, we try to *solve* the differential equation, that is, we look for a function $s = \phi(t)$ that reduces the equation to an identity upon substitution in the equation:

$$\frac{d}{dt}\phi(t) = k\phi(t) \qquad \text{for all } t \text{ in the domain}$$

Such a function is called a *solution* of the equation.

□ **Example 1**   The equation $ds/dt = ks$ is hard; we will be unable to solve it until exponential functions are discussed in the next chapter. Anticipating the result, we can tell you that $\phi(t)$ turns out to have the form

$$s = Ca^{kt}, \text{ where } C \text{ and } a \text{ are constants}$$

Assuming that our clock is turned on when the body starts to fall, we know that $s = 0$ when $t = 0$. (This information is called an *initial condition.*) Substituting into the equation $s = Ca^{kt}$, we find $0 = Ca^0$, from which $C = 0$. But this means that the solution $s = Ca^{kt}$ reduces to $s = 0$ for all t, which is absurd. (The falling body would not fall!) Hence Galileo's conjecture must be abandoned.   □

Galileo later decided that the velocity of a falling body is proportional to the *time*. In other words,

$$\frac{ds}{dt} = kt$$

where k is a constant (not the same k as before). Observe that this differential equation is a simple statement about the derivative of a function of t in *terms of* t (unlike $ds/dt = ks$, which involves the unknown function on the right side as well as its derivative on the left). To solve the equation, we need only find an antiderivative of kt:

$$s = \tfrac{1}{2}kt^2 + C$$

where C is a constant to be evaluated by making use of whatever additional information is available about $s = \phi(t)$. Since $s = 0$ when $t = 0$, we find $0 = \tfrac{1}{2}k \cdot 0 + C$, from which $C = 0$. Hence the law of motion implied by Galileo's second conjecture is $s = \tfrac{1}{2}kt^2$.

Most calculus students will recognize the form of this equation, which is usually written $s = \tfrac{1}{2}gt^2$. (The constant g is the acceleration due to gravity, approximately 32 ft/sec² in the foot-pound-second system of units.) Galileo's second conjecture, in other words, leads to a familiar formula. That doesn't make it correct, but at least it fits the facts in some rough fashion. A falling body that encounters little or no air resistance does travel a distance approximately equal to $s = \tfrac{1}{2}gt^2$ in t seconds. Hence Galileo had the right idea (long before Newton's calculus showed how to formulate and solve a differential equation).

The differential equation $ds/dt = kt$ (with $s = 0$ when $t = 0$) is an illustration of a *first-order initial value problem*. The term **first-order** refers to the fact that only the first derivative of the unknown function appears in the equation, while **initial value** is a phrase borrowed from physics (in which it is common for the value of a function, and sometimes its derivatives, to be specified at an initial time $t = 0$).

Remark   More generally, the **order** of a differential equation is the order of the highest derivative that occurs. Second-order equations are particularly important in mechanics, because Newton's Second Law involves a second derivative of position with respect to time. Hooke's Law, for example, says that the displacement x of a spring is related to the restoring force by the equation $F = -kx$ (where k is a positive constant). (See Section 7.5.) Upon substitution of mass times acceleration for the force, this becomes a second-order differential equation satisfied by x as a function of time:

$$m\frac{d^2x}{dt^2} + kx = 0$$

You can check that the function

$$x = \sin at \qquad (a = \sqrt{k/m})$$

satisfies the equation (as well as the initial conditions $x = 0$ and $dx/dt = a$ at $t = 0$), but it is not a simple matter to explain how such a solution is found. The subject of differential equations is vast and complicated, and has been the focus of extensive research for several hundred years (up to and including the present day). We barely scratch the surface in this section, but even so it is remarkable how soon the calculus student who is interested in applications runs up against differential equations. As you encounter them scattered through the rest of the book, you should try to impose some sort of order on the types and methods of solution that we present. We will summarize and extend the basic ideas in the íast chapter of the book.

□ **Example 2**   The slope of a curve $y = \phi(x)$ at each point (x, y) is known to be $\sin x$, and the curve passes through $(0,2)$. What is its equation?

Solution: We are told that $dy/dx = \sin x$ for all x (with $y = 2$ at $x = 0$). It follows that $dy = \sin x \, dx$ (a form in which it is easy to see why the term *differential equation* is used). Hence

$$\int dy = \int \sin x \, dx$$

Carrying out the indicated integrations, we find

$$y = -\cos x + C \qquad \text{(where } C \text{ is a constant)}$$

This result represents the *general solution* of the differential equation, because (according to Theorem 2 of Section 4.4) it includes all functions whose derivative is $\sin x$. In this case we are also told that $y = 2$ when $x = 0$. Hence $2 = -\cos 0 + C$, from which $C = 3$. Thus the equation of the curve is $y = 3 - \cos x$.   □

□ **Example 3**   Suppose that a drop of water (in the form of a sphere) evaporates at a rate numerically equal to its surface area. If its initial radius is 0.2 cm, how long will it take for the drop to completely evaporate?

Solution: Let V and A be the volume and surface area of the drop at time t. The differential equation describing the situation is $dV/dt = -A$. (The rate of evaporation is A, but since the volume is decreasing, its rate of change is $-A$.) The trouble with this equation is that it involves two unknown functions of the independent variable t, namely V and A. Both, however, can be expressed in terms of the radius r, by means of the familiar geometrical formulas

$$V = \tfrac{4}{3}\pi r^3 \qquad \text{and} \qquad A = 4\pi r^2$$

Hence our differential equation becomes

$$\frac{d}{dt}\left(\frac{4}{3}\pi r^3\right) = -4\pi r^2$$

or (by the Chain Rule)

$$4\pi r^2 \frac{dr}{dt} = -4\pi r^2$$

This reduces to $dr/dt = -1$ (or $dr = -dt$), a simple differential equation which yields $r = -t + C$. Since $r = 0.2$ when $t = 0$, we find $C = 0.2$ and hence $r = 0.2 - t$. The evaporation will be complete when $r = 0$, that is, when $t = 0.2$.   □

□ **Example 4**   Suppose that a certain population increases at a rate numerically equal to the square root of the population. If the initial population is 10,000 and time is measured in years, how long will it take for the population to reach 16,000?

Solution: Let P be the population at time t. We are told that

$$\frac{dP}{dt} = \sqrt{P} \qquad \text{(with } P = 10{,}000 \text{ at } t = 0)$$

Like Galileo's first conjecture about falling bodies ($ds/dt = ks$), this differential equation involves the unknown function on both sides. Simple antidifferentiation (as in Examples 2 and 3) is inadequate. Suppose, however, that we *separate the variables* by multiplying by dt and dividing by $\sqrt{P}$. Then the equation takes the form

$$\frac{dP}{\sqrt{P}} = dt$$

in which the left side involves only P and the right side only t. Integration now makes sense on each side:

$$\int \frac{dP}{\sqrt{P}} = \int dt$$

$$\int P^{-1/2}\, dP = \int dt$$

$$2P^{1/2} = t + C$$

Since $P = 10{,}000$ when $t = 0$, we find $C = 200$ and hence

$$2\sqrt{P} = t + 200$$

$$\sqrt{P} = \frac{1}{2}(t + 200)$$

$$P = \frac{1}{4}(t + 200)^2$$

This is a formula for P at any time t. To find when the population has reached 16,000, we need not have gone quite this far with our analysis; it is easier to return to the equation $2\sqrt{P} = t + 200$ and solve for t when $P = 16{,}000$:

$$2\sqrt{16{,}000} = t + 200$$
$$t = 80\sqrt{10} - 200 \approx 53 \text{ years}$$ □

Remark   In view of our success in Example 4, you may wonder what is hard about Galileo's first conjecture, $ds/dt = ks$. Why not separate the variables and integrate?

$$\frac{ds}{s} = k \, dt$$

$$\int \frac{ds}{s} = \int k \, dt = kt + C$$

The trouble is, we are stuck on the left side. For we have not encountered any function to put down as an antiderivative of $1/s$. The Power Rule says that

$$\int x^n \, dx = \frac{x^{n+1}}{n+1}$$

but that is no help when $n = -1$. The problem

$$\int x^{-1} \, dx = \int \frac{dx}{x}$$

is as yet beyond us; solving it will be the first order of business in the next chapter.

□ **Example 5**   A ball is thrown straight upward (with a speed of 16 ft/sec) from the top of a building 96 ft high. Find its height above the ground t seconds later.

Solution: Choose a coordinate axis with origin at ground level and positive direction upward. The problem is to find s (the coordinate of the ball) as a function of the time t. We are not given any differential equation involving s, but Newton's Second Law supplies one. For it says that the force on the ball is the product of mass and acceleration. Neglecting all forces except gravity (air resistance, for example), we assume that the only force on the ball is its weight,

$$F = -mg \qquad \text{(where } g \approx 32 \text{ in the foot-pound-second system)}$$

The negative sign is used because we have directed the s axis upward, whereas gravitational force acts downward. Since $F = ma$ (where a is the acceleration of the ball), we have

$$ma = -32m$$

and hence $a = -32$. The acceleration, however, is the second derivative of position, so Newton's Law is equivalent to the second-order differential equation

$$\frac{d^2s}{dt^2} = -32 \qquad \text{(with initial conditions } s = 96 \text{ and } \frac{ds}{dt} = 16 \text{ at } t = 0)$$

By recalling that velocity is $v = ds/dt$, we may convert our equation to the first-order initial value problem

$$\frac{dv}{dt} = -32 \qquad \text{(with } v = 16 \text{ at } t = 0)$$

It follows that $v = -32t + C_1$, where $C_1 = 16$ because $v = 16$ when $t = 0$. Replacing v by ds/dt, we now have another first-order problem, namely

$$\frac{ds}{dt} = 16 - 32t \qquad \text{(with } s = 96 \text{ at } t = 0)$$

This yields $s = 16t - 16t^2 + C_2$, where $C_2 = 96$ because $s = 96$ when $t = 0$. Thus the height of the ball at time t is

$$s = 96 + 16t - 16t^2 \qquad \qquad \square$$

□ **Example 6**   The equation $y = kx^2$ (where k is an arbitrary nonzero constant) describes a family of parabolas. Find their **orthogonal trajectories,** that is, the curves that intersect the parabolas at right angles. (See Figure 1.)

Figure 1
The family $y = kx^2$ and its orthogonal trajectories

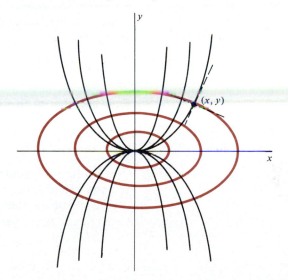

Solution: Let (x, y) be any point of the plane (not on a coordinate axis). One of the parabolas $y = kx^2$ passes through this point, with slope $m_1 = 2kx$. So does a member of the family of orthogonal trajectories, with slope m_2 satisfying the perpendicularity condition $m_1 m_2 = -1$. We do not know the equation of this curve, but whatever it is, differentiation must produce

$$\frac{dy}{dx} = -\frac{1}{2kx} \qquad \qquad (1)$$

It is easy to make a mistake at this point, regarding Equation (1) as a differential equation in which k is constant and y is to be found as a function of x. However, k depends on the original choice of the point (x, y). It is really a parameter whose value specifies which member of the family of parabolas

$y = kx^2$ we are talking about. Since that complicates life, we eliminate k by substituting $k = y/x^2$ in Equation (1):

$$\frac{dy}{dx} = -\frac{x}{2y}$$

This is a differential equation that describes the family of orthogonal trajectories. Separating the variables, we have

$$2y\,dy = -x\,dx$$

$$\int 2y\,dy = -\int x\,dx$$

$$y^2 = -\frac{x^2}{2} + C$$

$$2y^2 = -x^2 + 2C$$

$$x^2 + 2y^2 = 2C$$

Replacing $2C$ (an arbitrary positive constant) by the symbol a^2, we conclude that the family of orthogonal trajectories is defined by the equation $x^2 + 2y^2 = a^2$,

or $$\frac{x^2}{a^2} + \frac{y^2}{b^2} = 1 \qquad \text{(where } b^2 = a^2/2\text{)}$$

Each member of this family is an ellipse (as you may have guessed from Figure 1).   □

Given a differential equation involving an independent variable x, an unknown function of x, and one or more derivatives of this function, we find it convenient for theoretical puposes to express the highest derivative in terms of the other variables that may occur. In the equation

$$\frac{1}{x}\frac{dy}{dx} - y\sin x = x^2$$

for example, we may solve for dy/dx to obtain

$$\frac{dy}{dx} = x^3 + xy\sin x = \text{expression in } x \text{ and } y$$

Similarly, the second-order equation $y'' + 3y' + y = x$ may be written in the form

$$y'' = x - y - 3y' = \text{expression in } x, y, \text{ and } y'$$

Thus the general first-order equation has the form

$$y' = f(x,y) = \text{expression in } x \text{ and } y$$

where f is a function whose domain is a region in $\Re^2$. If f is reasonably well-behaved in this region, and (x_0, y_0) is an interior point of the region, we can guarantee the existence of a solution $y = \phi(x)$ satisfying the initial condition $\phi(x_0) = y_0$ and valid in an interval I containing x_0. Moreover, no other such solution exists, that is, ϕ is unique. This important statement, called the *Existence and Uniqueness Theorem,* will be especially useful in Chapter 13. For a complete statement of it, see Section 20.1.

□ **Example 7**   Show that the function $\phi(x) = -\cos x$ is a solution of the first-order equation $y' = \sqrt{1 - y^2}$ satisfying the initial condition $\phi(0) = -1$. Name an interval (containing 0) in which the solution is valid.

Solution: Since $\phi(0) = -\cos 0 = -1$, the required initial condition holds. To see whether ϕ is a solution of the differential equation, we substitute $y = -\cos x$ and $y' = \sin x$ in $y' = \sqrt{1 - y^2}$:

$$\sin x \overset{?}{=} \sqrt{1 - \cos^2 x} = \sqrt{\sin^2 x} = |\sin x|$$

This is an identity in any interval where $\sin x \geq 0$; we choose $I = [0, \pi]$ because $0 \in I$.

It is worth noting that the constant function $\psi(x) = -1$ is also a solution of $y' = \sqrt{1 - y^2}$ satisfying $\psi(0) = -1$. (Confirm!) This appears to contradict the Uniqueness Theorem. Observe, however, that the domain of $f(x, y) = \sqrt{1 - y^2}$ is the horizontal strip

$$D = \{(x, y): -1 \leq y \leq 1\}$$

and the initial point $(x_0, y_0) = (0, -1)$ is on the boundary of D. The Existence and Uniqueness Theorem requires (x_0, y_0) to be an *interior* point of the region.   □

The *general solution* of a differential equation is a relation between x and y (involving one or more arbitrary constants) that represents the class of all solutions of the equation. When initial conditions are imposed, the constants appearing in this relation are evaluated to obtain the unique solution satisfying the given conditions. See Example 5, where the general solution of the second-order equation $d^2s/dt^2 = -32$ is $s = -16t^2 + C_1 t + C_2$. The initial conditions $s = 96$ and $ds/dt = 16$ at $t = 0$ enable us to find $C_1 = 16$ and $C_2 = 96$, the unique solution satisfying these conditions being $s = -16t^2 + 16t + 96$.

The first-order equation $y' = f(x, y)$ is said to be **separable** if f has the form $g(x)h(y)$, where g is a function of x alone and h is a function of y alone:

$$\frac{dy}{dx} = g(x)h(y)$$

In an interval where $h(y) \neq 0$, we may convert this to the separated form

$$\frac{dy}{h(y)} = g(x)\,dx$$

and integrate both sides to arrive at a solution:

$$\int \frac{dy}{h(y)} = \int g(x)\,dx$$

Separable differential equations are the only ones we expect you to be able to handle now.

Problem Set 7.7

In each of the following, confirm that the function $y = \phi(x)$ is a solution of the given differential equation.

1. $xy' = \sqrt{1 - y^2}$; $\phi(x) = 1$

2. $xy' = y$; $\phi(x) = kx$

3. $y' = 1 + y^2$; $\phi(x) = \tan x$

4. $y' + \dfrac{y}{x} = 3x$; $\phi(x) = x^2 - \dfrac{3}{x}$

5. $xy' + 2y = 0$; $\phi(x) = \dfrac{4}{x^2}$

6. $y' = (y - 1)^{3/2}$; $\phi(x) = 1 + \dfrac{4}{(x - 3)^2}$

7. $y' = \dfrac{y^2 - 2xy}{x^2}$; $\phi(x) = \dfrac{6x}{x^3 + 2}$

8. $y'' + y = x^2$; $\phi(x) = x^2 - 2$

9. $y'' + y = 0$; $\phi(x) = A \cos x + B \sin x$

10. $y'' - 2y = 5 \cos x$; $\phi(x) = 1 - \cos x - 2 \sin x$

11. $(x^2 + 1)y'' - 2xy' + 2y = 0$; $\phi(x) = x^2 - 1$

12. $xy'' - (x + 1)y' + y = x^2$; $\phi(x) = -(x^2 + 2x + 2)$

Use the method of separation of variables to solve each of the following differential equations (subject to the given initial condition).

13. $\dfrac{dy}{dx} = y^{2/3}$; $y = 1$ when $x = 0$.

14. $\dfrac{dy}{dx} = 2xy^2$; $y = 1$ when $x = 0$.

15. $\dfrac{dy}{dx} = \sqrt{y - 1}$; $y = 2$ when $x = 1$.

16. $\dfrac{dy}{dx} = \sqrt{y} \cos x$; $y = 0$ when $x = 0$.

17. $\dfrac{dy}{dx} = y^2 \sin x$; $y = 1$ when $x = 0$.

18. $\dfrac{dy}{dx} = 2x \sqrt{1 - y}$; $y = 0$ when $x = 1$.

19. Use implicit differentiation to confirm that the curves $y^3 - 3y = 3x^2 + k$ are solutions of the differential equation

$$\frac{dy}{dx} = \frac{2x}{y^2 - 1}$$

20. Suppose that you were not told the solution curves of

$$\frac{dy}{dx} = \frac{2x}{y^2 - 1}$$

(as in the preceding problem). Find them by the method of separation of variables. Which one passes through the point $(0,2)$?

21. Show that the general solution of $dy/dx = x/y$ is a family of hyperbolas (together with their asymptotes as a special case).

22. Find the general solution of $dy/dx = 2\sqrt{y}$. What special solution is not included in your answer? *Hint:* Separation of variables is legitimate only if $y > 0$ in some interval.

23. Solve the initial value problem $dy/dx = y^2$, where $y = 1$ at $x = 0$.

24. Solve the initial value problem $dy/dx = xy^3$, where $y = 1$ at $x = 0$.

25. The slope at each point (x, y) of a certain curve is $\sec^2 x$. If the curve passes through the point $(\pi/4, 0)$, what is its equation?

26. Confirm that the function $\phi(x) = \frac{1}{4}(x - a)^2 + 1$ is a solution of the differential equation $y' = \sqrt{y - 1}$ in the interval $I = [a, \infty)$. (Note that neither the differential equation nor its solution contains any clue that I must be restricted like this; one might expect the interval to be the whole coordinate line. This illustrates the fact that in general the only way to discover the domain of a solution is to test it in the equation.)

27. Suppose that you were not told the solutions of $y' = \sqrt{y-1}$ (as in the preceding problem).

(a) Use the method of separation of variables to derive the general solution

$$\phi(x) = \tfrac{1}{4}(x-a)^2 + 1$$

(where a is an arbitrary constant).

(b) The general solution in part (a) supplies particular solutions satisfying initial conditions of the form $\phi(x_0) = y_0$, where x_0 is arbitrary and $y_0 \geq 1$. Find the one satisfying $\phi(1) = 2$. *Hint:* Of the two possible values of a that occur, only one can be chosen. Use the fact (from Problem 26) that the correct domain is $I = [a, \infty)$ and that I must contain the point $x_0 = 1$.

(c) Find the solution satisfying $\phi(0) = 1$. Then confirm that the constant function $\phi(x) = 1$ is also a solution satisfying this initial condition. [The existence of two such solutions would appear to contradict the Uniqueness Theorem mentioned in the text. Note, however, that the domain of $f(x, y) = \sqrt{y-1}$ (the right side of the differential equation) is $\{(x, y): y \geq 1\}$ and the point $(0, 1)$ is on the boundary. The uniqueness theorem applies only to interior points of the domain of f.]

(d) Why is it unreasonable to ask for a solution satisfying the initial condition $\phi(0) = 0$?

(e) Each of the solutions $y = \tfrac{1}{4}(x-a)^2 + 1$, $x \geq a$, is a half-parabola with vertex $(a, 1)$. If these are all graphed (together with the special solution $y = 1$), what part of the plane do they fill?

28. It can be proved that every solution of $y'' + y = 0$ has the form

$$y = A \cos x + B \sin x$$

(a) Find the solution satisfying the initial conditions $y = 2$ and $y' = -1$ at $x = 0$.

(b) There are infinitely many solutions satisfying the "boundary conditions" $y = 2$ at $x = 0$ and $y' = -2$ at $x = \pi/2$. Find them all.

[Note the distinction between *initial conditions* (values of y and y' are specified at the same point) and *boundary conditions* (specifications are given at the endpoints of an interval). This turns out to be important in the theory of differential equations.]

29. The differential equation $dy/dx + y/x = 2$ cannot be solved by separating the variables. (Why not?) Solve it instead by multiplying each side by the "integrating factor" x, as follows.

(a) Confirm that the effect of this multiplication is to convert the equation to the form

$$\frac{d}{dx}(xy) = 2x$$

(b) Integrate the equation in part (a) to obtain the general solution $y = x + (C/x)$.

(c) What solution satisfies the initial condition $y = 0$ at $x = 1$?

30. Use the integrating factor x^{-3} to find the general solution of

$$y' - \frac{3y}{x} = x^3$$

Hint: Upon multiplication by x^{-3}, the left side becomes

$$\frac{d}{dx}(x^{-3}y)$$

31. A falling body takes 5 seconds to hit the ground. Neglecting air resistance, find the altitude from which it was dropped.

32. A ball is thrown directly upward from the ground with a speed of 24 ft/sec. How high does it rise?

33. A girl in a helicopter (rising vertically at 24 ft/sec) drops her camera when she is 160 ft above the ground. How long will it take the camera to fall? With what speed will it hit the ground?

34. The speed of a ball rolling along the ground decreases at the rate of 4 ft/sec² due to friction. If its initial speed is 12 ft/sec, how far does it roll?

35. The brakes of a car traveling 60 mph decelerate the car at the rate of 22 ft/sec². If the car is 175 ft from a barrier when the brakes are applied, will it hit the barrier?

36. A ball rolls down a roof with a constant acceleration of 3 ft/sec². If it started from rest 15 ft from the edge of the roof, when will it fall off?

37. Suppose that the high-jump record for fully equipped astronauts is 4 ft on earth. What would it be on the moon? (The acceleration due to gravity on the moon is about 5 ft/sec², while on earth it is about 32 ft/sec².)

38. Newton's First Law says that in the absence of any force a body at rest will remain at rest and a body in motion suffers no change in the speed or direction of its motion. Why does this follow from his Second Law in the case of straight-line motion?

39. Hooke's Law for the displacement of a spring (together with Newton's Second Law) yields the differential equation

$$m\frac{d^2x}{dt^2} + kx = 0$$

where x is the displacement at time t.

(a) Confirm that the function

$$x = \sin at \qquad (a = \sqrt{k/m})$$

satisfies the initial value problem consisting of the differential equation and the conditions $x = 0$ and $dx/dt = a$ at $t = 0$. [These conditions correspond to our starting the spring in motion from its natural (unstretched) position with an initial velocity a. The spring then oscillates according to the law of motion $x = \sin at$.]

(b) The general solution of Hooke's equation is

$$x = A\cos at + B\sin at$$

What is the law of motion of a spring released from a stretched position of $x = 1$?

(c) What is the law of motion of a spring with an initial velocity $dx/dt = a$ from the compressed position $x = -1$?

40. A snowball melts at a rate proportional to its surface area. If it takes an hour for its radius to decrease from 5 cm to 2 cm, how long will it take (from the time its radius was 5 cm) to melt altogether?

41. A conical tank full of water is 20 ft deep and 12 ft across the top. After ten days of evaporation (at a rate proportional to the exposed surface area of the water) the water level has fallen to 4 ft. Find the depth at the end of two more days.

42. A certain population increases at a rate proportional to the square root of the population. If the population goes from 2500 to 3600 in five years, what is it at the end of t years?

43. A certain curve has the property that the normal line at (x,y) always passes through the point $(2,0)$. If the curve contains the point $(2,3)$, find its equation by **(a)** using elementary geometry, and by **(b)** solving an appropriate differential equation.

44. A certain curve has the property that each of its tangents intersects the coordinate axes in two points P and Q such that the midpoint of PQ is the point of tangency.

(a) What differential equation describes this curve?

(b) What difficulty stands in the way of our solving the differential equation?

(c) Despite part (b), can you guess what curve we are talking about?

45. It is geometrically apparent that the orthogonal trajectories of the family of straight lines through the origin are the circles $x^2 + y^2 = a^2$. Confirm this by formulating and solving an appropriate differential equation.

46. Find the orthogonal trajectories of the cubic curves $y = kx^3$.

True–False Quiz

1. The area of the region bounded by the curves $y = x^2$ and $y = 2 - x^2$ is

$$A = 4\int_0^1 (1 - x^2)\, dx$$

2. The area of the region bounded by the curve $y = \cos x$, the x axis, and the lines $x = 0$ and $x = \pi$ is

$$A = \int_0^\pi \cos x\, dx$$

3. The area of the region bounded by the curve $y = \sqrt{x}$, the x axis, and the line $x = 4$ is

$$A = \int_0^2 (4 - y^2)\, dy$$

4. If the region bounded by the line $y = hx/r$, the y axis, and the line $y = h$ is rotated about the y axis, the volume generated is

$$V = \frac{r^2}{h^2}\int_0^h \pi y^2\, dy$$

5. If the region in the first quadrant bounded by the curve $y = x^2$, the y axis, and the line $y = 1$ is rotated about the x axis, the volume generated is

$$V = \pi\int_0^1 (1 - x^2)^2\, dx$$

6. The volume in the preceding problem may be computed in the form

$$V = 2\pi\int_0^1 y^{3/2}\, dy$$

7. The curves

$$C_1: x = t, \ y = t^2 \qquad -1 \leq t \leq 1$$

and $\quad C_2: x = \cos t, \ y = \cos^2 t \qquad 0 \leq t \leq 2\pi$

have the same track.

8. The length of the curve $x = t^2, \ y = 1 - t, \ 0 \leq t \leq 1$, is

$$s = \int_0^1 \sqrt{4t^2 + 1} \ dt$$

9. The length of the graph of $xy = 2$ between $(1,2)$ and $(2,1)$ is

$$s = \int_1^2 \frac{\sqrt{x^4 + 4}}{x^2} \ dx$$

10. If the graph of $y = \cos x, \ 0 \leq x \leq \pi/2$, is rotated about the y axis, the area of the generated surface is

$$\int_0^{\pi/2} 2\pi x \sqrt{1 + \sin^2 x} \ dx$$

11. If the force required to hold a spring stretched x units from its natural length is $F(x) = 2x$, the work done in stretching the spring from $x = 2$ to $x = 5$ is 21 units.

12. A tank in the form of an inverted cone is full of liquid weighing 9 lb/ft^3. If the tank is 2 ft across the top and 3 ft deep, the work done in pumping the liquid over the rim is

$$W = \int_0^3 \pi h^2 (3 - h) \ dh$$

13. If the density of a straight wire is proportional to the distance from its midpoint, its center of mass is at the midpoint.

14. If the first-quadrant region inside the unit circle is covered by a piece of tinfoil of constant density 1, its moment relative to the x axis is

$$M_x = \int_0^1 (1 - x^2) \ dx$$

15. The center of mass of a wire in the shape of a smooth curve lies on the wire.

16. If the brakes of a truck traveling 30 mph decelerate the truck at the rate of 11 ft/sec^2, and are applied when the truck is 90 ft from a barrier, the truck will hit the barrier.

17. The differential equation $y' = 2xy^2$ can be solved by separating the variables.

Additional Problems

In each of the following, find the area of the region bounded by the given curves.

1. $y^2 = x$ and $y = x^3$
2. $y = x^2$ and $y = 2x - x^2$
3. $y = 2 - x^2, \ y = x^3$, and the y axis
4. $x + y = 2, \ x = y^3$, and the x axis. *Hint:* The first two curves intersect at $(1,1)$.
5. $y = \sqrt{x}, \ x + y = 2$, and the y axis
6. $y^2 = x^3$ and $x = 1$
7. $y^2 = x + 5$ and $x - y = 1$
8. $y = 4 - x^2$ and $y = 3/x^2$
9. $y = 1 - |x - 1|$ and the x axis
10. $y = |x^2 - 4|$ and $y = 5$
11. $y = \sec^2 x, \ y = \cos x, \ x = 0$, and $x = \pi/4$
12. $y = \sin x, \ y = x^2 + 1$, and $x = \pm 3$

In each of the following, the region bounded by the given curves is rotated about the given line. Find the volume of the resulting solid of revolution.

13. $y = \sqrt{1 - x^2}$ and the x axis (about the x axis)
14. $y = x^3, \ x = 1$, and $y = 0$ (about the x axis)
15. $y = x^{2/3}$ and $y = 1$ (about the y axis)
16. $y^2 = x^2(5 - x)$ (about the x axis)
17. $y = 1/\sqrt{x}, \ x = 1, \ x = 4$, and the x axis (about the y axis)
18. $y = 1/\sqrt{1 + x^2}, \ x = 1$, and the coordinate axes (about the y axis)
19. $|x| + |y| = 1$ (about the x axis)
20. $\sqrt{x} + \sqrt{y} = 1$ and the coordinate axes (about the y axis)
21. $y = \sqrt{2x - x^2}$ and the x axis (about the x axis)
22. $y = 4 - x^2$ and the x axis (about the y axis)
23. $y = \cos x, \ x = \pi/2$, and the coordinate axes (about the x axis)
24. $y = \cot x, \ x = \pi/4$, and the x axis (about the x axis)
25. A solid with a circular base of radius 3 has square cross sections perpendicular to a diameter of the base. Find its volume.
26. The base of a certain solid is circular, with radius 4. Cross sections of the solid perpendicular to a diameter of the

base are isosceles right triangles whose legs lie in the base and are perpendicular to the base, respectively. Find the volume of the solid.

Find the length of each of the following curves.

27. $x = t \cos t - \sin t$, $y = t \sin t + \cos t$, $0 \leq t \leq \pi$

28. $x = t^2 - t$, $y = t - t^2$, $0 \leq t \leq 1$

29. $x = 2t^{3/2} + 1$, $y = 2t - 1$, $0 \leq t \leq 1$

30. $x = 3 \sin t - 1$, $y = 2 - 3 \cos t$, $0 \leq t \leq 2\pi$

31. $x = 2t - 1$, $y = |t|$, $-1 \leq t \leq 1$

32. The position of a particle at time t is given by

$$x = t^2, \quad y = \tfrac{1}{3}(t^3 - 3t)$$

(a) Explain why the path of the particle intersects the x axis three times.

(b) Show that $dy/dx = (t^2 - 1)/2t$ and use the result to find where the path has horizontal and vertical tangents. In what order are these points encountered?

(c) Sketch the path and indicate the direction of travel.

(d) Find the length of the loop.

Find the length of each of the following graphs.

33. $y = \tfrac{3}{2}x^{2/3}$, $1 \leq x \leq 8$

34. $y = \tfrac{1}{3}(x^2 - 2)^{3/2}$, $2 \leq x \leq 5$

35. $y = \tfrac{2}{3}(x^2 + 1)^{3/2}$, $0 \leq x \leq 1$

36. $y = (4 - x^{2/3})^{3/2}$, $1 \leq x \leq 8$

Find the area of the surface generated by rotating the given curve about the indicated axis.

37. $x = 1 - \sin t$, $y = \cos t$, $-\pi/2 \leq t \leq \pi/2$ (about the x axis)

38. $x = t^3$, $y = t^2$, $0 \leq t \leq 1$ (about the x axis)

39. $x = t - 1$, $y = \tfrac{1}{2}t^2 - t$, $1 \leq t \leq 3$ (about the y axis)

40. $y = \sin x$, $0 \leq x \leq \pi$ (about the x axis)

41. $y = \tfrac{2}{3}(x^2 + 1)^{3/2}$, $0 \leq x \leq 1$ (about the y axis)

42. $y = x^{3/2}$, $0 \leq x \leq 4$ (about the y axis)

43. A certain spring exerts a restoring force of 0.8 newtons when it is compressed 0.2 meters. How much work is done in stretching the spring 0.3 meters?

44. In Problem 43 find how much work is done in stretching the spring from a length of 0.1 meters to a length of 0.4 meters.

45. A bucket of coal weighing 75 lb is hauled up a mine shaft 100 ft deep. If the cable hauling it up weighs 2 lb/ft, how much work is done?

46. Find the work done against gravity when a satellite weighing $\frac{1}{2}$ ton on the surface of the earth is lifted to an altitude of 80 miles. (Assume that the radius of the earth is 4000 miles.)

47. A conical tank is 20 ft across the top and 5 ft deep, and it is filled to within 2 ft of the top by oil weighing 50 lb/ft³. Find the work done in pumping the oil over the rim.

48. In Problem 47 find the work done in pumping the oil to an outlet pipe 2 ft below the top of the tank.

49. The density of a straight wire of length 2 is proportional to the distance from its center, having the value 3 at each endpoint. Find the mass and center of mass.

50. A straight wire extending from $x = -2$ to $x = 2$ has density $\delta(x) = |x|$ at point x. Find its center of mass.

Find the centroid of each of the following regions (bounded by the given curves).

51. $y = 1 - x^2$ and the x axis

52. $y = 2x - x^2$ and the x axis

53. $y = 1 - x^{2/3}$ and the x axis

54. $y = \cos x$, $y = 0$, and $x = \pm\pi/3$

55. The region in Problem 51 is covered by a thin material whose density at (x, y) is $\delta(x, y) = 2y$. Find the center of mass.

56. Find the centroid of the circular arc

$$x = \cos t, \quad y = \sin t, \quad 0 \leq t \leq \pi/2$$

57. Solve the initial value problem $dy/dx = 2x\sqrt{y}$, where $y = 1$ when $x = 0$.

58. A certain population grows at a rate proportional to the square of the population. If the population increases from 10,000 to 15,000 in three years, what is it at the end of t years? How long can it grow at this rate before exploding beyond all bounds?

59. Find the general solution of the differential equation $y' + 2xy^2 = 0$. What special solution is not included in your answer?

60. Find the orthogonal trajectories of the hyperbolas $xy = k$.

8 | Exponential and Logarithmic Functions

I never used a logarithm in my life, and could not undertake to extract the square root of four without misgivings.
GEORGE BERNARD SHAW (1856–1950)

The moving power of mathematical invention is not reasoning but imagination.
AUGUSTUS DE MORGAN (1806–1871)

The age of chivalry is gone. That of sophisters, economists, and calculators has succeeded.
EDMUND BURKE (1729–1797)

Most of the functions occurring in previous chapters (with the exception of the trigonometric functions) are in a class called **algebraic.** Loosely speaking, this means that their definition involves nothing more complicated than addition, subtraction, multiplication, division, and extraction of roots (the so-called *algebraic operations*). In this chapter and the next, we are going to discuss a number of *nonalgebraic* functions, which (together with the algebraic and trigonometric functions already introduced) are often called the **elementary functions of analysis.** Some books use the term *transcendental* for "nonalgebraic" because such functions transcend the ordinary processes of algebra.

These functions have been singled out from the vast class of real functions (most of which do not even have names) because they are useful. They occur, for example, in certain integration problems, including many types of differential equations. Some of them seem to arise inevitably from the analysis of certain natural phenomena, as if they were built into the scheme of things in the real world. Hence they have been given names, their values have been tabulated (more recently, programmed into calculators and computers), and their analytic properties have been intensively studied.

As you learn about these functions, it may not always be clear why they are important (because we cannot present more than a sampling of their applications). By the time you have finished this chapter and the next, however, you will have acquired an arsenal of real functions that will supply ammunition for many things to come.

8.1
THE NATURAL LOGARITHM

As you know, the power formula

$$\int x^n \, dx = \frac{x^{n+1}}{n+1} + C$$

has a gap in it due to the restriction $n \neq -1$. Moreover, no familiar function is available as a solution of the antidifferentiation problem

$$\int x^{-1} \, dx = \int \frac{dx}{x}$$

In Section 6.4 we observed that the first part of the Fundamental Theorem of Calculus supplies a theoretical answer, namely the function

$$F(x) = \int_1^x \frac{dt}{t} \qquad (x > 0)$$

For we know that

$$F'(x) = \frac{d}{dx} \int_1^x \frac{dt}{t} = \frac{1}{x}$$

that is, $F(x)$ is an antiderivative of $f(x) = 1/x$ in the interval $(0,\infty)$.

If this were the only virtue of the function F, we might not pay much attention. After all, we can do the same thing with any (continuous) function f. Simply let

$$F(x) = \int_a^x f(t) \, dt$$

and we have $F'(x) = f(x)$. The function

$$F(x) = \int_1^x \frac{dt}{t}$$

however, has some remarkable properties. Among other things, we are going to prove that if a and b are positive,

$$F(ab) = F(a) + F(b) \qquad (F \text{ turns multiplication into addition})$$
$$F(a/b) = F(a) - F(b) \qquad (F \text{ turns division into subtraction})$$

These statements strongly suggest the *laws of logarithms*. Recall from previous coursework in mathematics that

$$\log ab = \log a + \log b \qquad \text{and} \qquad \log \frac{a}{b} = \log a - \log b$$

where the symbol $\log x$ $(x > 0)$ means *common logarithm of x* (or *logarithm of x to the base ten*) and is defined to be the exponent of the power of ten that equals x:

$$y = \log x \Longleftrightarrow x = 10^y$$

More generally, if a is any positive constant ($a \neq 1$), then

$$y = \log_a x \Leftrightarrow x = a^y$$

(The symbol $\log_a x$ is read *logarithm of x to the base a.*)

There are difficulties with this definition. When we say, for example, that

$$y = \log_3 17 \Leftrightarrow 17 = 3^y$$

what y do we have in mind? Since $3^2 = 9$ and $3^3 = 27$, it may be clear that y is somewhere between 2 and 3 (if it exists at all!). But there is no rational value of y satisfying $3^y = 17$. Since irrational exponents are rarely defined in precalculus mathematics, the usual procedure is to tell the student that they *can* be defined and then assume that they have been. This includes the (unproved) statement that the usual laws of exponents hold even when the exponents are irrational. One of our objectives in this chapter is to supply a reasonable definition of such exponents and to prove the usual laws.

While the statements $F(ab) = F(a) + F(b)$ and $F(a/b) = F(a) - F(b)$ suggest the laws of logarithms (ordinarily derived from the corresponding laws of exponents), their proof as given here will not involve exponents, nor will any base be mentioned. What we propose to do in this section is to use F as the logical starting point of a discussion that may strike you as backwards. For we intend to derive the properties of logarithms first, and then introduce exponential functions in terms of what we have learned about logarithms. That is the reverse of what is usually done in precalculus mathematics.

We begin with the following definition.

For each $x > 0$ we define the **natural logarithm** *of* x to be the number

$$\ln x = \int_1^x \frac{dt}{t}$$

The term *natural* is used because (as we will see in Section 8.4) there are many different logarithm functions, each with its own base; this one has a base that arises from a "natural" exponential function with which it is (eventually) associated. Note, however, that none of that is mentioned in the definition. At this stage of our discussion the function $F(x) = \ln x$ is a "baseless" logarithm having no apparent connection with exponents.

The above definition yields one important result immediately, namely a new differentiation formula:

$$D_x \ln x = \frac{1}{x}$$

As usual when a new derivative is developed, we build it into the Chain Rule:

$$\frac{d}{dx}(\ln u) = \frac{1}{u}\frac{du}{dx}$$

□ **Example 1**   Find the derivative of

(a) $f(x) = \ln 3x$ (b) $f(x) = \ln(x^2 + 1)$ (c) $f(x) = \ln \sin x$

Solution:

(a) $f'(x) = \dfrac{1}{3x}\dfrac{d}{dx}(3x) = \dfrac{3}{3x} = \dfrac{1}{x}$

(b) $f'(x) = \dfrac{1}{x^2 + 1}\dfrac{d}{dx}(x^2 + 1) = \dfrac{2x}{x^2 + 1}$

(c) $f'(x) = \dfrac{1}{\sin x}\dfrac{d}{dx}(\sin x) = \dfrac{\cos x}{\sin x} = \cot x$ □

We will have more to say about differentiation (and integration) involving the natural logarithm in Section 8.5. For the present, however, the most important observation is that the function $\ln ax$ has the same derivative as $\ln x$:

$$\frac{d}{dx}(\ln ax) = \frac{1}{ax} \cdot a = \frac{1}{x}\qquad \text{(where } a \text{ is a positive constant)}$$

This leads directly to the *laws of logarithms*.

□ **Theorem 1**   If a and b are positive, $\ln ab = \ln a + \ln b$.

Proof: Since $\ln ax$ and $\ln x$ have the same derivative in the interval $(0, \infty)$, they must differ by a constant:

$$\ln ax - \ln x = C\qquad \text{for all } x > 0$$

Let $x = 1$ to find

$$C = \ln a - \ln 1 = \ln a\qquad \left(\text{because } \ln 1 = \int_1^1 \frac{dt}{t} = 0\right)$$

Hence $\ln ax - \ln x = \ln a$ for all $x > 0$. Let $x = b$ to obtain

$$\ln ab = \ln a + \ln b\qquad □$$

You have probably seen Theorem 1 proved in terms of the exponential law

$$x^u x^v = x^{u+v}$$

Note that the proof given here has nothing to do with exponents. It is a different kind of argument altogether, involving many of the basic ideas of calculus that we have developed in previous chapters. The list is impressive:

- The integral (in terms of which $\ln x$ is defined)
- The derivative (on which the whole proof depends)
- The Fundamental Theorem of Calculus (which enables us to say that the derivative of $\ln x$ is $1/x$)
- The Chain Rule (needed to show that the derivative of $\ln ax$ is also $1/x$)
- The Mean Value Theorem (which lies behind the statement that two functions with the same derivative differ by a constant)

That is one reason the natural logarithm is interesting. It may be the first function you have encountered whose definition has its source in calculus and whose properties are unintelligible without calculus.

□ **Theorem 2**   If a and b are positive, $\ln (a/b) = \ln a - \ln b$.

Proof: All that is needed is an algebraic device to reduce this law to Theorem 1:

$$\ln a = \ln \left(b \cdot \frac{a}{b} \right) = \ln b + \ln \frac{a}{b}$$

Hence $\ln (a/b) = \ln a - \ln b$.   □

□ **Theorem 3**   If r is rational, $\ln x^r = r \ln x$ $(x > 0)$.

Proof: By the Chain Rule,

$$\frac{d}{dx}(\ln x^r) = \frac{1}{x^r} \frac{d}{dx}(x^r) = \frac{1}{x^r} \cdot r x^{r-1} = \frac{r}{x}$$

Since

$$\frac{d}{dx}(r \ln x) = \frac{r}{x}$$

the functions $\ln x^r$ and $r \ln x$ differ by a constant:

$$\ln x^r - r \ln x = C \qquad \text{for all } x > 0$$

Let $x = 1$ to find

$$C = \ln 1 - r \ln 1 = 0$$

Hence $\ln x^r = r \ln x$ for all $x > 0$.   □

Note that in the proof of Theorem 3 we used the familiar Power Rule

$$D_x(x^r) = r x^{r-1}$$

It is important to recall that this was proved (in Section 3.2) only for rational exponents. We have assumed throughout the book that irrational exponents are as yet undefined; it is one of our present objectives to properly define them. When we do, the formula

$$\ln x^r = r \ln x \qquad \text{(proved in Theorem 3 for rational values of } r\text{)}$$

will serve as the idea behind the definition.

To get a better idea of the behavior of $F(x) = \ln x$, we study its graph. Since

$$\ln x = \int_1^x \frac{dt}{t}$$

it follows that

$\ln x$ is undefined when $x \leq 0$.

$\ln x$ is positive, zero, or negative depending on whether $x > 1$, $x = 1$, or $0 < x < 1$, respectively.

These statements may not be obvious, however. To see why $\ln x$ is undefined when $x \leq 0$, observe that

$$\int_1^x \frac{dt}{t}$$

does not exist if the domain of integration includes the origin (because the integrand $1/t$ is unbounded in every neighborhood of 0).

When $x > 1$ we have

$$\ln x = \int_1^x \frac{dt}{t} > 0$$

because the integrand is positive and the limits of integration are in the normal order. (See Figure 1 for a geometric interpretation. But do not confuse this figure with the graph of $y = \ln x$!)

When $0 < x < 1$ the limits of integration are reversed from the normal order, as in

$$\ln 0.5 = \int_1^{0.5} \frac{dt}{t}$$

Hence the integral is negative. (See Figure 2.) Of course when $x = 1$ we get zero, that is,

$$\ln 1 = \int_1^1 \frac{dt}{t} = 0$$

Since

$$F(x) = \int_1^x \frac{dt}{t}$$

Figure 1

$\ln x$ as area under the graph of $y = \dfrac{1}{t}$ $(x > 1)$

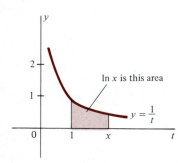

$\ln x$ is this area

$y = \dfrac{1}{t}$

Figure 2

$\ln x$ as the negative of area under the graph of $y = \dfrac{1}{t}$ $(0 < x < 1)$

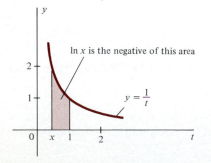

$\ln x$ is the negative of this area

$y = \dfrac{1}{t}$

Figure 3
Graph of the natural logarithm function

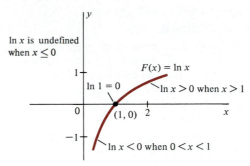

we have (for all $x > 0$) $F'(x) = 1/x > 0$ and $F''(x) = -1/x^2 < 0$. Hence the graph of F is continuous, rising, and concave down, as shown in Figure 3.

It looks very much from this figure as though $y = \ln x$ decreases without bound when $x \to 0$ (because the graph apparently has the negative y axis as an asymptote). It is not so clear that $\ln x$ *increases* without bound when $x \to \infty$. For the graph appears to flatten out, as though it has a horizontal asymptote. If so, $\ln x$ would be bounded for large x. It is important to show that such is not the case!

□ **Theorem 4**   The range of $F(x) = \ln x$ is $(-\infty, \infty)$, that is, $\ln x$ takes on all real values.

Proof: Let y be any real number. The problem is to demonstrate the existence of a number x satisfying $F(x) = y$. To do that, we propose to flank y by two values of F, say

$$F(a) < y < F(b)$$

Since F is continuous (why?), the Intermediate Value Theorem (Section 4.2) guarantees the existence of x satisfying $F(x) = y$.

The easiest way to name a and b is to find a positive integer n large enough to satisfy

$$\ln 2^{-n} < y < \ln 2^n$$

For these inequalities are equivalent to

$$-n \ln 2 < y < n \ln 2 \qquad \text{(Theorem 3)}$$

or (since $\ln 2$ is positive), $|y| < n \ln 2$. But this is equivalent to

$$n > \frac{|y|}{\ln 2}$$

so we need only choose n larger than $|y|/\ln 2$ and we have y flanked as advertised: $F(2^{-n}) < y < F(2^n)$.   □

Problem Set 8.1

1. Find an approximate value of ln 2 as follows.

(a) Express ln 2 as an integral.

(b) Compute a Riemann sum approximating the integral by dividing the interval of integration into four subintervals of equal length and evaluating the integrand at the midpoint of each interval. Compare with the value ln 2 ≈ 0.693 obtained from a table of natural logarithms (or a calculator).

2. Repeat Problem 1 for ln 3, using six subintervals. Compare with the calculator value ln 3 ≈ 1.099.

Given ln 2 ≈ 0.69 and ln 3 ≈ 1.10, compute an approximate value of each of the following. (No tables or calculators allowed!)

3. ln 18   *Hint:* $18 = 2 \cdot 3^2$

4. ln 24   **5.** ln 1.5   **6.** ln $\frac{1}{9}$

7. ln 4 $\sqrt[3]{2}$ **8.** ln 9 $\sqrt{2}$

Let x, y, and z be positive numbers whose natural logarithms are presumed known. Write each of the following in terms of ln x, ln y, and ln z. (Example: $\ln x^2 yz = 2 \ln x + \ln y + \ln z$.)

9. $\ln xyz^3$ **10.** $\ln x^2 y^4 z$ **11.** $\ln \dfrac{xy^2}{z}$

12. $\ln \dfrac{xy}{z^2}$ **13.** $\ln x \sqrt{yz}$ **14.** $\ln xy \sqrt[3]{z}$

15. Since the natural logarithm function turns multiplication into addition, we might expect that $\ln 0 = \ln(0 \cdot 2) = \ln 0 + \ln 2$, from which it would follow that $\ln 2 = 0$. Find the mistake.

Use your knowledge of the graph of $y = \ln x$ to sketch the graph of each of the following.

16. $y = \ln 2x$ *Hint:* $\ln 2x = \ln x + \ln 2$

17. $y = \ln \dfrac{x}{2}$ **18.** $y = \ln \dfrac{1}{x}$ **19.** $y = \ln |x|$

20. Describe the graph of the equation

$$\ln x + \ln y = 0$$

(Note that both x and y must be positive.)

Find the derivative of each of the following functions.

21. $f(x) = \ln(2x - 5)$
22. $f(x) = \ln(x^3 - 1)$
23. $f(x) = \ln \cos x$

24. $f(x) = \ln \sec x$ Why is the result the negative of the derivative in Problem 23?

25. $f(x) = (\ln x)^2$ **26.** $f(x) = x \ln x$

27. $f(x) = \dfrac{\ln x}{x}$ **28.** $f(x) = \sin(\ln x)$

In each of the following, find the derivative of the function as given. Then check by using laws of logarithms to simplify the function before differentiating. (Assume that both versions of the function are restricted to the same domain.)

29. ln 5x   *Hint:* The "simpler" version of the function is ln 5 + ln x.

30. ln ax (where a is a positive constant)

31. $\ln x^3$ **32.** $\ln x^2$ **33.** $\ln x \sin x$

34. $\ln x \tan x$ **35.** $\ln \sqrt{x^2 + 1}$ **36.** $\ln \sqrt{1 - x^2}$

37. $\ln \dfrac{1}{x^2 + 1}$ **38.** $\ln \dfrac{x}{1 + x^2}$

39. Two students are asked to graph the function

$$f(x) = \ln \tan x + \ln \cot x$$

(a) The first student decides to investigate f by differentiating it as it stands. What conclusion may be drawn from this approach?

(b) The second student reaches the same conclusion in a few seconds by a different approach. How?

(c) Describe the graph of f. (Pay attention to the domain!)

40. Show that $D_x \ln |x| = 1/x$ in two ways.

(a) Use the formula $D_x |x| = |x|/x$ from Section 3.3 (together with the Chain Rule).

(b) Consider the cases $x > 0$ and $x < 0$ separately. *Note:* This result (which extends the formula $D_x \ln x = 1/x$ to a larger domain) is important, particularly when we derive integration formulas involving the natural logarithm.

41. In Problem 32 we may write $f(x) = \ln x^2 = 2 \ln x$ if $x > 0$, the derivative of each version of the function being $f'(x) = 2/x$. If the domain is enlarged to $x \neq 0$, explain why $\ln x^2 = 2 \ln |x|$ and show that the derivative is still $f'(x) = 2/x$.

42. Use the identity $\sin 2x = 2 \sin x \cos x$ to differentiate

$$f(x) = \ln |\sin 2x|$$

in two ways and reconcile the answers. (You will also need the identity $\cos 2x = \cos^2 x - \sin^2 x$.)

43. Use implicit differentiation (Section 3.4) to find y' from the relation

$$\ln xy - 2x - y = 5$$

44. The equation $\ln y = x$ defines y (implicitly) as a function of x. Show that $dy/dx = y$. (This function has the remarkable property that it is its own derivative. It is the subject of Section 8.3.)

45. Reasoning from the graph of $F(x) = \ln x$, explain why

$$\ln a = \ln b \Rightarrow a = b$$

Is it true of every function f that if $f(a) = f(b)$, then $a = b$? If so, prove it; if not, give a counterexample.

46. Find all local extreme values of the function

$$f(x) = 2x^2 \ln x - x^2$$

(Pay attention to the domain!)

47. Sketch the graph of $y = x^2 - 2 \ln x$ after answering the following questions.
   (a) What is the domain?
   (b) How does y behave when $x \to 0$? *Hint:* $\ln x \to -\infty$ when $x \to 0$. (Why?)
   (c) Find all local extreme values.
   (d) Discuss concavity of the graph.

48. Sketch the graph of

$$y = \ln \frac{1 + x^2}{1 - x^2}$$

after discussing domain, asymptotes, and extreme values. *Hint:* Use Theorem 2 before differentiating.

49. Draw any tangent to the graph of $y = \ln x$ and let A be its point of intersection with the y axis. Then draw the horizontal line through the point of tangency, intersecting the y axis at B. Prove that AB is always 1.

50. Sketch the graphs of $y = \ln x$ and $y = -x$ to explain why the equation $x + \ln x = 0$ has exactly one root. Then use Newton's Method (Section 5.4) to approximate the root.

51. In our definition of

$$\ln x = \int_1^x \frac{dt}{t}$$

we chose the lower limit to be 1. The function

$$f(x) = \int_a^x \frac{dt}{t} \qquad \text{(where } a \text{ is a positive constant)}$$

is not much different. Find a formula for f in terms of ln.

52. Give a different proof of Theorem 1 as follows.
   (a) Make the substitution $u = at$ to show that

$$\int_1^b \frac{dt}{t} = \int_a^{ab} \frac{du}{u}$$

   (b) Explain why

$$\int_a^{ab} \frac{du}{u} = \ln ab - \ln a$$

and use the result to finish the proof.

53. Give a different proof of Theorem 2 by using Theorems 1 and 3. (First convince yourself that our proof of Theorem 3 does not depend on Theorem 2!)

54. In our proof of Theorem 4, which says that the range of $F(x) = \ln x$ is $(-\infty, \infty)$, we appealed to the fact that F is continuous. What justifies this assertion?

55. Use the continuity of ln to explain why

$$\lim_{x \to 0} \ln \frac{\sin x}{x} = 0$$

56. Use the area interpretation of

$$\ln x = \int_1^x \frac{dt}{t}$$

to argue that for all $x > 1$, $1 - 1/x < \ln x < x - 1$.

8.2
INVERSE FUNCTIONS

We suggested in the last section that the natural logarithm is related to exponents. To investigate this relation, we are going to reverse the function $F(x) = \ln x$, thus defining a new function G with familiar exponential properties. This new function will be discussed more fully in the next section; our purpose now is to explain what is meant by "reversing" a function.

The graph of $F(x) = \ln x$ is shown in Figure 1. Like every function, F converts each number x in the domain into a definite y in the range. This is indicated by the arrows in Figure 1, which show that x comes first and then $y = \ln x$ is determined. Unlike some functions, $\ln x$ is continuous and in-

Figure 1
Conversion of x into y by the function F

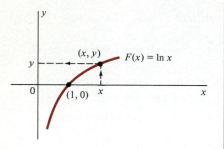

creasing, which means that *no horizontal line intersects its graph more than once*. In such circumstances it is clear that the rule of correspondence $x \to y$ may be turned around to read $y \to x$. In other words the arrows in Figure 1 may be reversed, with y coming first and x being determined by y.

Of course that creates a new function, say G, defined by the statement that

$$x = G(y) \Longleftrightarrow y = F(x)$$

The independent variable in this new function is y, its domain (the set of values of y) being the range of the original function F. The range of G (the set of values of the dependent variable x) is the domain of F.

Since the letters used to represent variables are immaterial, we can interchange x and y to define G by the statement

$$y = G(x) \Longleftrightarrow x = F(y)$$

This has the advantage of displaying the points of the graph of G in the usual order (x, y), x now being independent and y dependent. Hence G may be graphed in the same coordinate plane with F. If (a, b) is a point of its graph, then $b = G(a)$ and $a = F(b)$, so the corresponding point of the graph of F is (b, a). Thus the graph of G is obtained by reversing coordinates of the points of the graph of F. In other words (see Section 1.3):

The graph of $y = G(x)$ is the reflection of the graph of $y = F(x)$ in the line $y = x$. (See Figure 2.)

The above discussion need not be confined to the function $F(x) = \ln x$, but applies to any function with the property that no horizontal line intersects its graph more than once. The following examples will show what we mean.

Figure 2
Reflection of F in the line $y = x$

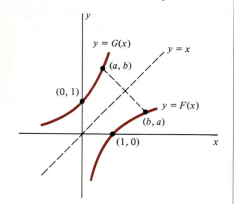

□ **Example 1**   The graph of $f(x) = 2x - 1$ is the straight line of slope 2 shown in Figure 3. The reverse function g in this case is defined by

$$y = g(x) \Longleftrightarrow x = f(y) = 2y - 1$$

Figure 3
Functions reflected about the line $y = x$

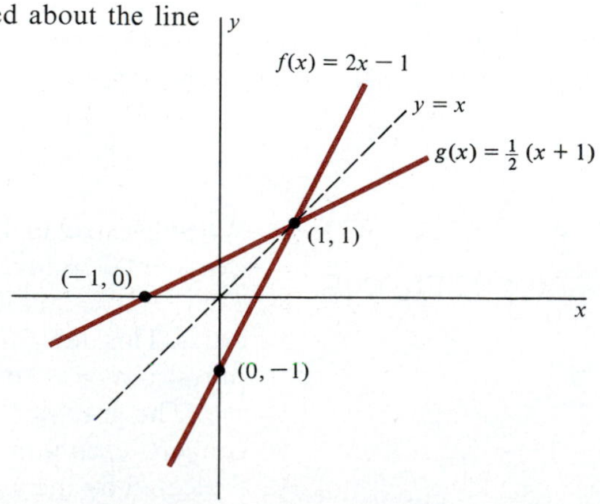

$f(x) = 2x - 1$

$y = x$

$g(x) = \frac{1}{2}(x + 1)$

$(1, 1)$

$(-1, 0)$

x

$(0, -1)$

To find an explicit formula for $g(x)$, we solve for y in terms of x in the equation $x = 2y - 1$ (which is merely a matter of reversing the indicated operations):

$$x = 2y - 1$$
$$x + 1 = 2y$$
$$y = \tfrac{1}{2}(x + 1)$$

Hence $g(x) = \tfrac{1}{2}(x + 1)$, the graph of which is the straight line of slope $\tfrac{1}{2}$ shown in Figure 3. Note that the graphs of

$$f(x) = 2x - 1 \quad \text{and} \quad g(x) = \tfrac{1}{2}(x + 1)$$

are, as advertised, reflections of one another in the line $y = x$.   □

When two functions f and g are related in this way, each is called the *inverse* of the other. The domain of one is the range of the other; moreover, they reverse each other's effect:

$$g[f(x)] = x \quad \text{for each } x \text{ in the domain of } f$$
$$f[g(x)] = x \quad \text{for each } x \text{ in the domain of } g$$

To see why, recall that the equations $x = g(y)$ and $y = f(x)$ are equivalent. Substituting from the second into the first, we have

$$x = g(y) = g[f(x)] \quad \text{for each } x \text{ in the domain of } f$$

Similarly, since $y = g(x) \Leftrightarrow x = f(y)$, we obtain

$$x = f(y) = f[g(x)] \quad \text{for each } x \text{ in the domain of } g$$

These identities should help explain the idea of inverse function, particularly if they are translated into words and pictures:

> *An input that goes into one function and then into its inverse is returned unchanged.*

(See Figure 4.) In other words whatever one function does to the input, the inverse function "undoes." They reverse each other's operations. To see this working in Example 1, observe that

$$g[f(x)] = g(2x - 1) = \tfrac{1}{2}[(2x - 1) + 1] = \tfrac{1}{2}(2x) = x$$
$$f(g(x)) = f[\tfrac{1}{2}(x + 1)] = 2[\tfrac{1}{2}(x + 1)] - 1 = (x + 1) - 1 = x$$

To indicate the inverse relationship between f and g, we often use the symbol f^{-1} for g, in which case the above identities read

$$f^{-1}[f(x)] = x \quad \text{for each } x \text{ in the domain of } f$$
$$f[f^{-1}(x)] = x \quad \text{for each } x \text{ in the domain of } f^{-1}$$

Figure 4
Inverse function reversing the effect of function

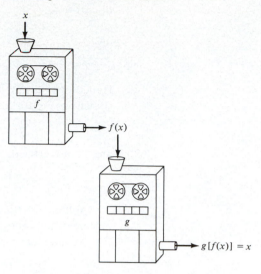

(a) *g* reverses the effect of *f*

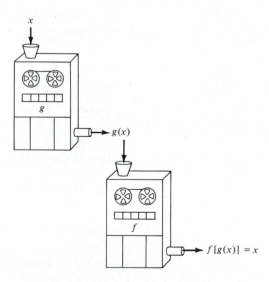

(b) *f* reverses the effect of *g*

The symbol f^{-1} should not be interpreted as a power (although of course its use is suggested by analogy with the reciprocal relation $a \cdot a^{-1} = 1$ in arithmetic). It is not the reciprocal of *f* we are talking about, but the inverse. To see the distinction in Example 1, note that

$$f^{-1}(x) = \frac{1}{2}(x + 1), \text{ while } \left(\frac{1}{f}\right)(x) = \frac{1}{2x - 1}$$

Figure 5
Horizontal line intersecting the graph of $f(x) = x^2$ more than once

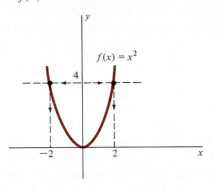

□ **Example 2**   Discuss the inverse (if it exists) of the function $f(x) = x^2$.

Solution: The graph of f in Figure 5 shows that although each x determines a unique y, we cannot reverse the correspondence to read $y \to x$. For there are horizontal lines that intersect the graph of f more than once. If $y = 4$, what is the corresponding x? Since the answer is ambiguous ($x = \pm 2$), the set of points obtained from the graph of f by reversing coordinates of its points is not the graph of a function. See Figure 6, which shows this set to be the graph of the *relation* $x = y^2$. Solving for y in terms of x yields the ambiguous formula $y = \pm \sqrt{x}$, which does not define a function. Thus we conclude that the inverse of f does not exist. (The graph of $x = y^2$ is still the reflection of the graph of $y = x^2$ in the line $y = x$, however. Hence we call $x = y^2$ and $y = x^2$ **inverse relations.**)

Figure 6
Inverse relations

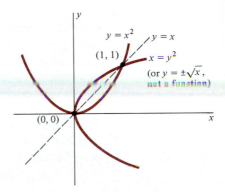

Suppose now that we restrict the domain of f to the interval $[0, \infty)$, thus cutting off the left-hand half of the parabola $y = x^2$. This new function has an inverse, namely

$$x = f^{-1}(y) \quad \text{defined by } y = f(x) = x^2 \quad (x \geq 0)$$

Or, interchanging x and y, we have

$$y = f^{-1}(x) \quad \text{defined by } x = f(y) = y^2 \quad (y \geq 0)$$

Solving for y in the equation $x = y^2$ no longer involves the ambiguous double sign (because $y \geq 0$). Hence the inverse function is

$$y = f^{-1}(x) = \sqrt{x}$$

as shown in Figure 7.   □

Figure 7
The square and square root functions as inverses

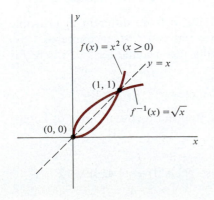

Examples 1 and 2 are misleading in one sense. In each case the function f is defined by an algebraic formula that is easy to solve for one variable in terms of the other. We cannot ordinarily expect such good fortune.

Figure 8

Graph of $y = x^3 + x - 2$

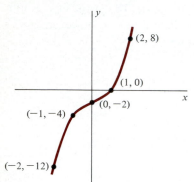

□ **Example 3**   Discuss the inverse (if it exists) of the function $f(x) = x^3 + x - 2$.

Solution: The graphical test for existence of the inverse of f is that no horizontal line intersects the graph of f more than once. In the present case f is continuous, and since

$$f'(x) = 3x^2 + 1 > 0 \qquad \text{for all } x$$

the graph of f is always rising. (See Figure 8.) Hence f^{-1} exists.

Knowing that the inverse exists and finding a formula for it are two different matters! To find $x = f^{-1}(y)$, we must solve the cubic equation $y = x^3 + x - 2$ for x in terms of y. Or, if we want the result to read $y = f^{-1}(x)$, we must solve $x = y^3 + y - 2$ for y in terms of x. In any case it is algebraically difficult to finish the problem. We leave it as an illustration of the fact that an inverse function may exist without our knowing how to write a formula for it.   □

As we observed at the beginning of this section, the function $y = \ln x$ has an inverse. Unlike Example 3 (in which it is merely difficult to find x in terms of y) this is impossible. No algebraic procedure exists for solving the equation.

It is precisely the impossible cases that are interesting. *The inverse of the natural logarithm function exists, but we cannot give you an algebraic formula for it.* Hence we are going to have to invent a name for it. An obvious candidate is $\ln^{-1}$ (because we have agreed that the inverse of a function f may be denoted by f^{-1}). Using this notation, we can solve the equation $y = \ln x$ by writing $x = \ln^{-1} y$. Or, interchanging x and y, we have

$$y = \ln^{-1} x \Leftrightarrow x = \ln y$$

Values of this new function are found indirectly in terms of known values of the natural logarithm. For example,

$$\ln^{-1} 0 = 1 \qquad \text{because } \ln 1 = 0 \quad \text{(Figure 2)}$$
$$\ln^{-1} 0.69 \approx 2 \qquad \text{because } \ln 2 \approx 0.69 \quad \text{(Problem 1, Section 8.1)}$$
$$\ln^{-1} 1.10 \approx 3 \qquad \text{because } \ln 3 \approx 1.10 \quad \text{(Problem 2, Section 8.1)}$$

This procedure may seem cumbersome, but it is no different in principle from the kind of thing you had to do when you first encountered (say) square roots. The radical sign $\sqrt{\ }$ is really just a name for f^{-1}, where f is the square function. In other words

$$y = f^{-1}(x) = \sqrt{x} \Leftrightarrow x = f(y) = y^2 \qquad (y \geq 0)$$

Values of the square root function are found indirectly in terms of known squares. For example,

$$f^{-1}(9) = \sqrt{9} = 3 \qquad \text{because } f(3) = 3^2 = 9$$
$$f^{-1}(2) = \sqrt{2} \approx 1.414 \qquad \text{because } f(1.414) = (1.414)^2 \approx 2$$

and so on.

In the next section we will introduce another symbol for $\ln^{-1}$ and develop its properties in more detail. It is an important function in its own right, having acquired a name that is independent of $\ln$ for the same reason that $\sqrt{}$ is used instead of f^{-1} when f is the square function.

We end this section by investigating how inverse functions are differentiated. As an example, consider the derivative of $y = \sqrt{x}$. Since

$$y = \sqrt{x} \Leftrightarrow x = y^2 \qquad (y \geq 0)$$

we can find dy/dx by implicit differentiation in the equation $x = y^2$:

$$1 = 2y \frac{dy}{dx}$$

$$\frac{dy}{dx} = \frac{1}{2y} = \frac{1}{2\sqrt{x}}$$

Of course, in this case it is easier to differentiate directly by using the Power Rule:

$$y = \sqrt{x} = x^{1/2}$$

$$\frac{dy}{dx} = \frac{1}{2}x^{-1/2} = \frac{1}{2\sqrt{x}}$$

The point is that we do not need the direct method. If we know how to differentiate the square function, the derivative of its inverse can be found by implicit differentiation.

There is no reason why we cannot apply this idea in general. Suppose that f is a differentiable function with an inverse. To find the derivative of $y = f^{-1}(x)$, we differentiate implicitly in the equivalent equation $x = f(y)$:

$$1 = f'(y) \frac{dy}{dx}$$

$$\frac{dy}{dx} = \frac{1}{f'(y)} \qquad [\text{provided that } f'(y) \neq 0]$$

Derivative of an Inverse Function

Suppose that f is a differentiable function with an inverse function $y = f^{-1}(x)$. Then

$$\frac{dy}{dx} = \frac{1}{f'(y)} \qquad \text{if } f'(y) \neq 0$$

□ **Example 4**   Use the above formula to find the derivative of the inverse of $f(x) = x^3$.

Solution: Let $y = f^{-1}(x)$. Since $f'(x) = 3x^2$, we find

$$\frac{dy}{dx} = \frac{1}{f'(y)} = \frac{1}{3y^2}$$

If we don't have a formula for y in terms of x, this is as far as we can go. In the present case, however, we know that $y = \sqrt[3]{x}$. Hence

$$\frac{dy}{dx} = \frac{1}{3(\sqrt[3]{x})^2} = \frac{1}{3x^{2/3}}$$

Note that this checks with the result of direct differentiation:

$$y = \sqrt[3]{x} = x^{1/3}$$

$$\frac{dy}{dx} = \frac{1}{3}x^{-2/3} = \frac{1}{3x^{2/3}}$$

□ **Example 5**   If $f(x) = x^3 + x - 2$, find the slope of the graph of $y = f^{-1}(x)$ at the point $(0,1)$.

Solution: First we check whether the question even makes sense. Does f^{-1} exist and is $(0,1)$ a point of its graph? The answer is supplied by Example 3, where we showed that f has an inverse but that it is hard to find an explicit formula for f^{-1}. The point $(0,1)$ is on the graph of $y = f^{-1}(x)$ because $(1,0)$ is on the graph of $f(x) = x^3 + x - 2$.

Note that direct differentiation of f^{-1} is out of the question unless we know a formula for $y = f^{-1}(x)$. We do know, however, that $f'(x) = 3x^2 + 1$. Hence

$$\frac{dy}{dx} = \frac{1}{f'(y)} = \frac{1}{3y^2 + 1}$$

and the slope of the graph of $y = f^{-1}(x)$ at $(0,1)$ is $1/4$.   □

It is worth observing in Example 5 that the derivative of $y = f^{-1}(x)$ at $(0,1)$ and the derivative of $y = f(x)$ at $(1,0)$ are reciprocals ($\frac{1}{4}$ and 4, respectively). That this is true in general may be seen by writing the formula

$$\frac{dy}{dx} = \frac{1}{f'(y)}$$

in a more elegant version. Since the equations $y = f^{-1}(x)$ and $x = f(y)$ are equivalent, and since differentiation of $x = f(y)$ with respect to y yields $dx/dy = f'(y)$, we have

$$\frac{dy}{dx} = \frac{1}{f'(y)} = \frac{1}{dx/dy}$$

This result might have been predicted because of the remarkable way in which Leibniz symbols behave like fractions. The equation

$$\frac{dy}{dx} = \frac{1}{dx/dy}$$

looks obvious. Nevertheless it is dangerous to take it too literally. A common error is to reason that since dy/dx and dx/dy are reciprocals, it follows that

$$\frac{dy}{dx} = \frac{1}{f'(x)} \qquad \left(\text{instead of the correct formula } \frac{dy}{dx} = \frac{1}{f'(y)}\right)$$

You can see that this is wrong by looking at Example 5, in which the derivative of $y = f^{-1}(x)$ at $x = 0$ is $\frac{1}{4}$. This is not the reciprocal of $f'(x) = 3x^2 + 1$ at $x = 0$ (which is 1), but the reciprocal of $f'(y) = 3y^2 + 1$ at $y = 1$ (namely 4). In other words the derivative of $f^{-1}(x)$ is the reciprocal of the derivative of f *evaluated at* $y = f^{-1}(x)$. The formula

$$\frac{d}{dx}[f^{-1}(x)] = \frac{1}{f'[f^{-1}(x)]}$$

is clumsy but precise!

In practice, however, we will not use it often. When the derivative of a specific inverse function $y = f^{-1}(x)$ is wanted, it is usually easier to start from scratch, differentiating implicitly in the equivalent equation $x = f(y)$ as if the theoretical formula were not known.

Optional Note (*on differentiation of inverse functions*)   When implicit differentiation is used in the equation $x = f(y)$ to write

$$1 = f'(y)\frac{dy}{dx}$$

we are assuming not only that f is differentiable, but that $y = f^{-1}(x)$ is. (Otherwise dy/dx would not exist.) It is easy to argue geometrically (harder to prove analytically) that this assumption is valid. For if we know that f is differentiable, the curve $y = f(x)$ has a tangent at each point. Reflection of this curve in the line $y = x$ yields the graph of $y = f^{-1}(x)$; obviously the curves are congruent. Hence there is a tangent at each point of the curve $y = f^{-1}(x)$.

In other words f^{-1} is differentiable when f is. The only exception to this statement occurs when f' is zero. For then the tangent to the graph of f is horizontal and hence the tangent to the graph of f^{-1} is vertical. (Why?) In this case the derivative of f^{-1} does not exist; that fact is reflected by the formula

$$\frac{dy}{dx} = \frac{1}{f'(y)}$$

which is meaningless when $f'(y) = 0$.

Problem Set 8.2

In each of the following, sketch the graph of f and decide whether f^{-1} exists. When it does, find a formula for $y = f^{-1}(x)$ and sketch the graph of f^{-1} on the same coordinate plane with your graph of f.

1. $f(x) = 2x$
2. $f(x) = \frac{1}{2}x - 1$
3. $f(x) = \sqrt[3]{x}$
4. $f(x) = \sqrt{x - 1}$
5. $f(x) = x^2 + 1$
6. $f(x) = 1 - x^2$
7. $f(x) = x^2 + 1, x \geq 0$
8. $f(x) = 1 - x^2, x \geq 0$
9. $f(x) = x^3 + 1$
10. $f(x) = x^3 - 1$
11. $f(x) = \frac{1}{x}$
12. $f(x) = \frac{1}{x - 1}$

13. $f(x) = \frac{x}{x - 1}$
14. $f(x) = \frac{1}{x^2 + 1}, x \geq 0$

In each of the following, a function that has an inverse is given. (See the corresponding exercise in Problems 1 through 14.) Confirm the identities

$$f^{-1}[f(x)] = x \quad \text{and} \quad f[f^{-1}(x)] = x$$

15. $f(x) = 2x$
16. $f(x) = \frac{1}{2}x - 1$
17. $f(x) = \sqrt[3]{x}$
18. $f(x) = \sqrt{x - 1}$
19. $f(x) = x^2 + 1, x \geq 0$
20. $f(x) = 1 - x^2, x \geq 0$
21. $f(x) = x^3 + 1$
22. $f(x) = x^3 - 1$

23. $f(x) = \dfrac{1}{x}$ **24.** $f(x) = \dfrac{1}{x-1}$

25. $f(x) = \dfrac{x}{x-1}$ **26.** $f(x) = \dfrac{1}{x^2+1}$, $x \geq 0$

In each of the following, find the derivative of $y = f^{-1}(x)$

(a) by differentiating $f^{-1}(x)$ directly;

(b) by using the formula $\dfrac{dy}{dx} = \dfrac{1}{f'(y)}$.

(See the corresponding exercise in Problems 1 through 14.)

27. $f(x) = 2x$ **28.** $f(x) = \frac{1}{2}x - 1$
29. $f(x) = \sqrt[3]{x}$ **30.** $f(x) = \sqrt{x-1}$
31. $f(x) = x^2 + 1$, $x \geq 0$ **32.** $f(x) = 1 - x^2$, $x \geq 0$
33. $f(x) = x^3 + 1$ **34.** $f(x) = x^3 - 1$

35. $f(x) = \dfrac{1}{x}$ **36.** $f(x) = \dfrac{1}{x-1}$

37. $f(x) = \dfrac{x}{x-1}$ **38.** $f(x) = \dfrac{1}{x^2+1}$, $x \geq 0$

Sketch the graph of each of the following functions and determine whether the inverse exists. (Do not try to find an algebraic formula for it!)

39. $f(x) = \sin x$, $0 \leq x \leq 2\pi$
40. $f(x) = \sin x$, $-\pi/2 \leq x \leq \pi/2$
41. $f(x) = \cos x$, $0 \leq x \leq \pi$
42. $f(x) = \cos x$, $-\pi/2 \leq x \leq \pi/2$
43. $f(x) = \tan x$, $-\pi/2 < x < \pi/2$
44. $f(x) = \tan x$, $-\pi/2 < x < 3\pi/2$ ($x \neq \pi/2$)
45. Explain why the inverse of $f(x) = x^3 + 2x - 1$ exists and why $(-1,0)$ is a point of the graph of $y = f^{-1}(x)$. Find the slope at this point and confirm that it is the reciprocal of the slope of the graph of $y = f(x)$ at $(0,-1)$.
46. Find the slope of the graph of $y = \ln^{-1} x$ at $(0,1)$ and confirm that it is the reciprocal of the slope of the graph of $y = \ln x$ at $(1,0)$.
47. The inverse of $f(x) = \sin x$, $-\pi/2 \leq x \leq \pi/2$, is written $f^{-1}(x) = \sin^{-1} x$. Explain why $(0,0)$ and $(1,\pi/2)$ are points of its graph and discuss the slope at these points. Compare with the slope of the graph of f at $(0,0)$ and $(\pi/2,1)$.
48. The inverse of $f(x) = \tan x$, $-\pi/2 < x < \pi/2$, is written $f^{-1}(x) = \tan^{-1} x$. Explain why $(0,0)$ and $(1,\pi/4)$ are points of its graph and find the slope at these points. Compare with the slope of the graph of f at $(0,0)$ and $(\pi/4,1)$.

49. If $f(x) = x^2$, then $f(1) = f(-1)$. Applying f^{-1} to each side, we have

$$f^{-1}[f(1)] = f^{-1}[f(-1)]$$

from which $1 = -1$. Find the mistake.
50. Suppose that the function f is its own inverse, that is, $f^{-1} = f$. (See Problems 35 and 37 for examples.) What can you say about the graph of f?
51. Suppose that f and g are inverse functions. Why is it geometrically apparent that if f is continuous, g is, too?
52. Although the function $f(x) = x^3$ has an inverse, the formula

$$\frac{dy}{dx} = \frac{1}{dx/dy}$$

for the derivative of $y = f^{-1}(x)$ does not apply at $x = 0$. Why? What is the geometrical interpretation?
53. Differentiate implicitly in the equation $x^2 + y^2 = 1$ to find dy/dx and dx/dy. Then check the formula

$$\frac{dy}{dx} = \frac{1}{dx/dy}$$

54. If the graph of $y = f(x)$ is a straight line of slope $m \neq 0$, show without calculus that the graph of $y = f^{-1}(x)$ is a straight line of slope $1/m$. How is this confirmed by calculus?

The remaining problems in this set develop properties of the function $\ln^{-1}$. They are good practice in working with the ideas of this section. The properties themselves, however, will be discussed in the next section in terms of different notation for $\ln^{-1}$.

55. Explain why the domain of $y = \ln^{-1} x$ is $\mathfrak{R}$ and the range is $(0, \infty)$.
56. Show that

$$\frac{d}{dx}(\ln^{-1} x) = \ln^{-1} x$$

(Thus the inverse of the natural logarithm function has the remarkable property that it is its own derivative.)
57. Use Theorem 1, Section 8.1, to show that

$$\ln^{-1}(u + v) = (\ln^{-1} u)(\ln^{-1} v)$$

58. Use Theorem 2, Section 8.1, to show that

$$\ln^{-1}(u - v) = \frac{\ln^{-1} u}{\ln^{-1} v}$$

59. Use Theorem 3, Section 8.1, to show that if r is rational,

$$(\ln^{-1} x)^r = \ln^{-1}(rx)$$

8.3
THE NATURAL EXPONENTIAL FUNCTION

In the last section we observed that since the function $F(x) = \ln x$ is continuous and increasing, it has an inverse, defined by

$$y = \ln^{-1} x \Leftrightarrow x = \ln y$$

The graph of $y = \ln^{-1} x$ is the reflection of the graph of $y = \ln x$ in the line $y = x$, as shown in Figure 1. If you worked Problems 57 through 59 in Section 8.2, you know that

$$\ln^{-1}(u + v) = (\ln^{-1} u)(\ln^{-1} v) \quad \text{and} \quad \ln^{-1}(u - v) = \frac{\ln^{-1} u}{\ln^{-1} v}$$

and (for rational values of r)

$$(\ln^{-1} x)^r = \ln^{-1}(rx)$$

These properties (which we will derive in this section) should remind you of the laws of exponents

$$a^{u+v} = (a^u)(a^v) \qquad a^{u-v} = \frac{a^u}{a^v} \qquad (a^x)^r = a^{rx}$$

with which you have long been familiar. It appears that the function $y = \ln^{-1} x$ has an exponential character. Hence we denote $\ln^{-1}$ by the more suggestive symbol *exp*, as in the following definition.

Figure 1
Graphs of ln and $\ln^{-1}$

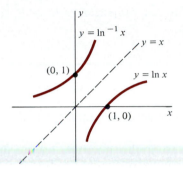

The Natural Exponential Function

The inverse of ln is called **exp.** It is defined by the statement that

$$y = \exp x \Leftrightarrow x = \ln y$$

Since the domain of ln is $(0, \infty)$ and the range is $\mathfrak{R}$, the function $y = \exp x$ is defined for all x and takes on all positive values. Its graph is the reflection of the graph of $y = \ln x$ in the line $y = x$. Since ln and exp are inverses,

$$\exp(\ln x) = x \text{ for all } x > 0 \quad \text{and} \quad \ln(\exp x) = x \text{ for all } x$$

To derive the exponential properties of exp, we use the logarithmic laws

$$\ln ab = \ln a + \ln b, \ln \frac{a}{b} = \ln a - \ln b, \ln x^r = r \ln x$$

(Theorems 1–3 in Section 8.1).

□ **Theorem 1**   For all u and v,

$$\exp(u + v) = \exp u \cdot \exp v \quad \text{and} \quad \exp(u - v) = \frac{\exp u}{\exp v}$$

Proof: Starting with the left side, we replace u and v by $\ln(\exp u)$ and $\ln(\exp v)$, respectively:

$$\exp(u + v) = \exp[\ln(\exp u) + \ln(\exp v)]$$

Since $\ln a + \ln b = \ln ab$, this becomes

$$\exp(u + v) = \exp[\ln(\exp u \cdot \exp v)]$$
$$= \exp u \cdot \exp v \qquad (\text{because exp and ln are inverses})$$

The second formula is proved in the same way.   □

□ **Theorem 2**   If r is rational, then $(\exp x)^r = \exp rx$ for all x.

Proof: As in Theorem 1, we use the inverse properties of $\ln$ and $\exp$:

$$\exp rx = \exp[r \ln(\exp x)]$$
$$= \exp[\ln(\exp x)^r] \qquad (\text{why})?$$
$$= (\exp x)^r$$ □

Having established that exp behaves like an exponential function, we come to a critical question. What is the base? That is, if $\exp x = a^x$ for some constant a, what is a? When we try to answer this question, we run up against the fact that we do not really know any exponential function that is defined for all x. Our point of view from the start has been that the symbol a^x is defined only for rational values of x; irrational exponents remain to be explained.

Let us rephrase the question. Is there a number a such that $\exp x = a^x$ for all *rational* x? If so, it would be natural to extend the definition of a^x to irrational values of x by agreeing that $a^x = \exp x$ for *all* x. (This would be a legitimate agreement because $\exp x$ is already defined for all x.)

Thus we are poised for a discovery that will kill two birds with one stone. Naming the base a will enable us to regard exp as an exponential function as understood in precalculus mathematics; it will simultaneously extend the definition of exponents to irrational values.

The clue, while not obvious, is nevertheless close at hand. Look at Theorem 2, which says that

$$\exp rx = (\exp x)^r \qquad (\text{provided that } r \text{ is rational})$$

Taking $x = 1$, we may reduce this to $\exp r = (\exp 1)^r$. Evidently the base we seek is the number $\exp 1$. It is such an important number that a letter of the alphabet has been taken out of circulation to name it, as in the following definition:

$$\boxed{e = \exp 1}$$

Since $e = \exp 1 \Leftrightarrow \ln e = 1$, we may describe e as the number whose natural logarithm is 1 (guaranteed to exist because $\ln x$ takes on all real values). We shall also find an expression for it as a certain limit (and in Chapter 12 as the sum of a certain infinite series). It can be proved that e is irrational (like π); its first few decimal places are given by

$$e = 2.7182818284\ldots$$

While this may seem a strange number to choose as the base of an exponential function, it is not out of order to call it "natural." For now we can write $\exp x = e^x$ (for rational x). No other choice of base enables us to replace $\exp x$ by an exponential symbol.

One loose end remains. The equation $\exp x = e^x$ is true only when x is rational (because the symbol e^x is still meaningless when x is irrational). Hence we propose the following definition.

When x is irrational, we define $e^x = \exp x$.

As an example, consider the symbol $e^{\sqrt{2}}$ (hitherto undefined). Having agreed to the definition $e^{\sqrt{2}} = \exp \sqrt{2}$ and knowing that

$$y = \exp \sqrt{2} \Leftrightarrow \ln y = \sqrt{2}$$

we may say that $e^{\sqrt{2}}$ is the number whose natural logarithm is $\sqrt{2}$. That gives us a handle for computing it (if we assume that values of the natural logarithm are available in tables or programmed into calculators). In fact, however, the function e^x is so important that its values are also tabulated. Hence in practice we compute $e^{\sqrt{2}}$ directly (without reference to the natural logarithm).

With the above agreements in hand, we shall abandon the notation $\exp x$, replacing it by e^x for all x. This means, for example, that the inverse relation between the natural exponential function and the natural logarithm may now be written

$$y = e^x \Leftrightarrow x = \ln y$$

and Theorems 1 and 2 may be rendered in the form

$$e^{u+v} = e^u e^v, \quad e^{u-v} = \frac{e^u}{e^v}, \quad (e^x)^r = e^{rx} \qquad \text{(if } r \text{ is rational)}$$

Moreover, the identities $\exp(\ln x) = x$ and $\ln(\exp x) = x$ now read

$$e^{\ln x} = x \text{ for all } x > 0$$

and

$$\ln e^x = x \text{ for all } x$$

If you worked Problem 56, Section 8.2, you know that $D_x \ln^{-1} x = \ln^{-1} x$, or in our present notation, $D_x e^x = e^x$. We derive this formula by letting $y = e^x$ and using implicit differentiation in the equivalent equation $x = \ln y$:

$$x = \ln y$$

$$1 = \frac{1}{y}\frac{dy}{dx}$$

$$\frac{dy}{dx} = y = e^x$$

Thus the natural exponential function has the remarkable property that it is its own derivative. In other words, it is unchanged by differentiation:

$$\boxed{D_x e^x = e^x}$$

Remark   To differentiate $y = e^x$, we could have used the formula

$$\frac{dy}{dx} = \frac{1}{f'(y)}$$

for the derivative of the inverse of f (Section 8.2). In this case $f(x) = \ln x$ and $f'(x) = 1/x$, so if $y = f^{-1}(x) = e^x$, we have

$$\frac{dy}{dx} = \frac{1}{f'(y)} = \frac{1}{1/y} = y = e^x$$

As noted in Section 8.2, however, it is easier to ignore this formula, differentiating implicitly in the equation $x = \ln y$ instead.

As usual when a new differentiation formula is developed, we build it into the Chain Rule:

$$\boxed{\frac{d}{dx}(e^u) = e^u \frac{du}{dx}}$$

□ **Example 1**   Find the derivative of

(a) $f(x) = e^{2x}$ (b) $f(x) = e^{x^2+1}$ (c) $f(x) = e^{\sin x}$

Solution:

(a) $f'(x) = e^{2x}\frac{d}{dx}(2x) = 2e^{2x}$

(b) $f'(x) = e^{x^2+1}\frac{d}{dx}(x^2 + 1) = 2xe^{x^2+1}$

(c) $f'(x) = e^{\sin x}\frac{d}{dx}(\sin x) = e^{\sin x}\cos x$ □

We end this section with two examples of how our new exponential function is used in practice.

□ **Example 2**   Find the inverse of

(a) $f(x) = \frac{1}{2}\ln(x - 1)$ (b) $f(x) = \dfrac{e^{2x} - 1}{e^{2x} + 1}$

Solution:

(a) Since $y = f^{-1}(x)$ is equivalent to $x = f(y)$, we solve for y in the equation $x = \frac{1}{2}\ln(y - 1)$:

$$2x = \ln(y - 1)$$

$$e^{2x} = e^{\ln(y-1)} = y - 1$$

$$y = e^{2x} + 1$$

(b) Solve for y in the equation $x = f(y)$:

$$x = \frac{e^{2y} - 1}{e^{2y} + 1}$$

$$x(e^{2y} + 1) = e^{2y} - 1$$

$$xe^{2y} + x = e^{2y} - 1$$

$$1 + x = e^{2y} - xe^{2y} = (1 - x)e^{2y}$$

$$e^{2y} = \frac{1 + x}{1 - x}$$

$$2y = \ln\frac{1 + x}{1 - x}$$

$$y = \frac{1}{2}\ln\frac{1 + x}{1 - x}$$ □

□ **Example 3**   Discuss and sketch the graph of $f(x) = x^2 e^{-x}$.

Solution: We use the Product Rule to write

$$f'(x) = x^2 D_x(e^{-x}) + e^{-x}D_x(x^2) = x^2(-e^{-x}) + e^{-x}(2x) = xe^{-x}(2 - x)$$

Since e^{-x} is always positive (why?), the critical values of f are $x = 0$ and $x = 2$. The sign change diagram in Figure 2 shows that the graph of f is falling in the intervals $(-\infty, 0)$ and $(2, \infty)$, and rising in $(0, 2)$. Hence $f(0) = 0$ is a local minimum and $f(2) = 4e^{-2} \approx 0.54$ is a local maximum. (We used a calculator to find e^{-2}.)

     When $x \to -\infty$, $e^{-x} \to \infty$ (why?), so $f(x) = x^2 e^{-x} \to \infty$. When $x \to \infty$, it is not so clear what $x^2 e^{-x}$ does. The factor x^2 increases without bound, but at the same time e^{-x} approaches zero. (Why?) It turns out (as we will show in Chapter 11) that the effect of the exponential factor is decisive, so $f(x) \to 0$ as $x \to \infty$. We can get an intuitive feeling for why that should be so by observing that $x^2 e^{-x}$ is positive and decreasing for $x > 2$. Hence the graph falls (but never hits the x axis) as x increases beyond 2. It therefore appears that the x axis is an asymptote. (See Figure 3.)   □

Figure 2
Sign of $f'(x) = xe^{-x}(2 - x)$

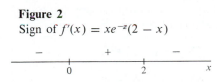

Figure 3
Graph of $y = x^2 e^{-x}$

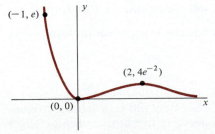

Problem Set 8.3

1. If you have a calculator, display any (reasonably small) number. Then punch the e^x key followed by the $\ln x$ key. What happens and why?

2. Display any positive number on your calculator. Then punch the $\ln x$ key followed by the e^x key. What happens and why?

What is the distinction (if any) between the functions in each of the following?

3. $f_1(x) = e^{\ln x}$ and $f_2(x) = \ln e^x$

4. $f_1(x) = e^{2\ln x}$ and $f_2(x) = x^2$

5. $f_1(x) = e^{3\ln x}$ and $f_2(x) = x^3$

6. $f_1(x) = e^{\ln 2 + \ln x}$ and $f_2(x) = 2x$

7. $f_1(x) = e^{x + \ln x}$ and $f_2(x) = xe^x$

8. $f_1(x) = e^{\ln x - x}$ and $f_2(x) = xe^{-x}$

In each of the following, find a formula for $f^{-1}(x)$.

9. $f(x) = e^{-x}$

10. $f(x) = e^x - 1$

11. $f(x) = \dfrac{e^{2x} + 1}{e^{2x} - 1}$

12. $f(x) = \dfrac{e^x - e^{-x}}{e^x + e^{-x}}$

13. $f(x) = \ln 2x$

14. $f(x) = \ln \dfrac{1}{x}$

15. $f(x) = 2\ln(1 - x)$

16. $f(x) = \frac{1}{2}\ln(x + 1)$

Find the derivative of each of the following functions.

17. $y = e^{-x}$

18. $y = e^{1-x}$

19. $y = e^{-x^2}$

20. $y = e^{x^2}$

21. $y = e^{\cos x}$

22. $y = e^{\tan x}$

23. $y = xe^{2x}$

24. $y = xe^{-x}$

25. $y = e^{-x}\sin x$

26. $y = e^{-x}\cos x$

27. $y = \frac{1}{2}(e^x - e^{-x})$

28. $y = \frac{1}{2}(e^x + e^{-x})$

29. $y = \dfrac{e^x - e^{-x}}{e^x + e^{-x}}$

30. $y = \dfrac{e^x + e^{-x}}{e^x - e^{-x}}$

31. Find x if $\displaystyle\int_1^x \frac{dt}{t} = 1$.

32. Find x if $\displaystyle\int_1^x \frac{dt}{t} = -1$.

33. Reasoning from the graph of $y = e^x$, explain why $a < b \Rightarrow e^a < e^b$. In view of the facts that $\ln 2 < 1$ and $\ln 3 > 1$ (Problems 1 and 2, Section 8.1), why does it follow that $2 < e < 3$?

34. While the graph of $f(x) = e^x$ is easily sketched as the reflection of the graph of $y = \ln x$ in the line $y = x$, it may also be discussed more directly, as follows.

   (a) Explain why the domain and range of f are $(-\infty, \infty)$ and $(0, \infty)$, respectively.

   (b) Use the derivative of e^x to show that the graph of f is always rising and concave up.

   (c) Why does the graph intersect the y axis at $(0,1)$?

   (d) Why does e^x increase without bound when x does, and approach 0 when $x \to -\infty$?

35. Discuss and sketch the graph of $f(x) = e^{-x}$, as follows.

   (a) Explain why the domain and range of f are $(-\infty, \infty)$ and $(0, \infty)$, respectively.

   (b) Use the derivative of e^{-x} to show that the graph of f is always falling and concave up.

   (c) Why does the graph intersect the y axis at $(0,1)$?

   (d) Why does e^{-x} approach 0 when $x \to \infty$ and increase without bound when $x \to -\infty$?

Use your knowledge of the graph of $y = e^x$ to sketch the graph of each of the following.

36. $y = e^{2x}$

37. $y = e^{\frac{1}{2}x}$

38. $y = e^{-x} + 1$

39. $y = -e^x$

40. $y = e^{x-1}$

41. $y = e^{1-x}$

Sketch the graph of each of the following by discussing domain, symmetry, intercepts, behavior of y for large $|x|$, asymptotes, extreme values, and concavity.

42. $y = \frac{1}{2}(e^x - e^{-x})$

43. $y = \frac{1}{2}(e^x + e^{-x})$

44. $y = \dfrac{e^x - e^{-x}}{e^x + e^{-x}}$

45. $y = \dfrac{e^x + e^{-x}}{e^x - e^{-x}}$

46. $y = \dfrac{2}{e^x + e^{-x}}$

47. $y = \dfrac{2}{e^x - e^{-x}}$

(The functions in Problems 42 through 47 are called the *hyperbolic functions*. We will discuss them in Chapter 9.)

48. $y = e^{-x^2}$ (This function is important in the theory of probability.)

49. $y = 1 - e^{-x^2}$

Sketch the graph of each of the following by discussing what seems relevant. (Certain questions will have to be left to intuition; for example, how does $y = xe^{-x}$ behave when x is large? Fill in the gaps as best you can by using other information.)

50. $y = xe^x$ **51.** $y = xe^{-x}$

52. $y = \dfrac{e^x}{x}$ **53.** $y = e^x - x$

54. $y = x \ln x$ **55.** $y = x \ln x - x$

56. $y = \dfrac{\ln x}{x}$ **57.** $y = \dfrac{x}{\ln x}$

58. Discuss the graph of $f(x) = e^{-x} \sin x$ as follows.

(a) Why does the curve lie between the graphs of $y = e^{-x}$ and $y = -e^{-x}$? Where does it intersect these graphs?

(b) Where does the curve intersect the x axis?

(c) Show that the extreme values of f occur when $\tan x = 1$.

(d) How does $f(x)$ behave for large x?

(e) Why is it appropriate to call the graph a "damped sine wave"?

59. Discuss the graph of $f(x) = e^{-x} \cos x$ as in Problem 58.

60. Show that the minimum value of $f(x) = e^x - \ln x$ occurs when $e^{-x} = x$. Then use Newton's Method to approximate this critical value of x.

61. Show that if a, b, k are positive constants, the minimum value of $f(x) = ae^{kx} + be^{-kx}$ is $2\sqrt{ab}$.

62. *Stirling's Formula* says that when n is large, the value of $n!$ ("n factorial," defined to be the product $1 \cdot 2 \cdot 3 \cdots n$) is closely approximated by the formula

$$n! \approx \sqrt{2\pi n}\left(\frac{n}{e}\right)^n$$

(a) Compute $10! = 1 \cdot 2 \cdot 3 \cdots 10$ and compare with the approximation given by Stirling's Formula. (Use a calculator!)

(b) Do the same with 50! (One fascinating aspect of this formula is that $n!$ is a function of positive integers, while its approximation involves the irrational numbers π and e.)

63. Prove the second part of Theorem 1,

$$\exp(u - v) = \frac{\exp u}{\exp v}$$

in two ways, as follows.

(a) Imitate the proof of the first part given in the text.

(b) Use the first part together with Theorem 2. (Note that Theorem 2 is proved in the text independently of Theorem 1.)

64. Give a different proof of the first part of Theorem 1 ($e^{u+v} = e^u e^v$) as follows.

(a) Let $a = e^u$ and $b = e^v$ and write logarithmic equations equivalent to these.

(b) Explain why $\ln ab = u + v$ and convert this equation to exponential form to finish the proof.

65. Show that if $a > 0$ and r is rational,

$$a^r = e^{r \ln a}$$

(In the next section we will use this formula to *define* a^r when r is irrational.)

66. What is the derivative of e^x at $x = 0$? Use the answer to prove that

$$\lim_{h \to 0} \frac{e^h - 1}{h} = 1$$

67. In the text we derived the formula $D_x e^x = e^x$ from the known derivative of $\ln$. What if the derivative of exp had come first? Show how it would follow that $D_x \ln x = 1/x$.

8.4 GENERAL EXPONENTIAL AND LOGARITHMIC FUNCTIONS

Although we now know what e^x means for all x (irrational as well as rational), the more general symbol a^x remains undefined for irrational x. The means are at hand, however, for defining it. For if x is rational, we can use Theorem 3 of Section 8.1 to write

$$a^x = e^{\ln a^x} \quad \text{(because exp and ln are inverses)}$$
$$= e^{x \ln a}$$

The expression $e^{x \ln a}$ makes sense for any x, so it is reasonable to adopt the following definition:

> Let a be any positive number. If x is irrational, we define
> $$a^x = e^{x \ln a}$$

Thus the formula $a^x = e^{x \ln a}$ is now true for all x. For the first time in this book we can make sense of arbitrary exponents and arbitrary (positive) bases.

□ **Example 1**   If we already know how to find values of ln and exp, a number like 2^π is (theoretically) a simple matter:

$$2^\pi = e^{\pi \ln 2} = e^{(3.14159\cdots)(0.69314\cdots)} = e^{2.17758\cdots} = 8.82497\cdots$$

These figures were obtained from a calculator, but our intent (for the moment) is to regard them as infinite decimals giving a unique (not merely approximate) answer. Of course many calculators have a y^x key that will give 2^π directly; in practice the formula $2^\pi = e^{\pi \ln 2}$ is unnecessary. Our point is the purely mathematical observation that the previously meaningless symbol 2^π is now adequately defined in terms of exp and ln.   □

□ **Example 2** (*Comedy Routine*)   Use our new formula for arbitrary exponentials to compute 3^2.

Solution: Using a calculator, we find

$$3^2 = e^{2 \ln 3} = e^{2(1.09861\cdots)} = e^{2.19722\cdots} = 9$$ □

□ **Example 3**   Determine which of the numbers e^π and π^e is smaller.

Solution: A calculator gives

$$e^\pi = 23.14069\cdots \qquad \text{(directly from the } e^x \text{ key)}$$
$$\pi^e = e^{e \ln \pi} = e^{(2.71828\cdots)(1.14472\cdots)} = e^{3.11169\cdots} = 22.45915\cdots$$

Hence $\pi^e < e^\pi$. It is more interesting, however, to discuss the question without depending on a calculator. We will compare the exponents in the two expressions e^π and $\pi^e = e^{e \ln \pi}$, that is, we will determine whether π or $e \ln \pi$ is smaller. To do that, look at the ratio $(\ln \pi)/\pi$. Since it is a value of the function $f(x) = (\ln x)/x$, we can say something about its size by finding the extreme values of f. Compute

$$f'(x) = \frac{(x)(1/x) - (\ln x)(1)}{x^2} = \frac{1 - \ln x}{x^2}$$

and note that $f'(x) = 0$ when $\ln x = 1$, that is, when $x = e$. Since $f'(x)$ changes from positive to negative as x increases through the critical value e (why?), we have located the global maximum of f, namely

$$f(e) = \frac{\ln e}{e} = \frac{1}{e}$$

In other words, $f(x) \leq 1/e$ for all $x > 0$ (the inequality being strict when $x \neq e$). Hence (in particular)

$$f(\pi) < \frac{1}{e}$$

that is, $(\ln \pi)/\pi < 1/e$. It follows that

$$e \ln \pi < \pi$$

and hence (applying exp to each side) $e^{e \ln \pi} < e^{\pi}$, that is, $\pi^e < e^{\pi}$.   □

It is worth observing that when $a = 1$, the formula $a^x = e^{x \ln a}$ reduces to

$$1^x = e^{x \ln 1} = e^{x \cdot 0} = e^0 = 1 \qquad \text{(for all } x\text{)}$$

Since this is merely a constant function (already classified as a polynomial in Section 1.5), we exclude it from the class of exponential functions. To be precise, we offer the following summary of exponentials developed in this section and the last.

Let a be any positive number except 1. The *exponential function of base a* is

$$a^x = e^{x \ln a} \qquad \text{(defined for all } x\text{)}$$

When $a = e$ this reduces to e^x and is called the *natural* exponential function.

When $a \neq e$ the function $a^x = e^{x \ln a}$ is much like the simpler e^x, but it has the constant factor $\ln a$ in the exponent to compensate for the "unnatural" choice of base. From that point of view the exponential functions you may have studied in precalculus mathematics (like 2^x) are, despite appearances, more complicated than the natural exponential e^x. In calculus and its applications they are seldom used.

□ **Example 4**   Find the derivative of the exponential function a^x.

Solution: Since we already know that $D_x e^x = e^x$, we need only use the Chain Rule:

$$\frac{d}{dx}(a^x) = \frac{d}{dx}(e^{x \ln a}) = e^{x \ln a} \frac{d}{dx}(x \ln a) = a^x \ln a$$

$$\boxed{D_x a^x = a^x \ln a}$$

Although we put this formula in a box (for reference purposes), it is hardly worth learning. When $a = e$ it reduces to the familiar rule $D_x e^x = e^x$; otherwise the compensating factor $\ln a$ is attached. (For example, $D_x 2^x = 2^x \ln 2$.) On those rare occasions when you need to differentiate a^x ($a \neq e$), it is probably easier to rederive the formula.   □

□ **Example 5**   Prove that the familiar power rule $D_x x^r = r x^{r-1}$ (derived in Section 3.2 for rational values of r) holds for any real number r.

Solution: Keeping in mind that r is constant (the variable being x), we have

$$D_x x^r = D_x e^{r \ln x} = e^{r \ln x} D_x(r \ln x) = x^r \cdot \frac{r}{x} = rx^{r-1}$$

Thus, for example, $D_x x^\pi = \pi x^{\pi - 1}$.   □

There is a gap in the argument in Example 5. For in writing

$$x^r \cdot \frac{r}{x} = rx^{r-1}$$

we are using a law of exponents ($a^u/a^v = a^{u-v}$) that is not yet proved for arbitrary exponentials. It is not hard to show, however, that all the usual laws are true:

$$
\begin{array}{ll}
a^u a^v = a^{u+v} & (ab)^u = a^u b^u \\[2mm]
(a^u)^v = a^{uv} & \left(\dfrac{a}{b}\right)^u = \dfrac{a^u}{b^u} \\[2mm]
\dfrac{a^u}{a^v} = a^{u-v} &
\end{array}
$$

We will prove the first one and leave the others for the problem set:

$$a^u a^v = e^{u \ln a} e^{v \ln a} = e^{u \ln a + v \ln a} \qquad \text{(Theorem 1, Section 8.3)}$$
$$= e^{(u+v) \ln a} = a^{u+v}$$

Another loose end in our discussion is that $\ln x^r = r \ln x$ (Theorem 3, Section 8.1) is still restricted to rational values of r. Now that we have defined irrational exponents, however, we can write (for arbitrary r)

$$\ln x^r = \ln (e^{r \ln x}) = r \ln x \qquad \text{(because ln and exp are inverses)}$$

This result is used in the next example.

□ **Example 6**   Find the derivative of $f(x) = x^x$ ($x > 0$).

Solution: This function is of a type not previously encountered. Unlike the power function x^r (in which the exponent is constant) and the exponential function a^x (in which the base is constant), the function x^x involves a variable in both the exponent and the base. We find its derivative by a process called **logarithmic differentiation** (first used in 1697 by Johann Bernoulli). The idea is to apply the natural logarithm before differentiating:

$$\ln f(x) = \ln x^x = x \ln x$$

This effectively removes the difficulty; we need only use familiar rules of differentiation to finish the problem:

$$\frac{1}{f(x)} \cdot f'(x) = x \cdot \frac{1}{x} + \ln x = 1 + \ln x$$

$$f'(x) = f(x)(1 + \ln x) = x^x(1 + \ln x) \qquad \square$$

□ **Example 7** Logarithmic differentiation is effective in certain problems that have no apparent connection with logarithms. To find the derivative of

$$y = \frac{x^3\sqrt{x^2 + 1}}{(x + 2)^4}$$

for example, we would normally use the Quotient Rule. As you can see by trying it, however, that promises to be gruesome. We simplify the problem considerably by taking logarithms first. (We also use absolute values to avoid difficulties with the domain. See the remark following this example.)

$$\ln |y| = \ln \left| \frac{x^3\sqrt{x^2 + 1}}{(x + 2)^4} \right| = \ln |x^3\sqrt{x^2 + 1}| - \ln |x + 2|^4$$

$$= \ln |x|^3 + \ln \sqrt{x^2 + 1} - \ln |x + 2|^4$$

$$= 3 \ln |x| + \tfrac{1}{2} \ln (x^2 + 1) - 4 \ln |x + 2|$$

This is easy to differentiate:

$$\frac{1}{y} \frac{dy}{dx} = \frac{3}{x} + \frac{x}{x^2 + 1} - \frac{4}{x + 2}$$

Multiplying each side by y, we find

$$\frac{dy}{dx} = \frac{x^3\sqrt{x^2 + 1}}{(x + 2)^4} \left(\frac{3}{x} + \frac{x}{x^2 + 1} - \frac{4}{x + 2} \right) \qquad \square$$

Remark In Example 7 note that y is negative when $x < 0$. If we did not use absolute values, the equation

$$\ln y = \ln \frac{x^3\sqrt{x^2 + 1}}{(x + 2)^4}$$

would require the restriction $x > 0$. Of course you may feel that the derivative is complicated by the introduction of absolute value, but in the case of the natural logarithm there is no problem. Since

$$D_x \ln |x| = \frac{1}{|x|} D_x|x| = \frac{1}{|x|} \cdot \frac{|x|}{x} = \frac{1}{x}$$

the functions $\ln x$ and $\ln |x|$ have the same derivative in their common domain. The difference is that $\ln |x|$ has a larger domain ($x \neq 0$ instead of $x > 0$).

Another bonus that we can collect from the exponentials defined in this section is an important expression for e as a limit.

□ **Example 8**   The derivative of $f(x) = \ln x$ is $f'(x) = 1/x$, from which $f'(1) = 1$. Using the definition of derivative, we can write this in the form

$$\lim_{h \to 0} \frac{f(1 + h) - f(1)}{h} = \lim_{h \to 0} \frac{\ln (1 + h) - \ln 1}{h}$$

$$= \lim_{h \to 0} \frac{1}{h} \ln (1 + h) = \lim_{h \to 0} \ln (1 + h)^{1/h} = 1$$

Since ln is continuous, this can be changed to

$$\ln \left[\lim_{h \to 0} (1 + h)^{1/h} \right] = 1$$

from which

$$\boxed{\lim_{h \to 0} (1 + h)^{1/h} = e}$$

Many textbooks *define* e by this limit. You might be interested in using it (with the help of a calculator) to find approximate values of e. For example, take $h = 0.00001$. Then

$$(1 + h)^{1/h} = (1.00001)^{100,000} = 2.718268\cdots$$

This is correct to four decimal places.   □

Now we turn to the last question we want to raise about exponents and logarithms. We know that exp and ln are inverse functions, that is,

$$y = e^x \Leftrightarrow x = \ln y$$

Having introduced the more general exponential function $y = a^x$, we might naturally ask, what is its inverse? The question is easily answered by using the natural logarithm to solve for x in terms of y:

$$y = a^x$$
$$\ln y = \ln a^x = x \ln a$$

Since $a \neq 1$ (we excluded that base earlier!), $\ln a \neq 0$ and hence

$$x = \frac{\ln y}{\ln a}$$

Interchanging x and y, we conclude that

$$y = \frac{\ln x}{\ln a} \Leftrightarrow x = a^y$$

We have come almost full circle from our remarks in Section 8.1 about logarithms and exponents. We said then that in precalculus mathematics the function $y = \log_a x$ is usually defined by writing

$$y = \log_a x \Leftrightarrow x = a^y$$

In view of the above conclusion about the inverse of $x = a^y$, the next definition suggests itself.

Let a be any positive number except 1. The **logarithmic function of base a** is defined by

$$\log_a x = \frac{\ln x}{\ln a} \qquad (x > 0)$$

It is the inverse of the exponential function of base a, that is,

$$y = \log_a x \Leftrightarrow x = a^y$$

Moreover, when $a = e$ it is the same as the natural logarithm:

$$\log_e x = \ln x$$

When the natural logarithm was defined in Section 8.1, we pointed out that it is "baseless." Now, however, we have identified it with the logarithmic function of base e. As in the case of e^x (which is the simplest exponential), $\ln x = \log_e x$ is the simplest logarithm. All others involve the compensating factor $\ln a$. The definition

$$\log_a x = \frac{\ln x}{\ln a}$$

is essentially a "change of base" formula enabling us to express the general logarithm in terms of the natural logarithm. If $a = 10$, for example, we have a conversion formula for the common logarithm:

$$\log x = \frac{\ln x}{\ln 10} \approx \frac{\ln x}{2.3026} \approx 0.4343 \ln x$$

□ **Example 9**   Find the derivative of the logarithmic function $\log_a x$.

Solution: Since $\log_a x = \ln x / \ln a$, we have

$$\boxed{D_x \log_a x = \frac{1}{x \ln a}}$$

As in the case of the formula $D_x a^x = a^x \ln a$, this derivative is not worth learning; we will rarely use it. Note that when $a = e$ it reduces to the familiar rule

$$D_x \ln x = \frac{1}{x} \qquad\qquad \square$$

Problem Set 8.4

Use the formula $a^x = e^{x \ln a}$ (and a calculator) to find each of the following.

1. 3^π **2.** π^π

3. $3^{\sqrt{2}}$ **4.** $(\ln 2)^e$

5. Sketch the graph of $y = 2^x$ by plotting the points corresponding to $x = -2, -1, 0, 1, 2$ and drawing a smooth curve through the points. (What properties of the exponential function justify this?) Then sketch the inverse function $y = \log_2 x$.

6. Repeat Problem 5 for $y = 3^x$.

7. Repeat Problem 5 for $y = (\frac{1}{2})^x$.

8. Since $(\frac{1}{2})^x = 2^{-x}$, the graph in Problem 7 may be obtained by reference to the graph of $y = 2^x$ in Problem 5. Explain how.

Find the derivative of each of the following functions.

9. $y = 3^x$ **10.** $y = \pi^x$

11. $y = 2^{x^2}$ **12.** $y = 5^{\sin x}$

13. $y = x^e$ **14.** $y = x^{\sqrt{2}}$

15. $y = (\sin x)^\pi$ **16.** $y = (\tan x)^{\sqrt{3}}$

17. In Example 6 we found the derivative of $f(x) = x^x$ by "logarithmic differentiation." We could have written $f(x) = e^{x \ln x}$, however, and then differentiated directly. Show that the same result is obtained.

Find the derivative of each of the following functions.

18. $y = |x|^x$ **19.** $y = x^{\sin x}$ **20.** $y = (\sin x)^x$

21. $y = x^{\ln x}$ **22.** $y = (2x)^{1/x}$

23. If $f(x) = (\sin x)^{\cos x}$, find $f'(\pi/2)$. What can you say about $f'(3\pi/2)$?

24. If $f(x) = (\tan x)^x$, find $f'(\pi/4)$. What can you say about $f'(0)$?

Use "logarithmic differentiation" to find the derivative of each of the following functions.

25. $y = x^2 \sqrt{1 - x}$ **26.** $y = x^3(2x - 1)^4(x + 5)^2$

27. $y = \dfrac{x(x^2 - 1)^3}{\sqrt{x^2 + 1}}$ **28.** $y = \dfrac{x^2 \sqrt{1 - x^2}}{(1 + x^2)^3}$

Find the derivative of each of the following functions. *Note:* The symbol $\log x$ means logarithm of x to the base 10 (the *common logarithm* of x).

29. $y = \log_2 x$ **30.** $y = \log x$

31. $y = x \log_3 x$ **32.** $y = \dfrac{\log x}{x}$

33. $y = \sin (\log_2 x)$ **34.** $y = \log (\sin x)$

35. In defining the symbol a^x for all x, why don't we allow the base to be negative? There is nothing wrong (for example) with the symbol $(-1)^2$. Why not define $(-1)^x$ for all x?

36. Explain why the definition $a^x = e^{x \ln a}$ does not apply when the base is 0.

37. Use a calculator to compute $(1 + h)^{1/h}$ when $h = 10^{-6}$ and when $h = -10^{-6}$ and compare with the value of e.

38. In Example 8 we used the fact that $D_x \ln x = 1/x$ to prove that

$$e = \lim_{h \to 0} (1 + h)^{1/h}$$

Suppose, however, that we had *defined* e to be this limit, and had followed the precalculus pattern of developing the properties of $\ln x$ as the inverse of e^x. Derive the formula $D_x \ln x = 1/x$ by using the definition of derivative,

$$D_x \ln x = \lim_{h \to 0} \frac{\ln (x + h) - \ln x}{h}$$

Hint: Fix $x > 0$ and transform the limit to

$$\lim_{t \to 0} \frac{1}{x} \ln (1 + t)^{1/t} \qquad (t = h/x)$$

Then use the continuity of $\ln$ and the above definition of e to finish.

39. Theorem 2 of Section 8.3 says that if r is rational, $(e^x)^r = e^{rx}$ for all x. The more general formula $(e^u)^v = e^{uv}$ (for all u and v) was left for this section. Prove it.

Suppose that a and b are positive. Prove the following.

40. $a^x > 0$ (for all x)

41. $a^{-x} = 1/a^x$ (for all x)

42. $(a^u)^v = a^{uv}$ (for all u and v)

43. $a^u/a^v = a^{u-v}$ (for all u and v)

44. $(ab)^x = a^x b^x$ (for all x)

45. $(a/b)^x = a^x/b^x$ (for all x)

46. Why is the symbol $\log_a x$ meaningless if $a = 1$?

Suppose that a is any positive number except 1. Prove the following.

47. $\log_a 1 = 0$ **48.** $\log_a a = 1$

49. $\log_a uv = \log_a u + \log_a v$ (for all positive u and v)

50. $\log_a (u/v) = \log_a u - \log_a v$ (for all positive u and v)

51. $\log_a u^v = v \log_a u$ (for all $u > 0$ and all v)

52. $\log_a a^x = x$ (for all x)

53. $a^{\log_a x} = x$ (for all positive x)

54. Let a and b be any positive numbers except 1. Explain why

$$(\log_a b)(\log_b a) = 1$$

55. Let a and b be any positive numbers except 1. Derive the general "change of base" formula

$$\log_b x = \frac{\log_a x}{\log_a b} \quad (x > 0)$$

Thus, for example, base three logarithms can be expressed in terms of base two logarithms by writing

$$\log_3 x = \frac{\log_2 x}{\log_2 3} \quad (x > 0)$$

8.5

INTEGRATION INVOLVING EXPONENTIAL AND LOGARITHMIC FUNCTIONS

In Section 6.4 we began a list of antidifferentiation formulas. The list was not very long, and the omissions were conspicuous:

$$\int x^n \, dx = \frac{x^{n+1}}{n+1} + C \quad (n \neq -1)$$

$$\int \sin x \, dx = -\cos x + C$$

$$\int \cos x \, dx = \sin x + C$$

$$\int \sec^2 x \, dx = \tan x + C$$

$$\int \csc^2 x \, dx = -\cot x + C$$

$$\int \sec x \tan x \, dx = \sec x + C$$

$$\int \csc x \cot x \, dx = -\csc x + C$$

Now we are in a position to fill in the gaps. Each formula we add to the list will be based on the fact that

$$D_x \ln |x| = \frac{1}{x}$$

(See the remark following Example 7 in the last section; also see Problem 40, Section 8.1.) This implies the important formula

$$\int \frac{dx}{x} = \ln |x| + C$$

which already provides an answer for the missing part of the Power Rule (when $n = 1$).

□ **Example 1**   Evaluate $\int_3^6 \dfrac{dx}{x}$.

Solution:

$$\int_3^6 \frac{dx}{x} = \ln x \Big|_3^6 \qquad \text{(absolute value omitted because } x > 0\text{)}$$

$$= \ln 6 - \ln 3 = \ln \frac{6}{3} \qquad \text{(Theorem 2, Section 8.1)}$$

$$= \ln 2 \approx 0.693 \qquad\qquad\qquad □$$

□ **Example 2**   Evaluate $\int_0^1 \dfrac{x\, dx}{x^2 - 4}$.

Solution: Make the substitution $u = x^2 - 4$, $du = 2x\, dx$. Then

$$\int_0^1 \frac{x\, dx}{x^2 - 4} = \frac{1}{2} \int_{-4}^{-3} \frac{du}{u} = \frac{1}{2} \ln |u| \Big|_{-4}^{-3} = \frac{1}{2}(\ln 3 - \ln 4)$$

$$= \frac{1}{2} \ln 0.75 \approx -0.144$$

Note that omission of absolute value in Example 2 leads to the meaningless equation

$$\frac{1}{2} \int_{-4}^{-3} \frac{du}{u} = \frac{1}{2} \ln u \Big|_{-4}^{-3} = \frac{1}{2}[\ln(-3) - \ln(-4)]$$

Some students (resourceful, to be sure!) will argue that this is all right, proposing that

$$\frac{1}{2}[\ln(-3) - \ln(-4)] = \frac{1}{2} \ln \frac{-3}{-4} = \frac{1}{2} \ln 0.75$$

Perhaps the only way to deal with this proposal is to play similar games, arguing (for example) that

$$\ln 0 = \ln (0 \cdot 2) = \ln 0 + \ln 2$$

and hence $\ln 2 = 0$. (See Problem 15, Section 8.1.)   □

□ **Example 3**   Find a formula for $\int \sec x\, dx$.

Solution: The device used here is not obvious. You should have no trouble seeing how it works, however:

$$\int \sec x\, dx = \int \frac{(\sec x + \tan x)\sec x\, dx}{\sec x + \tan x}$$

$$= \int \frac{(\sec x \tan x + \sec^2 x)\, dx}{\sec x + \tan x}$$

$$= \int \frac{du}{u} \qquad [u = \sec x + \tan x, \; du = (\sec x \tan x + \sec^2 x)\,dx]$$

$$= \ln |u| + C = \ln |\sec x + \tan x| + C \qquad \qquad \Box$$

Similar substitution techniques (we leave the details for the problem set) can be used to obtain antiderivatives of the trigonometric functions still missing from our list. For reference we give them all here:

$$\int \tan x \, dx = \ln |\sec x| + C = -\ln |\cos x| + C$$

$$\int \cot x \, dx = -\ln |\csc x| + C = \ln |\sin x| + C$$

$$\int \sec x \, dx = \ln |\sec x + \tan x| + C$$

$$\int \csc x \, dx = \ln |\csc x - \cot x| + C$$

□**Example 4**   Evaluate $\int_0^{\pi/6} \tan 2t \, dt$.

Solution: Let $u = 2t$, $du = 2 \, dt$. Then

$$\int_0^{\pi/6} \tan 2t \, dt = \frac{1}{2} \int_0^{\pi/3} \tan u \, du$$

$$= \frac{1}{2} \ln \sec u \Big|_0^{\pi/3} \qquad \text{(Why isn't absolute value needed?)}$$

$$= \frac{1}{2}(\ln 2 - \ln 1) = \frac{1}{2} \ln 2 \approx 0.347 \qquad \Box$$

Another simple (but important) integration formula can be added to our table by recalling from Section 8.3 that $D_x e^x = e^x$:

$$\int e^x \, dx = e^x + C$$

□**Example 5**   Find a formula for $\int x e^{-x^2} \, dx$.

Solution: Let $u = -x^2$, $du = -2x \, dx$. Then

$$\int x e^{-x^2} \, dx = -\tfrac{1}{2} \int e^u \, du = -\tfrac{1}{2} e^u + C = -\tfrac{1}{2} e^{-x^2} + C \qquad \Box$$

Figure 1
Infinite region under the graph of
$y = \dfrac{1}{x}$

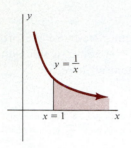

□ **Example 6** (*The Paradox of Gabriel's Horn*)   Let R be the infinite region to the right of $x = 1$ between the curve $y = 1/x$ and the x axis. (See Figure 1.) Investigate the following statements.

(a) No amount of paint is sufficient to cover R.
(b) The infinite solid of revolution generated by revolving R about the x axis can be filled with a finite amount of paint.
(c) The surface of this solid cannot be painted.

Solution:

(a) The question is whether R has a finite area. We have no way of investigating that question directly, but we *can* compute the area of the finite region bounded by the curve $y = 1/x$, the x axis, and the vertical lines $x = 1$ and $x = r \, (r > 1)$. This area is

$$A = \int_1^r y \, dx = \int_1^r \frac{dx}{x} = \ln x \Big|_1^r = \ln r$$

If the area of R exists at all, it must be the limiting value of A as the boundary $x = r$ is moved to the right. When $r \to \infty$, however, $\ln r$ increases without bound; no such limiting value exists. Hence R cannot be painted.

(b) As in part (a), consider the volume of the solid generated by rotating about the x axis the finite region below the curve $y = 1/x$, $1 \le x \le r$. This volume (Section 7.2) is

$$V = \int_1^r \pi y^2 \, dx = \pi \int_1^r x^{-2} \, dx = -\pi x^{-1} \Big|_1^r = \pi - \frac{\pi}{r}$$

When $r \to \infty$, V approaches the limiting value π, which suggests that the infinite solid of revolution generated by rotating R about the x axis has finite volume π. This solid is called "Gabriel's Horn." If you can imagine trying to fill it with paint, it appears to require only a finite amount (π cubic units). Yet its cross section in the xy plane cannot be painted!

(c) Rotate the curve $y = 1/x$, $1 \le x \le r$, about the x axis to generate a surface of revolution. The area of this surface (Section 7.4) is

$$S = \int_{x=1}^{x=r} 2\pi y \, ds = 2\pi \int_1^r y \sqrt{1 + y'^2} \, dx$$

Since

$$y' = \frac{1}{x^2} \quad \text{and} \quad 1 + y'^2 = 1 + \frac{1}{x^4} = \frac{x^4 + 1}{x^4}$$

we find

$$S = 2\pi \int_1^r \frac{\sqrt{x^4 + 1}}{x^3} \, dx$$

This integral is hard to evaluate. Observe, however, that

$$\frac{\sqrt{x^4 + 1}}{x^3} > \frac{\sqrt{x^4}}{x^3} = \frac{1}{x}$$

from which it follows that

$$S > 2\pi \int_1^r \frac{dx}{x} = 2\pi \ln r$$

Since $\ln r$ increases without bound as $r \to \infty$, S has no limit as the boundary $x = r$ is moved to the right. Hence the surface of Gabriel's Horn cannot be painted. On the other hand, wouldn't it automatically get painted (on its interior side) when the horn is filled with π cubic units of paint?

We have no resolution of this paradox to offer. It is an illustration of the difficulties encountered when we try to use intuitive geometric ideas in connection with infinite regions of the plane or in three-dimensional space.

□

Problem Set 8.5

Find a formula for each of the following antiderivatives.

1. $\int \dfrac{dx}{x + 3}$ **2.** $\int \dfrac{dx}{1 - x}$ **3.** $\int \dfrac{dx}{2x + 1}$

4. $\int \dfrac{dx}{3x - 2}$ **5.** $\int \dfrac{x\,dx}{x^2 + 1}$ **6.** $\int \dfrac{x\,dx}{1 - x^2}$

7. $\int \dfrac{x^2\,dx}{x^3 - 1}$ **8.** $\int \dfrac{x^2\,dx}{2x^3 + 1}$ **9.** $\int e^{2x}\,dx$

10. $\int e^{-x}\,dx$ **11.** $\int xe^{x^2}\,dx$ **12.** $\int xe^{1-x^2}\,dx$

13. $\int e^{\cos t} \sin t\,dt$ **14.** $\int e^{\sin 2t} \cos 2t\,dt$

15. $\int \dfrac{dx}{x \ln x}$ *Hint: Let $u = \ln x$*

16. $\int \dfrac{\ln x}{x}\,dx$ **17.** $\int \dfrac{\cos 2t\,dt}{1 + \sin 2t}$

18. $\int \dfrac{\sin t\,dt}{2 \cos t - 5}$ **19.** $\int \dfrac{\sec^2 t\,dt}{\tan t + 2}$

20. $\int \dfrac{\csc^2 3t\,dt}{1 - \cot 3t}$ **21.** $\int \dfrac{\sec(t/2)\tan(t/2)\,dt}{4 - \sec(t/2)}$

22. $\int \dfrac{\csc t \cot t\,dt}{\csc t + 1}$ **23.** $\int \tan 2t\,dt$

24. $\int \cot(t/2)\,dt$ **25.** $\int \dfrac{1 + \sin t}{\cos t}\,dt$

26. $\int \dfrac{1 - \cos t}{\sin t}\,dt$ **27.** $\int \dfrac{dt}{\sin 2t}$

28. $\int \dfrac{t\,dt}{\cos(t^2)}$

Evaluate each of the following.

29. $\int_2^6 \dfrac{du}{u}$ **30.** $\int_2^5 \dfrac{dx}{2x - 1}$

31. $\int_0^1 \dfrac{x\,dx}{x^2 + 4}$ **32.** $\int_2^3 \dfrac{x^3\,dx}{x^4 - 1}$

33. $\int_0^1 \dfrac{x^2\,dx}{x^3 - 8}$ **34.** $\int_1^2 \dfrac{x\,dx}{x^2 - 9}$

35. $\int_0^{\ln 2} e^x\,dx$ **36.** $\int_{\ln 2}^{\ln 3} e^{2x}\,dx$

37. $\int_0^2 e^{x/2}\,dx$ **38.** $\int_{-1}^0 e^{-3x}\,dx$

39. $\int_0^1 x^2 e^{x^3}\,dx$ **40.** $\int_1^2 xe^{x^2}\,dx$

41. $\int_{\pi/3}^{\pi/2} \dfrac{\sin t\,dt}{1 - \cos t}$ **42.** $\int_0^{\pi/2} \dfrac{\cos t\,dt}{\sin t + 1}$

43. $\int_0^{\pi/4} \sec t\,dt$ **44.** $\int_{\pi/2}^{5\pi/6} \csc t\,dt$

45. $\int_{\pi/12}^{\pi/4} \cot 2t\,dt$ **46.** $\int_0^{\pi/2} \tan(t/2)\,dt$

47. Find the average value of $f(x) = 1/x$ in the interval $[1,3]$.

48. What is the average value of $\tan x$ in the interval $[0, \pi/3]$?

49. Find the average value of e^x in the interval $[0,1]$.

50. Find the area under the curve

$$y = \frac{x}{x^2 - 1} \qquad 2 \le x \le 5$$

51. Find the area of the region bounded by the curves $xy = 2$ and $x + y = 3$.

52. Find the area of the region bounded by the graphs of $y = e^x$, $y = e^{-x}$, and $x = 1$.

53. The region under the curve $y = 1/\sqrt{x}$, $1 \le x \le 4$, is rotated about the x axis. Find the volume of the solid generated.

54. Find the volume of the solid generated by rotating the region under the curve

$$y = \frac{1}{1 + x^2} \qquad 0 \le x \le 1$$

about the y axis.

55. The region in the first quadrant bounded by $y = e^{-x^2}$ and $x = 1$ is rotated about the y axis. What is the volume of the solid generated?

56. A rectangle with base along the x axis has two vertices on the curve $y = e^{-x^2}$. Show that the largest area of such a rectangle occurs when these vertices are points of inflection of the curve.

57. Find the length of the curve $y = \ln \sec x$, $0 \le x \le \pi/3$.

58. Let R be the infinite region in the first quadrant under the curve $y = e^{-x}$.

(a) Can R be painted? If so, how much area must be covered?

(b) Can the infinite solid generated by revolving R about the x axis be filled with paint? If so, how many cubic units of paint are needed?

59. Derive the formulas

$$\int \tan x \, dx = \ln |\sec x| + C = -\ln |\cos x| + C$$

as follows.

(a) Make the substitution $u = \cos x$ in order to obtain $-\ln |\cos x| + C$.

(b) Write

$$\int \tan x \, dx = \int \frac{\sec x \tan x \, dx}{\sec x}$$

and let $u = \sec x$ to obtain $\ln |\sec x| + C$.

(c) Different methods of antidifferentiation yield answers that differ by a constant. What can you say about the answers obtained in parts (a) and (b)?

60. Derive the formulas

$$\int \cot x \, dx = -\ln |\csc x| + C = \ln |\sin x| + C$$

61. Derive the formula

$$\int \csc x \, dx = \ln |\csc x - \cot x| + C$$

by imitating Example 3.

8.6
APPLICATIONS OF EXPONENTIAL AND LOGARITHMIC FUNCTIONS

In this section we present a few examples of applications of exponential and logarithmic functions outside of mathematics. These examples (and similar questions in the problem set) may seem harder than they are, because of the context in which they arise. You should be impressed, however, by the power of calculus to solve them. There is some truth in the statement that we cannot claim to understand mathematics unless we are capable of applying it to real problems. These will give you an opportunity to test whether you have come to grips with exponents and logarithms.

□ **Example 1**   Suppose that a certain population grows at a rate proportional to the population. If the initial population is 100 and has grown to 5000 at the end of six months, find the population as a function of time.

Solution: We are told that $dP/dt = kP$, where P is the population at time t (in months) and k is the constant of proportionality. Writing this differential equation in the form

$$\frac{dP}{P} = k \, dt$$

we integrate each side to obtain

$$\int \frac{dP}{P} = \int k \, dt$$

that is,

$$\ln P = kt + C \quad \text{(absolute value not needed because } P > 0\text{)} \quad (1)$$

Since the population is 100 at $t = 0$, we evaluate C by writing

$$\ln 100 = k \cdot 0 + C$$

from which $C = \ln 100$. Hence Equation (1) becomes

$$\ln P = kt + \ln 100$$

$$\ln P - \ln 100 = kt$$

$$\ln \frac{P}{100} = kt \qquad (2)$$

$$\frac{P}{100} = e^{kt} \quad \text{(Why?)}$$

$$P = 100e^{kt} \qquad (3)$$

To evaluate k, use the information that $P = 5000$ when $t = 6$. The easiest place to substitute this is in Equation (2):

$$\ln \frac{5000}{100} = k \cdot 6$$

from which

$$k = \frac{1}{6} \ln 50$$

$$kt = \frac{t}{6} \ln 50 = \ln (50)^{t/6}$$

$$e^{kt} = (50)^{t/6}$$

Hence Equation (3) becomes $P = 100(50)^{t/6}$.

This formula may be used to find (for example) the population at the end of one year:

$$P = 100(50)^2 = 250,000$$

Of course a student knowing nothing of calculus might be able to predict this result. Since the population has grown by a factor of 50 in the first six months (from 100 to 5000), it seems reasonable to guess that it grows by a factor of 50 in the next six months (from 5000 to 250,000). The method used in Example 1, however, is not guesswork.   □

Radioactive decay is a process in which a substance disintegrates by conversion of its mass into radiation. The rate of decay is proportional to the mass, that is, $dm/dt = -km$, where m is the mass at time t and k is a positive constant. The *half-life* of the substance is the time required for half the original mass to disappear. If m_0 is the original mass (the value of m when $t = 0$), we may solve the above differential equation by writing

$$\frac{dm}{m} = -k\,dt$$

$$\int \frac{dm}{m} = -k \int dt$$

$$\ln m = -kt + C = -kt + \ln m_0$$

$$\ln m - \ln m_0 = -kt$$

$$\ln \frac{m}{m_0} = -kt$$

$$\frac{m}{m_0} = e^{-kt}$$

$$m = m_0 e^{-kt}$$

Our next example illustrates how these formulas for radioactive decay are used to determine the age of fossils. The method is called "carbon dating," and won a Nobel Prize for its discoverer, Willard Libby of UCLA.

□ **Example 2**   Living tissue contains two isotopes of carbon, one radioactive and the other stable (the ratio of the two being constant). When the tissue dies the stable isotope remains, but the radioactive one decays (with a half-life of about 5500 years). Use these facts to determine the age of a fossil in which the radioactive isotope has decayed to 20% of its original amount. (The percentage is determined by comparing the present ratio of isotopes in the fossil to the known ratio in living tissue.)

Solution: The fossil was once living tissue; take $t = 0$ to be the time when it died, and let m be the amount of radioactive carbon in the fossil t years later. If m_0 is the initial amount, we have $m = 0.20 m_0 = \frac{1}{5} m_0$ today. The problem is to find the corresponding value of t. From the above discussion of radioactive decay we know that

$$\ln \frac{m}{m_0} = -kt$$

Since $m = \frac{1}{2} m_0$ when $t = 5500$, we find

$$\ln \tfrac{1}{2} = -k(5500)$$

$$-\ln 2 = -5500k$$

$$k = \frac{\ln 2}{5500}$$

Hence
$$\ln \frac{m}{m_0} = \frac{-t \ln 2}{5500}$$

from which
$$t = \frac{-5500 \ln (m/m_0)}{\ln 2}$$

When $m = \frac{1}{5}m_0$ we have

$$t = \frac{-5500 \ln \frac{1}{5}}{\ln 2} = \frac{5500 \ln 5}{\ln 2} \approx 12{,}771 \text{ years} \qquad \square$$

□ **Example 3**   According to Newton's Law of Cooling, a hot object placed in surroundings at constant temperature cools at a rate proportional to the difference between the temperature of the object and the temperature of the surroundings. Suppose that an object at 120°C is brought into a room at 20°C. If it has cooled to 80°C ten minutes later, what will its temperature be at the end of an hour? What happens to its temperature as time goes on?

Solution: Newton's Law of Cooling can be written as a differential equation,

$$\frac{dT}{dt} = -k(T - 20)$$

where T is the temperature of the object t minutes after it is brought into the room. Writing this in the form

$$\frac{dT}{T - 20} = -k \, dt$$

we integrate both sides to obtain

$$\int \frac{dT}{T - 20} = -k \int dt$$

$$\ln (T - 20) = -kt + C \quad \text{(no absolute value because } T > 20) \quad (1)$$

Since $T = 120$ when $t = 0$, we find $C = \ln 100$ and Equation (1) becomes

$$\ln (T - 20) - \ln 100 = -kt$$

$$\ln \frac{T - 20}{100} = -kt \qquad (2)$$

$$\frac{T - 20}{100} = e^{-kt}$$

$$T - 20 = 100e^{-kt}$$

$$T = 20 + 100e^{-kt}$$

Since k is positive, $e^{-kt} \to 0$ as $t \to \infty$, so the temperature of the object approaches the room temperature 20°C as time goes on. Its value at the end of an hour ($t = 60$) is

$$T = 20 + 100e^{-60k}$$

so we need the value of k. We are told that $T = 80$ when $t = 10$. Substituting this information in Equation (2), we have

$$\ln 0.6 = -k(10)$$

$$k = -\frac{1}{10}\ln 0.6$$

$$-60k = 6\ln 0.6 = \ln (0.6)^6$$

$$e^{-60k} = (0.6)^6$$

Hence when $t = 60$ we find

$$T = 20 + 100(0.6)^6 \approx 24.7°\text{C} \qquad \text{(from a calculator)} \qquad \square$$

Remark   Sometimes students point out that analysis like Example 3 does not fit the facts. For the equation $T = 20 + 100e^{-kt}$ says that $T > 20$ for all t (because the exponential term is always positive). Surely the hot object eventually cools to 20°! Without arguing that point (who knows what the "facts" really are?) we agree with the complaint. Keep in mind that even the great Newton was only imitating reality with his Law of Cooling. It is a *model,* an approximation, of the real thing. No room exists in which the air is at truly constant temperature (not to mention the ventilation required to keep it that way when a hot object is brought into it).

The differential equation $dT/dt = -k(T - 20)$ is a further step away from reality (a *mathematical* model), for it assumes that Newton's statement has reference to a function of time whose derivative exists and behaves as advertised. In reality the temperature involves other variables besides time, the relations being so complicated that nobody knows what they are (nor whether they are susceptible to mathematical analysis at all). Even the notion of cause and effect (one variable depending on others) is a controversial one when philosophers talk about "reality."

Thus when a mathematical model leads to conclusions that do not seem precisely in accord with the "facts," there is no cause for alarm (unless the divergence is so wild that it suggests a poorly conceived model, misunderstanding of the facts, or both). A better model may improve matters, but nobody should expect to imitate reality in all its detail.

Mathematicians in particular are aware that what they do takes place in an imaginary world of their own, some of them even rejoicing in that fact. Others (especially students who grow impatient with theory) often accuse them of irrelevance. The anomaly that abstract ("useless") mathematics often leads to concrete ("useful") results is elegantly expressed in the following sonnet by Clarence R. Wylie, Jr. (used by permission of the author and of *Science*).

Paradox

Not truth, nor certainty. These I forswore
In my novitiate, as young men called
To holy orders must abjure the world.
"If . . . , then . . . ," this only I assert;
And my successes are but pretty chains
Linking twin doubts, for it is vain to ask
If what I postulate be justified,
Or what I prove possess the stamp of fact.

Yet bridges stand, and men no longer crawl
In two dimensions. And such triumphs stem

In no small measure from the power this game
Played with the thrice-attenuated shades
Of things, has over their originals.
How frail the wand, but how profound the spell!

□ **Example 4**   The pressure and volume of a gas confined in a cylinder at constant temperature are related by Boyle's Law, $PV = k$, where k is a constant. If the pressure is 20 lb/in² when the volume is 8 in³, find the work done by a piston that compresses the gas from 10 in³ to 5 in³.

Solution: Since $P = 20$ when $V = 8$, we find

$$k = PV = (20)(8) = 160$$

Hence the formula relating P and V is $PV = 160$. According to Problem 31, Section 7.5, the work done by a gas in expanding from V_1 in³ to V_2 in³ is

$$W = \int_{V_1}^{V_2} P \, dV$$

The work done by a piston in compressing the gas from $V_2 = 10$ to $V_1 = 5$ is numerically the same:

$$W = \int_5^{10} P \, dV = \int_5^{10} \frac{160}{V} \, dV \qquad \text{(because } PV = 160\text{)}$$

$$= 160 \ln V \Big|_5^{10} = 160(\ln 10 - \ln 5) = 160 \ln 2 \approx 110.9 \text{ (inch-pounds)} \ \square$$

Compound interest (on money deposited in a savings account, for example) is computed by periodically converting interest earned into new principal, which earns interest along with the original deposit. As an illustration, suppose that we invest P dollars in an account earning 6% interest compounded semiannually. At the end of the first six months the interest will be

$$\text{(Principal)(Interest Rate)(Time)} = P(0.06)(\tfrac{1}{2}) = 0.03P$$

This is added to the account to form the new principal

$$S_1 = P + 0.03P = P(1.03)$$

At the end of the second six-month period the interest on S_1 will be $0.03S_1$ and the new principal will be

$$S_2 = S_1 + 0.03S_1 = S_1(1.03) = P(1.03)^2$$

By continuing this process (formally, by mathematical induction) we conclude that after t years (or $2t$ six-month periods) the original investment will have grown to the amount

$$S_{2t} = P(1.03)^{2t}$$

This is an example of the *compound interest formula,* which says that an investment of P dollars grows to the amount

$$S_{nt} = P\left(1 + \frac{r}{n}\right)^{nt}$$

in t years, where r is the (annual) interest rate and n is the number of times per year the interest is compounded.

Assuming that P, r, and t are fixed, we would certainly prefer a bank that compounds interest often, that is, we would like n to be large. Suppose, for example, that $P = 1000$, $r = 0.06$, and $t = 10$. The following figures indicate the effect of n on the final amount:

$n = 2$ (semiannual): $S_{20} = 1000(1.03)^{20} = \$1806.11$
$n = 4$ (quarterly): $S_{40} = 1000(1.015)^{40} = \$1814.02$
$n = 12$ (monthly): $S_{120} = 1000(1.005)^{120} = \$1819.40$
$n = 360$ (approximately daily): $S_{3600} = 1000(1.000\overline{16})^{3600} = \$1822.03$

It is interesting to inquire what would happen if n were made to increase indefinitely, so that interest would be compounded hourly, every minute, every second, and so on. What would happen, in fact, if interest were compounded *continuously?*

□ **Example 5**   Suppose that P dollars are invested in an account earning interest at the annual rate r. Show that if interest is continuously compounded, the investment grows to the amount $S = Pe^{rt}$ in t years.

Solution: If interest is compounded n times per year, the amount after t years is

$$S_{nt} = P\left(1 + \frac{r}{n}\right)^{nt}$$

To find S (the amount when interest is continuously compounded), we evaluate the limit of S_{nt} as $n \to \infty$ (keeping t fixed). Letting $h = r/n$, and observing that $h \to 0$ when $n \to \infty$, we find the limit to be

$$S = \lim_{h \to 0} P(1 + h)^{rt/h} = P \lim_{h \to 0} [(1 + h)^{1/h}]^{rt} = P[\lim_{h \to 0} (1 + h)^{1/h}]^{rt} \qquad \text{(why?)}$$

$$= Pe^{rt} \qquad \text{(Example 8, Section 8.4)}$$

This result is remarkable both because of its simplicity and the occurrence of e. (Who would expect that number to pop up in finance theory?) Note that when 6% interest is compounded continuously for ten years, $1000 grows to the amount

$$S = 1000e^{(0.06)(10)} = 1000e^{0.6} = \$1822.12$$

(only nine cents more than when interest is compounded daily). The formula $S = Pe^{rt}$ is a good approximation to the common practice of using a daily rate.   □

Problem Set 8.6

1. If $dP/dt = 2P$ and $P = 50$ when $t = 0$, find P as a function of t. (Assume that $P > 0$ for all t.)

2. If $dm/dt = -m$ and $m = 2$ when $t = 0$, find m as a function of t. (Assume that $m > 0$ for all t.)

3. If $dT/dt = -2(T - 30)$ and $T = 90$ when $t = 0$, find T as a function of t. (Assume that $T > 30$ for all t.) What limiting value does T approach as t increases without bound?

4. If $dV/dt = 3(100 - V)$ and $V = 25$ when $t = 0$, find V as a function of t. (Assume that $V < 100$ for all t.) What limiting value does V approach as t increases without bound?

5. Suppose that $dm/dt = -km$, where $m > 0$ for all t and k is constant. If $m = 10$ when $t = 0$ and $m = 5$ when $t = 1$, find m as a function of t.

6. Suppose that $dP/dt = kP$, where $P > 0$ for all t and k is constant. If $P = 100$ when $t = 0$ and $P = 500$ when $t = 2$, find P as a function of t.

7. The number of bacteria in a certain culture increases at a rate proportional to the population. The initial number is 20,000 and this grows to 48,000 in 3 hours.

   (a) Find the population as a function of time.

   (b) What is the population after 7 hours?

   (c) How long does it take for the population to reach 1 million?

8. The population of a certain boom town grows at a rate proportional to the population. The initial population is 1000 and this grows to 15,000 in 5 years.

   (a) Find the population as a function of time.

   (b) What is the population after 3 years?

   (c) How long does it take for the population to reach 20,000?

9. A puppy is growing at a rate proportional to its weight. If it weighed 3 lb at birth and 5 lb one month later, how much will it weigh when it is 6 months old? How old will it be when it weighs 90 lb?

10. Atmospheric pressure changes with altitude at a rate proportional to the pressure (if the temperature is constant). If the pressure at sea level is 30 and it is 29 at 1000 ft, find the pressure at 1700 ft.

11. The rate at which a certain chemical dissolves in water is proportional to the amount still undissolved. If 8 grams of the chemical are placed in water and 3 grams dissolve in 5 minutes, when will the chemical be 99% dissolved?

12. An object at 200°C is placed in a room at 25°C, cooling to 150°C in 1 hr. According to Newton's Law of Cooling, what is the temperature of the object t hours after being placed in the room? How does the temperature behave as time goes on?

13. A machine costing $50,000 is assumed to depreciate at a rate proportional to the difference between its value and its scrap value of $1000.

   (a) If the machine is worth $30,000 at the end of 2 years, what is its value when it is 10 years old?

   (b) Find the value of the machine as a function of time. How does the value behave as time goes on?

14. Suppose that $10,000 is invested for 5 years at 14%. Find its value if interest is compounded (a) semiannually, (b) quarterly, (c) monthly, (d) 360 times per year, and (e) continuously. (Use a calculator!)

15. Repeat Problem 14 for $1000 invested for 6 years at 12%.

16. According to the legend, Manhattan Island was purchased from Indians for $24 in 1626. Suppose that the Indians had regarded the price as ludicrous and Peter Minuit had deposited his $24 in a bank paying 5% compounded continuously. The present-day heirs of the Indians would then own Manhattan, while Minuit's heirs would have the bank account. How much would be in the account in 1984? [The answer is in excess of one billion dollars. While it is true that interest rates have not averaged 5% historically, it is still necessary to compare the modern value of Manhattan land with the 1984 value of $24 paid in 1626 (not the original $24).]

17. An investment of P dollars grows at a rate proportional to its value. Show that this statement leads to the formula for continuous compounding of interest, $S = Pe^{rt}$ (where r is the given proportionality constant).

18. Explain why the number e may be described as the amount to which one dollar grows in one year at an interest rate of 100% compounded continuously.

19. The half-life of radium is about 1600 years. If m_0 is the amount present when $t = 0$, find a formula for the amount m after t years.

20. Forty per cent of a radioactive substance disappears in 100 years. What is the half-life of the substance?

21. Use the method of carbon dating (Example 2) to determine the age of a fossil in which the radioactive carbon has decayed to 15% of its original amount.

22. Show that the half-life of a radioactive substance is independent of the original amount present.

23. At one time Galileo thought that the velocity of a body falling from rest is proportional to the distance fallen. Why is this impossible? *Hint:* If x is the distance fallen after t seconds, then $x = 0$ when $t = 0$.

24. Galileo later speculated that the velocity of a body falling from rest is proportional to the *time* (not the distance, as in Problem 23). What law of motion does this imply?

25. The speed of transmission in a telegraph cable is directly proportional to $x^2 \ln(1/x)$, where x is the ratio of the radius of the core to the thickness of the insulation $(0 < x < 1)$. To achieve maximum speed of transmission, what value of x should be chosen?

26. A certain electric circuit has resistance R and inductance L. When the current is I_0, its source is removed from the circuit, causing the current to die out according to the law

$$I = I_0 e^{-Rt/L} \quad \text{(where } I \text{ is the current at time } t)$$

(a) What is the value of I when $t = 0$?

(b) Show that the current dies at a rate proportional to the current.

(c) Explain why the graph of I as a function of t is always concave up.

(d) How does I behave as time goes on?

(e) Sketch the graph.

27. A battery drives a current I_0 through a circuit with resistance R and inductance L. When the battery is switched out of the circuit, the current behaves according to the differential equation

$$L\frac{dI}{dt} + RI = 0$$

Find I as a function of t.

28. A battery supplying E volts is switched into a circuit with resistance R and inductance L, causing a current I that satisfies the differential equation

$$L\frac{dI}{dt} + RI = E \quad \text{(with } I = 0 \text{ when } t = 0)$$

(a) Solve for I as a function of t to obtain

$$I = \frac{E}{R}(1 - e^{-Rt/L})$$

(b) Explain why Ohm's Law, $I = E/R$, describes the "steady state" approached by I as time goes on.

(c) Sketch the graph of I as a function of t.

29. Show that if $y = Ce^{kx}$ (where C and k are constants), then $dy/dx = ky$, that is, y changes at a rate proportional to itself.

30. To prove the converse of Problem 29, suppose that $y = f(x)$ is an unknown function whose rate of change is proportional to itself, that is, $dy/dx = ky$ (where k is a constant). Show that f must have the exponential form $f(x) = Ce^{kx}$, as follows.

(a) Let $g(x) = f(x)e^{-kx}$. Show that $g'(x) = 0$.

(b) Why does it follow that $f(x) = Ce^{kx}$ for some constant C?

(c) The zero function has the property described above. Does that contradict the conclusion?

31. A more straightforward way to do Problem 30 is to write the given differential equation in the form $dy/y = k\,dx$ and integrate to find y in terms of x.

(a) What special case is overlooked by this method?

(b) Leaving aside the special case, observe that we do not know the sign of $y = f(x)$. Integration therefore requires absolute value, namely

$$\ln|y| = kx + A \quad \text{(where } A \text{ is an arbitrary constant)}$$

Show that this yields $|y| = Be^{kx}$ (where B is a positive constant) and hence $y = Ce^{kx}$ (where C is a nonzero constant).

(c) Explain how the special case we left aside can be incorporated into the formula $y = Ce^{kx}$. (Thus in the end all is well. But perhaps this method is not so straightforward as advertised! Problem 30 involves fewer difficulties.)

32. Show that $y = e^x$ and $y = xe^x$ are both solutions of the differential equation $y'' - 2y' + y = 0$.

33. Show that $y = e^x \cos x$ and $y = e^x \sin x$ are both solutions of the differential equation $y'' - 2y' + 2y = 0$.

34. Suppose that r is a (real) root of the quadratic equation $ax^2 + bx + c = 0$. Show that the function $y = e^{rx}$ is a solution of the differential equation $ay'' + by' + cy = 0$.

In each of the following, use Problem 34 to find two solutions (neither of which is a constant multiple of the other) of the given differential equation.

35. $y'' - y = 0$ **36.** $y'' - 3y' + 2y = 0$

37. $y'' - y' = 0$

38. Explain why Problem 34 yields one (but not both) of the solutions of $y'' - 2y' + y = 0$ given in Problem 32. (This happens whenever r is a double root of $ax^2 + bx + c = 0$. To obtain a second solution that is not a constant multiple of the first, we need other methods.)

39. Explain why Problem 34 does not apply at all to the equation $y'' - 2y' + 2y = 0$ in Problem 33. (However, see the next problem.)

40. A famous formula for imaginary exponents (due to Euler) says that

$$e^{ix} = \cos x + i \sin x$$

where $i = \sqrt{-1}$ is the familiar "imaginary unit."

(a) One of the roots of $x^2 - 2x + 2 = 0$ (which arises in applying Problem 34 to $y'' - 2y' + 2y = 0$) is $r = 1 + i$. Assuming that Euler's formula is consistent with the usual laws of exponents (you might be interested in the trigonometry involved in proving this!), show that

$$e^{rx} = e^x \cos x + ie^x \sin x$$

(b) A "complex-valued" function has the form

$$f(x) = g(x) + ih(x)$$

where g and h are real functions. Its derivative is defined to be

$$f'(x) = g'(x) + ih'(x)$$

Show that if f satisfies the differential equation $ay'' + by' + cy = 0$, so do g and h.

(c) Assuming that the formula $D_x e^{rx} = re^{rx}$ applies when r is complex [this can be proved using the definition of derivative in part (b)], show that if $r = 1 + i$, then e^{rx} satisfies $y'' - 2y' + 2y = 0$.

(d) Now explain where we got the solutions of $y'' - 2y' + 2y = 0$ proposed in Problem 33.

41. Put $x = \pi$ in Euler's formula (Problem 40) to obtain $e^{\pi i} + 1 = 0$. (This remarkable equation involves the five most important numbers of mathematics, namely 0, 1, π, e, and i. Euler's formula is clearly worth discussing! We'll return to it in Chapter 13.)

True–False Quiz

1. $\ln \dfrac{1}{2} = -\displaystyle\int_1^2 \dfrac{dt}{t}$.   **2.** $D_x \ln 2x = \dfrac{1}{x}$.

3. A number M exists such that $\ln x \leq M$ for all $x > 0$.

4. $\displaystyle\int_2^8 \dfrac{dx}{x} = 2 \ln 2$.

5. $\displaystyle\lim_{x \downarrow 0} (\ln \sin x + \ln \cos x - \ln x) = 0$.

6. $\displaystyle\int \dfrac{x\,dx}{x^2 + 1} = \ln \sqrt{x^2 + 1} + C$.

7. The graph of an invertible function is intersected at most once by any horizontal line.

8. If (a,b) is a point of the graph of the invertible function f, then (b,a) is a point of the graph of f^{-1}.

9. If $f(x) = x^3 + 2$, then $f^{-1}(x) = \sqrt[3]{x - 2}$.

10. If $f(x) = x^2$ ($x \geq 0$), then $D_x f^{-1}(x) = 1/(2x)$.

11. $\exp(-x) = -\exp x$ for all x.

12. $\displaystyle\int_1^{e^x} \dfrac{dt}{t} = x$.   **13.** $\displaystyle\int_0^{\ln 5} e^{2x}\,dx = 12$.

14. The graph of $y = \int_0^x e^{-t^2}\,dt$ is concave up in the first quadrant.

15. $2^\pi = e^{2 \ln \pi}$.   **16.** $D_x 2^x = 2^x$.

17. $\log e = 1/\ln 10$.

18. The graph of $y = \log_2 x$ is the reflection in the line $y = x$ of the graph of $y = 2^x$.

19. $\displaystyle\int_0^{\pi/2} \tan \dfrac{x}{2}\,dx = \ln 2$.

20. If the population of a bacteria culture increases at a rate proportional to itself, and is 1000 when $t = 0$, its size at time $t \geq 0$ is $P = 1000e^{kt}$, where k is the constant of proportionality.

Additional Problems

1. Find an approximate value of $\ln 5$ as follows.

   (a) Express $\ln 5$ as an integral.

   (b) Compute a Riemann sum approximating the integral by dividing the interval of integration into four subintervals of equal length and evaluating the integrand at the midpoint of each subinterval. Compare with the calculator value $\ln 5 \approx 1.609$.

2. Given $\ln 2 \approx 0.69$, $\ln 3 \approx 1.10$, and $\ln 5 \approx 1.61$, find an approximate value of $\ln \sqrt{75/8}$.

3. Find $D_x \ln \sqrt{x^2 - 1}$.

4. Find $f'(x)$ if $f(x) = e^{1-x^2}$.

5. If $f(x) = \ln \tan x$, show that $f'(x) = 2 \csc 2x$.

Find the critical points of each of the following functions.

6. $f(x) = \dfrac{x}{\ln x}$ 7. $f(x) = e^x \cos x$

8. $f(x) = xe^{-x^2}$ 9. $f(x) = x - \ln x$

10. Draw the graph of $y = \ln \cos x + \ln \sec x$.

11. Find the coordinates of the lowest point of the graph of $y = x \ln 2x$ and sketch the graph.

12. Explain why the graph of $y = e^{\tan x}$, $-\pi/2 < x < \pi/2$, is always rising, and sketch the graph.

13. Sketch the graph of $y = 3xe^{-x}$ after answering the following questions.

   (a) For what values of x is y positive? negative?
   (b) Where is the graph rising? falling?
   (c) Where is the graph concave up? concave down?
   (d) Make an intelligent guess about the behavior of y when $x \to \infty$. What happens when $x \to -\infty$?

14. Does the function $f(x) = \cot x$, $0 < x < \pi$, have an inverse? Explain.

15. Find $f^{-1}(x)$ if $f(x) = \sqrt{x - 1}$ and sketch the graphs of f and f^{-1} on the same coordinate plane.

In each of the following, find a formula for $f^{-1}(x)$.

16. $f(x) = \sqrt[4]{x}$

17. $f(x) = \dfrac{x + 3}{x}$

18. $f(x) = \frac{1}{2} \ln (x - 2) + 1$

19. $f(x) = 2e^{1-x}$

20. $f(x) = e^{x^3}$

21. $f(x) = 1 - \ln \sqrt{x + 2}$

22. If $f(x) = x^3 + x + 2$, find the slope of the graph of $y = f^{-1}(x)$ at the point $(2,0)$.

23. To compute $\log_2 5$ on a calculator giving natural logarithms, what formula would you use? What if the calculator gives common logarithms?

24. A bug moves from left to right along the graph of $y = (1 + x)^{1/x}$. As the bug crosses the y axis, it encounters a hole in the graph. What are the coordinates of the hole?

Use differentiation to confirm each of the following integration formulas. (These appear to come from nowhere! Later we will discuss methods for *discovering* them instead of merely checking that they are correct.)

25. $\displaystyle\int \ln x \, dx = x \ln x - x + C$

26. $\displaystyle\int \frac{dx}{\sqrt{x^2 + a^2}} = \ln |x + \sqrt{x^2 + a^2}| + C$

27. $\displaystyle\int \frac{dx}{\sqrt{x^2 - a^2}} = \ln |x + \sqrt{x^2 - a^2}| + C$

28. $\displaystyle\int \frac{dx}{a^2 - x^2} = \frac{1}{2a} \ln \left| \frac{a + x}{a - x} \right| + C$

Find each of the following antiderivatives.

29. $\displaystyle\int xe^{-x^2} \, dx$ 30. $\displaystyle\int \frac{x \, dx}{9x^2 - 4}$

31. $\displaystyle\int \frac{\sin t \, dt}{1 - \sin^2 t}$ 32. $\displaystyle\int \frac{dt}{1 - \sec^2 t}$

33. $\displaystyle\int \frac{e^{2x} \, dx}{e^{2x} + 1}$ 34. $\displaystyle\int \frac{x}{x - 2} \, dx$

35. A certain curve has slope $2xe^{-y}$ at each point (x,y) and contains the point $(2,0)$. Find an equation of the curve (giving y in terms of x).

Evaluate each of the following integrals.

36. $\displaystyle\int_0^1 \frac{x \, dx}{x^2 - 3}$ 37. $\displaystyle\int_{\pi/4}^{\pi/2} \cot x \, dx$

38. $\displaystyle\int_0^{\ln 2} e^{-3x} \, dx$ 39. $\displaystyle\int_1^4 \frac{2x + 1}{x^2 + x + 1} \, dx$

40. $\displaystyle\int_0^{\pi/6} \frac{\cos t \, dt}{\sin t - 1}$ 41. $\displaystyle\int_0^1 (e^{2x} + e^{-2x}) \, dx$

42. Find the average value of $\sec x$ in the interval $[0, \pi/4]$.

43. Find the length of the curve $y = \ln \csc x$, $\pi/4 \le x \le \pi/2$.

44. Find the area of the region bounded by the curve $y = x/(1 + x^2)$, the coordinate axes, and the line $x = 1$.

45. Find the area of the region bounded by the curves $y = x^2$, $xy = 1$, and $x = 2$.

46. The region under the curve $y = e^{-x}$, $0 \le x \le 1$, is rotated about the x axis. Find the volume generated.

47. The region under the curve $y = e^{-x^2}$, $0 \le x \le 1$, is rotated about the y axis. Find the volume generated.

48. Find $D_x x^{2x}$.

49. Find the tangent to the curve $y = (\cos x)^{\sin x}$ at $(0,1)$.

50. A biologist using the population formula $P = P_0 e^{3t}$ computes P when $t = 2$. If the measurement of t is subject to a relative error of 2%, approximately what per cent error should be expected in the computation of P?

51. The charge on a capacitor decreases at a rate proportional to the charge. If the original charge is 10 volts and is 5 volts 1 second later, find the charge as a function of time. How long does it take for the charge to decrease from 10 volts to 1 volt?

52. Use the method of carbon dating (Example 2, Section 8.6) to find the age of a fossil in which the radioactive carbon has decayed to 6% of its original amount.

53. An auto costing $7000 depreciates at a rate proportional to the difference between its value and its scrap value of $200. If it is worth $5000 1 year after purchase, what is its value 5 years after purchase?

54. For each positive integer n, let $p(n)$ be the number of primes $(2, 3, 5, 7, 11, \ldots)$ less than n. Thus $p(1) = 0$, $p(2) = 0, p(3) = 1, p(4) = 2$, and so on. The *Prime Number Theorem* says that the functions

$$p(n) \quad \text{and} \quad f(n) = \frac{n}{\ln n}$$

are "asymptotic" for large n. Compare the values of $p(50)$ and $f(50)$, using $\ln 50 \approx 3.9$.

55. Let $f(x) = \int_0^x \frac{dt}{1 + t^2}$.

(a) Explain why f has an inverse g.

(b) Show that $y = g(x)$ satisfies the differential equation $y' = 1 + y^2$ and the initial condition $g(0) = 0$. *Hint:* To find y' from $y = g(x)$, differentiate implicitly in the equivalent equation $x = f(y)$.

(c) Confirm that the function $y = \tan x$ also satisfies $y' = 1 + y^2$ and the initial condition $y = 0$ at $x = 0$.

(d) It follows from a uniqueness theorem in differential equations (Section 7.7) that g and tan are the same function in a neighborhood of 0. What is a formula for $\tan^{-1} x$ in the domain corresponding to this neighborhood?

56. Let $f(x) = \int_0^x \frac{dt}{\sqrt{1 - t^2}}$.

(a) Explain why f has an inverse g.

(b) Show that $y = g(x)$ satisfies the differential equation $y'' + y = 0$ and the initial conditions $g(0) = 0$, $g'(0) = 1$.

(c) Confirm that the function $y = \sin x$ also satisfies $y'' + y = 0$ and the initial conditions $y = 0$ and $y' = 1$ at $x = 0$.

(d) It follows from a uniqueness theorem in differential equations (Section 7.7) that g and sin are the same

function in a neighborhood of 0. What is a formula for $\sin^{-1} x$ in the domain corresponding to this neighborhood?

Problems 55 and 56 suggest a new way of defining trigonometric functions. If we had never heard of the sine, for example, it could be introduced as the inverse of f in Problem 56, in much the same way that exp is defined as the inverse of

$$F(x) = \int_1^x \frac{dt}{t}.$$

The advantages of this approach are considerable. (For example, continuity and differentiability of the sine are easy consequences of the corresponding properties of f.) Since we have already studied the trigonometric functions, however, Problems 55 and 56 provide formulas for $\tan^{-1}$ and $\sin^{-1}$ (to be introduced in Section 9.1).

57. Derive an approximation formula for

$$\ln x = \int_1^x \frac{dt}{t}$$

as follows.

(a) Make the substitution $u = t - 1$ to obtain

$$\ln x = \int_0^{x-1} \frac{du}{1 + u}.$$

(b) Divide 1 by $1 + u$ (stopping at the quadratic term in the quotient) to express the integral in part (a) in the form

$$\int_0^{x-1} \left(1 - u + u^2 - \frac{u^3}{1 + u} \right) du$$

thus obtaining

$$\ln x = (x - 1) - \frac{1}{2}(x - 1)^2 + \frac{1}{3}(x - 1)^3$$

$$- \int_0^{x-1} \frac{u^3 \, du}{1 + u}$$

(Note that the division could be continued, or stopped earlier. See the next problem.)

(c) Assuming that $x \geq 1$, explain why

$$0 \leq \int_0^{x-1} \frac{u^3 \, du}{1 + u} \leq \int_0^{x-1} u^3 \, du = \frac{1}{4}(x - 1)^4$$

and conclude that

$$\ln x \approx (x - 1) - \frac{1}{2}(x - 1)^2 + \frac{1}{3}(x - 1)^3$$

with an error that is no larger than $\frac{1}{4}(x - 1)^4$.

(d) Use this formula to find an approximate value of ln 1.1, retaining as many decimal places as justified by the error bound. Compare with the calculator value ln 1.1 ≈ 0.095310.

58. Referring to the preceding problem, assume again that $x \geq 1$.

(a) Derive the formula

$$\ln x \approx (x - 1) - \tfrac{1}{2}(x - 1)^2 + \tfrac{1}{3}(x - 1)^3 - \tfrac{1}{4}(x - 1)^4$$

with an error that is no larger than $\tfrac{1}{5}(x - 1)^5$.

(b) Use this formula to find an approximate value of ln 1.1, retaining as many decimal places as justified by the error bound. Compare with the approximation found in Problem 57, and with the calculator value given there.

(c) The formula does not apply directly to ln 0.8. (Why?) Nevertheless it can be used. Explain how, and compute ln 0.8 correct to as many decimal places as justified by the error bound. Compare with the calculator value ln 0.8 ≈ −0.22314.

59. Referring to the preceding two problems, assume again that $x \geq 1$.

(a) Explain why it is reasonable to expect that ln x is *exactly* given by the infinite series

$$\ln x = (x - 1) - \tfrac{1}{2}(x - 1)^2$$
$$+ \tfrac{1}{3}(x - 1)^3 - \tfrac{1}{4}(x - 1)^4 + \cdots$$

in the sense that it is the limit (as n increases) of approximating polynomials with n terms. (Even if this is not altogether clear, you are in a better position than the pioneers of calculus! Their understanding of infinite series, while based on superb intuition, suffered from imprecise definitions. We will discuss the subject in Chapter 12.)

(b) If this series is cut off after n terms to obtain an approximation formula, what upper bound on the error would you expect?

(c) Explain why $\ln 2 = 1 - \tfrac{1}{2} + \tfrac{1}{3} - \tfrac{1}{4} + \cdots$. (This famous formula is remarkable to contemplate! Apparently we can find ln 2 to any desired accuracy by doing nothing more complicated than adding and subtracting ordinary reciprocals.)

(d) To bring us back to reality, explain why 200,000 terms of the series in part (c) are needed to compute ln 2 to five-place accuracy. *Hint:* The error must be forced to be less than $\tfrac{1}{2} \times 10^{-5}$. (Why?)

9 | Inverse Trigonometric, Hyperbolic, and Inverse Hyperbolic Functions

[Calculus is] the outcome of a dramatic intellectual struggle which has lasted for twenty-five hundred years . . .
RICHARD COURANT

Mathematics is the subject in which we never know what we are talking about, nor whether what we are saying is true.
BERTRAND RUSSELL

Do I contradict myself?
Very well then I contradict myself.
(I am large, I contain multitudes.)
WALT WHITMAN (1819–1892)

This chapter continues the development of nonalgebraic ("transcendental") functions begun in Chapter 8. In the first half we discuss the *inverse trigonometric functions*, singling out three that are important for purposes of integration. Then we turn to certain combinations of exponentials called *hyperbolic functions*, which are remarkably analogous to the familiar trigonometric functions (and easier to discuss in some respects). They are also important in applications. Finally we derive logarithmic formulas for the *inverse hyperbolic functions*, which lead to integration formulas like those involving the inverse trigonometric functions. At that point you will have a substantial list of "standard forms" to take into the next chapter (which is devoted to techniques of integration). More important (in the long run), you will have learned all the "elementary functions of analysis," which are basic working tools of mathematics and its applications.

9.1

INVERSE TRIGONOMETRIC FUNCTIONS

We began Chapter 8 by seeking a function that would serve as an answer to the antidifferentiation problem

$$\int \frac{dx}{x} = ?$$

Since there seemed to be no other way to proceed, we simply gave a name to the function

$$F(x) = \int_1^x \frac{dt}{t}$$

(the natural logarithm) and then used it to solve our problem by writing

$$\int \frac{dx}{x} = \ln x + C \qquad (x > 0)$$

As we pointed out at the time, *any* missing antiderivative, say

$$\int f(x)\,dx = ?$$

can be supplied in this way, by defining

$$F(x) = \int_a^x f(t)\,dt$$

and observing (by the Fundamental Theorem of Calculus) that $F'(x) = f(x)$.

Ordinarily this is not a profitable thing to do. But as you have seen in Chapter 8 the natural logarithm (and its exponential inverse) have many useful properties that justify our singling them out.

In this section we are going to introduce functions that supply other important missing antiderivatives. One of them, for example, is an answer to the problem

$$\int \frac{dx}{1 + x^2} = ?$$

We could proceed as in the case of the natural logarithm by writing

$$F(x) = \int_0^x \frac{dt}{1 + t^2}$$

Then $F'(x) = 1/(1 + x^2)$ and our problem is (theoretically) solved. Give F a name, tabulate its values, study its properties (including the question of what its inverse is like), and soon it would become a familiar function in much the same way the logarithm has been added to our repertoire. (See Additional Problems 55 and 56 at the end of Chapter 8.)

The reason we do not take this route is that it is unnecessary. For it happens that F is the inverse of a function that is already adequately defined and well known, namely the tangent. It is therefore more natural (although not any easier from a theoretical point of view) to begin with the tangent and then introduce F as its inverse.

These remarks should help you understand why we now investigate what the inverse trigonometric functions are like. It is not because we have suffered an attack of renewed interest in trigonometry, but because of the important role these functions play in calculus.

We begin with the inverse sine. It may seem perverse (after this preamble) to point out that the sine does not even have an inverse! For its graph fails the horizontal line test; given a number y in the range of $y = \sin x$, there are infinitely many values of x in its domain such that $\sin x = y$. (See Figure 1.)

Figure 1
Failure of the sine to meet the horizontal line test

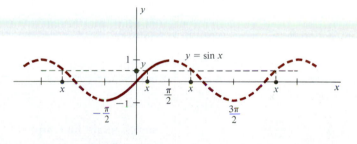

This is no problem, however. As in the case of other functions without an inverse, we simply restrict the domain in such a way that $\sin x$ takes each value in its range exactly once. Figure 1 shows that the most natural choice is the domain $[-\pi/2, \pi/2]$. The new sine function (the solid portion of the graph) does have an inverse, namely

$$x = \sin^{-1} y \qquad \text{defined by } y = \sin x, \quad -\pi/2 \le x \le \pi/2$$

As usual when dealing with an inverse function, we interchange x and y in order to discuss the new function with its variables labeled conventionally. Hence our formal definition of the inverse sine is as follows.

The **inverse sine** function is given by

$$y = \sin^{-1} x \Leftrightarrow x = \sin y \qquad -\pi/2 \le y \le \pi/2$$

It is defined for $-1 \le x \le 1$, while its range (the domain of the restricted sine) is $[-\pi/2, \pi/2]$.

Figure 2
Graph of the inverse sine

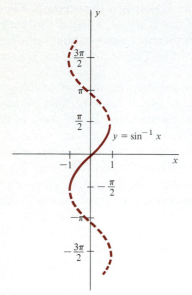

$y = \sin^{-1} x$

The graph of the inverse sine (the reflection of the restricted sine in the line $y = x$) is shown in Figure 2.

Remark   Some books use the notation arcsin x in place of $\sin^{-1} x$. The idea is that $y = $ arcsin x may be read "y is the arc whose sine is x," that is, $\sin y = x$. (This makes sense in view of the unit circle definitions of the trigonometric functions, where the input is often interpreted as an arc of the circle.) There is no harm in reading $y = \sin^{-1} x$ as "y is the *angle* whose sine is x," provided that you understand what angle is meant (and that it must be measured in radians to match the numerical output of the inverse sine). The notation and its verbal translation are not important; the essential thing is to know what the inverse sine is. Note particularly that it is not the reciprocal of sine, that is, $\sin^{-1} x \neq (\sin x)^{-1}$.

□ **Example 1**   Find each of the following.
(a) $\sin^{-1} 1$ (b) $\sin^{-1} \frac{1}{2}$ (c) $\sin^{-1}(-\sqrt{3}/2)$ (d) $\sin^{-1} 0.8$
(e) $\sin^{-1} 2$

Solution:

(a) The equation $y = \sin^{-1} 1$ is equivalent to

$$\sin y = 1 \qquad -\pi/2 \leq y \leq \pi/2$$

The only number in $[-\pi/2, \pi/2]$ whose sine is 1 is $y = \pi/2$, so $\sin^{-1} 1 = \pi/2$.

(b) $\sin^{-1} \frac{1}{2} = \pi/6$, because $\sin(\pi/6) = \frac{1}{2}$ and $\pi/6$ is between $-\pi/2$ and $\pi/2$.

(c) $\sin^{-1}(-\sqrt{3}/2) = -\pi/3$, because $\sin(-\pi/3) = -\sqrt{3}/2$ and $-\pi/3$ is between $-\pi/2$ and $\pi/2$. Note that it is incorrect to write

$$\sin^{-1}(-\sqrt{3}/2) = 4\pi/3$$

as you might be tempted to do because of your experience in trigonometry. For although it is true that $\sin 4\pi/3 = -\sqrt{3}/2$, the number $4\pi/3$ is not in the range of the inverse sine. Worse yet, do not write

$$\sin^{-1}(-\sqrt{3}/2) = 240° \text{ (or even } -60°)$$

The inverse sine (like the sine) is a function with numerical inputs and outputs; angles (except in radian measure) only muddy the water.

(d) To find $y = \sin^{-1} 0.8$ (equivalent to $\sin y = 0.8$, $-\pi/2 \leq y \leq \pi/2$), we need a table or a calculator. The latter is simplest, for it is programmed to give a numerical answer directly from the inverse sine key. (Be sure, however, to set the angle indicator on radians!) Thus $\sin^{-1} 0.8 = 0.92729 \cdots$.

(e) The equation $y = \sin^{-1} 2$ is equivalent to

$$\sin y = 2 \qquad -\pi/2 \leq y \leq \pi/2$$

Since the range of sine is $[-1,1]$, no such y exists; $\sin^{-1} 2$ is undefined.

□

□ **Example 2**   Discuss the distinction between the functions

$$f(x) = \sin(\sin^{-1} x) \qquad \text{and} \qquad g(x) = \sin^{-1}(\sin x)$$

Figure 3
Graph of $f(x) = \sin(\sin^{-1} x)$

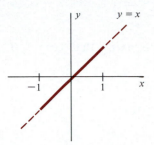

Solution: Since the sine and inverse sine are inverse functions, we know that $\sin(\sin^{-1} x) = x$ for all x in the domain of $\sin^{-1}$. (See Section 8.2.) This domain is the closed interval $[-1, 1]$, so the graph of f is as shown in Figure 3 (the solid part of the line $y = x$).

The function $g(x) = \sin^{-1}(\sin x)$, on the other hand, is defined for all x. (Why?) It is easy to make the mistake of writing $\sin^{-1}(\sin x) = x$ for all x, in which case the graph of g would be the line $y = x$. The identity holds, however, only in the domain of the restricted sine, that is,

$$\sin^{-1}(\sin x) = x \qquad \text{for } -\pi/2 \le x \le \pi/2$$

When x is outside this domain, things are not so simple. For example,

$$\sin^{-1}(\sin \pi) = \sin^{-1} 0 = 0 \qquad (\text{not } \pi)$$

You should be able to figure out that

$$\sin^{-1}(\sin x) = x \qquad \text{if } -\pi/2 \le x \le \pi/2$$
$$\sin^{-1}(\sin x) = \pi - x \qquad \text{if } \pi/2 \le x \le 3\pi/2$$
$$\sin^{-1}(\sin x) = x - 2\pi \qquad \text{if } 3\pi/2 \le x \le 5\pi/2$$

and so on. The graph is shown in Figure 4.   □

Figure 4
Graph of $y = \sin^{-1}(\sin x)$

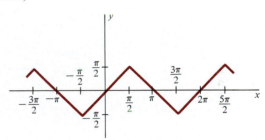

□ **Example 3**   Find the derivative of $y = \sin^{-1} x$.

Solution: Differentiate implicitly in the equivalent equation

$$\sin y = x \qquad -\pi/2 \le y \le \pi/2$$

to obtain

$$\frac{d}{dx}(\sin y) = 1$$

$$\cos y \cdot \frac{dy}{dx} = 1$$

$$\frac{dy}{dx} = \frac{1}{\cos y}$$

To express this result in terms of x (remembering that $x = \sin y$), we need a relation between $\sin y$ and $\cos y$. It is not hard to dig one up:

$$\sin^2 y + \cos^2 y = 1$$
$$\cos^2 y = 1 - \sin^2 y = 1 - x^2$$
$$\cos y = \pm\sqrt{1 - x^2}$$

The ambiguous sign can be settled by observing that $\cos y \geq 0$ when $-\pi/2 \leq y \leq \pi/2$. Hence we conclude that

$$\boxed{D_x \sin^{-1} x = \frac{1}{\sqrt{1 - x^2}}}$$

As usual, this formula should be built into the Chain Rule:

$$\frac{d}{dx}(\sin^{-1} u) = \frac{1}{\sqrt{1 - u^2}}\frac{du}{dx}$$

Thus (for example)

$$\frac{d}{dx}(\sin^{-1} x^2) = \frac{1}{\sqrt{1 - (x^2)^2}}\frac{d}{dx}(x^2) = \frac{2x}{\sqrt{1 - x^4}} \qquad\qquad \square$$

Now we turn to the inverse tangent, which is defined as follows. (See Figure 5.)

Figure 5
Graph of the restricted tangent

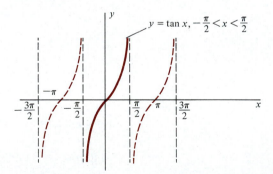

The **inverse tangent** function is given by

$$y = \tan^{-1} x \Leftrightarrow x = \tan y \qquad -\pi/2 < y < \pi/2$$

It is defined for all x (because the range of tangent is $\mathcal{R}$), while its range is the domain of the restricted tangent, namely $(-\pi/2, \pi/2)$.

Figure 6
Graph of the inverse tangent

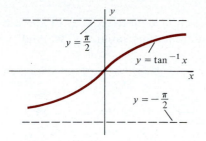

The graph of the inverse tangent (the reflection of the restricted tangent in the line $y = x$) is shown in Figure 6.

□ **Example 4**   Find the derivative of the inverse tangent.

Solution: If $y = \tan^{-1} x$, implicit differentiation in $\tan y = x$ gives

$$\sec^2 y \cdot \frac{dy}{dx} = 1$$

$$\frac{dy}{dx} = \frac{1}{\sec^2 y} = \frac{1}{1 + \tan^2 y} = \frac{1}{1 + x^2}$$

$$\boxed{D_x \tan^{-1} x = \frac{1}{1 + x^2}}$$

Using this formula with the Chain Rule, we find (for example)

$$\frac{d}{dx}(\tan^{-1} e^{-x}) = \frac{1}{1 + (e^{-x})^2} \frac{d}{dx}(e^{-x})$$

$$= \frac{-e^{-x}}{1 + e^{-2x}} = \frac{-e^x}{e^{2x} + 1}$$ □

When we come to the inverse secant, the domain to be chosen is not so apparent as in the preceding cases. Look at Figure 7 to see why. There is no single interval in which secant takes on its values exactly once; no matter how we do it, our domain is going to be in two pieces. For the moment, let's postpone a decision.

Figure 7
Graph of the restricted secant

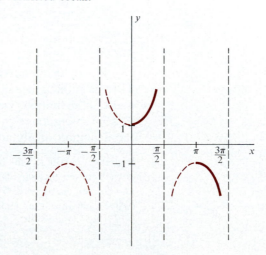

□ **Example 5**   Assuming that some decision has been made about the restricted secant, discuss the derivative of its inverse.

Solution: The equation $y = \sec^{-1} x$ is equivalent to $\sec y = x$ (where y lies in the domain not yet specified). Differentiating implicitly, we find

$$\sec y \tan y \cdot \frac{dy}{dx} = 1$$

$$\frac{dy}{dx} = \frac{1}{\sec y \tan y} = \frac{1}{x \tan y} \qquad \text{(because } \sec y = x\text{)}$$

To express this result entirely in terms of x, we use the identity $\sec^2 y - \tan^2 y = 1$:

$$\tan^2 y = \sec^2 y - 1 = x^2 - 1$$
$$\tan y = \pm \sqrt{x^2 - 1}$$

The ambiguous sign cannot be settled (like it was in Example 3) until we know the domain of the restricted secant. *We choose the domain to make the tangent nonnegative.* Certainly the interval $[0, \pi/2)$ should be part of it; the other part should be an interval in which secant takes its remaining values and tangent is never negative. We select $[\pi, 3\pi/2)$.

The **inverse secant** function is given by

$$y = \sec^{-1} x \Leftrightarrow x = \sec y \qquad 0 \leq y < \pi/2 \text{ or } \pi \leq y < 3\pi/2$$

It is defined for $x \geq 1$ or $x \leq -1$ and its range is the union of the intervals $[0, \pi/2)$ and $[\pi, 3\pi/2)$.

With this definition in hand, we can finish Example 5 by writing

$$D_x \sec^{-1} x = \frac{1}{x \sqrt{x^2 - 1}}$$ □

It may seem underhanded to fix things up (in Example 5) after the fact. There is nothing illegal, immoral, or fattening about it, however, since it is a matter of definition. We could have given the definition first (and then settled the ambiguous sign as in Example 3), but it is better to offer a reason for the definition finally adopted.

Remark   You should be aware that the domain of the restricted secant is not always chosen this way. Sometimes the interval $[\pi, 3\pi/2)$ is replaced by $[-\pi, -\pi/2)$ and sometimes by $(\pi/2, \pi]$. The former choice leads to the same derivative as above (why?), but the latter yields

$$D_x \sec^{-1} x = \frac{1}{|x| \sqrt{x^2 - 1}}$$

The absolute value is awkward (and leads to an ambiguous integration formula later), which is why we have chosen a different domain. You must read other books carefully on this point (to avoid confusion due to alternate definitions). One thing, at least, is common to every book, namely the choice of $(0,\pi/2)$ as part of the domain of each restricted trigonometric function f. Hence the evaluation of $f^{-1}(x)$ for $x > 0$ (the situation most often encountered in applications) is no problem.

The only remaining question (which you may have been wondering about) is what happened to $\cos^{-1}$, $\cot^{-1}$, and $\csc^{-1}$? The reason we have left them for last is that in calculus they are superfluous. To see why, consider (for example) the inverse cosine. Figure 8 shows that a good definition is

$$y = \cos^{-1} x \Leftrightarrow x = \cos y \qquad 0 \le y \le \pi$$

Figure 8
Graph of the restricted cosine

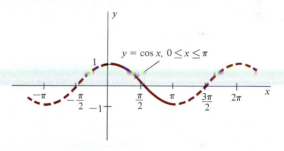

By imitating Example 3, you should be able to prove that

$$D_x \cos^{-1} x = \frac{-1}{\sqrt{1 - x^2}}$$

Since this is the negative of the derivative of $\sin^{-1} x$, it is of no interest in antidifferentiation. (See Section 9.2.) Moreover, the inverse cosine can be written in terms of the inverse sine, as the following example shows.

□**Example 6**   Explain why $\cos^{-1} x = \pi/2 - \sin^{-1} x$, $-1 \le x \le 1$.

Solution: We know that $\sin^{-1} x$ and $-\cos^{-1} x$ have the same derivative in the open interval $(-1,1)$. Hence they differ by a constant:

$$\sin^{-1} x - (-\cos^{-1} x) = C \qquad \text{(that is, } \sin^{-1} x + \cos^{-1} x = C\text{)}$$

Put $x = 0$ in this identity to find $C = \pi/2$. It follows that

$$\cos^{-1} x = \frac{\pi}{2} - \sin^{-1} x \qquad \text{for } -1 < x < 1$$

To extend the formula to the endpoints of the interval, we check $x = \pm 1$ directly:

$$\sin^{-1} 1 = \frac{\pi}{2} \quad \text{and} \quad \cos^{-1} 1 = 0$$

so $\cos^{-1} 1 = \pi/2 - \sin^{-1} 1$; and

$$\sin^{-1}(-1) = -\frac{\pi}{2} \quad \text{and} \quad \cos^{-1}(-1) = \pi$$

so $\cos^{-1}(-1) = \pi/2 - \sin^{-1}(-1)$.   □

Remark   The inverse cosine, while superfluous in calculus, is used to find the angle between two vectors. (See Section 15.1.) Hence it is worth remembering.

The inverse cotangent is defined by

$$y = \cot^{-1} x \Leftrightarrow x = \cot y \quad 0 < y < \pi$$

(Draw the graph of cotangent to see why.) Its derivative is

$$D_x \cot^{-1} x = \frac{-1}{1 + x^2}$$

which is the negative of the derivative of $\tan^{-1} x$. Moreover (as in Example 6), it can be shown that

$$\cot^{-1} x = \frac{\pi}{2} - \tan^{-1} x \quad \text{for all } x$$

It might help your understanding of this section to work through the reasons for these statements.

The inverse cosecant is hardly worth mentioning. For the record, however, we define it by

$$y = \csc^{-1} x \Leftrightarrow x = \csc y \quad 0 < y \le \pi/2 \text{ or } \pi < y \le 3\pi/2$$

Its derivative is

$$D_x \csc^{-1} x = \frac{-1}{x\sqrt{x^2 - 1}}$$

which is the negative of the derivative of $\sec^{-1} x$.

Problem Set 9.1

Find each of the following in exact form (no approximations).

1. $\sin^{-1}\frac{1}{2}$

2. $\cos^{-1}(-1)$

3. $\tan^{-1}\sqrt{3}$

4. $\cot^{-1} 0$

5. $\sec^{-1} 1$

6. $\cos^{-1} 0$

7. $\cos^{-1} 2$

8. $\sin^{-1}(-\frac{1}{2})$

9. $\cot^{-1}(-\sqrt{3})$

10. $\sec^{-1}\sqrt{2}$

11. $\tan^{-1}(-1/\sqrt{3})$

12. $\sec^{-1}\frac{1}{2}$

13. $\tan(\tan^{-1} 2)$

14. $\cos^{-1}(\cos 3\pi/2)$

15. $\cos(\sec^{-1} 3)$

16. $\sin(2 \tan^{-1} 3)$ *Hint:* Let $t = \tan^{-1} 3$ and use a formula for $\sin 2t$.

17. $\cos(\sin^{-1}\frac{3}{5} + \cos^{-1}\frac{5}{13})$ *Hint:* Use a formula for $\cos(u + v)$.

18. $\tan(\frac{1}{2}\sin^{-1}\frac{3}{5})$

Find the derivative of each of the following functions.

19. $y = \sin^{-1}(x/2)$ **20.** $y = \tan^{-1} 2x$

21. $y = \cos^{-1}(1/x^2)$ **22.** $y = \sec^{-1} x^2$

23. $y = \cot^{-1}(1 - x)$ **24.** $y = \tan^{-1} \sqrt{x}$

25. $y = \sin^{-1} x + \sqrt{1 - x^2}$

26. $y = x^2 \tan^{-1} x$

27. $y = x \sin^{-1} x + \sqrt{1 - x^2}$

28. $y = \sin^{-1} x + x\sqrt{1 - x^2}$

29. $y = \tan^{-1}\left(\dfrac{x - 1}{x + 1}\right)$ **30.** $y = \sin^{-1}\left(\dfrac{1 - x}{1 + x}\right)$

31. $y = \cot^{-1}(\tan x)$ **32.** $y = \sec^{-1}(\csc x)$

33. $y = \sin^{-1}(\cos x)$ **34.** $y = \cos^{-1}(\sin x)$

35. $y = \tan^{-1} x + \frac{1}{2} \ln(1 + x^2)$

36. $y = x \tan^{-1} x - \frac{1}{2} \ln(1 + x^2)$

37. Find dy/dx from the relation

$$\tan^{-1}\frac{y}{x} = \ln \sqrt{x^2 + y^2}$$

38. Some students expect the inverse trigonometric functions to satisfy identities analogous to familiar trigonometric formulas, for example,

$$\tan^{-1} x = \frac{\sin^{-1} x}{\cos^{-1} x}$$

Give a numerical example showing this formula to be false.

39. Give a numerical example disproving the formula $\sin^{-1} x = (\sin x)^{-1}$.

40. The formula $\cos^{-1} x = \pi/2 - \sin^{-1} x$ $(-1 \le x \le 1)$ was derived in Example 6 by using calculus. Prove it directly from the definitions, as follows.

  (a) Let $y = \sin^{-1} x$. Explain why $x = \cos(\pi/2 - y)$.
  Hint: Recall the cofunction identities from trigonometry.

  (b) In view of the fact that $-\pi/2 \le y \le \pi/2$, explain why the equation $x = \cos(\pi/2 - y)$ is equivalent to $\cos^{-1} x = \pi/2 - y$.

41. Show that $\cot^{-1} x = \pi/2 - \tan^{-1} x$ for all x.

42. Explain why the graphs of $y = \sin(\sin^{-1} x)$ and $y = \cos(\cos^{-1} x)$ are identical segments of the line $y = x$.

43. Confirm that

$$D_x \sin^{-1}(\sin x) = \frac{\cos x}{|\cos x|}$$

and use the result to check the graph of $y = \sin^{-1}(\sin x)$ in Figure 4.

44. Confirm that

$$D_x \cos^{-1}(\cos x) = \frac{\sin x}{|\sin x|}$$

and use the result to help sketch the graph of $y = \cos^{-1}(\cos x)$.

45. Confirm that $D_x \tan^{-1}(\cot x) = -1$ and use the result to help sketch the graph of $y = \tan^{-1}(\cot x)$. Watch the domain!

46. Sketch the graph of the inverse cosine.

47. Sketch the graph of the cotangent and explain why it is natural to restrict its domain to $(0, \pi)$ in order to guarantee an inverse.

48. Sketch the graph of the inverse cotangent.

49. Use the formula

$$\frac{dy}{dx} = \frac{1}{dx/dy} \qquad \text{(Section 8.2)}$$

to find the derivative of $y = \sin^{-1} x$, as follows.

  (a) Explain why

$$\frac{dy}{dx} = \frac{1}{\cos y} = \frac{1}{\cos(\sin^{-1} x)}$$

  (b) Show that $\cos(\sin^{-1} x) = \sqrt{1 - x^2}$.

50. Use the method of Problem 49 to find the derivative of $y = \tan^{-1} x$.

51. Use differentiation to show that

$$\sin^{-1}\frac{x}{\sqrt{1 + x^2}} = \tan^{-1} x \qquad \text{for all } x$$

52. Use differentiation to show that

$$\tan^{-1}\frac{x}{\sqrt{1 - x^2}} = \sin^{-1} x \qquad \text{if } -1 < x < 1$$

53. Verify the formula

$$\int \sqrt{1 - x^2}\, dx = \frac{1}{2} x \sqrt{1 - x^2} + \frac{1}{2} \sin^{-1} x + C$$

54. Derive the formula $D_x \cos^{-1} x = -1/\sqrt{1 - x^2}$.

55. Derive the formula $D_x \cot^{-1} x = -1/(1 + x^2)$.

56. Since tangent and cotangent are reciprocals, it seems reasonable to expect that $\cot^{-1}(1/x) = \tan^{-1} x$. Investigate the validity of this formula as follows.

  (a) Confirm that $D_x \cot^{-1}(1/x) = D_x \tan^{-1} x$ for $x \ne 0$.

  (b) It would seem to follow that $\cot^{-1}(1/x) = \tan^{-1} x + C$. Put $x = 1$ to obtain $C = 0$, which apparently establishes the desired formula.

  (c) Put $x = -1$ in the formula $\cot^{-1}(1/x) = \tan^{-1} x$ to show that it is false. (!)

(d) Look up Theorem 2, Section 4.4, to discover what went wrong. When is the formula $\cot^{-1}(1/x) = \tan^{-1} x$ correct?

(e) Explain why $\cot^{-1}(1/x) = \pi + \tan^{-1} x$ if $x < 0$.

57. Use differentiation to prove that

$$\sec^{-1}\frac{1}{x} = \begin{cases} \cos^{-1} x & \text{if } 0 < x \leq 1 \\ 2\pi - \cos^{-1} x & \text{if } -1 \leq x < 0 \end{cases}$$

58. The function $y = \cos^{-1} x - \sec^{-1} x$ has the same value for all x in its domain. But its derivative, far from being zero, does not exist for any value of x. Explain.

59. We suggested in Problems 55 and 56 at the end of Chapter 8 that the inverse sine and inverse tangent could have been defined as integrals,

$$\sin^{-1} x = \int_0^x \frac{dt}{\sqrt{1 - t^2}} \qquad \text{and} \qquad \tan^{-1} x = \int_0^x \frac{dt}{1 + t^2}$$

Having given different definitions, however, we must regard these formulas as unproved. Why are they true?

9.2

INTEGRATION INVOLVING INVERSE TRIGONOMETRIC FUNCTIONS

The most important fact about the inverse trigonometric functions is that they supply powerful new integration techniques. For example, we know that

$$D_x \sin^{-1} x = \frac{1}{\sqrt{1 - x^2}}$$

from which it follows that

$$\int \frac{dx}{\sqrt{1 - x^2}} = \sin^{-1} x + C$$

A slightly broader version of this formula is more useful (and is the one you should learn):

$$\boxed{\int \frac{dx}{\sqrt{a^2 - x^2}} = \sin^{-1}\frac{x}{a} + C}$$

(In all integration formulas of this type, involving a constant a^2, we assume that $a > 0$. Otherwise it is sometimes necessary to write $|a|$, which is annoying.)

The above formula can be confirmed by differentiation, but only if the answer is known in advance. To prove it directly, write

$$\sqrt{a^2 - x^2} = \sqrt{a^2(1 - x^2/a^2)} = a\sqrt{1 - x^2/a^2} \qquad \text{(because } a > 0\text{)}$$

Hence

$$\int \frac{dx}{\sqrt{a^2 - x^2}} = \int \frac{dx}{a\sqrt{1 - x^2/a^2}} = \int \frac{du}{\sqrt{1 - u^2}} \qquad \left[u = \frac{x}{a}, \ du = \frac{1}{a}dx \right]$$

$$= \sin^{-1} u + C = \sin^{-1}\frac{x}{a} + C$$

□ **Example 1**   Compute the value of $\displaystyle\int_0^1 \frac{dx}{\sqrt{9 - 4x^2}}$.

Solution: Let $u = 2x$, $du = 2\,dx$. Then

$$\int_0^1 \frac{dx}{\sqrt{9 - 4x^2}} = \frac{1}{2}\int_0^2 \frac{du}{\sqrt{9 - u^2}} = \frac{1}{2}\sin^{-1}\frac{u}{3}\Big|_0^2$$

$$= \frac{1}{2}\left(\sin^{-1}\frac{2}{3} - \sin^{-1}0\right)$$

$$= \frac{1}{2}\sin^{-1}\frac{2}{3} \approx 0.365 \qquad \text{(from a calculator)} \qquad □$$

No worthwhile integration formula is associated with the inverse cosine. For although the equation

$$D_x \cos^{-1} x = \frac{-1}{\sqrt{1 - x^2}}$$

implies that

$$\int \frac{dx}{\sqrt{1 - x^2}} = -\cos^{-1} x$$

we already know that

$$\int \frac{dx}{\sqrt{1 - x^2}} = \sin^{-1} x$$

(with an arbitrary constant added in each case). Hence we only complicate life by adding a new formula to our table of integrals. The derivative of the inverse tangent, on the other hand,

$$D_x \tan^{-1} x = \frac{1}{1 + x^2}$$

yields the formula

$$\int \frac{dx}{1 + x^2} = \tan^{-1} x + C$$

More generally (as you may confirm)

$$\boxed{\int \frac{dx}{a^2 + x^2} = \frac{1}{a}\tan^{-1}\frac{x}{a} + C}$$

□ **Example 2**   Find the area under the curve $y = 1/(x^2 + 4)$, $-2 \le x \le 2$.

Solution: The area is

$$\int_{-2}^{2} \frac{dx}{x^2 + 4} = 2 \int_{0}^{2} \frac{dx}{x^2 + 4} \qquad \text{(why?)}$$

$$= 2 \cdot \frac{1}{2} \tan^{-1} \frac{x}{2} \Big|_{0}^{2} = \tan^{-1} 1 - \tan^{-1} 0 = \frac{\pi}{4} \qquad \square$$

The inverse cotangent, like the inverse cosine, is not useful for integration (because its derivative is merely the negative of the derivative of the inverse tangent). Hence we turn to the inverse secant, with derivative

$$D_x \sec^{-1} x = \frac{1}{x\sqrt{x^2 - 1}}$$

The corresponding integral formula is

$$\int \frac{dx}{x\sqrt{x^2 - 1}} = \sec^{-1} x + C$$

or (more generally)

$$\boxed{\int \frac{dx}{x\sqrt{x^2 - a^2}} = \frac{1}{a} \sec^{-1} \frac{x}{a} + C}$$

□ **Example 3**   Evaluate $\displaystyle\int_{1}^{2} \frac{dx}{x\sqrt{16x^2 - 5}}$.

Solution: Let $u = 4x$, $du = 4\, dx$. Then

$$\int_{1}^{2} \frac{dx}{x\sqrt{16x^2 - 5}} = \int_{1}^{2} \frac{4\, dx}{4x\sqrt{16x^2 - 5}} = \int_{4}^{8} \frac{du}{u\sqrt{u^2 - 5}} = \frac{1}{\sqrt{5}} \sec^{-1} \frac{u}{\sqrt{5}} \Big|_{4}^{8}$$

$$= \frac{1}{\sqrt{5}} \left(\sec^{-1} \frac{8}{\sqrt{5}} - \sec^{-1} \frac{4}{\sqrt{5}} \right)$$

$$= \frac{1}{\sqrt{5}} \left(\cos^{-1} \frac{\sqrt{5}}{8} - \cos^{-1} \frac{\sqrt{5}}{4} \right) \qquad \text{(why?)}$$

$$\approx 0.14 \qquad \text{(from a calculator)} \qquad \square$$

Remark   We changed from $\sec^{-1}$ to $\cos^{-1}$ in Example 3 because most calculators do not have an inverse secant key. You should convince yourself that $\sec^{-1}(1/x) = \cos^{-1} x$ when $0 < x \leq 1$, while noting that when $-1 \leq x < 0$ the formula is false. (See Problem 57, Section 9.1.)

Our last example shows the usefulness of an inverse trigonometric function in an impressive application.

□ **Example 4**   Hooke's Law says that the restoring force exerted by a spring displaced x units from its natural position is $F = -kx$ ($k > 0$). Newton's Second Law (force equals mass times acceleration) converts this equation to the form $m(d^2x/dt^2) = -kx$, where t is time. Thus the motion of the spring (with displacement x at time t) is described by the second-order differential equation

$$\frac{d^2x}{dt^2} = -a^2x \qquad (a = \sqrt{k/m}\,)$$

Suppose that the motion starts with $x = 0$ and $v = dx/dt = a$ at $t = 0$. (The spring moves from its natural, unstretched position with initial velocity a.) What is the law of motion?

Solution: Reduce the equation to first-order by writing $dv/dt = -a^2x$. The awkward presence of t as the independent variable (when the right side involves x) can be cured by using a device due to Newton:

$$\frac{dv}{dt} = \frac{dv}{dx}\frac{dx}{dt} = v\frac{dv}{dx} \qquad \text{(by the Chain Rule)}$$

Hence our equation becomes $v\,(dv/dx) = -a^2x$. Separate the variables by writing $v\,dv = -a^2x\,dx$ and integrate:

$$\frac{v^2}{2} = -\frac{a^2x^2}{2} + C_1$$

Since $v = a$ when $x = 0$, we find $C_1 = a^2/2$ and hence $v^2 = a^2(1 - x^2)$. The spring starts out with a positive velocity (which persists during an interval after $t = 0$), so we choose the positive square root when solving for v:

$$v = a\sqrt{1 - x^2}, \text{ that is, } \frac{dx}{dt} = a\sqrt{1 - x^2}$$

Now separate the variables again and integrate:

$$\frac{dx}{\sqrt{1 - x^2}} = a\,dt$$

$$\int \frac{dx}{\sqrt{1 - x^2}} = \int a\,dt$$

$$\sin^{-1} x = at + C_2$$

Since $x = 0$ when $t = 0$, we find $C_2 = 0$ and hence

$$\sin^{-1} x = at$$
$$x = \sin at$$

This is the law of motion we mentioned in Section 7.7 (with the remark that it is not easy to explain). Note the role of the inverse sine in finding it.

□

Remark   In Chapter 20 we will explain why the general solution of the differential equation

$$\frac{d^2x}{dt^2} + a^2x = 0$$

is $x = A \cos at + B \sin at$. (See Problem 39, Section 7.7, where the same statement is made.) If we assume that fact, Example 4 requires considerably less effort. For then we can differentiate the general solution to obtain

$$\frac{dx}{dt} = -aA \sin at + aB \cos at$$

and the initial conditions $x = 0$ and $dx/dt = a$ at $t = 0$ yield

$$0 = A \cos 0 + B \sin 0 \qquad \text{and} \qquad a = -aA \sin 0 + aB \cos 0$$

It follows that $A = 0$ and $B = 1$, so the solution of Example 4 is (as before) $x = \sin at$.

Problem Set 9.2

Find a formula for each of the following.

1. $\displaystyle\int \frac{dx}{\sqrt{25 - x^2}}$

2. $\displaystyle\int \frac{dx}{\sqrt{1 - 9x^2}}$

3. $\displaystyle\int \frac{x\,dx}{\sqrt{9 - x^4}}$ *Hint:* Let $u = x^2$.

4. $\displaystyle\int \frac{dx}{x^2 + 12}$

5. $\displaystyle\int \frac{dx}{9 + 16x^2}$

6. $\displaystyle\int \frac{e^x\,dx}{1 + e^{2x}}$

7. $\displaystyle\int \frac{dx}{x\sqrt{9x^2 - 1}}$

8. $\displaystyle\int \frac{dx}{x\sqrt{16x^2 - 9}}$

9. $\displaystyle\int \frac{dx}{x^2 - 2x + 5}$ *Hint:* Complete the square to write $x^2 - 2x + 5 = (x - 1)^2 + 4$.

10. $\displaystyle\int \frac{dx}{\sqrt{6x - x^2}}$ *Hint:* Complete the square.

Evaluate each of the following.

11. $\displaystyle\int_{-2}^{2} \frac{dx}{\sqrt{16 - x^2}}$

12. $\displaystyle\int_{0}^{1} \frac{x\,dx}{\sqrt{4 - x^4}}$

13. $\displaystyle\int_{0}^{\pi/2} \frac{\sin x\,dx}{\sqrt{4 - \cos^2 x}}$

14. $\displaystyle\int_{1}^{3} \frac{dx}{x^2 + 3}$

15. $\displaystyle\int_{0}^{1} \frac{x + 1}{x^2 + 1}\,dx$ *Hint:* Express the integrand as a sum.

16. $\displaystyle\int_{0}^{\pi/2} \frac{\cos x\,dx}{9 + \sin^2 x}$

17. $\displaystyle\int_{3}^{4} \frac{dx}{x\sqrt{x^2 - 4}}$

18. $\displaystyle\int_{2}^{5} \frac{dx}{x\sqrt{9x^2 - 16}}$

19. $\displaystyle\int_{3}^{5} \frac{dx}{x^2 - 6x + 13}$

20. $\displaystyle\int_{0}^{1} \frac{dx}{\sqrt{3 + 2x - x^2}}$

21. Find the area of the region bounded by the graphs of $y = 1/\sqrt{1 - x^2}$ and $y = 2$.

22. Find the area under the curve $y = 9/(9 + x^2)$, $-3 \le x \le 3$.

23. Find the volume of the solid generated by rotating the region under the curve

$$y = \frac{1}{\sqrt{9 + x^2}} \qquad 0 \le x \le 3$$

about the x axis.

24. Repeat Problem 23 for rotation about the y axis.

25. Use calculus to find the length of the curve

$$y = \sqrt{r^2 - x^2} \qquad -r \le x \le r$$

How could the result have been predicted?

26. Solve the initial value problem $dy/dx = 16 + y^2$, where $y = 0$ when $x = 0$.

27. The motion of a spring with displacement x at time t is described by $d^2x/dt^2 = -a^2x$. If the spring is released from a stretched position of $x = 1$ at $t = 0$, find its law of motion. *Hint:* "Released" means that the velocity is zero at $t = 0$. Immediately thereafter it is negative (which will tell you which square root to choose when the time comes).

28. The beacon of a lighthouse 1 km from a straight shore revolves five times per minute. Find the speed of its beam along the shore in two ways, as follows. (See Figure 1.)

(a) Use the relation $x = \tan \theta$ to show that

$$\frac{dx}{dt} = 600\pi \sec^2 \theta \text{ km/hr}$$

(Compare with Problem 23, Section 5.1.)

(b) Use the relation $\theta = \tan^{-1} x$ to show that

$$\frac{dx}{dt} = 600\pi(1 + x^2) \text{ km/hr}$$

(c) Reconcile the answers. (Note that the use of an inverse trigonometric function offers no advantage, although it is a refreshing change. Most applications involving trigonometry can be treated either way.)

Figure 1
Lighthouse beam moving along a shore

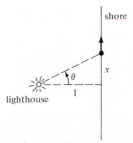

29. A camera located 10 meters from the finish line is televising a race. (See Figure 2.) When the runners are 10 meters from the finish line they are going 9 m/sec. How fast is the camera turning at that instant?

30. A painting 5 ft high is hung on a wall so that its lower edge is 1 ft above eye level. How far from the wall should an observer stand for the best view? ("Best view" means maximum angle between the lines of sight to the top and bottom of the painting.) *Hint:* If θ is the angle and x is the distance from the wall, then

$$\theta = \cot^{-1} \frac{x}{6} - \cot^{-1} x \quad \text{(Draw a picture!)}$$

Figure 2
Televising a race

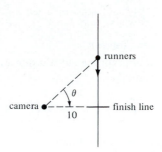

31. A diver is descending vertically from the center of a hemispherical tank (Figure 3) at the rate of 2 ft/sec. A light at the edge of the tank throws her shadow on the curved surface of the tank. How fast is her shadow moving along the tank when she is halfway down? *Hint:* First verify that $\alpha = \tan^{-1}(h/r) = \theta/2$. Also note (from trigonometry) that $s = r\theta$.

Figure 3
Shadow moving along a swimming tank

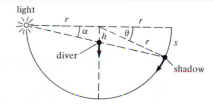

32. Derive the formula

$$\int \frac{dx}{\sqrt{a^2 - x^2}} = \sin^{-1} \frac{x}{a} + C$$

by differentiating the right side. (Note the role of the assumption $a > 0$.)

33. Derive the formula

$$\int \frac{dx}{a^2 + x^2} = \frac{1}{a} \tan^{-1} \frac{x}{a} + C$$

in two ways, as follows.

(a) Make an appropriate substitution to reduce it to the known formula

$$\int \frac{dx}{1 + x^2} = \tan^{-1} x + C$$

(b) Confirm it by differentiation.

34. Derive the formula

$$\int \frac{dx}{x\sqrt{x^2 - a^2}} = \frac{1}{a}\sec^{-1}\frac{x}{a} + C$$

in two ways, as follows.

(a) Make an appropriate substitution to reduce it to the known formula

$$\int \frac{dx}{x\sqrt{x^2 - 1}} = \sec^{-1}x + C$$

(b) Confirm it by differentiation.

Use differentiation to confirm each of the following integration formulas. (Later we will show how to discover them, rather than merely checking them.)

35. $\int \sin^{-1}x \, dx = x\sin^{-1}x + \sqrt{1 - x^2} + C$

36. $\int \tan^{-1}x \, dx = x\tan^{-1}x - \ln\sqrt{1 + x^2} + C$

37. $\int \sec^{-1}x \, dx = x\sec^{-1}x - \ln|x + \sqrt{x^2 - 1}| + C$

38. $\int \sqrt{a^2 - x^2}\, dx = \frac{x}{2}\sqrt{a^2 - x^2} + \frac{a^2}{2}\sin^{-1}\frac{x}{a} + C$

39. In previous chapters we have often computed

$$\int_{-a}^{a}\sqrt{a^2 - x^2}\,dx = \frac{\pi a^2}{2}$$

by using the area interpretation of integral. Confirm this result by using Problem 38.

40. The integrand of

$$\int_{0}^{2}\frac{dx}{\sqrt{4 - x^2}}$$

is unbounded in the domain of integration, so the integral does not exist. (See Section 6.2.) We can, however, compute

$$\int_{0}^{t}\frac{dx}{\sqrt{4 - x^2}}$$

for values of t close to 2 (and less than 2). What is the limit of the result as $t \to 2$? (The original integral may be assigned this value; we will discuss such "improper integrals" later. Note that in this example it would be easy to overlook the difficulty and compute the integral directly in terms of the inverse sine. That does not always work, however.)

41. What does the area under the curve

$$y = \frac{1}{1 + x^2} \qquad 0 \le x \le b$$

approach as b increases without bound?

9.3
HYPERBOLIC
FUNCTIONS

As you know from trigonometry, the coordinates of any point (x, y) on the unit circle can be written in the form $x = \cos t$, $y = \sin t$, where t is the measure of the arc from $(1,0)$ to (x, y). (See Figure 1.) Since $t = \theta$ (in radians) and since the area of a circular sector of radius r and central angle θ is $\frac{1}{2}r^2\theta$, we may also interpret t as twice the area of the shaded region in Figure 1:

$$2(\text{area of sector}) = 2 \cdot \tfrac{1}{2} \cdot 1^2\theta = t$$

This fact leads directly to the subject of this section. For suppose we try the same idea in connection with the unit hyperbola (the curve $x^2 - y^2 = 1$ in Figure 2). Letting t be twice the area of the shaded region (drawn by analogy to Figure 1), we propose to find x and y in terms of t. The resulting functions are called the *hyperbolic* cosine and sine, respectively; their similarity to the *circular* functions $x = \cos t$ and $y = \sin t$ is striking. Moreover,

Figure 1
Unit circle definition of sine and cosine

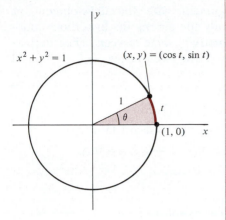

they turn out to be useful in unexpected ways (which is why they are included in this chapter).

Anticipating the solution (which is given in an optional note at the end of this section), we assert that

$$x = \tfrac{1}{2}(e^t + e^{-t}) \quad \text{and} \quad y = \tfrac{1}{2}(e^t - e^{-t})$$

These remarkable formulas (one wonders if there are similar exponential expressions for $x = \cos t$ and $y = \sin t$!) provide the starting point of our discussion.

The Hyperbolic Functions

The **hyperbolic** sine, cosine, tangent, . . . are defined by

$$\sinh t = \tfrac{1}{2}(e^t - e^{-t}) \qquad\qquad \coth t = \frac{\cosh t}{\sinh t} = \frac{e^t + e^{-t}}{e^t - e^{-t}}, \; t \neq 0$$

$$\cosh t = \tfrac{1}{2}(e^t + e^{-t}) \qquad\qquad \operatorname{sech} t = \frac{1}{\cosh t} = \frac{2}{e^t + e^{-t}}$$

$$\tanh t = \frac{\sinh t}{\cosh t} = \frac{e^t - e^{-t}}{e^t + e^{-t}} \qquad\qquad \operatorname{csch} t = \frac{1}{\sinh t} = \frac{2}{e^t - e^{-t}}, \; t \neq 0$$

The fundamental identity for hyperbolic functions (like $\cos^2 t + \sin^2 t = 1$ in trigonometry) is

$$\boxed{\cosh^2 t - \sinh^2 t = 1}$$

This follows from the fact that the point $(x, y) = (\cosh t, \sinh t)$ lies on the unit hyperbola $x^2 - y^2 = 1$. Other identities can be derived from this one, such as

$$\tanh^2 t + \operatorname{sech}^2 t = 1 \quad \text{and} \quad \coth^2 t - \operatorname{csch}^2 t = 1$$

but they are easily confused with similar formulas in trigonometry; we don't recommend your trying to learn them. The points of similarity and difference between the circular and hyperbolic functions are so unpredictable that you should take nothing for granted. For example, hyperbolic functions are even and odd in the same pattern as the trigonometric functions:

$$\sinh(-t) = -\sinh t \qquad \cosh(-t) = \cosh t \qquad \tanh(-t) = -\tanh t$$

and so on. (Why?) On the other hand, none of the hyperbolic functions is periodic, as you can see from their definition.

Figure 2
Unit hyperbola definition of hyperbolic sine and cosine

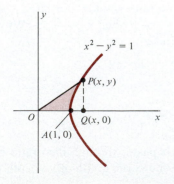

□ **Example 1**   Find the derivative of $y = \sinh x$.

Solution:

$$D_x \sinh x = \tfrac{1}{2} D_x (e^x - e^{-x}) = \tfrac{1}{2}(e^x + e^{-x}) = \cosh x$$

Compare the simplicity of this argument with the development of $D_x \sin x = \cos x$ in Section 2.3! Evidently life among the hyperbolic functions is going to be easier than trigonometry in some respects. Their derivatives (which are left for the problem set) are given below. □

$$D_x \sinh x = \cosh x \qquad\qquad D_x \coth x = -\operatorname{csch}^2 x$$
$$D_x \cosh x = \sinh x \qquad\qquad D_x \operatorname{sech} x = -\operatorname{sech} x \tanh x$$
$$D_x \tanh x = \operatorname{sech}^2 x \qquad\qquad D_x \operatorname{csch} x = -\operatorname{csch} x \coth x$$

□ **Example 2**   Discuss the graph of $y = \sinh x$.

Solution: As noted earlier, sinh is an odd function, so its graph is symmetric about the origin. It is defined for all x, and for large x its values are close to $\tfrac{1}{2} e^x$. (Why?) Since

$$D_x \sinh x = \cosh x = \tfrac{1}{2}(e^x + e^{-x}) > 0 \qquad \text{for all } x$$

the graph of sinh is always rising. Moreover,

$$D_x^2 \sinh x = \sinh x = \tfrac{1}{2}(e^x - e^{-x}) > 0 \qquad \text{for } x > 0$$

so the graph is concave up in $(0, \infty)$. It passes through the origin because

$$\sinh 0 = \tfrac{1}{2}(e^0 - e^0) = 0$$

Hence it has the appearance shown in Figure 3.

We leave it to you to discuss the graph of $y = \cosh x$ (also shown in Figure 3). □

Figure 3
Graphs of sinh and cosh

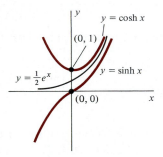

□ **Example 3**   Discuss the graph of $y = \tanh x$.

Solution: Since $\tanh x = (e^x - e^{-x})/(e^x + e^{-x})$, the domain is $\mathfrak{R}$. The graph is symmetric about the origin (because tanh is odd). For large x the values of $\tanh x$ are close to 1, while $\tanh x \to -1$ when $x \to -\infty$. (Why?) Hence the lines $y = \pm 1$ are asymptotes. Since

$$D_x \tanh x = \operatorname{sech}^2 x > 0 \qquad \text{for all } x$$

the graph is always rising. Moreover,

$$D_x^2 \tanh x = 2 \operatorname{sech} x (-\operatorname{sech} x \tanh x)$$
$$= -2 \operatorname{sech}^2 x \tanh x < 0 \qquad \text{for } x > 0$$

Thus the graph is concave down in $(0, \infty)$. It passes through the origin because $\tanh 0 = 0$. (See Figure 4, which also shows the graph of $y = \coth x$.)

Figure 4
Graphs of tanh and coth

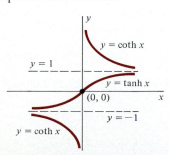

Figure 5
Graph of sech

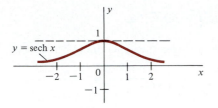

The graphs of $y = \text{sech } x$ and $y = \text{csch } x$ (which you should verify along with the graph of $y = \coth x$) are shown in Figures 5 and 6. □

□ **Example 4**   In 1700 Jakob Bernoulli proved that a flexible chain or cable hanging from its ends (and supporting no other weight but its own) takes the shape of the curve $y = a \cosh (x/a)$ (called a *catenary* from the Latin word for chain). Find the length of the catenary $y = \cosh x$ between $x = -1$ and $x = 1$. (See the graph of cosh in Figure 3.)

Solution: We use the formula

$$s = \int_{-1}^{1} \sqrt{1 + y'^2} \, dx$$

from Section 7.3. Since $y' = \sinh x$, we find

$$\sqrt{1 + y'^2} = \sqrt{1 + \sinh^2 x} = \sqrt{\cosh^2 x} \quad \text{(why?)}$$
$$= \cosh x \quad \text{(why?)}$$

Figure 6
Graph of csch

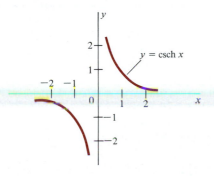

Hence $$s = \int_{-1}^{1} \cosh x \, dx = 2 \int_{0}^{1} \cosh x \, dx \quad \text{(why?)}$$

$$= 2 \sinh x \Big|_{0}^{1} \quad \text{(why?)}$$

$$= 2(\sinh 1 - \sinh 0) = 2 \sinh 1 \quad \text{(because } \sinh 0 = 0)$$

Some calculators have a $\sinh x$ key, and give $\sinh 1 \approx 1.1752$ directly. We can also write

$$\sinh 1 = \tfrac{1}{2}(e - e^{-1}) \approx 1.1752$$

In any case the length of the curve is approximately $s \approx 2.35$.   □

□ **Example 5**   Compute $\int_{0}^{\ln 2} \sinh t \cosh^2 t \, dt$.

Solution: Make the substitution $u = \cosh t$. Then $du = \sinh t \, dt$ and the integral takes the form $\int u^2 \, du$. The new limits correspond to $t = 0$ and $t = \ln 2$, respectively:

$$t = 0, \, u = \cosh 0 = 1$$

$$t = \ln 2, \, u = \cosh (\ln 2) = \tfrac{1}{2}(e^{\ln 2} + e^{-\ln 2}) = \tfrac{1}{2}(2 + \tfrac{1}{2}) = \tfrac{5}{4}$$

Hence

$$\int_{0}^{\ln 2} \sinh t \cosh^2 t \, dt = \int_{1}^{5/4} u^2 \, du = \frac{61}{192}$$ □

Optional Note (*on finding $x = \cosh t$ and $y = \sinh t$ in terms of t*)   In Figure 2 we show a point $P(x, y)$ on the unit hyperbola $x^2 - y^2 = 1$. Since t is twice the area shaded in this figure, we have

$$\frac{t}{2} = (\text{area of triangle } OPQ) - (\text{area of region } APQ)$$

The region APQ is bounded by the hyperbola, the x axis, and the vertical lines through A and Q, so we find it by integration:

$$\text{area of region } APQ = \int_1^x \sqrt{u^2 - 1} \, du$$

(We use u in the integrand to avoid confusion with x in the upper limit.) The area of triangle OPQ is of course

$$\tfrac{1}{2}(\text{base})(\text{altitude}) = \tfrac{1}{2}xy = \tfrac{1}{2}x\sqrt{x^2 - 1}$$

Hence

$$t = x\sqrt{x^2 - 1} - 2\int_1^x \sqrt{u^2 - 1} \, du$$

This looks hopeless to solve for x in terms of t. We can get rid of the integral, however, by differentiation (using the Fundamental Theorem of Calculus):

$$\frac{dt}{dx} = x \cdot \frac{x}{\sqrt{x^2 - 1}} + \sqrt{x^2 - 1} - 2\sqrt{x^2 - 1} = \frac{1}{\sqrt{x^2 - 1}}$$

Now separate the variables and integrate:

$$\frac{dx}{\sqrt{x^2 - 1}} = dt$$

$$\int \frac{dx}{\sqrt{x^2 - 1}} = \int dt \qquad (1)$$

It may appear that we are going in circles (first differentiating and then integrating). The integral on the left, however, may be found in Additional Problem 27 at the end of Chapter 8:

$$\int \frac{dx}{\sqrt{x^2 - a^2}} = \ln|x + \sqrt{x^2 - a^2}| + C$$

Of course we do not expect you to know this formula; not surprisingly, we will derive it by using hyperbolic functions (in the next section). Meanwhile, simply accept it as one that can be checked by differentiation (independently of hyperbolic functions). Then (1) becomes

$$\ln|x + \sqrt{x^2 - 1}| + C = t$$

Since $x \geq 1$ (see Figure 2), we may drop the absolute value. Moreover, $C = 0$ because $t = 0$ when $x = 1$. (Why?) Hence

$$\ln(x + \sqrt{x^2 - 1}) = t$$

Now it is relatively easy to solve for x:

$$x + \sqrt{x^2 - 1} = e^t$$

$$\sqrt{x^2 - 1} = e^t - x$$

$$x^2 - 1 = (e^t - x)^2 = e^{2t} - 2xe^t + x^2$$

$$2xe^t = e^{2t} + 1$$

$$x = \frac{e^{2t} + 1}{2e^t} = \frac{1}{2}(e^t + e^{-t})$$

To find y, use the fact that $x^2 - y^2 = 1$:

$$y^2 = x^2 - 1 = \tfrac{1}{4}(e^t + e^{-t})^2 - 1 = \tfrac{1}{4}(e^{2t} + 2 + e^{-2t}) - 1$$
$$= \tfrac{1}{4}(e^{2t} - 2 + e^{-2t}) = \tfrac{1}{4}(e^t - e^{-t})^2$$

Since $y \geq 0$, we have

$$y = \tfrac{1}{2}|e^t - e^{-t}| = \tfrac{1}{2}(e^t - e^{-t}) \qquad \text{(because } t \geq 0\text{)}$$

Problem Set 9.3

1. What curve is described by the parametric equations $x = \cosh t$, $y = \sinh t$? As t increases through its domain $(-\infty, \infty)$, how does the point (x, y) move along this curve?

2. Explain why sinh is an odd function.

3. Explain why cosh is an even function.

4. Use Problems 2 and 3 to show that tanh, coth, and csch are odd, while sech is even.

5. Use the formula $\cosh^2 t - \sinh^2 t = 1$ to derive the identity $\tanh^2 t + \text{sech}^2 t = 1$.

6. Use the formula $\cosh^2 t - \sinh^2 t = 1$ to derive the identity $\coth^2 t - \text{csch}^2 t = 1$.

7. Given the value $\tanh t = -\tfrac{3}{4}$, find the value of each of the remaining hyperbolic functions. Include an explanation of why each value is unique.

8. Repeat Problem 7 given the value $\sinh t = 2$.

Use the definitions of sinh and cosh to derive the following formulas.

9. $\sinh(u + v) = \sinh u \cosh v + \cosh u \sinh v$
10. $\cosh(u + v) = \cosh u \cosh v + \sinh u \sinh v$

Use Problems 9 and 10 to derive the following formulas.

11. $\sinh(u - v) = \sinh u \cosh v - \cosh u \sinh v$
12. $\cosh(u - v) = \cosh u \cosh v - \sinh u \sinh v$
13. $\sinh 2t = 2 \sinh t \cosh t$
14. $\cosh 2t = \cosh^2 t + \sinh^2 t$

Use Problem 14 (together with $\cosh^2 t - \sinh^2 t = 1$) to derive the following formulas.

15. $\sinh^2 t = \tfrac{1}{2}(\cosh 2t - 1)$
16. $\cosh^2 t = \tfrac{1}{2}(\cosh 2t + 1)$
17. Use the definition of cosh to prove that

$$D_x \cosh x = \sinh x$$

Use the derivatives of sinh and cosh to prove the following.

18. $D_x \tanh x = \text{sech}^2 x$ 19. $D_x \coth x = -\text{csch}^2 x$
20. $D_x \text{sech } x = -\text{sech } x \tanh x$
21. $D_x \text{csch } x = -\text{csch } x \coth x$

The graph of each of the following functions is shown in the text. Confirm that it is correct by discussing domain, symmetry, asymptotes, extreme values, and concavity.

22. $y = \cosh x$ 23. $y = \coth x$
24. $y = \text{sech } x$ 25. $y = \text{csch } x$
26. Show that the function $y = A \cosh x + B \sinh x$ satisfies the differential equation $y'' - y = 0$. (Compare with the fact that $y = A \cos x + B \sin x$ satisfies $y'' + y = 0$.)

Find the derivative of each of the following functions.

27. $y = \sinh 2x$ 28. $y = \cosh(1 - x)$
29. $y = \tanh x^2$ 30. $y = \sinh t + \cosh t$
31. $y = \sinh t - \cosh t$ 32. $y = \cosh^2 t - \sinh^2 t$
33. $y = x \sinh x$ 34. $y = \dfrac{\tanh x}{x}$
35. $y = \ln \cosh x$ 36. $y = \ln \sinh x$
37. $y = e^{-x} \sinh x$ 38. $y = e^{2x} \cosh x$
39. For what value of x does the function $y = \sinh x + 2 \cosh x$ have its minimum value?
40. Repeat Problem 39 for the function $y = 3 \cosh x - 2 \sinh x$.

Evaluate each of the following integrals.

41. $\displaystyle\int_0^{\ln 2} \sinh x \, dx$ (The answer is a rational number.)

42. $\displaystyle\int_0^{\ln 3} \cosh 2x \, dx$ (The answer is a rational number.)

43. $\int_0^1 \text{sech}^2 x \, dx$

44. $\int_0^2 \text{sech} \frac{x}{2} \tanh \frac{x}{2} \, dx$

45. $\int_0^1 \tanh x \, dx$ *Hint:* Make the substitution $u = \cosh x$.

46. $\int_1^2 \coth x \, dx$ [Can you reduce the answer to the form $\ln(e^2 + 1) - 1$?]

47. Find the area of the region bounded by the curves $y = \sinh x$ and $y = \cosh x$, the y axis, and the line $x = 1$. (Can you reduce the answer to $1 - e^{-1}$?)

48. Find the area of the region bounded by the curve $y = \tanh x$, the y axis, and the lines $x = 1$ and $y = 1$. (Use the hint in Problem 45.)

49. The region under the curve $y = \tanh x$, $0 \le x \le 1$, is rotated about the x axis. Find the volume of the resulting solid of revolution. *Hint:* Use Problem 5.

50. The region under the curve $y = \cosh x$, $0 \le x \le 1$, is rotated about the x axis. Find the volume of the resulting solid of revolution. *Hint:* Use Problem 16.

51. Find the surface area of the solid of revolution in Problem 50.

52. Find the length of the catenary

$$y = a \cosh(x/a)$$

between $x = -a$ and $x = a$.

53. Show that the length of the catenary

$$y = a \cosh(x/a)$$

from $(0,a)$ to (x,y) is $s = a \sinh(x/a)$.

54. Why is it meaningless to ask for the derivative of $y = \sin^{-1}(\cosh x)$?

55. Euler's formula for imaginary exponents (Problem 40, Section 8.6) says that $e^{ix} = \cos x + i \sin x$.

(a) Explain why $e^{-ix} = \cos x - i \sin x$.

(b) Show that

$$\cos x = \frac{1}{2}(e^{ix} + e^{-ix})$$

and

$$\sin x = \frac{1}{2i}(e^{ix} - e^{-ix})$$

(These remarkable formulas put the sine and cosine in the context of complex-valued functions. From the point of view of exponentials, the real functions sinh and cosh are simpler.)

56. A formula due to Abraham De Moivre (1667–1754) says that if n is any positive integer, then

$$(\cos t + i \sin t)^n = \cos nt + i \sin nt$$

(a) Use mathematical induction, together with the formulas for $\sin(u + v)$ and $\cos(u + v)$, to prove De Moivre's theorem.

(b) Assuming that complex exponentials obey the usual rules, give a simpler argument based on Euler's formula.

(c) Prove the hyperbolic formula

$$(\cosh t + \sinh t)^n = \cosh nt + \sinh nt$$

noting how uncomplicated it is compared to De Moivre's theorem.

9.4
INVERSE HYPERBOLIC FUNCTIONS

We saw in Section 9.2 that the inverse trigonometric functions are important for purposes of integration. Because of the analogy between trigonometric ("circular") functions and hyperbolic functions, we expect the same thing to be true of the inverse hyperbolic functions. You may be pleased to learn that these functions are ripe for the picking. Unlike the inverse trigonometric functions they involve nothing essentially new, but can be expressed in terms of the natural logarithm.

To see what we mean, consider the inverse hyperbolic sine. You might expect that we would proceed as so often before, observing from Figure 3 in the last section that $f(x) = \sinh x$ has an inverse because it is continuous and increasing. We define it by writing

$$y = \sinh^{-1} x \Leftrightarrow x = \sinh y$$

While this is reasonable enough, a second thought should occur to us. Since

$$x = \sinh y = \tfrac{1}{2}(e^y - e^{-y})$$

why not find the inverse by solving for y in terms of x? This would make the symbol $\sinh^{-1}$ superfluous (because an explicit formula for it exists).

□ **Example 1**   Find a formula for $y = \sinh^{-1} x$.

Solution: As suggested above, we solve for y in the equation

$$x = \tfrac{1}{2}(e^y - e^{-y})$$

The algebra is interesting:

$$x = \tfrac{1}{2}(e^y - e^{-y})$$
$$2x = e^y - e^{-y}$$
$$2xe^y = e^{2y} - 1$$

Regarding this equation as quadratic in e^y, $(e^y)^2 - 2xe^y - 1 = 0$, we use the quadratic formula (with $a = 1$, $b = -2x$, $c = -1$):

$$e^y = \frac{2x \pm \sqrt{4x^2 + 4}}{2} = x \pm \sqrt{x^2 + 1}$$

Since e^y is always positive, and $\sqrt{x^2 + 1} > x$ for all x, the ambiguous sign must be plus. Hence

$$e^y = x + \sqrt{x^2 + 1}$$
$$y = \ln(x + \sqrt{x^2 + 1})$$

In other words, the inverse hyperbolic sine is really just a logarithm:

$$\boxed{\sinh^{-1} x = \ln(x + \sqrt{x^2 + 1})}$$

□

□ **Example 2**   Find the derivative of $y = \sinh^{-1} x$.

Solution: The most direct procedure is to differentiate the above logarithm. It is easier, however, to differentiate implicitly in the equation $\sinh y = x$:

$$\cosh y \cdot \frac{dy}{dx} = 1$$

$$\frac{dy}{dx} = \frac{1}{\cosh y}$$

Since $\cosh^2 y - \sinh^2 y = 1$, we find

$$\cosh y = \sqrt{\sinh^2 y + 1} = \sqrt{x^2 + 1}$$

(Why is the positive square root chosen?) Hence

$$\boxed{D_x \sinh^{-1} x = \frac{1}{\sqrt{x^2 + 1}}}$$

□

The integration formula corresponding to this result is

$$\int \frac{dx}{\sqrt{x^2 + 1}} = \sinh^{-1} x + C$$

or (more generally)

$$\int \frac{dx}{\sqrt{x^2 + a^2}} = \sinh^{-1} \frac{x}{a} + C$$

(Confirm!) While this is good enough as it stands, it is often more useful to write it as a logarithm (using the formula for $\sinh^{-1}$):

$$\sinh^{-1} \frac{x}{a} = \ln \left[\frac{x}{a} + \sqrt{(x/a)^2 + 1} \right] = \ln \left(\frac{x + \sqrt{x^2 + a^2}}{a} \right)$$
$$= \ln (x + \sqrt{x^2 + a^2}) - \ln a$$

Since $-\ln a$ is a constant, we may drop it in the integration formula, obtaining

$$\int \frac{dx}{\sqrt{x^2 + a^2}} = \ln (x + \sqrt{x^2 + a^2}) + C$$

The inverse hyperbolic cosine is less useful (because the domain of cosh must be artificially restricted before the inverse can be said to exist). (See Figure 3, Section 9.3.) We define it by

$$y = \cosh^{-1} x \Leftrightarrow x = \cosh y \qquad y \geq 0$$

Proceeding as in Example 1, we find

$$\cosh^{-1} x = \ln (x + \sqrt{x^2 - 1}) \qquad (x \geq 1)$$

and (as in Example 2)

$$D_x \cosh^{-1} x = \frac{1}{\sqrt{x^2 - 1}} \qquad (x > 1)$$

This leads to

$$\int \frac{dx}{\sqrt{x^2 - a^2}} = \cosh^{-1} \frac{x}{a} + C \qquad (x > a)$$

or, equivalently,

$$\int \frac{dx}{\sqrt{x^2 - a^2}} = \ln (x + \sqrt{x^2 - a^2}) + C \qquad (x > a)$$

Since these formulas involve the restriction $x > a$ (whereas the integrand is defined for $x < -a$ as well as for $x > a$), it is desirable to find an unrestricted formula. This is not hard to do. Simply forget about the inverse hyperbolic cosine (it has served its purpose by leading us to the formula) and use an absolute value sign:

$$\int \frac{dx}{\sqrt{x^2 - a^2}} = \ln |x + \sqrt{x^2 - a^2}| + C$$

This holds in the domain $x < -a$ as well as $x > a$, as you can check.

□ **Example 3**   Find the inverse of the hyperbolic tangent.

Solution: Figure 4 in Section 9.3 shows that no restriction on tanh is necessary to guarantee an inverse. Hence

$$y = \tanh^{-1} x \Leftrightarrow x = \tanh y$$

To find the inverse, solve for y:

$$x = \frac{e^y - e^{-y}}{e^y + e^{-y}} = \frac{e^{2y} - 1}{e^{2y} + 1}$$

$$xe^{2y} + x = e^{2y} - 1$$

$$1 + x = e^{2y}(1 - x)$$

$$e^{2y} = \frac{1 + x}{1 - x}$$

$$2y = \ln \frac{1 + x}{1 - x}$$

$$y = \frac{1}{2} \ln \frac{1 + x}{1 - x}$$

Thus we have derived the formula

$$\tanh^{-1} x = \frac{1}{2} \ln \frac{1 + x}{1 - x} \qquad (|x| < 1)$$

□

To differentiate this, write it in the simpler form

$$y = \frac{1}{2}[\ln (1 + x) - \ln (1 - x)]$$

Then $$\frac{dy}{dx} = \frac{1}{2}\left(\frac{1}{1 + x} + \frac{1}{1 - x}\right) = \frac{1}{2} \cdot \frac{2}{1 - x^2} = \frac{1}{1 - x^2}$$

and hence

$$D_x \tanh^{-1} x = \frac{1}{1 - x^2} \qquad (|x| < 1)$$

A similar discussion of the inverse hyperbolic cotangent shows that

$$\coth^{-1} x = \frac{1}{2} \ln \frac{x+1}{x-1} \qquad (|x| > 1)$$

and

$$D_x \coth^{-1} x = \frac{1}{1-x^2} \qquad (|x| > 1)$$

Note that $\tanh^{-1} x$ and $\coth^{-1} x$ have the same derivative, but in different domains. There are two corresponding integration formulas:

$$\int \frac{dx}{1-x^2} = \tanh^{-1} x + C \qquad (|x| < 1)$$

and

$$\int \frac{dx}{1-x^2} = \coth^{-1} x + C \qquad (|x| > 1)$$

or (more generally)

$$\int \frac{dx}{a^2 - x^2} = \begin{cases} \dfrac{1}{a} \tanh^{-1} \dfrac{x}{a} + C & (|x| < a) \\ \dfrac{1}{a} \coth^{-1} \dfrac{x}{a} + C & (|x| > a) \end{cases}$$

A single version of these formulas is sometimes more useful, and may be obtained from the logarithmic formulas for $\tanh^{-1} x$ and $\coth^{-1} x$ given above:

$$\int \frac{dx}{a^2 - x^2} = \frac{1}{2a} \ln \left| \frac{a+x}{a-x} \right| + C$$

We will confirm this in the case $|x| > a$ and leave the case $|x| < a$ for you:

$$\int \frac{dx}{a^2 - x^2} = \frac{1}{a} \coth^{-1} \frac{x}{a} + C = \frac{1}{a} \cdot \frac{1}{2} \ln \left(\frac{x/a + 1}{x/a - 1} \right) + C$$

$$= \frac{1}{2a} \ln \frac{x+a}{x-a} + C = \frac{1}{2a} \ln \left| \frac{a+x}{a-x} \right| + C$$

$$\left(\text{because } \left| \frac{a+x}{a-x} \right| = \frac{x+a}{x-a} \text{ when } |x| > a \right).$$

The inverse hyperbolic secant and cosecant are relatively unimportant, so we will not dwell on them. They are defined by

$$y = \operatorname{sech}^{-1} x \Leftrightarrow x = \operatorname{sech} y \qquad y \geq 0$$
$$y = \operatorname{csch}^{-1} x \Leftrightarrow x = \operatorname{csch} y$$

(See Figures 5 and 6, Section 9.3, to understand why a restriction is needed in the first case but not the second.) It follows that

$$\text{sech}^{-1} x = \cosh^{-1} \frac{1}{x} \quad \text{and} \quad \text{csch}^{-1} x = \sinh^{-1} \frac{1}{x} \quad \text{(why?)}$$

which means that both functions can be differentiated by formulas already developed. The results are

$$D_x \text{sech}^{-1} x = \frac{-1}{x\sqrt{1-x^2}} \quad (0 < x < 1)$$

$$D_x \text{csch}^{-1} x = \frac{-1}{|x|\sqrt{1+x^2}} \quad (x \neq 0)$$

We will spare you the details of developing integration formulas based on these derivatives, and simply state the results:

$$\int \frac{dx}{x\sqrt{a^2-x^2}} = -\frac{1}{a}\cosh^{-1}\frac{a}{|x|} + C = \frac{1}{a}\ln\left|\frac{a-\sqrt{a^2-x^2}}{x}\right| + C$$

$$\int \frac{dx}{x\sqrt{a^2+x^2}} = -\frac{1}{a}\sinh^{-1}\frac{a}{|x|} + C = \frac{1}{a}\ln\left|\frac{a-\sqrt{a^2+x^2}}{x}\right| + C$$

Our last example illustrates the usefulness of hyperbolic functions and their inverses in a concrete setting.

□ **Example 4**   A body of mass m falls from rest, encountering air resistance that is proportional to the square of the speed. What is the law of motion?

Solution: It is convenient to direct the line of motion downward (so that the velocity is positive). Take the origin at the point where the body starts to fall (and turn on the clock). If s is the position of the body at time t, then $s = 0$ and $v = ds/dt = 0$ at $t = 0$.

The downward force of gravity (mg) and the oppositely directed force of air resistance $(-kv^2$, where $k > 0)$ combine to produce a force $F = mg - kv^2$ on the body. Newton's Second Law (force equals mass times acceleration) reduces this to the differential equation

$$m\frac{dv}{dt} = mg - kv^2$$

Separate the variables and integrate:

$$\frac{m\,dv}{mg - kv^2} = dt$$

$$m\int \frac{dv}{mg - kv^2} = \int dt = t + C_1 \tag{1}$$

The force $F = mg - kv^2$ is always positive (directed downward), so we have

$$kv^2 < mg$$
$$v^2 < a^2 \qquad \text{(where } a = \sqrt{mg/k}\text{)}$$
$$|v| < a$$

Hence Equation (1) becomes

$$\frac{m}{k} \int \frac{dv}{a^2 - v^2} = \frac{m}{k} \cdot \frac{1}{a} \tanh^{-1} \frac{v}{a} = t + C_1$$

Since $v = 0$ when $t = 0$, we find $C_1 = 0$ and thus

$$\frac{m}{ka} \tanh^{-1} \frac{v}{a} = t$$

$$\tanh^{-1} \frac{v}{a} = bt \qquad \text{(where } b = ka/m\text{)}$$

$$\frac{v}{a} = \tanh bt$$

$$v = a \tanh bt$$

Since $v = ds/dt$, we integrate again:

$$s = \int a \tanh bt \, dt = a \int \frac{\sinh bt \, dt}{\cosh bt}$$

$$= \frac{a}{b} \int \frac{du}{u} \qquad (u = \cosh bt, \; du = b \sinh bt \, dt)$$

$$= \frac{a}{b} \ln |u| + C_2 = \frac{a}{b} \ln \cosh bt + C_2 \qquad (\cosh bt \text{ is positive})$$

Since $s = 0$ when $t = 0$, we find $C_2 = 0$ and hence the desired law of motion is

$$s = \frac{a}{b} \ln \cosh bt = \frac{m}{k} \ln (\cosh \sqrt{kg/m} \, t)$$

Recall from Section 7.7 that Galileo's law of motion (for a body falling from rest in a vacuum) is simply $s = \frac{1}{2} g t^2$. The hypothesis that air resistance is proportional to the square of the speed is not unreasonable, but observe how it complicates matters! The heavy artillery involved in the solution of Example 4 is impressive:

- Newton's Second Law
- Separation of variables in a differential equation
- Integration involving the inverse hyperbolic tangent and the natural logarithm
- The hyperbolic sine, cosine, and tangent (which in turn involve the natural exponential function)   □

Problem Set 9.4

Find each of the following in exact form (no approximations).

1. $\sinh^{-1} 0$ **2.** $\cosh^{-1} 1$

3. $\tanh^{-1} 0$ **4.** $\sinh^{-1}(-1)$

5. $\tanh^{-1} \frac{1}{2}$ **6.** $\cosh^{-1} 0$

7. $\tanh^{-1} 2$

8. If $\sinh x = 2$, find x in terms of logarithms.

9. If $\cosh 2x = 3$, find x in terms of logarithms. Why are there two answers?

10. If $\tanh \sqrt{x} = \frac{2}{5}$, find x in terms of logarithms.

11. Show that the curves $y = \tanh x$ and $y = \mathrm{sech}\, x$ intersect at (x, y), where $x = \sinh^{-1} 1$ and $y = \tanh(\sinh^{-1} 1) = 1/\sqrt{2}$.

12. If $f(x) = \frac{1}{2} \sinh(x - 1)$, find a formula for $f^{-1}(x)$.

13. If $f(x) = 2 \tanh(x/2)$, find a formula for $f^{-1}(x)$. What is the domain of f^{-1}?

14. Reflect the graph of $y = \sinh x$ (Section 9.3) in the line $y = x$ to obtain the graph of $y = \sinh^{-1} x$.

15. Reflect the graph of $y = \cosh x$ (Section 9.3) in the line $y = x$ to obtain the graph of the inverse relation. What part of this graph is the graph of $y = \cosh^{-1} x$?

16. Reflect the graph of $y = \coth x$ (Section 9.3) in the line $y = x$ to obtain the graph of $y = \coth^{-1} x$.

Find the derivative of each of the following functions.

17. $f(x) = 3 \sinh^{-1} 2x$ **18.** $f(x) = \cosh^{-1} \sqrt{x}$

19. $f(x) = \tanh^{-1} \frac{x}{2}$ **20.** $f(x) = \coth^{-1}(1/x)$

21. $f(x) = x \sinh^{-1} x - \sqrt{x^2 + 1}$

22. $f(x) = x \tanh^{-1} x + \frac{1}{2} \ln(1 - x^2)$

23. $f(x) = \sinh^{-1}(\tan x)$

24. $f(x) = \tanh^{-1}(\sin x)$

Evaluate each of the following integrals.

25. $\displaystyle\int_0^2 \frac{dx}{9 - x^2}$ **26.** $\displaystyle\int_4^6 \frac{dx}{9 - x^2}$

27. $\displaystyle\int_0^3 \frac{dx}{x^2 - 25}$ **28.** $\displaystyle\int_1^2 \frac{dx}{4x^2 - 1}$

29. $\displaystyle\int_0^4 \frac{dx}{\sqrt{x^2 + 4}}$ **30.** $\displaystyle\int_0^1 \frac{dx}{\sqrt{9x^2 + 25}}$

31. $\displaystyle\int_2^7 \frac{dx}{\sqrt{x^2 - 1}}$ **32.** $\displaystyle\int_2^4 \frac{dx}{\sqrt{9x^2 - 16}}$

33. $\displaystyle\int_1^2 \frac{dx}{x \sqrt{9 - x^2}}$ **34.** $\displaystyle\int_2^3 \frac{dx}{x \sqrt{16 + x^2}}$

35. When a body falls from rest (encountering air resistance proportional to the square of its speed), its velocity at time t is

$$v = a \tanh bt \qquad \text{(where } a \text{ and } b \text{ are positive constants)}$$

(See Example 4.)

(a) Explain why v increases with t, but is bounded. Name its least upper bound.

(b) How could this "terminal velocity" have been predicted before the formula for v was found? (Look closely at Example 4!)

36. A sky diver is falling at the rate of 20 ft/sec when her parachute opens; t seconds later her velocity satisfies the equation

$$\frac{dv}{dt} = 32 - \frac{v^2}{50}$$

(a) Solve this initial value problem to show that

$$t = \frac{5}{4}\left(\tanh^{-1} \frac{v}{40} - \tanh^{-1} \frac{1}{2}\right)$$

(b) What terminal velocity does the sky diver approach as time goes on?

(c) How long does it take her to reach 99% of her terminal velocity? (Use a calculator.)

37. Find the derivative of $\sinh^{-1} x = \ln(x + \sqrt{x^2 + 1})$ and compare with Example 2.

38. Derive the formula

$$\int \frac{dx}{\sqrt{x^2 + a^2}} = \sinh^{-1} \frac{x}{a} + C \qquad (a > 0)$$

by making a substitution that reduces it to the known formula

$$\int \frac{dx}{\sqrt{x^2 + 1}} = \sinh^{-1} x + C$$

39. Derive the formula $\cosh^{-1} x = \ln(x + \sqrt{x^2 - 1})$ as follows.

(a) Write the equation $y = \cosh^{-1} x$ in equivalent hyperbolic form and express the result as an equation that is quadratic in e^y.

(b) Use the quadratic formula to solve for e^y. Include a defense of your choice of the ambiguous sign.

(c) Solve for y.

40. Derive the formula $D_x \cosh^{-1} x = 1/\sqrt{x^2 - 1}$ in two ways, as follows.

(a) Use the formula in Problem 39.

(b) Let $y = \cosh^{-1} x$ and differentiate implicitly in the equivalent hyperbolic equation. Include a defense of your choice of the ambiguous sign.

41. Derive the formula

$$\int \frac{dx}{\sqrt{x^2 - a^2}} = \cosh^{-1} \frac{x}{a} + C \qquad (x > a)$$

from the known formula

$$\int \frac{dx}{\sqrt{x^2 - 1}} = \cosh^{-1} x + C \qquad (x > 1)$$

42. Show how the formula

$$\int \frac{dx}{\sqrt{x^2 - a^2}} = \ln(x + \sqrt{x^2 - a^2}) + C \qquad (x > a)$$

follows from

$$\int \frac{dx}{\sqrt{x^2 - a^2}} = \cosh^{-1} \frac{x}{a} + C \qquad (x > a)$$

43. Derive the formula

$$\coth^{-1} x = \frac{1}{2} \ln \frac{x + 1}{x - 1}$$

as follows.

(a) Write the equation $y = \coth^{-1} x$ in equivalent hyperbolic form and solve for e^{2y}.

(b) Find y.

44. Derive the formula $D_x \coth^{-1} x = 1/(1 - x^2)$ by using Problem 43.

45. Derive the formulas

$$\int \frac{dx}{a^2 - x^2} = \begin{cases} \dfrac{1}{a} \tanh^{-1} \dfrac{x}{a} + C & (|x| < a) \\[2mm] \dfrac{1}{a} \coth^{-1} \dfrac{x}{a} + C & (|x| > a) \end{cases}$$

from the known formulas

$$\int \frac{dx}{1 - x^2} = \begin{cases} \tanh^{-1} x + C & (|x| < 1) \\ \coth^{-1} x + C & (|x| > 1) \end{cases}$$

46. Confirm the formula

$$\int \frac{dx}{a^2 - x^2} = \frac{1}{2a} \ln \left| \frac{a + x}{a - x} \right| + C$$

in the case $|x| < a$. (We did the case $|x| > a$ in the text.)

47. Show that

$$\operatorname{sech}^{-1} x = \ln \frac{1 + \sqrt{1 - x^2}}{x} \qquad 0 < x \le 1$$

48. Derive the formula

$$D_x \operatorname{sech}^{-1} x = \frac{-1}{x \sqrt{1 - x^2}}$$

in two ways, as follows.

(a) Use Problem 47.

(b) Use the formula $\operatorname{sech}^{-1} x = \cosh^{-1}(1/x)$.

49. Use the formula $\operatorname{csch}^{-1} x = \sinh^{-1}(1/x)$ to show that

$$D_x \operatorname{csch}^{-1} x = \frac{-1}{|x| \sqrt{1 + x^2}}$$

50. Derive the formula

$$\int \frac{dx}{x \sqrt{a^2 - x^2}} = -\frac{1}{a} \cosh^{-1} \frac{a}{|x|} + C$$
$$= \frac{1}{a} \ln \left| \frac{a - \sqrt{a^2 - x^2}}{x} \right| + C$$

51. Derive the formula

$$\int \frac{dx}{x \sqrt{a^2 + x^2}} = -\frac{1}{a} \sinh^{-1} \frac{a}{|x|} + C$$
$$= \frac{1}{a} \ln \left| \frac{a - \sqrt{a^2 + x^2}}{x} \right| + C$$

9.5

THE CATENARY
(Optional)

In Example 4, Section 9.3, we stated that a flexible chain hanging from its ends (and supporting no other weight but its own) takes the shape of the curve $y = a \cosh(x/a)$. This curve is called a **catenary** (from the Latin word for chain). If you are interested in physics, you should enjoy seeing how the equation of a catenary is derived.

Figure 1
Equilibrium of forces on a hanging chain

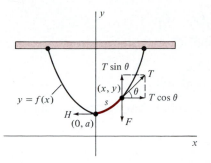

Figure 1 shows a chain hanging from two points of support on a horizontal beam. We have chosen the coordinate system so that the lowest point of the chain is at $(0, a)$, where a is a positive constant to be specified later. The point (x, y) is any other point of the chain, the relation between x and y being $y = f(x)$. The problem is to find a formula for $f(x)$.

We are assuming that the chain is of uniform density δ, that is, it weighs δ lb per ft. If s is the length of the portion from $(0, a)$ to (x, y), the weight of that portion is $F = \delta s$. This downward force is shown as a vector in Figure 1, with initial point at (x, y). The chain hangs motionless because the forces acting on it are in equilibrium. Hence there must be an upward force balancing the downward force F. Imagine two people holding onto the end-points of the portion shown in Figure 1 (as though the rest of the chain were removed). Each person exerts a tangential pull (called *tension*); these forces, together with the downward force F, keep the chain in the shape shown.

The tension T at (x, y) has horizontal and vertical components $T \cos \theta$ and $T \sin \theta$, respectively, whereas the tension H at $(0, a)$ is entirely horizontal. The upward force balancing F is therefore $T \sin \theta$, while the horizontal forces H and $T \cos \theta$ must cancel each other's effect. In other words, $F = T \sin \theta$ and $H = T \cos \theta$.

A student with no background in physics may not be at home with these ideas. However, we can now forget about physics and concentrate on calculus. The tension T is unknown to us, but we can eliminate it by using the slope m at (x, y):

$$m = \tan \theta = \frac{\sin \theta}{\cos \theta} = \frac{F/T}{H/T} = \frac{F}{H} = \left(\frac{\delta}{H} \right) s$$

Let $k = \delta/H$ (constant because the density δ and the tension H at the lowest point of the chain are constant). Then $m = ks$. Since $m = dy/dx$, this is a differential equation involving our unknown function $y = f(x)$. It is awkward, however, because s is an integral:

$$s = \int_0^x \sqrt{1 + f'(t)^2}\, dt$$

(See Section 7.3.) We can eliminate the integral by differentiating with respect to x in the equation $m = ks$:

$$\frac{dm}{dx} = k \frac{ds}{dx} = k \sqrt{1 + f'(x)^2}$$

(See the first part of the Fundamental Theorem of Calculus, Section 6.4.) Since $m = f'(x)$, we have $dm/dx = k \sqrt{1 + m^2}$ and things should begin to take shape. For this is a differential equation that can be solved by separating the variables:

$$\frac{dm}{\sqrt{1 + m^2}} = k\, dx$$

$$\int \frac{dm}{\sqrt{1 + m^2}} = k \int dx$$

You can also see why the catenary is discussed after a section on inverse hyperbolic functions! The integral on the left side fits one of our formulas in Section 9.4; we use it to write

$$\sinh^{-1} m = kx + C_1$$

Since $m = 0$ when $x = 0$ (why?), we find $C_1 = 0$ and hence

$$\sinh^{-1} m = kx$$
$$m = \sinh kx$$

Replacing m by dy/dx, we have another differential equation to solve:

$$\frac{dy}{dx} = \sinh kx$$
$$dy = \sinh kx\, dx$$
$$\int dy = \int \sinh kx\, dx$$
$$y = \frac{1}{k} \cosh kx + C_2$$

Since $y = a$ when $x = 0$, we find $C_2 = a - (1/k)$ and hence

$$y = \frac{1}{k} \cosh kx + \left(a - \frac{1}{k}\right)$$

Our choice of coordinate system in Figure 1 did not specify the value of a. Evidently we can make this number anything we like by moving the x axis up or down. The most convenient choice is the one making $a = 1/k$, in which case our equation for the catenary reduces to $y = a \cosh (x/a)$.

An interesting corollary of the above discussion is that the tension at (x, y) in Figure 1 is $T = \delta y$. (We ask you to show why in the problem set.) Since δy is the weight of y ft of the chain, we could remove one of the upper supports and let y ft hang from a nail driven into the wall at (x, y). (See Figure 2.) The chain would remain in place without slipping over the nail because the tension at (x, y) has been replaced by the weight of the vertical section of length y.

Figure 2
Chain hanging over a nail

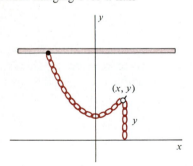

Problem Set 9.5

1. A rope 50 ft long is strung across a crevasse 40 ft wide, each end being fastened at the same elevation. Find the sag in the rope at its center as follows.

   (a) Using the coordinate system in Figure 1, we know that the rope takes the shape of the catenary $y = a \cosh (x/a)$. Explain why the sag at the center is

   $$a\left(\cosh \frac{20}{a} - 1\right)$$

   (b) Use Problem 53, Section 9.3, to show that a satisfies the equation $\sinh (20/a) = 25/a$.

   (c) To solve the equation in part (b), let $t = 20/a$. Confirm that the equation becomes $\sinh t = 1.25t$ and use Newton's Method (Section 5.4) to compute an approximate value of t. *Hint:* A reasonable first guess can be obtained from a sketch of the graphs of $y = \sinh t$ and $y = 1.25t$ on the same coordinate plane. (Why?) Use a calculator to improve it.

(d) The answer in part (c) is $t \approx 1.18$. Use this value to compute the sag.

2. Repeat Problem 1 for a rope 120 ft long strung across a crevasse 100 ft wide.

3. Prove that the tension at (x, y) in Figure 1 is $T = \delta y$, as follows.

(a) Show that $\tan \theta = \sinh (x/a)$.

(b) Why does it follow that $\sec \theta = \cosh (x/a)$?

(c) We know from the text that $H = T \cos \theta$. Explain why this gives

$$T = \delta a \cosh \frac{x}{a} = \delta y$$

4. If the rope in Problem 1 weighs 0.2 lb/ft, what is the tension at its center? at its ends?

5. If the rope in Problem 2 weighs 0.3 lb/ft, what is the tension at its center? at its ends?

6. Suppose that a rope is in the shape of the catenary $y = a \cosh (x/a)$, $-b \leq x \leq b$, and has length $2s$. Show that the sag in the rope at its center is $\sqrt{a^2 + s^2} - a$.

7. In Problem 1 we have $s = 25$ and $a = 20/t \approx 17$. Use these figures in Problem 6 to compute the sag and compare with the answer in Problem 1(d).

8. In Problem 2 we have $s = 60$ and $a = 50/t \approx 47$. Use these figures in Problem 6 to compute the sag and compare with the answer in Problem 2(d).

True–False Quiz

1. $\tan^{-1}(-1) = 3\pi/4$.

2. $\cos(\sin^{-1} x) \geq 0$.

3. $\tan^{-1} x = \sin^{-1} x / \cos^{-1} x$.

4. $\cos(\pi/2 - \sin^{-1} x) = x$.

5. $\tan^{-1}(\tan x) = x$ for all x in the domain of tangent.

6. $\sec^{-1} 5 = \cos^{-1} \frac{1}{5}$.

7. If $f(x) = \sin^{-1}(x/2)$, then $f'(1) = 2/\sqrt{3}$.

8. If $f(x) = \tan^{-1} \sqrt{x}$, then $f'(1) = \frac{1}{2}$.

9. If $f(x) = \sin^{-1}(\cos x)$, then $f'(4) = -1$.

10. $\displaystyle\int_{-2}^{2} \frac{dx}{4 + x^2} = \frac{\pi}{2}$.

11. If $f(x) = \displaystyle\int_{0}^{x} \frac{dt}{\sqrt{1 - t^2}}$, then $f^{-1}(\pi/2) = 1$.

12. $\displaystyle\int_{\sqrt{2}}^{2} \frac{dx}{x\sqrt{x^2 - 1}} = \frac{\pi}{12}$.

13. $\sinh(\ln 3) = \frac{4}{3}$.

14. The range of cosh is $\mathcal{R}$.

15. The domain of $f(x) = \sin^{-1}(\cosh x)$ is $\mathcal{R}$.

16. The hyperbolic tangent is an even function.

17. $0 < \operatorname{sech} x \leq 1$ for all x.

18. The function $y = \sinh x$ satisfies the differential equation $y'' + y = 0$.

19. $D_x \ln(\sinh x) = \coth x$.

20. $\displaystyle\int \frac{dx}{(e^x + e^{-x})^2} = \frac{1}{4} \tanh x + C$

21. The hyperbolic cosine has an inverse.

22. $\cosh^{-1} 0 = 1$.

23. $D_x \tanh^{-1}(\cos x) = -\csc x$.

24. $\displaystyle\int_{2}^{3} \frac{dx}{1 - x^2} = \tanh^{-1} 3 - \tanh^{-1} 2$.

25. $\displaystyle\int_{0}^{3} \frac{dx}{\sqrt{x^2 + 16}} = \ln 2$.

Additional Problems

Find each of the following in exact form. (Do not give approximations.)

1. $\tan^{-1}(-1)$ **2.** $\cos^{-1} \frac{1}{2}$

3. $\sin^{-1}\left(\sin \frac{3\pi}{4}\right)$ **4.** $\sec^{-1} 1$

5. $\cos(\cos^{-1} 0)$ **6.** $\sin^{-1} \frac{2}{3} + \cos^{-1} \frac{2}{3}$

7. $\sin^{-1} 2$ **8.** $\tanh 0$

9. $\cosh 0$ **10.** $\sinh (\ln 2)$

11. $\tanh^{-1} 0$ **12.** $\cosh^{-1} 0$

13. When the number 1.5 is entered in a calculator and the inverse sine key is pressed, the display panel flashes (or otherwise indicates nonsense). Why?

14. Most calculators do not have an inverse cotangent key. How would you use one to find $\cot^{-1} 2$?

Find the derivative of each of the following functions.

15. $y = \sin^{-1}(1 - x)$ **16.** $y = \tan^{-1}|x|$

17. $y = \sec^{-1} 2x$ **18.** $y = \cos^{-1}(1/x)$

19. $y = \cot^{-1}(x^2 - 1)$ **20.** $y = \tan^{-1} x + \cot^{-1} x$

21. $y = x \sin^{-1} x$ **22.** $y = (\tan^{-1} x)^2$

23. $y = x \tan^{-1} x - \ln \sqrt{1 + x^2}$

24. $y = \cosh x^2$ **25.** $y = \tanh \sqrt{x}$

26. $y = e^{2x} \sinh x$ **27.** $y = \ln \operatorname{sech} x$

28. $y = \tanh^{-1} x^2$ **29.** $y = \cosh^{-1} 2x$

30. Explain why the formula

$$\int \frac{dx}{\sqrt{a^2 - x^2}} = -\cos^{-1} \frac{x}{a} + C$$

is correct and use it to evaluate

$$\int_0^1 \frac{dx}{\sqrt{4 - x^2}}$$

Check by means of the standard formula.

Evaluate each of the following.

31. $\displaystyle\int_0^2 \frac{dx}{\sqrt{16 - x^2}}$ **32.** $\displaystyle\int_0^{\sqrt{3}} \frac{dx}{1 + x^2}$

33. $\displaystyle\int \frac{x\,dx}{\sqrt{1 - x^4}}$ **34.** $\displaystyle\int_5^6 \frac{dx}{x \sqrt{x^2 - 16}}$

35. $\displaystyle\int \frac{dx}{x^2 + 6x}$ *Hint:* Complete the square.

36. $\displaystyle\int_0^{\pi/2} \frac{\cos x\,dx}{\sqrt{9 - \sin^2 x}}$

37. $\displaystyle\int_0^2 \sinh \tfrac{1}{2}x\,dx$

38. $\displaystyle\int_0^{\ln 3} \sqrt{\sinh t}\,\cosh t\,dt$

39. Verify the formula

$$\int \operatorname{sech} x\,dx = \tan^{-1}(\sinh x) + C$$

by differentiating the right side.

40. If $dy/dx = 1 + y^2$ and $y = 1$ when $x = 0$, find y as a function of x.

41. If $dy/dx = \sqrt{1 - y^2}$ and $y = 0$ when $x = 1$, find y as a function of x.

42. If $dy/dx = 25 + y^2$ and $y = 5$ when $x = 0$, find y as a function of x.

43. Find the area of the region bounded by the graph of $y = \dfrac{1}{x^2 + 9}$, the x axis, and the lines $x = \pm 3$.

44. Find the area of the region bounded by the graph of $y = \dfrac{1}{\sqrt{4 - x^2}}$ and the line $y = 2$.

45. Find the volume of the solid generated by rotating the region under the curve $y = 1/\sqrt{9 - x^2}, 0 \le x \le 2$, about the x axis.

46. Repeat Problem 45 for rotation about the y axis.

47. If $f(x) = 2 \sinh \sqrt{x}$, find $f^{-1}(x)$.

48. The motion of a spring with displacement x at time t is given by $d^2x/dt^2 = -a^2x$. If the spring is released from a compressed position of $x = -1$ at $t = 0$, find its law of motion. *Hint:* If $v = dx/dt$, then $dv/dt = v(dv/dx)$.

49. What is the domain of the function $f(x) = \sin^{-1} x + \cos^{-1} x$? By examining $f'(x)$ and drawing appropriate conclusions, write $f(x)$ in simpler form.

50. For what values of x is it true to say that

$$D_x (\sin^{-1} x + \sec^{-1} x) = \frac{1}{\sqrt{1 - x^2}} + \frac{1}{x \sqrt{x^2 - 1}}?$$

51. Show that the function $y = 3 \sinh x - 2 \cosh x$ is always increasing.

52. Find $F'(x)$ if

$$F(x) = x \sqrt{1 - x^2} - 2 \int_1^x \sqrt{1 - t^2}\,dt$$

What is a simpler formula for $F(x)$?

53. Show that

$$D_x \cos^{-1} \frac{1}{x} = \frac{1}{|x| \sqrt{x^2 - 1}}$$

and use the result to explain why

$$\sec^{-1} x = \begin{cases} \cos^{-1} \dfrac{1}{x} & \text{if } x \ge 1 \\[2mm] 2\pi - \cos^{-1} \dfrac{1}{x} & \text{if } x \le -1 \end{cases}$$

54. Show that

$$\tanh\left(\frac{1}{2}\ln x\right) = \frac{x-1}{x+1}$$

55. Prove that the graphs of $y = \sinh x$ and $y = \operatorname{csch} x$ do not intersect at right angles.

56. Find the length of the curve $y = \cosh x$, $0 \le x \le 2$.

57. Find the centroid of the curve $y = \cosh x$, $-1 \le x \le 1$. *Hint:* Use the identity $\cosh^2 t = \frac{1}{2}(\cosh 2t + 1)$ in Problem 16, Section 9.3. Also see Example 4, Section 9.3, where the length of the curve is found to be $2\sinh 1$.

58. Find the area of the surface generated by rotating the catenary

$$y = a\cosh(x/a) \qquad -a \le x \le a$$

about the x axis. (It can be shown that no other curve with the same endpoints generates a smaller surface area when it is rotated about the x axis.)

59. The region bounded by the curves $y = \tanh x$ and $y = \operatorname{sech} x$ and the y axis is rotated about the x axis. Find the volume of the resulting solid of revolution as follows.

(a) Explain why the volume is given by

$$V = \pi(2\tanh c - c)$$

where c is the x coordinate of the point of intersection of the curves $y = \tanh x$ and $y = \operatorname{sech} x$.

(b) Show that $c = \ln(1 + \sqrt{2})$.

(c) Conclude that $V = \pi[\sqrt{2} - \ln(1 + \sqrt{2})]$.

60. The graph of a certain function (defined for all x) contains the point $(0,1)$ and at each point (x,y) the square of its slope is $y^2 - 1$. Find a formula for the function.

61. Given $t \in \mathbb{R}$, explain why there is a number x between $-\pi/2$ and $\pi/2$ such that $\sinh t = \tan x$. Then show that

$$\cosh t = \sec x$$
$$\tanh t = \sin x$$
$$\coth t = \csc x \qquad (t \ne 0)$$
$$\operatorname{sech} t = \cos x$$
$$\operatorname{csch} t = \cot x \qquad (t \ne 0)$$

62. The equation $\sinh t = \tan x$, $-\pi/2 < x < \pi/2$, in Problem 61 defines x as a function of t.

(a) What is the formula for x?

(b) Show that $dx/dt = 1/\sec x$.

(c) Separate the variables and integrate in part (b) to obtain

$$\int \sec x \, dx = \sinh^{-1}(\tan x) + C \qquad -\pi/2 < x < \pi/2$$

(d) Use the logarithmic formula for $\sinh^{-1}$ to find

$$\int \sec x \, dx = \ln(\sec x + \tan x) + C \qquad -\pi/2 < x < \pi/2$$

and compare with Example 3, Section 8.5.

63. Show that $\tan^{-1}(\sinh t) = \sin^{-1}(\tanh t)$ for all t in two ways, as follows.

(a) Use Problem 61.

(b) Use differentiation.

64. Show that $\sinh^{-1}(\tan x) = \tanh^{-1}(\sin x)$, $-\pi/2 < x < \pi/2$, in two ways, as follows.

(a) Use Problem 61.

(b) Use differentiation.

10 | Techniques of Integration

This chapter is devoted to what might be called the *art of integration*. We say art (rather than science) because it is hard to say anything very systematic about the methods used to tackle a given integral. Differentiation is straightforward; antidifferentiation often requires some imagination.

There are really only two general approaches to integration (although it may not look that way to the beginner). One is to make a substitution that reduces the integral to a familiar standard form; the other is *integration by parts* (the first topic of this chapter). From your experience with substitution in the last four chapters, you are undoubtedly aware of the central problem. *It is essential to perceive the standard form lurking behind an integral in order to know what substitution to try.* That is why we have been collecting integration formulas as though they were rare jewels.

You will find a list of previously derived results on the inside back cover of the book; they will be needed from here on out. Whether they should be memorized (in whole or in part) is a question we prefer to leave open. One can certainly argue that they need not be, on the grounds that we can always look them up. On the other hand, how do we recognize a good substitution if we are ignorant of the standard forms? It is no accident that most students who master the art of integration seem to know these formulas pretty well (whether required to learn them or not).

You will also find a list of hyperbolic integrals on the inside back cover (not given before) that are analogous to familiar trigonometric formulas. While they are not so often used, they are worth listing for reference (and are easily checked). Two of them (the integrals of sech and csch) are surprising; you might enjoy figuring out how they are obtained.

Although this list is a good start on a table of integrals, you should realize that no table is adequate for all integration problems. A simple example is

$$\int_0^1 e^{-x^2}\, dx = ?$$

No elementary function exists whose derivative is e^{-x^2}; in such a case it is necessary to resort to approximation techniques. We will take up that question at the end of the chapter.

10.1 INTEGRATION BY PARTS

The Product Rule for derivatives says that if u and v are functions of x, then

$$\frac{d}{dx}(uv) = u\frac{dv}{dx} + v\frac{du}{dx}$$

or (in differential form)

$$d(uv) = u\, dv + v\, du$$

Integrating both sides of this equation, we have $uv = \int u\, dv + \int v\, du$, from which

$$\int u\, dv = uv - \int v\, du$$

This innocent-looking formula is the source of a powerful method of integration known as *integration by parts*. A few examples will indicate how useful it is.

□ **Example 1**   Find $\int xe^x\, dx$.

Solution: We think of the integrand $xe^x\, dx$ as having two "parts," naming one part u and the other dv. This may be done in many ways, one of which is $u = x$ and $dv = e^x\, dx$. Calculating

$$du = dx \quad \text{and} \quad v = \int e^x\, dx = e^x \quad \text{(arbitrary constant omitted)}$$

we find

$$\int xe^x\,dx = \int u\,dv = uv - \int v\,du = xe^x - \int e^x\,dx = xe^x - e^x + C$$

Another choice of parts in this problem is $u = e^x$ and $dv = x\,dx$, from which $du = e^x\,dx$ and $v = x^2/2$. The parts formula gives

$$\int xe^x\,dx = \frac{1}{2}x^2e^x - \frac{1}{2}\int x^2e^x\,dx$$

While this is certainly correct, it is not going anywhere. The new integral (on the right side) is harder than the original one. This illustrates the fact that integration by parts is not a routine procedure; it requires some judgment (based on experience) of what choices of u and dv are good ones.   □

Remark   It is worth noting that once u is chosen in the method of integration by parts, the choice of dv is automatic. Thus in Example 1 we chose $u = x$; then we had no choice but to write $dv = e^x\,dx$. The calculation of $du = dx$ (by differentiation of $u = x$) and of $v = e^x$ (by integration of $dv = e^x\,dx$) is also automatic. In view of these facts the choice of u is clearly crucial. There are two critical considerations to keep in mind when this choice is made:

1. We must be able to evaluate $\int dv$ (to obtain v from dv).
2. The new integral $\int v\,du$ must be easier to evaluate than the original integral $\int u\,dv$. (See Example 7 for an exception to this statement, however.)

□ **Example 2**   You may have already encountered the formula

$$\int \ln x\,dx = x \ln x - x + C \qquad \text{(Additional Problem 25, Chapter 8)}$$

To see where it comes from, let $u = \ln x$ and $dv = dx$. Then $du = dx/x$ and $v = x$, from which

$$\int \ln x\,dx = uv - \int v\,du = x \ln x - \int dx = x \ln x - x + C \qquad □$$

□ **Example 3**   To find $\int \sin^{-1} x\,dx$ (Problem 35, Section 9.2), let $u = \sin^{-1} x$ and $dv = dx$. Then

$$du = \frac{dx}{\sqrt{1 - x^2}} \qquad \text{and} \qquad v = x$$

from which $\displaystyle \int \sin^{-1} x\,dx = x \sin^{-1} x - \int \frac{x\,dx}{\sqrt{1 - x^2}}$

The new integral may be found by making a substitution, namely $u = 1 - x^2$, $du = -2x\,dx$. Thus

$$\int \frac{x\,dx}{\sqrt{1 - x^2}} = -\frac{1}{2}\int u^{-1/2}\,du = -\frac{1}{2} \cdot 2u^{1/2} = -\sqrt{1 - x^2}$$

and we find

$$\int \sin^{-1} x \, dx = x \sin^{-1} x + \sqrt{1 - x^2} + C \qquad \qquad \square$$

Remark   The letter u occurs in Example 3 with different meanings. Such repetition is convenient (if the earlier u is no longer part of the problem), but of course it should not be done if there is any danger of confusion.

□ **Example 4**   To find

$$\int_0^{\pi/2} x^2 \sin x \, dx$$

let $u = x^2$ and $dv = \sin x \, dx$. Then $du = 2x \, dx$ and $v = -\cos x$, from which

$$\int x^2 \sin x \, dx = -x^2 \cos x + 2 \int x \cos x \, dx$$

While the problem is not solved, the new integral is simpler than the original one; we find it by using integration by parts again:

$$u = x, \, dv = \cos x \, dx \qquad (\text{hence } du = dx, \, v = \sin x)$$

Then

$$\int x \cos x \, dx = x \sin x - \int \sin x \, dx = x \sin x + \cos x$$

The original integral is

$$\int_0^{\pi/2} x^2 \sin x \, dx = (-x^2 \cos x + 2x \sin x + 2 \cos x) \Big|_0^{\pi/2} = \pi - 2 \qquad \square$$

Repeated integration by parts (as in Example 4) occurs so often that it is worthwhile to develop a formula for it. Suppose that we want to find

$$\int f(x) g(x) \, dx$$

where f is a polynomial of degree n and g is a function that can be integrated repeatedly. Letting

$$u = f(x) \qquad \text{and} \qquad dv = g(x) \, dx$$

we have $du = f'(x) \, dx$ and $v = G_1(x)$ (where G_1 is an antiderivative of g). The parts formula gives

$$\int f(x) g(x) \, dx = f(x) G_1(x) - \int f'(x) G_1(x) \, dx$$

Since f is a polynomial, so is f', and its degree is one less than the degree of f. Thus the new integral may be expected to be simpler than the original one (assuming that G_1 is no harder to integrate than g). Use the parts formula again, this time with

$$u = f'(x) \qquad \text{and} \qquad dv = G_1(x) \, dx$$

Then $du = f''(x)\,dx$ and $v = G_2(x)$ (where G_2 is an antiderivative of G_1). This gives

$$\int f(x)g(x)\,dx = f(x)G_1(x) - \left[f'(x)G_2(x) - \int f''(x)G_2(x)\,dx\right]$$

$$= f(x)G_1(x) - f'(x)G_2(x) + \int f''(x)G_2(x)\,dx$$

We will carry out one more step to clarify the general pattern. Let

$$u = f''(x) \quad \text{and} \quad dv = G_2(x)\,dx$$

Then $du = f'''(x)\,dx$ and $v = G_3(x)$ (where G_3 is an antiderivative of G_2) and we have

$$\int f(x)g(x)\,dx = f(x)G_1(x) - f'(x)G_2(x) + f''(x)G_3(x) - \int f'''(x)G_3(x)\,dx$$

This process stops in n steps (when the polynomial has been differentiated down to a constant). Hence the formula we are seeking reads as follows.

Repeated Integration by Parts

$$\int fg = fG_1 - f'G_2 + f''G_3 - f'''G_4 + \cdots + (-1)^n f^{(n)} G_{n+1} + C$$

where f is a polynomial of degree n and $G_1, G_2, \cdots, G_{n+1}$ are successive antiderivatives of g.

□ **Example 5**   Use the formula for repeated integration by parts to confirm the result in Example 4.

Solution: With $f(x) = x^2$ and $g(x) = \sin x$, our formula reads

$$\int x^2 \sin x\,dx = (x^2)(-\cos x) - (2x)(-\sin x) + (2)(\cos x) + C$$

$$= -x^2 \cos x + 2x \sin x + 2 \cos x + C$$

As you can see, this is the same antiderivative we found in Example 4.
                                                                        □

□ **Example 6**   Find $\int x^4 e^{2x}\,dx$.

Solution: This would take a while using the ordinary parts formula repeatedly. The above formula, however, reduces it to an easy problem:

$$\int x^4 e^{2x}\,dx = (x^4)(\tfrac{1}{2}e^{2x}) - (4x^3)(\tfrac{1}{4}e^{2x}) + (12x^2)(\tfrac{1}{8}e^{2x})$$

$$- (24x)(\tfrac{1}{16}e^{2x}) + (24)(\tfrac{1}{32}e^{2x}) + C$$

$$= \tfrac{1}{4}e^{2x}(2x^4 - 4x^3 + 6x^2 - 6x + 3) + C \qquad □$$

□ **Example 7**   To find $\int e^x \cos x\,dx$, let

$$u = e^x \quad \text{and} \quad dv = \cos x\,dx$$

Then $du = e^x\, dx$ and $v = \sin x$, from which

$$\int e^x \cos x\, dx = e^x \sin x - \int e^x \sin x\, dx$$

In the new integral let

$u = e^x$ and $dv = \sin x\, dx$ (hence $du = e^x\, dx$ and $v = -\cos x$)

This gives $\displaystyle \int e^x \cos x\, dx = e^x \sin x - \left(-e^x \cos x + \int e^x \cos x\, dx\right)$

$$= e^x \sin x + e^x \cos x - \int e^x \cos x\, dx$$

While it may appear that we are going in circles, in fact we are not. Simply solve for the original integral to obtain

$$2 \int e^x \cos x\, dx = e^x \sin x + e^x \cos x \qquad \text{(arbitrary constant omitted)}$$

$$\int e^x \cos x\, dx = \tfrac{1}{2}e^x(\sin x + \cos x) + C \qquad \qquad \square$$

□ **Example 8**   Derive a "reduction formula" for $\int \sin^n x\, dx$ (where n is a positive integer greater than 1).

Solution: Let $u = \sin^{n-1} x$ and $dv = \sin x\, dx$. Then

$$du = (n-1) \sin^{n-2} x \cos x\, dx \qquad \text{and} \qquad v = -\cos x$$

from which $\displaystyle \int \sin^n x\, dx = -\sin^{n-1} x \cos x + (n-1) \int \sin^{n-2} x \cos^2 x\, dx$

Replacing $\cos^2 x$ by $1 - \sin^2 x$ in the new integral, we have

$$\int \sin^n x\, dx = -\sin^{n-1} x \cos x + (n-1) \int \sin^{n-2} x\, dx - (n-1) \int \sin^n x\, dx$$

from which $\displaystyle n \int \sin^n x\, dx = -\sin^{n-1} x \cos x + (n-1) \int \sin^{n-2} x\, dx$

Hence $\displaystyle \int \sin^n x\, dx = \frac{-\sin^{n-1} x \cos x}{n} + \frac{n-1}{n} \int \sin^{n-2} x\, dx \quad (n > 1)$ □

A special case of this result is worth recording for future use, along with a similar formula we ask you to derive in the problem set:

If n is a positive integer greater than 1, then

$$\int_0^{\pi/2} \sin^n x\, dx = \frac{n-1}{n} \int_0^{\pi/2} \sin^{n-2} x\, dx$$

$$\int_0^{\pi/2} \cos^n x\, dx = \frac{n-1}{n} \int_0^{\pi/2} \cos^{n-2} x\, dx$$

For example, when $n = 6$ we have

$$\int_0^{\pi/2} \sin^6 x \, dx = \frac{5}{6} \int_0^{\pi/2} \sin^4 x \, dx$$

$$= \frac{5}{6} \cdot \frac{3}{4} \int_0^{\pi/2} \sin^2 x \, dx \quad \text{(using the formula with } n = 4)$$

$$= \frac{5}{6} \cdot \frac{3}{4} \cdot \frac{1}{2} \int_0^{\pi/2} dx \quad \text{(using it again with } n = 2)$$

$$= \frac{5}{6} \cdot \frac{3}{4} \cdot \frac{1}{2} \cdot \frac{\pi}{2} = \frac{5\pi}{32}$$

Problem Set 10.1

Find a formula for each of the following antiderivatives.

1. $\int x e^{2x} \, dx$

2. $\int (3 - x)e^{3x} \, dx$

3. $\int x \sin x \, dx$

4. $\int x \cos 2x \, dx$

5. $\int x(x + 1)^4 \, dx$

6. $\int (x - 2)(2x - 1)^3 \, dx$

7. $\int (x - 1)\csc^2 x \, dx$

8. $\int x \sec^2 x \, dx$

9. $\int x \sqrt{x - 1} \, dx$

10. $\int x \sqrt{2 - x} \, dx$

11. $\int x \ln x \, dx$

12. $\int \ln (x^2 + 1) \, dx$

13. $\int x^2 e^x \, dx$

14. $\int x^2 e^{2x} \, dx$

15. $\int e^x \sin x \, dx$

16. $\int e^{2x} \cos 3x \, dx$

17. $\int \sin (\ln x) \, dx$

18. $\int \cos (\ln x^2) \, dx$

19. $\int \cos^{-1} x \, dx$

20. $\int \tan^{-1} x \, dx$

21. $\int \sinh^{-1} x \, dx$

22. $\int x \tan^{-1} x \, dx$

23. $\int \sin x \sin 3x \, dx$

24. $\int \cos 2x \cos 3x \, dx$

25. $\int \sec^3 x \, dx$ *Hint:* Let $u = \sec x$ and $dv = \sec^2 x \, dx$. Then replace $\tan^2 x$ by $\sec^2 x - 1$ in the new integral.

26. $\int \csc^3 x \, dx$

Evaluate each of the following.

27. $\int_0^1 x e^{-x} \, dx$

28. $\int_0^1 x \sinh x \, dx$

29. $\int_0^1 x(2x - 1)^3 \, dx$

30. $\int_{\pi/6}^{\pi/2} x \csc^2 x \, dx$

31. $\int_1^2 x^2 \ln x \, dx$

32. $\int_1^2 \ln (x^2 + 4) \, dx$

33. $\int_0^\pi e^{-x} \sin x \, dx$

34. $\int_{e^{-\pi}}^{e^\pi} \cos (\ln x) \, dx$

35. $\int_{\pi/4}^{\pi/2} \csc^3 x \, dx$

36. $\int_1^2 \sec^{-1} x \, dx$

37. $\int_0^{1/2} \tanh^{-1} x \, dx$

38. $\int_0^{\pi/6} \sin 2x \cos 3x \, dx$

39. Find $\int e^x \sinh x \, dx$ without using integration by parts.

40. Find $\int e^{-x} \cosh x \, dx$ without using integration by parts.

41. Find the area bounded by the curve $y = \ln x$, the x axis, and the line $x = e$.

42. Find the area bounded by the curve $y = \tan^{-1} x$, the x axis, and the line $x = 1$.

43. The region under the curve $y = \sin x$, $0 \le x \le \pi/2$, is rotated about the y axis. Find the volume generated.

44. The region under the curve $y = e^{-2x}$, $0 \leq x \leq 1$, is rotated about the y axis. Find the volume generated.

45. Find the centroid of the region under the curve $y = \cos x$, $0 \leq x \leq \pi/2$.

46. Find the centroid of the region under the curve $y = e^x$, $0 \leq x \leq 1$.

Use the formula for repeated integration by parts to find each of the following.

47. $\displaystyle\int x^2 e^{-x}\, dx$

48. $\displaystyle\int x^3 e^{2x}\, dx$

49. $\displaystyle\int_0^1 (1-x)^3 e^x\, dx$

50. $\displaystyle\int (2-x)^2 e^{-x}\, dx$

51. $\displaystyle\int (x^2+1)\sin x\, dx$

52. $\displaystyle\int (1-x^2)\cos x\, dx$

53. $\displaystyle\int (x^2-x+1)\cos x\, dx$

54. $\displaystyle\int_0^{\pi/4} x^4 \sin 2x\, dx$

55. $\displaystyle\int x^3(x-1)^4\, dx$

56. $\displaystyle\int (x-2)^2(x+2)^3\, dx$

57. Show that if $b \neq 0$,

$$\int e^{ax} \sin bx\, dx = \frac{e^{ax}(a \sin bx - b \cos bx)}{a^2 + b^2} + C$$

What if $b = 0$ (and $a \neq 0$)?

58. Show that if $b \neq 0$,

$$\int e^{ax} \cos bx\, dx = \frac{e^{ax}(b \sin bx + a \cos bx)}{a^2 + b^2} + C$$

What if $b = 0$ (and $a \neq 0$)?

59. Derive the reduction formula

$$\int \cos^n x\, dx = \frac{\cos^{n-1} x \sin x}{n} + \frac{n-1}{n}\int \cos^{n-2} x\, dx$$

where n is a positive integer greater than 1.

60. Use Problem 59 to explain why

$$\int_0^{\pi/2} \cos^n x\, dx = \frac{n-1}{n}\int_0^{\pi/2} \cos^{n-2} x\, dx$$

Evaluate each of the following integrals.

61. $\displaystyle\int_0^{\pi/2} \cos^6 x\, dx$

62. $\displaystyle\int_0^{\pi/2} \sin^8 x\, dx$

63. $\displaystyle\int_0^{\pi} \sin^5 \frac{x}{2}\, dx$

64. $\displaystyle\int_0^{\pi} \cos^7 2x\, dx$

65. Derive the famous *Wallis formulas*

$$\int_0^{\pi/2} \sin^{2k} x\, dx = \frac{1 \cdot 3 \cdot 5 \cdot (2k-1)}{2 \cdot 4 \cdot 6 \cdot (2k)} \cdot \frac{\pi}{2}$$

$$\int_0^{\pi/2} \sin^{2k+1} x\, dx = \frac{2 \cdot 4 \cdot 6 \cdot (2k)}{1 \cdot 3 \cdot 5 \cdot (2k+1)}$$

where k is any positive integer.

66. It is not hard to prove that the ratio of the integrals in Problem 65 approaches 1 as k increases.

(a) Explain why it follows that

$$\frac{\pi}{2} = \frac{2}{1} \cdot \frac{2}{3} \cdot \frac{4}{3} \cdot \frac{4}{5} \cdot \frac{6}{5} \cdot \frac{6}{7}$$

in the sense that $\pi/2$ may be approximated as closely as desired by using sufficiently many factors of this "infinite product."

(b) Show that part (a) can be written in the form

$$\pi = 2 \cdot \frac{4}{3} \cdot \frac{16}{15} \cdot \frac{36}{35} \cdot \frac{64}{63}$$

(c) Use the formula in part (b) (and a calculator) to get an idea of how fast the product converges to π. (Punch 2 times 4 divided by 3 times 16 divided by 15 times 36···.)

10.2

**INTEGRALS
INVOLVING
TRIGONOMETRIC
FUNCTIONS**

In the next section we will discuss an important class of substitutions that have the effect of transforming certain algebraic integrals into integrals involving trigonometric functions. If we can handle the latter, we are in a position to dispose of many previously intractable problems. Our purpose in this section is to discuss the trigonometric integrals that are most useful in this connection. They are of three types, namely

1. $\displaystyle\int \sin^m x \cos^n x\, dx$ 2. $\displaystyle\int \tan^m x \sec^n x\, dx$ 3. $\displaystyle\int \cot^m x \csc^n x\, dx$

Rather than considering a systematic listing of cases under these three types, we simply present examples to show you the techniques that are needed for dealing with trigonometric integrals.

□ **Example 1**   Find $\int \sin^3 x \sqrt{\cos x}\, dx$.

Solution: The odd exponent ($m = 3$) allows us to break off $\sin x\, dx$ and change the remaining even power ($\sin^2 x$) to an expression involving cosine:

$$\int \sin^3 x \sqrt{\cos x}\, dx = \int \sin^2 x \sqrt{\cos x} \cdot \sin x\, dx$$

$$= \int (1 - \cos^2 x) \sqrt{\cos x} \cdot \sin x\, dx$$

Now let $u = \cos x$ and $du = -\sin x\, dx$ to obtain

$$\int \sin^3 x \sqrt{\cos x}\, dx = -\int (1 - u^2) \sqrt{u}\, du = \int (u^{5/2} - u^{1/2})\, du$$

$$= \frac{2}{7} u^{7/2} - \frac{2}{3} u^{3/2} + C = \frac{2}{7}(\cos x)^{7/2} - \frac{2}{3}(\cos x)^{3/2} + C \quad □$$

□ **Example 2**   Find $\displaystyle\int_0^\pi \sin^2 \frac{t}{2} \cos^5 \frac{t}{2}\, dt$.

Solution: This time the odd exponent ($n = 5$) suggests breaking off $\cos \dfrac{t}{2}\, dt$ and expressing what remains in terms of $\sin \dfrac{t}{2}$:

$$\int_0^\pi \sin^2 \frac{t}{2} \cos^5 \frac{t}{2}\, dt = \int_0^\pi \sin^2 \frac{t}{2} \cos^4 \frac{t}{2} \cdot \cos \frac{t}{2}\, dt$$

$$= \int_0^\pi \sin^2 \frac{t}{2} \left(1 - \sin^2 \frac{t}{2}\right)^2 \cdot \cos \frac{t}{2}\, dt$$

$$= 2 \int_0^1 u^2 (1 - u^2)^2\, du \qquad \left(u = \sin \frac{t}{2},\, du = \frac{1}{2} \cos \frac{t}{2}\, dt\right)$$

$$= 2 \int_0^1 (u^6 - 2u^4 + u^2)\, du$$

$$= 2 \left(\frac{u^7}{7} - \frac{2u^5}{5} + \frac{u^3}{3}\right) \Big|_0^1 = \frac{16}{105} \qquad □$$

□ **Example 3**   Find $\int \sin^2 t \cos^2 t\, dt$.

Solution: The even exponents suggest the multiplication formulas

$$\sin^2 t = \tfrac{1}{2}(1 - \cos 2t) \qquad \text{and} \qquad \cos^2 t = \tfrac{1}{2}(1 + \cos 2t)$$

(See Section 2.3.) We use them to write

$$\int \sin^2 t \cos^2 t \, dt = \frac{1}{4} \int (1 - \cos^2 2t) \, dt = \frac{1}{4} \int dt - \frac{1}{4} \int \cos^2 2t \, dt$$

In the last integral we use a multiplication formula again, in the form

$$\cos^2 2t = \frac{1}{2}(1 + \cos 4t)$$

This yields

$$\int \sin^2 t \cos^2 t \, dt = \frac{1}{4} \int dt - \frac{1}{8} \int (1 + \cos 4t) \, dt = \frac{1}{8} \int dt - \frac{1}{8} \int \cos 4t \, dt$$

$$= \frac{1}{8} t - \frac{1}{32} \sin 4t + C \qquad\qquad \square$$

□ **Example 4**   Find $\int \dfrac{\sec^4 t}{\tan t} \, dt$.

Solution: The fact that $\sec^2 t$ is the derivative of $\tan t$ suggests that we break off $\sec^2 t \, dt$:

$$\int \frac{\sec^4 t}{\tan t} \, dt = \int \frac{\sec^2 t}{\tan t} \cdot \sec^2 t \, dt$$

The identity $\sec^2 t - \tan^2 t = 1$ enables us to write

$$\int \frac{\sec^4 t}{\tan t} \, dt = \int \frac{\tan^2 t + 1}{\tan t} \cdot \sec^2 t \, dt$$

Now let $u = \tan t$, $du = \sec^2 t \, dt$ to obtain

$$\int \frac{\sec^4 t}{\tan t} \, dt = \int \frac{u^2 + 1}{u} \, du = \int \left(u + \frac{1}{u} \right) du = \frac{u^2}{2} + \ln |u| + C$$

$$= \frac{1}{2} \tan^2 t + \ln |\tan t| + C \qquad\qquad \square$$

□ **Example 5**   Find $\displaystyle\int_0^{\pi/3} \tan^3 t \sec t \, dt$.

Solution: This time we break off $\sec t \tan t \, dt$, having in mind the derivative of $\sec t$ and intending to express what remains in terms of $\sec t$:

$$\int_0^{\pi/3} \tan^3 t \sec t \, dt = \int_0^{\pi/3} \tan^2 t \cdot \sec t \tan t \, dt$$

$$= \int_0^{\pi/3} (\sec^2 t - 1) \cdot \sec t \tan t \, dt$$

$$= \int_1^2 (u^2 - 1)\, du \qquad (u = \sec t,\ du = \sec t \tan t\, dt)$$

$$= \frac{4}{3}$$
□

□ **Example 6**   Find $\int \cot^4 t\, dt$.

Solution: Use the identity $\csc^2 t - \cot^2 t = 1$ to write

$$\int \cot^4 t\, dt = \int \cot^2 t\, (\csc^2 t - 1)\, dt = \int \cot^2 t \cdot \csc^2 t\, dt - \int \cot^2 t\, dt$$

$$= \int \cot^2 t \cdot \csc^2 t\, dt - \int (\csc^2 t - 1)\, dt$$

$$= -\frac{1}{3} \cot^3 t + \cot t + t + C$$

These examples and the problem set should give you the idea. We will put the results to work in the next section.
□

Problem Set 10.2

Find each of the following integrals.

1. $\displaystyle\int_0^{\pi/2} \sin^4 x \cos^3 x\, dx$ **2.** $\displaystyle\int_0^{\pi/2} \cos x \sqrt{\sin x}\, dx$

3. $\displaystyle\int \cos^3 x \sqrt[3]{\sin x}\, dx$ **4.** $\displaystyle\int \cos^3 t \sin^2 t\, dt$

5. $\displaystyle\int \sin^3 x \cos^3 x\, dx$ **6.** $\displaystyle\int_0^{\pi/6} \frac{\sin^3 t}{\cos^2 t}\, dt$

7. $\displaystyle\int \sin^3 x \cos^4 x\, dx$ **8.** $\displaystyle\int \frac{\cos^4 t}{\sin^2 t}\, dt$

9. $\displaystyle\int_0^{\pi} \cos^4 t\, dt$ **10.** $\displaystyle\int \sin^5 t\, dt$

11. $\displaystyle\int_0^{\pi/2} \cos^3 x\, dx$ **12.** $\displaystyle\int \sin^2 2x\, dx$

13. $\displaystyle\int \frac{\sin^5 x}{\cos^5 x}\, dx$ **14.** $\displaystyle\int \sin^3 x\, dx$

15. $\displaystyle\int_0^{\pi/4} \frac{\sin^4 x}{\cos^2 x}\, dx$ **16.** $\displaystyle\int \sin t \cot t\, dt$

17. $\displaystyle\int \cos^2 t \tan t\, dt$ **18.** $\displaystyle\int \cot^2 t \sin t\, dt$

19. $\displaystyle\int_0^{\pi/6} \sec^3 t \sin t\, dt$ **20.** $\displaystyle\int \sin x \tan^2 x\, dx$

21. $\displaystyle\int \frac{dt}{\sin t \cos t}$ **22.** $\displaystyle\int \tan^4 x\, dx$

23. $\displaystyle\int \tan x \sec^3 x\, dx$ **24.** $\displaystyle\int \tan^2 x \sec^4 x\, dx$

25. $\displaystyle\int_{\pi/4}^{\pi/3} \tan^4 t \sec^4 t\, dt$ **26.** $\displaystyle\int_0^{\pi/4} \sqrt{\tan t}\, \sec^4 t\, dt$

27. $\displaystyle\int \tan^3 x\, dx$ **28.** $\displaystyle\int (\tan x + \cot x)^2\, dx$

29. $\displaystyle\int_0^{\pi/4} \sec^6 t\, dt$ **30.** $\displaystyle\int \tan^6 t\, dt$

31. $\int \tan^3 x (\sec x)^{3/2} \, dx$ **32.** $\int \dfrac{\sec^4 t}{\tan^2 t} \, dt$

33. $\int_{\pi/4}^{\pi/2} \cot^2 t \, dt$ **34.** $\int \cot^3 t \, dt$

35. $\int \csc^4 t \cot^2 t \, dt$ **36.** $\int_{\pi/4}^{\pi/2} \cot^3 t \csc t \, dt$

37. $\int_{\pi/4}^{\pi/2} \csc^4 x \, dx$ **38.** $\int \sec^4 x \csc^4 x \, dx$

39. $\int_{\pi/6}^{\pi/2} \dfrac{\cot t}{\csc^3 t} \, dt$

40. Work Example 3 by using the formula $\sin 2t = 2 \sin t \cos t$.

10.3
TRIGONOMETRIC SUBSTITUTIONS

Many important integration problems involve expressions of the form

$$\sqrt{a^2 - x^2} \qquad \sqrt{a^2 + x^2} \qquad \sqrt{x^2 - a^2}$$

where a is a (positive) constant. In such cases a *trigonometric substitution* is often helpful, of the type

$$x = a \sin t \qquad x = a \tan t \qquad x = a \sec t$$

respectively. To see what effect these substitutions have on the corresponding expressions, observe that

$$a^2 - x^2 = a^2 - a^2 \sin^2 t = a^2(1 - \sin^2 t) = a^2 \cos^2 t$$
$$a^2 + x^2 = a^2 + a^2 \tan^2 t = a^2(1 + \tan^2 t) = a^2 \sec^2 t$$
$$x^2 - a^2 = a^2 \sec^2 t - a^2 = a^2(\sec^2 t - 1) = a^2 \tan^2 t$$

respectively. If we assume that our substitutions are invertible (that is, only restricted trigonometric functions having inverses are used), then

$\sqrt{a^2 - x^2} = a|\cos t| = a \cos t$ (because $-\pi/2 \le t \le \pi/2$)
$\sqrt{a^2 + x^2} = a|\sec t| = a \sec t$ (because $-\pi/2 < t < \pi/2$)
$\sqrt{x^2 - a^2} = a|\tan t| = a \tan t$ (because $0 \le t < \pi/2$ or $\pi \le t < 3\pi/2$)

In other words, each substitution eliminates the radical by replacing it with a trigonometric function. This has a remarkable effect on many otherwise difficult integrals, as you will see.

Remark   A review of Section 9.1 may be needed here. What we are saying is that the substitutions

$$x = a \sin t \qquad x = a \tan t \qquad x = a \sec t$$

are equivalent to

$$t = \sin^{-1} \frac{x}{a} \qquad t = \tan^{-1} \frac{x}{a} \qquad t = \sec^{-1} \frac{x}{a}$$

respectively, provided that sine, tangent, and secant are restricted as described in Section 9.1. You may also want to look at Section 6.5 again, particularly the two substitution theorems stated there (and Example 7). These theorems call for a "change of variable" function $t = h(x)$ that gives the new variable in terms of the old. If the function is invertible, however, it makes no difference, since either x or t can be expressed in terms of the other.

□ **Example 1**   An important standard form (first stated in Problem 38, Section 9.2, but as yet unexplained) is

$$\int \sqrt{a^2 - x^2}\, dx = \frac{x}{2}\sqrt{a^2 - x^2} + \frac{a^2}{2}\sin^{-1}\frac{x}{a} + C$$

To see where it comes from, make the substitution

$$x = a \sin t \qquad (-\pi/2 \le t \le \pi/2)$$

Then $dx = a \cos t\, dt$ and (as explained above) $\sqrt{a^2 - x^2} = a \cos t$. Our integral becomes

$$\int \sqrt{a^2 - x^2}\, dx = \int a \cos t \cdot a \cos t\, dt = a^2 \int \cos^2 t\, dt$$

$$= \frac{a^2}{2}\int (1 + \cos 2t)\, dt$$

$$= \frac{a^2}{2}\left(t + \frac{1}{2}\sin 2t\right) + C$$

$$= \frac{a^2}{2}(t + \sin t \cos t) + C \qquad (\text{because } \sin 2t = 2 \sin t \cos t)$$

Since our change of variable equation ($x = a \sin t$) is equivalent to

$$t = \sin^{-1}\frac{x}{a}$$

and since $\qquad \sin t = \dfrac{x}{a} \qquad$ and $\qquad \cos t = \dfrac{\sqrt{a^2 - x^2}}{a}$

we find $\qquad \displaystyle\int \sqrt{a^2 - x^2}\, dx = \frac{a^2}{2}\left(\sin^{-1}\frac{x}{a} + \frac{x}{a}\cdot\frac{\sqrt{a^2 - x^2}}{a}\right) + C$

$$= \frac{x}{2}\sqrt{a^2 - x^2} + \frac{a^2}{2}\sin^{-1}\frac{x}{a} + C$$

as advertised.   □

□ **Example 2**   To find

$$\int_0^3 \frac{x^3\, dx}{\sqrt{9 + x^2}}$$

make the substitution

$$x = 3 \tan t \qquad (-\pi/2 < t < \pi/2)$$

Then $dx = 3 \sec^2 t\, dt$ and

$$\sqrt{9 + x^2} = \sqrt{9 + 9 \tan^2 t} = 3\sqrt{1 + \tan^2 t} = 3\sqrt{\sec^2 t}$$

$$= 3 \sec t \qquad (-\pi/2 < t < \pi/2)$$

Since $x = 0 \Rightarrow t = 0$ and $x = 3 \Rightarrow t = \pi/4$ (why?), we find

$$\int_0^3 \frac{x^3\, dx}{\sqrt{9 + x^2}} = \int_0^{\pi/4} \frac{27 \tan^3 t \cdot 3 \sec^2 t\, dt}{3 \sec t} = 27 \int_0^{\pi/4} \sec t \tan^3 t\, dt$$

At this point we are faced with a new problem (not necessarily easier than the original one, but let's hope for the best). The appearance of secant and tangent together in the integrand is a good sign, because we know an identity ($\sec^2 t - \tan^2 t = 1$) and several differentiation and integration formulas involving these functions.

You may have to stare at the problem for a while to see what to do. Let's break off a tangent (to go with secant) and see what happens:

$$\int \sec t \tan^3 t\, dt = \int \tan^2 t \cdot \sec t \tan t\, dt$$

The idea behind this move is that $\sec t \tan t$ is the derivative of $\sec t$; if we can express the rest of the integrand in terms of $\sec t$, the substitution

$$u = \sec t \qquad du = \sec t \tan t\, dt$$

will do the trick. To this end, write $\tan^2 t = \sec^2 t - 1$. Then

$$\int \sec t \tan^3 t\, dt = \int (\sec^2 t - 1) \sec t \tan t\, dt = \int (u^2 - 1)\, du$$

Now it is easy to finish:

$$\int_0^3 \frac{x^3\, dx}{\sqrt{9 + x^2}} = 27 \int_0^{\pi/4} \sec t \tan^3 t\, dt$$

$$= 27 \int_1^{\sqrt{2}} (u^2 - 1)\, du \quad (t = 0 \Rightarrow u = 1 \text{ and } t = \frac{\pi}{4} \Rightarrow u = \sqrt{2})$$

$$= 27 \left(\frac{u^3}{3} - u\right)\Big|_1^{\sqrt{2}} = 9u(u^2 - 3)\Big|_1^{\sqrt{2}}$$

$$= 9[\sqrt{2}(2 - 3) - (1 - 3)] = 9(2 - \sqrt{2}) \qquad \square$$

Remark   There is an easier way to do Example 2, based on the substitution $u = \sqrt{9 + x^2}$. (See the problem set.) Trigonometric substitution is an important technique of integration, but it is not guaranteed to give the quickest results in all cases that appear to call for it.

□ **Example 3**   Find a formula for $\int \sqrt{x^2 - a^2}\, dx$.

Solution: Let

$$x = a \sec t \qquad (0 \le t < \pi/2 \text{ or } \pi \le t < 3\pi/2)$$

Then $dx = a \sec t \tan t\, dt$ and (as already explained) $\sqrt{x^2 - a^2} = a \tan t$. Our integral becomes

$$\int \sqrt{x^2 - a^2}\, dx = \int a \tan t \cdot a \sec t \tan t\, dt = a^2 \int \sec t \tan^2 t\, dt$$

The device used in Example 2 (breaking off $\sec t \tan t$ in the hope of expressing the rest of the integrand in terms of $\sec t$) does not work this time. (Why?) Instead we write

$$\int \sec t \tan^2 t \, dt = \int \sec t \, (\sec^2 t - 1) \, dt = \int \sec^3 t \, dt - \int \sec t \, dt$$

The second integral on the right appears as a standard form in our table of integrals. The first is not easy, but it can be handled by integration by parts. In fact it has already appeared in Problem 25 of Section 10.1, the result of which is sufficiently useful to be recorded:

$$\int \sec^3 x \, dx = \frac{1}{2} \sec x \tan x + \frac{1}{2} \ln |\sec x + \tan x| + C$$

Using these results, we find

$$\int \sec t \tan^2 t \, dt = \frac{1}{2} \sec t \tan t + \frac{1}{2} \ln |\sec t + \tan t| - \ln |\sec t + \tan t| + C$$

$$= \frac{1}{2} \sec t \tan t - \frac{1}{2} \ln |\sec t + \tan t| + C$$

$$= \frac{1}{2} \cdot \frac{x}{a} \cdot \frac{\sqrt{x^2 - a^2}}{a} - \frac{1}{2} \ln \left| \frac{x}{a} + \frac{\sqrt{x^2 - a^2}}{a} \right| + C$$

$$= \frac{x}{2a^2} \sqrt{x^2 - a^2} - \frac{1}{2} \ln \left| \frac{x + \sqrt{x^2 - a^2}}{a} \right| + C$$

$$= \frac{x}{2a^2} \sqrt{x^2 - a^2} - \frac{1}{2} \ln |x + \sqrt{x^2 - a^2}| + \frac{1}{2} \ln a + C$$

The constant $\frac{1}{2} \ln a$ may be dropped. (Why?) Hence the original integral is

$$\int \sqrt{x^2 - a^2} \, dx = \frac{x}{2} \sqrt{x^2 - a^2} - \frac{a^2}{2} \ln |x + \sqrt{x^2 - a^2}| + C \qquad \square$$

We leave it to you (in the problem set) to derive the companion formula

$$\int \sqrt{x^2 + a^2} \, dx = \frac{x}{2} \sqrt{x^2 + a^2} + \frac{a^2}{2} \ln (x + \sqrt{x^2 + a^2}) + C$$

$\square$ **Example 4**   The semicircular region $R = \{(x,y): x^2 + y^2 \le a^2, y \ge 0\}$ is covered by a thin material whose density at (x,y) is $\delta(x,y) = y$. Find the center of mass of the material.

Figure 1
Center of mass by horizontal strips

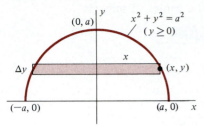

Solution: See Figure 1, in which we have used a horizontal strip as the element of area (because the density is essentially constant in such a strip if it is thin). The mass of the strip is

$$\Delta m \approx \delta \, \Delta A \approx y \cdot 2x \, \Delta y = 2xy \, \Delta y$$

so the mass of the material covering R is

$$m = \int dm = 2 \int_0^a y \sqrt{a^2 - y^2} \, dy$$

$$= -\int_{a^2}^0 u^{1/2} \, du \qquad (u = a^2 - y^2, \, du = -2y \, dy)$$

$$= \frac{2}{3} u^{3/2} \Big|_0^{a^2} = \frac{2}{3} a^3$$

The moment of the strip relative to the x axis is

$$\Delta M_x \approx y \, \Delta m \approx 2xy^2 \, \Delta y$$

so the total moment is

$$M_x = 2 \int_0^a y^2 \sqrt{a^2 - y^2} \, dy$$

This integral calls for the trigonometric substitution

$$y = a \sin t \qquad (-\pi/2 \le t \le \pi/2)$$

Since $dy = a \cos t \, dt$ and

$$\sqrt{a^2 - y^2} = \sqrt{a^2 - a^2 \sin^2 t} = a \cos t$$

we find

$$M_x = 2 \int_0^{\pi/2} a^2 \sin^2 t \cdot a \cos t \cdot a \cos t \, dt = 2a^4 \int_0^{\pi/2} \sin^2 t \cos^2 t \, dt$$

The device to be used at this point is not obvious, but you can see how it works. We use the identity

$$\sin 2t = 2 \sin t \cos t \qquad \text{(from which } \sin^2 t \cos^2 t = \tfrac{1}{4} \sin^2 2t\text{)}$$

to write

$$M_x = \frac{a^4}{2} \int_0^{\pi/2} \sin^2 2t \, dt$$

Then the identity

$$\sin^2 t = \tfrac{1}{2}(1 - \cos 2t) \qquad \text{(with } t \text{ replaced by } 2t\text{)}$$

yields

$$M_x = \frac{a^4}{4} \int_0^{\pi/2} (1 - \cos 4t) \, dt = \frac{a^4}{4}\left(t - \frac{1}{4} \sin 4t\right)\Big|_0^{\pi/2} = \frac{\pi a^4}{8}$$

The y coordinate of the center of mass is therefore

$$\bar{y} = \frac{M_x}{m} = \frac{\pi a^4}{8} \cdot \frac{3}{2a^3} = \frac{3}{16} \pi a$$

The x coordinate is 0 because of the symmetry of the region. (Note that this is unaffected by the variable density, since $\delta = y$. The mass is distributed symmetrically about the y axis.) Thus the center of mass is

$$(\bar{x},\bar{y}) = (0,\tfrac{3}{16}\pi a)$$

□ **Example 5**   To compute

$$\int_2^6 \frac{dx}{x^2\sqrt{4+x^2}}$$

make the substitution

$$x = 2\tan t \qquad (-\pi/2 < t < \pi/2)$$

Then $dx = 2\sec^2 t\, dt$ and

$$\sqrt{4+x^2} = \sqrt{4+4\tan^2 t} = 2\sec t$$

from which

$$\int \frac{dx}{x^2\sqrt{4+x^2}} = \int \frac{2\sec^2 t\, dt}{4\tan^2 t \cdot 2\sec t} = \frac{1}{4}\int \frac{\sec t\, dt}{\tan^2 t}$$

Since there is no obvious way to work this out in terms of secant and tangent, we change to sine and cosine:

$$\int \frac{dx}{x^2\sqrt{4+x^2}} = \frac{1}{4}\int \frac{\cos t\, dt}{\sin^2 t}$$

$$= \frac{1}{4}\int \frac{du}{u^2} \qquad (u = \sin t,\ du = \cos t\, dt)$$

$$= -\frac{1}{4}u^{-1} = -\frac{1}{4}\csc t \qquad \text{(arbitrary constant omitted)}$$

At this point we have two options. One is to change limits and write

$$\int_2^6 \frac{dx}{x^2\sqrt{4+x^2}} = -\frac{1}{4}\csc t \Big|_{\pi/4}^{\tan^{-1}3} = \frac{1}{4}[\sqrt{2} - \csc(\tan^{-1}3)]$$

while the other is to return to the original variable (expressing $-\frac{1}{4}\csc t$ in terms of x).

The first option requires us to find $\csc(\tan^{-1}3)$. Letting $\theta = \tan^{-1}3$ (from which $\tan\theta = 3$), we observe from Figure 2 that

$$\csc(\tan^{-1}3) = \csc\theta = \frac{\sqrt{10}}{3}$$

Hence

$$\int_2^6 \frac{dx}{x^2\sqrt{4+x^2}} = \frac{1}{4}\left(\sqrt{2} - \frac{\sqrt{10}}{3}\right) = \frac{1}{12}(3\sqrt{2} - \sqrt{10})$$

Figure 2
Finding $\csc(\tan^{-1}3)$

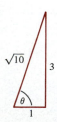

Figure 3
Finding csc t

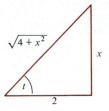

The second option requires that we find csc t in terms of $x = 2 \tan t$. Using Figure 3, we observe that $\csc t = \sqrt{4 + x^2}/x$. Hence

$$\int_2^6 \frac{dx}{x^2\sqrt{4 + x^2}} = -\frac{1}{4}\csc t\Big|_{x=2}^{x=6} = -\frac{1}{4}\cdot\frac{\sqrt{4 + x^2}}{x}\Big|_2^6 = \frac{1}{4}\left(\frac{\sqrt{8}}{2} - \frac{\sqrt{40}}{6}\right)$$

$$= \frac{1}{4}\left(\sqrt{2} - \frac{\sqrt{10}}{3}\right) = \frac{1}{12}(3\sqrt{2} - \sqrt{10})$$

as before.

As you can see, there is not much to choose between these options. Note, however, the usefulness of the triangles. Drawing such figures is often helpful in problems involving trigonometric substitutions.   □

Triangles that may be used in connection with our three trigonometric substitutions are shown in Figure 4. Because of our choice of domain for each (invertible) trigonometric function, these triangles apply when $x < 0$ as well as $x > 0$. They are drawn for the case $0 < t < \pi/2$ (t is an acute

Figure 4
Triangles for trigonometric substitutions

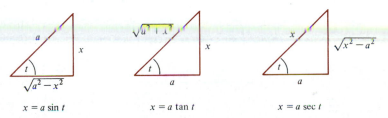

$x = a \sin t$ $x = a \tan t$ $x = a \sec t$

angle), in which case x is positive. Consider the third one, however, in the case $\pi < t < 3\pi/2$ (the other part of the range of sec^{-1}). The proper way to draw it is shown in Figure 5. The legs are directed distances (both negative in this case), while the hypotenuse, as always, is an ordinary positive distance (because $x < 0$). These precautions are unnecessary, however, because the other functions come out just as they do in the acute case:

$$\sin t = \frac{-\sqrt{x^2 - a^2}}{-x} = \frac{\sqrt{x^2 - a^2}}{x}$$

$$\cos t = \frac{-a}{-x} = \frac{a}{x}$$

$$\tan t = \frac{-\sqrt{x^2 - a^2}}{-a} = \frac{\sqrt{x^2 - a^2}}{a}$$

Figure 5
Triangle for $x = a \sec t$
($\pi < t < 3\pi/2$)

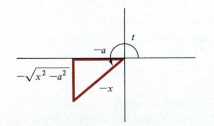

and so on.

It is worth noting that these trouble-free triangles are no accident! They are a consequence of our choice of domains in Section 9.1 (particularly in the case of sec^{-1}, which as we noted at the time is not always defined the same way).

Problem Set 10.3

Find a formula for each of the following.

1. $\displaystyle\int \frac{dx}{x^2 \sqrt{4 - x^2}}$ **2.** $\displaystyle\int \frac{x^2\, dx}{\sqrt{25 - x^2}}$

3. $\displaystyle\int \frac{\sqrt{1 + x^2}}{x^2}\, dx$ *Hint:* $\displaystyle\int \frac{\sec^3 t}{\tan^2 t}\, dt$ can be written as a sum of manageable integrals by using $\sec^2 t = \tan^2 t + 1$.

4. $\displaystyle\int \frac{dx}{(9 + x^2)^2}$ **5.** $\displaystyle\int \frac{x^2\, dx}{\sqrt{x^2 - 1}}$

6. $\displaystyle\int \frac{dx}{x^4 \sqrt{x^2 - 1}}$ *Hint:* $\displaystyle\int \cos^3 t\, dt = \int (1 - \sin^2 t) \cos t\, dt$

7. $\displaystyle\int \frac{dx}{(1 - x^2)^{3/2}}$ **8.** $\displaystyle\int x^2 \sqrt{4 - x^2}\, dx$

9. $\displaystyle\int \frac{dx}{(4 + 9x^2)^{3/2}}$ *Hint:* Let $3x = 2 \tan t$

10. $\displaystyle\int \frac{(1 + x^2)^{3/2}\, dx}{x^6}$

11. $\displaystyle\int \frac{\sqrt{1 - 4x^2}}{x^2}\, dx$ *Hint:* $\displaystyle\int \cot^2 t\, dt = \int (\csc^2 t - 1)\, dt$

12. $\displaystyle\int \frac{\sqrt{x^2 - 4}}{x^2}\, dx$ **13.** $\displaystyle\int \frac{x^2\, dx}{\sqrt{9 + x^2}}$

14. $\displaystyle\int \frac{\sqrt{9 + x^2}}{x^4}\, dx$ **15.** $\displaystyle\int \frac{\sqrt{x^2 - 9}}{x^4}\, dx$

16. $\displaystyle\int \frac{x^4\, dx}{\sqrt{1 - x^2}}$ **17.** $\displaystyle\int x \sin^{-1} x\, dx$

18. $\displaystyle\int x \sinh^{-1} x\, dx$

Evaluate each of the following.

19. $\displaystyle\int_2^4 \frac{\sqrt{16 - x^2}}{x^2}\, dx$ **20.** $\displaystyle\int_0^3 x^2 \sqrt{9 - x^2}\, dx$

21. $\displaystyle\int_2^{2\sqrt{3}} \frac{dx}{x(x^2 + 4)}$ **22.** $\displaystyle\int_0^2 \frac{x^2\, dx}{\sqrt{4 + x^2}}$

23. $\displaystyle\int_0^1 \frac{x^2\, dx}{(4 - x^2)^{3/2}}$

24. $\displaystyle\int_2^3 \frac{dx}{x^4 \sqrt{16 - x^2}}$

Hint: $\displaystyle\int \csc^4 t\, dt = \int (\cot^2 t + 1) \csc^2 t\, dt$

25. $\displaystyle\int_1^2 \frac{\sqrt{1 + 4x^2}}{x^4}\, dx$

26. $\displaystyle\int_3^6 \frac{\sqrt{x^2 - 9}}{x^2}\, dx$

27. $\displaystyle\int_4^8 \frac{(x^2 - 16)^{3/2}}{x^2}\, dx$

28. $\displaystyle\int_0^2 \frac{x^2\, dx}{(x^2 + 4)^{3/2}}$

29. Find $\displaystyle\int \frac{x\, dx}{x^2 + 9}$

in two ways, as follows. **(a)** Make an appropriate algebraic substitution; **(b)** make a trigonometric substitution; and **(c)** reconcile the results.

30. Find $\displaystyle\int \frac{x\, dx}{\sqrt{x^2 - 1}}$

in two ways, as follows. **(a)** Make an appropriate algebraic substitution; **(b)** make a trigonometric substitution; and **(c)** reconcile the results.

31. Find $\displaystyle\int_0^3 \frac{x^3\, dx}{\sqrt{9 + x^2}}$

by making the substitution $u = \sqrt{9 + x^2}$. *Hint:* $u^2 = 9 + x^2$ and $u\, du = x\, dx$.

32. Find $\displaystyle\int_1^2 \frac{\sqrt{4 - x^2}}{x}\, dx$

in two ways, as follows. **(a)** Make the substitution $u = \sqrt{4 - x^2}$; and **(b)** make a trigonometric substitution.

Problems 31 and 32 raise a question. If substitution of u for the radical works in these problems, why not in all the earlier ones? The answer is that $x\, dx$ is called for in the integrand; only if there is an *odd* power of x already present (becoming *even* when the needed x is supplied) is the substitution effective.

33. Find the length of the parabolic arc $y = \frac{1}{2}x^2$, $0 \le x \le 2$.

34. Find the length of the curve $y = \ln x$, $1 \le x \le 2$.

35. Solve the initial value problem

$$\frac{dy}{dx} = \frac{\sqrt{x^2 - 1}}{x^2} \qquad y = 0 \text{ when } x = 1$$

Each of the following standard forms has already been derived. Use a trigonometric substitution to confirm it. (Note the comparative simplicity of the argument, with the possible exception of the last formula, which requires some manipulation.)

36. $\int \dfrac{dx}{\sqrt{a^2 - x^2}} = \sin^{-1}\dfrac{x}{a} + C$

37. $\int \dfrac{dx}{a^2 + x^2} = \dfrac{1}{a}\tan^{-1}\dfrac{x}{a} + C$

38. $\int \dfrac{dx}{x\sqrt{x^2 - a^2}} = \dfrac{1}{a}\sec^{-1}\dfrac{x}{a} + C$

39. $\int \dfrac{dx}{\sqrt{x^2 + a^2}} = \ln\left(x + \sqrt{x^2 + a^2}\right) + C$

40. $\int \dfrac{dx}{\sqrt{x^2 - a^2}} = \ln\left|x + \sqrt{x^2 - a^2}\right| + C$

41. $\int \dfrac{dx}{x\sqrt{a^2 - x^2}} = \dfrac{1}{a}\ln\left|\dfrac{a - \sqrt{a^2 - x^2}}{x}\right| + C$

42. $\int \dfrac{dx}{x\sqrt{a^2 + x^2}} = \dfrac{1}{a}\ln\left|\dfrac{a - \sqrt{a^2 + x^2}}{x}\right| + C$

43. $\int \dfrac{dx}{a^2 - x^2} = \dfrac{1}{2a}\ln\left|\dfrac{a + x}{a - x}\right| + C$

44. Derive the formula

$$\int \sqrt{x^2 + a^2}\, dx =$$

$$\dfrac{x}{2}\sqrt{x^2 + a^2} + \dfrac{a^2}{2}\ln\left(x + \sqrt{x^2 + a^2}\right) + C$$

45. The triangles in Figure 4 are drawn for the case $0 < t < \pi/2$, when

$$x = a\sin t \qquad x = a\tan t \qquad \text{or} \qquad x = a\sec t$$

is positive. In Figure 5 we confirmed that the third one works out when x is negative. Do the same thing with the first and second triangles.

Hyperbolic substitutions work as well as trigonometric ones, the techniques being much the same. The following problems suggest the possibilities (and connections with earlier results).

46. Derive the formula

$$\int \dfrac{dx}{\sqrt{a^2 + x^2}} = \sinh^{-1}\dfrac{x}{a} + C$$

by substituting $x = a\sinh t$ and using the identity $\cosh^2 t - \sinh^2 t = 1$.

47. Derive the formula

$$\int \dfrac{dx}{a^2 + x^2} = \dfrac{1}{a}\tan^{-1}\dfrac{x}{a} + C$$

by substituting $x = a\sinh t$ and using the standard form

$$\int \operatorname{sech} t\, dt = \tan^{-1}(\sinh t) + C$$

48. Derive the formula

$$\int \dfrac{dx}{\sqrt{a^2 - x^2}} = \sin^{-1}\dfrac{x}{a} + C$$

as follows.

(a) Let $x = a\tanh t$ and use the identity $\tanh^2 t + \operatorname{sech}^2 t = 1$ (together with the integral of sech) to obtain

$$\int \dfrac{dx}{\sqrt{a^2 - x^2}} = \tan^{-1}(\sinh t) + C$$

(b) Use the identity

$$\tan^{-1}(\sinh t) = \sin^{-1}(\tanh t)$$

given in Additional Problem 63, Chapter 9.

49. Use the substitution $x = 2\sinh t$ to find

$$\int_2^6 \dfrac{dx}{x^2\sqrt{4 + x^2}}$$

and compare with Example 5.

50. Show that if $x = 2\tan^{-1} u$, then

$$dx = \dfrac{2\,du}{1 + u^2} \qquad \sin x = \dfrac{2u}{1 + u^2} \qquad \cos x = \dfrac{1 - u^2}{1 + u^2}$$

Use the substitution in Problem 50 to find the following integrals.

51. $\int \dfrac{dx}{1 + \cos x}$

52. $\int_0^{\pi/3} \dfrac{dx}{1 - \sin x}$

53. $\int \dfrac{dx}{\sin x + \tan x}$

54. $\int \dfrac{dx}{\tan x - \sin x}$

10.4
DECOMPOSITION OF RATIONAL FUNCTIONS INTO PARTIAL FRACTIONS

The idea of this section is that *rational functions* (quotients of polynomials) can be integrated by breaking them into simpler parts. We begin with an example that leads to such an integration problem.

□ **Example 1**   Suppose that the earth can support no more than 10 billion people. The *Law of Inhibited Growth* is a model of the situation in which we assume that the population x grows at a rate jointly proportional to x itself and $10 - x$ (the difference between x and its upper limit, measured in billions). In other words,

$$\frac{dx}{dt} = kx(10 - x) \qquad \text{(where } t \text{ is time and } k \text{ is a constant)}$$

To solve this differential equation, we separate the variables and integrate:

$$\frac{dx}{x(10 - x)} = k\, dt$$

$$\int \frac{dx}{x(10 - x)} = \int k\, dt = kt + C$$

The integral on the left side is of the type we propose to discuss (because the integrand is a rational function). The idea is to "decompose" the function into a sum of manageable fractions (called "partial fractions"). When this has been done, the population may be found as a function of time by carrying out the integration and solving for x. (See the problem set. For an interesting account of current research involving such analysis, see *Human Population Growth: Stability or Catastrophe?* by David A. Smith, in the September, 1977 issue of *Mathematics Magazine*.)   □

In Example 1 it is not hard to see which decomposition we should try:

$$\frac{1}{x(10 - x)} = \frac{A}{x} + \frac{B}{10 - x} \tag{1}$$

where A and B are constants to be specified later. We expect such a decomposition to work because the denominator $x(10 - x)$ on the left side is the common denominator of the fractions on the right. When these fractions are recombined, we should be able to make the numerators on each side the same by an appropriate choice of A and B.

To see what this choice is, multiply each side of Equation (1) by $x(10 - x)$, obtaining the identity

$$1 = A(10 - x) + Bx \tag{2}$$

This equation is true for all values of x, even $x = 0$ and $x = 10$. [These values are excluded in Equation (1), but they are legitimate in Equation (2) because both sides are continuous.] *Hence we may substitute any values of x we choose in Equation* (2). Two substitutions will enable us to evaluate the constants A and B. (Why?)

The most convenient substitutions in Equation (2) are $x = 0$ and

$x = 10$, because they cause one term or the other (on the right side) to drop out. Letting $x = 0$, we find $1 = 10A$ (from which $A = \frac{1}{10}$), while substitution of $x = 10$ yields $1 = 10B$ (and hence $B = \frac{1}{10}$).

Thus we have discovered that

$$\frac{1}{x(10 - x)} = \frac{1}{10} \cdot \frac{1}{x} + \frac{1}{10} \cdot \frac{1}{10 - x}$$

and our integration problem in Example 1 is solved:

$$\int \frac{dx}{x(10 - x)} = \frac{1}{10} \int \frac{dx}{x} + \frac{1}{10} \int \frac{dx}{10 - x}$$

$$= \frac{1}{10} \ln x - \frac{1}{10} \ln (10 - x) \qquad \text{(arbitrary constant omitted)}$$

$$= \frac{1}{10} \ln \frac{x}{10 - x}$$

(We left out the usual absolute values because the population limitations in Example 1 put x between 0 and 10. Hence x and $10 - x$ are positive.)

The substance of this section is that any rational function can be integrated as in the preceding example (by decomposition followed by use of standard formulas already developed). Rather than dwelling on the theorems from algebra that justify this statement, we present a number of examples to show you what we mean. The algebraic theory will emerge as we proceed.

□ **Example 2**   Find $\int \frac{x^3 + 2}{x^2 - x - 6} \, dx$.

Solution: The integrand is an *improper* rational fraction, meaning that the degree of the numerator is at least as large as the degree of the denominator. In such circumstances, long division reduces the fraction to *the sum of a polynomial and a proper fraction:*

$$\frac{x^3 + 2}{x^2 - x - 6} = (x + 1) + \frac{7x + 8}{x^2 - x - 6}$$

(Confirm!) Since we can certainly integrate a polynomial, our discussion of decomposition may be confined to proper fractions.

Factoring the denominator of our proper fraction, we have

$$\frac{7x + 8}{x^2 - x - 6} = \frac{7x + 8}{(x - 3)(x + 2)}$$

If this is going to break up into a sum, the only (simple) possibility is

$$\frac{7x + 8}{(x - 3)(x + 2)} = \frac{A}{x - 3} + \frac{B}{x + 2}$$

(for reasons explained in the discussion following Example 1). Since this equation implies that

$$7x + 8 = A(x + 2) + B(x - 3) \qquad \text{for all } x$$

we may substitute any values of x that seem convenient. Putting $x = 3$, we find $29 = 5A$ (or $A = \frac{29}{5}$), while $x = -2$ gives $-6 = -5B$ (or $B = \frac{6}{5}$). Thus

$$\frac{7x + 8}{(x - 3)(x + 2)} = \frac{29}{5} \cdot \frac{1}{x - 3} + \frac{6}{5} \cdot \frac{1}{x + 2}$$

and hence

$$\int \frac{x^3 + 2}{x^2 - x - 6}\, dx = \int (x + 1)\, dx + \frac{29}{5} \int \frac{dx}{x - 3} + \frac{6}{5} \int \frac{dx}{x + 2}$$

$$= \frac{1}{2}(x + 1)^2 + \frac{29}{5} \ln |x - 3| + \frac{6}{5} \ln |x + 2| + C \quad \square$$

Remark   The substitution of $x = 3$ and $x = -2$ in Example 2 bypasses a longer method which is sometimes needed. The identity

$$7x + 8 = A(x + 2) + B(x - 3)$$

can be written in the form

$$7x + 8 = (A + B)x + (2A - 3B)$$

Equal polynomials have equal coefficients (of corresponding terms), which implies in this case that

$$A + B = 7 \qquad \text{and} \qquad 2A - 3B = 8$$

This is a system of linear equations in A and B, yielding $A = \frac{29}{5}$ and $B = \frac{6}{5}$ (as you can check). However, you should avoid this approach (or at least cut it down) whenever the substitution of special values of x promises to yield one or more of the unknown constants.

Example 2 works out nicely because the denominator $x^2 - x - 6$ is easily factored into $(x - 3)(x + 2)$. All the examples and problems in this section are chosen to avoid difficult factoring, but nevertheless you should be aware of a general theorem about the question. It is proved in algebra that *any polynomial with real coefficients can be expressed as a product of real factors of degree no higher than 2.* For example,

$$x^2 - x - 6 = (x - 3)(x + 2)$$

(a product of linear factors alone), while

$$x^4 - 16 = (x^2 - 4)(x^2 + 4) = (x - 2)(x + 2)(x^2 + 4)$$

(a product of both linear and quadratic factors). The point of this theorem (in the present context) is that the *denominator* of any rational function we encounter may always be factored this way. The quadratic factors (if any) are understood to be *irreducible* (that is, they cannot be factored further using real coefficients). For example, $x^2 - 4$ is *reducible* because it factors into $(x - 2)(x + 2)$, but $x^2 + 4$ is irreducible.

$\square$ **Example 3**   Find $\displaystyle \int \frac{x + 5}{x^3 - 2x^2 + x}\, dx$.

Solution: The integrand is already proper, so we factor its denominator:

$$x^3 - 2x^2 + x = x(x^2 - 2x + 1) = x(x - 1)^2$$

Thus the problem is to decompose

$$\frac{x + 5}{x(x - 1)^2}$$

It is not so obvious this time what form the decomposition should take. A little thought, however, should convince you that partial fractions of the type

$$\frac{A}{x} \qquad \frac{B}{x - 1} \qquad \frac{C}{(x - 1)^2}$$

may all be involved in a sum that gives back the original fraction when its terms are recombined. Moreover, no other fractions *need* be involved. Hence our decomposition has the form

$$\frac{x + 5}{x(x - 1)^2} = \frac{A}{x} + \frac{B}{x - 1} + \frac{C}{(x - 1)^2}$$

Since the least common denominator of the fractions on the right side is $x(x - 1)^2$, we arrive at the equation

$$x + 5 = A(x - 1)^2 + Bx(x - 1) + Cx \qquad \text{for all } x$$

The useful special values of x this time are $x = 0$ and $x = 1$, yielding $A = 5$ and $C = 6$ (respectively) when they are substituted. To find B, rewrite the identity in the form

$$x + 5 = A(x^2 - 2x + 1) + B(x^2 - x) + Cx$$
$$= (A + B)x^2 + (-2A - B + C)x + A$$

and equate the coefficients of x^2 on each side:

$$0 = A + B \qquad \text{(from which } B = -5 \text{ because } A = 5\text{)}$$

(In reasonably simple cases, like this one, you can probably put down $0 = A + B$ by inspection, without rewriting the identity.)

Thus we have found the decomposition

$$\frac{x + 5}{x(x - 1)^2} = \frac{5}{x} - \frac{5}{x - 1} + \frac{6}{(x - 1)^2}$$

and our integral is

$$\int \frac{x + 5}{x^3 - 2x^2 + x}\,dx = 5 \int \frac{dx}{x} - 5 \int \frac{dx}{x - 1} + 6 \int \frac{dx}{(x - 1)^2}$$

$$= 5 \ln |x| - 5 \ln |x - 1| - \frac{6}{x - 1} + C$$

$$= 5 \ln \left| \frac{x}{x - 1} \right| - \frac{6}{x - 1} + C \qquad\qquad \square$$

Examples 2 and 3 suggest what happens in general when the denominator of our fraction involves linear factors:

- A linear factor $ax + b$ that occurs only once calls for a term of the form $A/(ax + b)$ in the decomposition.
- A repeated linear factor, say $(ax + b)^k$, calls for a sum of terms of the form

$$\frac{A_1}{ax + b} + \frac{A_2}{(ax + b)^2} + \cdots + \frac{A_k}{(ax + b)^k}$$

(The second statement includes the first as a special case by taking $k = 1$.)

Now we turn to problems in which (irreducible) quadratic factors are allowed as well.

□ **Example 4**   Find $\int \dfrac{x^2 - x + 5}{x(x^2 + 1)} \, dx$.

Solution: The integrand is proper and the denominator is factored, so the preliminaries are already out of the way. The decomposition takes the form

$$\frac{x^2 - x + 5}{x(x^2 + 1)} = \frac{A}{x} + \frac{Bx + C}{x^2 + 1}$$

Note the *linear* (not constant) numerator in the second fraction. If we wrote only

$$\frac{A}{x} + \frac{C}{x^2 + 1} \qquad \left(\text{leaving out } \frac{Bx}{x^2 + 1}\right)$$

we would not be putting down all the fractions that might contribute.
Using the least common denominator as before, we find

$$x^2 - x + 5 = A(x^2 + 1) + x(Bx + C) \qquad \text{for all } x$$

The only special value of x of any interest this time is $x = 0$, substitution of which gives $A = 5$. That leaves two unknowns; we need two equations to determine their values. Hence we equate coefficients of both x^2 and x:

$$1 = A + B \qquad \text{and} \qquad -1 = C$$

Since $A = 5$, we find $B = -4$, and our decomposition reads

$$\frac{x^2 - x + 5}{x(x^2 + 1)} = \frac{5}{x} + \frac{(-4x - 1)}{x^2 + 1} = \frac{5}{x} - \frac{4x}{x^2 + 1} - \frac{1}{x^2 + 1}$$

The integral is

$$\int \frac{x^2 - x + 5}{x(x^2 + 1)} \, dx = 5 \int \frac{dx}{x} - 4 \int \frac{x \, dx}{x^2 + 1} - \int \frac{dx}{x^2 + 1}$$

$$= 5 \ln |x| - 2 \ln (x^2 + 1) - \tan^{-1} x + C \qquad \square$$

□ **Example 5**   Find $\int \dfrac{2x^2 + x + 7}{(x^2 + 4)^2}\, dx$.

Solution: The repeated quadratic factor calls for a decomposition of the form

$$\frac{2x^2 + x + 7}{(x^2 + 4)^2} = \frac{Ax + B}{x^2 + 4} + \frac{Cx + D}{(x^2 + 4)^2}$$

(as in the linear case). We find

$$2x^2 + x + 7 = (x^2 + 4)(Ax + B) + (Cx + D)$$

No special value of x is of any help this time, so we equate coefficients of corresponding powers of x.

$$x^3: 0 = A \qquad x^2: 2 = B \qquad x^1: 1 = 4A + C \qquad x^0: 7 = 4B + D$$

Using the values $A = 0$ and $B = 2$ in the last two equations, we find $C = 1$ and $D = -1$. Our decomposition is

$$\frac{2x^2 + x + 7}{(x^2 + 4)^2} = \frac{2}{x^2 + 4} + \frac{x - 1}{(x^2 + 4)^2} = \frac{2}{x^2 + 4} + \frac{x}{(x^2 + 4)^2} - \frac{1}{(x^2 + 4)^2}$$

and our integral is

$$\int \frac{2x^2 + x + 7}{(x^2 + 4)^2}\, dx = 2\int \frac{dx}{x^2 + 4} + \int \frac{x\, dx}{(x^2 + 4)^2} - \int \frac{dx}{(x^2 + 4)^2}$$

$$= \tan^{-1}\frac{x}{2} - \frac{1}{2(x^2 + 4)} - \;?$$

The question mark is inserted because the third integral is not easy enough to put down by inspection. (We hope the first two are!) Make the substitution

$$x = 2 \tan t \qquad (-\pi/2 < t < \pi/2)$$

in this integral. Then $dx = 2 \sec^2 t\, dt$ and $(x^2 + 4)^2 = 16 \sec^4 t$, from which

$$\int \frac{dx}{(x^2 + 4)^2} = \int \frac{2 \sec^2 t\, dt}{16 \sec^4 t} = \frac{1}{8}\int \cos^2 t\, dt = \frac{1}{16}\int (1 + \cos 2t)\, dt$$

$$= \frac{1}{16}\left(t + \frac{1}{2}\sin 2t\right) = \frac{1}{16}(t + \sin t \cos t)$$

$$= \frac{1}{16}\left(\tan^{-1}\frac{x}{2} + \frac{x}{\sqrt{x^2 + 4}} \cdot \frac{2}{\sqrt{x^2 + 4}}\right)$$

$$= \frac{1}{16}\tan^{-1}\frac{x}{2} + \frac{x}{8(x^2 + 4)} \qquad \text{(arbitrary constant omitted)}$$

(See Figure 1.) Hence the original integral is

$$\int \frac{2x^2 + x + 7}{(x^2 + 4)^2}\, dx = \tan^{-1}\frac{x}{2} - \frac{1}{2(x^2 + 4)} - \frac{1}{16}\tan^{-1}\frac{x}{2} - \frac{x}{8(x^2 + 4)} + C$$

$$= \frac{15}{16}\tan^{-1}\frac{x}{2} - \frac{x + 4}{8(x^2 + 4)} + C \qquad\qquad\qquad □$$

Figure 1
Finding $\sin t$ and $\cos t$

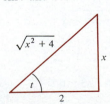

Examples 4 and 5 indicate that the rule for (irreducible) quadratic factors is the same as for linear factors, except that the numerators are linear instead of constant. None of these statements, incidentally, has been proved! We are relying on your intuition of what a reasonable decomposition should look like.

□ **Example 6**   Find $\displaystyle\int_2^3 \frac{dx}{x^3 - 1}$.

Solution: Since $x^3 - 1 = (x - 1)(x^2 + x + 1)$ (the second factor being irreducible), we write

$$\frac{1}{x^3 - 1} = \frac{A}{x - 1} + \frac{Bx + C}{x^2 + x + 1}$$

We leave it to you to confirm that $A = \frac{1}{3}$, $B = -\frac{1}{3}$, $C = -\frac{2}{3}$. Hence

$$\int \frac{dx}{x^3 - 1} = \frac{1}{3}\int \frac{dx}{x - 1} - \frac{1}{3}\int \frac{(x + 2)\,dx}{x^2 + x + 1} = \frac{1}{3}\ln|x - 1| - \frac{1}{3}(?)$$

To find

$$\int \frac{(x + 2)\,dx}{x^2 + x + 1}$$

let $u = x^2 + x + 1$, $du = (2x + 1)\,dx$. Since the numerator is $x + 2$ (when we would like it to be $2x + 1$), we fix it up by writing

$$x + 2 = \frac{1}{2}(2x + 1) + \frac{3}{2}$$

Then $\displaystyle\int \frac{(x + 2)\,dx}{x^2 + x + 1} = \frac{1}{2}\int \frac{(2x + 1)\,dx}{x^2 + x + 1} + \frac{3}{2}\int \frac{dx}{x^2 + x + 1}$

$$= \frac{1}{2}\int \frac{du}{u} + \frac{3}{2}\int \frac{dx}{x^2 + x + 1}$$

The first integral on the right is a logarithm. The second is an inverse tangent, as you can see by completing the square in the denominator:

$$x^2 + x + 1 = (x^2 + x + \tfrac{1}{4}) + \tfrac{3}{4} = (x + \tfrac{1}{2})^2 + \tfrac{3}{4}$$

This prepares for integration using the standard form

$$\int \frac{dx}{a^2 + x^2} = \frac{1}{a}\tan^{-1}\frac{x}{a} + C$$

Hence $\displaystyle\int \frac{(x + 2)\,dx}{x^2 + x + 1} = \frac{1}{2}\int \frac{du}{u} + \frac{3}{2}\int \frac{dx}{(x + \frac{1}{2})^2 + \frac{3}{4}}$

$$= \frac{1}{2}\ln|u| + \frac{3}{2}\cdot\frac{2}{\sqrt{3}}\tan^{-1}\left(\frac{2x + 1}{\sqrt{3}}\right)$$

$$= \frac{1}{2}\ln(x^2 + x + 1) + \sqrt{3}\tan^{-1}\left(\frac{2x + 1}{\sqrt{3}}\right)$$

Thus

$$\int_2^3 \frac{dx}{x^3 - 1} = \left[\frac{1}{3} \ln |x - 1| - \frac{1}{6} \ln (x^2 + x + 1) - \frac{\sqrt{3}}{3} \tan^{-1} \left(\frac{2x + 1}{\sqrt{3}} \right) \right] \Big|_2^3$$

$$= \frac{1}{3} \ln 2 - \frac{1}{6} \ln 13 + \frac{1}{6} \ln 7 - \frac{\sqrt{3}}{3} \tan^{-1} \left(\frac{7}{\sqrt{3}} \right)$$

$$+ \frac{\sqrt{3}}{3} \tan^{-1} \left(\frac{5}{\sqrt{3}} \right)$$

$$\approx 0.075 \quad \text{(from a calculator)} \qquad \square$$

Problem Set 10.4

Use the methods of this section to find each of the following.

1. $\int \dfrac{dx}{x^2 + 2x - 3}$

2. $\int \dfrac{x \, dx}{x^2 - 5x + 6}$

3. $\int \dfrac{x^2 + 1}{x(x^2 - 1)} \, dx$

4. $\int \dfrac{(x - 3) \, dx}{x(x^2 + x - 2)}$

5. $\int \dfrac{dx}{x^3 - x^2}$

6. $\int \dfrac{dx}{x(x - 1)^2}$

7. $\int \dfrac{dx}{x(x^2 + 4)}$

8. $\int \dfrac{x^2 \, dx}{(x - 1)(x^2 + 1)}$

9. $\int \dfrac{dx}{x^2(x^2 + 9)}$

10. $\int \dfrac{dx}{x^4 - 1}$

11. $\int \dfrac{x^3 \, dx}{(x^2 + 1)^2}$

12. $\int \dfrac{x^2 - x + 1}{(x^2 + 4)^2} \, dx$

13. $\int \dfrac{dx}{x^3 - 8}$

14. $\int \dfrac{dx}{x^3 - 27}$

Evaluate each of the following.

15. $\int_0^1 \dfrac{x \, dx}{x^2 - x - 2}$

16. $\int_1^5 \dfrac{dx}{x^2 + 5x}$

17. $\int_2^3 \dfrac{x \, dx}{(x - 1)^2}$

18. $\int_1^2 \dfrac{dx}{x^3(x + 2)}$

19. $\int_1^2 \dfrac{dx}{x^3 + x}$

20. $\int_0^1 \dfrac{x^2 \, dx}{16 - x^4}$

21. $\int_2^4 \dfrac{dx}{x^2(x^2 + 4)}$

22. $\int_2^3 \dfrac{dx}{x^3(x^2 - 2x + 1)}$

23. $\int_0^1 \dfrac{dx}{x^3 + 1}$

24. $\int_0^2 \dfrac{dx}{x^3 + 8}$

25. To find

$$\int \frac{x^3 + 1}{x(x^2 + 1)} \, dx$$

suppose we write

$$\frac{x^3 + 1}{x(x^2 + 1)} = \frac{A}{x} + \frac{Bx + C}{x^2 + 1}$$

(a) Show that this implies $A = 1$, $B = -1$, $C = 0$ and hence

$$\frac{x^3 + 1}{x(x^2 + 1)} = \frac{1}{x} - \frac{1}{x^2 + 1}$$

(b) When $x = 1$ the "identity" in part (a) yields $1 = \frac{1}{2}$. What went wrong?

(c) Find the integral.

26. To find

$$\int \frac{dx}{(x^2 + 1)^2}$$

suppose we write

$$\frac{1}{(x^2 + 1)^2} = \frac{Ax + B}{x^2 + 1} + \frac{Cx + D}{(x^2 + 1)^2}$$

(a) What is the result and how could it have been foreseen?

(b) Find the integral.

27. Find

$$\int \frac{dx}{x(10 - x)}$$

by completing the square in the denominator and using a standard integration formula. (Assume that $0 < x < 10$, as in the opening example of this section.)

28. Find

$$\int \frac{dx}{x^2 + 2x - 3}$$

by completing the square. (Compare with Problem 1.)

29. Use the ideas of this section to derive the standard form

$$\int \frac{dx}{a^2 - x^2} = \frac{1}{2a} \ln \left| \frac{a + x}{a - x} \right| + C$$

30. Derive the formula

$$\int \frac{dx}{x(ax + b)} = \frac{1}{b} \ln \left| \frac{x}{ax + b} \right| + C$$

31. Derive the formula

$$\int \frac{dx}{x^2(ax + b)} = -\frac{1}{bx} + \frac{a}{b^2} \ln \left| \frac{ax + b}{x} \right| + C$$

32. The *Law of Mass Action* in chemistry leads to the differential equation

$$\frac{dx}{dt} = k(a - x)(b - x) \qquad (k, a, b > 0)$$

where x is the amount at time t of a substance being formed from the reaction of two others ($x = 0$ at $t = 0$).

(a) Assuming that $a \neq b$, solve this initial value problem to obtain

$$\frac{a - x}{b - x} = \frac{a}{b} e^{(a-b)kt}$$

(b) Explain why x approaches the smaller of a and b as time goes on.

(c) Taking $a = 3$ and $b = 6$, suppose that 1 gram of the substance is formed in 10 minutes. How many grams are present 10 minutes later?

(d) Solve the differential equation in the case $a = b$.

33. Suppose that the upper limit of world population is 10 billion and that there were 2 billion people in 1920 and 4 billion in 1980.

(a) Use the Law of Inhibited Growth to show that the population t years after 1920 is

$$x = \frac{10}{1 + 4e^{-10kt}} \qquad \text{where } k = \frac{1}{600} \ln \frac{8}{3}$$

What does x approach as time goes on?

(b) When will the population be 8 billion?

(c) If the Law of Exponential Growth ($x = x_0 e^{ct}$) is used instead, when will the population be 8 billion?

34. Suppose that the upper limit of world population is 8 billion, and assume the Law of Inhibited Growth (together

with the statistics of 2 billion people in 1920 and 4 billion in 1980).

(a) Show that the population t years after 1920 is

$$x = \frac{8}{1 + 3e^{-8kt}} \qquad \text{where } k = \frac{1}{480} \ln 3$$

What does x approach as time goes on?

(b) Confirm that

$$x = \frac{8}{1 + 3^{1-(t/60)}}$$

and use the result to find when the population will be 6 billion.

(c) If the Law of Exponential Growth is used instead, when will the population be 6 billion?

35. A law of population growth that applies in some circumstances is

$$\frac{dx}{dt} = kx + ax^2 \qquad \text{(where } k \text{ and } a \text{ are positive)}$$

(a) Solve this differential equation to obtain

$$x = \frac{k}{(a + k/x_0)e^{-kt} - a}$$

where x_0 is the initial population. What would this become if a were 0? *Hint:* Use Problem 30.

(b) Explain why it is unreasonable to use this model of population growth over a long period of time.

(c) More precisely, show that the population grows without bound in a finite time. When is doomsday?

36. Use the substitution $x = 2 \tan^{-1} u$ (Problem 50, Section 10.3) to find

$$\int \frac{dx}{1 + \sin x - \cos x}$$

(This substitution changes rational functions of $\sin x$ and $\cos x$ to rational functions of u.)

37. Repeat Problem 36 in the case of $\displaystyle\int \frac{dx}{1 + \cos x - \sin x}$.

10.5

**MISCELLANEOUS
INTEGRATION
PROBLEMS**

As we mentioned at the beginning of this chapter, there are only two general methods of integration. One is *substitution* (including many special devices not covered in this book); the other is *integration by parts*. (Decomposition of rational functions is not really a method of integration, but an algebraic technique for breaking up a fraction into a sum.)

Nothing new is offered in this section (except for fuller explanation of some substitutions that have previously occurred only in the problem sets). Our purpose is to help you develop more confidence in your mastery of technique by presenting miscellaneous examples and problems.

□ **Example 1**   Find $\int \dfrac{\sqrt{x}\, dx}{x^2 - 1}$.

Solution: The substitution needed here was first suggested in Example 6, Section 6.5, but has not appeared too often since. Because the radical involves a *linear* expression, nothing so fancy as a trigonometric substitution is called for; simply let $u = \sqrt{x}$, $u^2 = x$, $2u\, du = dx$. Then

$$\int \frac{\sqrt{x}\, dx}{x^2 - 1} = \int \frac{u(2u\, du)}{u^4 - 1} = \int \frac{2u^2\, du}{u^4 - 1}$$

an integral which calls for a decomposition of the type described in the last section. We leave it to you to confirm that the integrand is

$$\frac{2u^2}{(u - 1)(u + 1)(u^2 + 1)} = \frac{1}{2} \cdot \frac{1}{u - 1} - \frac{1}{2} \cdot \frac{1}{u + 1} + \frac{1}{u^2 + 1}$$

Hence

$$\int \frac{\sqrt{x}\, dx}{x^2 - 1} = \frac{1}{2} \ln |u - 1| - \frac{1}{2} \ln |u + 1| + \tan^{-1} u + C$$

$$= \frac{1}{2} \ln \left| \frac{\sqrt{x} - 1}{\sqrt{x} + 1} \right| + \tan^{-1} \sqrt{x} + C$$ □

□ **Example 2**   Find $\displaystyle\int_0^1 x^2 \sin^{-1} x\, dx$.

Solution: Integration by parts seems a good way to start. Let

$$u = \sin^{-1} x \qquad \text{and} \qquad dv = x^2\, dx$$

Then $du = dx / \sqrt{1 - x^2}$ and $v = x^3/3$, from which

$$\int x^2 \sin^{-1} x\, dx = \frac{x^3}{3} \sin^{-1} x - \frac{1}{3} \int \frac{x^3\, dx}{\sqrt{1 - x^2}}$$

The new integral can be handled by a trigonometric substitution. In view of the odd power in the numerator, however (see Problems 31 and 32, Section 10.3), we will let

$$u = \sqrt{1 - x^2} \qquad u^2 = 1 - x^2 \qquad u\,du = -x\,dx$$

Then

$$\int \frac{x^3\,dx}{\sqrt{1 - x^2}} = \int \frac{(1 - u^2)(-u\,du)}{u}$$

$$= \int (u^2 - 1)\,du = \frac{u^3}{3} - u = \frac{u}{3}(u^2 - 3)$$

(We omitted the arbitrary constant.) The original integral is

$$\int_0^1 x^2 \sin^{-1} x\,dx = \frac{x^3}{3} \sin^{-1} x \Big|_0^1 - \frac{u}{9}(u^2 - 3)\Big|_1^0 = \frac{\pi}{6} - \frac{2}{9} \qquad \square$$

Remark   In Example 2 we delayed inserting the limits of integration until the end. A funny thing happens if they are inserted early:

$$\int_0^1 x^2 \sin^{-1} x\,dx = \frac{x^3}{3} \sin^{-1} x \Big|_0^1 - \frac{1}{3} \int_0^1 \frac{x^3\,dx}{\sqrt{1 - x^2}}$$

The original integrand is continuous in $[0,1]$, so the integral certainly exists. The new integral, however, is *improper* (see Problem 40, Section 9.2) because $x^3/\sqrt{1 - x^2}$ is unbounded in the domain of integration. We cannot integrate from 0 to 1, but only from 0 to t (where t is close to, but less than, 1). The result is

$$\int_0^t \frac{x^3\,dx}{\sqrt{1 - x^2}} = \frac{u}{3}(u^2 - 3)\Big|_1^{\sqrt{1-t^2}} \qquad (u = \sqrt{1 - x^2})$$

$$= \frac{2}{3} - \frac{1}{3}\sqrt{1 - t^2}(2 + t^2)$$

which approaches $\frac{2}{3}$ as $t \to 1$. The improper integral may be assigned this value and all is well. Since we have not yet formally discussed the subject, however, this approach is a little tricky! In our first solution the difficulty evaporated when u cancelled in the step

$$\int \frac{(1 - u^2)(-u\,du)}{u} = \int (u^2 - 1)\,du$$

It is easy to overlook improper integrals that arise in this way. (If you worked Problem 36 in Section 10.1, for example, you may have sailed right by one!)

□ **Example 3**   Find $\displaystyle \int_0^\pi \frac{\sin x\,dx}{4 + \cos^2 x}$.

Solution: This becomes an inverse tangent upon substitution of $u = \cos x$, $du = -\sin x\,dx$:

$$\int_0^\pi \frac{\sin x\, dx}{4 + \cos^2 x} = -\int_1^{-1} \frac{du}{4 + u^2} = 2\int_0^1 \frac{du}{4 + u^2} \qquad \text{(why?)}$$

$$= \tan^{-1} \frac{u}{2}\Big|_0^1 = \tan^{-1} \frac{1}{2} \approx 0.46 \qquad \square$$

□ **Example 4**   Find $\displaystyle\int \frac{dx}{1 - \sin x + \cos x}$.

Solution: The integrand is a rational function of $\sin x$ and $\cos x$, which calls for a substitution we have mentioned only in the problem sets. (See Problems 50 through 54, Section 10.3, and Problems 36 and 37, Section 10.4.) It is not an obvious device, but one of those clever ideas that has been around for a long time (and is worth understanding). Let

$$x = 2\tan^{-1} u, \; dx = \frac{2\, du}{1 + u^2}$$

Since $\tan(x/2) = u$, we may use Figure 1 to find

$$\sin x = 2\sin \frac{x}{2} \cos \frac{x}{2} = 2 \cdot \frac{u}{\sqrt{1 + u^2}} \cdot \frac{1}{\sqrt{1 + u^2}} = \frac{2u}{1 + u^2}$$

$$\cos x = \cos^2 \frac{x}{2} - \sin^2 \frac{x}{2} = \frac{1}{1 + u^2} - \frac{u^2}{1 + u^2} = \frac{1 - u^2}{1 + u^2}$$

With these results in hand, we can write

$$\frac{dx}{1 - \sin x + \cos x} = \frac{1}{1 - \dfrac{2u}{1 + u^2} + \dfrac{1 - u^2}{1 + u^2}} \cdot \frac{2\, du}{1 + u^2} = \frac{du}{1 - u}$$

from which

$$\int \frac{dx}{1 - \sin x + \cos x} = \int \frac{du}{1 - u} = -\ln|1 - u| + C$$

$$= -\ln \left|1 - \tan \frac{x}{2}\right| + C \qquad \square$$

The substitution in Example 4 is worth emphasizing:

To find the integral of a rational function of $\sin x$ and $\cos x$, make the substitution

$$x = 2\tan^{-1} u, \qquad dx = \frac{2\, du}{1 + u^2}$$

$$\sin x = \frac{2u}{1 + u^2}, \qquad \cos x = \frac{1 - u^2}{1 + u^2}$$

Figure 1
Finding $\sin x$ and $\cos x$

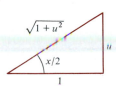

□ **Example 5**   Find $\displaystyle\int_3^4 \frac{x\,dx}{x^4 - 16}$.

Solution: This appears to call for decomposition of the rational integrand. It is easier, however, to make the substitution $u = x^2$, $du = 2x\,dx$. Then

$$\int_3^4 \frac{x\,dx}{x^4 - 16} = \frac{1}{2}\int_9^{16} \frac{du}{u^2 - 16} = -\frac{1}{2}\int_9^{16} \frac{du}{16 - u^2} = -\frac{1}{2}\cdot\frac{1}{4}\coth^{-1}\frac{u}{4}\Big|_9^{16}$$

$$= \frac{1}{8}\left(\coth^{-1}\frac{9}{4} - \coth^{-1}4\right)$$

$$= \frac{1}{8}\left(\tanh^{-1}\frac{4}{9} - \tanh^{-1}\frac{1}{4}\right)\qquad [\text{because } \coth^{-1}x = \tanh^{-1}(1/x)]$$

$$\approx 0.028\qquad (\text{from a calculator with a } \tanh^{-1} \text{ key})$$

The problem can also be done in terms of logarithms:

$$\int_3^4 \frac{x\,dx}{x^4 - 16} = \frac{1}{2}\int_9^{16} \frac{du}{u^2 - 16} = -\frac{1}{2}\int_9^{16} \frac{du}{16 - u^2} = -\frac{1}{16}\ln\left|\frac{4 + u}{4 - u}\right|\Big|_9^{16}$$

$$= -\frac{1}{16}\ln\frac{5}{3} + \frac{1}{16}\ln\frac{13}{5} = \frac{1}{16}\ln\frac{39}{25} \approx 0.028 \qquad\qquad □$$

□ **Example 6**   Find $\displaystyle\int \frac{x\,dx}{\sqrt[3]{x + 1}}$.

Solution: As in Example 1, we substitute for the radical:

$$u = \sqrt[3]{x + 1}, \; u^3 = x + 1, \; 3u^2\,du = dx$$

Then

$$\int \frac{x\,dx}{\sqrt[3]{x + 1}} = \int \frac{(u^3 - 1)(3u^2\,du)}{u}$$

$$= \int (3u^4 - 3u)\,du = \frac{3}{5}u^5 - \frac{3}{2}u^2 + C = \frac{3}{10}u^2(2u^3 - 5) + C$$

$$= \frac{3}{10}(x + 1)^{2/3}(2x - 3) + C \qquad\qquad □$$

□ **Example 7**   Find $\displaystyle\int \frac{x^4\,dx}{\sqrt{x^2 - 4}}$.

Solution: Make the trigonometric substitution

$$x = 2\sec t \qquad (0 \le t < \pi/2 \text{ or } -\pi \le t < -\pi/2)$$

Then $dx = 2 \sec t \tan t \, dt$ and $\sqrt{x^2 - 4} = \sqrt{4 \sec^2 t - 4} = 2 \tan t$, from which

$$\int \frac{x^4 \, dx}{\sqrt{x^2 - 4}} = \int \frac{16 \sec^4 t \cdot 2 \sec t \tan t \, dt}{2 \tan t} = 16 \int \sec^5 t \, dt$$

Now use integration by parts, letting $u = \sec^3 t$ and $dv = \sec^2 t \, dt$. Then

$$du = 3 \sec^3 t \tan t \, dt \qquad \text{and} \qquad v = \tan t$$

from which

$$\int \sec^5 t \, dt = \sec^3 t \tan t - 3 \int \sec^3 t \tan^2 t \, dt$$

$$= \sec^3 t \tan t - 3 \int \sec^3 t \, (\sec^2 t - 1) \, dt$$

$$= \sec^3 t \tan t - 3 \int \sec^5 t \, dt + 3 \int \sec^3 t \, dt$$

Solving for $\int \sec^5 t \, dt$, and recalling from Section 10.3 that

$$\int \sec^3 t \, dt = \frac{1}{2} \sec t \tan t + \frac{1}{2} \ln |\sec t + \tan t|$$

(arbitrary constant omitted)

we find

$$\int \frac{x^4 \, dx}{\sqrt{x^2 - 4}} = 4 \sec^3 t \tan t + 6 \sec t \tan t + 6 \ln |\sec t + \tan t| + C$$

$$= 2 \sec t \tan t \, (2 \sec^2 t + 3) + 6 \ln |\sec t + \tan t| + C$$

$$= 2 \cdot \frac{x}{2} \cdot \frac{\sqrt{x^2 - 4}}{2} \left(2 \cdot \frac{x^2}{4} + 3 \right) + 6 \ln \left| \frac{x}{2} + \frac{\sqrt{x^2 - 4}}{2} \right| + C$$

$$= \frac{1}{4} x \sqrt{x^2 - 4} \, (x^2 + 6) + 6 \ln |x + \sqrt{x^2 - 4}| + C$$

(We dropped the constant term $-6 \ln 2$.) □

□ **Example 8**   Find $\int \dfrac{dx}{\sqrt{6x - x^2}}$.

Solution: We complete the square by writing

$$6x - x^2 = -(x^2 - 6x + 9) + 9 = 9 - (x - 3)^2$$

Then $\displaystyle \int \frac{dx}{\sqrt{6x - x^2}} = \int \frac{dx}{\sqrt{9 - (x - 3)^2}} = \sin^{-1} \left(\frac{x - 3}{3} \right) + C$ □

Problem Set 10.5

Find each of the following integrals.

1. $\displaystyle\int_1^3 \frac{dx}{x\sqrt{4-x}}$

2. $\displaystyle\int \frac{dx}{x^2+8x+20}$

21. $\displaystyle\int \sec^6 t \, dt$

22. $\displaystyle\int \frac{x^2 \, dx}{\sqrt[3]{x+2}}$

3. $\displaystyle\int \sin\sqrt{x}\,dx$

4. $\displaystyle\int_0^1 \frac{e^x \, dx}{e^{2x}+1}$

23. $\displaystyle\int \frac{x^3+8}{x-2}\,dx$

24. $\displaystyle\int \frac{dx}{x^2\sqrt{x^2+9}}$

5. $\displaystyle\int_0^{\pi/6} \frac{\sin x \cos x}{1-\sin x}\,dx$

6. $\displaystyle\int x^2 \tan^{-1} x \, dx$

25. $\displaystyle\int \frac{x^2+4}{x^3-x}\,dx$

26. $\displaystyle\int_0^1 \sqrt{x^2-x^4}\,dx$

7. $\displaystyle\int \frac{dx}{x^2-2x}$

8. $\displaystyle\int \frac{dx}{(a^2-x^2)^{3/2}}$

27. $\displaystyle\int \frac{e^t \, dt}{e^t-1}$

28. $\displaystyle\int x^3 \sin x \, dx$

9. $\displaystyle\int_0^1 \frac{x^2 \, dx}{(x^2+1)^3}$

10. $\displaystyle\int_0^{\pi/4} \frac{\tan^2 x + 1}{\tan x + 1}\,dx$

29. $\displaystyle\int_0^1 \frac{\tan^{-1} x}{1+x^2}\,dx$

30. $\displaystyle\int_4^9 \frac{dx}{\sqrt{x}(\sqrt{x}-1)}$

11. $\displaystyle\int_0^{\pi/2} \frac{dx}{1+\sin x}$

12. $\displaystyle\int \frac{x^2 \, dx}{(x-1)^3}$

31. $\displaystyle\int \frac{x \, dx}{\sqrt{6x-x^2}}$

32. $\displaystyle\int_0^{\pi/2} \frac{\cos t \, dt}{1+\sin^2 t}$

13. $\displaystyle\int \frac{x^4+1}{x^4-1}\,dx$

14. $\displaystyle\int \frac{dx}{1+\sec x}$

33. $\displaystyle\int \frac{\sqrt{4-x^2}}{x^3}\,dx$

34. $\displaystyle\int \frac{x \, dx}{x^4+1}$

15. $\displaystyle\int_0^1 \sqrt{4x-x^2}\,dx$

16. $\displaystyle\int_0^{\pi/4} \sec^4 x \, dx$

35. $\displaystyle\int \frac{\cos t \, dt}{3+\cos^2 t}$

36. $\displaystyle\int_0^3 \frac{(2x-3)\,dx}{x^2-3x+5}$

17. $\displaystyle\int (x^3+1)e^x \, dx$

18. $\displaystyle\int x \sec x \tan x \, dx$

37. $\displaystyle\int e^x \cos 2x \, dx$

38. $\displaystyle\int_{\pi/4}^{\pi/3} \csc^4 t \, dt$

19. $\displaystyle\int_0^1 x^3\sqrt{x^2+1}\,dx$

20. $\displaystyle\int_0^1 \frac{dx}{2-\sqrt[3]{x}}$

39. $\displaystyle\int x^3 e^{-2x} \, dx$

40. $\displaystyle\int \frac{dx}{x^2-4x}$

10.6

NUMERICAL INTEGRATION

As we have pointed out before, a table of integrals (no matter how extensive) cannot touch some problems. A simple example is

$$\int_0^1 \sqrt{x^3+1}\,dx$$

which cannot be evaluated by any of the methods we have described. In fact no elementary function exists whose derivative is $\sqrt{x^3+1}$. Moreover, we tend to forget that our "evaluation" of many integrals is in name only. A result such as

$$\int_0^1 \cos x \, dx = \sin 1$$

has to be converted to a decimal approximation to be useful. While tables and calculators give the values of common functions, there are situations in which we have to do better.

In this section we develop two formulas which enable us to compute

$$\int_a^b f(x)\, dx$$

as accurately as we please (provided of course that f itself is known to within the needed degree of precision). They are based on linear and quadratic approximation of $f(x)$, respectively.

Let $I = [a,b]$ and suppose that $P = \{x_0, x_1, \ldots, x_n\}$ is a partition of I into n subintervals of equal length. In Figure 1 we show the graph of f in the typical subinterval, together with its linear approximation by a line segment joining its endpoints. The area of the shaded trapezoid (from geometry) is

$$(\text{average base})(\text{altitude}) = \tfrac{1}{2}(y_{k-1} + y_k)\,\Delta x$$

To obtain an approximation of the integral of f from a to b, we need only add up these trapezoidal areas:

$$\int_a^b f(x)\, dx \approx \sum_{k=1}^n \frac{1}{2}(y_{k-1} + y_k)\,\Delta x$$

When this sum is written out, it becomes

$$\Delta x\left[\frac{1}{2}(y_0 + y_1) + \frac{1}{2}(y_1 + y_2) + \cdots + \frac{1}{2}(y_{n-2} + y_{n-1}) + \frac{1}{2}(y_{n-1} + y_n)\right]$$

$$= \Delta x\left(\frac{1}{2}y_0 + y_1 + y_2 + \cdots + y_{n-1} + \frac{1}{2}y_n\right)$$

Thus we have arrived at the following result.

Figure 1

Trapezoidal approximation of area under a curve

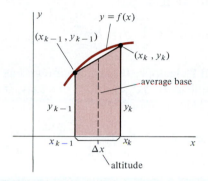

The Trapezoidal Rule

Let $\{x_0, x_1, \ldots, x_n\}$ be a partition of $[a,b]$ into subintervals of equal length, $\Delta x = (b - a)/n$. If

$$y_k = f(x_k) \qquad k = 0, 1, \ldots, n$$

then

$$\int_a^b f(x)\, dx \approx \Delta x\left(\frac{1}{2}y_0 + y_1 + y_2 + \cdots + y_{n-1} + \frac{1}{2}y_n\right)$$

We do not claim to have proved anything! How could we, when no hypotheses are given concerning f and no precision is imposed on the approximation? Not until an upper bound is given for the error does the rule acquire any mathematical substance beyond the intuitive ideas of area that lead to it. Note, however, that if f is integrable (which of course we assume when we write its integral), the approximation does approach the right answer as n increases. To see why, let

$$T_n = \Delta x \left(\frac{1}{2} y_0 + y_1 + y_2 + \cdots + y_{n-1} + \frac{1}{2} y_n \right)$$

$$= (y_1 + y_2 + \cdots + y_n) \Delta x - \frac{1}{2}(y_n - y_0) \Delta x$$

$$= \sum_{k=1}^{n} f(x_k) \Delta x - \frac{1}{2} \Big[f(b) - f(a) \Big] \Delta x$$

When n increases (which forces $\Delta x \to 0$) we find

$$\lim_{\Delta x \to 0} T_n = \lim_{\Delta x \to 0} \sum_{k=1}^{n} f(x_k) \Delta x - 0 = \int_a^b f(x)\, dx$$

This guarantees that the Trapezoidal Rule approximates the integral as closely as we please (when n is taken sufficiently large). More precisely, it can be proved that if f'' exists in I, and M is an upper bound for $|f''(x)|$ in I, that is,

$$|f''(x)| \le M \qquad \text{for } a \le x \le b$$

then the error in the Trapezoidal Rule is

$$E_n = \left| \int_a^b f - T_n \right| \le \frac{1}{12}(b-a)M(\Delta x)^2$$

□ **Example 1**   Use the Trapezoidal Rule with $n = 4$ to estimate $\displaystyle\int_1^3 \frac{dx}{x}$.

Solution: Since $\Delta x = \frac{1}{2}$, the points of subdivision are $x_0 = 1$, $x_1 = \frac{3}{2}$, $x_2 = 2$, $x_3 = \frac{5}{2}$, $x_4 = 3$. The corresponding functional values ($y_k = 1/x_k$) are $y_0 = 1$, $y_1 = \frac{2}{3}$, $y_2 = \frac{1}{2}$, $y_3 = \frac{2}{5}$, $y_4 = \frac{1}{3}$. Hence

$$\int_1^3 \frac{dx}{x} \approx \frac{1}{2} \left(\frac{1}{2} + \frac{2}{3} + \frac{1}{2} + \frac{2}{5} + \frac{1}{6} \right) \approx 1.12$$

Since

$$\int_1^3 \frac{dx}{x} = \ln 3 \approx 1.10$$

our approximation is correct in the first decimal place.   □

□ **Example 2**   Use the Trapezoidal Rule with $n = 4$ to estimate

$$\int_0^1 \sqrt{x^3 + 1}\, dx$$

Solution: Since $\Delta x = 0.25$, the points of subdivision are $x_0 = 0$, $x_1 = 0.25$, $x_2 = 0.5$, $x_3 = 0.75$, $x_4 = 1$. The corresponding values of the integrand (rounded off from a calculator) are $y_0 = 1$, $y_1 = 1.0078$, $y_2 = 1.0607$, $y_3 = 1.1924$, $y_4 = 1.4142$. Hence

$$\int_0^1 \sqrt{x^3 + 1}\, dx \approx 0.25(0.5 + 1.0078 + 1.0607 + 1.1924 + 0.7071) = 1.117$$

This time (unlike Example 1) we have no answer to serve as a check. We can, however, look at the error. If $f(x) = \sqrt{x^3 + 1}$, then

$$f'(x) = \frac{3x^2}{2\sqrt{x^3 + 1}} \qquad \text{and} \qquad f''(x) = \frac{3x(x^3 + 4)}{4(x^3 + 1)^{3/2}}$$

An upper bound of $|f''(x)|$ for $0 \le x \le 1$ is $M = \frac{15}{4}$ (why?), so the error is

$$E_4 \le \tfrac{1}{12}(b - a)M(\Delta x)^2 = \tfrac{1}{12}(1)(\tfrac{15}{4})(0.25)^2 < 0.02$$

Our approximation is therefore accurate to at least one decimal place.   □

Figure 2
Parabolic approximation to a curve

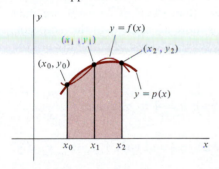

Now we turn to a numerical integration formula based on quadratic approximation of $f(x)$. Instead of joining consecutive points by straight line segments (as in Figure 1), we pass *parabolic arcs* through three points at a time. (Figure 2 shows the procedure in the first two subintervals of the partition.) In the problem set we ask you to prove that the area of the shaded region (under the parabola $y = p(x)$ which passes through the three points) is

$$\int_{x_0}^{x_2} p(x)\, dx = \frac{\Delta x}{3}(y_0 + 4y_1 + y_2)$$

Similarly, the area under the next parabola (from x_2 to x_4) is

$$\frac{\Delta x}{3}(y_2 + 4y_3 + y_4)$$

Assuming that n is even, we may continue in this way, the area under the last parabola (from x_{n-2} to x_n) being

$$\frac{\Delta x}{3}(y_{n-2} + 4y_{n-1} + y_n)$$

The sum of these areas is

$$\frac{\Delta x}{3}[(y_0 + 4y_1 + y_2) + (y_2 + 4y_3 + y_4) + \cdots + (y_{n-2} + 4y_{n-1} + y_n)]$$

$$= \frac{\Delta x}{3}(y_0 + 4y_1 + 2y_2 + 4y_3 + 2y_4 + \cdots + 2y_{n-2} + 4y_{n-1} + y_n)$$

Note that with the exception of the first and last terms (which have coefficient 1) the coefficients are alternately 4 and 2. We summarize this result as follows.

Simpson's Parabolic Rule

Let $\{x_0, x_1, \ldots, x_n\}$ be a partition of $[a,b]$ into an even number of subintervals of equal length $\Delta x = (b - a)/n$. If

$$y_k = f(x_k) \qquad k = 0, 1, \ldots, n$$

then

$$\int_a^b f(x)\,dx \approx \frac{\Delta x}{3}(y_0 + 4y_1 + 2y_2 + 4y_3 + \cdots + 2y_{n-2} + 4y_{n-1} + y_n)$$

Let S_n be the right side of Simpson's formula. It can be shown that if $f^{(4)}$ exists in I, and N is an upper bound of $|f^{(4)}(x)|$ for $a \leq x \leq b$, then the error is

$$E_n = \left| \int_a^b f - S_n \right| \leq \frac{1}{180}(b - a)N(\Delta x)^4$$

Since $(\Delta x)^4$ approaches zero much faster than $(\Delta x)^2$ as $\Delta x \to 0$, we may reasonably expect that Simpson's Rule is more accurate than the Trapezoidal Rule. (This is not always the case, however, because M and N are not the same.)

□ **Example 3**   Use Simpson's Rule with $n = 4$ to estimate $\int_1^3 \frac{dx}{x}$.

Solution: As in Example 1, we have $\Delta x = \frac{1}{2}$ and $y_0 = 1, y_1 = \frac{2}{3}, y_2 = \frac{1}{2}$, $y_3 = \frac{2}{5}, y_4 = \frac{1}{3}$. Hence

$$\int_1^3 \frac{dx}{x} \approx \frac{1}{6}\left(1 + \frac{8}{3} + 1 + \frac{8}{5} + \frac{1}{3}\right) = 1.100$$

The true value of the integral is $\ln 3 = 1.0986 \cdots$, so our approximation is accurate to two places. (Note the improvement over the Trapezoidal Rule.)
□

□ **Example 4**   Use Simpson's Rule with $n = 4$ to estimate

$$\int_0^1 \sqrt{x^3 + 1}\,dx$$

Solution: As in Example 2, we have $\Delta x = 0.25$ and $x_0 = 0, x_1 = 0.25$, $x_2 = 0.5, x_3 = 0.75, x_4 = 1$. Rounding off from a calculator, we find

$$\int_0^1 \sqrt{x^3 + 1}\,dx \approx \frac{0.25}{3}(1 + 4.0311 + 2.1213 + 4.7697 + 1.4142) = 1.1114$$

It is a nasty chore to estimate the error, because we need the fourth derivative of $f(x) = \sqrt{x^3 + 1}$ and an upper bound for its absolute value in the interval $[0,1]$. In any case, however, the error is

$$E_4 \leq \frac{N}{180}(0.25)^4$$

The fourth power of 0.25 is considerably smaller than the second power (and of course $\frac{1}{180}$ is more than ten times smaller than $\frac{1}{12}$). This suggests that the error is less than in the Trapezoidal Rule, but we cannot be sure without finding N. $\square$

All things considered, Simpson's Rule is superior to the Trapezoidal Rule (since parabolic arcs usually fit a curve better than line segments). It is no harder to compute, and the error is generally smaller. If you feel that the application of these rules is tedious, remember that to a computer they look easy. Computer-assisted numerical integration is the most practical way to find all but the simplest integrals.

Problem Set 10.6

Use the Trapezoidal Rule (with the given value of n) to compute an approximate value of each of the following integrals. When possible, compare with the true value of the integral.

1. $\int_1^2 \frac{dx}{x}$ $(n = 4)$

2. $\int_3^6 \frac{dx}{x - 1}$ $(n = 4)$

3. $\int_0^1 e^x \, dx$ $(n = 6)$

4. $\int_0^2 e^{-x} \, dx$ $(n = 6)$

5. $\int_0^{\pi} \sin x \, dx$ $(n = 6)$

6. $\int_0^{\pi/2} (1 - \cos x) \, dx$ $(n = 6)$

7. $\int_0^1 \frac{dx}{1 + x^2}$ $(n = 4)$

8. $\int_0^1 \frac{dx}{4 - x^2}$ $(n = 4)$

9. $\int_0^1 \sqrt{1 - x^2} \, dx$ $(n = 4)$

10. $\int_0^{\pi} \frac{\sin x}{x} \, dx$ $(n = 4)$ *Hint:* The integral is not improper. Define the integrand at $x = 0$ so as to make it continuous.

11. $\int_0^1 \sqrt{1 + x^4} \, dx$ $(n = 4)$

12. $\int_0^2 e^{-x^2} \, dx$ $(n = 4)$

13.–24. Use Simpson's Rule (with the given value of n) to compute an approximate value of each of the integrals in Problems 1 through 12. When possible, compare with the true value of the integral.

25.–30. Use the error bound given in the text to estimate your accuracy in Problems 1 through 6.

31.–36. Use the error bound given in the text to estimate your accuracy in Problems 13 through 18.

37. Use your answer to Problem 7 to obtain an approximation of π.

38. Use your answer to Problem 19 to obtain an approximation of π.

39. The width (at 2-inch intervals) of an irregularly shaped piece of material is shown in Figure 3. Use the Trapezoidal Rule to estimate the area of the material.

Figure 3
Irregular region

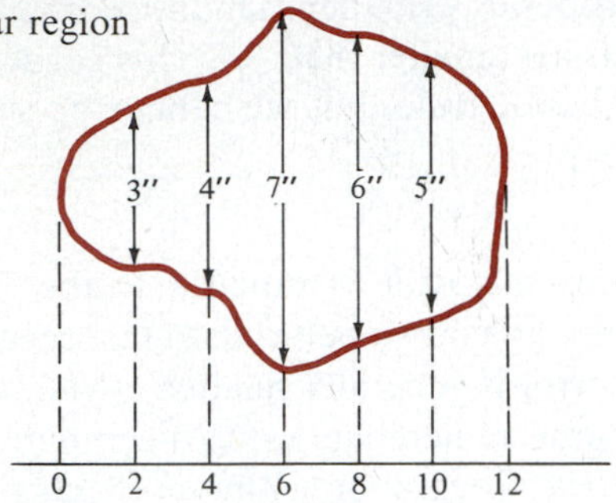

40. Repeat Problem 39 using Simpson's Rule.

41. The velocity of an object moving in a straight line was measured at regular intervals, with the following results:

t (sec)	0	1	2	3	4
v (m/sec)	2.6	3.0	3.2	4.0	4.4

Use the Trapezoidal Rule to estimate how far the object traveled from $t = 0$ to $t = 4$.

42. Repeat Problem 41 using Simpson's Rule.

43. The force applied to an object to move it 2 meters was measured at $\frac{1}{2}$-meter intervals (starting at the origin of the motion) and was found to be 15, 18, 20, 16, 18 newtons, respectively. Use the Trapezoidal Rule to estimate the work done.

44. Repeat Problem 43 using Simpson's Rule.

45. Use Simpson's Rule (with $n = 4$) to compute the length of the curve $y = \sin x$, $0 \le x \le \pi$.

46. Suppose that we want to compute

$$\int_1^2 \frac{dx}{x}$$

correct to five decimal places. If the Trapezoidal Rule is used, how many subintervals are needed to guarantee this accuracy? *Hint:* The error must be less than $\frac{1}{2} \times 10^{-5}$. (Why?)

47. Repeat Problem 46 using Simpson's Rule.

48. Suppose that $f(x)$ and $f''(x)$ are both positive in $[a,b]$. Use geometric reasoning to argue that the Trapezoidal Rule overestimates $\int_a^b f$. What if $f(x)$ is positive and $f''(x)$ is negative?

49. Suppose that $f(x)$ is a polynomial of degree no higher than 3. Why does Simpson's Rule give the exact value of $\int_a^b f$?

50. To complete the argument in the text for Simpson's Rule, we must prove that the area of the shaded region in Figure 2 is $\frac{1}{3}\Delta x(y_0 + 4y_1 + y_2)$. Do this as follows.

  (a) Let $p(x) = Ax^2 + Bx + C$ be an equation of the parabola passing through (x_0, y_0), (x_1, y_1), (x_2, y_2) and let $r = x_0$, $s = x_2$. Show that

  $$\int_r^s p(x)\, dx =$$

  $$\frac{1}{6}(s - r)[2A(s^2 + sr + r^2) + 3B(s + r) + 6C]$$

  (b) Noting that $x_1 = \frac{1}{2}(r + s)$, show that $y_0 + 4y_1 + y_2$ is the expression in brackets in part (a).

  (c) Combine parts (a) and (b) to finish the proof.

Additional Problems

Find each of the following integrals.

1. $\displaystyle\int \frac{x\, dx}{2 + x}$

2. $\displaystyle\int_3^5 \frac{dx}{x^2 - 6x + 13}$

3. $\displaystyle\int \frac{1 + \sin x}{1 - \sin x}\, dx$

4. $\displaystyle\int \frac{dx}{x^3 + x}$

5. $\displaystyle\int \frac{dx}{x^2 \sqrt{x^2 + 1}}$

6. $\displaystyle\int \cos \sqrt{x}\, dx$

7. $\displaystyle\int \frac{2 - \cos t}{\sin t}\, dt$

8. $\displaystyle\int \frac{2x^2 + 1}{x^3 + x}\, dx$

9. $\displaystyle\int e^{\sqrt{x}}\, dx$

10. $\displaystyle\int_0^1 \tan^{-1} \sqrt{x}\, dx$

11. $\displaystyle\int_0^1 \ln(x^2 + 1)\, dx$

12. $\displaystyle\int \frac{x^3\, dx}{\sqrt{x^2 - 25}}$

13. $\displaystyle\int \frac{e^{1/x}}{x^2}\, dx$

14. $\displaystyle\int \frac{e^x\, dx}{e^{2x} + 1}$

15. $\displaystyle\int x \cosh x\, dx$

16. $\displaystyle\int \frac{dx}{(4 - x^2)^{3/2}}$

17. $\displaystyle\int \tan^5 x \sec x\, dx$

18. $\displaystyle\int \frac{x + 4}{x^2 + 4}\, dx$

19. $\displaystyle\int \frac{x\, dx}{x^2 - 8x + 15}$

20. $\displaystyle\int \frac{dx}{3 \sin x + 4 \cos x}$

21. $\displaystyle\int_0^{\ln 3} \frac{dx}{1 + e^x}$

22. $\displaystyle\int \frac{x^2\, dx}{\sqrt{x + 2}}$

23. $\displaystyle\int_0^{\pi/2} e^{-x} \cos x\, dx$

24. $\displaystyle\int_2^4 \frac{dx}{\sqrt{-x^2 + 6x - 5}}$

25. $\displaystyle\int_{-1}^2 \frac{dx}{x^2 + 2x + 10}$

26. $\displaystyle\int_0^{\pi/2} \sin^5 t \cos^3 t\, dt$

27. $\displaystyle\int \frac{dx}{\sqrt{9x^2 - 16}}$

28. $\displaystyle\int \frac{dx}{\cos x + \cot x}$

29. $\displaystyle\int_0^2 \frac{x^2\, dx}{\sqrt{16 - x^2}}$

30. $\displaystyle\int \frac{dx}{x^3 - 4x}$

31. $\displaystyle\int_1^4 \frac{dx}{x + 2\sqrt{x}}$

32. $\displaystyle\int_0^1 \frac{x^3\, dx}{9 + x^8}$

33. $\displaystyle\int x \sin^{-1} x^2\, dx$

34. $\displaystyle\int x \sqrt{1 - x}\, dx$

35. $\displaystyle\int_0^{\ln 2} \frac{e^x\, dx}{9 - e^{2x}}$

36. $\displaystyle\int \frac{dx}{9x^2 + 4}$

37. $\displaystyle\int (\cos t + \sec t)^2\, dt$

38. $\displaystyle\int_0^5 \frac{x\, dx}{\sqrt{9 - x}}$

39. $\displaystyle\int \frac{dx}{\sqrt{x^2 - 4}}$

40. $\displaystyle\int x \sinh^{-1} x\, dx$

41. $\displaystyle\int_1^4 \frac{dx}{\sqrt{x}(1 + \sqrt{x})^4}$

42. $\displaystyle\int_1^2 x \ln x\, dx$

43. $\displaystyle\int x \sec^2 x\, dx$

44. $\displaystyle\int_0^1 x^3 e^{-x^2}\, dx$

45. $\displaystyle\int_5^{10} \frac{\sqrt{x^2 - 25}}{x}\, dx$

46. $\displaystyle\int \frac{\sqrt{x^2 + 16}}{x^3}\, dx$

47. $\displaystyle\int \frac{x^2 - 1}{2x - 5}\, dx$

48. $\displaystyle\int \frac{x^2\, dx}{x^4 - 1}$

49. $\displaystyle\int \frac{x\, dx}{\sqrt{4 + x^2}}$

50. $\displaystyle\int_0^4 x^3 \sqrt{16 - x^2}\, dx$

51. $\displaystyle\int_0^{\pi/4} \sec^2 x \tan^2 x\, dx$

52. $\displaystyle\int \frac{x^2 - 2}{x^3 - x^2}\, dx$

53. $\displaystyle\int \frac{x\, dx}{\sqrt{4 - x^4}}$

54. $\displaystyle\int \sec^4 x\, dx$

55. $\displaystyle\int_0^3 \frac{dx}{(x^2 + 9)^{3/2}}$

56. $\displaystyle\int_0^{\pi/2} \frac{\sin t}{1 + \cos^2 t}\, dt$

57. $\displaystyle\int \frac{x^2 + 1}{x^3 + 1}\, dx$

58. $\displaystyle\int_0^1 x(2x - 1)^7\, dx$

59. $\displaystyle\int_0^{\pi/2} \frac{dx}{2 + \sin x}$

60. $\displaystyle\int (x - \cos x)^2\, dx$

61. $\displaystyle\int \frac{dx}{x^2(x^2 + 1)}$

62. $\displaystyle\int_2^5 \frac{\sqrt{4 + x^2}}{x^4}\, dx$

63. $\displaystyle\int_0^{\pi/6} \frac{dx}{\cos 2x}$

64. $\displaystyle\int x \ln 2x\, dx$

65. $\displaystyle\int \frac{2x - 5}{\sqrt{x^2 - 5x + 4}}\, dx$

66. $\displaystyle\int \frac{x\, dx}{x^3 - 1}$

67. $\displaystyle\int x^3 e^{-x}\, dx$

68. $\displaystyle\int \frac{dx}{9x^2 - 18x + 13}$

69. $\displaystyle\int_0^\pi e^x \sin x\, dx$

70. $\displaystyle\int_3^5 \frac{dx}{x^2 - 6x + 13}$

71. $\displaystyle\int_0^1 x^3 \cosh x\, dx$

72. $\displaystyle\int \frac{\tan^{-1} x}{x^2}\, dx$

73. $\displaystyle\int x \sqrt{x^4 - 16}\, dx$

74. $\displaystyle\int_0^1 \frac{e^x\, dx}{e^x + 1}$

75. $\displaystyle\int_0^1 \frac{x^2\, dx}{x + 1}$

76. $\displaystyle\int \frac{x\, dx}{(x - 2)^3}$

77. $\displaystyle\int_1^2 x^3 \ln x\, dx$

78. $\displaystyle\int x^5 \sin x\, dx$

79. $\displaystyle\int_0^1 x(x - 1)^4\, dx$

80. $\displaystyle\int \frac{dx}{5 + 4 \cos x}$

81. $\displaystyle\int \frac{\sin x\, dx}{1 + \sin x}$

82. Find

$$\int_0^4 \frac{x^3\, dx}{\sqrt{x^2 + 9}}$$

by making an algebraic substitution.

83. Do Problem 82 by making a trigonometric substitution.

84. Use the identity $\sin^2(t/2) = \frac{1}{2}(1 - \cos t)$ to find

$$\int \frac{dt}{1 - \cos t}$$

85. Do Problem 84 by making the substitution $t = 2 \tan^{-1} u$. (See Example 4, Section 10.5.)

86. Use the identity $\cos^2(t/2) = \frac{1}{2}(1 + \cos t)$ to find

$$\int \frac{dt}{1 + \cos t}$$

87. Do Problem 86 by making the substitution $t = 2 \tan^{-1} u$.

Use the reduction formulas in Section 10.1 to evaluate each of the following integrals.

88. $\displaystyle\int_0^{\pi/2} \sin^5 x \, dx$ **89.** $\displaystyle\int_0^{\pi/2} \cos^7 x \, dx$

90. $\displaystyle\int_0^{\pi} \sin^6 \frac{x}{2} \, dx$ **91.** $\displaystyle\int_0^{\pi} \cos^8 \frac{x}{2} \, dx$

92. Show that

$$\int_0^1 \sinh^{-1} x \, dx = 1 - \sqrt{2} + \ln(1 + \sqrt{2})$$

93. Find the area of the region bounded by the curve $y = \tan^{-1} x$, the x axis, and the line $x = 1$.

94. The region enclosed by the curve $x = a \cos t$, $y = b \sin t$ is rotated about the x axis. Find the volume generated, as follows.

    (a) Use the parametric equations as they stand to show that

$$V = 2\pi a b^2 \int_0^{\pi/2} \sin^3 t \, dt$$

and evaluate the integral.

    (b) Eliminate the parameter before setting up an integral for V.

    (c) What is V if $a = b$?

95. In an ac circuit the *power* at time t is the product of current I and voltage V. Find the average power during one cycle in each of the following situations.

    (a) The current and voltage are in phase, that is,

$$I = I_0 \cos(\omega t) \quad \text{and} \quad V = V_0 \cos(\omega t)$$

    (b) The current and voltage are out of phase, that is,

$$I = I_0 \cos(\omega t) \quad \text{and} \quad V = V_0 \cos(\omega t + \alpha)$$

    where $0 < \alpha < \pi/2$.

In each of the following, use the given value of n in the Trapezoidal Rule to obtain an approximation to the integral.

96. $\displaystyle\int_0^{\pi} e^{\sin x} \, dx$ $(n = 4)$

97. $\displaystyle\int_0^2 \frac{dx}{x^3 + 2}$ $(n = 6)$

98. $\displaystyle\int_0^1 \frac{\tan x}{x} \, dx$ $(n = 4)$

99. Do Problem 96 by using Simpson's Rule.

100. Do Problem 97 by using Simpson's Rule.

101. Do Problem 98 by using Simpson's Rule.

11 More About Limits

Mathematics is the science which draws necessary conclusions.
BENJAMIN PEIRCE (1809–1880)

A chess player may offer the sacrifice of a pawn or even a piece, but a mathematician the game.
G. H. HARDY (1877–1947)
(commenting on the method of proof by contradiction)

In this chapter we are interested in four topics. First we enlarge the idea of limit (of a real function) to allow the independent variable or the functional value (or both) to "go to infinity." Then we discuss an elegant device (called L'Hôpital's Rule) for evaluating limits. Next we broaden the scope of integration by defining *improper integrals* (in terms of limits). Finally we develop a generalization of the Mean Value Theorem known as Taylor's Formula (for the approximation of functions by polynomials of increasing degree). Each of these topics is important in the upcoming chapters on infinite series.

11.1 LIMITS INVOLVING INFINITY

The idea of limit lies at the heart of calculus. It is easy to lose sight of that fact, for although derivatives and integrals are defined in terms of limits, we spend so much time on techniques and applications of differentiation and integration that we tend to forget their source.

It is important at this time to remember what a limit is. A reading of Section 2.4 may be profitable (although it appears in Chapter 2 as an optional assignment). Section 2.5, in which the properties of limits are discussed, is also worth reviewing. The purpose of this section is to discuss limits involving infinity. We have been evaluating such limits informally for a long time, using language that was designed to appeal to intuition. The

465

Figure 1
Unbounded behavior of ln x

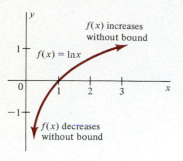

function $f(x) = \ln x$, for example, can be made arbitrarily large by choosing x to be large enough, and it can be made arbitrarily large negatively by choosing x to be sufficiently close to 0. (See Figure 1, and Theorem 4 in Section 8.1.) Symbolic shorthand for these statements is

$$\lim_{x \to \infty} \ln x = \infty \qquad \text{and} \qquad \lim_{x \to 0} \ln x = -\infty$$

To define their meaning, we use the "challenge and response" idea of an ordinary limit, suitably restated so as to make sense.

□ **Example 1**   Devise a definition of the statement $\lim_{x \to \infty} \ln x = \infty$.

Solution: The idea that $\ln x$ can be made arbitrarily large by choosing x to be large enough is made precise by imagining a critic who specifies what "arbitrarily large" means, and by responding to the critic with a specification of "large enough." The critic (who presumably doubts that $\ln x$ can be made arbitrarily large) names a number B and challenges us to force $\ln x > B$. We have control of x; the problem is to name a number A such that

$$x > A \Rightarrow \ln x > B$$

Figure 2 suggests that when B is given by our critic, we can specify A by solving the equation $\ln A = B$. Since this yields $A = e^B$, we satisfy our critic by arguing that

$$x > A \Rightarrow x > e^B \Rightarrow \ln x > \ln e^B \qquad \text{(Why?)}$$
$$\Rightarrow \ln x > B$$

Figure 2
$\lim_{x \to \infty} \ln x = \infty$

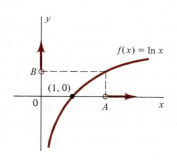

Thus our critic must admit that $\ln x$ can be made larger than any number B that is named, simply by choosing x to be sufficiently large (greater than e^B).

□

It is not hard to convert Example 1 to a general definition:

The statement

$$\lim_{x \to \infty} f(x) = \infty$$

means that given any number B, there is a number A such that $x > A \Rightarrow f(x) > B$.

□ **Example 2**   Give a definition of the statement $\lim_{x \to 0} \ln x = -\infty$.

Solution: This time $\ln x$ is supposed to be arbitrarily large negatively if x is sufficiently close to 0. Our critic defines the meaning of "arbitrarily large negatively" by proposing a negative number $-B$ and challenging us to force $\ln x < -B$. We respond by showing that if x is sufficiently close to 0 (and positive), then the critic's desired inequality will follow. More precisely, we name a (right) neighborhood of 0, say $M = (0, \delta)$, such that

$$x \in M \Rightarrow \ln x < -B$$

Figure 3 suggests that δ should satisfy $\ln \delta = -B$, that is, $\delta = e^{-B}$. Our critic will be convinced that the above implication is correct when we write

$$x \in M \Rightarrow 0 < x < e^{-B} \Rightarrow \ln x < \ln e^{-B}$$
$$\Rightarrow \ln x < -B \qquad \qquad \square$$

Example 2 suggests the following more general definition:

The statement

$$\lim_{x \to a} f(x) = -\infty$$

means that given any negative number $-B$, there is a neighborhood of a, say M, such that $x \in M \Rightarrow f(x) < -B$. ($M$ may have to be punctured, or a right or left neighborhood of a.)

□ **Example 3**   To sketch the graph of $f(x) = x/(x-2)$, we observe that

• $f(x)$ approaches 1 when x increases or decreases without bound.
• $f(x)$ increases without bound when x approaches 2 from the right.
• $f(x)$ decreases without bound when x approaches 2 from the left.

(See Figure 4.) This mouthful is avoided by using crisp notation:

$$\lim_{x \to \infty} \frac{x}{x-2} = 1 \quad \text{and} \quad \lim_{x \to -\infty} \frac{x}{x-2} = 1$$

$$\lim_{x \downarrow 2} \frac{x}{x-2} = \infty$$

$$\lim_{x \uparrow 2} \frac{x}{x-2} = -\infty$$

The question is whether these limits can be properly defined. We will choose the first one as further illustration of the challenge and response terminology that is needed for a good definition.

Suppose that a critic denies that

$$\lim_{x \to \infty} \frac{x}{x-2} = 1$$

meaning that we cannot force $f(x)$ to be close to $L = 1$ by making x large. Naming a neighborhood of 1, say $N = (1 - \varepsilon, 1 + \varepsilon)$, the critic challenges us to confine $f(x)$ to this neighborhood. To respond, we solve the inequality

$$1 - \varepsilon < f(x) < 1 + \varepsilon, \text{ that is, } |f(x) - 1| < \varepsilon$$

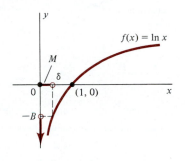

Figure 3
$\displaystyle \lim_{x \to 0} \ln x = -\infty$

Figure 4

Behavior of $f(x) = \dfrac{x}{x - 2}$

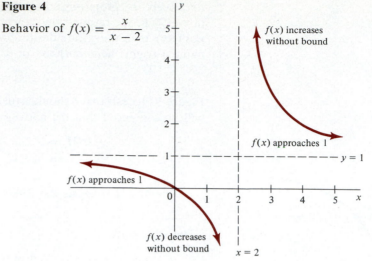

Rather than depending on the figure, we will proceed algebraically:

$$\left| \frac{x}{x - 2} - 1 \right| < \varepsilon$$

$$\left| \frac{2}{x - 2} \right| < \varepsilon$$

$$\left| \frac{x - 2}{2} \right| > \frac{1}{\varepsilon}$$

$$|x - 2| > \frac{2}{\varepsilon}$$

Since we claim that the inequality can be satisfied when x is large, we assume (at least) that $x > 2$, thus dropping the absolute value:

$$x - 2 > \frac{2}{\varepsilon}$$

$$x > \frac{2}{\varepsilon} + 2$$

Now we are ready to answer with precision. We name

$$A = \frac{2}{\varepsilon} + 2 \qquad \text{(a large number if } \varepsilon \text{ is small)}$$

and assert that $x > A \Longrightarrow f(x) \in N$. To verify the assertion, our critic need only reverse the above steps.   □

Remark   It is worth adding in Example 3 that if we were trying to explain the meaning of

$$\lim_{x \to -\infty} \frac{x}{x - 2} = 1$$

the procedure would be much the same. Our critic names $N = (1 - \varepsilon, 1 + \varepsilon)$ as before, and we solve the inequality

$$1 - \varepsilon < f(x) < 1 + \varepsilon$$

to the stage where $|x - 2| > 2/\varepsilon$. This time, however, we are interested in x decreasing without bound. Thus we assume that $x < 2$, which implies $|x - 2| = 2 - x$ and hence

$$2 - x > \frac{2}{\varepsilon}$$

$$x - 2 < -\frac{2}{\varepsilon}$$

$$x < -\frac{2}{\varepsilon} + 2$$

Naming $-A = (-2/\varepsilon) + 2$, we assert that

$$x < -A \Longrightarrow f(x) \in N$$

thus satisfying our critic that $f(x)$ is close to 1 when x is negatively large.

As you can see, there are several types of limits involving infinity. **Finite limits** [of the type $\lim f(x) = L$] may occur when the independent variable increases or decreases without bound ($x \to \infty$ or $x \to -\infty$). Graphically, these limits correspond to a horizontal asymptote ($y = L$) of the curve $y = f(x)$. **Infinite limits** [of the type $\lim f(x) = \infty$ or $\lim f(x) = -\infty$] may occur when the independent variable behaves in any of the modes $x \to a$, $x \downarrow a$, $x \uparrow a$, $x \to \infty$, $x \to -\infty$. (The first three correspond to a vertical asymptote $x = a$ of the graph of f.)

Rather than developing definitions of all these types, we will continue to rely on your intuition. Just as in the case of ordinary limits, technical definitions are rarely needed in practice; you can evaluate most of the limits that occur in this book by a judicious combination of theorems about limits and "inspection."

Before proceeding to the theorems, we need to discuss a controversial question. It is no problem to give technical definitions of infinite limits; for example, the statement

$$\lim_{x \to 0} \ln x = -\infty$$

has a definite meaning as explained in Example 2. Do we therefore say that the limit *exists?* If we do, several theorems about limits need qualifying. Consider, for instance, the statement that if $\lim f(x)$ and $\lim g(x)$ exist, then

$$\lim f(x) g(x) = [\lim f(x)][\lim g(x)]$$

(See Theorem 1, Section 2.5, on the "algebra of limits.") As you can see by taking $f(x) = x^2$ and $g(x) = 1/x^2$ (and letting $x \to 0$), the statement is false. For it reads

$$\lim_{x \to 0} \left(x^2 \cdot \frac{1}{x^2} \right) = \left[\lim_{x \to 0} x^2 \right] \left[\lim_{x \to 0} \frac{1}{x^2} \right]$$

or $1 = 0 \cdot \infty$.

In view of this nonsense, *we will avoid saying that an infinite limit "exists."* We do not mean that its definition lacks precision, but merely that this decision will keep our theorems simple.

In the case of *finite* limits involving infinity, the theorems stated in Section 2.5 apply without any essential changes. Thus the *algebra of limits* still holds:

If $\lim f(x)$ and $\lim g(x)$ exist, then

$$\lim (f + g)(x) = \lim f(x) + \lim g(x)$$
$$\lim (f - g)(x) = \lim f(x) - \lim g(x)$$
$$\lim (fg)(x) = [\lim f(x)][\lim g(x)]$$
$$\lim (f/g)(x) = \frac{\lim f(x)}{\lim g(x)}, \text{ provided that } \lim g(x) \neq 0$$

◻ **Example 4**

$$\lim_{x \to \infty} \left(\frac{x - 1}{x} + e^{-x} \right) = \lim_{x \to \infty} \frac{x - 1}{x} + \lim_{x \to \infty} e^{-x} = 1 + 0 = 1$$

In Example 4 we used the algebra of limits only in the first step (to break up the sum). The evaluation of

$$\lim_{x \to \infty} \frac{x - 1}{x}$$

cannot be done by the rule for a quotient, because $\lim_{x \to \infty} (x - 1)$ and $\lim_{x \to \infty} x$ do not exist. (See the preceding note about "existence." Each of these limits is infinite.) We can, however, write

$$\lim_{x \to \infty} \frac{x - 1}{x} = \lim_{x \to \infty} \left(1 - \frac{1}{x} \right) = \lim_{x \to \infty} 1 - \lim_{x \to \infty} \frac{1}{x} = 1 - 0 = 1$$

(the last limit being evaluated "by inspection"). Similarly,

$$\lim_{x \to \infty} e^{-x} = \lim_{x \to \infty} \frac{1}{e^x} = 0$$

"by inspection." (If a critic objects to doing limits by inspection, we are forced back to the definitions!)   ◻

◻ **Example 5**

$$\lim_{x \to -\infty} \frac{x \tan^{-1} x}{2 - x} = \left[\lim_{x \to -\infty} \frac{x}{2 - x} \right]\left[\lim_{x \to -\infty} \tan^{-1} x \right] = (-1)\left(-\frac{\pi}{2} \right) = \frac{\pi}{2}$$

In Example 5 (as in Example 4) the individual limits are not evaluated by the algebra of limits, but directly:

$$\lim_{x \to -\infty} \frac{x}{2-x} = \lim_{x \to -\infty} \frac{1}{\dfrac{2}{x} - 1} = -1 \qquad \text{("by inspection")}$$

$$\lim_{x \to -\infty} \tan^{-1} x = -\frac{\pi}{2} \qquad \text{(from our study of the inverse tangent in Section 9.1)} \qquad \square$$

Like the algebra of limits, the *Composite Function Theorem* still holds for finite limits:

> If $\lim g(x) = u_0$ exists and f is continuous at u_0, then
> $$\lim f[g(x)] = f[\lim g(x)]$$

□ **Example 6**

$$\lim_{x \to \infty} \cos \frac{1}{x} = \cos \left(\lim_{x \to \infty} \frac{1}{x} \right) = \cos 0 = 1 \qquad \square$$

□ **Example 7**

$$\lim_{x \to -\infty} \ln \frac{2x^2}{1+x^2} = \ln \left(\lim_{x \to -\infty} \frac{2x^2}{1+x^2} \right) = \ln 2 \qquad \square$$

The *Squeeze Play Theorem* also still applies:

> If $f(x)$ is between $g(x)$ and $h(x)$, and $\lim g(x) = \lim h(x) = L$, then $\lim f(x) = L$.

□ **Example 8**   Since $f(x) = e^{-x} \sin x$ is between $g(x) = e^{-x}$ and $h(x) = -e^{-x}$ for all x, and since

$$\lim_{x \to \infty} e^{-x} = \lim_{x \to \infty} (-e^{-x}) = 0$$

it follows that

$$\lim_{x \to \infty} (e^{-x} \sin x) = 0 \qquad \square$$

Infinite limits do not obey these rules. Nevertheless some rules can be stated. Suppose, for example, that $\lim f(x)$ is positive (or ∞) and $\lim g(x) = \infty$. Then it is intuitively evident that

$$\lim (fg)(x) = \infty$$

As an illustration, consider the statement that

$$\lim_{x \to \infty} \left(\frac{x - 1}{x} \ln x \right) = \infty$$

The idea is that since $f(x) = (x - 1)/x$ is close to 1 for large x, while $g(x) = \ln x$ increases without bound, the product

$$f(x)g(x) = \frac{x - 1}{x} \ln x$$

also increases without bound. Change the first function to $f(x) = (1 - x)/x$ (which approaches -1 as $x \to \infty$) and the product $f(x)g(x)$ goes to $-\infty$. (Why?) In practice we can usually figure out this sort of thing by inspection. (See the problem set.)

Despite the proliferation of definitions and theorems implicit in this section, we are not much interested in proofs. Intuition is sufficient for most limits involving infinity (as in the case of ordinary limits). In doubtful cases, however, an appeal to the definitions is the way to resolve the difficulty.

Problem Set 11.1

Evaluate each of the following limits.

1. $\lim\limits_{x \to \infty} \dfrac{x^2 - x + 1}{x^2 + x + 1}$

2. $\lim\limits_{x \to -\infty} \dfrac{x^3}{x^2 + 1}$

3. $\lim\limits_{x \to -\infty} x \sqrt{3 - x}$

4. $\lim\limits_{x \to \infty} \left(\dfrac{1}{x} - \dfrac{1}{x^2} \right)$

5. $\lim\limits_{x \to 0} \left(x^2 - \dfrac{1}{x^2} \right)$

6. $\lim\limits_{x \to \infty} (x - \sqrt{x^2 + 1})$

7. $\lim\limits_{x \to -\infty} (x - \sqrt{x^2 + 1})$

8. $\lim\limits_{x \downarrow 0} \dfrac{\cos x}{x}$

9. $\lim\limits_{x \downarrow 0} \dfrac{\ln x}{x}$

10. $\lim\limits_{x \downarrow 0} \left(\ln x - \dfrac{1}{x} \right)$

11. $\lim\limits_{x \to \infty} \sin \dfrac{1}{x}$

12. $\lim\limits_{x \to \infty} x \sin \dfrac{1}{x}$ *Hint:* Let $t = 1/x$.

13. $\lim\limits_{x \to \infty} x \cos \dfrac{1}{x}$

14. $\lim\limits_{x \to \infty} x \sinh x$

15. $\lim\limits_{x \to \infty} \dfrac{2^x}{5^x}$

16. $\lim\limits_{x \to \infty} \dfrac{5^x}{2^x}$

17. $\lim\limits_{x \downarrow 0} e^{1/x}$

18. $\lim\limits_{x \uparrow 0} e^{1/x}$

19. $\lim\limits_{x \uparrow 1} \ln \dfrac{1 + x}{1 - x}$

Each of the preceding problems has a definite answer (real number, ∞, or $-\infty$). Explain why the following do not.

20. $\lim\limits_{x \to 0} \dfrac{1}{x}$

21. $\lim\limits_{x \to \pi/2} \tan x$

22. $\lim\limits_{x \to \infty} \sin x$

23. $\lim\limits_{x \to \infty} x \sin x$

24. Explain why it does not make sense to investigate

$$\lim_{x \to -\infty} (1 - \sqrt{x})$$

25. Show that

$$\lim_{x \to \infty} \left(1 + \frac{1}{x} \right)^x = e$$

Hint: Let $h = 1/x$.

26. Use Problem 25 to find

$$\lim_{x \to \infty} \left(1 + \frac{1}{x} \right)^{2x}$$

27. The value of $\lim\limits_{n \to \infty} x^n$ depends on x. What is it when $0 \le x < 1$? when $x = 1$? when $x > 1$? Why don't we investigate it when $x < 0$?

28. Sketch the graph of

$$f(x) = \lim_{n \to \infty} \frac{2x^n}{x^n + 1} \qquad x \geq 0$$

Each of the following statements is a theorem about infinite limits. Explain it on an intuitive basis. (When more than one double sign appears, read the top signs together, or the bottom signs.)

29. $\lim f(x) = L$ (or $\pm\infty$) and $\lim g(x) = \pm\infty \Rightarrow$
$\lim (f + g)(x) = \pm\infty$

30. $\lim f(x) > 0$ (or ∞) and $\lim g(x) = \pm\infty \Rightarrow$
$\lim (fg)(x) = \pm\infty$

31. $\lim f(x) < 0$ (or $-\infty$) and $\lim g(x) = \pm\infty \Rightarrow$
$\lim (fg)(x) = \mp\infty$

32. $\lim f(x) = L$ and $\lim g(x) = \pm\infty \Rightarrow \lim (f/g)(x) = 0$

33. $\lim f(x) = \pm\infty$ and $\lim g(x) > 0 \Rightarrow \lim (f/g)(x) = \pm\infty$

34. $\lim f(x) = \pm\infty$ and $\lim g(x) < 0 \Rightarrow \lim (f/g)(x) = \mp\infty$

35. $\lim f(x) > 0$ (or ∞) and $g(x) \downarrow\uparrow 0 \Rightarrow \lim (f/g)(x) = \pm\infty$

36. $\lim f(x) < 0$ (or $-\infty$) and $g(x) \downarrow\uparrow 0 \Rightarrow \lim (f/g)(x) = \mp\infty$

Give a counterexample disproving each of the following.

37. $\lim f(x) = \infty$ and $\lim g(x) = \infty \Rightarrow \lim (f - g)(x) = 0$

38. $\lim f(x) = 0$ and $\lim g(x) = 0 \Rightarrow \lim (f/g)(x) = 1$

39. $\lim f(x) = \infty$ and $\lim g(x) = \infty \Rightarrow \lim (f/g)(x) = 1$

The remaining problems test your ability to state and use definitions of limits. You can omit them altogether without missing anything later; on the other hand, they will sharpen your perception of what a limit really is. (In Problems 40–44 the letters B and ε represent arbitrary positive numbers.)

40. Explain why the implication $x > B^2 \Rightarrow \sqrt{x} > B$ is correct and why it proves that

$$\lim_{x \to \infty} \sqrt{x} = \infty$$

41. Explain why the implication $x < -\sqrt{B} \Rightarrow x^2 > B$ is correct and why it proves that

$$\lim_{x \to -\infty} x^2 = \infty$$

42. Explain why the implication

$$0 < x < \frac{1}{B + 1} \Rightarrow 1 - \frac{1}{x} < -B$$

is correct and why it proves that

$$\lim_{x \downarrow 0} \left(1 - \frac{1}{x}\right) = -\infty$$

43. Explain why the implication

$$x > \frac{2}{\varepsilon} \Rightarrow \left|\frac{2}{x}\right| < \varepsilon$$

is correct and why it proves that

$$\lim_{x \to \infty} \frac{2}{x} = 0$$

44. Explain why the implication

$$x > \frac{1}{\varepsilon} + 1 \Rightarrow \left|\frac{x}{x - 1} - 1\right| < \varepsilon$$

is correct and why it proves that

$$\lim_{x \to \infty} \frac{x}{x - 1} = 1$$

Find each of the following limits and use an appropriate definition to prove that your answer is correct.

45. $\displaystyle\lim_{x \to \infty} (1 - \sqrt{x})$ **46.** $\displaystyle\lim_{x \downarrow 0} \frac{1}{x}$ **47.** $\displaystyle\lim_{x \to \infty} e^{-x}$

48. $\displaystyle\lim_{x \uparrow 1} \frac{1}{\sqrt{1 - x^2}}$ **49.** $\displaystyle\lim_{x \to \infty} \frac{x^2}{x^2 + 1}$

11.2
L'HÔPITAL'S RULE AND INDETERMINATE FORMS

We have evaluated many limits of the indeterminate type 0/0. Familiar examples are

$$\lim_{x \to 3} \frac{x^2 - x - 6}{x - 3} = \lim_{x \to 3} \frac{(x - 3)(x + 2)}{x - 3} = \lim_{x \to 3} (x + 2) = 5$$

(in which we execute an algebraic change of form to avoid 0/0), and

$$\lim_{x \to 0} \frac{\sin x}{x} = 1$$

(in which we use the Squeeze Play Theorem). Every derivative is of this type (which is one reason why such limits are important):

$$f'(x) = \lim_{z \to x} \frac{f(z) - f(x)}{z - x}$$

Limits of this kind are often hard to evaluate, as you can see by trying to find

$$\lim_{x \to 0} \frac{x - \sin x}{1 - \cos x}$$

Substitution of $x = 0$ merely gives $0/0$ (and there is no obvious maneuver that will avoid it). In this section we develop a remarkable device for solving such problems, known as **L'Hôpital's Rule** (pronounced low-pee-tăl). It is easy to give the bare bones of the rule (somewhat harder to flesh out the details):

> Suppose that f and g are differentiable in a (punctured) neighborhood of a. If the limit (as $x \to a$) of $f(x)/g(x)$ is indeterminate of the type $0/0$ or ∞/∞, then
>
> $$\lim_{x \to a} \frac{f(x)}{g(x)} = \lim_{x \to a} \frac{f'(x)}{g'(x)}$$
>
> provided that the latter limit is definite (either a real number or $\pm\infty$).

Remark   Sometimes students are unconvinced that $0/0$ is indeterminate. Consider (as examples)

$$\lim_{x \to 0} \frac{2x}{x} \quad \text{and} \quad \lim_{x \to 0} \frac{3x}{x}$$

Both limits are indeterminate of the type $0/0$; their values are 2 and 3, respectively. More generally, if c is an arbitrary real number,

$$\lim_{x \to 0} \frac{cx}{x} = c$$

□ **Example 1**   Since $x - \sin x$ and $1 - \cos x$ both approach 0 as $x \to 0$, L'Hôpital's Rule says that

$$\lim_{x \to 0} \frac{x - \sin x}{1 - \cos x} = \lim_{x \to 0} \frac{1 - \cos x}{\sin x}$$

(provided that the new limit is definite). Writing

$$\frac{1 - \cos x}{\sin x} = \tan \frac{x}{2} \qquad \text{(a trigonometric identity)}$$

we find

$$\lim_{x \to 0} \frac{x - \sin x}{1 - \cos x} = \lim_{x \to 0} \tan \frac{x}{2} = 0$$

An alternate approach (if the identity does not come to mind) is to observe that the new limit is also of the type 0/0. Hence we apply the rule again:

$$\lim_{x \to 0} \frac{x - \sin x}{1 - \cos x} = \lim_{x \to 0} \frac{1 - \cos x}{\sin x} = \lim_{x \to 0} \frac{\sin x}{\cos x} = \frac{0}{1} = 0 \qquad \square$$

In its simplest version, L'Hôpital's Rule is not hard to understand. Suppose that f and g are differentiable at a, with $g'(a) \neq 0$, and assume that $f(a) = g(a) = 0$. Then for x sufficiently near a we have

$$\frac{f(x)}{g(x)} = \frac{f(x) - f(a)}{g(x) - g(a)} = \frac{\dfrac{f(x) - f(a)}{x - a}}{\dfrac{g(x) - g(a)}{x - a}}$$

Letting $x \to a$, we find

$$\lim_{x \to a} \frac{f(x)}{g(x)} = \frac{f'(a)}{g'(a)}$$

Thus in this case the limit of $f(x)/g(x)$ does not require the evaluation of the limit of $f'(x)/g'(x)$, but is simply the quotient $f'(a)/g'(a)$.

In general the rule is not this easy. To avoid a digression, we have put an outline of its proof in the problem set. Meanwhile the examples are designed to show you how it works, and in what circumstances.

□ **Example 2**   Evaluate $\lim_{x \to 0} \sqrt{x}/\sin x$.

Solution: Since $\sqrt{x}$ and $\sin x$ both approach 0 as $x \to 0$, L'Hôpital's Rule applies (provided that the final limit is definite):

$$\lim_{x \to 0} \frac{\sqrt{x}}{\sin x} = \lim_{x \to 0} \frac{1/(2\sqrt{x})}{\cos x} = \lim_{x \to 0} \frac{1}{2\sqrt{x} \cos x} = \infty$$

Note that the final limit does not have to be a number; we allow ∞ or $-\infty$ as well. $\qquad \square$

L'Hôpital's Rule also applies when $x \to \infty$ or $x \to -\infty$. To see why, consider the problem of finding

$$\lim_{x \to \infty} \frac{f(x)}{g(x)} \qquad \text{where } \lim_{x \to \infty} f(x) = 0 \text{ and } \lim_{x \to \infty} g(x) = 0$$

We change the limit to one we can handle by letting $t = 1/x$. Since $t \downarrow 0$ when $x \to \infty$, we have

$$\lim_{x \to \infty} \frac{f(x)}{g(x)} = \lim_{t \downarrow 0} \frac{f(1/t)}{g(1/t)} = \lim_{t \downarrow 0} \frac{F(t)}{G(t)} \qquad [F(t) = f(1/t),\ G(t) = g(1/t)]$$

L'Hôpital's Rule applies to this limit because $F(t)$ and $G(t)$ both approach 0 as $t \downarrow 0$. Since

$$\frac{F'(t)}{G'(t)} = \frac{f'(1/t)(-1/t^2)}{g'(1/t)(-1/t^2)} \qquad \text{(Chain Rule)}$$

$$= \frac{f'(1/t)}{g'(1/t)} = \frac{f'(x)}{g'(x)}$$

we find

$$\lim_{x \to \infty} \frac{f(x)}{g(x)} = \lim_{t \downarrow 0} \frac{F'(t)}{G'(t)} = \lim_{x \to \infty} \frac{f'(x)}{g'(x)}$$

Similar reasoning is used in the case $x \to -\infty$ (replacing $t \downarrow 0$ by $t \uparrow 0$).

□ **Example 3**   Find $\lim_{x \to \infty} x \sin (1/x)$.

Solution: As it stands, this limit is indeterminate of the type $\infty \cdot 0$. By writing

$$x \sin \frac{1}{x} = \frac{\sin (1/x)}{1/x}$$

however, we change the form to $0/0$. Hence L'Hôpital's Rule applies:

$$\lim_{x \to \infty} x \sin \frac{1}{x} = \lim_{x \to \infty} \frac{\sin (1/x)}{1/x} = \lim_{x \to \infty} \frac{(-1/x^2) \cos (1/x)}{-1/x^2}$$

$$= \lim_{x \to \infty} \cos \frac{1}{x} = 1$$

Note that we could have avoided L'Hôpital's Rule altogether by letting $t = 1/x$:

$$\lim_{x \to \infty} x \sin \frac{1}{x} = \lim_{t \downarrow 0} \frac{\sin t}{t} = 1 \qquad \text{(Section 2.3)}$$

Elementary methods need not be discarded just because we have an elegant new rule!   □

□ **Example 4**   Find $\lim_{x \to \infty} \left(1 + \frac{3}{x}\right)^{2x}$

Solution: This limit takes the indeterminate form 1^∞ (*not* equal to 1, as you will see). To get rid of the exponent, let

$$y = \left(1 + \frac{3}{x}\right)^{2x}$$

and use logarithms (a standard device for evaluating limits of this type):

$$\ln y = \ln \left(1 + \frac{3}{x}\right)^{2x} = 2x \ln \left(1 + \frac{3}{x}\right)$$

Since the limit of this expression is still indeterminate (of the form $\infty \cdot 0$), we proceed as in Example 3:

$$\lim_{x \to \infty} \ln y = \lim_{x \to \infty} \frac{2 \ln (1 + 3/x)}{1/x} \qquad (0/0 \text{ type})$$

$$= \lim_{x \to \infty} \frac{\dfrac{2(-3/x^2)}{1 + 3/x}}{-1/x^2} = \lim_{x \to \infty} \frac{6x}{x + 3} = 6$$

Interchanging lim and ln (by the Composite Function Theorem for limits), we find

$$\ln \left(\lim_{x \to \infty} y\right) = 6$$

Hence the original limit is $\lim_{x \to \infty} y = e^6$.   □

So far we have confined the application of L'Hôpital's Rule to indeterminate forms of the type $0/0$. Although it is harder to prove, the rule also applies to forms of the type ∞/∞. Since the proof is more painful than enlightening, we omit it.

We use the symbol ∞/∞ in connection with $\lim f(x)/g(x)$ if $f(x)$ and $g(x)$ are unbounded in any way whatever x does. For example, we classify

$$\lim_{x \to 0} x \cot x = \lim_{x \to 0} \frac{\cot x}{1/x}$$

as an ∞/∞ type, although

$$\lim_{x \to 0} \cot x \neq \infty \qquad \text{and} \qquad \lim_{x \to 0} \frac{1}{x} \neq \infty \qquad \text{(Why?)}$$

But $\cot x$ and $1/x$ are both unbounded as $x \to 0$. This convention is also used in other indeterminate forms involving the symbol ∞.

□ **Example 5**   Find $\lim_{x \to \infty} x^2/e^x$.

Solution: Since x^2 and e^x both become infinite as x does, we may use L'Hôpital's Rule:

$$\lim_{x \to \infty} \frac{x^2}{e^x} = \lim_{x \to \infty} \frac{2x}{e^x}$$

This is still of the type ∞/∞, so we use the rule again:

$$\lim_{x \to \infty} \frac{x^2}{e^x} = \lim_{x \to \infty} \frac{2}{e^x} = 0 \qquad\qquad □$$

Example 5 illustrates a remarkable property of e^x. It grows so fast (as $x \to \infty$) that it dominates every power of x! For example,

$$\lim_{x \to \infty} \frac{x^{1984}}{e^x} = 0$$

(by applying L'Hôpital's Rule 1984 times). Both x^{1984} and e^x increase without bound as $x \to \infty$, but the limit is 0 because e^x grows faster.

□ **Example 6**   The most straightforward way to find $\lim_{x \to 0} x \cot x$ (of the type $0 \cdot \infty$) is to write

$$\lim_{x \to 0} x \cot x = \lim_{x \to 0} (\cos x)\left(\frac{x}{\sin x}\right) = \left[\lim_{x \to 0} \cos x\right]\left[\lim_{x \to 0} \frac{x}{\sin x}\right]$$
$$= (1)(1) = 1 \qquad \text{(Section 2.3)}$$

However, L'Hôpital's Rule may also be used:

$$\lim_{x \to 0} x \cot x = \lim_{x \to 0} \frac{x}{\tan x} \qquad \text{(of the type 0/0)}$$

$$= \lim_{x \to 0} \frac{1}{\sec^2 x} = 1$$

An alternate use of the rule is to write

$$\lim_{x \to 0} x \cot x = \lim_{x \to 0} \frac{\cot x}{1/x} \qquad \text{(of the type } \infty/\infty)$$

$$= \lim_{x \to 0} \frac{-\csc^2 x}{-1/x^2} = \lim_{x \to 0} \left(\frac{x}{\sin x}\right)^2 = 1$$

This method illustrates the fact that L'Hôpital's Rule (sometimes overrated by students) can make a simple problem hard.   □

□ **Example 7**   Find $\lim_{x \to \infty} x^{1/x}$.

Solution: The indeterminate form this time is of the type ∞^0. To evaluate it, write $y = x^{1/x}$. Then

$$\ln y = \frac{1}{x} \ln x$$

$$\lim_{x \to \infty} \ln y = \lim_{x \to \infty} \frac{\ln x}{x} \qquad \text{(of the type } \infty/\infty)$$

$$= \lim_{x \to \infty} \frac{1/x}{1} \qquad \text{(L'Hôpital's Rule)}$$

$$= 0$$

Hence $\lim_{x \to \infty} y = e^0 = 1$.   □

□ **Example 8**   Find $\lim_{x \downarrow 0} (\sin x)^{\tan x}$.

Solution: This is of the indeterminate type 0^0. To evaluate it, take logarithms (as in Example 4):

$$y = (\sin x)^{\tan x}$$

$$\ln y = \tan x \ln (\sin x)$$

$$\lim_{x \downarrow 0} \ln y = \lim_{x \downarrow 0} \tan x \ln (\sin x) \qquad \text{(of the type } 0 \cdot \infty)$$

$$= \lim_{x \downarrow 0} \frac{\ln (\sin x)}{\cot x} \qquad \text{(of the type } \infty/\infty)$$

$$= \lim_{x \downarrow 0} \frac{\cot x}{-\csc^2 x} \qquad \text{(L'Hôpital's Rule)}$$

$$= \lim_{x \downarrow 0} (-\sin x \cos x) \qquad \text{(simplification, } not \text{ L'Hôpital's Rule)}$$

$$= 0$$

Hence $\lim_{x \downarrow 0} y = e^0 = 1$. □

□ **Example 9**   Find $\lim_{x \to \infty} (x - \sinh x)$.

Solution: This is of the indeterminate type $\infty - \infty$. Writing

$$\lim_{x \to \infty} (x - \sinh x) = \lim_{x \to \infty} x \left(1 - \frac{\sinh x}{x} \right)$$

observe that

$$\lim_{x \to \infty} \frac{\sinh x}{x}$$

is of the type ∞/∞. L'Hôpital's Rule changes it to

$$\lim_{x \to \infty} \frac{\cosh x}{1} = \infty$$

so we conclude that

$$\lim_{x \to \infty} x \left(1 - \frac{\sinh x}{x} \right)$$

is *not* an indeterminate type. Instead it is the limit of a product of two factors that both become infinite as $x \to \infty$ (one positive, the other negative). Hence

$$\lim_{x \to \infty} (x - \sinh x) = -\infty \qquad □$$

In summary, we conclude from these examples that L'Hôpital's Rule applies directly to indeterminate forms of the type $0/0$ and ∞/∞, and indirectly (after a change to one of these two) to limits of the type $\infty - \infty$, $0 \cdot \infty$, 1^∞, 0^0, ∞^0 (each of which is indeterminate). Three things to remember:

1. Be sure you have a $0/0$ or ∞/∞ type before using the rule.

2. Simplify at every opportunity, looking for ways to evaluate the limit directly. (Otherwise L'Hôpital's Rule may make things worse instead of better.)

3. Remember that the rule is validated only when the new limit is definite. (See the problem set for an example in which failure to meet this hypothesis leads to nonsense.)

Problem Set 11.2

Evaluate each of the following limits in two ways (with and without L'Hôpital's Rule).

1. $\lim_{x \to 0} \dfrac{\sin^2 x}{1 - \cos x}$ **2.** $\lim_{x \to 0} \dfrac{x + \tan x}{\sin x}$

3. $\lim_{x \to \infty} \dfrac{x - 2}{x^2 - 4}$ **4.** $\lim_{x \to \infty} \dfrac{2x^3 - 5x^2 + 1}{x^3 - 1}$

Evaluate each of the following limits by any method.

5. $\lim_{x \to 0} \dfrac{x - \sin x}{\tan x}$ **6.** $\lim_{x \to \infty} \dfrac{\sqrt{x}}{x^2 - 1}$

7. $\lim_{x \to 0} \dfrac{e^x - 1}{xe^x}$ **8.** $\lim_{x \to 0} \dfrac{\sinh x}{\sin x}$

9. $\lim_{x \to 0} \dfrac{\ln \cos x}{x}$ **10.** $\lim_{x \to 0} x \csc x$

11. $\lim_{x \to 0} \dfrac{\ln (1 + x)}{x}$ **12.** $\lim_{x \to 1} \dfrac{x^2 - 1}{\ln x}$

13. $\lim_{x \to \pi/2} \dfrac{\sec x + 1}{\tan x - 1}$ **14.** $\lim_{x \downarrow 0} \dfrac{x^2}{\sin x - x}$

15. $\lim_{x \to 0} \dfrac{\tan x}{\tan 2x}$ **16.** $\lim_{x \to 0} \dfrac{1 - \cos x}{x^2}$

17. $\lim_{x \to 1} \dfrac{x^2 - 1}{\sqrt[3]{x} - 1}$ **18.** $\lim_{x \to \pi/2} (\sec x - \tan x)$

19. $\lim_{x \to 0} (\csc x - \cot x)$ **20.** $\lim_{x \to 0} \left(\csc x - \dfrac{1}{x}\right)$

21. $\lim_{x \to \infty} (x - \cosh x)$ **22.** $\lim_{x \to -\infty} (x^2 - \sinh x)$

23. $\lim_{x \to \infty} x^3 e^{-x}$ **24.** $\lim_{x \to \infty} \dfrac{e^x + 1}{x^2}$

25. $\lim_{x \to 0} x^2 \ln x$ **26.** $\lim_{x \downarrow 0} xe^{1/x}$

27. $\lim_{x \to 0} (1 - x)^{1/x}$ **28.** $\lim_{x \to \infty} (x^2 + 1)^{1/x}$

29. $\lim_{x \downarrow 0} x^{\sin x}$ **30.** $\lim_{x \uparrow \pi/2} (\tan x)^{\cos x}$

The instructions in Problems 50–56, Section 8.3, called for sketching certain graphs without all the relevant information. Fill in the gaps as follows, and see if you guessed correctly when you drew the graphs. (Not all these questions involve indeterminate forms; you probably answered some of them before.)

31. $y = xe^x$ How does y behave when $x \to \infty$ and $x \to -\infty$?

32. $y = xe^{-x}$ How does y behave when $x \to \infty$ and $x \to -\infty$?

33. $y = \dfrac{e^x}{x}$ How does y behave when $x \downarrow 0$ and $x \uparrow 0$? when $x \to \infty$ and $x \to -\infty$?

34. $y = e^x - x$ How does y behave when $x \to \infty$ and $x \to -\infty$?

35. $y = x \ln x$ How does y behave when $x \downarrow 0$? when $x \to \infty$?

36. $y = x \ln x - x$ How does y behave when $x \downarrow 0$? when $x \to \infty$?

37. $y = \dfrac{\ln x}{x}$ How does y behave when $x \downarrow 0$? when $x \to \infty$?

38. One student writes

$$\lim_{x \to 1} \frac{x^3 - x^2}{x^3 - 1} = \lim_{x \to 1} \frac{x^2}{x^2 + x + 1} = \frac{1}{3}$$

and another writes

$$\lim_{x \to 1} \frac{x^3 - x^2}{x^3 - 1} = \lim_{x \to 1} \frac{3x^2 - 2x}{3x^2} = \lim_{x \to 1} \frac{6x - 2}{6x} = \frac{2}{3}$$

Which one is right? Or are they both wrong?

39. What happens when you try to find

$$\lim_{x \to \infty} \frac{\sqrt{1 + x^2}}{x}$$

by L'Hôpital's Rule? What is the value of the limit?

40. Explain why

$$\lim_{x \to \infty} \frac{x - \sin x}{x} = 1$$

Why can't L'Hôpital's Rule be used to evaluate this limit?

41. In Section 2.3 we went to considerable trouble to show that

$$\lim_{x \to 0} \frac{\sin x}{x} = 1$$

(a) Use L'Hôpital's Rule to evaluate this limit.

(b) If this rule had been available to us in Section 2.3, could we have used it?

42. Sometimes students argue that the forms 0^0 and ∞^0 are not indeterminate but should equal 1. To see why this is wrong, confirm that

$$\lim_{x \to \infty} (e^{-x})^{1/x} \quad \text{and} \quad \lim_{x \to \infty} (e^x)^{1/x}$$

are of the type 0^0 and ∞^0, respectively, and evaluate them.

43. Let x be any real number. Show that

$$\lim_{h \to 0} (1 + xh)^{1/h} = e^x$$

44. We have classified $0/0$, ∞/∞, $\infty - \infty$, $0 \cdot \infty$, 1^∞, 0^0, ∞^0 as indeterminate forms. There are many other possibilities, like $0/\infty$, $\infty/0$, $\infty + \infty$, $\infty \cdot \infty$, ∞^1, 0^∞, ∞^∞. Why aren't these indeterminate? (*Note:* Some confusion is possible here, because we said in the text that when ∞ is involved in an indeterminate form we mean to include any kind of unbounded behavior. In this question assume that ∞ refers as usual to unbounded *increase*.)

45. Use a calculator to check up on Example 9,

$$\lim_{x \to \infty} (x - \sinh x) = -\infty$$

46. Define f by the rule

$$f(x) = \begin{cases} e^{-1/x^2} & \text{if } x \neq 0 \\ 0 & \text{if } x = 0 \end{cases}$$

(a) Explain why

$$f'(0) = \lim_{x \to 0} \frac{1}{x} e^{-1/x^2}$$

(b) Evaluate the limit to show that $f'(0) = 0$.   *Hint:* Let $t = 1/x$.

47. For a fixed value of $x > 0$ we evaluate

$$\int_1^x t^n \, dt$$

by the power rule if $n \neq -1$, whereas the case $n = -1$ requires the separate formula

$$\int_1^x \frac{dt}{t} = \ln x$$

Show that these cases are not all that different by proving that

$$\lim_{n \to -1} \int_1^x t^n \, dt = \ln x$$

Hint: The variable is n, not x.

48. The current flowing in a circuit with resistance R and inductance L (t seconds after a source of E volts is switched into the circuit) is a function of time, namely

$$I = \frac{E}{R}(1 - e^{-Rt/L})$$

(a) Keeping R, E, and t fixed, explain why

$$\lim_{L \downarrow 0} I = \frac{E}{R}$$

(b) Keeping L, E, and t fixed, explain why

$$\lim_{R \downarrow 0} I = \frac{E}{L} t$$

(c) Each of these limits is a function of t (for although we fixed t, its value was arbitrary). Contrast the behavior of I as a function of t in a circuit with no inductance to its behavior in a circuit with no resistance.

49. In Figure 1 we show the line joining $P(\cos t, \sin t)$ and $Q(1 - t, 0)$, where t is a small positive number.

Figure 1
What does B approach as $t \to 0$?

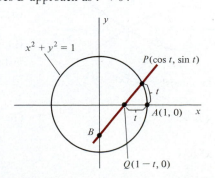

(a) What do P and Q approach as $t \to 0$? Can you guess what B approaches?

(b) Show that the y coordinate of B is

$$\frac{(t-1)\sin t}{(t-1) + \cos t}$$

and use the result to find what B approaches as $t \to 0$. Was your guess right?

50. Formally stated, L'Hôpital's Rule says that

$$\lim_{x \to a} \frac{f(x)}{g(x)} = \lim_{x \to a} \frac{f'(x)}{g'(x)}$$

provided that f and g are differentiable in a (punctured) neighborhood of a in which $g'(x)$ is never 0, each has limit 0 as $x \to a$, and the limit of $f'(x)/g'(x)$ as $x \to a$ is definite (a real number or $\pm\infty$). Prove it as follows.

(a) If N is the given (punctured) neighborhood of a, define the functions

$$F(x) = \begin{cases} f(x) & \text{if } x \in N \\ 0 & \text{if } x = a \end{cases}$$

and

$$G(x) = \begin{cases} g(x) & \text{if } x \in N \\ 0 & \text{if } x = a \end{cases}$$

and let $x \in N$. Look up Cauchy's Mean Value Theorem (Problem 31, Section 4.3) and explain why F and G satisfy its hypotheses in the interval with endpoints a and x.

(b) Why does it follow that there is a point c between a and x such that $f(x)/g(x) = f'(c)/g'(c)$?

(c) Use part (b) to finish the proof.

11.3

IMPROPER INTEGRALS

Suppose that we want to find the length of the loop of the curve $9y^2 = x(x-3)^2$. (See Figure 1.) Since the loop is symmetric about the x axis, its length (Section 7.3) is

$$s = 2 \int_0^3 \sqrt{1 + y'^2}\, dx$$

where y is either one of the functions indicated in the figure.

We propose this problem because it is an interesting example of how an *improper integral* can turn up in unexpected places. In Section 6.2 we said that a function f is *integrable* over the interval $I = [a,b]$ if

$$\int_a^b f(x)\, dx \quad \text{exists}$$

and we stated theorems to the effect that f is integrable if it is continuous (or if it is bounded and has only a finite number of discontinuities in I). We will see that the integrand occurring in the above length problem fails to meet these criteria. Yet the loop in Figure 1 looks simple enough to have a definite length; there ought to be a natural extension of our definition of integral that will enable us to compute it. (See Example 3 for the calculation.)

To arrive at this more general definition, suppose that f is continuous in the half-closed interval $[a,b)$ and *unbounded* as $x \uparrow b$. According to the above criteria,

$$\int_a^b f(x)\, dx \quad \text{does not exist}$$

On the other hand, we can certainly find

$$\int_a^t f(x)\, dx$$

Figure 1
Graph of $9y^2 = x(x-3)^2$

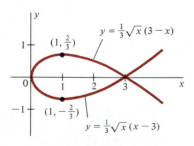

if t is any number close to, but less than, b. It seems reasonable to compute this integral and then find its limit as $t \uparrow b$. If the limit exists, we assign its value (by definition) to the integral of f from a to b.

> If f is continuous in $[a,b)$ and unbounded as $x \uparrow b$, we define
>
> $$\int_a^b f(x)\,dx = \lim_{t \uparrow b} \int_a^t f(x)\,dx$$
>
> provided that the limit exists.

Such an integral is called **improper,** not because there is anything illogical about its definition, but to distinguish it from the ordinary "proper" integral in Section 6.2. If the limit in the definition exists, we say that the integral **converges;** otherwise it is said to **diverge.**

There are several types of improper integrals, as we will see. For the moment, however, we will concentrate on this one.

□ **Example 1**   Find

$$\int_0^1 \frac{dx}{\sqrt{1-x^2}}$$

(or determine that it diverges).

Solution: The integrand is continuous in $[0,1)$ and unbounded as $x \uparrow 1$. Hence we compute

$$\int_0^t \frac{dx}{\sqrt{1-x^2}} = \sin^{-1} x \ \Big|_0^t = \sin^{-1} t$$

and find its limit as $t \uparrow 1$:

$$\int_0^1 \frac{dx}{\sqrt{1-x^2}} = \lim_{t \uparrow 1} \sin^{-1} t = \sin^{-1} 1 = \frac{\pi}{2}$$ □

□ **Example 2**   Find

$$\int_0^1 \frac{dx}{1-x}$$

(or determine that it diverges).

Solution: As in Example 1, the integral is improper because the integrand becomes infinite as $x \uparrow 1$. Hence we compute

$$\int_0^t \frac{dx}{1-x} = -\ln|1-x| \ \Big|_0^t = -\ln(1-t) \qquad (0 < t < 1)$$

The original integral is

$$\int_0^1 \frac{dx}{1-x} = -\lim_{t\uparrow 1} \ln(1-t) = -(-\infty) = \infty$$

Hence it diverges.   □

□ **Example 3**   Find the length of the loop of the curve $9y^2 = x(x-3)^2$.

Solution: This is the problem with which we began. Nothing about the formula

$$s = 2\int_0^3 \sqrt{1+y'^2}\, dx$$

suggests any difficulty, but let's examine the details. To find y', differentiate implicitly in the equation

$$9y^2 = x(x-3)^2 = x^3 - 6x^2 + 9x$$

This gives

$$18yy' = 3x^2 - 12x + 9 = 3(x^2 - 4x + 3) = 3(x-1)(x-3)$$

$$y' = \frac{(x-1)(x-3)}{6y}$$

Hence

$$1 + y'^2 = 1 + \frac{(x-1)^2(x-3)^2}{36y^2} = 1 + \frac{(x-1)^2(x-3)^2}{4x(x-3)^2} = 1 + \frac{(x-1)^2}{4x}$$

$$= \frac{x^2 + 2x + 1}{4x} = \frac{(x+1)^2}{4x}$$

and

$$\sqrt{1+y'^2} = \frac{|x+1|}{2\sqrt{x}} = \frac{x+1}{2\sqrt{x}}$$

(because $x + 1 > 0$ in the domain of integration).

At this point the difficulty becomes apparent. We cannot integrate from 0 to 3 because $(x+1)/2\sqrt{x}$ is unbounded as $x \downarrow 0$. Let's agree to extend the formula for length to include improper integrals (if they converge). Then

$$s = \int_0^3 \frac{x+1}{\sqrt{x}}\, dx = \int_0^3 (x^{1/2} + x^{-1/2})\, dx$$

provided the integral exists.

Although the difficulty with this integral is at its lower limit (whereas our first definition involves the upper limit), it should be clear how to proceed. Integrate from t to 3 (where t is close to, but greater than, 0) and find the limit as $t \downarrow 0$. Thus we compute

$$\int_t^3 (x^{1/2} + x^{-1/2})\, dx = \left(\frac{2}{3}x^{3/2} + 2x^{1/2}\right)\Big|_t^3 = \frac{2}{3}\sqrt{x}(x+3)\Big|_t^3$$

$$= \frac{2}{3}[6\sqrt{3} - \sqrt{t}(t+3)]$$

Letting $t \downarrow 0$, we find

$$s = \lim_{t \downarrow 0} \tfrac{2}{3}[6\sqrt{3} - \sqrt{t}(t + 3)] = 4\sqrt{3} \qquad \square$$

If f is continuous in $(a,b]$ and unbounded as $x \downarrow a$, we define

$$\int_a^b f(x)\, dx = \lim_{t \downarrow a} \int_t^b f(x)\, dx$$

provided that the limit exists.

$\square$ **Example 4**   Find $\int_0^\pi \sec^2 x\, dx$.

Solution: It would be easy to come up with the clinker

$$\int_0^\pi \sec^2 x\, dx = \tan x \Big|_0^\pi = 0$$

This cannot be right, however, because the integrand is positive. What has been overlooked is that $\sec^2 x$ is undefined at $x = \pi/2$ (and indeed unbounded as $x \to \pi/2$). Hence we are dealing with an improper integral (of a type not yet defined). A reasonable definition would be to break up the integral by writing

$$\int_0^\pi \sec^2 x\, dx = \int_0^{\pi/2} \sec^2 x\, dx + \int_{\pi/2}^\pi \sec^2 x\, dx$$

We investigate each of these improper integrals separately. If they both converge, the original integral is their sum; otherwise it diverges.

We look first at

$$\int_0^{\pi/2} \sec^2 x\, dx = \lim_{t \uparrow \pi/2} \int_0^t \sec^2 x\, dx = \lim_{t \uparrow \pi/2} \tan t = \infty$$

Since this diverges, we don't bother to look at the other one; the original integral diverges by definition. $\square$

Let c be an interior point of $I = [a,b]$ and suppose that f is unbounded as $x \to c$ (but continuous elsewhere in I). We define

$$\int_a^b f(x)\, dx = \int_a^c f(x)\, dx + \int_c^b f(x)\, dx$$

provided that both terms of the sum exist.

Figure 2
Do R_1 and R_2 have equal areas?

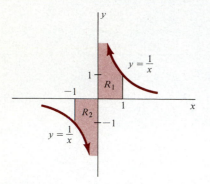

□ **Example 5**   Perhaps the most controversial question in elementary calculus (to judge by the experience of the author with his students) is the value of

$$\int_{-1}^{1} \frac{dx}{x}$$

The graph of $y = 1/x$ (Figure 2) suggests that the answer is zero, since the area interpretation of integral seems to say that

$$\int_{-1}^{1} \frac{dx}{x} = \int_{-1}^{0} \frac{dx}{x} + \int_{0}^{1} \frac{dx}{x} = -(\text{Area of } R_2) + (\text{Area of } R_1) = 0$$

The question is whether our definition of improper integral squares with this result. The function $1/x$ is unbounded as $x \to 0$, so we write

$$\int_{-1}^{1} \frac{dx}{x} = \int_{-1}^{0} \frac{dx}{x} + \int_{0}^{1} \frac{dx}{x}$$

Since

$$\int_{0}^{1} \frac{dx}{x} = \lim_{t \downarrow 0} \int_{t}^{1} \frac{dx}{x} = \lim_{t \downarrow 0} (-\ln t) = \infty$$

we conclude that the original integral does not exist.

The trouble with the area interpretation is that neither R_1 nor R_2 is a bounded region. Even when a finite area can be assigned to such a region (not the case in this example!) we must not expect too much from geometric intuition. For an illustration of what we mean, see Example 6, Section 8.5. An unbounded region with infinite area generates a finite volume and infinite surface area when it is rotated about the x axis! In view of that kind of nonsense, some caution is called for in connection with unbounded regions.

Good students die hard, however. Consider the following argument that the answer is zero:

$$\int_{-1}^{1} \frac{dx}{x} = \int_{-1}^{0} \frac{dx}{x} + \int_{0}^{1} \frac{dx}{x} = \lim_{t \to 0} \int_{-1}^{t} \frac{dx}{x} + \lim_{t \to 0} \int_{t}^{1} \frac{dx}{x}$$
$$= \lim_{t \to 0} \ln |t| + \lim_{t \to 0} (-\ln |t|) = \lim_{t \to 0} (\ln |t| - \ln |t|) = 0$$

We leave it to you to find the mistakes in this argument. (There are two.)

□

Problem Set 11.3

Evaluate each of the following improper integrals (or conclude that it diverges).

1. $\int_{0}^{2} \frac{dx}{\sqrt{2 - x}}$ **2.** $\int_{0}^{1} \frac{x \, dx}{\sqrt{1 - x^2}}$

3. $\int_{0}^{1} \frac{dx}{x^2}$ **4.** $\int_{0}^{1} \frac{dx}{\sqrt{x}}$

5. $\int_{0}^{\pi} \sec t \tan t \, dt$ **6.** $\int_{0}^{\pi} \sec t \, dt$

7. $\int_0^4 \dfrac{dx}{4 - x^2}$

8. $\int_0^5 \dfrac{x\, dx}{9 - x^2}$

9. $\int_0^2 \dfrac{dx}{\sqrt{4 - x^2}}$

10. $\int_0^1 \dfrac{dx}{\sqrt{1 - x}}$

11. $\int_0^1 \dfrac{dx}{x^3}$

12. $\int_0^3 \dfrac{dx}{x - 3}$

13. $\int_0^2 \dfrac{dx}{(x - 1)^2}$

14. $\int_0^\pi \tan x\, dx$

15. $\int_{-1}^1 \dfrac{dx}{\sqrt[3]{x}}$

16. $\int_0^1 \ln x\, dx$

17. $\int_0^{\pi/2} \dfrac{dx}{1 - \sin x}$ *Hint:* Let $x = 2 \tan^{-1} u$ (Example 4, Section 10.5).

18. $\int_0^{\pi/6} \dfrac{\cos x}{1 - 4 \sin^2 x}\, dx$

19. $\int_0^2 \dfrac{dx}{\sqrt{2x - x^2}}$ *Hint:* Complete the square.

20. $\int_0^4 \dfrac{dx}{x^2 - 4x}$

21. $\int_0^1 \dfrac{dx}{e^x - e^{-x}}$

22. $\int_0^1 \dfrac{\ln x}{x}\, dx$

23. $\int_0^1 \coth x\, dx$

24. $\int_0^{\pi/2} \csc^2 t\, dt$

25. $\int_1^2 \dfrac{dx}{\sqrt{x^2 - 1}}$

26. $\int_2^4 \dfrac{dx}{x\sqrt{x^2 - 4}}$

27. $\int_0^3 \dfrac{x^2\, dx}{\sqrt{9 - x^2}}$

28. $\int_0^2 \dfrac{x^3\, dx}{\sqrt{4 - x^2}}$

29. $\int_0^3 \dfrac{x\, dx}{x^2 - x - 6}$

30. $\int_0^{\pi/2} \csc t \cot t\, dt$

31. Find the area of the region bounded by the curve $y = x \ln x$ and the x axis. (Draw a picture! The graph is discussed in Problem 54, Section 8.3, and Problem 35, Section 11.2.)

32. Show that the unbounded region below the curve

$$y = \dfrac{1}{\sqrt{x}} \qquad 0 < x \le 1$$

has a finite area, and find the area.

33. The region in the preceding problem is rotated about the x axis. Can a volume be assigned to the unbounded solid that is generated?

34. Find the area in the first and third quadrants between the curve $y^2 = x^2/(1 - x^2)$ and its asymptotes (or conclude that it does not exist).

35. Find the length of the loop of the curve

$$15y^2 = x(x - 5)^2$$

36. Evaluate the improper integral

$$\int_{-1}^1 \sqrt{\dfrac{1 + x}{1 - x}}\, dx$$

(which occurs in aerodynamics) as follows.
(a) Explain why

$$\sqrt{\dfrac{1 + x}{1 - x}} = \dfrac{1 + x}{\sqrt{1 - x^2}} \qquad (-1 < x < 1)$$

(b) Use part (a) to find a formula for

$$\int \sqrt{\dfrac{1 + x}{1 - x}}\, dx$$

and conclude that the original integral has value π.

37. Find x satisfying $\int_0^x \ln t\, dt = 0$.

38. Explain why $\int_0^1 (\sin x)/x\, dx$ is proper.

39. Show that

$$\int_0^1 \dfrac{dx}{x^p}$$

converges if $p < 1$ and diverges if $p \ge 1$. (Consider the cases $p < 1$, $p = 1$, and $p > 1$ separately.)

40. Show that

$$\int_0^{\pi/2} (\sec^2 x - \sec x \tan x)\, dx = 1$$

while

$$\int_0^{\pi/2} \sec^2 x\, dx - \int_0^{\pi/2} \sec x \tan x\, dx$$

is meaningless. (Thus the formula

$$\int [af(x) + bg(x)]\, dx = a \int f(x)\, dx + b \int g(x)\, dx$$

must be used with caution in the case of improper integrals! The left side may exist when the terms on the right do not.)

41. If f is continuous in $[a,b]$, then $\int_a^b f(x)\,dx$ is an ordinary (proper) integral. Suppose that we mistakenly classify it as improper (because of an apparent difficulty at b) and write

$$\int_a^b f(x)\,dx = \lim_{t\uparrow b}\int_a^t f(x)\,dx$$

Show that no harm is done.

42. Criticize the reasoning

$$\int_{-1}^1 \frac{dx}{x} = \ln|x|\;\Big|_{-1}^1 = \ln 1 - \ln 1 = 0$$

How about the claim that the integral is zero because the integrand is odd? (See Problem 44, Section 6.5.)

43. Find the errors in the argument

$$\int_{-1}^1 \frac{dx}{x} = \int_{-1}^0 \frac{dx}{x} + \int_0^1 \frac{dx}{x}$$

$$= \lim_{t\to 0}\int_{-1}^t \frac{dx}{x} + \lim_{t\to 0}\int_t^1 \frac{dx}{x}$$

$$= \lim_{t\to 0}\ln|t| + \lim_{t\to 0}(-\ln|t|)$$

$$= \lim_{t\to 0}(\ln|t| - \ln|t|) = 0$$

(There are two.)

11.4 OTHER TYPES OF IMPROPER INTEGRALS

The integrals in Section 11.3 are improper because the integrand becomes infinite in the domain of integration. The domain itself has been kept finite. Now we are going to consider integrals that are improper because the interval of integration is unbounded.

□ **Example 1**   A space capsule weighing 1000 pounds on the surface of the earth is propelled out of the earth's gravitational field. How much work is done against the force of gravity?

Solution: In Example 4, Section 7.5, we found the work done in propelling the capsule to an altitude of 100 miles. Our analysis required Newton's *Inverse-Square Law*, which says that the force exerted by the earth on the capsule (the force of gravity) is $f(x) = -k/x^2$, where x is the distance of the capsule from the center of the earth and k is a positive constant. The lifting force (which opposes gravity) is therefore $F(x) = -f(x) = k/x^2$. It is this force that does the work on the capsule.

To evaluate k, we assume that the radius of the earth is 4000 miles. Then $F(4000) = 1000$ because the capsule weighs 1000 pounds on the surface of the earth. This yields $k = 16 \times 10^9$ and hence the lifting force is

$$F(x) = \frac{16 \times 10^9}{x^2}$$

The work done in lifting the capsule from $x = 4000$ (the surface of the earth) to $x = R$ (an arbitrary distance from the center of the earth) is

$$\int_{4000}^R F(x)\,dx$$

In Section 7.5 we took $R = 4100$ (100 miles above the surface of the earth), but now we are interested in what happens as $R \to \infty$. Obvious notation for the answer is

$$\int_{4000}^{\infty} F(x)\, dx$$

where the use of ∞ simply means that we evaluate

$$\int_{4000}^{R} F(x)\, dx$$

and then find the limit as $R \to \infty$. Such an integral is improper, not because of any difficulty with the integrand, but because the domain of integration is the unbounded interval $[4000, \infty)$. Since

$$\int_{4000}^{R} F(x)\, dx = \int_{4000}^{R} \frac{16 \times 10^9}{x^2}\, dx$$

$$= -\frac{16 \times 10^9}{x} \Big|_{4000}^{R} = 16 \times 10^9 \left(\frac{1}{4000} - \frac{1}{R} \right)$$

the work done in sending the capsule out of the earth's gravitational field is

$$\int_{4000}^{\infty} F(x)\, dx = 16 \times 10^9 \lim_{R \to \infty} \left(\frac{1}{4000} - \frac{1}{R} \right) = 16 \times 10^9 \left(\frac{1}{4000} \right)$$

$$= 4 \times 10^6 \text{ (mile-pounds)} \qquad \square$$

Example 1 suggests the following definitions.

Suppose that $f(x)$ is continuous for all $x \geq a$. We define

$$\int_{a}^{\infty} f(x)\, dx = \lim_{b \to \infty} \int_{a}^{b} f(x)\, dx$$

provided that the limit exists. Similarly (if f is continuous for all $x \leq b$) we define

$$\int_{-\infty}^{b} f(x)\, dx = \lim_{a \to -\infty} \int_{a}^{b} f(x)\, dx$$

provided that the limit exists. If f is continuous for all x, we define

$$\int_{-\infty}^{\infty} f(x)\, dx = \int_{-\infty}^{0} f(x)\, dx + \int_{0}^{\infty} f(x)\, dx$$

provided that both terms of the sum exist.

$\square$ **Example 2**   Can a finite area be assigned to the unbounded region between the curve $y = 1/(1 + x^2)$ and the x axis? If so, find it.

Figure 1
Unbounded region between curve
and its asymptote

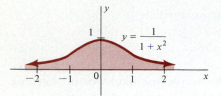

Solution: The region is shown in Figure 1. Its "area" is not defined by anything we have said before, so we are free to make whatever definition seems reasonable. Since the area under the curve from $x = 0$ to $x = b$ is

$$\int_0^b \frac{dx}{1 + x^2} = \tan^{-1} b$$

it is not surprising that we define the area in the first quadrant to be

$$\int_0^\infty \frac{dx}{1 + x^2} = \lim_{b \to \infty} \tan^{-1} b = \frac{\pi}{2}$$

The area of the congruent second-quadrant region should surely be defined to be the same, so the total area is

$$A = 2 \int_0^\infty \frac{dx}{1 + x^2} = \pi$$

Unless our intuition deceives us, this should also be given by

$$\int_{-\infty}^\infty \frac{dx}{1 + x^2} = \int_{-\infty}^0 \frac{dx}{1 + x^2} + \int_0^\infty \frac{dx}{1 + x^2}$$

Since

$$\int_{-\infty}^0 \frac{dx}{1 + x^2} = \lim_{a \to -\infty} \int_a^0 \frac{dx}{1 + x^2} = \lim_{a \to -\infty} (-\tan^{-1} a) = -\left(-\frac{\pi}{2} \right) = \frac{\pi}{2}$$

we see that it is.   □

□ **Example 3**   Find the area of the region between the curve $y = 2x/(1 + x^2)$ and its asymptote (if it exists).

Solution: The region is shown in Figure 2. The area in the first quadrant is

$$\int_0^\infty \frac{2x \, dx}{1 + x^2} = \lim_{b \to \infty} \int_0^b \frac{2x \, dx}{1 + x^2} = \lim_{b \to \infty} \ln (1 + b^2) = \infty$$

Figure 2
Unbounded region between curve and its asymptote

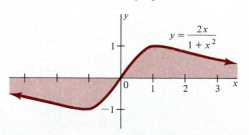

More properly (in view of the result), we should say that the region cannot be assigned an area.   □

The results in Examples 2 and 3 may puzzle you. Both curves approach the x axis asymptotically as $x \to \infty$. Why should the region have a finite area in one case and not the other? The "answer" is not too satisfying; about all we can say is that $y = 1/(1 + x^2)$ approaches 0 faster than $y = 2x/(1 + x^2)$ as $x \to \infty$. After integrating from 0 to b, we have too much area left over in the second case; the limit as $b \to \infty$ does not exist. In the first case the area from 0 to b, while increasing with b, does not increase without bound.

□ **Example 4**   If interest is continuously compounded, an investment of P dollars at an annual rate r will grow to the amount $S = Pe^{rt}$ in t years. (See Example 5, Section 8.6.) By solving this equation for P, we can say that the *present value* of S dollars (due to be acquired t years from now) is $P = Se^{-rt}$.

Suppose that a company expects its annual profit t years from now to be $f(t)$ dollars, where f is assumed to be continuous. During a short interval of time (from t to $t + \Delta t$) its profit is approximately $f(t) \Delta t$; according to the above formula, the present value of that amount is

$$e^{-rt}f(t) \, \Delta t$$

Hence the present value of all profits between $t = 0$ and $t = b$ is

$$\int_0^b e^{-rt}f(t) \, dt$$

The *present value of all future profits* is therefore

$$\int_0^\infty e^{-rt}f(t) \, dt$$

a formula that enables a prospective investor to assess the worth of the company (assuming that its assets and liabilities are known, and that the function f is a reasonable projection of future profit).

As an illustration of the use of this formula, suppose that the interest rate is 8% and the company expects a constant profit of \$10,000 per year. Then $f(t) = 10,000$. If the company does business forever, the total profit it will make is infinite. The *present value* of all future profits, however, is finite, namely

$$\int_0^\infty e^{-0.08t}(10,000) \, dt = \$125,000 \qquad \text{(Confirm!)}$$

This is the amount that an investor would consider the future profits to be worth today.   □

Remark   In Example 4 the present value depends on the interest rate r, that is,

$$P(r) = \int_0^\infty e^{-rt} f(t)\, dt$$

is a function of r. This equation may be regarded as a transformation of the function f into a new function P and is important in the subject of differential equations. P is called the **Laplace Transform** of f.

The definitions given so far do not cover the case of an integral that is improper in two ways (because its integrand is unbounded in the neighborhood of some point and because the domain of integration is also unbounded). An example is

$$\int_0^\infty \frac{dx}{2 - x}$$

which is improper because $1/(2 - x)$ is unbounded as $x \to 2$ and also because the interval of integration is infinite. We define such an integral by breaking up the domain so that only one kind of improper integral is considered at a time. In the present case we may write

$$\int_0^\infty \frac{dx}{2 - x} = \int_0^2 \frac{dx}{2 - x} + \int_2^3 \frac{dx}{2 - x} + \int_3^\infty \frac{dx}{2 - x}$$

and agree that the original integral converges if each term of the sum does. (The choice of the point 3 is not important; it is easy to prove that any choice gives the same result.) We leave it to you to show that this example diverges.

Problem Set 11.4

Evaluate each of the following improper integrals (or conclude that it diverges).

1. $\displaystyle\int_0^\infty \frac{dx}{4 + x^2}$

2. $\displaystyle\int_0^\infty \frac{dx}{\sqrt{4 + x^2}}$

3. $\displaystyle\int_0^\infty e^{-x}\, dx$

4. $\displaystyle\int_0^\infty x e^{-x}\, dx$

5. $\displaystyle\int_{-\infty}^\infty x e^{-x^2}\, dx$

6. $\displaystyle\int_{-\infty}^\infty \frac{dx}{x^2 + 9}$

7. $\displaystyle\int_{-\infty}^\infty \sin x\, dx$

8. $\displaystyle\int_{-\infty}^\infty \cos x\, dx$

9. $\displaystyle\int_{-\infty}^\infty \operatorname{sech} x\, dx$

10. $\displaystyle\int_0^\infty e^{-x} \sin x\, dx$

11. $\displaystyle\int_0^\infty e^{-x} \cos x\, dx$

12. $\displaystyle\int_2^\infty \frac{x\, dx}{x^2 - 1}$

13. $\displaystyle\int_2^\infty \frac{dx}{x^2 - 1}$

14. $\displaystyle\int_2^\infty \frac{dx}{(x - 1)^2}$

15. $\displaystyle\int_1^\infty \frac{\sqrt{x^2 - 1}}{x^2}\, dx$

16. $\displaystyle\int_3^\infty \frac{dx}{x^2 + x - 6}$

17. $\displaystyle\int_2^\infty \frac{x\, dx}{x^3 - 1}$

18. $\displaystyle\int_1^\infty \frac{dx}{x(x^2 + 1)}$

19. $\displaystyle\int_0^\infty \frac{dx}{\sqrt{x}(1 + x)}$

20. $\displaystyle\int_1^\infty \frac{dx}{x\sqrt{x^2 - 1}}$

21. $\displaystyle\int_0^\infty \frac{dx}{2 - x}$

22. $\displaystyle\int_0^\infty \frac{dx}{(x - 1)^2}$

23. Show that $\int_1^\infty dx/x^p$ converges if $p > 1$ and diverges if $p \leq 1$.

24. In view of the preceding problem and Problem 39, Section 11.3, what can you say about

$$\int_0^\infty \frac{dx}{x^p}$$

for every value of p?

25. Find $\int_1^\infty dx/x(x + 1)$ as follows.

   (a) Use decomposition of rational functions (Section 10.4) to show that

$$\int \frac{dx}{x(x + 1)} = \ln\frac{x}{x + 1} + C \qquad (x > 0)$$

   (b) Conclude that

$$\int_1^\infty \frac{dx}{x(x + 1)} = \ln 2$$

26. The decomposition called for in the preceding problem is

$$\frac{1}{x(x + 1)} = \frac{1}{x} - \frac{1}{x + 1}$$

Explain why

$$\int_1^\infty \frac{dx}{x(x + 1)} \neq \int_1^\infty \frac{dx}{x} - \int_1^\infty \frac{dx}{x + 1}$$

27. Does the region between the curves $y = e^{-x}$ and $y = -e^{-x}$ (to the right of the y axis) have a finite area? If so, find it.

28. The region between the curve $y = 1/(1 - x^2)$ and its asymptotes is in three parts. (Draw a picture!) Can a finite area be assigned to any of these parts? If so, find it.

29. The region in Problem 27 is rotated about the x axis. Does the resulting solid of revolution have a finite volume? If so, find it.

30. The region under the curve $y = 1/x$, $x \geq 1$, does not have a finite area. Does the solid of revolution obtained by rotating this region about the x axis have a finite volume? If so, find it.

31. Suppose that the universe had been made differently, so that Newton's Inverse-Square Law for the force of gravity were an inverse proportion instead. Then the lifting force on an object x miles from the center of the earth (Example 1) would be $F(x) = k/x$. Show that this state of affairs would impose a quarantine on earthlings (and others for that matter); it would be impossible to send *anything* out of the gravitational field of a planet.

32. Suppose that the annual profit of a company t years from

now is $f(t) = k$ (where k is a constant). Show that the present value of all future profits is $P(r) = k/r$, where r is the annual interest rate (continuously compounded).

33. Explain why the present value of a company's future profits goes *down* as interest rates go up. (Thus investors tend to find savings accounts more attractive than future profits of a company when rates go up.)

34. We defined

$$\int_{-\infty}^\infty f(x)\,dx = \int_{-\infty}^0 f(x)\,dx + \int_0^\infty f(x)\,dx$$

provided that each term of the sum exists. Show that the same result is obtained by using an arbitrary point c in place of 0.

35. An alternate definition of

$$\int_{-\infty}^\infty f(x)\,dx$$

that might occur to you is

$$\int_{-\infty}^\infty f(x)\,dx = \lim_{b \to \infty} \int_{-b}^b f(x)\,dx$$

Show, however, that this is not equivalent to our definition by considering the example

$$\int_{-\infty}^\infty \frac{x\,dx}{1 + x^2}$$

36. Show that $\int_1^\infty e^{-x^2}\,dx$ converges as follows.

   (a) Explain why $0 < e^{-x^2} \leq e^{-x}$ for $x \geq 1$ and hence

$$0 \leq \int_1^b e^{-x^2}\,dx \leq \int_1^b e^{-x}\,dx \qquad \text{for } b \geq 1$$

   (See Theorem 5, Section 6.3.)

   (b) Why does

$$\int_1^b e^{-x^2}\,dx$$

increase with b? Does it increase without bound?

   (c) Why do the answers in part (b) imply that

$$\int_1^\infty e^{-x^2}\,dx$$

converges? What is an upper bound for its value?

37. The *gamma function* is defined for all $x \geq 1$ by the rule

$$\Gamma(x) = \int_0^\infty t^{x-1}e^{-t}\,dt$$

   (a) Show that $\Gamma(1) = 1$ and $\Gamma(2) = 1$.

(b) Use integration by parts to prove that

$$\Gamma(x + 1) = x\Gamma(x) \qquad \text{for all } x \geq 1$$

(c) If we agree to the convention that $0! = 1$, explain why $\Gamma(n) = (n - 1)!$ for every positive integer n.

(Thus the gamma function is an ordinary factorial when $x = n$. It is also defined for values like $x = \frac{3}{2}$ or $x = \pi$, when the factorial notation does not make sense.)

The *Laplace Transform* of a function f is the function L defined by

$$L(x) = \int_0^\infty e^{-xt} f(t) \, dt$$

(See Example 4.) Find the Laplace Transform of each of the following functions.

38. $f(t) = 1$ **39.** $f(t) = t$ **40.** $f(t) = \sin t$

41. $f(t) = \cos t$

11.5
TAYLOR POLYNOMIALS

In Section 4.6 we discussed the linear approximation formula

$$f(x) \approx f(a) + f'(a)(x - a)$$

where a is a *base point* at which we know how to compute f and f', and x is kept close to a in order for the approximation to be accurate. The accuracy is estimated by means of the exact formula

$$f(x) = f(a) + f'(a)(x - a) + \tfrac{1}{2}f''(c)(x - a)^2 \qquad (c \text{ is between } a \text{ and } x)$$

the error being the absolute value of $\tfrac{1}{2}f''(c)(x - a)^2$.

Linear approximation is a special case of a more general idea, which we will introduce by means of an example. Suppose that we are interested in computing values of $f(x) = \cos x$ in a neighborhood of the base point $a = 0$. A natural way to do this is to use polynomials (whose values are easy to compute because only addition and multiplication are required). These have the form

$$P_n(x) = c_0 + c_1 x + c_2 x^2 + \cdots + c_n x^n$$

where n is a nonnegative integer and the coefficients $c_0, c_1, \ldots, c_n$ are chosen to make the graph of $y = P_n(x)$ fit the graph of $f(x) = \cos x$ near the point $(0,1)$. (See Figure 1.)

Figure 1
Polynomials fitting $f(x) = \cos x$ near $(0,1)$

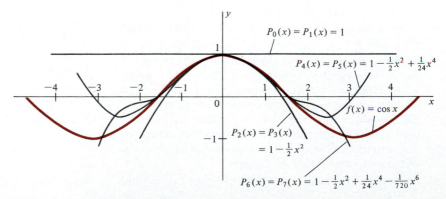

$P_0(x) = P_1(x) = 1$

$P_4(x) = P_5(x) = 1 - \frac{1}{2}x^2 + \frac{1}{24}x^4$

$f(x) = \cos x$

$P_2(x) = P_3(x)$
$= 1 - \frac{1}{2}x^2$

$P_6(x) = P_7(x) = 1 - \frac{1}{2}x^2 + \frac{1}{24}x^4 - \frac{1}{720}x^6$

The meaning of "fit" may be a little vague, but certainly it includes these geometric ideas:

- The graph of $y = P_n(x)$ should pass through the point $(0,1)$.
- It should have the same slope at $(0,1)$ as does the graph of $f(x) = \cos x$, namely $f'(0) = -\sin 0 = 0$.
- It should have the same concavity at $(0,1)$ as does the graph of $f(x) = \cos x$, namely $f''(0) = -\cos 0 = -1$.

In other words, we select the coefficients of $P_n(x)$ in such a way as to force $P_n(x)$ and its first two derivatives to match $\cos x$ and its first two derivatives at the base point $a = 0$:

$$P_n(0) = f(0) = 1, \ P_n'(0) = f'(0) = 0, \ P_n''(0) = f''(0) = -1$$

A geometric interpretation of higher-order derivatives is not available to us, but the analytic idea ought to be clear. To ensure a close fit, why not make *all* the derivatives match at $a = 0$? Of course the derivatives of $P_n(x)$ peter out after a while (depending on what n is), but as long as we have control of them, let's agree to make

$$P_n^{(k)}(0) = f^{(k)}(0) \qquad k = 0, 1, \ldots, n$$

These choices determine a definite polynomial $P_n(x)$, called the nth-order **Taylor polynomial** associated with $f(x) = \cos x$ at the point $a = 0$.

□ **Example 1**   Find the fourth-order Taylor polynomial associated with $f(x) = \cos x$ at the base point $a = 0$.

Solution: Since $n = 4$, our polynomial has the form

$$P_4(x) = c_0 + c_1 x + c_2 x^2 + c_3 x^3 + c_4 x^4$$

Its successive derivatives are

$$P_4'(x) = c_1 + 2c_2 x + 3c_3 x^2 + 4c_4 x^3$$
$$P_4''(x) = 2c_2 + 6c_3 x + 12c_4 x^2$$
$$P_4'''(x) = 6c_3 + 24c_4 x$$
$$P_4^{(4)}(x) = 24c_4$$

(All derivatives after the fourth are zero.) The first four derivatives of $f(x) = \cos x$ are

$$f'(x) = -\sin x, \ f''(x) = -\cos x, \ f'''(x) = \sin x, \ f^{(4)}(x) = \cos x$$

so the conditions

$$P_n^{(k)}(0) = f^{(k)}(0) \qquad k = 0, 1, 2, 3, 4$$

yield $c_0 = 1, c_1 = 0, 2c_2 = -1, 6c_3 = 0, 24c_4 = 1$. The coefficients of $P_4(x)$ are therefore $c_0 = 1, c_1 = 0, c_2 = -\frac{1}{2}, c_3 = 0, c_4 = \frac{1}{24}$, and hence

$$P_4(x) = 1 - \tfrac{1}{2}x^2 + \tfrac{1}{24}x^4$$

It is reasonable to expect that this serves as a good approximation of $\cos x$ near the base point $a = 0$, that is,

$$\cos x \approx 1 - \tfrac{1}{2}x^2 + \tfrac{1}{24}x^4 \qquad \text{for } x \text{ close to } 0$$

Taking $x = 0.2$, for example, we have

$$\cos 0.2 \approx 1 - \tfrac{1}{2}(0.2)^2 + \tfrac{1}{24}(0.2)^4 = 0.980066\cdots$$

The correct five-place value (from a table or calculator) is $\cos 0.2 = 0.98007$, so our formula gives the right answer to at least five places. □

The preceding discussion can be generalized to any function f and any base point a (in a neighborhood of which the successive derivatives of f exist). For convenience we now write our polynomials in powers of $x - a$:

$$P_n(x) = c_0 + c_1(x - a) + c_2(x - a)^2 + \cdots + c_n(x - a)^n$$

(Any polynomial can be written this way; see the problem set.) To make $P_n(x)$ fit $f(x)$ near a, we impose the conditions

$$P_n^{(k)}(a) = f^{(k)}(a) \qquad k = 0, 1, \ldots, n$$

The successive derivatives of $P_n(x)$ are

$$P_n'(x) = c_1 + 2c_2(x - a) + 3c_3(x - a)^2 + 4c_4(x - a)^3 + \cdots$$
$$P_n''(x) = 2c_2 + 2 \cdot 3c_3(x - a) + 3 \cdot 4c_4(x - a)^2 + \cdots$$
$$P_n'''(x) = 2 \cdot 3c_3 + 2 \cdot 3 \cdot 4c_4(x - a) + \cdots$$
$$P_n^{(4)}(x) = 2 \cdot 3 \cdot 4c_4 + \cdots$$

and so on. Introducing the symbol $n! = 1 \cdot 2 \cdot 3 \cdots n$ (read "n factorial"), and agreeing to the special convention that $0! = 1$, we see that

$$P_n^{(k)}(x) = k!c_k + (\text{terms with the factor } x - a)$$

Since all terms with the factor $x - a$ drop out when $x = a$, we have

$$P_n^{(k)}(a) = k!c_k$$

Hence the conditions that fit $P_n(x)$ to $f(x)$ are

$$k!c_k = f^{(k)}(a) \qquad k = 0, 1, \ldots, n$$

We can find $f^{(k)}(a)$ by successive differentiation of f (evaluating each derivative at the base point a), so we have discovered how to fit a polynomial of arbitrary degree to a given function in the neighborhood of a given point. Simply choose its coefficients to be

$$c_k = \frac{f^{(k)}(a)}{k!} \qquad k = 0, 1, \ldots, n$$

and the polynomial is determined.

This result is historic (and has a colorful background). It is named after Brook Taylor (1685–1731), who was one of several mathematicians to have figured it out.

Let a be a point in the domain of the function f and suppose that $f^{(n)}(a)$ exists (where n is a nonnegative integer). Then

$$P_n(x) = f(a) + f'(a)(x - a) + \frac{f''(a)}{2!}(x - a)^2 + \cdots + \frac{f^{(n)}(a)}{n!}(x - a)^n$$

is called the nth-order **Taylor polynomial** associated with f at a.

□ **Example 2**   Let $f(x) = 1/x$ and take $a = 1$. (Why can't we choose the base point to be 0?) Successive derivatives of f (and their values at the base point) are

$$
\begin{aligned}
f(x) &= x^{-1} & f(1) &= 1 \\
f'(x) &= -x^{-2} & f'(1) &= -1 \\
f''(x) &= 2x^{-3} & f''(1) &= 2 \\
f'''(x) &= -3!x^{-4} & f'''(1) &= -3! \\
f^{(4)}(x) &= 4!x^{-5} & f^{(4)}(1) &= 4! \\
&\vdots & &\vdots \\
f^{(k)}(x) &= (-1)^k k! x^{-(k+1)} & f^{(k)}(1) &= (-1)^k k!
\end{aligned}
$$

Hence

$$c_k = \frac{f^{(k)}(1)}{k!} = (-1)^k \qquad k = 0, 1, \ldots, n$$

and the nth-order Taylor polynomial at $a = 1$ is

$$P_n(x) = 1 - (x - 1) + (x - 1)^2 - \cdots + (-1)^n(x - 1)^n$$

To try this out as an approximation to $f(x) = 1/x$, take $n = 3$ and write

$$\frac{1}{x} \approx 1 - (x - 1) + (x - 1)^2 - (x - 1)^3$$

This should give good results for x near 1. For example,

$$\frac{1}{1.3} \approx 1 - (0.3) + (0.3)^2 - (0.3)^3 = 0.763$$

The correct reciprocal of $x = 1.3$ (to three places) is 0.769, so the error is 0.006 and the relative error is $0.006/0.769 \approx 0.008$, or about $\frac{8}{10}$ of one per cent.   □

To discuss the question of accuracy in general, we need an estimate of the difference between $f(x)$ and each of its Taylor polynomials. To start with, let's give it a name:

$$R_n(x) = f(x) - P_n(x)$$

the **remainder** after the approximation of $f(x)$ by its nth-order Taylor polynomial. The following theorem tells us what it looks like.

Taylor's Formula

Let a be a point in the domain of the function f and suppose that $f^{(n+1)}$ exists in a neighborhood of a (where n is a nonnegative integer). For each x in this neighborhood,

$$f(x) = f(a) + f'(a)(x - a) + \frac{f''(a)}{2!}(x - a)^2$$

$$+ \cdots + \frac{f^{(n)}(a)}{n!}(x - a)^n + \frac{f^{(n+1)}(c)}{(n + 1)!}(x - a)^{n+1}$$

where c is between a and x.

In other words, the remainder after approximation of $f(x)$ by its nth-order Taylor polynomial is

$$R_n(x) = \frac{f^{(n+1)}(c)}{(n + 1)!}(x - a)^{n+1} \qquad \text{where } c \text{ is between } a \text{ and } x$$

This is called the **Lagrange form** of the remainder. (There are several others.) *It is the same as the next term in the Taylor approximation formula,* except that the derivative is not evaluated at the base point a but at a point c between a and x. The substance of Taylor's Formula is that the approximation is exact when the remainder is included. Note that the Mean Value Theorem,

$$f(x) = f(a) + f'(c)(x - a) \qquad \text{(where } c \text{ is between } a \text{ and } x\text{)}$$

is just Taylor's Formula when $n = 0$.

The proof of Taylor's Formula is not too hard to follow, but it is a whopper to think up. We have put an outline of it in the problem set.

□ **Example 3**   Apply Taylor's Formula to $f(x) = \cos x$ at the base point $a = 0$.

Solution: The first few derivatives of f are worked out in Example 1. In general we have

$$f^{(k)}(x) = \cos\left(x + \frac{k\pi}{2}\right) \qquad k = 0, 1, 2, \ldots$$

a formula you can check by noting that the sequence of derivatives

$$f^{(0)}(x) = \cos x = \cos\left(x + \frac{0\pi}{2}\right)$$

$$f^{(1)}(x) = -\sin x = \cos\left(x + \frac{1\pi}{2}\right)$$

$$f^{(2)}(x) = -\cos x = \cos\left(x + \frac{2\pi}{2}\right)$$

$$f^{(3)}(x) = \sin x = \cos\left(x + \frac{3\pi}{2}\right)$$

repeats. (Also see Example 2 and Problem 28, Section 4.5.) The nth-order Taylor polynomial associated with f at $a = 0$ has coefficients

$$c_k = \frac{f^{(k)}(0)}{k!} = \frac{1}{k!}\cos\frac{k\pi}{2} \qquad k = 0, 1, \ldots, n$$

so Taylor's Formula reads

$$\cos x = 1 + 0x - \frac{x^2}{2!} + 0x^3 + \frac{x^4}{4!} + \cdots + \left(\frac{1}{n!}\cos\frac{n\pi}{2}\right)x^n + R_n(x)$$

where

$$R_n(x) = \frac{1}{(n+1)!}\cos\left[c + \frac{(n+1)\pi}{2}\right]x^{n+1} \qquad (c \text{ between } 0 \text{ and } x)$$

This looks more complicated than it is. The error involved in using the formula without the remainder is

$$|\cos x - P_n(x)| = |R_n(x)| \le \frac{|x|^{n+1}}{(n+1)!}$$

the inequality being justified by the fact that $|\cos t| \le 1$ for all t. Note that the value of c (somewhere between 0 and x) is irrelevant in this estimate; we simply ignore it.

When n is even, say $n = 2m$ ($m = 0, 1, 2, \ldots$), the coefficient of the last term of $P_n(x)$ is

$$\frac{1}{(2m)!}\cos m\pi = \frac{(-1)^m}{(2m)!}$$

and the polynomial simplifies to

$$P_{2m}(x) = 1 - \frac{x^2}{2!} + \frac{x^4}{4!} - \cdots + \frac{(-1)^m}{(2m)!}x^{2m}$$

If n is odd, say $n = 2m + 1$, the last term is 0 and we have $P_{2m+1}(x) = P_{2m}(x)$.

Thus the best use of Taylor's Formula for $\cos x$ is when n is odd. For then it reduces to

$$\cos x = 1 - \frac{x^2}{2!} + \frac{x^4}{4!} - \cdots + \frac{(-1)^m}{(2m)!}x^{2m} + 0x^{2m+1} + R_{2m+1}(x)$$

where

$$|R_{2m+1}(x)| \le \frac{|x|^{2m+2}}{(2m+2)!}$$

If we take n to be even, the Taylor polynomial is the same, but the error is

$$|R_{2m}(x)| \le \frac{|x|^{2m+1}}{(2m+1)!}$$

The upper bound is less accurate in this case, since

$$\frac{1}{(2m+1)!} \text{ is not as small as } \frac{1}{(2m+2)!}$$

and (for values of x close to 0)

$$|x|^{2m+1} \text{ is not as small as } |x|^{2m+2} \qquad \square$$

□ **Example 4**   Taking $n = 5$ in the Taylor approximation of $\cos x$, we find

$$\cos x \approx 1 - \tfrac{1}{2}x^2 + \tfrac{1}{24}x^4$$

the error being

$$|R_5(x)| \le \frac{|x|^6}{720}$$

In Example 1 we used this approximation to write

$$\cos 0.2 \approx 1 - \tfrac{1}{2}(0.04) + \tfrac{1}{24}(0.0016) = 0.98006\overline{6}$$

Now we can say something intelligible about the accuracy:

$$|R_5(0.2)| \le \frac{(0.2)^6}{720} = \frac{64 \times 10^{-6}}{720} = \frac{8}{9} \times 10^{-7} = 0.\overline{8} \times 10^{-7}$$

Thus in writing $\cos 0.2 \approx 0.98006\overline{6}$, we are actually saying that

$$0.98006\overline{6} - 0.\overline{8} \times 10^{-7} \le \cos 0.2 \le 0.98006\overline{6} + 0.\overline{8} \times 10^{-7}$$

that is,

$$0.9800665\overline{7} \le \cos 0.2 \le 0.9800667\overline{5}$$

Hence $\cos 0.2 = 0.980067$ correct to six places.

The same approximation formula with $x = 1$ gives

$$\cos 1 \approx 1 - \frac{1}{2} + \frac{1}{24} = 0.541\overline{6}$$

with error

$$|R_5(1)| \le \frac{1}{720} = 0.0013\overline{8}$$

Hence

$$0.5402\overline{7} \le \cos 1 \le 0.5430\overline{5}$$

This time the best we can do is $\cos 1 = 0.54$ correct to two places. A five-place table gives $\cos 1 = 0.54030$; to achieve this accurary with Taylor polynomials, we must choose a larger n. □

Problem Set 11.5

1. Find the fourth-order Taylor polynomial associated with $f(x) = 5 - 2x + x^2 + 6x^3 + 2x^4$ at $a = 0$ and show that the remainder is zero. Can you generalize the result?

2. Use the ideas of this section to write $f(x) = 2 + x^2 - 5x^3$ in powers of $x - 1$.

3. Show that

$$e^x \approx 1 + x + \frac{x^2}{2!} + \frac{x^3}{3!} + \cdots + \frac{x^n}{n!}$$

4. Explain why the error in Problem 3 is

$$|R_n(x)| \le \frac{e^{|x|}}{(n+1)!}|x|^{n+1}$$

5. Use $n = 4$ in Problem 3 to compute $e^{0.2}$ as accurately as the error estimate in Problem 4 allows. (*Note:* The error estimate involves the very number you are trying to compute! It is not in the spirit of the question to use a calculator to find it. Instead use what you know about exponentials to find an upper bound for it.)

6. The value of e may be computed by taking $x = 1$ in Problems 3 and 4.

(a) Using the fact that $e < 3$, explain why the error is

$$|R_n(1)| < \frac{3}{(n + 1)!}$$

(b) What value of n will guarantee three-place accuracy? *Hint:* The error must be less than $\frac{1}{2} \times 10^{-3}$.

(c) Compute e correct to three decimal places.

7. Show that

$$e^{-x} \approx 1 - x + \frac{x^2}{2!} - \frac{x^3}{3!} + \cdots + (-1)^n \frac{x^n}{n!}$$

by applying Taylor theory to $f(x) = e^{-x}$.

8. Replace x by $-x$ in Problem 3 to confirm the formula in Problem 7.

9. Show that

$$\cosh x \approx 1 + \frac{x^2}{2!} + \frac{x^4}{4!}$$

by applying Taylor theory to $f(x) = \cosh x$. (Note the remarkable similarity to the Taylor polynomial for $\cos x$!)

10. Recalling that $\cosh x = \frac{1}{2}(e^x + e^{-x})$, confirm the formula in Problem 9 by combining formulas for e^x and e^{-x}.

11. Show that

$$\sinh x \approx x + \frac{x^3}{3!} + \frac{x^5}{5!}$$

by applying Taylor theory to $f(x) = \sinh x$.

12. Confirm the formula in Problem 11 by combining formulas for e^x and e^{-x}.

13. Show that the nth-order Taylor polynomial associated with $f(x) = \sin x$ at $a = 0$ is

$$P_n(x) = 0 + x + 0x^2 - \frac{x^3}{3!} + 0x^4$$

$$+ \frac{x^5}{5!} + \cdots + \left(\frac{1}{n!} \sin \frac{n\pi}{2}\right)x^n$$

Hint: In Example 2, Section 4.5, we showed that

$$f^{(k)}(x) = \sin\left(x + \frac{k\pi}{2}\right), \quad k = 0, 1, 2, \ldots$$

14. Explain why the error in writing $\sin x \approx P_n(x)$ is

$$|R_n(x)| \leq \frac{|x|^{n+1}}{(n + 1)!}$$

15. Explain why the best use of $\sin x \approx P_n(x)$ is when n is even, say $n = 2m$ ($m = 1, 2, 3, \ldots$) and that in this case

$$\sin x \approx x - \frac{x^3}{3!} + \frac{x^5}{5!} - \cdots + \frac{(-1)^{m-1}}{(2m - 1)!}x^{2m-1}$$

16. Use the formula $\sin x \approx x - \frac{1}{6}x^3$ (taking $n = 4$ in Problems 13 and 14) to confirm that

$$0.825\overline{0} \leq \sin 1 \leq 0.841\overline{6}$$

What is the best value of $\sin 1$ you can report from these figures?

17. Use the formula in Problem 16 to confirm that

$$0.1986640 \leq \sin 0.2 \leq 0.1986693$$

What is the best value of $\sin 0.2$ you can report from these figures?

18. Explain why the formula

$$\sin x \approx x - \frac{x^3}{3!} + \frac{x^5}{5!} - \frac{x^7}{7!} + \frac{x^9}{9!}$$

can be used to compute the entries in a five-place table of sines. *Hint:* We may restrict x to the interval $(0, \pi/2)$. Why?

19. Explain why $|\sin x - x| \leq \frac{1}{6}|x|^3$ for all x, and hence

$$\left|\frac{\sin x}{x} - 1\right| \leq \frac{1}{6}|x|^2 \qquad \text{for all } x \neq 0$$

Why does it follow that

$$\lim_{x \to 0} \frac{\sin x}{x} = 1?$$

20. If the formula $\cos x \approx 1 - x^2/2!$ is used in the interval $[-0.3, 0.3]$, how many decimal places can be guaranteed? *Hint:* Regard $1 - x^2/2!$ as the *third*-order Taylor polynomial associated with $\cos x$ at $a = 0$.

21. Show that

$$\left|\frac{1 - \cos x}{x}\right| \leq \frac{1}{2}|x| \qquad \text{for all } x \neq 0$$

Why does it follow that $\lim_{x \to 0} (1 - \cos x)/x = 0$?

22. Use $n = 5$ in Taylor's Formula for $\cos x$ at $a = 0$ to compute $\cos 0.1$ as accurately as the error estimate allows.

23. Use $n = 7$ in Taylor's Formula for $\cos x$ at $a = 0$ to compute $\cos 0.2$ as accurately as the error estimate allows.

24. Show that the first few Taylor approximations of $\tan x$ in a neighborhood of 0 are (a) $\tan x \approx x$; (b) $\tan x \approx x + \frac{1}{3}x^3$; and (c) $\tan x \approx x + \frac{1}{3}x^3 + \frac{2}{15}x^5$.

25. Considerable labor is involved in part (c) of the preceding problem. Obtain the same result by division of

$$\sin x \approx x - \frac{x^3}{3!} + \frac{x^5}{5!} \quad \text{by} \quad \cos x \approx 1 - \frac{x^2}{2!} + \frac{x^4}{4!}$$

26. Show that

$$|\tan x - x| \le \tfrac{8}{9}|x|^3 \qquad \text{if } -\pi/6 \le x \le \pi/6$$

Why does it follow that $\lim_{x \to 0} (\tan x)/x = 1$?

27. Show that if $\pi/4 < x < \pi/3$, then

$$\tan x = 1 + 2\left(x - \frac{\pi}{4}\right) + 2\left(x - \frac{\pi}{4}\right)^2 + R_2(x)$$

where

$$0 \le R_2(x) \le \frac{40}{3}\left(x - \frac{\pi}{4}\right)^3$$

28. Use Problem 27 to compute $\tan 0.8$ as accurately as the error estimate allows. (Compare with the value 1.0292 found by linear approximation in Example 5, Section 4.6.)

29. Find the second-order Taylor polynomial associated with $\sqrt{x}$ at $a = 1$ and use it to compute $\sqrt{0.98}$. (Compare with the value 0.99 found by linear approximation in Example 2, Section 4.6.)

30. Find a quadratic approximation formula for $\sec x$ in a neighborhood of 0 and use it to compute $\sec 0.2$.

31. Show that the nth-order Taylor polynomial associated with $\ln x$ at $a = 1$ is

$$P_n(x) = (x - 1) - \frac{(x - 1)^2}{2} + \frac{(x - 1)^3}{3}$$

$$- \cdots + (-1)^{n-1}\frac{(x - 1)^n}{n}$$

32. Show that the remainder in Problem 31 is

$$R_n(x) = \frac{(-1)^n}{n + 1}\left(\frac{x - 1}{c}\right)^{n+1} \qquad (c \text{ is between 1 and } x)$$

33. Use $n = 4$ in Problems 31 and 32 to confirm that $0.0953063 \le \ln 1.1 \le 0.0953103$. What is the best value of $\ln 1.1$ you can report from these figures?

34. Show that the nth-order Taylor polynomial associated with $1/(1 - x)$ at $a = 0$ is

$$P_n(x) = 1 + x + x^2 + \cdots + x^n$$

35. Referring to the preceding problem, explain why $P_n(x) - xP_n(x) = 1 - x^{n+1}$ and hence (if $x \ne 1$)

$$1 + x + x^2 + \cdots + x^n = \frac{1 - x^{n+1}}{1 - x}$$

(This is a formula for the sum of consecutive terms of a geometric progression, often derived in intermediate algebra books.)

36. The remainder in the formula

$$\frac{1}{1 - x} = 1 + x + x^2 + \cdots + x^n + R_n(x)$$

may be expected to involve an unknown point c between 0 and x. Use Problem 35, however, to show that

$$R_n(x) = \frac{x^{n+1}}{1 - x}$$

(Thus in this case the remainder is known in an exact form not involving c.)

37. Explain why it follows from Problem 36 that if $|x| < 1$, $\lim_{n \to \infty} R_n(x) = 0$. [This justifies the infinite series

$$\frac{1}{1 - x} = 1 + x + x^2 + x^3 + \cdots \qquad (-1 < x < 1)$$

in the sense that $1/(1 - x)$ may be approximated arbitrarily closely by using enough terms of the series.]

38. Replacing n by $n - 1$ in Problem 36 (and x by t), we have

$$\frac{1}{1 - t} \approx 1 + t + t^2 + \cdots + t^{n-1}$$

(a) Explain why it follows (for $-1 < x < 1$) that

$$\int_0^x \frac{dt}{1 - t} \approx x + \frac{x^2}{2} + \frac{x^3}{3} + \cdots + \frac{x^n}{n}$$

(b) Use part (a) to obtain (for $-1 < x < 1$)

$$\ln(1 - x) \approx -x - \frac{x^2}{2} - \frac{x^3}{3} - \cdots - \frac{x^n}{n}$$

(c) Replace x by $1 - x$ in part (b) to find

$$\ln x \approx (x - 1) - \frac{(x - 1)^2}{2} + \frac{(x - 1)^3}{3} - \cdots$$

$$+ (-1)^{n-1}\frac{(x - 1)^n}{n} \qquad (0 < x < 2)$$

and compare with Problem 31.

39. Show that

$$\tan^{-1} x = x - \frac{cx^2}{(1 + c^2)^2}$$

where c is between 0 and x.

40. Suppose that the formula in Problem 39 is used to assert that $\tan^{-1} 0.2 \approx 0.2$. Why is this accurate to at least one decimal place?

41. It is gruesome to find Taylor polynomials associated with $\tan^{-1} x$ by a direct attack. Use the following device instead.

(a) Replace x by $-t^2$ in the formula

$$\frac{1}{1-x} \approx 1 + x + x^2 + \cdots + x^{n-1}$$

to obtain

$$\frac{1}{1+t^2} \approx 1 - t^2 + t^4 - \cdots + (-1)^{n-1}t^{2n-2}$$

(b) Integrate from 0 to x to obtain

$$\tan^{-1} x \approx x - \frac{x^3}{3} + \frac{x^5}{5} - \cdots + (-1)^{n-1}\frac{x^{2n-1}}{2n-1}$$

42. Use the formula $\tan^{-1} x \approx x - x^3/3$ to compute $\tan^{-1} 0.2$.

43. Sketch the graphs of $y = x^2$ and $y = \sin x$ to convince yourself that the equation $\sin x = x^2$ has one nonzero root. Use the formula

$$\sin x \approx x - \tfrac{1}{6}x^3$$

to show that this root is approximately

$$\sqrt{15} - 3 \approx 0.87$$

(In Example 3, Section 5.4, we used Newton's Method to find the value 0.8767.)

44. Solve the equation $e^x = 1 + 2x$ as follows.

(a) Sketch the graphs of $y = e^x$ and $y = 1 + 2x$ to convince yourself that the equation has one positive root. What is the only other root?

(b) Use the formula

$$e^x \approx 1 + x + \frac{x^2}{2!} + \frac{x^3}{3!}$$

to show that the positive root is approximately

$$\tfrac{1}{2}(-3 + \sqrt{33}) \approx 1.37$$

(c) Use 1.37 as a first guess in Newton's Method (Section 5.4) to improve the approximation.

45. Compute $\int_0^{0.5} e^{-x^2}\, dx$ in two ways, as follows.
(a) Replace x by $-x^2$ in the formula

$$e^x \approx 1 + x + \frac{x^2}{2!} + \frac{x^3}{3!}$$

and integrate the right side from 0 to 0.5.

(b) Use Simpson's Rule (Section 10.6) with four subintervals.

46. Compute $\int_0^1 (\sin x)/x\, dx$ in two ways, as follows.
(a) Divide by x in the formula

$$\sin x \approx x - \frac{x^3}{3!} + \frac{x^5}{5!}$$

and integrate the right side from 0 to 1.

(b) Use Simpson's Rule (Section 10.5) with four subintervals. *Note:* The integrand is understood to be the function f defined by

$$\begin{cases} \dfrac{\sin x}{x} & \text{if } x \neq 0 \\[2mm] 1 & \text{if } x = 0 \end{cases}$$

The integral (despite appearances) is proper.

47. Prove Taylor's Formula as follows.

(a) Why is there nothing to prove if $x = a$? Assuming that x is different from a, relabel it b, and let $R_n(b) = f(b) - P_n(b)$. Explain why the problem is to demonstrate the existence of a point c between a and b such that

$$R_n(b) = \frac{f^{(n+1)}(c)}{(n+1)!}(b-a)^{n+1}$$

(b) Regarding b as fixed, define the function

$$F(x) = \left[f(x) + f'(x)(b-x) + \frac{f''(x)}{2!}(b-x)^2 \right.$$
$$\left. + \cdots + \frac{f^{(n)}(x)}{n!}(b-x)^n \right]$$
$$+ \frac{R_n(b)}{(b-a)^{n+1}}(b-x)^{n+1} - f(b)$$

Confirm that $F(a) = F(b) = 0$ and show that

$$F'(x) = \frac{f^{(n+1)}(x)}{n!}(b-x)^n - \frac{(n+1)R_n(b)}{(b-a)^{n+1}}(b-x)^n$$

Hint: The derivative of the expression in brackets in part (b) is a "telescoping sum."

(c) Apply Rolle's Theorem (Section 4.3) to F in the interval with endpoints a and b to complete the proof.

True–False Quiz

1. The implication

$$x > \ln \frac{1}{\varepsilon} \Rightarrow e^{-x} < \varepsilon \qquad \text{(for each } \varepsilon > 0\text{)}$$

shows that $\lim_{x \to \infty} e^{-x} = 0$.

2. $\lim_{x \to \infty} x e^{1/x} = \infty$.   **3.** $\lim_{x \to \infty} \frac{1984 x^3}{x^4 + 1} = \infty$.

4. If $\lim_{x \to a} f(x) = \lim_{x \to a} g(x) = \infty$, then $\lim_{x \to a} \frac{f(x)}{g(x)} = 1$.

5. If $\lim_{x \to a} f(x) = -1$ and $\lim_{x \to a} g(x) = \infty$, then

$$\lim_{x \to a} f(x) g(x) = -\infty$$

6. $\lim_{x \to \infty} \frac{x}{\ln x} = 0$.

7. $\lim_{x \to 1} \frac{x^2 - x}{x^2 + 1} = \lim_{x \to 1} \frac{2x - 1}{2x}$.

8. $\int_0^1 \frac{x \, dx}{1 - x}$ converges.

9. If f is an odd function and $\int_0^\infty f(x) \, dx$ converges, then

$$\int_{-\infty}^\infty f(x) \, dx = 0.$$

10. $\int_0^\infty x^2 e^{-x} \, dx$ converges.

11. The second-order Taylor polynomial associated with $f(x) = x^2$ at the base point 1 is $1 + 2(x - 1) + (x - 1)^2$.

12. The second-order Taylor polynomial associated with $f(x) = e^{-x^2}$ at the base point 0 is $1 - x^2$.

Additional Problems

Evaluate each of the following limits.

1. $\lim_{x \to \infty} \frac{x + 1}{x^2 + 1}$ **2.** $\lim_{x \to \infty} \frac{2x - 1}{x + 1}$

3. $\lim_{x \to \infty} \left(1 - \frac{1}{x}\right)^3$ **4.** $\lim_{x \to \infty} (x^2 - 2x)$

5. $\lim_{x \downarrow 0} \cot x$ **6.** $\lim_{x \uparrow 0} \cot x$

7. $\lim_{x \to \pi/2} \sqrt{1 + \tan^2 x}$ **8.** $\lim_{x \to 0} \tan x \cot x$

9. $\lim_{x \to \infty} \tan \frac{1}{x}$ **10.** $\lim_{x \to 0} x \sin \frac{1}{x}$

11. $\lim_{x \to \infty} x^2 \sin \frac{1}{x}$ **12.** $\lim_{x \to \infty} \tanh x$

13. $\lim_{x \downarrow 0} \frac{e^x + 1}{e^x - 1}$ **14.** $\lim_{x \to \infty} \tan^{-1}(2 - x)$

15. $\lim_{x \to \infty} (2 - \sin x) \ln x$ **16.** $\lim_{x \to \infty} \frac{e^x - \ln x}{e^x}$

17. To prove that

$$\lim_{x \to \infty} e^{2x} = \infty$$

let B be given. Name A such that $x > A \Rightarrow e^{2x} > B$.

18. To prove that

$$\lim_{x \to 0} \frac{1}{x^2} = \infty$$

let B be given. Name δ such that $0 < |x| < \delta \Rightarrow 1/x^2 > B$.

19. To prove that

$$\lim_{x \uparrow 1} \frac{1}{1 - x} = \infty$$

let B be given. Name δ such that $0 < 1 - x < \delta \Rightarrow 1/(1 - x) > B$.

20. To prove that

$$\lim_{x \to -\infty} \frac{2}{x} = 0$$

let ε be given. Name A such that $x < -A \Rightarrow |2/x| < \varepsilon$.

21. To prove that

$$\lim_{x \uparrow 2} \frac{x}{x - 2} = -\infty$$

let B be given. Name δ such that $0 < 2 - x < \delta \Rightarrow x/(x - 2) < -B$.

Find each of the following limits and use an appropriate definition to prove that your answer is correct.

22. $\lim\limits_{x \to \infty} \dfrac{x-1}{x}$ **23.** $\lim\limits_{x \to -\infty} (2-x)$

24. $\lim\limits_{x \uparrow 0} \dfrac{1}{x}$ **25.** $\lim\limits_{x \downarrow 0} \left(2 - \dfrac{1}{x}\right)$

26. $\lim\limits_{x \to \infty} (1 - \ln x)$ **27.** $\lim\limits_{x \to \infty} \dfrac{\sin x}{x}$

28. Evaluate

$$\lim_{x \to 0} \frac{1 - \cos x}{\sin^2 x}$$

without using L'Hôpital's Rule.

29. Do Problem 28 by using L'Hôpital's Rule.

Using L'Hôpital's Rule where it is appropriate, evaluate each of the following limits.

30. $\lim\limits_{x \to 1} \dfrac{x^4 - 1}{x^3 - 1}$ **31.** $\lim\limits_{x \to 0} \dfrac{\sin^{-1} x}{\tan^{-1} x}$

32. $\lim\limits_{x \to 0} \dfrac{\sec x - \tan x}{\cos x}$ **33.** $\lim\limits_{x \to 0} (\cos x)^{1/x}$

34. $\lim\limits_{x \to 0} \dfrac{\sin 4x}{\sin 2x}$ **35.** $\lim\limits_{x \to \infty} \dfrac{x^4}{e^x}$

36. $\lim\limits_{x \to 0} \dfrac{x - \sin x}{x^2}$ **37.** $\lim\limits_{x \to \infty} \dfrac{\tan^{-1} x}{\tanh x}$

38. $\lim\limits_{x \downarrow 0} x^2 \ln x$ **39.** $\lim\limits_{x \to 0} \dfrac{x + \tan x}{x + \sin x}$

40. $\lim\limits_{x \to \infty} \dfrac{x}{\sqrt{1 + x^2}}$ **41.** $\lim\limits_{x \downarrow 0} \dfrac{x^2 - \tan x}{1 - \cos x}$

42. $\lim\limits_{x \to 0} (1 - \sin x)^{1/x}$

43. Explain why L'Hôpital's Rule does not apply to

$$\lim_{x \to \infty} \frac{x - \cos x}{x}$$

What is the value of the limit?

44. Sketch the graph of $y = \ln x + (1/x)$ after answering the following questions.

(a) What is the domain?

(b) How does y behave when $x \downarrow 0$?

(c) How does y behave when $x \to \infty$?

(d) Find all local extreme values.

(e) Find all points of inflection.

Evaluate each of the following improper integrals (or conclude that it diverges).

45. $\displaystyle\int_1^2 \dfrac{dx}{\sqrt{x - 1}}$ **46.** $\displaystyle\int_2^\infty \dfrac{dx}{x(\ln x)^2}$

47. $\displaystyle\int_1^\infty \dfrac{dx}{x^{3/2}}$ **48.** $\displaystyle\int_0^3 \dfrac{dx}{\sqrt{9 - x^2}}$

49. $\displaystyle\int_0^{\pi/2} \sec x \tan x \, dx$

50. $\displaystyle\int_0^5 \dfrac{dx}{(x - 2)^{2/3}}$ **51.** $\displaystyle\int_0^\infty e^{-2x} \, dx$

52. $\displaystyle\int_{-\infty}^\infty \dfrac{x \, dx}{x^4 + 16}$ **53.** $\displaystyle\int_0^4 \dfrac{dx}{\sqrt{4 - x}}$

54. $\displaystyle\int_0^1 \ln x^2 \, dx$ **55.** $\displaystyle\int_3^5 \dfrac{dx}{x\sqrt{x^2 - 9}}$

56. $\displaystyle\int_0^\infty e^{-x} \sin x \, dx$ **57.** $\displaystyle\int_0^2 \dfrac{dx}{1 - x^2}$

58. $\displaystyle\int_0^\infty \tan^{-1} x \, dx$ **59.** $\displaystyle\int_0^\infty \dfrac{x \, dx}{x^8 + 1}$

60. Find the area of the region in the fourth quadrant between the curve $y = \ln x$ and the y axis (or conclude that it does not exist).

61. Find the area of the region below the curve $y = 1/\sqrt{1 - x^2}$, above the x axis, and between the lines $x = \pm 1$ (or conclude that it does not exist).

62. Find the third-order Taylor polynomial associated with $f(x) = 1/x^2$ at the base point 1.

63. Use the Taylor polynomial in Problem 62 to compute $1/(1.1)^2$ as accurately as the error estimate allows. Then compare with the exact value.

64. Find the third-order Taylor polynomial associated with $f(x) = \sin^{-1} x$ at the base point 0.

65. Use the Taylor polynomial in Problem 64 to find an approximate value of $\sin^{-1} 0.2$. Compare with the actual value $\sin^{-1} 0.2 = 0.201357 \cdots$.

66. Find the fourth-order Taylor polynomial associated with $f(x) = \sqrt{x}$ at the base point 1.

67. Use the Taylor polynomial in Problem 66 to find an approximate value of $\sqrt{1.2}$. Compare with the actual value $\sqrt{1.2} = 1.095445 \cdots$.

68. Find the nth-order Taylor polynomial associated with $f(x) = \ln(1 + x)$ at the base point 0.

12 | Infinite Series

The Taylor polynomials introduced in the last chapter strongly suggest the idea of *infinite series*. For example, the approximation

$$\cos x \approx 1 - \frac{x^2}{2!} + \frac{x^4}{4!} - \cdots + \frac{(-1)^m}{(2m)!} x^{2m} \qquad (m = 0, 1, 2, \ldots)$$

is supposed to improve with increasing m (and does if the remainder approaches zero when $m \to \infty$). If this has been established, it is natural to write

$$\cos x = 1 - \frac{x^2}{2!} + \frac{x^4}{4!} - \cdots$$

where the approximation sign has been replaced by an equals sign and the series on the right side is understood to be unending. The idea is that $\cos x$ (a number that seems simple until one gets down to the question of computing it) is not just approximated by the series; it *is* the series.

We can try to make this idea reasonable by saying that $\cos x$ may be approximated arbitrarily closely by taking enough terms of the series. When you think about it, what else could we say? Nobody can add infinitely many numbers. But any sum with a finite number of terms can be computed (in principle). What we must mean by an infinite series is the *limit* of such a sum as the number of terms increases.

To a student who understands what a limit is, this notion may not seem very deep. Like most great ideas, it is in fact pretty simple! Its implications for calculus, however, are profound (both in theory and in applications to real problems). This chapter is an introduction, the idea being to learn what an infinite series is and how to determine when it makes sense. In Chapter 13 we will put the results to work.

12.1
A BRIEF LOOK AT SEQUENCES

An important preliminary to the study of infinite series is the topic of sequences and their limits. A *sequence* is nothing more complicated than a listing of numbers in order, such as

$$\frac{1}{2}, \frac{2}{3}, \frac{3}{4}, \frac{4}{5}, \ldots, \frac{n}{n+1}, \ldots$$

Its *limit* (if there is one) is the number approached by the *n*th term of the sequence as $n \to \infty$. The typical term of the preceding sequence is

$$s_n = \frac{n}{n+1}, \quad n = 1, 2, 3, \ldots$$

and its limit, as you can see by inspection, is

$$\lim_{n \to \infty} \frac{n}{n+1} = 1$$

More formal definitions of these ideas are as follows.

A *sequence* is a function whose domain is the set of positive integers 1, 2, 3, 4, ... If the function is *f*, the *terms* of the sequence are

$$s_1 = f(1), \; s_2 = f(2), \; s_3 = f(3), \; s_4 = f(4), \ldots$$

Sometimes the sequence is denoted by $\{s_n\}$, but more often it is given by simply listing the terms in order,

$$s_1, s_2, s_3, \ldots, s_n, \ldots$$

(This is the source of the idea of a sequence as a listing of numbers in order.)

If a number *L* exists with the property that s_n can be forced arbitrarily close to *L* by making *n* sufficiently large, we call *L* the *limit of the sequence* and write

$$\lim_{n \to \infty} s_n = L$$

Note that since $s_n = f(n)$, the only difference between the preceding definition of limit and our definition of $\lim_{x \to \infty} f(x) = L$ in Chapter 11 is that the variable *n* is restricted to the domain of positive integers. Thus there is not much to say about the limit of a sequence that would not be repetitive.

We trust your intuition. Perhaps one thing, however, is worth pointing out. In finding limits of sequences, we will feel free to use such devices as L'Hôpital's Rule, even though it is stated for functions with continuous domains. The limit of the sequence

$$\ln 1, \frac{\ln 2}{2}, \frac{\ln 3}{3}, \ldots, \frac{\ln n}{n}, \ldots$$

for example, is

$$\lim_{n \to \infty} \frac{\ln n}{n} = \lim_{n \to \infty} \frac{1/n}{1} = 0$$

The reason this works is that the limit of the function

$$f(x) = \frac{\ln x}{x} \qquad \text{(defined for all } x > 0)$$

is 0. Since the terms of the sequence are $f(1), f(2), f(3), \ldots$, they must approach the same limit.

□ **Example 1**   Find the limit of each of the following sequences (or determine that it does not exist).

(a) $1, 1/2, 1/4, 1/8, \ldots, 1/2^{n-1}, \ldots$

(b) $1, -1, 1, -1, \ldots, (-1)^{n-1}, \ldots$

(c) $\{n \sin (1/n)\}$

(d) $\ln 2, \ln (3/2), \ln (4/3), \ldots, \ln (n + 1)/n, \ldots$

(e) $3/2, (3/2)^2, (3/2)^3, \ldots, (3/2)^n, \ldots$

(f) $\{(-1)^{n-1}/3^{n-1}\}$

Solution:

(a) Since 2^{n-1} increases without bound as $n \to \infty$, the limit of the sequence is 0.

(b) The terms alternate between 1 and -1, never approaching anything definite. The sequence does not have a limit.

(c) The limit of the typical term is indeterminate of the form $\infty \cdot 0$. By rewriting it as

$$\lim_{n \to \infty} \frac{\sin (1/n)}{1/n}$$

we change it to the form $0/0$ and can use L'Hôpital's Rule. It is easier, however, to let $t = 1/n$. Then

$$\lim_{n \to \infty} \frac{\sin (1/n)}{1/n} = \lim_{t \to 0} \frac{\sin t}{t} = 1 \qquad \text{(from Section 2.3)}$$

(d) The limit of the sequence is

$$\lim_{n \to \infty} \ln \frac{n + 1}{n} = \ln \left(\lim_{n \to \infty} \frac{n + 1}{n} \right) = \ln 1 = 0 \qquad \text{(Composite Function Theorem, Section 11.1)}$$

(e) Perhaps you can see by inspection that $(\frac{3}{2})^n \to \infty$ as $n \to \infty$. To make sure, however, let $y = (\frac{3}{2})^n$, from which

$$\ln y = \ln (\tfrac{3}{2})^n = n \ln (\tfrac{3}{2})$$

Since $\ln (\frac{3}{2})$ is positive, it is clear that

$$\lim_{n \to \infty} \ln y = \lim_{n \to \infty} n \ln (\tfrac{3}{2}) = \infty$$

Hence

$$\lim_{n \to \infty} y = \lim_{n \to \infty} (\tfrac{3}{2})^n = \infty$$

(f) The terms of this sequence are

$$1, \ -\frac{1}{3}, \ \frac{1}{3^2}, \ -\frac{1}{3^3}, \ \ldots$$

which alternate in sign while approaching 0 in value. Hence the limit of the sequence is 0. $\square$

Usually the limit of a sequence (of the kind considered in this book) can be found without resorting to the definition of limit. There is one important type of sequence, however, whose limit is not obvious.

$\square$ **Example 2**   Find the limit of the sequence $1, x, x^2, x^3, \ldots$, where x is a number with the property that $|x| < 1$.

Solution: Example 1(a) (the special case corresponding to $x = \frac{1}{2}$) suggests that the answer is 0. It is legitimate to doubt it, however, in cases like $x = 0.9$. We need the definition of limit to settle the question. The typical term of our sequence is x^{n-1}. By asserting that its limit is 0, we are saying that x^{n-1} can be forced arbitrarily close to 0 by making n sufficiently large. In other words, given a neighborhood of 0, say $N = (-\varepsilon, \varepsilon)$, we can name a number A such that $n > A \Rightarrow x^{n-1} \in N$. The following implications will tell us what A should be:

$$x^{n-1} \in N \Leftrightarrow -\varepsilon < x^{n-1} < \varepsilon \Leftrightarrow |x|^{n-1} < \varepsilon \Leftrightarrow \ln |x|^{n-1} < \ln \varepsilon$$

$$\Leftrightarrow (n-1) \ln |x| < \ln \varepsilon \Leftrightarrow n - 1 > \frac{\ln \varepsilon}{\ln |x|}$$

(The sense of the last inequality was reversed because $|x| < 1$ and hence $\ln |x|$ is negative. We also assume $x \neq 0$, since otherwise $\ln |x|$ is undefined. Note that when $x = 0$ the original sequence is $1, 0, 0, 0, \ldots$, with obvious limit 0.)

Now it is clear that we can force x^{n-1} into the given neighborhood N by choosing

$$n > \frac{\ln \varepsilon}{\ln |x|} + 1$$

Letting

$$A = \frac{\ln \varepsilon}{\ln |x|} + 1$$

we see that

$$n > A \Rightarrow x^{n-1} \in N$$

which is what we had to prove.   □

If $|x| < 1$, the sequence $1, x, x^2, x^3, \ldots$ has limit 0.

Figure 1
An increasing bounded sequence

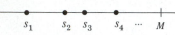

Our next theorem is more abstract. Suppose that the sequence s_1, s_2, s_3, s_4, . . . is *increasing* and has an *upper bound*. Geometrically, this means that the points s_1, s_2, s_3, s_4, . . . move from left to right on the number line (as shown in Figure 1) and that there is a number M beyond which they do not go. In other words,

$$s_1 < s_2 < s_3 < s_4 < \cdots$$

and $$s_n \leq M \text{ for all } n$$

We have shown only four points in Figure 1, but imagine plotting a hundred, or a thousand! Because they move from left to right, but never go beyond M, there is no room for them to do anything else but cluster toward a limit.

This theorem, that an increasing sequence with an upper bound has a limit, is not really obvious (despite the geometrical evidence). It is proved only by appealing to one of the basic axioms governing $\Re$, an axiom that we have chosen not to discuss. We have stated the axiom in the problem set, however, and have outlined a short argument that is not hard. You should have no trouble believing it if you believe the axiom on which it is based.

We have said that a sequence s_1, s_2, s_3, . . . has an *upper bound* if $s_n \leq M$ for all n. More generally, we call the sequence *bounded* if $|s_n| \leq M$ for all n. Since $|s_n| \leq M$ is equivalent to $-M \leq s_n \leq M$, a bounded sequence is one whose terms are boxed in between an upper bound M and a lower bound $-M$. In other words, the terms do not "go off to infinity" (in either direction).

It is not hard to show that a sequence with a limit must be bounded (see the problem set). The final version of our theorem about increasing sequences therefore reads as follows.

An increasing sequence has a (finite) limit if and only if it is bounded.

Problem Set 12.1

In each of the following, determine whether the sequence has a limit. Find the limit when it exists.

1. $1, 2/3, (2/3)^2, \ldots, (2/3)^{n-1}, \ldots$

2. $\cos \pi, \cos 2\pi, \cos 3\pi, \ldots, \cos n\pi, \ldots$

3. $2.9, 2.99, 2.999, 2.9999, \ldots$ (The nth term has n nines following the decimal point.)

4. $\{(1 + 1/n)^n\}$ *Hint:* Recall the formula for e in Section 8.4.

5. $-2, 4, -8, 16, \ldots, (-2)^n, \ldots$

6. $\{1 + (-1)^n\}$

7. $\cos 1, \cos \dfrac{1}{2}, \cos \dfrac{1}{3}, \ldots, \cos \dfrac{1}{n}, \ldots$

8. $1.01, 1.001, 1.0001, 1.00001, \ldots$ (The nth term has n zeros following the decimal point.)

9. $\dfrac{1}{e}, \dfrac{2}{e^2}, \dfrac{3}{e^3}, \ldots, \dfrac{n}{e^n}, \ldots$

10. $\{(0.98)^n\}$

11. $\{(1.02)^n\}$

12. $\dfrac{\sin 1}{1}, \dfrac{\sin 2}{2}, \dfrac{\sin 3}{3}, \ldots, \dfrac{\sin n}{n}, \ldots$

13. $s_n = n \sin \dfrac{n\pi}{2}$

14. $s_n = 1 + (-2/3)^n$

15. $s_n = \dfrac{n-1}{n+1}$

16. $s_n = e^n/n^3$

17. A machine purchased for \$1000 depreciates 10% each year.

(a) If s_n is the value of the machine at the end of n years, find a formula for s_n.

(b) What does the value of the machine approach as time goes on?

18. Prove that a sequence with a limit must be bounded, as follows.

(a) If L is the limit, explain why all but a finite number of terms of the sequence belong to the neighborhood $N = (L - 1, L + 1)$.

(b) Why does it follow that the sequence is bounded?

19. Does a bounded sequence necessarily have a limit? Explain.

20. To prove that a bounded increasing sequence $s_1, s_2, s_3, \ldots$ has a limit, we appeal to a basic property of $\mathcal{R}$ called the *Axiom of Completeness*, which says that any set of real numbers with an upper bound has a *least* upper bound. Let L be the least upper bound of the given sequence. Prove that $\lim_{n\to\infty} s_n = L$ as follows.

(a) Given $\varepsilon > 0$, explain why there is a member of the sequence (say s_N) that is larger than $L - \varepsilon$. Why is every term after s_N also larger than $L - \varepsilon$? Why is every term of the sequence less than $L + \varepsilon$?

(b) From part (a) we have

$$n > N \Rightarrow L - \varepsilon < s_n < L + \varepsilon$$

Why does this prove that $\lim_{n\to\infty} s_n = L$?

21. The Axiom of Completeness (Problem 20) also says that any set of real numbers with a lower bound has a *greatest* lower bound. Use that fact to prove that a bounded decreasing sequence has a limit.

12.2
INFINITE SERIES

Suppose that a ball falling from a given height above the floor always bounces back halfway. If it is dropped from a height of 1 ft, how far does it travel?

This question (or its equivalent) has been a historical source of controversy. If the ball bounces infinitely many times (as it must if we accept the hypothesis that it always rebounds halfway), how can it travel a finite distance in a finite time? Yet anybody who has watched a ping pong ball knows that it does.

Let's analyze the problem using what we know about limits. To start with, the ball falls 1 ft. It rebounds $\frac{1}{2}$ ft and falls $\frac{1}{2}$ ft, rebounds $\frac{1}{4}$ ft and falls $\frac{1}{4}$ ft, and so on. Apparently the distance traveled (if such a number exists) is the "sum"

$$s = 1 + (\tfrac{1}{2} + \tfrac{1}{2}) + (\tfrac{1}{4} + \tfrac{1}{4}) + (\tfrac{1}{8} + \tfrac{1}{8}) + (\tfrac{1}{16} + \tfrac{1}{16}) + \cdots$$
$$= 1 + 1 + \tfrac{1}{2} + \tfrac{1}{4} + \tfrac{1}{8} + \cdots$$
$$= 1 + 1 + \tfrac{1}{2} + (\tfrac{1}{2})^2 + (\tfrac{1}{2})^3 + \cdots$$

To make sense of this, cut it off at the term $(1/2)^{n-1}$. (We choose $n - 1$ instead of n for convenience in the formulas to follow.) Also ignore the first term (for the moment). In other words, look at the finite sum

$$1 + \tfrac{1}{2} + (\tfrac{1}{2})^2 + \cdots + (\tfrac{1}{2})^{n-1} \qquad \text{(which has } n \text{ terms)}$$

You may recognize this as a *geometric progression* of the type

$$S_n = 1 + x + x^2 + \cdots + x^{n-1}$$

A formula for S_n (derived in algebra courses and also Problem 35, Section 11.5) is

$$S_n = \frac{1 - x^n}{1 - x} \qquad (x \neq 1)$$

so our sum (obtained by taking $x = \frac{1}{2}$) is

$$1 + \tfrac{1}{2} + (\tfrac{1}{2})^2 + \cdots + (\tfrac{1}{2})^{n-1} = \frac{1 - (\tfrac{1}{2})^n}{1 - \tfrac{1}{2}} = 2[1 - (\tfrac{1}{2})^n]$$

Thus in hitting the floor $n + 1$ times ($n = 1, 2, 3, \ldots$) the ball has traveled a distance

$$s_n = 1 + [1 + \tfrac{1}{2} + (\tfrac{1}{2})^2 + \cdots (\tfrac{1}{2})^{n-1}] = 1 + 2[1 - (\tfrac{1}{2})^n]$$

Any calculus student knows what to do next! Let $n \to \infty$ and find the limit (if it exists). Since $(\frac{1}{2})^n$ approaches 0 as $n \to \infty$, we find

$$s = \lim_{n \to \infty} s_n = 1 + 2 = 3$$

Hence the ball travels a total distance of 3 ft.

Remark   Since the ball bounces infinitely often, the preceding statement should be slightly qualified to make sense. Actually there is no total distance, but rather a least upper bound (namely 3) of a sequence of "partial distances." When we ask how far the ball travels, we are setting up a paradox that rightly baffles people. (The ancients, after all, were no less intelligent than we are. If a question bothered them, we should not assume that they were stupid, or even ignorant. Instead we should discover what makes the question deep.) The true solution of the paradox is to ask a different question! Does the ball travel infinitely far? If not, what is the least upper bound of the distance traveled? The answers are "no" and "3 ft," respectively.

Of course it is easier to shorten this by saying that the ball travels 3 ft. It does this, incidentally, in a finite time (despite the fact that it bounces infinitely often). In the problem set we ask you to show that the time is a little less than 1.5 seconds.

This example suggests that an *infinite series* $a_1 + a_2 + a_3 + \cdots$ may have a sensible interpretation. Rather than taking it at face value (it *looks* like an addition of infinitely many terms, which is impossible), we proceed as in the case of the bouncing ball. Cut it off at (say) the term a_n, computing the finite sum

$$S_n = a_1 + a_2 + \cdots + a_n$$

Then find the limit of this (if it exists) as $n \to \infty$. The result is a *number,*

$$S = \lim_{n \to \infty} S_n$$

which we define to be the value (or "sum," if you will forgive the expression) of the series. The series, after all, is meaningless until a definition is given. Having decided to assign this meaning to it, we can write

$$S = a_1 + a_2 + a_3 + \cdots$$

with a straight face.

Let $a_1, a_2, a_3, \ldots$ be a sequence of real numbers. The expression

$$a_1 + a_2 + a_3 + \cdots = \sum_{k=1}^{\infty} a_k$$

is called the **infinite series** associated with the sequence. Its **partial sums** are the numbers

$$
\begin{aligned}
S_1 &= a_1 \\
S_2 &= a_1 + a_2 \\
S_3 &= a_1 + a_2 + a_3 \\
&\;\;\vdots \\
S_n &= a_1 + a_2 + \cdots + a_n = \sum_{k=1}^{n} a_k \\
&\;\;\vdots
\end{aligned}
$$

If the sequence of partial sums has a limit, that is, if

$$S = \lim_{n \to \infty} S_n$$

exists, we call S the **sum** of the series and write

$$S = \sum_{k=1}^{\infty} a_k = a_1 + a_2 + a_3 + \cdots$$

In that case the series is said to **converge** to S; otherwise it is said to **diverge.**

Remarks

1. There is a distinction between a *sequence* and a *series,* the former being no more than a listing of numbers in order, the latter being an indicated sum. Thus the sequence

$$1, \tfrac{1}{2}, \tfrac{1}{4}, \tfrac{1}{8}, \ldots, (\tfrac{1}{2})^{n-1}, \ldots$$

is one thing (having nothing to do with addition); the series

$$1 + \tfrac{1}{2} + \tfrac{1}{4} + \tfrac{1}{8} + \cdots + (\tfrac{1}{2})^{n-1} + \cdots$$

is another.

2. There are two sequences involved in every series. The first is the sequence of *terms*, $a_1, a_2, a_3, \ldots$. The second is the sequence of *partial sums*, $S_1, S_2, S_3, \ldots$. The sum of the series (when it exists) is not the limit of $a_1, a_2, a_3, \ldots$, but the limit of $S_1, S_2, S_3, \ldots$. For example, the sum of the series

$$1 + \frac{1}{2} + \frac{1}{4} + \frac{1}{8} + \cdots = \sum_{k=1}^{\infty} \left(\frac{1}{2}\right)^{k-1}$$

is not $\lim_{k \to \infty} (\tfrac{1}{2})^{k-1} = 0$ but rather

$$S = \lim_{n \to \infty} S_n = \lim_{n \to \infty} 2[1 - (\tfrac{1}{2})^n] = 2 \qquad \text{(as we saw earlier)}$$

3. An infinite series need not converge to be a conceptual reality. The series

$$1 + 3 + 5 + 7 + \cdots = \sum_{k=1}^{\infty} (2k - 1)$$

for example, has a definite sequence of partial sums, namely

$$S_1 = 1$$
$$S_2 = 1 + 3 = 4$$
$$S_3 = 1 + 3 + 5 = 9$$

$$S_n = 1 + 3 + 5 + \cdots + (2n - 1) = n^2$$

Since $\lim_{n \to \infty} S_n = \lim_{n \to \infty} n^2 = \infty$, the series does not converge (and does not have a sum). But it is nonetheless a mathematical entity that we can discuss. A source of legitimate confusion in connection with this remark is that in our definition we wrote $S = a_1 + a_2 + a_3 + \cdots$, thus identifying the sum of a series with the series itself. Strictly speaking, this should not be done; nevertheless everybody does it. Note that it makes sense only in the case of a convergent series.

4. We will often designate a series by the imprecise symbol $\sum a_k$, particularly when we are discussing its convergence without reference to its sum. One reason for this (aside from the obvious saving of time) is that the index k does not always start with 1. For example, the series

$$\tfrac{1}{4} + \tfrac{1}{8} + \tfrac{1}{16} + \cdots$$

may be written

$$\sum_{k=2}^{\infty} \left(\frac{1}{2}\right)^k$$

This is certainly different from the series

$$\sum_{k=0}^{\infty} \left(\frac{1}{2}\right)^k = 1 + \frac{1}{2} + \frac{1}{4} + \frac{1}{8} + \cdots$$

but if we are merely interested in convergence, it is worth observing that both series must do the same thing. (They either both converge or both diverge.) The reason ought to be apparent. The only difference between them is the sum $1 + \frac{1}{2}$ at the beginning; it is the behavior of partial sums S_n for large n that determines convergence or divergence. The symbol $\sum (\frac{1}{2})^k$ is ambiguous if we want to find the sum, but sufficiently precise if we are simply talking about convergence.

With these preliminaries out of the way, let's look at some examples and a theorem.

□ **Example 1**   Find the sum of the series

$$\sum_{k=1}^{\infty} \frac{1}{k(k+1)} = \frac{1}{1 \cdot 2} + \frac{1}{2 \cdot 3} + \frac{1}{3 \cdot 4} + \frac{1}{4 \cdot 5} + \cdots$$

Solution: By brute force we can discover a pattern in the sequence of partial sums:

$$S_1 = \frac{1}{1 \cdot 2} = \frac{1}{2}$$

$$S_2 = S_1 + \frac{1}{2 \cdot 3} = \frac{1}{2} + \frac{1}{2 \cdot 3} = \frac{2}{3}$$

$$S_3 = S_2 + \frac{1}{3 \cdot 4} = \frac{2}{3} + \frac{1}{3 \cdot 4} = \frac{3}{4}$$

$$S_4 = S_3 + \frac{1}{4 \cdot 5} = \frac{3}{4} + \frac{1}{4 \cdot 5} = \frac{4}{5}$$

A reasonable guess for the general case is $S_n = n/(n+1)$. A less gruesome approach is to decompose the typical term of the series by writing

$$a_k = \frac{1}{k(k+1)} = \frac{1}{k} - \frac{1}{k+1}$$

The nth partial sum then telescopes:

$$S_n = a_1 + a_2 + \cdots + a_n$$
$$= \left(1 - \frac{1}{2}\right) + \left(\frac{1}{2} - \frac{1}{3}\right) + \left(\frac{1}{3} - \frac{1}{4}\right) + \cdots + \left(\frac{1}{n} - \frac{1}{n+1}\right)$$
$$= 1 - \frac{1}{n+1} = \frac{n}{n+1}$$

In any case, the sum of the series is $\lim_{n \to \infty} S_n = \lim n/(n+1) = 1$.   □

Such good fortune (discovering a pattern in the sequence of partial sums) is not to be expected in very many cases. That is why we should treasure the results we do get, filing them away for future reference.

You should notice how we computed partial sums in Example 1. Each sum is the preceding sum plus the next term of the series, that is,

$$S_n = S_{n-1} + a_n$$

This observation leads to a simple (but important) theorem about convergence. Since the typical term of the series is $a_n = S_n - S_{n-1}$, its limit (as $n \to \infty$) must be zero if the series converges:

$$\lim_{n \to \infty} a_n = \lim_{n \to \infty} (S_n - S_{n-1}) = \lim_{n \to \infty} S_n - \lim_{n \to \infty} S_{n-1} = S - S = 0$$

(where S is the sum of the series).

Necessary Condition for Convergence (Test for Divergence)

If $\Sigma\, a_n$ converges, then $\lim a_n = 0$. Equivalently, if $\lim a_n \neq 0$, then $\Sigma\, a_n$ diverges.

Warning: The converse of this theorem is false. If $\lim a_n = 0$, it does not follow that $\Sigma\, a_n$ converges. (See Example 4.)

□ **Example 2**   The series

$$1 - 1 + 1 - 1 + \cdots = \sum_{n=1}^{\infty} (-1)^{n-1}$$

diverges because its typical term (1 or -1) does not approach 0. One might think that the series converges because each 1 is cancelled by the next -1 and hence the partial sums do not become infinite. While it is true that they are bounded, they still do not have a limit:

$$S_1 = 1$$
$$S_2 = 1 - 1 = 0$$
$$S_3 = 1 - 1 + 1 = 1$$
$$S_4 = 1 - 1 + 1 - 1 = 0$$
$$\vdots$$

In other words, the sequence of partial sums is $1, 0, 1, 0, \ldots$, which does not have a limit.   □

□ **Example 3**   The series

$$\sum_{n=1}^{\infty} n \sin \frac{1}{n} = \sin 1 + 2 \sin \frac{1}{2} + 3 \sin \frac{1}{3} + \cdots$$

diverges because its typical term does not approach zero:

$$\lim_{n \to \infty} n \sin \frac{1}{n} = \lim_{n \to \infty} \frac{\sin (1/n)}{1/n} = \lim_{t \downarrow 0} \frac{\sin t}{t} \qquad (t = 1/n)$$

$$= 1 \neq 0$$

□

◻ **Example 4**   The typical term of the series

$$\sum_{n=1}^{\infty} \ln \frac{n+1}{n} = \ln 2 + \ln \frac{3}{2} + \ln \frac{4}{3} + \cdots$$

is $a_n = \ln (n + 1)/n$, which approaches 0 as $n \to \infty$. What conclusion can be drawn?

Solution: None! Our necessary condition for convergence is just that; it is not a sufficient condition. Only when the typical term of the series does not approach zero can we use this theorem to decide anything. (The series diverges in that case.) When the typical term approaches zero, the question of convergence or divergence is still open, and must be decided in some other way. In the present case look at the nth partial sum:

$$S_n = a_1 + a_2 + \cdots + a_n = \ln \frac{2}{1} + \ln \frac{3}{2} + \ln \frac{4}{3} + \cdots + \ln \frac{n+1}{n}$$

$$= (\ln 2 - \ln 1) + (\ln 3 - \ln 2) + (\ln 4 - \ln 3) + \cdots$$
$$+ [\ln (n + 1) - \ln n]$$

$$= \ln (n + 1) \quad \text{(after telescoping)}$$

Since $\lim_{n\to\infty} S_n = \lim_{n\to\infty} \ln (n + 1) = \infty$, the series diverges. ◻

◻ **Example 5**   The series

$$1 + x + x^2 + \cdots = \sum_{k=0}^{\infty} x^k$$

is called a *geometric series*. Its nth partial sum is

$$S_n = 1 + x + x^2 + \cdots + x^{n-1} = \frac{1 - x^n}{1 - x} \quad (x \neq 1)$$

and its sum (if it exists) is $S = \lim_{n\to\infty} S_n$. We analyze the question of convergence in cases:

(a) If $|x| < 1$, then $x^n \to 0$ as $n \to \infty$ (Section 12.1). Therefore the series converges to

$$S = \lim_{n\to\infty} \frac{1 - x^n}{1 - x} = \frac{1}{1 - x}$$

(b) If $x > 1$, then $x^n \to \infty$ as $n \to \infty$ and $\lim S_n$ does not exist. Hence the series diverges.

(c) If $x < -1$, then x^n bounces back and forth between positive and negative values, its absolute value increasing without bound. Again $\lim S_n$ does not exist; the series diverges.

(d) If $x = -1$, the series is $1 - 1 + 1 - 1 + \cdots$, which diverges. (See Example 2.)

(e) If $x = 1$, the series is $1 + 1 + 1 + \cdots$, with partial sum $S_n = n$. Hence $\lim S_n$ does not exist and the series diverges. ◻

We summarize these results as follows.

The geometric series

$$1 + x + x^2 + \cdots = \sum_{k=0}^{\infty} x^k$$

converges to $\qquad \dfrac{1}{1-x} \qquad$ if $|x| < 1$

Otherwise it diverges.

□ **Example 6**   Nonterminating decimals are *defined* by infinite series. An illustration is $0.4999\cdots = 0.4 + 0.09 + 0.009 + 0.0009 + \cdots$. What rational number is represented by this decimal?

Solution: The given series is geometric in form after the first term. For we can rewrite it as

$$0.4 + 0.09(1 + 0.1 + 0.01 + \cdots) = \frac{4}{10} + \frac{9}{100}\left(1 + \frac{1}{10} + \frac{1}{10^2} + \cdots\right)$$

$$= \frac{4}{10} + \frac{9}{100} \cdot \frac{1}{1 - 1/10}$$

$$= \frac{4}{10} + \frac{9}{100} \cdot \frac{10}{9} = \frac{4}{10} + \frac{1}{10} = \frac{1}{2}$$

Sometimes students argue with this result, not realizing that it is a matter of definition. Another way to look at it is to let $x = 0.4999\cdots$. Then

$$10x = 4.999\cdots$$
$$100x = 49.999\cdots$$

and hence (by subtraction) $90x = 45$ and $x = \frac{1}{2}$. Of course this argument (often given in algebra courses) presumes that nonterminating decimals are already understood.   □

It is too restrictive to have to start every geometric series with 1 (as in Example 6). Hence we broaden our formula:

The geometric series

$$a + ar + ar^2 + \cdots = \sum_{k=0}^{\infty} ar^k = a \sum_{k=0}^{\infty} r^k$$

converges to $\qquad \dfrac{a}{1-r} \qquad$ if $|r| < 1$

Otherwise it diverges.

This result is important to learn; geometric series (as a class) serve as a source of known results with which other series can be compared.

Problem Set 12.2

In each of the following, determine whether the series converges by examining its sequence of partial sums. If it converges, find its sum.

1. $\displaystyle\sum_{k=0}^{\infty} \cos k\pi$ **2.** $\displaystyle\sum_{k=1}^{\infty} \sin \frac{k\pi}{2}$

3. $\displaystyle\sum_{k=1}^{\infty} \ln \frac{1}{k}$ **4.** $\displaystyle\sum_{k=1}^{\infty} k$

5. $\displaystyle\sum_{k=1}^{\infty} \frac{1}{(k+1)(k+2)}$ *Hint:* Decompose the typical term into a sum of fractions.

6. $\displaystyle\sum_{k=1}^{\infty} \frac{1}{4k^2 - 1}$ **7.** $\displaystyle\sum_{k=1}^{\infty} \int_{k}^{k+1} \frac{dx}{x^2}$

In each of the following, find the sum of the series (or determine that it diverges).

8. $1 + \frac{1}{3} + \frac{1}{9} + \frac{1}{27} + \cdots$ **9.** $1 - \frac{1}{3} + \frac{1}{9} - \frac{1}{27} + \cdots$

10. $1 + 1.01 + (1.01)^2 + (1.01)^3 + \cdots$

11. $\frac{1}{2} + \frac{1}{4} + \frac{1}{8} + \frac{1}{16} + \cdots$

12. $1 - \frac{1}{2} + \frac{1}{4} - \frac{1}{8} + \cdots$

13. $2 + 1 + \frac{1}{2} + \frac{1}{4} + \cdots$

14. $\displaystyle\sum_{k=1}^{\infty} 10^{-k} = 0.1 + 0.01 + 0.001 + \cdots = 0.\overline{1}$

15. $\displaystyle\sum_{k=0}^{\infty} \frac{2^k}{3^k}$ **16.** $\displaystyle\sum_{k=1}^{\infty} e^{-k}$

17. $\displaystyle\sum_{k=0}^{\infty} 2^k$ **18.** $\displaystyle\sum_{k=0}^{\infty} \left(\frac{\pi}{3}\right)^k$

19. $\displaystyle\sum_{k=0}^{\infty} \left(\frac{3}{\pi}\right)^k$

20. $\displaystyle\sum_{k=0}^{\infty} \sin^{2k} x$ where $-\pi/2 < x < \pi/2$

21. $\displaystyle\sum_{k=1}^{\infty} \cos^{2k} x$ where $0 < x < \pi$

22. $\displaystyle\sum_{k=0}^{\infty} \left(\frac{1}{1+x^2}\right)^k$ where $x \neq 0$

23. $\frac{1}{2} + \frac{2}{3} + \frac{3}{4} + \frac{4}{5} + \cdots$

24. $1 + \frac{2}{3} + \frac{3}{5} + \frac{4}{7} + \frac{5}{9} + \cdots$

25. $\displaystyle\sum_{k=1}^{\infty} \frac{\sinh k}{k}$

26. $1 + \dfrac{1}{\sqrt{2}} + \dfrac{1}{\sqrt[3]{3}} + \dfrac{1}{\sqrt[4]{4}} + \cdots$

27. Show that the series

$$1 + \frac{1}{\sqrt{2}} + \frac{1}{\sqrt{3}} + \frac{1}{\sqrt{4}} + \cdots$$

diverges, as follows.

(a) Explain why each term of the nth partial sum of the series is $\geq 1/\sqrt{n}$ and hence $S_n \geq \sqrt{n}$.

(b) What is $\lim_{n\to\infty} S_n$?

28. Show that the *harmonic series*

$$1 + \frac{1}{2} + \frac{1}{3} + \frac{1}{4} + \cdots + \frac{1}{n} + \cdots$$

diverges, as follows.

(a) Explain why partial sums with indices that are powers of 2 satisfy the inequalities

$$S_1 = 1 = 1 + \tfrac{1}{2}(0)$$

$$S_2 = S_1 + \tfrac{1}{2} = 1 + \tfrac{1}{2}(1)$$

$$S_4 = S_2 + \tfrac{1}{3} + \tfrac{1}{4} > S_2 + \tfrac{2}{4} = 1 + \tfrac{1}{2}(2)$$

$$S_8 = S_4 + \tfrac{1}{5} + \tfrac{1}{6} + \tfrac{1}{7} + \tfrac{1}{8} > S_4 + \tfrac{4}{8} > 1 + \tfrac{1}{2}(3)$$

$$S_{16} = S_8 + \tfrac{1}{9} + \tfrac{1}{10} + \tfrac{1}{11} + \tfrac{1}{12} + \tfrac{1}{13} + \tfrac{1}{14} + \tfrac{1}{15} + \tfrac{1}{16} >$$
$$S_8 + \tfrac{8}{16} > 1 + \tfrac{1}{2}(4)$$

and (in general) $S_n > 1 + \tfrac{1}{2}r$ if $n = 2^r$.

(b) Why does it follow that $\lim_{n\to\infty} S_n = \infty$?

29. For what values of x does the series

$$\sum_{k=0}^{\infty} (x-2)^k = 1 + (x-2) + (x-2)^2 + \cdots$$

converge? What is its sum when it does?

Using the fact that

$$\frac{1}{1-x} = 1 + x + x^2 + \cdots \qquad (-1 < x < 1)$$

express each of the following functions as an infinite series. State the domain in each case.

30. $f(x) = \dfrac{1}{1 - x^2}$ **31.** $f(x) = \dfrac{1}{1 + x}$

32. $f(x) = \dfrac{1}{2 - x}$ **33.** $f(x) = \dfrac{x}{1 - x}$

34. Use a geometric series to find the rational number represented by $0.\overline{36}$.

35. It is not hard to see that every rational number is represented by a periodic decimal. (See Problem 51, Section 1.1.) The converse (that every periodic decimal represents a rational number) requires the theory of geometric series. Outline the argument in general terms. (The details involve somewhat messy notation.)

36. A ball that rebounds 60% of the distance fallen is dropped from a height of 5 ft. How far does it travel?

37. Each time a ball falls h ft it rebounds rh ft, where $0 < r < 1$. If it is dropped from a height of a ft, how far does it travel?

38. The ball described at the beginning of this section will bounce infinitely often, but will it bounce forever? If not, how long will it bounce? *Hint:* The time t required for an object to rise (or fall) a distance s is $t = \sqrt{2s/g}$ (because $s = \frac{1}{2}gt^2$). Take $g = 32$.

39. Some bacteria reproduce themselves, others do not. If 80% of a certain culture is always reproductive, and there are no deaths, show that the original population is eventually multiplied by 5.

40. Suppose that 40 cents of every dollar spent in a small town winds up deposited in the local bank, which lends it immediately to others to spend. How much in total deposits is generated by a visiting professor who spends $100 in one night on the town?

41. Suppose that 75 cents of every dollar spent in the U.S. is spent again in the U.S. If the federal government pumps an extra billion into the economy, how much spending occurs as a result?

42. Suppose that p cents of every dollar loaned by a bank ends up deposited in the bank ($p < 100$). If it starts with D dollars in deposits, and is required by law to keep at least d dollars in deposits ($D > d$), how much money can it ultimately loan out? *Hint:* The bank loans $D - d$ dollars immediately, and continues to loan as much as the law allows. Assume that no other deposits are made.

43. A humane version of Russian roulette involves two people who alternate in rolling a six-sided die until one of them loses by rolling a 1. The probability that you win if you go first is

$$\frac{5}{6^2} + \frac{5^3}{6^4} + \frac{5^5}{6^6} + \cdots$$

(a formula that you might enjoy figuring out if you know a little about probability). What is this probability?

44. Suppose that each term of the series $\sum a_k$ is multiplied by the constant $c \neq 0$ to obtain the series $\sum ca_k$. Explain why this does not affect convergence or divergence, that is,

$$\sum a_k \text{ converges} \Leftrightarrow \sum ca_k \text{ converges}$$

Show that $\sum ca_k = c \sum a_k$ when $\sum a_k$ converges.

45. Show that if $\sum a_k$ and $\sum b_k$ converge, so does $\sum (a_k + b_k)$, and

$$\sum (a_k + b_k) = \sum a_k + \sum b_k$$

46. Give an example showing that convergence of $\sum (a_k + b_k)$ does not imply convergence of $\sum a_k$ and $\sum b_k$.

47. The series $1 - 1 + 1 - 1 + \cdots$ diverges, as we have seen. Explain why the series

$$(1 - 1) + (1 - 1) + (1 - 1) + \cdots$$

converges. (Thus the associative law of ordinary algebra does not apply to series.)

48. Explain why convergence of the series

$$\sum_{k=1}^{\infty} (a_{k+1} - a_k)$$

is equivalent to the existence of the limit of the sequence $a_1, a_2, a_3, \ldots.$

49. Show that if $\sum a_k$ converges, then $\sum \cos a_k$ diverges.

12.3
THE INTEGRAL TEST

Finding the sum of an infinite series is often difficult, despite the impression that may have been created in the last section (where our success in computing sums was due to the fact that most of the series considered were geometric). The rest of this chapter is devoted primarily to the more fundamental question of whether a given series converges at all. This is critical both for theoretical reasons and because it is clearly desirable to establish convergence before spending much time looking for a sum that may not exist. A computer can be programmed to approximate the sum of a convergent series, but it would be embarrassing (and expensive) to run a program for a series that turns out to diverge!

In this section (and the next) we restrict our attention to **positive series,** meaning series in which each term is positive. Of course in practice it does not matter if a finite number of terms are not positive, as in the "ultimately positive" series

$$\sum_{n=1}^{\infty} \frac{n-2}{n^2-10} = \frac{1}{9} + 0 - 1 + \frac{1}{3} + \frac{1}{5} + \frac{2}{13} + \cdots$$

For we can discard the first three terms (which are mixed) without affecting the convergence or divergence of the series.

We also assume that the typical term of our series approaches zero (since otherwise the series diverges, as we saw in Section 12.2). In most cases of practical interest the terms not only approach zero but are monotonically decreasing, that is, each term is smaller than the preceding one. When that happens we can develop an intuitively appealing test for convergence based on the idea of area under a curve.

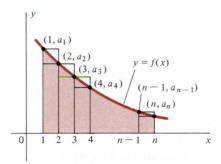

Figure 1
The Integral Test

Let $\sum a_n$ be the series in question, and let f be a continuous and decreasing function with the property that $f(n) = a_n$, $n = 1, 2, 3, \ldots$. There are many such functions (one of which is shown in Figure 1); in practice we usually choose the one suggested by the formula for a_n. Thus if the series is $\sum 1/n$, we let $f(x) = 1/x$.

It is clear from Figure 1 that the area under the graph of f from 1 to n is underestimated by the rectangles with altitudes $a_2, a_3, \ldots, a_n$ and overestimated by the ones with altitudes $a_1, a_2, \ldots, a_{n-1}$. Since the base of each rectangle is of length 1, we have

$$\sum_{k=2}^{n} a_k \leq \int_1^n f(x)\, dx \leq \sum_{k=1}^{n-1} a_k$$

Neither of these sums is S_n, but a slight reworking of the inequalities will box it in. Adding a_1 to each side of the first one yields

$$\sum_{k=1}^{n} a_k \leq a_1 + \int_1^n f(x)\, dx$$

whereas the second one implies that

$$\int_1^n f(x)\, dx \leq \sum_{k=1}^{n} a_k \qquad \left(\text{because } \sum_{k=1}^{n-1} a_k \leq \sum_{k=1}^{n} a_k\right)$$

Combining the results, we have

$$\int_1^n f(x)\,dx \le S_n \le a_1 + \int_1^n f(x)\,dx, \quad n = 1, 2, 3, \ldots$$

These inequalities strongly suggest a connection between the series $\Sigma\, a_n$ and the improper integral

$$\int_1^\infty f(x)\,dx$$

To be precise, observe that since the terms of $\Sigma\, a_n$ are positive, its partial sums are increasing:

$$\begin{aligned}
S_1 &= a_1 \\
S_2 &= S_1 + a_2 > S_1 \\
S_3 &= S_2 + a_3 > S_2 \\
&\ \ \vdots \\
S_n &= S_{n-1} + a_n > S_{n-1} \\
&\ \ \vdots
\end{aligned}$$

In such circumstances there are only two possibilities.

1. The sequence $S_1, S_2, S_3, \ldots$ increases without bound and the series $\Sigma\, a_n$ diverges. Then the inequality

$$S_n \le a_1 + \int_1^n f(x)\,dx$$

implies that

$$\lim_{n\to\infty} \int_1^n f(x)\,dx = \infty$$

that is, $\int_1^\infty f(x)\,dx$ diverges.

2. The sequence $S_1, S_2, S_3, \ldots$ is bounded. Since it is increasing, we know from Section 12.1 that it has a limit and hence the series $\Sigma\, a_n$ converges. The inequality

$$\int_1^n f(x)\,dx \le S_n$$

implies that the sequence of numbers

$$I_n = \int_1^n f(x)\,dx \qquad n = 1, 2, 3, \ldots$$

is bounded. Since it is also increasing, it has a limit. (See Section 12.1.) Hence $\int_1^\infty f(x)\,dx$ converges.

Thus we have proved the following theorem.

The Integral Test

Suppose that f is a continuous function that is positive and decreasing for $x \geq 1$, and $f(n) = a_n$, $n = 1, 2, 3, \ldots$. Then

$$\sum_{n=1}^{\infty} a_n \text{ converges} \iff \int_{1}^{\infty} f(x)\, dx \text{ converges}$$

In applications of this test neither the series nor the integral has to start with 1. Students sometimes think that the integral is being used to approximate the sum of the series, but notice that the test does not read that way. All it says is that convergence of one implies convergence of the other. Leaving off N terms of the series (or integrating from $N + 1$ to ∞) does not affect convergence.

□ **Example 1**   In Problem 28 of the last section we asked you to prove that the **harmonic series**

$$\sum_{n=1}^{\infty} \frac{1}{n} = 1 + \frac{1}{2} + \frac{1}{3} + \frac{1}{4} + \frac{1}{5} + \cdots$$

diverges. The suggested argument, while not difficult to follow, was hardly obvious! The Integral Test, however, is easy. Let $f(x) = 1/x$, a function with all the properties called for in the test. Since

$$\int_{1}^{\infty} \frac{dx}{x} = \lim_{b \to \infty} \int_{1}^{b} \frac{dx}{x} = \lim_{b \to \infty} \ln b = \infty$$

the series diverges.   □

□ **Example 2**   An important class of series called the **p-series** is

$$\sum_{n=1}^{\infty} \frac{1}{n^p} = 1 + \frac{1}{2^p} + \frac{1}{3^p} + \frac{1}{4^p} + \cdots$$

where p is a positive constant. (The harmonic series is the special case corresponding to $p = 1$.) To test for convergence, let

$$f(x) = \frac{1}{x^p} \qquad \text{(Why is } f \text{ a decreasing function for } x \geq 1?)$$

We leave it to you to show that

$$\int_{1}^{\infty} \frac{dx}{x^p}$$

converges if $p > 1$ and diverges if $p \leq 1$. (Also see Problem 23, Section 11.4.) Assuming this to be a known result, we get a quick answer to our problem:

The *p*-series

$$\sum_{n=1}^{\infty} \frac{1}{n^p} = 1 + \frac{1}{2^p} + \frac{1}{3^p} + \frac{1}{4^p} + \cdots$$

converges if $p > 1$ and diverges if $p \leq 1$.

For example, the series

$$\sum_{n=1}^{\infty} \frac{1}{n^{1/2}} = 1 + \frac{1}{\sqrt{2}} + \frac{1}{\sqrt{3}} + \frac{1}{\sqrt{4}} + \cdots$$

(which we brought up in Problem 27 of the last section) diverges because $p = \frac{1}{2} < 1$. On the other hand, the series

$$\sum_{n=1}^{\infty} \frac{1}{n^2} = 1 + \frac{1}{2^2} + \frac{1}{3^2} + \frac{1}{4^2} + \cdots$$

converges because $p = 2 > 1$. $\qquad\qquad\qquad\qquad\qquad\qquad\qquad\square$

The Integral Test is evidently a powerful tool when it applies. Its usefulness should not be overrated, however; the convergence or divergence of an improper integral is often tough to decide.

The argument that led to the Integral Test can be generalized to provide a clever estimate of the sum of a positive series. Any partial sum is of course an approximation, that is, $S \approx S_N$, where S is the sum of the series and N is the number of terms in the partial sum. More precisely,

$$S = S_N + \sum_{n=N+1}^{\infty} a_n$$

that is, the sum of a convergent series is any partial sum plus the "tail end" of the series. The idea is to estimate this remainder by

$$\int_{N+1}^{\infty} f(x)\, dx$$

where f is a function with the properties described in the Integral Test. (In other words replace addition by integration after the Nth term.) It is reasonable to expect that the resulting approximation,

---

$$S \approx S_N + \int_{N+1}^{\infty} f(x)\, dx$$

---

is more accurate than $S \approx S_N$. The error, in fact, is no larger than the next term of the series:

$$0 \leq S - \left(S_N + \int_{N+1}^{\infty} f(x)\, dx \right) \leq a_{N+1}$$

To see where this comes from, replace Figure 1 by a graph that runs from $N + 1$ to $N + n$ (instead of from 1 to n). The same argument as before leads to the inequalities

$$\int_{N+1}^{N+n} f(x)\, dx \leq S_{N+n} - S_N \leq a_{N+1} + \int_{N+1}^{N+n} f(x)\, dx$$

Letting $n \to \infty$, we obtain

$$\int_{N+1}^{\infty} f(x)\, dx \leq S - S_N \leq a_{N+1} + \int_{N+1}^{\infty} f(x)\, dx$$

from which the result follows by subtracting the integral from each side.

□ **Example 3**   How large should N be in order to compute

$$S = \sum_{n=1}^{\infty} e^{-n}$$

correct to five decimal places?

Solution: Let $f(x) = e^{-x}$ (which decreases for all x). Then

$$S \approx \sum_{n=1}^{N} e^{-n} + \int_{N+1}^{\infty} e^{-x}\, dx$$

the error being no larger than $a_{N+1} = e^{-(N+1)}$. To obtain five-place accuracy, we choose N satisfying

$$e^{-(N+1)} < \tfrac{1}{2} \times 10^{-5}$$
$$2 \times 10^{5} < e^{N+1}$$
$$\ln (2 \times 10^{5}) < N + 1$$
$$N > \ln (2 \times 10^{5}) - 1 \approx 11.2$$

Hence N should be at least 12. Our approximation is

$$S \approx \sum_{n=1}^{12} e^{-n} + \int_{13}^{\infty} e^{-x}\, dx = (e^{-1} + e^{-2} + \cdots + e^{-12}) + e^{-13}$$

$$= \frac{e^{-1}(1 - e^{-14})}{1 - e^{-1}}   \text{(using the formula for a geometric progression)}$$

$$\approx 0.58198   \text{(correct to five places)}$$

As a check, note that the series is geometric. Hence

$$S = e^{-1} + e^{-2} + e^{-3} + \cdots = \frac{e^{-1}}{1 - e^{-1}} = 0.58197670\cdots$$

Ordinarily, of course, the true sum of the series is not available as a check; this example is intended to give you a feeling of security! □

Problem Set 12.3

Decide which of the following series converge. (*Note:* Before applying the Integral Test, you must be sure that your function *f* is decreasing. In some cases that is most easily confirmed by using the derivative.)

1. $1 + \dfrac{1}{2\sqrt{2}} + \dfrac{1}{3\sqrt{3}} + \dfrac{1}{4\sqrt{4}} + \cdots$

2. $1 + \dfrac{1}{\sqrt[3]{2}} + \dfrac{1}{\sqrt[3]{3}} + \dfrac{1}{\sqrt[3]{4}} + \cdots$

3. $\displaystyle\sum \dfrac{1}{n^{1.01}}$

4. $\displaystyle\sum \dfrac{1}{n^{0.99}}$

5. $1 + \tfrac{1}{3} + \tfrac{1}{5} + \tfrac{1}{7} + \cdots$

6. $\tfrac{1}{2} + \tfrac{1}{4} + \tfrac{1}{6} + \tfrac{1}{8} + \cdots$

7. $\displaystyle\sum \dfrac{1}{(2n-1)^2}$

8. $\displaystyle\sum \dfrac{1}{(2n)^2}$

9. $\displaystyle\sum \dfrac{n}{n^2+1}$

10. $\displaystyle\sum \dfrac{1}{n^2+1}$

11. $\displaystyle\sum \dfrac{1}{\sqrt{n^2+1}}$

12. $\displaystyle\sum \dfrac{1}{\sqrt{n^2-1}}$

13. $\displaystyle\sum \dfrac{1}{n\sqrt{n^2-1}}$

14. $\displaystyle\sum \dfrac{1}{n\sqrt{n^2+1}}$

15. $\displaystyle\sum ne^{-n^2}$

16. $\displaystyle\sum ne^{-n}$

17. $\displaystyle\sum \dfrac{\ln n}{n}$

18. $\displaystyle\sum \dfrac{n}{\ln n}$

19. $\displaystyle\sum \dfrac{1}{n\ln n}$

20. $\displaystyle\sum \dfrac{\ln n}{n^2}$

21. $\displaystyle\sum \dfrac{1}{n(n+1)}$

22. $\displaystyle\sum \dfrac{n^2}{n^3+1}$

23. $\displaystyle\sum \operatorname{sech}^2 n$

24. Use the formula

$$S \approx S_N + \int_{N+1}^{\infty} f(x)\, dx$$

to find the sum of

$$\sum_{n=1}^{\infty} \frac{1}{n^2}$$

correct to one decimal place. How large should *N* be if you wanted five decimal places? (It is shown in the theory of Fourier series that $S = \pi^2/6 \approx 1.645$. Also see the Euler equation quoted at the beginning of this chapter.)

25. Use the formula in Problem 24 to find the sum of

$$\sum_{n=1}^{\infty} \frac{1}{n^2 + 1}$$

correct to two decimal places.

26. Use the formula in Problem 24 to find the sum of

$$\sum_{n=1}^{\infty} \operatorname{sech}^2 n$$

correct to four decimal places.

27. Use a picture similar to Figure 1 to show that for every positive integer *n*,

$$\ln(n+1) < 1 + \frac{1}{2} + \frac{1}{3} + \cdots + \frac{1}{n} < 1 + \ln n$$

28. In view of Problem 27, explain why it requires somewhere between 8103 and 22,027 terms of the harmonic series to obtain a sum larger than 10. How many terms are needed to obtain a sum larger than 100? (Use a calculator!)

29. After learning the Integral Test, students sometimes wonder why we don't write

$$\sum_{n=1}^{\infty} a_n \approx \int_{1}^{\infty} f(x)\, dx$$

where *f* is a suitable function with the property that $f(n) = a_n$. If we did use this approximation, how large might the error be?

30. Let S_n be the nth partial sum of the series

$$\sum_{n=1}^{\infty} a_n$$

and let

$$d_n = S_n - \int_1^n f(x)\, dx$$

where f is a continuous function that is positive and decreasing for $x \geq 1$, and $f(n) = a_n$, $n = 1, 2, 3, \ldots$. Explain why the sequence $d_1, d_2, d_3, \ldots$ has a limit if the given series is known to converge.

31. It is an interesting fact that the sequence $d_1, d_2, d_3, \ldots$ in Problem 30 has a limit even if the given series *diverges*. Prove this as follows.

(a) Explain why $d_n \geq 0$.

(b) Show that

$$d_n - d_{n+1} = \int_n^{n+1} f(x)\, dx - a_{n+1} \geq 0$$

(c) Why does it follow that $d_1, d_2, d_3, \ldots$ is a bounded decreasing sequence? (Thus it has a limit, by the same kind of reasoning we use to prove that a bounded increasing sequence has a limit. See Problems 20 and 21, Section 12.1.)

32. In the case of the harmonic series we may take $f(x) = 1/x$. What is d_n? (Its limit as $n \to \infty$ is known as **Euler's constant,** $\gamma = 0.57721\cdots$. Nobody knows whether it is rational or irrational; you can make a name for yourself by deciding the question!)

12.4
COMPARISON TESTS

The Integral Test for convergence of a positive series is difficult to apply in many cases. Even in those cases where it works, there are often easier ways. In this section we develop two simple tests that are based on the idea of "relative size" of an unknown series compared to a known one.

□ **Example 1**   We know that the series

$$\sum_{n=1}^{\infty} \frac{1}{n(n+1)} = \frac{1}{1 \cdot 2} + \frac{1}{2 \cdot 3} + \frac{1}{3 \cdot 4} + \cdots$$

converges to the sum 1. (See Example 1 in Section 12.2.) The series

$$\frac{1}{2^2} + \frac{1}{3^2} + \frac{1}{4^2} + \cdots$$

is term-by-term smaller, that is,

$$\frac{1}{2^2} < \frac{1}{1 \cdot 2}, \frac{1}{3^2} < \frac{1}{2 \cdot 3}, \frac{1}{4^2} < \frac{1}{3 \cdot 4}, \cdots$$

It seems apparent that the new series *converges by comparison* to the first one, and its sum is no greater than 1. To confirm this statement by the definition, we need only observe that the partial sums of the new series have two properties:

1. They are *increasing* (because the terms of the series are positive).
2. They are *bounded*. (The partial sums of the first series are bounded because they have a limit; each term of the new series is less than the corresponding term of the first series.)

Since a bounded increasing sequence has a limit, the convergence of the new series is established. It follows that the series

$$\sum_{n=1}^{\infty} \frac{1}{n^2} = 1 + \frac{1}{2^2} + \frac{1}{3^2} + \cdots$$

also converges, with a sum that is no larger than 2. (Why?) This is the series that we had in mind when we started this example, but we left off the first term in order to be correct in our inequalities. It is false that

$$\frac{1}{n^2} < \frac{1}{n(n+1)}$$

but a shift to the next term in $\sum 1/n^2$ makes this read

$$\frac{1}{(n+1)^2} < \frac{1}{n(n+1)}$$

which is true for all n.   □

Comparison Test

Let $\sum a_n$ and $\sum b_n$ be two series with the property that

$$0 \le a_n \le b_n \qquad \text{for large } n \text{ (all } n > N, \text{ where } N \text{ is fixed)}$$

If $\sum b_n$ converges, so does $\sum a_n$. (Equivalently, if $\sum a_n$ diverges, so does $\sum b_n$.)

We say "for large n" in this theorem (rather than "for all n") to allow for series in which the inequalities $0 \le a_n \le b_n$ do not hold in the early going. For example, the series $\sum 1/(n^2 - 5)$ converges by comparison to $\sum 2/n^2$ because

$$n^2 - 5 > \frac{n^2}{2} \qquad \text{for } n = 4, 5, 6, \ldots \qquad \text{(Why?)}$$

Hence for these values of n,

$$\frac{1}{n^2 - 5} < \frac{2}{n^2}$$

and the Comparison Test applies.

□ **Example 2**   Our comparison of

$$\sum a_n = \sum \frac{1}{n^2} \qquad \text{to} \qquad \sum b_n = \sum \frac{1}{n(n+1)}$$

required dropping the first term of $\sum a_n$. One wonders if such delicacy is really needed. The series are clearly alike, in the sense that

$$\frac{1}{n(n+1)} \approx \frac{1}{n^2} \qquad \text{for large } n$$

If one converges, it is reasonable to expect that the other does, too; fussing with inequalities seems a poor way to confirm it. Note that the ratio of a_n to b_n is

$$\frac{\dfrac{1}{n^2}}{\dfrac{1}{n(n+1)}} = \frac{n(n+1)}{n^2} = \frac{n+1}{n}$$

and

$$\lim_{n \to \infty} \frac{a_n}{b_n} = \lim_{n \to \infty} \frac{n+1}{n} = 1$$

The existence of this limit (its value need not be 1) is usually sufficient to guarantee that $\Sigma\, a_n$ and $\Sigma\, b_n$ behave alike. To see why, suppose that

$$\lim_{n \to \infty} \frac{a_n}{b_n} = L$$

Then for large n we have $a_n/b_n \approx L$, that is, $a_n \approx L b_n$. In other words, the typical term of $\Sigma\, a_n$ is (approximately) a constant multiple of the typical term of $\Sigma\, b_n$, which certainly suggests that the series behave alike. If $L > 0$, there is no doubt about it, for then we can also write

$$\frac{b_n}{a_n} \approx \frac{1}{L}, \text{ that is, } b_n \approx \frac{1}{L} a_n$$

and the constant multiple idea works both ways. In that case, convergence of either series implies convergence of the other. When $L = 0$ or $L = \infty$ the implication goes in only one direction, but even that fact is frequently useful. □

We summarize these observations in the following theorem, the proof of which is outlined in the problem set.

Limit Comparison Test

Suppose that $\Sigma\, a_n$ and $\Sigma\, b_n$ are positive series and

$$\lim_{n \to \infty} \frac{a_n}{b_n} = L$$

1. If $L > 0$, convergence of either series implies convergence of the other. (Equivalently, divergence of either series implies divergence of the other.) In other words, the series behave alike.
2. If $L = 0$, convergence of $\Sigma\, b_n$ implies convergence of $\Sigma\, a_n$. (Equivalently, divergence of $\Sigma\, a_n$ implies divergence of $\Sigma\, b_n$.)
3. If $L = \infty$, convergence of $\Sigma\, a_n$ implies convergence of $\Sigma\, b_n$. (Equivalently, divergence of $\Sigma\, b_n$ implies divergence of $\Sigma\, a_n$.)

□ **Example 3**   The series

$$\sum \frac{1}{2^n - 10}$$

is much like the geometric series $\sum 1/(2^n) = \sum (\frac{1}{2})^n$ (which we know is convergent). Hence it appears that the given series converges. To check it out, we look at the ratio

$$\frac{\dfrac{1}{2^n - 10}}{\dfrac{1}{2^n}} = \frac{2^n}{2^n - 10}$$

the limit of which (as $n \to \infty$) is $L = 1 > 0$. This confirms our guess. Note that the first three terms of the given series (assuming that it starts with $n = 1$) are negative. But for $n > 3$ the terms are positive, which is all that is required. (The Limit Comparison Test is stated for positive series, but since its conclusions refer to the limit of a_n/b_n as $n \to \infty$, we need only be concerned with "ultimately positive" series.)   □

□ **Example 4**   The series

$$\sum a_n = \sum \frac{2n + 1}{\sqrt{n^3 - 1}}$$

looks complicated. For large n, however,

$$\frac{2n + 1}{\sqrt{n^3 - 1}} \approx \frac{2n}{\sqrt{n^3}} = \frac{2}{\sqrt{n}}$$

which suggests a comparison with the divergent p-series

$$\sum b_n = \sum \frac{1}{n^{1/2}}$$

(See Example 2, Section 12.3.) We compute

$$\lim_{n\to\infty} \frac{a_n}{b_n} = \lim_{n\to\infty} \frac{\sqrt{n}(2n + 1)}{\sqrt{n^3 - 1}}$$

$$= \lim_{n\to\infty} \frac{2 + 1/n}{\sqrt{1 - 1/n^3}} \qquad \text{(dividing numerator and denominator by } n^{3/2}\text{)}$$

$$= 2$$

Since this is positive, the Limit Comparison Test says that our series diverges.

The ordinary Comparison Test is simpler in this example. For we can write

$$\frac{2n + 1}{\sqrt{n^3 - 1}} > \frac{2n}{\sqrt{n^3}} = \frac{2}{n^{1/2}} \qquad (n = 2, 3, 4, \ldots)$$

and conclude by a direct comparison with $\sum 2/(n^{1/2})$ that our series diverges. This suggests the use of some judgment in applying the tests. The Limit Comparison Test is the easier of the two in many cases, but not always. □

□ **Example 5**   The series

$$\sum \frac{2n - 1}{\sqrt{n^3 + 1}}$$

looks much like the series in Example 4. If we try to compare it with $\sum 2/n^{1/2}$, however (using the ordinary Comparison Test), we find that the inequality must be reversed:

$$\frac{2n - 1}{\sqrt{n^3 + 1}} < \frac{2n}{\sqrt{n^3}} = \frac{2}{n^{1/2}} \qquad (n = 1, 2, 3, \dots)$$

Since $\sum (2/n^{1/2})$ diverges, no conclusion follows. The Limit Comparison Test, on the other hand (with $\sum 1/n^{1/2}$ as the comparison series), applies as in Example 4. Hence the given series diverges. □

□ **Example 6**   Discuss the various ways of testing $\sum (\ln n)/n$.

Solution: One way is the Integral Test, since

$$\int_1^\infty \frac{\ln x}{x} \, dx$$

is easily shown to diverge. (See Problem 17 in the last section.) A quicker way, however (if you think of the right comparison series!) is to observe that

$$\frac{\ln n}{n} > \frac{1}{n} \qquad \text{for } n = 3, 4, 5, \dots$$

Since the harmonic series $\sum 1/n$ is known to diverge, it follows by the ordinary Comparison Test that $\sum (\ln n)/n$ diverges.

The Limit Comparison Test (with the harmonic series in the role of $\sum a_n$ and our given series in the role of $\sum b_n$) calls for the computation of

$$\lim_{n \to \infty} \frac{a_n}{b_n} = \lim_{n \to \infty} \frac{1}{\ln n} = 0$$

Statement (2) of the test then tells us that the given series diverges. Note that if the roles of the series are interchanged [$a_n = (\ln n)/n$ and $b_n = 1/n$],

$$\lim_{n \to \infty} \frac{a_n}{b_n} = \lim_{n \to \infty} \ln n = \infty$$

and Statement (3) of the test applies. □

The trouble with comparison tests is that one must choose wisely from among known series in order to reach a conclusion. That isn't always easy, as the next example shows.

□ **Example 7**   Use a comparison test to decide whether $\sum (\ln n)/n^2$ converges or diverges.

Solution: What series do we select for purposes of comparison? Our experience in Example 6 [where the harmonic series $\sum 1/n$ settled the question of how $\sum (\ln n)/n$ behaves] may suggest the use of $\sum 1/n^2$. Observing that

$$\frac{\ln n}{n^2} > \frac{1}{n^2}, \quad n = 3, 4, 5, \ldots$$

what can we conclude? Our comparison series *converges* (it is a p-series with $p = 2$). The fact that our given series dominates it is of no interest (as you can see by checking the statement of the ordinary Comparison Test). Nor does it help to use the Limit Comparison Test with $\sum 1/n^2$ in the role of either $\sum a_n$ or $\sum b_n$. (Try it!)

Evidently $\sum 1/n^2$ is a poor choice. Perhaps we should try the harmonic series. Letting $a_n = (\ln n)/n^2$ and $b_n = 1/n$, we find

$$\lim_{n \to \infty} \frac{a_n}{b_n} = \lim_{n \to \infty} \frac{\ln n}{n} = \lim_{n \to \infty} \frac{1/n}{1} \quad \text{(L'Hôpital's Rule)}$$

$$= 0$$

This leads to a conclusion provided that $\sum b_n$ converges. But $\sum 1/n$ *diverges*. Foiled again.

Rather than thrashing around, let's do some thinking. If $a_n = (\ln n)/n^2$, the choice of $b_n = 1/n^2$ is a decision to replace the function $\ln n$ by the constant function 1. That isn't reasonable; the functions are too unlike. The choice of

$$b_n = \frac{1}{n} = \frac{n}{n^2}$$

is a little better, since we are replacing $\ln n$ by the linear function n (which at least becomes infinite when $n \to \infty$, like $\ln n$). A glance at the graph of $y = \ln x$, however, shows that this is a poor choice, too; a curve that looks more like the logarithm is $y = \sqrt{x}$. Let's try it, that is, let's take

$$b_n = \frac{\sqrt{n}}{n^2} = \frac{1}{n^{3/2}}$$

This time we find

$$\lim_{n \to \infty} \frac{a_n}{b_n} = \lim_{n \to \infty} \frac{\ln n}{n^{1/2}} = \lim_{n \to \infty} \frac{1/n}{\frac{1}{2}n^{-1/2}} \quad \text{(L'Hôpital's Rule)}$$

$$= \lim_{n \to \infty} \frac{2}{\sqrt{n}} = 0$$

Since $\sum b_n$ is a convergent p-series ($p = \frac{3}{2}$), it follows from Statement (2) of the Limit Comparison Test that $\sum (\ln n)/n^2$ converges.   □

Problem Set 12.4

Each of the following series can be investigated by the Integral Test. (See Problem Set 12.3.) Use a comparison test instead.

1. $\sum \dfrac{1}{2n-1}$ **2.** $\sum \dfrac{1}{2n}$

3. $\sum \dfrac{1}{(2n-1)^2}$ **4.** $\sum \dfrac{1}{(2n)^2}$

5. $\sum \dfrac{n}{n^2+1}$ **6.** $\sum \dfrac{1}{n^2+1}$

7. $\sum \dfrac{1}{\sqrt{n^2+1}}$ **8.** $\sum \dfrac{1}{\sqrt{n^2-1}}$

9. $\sum \dfrac{1}{n\sqrt{n^2-1}}$ **10.** $\sum \dfrac{1}{n\sqrt{n^2+1}}$

Investigate each of the following series for convergence.

11. $\sum \dfrac{1}{n^2+2n}$ **12.** $\sum \dfrac{1}{\sqrt{n^2+n}}$

13. $\sum \dfrac{1}{\sqrt{n(n+2)}}$ **14.** $\sum \dfrac{1}{\sqrt[3]{n^2+1}}$

15. $\sum \dfrac{n-1}{n^2+n+1}$ **16.** $\sum \dfrac{1}{n^2-8}$

17. $\sum \dfrac{1}{\sqrt{n^2-8}}$ **18.** $\sum \dfrac{\sin^2 n}{n^2}$

19. $\sum \dfrac{1}{\ln n}$ **20.** $\sum \dfrac{1}{2^n+n}$

21. $\sum \dfrac{1}{e^n-n}$ **22.** $\sum \dfrac{2^n}{3^n-1000}$

23. $\sum \dfrac{n^2}{2^n}$ **24.** $\sum \operatorname{sech} n$

25. $\sum \left(\dfrac{n+1}{2n}\right)^n$

26. $\sum e^n \sin 2^{-n}$ *Hint:* $\dfrac{\sin 2^{-n}}{2^{-n}}$ approaches 1 as $n \to \infty$.

27. $\sum 2^n \sin e^{-n}$

28. Explain why the p-series

$$\sum \dfrac{1}{n^p} \qquad (p < 1)$$

diverges by comparison with the harmonic series.

29. Show that the series $\sum 1/n!$ converges by comparison with $\sum (\frac{1}{2})^{n-1}$.

30. Show that

$$\frac{1 \cdot 2 \cdot 3 \cdots n}{n \cdot n \cdot n \cdots n} \le \frac{2}{n^2} \qquad \text{if } n = 3, 4, 5, \ldots$$

Why does it follow that the series $\sum n!/n^n$ converges?

31. Explain why it follows from Problem 30 that $\lim_{n\to\infty} n!/n^n = 0$. (Thus n^n "dominates" $n!$ as $n \to \infty$.)

32. In Examples 6 and 7 we learned that $\sum (\ln n)/n$ diverges and $\sum (\ln n)/n^2$ converges.

(a) Explain why

$$\frac{\ln n}{n^2} < \frac{\ln n}{n^{3/2}} < \frac{\ln n}{n} \qquad (n = 2, 3, 4, \ldots)$$

(b) Why are these inequalities useless in the application of the ordinary comparison test to the series $\sum (\ln n)/n^{3/2}$?

(c) What does the series in part (b) do?

33. If $\sum a_n$ is a convergent series of positive terms, why does the series $\sum a_n^2$ converge?

34. If $\sum a_n$ is a convergent series of positive terms, why does the series $\sum \ln(1+a_n)$ converge?

35. Show that the positive series $\sum a_n$ must diverge if $\lim_{n\to\infty} na_n$ exists and is not 0.

36. Prove Statement (1) of the Limit Comparison Test as follows.

(a) Explain why there is a number N such that

$$\tfrac{1}{2}L < a_n/b_n < \tfrac{3}{2}L \qquad \text{for all } n > N$$

(b) The inequalities in part (a) are equivalent to

$$\tfrac{1}{2}Lb_n < a_n < \tfrac{3}{2}Lb_n \qquad \text{for all } n > N$$

Why? Explain why the inequality $a_n < \tfrac{3}{2}Lb_n$ proves that $\sum a_n$ converges when $\sum b_n$ does, and why the inequality $\tfrac{1}{2}Lb_n < a_n$ proves that $\sum b_n$ converges when $\sum a_n$ does.

37. Prove Statement (2) of the Limit Comparison Test. (Note that when $L = 0$ the argument in Problem 36 breaks down. Why?)

38. Prove Statement (3) of the Limit Comparison Test.

12.5
ALTERNATING SERIES

The convergence tests developed in the last two sections apply to positive series (and indirectly to negative series, since a factor of -1 may be removed from each term without affecting convergence or divergence). In this section we consider "mixed" series in which there are infinitely many terms of each sign. The simplest type is an **alternating series,** which (as the name suggests) has the form

$$a_1 - a_2 + a_3 - a_4 + \cdots = \sum_{n=1}^{\infty} (-1)^{n-1}a_n \qquad (a_n > 0 \text{ for each } n)$$

This is the only kind of mixed series we intend to discuss in detail; more general patterns of signs are both more difficult and (for our purposes) less important. (For an example of a mixed series that is not alternating, see Problem 28.)

Our first observation is that by replacing each term of a mixed series by its absolute value, we get a new series whose behavior can be investigated by the tests for positive series. *If this new series converges, so does the original series.*

□ **Theorem 1**   If $\sum |a_n|$ converges, then $\sum a_n$ converges.

An intuitive argument for Theorem 1 is that convergence of the positive series $\sum |a_n|$ must be due to the fact that its partial sums are bounded, which is entirely a matter of the size of the terms. They must approach zero so rapidly that the partial sums do not increase without bound. The series $\sum a_n$, in which there may be cancellations due to terms of mixed sign, has at least as good a chance to converge. (A formal proof is outlined in the problem set.)   □

□ **Example 1**   The alternating series

$$1 - \frac{1}{2^2} + \frac{1}{3^2} - \frac{1}{4^2} + \cdots = \sum_{n=1}^{\infty} \frac{(-1)^{n-1}}{n^2}$$

converges because the series of absolute values is

$$1 + \frac{1}{2^2} + \frac{1}{3^2} + \frac{1}{4^2} + \cdots = \sum_{n=1}^{\infty} \frac{1}{n^2}$$

This is a convergent *p*-series ($p = 2$).   □

□ **Example 2**   The series of absolute values associated with the alternating series

$$1 - \frac{1}{2} + \frac{1}{3} - \frac{1}{4} + \cdots = \sum_{n=1}^{\infty} \frac{(-1)^{n-1}}{n}$$

is the harmonic series

$$1 + \frac{1}{2} + \frac{1}{3} + \frac{1}{4} + \cdots = \sum_{n=1}^{\infty} \frac{1}{n}$$

Since this is known to diverge, Theorem 1 does not apply; convergence of the original series must be investigated in some other way. It turns out to converge (see Theorem 2), a fact that must be due to the mixture of signs in the partial sums:

$$S_1 = 1$$
$$S_2 = S_1 - \tfrac{1}{2} = \tfrac{1}{2} < S_1$$
$$S_3 = S_2 + \tfrac{1}{3} = \tfrac{5}{6} > S_2$$
$$S_4 = S_3 - \tfrac{1}{4} = \tfrac{7}{12} < S_3$$

These sums do not increase as in the case of a positive series (where the only possibility for convergence is that the partial sums increase slowly enough to be bounded). Instead they go up and down, which gives them an additional chance to approach a limit.   □

We will discuss the implications of this behavior shortly; what we want you to notice now is that Examples 1 and 2 exhibit different types of convergence. The series

$$1 - \frac{1}{2^2} + \frac{1}{3^2} - \frac{1}{4^2} + \cdots$$

"converges strongly," in the sense that its convergence survives a change to absolute values. In other words, the mixed signs in this series, while causing the partial sums to approach a limit they would otherwise not approach, are not responsible for the *existence* of a limit. For the partial sums of the series of absolute values also have a limit (which can happen only because they do not increase without bound).

On the other hand, the series $1 - \tfrac{1}{2} + \tfrac{1}{3} - \tfrac{1}{4} + \cdots$ does not survive a change to absolute values. Its convergence is "weak," in the sense that the partial sums of the series of absolute values are unbounded. Evidently the terms of the series do not approach zero fast enough to prevent this; the convergence of the original series must be due in part to the cancellation effect of the mixed signs.

Mathematicians call these kinds of convergence *absolute* and *conditional*, respectively, as in the following definition.

Suppose that $\Sigma\, a_n$ converges. Its convergence is called **absolute** if $\Sigma\, |a_n|$ also converges. If $\Sigma\, |a_n|$ diverges, however, then $\Sigma\, a_n$ is said to be **conditionally convergent.**

Now we turn to the main result of this section, a simple and powerful theorem about alternating series. We know from Section 12.2 that the typical term of any series must approach zero if the series is to converge. This is not a sufficient condition for convergence, however, since the harmonic series (for example) diverges even though its typical term approaches zero.

When the terms of a series alternate in sign, only one additional assumption is needed to assure convergence. Suppose that the series is

$$a_1 - a_2 + a_3 - a_4 + \cdots \qquad \text{(where each } a_n \text{ is positive)}$$

and suppose that the sequence $a_1, a_2, a_3, \ldots$ is decreasing. If the limit of this sequence is zero, the series converges. Moreover, the error in approximating the sum of the series by its nth partial sum is no larger than a_{n+1}.

□ **Theorem 2** (*Alternating Series Test*)   The alternating series

$$\sum_{n=1}^{\infty} (-1)^{n-1} a_n = a_1 - a_2 + a_3 - a_4 + \cdots \qquad (a_n > 0)$$

converges if the sequence $a_1, a_2, a_3, \ldots$ decreases to the limit 0. Moreover,

$$|S - S_n| \leq a_{n+1}, \; n = 1, 2, 3, \ldots$$

where S is the sum of the series and S_n is its nth partial sum.

The technical details of a proof of Theorem 2 are not very enlightening, but an informal argument is. See Figure 1, which shows the partial sums as points on a number line. The steps involved in computing these sums indicate why the figure has the appearance shown:

$$
\begin{array}{ll}
S_1 = a_1 & \text{(the first point plotted)} \\
S_2 = S_1 - a_2 & \text{(to the left of } S_1 \text{ because } a_2 > 0) \\
S_3 = S_2 + a_3 & \text{(to the right of } S_2 \\
& \text{but not as far as } S_1 \text{ because } 0 < a_3 < a_2) \\
S_4 = S_3 - a_4 & \text{(to the left of } S_3 \\
& \text{but not as far as } S_2 \text{ because } 0 < a_4 < a_3) \\
\vdots &
\end{array}
$$

As you can see, the odd-numbered sums $S_1, S_3, S_5, \ldots$ form a decreasing sequence (with a lower bound because the even-numbered sums are smaller). The even-numbered sums $S_2, S_4, S_6, \ldots$ form an increasing sequence (with an upper bound because the odd-numbered sums are larger). Each of these sequences has a limit, say S' and S'', respectively. (Why?) Moreover, S' and S'' are in fact the same number (contrary to the appearance in Figure 1, where they are shown as distinct). The reason for this is that the sequence $a_1, a_2, a_3, \ldots$ is not merely decreasing, but has limit 0. Hence the difference between consecutive partial sums,

$$S_n - S_{n-1} = a_n$$

approaches 0 as $n \to \infty$. Since one of these consecutive sums is odd-numbered and the other even-numbered, the sums approaching S' and the sums approaching S'' can be made arbitrarily close together by taking n suffi-

Figure 1
Partial sums of an alternating series

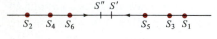

ciently large. This can be true only if $S' = S''$. If we call this common value S, then

$$\lim_{n\to\infty} S_n = S$$

that is, the series converges. Moreover, S_n differs from S by less than a_{n+1}, that is,

$$|S - S_n| \leq a_{n+1}$$

(because S is between S_n and S_{n+1} and they differ by a_{n+1}).   □

Theorem 2 fills in our discussion of Example 2, where we claimed that the series $1 - \frac{1}{2} + \frac{1}{3} - \frac{1}{4} + \cdots$ converges. As you can see, the hypotheses of the theorem are satisfied. The signs alternate and the absolute value of the typical term is $a_n = 1/n$, which decreases to 0 as $n \to \infty$.

□ **Example 3**   Test the series

$$\frac{\ln 1}{1} - \frac{\ln 2}{2} + \frac{\ln 3}{3} - \frac{\ln 4}{4} + \cdots = \sum_{n=1}^{\infty} (-1)^{n-1} \frac{\ln n}{n}$$

for convergence.

Solution: Strictly speaking, the first term should be ignored (since it is 0, which is neither positive nor negative). After that, however, the signs alternate and the absolute value of the typical term, $a_n = (\ln n)/n$, approaches 0 as $n \to \infty$. (Use L'Hôpital's Rule.) But does a_n decrease as $n \to \infty$? To find out, let $f(x) = (\ln x)/x$. Then

$$f'(x) = \frac{1 - \ln x}{x^2} < 0 \qquad \text{for } x > e \qquad \text{(Why?)}$$

This means that f is a decreasing function in the interval (e, ∞) and hence

$$a_{n+1} < a_n \qquad \text{for } n = 3, 4, 5, \ldots$$

The sequence

$$a_1 = 0, \ a_2 \approx 0.347, \ a_3 \approx 0.366, \ a_4 \approx 0.347, \ a_5 \approx 0.322, \ \ldots$$

is not decreasing for $n = 1, 2, 3, \ldots$, as you can see. But the sequence $a_3, a_4, a_5, \ldots$ is decreasing, which is all that is needed for Theorem 2 to work. (Leave out the first two terms of the series and apply the theorem to the rest.) Hence the series converges.   □

□ **Example 4**   How many terms of the series $1 - \frac{1}{2} + \frac{1}{3} - \frac{1}{4} + \cdots$ are needed to compute the sum correct to two decimal places?

Solution: According to Theorem 2, the error in the approximation $S \approx S_n$ is

$$|S - S_n| \leq a_{n+1}$$

In the present case, $a_n = 1/n$, so we choose n satisfying

$$\frac{1}{n+1} < \frac{1}{2} \times 10^{-2}$$

$$200 < n + 1$$

$$n > 199$$

Hence the approximation $S \approx S_{200}$ gives two-place accuracy. We will show later that the sum of this series is $\ln 2 \approx 0.693$, but as you can see, its convergence is slow.   □

□ **Example 5**   Compute the sum of the series

$$\frac{1}{0!} - \frac{1}{1!} + \frac{1}{2!} - \frac{1}{3!} + \cdots$$

correct to six decimal places.

Solution: The series converges by Theorem 2. (Its terms alternate in sign and their absolute values decrease to the limit 0.) The error in the approximation $S \approx S_n$ is no larger than

$$\frac{1}{a_{n+1}} = \frac{1}{n!}$$

so we choose n satisfying

$$\frac{1}{n!} < \frac{1}{2} \times 10^{-6}$$

$$2 \times 10^6 < n!$$

Rather than trying to solve for n algebraically, we simply compute factorials until we find the first one to exceed two million. This turns out to be 10!, so our approximation is

$$S \approx S_{10} = 1 - 1 + \left(\frac{1}{2!} - \frac{1}{3!} + \frac{1}{4!} - \cdots - \frac{1}{9!}\right) = 0.367879$$

(correct to six places). The true sum of this series (as we will show later) is $1/e = 0.367879441\cdots$.   □

Remark   It is important to note that the error bound we have stated for alternating series, namely $|S - S_n| \le a_{n+1}$, does not apply to series in general. Usually more complicated formulas are required, like

$$0 \le S - \left(S_N + \int_{N+1}^{\infty} f(x)\, dx\right) \le a_{N+1}$$

(given in Section 12.3 for positive series that satisfy the hypotheses of the Integral Test).

We end this section with some remarks about grouping and rearranging the terms of an infinite series. The series

$$1 - 1 + 1 - 1 + 1 - 1 + \cdots$$

diverges (because its partial sums are 1, 0, 1, 0, 1, 0, ..., which do not approach a limit). Grouping the terms of the series, however, we obtain

$$(1 - 1) + (1 - 1) + (1 - 1) + \cdots$$

an altogether different series which converges to the sum 0. Evidently the associative law of ordinary algebra does not apply to series!

Another example is the equation

$$S = 1 - \tfrac{1}{2} + \tfrac{1}{3} - \tfrac{1}{4} + \tfrac{1}{5} - \tfrac{1}{6} + \cdots$$

Can we rearrange this to read

$$S = (1 + \tfrac{1}{3} + \tfrac{1}{5} + \cdots) - (\tfrac{1}{2} + \tfrac{1}{4} + \tfrac{1}{6} + \cdots)?$$

Each of the series in parentheses diverges (by comparison with the harmonic series); in fact the partial sums of each become infinite as the number of terms increases. The result is $S = \infty - \infty$, and the less said the better.

A more interesting example is the following. Divide each side of

$$S = 1 - \tfrac{1}{2} + \tfrac{1}{3} - \tfrac{1}{4} + \tfrac{1}{5} - \tfrac{1}{6} + \cdots$$

by 2, obtaining

$$\tfrac{1}{2}S = \tfrac{1}{2} - \tfrac{1}{4} + \tfrac{1}{6} - \tfrac{1}{8} + \tfrac{1}{10} - \tfrac{1}{12} + \cdots$$

(This is legitimate because the series converges; see Problem 44, Section 12.2.) Adding the series, we find

$$\tfrac{3}{2}S = (1 + \tfrac{1}{2}) + (-\tfrac{1}{2} - \tfrac{1}{4}) + (\tfrac{1}{3} + \tfrac{1}{6}) + (-\tfrac{1}{4} - \tfrac{1}{8}) + (\tfrac{1}{5} + \tfrac{1}{10}) + (-\tfrac{1}{6} - \tfrac{1}{12}) + \cdots$$

So far, so good; we have done nothing that is not justified by theorems previously proved. Suppose, however, that we rearrange and regroup the terms as in ordinary algebra, writing

$$\tfrac{3}{2}S = 1 + (\tfrac{1}{2} - \tfrac{1}{2}) + (-\tfrac{1}{4} - \tfrac{1}{4}) + \tfrac{1}{3} + (\tfrac{1}{6} - \tfrac{1}{6}) + (-\tfrac{1}{8} - \tfrac{1}{8}) + \tfrac{1}{5} + (\tfrac{1}{10} - \tfrac{1}{10}) + (-\tfrac{1}{12} - \tfrac{1}{12}) + \tfrac{1}{7} + \cdots$$
$$= 1 - \tfrac{1}{2} + \tfrac{1}{3} - \tfrac{1}{4} + \tfrac{1}{5} - \tfrac{1}{6} + \tfrac{1}{7} - \cdots$$

If this were correct, it would follow that $\tfrac{3}{2}S = S$, or $3S = 2S$. Since $S \neq 0$, (why?), we have $3 = 2$. (!)

It can be proved that grouping and rearranging the terms of an *absolutely convergent* series is legitimate; neither its convergence nor its sum is affected. A conditionally convergent series, on the other hand (like the one just discussed), must be treated with caution. Remarkably enough, the terms of such a series can be rearranged to converge to any sum whatever, or even to diverge. (See the problem set.)

Problem Set 12.5

Classify each of the following series as absolutely convergent, conditionally convergent, or divergent.

1. $1 - \frac{1}{3} + \frac{1}{9} - \frac{1}{27} + \cdots$

2. $-1 - \frac{1}{2} - \frac{1}{4} - \frac{1}{8} - \frac{1}{16} - \cdots$

3. $-1 - \frac{1}{2} - \frac{1}{3} - \frac{1}{4} - \cdots$

4. $1 - \frac{1}{3} + \frac{1}{5} - \frac{1}{7} + \cdots$

5. $\frac{1}{2} - \frac{1}{4} + \frac{1}{6} - \frac{1}{8} + \cdots$

6. $1 - \dfrac{1}{\sqrt{2}} + \dfrac{1}{\sqrt{3}} - \dfrac{1}{\sqrt{4}} + \cdots$

7. $1 - \dfrac{1}{2\sqrt{2}} + \dfrac{1}{3\sqrt{3}} - \dfrac{1}{4\sqrt{4}} + \cdots$

8. $1 - \dfrac{1}{3^2} + \dfrac{1}{5^2} - \dfrac{1}{7^2} + \cdots$

9. $\dfrac{1}{2^2} - \dfrac{1}{4^2} + \dfrac{1}{6^2} - \dfrac{1}{8^2} + \cdots$

10. $\dfrac{2}{\ln 2} - \dfrac{3}{\ln 3} + \dfrac{4}{\ln 4} - \dfrac{5}{\ln 5} + \cdots$

11. $\sin 1 + \dfrac{\sin 4}{4} + \dfrac{\sin 9}{9} + \dfrac{\sin 16}{16} + \cdots$

12. $\displaystyle\sum_{n=1}^{\infty} \dfrac{(-1)^{n-1}n}{n^2 + 1}$ **13.** $\displaystyle\sum_{n=1}^{\infty} \dfrac{(-1)^{n-1}}{\sqrt{n^2 + 1}}$

14. $\displaystyle\sum_{n=1}^{\infty} \dfrac{(-1)^{n-1}}{n\sqrt{n^2 + 1}}$

15. In Example 3 we showed that the series

$$\frac{\ln 1}{1} - \frac{\ln 2}{2} + \frac{\ln 3}{3} - \frac{\ln 4}{4} + \cdots$$

converges. Does it converge absolutely?

Investigate the convergence of each of the following series.

16. $\displaystyle\sum_{n=1}^{\infty} (-1)^n \dfrac{n}{n^2 - 2}$ **17.** $\displaystyle\sum_{n=1}^{\infty} \dfrac{\cos \pi n}{n}$

18. $\displaystyle\sum_{n=2}^{\infty} \dfrac{(-1)^n}{\sqrt{n^2 - 1}}$ **19.** $\displaystyle\sum_{n=2}^{\infty} \dfrac{(-1)^n}{n\sqrt{n^2 - 1}}$

20. $\displaystyle\sum_{n=1}^{\infty} \dfrac{(-1)^{n-1}}{\sqrt{n^2 + n}}$ **21.** $\displaystyle\sum_{n=1}^{\infty} \dfrac{(-1)^{n-1}}{\sqrt[3]{n^2 + 1}}$

22. $\displaystyle\sum_{n=2}^{\infty} \dfrac{(-1)^n}{\ln n}$

23. $\frac{1}{2} - \frac{2}{3} + \frac{3}{4} - \frac{4}{5} + \cdots$

24. $\ln 1 - \ln \frac{1}{2} + \ln \frac{1}{3} - \ln \frac{1}{4} + \cdots$

25. $\dfrac{1}{e} - \dfrac{2}{e^2} + \dfrac{3}{e^3} - \dfrac{4}{e^4} + \cdots$

26. $\sin \dfrac{\pi}{2} - \sin \dfrac{\pi}{3} + \sin \dfrac{\pi}{4} - \sin \dfrac{\pi}{5} + \cdots$

27. $\cos \dfrac{\pi}{2} - \cos \dfrac{\pi}{3} + \cos \dfrac{\pi}{4} - \cos \dfrac{\pi}{5} + \cdots$

28. Explain why

$$\sin 1 + \frac{\sin 2}{2^2} + \frac{\sin 3}{3^2} + \frac{\sin 4}{4^2} + \cdots$$

is not an alternating series. Is it a positive series (or even "ultimately positive")? Does it converge?

29. Explain why the alternating p-series

$$1 - \frac{1}{2^p} + \frac{1}{3^p} - \frac{1}{4^p} + \cdots$$

converges for every $p > 0$. For what values of p is it absolutely convergent? conditionally convergent?

30. The Alternating Series Test can be used to investigate the series

$$1 - \frac{1}{2} + \frac{1}{4} - \frac{1}{8} + \cdots = \sum_{n=0}^{\infty} \frac{(-1)^n}{2^n}$$

and its sum can be approximated by $S \approx S_n$, with error less than a_{n+1}. Why is this a waste of time?

31. Explain why the geometric series $1 + x + x^2 + \cdots$ either converges absolutely or diverges, that is, it is never conditionally convergent.

32. How many terms of the series

$$1 - \frac{1}{\sqrt{2}} + \frac{1}{\sqrt{3}} - \frac{1}{\sqrt{4}} + \cdots$$

would you use to compute the sum correct to three decimal places?

33. Explain why the series

$$1 - \frac{1}{3!} + \frac{1}{5!} - \frac{1}{7!} + \cdots$$

converges and compute its sum correct to five decimal places. (We will show later that the sum is sin 1.)

34. Explain why the series

$$1 - \frac{1}{3 \cdot 2!} + \frac{1}{5 \cdot 3!} - \frac{1}{7 \cdot 4!} + \cdots$$

converges and find its sum correct to five decimal places.

35. Show that the alternating series

$$1 - 1 + \frac{1}{2} - \frac{1}{2} + \frac{1}{3} - \frac{1}{4} + \frac{1}{4} - \frac{1}{8} + \frac{1}{5} - \frac{1}{16}$$

$$+ \frac{1}{6} - \frac{1}{32} + \cdots + \frac{1}{n} - \frac{1}{2^{n-1}} + \cdots$$

diverges even though its nth term approaches zero as n increases. Why doesn't that contradict Theorem 2?

36. If $a_1 - a_2 + a_3 - a_4 + \cdots$ is an alternating series such that a_n decreases to the limit 0 as $n \to \infty$, why is its sum between a_1 and $a_1 - a_2$?

37. If $\Sigma\, a_n$ diverges, why does it follow that $\Sigma\, |a_n|$ diverges? (Thus there is no concept of absolute or conditional divergence, as there is for convergence.)

38. Give an example of convergent series $\Sigma\, a_n$ and $\Sigma\, b_n$ such that $\Sigma\, a_n b_n$ diverges.

39. In Chapter 13 we will show that

$$e = \frac{1}{0!} + \frac{1}{1!} + \frac{1}{2!} + \frac{1}{3!} + \cdots$$

and $\quad e^{-1} = \dfrac{1}{0!} - \dfrac{1}{1!} + \dfrac{1}{2!} - \dfrac{1}{3!} + \cdots$

Explain why it is legitimate to conclude that

$$\cosh 1 = \frac{1}{0!} + \frac{1}{2!} + \frac{1}{4!} + \cdots$$

40. Use five terms of the series in Problem 39 to compute an approximate value of $\cosh 1$.

41. In Problem 40 the first term not used is

$$\frac{1}{10!} = 0.000000275 \cdots < \frac{1}{2} \times 10^{-6}$$

The approximation obtained is $\cosh 1 \approx 1.54308035 \cdots$, whereas the true value is $1.54308063 \cdots$. Hence the approximation is correct to five places but not six. Does this contradict Theorem 2?

42. Explain what is wrong with the statement

$$1 - \frac{1}{\sqrt{2}} + \frac{1}{\sqrt{3}} - \frac{1}{\sqrt{4}} + \frac{1}{\sqrt{5}} - \frac{1}{\sqrt{6}} + \cdots$$

$$= \left(1 + \frac{1}{\sqrt{3}} + \frac{1}{\sqrt{5}} + \cdots\right) - \left(\frac{1}{\sqrt{2}} + \frac{1}{\sqrt{4}} + \frac{1}{\sqrt{6}} + \cdots\right)$$

43. Explain why it is legitimate to write

$$1 - \frac{1}{2^2} + \frac{1}{3^2} - \frac{1}{4^2} + \frac{1}{5^2} - \frac{1}{6^2} + \cdots$$

$$= \left(1 + \frac{1}{3^2} + \frac{1}{5^2} + \cdots\right) - \left(\frac{1}{2^2} + \frac{1}{4^2} + \frac{1}{6^2} + \cdots\right)$$

44. A conditionally convergent series, like

$$1 - \tfrac{1}{2} + \tfrac{1}{3} - \tfrac{1}{4} + \tfrac{1}{5} - \tfrac{1}{6} + \cdots$$

can be rearranged so as to converge to any number whatever, say π. Explain how this may be done, as follows.

(a) Why is it possible to add up enough positive terms $(1 + \tfrac{1}{3} + \tfrac{1}{5} + \tfrac{1}{7} + \cdots)$ to eventually exceed π?

(b) Supposing this to have been done (with no more terms than necessary), why is it possible to continue with negative terms $(-\tfrac{1}{2} - \tfrac{1}{4} - \tfrac{1}{6} - \cdots)$ until the sum in part (a) becomes less than π?

(c) Add more positive terms to get the sum above π again, then more negative terms to bring it below π, and continue in this way indefinitely (always using just enough terms to do the job). Do we run out of positive terms? of negative terms? Is every term of the original series eventually used? Why does this process produce sums that converge to π?

45. Explain why the series in Problem 44 can be rearranged to diverge, as follows.

(a) Why is it possible to add up enough positive terms to exceed 10? Why is it possible to continue with enough negative terms to bring the sum below -10^2?

(b) Add more positive terms to get the sum above 10^3, then more negative terms to bring it below -10^4, and continue in this way indefinitely. Do we run out of positive terms? of negative terms? Is every term of the original series eventually used? Why does this process produce sums that have no limit?

46. Prove Theorem 1 as follows.

(a) Confirm that $a_n = \tfrac{1}{2}(|a_n| + a_n) - \tfrac{1}{2}(|a_n| - a_n)$ and check the inequalities

$$0 \le \tfrac{1}{2}(|a_n| + a_n) \le |a_n|$$

and $\quad 0 \le \tfrac{1}{2}(|a_n| - a_n) \le |a_n|$

(b) Why do the inequalities in part (a) show that the series

$$\Sigma \tfrac{1}{2}(|a_n| + a_n) \qquad \text{and} \qquad \Sigma \tfrac{1}{2}(|a_n| - a_n)$$

converge? Why does it follow that $\Sigma\, a_n$ converges?

12.6
THE RATIO TEST
AND ROOT TEST

In this section we consider the last of the convergence tests of this chapter. They are the ones most often used in discussing the power series of Chapter 13; together with those already developed, they provide an effective battery of tests for investigating the series that commonly occur in calculus.

The ratio test arises from consideration of the ratios

$$\frac{|a_2|}{|a_1|}, \frac{|a_3|}{|a_2|}, \ldots, \frac{|a_{n+1}|}{|a_n|}, \ldots$$

where $\sum a_n$ is the series under investigation. These ratios measure the rate at which the terms of the series grow in absolute value as $n \to \infty$. In a geometric series $1 + r + r^2 + \cdots$ this rate is constant, namely $|r|$, and we know that the series converges when $|r| < 1$ and diverges when $|r| \geq 1$. In general the rate $|a_{n+1}|/|a_n|$ is not constant, but it is reasonable to expect that it is ultimately no greater than 1 if the series is to converge. (Otherwise how could the typical term approach zero?) This suggests that its limit,

$$\rho = \lim_{n \to \infty} \frac{|a_{n+1}|}{|a_n|}$$

can tell us something about convergence and divergence, and that $\rho = 1$ is the border line between the two. Hence the following theorem should not be surprising.

□ **Theorem 1** (*Ratio Test*)   Let $\sum a_n$ be a series of nonzero numbers and suppose that

$$\rho = \lim_{n \to \infty} \frac{|a_{n+1}|}{|a_n|}$$

either exists or is ∞. Then

$$\rho < 1 \Rightarrow \sum a_n \text{ converges absolutely}$$

$$\rho > 1 \text{ (or } \rho = \infty) \Rightarrow \sum a_n \text{ diverges}$$

When $\rho = 1$, the test fails to distinguish between convergence and divergence.

Proof: Suppose first that $\rho > 1$ (the easiest case). Then for sufficiently large n we have

$$\frac{|a_{n+1}|}{|a_n|} > 1, \text{ that is, } |a_{n+1}| > |a_n|$$

This means that (ultimately) the absolute values of the terms of the series are increasing. Therefore a_n cannot approach 0 as $n \to \infty$ and the series must diverge. (Note that this argument also applies when $\rho = \infty$.)

Next suppose that $\rho < 1$. Let r be any point between ρ and 1 ($\rho < r < 1$). We claim that $\sum |a_n|$ converges by comparison with $\sum r^n$ (a geometric series which converges because $0 < r < 1$). Our claim is based on another, namely that

$$\lim_{n \to \infty} \frac{|a_n|}{r^n} \text{ exists}$$

(in which case the Limit Comparison Test says that $\sum |a_n|$ converges because $\sum r^n$ does). To see why this limit exists, observe that since

$$\rho = \lim_{n \to \infty} \frac{|a_{n+1}|}{|a_n|}$$

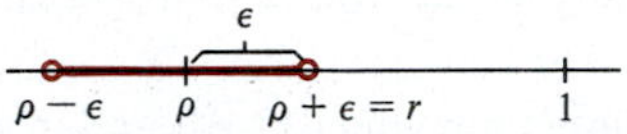

we can force the ratio $|a_{n+1}|/|a_n|$ into any interval $(\rho - \varepsilon, \rho + \varepsilon)$ by taking n sufficiently large. Choosing $\varepsilon = r - \rho$ (Figure 1), we know there is a number N such that

$$n > N \Rightarrow \rho - \varepsilon < \frac{|a_{n+1}|}{|a_n|} < \rho + \varepsilon = r$$

Thus (for $n > N$) we have $|a_{n+1}| < r|a_n|$ and (dividing by r^{n+1})

$$\frac{|a_{n+1}|}{r^{n+1}} < \frac{|a_n|}{r^n}$$

This means that the sequence whose typical term is $|a_n|/r^n$ decreases (at least for $n > N$). Its terms are positive (hence bounded below by 0). Therefore it has a limit, as we claimed. The series $\sum |a_n|$ converges and hence $\sum a_n$ converges absolutely.

To complete the proof, we must show that when $\rho = 1$ the test fails to distinguish between convergence and divergence. See Example 1. □

Figure 1
Proving the Ratio Test

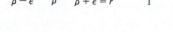

□ **Example 1**   The series

$$1 - \frac{1}{2} + \frac{1}{3} - \frac{1}{4} + \cdots = \sum_{n=1}^{\infty} \frac{(-1)^{n-1}}{n}$$

converges (by the Alternating Series Test), while its series of absolute values (the harmonic series) does not. In both cases the Ratio Test yields

$$\rho = \lim_{n \to \infty} \frac{|a_{n+1}|}{|a_n|} = \lim_{n \to \infty} \frac{\dfrac{1}{n+1}}{\dfrac{1}{n}} = \lim_{n \to \infty} \frac{n}{n+1} = 1$$

This shows that when $\rho = 1$ anything can happen; the Ratio Test is indecisive in such cases.

Example 1 is typical of a whole class of series in which the general term is an algebraic function of n (powers, roots, polynomials, quotients of polynomials). The Ratio Test is not worth trying because ρ turns out to be 1. □

□ **Example 2**   Investigate the series $\sum 2^n/n!$.

Solution: This kind of series (whose typical term involves repeated multiplication in an exponential function or factorial) is ideally suited for the Ratio Test. Since the terms are positive, absolute value is not needed; we compute the ratio

$$\frac{a_{n+1}}{a_n} = \frac{2^{n+1}}{(n+1)!} \cdot \frac{n!}{2^n} = \frac{2}{n+1} \qquad [\text{because } (n+1)! = n!(n+1)]$$

Hence

$$\rho = \lim_{n\to\infty} \frac{2}{n+1} = 0$$

Since this is less than 1, the Ratio Test says that the series converges.   □

Remark   In Example 5, Section 11.2 (and the note following it) we found that the exponential function e^x "dominates" every power of x as $x \to \infty$. Now (in Example 2) we have found something that dominates an exponential! For if the series $\sum 2^n/n!$ converges, its typical term must approach zero:

$$\lim_{n\to\infty} \frac{2^n}{n!} = 0$$

Thus although 2^n increases very rapidly as $n \to \infty$, the factorial grows even faster. In this connection you might also look up Problems 30 and 31, Section 12.4, in which it is shown that

$$\lim_{n\to\infty} \frac{n!}{n^n} = 0$$

Thus the function n^n dominates the factorial.

In fact we have developed an interesting hierarchy of functions, each of which goes to infinity with n, and each of which dominates its predecessor by growing more rapidly. The slowest (in the list we have in mind) is $\ln n$, since even the linear function dominates it:

$$\lim_{n\to\infty} \frac{\ln n}{n} = 0$$

Then we might list all positive integral powers of n, namely $n, n^2, n^3, \ldots$, each one dominating its predecessor. The exponential a^n ($a > 1$) dominates all these, $n!$ dominates a^n, and n^n dominates the factorial. This hierarchy is worth remembering when you are searching for ways to test a new series for convergence.

□ **Example 3**   A generalization of Example 2 is the series

$$1 + x + \frac{x^2}{2!} + \frac{x^3}{3!} + \cdots = \sum_{n=0}^{\infty} \frac{x^n}{n!}$$

where x is any real number. The series obviously converges if $x = 0$ (all its terms after the first are zero), so we assume that $x \neq 0$. (In Example 2, $x = 2$.) The typical term is $a_n = x^n/n!$, so

$$\frac{|a_{n+1}|}{|a_n|} = \frac{|x|^{n+1}}{(n+1)!} \cdot \frac{n!}{|x|^n} = \frac{|x|}{n+1}$$

Hence

$$\rho = \lim_{n\to\infty} \frac{|x|}{n+1} = |x| \lim_{n\to\infty} \frac{1}{n+1} = |x| \cdot 0 = 0$$

Since this is less than 1 (regardless of the value of x), we conclude that the series converges absolutely for every x. $\square$

□ **Example 4**   For each value of x the series in Example 3 converges; let $S(x)$ be its sum, that is,

$$S(x) = 1 + x + \frac{x^2}{2!} + \frac{x^3}{3!} + \cdots$$

Then S is a function with domain $\mathfrak{R}$. Assuming that it can be differentiated as though it were an ordinary polynomial, find $S'(x)$ and use the result to derive a "closed-form" expression for $S(x)$.

Solution: Letting $y = S(x)$, we find

$$\frac{dy}{dx} = \frac{d}{dx}\left(1 + x + \frac{x^2}{2!} + \frac{x^3}{3!} + \frac{x^4}{4!} + \cdots\right)$$

$$= 0 + 1 + \frac{2x}{2!} + \frac{3x^2}{3!} + \frac{4x^3}{4!} + \cdots$$

$$= 1 + x + \frac{x^2}{2!} + \frac{x^3}{3!} + \cdots = y$$

Thus $y = S(x)$ satisfies the differential equation $dy/dx = y$. By separating the variables and integrating, we find $y = Ce^x$ for some constant C. It is clear from the original series that $y = 1$ when $x = 0$, so $C = 1$ and hence $y = e^x$.

Thus we have discovered the remarkable formula

$$e^x = 1 + x + \frac{x^2}{2!} + \frac{x^3}{3!} + \cdots$$

In Chapter 13 we exploit ideas of this kind to discover many similar results. $\square$

□ **Example 5**   Find all values of x for which the series

$$(x - 1) - \frac{(x-1)^2}{2} + \frac{(x-1)^3}{3} - \cdots = \sum_{n=1}^{\infty} \frac{(-1)^{n-1}}{n}(x-1)^n$$

converges.

Solution: When $x = 1$ the series obviously converges (its terms are all zero). Assume that $x \neq 1$. The ratio called for in the Ratio Test is

$$\frac{|x-1|^{n+1}}{n+1} \cdot \frac{n}{|x-1|^n} = |x-1| \cdot \frac{n}{n+1}$$

and its limit is

$$\rho = |x - 1| \lim_{n \to \infty} \frac{n}{n + 1} = |x - 1|$$

This time (unlike Example 3) the limit depends on x. The Ratio Test says that the series converges absolutely or diverges depending on whether $\rho < 1$ or $\rho > 1$, whereas the issue is in doubt when $\rho = 1$.

Hence the series converges absolutely when

$$|x - 1| < 1$$
$$-1 < x - 1 < 1$$
$$0 < x < 2$$

and diverges outside this interval (when $x > 2$ or $x < 0$). The endpoints $x = 0$ and $x = 2$ correspond to the ambiguous case $\rho = 1$ and must be investigated separately. When $x = 0$ the original series is

$$-1 - \tfrac{1}{2} - \tfrac{1}{3} - \cdots$$

which diverges because it is the negative of the harmonic series. When $x = 2$ the series is

$$1 - \tfrac{1}{2} + \tfrac{1}{3} - \cdots$$

which converges by the Alternating Series Test. (Why is this conditional convergence?) The answer to the question is that the given series converges when $0 < x \le 2$ (absolutely except at $x = 2$, where the convergence is conditional).   □

We turn now to a convergence test that is similar to the Ratio Test. To some extent it duplicates the results of the Ratio Test (and is therefore less often used, being harder to apply in many cases). We leave its proof to you. (It is easier than the argument for the Ratio Test.)

□ **Theorem 2** (*Root Test*)   Let $\sum a_n$ be any series and suppose that

$$\rho = \lim_{n \to \infty} \sqrt[n]{|a_n|}$$

either exists or is ∞. Then

$$\rho < 1 \Rightarrow \sum a_n \text{ converges absolutely}$$

$$\rho > 1 \text{ (or } \rho = \infty) \Rightarrow \sum a_n \text{ diverges}$$

When $\rho = 1$, the test fails.   □

□ **Example 6**   Use the Root Test to investigate the series

$$\sum_{n=1}^{\infty} \left(\frac{n}{2n - 1} \right)^n$$

Solution: The root called for is

$$\sqrt[n]{\left(\frac{n}{2n-1}\right)^n} = \frac{n}{2n-1}$$

Its limit as $n \to \infty$ is $\rho = \frac{1}{2}$, so the series converges. The application of the Ratio Test to this series is much harder, as you will see if you try it.   □

□ **Example 7**   Apply the Root Test to the series in Example 5.

Solution: The root called for is

$$\left(\frac{|x-1|^n}{n}\right)^{1/n} = \frac{|x-1|}{n^{1/n}}$$

Its limit is

$$\rho = |x-1| \lim_{n \to \infty} n^{-1/n}$$

Since this is indeterminate of the type ∞^0 (see Section 11.2), we let $y = n^{-1/n}$ and take logarithms:

$$\ln y = -\frac{1}{n} \ln n = -\frac{\ln n}{n}$$

$$\lim_{n \to \infty} \ln y = -\lim_{n \to \infty} \frac{\ln n}{n} = 0 \quad \text{(L'Hôpital's Rule)}$$

Hence

$$\lim_{n \to \infty} y = e^0 = 1$$

and we find $\rho = |x-1|$. The rest of the analysis proceeds as in Example 5. As you can see, the Root Test is harder than the Ratio Test in this case.   □

Problem Set 12.6

Investigate each of the following series by using the Ratio Test or Root Test (unless they do not apply).

1. $\sum \dfrac{1}{n!}$

2. $\sum \dfrac{n^2}{n!}$

3. $\sum \dfrac{3^n}{n!}$

4. $\sum \dfrac{2^n}{(2n-1)!}$

5. $\sum \dfrac{(-2)^n}{(2n)!}$

6. $\sum \dfrac{n^2}{2^n}$

7. $\sum \dfrac{n}{3^n}$

8. $\sum \dfrac{n \cdot 2^n}{3^n}$

9. $\sum \dfrac{n!}{1 \cdot 3 \cdot 5 \cdots (2n-1)}$

10. $\sum \dfrac{2 \cdot 4 \cdot 6 \cdots (2n)}{n!}$

11. $\sum \dfrac{1}{n^2 - 2}$

12. $\sum \dfrac{(-1)^{n-1}}{\sqrt{n}}$

13. $\sum \dfrac{1}{n^n}$

14. $\sum \dfrac{2^n}{n^n}$

15. $\sum e^{-n^2}$

16. $\sum \dfrac{1}{(\ln n)^n}$

17. $\sum \dfrac{1}{1 + 2^n}$

18. $\sum \dfrac{n!}{n^n}$ *Hint:* Recall that $\lim_{h \to 0} (1+h)^{1/h} = e$.

19. $\sum \left(1 + \dfrac{1}{n}\right)^n$

20. $\sum \left(\dfrac{n}{n+1}\right)^n$

In each of the following, find all values of x for which the series converges. Distinguish between absolute and conditional convergence.

21. $\sum \dfrac{x^{2n}}{(2n)!}$

22. $\sum \dfrac{x^{2n-1}}{(2n-1)!}$

23. $\sum (-1)^{n-1} \dfrac{x^{2n-1}}{(2n-1)!}$

24. $\sum n! x^n$

25. $\sum x^n$ (The result should confirm what we said about geometric series in Section 12.2.)

26. $\sum n x^{n-1}$

27. $\sum n(n-1)x^{n-2}$

28. $\sum (x-1)^n$

29. $\sum \dfrac{x^n}{n}$

30. $\sum (-1)^{n-1} \dfrac{x^{2n-1}}{2n-1}$

31. $\sum \dfrac{(x-2)^n}{n^2}$

32. $\sum \dfrac{(x+2)^n}{2^n}$

33. $\sum 2^n x^n$

34. $\sum \dfrac{(x+1)^n}{\sqrt{n}}$

35. $\sum \dfrac{x^n}{n^n}$

36. $\sum \dfrac{n! x^n}{n^n}$

37. $\sum (x^2-1)^n$

38. $\sum \dfrac{2^n x^n}{n!}$

39. $\sum \dfrac{n! x^n}{3^n}$

40. $\sum \dfrac{x^n}{n\sqrt{n}}$

41. Suppose that

$$f(x) = (x-1) - \frac{(x-1)^2}{2} + \frac{(x-1)^3}{3} - \cdots$$

$$= \sum_{n=1}^{\infty} (-1)^{n-1} \frac{(x-1)^n}{n}$$

(a) Show that the domain of f is $(0,2]$.

(b) Assuming that $f'(x)$ can be found by differentiating the series as though it were a polynomial, show that

$$f'(x) = 1 - (x-1) + (x-1)^2 - \cdots$$

What is the domain of f'?

(c) Explain why the result in part (b) can be written in the form

$$f'(x) = \sum_{n=0}^{\infty} (1-x)^n$$

(d) We know from Section 12.2 that

$$\sum_{n=0}^{\infty} t^n = \frac{1}{1-t} \qquad (-1 < t < 1)$$

Use this to show that $f'(x) = 1/x$ ($0 < x < 2$). Does the domain $(0,2)$ check with that found in part (b)?

(e) The result in part (d) means that $y = f(x)$ satisfies the differential equation $dy/dx = 1/x$. Explain the initial condition $f(1) = 0$ and solve the equation to find $f(x)$.

(f) The result in part (e) is

$$f(x) = \ln x \qquad (0 < x < 2)$$

The endpoint $x = 2$ got lost in the shuffle, but assuming that it can be restored, explain why

$$\ln 2 = 1 - \frac{1}{2} + \frac{1}{3} - \frac{1}{4} + \cdots$$

(See Example 4, Section 12.5, where we stated this fact without proof. We still have not proved it, but at least you can see where it comes from.)

(g) Question for discussion: Does the discovery that $f(x) = \ln x$ really improve matters? Or is the definition of $f(x)$ as an infinite series adequate for practical purposes?

42. Suppose that $a_1, a_2, a_3, \ldots$ is a sequence with the property that

$$\lim_{n \to \infty} \left| \frac{a_{n+1}}{a_n} \right| < 1$$

Why does it follow that the limit of the sequence is 0?

43. Use Problem 42 to investigate whether the sequence

$$10, \frac{10^2}{2^{10}}, \frac{10^3}{3^{10}}, \ldots, \frac{10^n}{n^{10}}, \ldots$$

approaches 0.

44. Prove the Root Test as follows.

(a) Assuming that $\rho > 1$ (or $\rho = \infty$), explain why $|a_n| > 1$ for large n and why that implies divergence of $\sum a_n$.

(b) Assuming that $\rho < 1$, choose r between ρ and 1 and explain why $|a_n| < r^n$ for large n. Why does it follow that $\sum a_n$ converges absolutely?

(c) Give examples showing that when $\rho = 1$ the Root Test does not distinguish between convergence and divergence.

45. When $\rho < 1$ in the proof of both the Ratio Test and Root Test, we select a point r between ρ and 1 and then compare the series $\sum |a_n|$ with the geometric series $\sum r^n$. Why does the argument break down if $\rho = 1$?

46. Find the flaw in the following argument that when $\rho = 1$ the Root Test shows the series $\sum a_n$ to be divergent.

(a) If $\lim_{n \to \infty} \sqrt[n]{|a_n|} = 1$, then $\sqrt[n]{|a_n|}$ is close to 1 for large n.

(b) It follows that $|a_n|$ is close to 1 for large n. Hence a_n cannot approach 0 as $n \to \infty$ and $\sum a_n$ diverges.

True–False Quiz

1. If the series $\sum a_n$ diverges, its sequence of partial sums is unbounded.

2. The series $\cos 1 - \cos \frac{1}{2} + \cos \frac{1}{3} - \cos \frac{1}{4} + \cdots$ converges.

3. If $\lim_{n \to \infty} a_n = 0$, the series $\sum a_n$ converges.

4. The series $1 + 1.02 + (1.02)^2 + (1.02)^3 + \cdots$ converges.

5. The series $1 + \dfrac{1}{\sqrt[4]{2}} + \dfrac{1}{\sqrt[4]{3}} + \dfrac{1}{\sqrt[4]{4}} + \cdots$ converges.

6. The series $\sum n/(n^2 + 1)$ converges.

7. The series $\sum 1/n(n - 1)$ diverges.

8. If $0 \le u_n \le b_n$ for all n and $\sum b_n$ diverges, so does $\sum u_n$.

9. If $\sum a_n^2$ converges, so does $\sum a_n$ (provided that each a_n is positive).

10. The series $1 - \frac{1}{2} + \frac{1}{3} - \frac{1}{4} + \cdots$ converges absolutely.

11. The series $\sum_{n=2}^{\infty} (-1)^n/\ln n$ diverges.

12. The series $1 - \dfrac{1}{2^2} + \dfrac{1}{3^2} - \dfrac{1}{4^2} + \cdots$ is conditionally convergent.

13. The Ratio Test can be used to determine whether the harmonic series converges.

14. The series $\sum n!/2^n$ converges.

15. If $\sum u_n$ converges, then $\lim_{n \to \infty} |u_{n+1}|/|u_n| < 1$.

Additional Problems

Find the limit of each of the following sequences (or determine that it does not exist).

1. $1, 5/8, (5/8)^2, \ldots, (5/8)^{n-1}, \ldots$

2. $\left\{ \sin \dfrac{n\pi}{2} \right\}$

3. $-10/9, (-10/9)^2, (-10/9)^3, \ldots, (-10/9)^n, \ldots$

4. $\{n/\ln n\}$

5. The typical term of the sequence is the sum $S_n = \sum_{k=0}^{n-1} \left(\frac{1}{2}\right)^k$.

6. The typical term of the sequence is the sum $S_n = \sum_{k=1}^{n} 1/k$.

7. Find the sum of the series $1 - 1 + \frac{1}{2} - \frac{1}{2} + \frac{1}{3} - \frac{1}{3} + \frac{1}{4} - \frac{1}{4} + \cdots$ (or conclude that it diverges).

8. Find the sum of the series $\sum_{k=2}^{\infty} 1/(k^2 - 1)$ (or conclude that it diverges).

9. Find the sum of the series $\frac{1}{4} + \frac{2}{6} + \frac{3}{8} + \frac{4}{10} + \cdots$ (or conclude that it diverges).

10. Find the sum of the series $\frac{1}{2} - \frac{1}{4} + \frac{1}{8} - \frac{1}{16} + \cdots$ (or conclude that it diverges).

11. Show that $0.999 \cdots = 1$.

12. Write the periodic decimal $0.424242 \cdots$ as a geometric series and find what rational number it represents.

13. Find the sum of the series $\sum_{n=1}^{\infty} 5^{-n}$ (or conclude that it diverges).

14. Find the sum of the series $\sum_{n=1}^{\infty} n/(2n - 1)$ (or conclude that it diverges).

15. If $f(x) = 1 + x + x^2 + x^3 + \cdots$, find $f(\frac{2}{5})$. What can be said about $f(1)$?

16. Explain why

$$\sum_{n=1}^{\infty} x^n = \frac{x}{1 - x} \qquad \text{if } |x| < 1$$

17. Express the function $f(x) = 1/(1 - x^3)$ as an infinite series. In what domain is this legitimate?

18. A ball that rebounds $\frac{1}{3}$ of the distance fallen is dropped from a height of 3 ft. How far does it travel?

19. A ball that rebounds $\frac{2}{3}$ of the distance fallen is dropped from a height of 6 ft. How far does it travel?

20. Explain why the series

$$\sum_{n=1}^{\infty} \frac{1}{(2n-1)^2}$$

converges. Use the formula

$$S \approx S_N + \int_{N+1}^{\infty} f(x)\,dx \qquad \text{(Section 12.3)}$$

to compute its sum correct to two decimal places.

21. Repeat Problem 20 for the series $\sum_{n=1}^{\infty} 1/n^3$.

22. Investigate the convergence of $\sum 1/(n^2+4)$ by using **(a)** the Integral Test; **(b)** the Comparison Test; **(c)** the Limit Comparison Test; and **(d)** the Ratio Test. Which tests lead to a conclusion? What is the conclusion?

23. Repeat Problem 22 for the series $\sum 1/(n^2-4)$.

24. Explain why the series $1 - \frac{1}{4} + \frac{1}{9} - \frac{1}{16} + \cdots$ converges. To compute its sum with an error less than 0.005, how many terms will suffice?

25. Explain why the series $1 - \frac{1}{2!} + \frac{1}{4!} - \frac{1}{6!} + \cdots$ converges and compute its sum correct to five decimal places. Compare with the exact value $\cos 1$ (to be confirmed in Chapter 13).

26. Classify the series

$$\sum_{n=1}^{\infty} \frac{(-1)^{n-1}n}{n^2+4}$$

as absolutely convergent, conditionally convergent, or divergent.

27. Repeat Problem 26 for the series

$$\sum_{n=1}^{\infty} \frac{(-1)^{n-1}}{\sqrt{n+1}}$$

Investigate the convergence of each of the following series.

28. $\sum \dfrac{1}{\sqrt[3]{n^2}}$

29. $\sum n(\tfrac{7}{8})^n$

30. $\sum \dfrac{n}{n^3+1}$

31. $\sum \dfrac{\sin n}{n^2}$

32. $\sum \ln\left(1+\dfrac{2}{n}\right)$

33. $\sum \sin \dfrac{1}{n}$

34. $\sum \left(\dfrac{n-1}{n+1}\right)\left(\dfrac{3}{4}\right)^n$

35. $\sum \dfrac{10^{-n}}{n}$

36. $\sum \dfrac{5^n}{n!}$

37. $\sum \dfrac{n^3}{e^n}$

38. $\sum \dfrac{2^n}{n^n}$

39. $\sum \dfrac{(2n-1)!}{n \cdot 5^{2n-1}}$

In each of the following, find all values of x for which the series converges. Distinguish between absolute and conditional convergence.

40. $\sum \dfrac{x^n}{n^2+1}$

41. $\sum \dfrac{nx^n}{2^n}$

42. $\sum \dfrac{n!x^{2n}}{2^n}$

43. $\sum \dfrac{(x-2)^n}{n}$

44. $\sum \dfrac{(x+1)^n}{n \ln n}$

45. $\sum \dfrac{(x-1)^n}{n^2 3^n}$

46. Show that the series

$$\sum \frac{1}{n(\ln n)^p}$$

converges if $p > 1$ and diverges if $p \le 1$.

47. If $\sum a_n$ is a convergent series of positive terms, show that the series $\sum a_n^3$ also converges.

48. Given the positive series $\sum a_n$, suppose that $\lim_{n\to\infty} na_n > 0$. Show that the series diverges.

13 | Power Series

$$\cdots + \frac{1}{x^3} + \frac{1}{x^2} + \frac{1}{x} + 1 + x + x^2 + x^3 + \cdots = 0$$

LEONHARD EULER
(a false statement appearing in his *Introduction to Infinitesimal Analysis* of 1748, thus demonstrating that even genius has its limitations)

As for everything else, so for a mathematical theory: beauty can be perceived but not explained.
ARTHUR CAYLEY (1821–1895)

My work always tried to unite the true with the beautiful; but when I had to choose one or the other, I usually chose the beautiful.
HERMANN WEYL (1885–1955)
(as quoted by Freeman Dyson)

Polynomials are functions of the type

$$P(x) = c_0 + c_1 x + c_2 x^2 + \cdots + c_n x^n = \sum_{k=0}^{n} c_k x^k$$

where n is a nonnegative integer. They are continuous everywhere, have derivatives of all orders (which are easy to find), and are no problem to integrate. They serve as reasonable approximations to other functions (if their coefficients are properly chosen). In short, they are useful and easy to work with.

A **power series** (with base point 0) is an expression of the form

$$c_0 + c_1 x + c_2 x^2 + \cdots = \sum_{k=0}^{\infty} c_k x^k$$

It is an obvious generalization of the idea of polynomial, and turns out to have many of the same properties. In this chapter we study functions defined by series of this type, and we show how the elementary functions of analysis can be expressed in terms of such series. The results are both beautiful and useful. It is astonishing how many loose ends are tied up by an understanding of power series, not to mention the clarity of vision that comes from looking at elementary functions "as they really are" and the feeling of power one gets from realizing how they are actually computed.

Numerical tables, calculators, and computers lose some of their mystery when power series are studied. Approximation techniques (and the analysis of their errors) seem less forbidding. Hidden connections between different parts of mathematics emerge. In short, some light dawns!

We do not mean to suggest that utopia is at hand; many of these insights take time to mature. But the subject of power series does pretty well wrap up single-variable calculus. In some ways this chapter is the high point of its development.

13.1
FUNCTIONS DEFINED BY POWER SERIES

For each value of x the infinite series

$$1 - \frac{x^2}{2!} + \frac{x^4}{4!} - \frac{x^6}{6!} + \cdots = \sum_{n=0}^{\infty} \frac{(-1)^n x^{2n}}{(2n)!}$$

is absolutely convergent (as you can check by the Ratio Test). Hence it has a sum, the value of which depends on x. When $x = 0$, for example, the sum of the series is 1. (Why?) When $x = 1$ the sum is

$$1 - \frac{1}{2!} + \frac{1}{4!} - \frac{1}{6!} + \cdots$$

(a number we may not know, but which is nonetheless definite). Thus we are dealing with a function (say f) defined by the rule

$$f(x) = \sum_{n=0}^{\infty} \frac{(-1)^n x^{2n}}{(2n)!}$$

There is no difference in principle between this rule and the kind we have studied before. When we define $g(x) = \sin x$, for example, we may be dulled by familiarity into thinking that the rule is simple. It is not, however, as becomes apparent when we try to compute values of g. Is it any easier to find $g(1) = \sin 1$ than

$$f(1) = 1 - \frac{1}{2!} + \frac{1}{4!} - \frac{1}{6!} + \cdots ?$$

That may sound like a silly question in view of the availability of trigonometric tables and sine keys on calculators. But after all, tables and calculators give the values of sine *because it is hard to compute* (and because its values are needed in a variety of applications). Without their help, is our question so silly? As a matter of fact, the answer is that $f(1)$ is easier to find than $g(1)$, because we need only compute partial sums of the series to approximate its value as accurately as we please. Can we do that with sin 1?

The moral is clear. Functions defined by series are perfectly respectable, and from the computational point of view they are easier to deal with than many of the familiar functions of analysis. The problem is that they may intimidate us by *looking* complicated; a "closed-form" expression like

$g(x) = \sin x$ seems simpler because it is concise. However (as we will show in the next section) this expression is in reality nothing more than a name for a function defined by a series! One of the objectives of this chapter is to show that *every* (nonalgebraic) function we have studied can be represented by a series.

In this section we discuss the properties of functions defined by series of the type

$$f(x) = c_0 + c_1(x - a) + c_2(x - a)^2 + \cdots = \sum_{n=0}^{\infty} c_n(x - a)^n$$

where a is a number called the **base point** and $c_0, c_1, c_2, \ldots$ are the **coefficients** of the series. We call this expression a *power series* because each term involves a power of $x - a$. It is a generalization of the idea of polynomial.

Our first objective is to discover the domain of such a function, that is, the set of points for which the series converges. Here are four examples to give you an idea of what to expect.

□ **Example 1**   The domain of

$$f(x) = \sum_{n=0}^{\infty} n! x^n$$

consists of the single point $x = 0$. For if $x \neq 0$, we find

$$\rho = \lim_{n \to \infty} \frac{(n + 1)! |x|^{n+1}}{n! |x|^n} = |x| \lim_{n \to \infty} (n + 1) = \infty$$

and the Ratio Test says that the series diverges.   □

You can see that every power series

$$f(x) = c_0 + c_1(x - a) + c_2(x - a)^2 + \cdots$$

converges at the base point $x = a$. Example 1 shows that in some cases it may converge nowhere else (in which case it is a very dull function).

□ **Example 2**   The domain of

$$f(x) = \sum_{n=0}^{\infty} \frac{x^n}{n!}$$

is the entire number line, as we saw in Example 3, Section 12.6. This is the opposite extreme from Example 1, the series converging everywhere instead of nowhere but at the base point.   □

□ **Example 3**   The domain of

$$f(x) = \sum_{n=0}^{\infty} x^n$$

(a geometric series) is the interval $(-1,1)$, as we know from Section 12.2. Notice that the interval is symmetric about the base point $a = 0$, a property of the domain of a power series that turns out to be typical.   □

□ **Example 4**   The domain of

$$f(x) = (x - 1) - \frac{(x-1)^2}{2} + \frac{(x-1)^3}{3} - \cdots = \sum_{n=1}^{\infty} \frac{(-1)^{n-1}}{n}(x-1)^n$$

is the interval $(0,2]$, as we saw in Example 5, Section 12.6. For it turns out that $\rho = |x - 1|$ and hence the interval of convergence (not counting endpoints, which must be investigated separately) is the solution set of the inequality $|x - 1| < 1$. This is the source of symmetry about the base point. The solutions of the inequality are the points that are within 1 unit of the point $a = 1$, that is, they satisfy $0 < x < 2$.   □

With these examples in mind, let's use the Ratio Test to investigate the domain of

$$f(x) = \sum_{n=0}^{\infty} c_n (x - a)^n$$

When $x \neq a$ we find

$$\rho = \lim_{n \to \infty} \frac{|c_{n+1}||x - a|^{n+1}}{|c_n||x - a|^n} = |x - a| \lim_{n \to \infty} \left| \frac{c_{n+1}}{c_n} \right|$$

so the results of the test depend on the value of

$$L = \lim_{n \to \infty} \left| \frac{c_{n+1}}{c_n} \right| \qquad \text{(provided that } L \text{ is definite)}$$

Of course we are assuming that none of the coefficients is zero. To avoid that assumption, the Root Test may be used instead. But it is no loss of generality to omit all terms with zero coefficients, in which case the Ratio Test applies.

1. If $L > 0$ (not including $L = \infty$), then $\rho = L|x - a|$ and the Ratio Test says that the series converges absolutely when

$$L|x - a| < 1, \text{ that is, } |x - a| < r \qquad \text{(where } r = 1/L)$$

Not counting endpoints (which must be investigated separately) the domain of f is the interval $(a - r, a + r)$, with midpoint a and radius r.
2. If $L = 0$, then $\rho = 0$ and the series is absolutely convergent for all x.
3. If $L = \infty$, then $\rho = \infty$ and the series diverges for all $x \neq a$ (converging only at $x = a$).

Domain of a Power Series

The power series

$$\sum_{n=0}^{\infty} c_n(x-a)^n = c_0 + c_1(x-a) + c_2(x-a)^2 + \cdots$$

converges absolutely in an interval of one of the following types.

1. An open interval $(a-r, a+r)$ centered at the base point a and having radius $r > 0$ (called the **radius of convergence**). The series may also converge (absolutely or conditionally) at one or both endpoints.
2. The entire number line $(-\infty, \infty)$, in which case we say that $r = \infty$.
3. The single point $x = a$, in which case we say that $r = 0$.

Remark   In deriving this result we left out one possibility. What if L is nothing definite? Such limits can occur, for example $\lim_{n\to\infty} \sin n$, which is neither a number nor ∞. The way out of this difficulty is to introduce a type of limit we have not discussed. Thus our *proof* is defective, but the *result* is as we have stated it.

This theorem is more powerful than it might appear at first. Suppose, for example, that we have discovered *one* point at which the series

$$f(x) = c_0 + c_1 x + c_2 x^2 + \cdots$$

converges, say $x = 2$. Knowing nothing else about the series, we can say that it must converge (at least) for all x between 2 and -2. The reason for this is that we know the domain of f is not merely a scattered set of points, but an interval centered at the base point $(a = 0)$. Hence we lift ourselves by our bootstraps and deduce from a scrap of information (convergence at $x = 2$) a whole interval in which the series converges. Of course the interval may be larger; one way to state the theorem (when $a = 0$) is that the radius of convergence is the least upper bound of all $|x|$ for which the series converges (unless there is no upper bound, in which case the radius is $r = \infty$).

Now we turn to the question of differentiating a power series. If

$$f(x) = c_0 + c_1(x-a) + c_2(x-a)^2 + c_3(x-a)^3 + \cdots = \sum_{n=0}^{\infty} c_n(x-a)^n$$

it is reasonable to expect (by analogy with polynomials) that

$$f'(x) = c_1 + 2c_2(x-a) + 3c_3(x-a)^2 + \cdots = \sum_{n=1}^{\infty} nc_n(x-a)^{n-1}$$

The proof that this actually works is not easy, and we omit it. The result, however, is crucial for the development of this chapter.

Term-by-Term Differentiation of a Power Series

Suppose that the domain of

$$f(x) = \sum_{n=0}^{\infty} c_n(x-a)^n = c_0 + c_1(x-a) + c_2(x-a)^2 + \cdots$$

is not a single point. Then f is differentiable in the same domain (with the possible exception of its endpoints, if any). Moreover,

$$f'(x) = \sum_{n=1}^{\infty} nc_n(x-a)^{n-1} = c_1 + 2c_2(x-a) + 3c_3(x-a)^2 + \cdots$$

□ **Example 5**   The domain of

$$f(x) = (x-1) - \frac{(x-1)^2}{2} + \frac{(x-1)^3}{3} - \frac{(x-1)^4}{4} + \cdots$$

is $(0,2]$. (See Example 4.) According to our theorem on derivatives,

$$f'(x) = 1 - \frac{2(x-1)}{2} + \frac{3(x-1)^3}{3} - \frac{4(x-1)^4}{4} + \cdots$$
$$= 1 - (x-1) + (x-1)^2 - (x-1)^3 + \cdots$$

the domain of f' being at least $(0,2)$. (The endpoint $x=2$ must be excluded unless further investigation shows that $f'(2)$ exists.)

It is easy to overlook the power of the derivative theorem. The series

$$1 - (x-1) + (x-1)^2 - (x-1)^3 + \cdots$$

in Example 5 can be investigated by the Ratio Test, the result being that it converges in the interval $(0,2)$. But that says nothing about its sum. The derivative theorem tells us that the sum of the series is $f'(x)$, which is not a trivial result. It sounds obvious, but in fact it is not; without this theorem to lean on, we have no way of knowing what the sum is.   □

□ **Example 6**   We can do more with Example 5. Rewrite the series for the derivative in the form

$$f'(x) = 1 + (1-x) + (1-x)^2 + (1-x)^3 + \cdots$$

and fit it to the geometric series

$$1 + t + t^2 + t^3 + \cdots = \frac{1}{1-t} \qquad (-1 < t < 1)$$

Letting $t = 1 - x$, we have

$$f'(x) = \frac{1}{1-(1-x)} = \frac{1}{x} \qquad (0 < x < 2)$$

Hence $y = f(x)$ satisfies the differential equation $dy/dx = 1/x$. Solve the equation to obtain

$$y = \int \frac{dx}{x} = \ln x + C \qquad \text{(absolute value not needed because } 0 < x < 2)$$

Since $f(1) = 0$ (look at the series for f), we find $C = 0$ and hence

$$y = f(x) = \ln x$$

Thus we have discovered the formula

$$\ln x = (x - 1) - \frac{(x - 1)^2}{2} + \frac{(x - 1)^3}{3} - \frac{(x - 1)^4}{4} + \cdots \qquad (0 < x < 2)$$

$\square$

Remark   This formula might have been anticipated in Problem 31, Section 11.5, where the Taylor approximation of $\ln x$ (in a neighborhood of $a = 1$) is given as

$$\ln x \approx (x - 1) - \frac{(x - 1)^2}{2} + \frac{(x - 1)^3}{3} - \cdots + (-1)^{n-1} \frac{(x - 1)^n}{n}$$

If the error in this approximation approaches zero as $n \to \infty$, the formula can be turned into an infinite series for $\ln x$. In Example 6 we bypassed that question (which is not an easy one) by manipulating known series to obtain a formula for $\ln x$.

Our next theorem about functions defined by power series is that they are continuous. In view of the derivative theorem, this sounds obvious (since a differentiable function is automatically continuous). However, existence of the derivative is guaranteed only at interior points of the domain of the function, so the endpoints (if any) require special consideration. Since the details involve more advanced treatment of convergence, we omit the proof.

Continuity of Power Series

Suppose that the domain I of

$$f(x) = c_0 + c_1(x - a) + c_2(x - a)^2 + \cdots$$

is not a single point. Then f is continuous in I (including endpoints at which f is defined, if any). That is, if $x_0 \in I$,

$$\lim_{x \to x_0} f(x) = f(x_0)$$

$\square$ **Example 7**   The function

$$f(x) = (x - 1) - \frac{(x - 1)^2}{2} + \frac{(x - 1)^3}{3} - \frac{(x - 1)^4}{4} + \cdots$$

in Example 5 is defined for $0 < x \leq 2$. Hence it is continuous at $x = 2$ as well as for all x in the open interval $(0,2)$. This small detail of the theory enables us to prove a formula we have stated before, namely

$$\ln 2 = 1 - \tfrac{1}{2} + \tfrac{1}{3} - \tfrac{1}{4} + \cdots$$

For we know from Example 6 that $\ln x = f(x)$ if $0 < x < 2$. Since f is continuous at 2, we have

$$\lim_{x \uparrow 2} \ln x = \lim_{x \uparrow 2} f(x) = f(2)$$

that is, $\ln 2 = 1 - \tfrac{1}{2} + \tfrac{1}{3} - \tfrac{1}{4} + \cdots$.   □

Functions defined by power series can also be integrated term-by-term. More precisely, the following statement can be proved.

Term-by-Term Integration of a Power Series

Suppose that the domain of

$$f(t) = c_0 + c_1(t - a) + c_2(t - a)^2 + \cdots$$

is not a single point. If x is any point of the domain, then

$$\int_a^x f(t)\, dt = c_0(x - a) + \frac{1}{2} c_1(x - a)^2 + \frac{1}{3} c_2(x - a)^3 + \cdots$$

The proof of this theorem need not be deferred to a more advanced course. Let I be the domain of f and let

$$\begin{aligned} F(x) &= c_0(x - a) + \tfrac{1}{2} c_1(x - a)^2 + \tfrac{1}{3} c_2(x - a)^3 + \cdots \\ &= (x - a)[c_0 + \tfrac{1}{2} c_1(x - a) + \tfrac{1}{3} c_2(x - a)^2 + \cdots] \end{aligned}$$

The absolute value of each term of the series in brackets is no larger than the absolute value of the corresponding term of

$$f(x) = c_0 + c_1(x - a) + c_2(x - a)^2 + \cdots$$

so the series in brackets converges whenever the series for f does. Hence so does the series for F. Moreover, our theorem on derivatives says that

$$F'(x) = c_0 + c_1(x - a) + c_2(x - a)^2 + \cdots = f(x)$$

for all interior points of I. This makes F an antiderivative of f, so for every interior point $x \in I$ we have

$$\begin{aligned} \int_a^x f(t)\, dt &= F(x) - F(a) = F(x) \qquad \text{[because } F(a) = 0] \\ &= c_0(x - a) + \tfrac{1}{2} c_1(x - a)^2 + \tfrac{1}{3} c_2(x - a)^3 + \cdots \end{aligned}$$

Our theorem on continuity extends this formula to the endpoints of I (if any).

□ **Example 8**   Replacing x by $-t^2$ in the geometric series

$$\frac{1}{1-x} = 1 + x + x^2 + x^3 + \cdots \qquad (-1 < x < 1)$$

we have

$$\frac{1}{1+t^2} = 1 - t^2 + t^4 - t^6 + \cdots \qquad (-1 < t < 1)$$

If $x \in (-1,1)$, we may integrate from 0 to x to obtain

$$\int_0^x \frac{dt}{1+t^2} = \int_0^x (1 - t^2 + t^4 - t^6 + \cdots)\, dt$$

that is,

$$\tan^{-1} x = x - \frac{x^3}{3} + \frac{x^5}{5} - \frac{x^7}{7} + \cdots \qquad (-1 < x < 1)$$

In the problem set we ask you to explain why this equation is also valid at $x = 1$, the result being the famous Leibniz formula

$$\frac{\pi}{4} = 1 - \frac{1}{3} + \frac{1}{5} - \frac{1}{7} + \cdots$$

The preceding formula for $\tan^{-1} x$ is known as Gregory's series. It was discovered in 1671 by the Scottish mathematician James Gregory (who anticipated the publication of Taylor's series by more than forty years). See Problem 41, Section 11.5, in which the formula

$$\tan^{-1} x \approx x - \frac{x^3}{3} + \frac{x^5}{5} - \cdots + (-1)^{n-1}\frac{x^{2n-1}}{2n-1}$$

is given. By now it should be clear that there is a close connection between power series and Taylor polynomials. In this section we are not exploiting it; instead we are obtaining new series by differentiation and integration of known results. We will investigate Taylor's approach to series in the next section.   □

Problem Set 13.1

Find the domain of each of the following functions.

1. $f(x) = \sum_{n=0}^{\infty} \frac{(-1)^n x^n}{n!}$

2. $f(x) = \sum_{n=0}^{\infty} \frac{(-1)^n x^{2n}}{(2n)!}$

3. $f(x) = \sum_{n=1}^{\infty} \frac{(x-1)^n}{n^n}$

4. $f(x) = \sum_{n=1}^{\infty} \frac{2^n(x+1)^n}{n^n}$

5. $f(x) = \sum_{n=0}^{\infty} \frac{(-1)^n x^n}{2^n}$

6. $f(x) = \sum_{n=0}^{\infty} 3^n x^n$

7. $f(x) = \sum_{n=1}^{\infty} \frac{(x-1)^n}{n}$

8. $f(x) = \sum_{n=1}^{\infty} \frac{(-1)^{n-1}(x-2)^n}{\sqrt{n}}$

9. $f(x) = \sum_{n=1}^{\infty} \frac{(-1)^{n-1}(x+1)^{2n-1}}{2n-1}$

10. $f(x) = \sum_{n=1}^{\infty} \frac{(x+2)^n}{n^2}$

11. If f is the function in Problem 1, what is the value of $f(1)$ correct to three decimal places? Compare with the value of e^{-1} found from a table or calculator. [In the next section we will show that $f(x) = e^{-x}$.]

12. If f is the function in Problem 2, what is the value of $f(1)$ correct to three decimal places? Compare with the value of $\cos 1$ found from a table or calculator. [In the next section we will show that $f(x) = \cos x$.]

13. If f is the function in Problem 5, write $f(x)$ in "closed form," that is, as a finite expression not involving series. *Hint:* The series is geometric.

14. Repeat Problem 13 for the function in Problem 6.

15. Find the domain of

$$f(x) = 1 + (x - 2) + (x - 2)^2 + (x - 2)^3 + \cdots$$

and express $f(x)$ in closed form.

16. Repeat Problem 15 for the function

$$f(x) = 1 + (x - 1) + (x - 1)^2 + (x - 1)^3 + \cdots$$

17. Suppose that the domain of

$$f(x) = c_0 + c_1(x - 1) + c_2(x - 1)^2 + \cdots$$

includes $x = -1$. Does it necessarily include $x = 2$? $x = 3$? Explain.

18. Suppose that the domain of

$$f(x) = c_0 + c_1(x + 1) + c_2(x + 1)^2 + \cdots$$

includes $x = 1$. Does it necessarily include $x = 0$? $x = -3$? Explain.

19. The domain of the function

$$f(x) = (x - 1) - \tfrac{1}{2}(x - 1)^2 + \tfrac{1}{3}(x - 1)^3 - \tfrac{1}{4}(x - 1)^4 + \cdots$$

is (0,2]. (See Example 5.) Our theorem on derivatives guarantees that the domain of

$$f'(x) = 1 - (x - 1) + (x - 1)^2 - (x - 1)^3 + \cdots$$

is at least (0,2), but says nothing about the endpoint $x = 2$. Should that point be included? Explain.

20. Obtain a series formula for $1/(1 - x)^2$ by differentiating the geometric series

$$\frac{1}{1 - x} = 1 + x + x^2 + x^3 + \cdots$$

What is the domain of the result?

21.–30. Find $f'(x)$ and give its domain in Problems 1 through 10.

In each of the following, find $\int_a^x f(t)\, dt$, where a is the base point of the given series and x is any point of the domain of f.

31. $f(t) = 1 + t + \dfrac{t^2}{2!} + \dfrac{t^3}{3!} + \cdots$

32. $f(t) = t - \dfrac{t^3}{3!} + \dfrac{t^5}{5!} - \dfrac{t^7}{7!} + \cdots$

33. $f(t) = (t - 1) + (t - 1)^2 + (t - 1)^3 + \cdots$

34. $f(t) = 1 + \tfrac{1}{2}(t + 1)^2 + \tfrac{1}{4}(t + 1)^4 + \tfrac{1}{8}(t + 1)^6 + \cdots$

35. Derive the formula

$$\ln(1 + x) = x - \frac{x^2}{2} + \frac{x^3}{3} - \frac{x^4}{4} + \cdots \quad (-1 < x \le 1)$$

as follows.

(a) Show that the domain of the function

$$f(x) = x - \frac{x^2}{2} + \frac{x^3}{3} - \frac{x^4}{4} + \cdots$$

is the interval $(-1,1]$.

(b) Differentiate the series in part (a) to show that $y = f(x)$ satisfies the differential equation

$$\frac{dy}{dx} = \frac{1}{1 + x}$$

Is the domain of f' the same as the domain of f?

(c) Confirm that $y = 0$ when $x = 0$ and solve the above differential equation (with this initial condition) to show that

$$y = \ln(1 + x) \quad (-1 < x < 1)$$

Why does the result also hold at $x = 1$?

36. Do Problem 35 more directly by integrating the geometric series

$$\frac{1}{1 + t} = 1 - t + t^2 - t^3 + \cdots$$

from 0 to x $(-1 < x < 1)$. Why does the result also hold when $x = 1$?

37. In the text we derived the formula

$$\ln x = (x - 1) - \frac{(x - 1)^2}{2} + \frac{(x - 1)^3}{3} - \frac{(x - 1)^4}{4} + \cdots \quad (0 < x \le 2)$$

Replace x by $1 + x$ to confirm Problem 35.

38. Differentiate the series

$$\ln x = (x - 1) - \frac{(x - 1)^2}{2} + \frac{(x - 1)^3}{3}$$
$$- \frac{(x - 1)^4}{4} + \cdots \qquad (0 < x \le 2)$$

to obtain

$$\frac{1}{x} = 1 - (x - 1) + (x - 1)^2 - (x - 1)^3 + \cdots$$

Letting $x = 2$, we have $\frac{1}{2} = 1 - 1 + 1 - 1 + \cdots$. The series on the right, however, does not converge to $\frac{1}{2}$. (It diverges!) What went wrong?

39. Derive the formula

$$\frac{1}{2} \ln \frac{1 + x}{1 - x} = x + \frac{x^3}{3} + \frac{x^5}{5} + \cdots \qquad (-1 < x < 1)$$

as follows.

(a) Replace x by $-x$ in Problem 35 to obtain a series for $\ln (1 - x)$. What is the domain of the result?

(b) Subtract the series for $\ln (1 - x)$ from the series for $\ln (1 + x)$. Why doesn't the result hold at $x = 1$?

40. Use Problem 39 to explain why

$$\tanh^{-1} x = x + \frac{x^3}{3} + \frac{x^5}{5} + \cdots \qquad (-1 < x < 1)$$

and note the similarity to Gregory's series for $\tan^{-1} x$. (See Section 9.4.)

41. Confirm the series in Problem 40 as follows.

(a) Replace t by t^2 in the geometric series

$$\frac{1}{1 - t} = 1 + t + t^2 + t^3 + \cdots$$

and integrate from 0 to $x (-1 < x < 1)$.

(b) Use the formula

$$\int \frac{dt}{1 - t^2} = \tanh^{-1} t + C \qquad (-1 < t < 1)$$

from Section 9.4.

42. Derive the Leibniz formula

$$\frac{\pi}{4} = 1 - \frac{1}{3} + \frac{1}{5} - \frac{1}{7} + \cdots$$

by explaining why Gregory's series,

$$\tan^{-1} x = x - \frac{x^3}{3} + \frac{x^5}{5} - \frac{x^7}{7} + \cdots \qquad (-1 < x < 1)$$

holds at $x = 1$. Is the formula also valid at $x = -1$?

43. Derive the formula

$$\frac{\pi}{4} = 4 \tan^{-1} \frac{1}{5} - \tan^{-1} \frac{1}{239}$$

as follows.

(a) Letting $t = \tan^{-1} (\frac{1}{5})$, use trigonometric identities to show that

$$\tan 2t = \frac{5}{12}, \quad \tan 4t = \frac{120}{119}, \quad \tan \left(4t - \frac{\pi}{4}\right) = \frac{1}{239}$$

(b) Conclude that

$$\frac{\pi}{4} = 4 \tan^{-1} \frac{1}{5} - \tan^{-1} \frac{1}{239}$$

(With Gregory's series to compute the inverse tangents, this formula was used by John Machin in 1706 to find 100 decimal places of π.)

44. Use Problem 43 and Gregory's series to compute π correct to seven decimal places. *Hint:* Control the error by using the Alternating Series Test in Section 12.5.

45. Suppose that

$$\sum_{n=0}^{\infty} c_n (x - a)^n = 0$$

for all x in some neighborhood of the base point a. Show that every coefficient of the series must be zero by putting $x = a$ into the series and its successive derivatives.

46. Use Problem 45 to show that if

$$\sum_{n=0}^{\infty} a_n (x - a)^n = \sum_{n=0}^{\infty} b_n (x - a)^n$$

for all x in some neighborhood of a, then $a_n = b_n$ for every n.

47. Suppose that

$$f(x) = \sum_{n=0}^{\infty} c_n (x - a)^n$$

for all x in some neighborhood of a. Use Problem 46 to explain why no other power series (with base point a) can represent $f(x)$ in this neighborhood. (In other words the power series representation of a function is unique.)

48. Suppose that

$$f(x) = c_0 + c_1 x + c_2 x^2 + \cdots$$

is an even function, that is, $f(-x) = f(x)$ for all x in the domain. Show that

$$f(x) = c_0 + c_2 x^2 + c_4 x^4 + \cdots$$

49. Suppose that

$$f(x) = c_0 + c_1 x + c_2 x^2 + \cdots$$

is an odd function, that is, $f(-x) = -f(x)$ for all x in the domain. Show that

$$f(x) = c_1 x + c_3 x^3 + c_5 x^5 + \cdots$$

50. At the beginning of this chapter we gave Euler's "equation"

$$\cdots + \frac{1}{x^3} + \frac{1}{x^2} + \frac{1}{x} + 1 + x + x^2 + x^3 + \cdots = 0$$

Consider the following argument in defense of it.

(a) Replace x by $1/x$ in the geometric series

$$\frac{1}{1-x} = 1 + x + x^2 + x^3 + \cdots$$

and subtract 1 from each side of the result to obtain

$$\frac{1}{x-1} = \frac{1}{x} + \frac{1}{x^2} + \frac{1}{x^3} + \cdots$$

(b) Add the two series in part (a) to obtain Euler's equation. Why is this equation false and what is wrong with the "proof"?

13.2
TAYLOR SERIES

Let f be a function represented by a power series in a neighborhood of the base point a, say

$$f(x) = c_0 + c_1(x-a) + c_2(x-a)^2 + c_3(x-a)^3 + \cdots$$

Then we can differentiate term-by-term, obtaining

$$f'(x) = 1c_1 + 2c_2(x-a) + 3c_3(x-a)^2 + 4c_4(x-a)^3 + \cdots$$

in the same neighborhood of a. Similarly,

$$f''(x) = 1\cdot 2c_2 + 2\cdot 3c_3(x-a) + 3\cdot 4c_4(x-a)^2 + 4\cdot 5c_5(x-a)^3 + \cdots$$

$$f'''(x) = 1\cdot 2\cdot 3c_3 + 2\cdot 3\cdot 4c_4(x-a) + 3\cdot 4\cdot 5c_5(x-a)^2 + 4\cdot 5\cdot 6c_6(x-a)^3 + \cdots$$

and so on, each derivative being defined in the same neighborhood of a. Evaluating these derivatives at $x = a$, we find

$$f^{(0)}(a) = 0!c_0, \ f^{(1)}(a) = 1!c_1, \ f^{(2)}(a) = 2!c_2, \ f^{(3)}(a) = 3!c_3, \ \cdots$$

The general formula (which can be formally established by induction) is clear:

$$f^{(n)}(a) = n!c_n \qquad n = 0, 1, 2, \ldots$$

Hence we have the following theorem.

If
$$f(x) = c_0 + c_1(x-a) + c_2(x-a)^2 + \cdots$$
then
$$c_n = \frac{f^{(n)}(a)}{n!} \qquad n = 0, 1, 2, \ldots$$

Another way to state this theorem is that a function cannot be represented by two different power series with the same base point. In other words, *the power series representation of a function is unique.* What makes this statement so powerful is that it leaves us no choice in our attempts to represent known functions by power series (assuming that the base point is fixed). The following example shows what we mean.

▫ **Example 1**   If the function $f(x) = e^x$ can be represented by a power series with base point $a = 0$, what is the series?

Solution: Suppose that such a series exists, say

$$f(x) = c_0 + c_1 x + c_2 x^2 + \cdots$$

Since $f^{(n)}(x) = e^x$ for every n, we have

$$f^{(n)}(0) = 1 \qquad n = 0, 1, 2, \ldots$$

Our uniqueness theorem says that the coefficients of the above series *must* be

$$c_n = \frac{1}{n!} \qquad n = 0, 1, 2, \ldots$$

Hence if e^x can be written as a power series based at $a = 0$, no search is required; the series must be

$$e^x = 1 + x + \frac{x^2}{2!} + \frac{x^3}{3!} + \cdots \qquad \square$$

Given a function f (with derivatives of all orders in a neighborhood of the point a), we call the expression

$$\sum_{n=0}^{\infty} \frac{f^{(n)}(a)}{n!} (x - a)^n = f(a) + f'(a)(x - a)$$

$$+ \frac{f''(a)}{2!}(x - a)^2 + \frac{f'''(a)}{3!}(x - a)^3 + \cdots$$

the **Taylor series** associated with f at the base point a. Its partial sums,

$$P_n(x) = f(a) + f'(a)(x - a) + \frac{f''(a)}{2!}(x - a)^2 + \cdots + \frac{f^{(n)}(a)}{n!}(x - a)^n$$

($n = 0, 1, 2, \ldots$) are the *n*th-order *Taylor polynomials* defined in Section 11.5. When $a = 0$ the Taylor series is often called a **Maclaurin series** in honor of the Scottish mathematician Colin Maclaurin (1698–1746).

The principal question in connection with a Taylor series is not how it is found (that's usually the easy part!), but whether it represents the function with which it is associated.

□ **Example 2**   Does the Taylor series associated with $f(x) = e^x$ at $a = 0$ (its Maclaurin series) represent e^x?

Solution: In Example 1 we found the series to be

$$S(x) = 1 + x + \frac{x^2}{2!} + \frac{x^3}{3!} + \cdots$$

where we use $S(x)$ instead of $f(x)$ to name it because the question is whether $S(x) = f(x)$. In general this is not an easy question, but in the case of e^x we have already answered it. (See Example 4, Section 12.6.) What we did was to differentiate the series for $y = S(x)$ to obtain $dy/dx = y$. Separating the variables and integrating, we found $y = Ce^x$. Since $y = 1$ when $x = 0$ (why?), we obtained $C = 1$ and hence $y = e^x$, that is, $S(x) = f(x)$.   □

Example 2 illustrates what the general question (sometimes called the *representation problem*) is. Given a function f (with derivatives of all orders in a neighborhood of the base point a), we can construct its Taylor series at a by successive differentiation (as in Example 1). We can test the series for convergence, finding a domain (centered at a) in which the series has a sum. Let $S(x)$ be the sum. *The question is whether $f(x) = S(x)$.* Of course the question is real only if f is described by a formula (like e^x) which is independent of series. If f is *defined* by the rule

$$f(x) = c_0 + c_1(x - a) + c_2(x - a)^2 + \cdots$$

asking whether the series represents f is like asking whether Beethoven wrote Beethoven's symphonies.

One way to answer the question is to write $f(x) \approx P_n(x)$, $n = 0, 1, 2, \ldots,$ where $P_n(x)$ is the nth-order Taylor polynomial associated with f at a (and a partial sum of the Taylor series). The problem is that these approximations are not necessarily good ones, nor do they necessarily improve as n increases. What the question comes down to is whether the error is small, that is, whether (for each x in the domain)

$$\lim_{n \to \infty} R_n(x) = 0 \qquad \text{where } R_n(x) = f(x) - P_n(x)$$

In Section 11.5 we called $R_n(x)$ the *remainder* in the approximation of $f(x)$ by $P_n(x)$; the *Lagrange form* of it is

$$R_n(x) = \frac{f^{(n+1)}(c)}{(n + 1)!}(x - a)^{n+1} \qquad \text{where } c \text{ is between } a \text{ and } x$$

For each x in the domain of the Taylor series associated with f at the base point a, let $S(x)$ be the sum of the series. Then

$$f(x) = S(x) \Leftrightarrow \lim_{n \to \infty} R_n(x) = 0$$

where $R_n(x)$ is the remainder in the approximation of $f(x)$ by its nth-order Taylor polynomial.

Students are sometimes skeptical about the seriousness of the representation problem (because we seldom encounter functions that are not equal to their Taylor series). The following example shows that caution is needed.

□ **Example 3**   Define f by the rule

$$f(x) = \begin{cases} e^{-1/x^2} & \text{if } x \neq 0 \\ 0 & \text{if } x = 0 \end{cases}$$

It can be shown that this function has derivatives of all orders (defined for all x) and that $f^{(n)}(0) = 0$, $n = 0, 1, 2, \ldots$. Since it is tricky to compute these derivatives (and not very enlightening), we omit the proof. (See, however, Problem 46, Section 11.2.) But look at the implication! The Taylor series associated with f at $a = 0$ is

$$S(x) = 0 + 0x + 0x^2 + 0x^3 + \cdots = 0 \qquad \text{for all } x$$

This function differs from $f(x)$ for all $x \neq 0$. (!)   □

In view of Example 3, we cannot assume that a Taylor series found by successive differentiation of a given function is equal to the function; one way or another it must be proved. Fortunately we do not have to prove it by demonstrating that

$$\lim_{n \to \infty} R_n(x) = 0$$

This direct attack is often difficult, and in most cases of interest it is unnecessary. An illustration is Example 2, in which we proved that e^x is equal to its Maclaurin series by differentiating the series. Similar devices may be used in connection with most of the elementary functions of analysis. As a matter of fact, we have already accomplished something in that direction, as the following list of previously derived results shows.

$$\frac{1}{1-x} = 1 + x + x^2 + x^3 + \cdots \qquad (-1 < x < 1)$$

(Example 5, Section 12.2)

$$\frac{1}{(1-x)^2} = 1 + 2x + 3x^2 + 4x^3 + \cdots \qquad (-1 < x < 1)$$

(Problem 20, Section 13.1)

$$e^x = 1 + x + \frac{x^2}{2!} + \frac{x^3}{3!} + \cdots$$

(Example 2 in this section)

$$\ln(1+x) = x - \frac{x^2}{2} + \frac{x^3}{3} - \frac{x^4}{4} + \cdots \qquad (-1 < x \leq 1)$$

(Problem 35, Section 13.1)

$$\tan^{-1} x = x - \frac{x^3}{3} + \frac{x^5}{5} - \frac{x^7}{7} + \cdots \qquad (-1 \leq x \leq 1)$$

(Example 8, Section 13.1)

$$\tanh^{-1} x = x + \frac{x^3}{3} + \frac{x^5}{5} + \frac{x^7}{7} + \cdots \qquad (-1 < x < 1)$$

(Problem 40, Section 13.1)

□ **Example 4**   Find the Maclaurin series associated with $f(x) = \sin x$ and show that it represents $\sin x$.

Solution: Successive derivatives of f (and their values at $a = 0$) are

$$\begin{aligned}
f(x) &= \sin x & f(0) &= 0 \\
f'(x) &= \cos x & f'(0) &= 1 \\
f''(x) &= -\sin x & f''(0) &= 0 \\
f'''(x) &= -\cos x & f'''(0) &= -1 \\
f^{(4)}(x) &= \sin x & f^{(4)}(0) &= 0 \\
&\;\;\vdots & &\;\;\vdots
\end{aligned}$$

Hence the series we seek is

$$S(x) = 0 + 1x + 0x^2 + \frac{(-1)}{3!}x^3 + 0x^4 + \cdots$$

$$= x - \frac{x^3}{3!} + \frac{x^5}{5!} - \frac{x^7}{7!} + \cdots$$

To make a direct attack on the representation problem, we work on

$$R_n(x) = \frac{f^{(n+1)}(c)}{(n+1)!}x^{n+1} \qquad \text{(where } c \text{ is between 0 and } x\text{)}$$

Since every derivative of $f(x)$ is $\sin x$, $\cos x$, $-\sin x$, or $-\cos x$, we can be sure that

$$|f^{(n+1)}(c)| \le 1 \qquad \text{(regardless of the value of } c\text{)}$$

Hence $$|R_n(x)| \le \frac{|x|^{n+1}}{(n+1)!}$$

and the problem is solved if we can show that

$$\lim_{n \to \infty} \frac{|x|^{n+1}}{(n+1)!} = 0 \qquad \text{(Why?)}$$

This result is already known, however, because $\Sigma\, x^n/n!$ converges absolutely for all x and we know from Section 12.2 that the typical term of a convergent series must approach 0. It follows that

$$\lim_{n \to \infty} |R_n(x)| = 0 \quad \text{and therefore} \quad \lim_{n \to \infty} R_n(x) = 0 \qquad \text{(Why?)}$$

This proves that $f(x) = S(x)$, that is,

$$\sin x = x - \frac{x^3}{3!} + \frac{x^5}{5!} - \frac{x^7}{7!} + \cdots \qquad \text{(for all } x\text{)}$$

A way to avoid the direct attack is outlined in the problem set.   □

□ **Example 5**   Find the Maclaurin series associated with $\cos x$ and show that it represents $\cos x$.

Solution: There is no need to repeat the procedure of Example 4. Simply differentiate the Maclaurin series for $\sin x$:

$$\cos x = 1 - \frac{3x^2}{3!} + \frac{5x^4}{5!} - \frac{7x^6}{7!} + \cdots$$

$$= 1 - \frac{x^2}{2!} + \frac{x^4}{4!} - \frac{x^6}{6!} + \cdots \qquad \text{(for all } x\text{)} \qquad \square$$

□ **Example 6**   Find the Taylor series associated with $f(x) = 1/x$ at $a = 1$ and show that it represents $1/x$ in its domain.

Solution: The simplest way to do this is to write

$$\frac{1}{x} = \frac{1}{1 - (1 - x)}$$

and use the geometric series

$$\frac{1}{1 - t} = 1 + t + t^2 + t^3 + \cdots \qquad (-1 < t < 1)$$

Letting $t = 1 - x$ (which means that $0 < x < 2$), we find

$$\frac{1}{x} = 1 + (1 - x) + (1 - x)^2 + (1 - x)^3 + \cdots$$

$$= 1 - (x - 1) + (x - 1)^2 - (x - 1)^3 + \cdots$$

We are done! Remember that there is only one way to represent f by a power series based at $a = 1$, so this must be it. (It does not matter how we find it.) Moreover, there is no question that it converges to $f(x)$ in the interval $(0,2)$. The geometric series from which it came is a known result from Section 12.2.   □

Remark   Compare our solution of Example 6 with the direct attack. Successive derivatives of $f(x) = 1/x$ (and their values at $a = 1$) are

$$
\begin{aligned}
f(x) &= x^{-1} & f(1) &= 1 \\
f'(x) &= -x^{-2} & f'(1) &= -1 \\
f''(x) &= 2x^{-3} & f''(1) &= 2 \\
f'''(x) &= -3!x^{-4} & f'''(1) &= -3! \\
&\;\;\vdots & &\;\;\vdots \\
f^{(n)}(x) &= (-1)^n n! x^{-(n+1)} & f^{(n)}(1) &= (-1)^n n! \\
&\;\;\vdots & &\;\;\vdots
\end{aligned}
$$

Hence the Taylor series at 1 is

$$S(x) = f(1) + f'(1)(x - 1) + \frac{f''(1)}{2!}(x - 1)^2 + \frac{f'''(1)}{3!}(x - 1)^3 + \cdots$$

$$= 1 - (x - 1) + (x - 1)^2 - (x - 1)^3 + \cdots$$

To prove that $f(x) = S(x)$, we work on

$$|R_n(x)| = \frac{|f^{(n+1)}(c)|}{(n + 1)!}|x - 1|^{n+1} \qquad \text{(where } c \text{ is between 1 and } x\text{)}$$

$$= \frac{(n + 1)!|c|^{-(n+2)}}{(n + 1)!}|x - 1|^{n+1} = \frac{|x - 1|^{n+1}}{|c|^{n+2}}$$

We consider two cases.

1. Suppose that $x \geq 1$. Then $1 \leq c \leq x$ and hence

$$|R_n(x)| \leq (x - 1)^{n+1} \qquad \text{(Why?)}$$

We know that $(x - 1)^{n+1}$ approaches 0 as $n \to \infty$ if $|x - 1| < 1$, that is, if $0 < x < 2$. Combining this with $x \geq 1$, we conclude that

$$\lim_{n \to \infty} |R_n(x)| = 0 \qquad \text{if} \qquad 1 \leq x < 2$$

2. Suppose that $0 < x < 1$. Then $x \leq c \leq 1$ and hence

$$|R_n(x)| \leq \frac{(1 - x)^{n+1}}{x^{n+2}} = \frac{1}{x}\left(\frac{1 - x}{x}\right)^{n+1}$$

This approaches 0 as $n \to \infty$ if

$$\left|\frac{1 - x}{x}\right| < 1, \text{ that is, if } x > \frac{1}{2}$$

Combining this with $0 < x < 1$, we conclude that

$$\lim_{n \to \infty} |R_n(x)| = 0 \qquad \text{if} \qquad \tfrac{1}{2} < x < 1$$

Thus $f(x) = S(x)$ if $\frac{1}{2} < x < 2$. But what about the values of x between 0 and $\frac{1}{2}$? The value of c is not known to us, so for these values of x we are still in doubt whether the remainder approaches 0 as $n \to \infty$. It *may*, but a proof requires a different formula for the remainder. Rather than digressing to develop it, we simply observe that the direct attack on the representation problem has bogged down, and that our first solution of Example 6 is clearly superior.

These examples should convince you that the representation problem is real, that a direct attack on it is often hard, and that any devices we can find to avoid it are worthwhile. The exercises will provide additional practice in solving the problem, but you should not miss the forest for the trees! In the long run we are mainly interested in *results,* namely Taylor series that represent the elementary functions of analysis. Besides the ones listed earlier in this section, we now have

$$\sin x = x - \frac{x^3}{3!} + \frac{x^5}{5!} - \frac{x^7}{7!} + \cdots$$

$$\cos x = 1 - \frac{x^2}{2!} + \frac{x^4}{4!} - \frac{x^6}{6!} + \cdots$$

$$\frac{1}{x} = 1 - (x - 1) + (x - 1)^2 - (x - 1)^3 + \cdots \qquad 0 < x < 2$$

Two other important results (derived in the problem set) are

$$\sinh x = x + \frac{x^3}{3!} + \frac{x^5}{5!} + \frac{x^7}{7!} + \cdots$$

$$\cosh x = 1 + \frac{x^2}{2!} + \frac{x^4}{4!} + \frac{x^6}{6!} + \cdots$$

Note the striking resemblance to the series for $\sin x$ and $\cos x$!

Problem Set 13.2

In each of the following, use the direct approach (successive differentiation of the given function) to find the Maclaurin series associated with f. You need not prove that the series represents f.

1. $f(x) = e^x$ **2.** $f(x) = e^{-x}$

3. $f(x) = \cosh x$ **4.** $f(x) = \sinh x$

5. $f(x) = \cos x$ **6.** $f(x) = \ln(1 + x)$

7. $f(x) = \dfrac{1}{1 - x}$ **8.** $f(x) = \dfrac{1}{1 + x}$

9. $f(x) = \dfrac{1}{(1 - x)^2}$ **10.** $f(x) = \dfrac{1}{(1 + x)^2}$

11. $f(x) = (1 + x)^{1/2}$ **12.** $f(x) = (1 - x)^{1/2}$

In each of the following, use the direct approach to find the Taylor series associated with f at the base point a. You need not prove that the series represents f.

13. $f(x) = e^x$ at $a = 1$ **14.** $f(x) = \sin x$ at $a = \pi/4$

15. $f(x) = \cos x$ at $a = \pi$ **16.** $f(x) = 1/x^2$ at $a = 1$

17. $f(x) = \ln x$ at $a = 1$ **18.** $f(x) = \sqrt{x}$ at $a = 1$

19. Find the first three nonzero terms of the Maclaurin series for $\tan x$.

20. Find the first three nonzero terms of the Maclaurin series for $\sec x$.

21. Find the first four nonzero terms of the Maclaurin series for $e^x \cos x$.

22. Find the first four nonzero terms of the Maclaurin series for $e^{-x} \cos x$.

23. If you were asked to find $f^{(4)}(0)$ and $f^{(5)}(0)$ by successive differentiation of $f(x) = \tan^{-1} x$, you might object. Find these numbers by inspection of the series

$$\tan^{-1} x = x - \frac{x^3}{3} + \frac{x^5}{5} - \frac{x^7}{7} + \cdots$$

24. Find $f^{(4)}(0)$ and $f^{(5)}(0)$ by using the series

$$f(x) = \tanh^{-1} x = x + \frac{x^3}{3} + \frac{x^5}{5} + \frac{x^7}{7} + \cdots$$

25. We know from Example 2 that the Maclaurin series for e^x (Problem 1) actually equals e^x. Replace x by $-x$ in this series to derive the formula

$$e^{-x} = 1 - x + \frac{x^2}{2!} - \frac{x^3}{3!} + \cdots \qquad \text{(for all } x\text{)}$$

Why does this show that the Maclaurin series for e^{-x} (Problem 2) actually equals e^{-x}?

26. Show that the Maclaurin series for $f(x) = e^{-x}$ (Problem 2) actually equals e^{-x} by proving that $R_n(x)$ approaches 0 as $n \to \infty$.   *Hint:* As in Problem 4, Section 11.5, explain why

$$|R_n(x)| \leq \frac{e^{|x|}}{(n + 1)!} |x|^{n+1}$$

27. Use the Maclaurin series for e^x and e^{-x}, together with the formula

$$\cosh x = \tfrac{1}{2}(e^x + e^{-x})$$

to show that

$$\cosh x = 1 + \frac{x^2}{2!} + \frac{x^4}{4!} + \frac{x^6}{6!} + \cdots \qquad \text{(for all } x\text{)}$$

Why does this settle the representation problem associated with Problem 3?

28. Show that

$$\sinh x = x + \frac{x^3}{3!} + \frac{x^5}{5!} + \frac{x^7}{7!} + \cdots \qquad \text{(for all } x\text{)}$$

by proceeding as in Problem 27.

29. Confirm Problem 28 by differentiating the Maclaurin series for $\cosh x$.

30. Confirm Problem 28 by integrating the Maclaurin series for $\cosh t$ from 0 to x.

31. Confirm Problem 28 by a direct attack; that is, show that the Maclaurin series for $f(x) = \sinh x$ represents $\sinh x$ by proving that $R_n(x)$ approaches 0 as $n \to \infty$. *Hint:* If c is between 0 and x, then

$$|f^{(n+1)}(c)| \leq \cosh c \leq \cosh x$$

32. Show that

$$\cos x = 1 - \frac{x^2}{2!} + \frac{x^4}{4!} - \frac{x^6}{6!} + \cdots \qquad \text{(for all } x\text{)}$$

by integrating the Maclaurin series for $\sin t$. (We know from Example 4 that the latter series represents $\sin t$.)

33. Confirm Problem 32 by differentiating the Maclaurin series for $\sin x$.

34. Confirm Problem 32 by a direct attack; that is, show that the Maclaurin series for $f(x) = \cos x$ represents $\cos x$ by proving that $R_n(x)$ approaches 0 as $n \to \infty$.

35. The Maclaurin series for $f(x) = \sin x$ is

$$S(x) = x - \frac{x^3}{3!} + \frac{x^5}{5!} - \frac{x^7}{7!} + \cdots$$

Prove that $f(x) = S(x)$ as follows (thus avoiding the direct attack in Example 4).

(a) Show that $y = S(x)$ satisfies the differential equation $y'' + y = 0$ and the initial conditions $S(0) = 0$, $S'(0) = 1$.

(b) Confirm that $f(x) = \sin x$ satisfies the same differential equation and initial conditions. *Note:* This is sufficient to prove that $f(x) = S(x)$; there cannot be more than one function satisfying this kind of differential equation and initial conditions. (See Section 7.7, where this "uniqueness theorem" is mentioned.)

36. Let $S(x)$ be the Maclaurin series for $f(x) = \cos x$. Prove that $f(x) = S(x)$ by showing that both $y = S(x)$ and $f(x) = \cos x$ satisfy the differential equation $y'' + y = 0$ with initial conditions $y = 1$ and $y' = 0$ at $x = 0$.

37. The Taylor series associated with $f(x) = 1/x^2$ at $a = 1$ is found in Problem 16, but a direct attack on the corresponding representation problem is hard. Avoid it by differentiating the series

$$\frac{1}{x} = 1 - (x - 1) + (x - 1)^2 - (x - 1)^3 + \cdots$$

found in Example 6.

38. Use the fact that $e^x = e^a e^{x-a}$ to write e^x as a Taylor series with base point a.

39. Express 2^x as a Maclaurin series. *Hint:* $2^x = e^{x \ln 2}$.

40. A famous formula due to Euler says that

$$e^{ix} = \cos x + i \sin x$$

where $i = \sqrt{-1}$ is the imaginary unit. (See Problem 40, Section 8.6, and Problem 55, Section 9.3.) We have not discussed infinite series involving complex numbers, but you can still have some fun "deriving" Euler's formula, as follows.

(a) Replace x by ix in the series for e^x and group the result to obtain

$$\left(1 - \frac{x^2}{2!} + \frac{x^4}{4!} - \frac{x^6}{6!} + \cdots\right)$$

$$+ i\left(x - \frac{x^3}{3!} + \frac{x^5}{5!} - \frac{x^7}{7!} + \cdots\right)$$

Hint: $i^2 = -1$, $i^3 = -i$, $i^4 = 1$, and so on.

(b) Why does Euler's formula follow? (Try to imagine the state of mind of Euler and others who did this sort of thing for the first time. The joy of discovery must have been very great indeed, like "Wow! I'm beginning to see how these things hang together.")

13.3
MORE ABOUT POWER SERIES

In previous sections we have manipulated power series in various ways to derive new results. The Maclaurin series for cosh x, for example, is most easily found by adding the series for e^x and e^{-x} and multiplying the sum by $\frac{1}{2}$:

$$\cosh x = \frac{1}{2}(e^x + e^{-x})$$

$$= \frac{1}{2}\left[\left(1 + x + \frac{x^2}{2!} + \frac{x^3}{3!} + \cdots\right) + \left(1 - x + \frac{x^2}{2!} - \frac{x^3}{3!} + \cdots\right)\right]$$

$$= \frac{1}{2}\left(2 + \frac{2x^2}{2!} + \frac{2x^4}{4!} + \cdots\right) = 1 + \frac{x^2}{2!} + \frac{x^4}{4!} + \cdots$$

Such operations with convergent series are legitimate (see Problems 44 and 45, Section 12.2); moreover, since every power series whose domain is not a single point *converges absolutely* in an open interval centered at the base point, we are free to group and rearrange its terms as in ordinary algebra. (See the end of Section 12.5, where we stated that neither the convergence nor the sum of the series is affected by such manipulation.)

In this section we present a few miscellaneous examples of other things we can do with power series. The most important of these is multiplication, which proceeds in the same way as if power series were polynomials.

To see what we mean, suppose that

$$f(x) = \sum_{n=0}^{\infty} a_n x^n \quad \text{and} \quad g(x) = \sum_{n=0}^{\infty} b_n x^n$$

are two power series with base point $a = 0$ (and intervals of convergence that are not single points). Their product is the series

$$f(x)g(x) = (a_0 + a_1 x + a_2 x^2 + a_3 x^3 + \cdots)(b_0 + b_1 x + b_2 x^2 + b_3 x^3 + \cdots)$$
$$= a_0 b_0 + (a_0 b_1 + a_1 b_0)x + (a_0 b_2 + a_1 b_1 + a_2 b_0)x^2$$
$$+ (a_0 b_3 + a_1 b_2 + a_2 b_1 + a_3 b_0)x^3 + \cdots$$

which has the same pattern of coefficients as in ordinary algebra (when two polynomials are multiplied). It can be proved that this "product series" converges to $f(x)g(x)$ in an interval whose radius is at least as large as the smaller of the radii of convergence of the given series. For example, if the intervals of convergence of the original series are $(-1,1)$ and $(-2,2)$, the product series represents $f(x)g(x)$ in at least $(-1,1)$.

□ **Example 1**   The Maclaurin series associated with $f(x) = e^{-x} \sin x$ is not easy to find by successive differentiation. We know, however, that

$$e^{-x} = 1 - x + \frac{x^2}{2!} - \frac{x^3}{3!} + \frac{x^4}{4!} - \frac{x^5}{5!} + \cdots$$

$$\sin x = x - \frac{x^3}{3!} + \frac{x^5}{5!} - \cdots$$

It is often possible to multiply series by inspection (finding all terms involving a given power of x and adding their coefficients mentally). To illustrate our general formula, however, we label the coefficients of each series:

$$a_0 = 1, \, a_1 = -1, \, a_2 = \tfrac{1}{2}, \, a_3 = -\tfrac{1}{6}, \, a_4 = \tfrac{1}{24}, \, a_5 = -\tfrac{1}{120}, \, \ldots$$
$$b_0 = 0, \, b_1 = 1, \, b_2 = 0, \, b_3 = -\tfrac{1}{6}, \, b_4 = 0, \, b_5 = \tfrac{1}{120}, \, \ldots$$

Hence

$$e^{-x} \sin x = 0 + (1 + 0)x + (0 - 1 + 0)x^2$$
$$+ (-\tfrac{1}{6} + 0 + \tfrac{1}{2} + 0)x^3 + (0 + \tfrac{1}{6} + 0 - \tfrac{1}{6} + 0)x^4$$
$$+ (\tfrac{1}{120} + 0 - \tfrac{1}{12} + 0 + \tfrac{1}{24} + 0)x^5 + \cdots$$
$$= x - x^2 + \tfrac{1}{3}x^3 - \tfrac{1}{30}x^5 + \cdots$$

The result has no obvious pattern of coefficients. Nevertheless any desired term can be obtained by nothing more complicated than ordinary arithmetic.   □

□ **Example 2**   Find the value of $\displaystyle\int_0^1 e^{-x} \sin x \, dx$.

Solution: Integration by parts (which takes a while!) yields

$$\int_0^1 e^{-x} \sin x \, dx = -\frac{1}{2} e^{-x}(\sin x + \cos x) \Big|_0^1 = \frac{1}{2} - \frac{1}{2} e^{-1}(\sin 1 + \cos 1)$$

$$= 0.2458 \cdots$$

Why not use the series in Example 1 instead? This gives

$$\int_0^1 e^{-x} \sin x \, dx = \int_0^1 \left(x - x^2 + \frac{1}{3} x^3 - \frac{1}{30} x^5 + \cdots \right) dx$$

$$\approx \frac{1}{2} - \frac{1}{3} + \frac{1}{12} - \frac{1}{180} = 0.2444 \cdots$$

As you can see, the result is not far wrong (and can be made more accurate by using more terms of the series). □

□ **Example 3**   Compute $\int_0^1 \dfrac{\sin x}{x} \, dx$ correct to six decimal places.

Solution: No elementary antiderivative of $(\sin x)/x$ exists, so we are forced to use an approximation of some kind. Since

$$\frac{\sin x}{x} = \frac{1}{x} \left(x - \frac{x^3}{3!} + \frac{x^5}{5!} - \frac{x^7}{7!} + \cdots \right) = 1 - \frac{x^2}{3!} + \frac{x^4}{5!} - \frac{x^6}{7!} + \cdots$$

we have

$$\int_0^1 \frac{\sin x}{x} \, dx = 1 - \frac{1}{3(3!)} + \frac{1}{5(5!)} - \frac{1}{7(7!)} + \cdots$$

To obtain six-place accuracy, observe that

$$\frac{1}{9(9!)} < \frac{1}{2} \times 10^{-6}$$

so the first four terms are enough (by the Alternating Series Test). Using a calculator, we find

$$\int_0^1 \frac{\sin x}{x} \, dx = 0.946083 \qquad \text{(correct to six places)}$$

Simpson's Rule with four subintervals (Section 10.6) yields

$$\int_0^1 \frac{\sin x}{x} \, dx \approx 0.946087$$

which is less accurate and takes more work. □

□ **Example 4**   The Maclaurin series associated with $\tan x$ is hard to find directly. A shortcut is to divide

$$\sin x = x - \frac{x^3}{3!} + \frac{x^5}{5!} - \frac{x^7}{7!} + \cdots$$

by
$$\cos x = 1 - \frac{x^2}{2!} + \frac{x^4}{4!} - \frac{x^6}{6!} + \cdots$$

The work is shown below.

$$x + \frac{x^3}{3} + \frac{2x^5}{15} + \frac{17x^7}{315} + \cdots$$

$$1 - \frac{x^2}{2} + \frac{x^4}{24} - \frac{x^6}{720} + \cdots \Bigg| x - \frac{x^3}{6} + \frac{x^5}{120} - \frac{x^7}{5040} + \cdots$$

$$\underline{x - \frac{x^3}{2} + \frac{x^5}{24} - \frac{x^7}{720} + \cdots}$$

$$\frac{x^3}{3} - \frac{x^5}{30} + \frac{x^7}{840} - \cdots$$

$$\underline{\frac{x^3}{3} - \frac{x^5}{6} + \frac{x^7}{72} - \cdots}$$

$$\frac{2x^5}{15} - \frac{4x^7}{315} + \cdots$$

$$\underline{\frac{2x^5}{15} - \frac{x^7}{15} + \cdots}$$

$$\frac{17x^7}{315} - \cdots$$

Hence
$$\tan x = x + \frac{1}{3}x^3 + \frac{2}{15}x^5 + \frac{17}{315}x^7 + \cdots$$

It should be mentioned that this series cannot be expected to represent $\tan x$ in any interval larger than $(-\pi/2, \pi/2)$. (Why?) Yet the Maclaurin series for $\sin x$ and $\cos x$ are valid for all x. This illustrates the fact that quotient series are more complicated to discuss than product series. Our purpose in this section is not to get into details from advanced calculus, but simply to give you some additional ideas about power series and how they are used. □

Problem Set 13.3

1. Replace t by x^2 in the geometric series for $1/(1 - t)$ to derive the Maclaurin series representation of $1/(1 - x^2)$. What is the domain?

2. Confirm Problem 1 by differentiating the series

$$\tanh^{-1} x = x + \frac{x^3}{3} + \frac{x^5}{5} + \frac{x^7}{7} + \cdots (-1 < x < 1)$$

(See Section 13.2.)

3. Confirm Problem 1 by multiplying the Maclaurin series for $1/(1 - x)$ and $1/(1 + x)$.

4. Confirm Problem 1 by dividing 1 by $1 - x^2$.

5. In Example 6, Section 13.2, we derived the formula

$$\frac{1}{x} = 1 - (x - 1) + (x - 1)^2 - (x - 1)^3 + \cdots$$

$$(0 < x < 2)$$

Replace x by $2 - x$ to obtain the formula

$$\frac{1}{2 - x} = 1 + (x - 1) + (x - 1)^2 + (x - 1)^3 + \cdots$$

What is the domain?

6. Find the Taylor series associated with $1/[x(2 - x)]$ at

$a = 1$ by multiplying the series for $1/x$ and $1/(2 - x)$ in Problem 5.

7. Confirm Problem 6 by finding the decomposition

$$\frac{1}{x(2 - x)} = \frac{A}{x} + \frac{B}{2 - x}$$

and adding the Taylor series associated with A/x and $B/(2 - x)$ at $a = 1$.

8. Confirm Problem 6 by dividing 1 by $x(2 - x)$. *Hint:* To obtain a series with base point $a = 1$, write $x(2 - x) = 1 - (x - 1)^2$.

9. Find the Maclaurin series representation of $\sin^2 (x/2)$ by using the identity $\sin^2 (x/2) = \frac{1}{2}(1 - \cos x)$ together with the Maclaurin series for $\cos x$.

10. Find the Maclaurin series representation of $\cos^2 (x/2)$ by using the identity $\cos^2 (x/2) = \frac{1}{2}(1 + \cos x)$ together with the Maclaurin series for $\cos x$.

11. Confirm Problem 10 by writing $\cos^2 (x/2) = 1 - \sin^2 (x/2)$ and using Problem 9.

12. Use Problem 9 to show that

$$\sin^2 x = \sum_{n=1}^{\infty} \frac{(-1)^{n-1} 2^{2n-1} x^{2n}}{(2n)!}$$

13. Derive the formula

$$\sin (x + c) = \sin c + (\cos c)x - (\sin c)\frac{x^2}{2!}$$
$$- (\cos c)\frac{x^3}{3!} + (\sin c)\frac{x^4}{4!} + \cdots$$

where c is a constant.

14. Derive the formula

$$\cos (x + c) = \cos c - (\sin c)x - (\cos c)\frac{x^2}{2!}$$
$$+ (\sin c)\frac{x^3}{3!} + (\cos c)\frac{x^4}{4!} + \cdots$$

where c is a constant.

15. Rewrite the equation in Problem 13 in the form

$$\sin c (\qquad) + \cos c (\qquad)$$

and obtain the familiar addition formula for $\sin (x + c)$.

16. Rewrite the equation in Problem 14 in the form

$$\cos c (\qquad) - \sin c (\qquad)$$

and obtain the familiar addition formula for $\cos (x + c)$.

17. The function

$$E(x) = \frac{2}{\sqrt{\pi}} \int_0^x e^{-t^2} dt$$

is important in statistics. Show that

$$E(x) = \frac{2}{\sqrt{\pi}} \sum_{n=0}^{\infty} \frac{(-1)^n x^{2n+1}}{(2n + 1)n!}$$

18. In Problem 21, Section 13.2, we asked you to find the first four nonzero terms of the Maclaurin series for $e^x \cos x$. Confirm the results by multiplying the series for e^x and $\cos x$.

19. In Problem 22, Section 13.2, we asked you to find the first four nonzero terms of the Maclaurin series for $e^{-x} \cos x$. Confirm the results by multiplying the series for e^{-x} and $\cos x$.

20. Do Problem 19 by replacing x by $-x$ in the series for $e^x \cos x$ (Problem 18).

21. Use integration by parts to show that

$$\int_0^{0.5} e^x \cos x \, dx = 0.618664 \cdots$$

22. Approximate the result in Problem 21 by using Simpson's Rule with four subintervals. (See Section 10.6.)

23. Approximate the result in Problem 21 by integrating the first four nonzero terms of the Maclaurin series for $e^x \cos x$. (See Problem 18.)

24. Use integration by parts to show that

$$\int_0^{0.5} e^{-x} \cos x \, dx = 0.379252 \cdots$$

25. Approximate the result in Problem 24 by using Simpson's Rule with four subintervals.

26. Approximate the result in Problem 24 by integrating the first four nonzero terms of the Maclaurin series for $e^{-x} \cos x$. (See Problem 19.)

27. Find the first four nonzero terms of the Maclaurin series for $\sec x$ by division of 1 by the series for $\cos x$. (Compare with Problem 20, Section 13.2.)

28. In Example 4 we found

$$\tan x = x + \frac{1}{3}x^3 + \frac{2}{15}x^5 + \frac{17}{315}x^7 + \cdots$$

Confirm this as follows.

(a) Square the series

$$\sec t = 1 + \frac{1}{2}t^2 + \frac{5}{24}t^4 + \frac{61}{720}t^6 + \cdots$$

found in Problem 27 to obtain

$$\sec^2 t = 1 + t^2 + \tfrac{2}{3}t^4 + \tfrac{17}{45}t^6 + \cdots$$

(b) Integrate from 0 to x to find the series for $\tan x$.

29. Use series to compute

$$\int_0^1 e^{-x^2} \, dx$$

correct to two decimal places.

30. Use series to compute

$$\int_0^1 \sin \sqrt{x} \, dx$$

correct to three decimal places.

31. Use series to compute

$$\int_0^{0.5} \frac{dx}{1 + x^4}$$

correct to four decimal places.

32. The series

$$\ln \frac{1 + x}{1 - x} = 2\left(x + \frac{x^3}{3} + \frac{x^5}{5} + \frac{x^7}{7} + \cdots\right)$$

$(-1 < x < 1)$ is derived in Problem 39, Section 13.1. Explain why it can be used to find the logarithm of any positive number. *Hint:* Given $N > 0$, find x satisfying

$$N = \frac{1 + x}{1 - x}$$

and show that $-1 < x < 1$.

33. Use the first four terms of the series in Problem 32 to compute $\ln 2$.

34. Use L'Hôpital's Rule to evaluate

$$\lim_{x \to 0} \frac{\sin x - x + x^3/6}{x^5}$$

35. Do Problem 34 by replacing $\sin x$ by its Maclaurin series.

13.4
THE BINOMIAL SERIES

The familiar formulas

$$(a + b)^1 = a + b$$
$$(a + b)^2 = a^2 + 2ab + b^2$$
$$(a + b)^3 = a^3 + 3a^2b + 3ab^2 + b^3$$

are special cases of a more general statement called the **Binomial Theorem,** often proved (or at least presented) in high school algebra:

$$(a + b)^n = a^n + na^{n-1}b + \frac{n(n - 1)}{2}a^{n-2}b^2$$

$$+ \frac{n(n - 1)(n - 2)}{2 \cdot 3}a^{n-3}b^3 + \cdots + b^n$$

where n is a positive integer. (If you have not seen this formula before, don't worry about it. It is a special case of the binomial series, which we will develop from scratch.) The form in which this theorem is of interest to us is obtained by taking $a = 1$ and $b = x$, in which case it reads

$$(1 + x)^n = \sum_{k=0}^n c_k x^k$$

where

$$c_0 = 1, \; c_1 = n, \; c_2 = \frac{n(n - 1)}{2}, \; c_3 = \frac{n(n - 1)(n - 2)}{2 \cdot 3}, \ldots, \; c_n = 1$$

Although this theorem had been known for centuries before Newton, it remained for him to make the transition to fractional exponents (in which case the formula does not terminate, but becomes an infinite series). When $n = 1/2$, for example, it reads

$$(1 + x)^{1/2} = 1 + \tfrac{1}{2}x + \frac{(\tfrac{1}{2})(-\tfrac{1}{2})}{2}x^2 + \frac{(\tfrac{1}{2})(-\tfrac{1}{2})(-\tfrac{3}{2})}{2 \cdot 3}x^3 + \cdots$$

To make sense of this, Newton was forced to develop a coherent theory of infinite processes that led him to calculus; the binomial series was one of his major discoveries.

Our objective in this section is to discuss the Maclaurin series for

$$f(x) = (1 + x)^p$$

where p is any real number. We have an advantage over Newton! For we need not know anything about the Binomial Theorem to see how the coefficients are generated:

$$
\begin{aligned}
f(x) &= (1 + x)^p & f(0) &= 1 \\
f'(x) &= p(1 + x)^{p-1} & f'(0) &= p \\
f''(x) &= p(p - 1)(1 + x)^{p-2} & f''(0) &= p(p - 1) \\
f'''(x) &= p(p - 1)(p - 2)(1 + x)^{p-3} & f'''(0) &= p(p - 1)(p - 2) \\
&\;\;\vdots & &\;\;\vdots
\end{aligned}
$$

Thus the series we seek is

$$1 + px + \frac{p(p - 1)}{2!}x^2 + \frac{p(p - 1)(p - 2)}{3!}x^3 + \cdots = \sum_{n=0}^{\infty} c_n x^n$$

where
$$c_0 = 1$$
$$c_1 = p = (p)c_0$$
$$c_2 = \frac{p(p - 1)}{2} = \left(\frac{p - 1}{2}\right)c_1$$
$$c_3 = \frac{p(p - 1)(p - 2)}{2 \cdot 3} = \left(\frac{p - 2}{3}\right)c_2$$

and in general
$$c_{n+1} = \left(\frac{p - n}{n + 1}\right)c_n \qquad n = 0, 1, 2, \ldots$$

To test for convergence, we compute the ratio

$$\frac{|c_{n+1}x^{n+1}|}{|c_n x^n|} = \left|\frac{p - n}{n + 1}\right| |x|$$

Its limit as $n \to \infty$ is

$$|x| \lim_{n\to\infty} \left|\frac{p - n}{n + 1}\right| = |x| \qquad \text{(because } p \text{ is fixed)}$$

Hence the binomial series converges absolutely in the interval $(-1, 1)$.

To see whether the series converges to $f(x)$ (the "representation problem" of Section 13.2), we let

$$S(x) = \sum_{n=0}^{\infty} c_n x^n$$

and try to prove that $f(x) = S(x)$. This is hard to do directly (by a consideration of the remainder). We avoid a confrontation by the familiar device of differentiating the series.

Observe that since $f(x) = (1 + x)^p$ and $f'(x) = p(1 + x)^{p-1}$, we can write

$$(1 + x)f'(x) = p(1 + x)^p = pf(x)$$

Hence $y = f(x)$ satisfies the differential equation $(1 + x)y' = py$. Moreover, $f(1) = 0$. We claim that $S(x)$ satisfies the same differential equation and initial condition, in which case it follows that $f(x) = S(x)$. (See the uniqueness theorem mentioned in Section 7.7.)

Since $S(0) = c_0 = 1$, the initial condition is trivial. The problem therefore reduces to verification of the equation

$$(1 + x)S'(x) = pS(x) \tag{1}$$

Since

$$S(x) = \sum_{n=0}^{\infty} c_n x^n$$

we have

$$S'(x) = \sum_{n=1}^{\infty} nc_n x^{n-1}$$

Hence the left side of Equation (1) is

$$(1 + x)S'(x) = S'(x) + xS'(x) = \sum_{n=1}^{\infty} nc_n x^{n-1} + \sum_{n=1}^{\infty} nc_n x^n$$

The series on the right can be more easily combined if x^n appears in the typical term of each. Hence we replace n by $n + 1$ in the first series to obtain

$$(1 + x)S'(x) = \sum_{n=0}^{\infty} (n + 1)c_{n+1}x^n + \sum_{n=1}^{\infty} nc_n x^n$$

Break off the first term of the first series so that they both start with $n = 1$:

$$(1 + x)S'(x) = c_1 + \sum_{n=1}^{\infty} (n + 1)c_{n+1}x^n + \sum_{n=1}^{\infty} nc_n x^n$$

$$= c_1 + \sum_{n=1}^{\infty} [(n + 1)c_{n+1} + nc_n]x^n$$

Since $c_1 = p$ and $c_{n+1} = \left(\dfrac{p - n}{n + 1}\right)c_n$

we find

$$(1 + x)S'(x) = p + \sum_{n=1}^{\infty} [(p - n)c_n + nc_n]x^n = p + \sum_{n=1}^{\infty} pc_nx^n$$

$$= p\left(1 + \sum_{n=1}^{\infty} c_nx^n\right) = p\sum_{n=0}^{\infty} c_nx^n = pS(x)$$

This proves Equation (1) and our argument is complete.

The Binomial Series

If p is any real number, then

$$(1 + x)^p = 1 + px + \frac{p(p - 1)}{2!}x^2 + \frac{p(p - 1)(p - 2)}{3!}x^3 + \cdots$$

for $-1 < x < 1$.

□ **Example 1**   Find the Maclaurin series for $f(x) = \sqrt[3]{1 + x}$ and use it to compute $\sqrt[3]{30}$ correct to three places.

Solution: Taking $p = \frac{1}{3}$ in the binomial series, we find

$$f(x) = (1 + x)^{1/3}$$

$$= 1 + \frac{1}{3}x + \frac{(\frac{1}{3})(-\frac{2}{3})}{2!}x^2 + \frac{(\frac{1}{3})(-\frac{2}{3})(-\frac{5}{3})}{3!}x^3 + \cdots$$

$$= 1 + \frac{1}{3}x - \frac{2}{3^2 \cdot 2!}x^2 + \frac{2 \cdot 5}{3^3 \cdot 3!}x^3 - \frac{2 \cdot 5 \cdot 8}{3^4 \cdot 4!}x^4 + \cdots$$

To compute $\sqrt[3]{30}$, write

$$\sqrt[3]{30} = (27 + 3)^{1/3} = [27(1 + \tfrac{1}{9})]^{1/3} = 3(1 + \tfrac{1}{9})^{1/3} = 3f(\tfrac{1}{9})$$

$$= 3\left[1 + \frac{1}{3}\left(\frac{1}{9}\right) - \frac{2}{3^2 \cdot 2!}\left(\frac{1}{9}\right)^2 + \frac{2 \cdot 5}{3^3 \cdot 3!}\left(\frac{1}{9}\right)^3 - \cdots\right]$$

From the second term on, this is an alternating series. (Why?) Since

$$3\left[\frac{2 \cdot 5}{3^3 \cdot 3!}\left(\frac{1}{9}\right)^3\right] < 0.0003 \qquad \text{(from a calculator)}$$

we may use the first three terms to obtain three-place accuracy. The result is $\sqrt[3]{30} \approx 3.107$.

The reader may object that if we use a calculator to estimate the error, we might as well use it to find $\sqrt[3]{30}$ in the first place. In practice we do; the purpose of Example 1 is to show you the kind of formula that is programmed into the calculator to enable it to compute cube roots.   □

□ **Example 2**   Compute $\int_0^{0.7} \sqrt{1 + x^3}\, dx$ correct to three decimal places.

Solution: Let $p = \frac{1}{2}$ in the binomial series and replace x by x^3:

$$(1 + x^3)^{1/2} = 1 + \frac{1}{2}x^3 + \frac{(\frac{1}{2})(-\frac{1}{2})}{2!}x^6 + \frac{(\frac{1}{2})(-\frac{1}{2})(-\frac{3}{2})}{3!}x^9 + \cdots$$

$$= 1 + \frac{1}{2}x^3 - \frac{1}{8}x^6 + \frac{1}{16}x^9 - \cdots$$

Hence

$$\int_0^{0.7} \sqrt{1 + x^3}\, dx = 0.7 + \frac{(0.7)^4}{8} - \frac{(0.7)^7}{56} + \frac{(0.7)^{10}}{160} - \cdots$$

As in Example 1, three terms are enough, because the series alternates (after the first term) and

$$\frac{(0.7)^{10}}{160} < \frac{1}{2} \times 10^{-3}$$

The result (from a calculator) is 0.729.   □

□ **Example 3**   Find the Maclaurin series for $\sin^{-1} x$.

Solution: Since

$$\sin^{-1} x = \int_0^x \frac{dt}{\sqrt{1 - t^2}}$$

we can find the desired series by integrating the series for

$$\frac{1}{\sqrt{1 - t^2}} = (1 - t^2)^{-1/2}$$

Let $p = -\frac{1}{2}$ in the binomial series and replace x by $-t^2$:

$$(1 - t^2)^{-1/2} = 1 + (-\tfrac{1}{2})(-t^2) + \frac{(-\frac{1}{2})(-\frac{3}{2})}{2!}(-t^2)^2$$

$$+ \frac{(-\frac{1}{2})(-\frac{3}{2})(-\frac{5}{2})}{3!}(-t^2)^3 + \cdots$$

$$= 1 + \frac{1}{2}t^2 + \frac{1 \cdot 3}{2^2 \cdot 2!}t^4 + \frac{1 \cdot 3 \cdot 5}{2^3 \cdot 3!}t^6 + \cdots$$

$$= 1 + \frac{1}{2}t^2 + \frac{1 \cdot 3}{2 \cdot 4}t^4 + \frac{1 \cdot 3 \cdot 5}{2 \cdot 4 \cdot 6}t^6 + \cdots$$

$$[\text{because } 2^n \cdot n! = 2 \cdot 4 \cdot 6 \cdots (2n)]$$

This holds for $-1 < t < 1$ (why?), so

$$\sin^{-1} x = x + \frac{1}{2} \cdot \frac{x^3}{3} + \frac{1 \cdot 3}{2 \cdot 4} \cdot \frac{x^5}{5} + \frac{1 \cdot 3 \cdot 5}{2 \cdot 4 \cdot 6} \cdot \frac{x^7}{7} + \cdots \qquad (-1 < x < 1)$$

Since the domain of $\sin^{-1} x$ is the closed interval $[-1,1]$, one may expect this formula to hold when $x = \pm 1$. Can you prove that it does? When $x = 1$ we get another interesting (but slow to converge) expression for π, namely

$$\frac{\pi}{2} = 1 + \frac{1}{2} \cdot \frac{1}{3} + \frac{1 \cdot 3}{2 \cdot 4} \cdot \frac{1}{5} + \frac{1 \cdot 3 \cdot 5}{2 \cdot 4 \cdot 6} \cdot \frac{1}{7} + \cdots$$ □

Problem Set 13.4

In each of the following, use the binomial series to find the Maclaurin series representing f.

1. $f(x) = \dfrac{1}{1 + x}$

2. $f(x) = \dfrac{1}{(1 + x)^2}$

3. $f(x) = \sqrt{1 + x}$

4. $f(x) = (1 + x)^{2/3}$

5. $f(x) = \dfrac{1}{\sqrt{1 + x^2}}$

6. $f(x) = \sqrt[3]{1 - x^2}$

7. $f(x) = (1 + x)^4$

8. $f(x) = (1 + x)^6$

9. $f(x) = \sqrt{4 + x^2}$ What is the domain?

10. Use the binomial series to derive the geometric series

$$\frac{1}{1 - x} = 1 + x + x^2 + x^3 + \cdots \qquad (-1 < x < 1)$$

(Many of our earlier results are based on the geometric series. This shows that the binomial series is even more fundamental.)

11. Explain why the formula $\sqrt[5]{1 + x} \approx 1 + \frac{1}{5}x$ may be used in the interval $(-1,1)$ with an error no greater than $0.08x^2$.

12. Explain why the formula

$$(1 + x)^{-1/3} \approx 1 - \tfrac{1}{3}x + \tfrac{2}{9}x^2$$

may be used in the interval $(-1,1)$ with an error less than $0.18|x|^3$.

13. Use the formula in Problem 11 to compute $\sqrt[5]{35}$ correct to as many places as justified by the error estimate. *Hint:* $\sqrt[5]{35} = (32 + 3)^{1/5} = 2(1 + \frac{3}{32})^{1/5}$.

14. Use the formula in Problem 12 to compute $1/\sqrt[3]{25}$ correct to as many decimal places as justified by the error estimate. *Hint:* $25 = 27 - 2$.

15. Einstein's theory of relativity predicts that a body of mass m_0 (at rest) has mass

$$m = \frac{m_0}{\sqrt{1 - v^2/c^2}} \qquad (c = \text{velocity of light})$$

when its velocity is v. Use the binomial series to obtain the approximation $mc^2 \approx m_0 c^2 + \frac{1}{2}m_0 v^2$.

16. Einstein's famous mass-energy formula is $E = mc^2$, while the kinetic energy of a body of (rest) mass m_0 with velocity v is $\frac{1}{2}m_0 v^2$. What interpretation can you give of the approximation in Problem 15?

17. Use the first four nonzero terms of the Maclaurin series for $\sqrt{1 - x^3}$ to compute

$$\int_0^{0.5} \sqrt{1 - x^3}\, dx$$

18. Use the first four nonzero terms of the series for $\sin^{-1} \frac{1}{2}$ (Example 3) to compute π.

19. Use the formula

$$\sinh^{-1} x = \int_0^x \frac{dt}{\sqrt{1 + t^2}} \qquad \text{(Section 9.4)}$$

to find the Maclaurin series for $\sinh^{-1} x$.

20. Find the Taylor series associated with $f(x) = \sqrt{x}$ at $a = 1$ by using the binomial series. What is the domain? *Hint:* $\sqrt{x} = \sqrt{1 + (x - 1)}$.

21. Use the series in Problem 20 to compute $\sqrt{10}$ correct to four decimal places. *Hint:* $\sqrt{10} = 3\sqrt{10/9}$.

22. Use the binomial series to expand $(a + b)^5$, as follows.
(a) Find the series for $(1 + x)^5$, noting that it terminates (and hence converges for all x).

(b) Write $(a + b)^5 = a^5\left(1 + \dfrac{b}{a}\right)^5$.

23. Use the binomial series to explain why

$$(1 + h)^{1/h} = 1 + 1 + \frac{1 - h}{2!} + \frac{(1 - h)(1 - 2h)}{3!}$$
$$+ \frac{(1 - h)(1 - 2h)(1 - 3h)}{4!} + \cdots$$

for $-1 < h < 1$ ($h \neq 0$).

24. What series is obtained from the right side of the equation in Problem 23 when h is allowed to approach 0? Does this agree with what you know about $\lim_{h \to 0} (1 + h)^{1/h}$?

25. Compute $\int_0^1 \sqrt{4 + x^2}\, dx$ correct to four decimal places by using the formula

$$\int \sqrt{x^2 + a^2}\, dx = \tfrac{1}{2} x \sqrt{x^2 + a^2}$$
$$+ \tfrac{1}{2} a^2 \ln\left(x + \sqrt{x^2 + a^2}\right) + C$$

from Section 10.3.

26. Do Problem 25 by using the series found in Problem 9.

27. Let $f(x) = (1 + x)^p$ and suppose that $S(x)$ is the function defined by the binomial series.

(a) In the text we proved that $(1 + x)S'(x) = pS(x)$.

Multiply each side of this equation by $(1 + x)^{p-1}$ to obtain

$$f(x)S'(x) - f'(x)S(x) = 0$$

and conclude that

$$\frac{d}{dx}\left[\frac{S(x)}{f(x)}\right] = 0$$

(b) Why does it follow that $S(x) = Cf(x)$ for some constant C? Use the values of $f(x)$ and $S(x)$ at $x = 0$ to show that $C = 1$ and hence $S(x) = f(x)$. *Note:* This is an alternate proof that $f(x)$ and $S(x)$ are the same function. In the text we appealed to a uniqueness theorem from differential equations.

13.5
SERIES SOLUTIONS OF DIFFERENTIAL EQUATIONS (Optional)

In the last section we proved that the binomial series converges to

$$f(x) = (1 + x)^p$$

by labeling the series $S(x)$ and by showing that $y = S(x)$ satisfies the differential equation

$$(1 + x)y' = py$$

We did this by substituting the functions

$$S(x) = \sum_{n=0}^{\infty} c_n x^n \quad \text{and} \quad S'(x) = \sum_{n=1}^{\infty} n c_n x^{n-1}$$

for y and y' in the equation, manipulating the series until we could see that $(1 + x)y'$ equals py. Since we knew the binomial coefficients $c_0, c_1, c_2, \ldots$ ahead of time, our proof was really nothing more than checking that a given function (defined by a known series) actually satisfies the differential equation.

In this section we do the same sort of thing in reverse. Rather than substituting a series with known coefficients in a differential equation, we seek a solution in the form of a series with unknown coefficients. The idea is that by substituting this series in the equation, the coefficients (and hence the solution) can be found.

This description of what we are going to do is probably not very enlightening until we do it. The following examples illustrate what we mean.

□ **Example 1**   Use a power series based at $a = 0$ to find the solution of

$$y' = y$$

satisfying the initial condition $y = 1$ when $x = 0$.

Solution: It is probably obvious that the solution is $y = e^x$. Let's act ignorant, however, and assume nothing more than that the solution (whatever it is) can be written in the form

$$y = \sum_{n=0}^{\infty} c_n x^n$$

The problem is to determine the coefficients $c_0, c_1, c_2, \ldots$.

Our first step is to differentiate the series to obtain

$$y' = \sum_{n=1}^{\infty} n c_n x^{n-1}$$

(so that we may substitute series for y and y' in the given equation). The condition to be satisfied is then

$$\sum_{n=1}^{\infty} n c_n x^{n-1} = \sum_{n=0}^{\infty} c_n x^n$$

The second step is to rewrite this condition in the form of a single series equal to 0, say

$$\sum_{n=0}^{\infty} a_n x^n = 0$$

Then (thinking of the right side as the series $0 + 0x + 0x^2 + \cdots$) we may assert that every coefficient on the left side must be 0. That fact will enable us to discover the unknown coefficients $c_0, c_1, c_2, \ldots$.

To be specific, we proceed with the second step as follows:

$$\sum_{n=1}^{\infty} n c_n x^{n-1} - \sum_{n=0}^{\infty} c_n x^n = 0$$

Replace n by $n + 1$ in the first series (so that the typical term in each series will involve x^n):

$$\sum_{n=0}^{\infty} (n + 1)c_{n+1} x^n - \sum_{n=0}^{\infty} c_n x^n = 0$$

$$\sum_{n=0}^{\infty} [(n + 1)c_{n+1} - c_n] x^n = 0$$

It follows (as explained above) that for every n,

$$(n + 1)c_{n+1} - c_n = 0$$

that is,

$$c_{n+1} = \frac{c_n}{n + 1}, \quad n = 0, 1, 2, \ldots$$

This *recursion formula* tells us what each coefficient is in terms of the preceding one; if we know the first one, we know them all. Hence we turn to the initial condition $y = 1$ when $x = 0$. Since

$$y = c_0 + c_1 x + c_2 x^2 + \cdots$$

this condition yields $c_0 = 1$. The recursion formula then gives

$$c_1 = c_0/1 = 1$$

$$c_2 = c_1/2 = \frac{1}{2}$$

$$c_3 = c_2/3 = \frac{1}{3!}$$

$$c_4 = c_3/4 = \frac{1}{4!}$$

$$\vdots$$

The unknown function with which we started, namely

$$y = \sum_{n=0}^{\infty} c_n x^n$$

is now known; it is

$$y = \sum_{n=0}^{\infty} \frac{x^n}{n!} = 1 + x + \frac{x^2}{2!} + \frac{x^3}{3!} + \cdots$$

Thus we have solved the problem. It is unnecessary to go further and observe that the series is equal to e^x. *The series is the solution.* In all but the simplest cases we won't be able to replace it by a "closed-form" expression like e^x. In fact the point of solving differential equations by series is to handle problems that defy the usual approach. □

□ **Example 2**   Use series to find the general solution of

$$xy' = y$$

Solution: Assuming a solution in the form

$$y = \sum_{n=0}^{\infty} c_n x^n$$

we differentiate to find

$$y' = \sum_{n=1}^{\infty} n c_n x^{n-1}$$

and substitute in the equation $xy' = y$:

$$x \sum_{n=1}^{\infty} nc_n x^{n-1} = \sum_{n=0}^{\infty} c_n x^n$$

$$\sum_{n=1}^{\infty} nc_n x^n - \sum_{n=0}^{\infty} c_n x^n = 0$$

No need to shift the index of summation this time! To make each series start at the same point, however, break off the first term of the second series:

$$\sum_{n=1}^{\infty} nc_n x^n - \left(c_0 + \sum_{n=1}^{\infty} c_n x^n \right) = 0$$

$$-c_0 + \sum_{n=1}^{\infty} (nc_n - c_n)x^n = 0$$

Since every coefficient on the left side must be 0, we find $c_0 = 0$ and (for $n = 1, 2, 3, \ldots$)

$$nc_n - c_n = 0$$
$$(n - 1)c_n = 0$$

This implies that $c_n = 0$ for $n = 2, 3, 4, \ldots$, but when $n = 1$ it says nothing about c_1. Hence c_1 is arbitrary (call it C) and every other coefficient is 0. The solution

$$y = \sum_{n=0}^{\infty} c_n x^n = c_0 + c_1 x + c_2 x^2 + \cdots$$

collapses to

$$y = Cx$$

You can check that this is a solution of $xy' = y$ by substitution in the equation.   □

□ **Example 3**   Use series to solve the initial value problem

$$y'' + y = 0, \text{ where } y = 1 \text{ and } y' = 0 \text{ at } x = 0$$

Solution: This time we differentiate the trial solution

$$y = \sum_{n=0}^{\infty} c_n x^n$$

twice, obtaining

$$y' = \sum_{n=1}^{\infty} nc_n x^{n-1} \quad \text{and} \quad y'' = \sum_{n=2}^{\infty} n(n - 1)c_n x^{n-2}$$

Substitution in $y'' + y = 0$ yields

$$\sum_{n=2}^{\infty} n(n-1)c_n x^{n-2} + \sum_{n=0}^{\infty} c_n x^n = 0$$

Replace n by $n + 2$ in the first series to obtain

$$\sum_{n=0}^{\infty} (n+2)(n+1)c_{n+2} x^n + \sum_{n=0}^{\infty} c_n x^n = 0$$

$$\sum_{n=0}^{\infty} [(n+2)(n+1)c_{n+2} + c_n] x^n = 0$$

Since every coefficient must be 0, we find (for $n = 0, 1, 2, \ldots$)

$$(n+2)(n+1)c_{n+2} + c_n = 0$$

Hence our recursion formula is

$$c_{n+2} = \frac{-c_n}{(n+1)(n+2)}, \quad n = 0, 1, 2, \ldots$$

Put the initial condition $y = 1$ at $x = 0$ into

$$y = c_0 + c_1 x + c_2 x^2 + \cdots$$

to find $c_0 = 1$. Similarly, substitution of the condition $y' = 0$ at $x = 0$ in

$$y' = c_1 + 2c_2 x + 3c_3 x^2 + \cdots$$

yields $c_1 = 0$. The recursion formula now gives two sets of coefficients (by taking $n = 0, 2, 4, \ldots$ and $n = 1, 3, 5, \ldots$ in turn):

$$n = 0 \Rightarrow c_2 = \frac{-c_0}{(1)(2)} = -\frac{1}{2!}$$

$$n = 2 \Rightarrow c_4 = \frac{-c_2}{(3)(4)} = \frac{1}{4!}$$

$$n = 4 \Rightarrow c_6 = \frac{-c_4}{(5)(6)} = -\frac{1}{6!}$$

$$\vdots$$

$$n = 1 \Rightarrow c_3 = \frac{-c_1}{(2)(3)} = 0$$

$$n = 3 \Rightarrow c_5 = \frac{-c_3}{(4)(5)} = 0$$

$$\vdots$$

Hence the solution is

$$y = c_0 + c_1 x + c_2 x^2 + \cdots = 1 - \frac{x^2}{2!} + \frac{x^4}{4!} - \frac{x^6}{6!} + \cdots$$

In closed form this is $y = \cos x$ (which may have been obvious from the beginning). $\qquad\qquad\qquad\qquad\qquad\qquad\qquad\qquad\qquad\qquad\qquad$ □

□ **Example 4** $\qquad$ Use series to find the general solution of

$$y'' - xy = 0$$

Solution: $\qquad$ As in Example 3, substitute

$$y = \sum_{n=0}^{\infty} c_n x^n \qquad \text{and} \qquad y'' = \sum_{n=2}^{\infty} n(n-1)c_n x^{n-2}$$

in the given equation:

$$\sum_{n=2}^{\infty} n(n-1)c_n x^{n-2} - x \sum_{n=0}^{\infty} c_n x^n = 0$$

$$\sum_{n=2}^{\infty} n(n-1)c_n x^{n-2} - \sum_{n=0}^{\infty} c_n x^{n+1} = 0$$

To get x^n in the typical term of each series, shift the index down by 2 in the first series and up by 1 in the second series:

$$\sum_{n=0}^{\infty} (n+2)(n+1)c_{n+2} x^n - \sum_{n=1}^{\infty} c_{n-1} x^n = 0$$

Break off the first term of the first series:

$$2c_2 + \sum_{n=1}^{\infty} (n+2)(n+1)c_{n+2} x^n - \sum_{n=1}^{\infty} c_{n-1} x^n = 0$$

$$2c_2 + \sum_{n=1}^{\infty} [(n+2)(n+1)c_{n+2} - c_{n-1}]x^n = 0$$

Since each coefficient must be 0, we find $2c_2 = 0$ (or $c_2 = 0$) and

$$(n+2)(n+1)c_{n+2} - c_{n-1} = 0$$

$$c_{n+2} = \frac{c_{n-1}}{(n+1)(n+2)}, \quad n = 1, 2, 3, \ldots$$

The values $n = 1, 4, 7, \ldots$ yield

$$c_3 = \frac{c_0}{(2)(3)} = \frac{c_0}{3!} \qquad \text{(where } c_0 \text{ is arbitrary)}$$

$$c_6 = \frac{c_3}{(5)(6)} = \frac{4c_0}{6!}$$

$$c_9 = \frac{c_6}{(8)(9)} = \frac{4 \cdot 7c_0}{9!}$$

The values $n = 2, 5, 8, \ldots$ yield

$$c_4 = \frac{c_1}{(3)(4)} = \frac{2c_1}{4!} \qquad \text{(where } c_1 \text{ is arbitrary)}$$

$$c_7 = \frac{c_4}{(6)(7)} = \frac{2 \cdot 5 c_1}{7!}$$

$$c_{10} = \frac{c_7}{(9)(10)} = \frac{2 \cdot 5 \cdot 8 c_1}{10!}$$
$$\vdots$$

The values $n = 3, 6, 9, \ldots$ yield

$$c_5 = c_8 = c_{11} = \cdots = 0 \qquad \text{(because } c_2 = 0)$$

Hence the general solution is

$$
\begin{aligned}
y &= c_0 + c_1 x + c_2 x^2 + \cdots \\
&= (c_0 + c_3 x^3 + c_6 x^6 + \cdots) + (c_1 x + c_4 x^4 + c_7 x^7 + \cdots) \\
&\qquad\qquad\qquad\qquad + (c_2 x^2 + c_5 x^5 + c_8 x^8 + \cdots) \\
&= c_0\left(1 + \frac{x^3}{3!} + \frac{4x^6}{6!} + \frac{4 \cdot 7 x^9}{9!} + \cdots\right) + c_1\left(x + \frac{2x^4}{4!} + \frac{2 \cdot 5 x^7}{7!} + \cdots\right)
\end{aligned}
$$

This time it is not easy to identify any closed-form expression represented by the series. The solution is nonetheless definite. Note that it involves two arbitrary constants (c_0 and c_1), which is to be expected in view of the fact that the given equation is second-order.   □

The form of the general solution in Example 4 is

$$y = a_1 \phi_1(x) + a_2 \phi_2(x)$$

where a_1 and a_2 are arbitrary constants and $\phi_1(x)$ and $\phi_2(x)$ are the functions defined by the series in parentheses. This illustrates a theorem that we will discuss in Chapter 20 (on differential equations). Loosely stated, it says that a second-order equation of the type

$$ay'' + by' + cy = 0$$

has two solutions ϕ_1 and ϕ_2 (neither of which is a multiple of the other) in terms of which every solution can be written. That is, the general solution is a linear combination of the basic solutions ϕ_1 and ϕ_2, namely

$$y = a_1 \phi_1(x) + a_2 \phi_2(x)$$

Problem Set 13.5

Use series to solve each of the following initial value problems.

1. $y' - 2xy = 0$, $y = 1$ at $x = 0$
2. $y' - y = x$, $y = 1$ at $x = 0$
3. $y'' - y = 0$, $y = 0$ and $y' = 1$ at $x = 0$
4. $(1 - x^2)y'' + 2y = 0$, $y = 1$ and $y' = 0$ at $x = 0$
5. $y'' - 2xy = 0$, $y = 0$ and $y' = 1$ at $x = 0$
6. $xy'' + y' + xy = 0$, $y = 1$ at $x = 0$

7. $y'' - 2xy' = 0$, $y = 1$ and $y' = 1$ at $x = 0$

8. What function does the series solution of Problem 1 represent?

9. What function does the series solution of Problem 2 represent?

10. What function does the series solution of Problem 3 represent?

11. Use series to find the general solution of

$$y' - y = e^x$$

and identify the function it represents.

12. Use series to find the general solution of

$$xy' - 2y = 0$$

13. Show that the equation

$$x^2 y'' - y = 0$$

has only one solution that can be represented by a Maclaurin series. What is this solution?

14. Suppose that the series

$$y = \sum_{n=0}^{\infty} c_n x^n$$

is tried in the differential equation

$$x^2 y'' + xy' - y = 0$$

(a) Show that this trial solution turns out to be $y = Cx$, where C is an arbitrary constant.

(b) Confirm that another family of solutions of the equation is $y = C/x$. Why were none of these functions found in part (a)?

True-False Quiz

1. $\sin x + \sin 2x + \sin 3x + \cdots$ is a power series.

2. If the series $\sum c_n x^n$ diverges when $x = \frac{1}{3}$, it also diverges when $x = -\frac{2}{3}$.

3. The radius of convergence of the power series

$$\sum_{n=1}^{\infty} \frac{(x-3)^n}{n}$$

is 1.

4. The derivative of

$$f(x) = 1 + \frac{1}{2}x + \frac{2!}{2^2}x^2 + \frac{3!}{2^3}x^3 + \cdots$$

is

$$f'(x) = \frac{1}{2} + \frac{2(2!)}{2^2}x + \frac{3(3!)}{2^3}x^2 + \cdots$$

5. The function $\phi(x) = 1 + x + x^2 + x^3 + \cdots$ is a solution of the differential equation $y' = y^2$ in the interval $(-1,1)$.

6. If $f(x) = 1 + 2x + 3x^2 + 4x^3 + \cdots$, then $f'''(0) = 4!$.

7. If f has derivatives of all orders in a neighborhood of the point $x = a$, then

$$f(x) = f(a) + f'(a)(x-a) + \frac{f''(a)}{2!}(x-a)^2 + \cdots$$

for all x in the neighborhood.

8. $e = \sum_{n=0}^{\infty} \frac{1}{n!}$

9. The function $f(x) = \sqrt{x}$ can be represented by a power series in x.

10. The formula

$$\sqrt{1 + x^2} \approx 1 + \tfrac{1}{2}x^2$$

may be used in the interval $(-1,1)$ with an error no greater than $\frac{1}{8}x^4$.

Additional Problems

Find the domain of each of the following functions.

1. $f(x) = \sum \frac{2^n x^n}{n}$ 2. $f(x) = \sum \frac{x^n}{n(n+1)}$

3. $f(x) = \sum \frac{x^{2n-1}}{2n-1}$ 4. $f(x) = \sum \frac{(x-2)^n}{\sqrt{n}(3^n)}$

5. $f(x) = \sum \frac{(x+1)^n}{n^2(2^n)}$

6. A Maclaurin series is known to converge at $x = -4$. For what values of x must it converge?

7. The series

$$f(x) = \sum c_n(x - 1)^n$$

is known to converge at $x = 2$ and to diverge at $x = 0$. What is the domain of f?

In each of the following, use the direct approach (successive differentiation of the given function) to find the Taylor series associated with f at the base point a.

8. $f(x) = e^{2x}$ at $a = 0$

9. $f(x) = \sin x$ at $a = \pi$

10. $f(x) = \ln(1 - x)$ at $a = 0$

11. $f(x) = 1/x^3$ at $a = 1$

12. $f(x) = \sqrt{2 + x}$ at $a = 0$

13. Find the fourth-order Maclaurin polynomial approximation to $f(x) = \ln \cos x$.

14. Repeat Problem 13 for $f(x) = \ln \sin x$.

15. If $f(x) = 1/(1 + x^2)$, find $f^{(4)}(0)$. *Hint:* What is the Maclaurin series representation of $f(x)$?

16. If $f(x) = 1/(1 - x^3)$, find $f^{(6)}(0)$.

17. Find the Maclaurin series for $t/(1 + t^2)$.

18. Integrate the series in Problem 17 (from 0 to x) to find the Maclaurin series for $\ln(1 + x^2)$.

19. The Maclaurin series associated with e^{-x} is

$$S(x) = 1 - x + \frac{x^2}{2!} - \frac{x^3}{3!} + \cdots$$

Show that $e^{-x} = S(x)$ as follows.

(a) What is the domain of S?

(b) Show that $S'(x) = -S(x)$ for all x.

(c) Part (b) shows that $y = S(x)$ satisfies the differential equation $dy/dx = -y$. Solve this equation to find $y = Ce^{-x}$, where C is a constant.

(d) Explain why $S(0) = 1$ and hence $S(x) = e^{-x}$.

20. Derive the formula

$$\ln x = (x - 1) - \frac{(x - 1)^2}{2} + \frac{(x - 1)^3}{3} - \cdots$$

as follows.

(a) Replace x by $1 - t$ in the geometric series for $1/(1 - x)$ to obtain

$$\frac{1}{t} = 1 + (1 - t) + (1 - t)^2 + (1 - t)^3 + \cdots$$

(b) Integrate the series in part (a) from 1 to x. For what values of x is the result valid?

21. Derive the formula

$$\frac{1}{1 + x^2} = 1 - x^2 + x^4 - x^6 + \cdots \qquad (-1 < x < 1)$$

in three ways, as follows.

(a) Replace t by $-x^2$ in the geometric series for $1/(1 - t)$.

(b) Differentiate the series for $\tan^{-1} x$.

(c) Divide 1 by $1 + x^2$.

22. A function f satisfies the conditions

$$f(0) = 2, \; f'(0) = -1, \; f''(0) = 0, \; f'''(0) = 1$$

and $f^{(k)}(0) = 0$ for $k = 4, 5, 6, \ldots$. Find a formula for $f(x)$.

23. Differentiate the Maclaurin series for $1/(1 + x)$ to obtain the Maclaurin series for $1/(1 + x)^2$.

24. Find the Maclaurin series for $1/(1 + x)^2$ by squaring the Maclaurin series for $1/(1 + x)$.

25. Multiply the geometric series for $1/(1 - x)$ by the Maclaurin series for $\sin x$ to obtain the first four nonzero terms of the Maclaurin series for $(\sin x)/(1 - x)$.

26. Obtain the series in Problem 25 by division of the series for $\sin x$ by $1 - x$.

27. Find the first three nonzero terms of the Maclaurin series for $e^x \sin x$ in two ways, as follows.

(a) Use successive differentiation of $f(x) = e^x \sin x$.

(b) Multiply the series for e^x and $\sin x$.

28. Find $\int_0^{0.5} e^x \sin x \, dx$ in three ways, as follows.

(a) Using integration by parts, show that

$$\int_0^{0.5} e^x \sin x \, dx = \frac{1}{2} e^{0.5}(\sin 0.5 - \cos 0.5) + \frac{1}{2}$$

$$= 0.171775 \cdots$$

(b) Use the formula $e^x \sin x \approx x + x^2 + \frac{1}{3}x^3$ found in Problem 27.

(c) Use Simpson's Rule with four subintervals (Section 10.6).

29. Use series to compute $\int_0^1 x \cos x \, dx$ correct to three decimal places. Check by integration by parts.

30. Use the Maclaurin series for e^{-x^2} to compute

$$\int_0^{1/2} e^{-x^2} \, dx$$

correct to four decimal places.

31. Use series to compute

$$\int_0^1 \cos \sqrt{x} \, dx$$

correct to four decimal places.

32. Use series to compute e^{-1} correct to three decimal places.

33. Use a Maclaurin series to estimate $\sqrt{e}$.

34. The series

$$\ln\frac{1+x}{1-x} = 2\left(x + \frac{x^3}{3} + \frac{x^5}{5} + \cdots\right) \qquad (-1 < x < 1)$$

is given in Problem 39, Section 13.1. Use it to compute $\ln 3$.

35. The equation $e^x = 3x + 4$ has a positive root. Find it approximately by replacing e^x by the first three terms of its Maclaurin series.

36. Use series to prove that $e^x > 1 + x$ for $x > 0$.

37. Explain why

$$a^x = 1 + x \ln a + \frac{(x \ln a)^2}{2!} + \frac{(x \ln a)^3}{3!} + \cdots \text{ for all } x$$

38. Show that if $x \geq \frac{1}{2}$, then

$$\ln x = \frac{x-1}{x} + \frac{1}{2}\left(\frac{x-1}{x}\right)^2 + \frac{1}{3}\left(\frac{x-1}{x}\right)^3 + \cdots$$

39. Find the first three terms of the Maclaurin series for $f(x) = e^{\sin x}$ in two ways, as follows.

(a) Use successive differentiation of f.

(b) Replace x by $\sin x$ in the series for e^x and substitute the series for $\sin x$ and $\sin^2 x$. (The latter may be found in Problem 12, Section 13.3.)

40. Obtain the Maclaurin series for $1/(1+x)^2$ in two ways, as follows.

(a) Differentiate a known series for $1/(1+x)$.

(b) Apply the binomial series.

What is the domain of the result?

41. Use the binomial series to find the Maclaurin series representing $\sqrt[4]{1 + x^2}$.

42. Explain why the formula $\sqrt[3]{1 + x} \approx 1 + \frac{1}{3}x$ can be used in the interval $(-1,1)$ with an error no greater than $x^2/9$.

43. Explain why the formula $\sqrt[4]{1 + x} \approx 1 + \frac{1}{4}x - \frac{3}{32}x^2$ can be used in the interval $(-1,1)$ with an error no greater than $\frac{7}{128}|x|^3$.

44. Use the formula in Problem 43 to compute $\sqrt[4]{15}$ correct to as many decimal places as justified by the error estimate.

45. Use the binomial series to compute

$$\int_0^{1/2} \frac{dx}{\sqrt{1 - x^2}}$$

46. Use the first four nonzero terms of the Maclaurin series for $\sqrt{1 - x^4}$ to compute

$$\int_0^{1/2} \sqrt{1 - x^4}\, dx$$

47. Suppose that $y = \phi(x)$ is the solution of the differential equation $y' = 1 + y$ whose graph contains the point $(0,1)$.

(a) Use the given information and repeated differentiation to compute $\phi(0)$, $\phi'(0)$, $\phi''(0)$, $\phi'''(0)$, ...

(b) What is the Taylor series associated with ϕ at the origin? Use it to find a closed-form expression for $\phi(x)$.

(c) Use separation of variables to solve the given initial value problem for $\phi(x)$, obtaining the same result as in part (b).

48. Suppose that $y = \phi(x)$ is the solution of the differential equation $y' = x + y$ whose graph contains the origin.

(a) Use the given information and repeated differentiation to compute $\phi(0)$, $\phi'(0)$, $\phi''(0)$, $\phi'''(0)$, ...

(b) What is the Maclaurin series for ϕ? Use it to find a closed-form expression for $\phi(x)$.

(c) Rewrite the given equation in the form $y' - y = x$, multiply by the "integrating factor" e^{-x} to obtain $(ye^{-x})' = xe^{-x}$, and solve for $\phi(x)$, obtaining the same result as in part (b).

49. Use series to solve the initial value problem

$$y'' + y' = 0 \qquad \text{where } y = 0 \text{ and } y' = 1 \text{ at } x = 0$$

What function does the result represent?

50. Use series to solve the initial value problem

$$y'' - 2y' = 0 \qquad \text{where } y = 1 \text{ and } y' = 0 \text{ at } x = 0$$

What function does the result represent?

51. Use series to solve the initial value problem

$$y'' - y = 0 \qquad \text{where } y = 1 \text{ and } y' = -1 \text{ at } x = 0$$

What function does the result represent?

14

Geometry in the Plane

As far as the laws of mathematics refer to reality, they are not certain; and as far as they are certain, they do not refer to reality.
ALBERT EINSTEIN

The deep study of nature is the most fruitful source of mathematical discoveries.
JEAN-BAPTISTE-JOSEPH FOURIER (1768–1830)

There are two main topics in this chapter, *polar coordinates* and *conic sections*. The former is a new way to locate points in the coordinate plane, while the latter is the study of *second-degree curves* (circles, ellipses, parabolas, and hyperbolas). Except for some applications here and there, calculus is not much involved; most of this material could be (and often is) discussed in a precalculus course. That does not diminish its importance, as you will see when we put the results to work.

14.1
POLAR
COORDINATES

Many interesting and useful graphs have the property of looping or spiraling or otherwise revolving about some central point. Such a curve is not the graph of a function in a rectangular coordinate system (because it crosses vertical lines more than once), nor is its equation in x and y likely to be simple or enlightening. It is natural in such circumstances to locate the typical point of the curve in terms of its distance from the central point, together with some measure of rotation about that point.

More precisely, regard the central point as the origin O of a fixed ray (or half-line) in the plane, and let P be any point of the plane. **Polar coordinates** of P are numbers r and θ, written in the order (r,θ), such that $r = OP$

Figure 1
Polar coordinates of a point

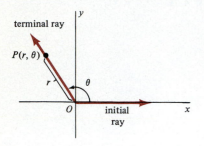

Figure 2
Determining a point from a pair (r,θ)

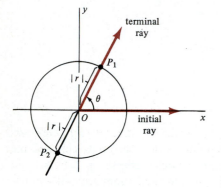

Figure 3
Points in polar coordinates

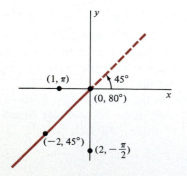

and θ is the radian measure of any angle whose initial side is the fixed ray and whose terminal side contains P. (See Figure 1, in which we have imposed this scheme on the usual rectangular coordinate plane.)

Since θ is not uniquely determined by the initial and terminal rays, a given point P has infinitely many pairs of polar coordinates. For example, $(2,60°)$, $(2,-300°)$, and $(2,420°)$ all represent the same point. Given a pair of real numbers (r,θ), however, we can find a unique point P having these numbers as polar coordinates, as follows. Regard $|r|$ as the radius of a circle with center O and regard θ as radian measure of the angle whose initial side is the positive x axis and whose terminal side is obtained by rotation of this ray through θ radians (counterclockwise or clockwise or not at all depending on whether $\theta > 0$ or $\theta < 0$ or $\theta = 0$). The line containing the terminal ray intersects the circle in two points P_1 and P_2 as shown in Figure 2. Take $P = P_1$ if $r > 0$ and $P = P_2$ if $r < 0$. (Of course $P = O$ if $r = 0$, regardless of the value of θ.)

This scheme associates a definite point P with each pair of real numbers (r,θ), that is, the correspondence $(r,\theta) \to P$ is a function

$$T: \Re^2 \to \Re^2 \text{ defined by } T(r,\theta) = P$$

If the rectangular coordinates of P are (x,y), then

$$x = r \cos \theta \qquad \text{and} \qquad y = r \sin \theta$$

as you can see by recalling the definition of sine and cosine. That is, T may be regarded as the function

$$T(r,\theta) = (x,y) = (r \cos \theta, r \sin \theta)$$

which sends polar coordinates into rectangular coordinates.

Since a given point P has infinitely many pairs of polar coordinates, the function T is not invertible, that is, the correspondence $P \to (r,\theta)$ is not a function. This is reflected by the fact that we cannot solve the equations $x = r \cos \theta$ and $y = r \sin \theta$ for r and θ in terms of x and y. The best we can do is to observe that

$$r^2 = x^2 + y^2 \qquad \text{and (if } x \neq 0) \qquad \tan \theta = \frac{y}{x}$$

▢ **Example 1**   Plot the points with polar coordinates $(1,\pi)$, $(2,-\pi/2)$, $(-2,45°)$, $(0,80°)$, and find their rectangular coordinates.

Solution: The points are shown in Figure 3. Their rectangular coordinates are

$$T(1,\pi) = (1 \cos \pi, 1 \sin \pi) = (-1,0)$$

$$T\left(2,-\frac{\pi}{2}\right) = \left(2 \cos\left(-\frac{\pi}{2}\right), 2 \sin\left(-\frac{\pi}{2}\right)\right) = (0,-2)$$

$$T(-2,45°) = (-2 \cos 45°, -2 \sin 45°) = (-\sqrt{2}, -\sqrt{2})$$
$$T(0,80°) = (0 \cos 80°, 0 \sin 80°) = (0,0)$$

respectively.   □

The equations

$$x = r \cos \theta, \ y = r \sin \theta, \ r^2 = x^2 + y^2, \ \tan \theta = \frac{y}{x}$$

provide the means for changing from one coordinate system to the other, as illustrated in the next three examples.

□ **Example 2**   The circle of radius 3 centered at the origin is represented by the equation $x^2 + y^2 = 9$ in rectangular coordinates. Transforming to polar coordinates, we have $r^2 = 9$ or simply $r = 3$. (Why does the other possibility, $r = -3$, represent the same circle?)   □

> The polar equation $r = a$ $(a \neq 0)$ represents a circle of radius $|a|$ and center at O.

□ **Example 3**   The equation $\theta = \pi/4$ in polar coordinates represents the straight line with equation $y = x$ in rectangular coordinates. For r is unrestricted by the equation; it may be any real number. When $r > 0$ we get the first-quadrant portion of the line, whereas the third-quadrant part corresponds to $r < 0$. The origin is included as the point $(0,\pi/4)$. To confirm this by our transformation equations, observe that if $x \neq 0$, the equation $y = x$ can be written $y/x = 1$, or $\tan \theta = 1$ in polar coordinates. One solution of this is $\theta = \pi/4$. The origin is excluded in this analysis, but since it satisfies $y = x$ in the rectangular form $(0,0)$ and satisfies $\theta = \pi/4$ in the polar form $(0,\pi/4)$, it is part of the graph.   □

> Polar equations of the form $\theta = k$ represent straight lines through O.

□ **Example 4**   Discuss the graph of the polar equation $r = 2a \cos \theta$ (where a is a positive constant).

Solution: Unless this can be transformed into something manageable in rectangular coordinates, we have little recourse but to plot points by constructing a table of values of r and θ. In many cases we will have to do just that, but when possible we should avoid it. In the present case, multiplication of each side by r yields an equation that is easily transformed into rectangular coordinates, namely

$$r^2 = 2ar \cos \theta$$

Replacing r^2 by $x^2 + y^2$ and $r \cos \theta$ by x, we have $x^2 + y^2 = 2ax$, which is the equation of a circle. To be precise, complete the square:

$$(x^2 - 2ax + a^2) + y^2 = a^2$$
$$(x - a)^2 + y^2 = a^2$$

Hence the circle has center $(a,0)$ and radius a. (See Figure 4.)

If we think of this circle as the collection of points (r,θ) satisfying $r = 2a \cos \theta$, it is important to note that it is generated when θ goes from 0 to π, not 0 to 2π. The upper half is obtained by letting θ run through the interval $[0,\pi/2]$, the corresponding values of r decreasing from $2a$ to 0. In the interval $(\pi/2,\pi)$ the values of r are negative and the lower half of the circle is generated. (Why?) When $\theta = \pi$, $r = -2a$ and we are back to the point $(2a,0)$ corresponding to $\theta = 0$.   □

We leave it to you to show that the graph of the polar equation $r = 2a \sin \theta$ is the circle of center $(0,a)$ (in rectangular coordinates) and radius a.

The polar equations

$$r = 2a \cos \theta \quad \text{and} \quad r = 2a \sin \theta \qquad (a > 0)$$

represent circles of radius a with centers $(a,0)$ and $(0,a)$, respectively (in rectangular coordinates).

□ **Example 5**   Discuss the graph of the polar equation $r = a(1 - \cos \theta)$ (where a is a positive constant).

Solution: This is the first example in which polar coordinates are truly worthwhile. Changing to rectangular coordinates is not easy, and is of little use (as you can see if you try it). Hence we will discuss the equation as it stands. Note first that since $\cos(-\theta) = \cos \theta$, the point $(r,-\theta)$ lies on the graph whenever (r,θ) does. Hence the graph is symmetric about the x axis. (Why?)

The following description of a table of values indicates how the curve should be sketched.

- When $\theta = 0$, $r = 0$ (the origin).
- As θ increases from 0 to $\pi/2$, r goes from 0 to a. (This gives the first-quadrant portion of Figure 5.)

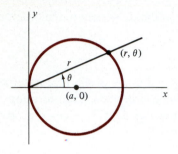

Figure 4
Graph of $r = 2a \cos \theta$

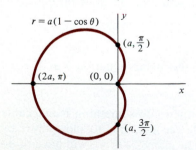

Figure 5
The cardioid $r = a(1 - \cos \theta)$

• As θ increases from $\pi/2$ to π, r increases from a to $2a$. (This gives the second-quadrant portion of Figure 5.)
• The same behavior in reverse occurs as θ goes from π to 2π. (We need not consider this, however, because of the symmetry about the x axis already mentioned.)

The graph is called a *cardioid* because it is heart-shaped. Similar curves are obtained from the equations

$$r = a(1 + \cos\theta), \ r = a(1 - \sin\theta), \ r = a(1 + \sin\theta)$$

(See the problem set.)   □

□ **Example 6**   Discuss the graph of the polar equation $r = \sin 2\theta$.

Solution: The easiest procedure is to make a table of values and plot points. To illustrate various ways of discussing symmetry, however, we make the following observations.

1. If θ is replaced by $-\theta$, the equation becomes

$$r = \sin(-2\theta) = -\sin 2\theta$$

Hence $(-r, -\theta)$ is on the graph whenever (r, θ) is. This implies symmetry about the y axis. (Why?)
2. If θ is replaced by $\theta + \pi$, the equation becomes

$$r = \sin 2(\theta + \pi) = \sin(2\theta + 2\pi) = \sin 2\theta$$

Hence $(r, \theta + \pi)$ is on the graph whenever (r, θ) is. This implies symmetry about the origin. (Why?)
3. Symmetry about the y axis and about the origin implies symmetry about the x axis. (Why?)

With these observations in hand, we need only find what the graph looks like in the first quadrant. Reflection about the coordinate axes will then produce the whole graph. The following description of a table of values suggests how to sketch the curve.

• When $\theta = 0$, $r = 0$ (the origin).
• As θ increases from 0 to $\pi/4$, r goes from 0 to 1 (its maximum value).
• As θ increases from $\pi/4$ to $\pi/2$, r goes from 1 back to 0. This is sufficient to give the first-quadrant loop shown in Figure 6, and hence the whole graph.

It is worthwhile to continue the description of the table, however:

• The fourth-quadrant loop is obtained as θ runs through the interval $(\pi/2, \pi)$, when $r < 0$.
• The third-quadrant loop is obtained as θ runs through $(\pi, 3\pi/2)$, when $r > 0$.

Figure 6
The four-leaf rose $r = \sin 2\theta$

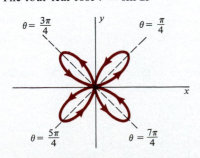

• The second-quadrant loop is obtained as θ runs through $(3\pi/2, 2\pi)$, when $r < 0$.

The arrows indicate the order in which the loops are traversed as θ goes from 0 to 2π. The curve is called a *four-leaf rose*. □

□ **Example 7**   Find the points of intersection of the circle $r = 6 \cos \theta$ and the cardioid $r = 2(1 + \cos \theta)$.

Solution: The curves are shown in Figure 7, from which it is clear that there are three points of intersection. The usual procedure for finding such points is to solve the equations simultaneously:

$$6 \cos \theta = 2(1 + \cos \theta)$$

$$3 \cos \theta = 1 + \cos \theta$$

$$2 \cos \theta = 1$$

$$\cos \theta = \frac{1}{2}$$

$$\theta = \pm \frac{\pi}{3} \qquad \text{(plus multiples of } 2\pi\text{)}$$

Hence the points $(3, \pi/3)$ and $(3, -\pi/3)$ are two points of intersection.
While this is good enough as far as it goes, you can see that we missed the origin. The reason is that the origin satisfies the equation $r = 6 \cos \theta$ in the form $(0, \pi/2)$ and the equation $r = 2(1 + \cos \theta)$ in the form $(0, \pi)$, but neither pair of polar coordinates satisfies both equations. One way to look at it is to imagine two bugs traversing the curves, both starting at $\theta = 0$. The bug on the circle reaches the origin when $\theta = \pi/2$, but the bug on the cardioid does not reach it until $\theta = \pi$. The origin is on both curves, but not "at the same time." □

There are other difficulties that can make points of intersection tricky; it is hard to give general rules covering all cases. A picture is always helpful (to get an idea of where the curves intersect), and of course we should solve the equations simultaneously. The origin often requires special consideration, as in Example 7. If it looks as if we missed other points, it is sometimes helpful to remember that (r, θ), $(r, \theta + 2n\pi)$, and $(-r, \theta + \pi)$ are all polar coordinates of the same point.

Figure 7
Intersection of cardioid and circle

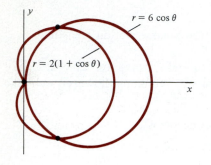

$r = 6 \cos \theta$

$r = 2(1 + \cos \theta)$

Problem Set 14.1

Each of the following points is given in polar coordinates. Find its rectangular coordinates.

1. $(3, \pi)$ **2.** $(1, \pi/2)$ **3.** $(-1, 3\pi)$
4. $(0, \pi)$ **5.** $(2, 135°)$ **6.** $(4, 30°)$
7. $(4, -30°)$ **8.** $(2, \pi/3)$ **9.** $(-2, 4\pi/3)$
10. $(-1, \pi/2)$

Each of the following points is given in rectangular coordinates. Find a set of polar coordinates.

11. $(3, 0)$ **12.** $(0, 2)$
13. $(1, 1)$ **14.** $(1, -1)$
15. $(\sqrt{3}, -1)$ **16.** $(-1, \sqrt{3})$
17. $(0, -1)$ **18.** $(-1, -1)$

Change each of the following polar equations to an equation in rectangular coordinates and describe the graph.

19. $r = 2$

20. $\theta = 3\pi/4$

21. $\tan \theta = 2$

22. $r = 2 \sec \theta$

23. $r = \dfrac{1}{\sin \theta + \cos \theta}$

24. $r = 2 \sin \theta$

25. $r = -6 \cos \theta$

26. $r = 2 \sin \theta + 4 \cos \theta$

27. $r^2 = 2 \csc 2\theta$

28. $r^2 = \sec 2\theta$

29. By changing to rectangular coordinates, show that the polar equation

$$r = \frac{1}{1 - \cos \theta}$$

represents a parabola.

30. Change the polar equation $r = \sin 2\theta$ to an equation in rectangular coordinates. Is the result much help in graphing the equation?

Graph each of the following cardioids.

31. $r = 2(1 + \cos \theta)$

32. $r = 3(1 - \sin \theta)$

33. $r = 1 + \sin \theta$

34. $r = 4(1 - \cos \theta)$

Sketch the graph of each of the following polar equations.

35. $r = -1$

36. $\theta = 3\pi$

37. $r = \cos 2\theta$

38. $r = \sin 3\theta$

39. $r = 2 - \cos \theta$

40. $r = 2 - \sin \theta$

41. $r = 1 - 2 \cos \theta$ *Hint:* The curve has an inside loop.

42. $r = 1 - 2 \sin \theta$

43. $r = \theta$ $(\theta \geq 0)$

44. $r = 1/\theta, \theta > 0$ is an asymptote.   *Hint:* First explain why the line $y = 1$

45. $r = e^\theta$

46. $r^2 = \cos 2\theta$

47. $r^2 = \sin 2\theta$

48. $r = 2 \sec \theta - 1$

Find the points of intersection of each of the following pairs of polar curves.

49. $r = 2 \cos \theta, r = 2 \sin \theta$

50. $r = 1 + \cos \theta, r = 2 \cos \theta$

51. $r = \cos \theta, r = \sin 2\theta$

52. $r = \sin \theta, r = \cos 2\theta$

53. How many points of intersection do the polar curves $r = \theta$ and $\theta = \pi/2$ have? Which of these do you find by solving the equations simultaneously?

54. Show that the distance between the points (r_1, θ_1) and (r_2, θ_2) in polar coordinates is

$$\sqrt{r_1^2 + r_2^2 - 2r_1 r_2 \cos (\theta_1 - \theta_2)}.$$

14.2
AREA AND LENGTH IN POLAR COORDINATES

Let D be the region bounded by two rays from the origin and by a curve with polar equation $r = f(\theta)$ that is intersected just once by each ray in between. (See Figure 1.) Restricting $r \geq 0$, we may represent the given rays by $\theta = \alpha$ and $\theta = \beta$, where $0 \leq \beta - \alpha \leq 2\pi$. (This restriction ensures natural choices of α and β. For it implies that $\alpha \leq \beta$ and that no more than one revolution is allowed in going from α to β.) Each point of the curved boundary may be represented by polar coordinates (r, θ), where

$$r = f(\theta) \geq 0 \qquad \alpha \leq \theta \leq \beta$$

Assuming that f is continuous, we propose to find the area of D by integration involving polar coordinates.

Our procedure is like that adopted so often before. (See Chapter 7.) Let

$$P = \{\theta_0, \theta_1, \ldots, \theta_n\}$$

be a partition of the interval $[\alpha, \beta]$. The rays $\theta = \theta_k$, $k = 0, 1, \ldots, n$, divide the region D into n "sectors," the typical sector being shown in Figure 2. Choose a number θ in the typical subinterval of $[\alpha, \beta]$ and let $r = f(\theta)$. The polar coordinates (r, θ) represent a point on the curved boundary of the sector; the area of the sector is

$$\Delta A \approx \tfrac{1}{2} r^2 \, \Delta \theta$$

Figure 1
Region in polar coordinates

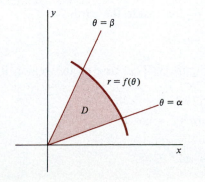

Figure 2

Typical subregion in polar coordinates

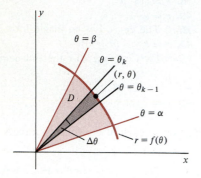

(If the boundary of the sector were a circular arc of radius r, this formula would be exact. Recall from trigonometry that the area of a circular sector of radius r and central angle θ is $\frac{1}{2}r^2\theta$.) The total area is

$$A = \sum \Delta A \approx \sum \tfrac{1}{2} r^2 \, \Delta\theta = \sum \tfrac{1}{2} f(\theta)^2 \, \Delta\theta$$

which is a Riemann sum of the function $\frac{1}{2} f(\theta)^2$ associated with the partition P. Hence

$$A = \lim_{\|P\|\to 0} \sum \frac{1}{2} r^2 \, \Delta\theta = \int_\alpha^\beta \frac{1}{2} r^2 \, d\theta = \frac{1}{2} \int_\alpha^\beta f(\theta)^2 \, d\theta$$

Before stating this result as a formal conclusion, we should point out a difficulty. Area in rectangular coordinates and area in polar coordinates might conceivably be different (because although both are computed by Riemann sums, the approximations leading to the sums are not the same). What is needed is a general definition of area in the plane (independent of coordinate systems), on the basis of which a proof can be given that rectangular and polar coordinates yield the same result. Such a definition is usually discussed in advanced calculus.

Area in Polar Coordinates

Let D be the region bounded by two rays from the origin and by a curve that is intersected just once by each ray in between. If the curve is represented by the polar equation

$$r = f(\theta) \qquad \alpha \le \theta \le \beta \text{ (where } 0 \le \beta - \alpha \le 2\pi)$$

and if f is continuous in $[\alpha,\beta]$, the area of D is

$$A = \frac{1}{2} \int_\alpha^\beta r^2 \, d\theta = \frac{1}{2} \int_\alpha^\beta f(\theta)^2 \, d\theta$$

Remark   In this statement of the area formula we have dropped the restriction that $r \ge 0$. The crucial step of the derivation ($\Delta A \approx \frac{1}{2} r^2 \, \Delta\theta$) works just as well if $r < 0$, so we allow situations in which a ray is the backward extension of the terminal side of θ. See Examples 2 and 4 for illustrations.

□ **Example 1**   Find the area of the region inside the cardioid

$$r = a(1 - \cos\theta)$$

Solution: The curve is shown in Figure 3. The area of the region it encloses is

$$A = \frac{1}{2} \int_0^{2\pi} r^2 \, d\theta = \frac{1}{2} \int_0^{2\pi} a^2 (1 - \cos\theta)^2 \, d\theta$$

$$= \frac{a^2}{2} \int_0^{2\pi} (1 - 2\cos\theta + \cos^2\theta) \, d\theta$$

Figure 3

The cardioid $r = a(1 - \cos\theta)$

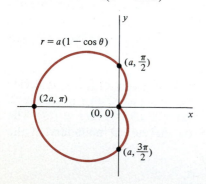

$$= \frac{a^2}{2} \int_0^{2\pi} \left[1 - 2\cos\theta + \frac{1}{2}(1 + \cos 2\theta) \right] d\theta$$

$$= \frac{a^2}{2} \left(\frac{3}{2}\theta - 2\sin\theta + \frac{1}{4}\sin 2\theta \right) \Big|_0^{2\pi} = \frac{3}{2}\pi a^2 \qquad \square$$

□ **Example 2**   Find the area of the region inside the polar curve

$$r = 2a\cos\theta$$

Solution: The graph is the circle of radius a shown in Figure 4. Hence the area is πa^2 by plane geometry. To confirm this by integration, recall from Example 4, Section 14.1, that the circle is generated when θ goes from 0 to π, not 0 to 2π. Hence the area is

$$A = \frac{1}{2} \int_0^\pi r^2 \, d\theta = \frac{1}{2} \int_0^\pi 4a^2 \cos^2\theta \, d\theta = a^2 \int_0^\pi (1 + \cos 2\theta) \, d\theta$$

$$= a^2 \left(\theta + \frac{1}{2}\sin 2\theta \right) \Big|_0^\pi = \pi a^2 \qquad \square$$

□ **Example 3**   Find the area of the region inside the four-leaf rose

$$r = \sin 2\theta$$

Solution: The curve is shown in Figure 5. One of the four loops is generated as θ goes from 0 to $\pi/2$, so the area is

$$A = 4 \cdot \frac{1}{2} \int_0^{\pi/2} r^2 \, d\theta = 2 \int_0^{\pi/2} \sin^2 2\theta \, d\theta = \int_0^{\pi/2} (1 - \cos 4\theta) \, d\theta$$

$$= \left(\theta - \frac{1}{4}\sin 4\theta \right) \Big|_0^{\pi/2} = \frac{\pi}{2} \qquad \square$$

□ **Example 4**   Find the area of the inside loop of the polar curve

$$r = 1 - 2\cos\theta$$

Solution: The curve is shown in Figure 6. The inside loop is generated as θ runs through the interval $[-\pi/3, \pi/3]$, while the rest of the curve corresponds to the interval $(\pi/3, 5\pi/3)$. (Confirm!) The lower half of the inside loop is obtained when θ goes from 0 to $\pi/3$, so the area is

$$A = 2 \cdot \frac{1}{2} \int_0^{\pi/3} (1 - 2\cos\theta)^2 \, d\theta = \int_0^{\pi/3} (1 - 4\cos\theta + 4\cos^2\theta) \, d\theta$$

$$= \int_0^{\pi/3} [1 - 4\cos\theta + 2(1 + \cos 2\theta)] \, d\theta = (3\theta - 4\sin\theta + \sin 2\theta) \Big|_0^{\pi/3}$$

$$= \pi - 4\sin\frac{\pi}{3} + \sin\frac{2\pi}{3} = \pi - \frac{3}{2}\sqrt{3} \qquad \square$$

Figure 4
Graph of $r = 2a\cos\theta$

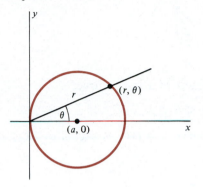

Figure 5
The four-leaf rose $r = \sin 2\theta$

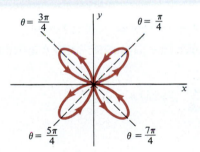

Figure 6
Graph of $r = 1 - 2 \cos \theta$

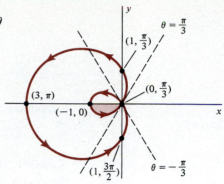

Now we turn to the question of length of a polar curve

$$r = f(\theta) \qquad \alpha \leq \theta \leq \beta$$

Our formula for length of a smooth curve $x = f(t)$, $y = g(t)$, $a \leq t \leq b$ (Section 7.3) is

$$s = \int_a^b \sqrt{(dx/dt)^2 + (dy/dt)^2} \, dt$$

To apply this to our polar curve, observe that the rectangular coordinates of a typical point (r, θ) of the curve are

$$x = r \cos \theta = f(\theta) \cos \theta, \; y = r \sin \theta = f(\theta) \sin \theta \qquad \alpha \leq \theta \leq \beta$$

These are parametric equations of the curve (the parameter being θ instead of t). We compute

$$\frac{dx}{d\theta} = -r \sin \theta + \cos \theta \, \frac{dr}{d\theta}, \; \frac{dy}{d\theta} = r \cos \theta + \sin \theta \, \frac{dr}{d\theta}$$

from which

$$\left(\frac{dx}{d\theta}\right)^2 + \left(\frac{dy}{d\theta}\right)^2 = r^2 \sin^2 \theta - 2r \sin \theta \cos \theta \, \frac{dr}{d\theta} + \cos^2 \theta \left(\frac{dr}{d\theta}\right)^2$$

$$+ r^2 \cos^2 \theta + 2r \sin \theta \cos \theta \, \frac{dr}{d\theta} + \sin^2 \theta \left(\frac{dr}{d\theta}\right)^2$$

$$= r^2 + \left(\frac{dr}{d\theta}\right)^2$$

Hence:

The length of the smooth polar curve $r = f(\theta)$, $\alpha \leq \theta \leq \beta$, is

$$s = \int_\alpha^\beta \sqrt{r^2 + (dr/d\theta)^2} \, d\theta$$

Remark   In Section 7.3 the curve $x = f(t)$, $y = g(t)$, $a \leq t \leq b$, is defined to be *smooth* if $(dx/dt)^2 + (dy/dt)^2 > 0$ for $a < t < b$ (all interior points of the domain). In the polar case this condition is $r^2 + (dr/d\theta)^2 > 0$ for $\alpha < \theta < \beta$. It is important to keep this hypothesis in mind, to avoid errors caused by inadvertent integration past cusps and other pathological points. See Example 5, in which there is a cusp at $\theta = 0$ (repeated at $\theta = 2\pi$). Note that we are careful not to integrate past this point.

□ **Example 5**   Find the length of the cardioid $r = a(1 - \cos \theta)$.

Solution: Since $dr/d\theta = a \sin \theta$, we have

$$r^2 + \left(\frac{dr}{d\theta}\right)^2 = a^2(1 - \cos \theta)^2 + a^2 \sin^2 \theta = a^2(1 - 2 \cos \theta + \cos^2 \theta + \sin^2 \theta)$$

$$= a^2(2 - 2 \cos \theta) = 2a^2(1 - \cos \theta)$$

$$= 4a^2 \sin^2 \frac{\theta}{2} \quad \left[\text{because } \sin^2 \frac{\theta}{2} = \frac{1}{2}(1 - \cos \theta)\right]$$

The cardioid is generated when θ runs through the interval $[0, 2\pi]$ and in that interval $\sin(\theta/2)$ is zero only at the endpoints. Hence our curve is smooth (from cusp back to cusp); its length is

$$s = \int_0^{2\pi} 2a \sin \frac{\theta}{2} \, d\theta \quad \left[\sqrt{\sin^2(\theta/2)} = \left|\sin \frac{\theta}{2}\right| = \sin \frac{\theta}{2} \text{ because } 0 \leq \frac{\theta}{2} \leq \pi\right]$$

$$= -4a \cos \frac{\theta}{2} \Big|_0^{2\pi} = 8a \qquad\qquad □$$

□ **Example 6**   Find the length of the polar curve $r = 2a \cos \theta$.

Solution: Since this is a circle of radius a, its length is the circumference $2\pi a$. To confirm this by our length formula, compute

$$\frac{dr}{d\theta} = -2a \sin \theta$$

and

$$r^2 + \left(\frac{dr}{d\theta}\right)^2 = 4a^2 \cos^2 \theta + 4a^2 \sin^2 \theta = 4a^2$$

Hence the length is

$$s = \int_0^{\pi} 2a \, d\theta = 2\pi a$$

(Note that the integration is from 0 to π, not 0 to 2π.)   □

□ **Example 7**   Find the length of the four-leaf rose $r = \sin 2\theta$.

Solution: Since $dr/d\theta = 2 \cos 2\theta$, we have

$$r^2 + \left(\frac{dr}{d\theta}\right)^2 = \sin^2 2\theta + 4 \cos^2 2\theta = 1 + 3 \cos^2 2\theta$$

Integrating from 0 to $\pi/2$ and multiplying by 4 (as in Example 3), we find

$$s = 4 \int_0^{\pi/2} \sqrt{1 + 3\cos^2 2\theta}\, d\theta$$

This cannot be evaluated in terms of elementary functions, so we leave it in this form. Simpson's Rule (for example) can be used to approximate it.

□

Problem Set 14.2

In each of the following, find the area swept out by a ray from the origin to the given polar curve in the given interval.

1. $r = \theta,\ 0 \le \theta \le 2\pi$

2. $r = e^\theta,\ 0 \le \theta \le 2\pi$

3. $r = \dfrac{1}{\sqrt{1 + \theta}},\ 0 \le \theta \le 2\pi$

4. $r = 2\sec\theta,\ 0 \le \theta \le \pi/4$ (Check by geometry.)

Use integration to find the area of the region bounded by each of the following polar curves.

5. $r = 2(1 + \cos\theta)$ **6.** $r = 3(1 + \sin\theta)$

7. $r = 2\sin\theta$ (Check by geometry.)

8. $r = \cos 2\theta$ **9.** $r = \sin 3\theta$

10. $r^2 = \cos 2\theta$ **11.** $r^2 \cos^4\theta - 2\cos^2\theta + 1 = 0$

12. $r = \dfrac{2}{1 - \cos\theta}$ and $\theta = \dfrac{\pi}{2}$

In each of the following, find the area of the region inside both of the given polar curves.

13. $r = 2a\cos\theta$ and $r = 2a\sin\theta$

14. $r = 2\sin\theta$ and $r = 2(1 - \sin\theta)$

15. $r = 2(1 + \cos\theta)$ and $r = 3$

16. Find the area of the inside loop of the polar curve $r = 1 + 2\cos\theta$.

17. Find the area of the smaller of the two regions bounded by the polar curves $r = 4\cos\theta$ and $r = 3\sec\theta$.

18. Find the area of the region outside the circle $r = 2\sin\theta$ and inside the cardioid $r = 1 + \sin\theta$.

19. Find the area of the region outside the cardioid $r = 2(1 + \cos\theta)$ and inside the circle $r = 6\cos\theta$.

20. The circular region of radius 2 centered at the origin has the property that any line through the origin intersects the region in a segment of length 4.

(a) Show that the same thing is true of the region inside the polar curve $r = 2 + \cos\theta$.

(b) Show that the area of the region in part (a) is not the same as the area of the circle.

(c) Question for discussion: Is the area of a region determined if the lengths of all "diameters" through a given interior point are known?

Use integration to find the length of each of the following polar curves.

21. $r = 6\sin\theta$ (Check by geometry.)

22. $r = 2\cos\theta$ (Check by geometry.)

23. $r = 2\sec\theta,\ 0 \le \theta \le \pi/4$ (Check by geometry.)

24. $r = 5\csc\theta,\ \pi/4 \le \theta \le 3\pi/4$ (Check by geometry.)

25. $r = e^\theta,\ 0 \le \theta \le 2\pi$ **26.** $r = \theta^2,\ 0 \le \theta \le 2\pi$

27. $r = 1 - \sin\theta$ **28.** $r = 2(1 + \cos\theta)$

29. $r = 3(1 - \cos\theta)$ **30.** $r = 2(1 + \sin\theta)$

31. The polar curve $r = e^{-\theta}$ ($\theta \ge 0$) spirals toward the origin as θ increases. What can you say about its total length? *Hint:* An improper integral is involved.

32. The polar curve $r = 1/\theta$ ($\theta > 0$) spirals toward the origin as θ increases. What can you say about the length of the part of this curve that is inside the unit circle $r = 1$?

14.3
CONICS:
THE PARABOLA

In Section 1.3 we investigated some elementary properties of circles, parabolas, ellipses, and hyperbolas (often called **conic sections** because they may be obtained from a cone by slicing it at various angles with a plane). Our purpose was modest and our discussion sketchy. We did not give general definitions of the parabola, ellipse, or hyperbola, but merely examined some equations whose graphs go by these names. Since then we have used these curves in so many parts of calculus that you are probably confident that you know what they are. For most practical purposes, you do. The next two sections are designed to fill in details that may be new to you. More precisely, our objective is to unify the treatment of these curves by considering them under the general heading of *conics* (a term that we will define in the next section after treating the parabola as a special case).

A **parabola** is the set of points that are equidistant from a given point and a given line. More precisely, if F is the point and L is a line not containing F, the parabola with **focus** F and **directrix** L is the set of points P such that $d(P,F) = d(P,L)$.

Figure 1
Horizontal parabola with focus F and directrix L

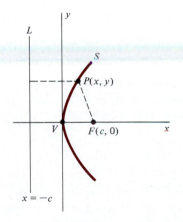

Since this definition makes no reference to a coordinate system, we cannot obtain equations of the parabola until we have decided where to locate it. One simple choice is to run the x axis through the focus perpendicular to the directrix, with the origin halfway in between. (See Figure 1.) Then F is the point $(c,0)$ and L is the line $x = -c$, where c is a constant (positive in Figure 1, negative if the focus is to the left of the origin). If S is the parabola and $P(x,y)$ is any point of the plane, our definition says that

$$P \in S \Leftrightarrow d(P,F) = d(P,L)$$
$$\Leftrightarrow \sqrt{(x - c)^2 + y^2} = \sqrt{(x + c)^2}$$
$$\Leftrightarrow (x - c)^2 + y^2 = (x + c)^2 \quad \text{(Why does the implication go both ways?)}$$
$$\Leftrightarrow x^2 - 2cx + c^2 + y^2 = x^2 + 2cx + c^2$$
$$\Leftrightarrow y^2 = 4cx$$

This is a familiar equation; the only thing new about it is that the constant $4c$ has a meaning not previously discussed:

$$|4c| = \text{twice the distance between the focus and directrix}$$

The point midway between the focus and directrix is called the **vertex** of the parabola. Hence

$$|c| = \text{distance between the vertex and focus (or directrix)}$$

Note that the sign of c determines the direction in which the parabola opens (to the right if $c > 0$, to the left if $c < 0$).

Had we run the y axis through the focus perpendicular to the directrix (as in Figure 2), a similar analysis would yield

$$P \in S \Leftrightarrow x^2 = 4cy$$

Figure 2
Vertical parabola with focus F and directrix L

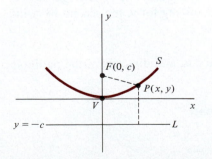

where c has the same meaning as before. This time the parabola opens upward if $c > 0$ and downward if $c < 0$.

A more general version of these remarks is obtained by taking the vertex to be some point other than the origin, say (h,k). This has the effect of translating the axes, which replaces x and y by $x - h$ and $y - k$, respectively. Otherwise the analysis is the same as before. In each case the **axis** of the parabola is the line through the focus and vertex perpendicular to the directrix; it is a line of symmetry of the parabola.

A parabola with vertex (h,k) and axis parallel to a coordinate axis may be represented by

$$4c(x - h) = (y - k)^2 \qquad \text{or} \qquad 4c(y - k) = (x - h)^2$$

depending on whether its axis is horizontal or vertical. The parabola opens in the positive direction if $c > 0$, in the negative direction if $c < 0$. In both cases $|c|$ is the distance between the vertex and focus (or directrix).

□ **Example 1**   Find the vertex, axis, focus, and directrix of the parabola $x^2 - 6x - 6y + 3 = 0$.

Solution: Perhaps the first question to raise is how we know this is a parabola. An examination of the standard forms reveals that in each case the equation is quadratic in one variable and linear in the other. Any equation of that type can be put in the standard form for a parabola by completing the square:

$$x^2 - 6x + \underline{\quad} = 6y - 3 + \underline{\quad}$$
$$x^2 - 6x + 9 = 6y - 3 + 9$$
$$(x - 3)^2 = 6(y + 1)$$

We recognize this as the equation of a vertical parabola with vertex $(3, -1)$, opening upward because $c = 3/2 > 0$. (See Figure 3.) The focus is c units above the vertex and the directrix c units below it. Hence $F = (3, \frac{1}{2})$ and L is the horizontal line $y = -\frac{5}{2}$. The axis is the vertical line $x = 3$.

Note the plotting of the points $(0, \frac{1}{2})$ and $(6, \frac{1}{2})$, designed to help sketch the curve. They are chosen for convenience, the first one by letting $x = 0$ in the equation $x^2 - 6x - 6y + 3 = 0$ and solving for y, the second by using symmetry about the axis.   □

□ **Example 2**   Find the vertex, axis, focus, and directrix of the parabola $y^2 + x - 2y = 0$ and sketch the graph.

Solution: First we complete the square:

$$y^2 - 2y + \underline{\quad} = -x + \underline{\quad}$$
$$y^2 - 2y + 1 = -x + 1$$
$$(y - 1)^2 = -(x - 1)$$

Figure 3
The parabola $6(y + 1) = (x - 3)^2$

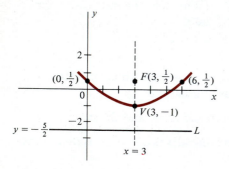

Figure 4
The parabola $-(x-1) = (y-1)^2$

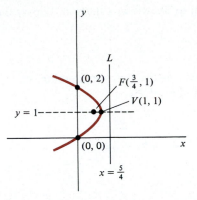

This is the equation of a horizontal parabola with vertex $(1,1)$, opening to the left because $c = -\frac{1}{4} < 0$. (See Figure 4.) The focus is $|c|$ units to the left of the vertex and the directrix $|c|$ units to the right of it. Hence $F = (\frac{3}{4},1)$ and L is the vertical line $x = \frac{5}{4}$. The axis is the horizontal line $y = 1$.   □

□ **Example 3**   In Figure 5 we show the parabola $y^2 = 4cx$ ($c > 0$), with focus $F(c,0)$ and typical point $P(x,y)$. Assuming that the parabola lies in a reflecting surface (like the housing of an auto headlight), prove that a light ray from the focus to P is reflected parallel to the axis.

Figure 5
Optical property of the parabola

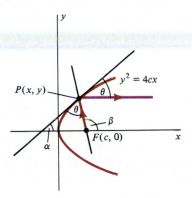

Solution: It is a property of reflected light that the angle of incidence is equal to the angle of reflection. The light ray is reflected at P as though it were hitting a flat mirror along the tangent. Hence we have marked the angle between the tangent and the incoming ray, and between the tangent and the reflected ray, by the same letter θ. The problem is to show that the reflected ray is horizontal, which is equivalent to proving that θ is the angle of inclination of the tangent. (Why?)

Let α and β be the inclination angles of the tangent line and the line PF, respectively. Then $\theta = \beta - \alpha$ and

$$\tan \theta = \tan(\beta - \alpha) = \frac{\tan \beta - \tan \alpha}{1 + \tan \alpha \tan \beta}$$

The slope of the tangent is found by implicit differentiation in $y^2 = 4cx$:

$$2y \frac{dy}{dx} = 4c$$

$$\frac{dy}{dx} = \frac{2c}{y} = \text{slope of the tangent} = \tan \alpha$$

The slope of the line PF is $y/(x-c) = \tan \beta$. Hence

$$\tan \theta = \frac{\dfrac{y}{x-c} - \dfrac{2c}{y}}{1 + \dfrac{2c}{y} \cdot \dfrac{y}{x-c}} = \frac{y^2 - 2c(x-c)}{y(x-c) + 2cy} = \frac{y^2 - 2cx + 2c^2}{xy - cy + 2cy}$$

$$= \frac{4cx - 2cx + 2c^2}{xy + cy} \qquad \text{(because } y^2 = 4cx)$$

$$= \frac{2cx + 2c^2}{y(x+c)} = \frac{2c(x+c)}{y(x+c)} = \frac{2c}{y} = \tan \alpha$$

Since θ and α are acute angles, it follows that $\theta = \alpha$, which is what we had to prove.   □

Figure 6

Parabola in polar coordinates

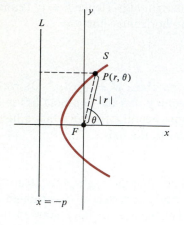

Polar coordinates are well suited for the representation of conics, as we now show in the case of the parabola. Place the focus at the origin and the directrix p units to the left of the origin, as shown in Figure 6. If S is the parabola, we have

$$P \in S \Leftrightarrow d(P,F) = d(P,L)$$

$$\Leftrightarrow |r| = |x + p| \quad \text{(where } x \text{ is the first rectangular coordinate of } P\text{)}$$

$$\Leftrightarrow r = \pm(x + p) \Leftrightarrow r = \pm(r \cos \theta + p)$$

$$\Leftrightarrow r \mp r \cos \theta = \pm p \Leftrightarrow r(1 \mp \cos \theta) = \pm p$$

$$\Leftrightarrow r = \frac{\pm p}{1 \mp \cos \theta}$$

Since the polar equations

$$r = \frac{p}{1 - \cos \theta} \quad \text{and} \quad r = \frac{-p}{1 + \cos \theta}$$

represent the same graph (replace r by $-r$ and θ by $\theta + \pi$ to obtain one from the other), we conclude that the parabola may be represented by the equation

$$r = \frac{p}{1 - \cos \theta}$$

If the parabola opens to the left (directrix $x = p > 0$), a similar analysis yields the equation

$$r = \frac{p}{1 + \cos \theta}$$

If it opens upward (directrix $y = -p$) or downward (directrix $y = p$), the forms obtained are

$$r = \frac{p}{1 - \sin \theta} \quad \text{or} \quad r = \frac{p}{1 + \sin \theta}$$

respectively.

Figure 7

The parabola $r = \dfrac{2}{1 - \sin \theta}$

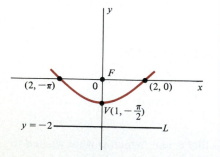

□ **Example 4**   Discuss the graph of the polar equation $r = 2/(1 - \sin \theta)$.

Solution: Assuming that we recognize this as a parabola with focus at the origin, it is unnecessary to memorize which type it is. Simply observe that the curve is swept out as θ runs through the interval $(-3\pi/2, \pi/2)$, with $r = 2$ at $\theta = -\pi$, $r = 1$ at $\theta = -\pi/2$, and $r = 2$ at $\theta = 0$. This orients the curve as shown in Figure 7. Since the vertex is 1 unit from the focus, the directrix is the horizontal line $y = -2$ (as may also be seen from the fact that $p = 2$ and the parabola opens upward).   □

Problem Set 14.3

Find the vertex, axis, focus, and directrix of each of the following parabolas, and sketch the graph.

1. $y = 4 - x^2$

2. $y = x^2 - 4$

3. $y^2 = 4(1 - x)$

4. $y^2 = 6(x - 2)$

5. $y = x^2 - 2x$

6. $y = 6x - x^2$

7. $y^2 - 2y - 8x + 1 = 0$

8. $x^2 - 4x + 4y - 8 = 0$

9. $4x^2 + 12x - 24y + 33 = 0$

10. $y^2 - 4x - 4y + 4 = 0$

Find an equation of each of the following parabolas.

11. Focus $(1,2)$, directrix $y = 0$.

12. Focus $(2,0)$, directrix $x = 0$.

13. Vertex $(0,1)$, focus $(0,\frac{1}{2})$.

14. Vertex $(-1,1)$, directrix $x = -\frac{5}{2}$.

15. Vertex $(0,0)$, axis $y = 0$, through the point $(1,4)$.

16. Vertical axis, through the points $(0,1)$, $(2,1)$, $(4,-1)$.

17. Horizontal axis, through the points $(-1,3)$, $(0,2)$, $(0,0)$.

18. Vertex $(1,1)$, axis $x = 1$, through the point $(2,3)$.

19. Let S be the parabola with focus $(0,c)$ and directrix $y = -c$. Use the definition of parabola to show that if $P(x,y)$ is any point of the plane,

$$P \in S \Leftrightarrow x^2 = 4cy$$

20. Use the definition of parabola to show that an equation of the parabola with vertex (h,k), focus $(h + c,k)$, and directrix $x = h - c$ is

$$(y - k)^2 = 4c(x - h)$$

21. What is an equation of the directrix of the parabola with vertex at the origin and focus $(1,-1)$?

22. Find an equation of the parabola with focus $(1,1)$ and directrix $x + y + 2 = 0$. *Hint:* The distance between the point (x_0,y_0) and the line $Ax + By + C = 0$ is

$$d = \frac{|Ax_0 + By_0 + C|}{\sqrt{A^2 + B^2}}$$

(See Problem 56, Section 1.2.)

Sketch the graph of each of the following polar equations and name the vertex, axis, focus, and directrix.

23. $r = \dfrac{4}{1 - \cos \theta}$

24. $r = \dfrac{2}{1 + \cos \theta}$

25. $r = \dfrac{1}{1 - \sin \theta}$

26. $r = \dfrac{3}{1 + \sin \theta}$

27. Explain why the polar equation

$$r = \frac{p}{2} \csc^2 \frac{\theta}{2}$$

represents a parabola.

28. Show that the parabola with focus at the origin and directrix $y = -p$ $(p > 0)$ may be represented by the polar equation

$$r = \frac{p}{1 - \sin \theta}$$

29. Show that the tangent to the parabola $y^2 = 4cx$ at (x_0,y_0) is $y_0 y = 2c(x + x_0)$.

30. The **latus rectum** of a parabola is the chord through the focus perpendicular to the axis. Show that its length is $4c$, where c is the distance between the vertex and focus.

31. Show that the tangents at the endpoints of the latus rectum of a parabola intersect on the directrix. What is the angle between them?

32. Find an equation of the circle that contains the vertex and endpoints of the latus rectum of the parabola $y^2 = 4cx$.

33. A chord through the focus of the parabola $y^2 = 4cx$ has endpoints (x_1,y_1) and (x_2,y_2). Show that its length is $x_1 + x_2 + 2c$. (Assume that $c > 0$.)

34. The normal to a parabola at P intersects the axis of the parabola at Q. Show that P and Q are the same distance from the focus. (Assume that P is not the vertex.)

35. Prove that the point of a parabola closest to the focus is the vertex. *Hint:* Use polar coordinates.

36. The path of a comet is a parabola with the sun at the focus. The ray from the sun through the comet makes an angle of $60°$ with the ray from the sun along the axis (away from the vertex) when the comet is 80,000,000 km from the sun. How close does the comet come to the sun?

14.4
CONICS:
THE ELLIPSE
AND HYPERBOLA

In the last section we defined the parabola as a set of points equidistant from a given point F and a given line L:

$$d(P,F) = d(P,L)$$

Another way of saying this is

$$\frac{d(P,F)}{d(P,L)} = 1$$

which serves as a clue to the general definition of conic.

> Let F, L, and e be a point, a line not containing the point, and a positive number, respectively. The **conic** with **focus** F, **directrix** L, and **eccentricity** e is the set of points P such that
>
> $$\frac{d(P,F)}{d(P,L)} = e$$
>
> It is called an **ellipse** if $e < 1$, a **parabola** if $e = 1$, and a **hyperbola** if $e > 1$. The line through the focus perpendicular to the directrix is called its **axis;** each point of the conic lying on the axis is called a **vertex.**

Figure 1
Vertices and center of conic ($e < 1$)

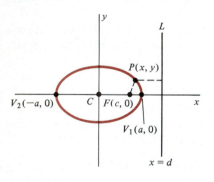

Figure 2
Vertices and center of conic ($e > 1$)

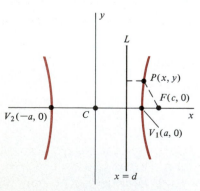

We saw in the last section that the term *parabola* (the case $e = 1$) corresponds to what we have been calling a parabola all along. We do not mean to assert (yet) that the same thing is true of *ellipse* and *hyperbola*. Of course it will turn out that way (otherwise we would not muddy the water by using familiar terminology to mean something new). For the moment, however, pretend that you have not heard the terms before.

A parabola has only one vertex, namely the point halfway between the focus and directrix. But the ellipse and the hyperbola have two. See Figures 1 and 2, in which we show a vertex V_1 between the focus and directrix, and another vertex V_2 that is not between them. Each of these points is on the conic, that is,

$$\frac{d(V_1,F)}{d(V_1,L)} = e \qquad (1)$$

and

$$\frac{d(V_2,F)}{d(V_2,L)} = e \qquad (2)$$

(We are assuming that $e \neq 1$, the case $e = 1$ having been covered in the last section.) The **center** of the conic is the point C midway between the vertices.

Imagine a coordinate system as shown in Figures 1 and 2, so that V_1 and V_2 have coordinates a and $-a$, respectively ($a > 0$) and F has coordinate $c > 0$. Let $x = d$ be the equation of the directrix ($d > 0$). Equation (1) yields

$$|c - a| = e|d - a|$$

which reduces to

$$a - c = e(d - a) \qquad (3)$$

because $a - c$ and $d - a$ have the same sign (positive in Figure 1, negative in Figure 2). Similarly, Equation (2) yields

$$a + c = e(d + a) \qquad (4)$$

Adding Equations (3) and (4), we find $2a = 2de$, from which $d = a/e$. Hence

$$\boxed{\frac{a}{e} = \text{distance from center to directrix}}$$

Subtraction of Equation (3) from Equation (4) gives $2c = 2ae$, from which

$$\boxed{e = \frac{c}{a} = \text{eccentricity}}$$

Now we are ready to derive an equation of the conic.

If S is the conic and $P(x,y)$ is any point of the plane, the definition says that

$$P \in S \Leftrightarrow \frac{d(P,F)}{d(P,L)} = e$$

$$\Leftrightarrow \sqrt{(x - c)^2 + y^2} = e \sqrt{(x - d)^2}$$

$$\Leftrightarrow (x - c)^2 + y^2 = \frac{c^2}{a^2}\left(x - \frac{a^2}{c}\right)^2 \qquad \left(e = \frac{c}{a}, \; d = \frac{a}{e} = \frac{a^2}{c}\right)$$

$$\Leftrightarrow a^2[(x - c)^2 + y^2] = (cx - a^2)^2$$

$$\Leftrightarrow a^2x^2 - 2a^2cx + a^2c^2 + a^2y^2 = c^2x^2 - 2a^2cx + a^4$$

$$\Leftrightarrow (a^2 - c^2)x^2 + a^2y^2 = a^2(a^2 - c^2)$$

Now we separate the discussion into two cases.

(a) Suppose that the conic is an ellipse. Then $e < 1$ and since $e = c/a$, we have $a > c$ and hence $a^2 - c^2 > 0$. It is therefore legitimate to define $b = \sqrt{a^2 - c^2}$ and our equation becomes

$$b^2x^2 + a^2y^2 = a^2b^2 \qquad \text{or} \qquad \frac{x^2}{a^2} + \frac{y^2}{b^2} = 1$$

(b) Suppose that the conic is a hyperbola. Then $e > 1$ and hence $a < c$ and $a^2 - c^2 < 0$. This time we cannot let $b = \sqrt{a^2 - c^2}$. (Why?) Instead we write our equation in the form

$$(c^2 - a^2)x^2 - a^2y^2 = a^2(c^2 - a^2)$$

and define $b = \sqrt{c^2 - a^2}$. Our equation becomes

$$b^2x^2 - a^2y^2 = a^2b^2 \qquad \text{or} \qquad \frac{x^2}{a^2} - \frac{y^2}{b^2} = 1$$

At this point we note that the whole discussion can be repeated with $(-c, 0)$ in place of F and $x = -d$ in place of L. The results are the same, as you can see by looking over the steps by which we derived our equations.

This means that the ellipse and hyperbola have two foci and two directrices, namely the points

$$F_1 = (c,0) \quad \text{and} \quad F_2 = (-c,0)$$

and the lines

$$L_1: x = d \quad \text{and} \quad L_2: x = -d$$

Either F_1 and L_1 or F_2 and L_2 can be used as the focus-directrix pair in the definition of conic.

Moreover, the whole discussion can be repeated with the foci on the y axis. This has the effect of interchanging x and y in the equations, so that the ellipse becomes

$$\frac{y^2}{a^2} + \frac{x^2}{b^2} = 1$$

and the hyperbola becomes

$$\frac{y^2}{a^2} - \frac{x^2}{b^2} = 1$$

Moreover, the center can be placed at the point (h,k) instead of at the origin. This translates the axes and replaces x and y by $x - h$ and $y - k$, respectively. We summarize as follows.

An ellipse with center (h,k) and axis parallel to a coordinate axis may be represented by

$$\frac{(x-h)^2}{a^2} + \frac{(y-k)^2}{b^2} = 1 \quad \text{or} \quad \frac{(y-k)^2}{a^2} + \frac{(x-h)^2}{b^2} = 1$$

depending on whether its axis is horizontal or vertical. In each case $a > b > 0$ and the foci are c units from the center, where

$$c^2 = a^2 - b^2 \quad (0 < c < a)$$

The eccentricity is $e = c/a < 1$. The directrices are perpendicular to the axis d units from the center, where $d = a/e > a$. Hence as we move out from the center in either direction, we encounter a focus, vertex, and directrix in that order.

A hyperbola with center (h,k) and axis parallel to a coordinate axis may be represented by

$$\frac{(x-h)^2}{a^2} - \frac{(y-k)^2}{b^2} = 1 \quad \text{or} \quad \frac{(y-k)^2}{a^2} - \frac{(x-h)^2}{b^2} = 1$$

depending on whether its axis is horizontal or vertical. In each case a and b are positive (but neither is necessarily greater than the other). The foci are c units from the center, where

$$c^2 = a^2 + b^2 \quad (0 < a < c)$$

The eccentricity is $e = c/a > 1$. The directrices are perpendicular to the axis d units from the center, where $d = a/e < a$. Hence as we move out from the center in either direction, we encounter a directrix, vertex, and focus in that order.

As you can see, there is a lot of information to absorb! None of it is hard, but it takes a while to get it all straight. The following examples should help.

□ **Example 1**   Discuss the ellipse $4x^2 + 9y^2 - 8x - 32 = 0$.

Solution: How do we know that the equation represents an ellipse? The answer is to be found in the standard forms, which are quadratic in x and y, the coefficients of x^2 and y^2 having *the same sign but different size*. To discover what ellipse we are dealing with, complete the square:

$$4(x^2 - 2x + \underline{\quad}) + 9y^2 = 32 + \underline{\quad}$$

$$4(x^2 - 2x + 1) + 9y^2 = 32 + 4 \qquad \text{(Add 4 to the right side, not 1!)}$$

$$4(x - 1)^2 + 9y^2 = 36$$

$$\frac{(x - 1)^2}{9} + \frac{y^2}{4} = 1$$

Hence $a = 3$, $b = 2$, and $c = \sqrt{a^2 - b^2} = \sqrt{5}$. The axis is horizontal (because the larger denominator in the standard form is associated with x). The center is $C = (1,0)$ and the vertices are $V_1 = (4,0)$ and $V_2 = (-2,0)$, each being $a = 3$ units from the center. The foci are $c = \sqrt{5} \approx 2.2$ units from the center, namely

$$F_1 = (1 + \sqrt{5},0) \qquad \text{and} \qquad F_2 = (1 - \sqrt{5},0)$$

The eccentricity is $e = c/a = \sqrt{5}/3 \approx 0.75$. The directrices are the vertical lines whose distance from the center is $a/e = a^2/c = 9/\sqrt{5} \approx 4$, namely

$$L_1: x = 1 + \frac{9}{\sqrt{5}} \qquad \text{and} \qquad L_2: x = 1 - \frac{9}{\sqrt{5}}$$

(See Figure 3.)

Figure 3

The ellipse $\dfrac{(x - 1)^2}{9} + \dfrac{y^2}{4} = 1$

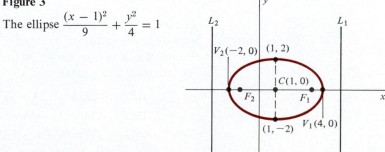

Note that we have also plotted the points $(1,2)$ and $(1,-2)$ in Figure 3. These are the endpoints of the *minor diameter* of the ellipse, just as the vertices are the endpoints of the *major diameter*. The lengths of the major and minor diameters are $2a$ and $2b$, respectively.   □

How much of the information developed in Example 1 is worthwhile? That depends on what we are trying to do. If the graph is all we care about, the center, vertices, and endpoints of the minor diameter are sufficient for a good sketch; there is no need to find the foci, directrices, or eccentricity. On the other hand, many important applications of conics involve this information. Kepler's First Law of Planetary Motion, for example, says that the orbit of a planet is an ellipse with the sun at one focus. That statement cannot even be understood without having studied conics. Another illustration is the optical property of the parabola (Example 3 in the last section), which involves the idea of focus. The ellipse and hyperbola have similar properties.

□ **Example 2**   Find an equation of the ellipse with vertices $(-1,0)$ and $(-1,4)$, and eccentricity $\sqrt{3}/2$.

Solution: The axis is the vertical line $x = -1$, so the standard equation is of the form

$$\frac{(y-k)^2}{a^2} + \frac{(x-h)^2}{b^2} = 1$$

We need the center (h,k) and the semidiameters a and b. Since the center is on the axis halfway between the vertices, it must be $C = (-1,2)$. The distance from the center to either vertex is $a = 2$. The eccentricity is $e = c/a = \sqrt{3}/2$, from which $c = a\sqrt{3}/2 = \sqrt{3}$. Hence $b^2 = a^2 - c^2 = 4 - 3 = 1$. The equation is

$$\frac{(y-2)^2}{4} + \frac{(x+1)^2}{1} = 1$$

(See Figure 4.)   □

Figure 4

The ellipse $\dfrac{(y-2)^2}{4} + \dfrac{(x+1)^2}{1} = 1$

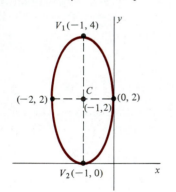

□ **Example 3**   Discuss the hyperbola $16x^2 - 9y^2 - 32x - 54y - 209 = 0$.

Solution: As in the case of the ellipse, the standard equations of the hyperbola are quadratic in x and y. The coefficients of x^2 and y^2 are *opposite in sign*, which is how we know the given equation represents a hyperbola. We complete the square:

$$16(x^2 - 2x + \underline{\quad}) - 9(y^2 + 6y + \underline{\quad}) = 209 + \underline{\quad}$$
$$16(x^2 - 2x + 1) - 9(y^2 + 6y + 9) = 209 + 16 - 81$$
$$16(x - 1)^2 - 9(y + 3)^2 = 144$$
$$\frac{(x-1)^2}{9} - \frac{(y+3)^2}{16} = 1$$

Hence $a = 3$, $b = 4$, and $c = \sqrt{a^2 + b^2} = 5$. The axis is horizontal (because the positive term in the standard form involves x). The center is $C = (1, -3)$ and the vertices ($a = 3$ units from the center) are $V_1 = (4, -3)$ and $V_2 = (-2, -3)$. The foci ($c = 5$ units from the center) are $F_1 = (6, -3)$ and $F_2 = (-4, -3)$. The eccentricity is $e = c/a = \frac{5}{3}$. The directrices are the vertical lines whose distance from the center is $a/e = a^2/c = \frac{9}{5}$, namely

$$L_1: x = \tfrac{14}{5} \qquad \text{and} \qquad L_2: x = -\tfrac{4}{5}$$

(See Figure 5.)

Figure 5
The hyperbola $\dfrac{(x-1)^2}{9} - \dfrac{(y+3)^2}{16} = 1$

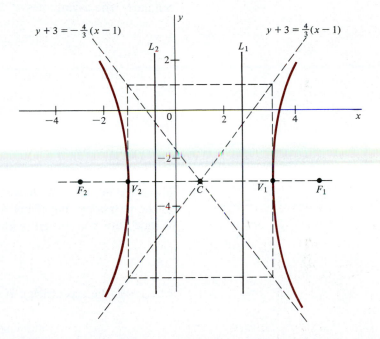

Note the asymptotes shown in Figure 5. These should be drawn first, because they serve as guidelines for sketching the hyperbola. They contain the diagonals of the rectangle centered at C, of length $2a = 6$ and altitude $2b = 8$. Hence their slopes are $\pm b/a = \pm\frac{4}{3}$ and their equations are

$$y + 3 = \pm\tfrac{4}{3}(x - 1)$$

Thus although there is no concept of major and minor diameter (as in the case of the ellipse), the numbers a and b still have geometric significance. Sometimes the segment with endpoints V_1 and V_2 is called the **transverse diameter** of the hyperbola.   □

In general the asymptotes of the hyperbola

$$\frac{(x-h)^2}{a^2} - \frac{(y-k)^2}{b^2} = 1$$

are the lines obtained from the standard form by replacing 1 by 0 on the right side:

$$\frac{(x-h)^2}{a^2} - \frac{(y-k)^2}{b^2} = 0$$

$$\frac{(x-h)^2}{a^2} = \frac{(y-k)^2}{b^2}$$

$$y - k = \pm\frac{b}{a}(x-h)$$

(See Problem 57, Section 1.3, and Problem 35, Section 5.3, where this statement is discussed in the context of a hyperbola with center at the origin.) Similarly, the asymptotes of the vertical hyperbola

$$\frac{(y-k)^2}{a^2} - \frac{(x-h)^2}{b^2} = 1$$

are obtained by replacing 1 by 0:

$$\frac{(y-k)^2}{a^2} = \frac{(x-h)^2}{b^2}$$

$$y - k = \pm\frac{a}{b}(x-h)$$

As we pointed out in the last section, polar coordinates simplify the discussion of conics. We leave it to you to show (almost exactly as in the case of the parabola) that if the focus is at the origin and the directrix is the vertical line $x = -p$ $(p > 0)$, a polar equation of the conic is

$$r = \frac{ep}{1 - e\cos\theta}$$

where e is the eccentricity. If the directrix is $x = p$, the equation is

$$r = \frac{ep}{1 + e\cos\theta}$$

whereas if it is $y = -p$ or $y = p$, the equations are

$$r = \frac{ep}{1 - e\sin\theta} \qquad \text{or} \qquad r = \frac{ep}{1 + e\sin\theta}$$

respectively.

□ **Example 4**   Discuss the graph of the polar equation

$$r = \frac{3}{1 + 2\cos\theta}$$

Solution: Since $e = 2 > 1$, the equation represents a hyperbola. The left branch is obtained when θ runs through the interval $(-2\pi/3, 2\pi/3)$, in

Figure 6

The hyperbola $r = \dfrac{3}{1 + 2\cos\theta}$

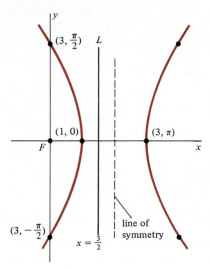

which case $r > 0$; we show three of its points (in polar coordinates) in Figure 6. The right branch corresponds to the interval $(2\pi/3, 4\pi/3)$, when $r < 0$; after plotting $(3, \pi)$ we can sketch the rest of it by symmetry with the left branch. Since $ep = 3$ and $e = 2$, we have $p = \frac{3}{2}$, so the directrix corresponding to the focus at the origin is the vertical line $x = \frac{3}{2}$. □

We end this section with some remarks about **second-degree curves,** which are defined by equations of the form

$$Ax^2 + Bxy + Cy^2 + Dx + Ey + F = 0 \qquad (A,\ B,\ C \text{ not all zero})$$

An examination of the standard forms for circles and conics with lines of symmetry parallel to the coordinate axes reveals that each is of this type with $B = 0$. Conversely (with the exception of degenerate cases) every equation of the form

$$Ax^2 + Cy^2 + Dx + Ey + F = 0 \qquad (A \text{ and } C \text{ not both zero})$$

represents

- an *ellipse* if $AC > 0$ (that is, if A and C have the same sign), unless $A = C$, in which case the graph is a circle
- a *parabola* if $AC = 0$ (that is, if $A = 0$ or $C = 0$)
- a *hyperbola* if $AC < 0$ (that is, if A and C have opposite signs)

Each of these curves has lines of symmetry parallel to the coordinate axes.

A more complete discussion of second-degree curves would include a study of *rotation of axes*. If you are interested in the details, see the Additional Problems at the end of this chapter. It turns out that when $B \neq 0$ in the general second-degree equation, the graph is still a conic, but its axis is "tilted." Rotating the coordinate axes to bring them into line with the lines of symmetry of the conic changes its equation to one with no xy term, and permits its expression in one of the standard forms already studied.

Problem Set 14.4

Each of the following equations represents an ellipse. Find the center, endpoints of the major and minor diameters, foci, directrices, and eccentricity, and sketch the graph.

1. $4x^2 + 9y^2 = 36$

2. $5x^2 + y^2 - 20x = 0$

3. $x^2 + 4y^2 - 4x - 8y - 8 = 0$

4. $4x^2 + y^2 - 24x + 32 = 0$

5. $16x^2 + 25y^2 - 64x + 50y - 311 = 0$

6. Despite appearances, the equation $x^2 + 4y^2 - 24y + 36 = 0$ does not represent an ellipse. What is its graph?

Find an equation of each of the following ellipses.

7. Foci $(0,\pm 2)$, major diameter of length 8.

8. Vertices $(3,1)$ and $(-1,1)$, eccentricity $\frac{2}{3}$.

9. Foci $(3,0)$ and $(1,0)$, passing through $(1,\frac{3}{2})$.

10. Center $(-3,5)$, focus $(-5,5)$, and corresponding directrix $x = -7$.

11. Directrices $y = 7$ and $y = -3$, endpoints of minor diameter $(1,2)$ and $(-3,2)$. (Two answers)

12. Describe the parametric path

$$x = 1 + 3\cos t, \ y = -2 + 5\sin t \qquad 0 \le t \le 2\pi$$

Each of the following equations represents a hyperbola. Find the center, vertices, foci, directrices, and eccentricity, and sketch the hyperbola with its asymptotes.

13. $16x^2 - 9y^2 = 144$

14. $9x^2 - 4y^2 - 18x - 16y - 43 = 0$

15. $16x^2 - 9y^2 - 96x + 128 = 0$

16. $9x^2 - 4y^2 + 36x - 8y + 68 = 0$

17. $12x^2 - 32y^2 - 12x + 96y + 27 = 0$

18. Despite appearances, the equation $9x^2 - 4y^2 - 18x + 8y + 5 = 0$ does not represent a hyperbola. Describe its graph.

Find an equation of each of the following hyperbolas.

19. Vertices $(0,\pm 2)$, asymptotes $y = \pm\frac{1}{2}x$.

20. Vertices $(1,2)$ and $(-3,2)$, eccentricity $\sqrt{2}$.

21. Foci $(7,0)$ and $(-1,0)$, containing $(6,\sqrt{15})$.

22. Center at the origin, focus $(0,3)$, and corresponding directrix $y = 2$.

23. Describe the parametric path

$$x = 2\cosh t, \ y = 2 + 3\sinh t$$

24. Explain why the hyperbolas $x^2 - y^2 = \pm 1$ have the same asymptotes. Sketch the graphs on the same coordinate system.

25. What is the eccentricity of a hyperbola whose asymptotes are perpendicular?

26. Why are the directrices of an ellipse outside the ellipse? Why are the directrices of a hyperbola between the vertices?

27. If F is a point not on the line L, what kind of curve is

$$S = \{P \in \mathbb{R}^2 : d(P,L) = 2 \cdot d(P,F)\}?$$

28. If $k \neq 1$, the equation

$$\frac{x^2}{1} + \frac{y^2}{1 - k} = 1$$

represents a conic section through the points $(\pm 1, 0)$.

(a) Describe the curve for $k < 0$ and how it changes as $k \uparrow 0$. What is it when $k = 0$?

(b) Describe the curve for $0 < k < 1$ and how it changes as $k \uparrow 1$.

(c) Describe the curve for $k > 1$.

Sketch the graph of each of the following polar equations, and name the center, foci, directrices, and eccentricity.

29. $r = \dfrac{16}{5 - 3\cos\theta}$

30. $r = \dfrac{6}{2 + \sin\theta}$

31. $r = \dfrac{3}{1 + 2\sin\theta}$

32. $r = \dfrac{6}{1 - 2\cos\theta}$

33. A **latus rectum** of a conic is a chord through a focus perpendicular to the axis. Show that the length of each latus rectum of the polar graph

$$r = \frac{ep}{1 - e \cos \theta}$$

is $2ep$.

34. Show that each latus rectum of an ellipse or hyperbola has length $2b^2/a$, where a and b have their usual meaning.

35. The tangents at the endpoints of the latus rectum of a parabola intersect on the directrix. (See Problem 31, Section 14.3.) Show that the same thing is true of an ellipse or hyperbola.

36. Show that the tangent to the ellipse $x^2/a^2 + y^2/b^2 = 1$ at (x_0, y_0) is

$$\frac{x_0 x}{a^2} + \frac{y_0 y}{b^2} = 1$$

37. Show that the tangent to the hyperbola $x^2/a^2 - y^2/b^2 = 1$ at (x_0, y_0) is

$$\frac{x_0 x}{a^2} - \frac{y_0 y}{b^2} = 1$$

38. The vertex of the parabola $y^2 = 4cx$ is the center of an ellipse and its focus is an endpoint of the minor diameter of the ellipse. If the parabola and ellipse intersect at right angles, what is the eccentricity of the ellipse?

39. Find an equation of the parabola with the same vertex and focus as the right-hand branch of the hyperbola $x^2/a^2 - y^2/b^2 = 1$.

40. The parabola in the preceding problem is

$$y^2 = 4(c - a)(x - a)$$

where $(c, 0)$ is the focus. The equation of the hyperbola can be written

$$y^2 = \frac{b^2}{a^2}(x^2 - a^2)$$

(a) Show that the ratio of y^2 for the hyperbola to y^2 for the parabola is

$$\frac{(1 + e)(x + a)}{4a}$$

where e is the eccentricity of the hyperbola.

(b) Explain why the ratio in part (a) is greater than 1. Why does it follow that the parabola is "inside" the hyperbola?

41. Repeat Problems 39 and 40 with the ellipse $x^2/a^2 + y^2/b^2 = 1$ ($a > b$) in place of the hyperbola, this time showing that the parabola is outside the ellipse.

42. The point of a parabola closest to the focus is the vertex. (See Problem 35, Section 14.3.) Show that the same thing is true of the ellipse and hyperbola (with the understanding that the vertex nearest the given focus is meant).

43. The earth's orbit is an ellipse with the sun at one focus, the eccentricity being approximately 0.0167 and the major diameter about 185.8 million miles. How close to the sun does the earth come? How far away? Find the average of these two figures.

44. On level ground an observer hears the sound of a gun and the sound of the bullet striking a target at the same time. Describe the set of possible positions of the observer. (Assume that the bullet travels faster than sound.)

True–False Quiz

1. (r, θ) and $(-r, \theta + \pi)$ are polar coordinates of the same point.

2. The graph of the polar equation $r \cos(\theta - \pi/2) = 3$ is a straight line.

3. The graph of the polar equation $r = \cos \theta$ is a circle.

4. A bug crawling clockwise along the graph of the polar equation $r = e^\theta$ approaches the origin.

5. The graph of the polar equation $r = \cos 2\theta$ is a four-leaf rose.

6. The graph of the polar equation

$$r = \frac{2}{1 - 2 \cos \theta}$$

is an ellipse.

7. The distance between the focus and directrix of the parabola $4y = x^2$ is 1.

8. The eccentricity of a parabola is 1.

9. The equation $9x^2 + 4y^2 - 8y - 32 = 0$ represents a vertical ellipse.

10. The directrices of an ellipse are outside the ellipse.

11. The eccentricity of the hyperbola $x^2/16 - y^2/9 = 1$ is $5/4$.

12. The vertex of the parabola $y^2 - 4x + 6y + 17 = 0$ is $(-2, 3)$.

13. A circle is a conic of eccentricity 0.

14. The line $y = 2x$ is an asymptote of the hyperbola $x^2 - 4y^2 = 4$.

15. The graph of $x^2 - y^2 = 0$ is a hyperbola.

Additional Problems

Sketch the graphs of the polar equations in each of the following, and find their points of intersection.

1. $r = 2(1 + \cos\theta)$ and $r = 2$

2. $r = 1 - \cos\theta$ and $r = 1 - \sin\theta$

3. $r = 6\cos\theta$ and $r = 6\sin 2\theta$

4. $r = 2\sin\theta$ and $r = 2\cos 2\theta$

5. Show that the polar equation

$$r = \frac{3}{1 + \sin\theta}$$

represents a parabola.

6. Find the area bounded by the parabola in Problem 5 and the x axis.

7. Find the area bounded by the polar curve $r = 3\csc\theta$ and the rays $\theta = \pi/4$ and $\theta = 3\pi/4$.

8. Find the area inside the cardioid $r = 1 - \sin\theta$ and outside the circle $r = 1$.

9. Find the area inside the circle $r = 3\cos\theta$ and outside the cardioid $r = 1 + \cos\theta$.

10. Find the area inside the rose $r = 3\cos 2\theta$.

11. Find the area inside the rose $r = \cos 3\theta$.

Find the length of each of the following polar curves.

12. $r = 4\cos\theta$ **13.** $r = 4\sec\theta,\ 0 \le \theta \le \pi/3$

14. $r = \sin^2\dfrac{\theta}{2}$ **15.** $r = e^{-2\theta},\ \theta \ge 0$

16. $r = 1 + \sin\theta$ **17.** $r = a(1 + \cos\theta)$

18. Find the vertex, axis, focus, and directrix of the parabola $y^2 + 4x + 4y = 0$ and sketch it.

19. Repeat Problem 18 for the parabola $x^2 - 6x + 3y = 0$.

20. Sketch the ellipse $4x^2 + 9y^2 - 40x + 64 = 0$, naming the center, endpoints of the major and minor diameters, foci, directrices, and eccentricity.

21. Repeat Problem 20 for the ellipse $16x^2 + 9y^2 + 64x - 54y + 1 = 0$.

22. Sketch the hyperbola $9x^2 - 4y^2 + 36 = 0$, naming the center, endpoints of the transverse diameter, foci, directrices, asymptotes, and eccentricity.

23. Repeat Problem 22 for the hyperbola $9x^2 - 16y^2 - 36x - 108 = 0$.

24. Show that a line through a focus of a hyperbola perpendicular to an asymptote intersects the asymptote on a directrix.

25. (*Optical Property of the Ellipse*) Show that the rays from the foci through a point P of the ellipse $x^2/a^2 + y^2/b^2 = 1$ make equal angles with the tangent at P, as follows.

   (a) Let θ_1 and θ_2 be the angles the rays from $F_1(c,0)$ and $F_2(-c,0)$ make with the tangent. Draw a picture showing that $\theta_1 = \alpha - \beta_1$ and $\theta_2 = \pi - \alpha + \beta_2$, where P is in the first quadrant and α, β_1, β_2 are inclination angles of the tangent and the rays from F_1 and F_2, respectively.

   (b) Show that $\theta_1 = \theta_2$ by showing that $\tan\theta_1$ and $\tan\theta_2$ are both equal to b^2/cy. Hint: $b^2x^2 + a^2y^2 = a^2b^2$ and $c^2 = a^2 - b^2$.

   (This optical property is also acoustical, and explains the phenomenon of "whispering galleries." These are elliptically shaped rooms in which a whisper at one focus can be clearly heard at the other.)

26. (*Optical Property of the Hyperbola*) Show that the rays from the foci through a point P of a hyperbola make equal angles with the tangent at P.

27. Show that an ellipse and a hyperbola with the same foci intersect at right angles. *Hint:* A direct attack is hard; use the optical properties of these curves to make it easy.

28. Explain why each member of the family of ellipses

$$\frac{x^2}{k^2} + \frac{y^2}{k^2 - 1} = 1 \qquad (k > 1)$$

and each member of the family of hyperbolas

$$\frac{x^2}{l^2} - \frac{y^2}{1 - l^2} = 1 \qquad (0 < l < 1)$$

has foci $(\pm 1, 0)$. Sketch a few members of each family on the same coordinate system, noting from Problem 27 that they intersect at right angles.

29. Use the definition of conic to find an equation of the ellipse with eccentricity $\frac{1}{2}$, focus $(2,1)$, and directrix $2x + y - 8 = 0$. *Hint:* Use the formula for distance between a point and a line (Problem 56, Section 1.2).

30. Show that an alternate definition of the ellipse with foci F_1 and F_2 and major diameter $2a$ is

$$S = \{P \in \mathcal{R}^2 : d(P,F_1) + d(P,F_2) = 2a\}$$

(This provides a mechanical way to draw an ellipse. Place thumbtacks at the foci and attach a string of length $2a$. A pencil held against the string and moved so as to keep it taut will trace out the ellipse.)

31. Show that an alternate definition of the hyperbola with foci F_1 and F_2 and transverse diameter $2a$ is

$$S = \{P \in \mathfrak{R}^2 : d(P,F_1) - d(P,F_2) = \pm 2a\}$$

32. Suppose that the coordinate axes are rotated through a counterclockwise angle α $(0 < \alpha < \pi/2)$. (See Figure 1.)

Figure 1
Rotation of axes

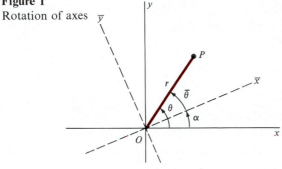

   (a) Letting (r,θ) be polar coordinates of P in the original system, explain why polar coordinates $(\overline{r},\overline{\theta})$ in the new system may be chosen satisfying $\overline{r} = r$ and $\overline{\theta} = \theta - \alpha$.

   (b) The new rectangular coordinates of P are

$$\overline{x} = \overline{r}\cos\overline{\theta} \quad \text{and} \quad \overline{y} = \overline{r}\sin\overline{\theta}$$

   Show that the old ones are

$$x = \overline{x}\cos\alpha - \overline{y}\sin\alpha \quad \text{and} \quad y = \overline{x}\sin\alpha + \overline{y}\cos\alpha$$

33. Given the equation $Ax^2 + Bxy + Cy^2 + Dx + Ey + F = 0$, show that a rotation of axes through the angle α (as in Problem 32) transforms the equation to

$$\overline{A}\overline{x}^2 + \overline{B}\overline{x}\,\overline{y} + \overline{C}\overline{y}^2 + \overline{D}\overline{x} + \overline{E}\overline{y} + \overline{F} = 0$$

   where

$$\overline{A} = A\cos^2\alpha + B\sin\alpha\cos\alpha + C\sin^2\alpha$$
$$\overline{B} = B\cos 2\alpha - (A - C)\sin 2\alpha$$
$$\overline{C} = A\sin^2\alpha - B\sin\alpha\cos\alpha + C\cos^2\alpha$$
$$\overline{D} = D\cos\alpha + E\sin\alpha$$
$$\overline{E} = -D\sin\alpha + E\cos\alpha$$
$$\overline{F} = F$$

34. If we want the new equation in Problem 33 to have no xy term, we must choose α to make $\overline{B} = 0$. Explain why the angle α satisfying

$$\cot 2\alpha = \frac{A - C}{B}$$

   will do the job.

35. Show that the quantity $A + C$ is invariant under rotation of axes, that is, $\overline{A} + \overline{C} = A + C$.

36. (*For a Dreary Day*) Show that the quantity $B^2 - 4AC$ is invariant under rotation of axes, that is, $\overline{B}^2 - 4\overline{A}\overline{C} = B^2 - 4AC$.

37. Use Problem 36 to explain why (except for degenerate cases) the graph of $Ax^2 + Bxy + Cy^2 + Dx + Ey + F = 0$ is an ellipse, parabola, or hyperbola depending on whether $B^2 - 4AC$ is negative, zero, or positive, respectively. (The quantity $B^2 - 4AC$ is called the *discriminant*.)

38. Use the results of the preceding problems to eliminate the xy term from the equation $x^2 + xy + y^2 = 1$, as follows.

   (a) What angle α should be chosen to make $\overline{B} = 0$?

   (b) Compute $\overline{A}$ using Problem 33 and $\overline{C}$ using Problem 35. Why are $\overline{D}$ and $\overline{E}$ zero?

   (c) Show that the new equation can be written in the form

$$\frac{\overline{x}^2}{\frac{2}{3}} + \frac{\overline{y}^2}{2} = 1$$

   and describe the graph.

In each of the following, use the discriminant $B^2 - 4AC$ to classify the curve. Then eliminate the xy term by an appropriate rotation of axes and confirm the identification.

39. $x^2 - 2xy + y^2 - 2x + 5 = 0$

40. $4x^2 - 4xy + y^2 - 5 = 0$ *Hint:* No need to find α! After determining $\cot 2\alpha$, use the identities

$$\sin^2\alpha = \tfrac{1}{2}(1 - \cos 2\alpha) \quad \text{and} \quad \cos^2\alpha = \tfrac{1}{2}(1 + \cos 2\alpha)$$

   to find $\sin\alpha$ and $\cos\alpha$.

41. $x^2 - xy + 2y^2 = 1$

42. Show that the equation $xy = 1$ represents a hyperbola with focus $(\sqrt{2}, \sqrt{2})$, directrix $x + y = \sqrt{2}$, and eccentricity $\sqrt{2}$. Sketch the hyperbola and name its center, vertices, and asymptotes.

In each of the following, use the discriminant $B^2 - 4AC$ to classify the curve. Then factor the left side of the equation to show that the graph is a degenerate "conic," and describe it.

43. $x^2 - x - 6 = 0$

44. $x^2 - xy - 2y^2 = 0$

45. $x^2 - 2xy + y^2 - 1 = 0$

46. Explain why the equation $Ax^2 + Bxy + Cy^2 + Dx + Ey + F = 0$ cannot represent a circle if $B \neq 0$.

15 | Vector Analysis in the Plane

God ever geometrizes.
PLATO (428–348 B.C.)

God ever arithmetizes.
K.G.J. JACOBI (1804–1851)

God is subtle, but he is not malicious.
ALBERT EINSTEIN

This chapter completes our study of single-variable calculus, and serves as a bridge to multivariable calculus. The vector methods presented here (together with the information in Chapter 14 on polar coordinates and conics) are involved in Newton's celebrated demonstration that the orbit of a planet must be an ellipse if the inverse-square law of gravitational attraction is assumed. That is still a first-class problem; by the time you have finished this chapter, you will have the mathematical tools for solving it. We will discuss it in an optional section at the end of the chapter.

15.1 VECTORS IN THE PLANE

In many parts of mathematics and its applications it is important to assign both a *numerical size* and a *direction* to certain quantities. We do this routinely in one dimension, because real numbers have both size and sign. To dramatize that fact, we could represent each real number by an arrow from the origin to a point on the number line (as shown in Figure 1). The length of the arrow is the absolute value of the number; its direction indicates the sign.

Such an arrow is often called a **vector** (although in one dimension the term is hardly worth mentioning). Velocity, for example, is a vector quantity; it tells us not only how fast an object is moving, but also in what direction. Another example is force.

Figure 1
Vectors in one-dimensional space

Figure 2
Velocity as an arrow in the plane

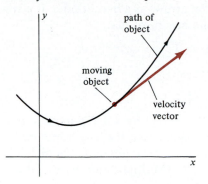

Figure 3
Vector as a number pair

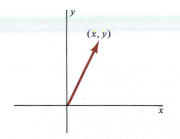

Figure 4
Addition of vectors

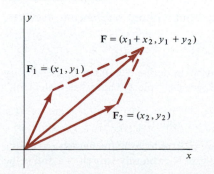

Figure 5
Multiplication of vectors by real numbers

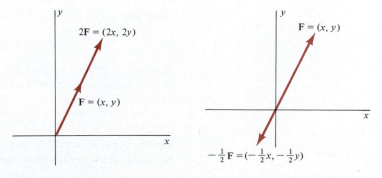

When we generalize these ideas to higher dimensions, some discussion is needed. Consider, for example, the notion of velocity in connection with an object moving along a curve. If it is to play the same role as in motion along a straight line, it must possess a *magnitude* (telling us how fast the object is moving) and a *direction* (telling us where the object is heading). Arrows ("vectors") seem ready-made for the purpose, as suggested in Figure 2. If we make the length of the arrow equal to the speed and make it point in the direction of motion, we have a vivid (and precise) description of velocity. Evidently some definitions are needed, however. What is "speed" in curvilinear motion? How do we determine direction? Is velocity the derivative of position, as in straight-line motion? What is "position?" We will answer these questions eventually, but in this section we concentrate on the primary concept of *vector* in two-dimensional space.

In Figure 2 we placed the initial point of the velocity vector at the position of the moving object. There is no real need for this; any other arrow with the same length and direction would carry the same information. For the moment let's agree that all arrows start at the origin, as shown in Figure 3. Then it is clear that every vector determines a pair of numbers, namely the coordinates (x, y) of its terminal point, and every pair of numbers determines a vector (starting at the origin).

In view of this fact, it is common practice to say that *a vector is a pair of numbers* (just as at the beginning of analytic geometry we agreed to identify a number pair with a point of the plane). While this may seem to load the notation (x, y) with too many possible meanings, it enables us to treat vectors algebraically, which is an enormous simplification (as you will see).

Unlike a point of the plane, a vector is more than just a pair of numbers. For we are going to *add* vectors and *multiply* them by real numbers, thus imposing an algebraic structure on the set of vectors that the set of points in the plane does not have. The reason for this is probably best understood by recalling the concept of force in physics. *Forces are added* to produce a "resultant" force; *they are multiplied by real numbers* to obtain larger or smaller forces in the same or opposite direction.

More precisely, consider the vectors in Figures 4 and 5. If you have studied physics, you know that two forces are added by the "Parallelogram Law," their sum being an arrow from a common initial point to the opposite vertex of a parallelogram having the given arrows as adjacent sides. It is not hard to see that if the initial point is the origin, so that

$$\mathbf{F}_1 = (x_1, y_1) \quad \text{and} \quad \mathbf{F}_2 = (x_2, y_2)$$

can be written as number pairs, the sum is

$$\mathbf{F} = (x_1 + x_2, y_1 + y_2)$$

Similarly, multiplication of a force $\mathbf{F} = (x, y)$ by a real number c produces another force $c\mathbf{F} = (cx, cy)$ whose magnitude is the length of $\mathbf{F}$ multiplied by $|c|$ and whose direction is the same as, or opposite to, the direction of $\mathbf{F}$ depending on whether $c > 0$ or $c < 0$. (Multiplication by $c = 0$ is a special case.)

For a graphic illustration of addition of vectors as shown in Figure 4, imagine a ping pong ball at the origin. If $\mathbf{F}_1$ represents the distance and direction traveled by the ball due to a jet of compressed air at the origin, and $\mathbf{F}_2$ represents the motion of the ball due to a second jet, then $\mathbf{F}$ shows the resultant motion of the ball under the simultaneous impulse of both jets.

These ideas from physics need not concern us now; we mention them only to help you see the reason for the following definition.

The Coordinate Plane as a Vector Space

The set $\mathfrak{R}^2 = \{(x, y): x \in \mathfrak{R}, y \in \mathfrak{R}\}$ is called a **vector space** (and its elements are called **vectors**) when the following definitions of addition and multiplication by a real number are imposed:

- If $\mathbf{x}_1 = (x_1, y_1)$ and $\mathbf{x}_2 = (x_2, y_2)$, then $\mathbf{x}_1 + \mathbf{x}_2 = (x_1 + x_2, y_1 + y_2)$.
- If $\mathbf{x} = (x, y)$ and $c \in \mathfrak{R}$, then $c\mathbf{x} = (cx, cy)$.

□ **Example 1**   If $\mathbf{x}_1 = (5, -8)$ and $\mathbf{x}_2 = (3, 4)$, then

$$\mathbf{x}_1 + \mathbf{x}_2 = (5, -8) + (3, 4) = (5 + 3, -8 + 4) = (8, -4)$$

and

$$2\mathbf{x}_1 = 2(5, -8) = (10, -16)$$

Note that these operations can be carried out without reference to pictures.

□

Remark   The above definition can be used to derive a number of elementary algebraic properties of $\mathfrak{R}^2$ as a vector space. For example, $\mathfrak{R}^2$ has a "zero element," namely $\mathbf{0} = (0, 0)$, since

$$(x, y) + (0, 0) = (x, y) \quad \text{for each } (x, y) \in \mathfrak{R}^2$$

Each vector $\mathbf{x} = (x, y)$ in $\mathfrak{R}^2$ has an "inverse" relative to addition, namely $(-x, -y)$, because $(x, y) + (-x, -y) = (0, 0)$. This inverse is ordinarily written $-\mathbf{x}$ and is the same as $(-1)\mathbf{x}$. (Why?) Such formulas as

$$(a + b)\mathbf{x} = a\mathbf{x} + b\mathbf{x}$$

(which look like the usual laws of algebra) are proved by appealing to what we already know about real numbers:

$$(a + b)\mathbf{x} = (a + b)(x,y) = ((a + b)x, (a + b)y) = (ax + bx, ay + by)$$
$$= (ax,ay) + (bx,by) = a(x,y) + b(x,y) = a\mathbf{x} + b\mathbf{x}$$

You should compare the simplicity of these remarks with the geometric arguments that would be needed if we had defined vectors as arrows in the plane. Our number pairs may be *interpreted* as arrows (and indeed that is one of their main uses in applications). But it is much easier to discuss the mathematical properties of vectors algebraically than geometrically.

Although the motivation for our definition of $\mathcal{R}^2$ as a vector space is geometric, the definition itself is purely algebraic. Hence we are free to adopt any geometrical interpretation of vectors as number pairs that seems convenient or useful. *One* such interpretation is that $\mathbf{x} = (x,y)$ is the arrow from the origin to the point (x,y). We are not bound by this, however. Another useful interpretation is that $\mathbf{x} = (x,y)$ is the arrow from (x_1,y_1) to (x_2,y_2), where the initial point is arbitrary and the terminal point is chosen to satisfy $x = x_2 - x_1$ and $y = y_2 - y_1$. (See Figure 6.) As you can see, the arrows in this figure have the same length and direction; by allowing this interpretation we are saying that a vector $\mathbf{x} = (x,y)$ is *any* arrow of this length and direction.

A more sophisticated version of this statement is that a vector (geometrically speaking) is a whole class of arrows, all with the same length and direction. Any member of the class is then a *representative* of the vector. In any case, we have a freedom of interpretation allowing us to place the initial point anywhere we like, depending on what application we have in mind. A **position vector,** for example (whose terminal point coincides with a particle moving in the plane), is considered to start at the origin. The *force* on the particle, on the other hand, is more naturally regarded as an arrow whose initial point coincides with the particle.

Figure 6
Two arrows representing the same vector

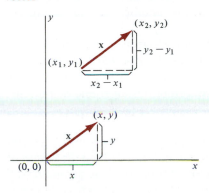

Figure 7
Vector from P_1 to P_2

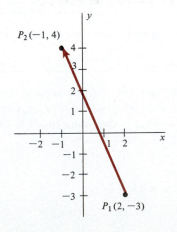

Let $P_1(x_1,y_1)$ and $P_2(x_2,y_2)$ be any points of the plane. The vector from P_1 to P_2 is

$$\overrightarrow{P_1P_2} = (x_2 - x_1,\ y_2 - y_1)$$

Note the pattern "terminal coordinate minus initial coordinate" in the formation of this number pair.

◻ **Example 2**   If $P_1(2,-3)$ and $P_2(-1,4)$ are two points (Figure 7), the vector from P_1 to P_2 is

$$\overrightarrow{P_1P_2} = (-1 - 2, 4 - (-3)) = (-3,7)$$ ◻

Figure 8
Addition of vectors

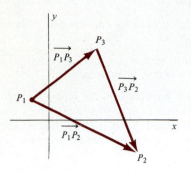

It is worth noting in this connection that the formula

$$\overrightarrow{P_1P_2} = \overrightarrow{P_1P_3} + \overrightarrow{P_3P_2}$$

is true regardless of the choice of points P_1, P_2, P_3. To see why, let $P_3 = (x_3, y_3)$ and use the definition of addition:

$$\overrightarrow{P_1P_3} + \overrightarrow{P_3P_2} = (x_3 - x_1, y_3 - y_1) + (x_2 - x_3, y_2 - y_3)$$
$$= (x_3 - x_1 + x_2 - x_3, y_3 - y_1 + y_2 - y_3)$$
$$= (x_2 - x_1, y_2 - y_1)$$
$$= \overrightarrow{P_1P_2}$$

(See Figure 8.) We may state the result in words by saying that two vectors are added by starting the second one where the first one terminates; the sum starts at the initial point of the first vector and ends at the terminal point of the second vector. As you can see, this is just another way of stating the Parallelogram Law.

Sometimes it is convenient to write vectors in terms of the *perpendicular unit vectors* $\mathbf{i} = (1,0)$ and $\mathbf{j} = (0,1)$. (See Figure 9.) The Parallelogram Law suggests that if $\mathbf{x} = (x,y)$, then $\mathbf{x} = x\mathbf{i} + y\mathbf{j}$. We do not have to depend on geometry to verify this, however; it follows from our algebraic definitions:

$$\mathbf{x} = (x,y) = (x,0) + (0,y) = x(1,0) + y(0,1) = x\mathbf{i} + y\mathbf{j}$$

Figure 9
Vector in terms of **i** and **j**

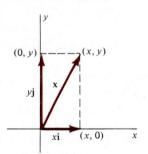

Note that we referred to $\mathbf{i}$ and $\mathbf{j}$ as "perpendicular unit vectors." Strictly speaking, these terms are premature, since we have not defined the length of a vector nor the angle between two vectors. Of course it is geometrically apparent that the arrows from the origin to $(1,0)$ and $(0,1)$ are of length 1 and are perpendicular. But it is enlightening to see what develops when we seek algebraic definitions.

The formula for length is obvious, being nothing more than the distance between the initial and terminal points of the arrow representing the vector:

If $\mathbf{x} = (x,y)$ is a vector in $\mathcal{R}^2$, its **length** (or *magnitude*) is $|\mathbf{x}| = \sqrt{x^2 + y^2}$.

Note that this formula makes sense whether the arrow representing $\mathbf{x} = (x,y)$ starts at the origin and terminates at (x,y) or starts at (x_1,y_1) and ends at (x_2,y_2). Since $x_2 - x_1 = x$ and $y_2 - y_1 = y$, the same result is obtained in any case.

□ **Example 3**   The length of the vector $\mathbf{x} = (5,-2)$ is

$$|\mathbf{x}| = \sqrt{25 + 4} = \sqrt{29}$$ □

Figure 10
Angle between vectors

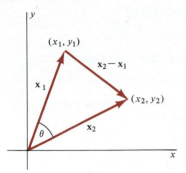

The formula for angle is not obvious. To see how it is obtained, consider Figure 10, in which we show two nonzero vectors $\mathbf{x}_1 = (x_1, y_1)$ and $\mathbf{x}_2 = (x_2, y_2)$ and the angle θ between them. The Law of Cosines (Problem 52, Section 2.3) says that

$$|\mathbf{x}_2 - \mathbf{x}_1|^2 = |\mathbf{x}_1|^2 + |\mathbf{x}_2|^2 - 2|\mathbf{x}_1||\mathbf{x}_2| \cos \theta$$

Since $\mathbf{x}_2 - \mathbf{x}_1 = (x_2 - x_1, y_2 - y_1)$, this becomes

$$(x_2 - x_1)^2 + (y_2 - y_1)^2 = (x_1^2 + y_1^2) + (x_2^2 + y_2^2) - 2|\mathbf{x}_1||\mathbf{x}_2| \cos \theta$$

Expanding the products and simplifying, we find

$$|\mathbf{x}_1||\mathbf{x}_2| \cos \theta = x_1 x_2 + y_1 y_2$$

from which
$$\cos \theta = \frac{x_1 x_2 + y_1 y_2}{|\mathbf{x}_1||\mathbf{x}_2|}$$

If we agree that the angle between two vectors satisfies $0 \leq \theta \leq \pi$, this formula determines it uniquely.

□ **Example 4**   Find the angle between $\mathbf{x}_1 = (2, -1)$ and $\mathbf{x}_2 = (3, 4)$.

Solution: No picture is needed; simply compute

$$\cos \theta = \frac{(2)(3) + (-1)(4)}{5\sqrt{5}} = \frac{2}{5\sqrt{5}}$$

Hence $\theta = \cos^{-1}(2/(5\sqrt{5})) \approx 1.391$ (or about $79.7°$).   □

The *dot product* of two vectors $\mathbf{x}_1$ and $\mathbf{x}_2$ is often defined in physics as

$$\mathbf{x}_1 \cdot \mathbf{x}_2 = |\mathbf{x}_1||\mathbf{x}_2| \cos \theta$$

(motivated by considerations of vector components of force involved in the concept of work). From our point of view, this definition is backwards (for it presupposes that we know what the angle between two vectors is). Instead, look at the formula

$$|\mathbf{x}_1||\mathbf{x}_2| \cos \theta = x_1 x_2 + y_1 y_2$$

we just derived. Taking our clue from physics (later we will have other good reasons), we define the dot product as follows.

Let $\mathbf{x}_1 = (x_1, y_1)$ and $\mathbf{x}_2 = (x_2, y_2)$ be vectors in $\mathfrak{R}^2$. Their **dot product** is the number

$$\mathbf{x}_1 \cdot \mathbf{x}_2 = x_1 x_2 + y_1 y_2$$

Note that this definition (like all our others) is purely algebraic. A geometric interpretation is that if $\mathbf{x}_1$ and $\mathbf{x}_2$ are nonzero vectors with the same initial point and θ is the angle between them, then

$$\mathbf{x}_1 \cdot \mathbf{x}_2 = |\mathbf{x}_1||\mathbf{x}_2| \cos \theta$$

This leads to our final formula for the angle between two vectors:

If $\mathbf{x}_1$ and $\mathbf{x}_2$ are nonzero vectors in $\mathcal{R}^2$, the **angle** between them is measured by the number θ $(0 \leq \theta \leq \pi)$ satisfying

$$\cos \theta = \frac{\mathbf{x}_1 \cdot \mathbf{x}_2}{|\mathbf{x}_1||\mathbf{x}_2|}$$

The vectors are said to be **perpendicular** if $\theta = \pi/2$ and **parallel** if $\theta = 0$ or $\theta = \pi$. They are said to have the **same direction** if $\theta = 0$, **opposite directions** if $\theta = \pi$.

Note that θ is defined only for nonzero vectors. The zero vector $\mathbf{0} = (0,0)$ can be regarded as an "arrow" beginning and terminating at the same point, but it is hardly possible to speak of its direction, or the angle it makes with another vector. For convenience in the statement of certain theorems, we adopt the convention that it has *any* direction. In particular, it is perpendicular to every vector (and parallel to every vector). This sounds strange, but it does not contradict our definition of angle (which excludes the zero vector). Moreover, it allows us to state the following elegant test for perpendicularity without making a special case of zero.

$$\mathbf{x}_1 \text{ and } \mathbf{x}_2 \text{ are perpendicular} \Leftrightarrow \mathbf{x}_1 \cdot \mathbf{x}_2 = 0$$

To see why this is true, assume first that $\mathbf{x}_1$ and $\mathbf{x}_2$ are nonzero. If θ is the angle between them, we have

$$\mathbf{x}_1 \text{ and } \mathbf{x}_2 \text{ are perpendicular} \Leftrightarrow \theta = \pi/2$$
$$\Leftrightarrow \cos \theta = 0 \qquad \text{(because } 0 \leq \theta \leq \pi\text{)}$$
$$\Leftrightarrow \mathbf{x}_1 \cdot \mathbf{x}_2 = 0 \qquad \text{(Why?)}$$

If either of the vectors is zero, then $\mathbf{x}_1 \cdot \mathbf{x}_2 = 0$ (why?) and by the convention just adopted they are perpendicular. Hence the test applies to all vectors.

□ **Example 5**   Show that $\mathbf{x}_1 = (2,-3)$ and $\mathbf{x}_2 = (6,4)$ are perpendicular.

Solution: Again no picture is needed; simply compute

$$\mathbf{x}_1 \cdot \mathbf{x}_2 = (2)(6) + (-3)(4) = 0 \qquad\qquad\qquad □$$

The dot product has several familiar-looking algebraic properties, each of which you should verify:

1. $x_1 \cdot x_2 = x_2 \cdot x_1$
2. $x_1 \cdot (x_2 + x_3) = x_1 \cdot x_2 + x_1 \cdot x_3$
3. $(cx_1) \cdot x_2 = c(x_1 \cdot x_2) = x_1 \cdot (cx_2)$
4. $x \cdot x = |x|^2$

□ **Example 6**   Show that if x is any vector and c is a real number, then $|cx| = |c||x|$.

Solution: By Property 4 we have

$$|cx|^2 = (cx) \cdot (cx) = c^2(x \cdot x) \quad \text{(Property 3)}$$
$$= c^2|x|^2 \quad \text{(Property 4 again)}$$

Hence $|cx| = |c||x|$. This formula can also be verified by letting $x = (x, y)$ and using the definition $|x| = \sqrt{x^2 + y^2}$, but the above "coordinate-free" argument is easier.   □

□ **Example 7**   Show that if $x \neq 0$, the vector

$$u = \frac{x}{|x|} \quad \left(\text{meaning } \frac{1}{|x|}x\right)$$

is a unit vector with the same direction as x.

Solution: Let $c = 1/|x|$. Then $u = cx$ and by Example 6 we have

$$|u| = |c||x| = \frac{|x|}{|x|} = 1$$

Hence u is a unit vector. It is geometrically apparent that u and x have the same direction (because $u = cx$ and $c > 0$). However, let's confirm it by the definition of the angle θ between them:

$$\cos \theta = \frac{u \cdot x}{|u||x|} = \frac{u \cdot x}{|x|} = u \cdot \frac{x}{|x|} \quad \text{(Property 3 of the dot product)}$$
$$= u \cdot u = |u|^2 = 1$$

This implies that $\theta = 0$ (because $0 \leq \theta \leq \pi$), so u and x have the same direction.

As an illustration of Example 7, take $x = (3, 4)$. Then $|x| = 5$ and hence a unit vector with the same direction as x is $u = \frac{1}{5}(3, 4) = (\frac{3}{5}, \frac{4}{5})$. (*Cookbook Recipe:* Divide x by its own length.)   □

Problem Set 15.1

If $x_1 = (6, -8)$ and $x_2 = (-9, 12)$, find the following.

1. $3x_1 + 2x_2$ **2.** $x_1 \cdot x_2$
3. The angle between x_1 and x_2

If $x_1 = (3, 0)$ and $x_2 = (-1, 1)$, find the following.

4. $x_1 - 2x_2$ **5.** $x_1 \cdot x_2$
6. The angle between x_1 and x_2

Given the points $P_1(3,4)$ and $P_2(6,-5)$, find the following.

7. $\overrightarrow{P_1P_2}$ **8.** $\overrightarrow{P_2P_1}$

9. $\overrightarrow{P_1P_2} + \overrightarrow{P_2P_1}$ **10.** $|\overrightarrow{P_1P_2}|$

11. The cosine of the angle between $\overrightarrow{P_1P_2}$ and $\mathbf{i}$

12. The cosine of the angle between $\overrightarrow{P_1P_2}$ and $\mathbf{j}$

13. The point two-thirds of the way from P_1 to P_2

Find a unit vector in the direction of each of the following vectors.

14. $(3,0)$ **15.** $-2\mathbf{j}$

16. $(4,-3)$ **17.** $5\mathbf{i} - 12\mathbf{j}$

18. (π,π)

19. Show that $\mathbf{u}_1 = (1/2, \sqrt{3}/2)$ and $\mathbf{u}_2 = (-\sqrt{3}/2, 1/2)$ are perpendicular unit vectors.

20. Find all unit vectors perpendicular to $\mathbf{x} = (2,1)$. (From a picture, how many do you expect?)

21. Let $\mathbf{x} = (x,y)$ be the position vector (initial point at the origin) of a point (x,y) moving in the plane, and suppose that $\mathbf{a} = (a,b)$ is not zero. What is the path of the point if $\mathbf{a} \cdot \mathbf{x} = 0$? What is the geometric relation of $\mathbf{a}$ to this path?

22. Show that the vector (a,b) is perpendicular to the straight line $ax + by + c = 0$.

23. Use vector methods to find the angles of the triangle with vertices $(-2,1)$, $(1,4)$, $(0,1)$.

24. Let $P_1(x_1,y_1)$ and $P_2(x_2,y_2)$ be two points of the plane and let $M(x,y)$ be the midpoint of the segment joining them.

   (a) Draw a picture to convince yourself that $\overrightarrow{P_1M} = \overrightarrow{MP_2}$.

   (b) Use the vector equation in part (a) to derive the midpoint formulas

$$x = \tfrac{1}{2}(x_1 + x_2), \quad y = \tfrac{1}{2}(y_1 + y_2)$$

25. Let A and B be endpoints of a diameter of a circle and let C be a third point of the circle. Use vector methods to prove that angle ACB is a right angle.

26. Show that the line segment joining the midpoints of two sides of a triangle is parallel to the third side and half as long, as follows.

   (a) Let P_1, P_2, and P_3 be the vertices of the triangle and let M_1 and M_2 be the midpoints of P_1P_2 and P_2P_3, respectively. Why is it sufficient to show that $\overrightarrow{M_1M_2} = \tfrac{1}{2}\overrightarrow{P_1P_3}$? Start by confirming that

$$\overrightarrow{P_1M_1} + \overrightarrow{M_1M_2} + \overrightarrow{M_2P_3} = \overrightarrow{P_1P_3}$$

   (b) Explain why $\overrightarrow{P_1M_1} = \tfrac{1}{2}\overrightarrow{P_1P_2}$ and $\overrightarrow{M_2P_3} = \tfrac{1}{2}\overrightarrow{P_2P_3}$, and substitute in part (a) to finish the proof.

27. Prove that the midpoints of the sides of any quadrilateral (four-sided figure) are the vertices of a parallelogram.

28. Let P be the point two-thirds of the way from one vertex of a triangle to the midpoint of the opposite side.

   (a) Use vector methods to find the coordinates of P in terms of the coordinates of the vertices.

   (b) Argue from part (a) that the medians of the triangle intersect at P.

29. Show that if $\mathbf{x}_1$ and $\mathbf{x}_2$ are any vectors in $\mathcal{R}^2$, then

$$(\mathbf{x}_1 + \mathbf{x}_2) \cdot (\mathbf{x}_1 - \mathbf{x}_2) = |\mathbf{x}_1|^2 - |\mathbf{x}_2|^2$$

Why does it follow that the diagonals of a square are perpendicular?

30. Prove that if a and b are real numbers and $\mathbf{x} \in \mathcal{R}^2$, then $a(b\mathbf{x}) = (ab)\mathbf{x}$.

31. Prove that if $c \in \mathcal{R}$ and $\mathbf{x}_1$ and $\mathbf{x}_2$ are in $\mathcal{R}^2$, then

$$c(\mathbf{x}_1 + \mathbf{x}_2) = c\mathbf{x}_1 + c\mathbf{x}_2$$

32. Let c be a real number and $\mathbf{x}$ a vector. Explain why

$$c\mathbf{x} = \mathbf{0} \Leftrightarrow c = 0 \text{ or } \mathbf{x} = \mathbf{0}$$

33. Verify the formula $\mathbf{x} \cdot \mathbf{x} = |\mathbf{x}|^2$.

34. Show that $\mathbf{x} \cdot \mathbf{x} = 0 \Leftrightarrow \mathbf{x} = \mathbf{0}$.

35. Which part of the double implication

$$\mathbf{x}_1 \cdot \mathbf{x}_2 = 0 \Leftrightarrow \mathbf{x}_1 = \mathbf{0} \text{ or } \mathbf{x}_2 = \mathbf{0}$$

is correct? Show by means of an example that the other part is incorrect.

Derive the following properties of the dot product.

36. $\mathbf{x}_1 \cdot \mathbf{x}_2 = \mathbf{x}_2 \cdot \mathbf{x}_1$

37. $\mathbf{x}_1 \cdot (\mathbf{x}_2 + \mathbf{x}_3) = \mathbf{x}_1 \cdot \mathbf{x}_2 + \mathbf{x}_1 \cdot \mathbf{x}_3$

38. $(c\mathbf{x}_1) \cdot \mathbf{x}_2 = c(\mathbf{x}_1 \cdot \mathbf{x}_2) = \mathbf{x}_1 \cdot (c\mathbf{x}_2)$

39. The algebraic properties of the dot product do not include the associative law

$$(\mathbf{x}_1 \cdot \mathbf{x}_2) \cdot \mathbf{x}_3 = \mathbf{x}_1 \cdot (\mathbf{x}_2 \cdot \mathbf{x}_3)$$

Why not?

40. Show that $\mathbf{x}$ and $c\mathbf{x}$ have the same direction if $c \geq 0$ and opposite directions if $c < 0$.

41. Let $\mathbf{x}_1$ and $\mathbf{x}_2$ be any vectors in the plane.

   (a) Assuming that $\mathbf{x}_1$ and $\mathbf{x}_2$ are not zero, show that the angle θ between them satisfies

$$\sin^2\theta = \frac{|\mathbf{x}_1|^2|\mathbf{x}_2|^2 - (\mathbf{x}_1 \cdot \mathbf{x}_2)^2}{|\mathbf{x}_1|^2|\mathbf{x}_2|^2}$$

(b) Explain why it follows that

$$\mathbf{x}_1 \text{ and } \mathbf{x}_2 \text{ are parallel} \Leftrightarrow |\mathbf{x}_1 \cdot \mathbf{x}_2| = |\mathbf{x}_1||\mathbf{x}_2|$$

42. The preceding problem shows that the formula $|\mathbf{x}_1 \cdot \mathbf{x}_2| = |\mathbf{x}_1||\mathbf{x}_2|$ is not true in general. Prove the famous *Cauchy-Schwarz Inequality*,

$$|\mathbf{x}_1 \cdot \mathbf{x}_2| \leq |\mathbf{x}_1||\mathbf{x}_2|$$

43. Draw a picture showing why the statement

$$|\mathbf{x}_1 + \mathbf{x}_2| \leq |\mathbf{x}_1| + |\mathbf{x}_2|$$

is called the *Triangle Inequality*. Then use the Cauchy-Schwarz Inequality to prove it. *Hint:*

$$|\mathbf{x}_1 + \mathbf{x}_2|^2 = (\mathbf{x}_1 + \mathbf{x}_2) \cdot (\mathbf{x}_1 + \mathbf{x}_2)$$

44. Derive the trigonometric identity

$$\cos(\alpha - \beta) = \cos\alpha\cos\beta + \sin\alpha\sin\beta$$

as follows.

(a) Explain why $\mathbf{u} = (\cos\alpha, \sin\alpha)$ and $\mathbf{v} = (\cos\beta, \sin\beta)$ are unit vectors.

(b) Explain why the angle between $\mathbf{u}$ and $\mathbf{v}$ is $\theta = \pm(\alpha - \beta) + 2\pi n$ for some integer n.

(c) Why does it follow that $\mathbf{u} \cdot \mathbf{v} = \cos(\alpha - \beta)$ and why does this prove the required identity?

45. The definitions of sine and cosine in trigonometry imply that any point of the unit circle can be written in the form $(\cos\theta, \sin\theta)$ for some θ.

(a) Use this fact to explain why any two perpendicular unit vectors can be written in the form

$$\mathbf{u}_1 = (\cos\theta, \sin\theta) \quad \text{and} \quad \mathbf{u}_2 = (-\sin\theta, \cos\theta)$$

for some θ.

(b) Given the unit vectors of part (a), show that every vector $\mathbf{x} \in \mathcal{R}^2$ can be expressed in terms of $\mathbf{u}_1$ and $\mathbf{u}_2$ by writing

$$\mathbf{x} = c_1\mathbf{u}_1 + c_2\mathbf{u}_2$$

where $c_1 = \mathbf{x} \cdot \mathbf{u}_1$ and $c_2 = \mathbf{x} \cdot \mathbf{u}_2$.

15.2
VECTOR FUNCTIONS

Recall from Section 7.3 that if $I = [a,b]$, the parametric equations

$$x = f(t), \quad y = g(t) \qquad (t \in I)$$

describe a curve in the plane. The curve is continuous if f and g are continuous in I, and "smooth" if dx/dt and dy/dt exist and are continuous in I, with

$$\left(\frac{dx}{dt}\right)^2 + \left(\frac{dy}{dt}\right)^2 > 0 \qquad \text{for all interior points of } I$$

As we pointed out when this definition was given, the term **smooth** means that the point (x,y) suffers no abrupt changes in direction as it moves along the path. In particular it never stops or reverses direction.

Our technical definition of "curve" in Section 7.3 was that the curve is actually the function

$$F: I \to \mathcal{R}^2 \text{ defined by } F(t) = (f(t), g(t)), \ t \in I$$

Now that we have discussed vectors in $\mathcal{R}^2$, you should recognize that the functional values

$$F(t) = (x,y) = (f(t), g(t))$$

are vectors; for that reason F is called a **vector function**. To be consistent with our notation in the last section, we will use boldface type, writing **F** instead of F to emphasize that we are dealing with a function whose outputs are number pairs (vectors).

The graph of **F**, namely the set of points

$$\{(x,y): x = f(t), y = g(t), t \in I\}$$

was called the *track* of the curve in Section 7.3. The vector interpretation is that $\mathbf{x} = \mathbf{F}(t) = (x, y)$ is the *position vector* of the typical point of the track. (See Figure 1.) As t runs through the domain I, this vector follows the point (x, y) in its motion, just as if we had a radar tracking system at the origin.

Our definition that **F** is continuous if f and g are continuous amounts to saying that the limit of a vector function is found "componentwise," that is,

$$\lim_{t \to c} \mathbf{F}(t) = \left(\lim_{t \to c} f(t), \lim_{t \to c} g(t)\right)$$

Continuity of f and g at c implies that

$$\lim_{t \to c} \mathbf{F}(t) = (f(c), g(c)) = \mathbf{F}(c)$$

which is the expected meaning of continuity of **F** at c.

The derivative of a vector function is defined by componentwise differentiation:

$$\mathbf{F}'(t) = (f'(t), g'(t))$$

Since limits are found componentwise, this is equivalent to the familiar-looking formula

$$\mathbf{F}'(t) = \lim_{z \to t} \frac{\mathbf{F}(z) - \mathbf{F}(t)}{z - t}$$

(See the problem set.)

It is enlightening to examine the geometric meaning of this formula. We assume that the curve $\mathbf{x} = \mathbf{F}(t)$ is smooth, so that

$$\frac{d\mathbf{x}}{dt} = \mathbf{F}'(t) = (dx/dt, dy/dt)$$

exists for $t \in I$ and

$$\left(\frac{dx}{dt}\right)^2 + \left(\frac{dy}{dt}\right)^2 > 0 \qquad \text{for all interior points of } I$$

Taking the formula for $\mathbf{F}'(t)$ step by step, we begin by fixing $t \in I$ and computing the vector $\mathbf{F}(t)$. (See Figure 2.) Then we choose a neighboring value $z \in I$ and compute $\mathbf{F}(z)$, obtaining a vector that is close to $\mathbf{F}(t)$ in both magnitude and direction. Hence we expect the vector $\mathbf{F}(z) - \mathbf{F}(t)$ to be short. This is compensated, however, by the lengthening effect of multiplying by $1/(z - t)$, because $z - t$ is small. Since the derivative exists, the compensating process causes the vector

$$\frac{\mathbf{F}(z) - \mathbf{F}(t)}{z - t}$$

to approach a definite limit, namely the vector $\mathbf{F}'(t) = (dx/dt, dy/dt)$.

We can say two things about this vector. First, its length is

$$|\mathbf{F}'(t)| = \sqrt{(dx/dt)^2 + (dy/dt)^2}$$

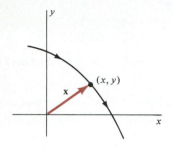

Figure 1
Position vector tracking a moving point

Figure 2
Computing the vector
derivative $\mathbf{F}'(t)$

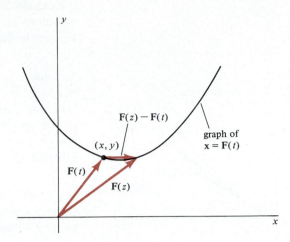

which is never zero (except possibly at the endpoints of I) because the curve is smooth. Hence $\mathbf{F}'(t)$ is not the zero vector. Second, $\mathbf{F}'(t)$ is parallel to the tangent at (x,y). For its slope is

$$\frac{dy/dt}{dx/dt} = \frac{dy}{dx}$$

(unless $dx/dt = 0$, in which case both $\mathbf{F}'(t)$ and the tangent are vertical).

The derivative of a smooth vector function is a nonzero vector parallel to the tangent to the graph of the function.

Figure 3
Graph of $\mathbf{F}(t) = (\cos t, \sin t)$

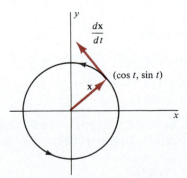
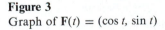

□ **Example 1**   The vector function $\mathbf{F}(t) = (\cos t, \sin t)$, $0 \leq t \leq 2\pi$, represents the unit circle shown in Figure 3, parametric equations being

$$x = \cos t, \ y = \sin t$$

The position vector $\mathbf{x} = \mathbf{F}(t) = (x,y)$ terminates at the typical point of the circle; its derivative

$$\frac{d\mathbf{x}}{dt} = \mathbf{F}'(t) = (dx/dt, dy/dt) = (-\sin t, \cos t)$$

is tangent to the circle at (x,y). We may confirm this statement by elementary geometry, because we know that the radius and the tangent are perpendicular. Hence the dot product of the position vector and tangent vector should be zero:

$$\mathbf{x} \cdot \frac{d\mathbf{x}}{dt} = (\cos t, \sin t) \cdot (-\sin t, \cos t) = (\cos t)(-\sin t) + (\sin t)(\cos t) = 0$$

□

□ **Example 2**   The flight of a bumblebee is given by the parametric equations $x = 2t^2$, $y = t^3$, $t \geq 0$ (where t is time). When $t = 1$ the bee spies a flower at the point (5,3) and flies off on a tangent. Will it reach the flower?

Solution: The vector function $\mathbf{F}(t) = (2t^2, t^3)$ gives the position of the bee at time t. Hence the bee is at the point $\mathbf{F}(1) = (2,1)$ when it sees the flower at (5,3) and flies off on a tangent. The question is whether the point (5,3) is on the tangent. (See Figure 4.) A vector with initial point (2,1) and lying in the tangent is $\mathbf{F}'(1)$. Since $\mathbf{F}'(t) = (4t, 3t^2)$, this vector is (4,3). On the other hand, the vector from $P_1(2,1)$ to $P_2(5,3)$ is

$$\overrightarrow{P_1 P_2} = (5 - 2, 3 - 1) = (3, 2)$$

If these vectors are parallel, the flower is on the tangent, otherwise not. The angle θ between them satisfies

$$\cos \theta = \frac{(4,3) \cdot (3,2)}{|(4,3)|\,|(3,2)|} = \frac{(4)(3) + (3)(2)}{5\sqrt{13}} = \frac{18}{5\sqrt{13}}$$

Since $\cos \theta \neq \pm 1$, θ is neither 0 nor π, so the vectors are not parallel. The bee will miss the flower. (A simpler way to verify this is that neither vector is a multiple of the other, so they cannot be parallel.)   □

Having defined what we mean by a vector function and its derivative, we need to know how it fits into the usual rules of differentiation. Suppose, for example, that we are interested in finding the derivative of

$$\mathbf{G}(t) = e^t(\cos t, \sin t)$$

which is the product of an ordinary real function $f(t) = e^t$ and the vector function $\mathbf{F}(t) = (\cos t, \sin t)$. (*Note:* In the context of vectors we often use "scalar" as a synonym for "real," because real numbers are located on a *scale,* the coordinate line. Hence we refer to f as a *scalar function,* in contrast to the *vector function* $\mathbf{F}$.) One way to do the problem is to carry out the multiplication first, writing

$$\mathbf{G}(t) = (e^t \cos t, e^t \sin t)$$

Using the Product Rule in the differentiation of each component, we find

$$\begin{aligned} \mathbf{G}'(t) &= (-e^t \sin t + e^t \cos t, \; e^t \cos t + e^t \sin t) \\ &= e^t(\cos t - \sin t, \cos t + \sin t) \end{aligned}$$

Another method that may occur to us, however, is to apply the Product Rule to the function as it stands. In other words, if $\mathbf{G}(t) = f(t)\mathbf{F}(t)$, why not write

$$\mathbf{G}'(t) = f(t)\mathbf{F}'(t) + f'(t)\mathbf{F}(t)?$$

Of course this requires proof, but let's try it in the present example:

$$\mathbf{G}(t) = e^t(\cos t, \sin t)$$

Figure 4
Flight of the bumblebee

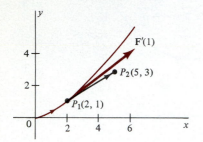

$$\mathbf{G}'(t) = e^t \frac{d}{dt}(\cos t, \sin t) + \frac{d}{dt}(e^t)(\cos t, \sin t)$$

$$= e^t(-\sin t, \cos t) + e^t(\cos t, \sin t)$$

$$= e^t(\cos t - \sin t, \cos t + \sin t)$$

As you can see, the result is the same either way, which suggests the following theorem. We leave the proof for you in the problem set.

Product Rule for Scalar Function Times Vector Function

If f is a scalar function and $\mathbf{F}$ is a vector function, both differentiable, then

$$\frac{d}{dt} f(t)\mathbf{F}(t) = f(t)\mathbf{F}'(t) + f'(t)\mathbf{F}(t)$$

Another situation in which a Product Rule is called for is the dot product of two vector functions, for example,

$$f(t) = (t^2, t^3) \cdot (\cosh t, \sinh t)$$

One way to differentiate this is to multiply first, obtaining

$$f(t) = t^2 \cosh t + t^3 \sinh t$$

Since this is an ordinary scalar function, nothing new is needed to find its derivative:

$$f'(t) = t^2 \sinh t + 2t \cosh t + t^3 \cosh t + 3t^2 \sinh t$$

The same result is obtained by using the Product Rule pattern in the original function:

$$f'(t) = (t^2, t^3) \cdot \frac{d}{dt}(\cosh t, \sinh t) + (\cosh t, \sinh t) \cdot \frac{d}{dt}(t^2, t^3)$$

$$= (t^2, t^3) \cdot (\sinh t, \cosh t) + (\cosh t, \sinh t) \cdot (2t, 3t^2)$$

$$= t^2 \sinh t + t^3 \cosh t + 2t \cosh t + 3t^2 \sinh t$$

Product Rule for a Dot Product of Vector Functions

If $\mathbf{F}$ and $\mathbf{G}$ are differentiable vector functions, then

$$\frac{d}{dt}\mathbf{F}(t) \cdot \mathbf{G}(t) = \mathbf{F}(t) \cdot \mathbf{G}'(t) + \mathbf{F}'(t) \cdot \mathbf{G}(t)$$

Finally, the Chain Rule applies in an example like $\mathbf{x} = (u^2, e^u)$, where $u = \sin t$. Again we have a choice of methods. Substitution of $u = \sin t$ in the vector function gives

$$\mathbf{x} = (\sin^2 t, e^{\sin t})$$

which may be differentiated by the ordinary Chain Rule applied to each component:

$$\frac{d\mathbf{x}}{dt} = (2 \sin t \cos t, e^{\sin t} \cos t)$$

The same result is obtained by applying the Chain Rule idea to the original function:

$$\frac{d\mathbf{x}}{dt} = \frac{d\mathbf{x}}{du}\frac{du}{dt} = \frac{d}{du}(u^2, e^u)\frac{d}{dt}(\sin t) = (2u, e^u) \cos t$$

$$= (2u \cos t, e^u \cos t) = (2 \sin t \cos t, e^{\sin t} \cos t)$$

Chain Rule for Vector Functions

If $\mathbf{x} = \mathbf{F}(u)$ is a vector function and $u = f(t)$ is a scalar function, both differentiable, then $\mathbf{x}$ is a differentiable function of t and

$$\frac{d\mathbf{x}}{dt} = \frac{d\mathbf{x}}{du}\frac{du}{dt}$$

Problem Set 15.2

In each of the following, find an equation in rectangular coordinates of the graph of the vector function. Sketch the graph and indicate the direction in which it is traversed by considering the behavior of the vector derivative for appropriate values of the parameter.

1. $\mathbf{F}(t) = (2t, 3t + 1)$
2. $\mathbf{F}(t) = (t^2, t)$
3. $\mathbf{F}(t) = (t, t^3)$
4. $\mathbf{F}(t) = (\cos t^2, \sin t^2)$, $t \geq 0$
5. $\mathbf{F}(t) = (a \cos t, b \sin t)$, $a > b > 0$
6. $\mathbf{F}(t) = (\cosh t, \sinh t)$
7. $\mathbf{F}(t) = (e^t, e^{-t})$
8. Explain why the graph of the vector function

$$\mathbf{F}(t) = (\tan t, \cos^2 t) \qquad -\pi/2 < t < \pi/2$$

is the curve $y = 1/(x^2 + 1)$. Sketch the curve and indicate the direction in which it is traversed.

9. Explain why the graph of $\mathbf{x} = (x, y)$ defined by

$$x = \frac{1}{\sqrt{1 + t^2}}, y = \frac{t}{\sqrt{1 + t^2}}$$

is the right-hand half of the unit circle $x^2 + y^2 = 1$ (excluding the endpoints). In what direction is it traversed?

10. Explain why the graph of

$$\mathbf{F}(t) = (t \cos t, t \sin t) \qquad t \geq 0$$

is the spiral whose polar equation is $r = \theta$, $\theta \geq 0$.

11. Explain why the graph of

$$\mathbf{F}(t) = (e^t \cos t, e^t \sin t)$$

is the spiral whose polar equation is $r = e^\theta$.

12. Graph the vector function

$$\mathbf{F}(t) = (\sin t, \sin 2t) \qquad 0 \leq t \leq 2\pi$$

by locating the points corresponding to $t = 0$, $\pi/4$, $\pi/2$, $3\pi/4$, π, $5\pi/4$, $3\pi/2$, $7\pi/4$, 2π, and drawing a tangent vector at each point.

13. Sketch the parametric path

$$x = a(\theta - \sin \theta), y = a(1 - \cos \theta)$$

as follows.

(a) Locate the points corresponding to $\theta = 0$, $\pi/2$, π, $3\pi/2$, 2π, and draw a tangent vector (if possible) at each point.

(b) The tangent vector at each multiple of 2π is zero.

(Why?) To discover the behavior of the graph at these points, show that

$$\frac{dy}{dx} = \cot\frac{\theta}{2}$$

and discuss the slope when θ is near each multiple of 2π. *Hint:* Use the identity

$$\tan\frac{\theta}{2} = \frac{1 - \cos\theta}{\sin\theta}$$

(c) This curve (called a *cycloid*) is the path of a point on the rim of a wheel of radius a rolling on the x axis. Look up Problem 37, Section 7.3, and see if you figured it out correctly.

14. The graph of $\mathbf{F}(t) = (e^t \cos t, e^t \sin t)$ is the spiral whose polar equation is $r = e^\theta$. (See Problem 11.)

(a) Show that the angle between the position vector and tangent vector is always 45°.

(b) Locate the points corresponding to $t = -\pi/2, -\pi/4, 0, \pi/4, \pi/2$, and draw a tangent vector at each point [using part (a)]. How does this kind of data help sketch the graph?

In each of the following, find $d\mathbf{x}/dt$ in two ways.

15. $\mathbf{x} = t(1, t^2)$ 16. $\mathbf{x} = t(\cos t, \sin t)$

17. $\mathbf{x} = e^t(\cos t, \sin t)$

18. $\mathbf{x} = (u^2, u^3)$, where $u = \sqrt{t}$

19. $\mathbf{x} = (\cos u, \sin u)$, where $u = t^2$

In each of the following, find $f'(t)$ in two ways.

20. $f(t) = (e^t, e^{-t}) \cdot (e^{-t}, e^t)$

21. $f(t) = (t^3, t^2) \cdot (\cos t, \sin t)$

22. $f(t) = (2t, t^2) \cdot (e^{2t}, e^{-t})$

23. Let $\mathbf{m} = (a, b)$ be a nonzero vector and $\mathbf{x}_0 = (x_0, y_0)$ the position vector of a fixed point. Let $\mathbf{x} = (x, y)$ be the position vector of any point.

(a) Draw a picture showing that the line through (x_0, y_0) parallel to $\mathbf{m}$ is the graph of the vector equation $\mathbf{x} - \mathbf{x}_0 = t\mathbf{m}$, $t \in \mathcal{R}$.

(b) Conclude that parametric equations of the line are $x = at + x_0$, $y = bt + y_0$.

(c) Find the derivative of the vector function $\mathbf{x} = \mathbf{F}(t) = t\mathbf{m} + \mathbf{x}_0$. What is the geometric interpretation?

24. Suppose that the derivative of the vector function $\mathbf{x} = \mathbf{F}(t)$ is $\mathbf{F}'(t) = \mathbf{m}$ for all t, where $\mathbf{m}$ is a nonzero constant vector. Explain why the graph of $\mathbf{F}$ is a straight line parallel to $\mathbf{m}$.

25. The slope of the parametric path $x = f(t)$, $y = g(t)$ is

$$\frac{dy}{dx} = \frac{dy/dt}{dx/dt} = \frac{g'(t)}{f'(t)}$$

How does this fit in with the fact that the derivative of $\mathbf{F}(t) = (f(t), g(t))$ is $\mathbf{F}'(t) = (f'(t), g'(t))$?

26. Show that if $\mathbf{x} = \mathbf{F}(t)$ is differentiable, then

$$\frac{d}{dt}|\mathbf{x}|^2 = 2\mathbf{x} \cdot \frac{d\mathbf{x}}{dt}$$

27. Suppose that the magnitude of the vector function $\mathbf{x} = \mathbf{F}(t)$ is $|\mathbf{F}(t)| = a$ for all t, where a is a positive constant.

(a) Give a geometric reason why the graph of $\mathbf{F}$ is the circle of radius a centered at the origin.

(b) Show without reference to geometry that

$$\mathbf{x} \cdot \frac{d\mathbf{x}}{dt} = 0 \qquad \text{for all } t$$

Why is this consistent with part (a)? *Hint:* Differentiate $|\mathbf{x}|^2 = a^2$.

28. Suppose that $(x_0, y_0) \neq (0, 0)$ is a point of the graph of $\mathbf{x} = \mathbf{F}(t)$ that is closest to the origin. Show that the position vector and tangent vector at (x_0, y_0) are perpendicular. *Hint:* $|\mathbf{x}|^2$ has its minimum value at (x_0, y_0). What does this imply about its derivative?

29. We defined the derivative of the vector function $\mathbf{F}(t) = (f(t), g(t))$ to be $\mathbf{F}'(t) = (f'(t), g'(t))$. Use the fact that limits of vector functions are found "componentwise" to show that

$$\mathbf{F}'(t) = \lim_{z \to t} \frac{\mathbf{F}(z) - \mathbf{F}(t)}{z - t}$$

30. Prove that if $\mathbf{F}$ and $\mathbf{G}$ are differentiable, then

$$\frac{d}{dt}[\mathbf{F}(t) + \mathbf{G}(t)] = \mathbf{F}'(t) + \mathbf{G}'(t)$$

31. Prove the Product Rule for a scalar function times a vector function:

$$\frac{d}{dt}f(t)\mathbf{F}(t) = f(t)\mathbf{F}'(t) + f'(t)\mathbf{F}(t)$$

32. Prove the Product Rule for a dot product of vector functions:

$$\frac{d}{dt}\mathbf{F}(t) \cdot \mathbf{G}(t) = \mathbf{F}(t) \cdot \mathbf{G}'(t) + \mathbf{F}'(t) \cdot \mathbf{G}(t)$$

33. Prove the Chain Rule for vector functions:

$$\frac{d\mathbf{x}}{dt} = \frac{d\mathbf{x}}{du}\frac{du}{dt}$$

15.3

**MOTION IN
THE PLANE**

Suppose that the parameter in the equations $x = f(t)$, $y = g(t)$, $a \leq t \leq b$, is *time*. Then the vector

$$\mathbf{x} = \mathbf{F}(t) = (x,y) = (f(t),g(t))$$

gives the position at time t of a point (x,y) moving in the plane. Assuming that $\mathbf{F}$ is a smooth vector function, we know from Section 7.3 that the distance traveled by the moving point from $t = a$ to $t = b$ is

$$s = \int_a^b \sqrt{(dx/dt)^2 + (dy/dt)^2}\, dt$$

If we replace the upper limit by the variable t ($a \leq t \leq b$), we have the distance traveled from the initial position $\mathbf{F}(a)$ to any position $\mathbf{F}(t)$, as shown in Figure 1. This is a function of t, namely

$$s(t) = \int_a^t \sqrt{[f'(u)]^2 + [g'(u)]^2}\, du$$

Figure 1
Distance traveled at time t

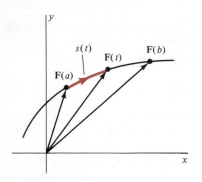

(The variable of integration is changed to u to avoid confusion with the upper limit t.) By the Fundamental Theorem of Calculus,

$$s'(t) = \sqrt{[f'(t)]^2 + [g'(t)]^2}$$

or simply $ds/dt = \sqrt{(dx/dt)^2 + (dy/dt)^2}$.

This is a formula for rate of change of distance with respect to time; it is a measure of *how fast* the point (x,y) is moving along its path. In other words it is *speed*.

In Section 2.1 (where we discussed straight line motion) speed was a creature of *velocity*, namely its absolute value. This raises the question of what velocity is in the case of motion in two dimensions. In straight line motion it is the derivative of position; is it the same thing here?

Position in two dimensions is a *vector*, namely $\mathbf{x} = \mathbf{F}(t) = (x,y)$. The derivative of this vector is

$$\frac{d\mathbf{x}}{dt} = \mathbf{F}'(t) = (dx/dt, dy/dt)$$

Figure 2
Vector derivative pointing in the direction of motion

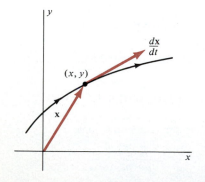

which (as we saw in the last section) is parallel to the tangent to the graph of $\mathbf{F}$ (the path of the moving point). If we place the initial point of $d\mathbf{x}/dt$ at (x,y), it lies in the tangent line and points in the direction of motion. (See Figure 2.) Moreover, its length is

$$|\mathbf{F}'(t)| = \sqrt{(dx/dt)^2 + (dy/dt)^2}$$

which (as we just saw) is speed. What more could we ask of a velocity vector? Since $d\mathbf{x}/dt$ specifies both the direction and speed of motion, we are justified in making the following definition.

If the position of a moving point at time t is given by the smooth vector function $\mathbf{x} = \mathbf{F}(t)$, the **velocity** of the point at time t is the vector $\mathbf{v} = d\mathbf{x}/dt$. The magnitude of this vector,

$$|\mathbf{v}| = ds/dt = \sqrt{(dx/dt)^2 + (dy/dt)^2}$$

(where s is distance measured along the path from some initial point) is called **speed.**

Since the derivative of a smooth vector function is never zero (except possibly at the endpoints of its domain), we can form the *unit tangent vector*

$$\mathbf{T} = \frac{\mathbf{F}'(t)}{|\mathbf{F}'(t)|}$$

If t is time, this vector is $\mathbf{T} = \mathbf{v}/|\mathbf{v}|$, from which we have the formula

$$\mathbf{v} = \frac{ds}{dt}\mathbf{T}$$

In other words, *velocity is speed multiplied by a unit vector in the direction of motion.*

□ **Example 1**   How fast and in what direction is the bumblebee of Example 2, Section 15.2, going when it flies off on a tangent?

Solution: The vector function giving the position of the bee at time t is

$$\mathbf{F}(t) = (2t^2, t^3) \qquad t \geq 0$$

The bee's velocity is $\mathbf{F}'(t) = (4t, 3t^2)$, which is the vector $\mathbf{v} = (4,3)$ at $t = 1$ (when the bee flies off on a tangent). Its speed is the magnitude of this vector, namely $|\mathbf{v}| = 5$. Its direction may be specified by the angle θ between $\mathbf{T} = (\frac{4}{5}, \frac{3}{5})$ and $\mathbf{i}$, which satisfies

$$\cos \theta = \mathbf{T} \cdot \mathbf{i} = \tfrac{4}{5}$$

Hence $\theta = \cos^{-1} 0.8 \approx 0.64$ (or about 37° from the positive x axis).   □

□ **Example 2**   A particle travels in the ellipse $x = 3 \cos t$, $y = 2 \sin t$, where t is time. Find the maximum and minimum values of the speed, and the points where they occur.

Solution: Since speed is ds/dt, we compute

$$\left(\frac{ds}{dt}\right)^2 = \left(\frac{dx}{dt}\right)^2 + \left(\frac{dy}{dt}\right)^2 = (-3 \sin t)^2 + (2 \cos t)^2 = 9 \sin^2 t + 4 \cos^2 t$$

$$= 5 \sin^2 t + 4$$

The maximum value of ds/dt is evidently 3 (when $\sin^2 t = 1$). This occurs when $\sin t = \pm1$ and $\cos t = 0$, that is, at the points $(0, \pm2)$ of the ellipse (the endpoints of the minor diameter). The minimum value of ds/dt is 2, occurring when $\sin t = 0$ and $\cos t = \pm1$, that is, at the vertices $(\pm3, 0)$ of the ellipse. (See Figure 3.)   □

Figure 3
The elliptical path $x = 3 \cos t$, $y = 2 \sin t$

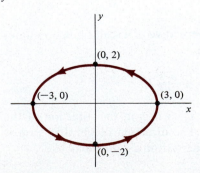

In view of the preceding discussion, it should come as no surprise that we define acceleration as the vector derivative of velocity.

If the position of a moving point at time t is given by the smooth vector function $\mathbf{x} = \mathbf{F}(t)$, the **acceleration** of the point at time t is

$$\mathbf{a} = \frac{d\mathbf{v}}{dt} = \frac{d^2\mathbf{x}}{dt^2}$$

where $\mathbf{v}$ is velocity.

□ **Example 3**   Show that the force keeping the particle of Example 2 in its elliptical path is directed toward the origin (assuming that its initial point coincides with the particle).

Solution: Newton's Second Law says that force is mass times acceleration, $\mathbf{F} = m\mathbf{a}$. The vector function giving the position of the particle at time t is

$$\mathbf{x} = (3 \cos t, 2 \sin t)$$

Hence the velocity is

$$\mathbf{v} = \frac{d\mathbf{x}}{dt} = (-3 \sin t, 2 \cos t)$$

and the acceleration is

$$\mathbf{a} = \frac{d\mathbf{v}}{dt} = (-3 \cos t, -2 \sin t) = -(3 \cos t, 2 \sin t) = -\mathbf{x}$$

Since $\mathbf{x}$ points from the origin to the moving particle, $\mathbf{a} = -\mathbf{x}$ points from the particle to the origin. Hence $\mathbf{F} = m\mathbf{a}$ is directed toward the origin. □

We turn now to an important formula for acceleration in terms of *tangential and normal components*. To see what we mean by this terminology, return to the equation

$$\mathbf{v} = \frac{ds}{dt} \mathbf{T}$$

for velocity. By the Product Rule for the derivative of a scalar function times a vector function (Section 15.2), we have

$$\mathbf{a} = \frac{d\mathbf{v}}{dt} = \frac{ds}{dt} \frac{d\mathbf{T}}{dt} + \frac{d^2s}{dt^2} \mathbf{T}$$

The second term is a scalar multiple of $\mathbf{T}$, so it is in the direction of the tangent (the *tangential component of* $\mathbf{a}$). The direction of the first term is harder to discuss; the question is what the vector $d\mathbf{T}/dt$ looks like.

For reasons that are not obvious (unless you have advance information from physics), we write

$$\frac{d\mathbf{T}}{dt} = \frac{d\mathbf{T}}{ds}\frac{ds}{dt}$$

(by the Chain Rule for vector functions stated in Section 15.2). Of course this formula makes sense only if $\mathbf{T}$ can be regarded as a function of s. We know that s is a function of the parameter t. Since $ds/dt > 0$ (because the curve is smooth), this function is increasing and hence invertible. (See Section 8.2.) Therefore we can turn it around and regard t as a function of s. Another way of saying the same thing is that arc length s, measured from some initial point of the curve, may be assigned as a "curvilinear coordinate" of the moving point, determining its position unambiguously because the motion is in one direction only. The parameter t (and hence all the variables occurring in our discussion) may therefore be regarded as functions of s.

Now use the Chain Rule again to write

$$\frac{d\mathbf{T}}{ds} = \frac{d\mathbf{T}}{d\phi}\frac{d\phi}{ds}$$

where ϕ is the angle measured counterclockwise from $\mathbf{i}$ to $\mathbf{T}$. (See Figure 4, which also shows a vector $\mathbf{n}$ to be described shortly.)

The reason for this strange-looking step is that both $d\mathbf{T}/d\phi$ and $d\phi/ds$ have geometric meanings that are not hard to figure out. We know that $\mathbf{T}$ can be written in the form

$$\mathbf{T} = (\cos\phi, \sin\phi)$$

(See Figure 5, in which we show $\mathbf{T}$ with initial point at the origin and terminal point on the unit circle.) It follows that the vector

$$\mathbf{n} = \frac{d\mathbf{T}}{d\phi} = (-\sin\phi, \cos\phi) = (\cos(\phi + \pi/2), \sin(\phi + \pi/2))$$

is a unit vector 90° counterclockwise from $\mathbf{T}$.

Remark   In straight-line motion the angle ϕ is constant and the "function" $\mathbf{T} = (\cos\phi, \sin\phi)$ cannot be differentiated with respect to ϕ. In that case $\mathbf{n} = d\mathbf{T}/d\phi$ does not make sense. We can make a separate definition, however, agreeing that $\mathbf{n}$ is still to be drawn as a unit vector 90° counterclockwise from $\mathbf{T}$ (which in straight-line motion has constant direction).

Before proceeding, let's summarize the steps taken so far:

$$\mathbf{a} = \frac{ds}{dt}\frac{d\mathbf{T}}{dt} + \frac{d^2s}{dt^2}\mathbf{T} = \frac{ds}{dt}\left(\frac{d\mathbf{T}}{ds}\frac{ds}{dt}\right) + \frac{d^2s}{dt^2}\mathbf{T} = \left(\frac{ds}{dt}\right)^2\left(\frac{d\mathbf{T}}{d\phi}\frac{d\phi}{ds}\right) + \frac{d^2s}{dt^2}\mathbf{T}$$

$$= \frac{d^2s}{dt^2}\mathbf{T} + \frac{d\phi}{ds}\left(\frac{ds}{dt}\right)^2\mathbf{n}$$

Figure 4
Unit tangent and normal vectors

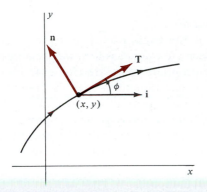

Figure 5
Unit tangent and normal vectors in terms of ϕ

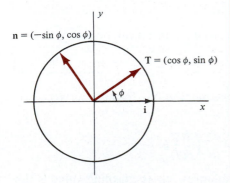

The two components of vector acceleration have now taken shape as

$$\text{tangential: } \frac{d^2s}{dt^2}\mathbf{T} \qquad \text{normal: } \frac{d\phi}{ds}\left(\frac{ds}{dt}\right)^2\mathbf{n}$$

Moreover, we can identify two of the three scalar functions that occur:

$$\frac{d^2s}{dt^2} = \frac{d}{dt}\left(\frac{ds}{dt}\right) = \text{rate of change of speed}$$

$$\left(\frac{ds}{dt}\right)^2 = \text{square of speed}$$

It remains to identify the scalar function $d\phi/ds$.

Think about it in words! Since ϕ is the angle from $\mathbf{i}$ to $\mathbf{T}$, and s is distance measured along the path, the derivative $d\phi/ds$ is the *rate of change of direction with respect to distance*. Appropriate units are (for example) radians per meter. If $d\phi/ds = 0$, then ϕ is constant and the path is straight; in general, $d\phi/ds$ is the rate at which the path turns as we move along it. Note that it has nothing to do with time, but is a purely geometric property of the path, defined in terms of direction angle and arc length. Also note that it is positive or negative depending on whether ϕ is increasing or decreasing as we move along the path in the direction of increasing s. Its absolute value is called *curvature*, as in the following definition.

Suppose that $\mathbf{x} = \mathbf{F}(t)$ is a smooth curve and $\mathbf{T} = \mathbf{F}'(t)/|\mathbf{F}'(t)|$ is the unit tangent vector. If ϕ is the counterclockwise angle from $\mathbf{i}$ to $\mathbf{T}$ and s is distance measured along the curve from some initial point, the **curvature** at t is

$$\kappa = \left|\frac{d\phi}{ds}\right|$$

(Note that the parameter t need not be time; in fact it can be shown that κ is independent of the parameter.)

Now we are ready for the final statement of our formula for acceleration, which we left at the stage

$$\mathbf{a} = \frac{d^2s}{dt^2}\mathbf{T} + \frac{d\phi}{ds}\left(\frac{ds}{dt}\right)^2\mathbf{n}$$

We have identified $d\phi/ds$ as a scalar quantity whose absolute value is the curvature κ; hence

$$\mathbf{a} = \frac{d^2s}{dt^2}\mathbf{T} + \kappa\left(\frac{ds}{dt}\right)^2\mathbf{N} \qquad \text{where } \mathbf{N} = \pm\mathbf{n}$$

the sign depending on whether $d\phi/ds$ is ≥ 0 or < 0. Recall that $\mathbf{n}$ is $90°$ counterclockwise from $\mathbf{T}$ (being one of the two unit normal vectors that can

Figure 6
$N = -n$ when $d\phi/ds < 0$

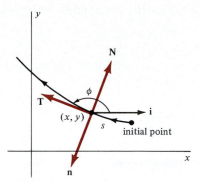

be drawn at a given point). By considering the ways in which a path can bend as it is traversed in one direction or the other, you should convince yourself that **N** always points "inward" (in the direction of bending), as shown in Figure 6. This figure is only one of several that may be drawn; it shows a case where ϕ is decreasing ($d\phi/ds < 0$) and hence $N = -n$. Reverse the direction of motion in this figure (which reverses **T** and **n**) and ϕ becomes an increasing function ($d\phi/ds > 0$). In that case $N = n$, but in both cases **N** is the same inward-directed normal vector.

Tangential and Normal Components of Acceleration

If the position of a moving point at time t is given by the smooth vector function $\mathbf{x} = \mathbf{F}(t)$, the acceleration is

$$\mathbf{a} = \frac{d^2s}{dt^2}\mathbf{T} + \kappa \left(\frac{ds}{dt}\right)^2 \mathbf{N}$$

where s is distance measured along the path from some initial point, κ is curvature, **T** is the unit tangent vector in the direction of motion, and **N** is the inward-directed unit normal vector.

□ **Example 4**   Discuss the force involved in the motion of a car on a curved road.

Solution: Since force is mass times acceleration, we have

$$\mathbf{F} = \left(m\frac{d^2s}{dt^2}\right)\mathbf{T} + m\kappa \left(\frac{ds}{dt}\right)^2 \mathbf{N}$$

The (scalar) tangential component of this force is

$$F_T = m\frac{d^2s}{dt^2} = m\frac{d}{dt}\left(\frac{ds}{dt}\right)$$

Since $\dfrac{d}{dt}\left(\dfrac{ds}{dt}\right) =$ rate of change of speed

this component of the force is what we usually have in mind when we speak of accelerating or decelerating the car (by pushing down or letting up on the "accelerator"). The (scalar) normal component of force is

$$F_N = m\kappa \left(\frac{ds}{dt}\right)^2$$

This is what keeps the car on the road (through friction between the tires and the surface). It is proportional to curvature (which is a property of the curve independent of the motion of the car). The greater the curvature, the more force is needed to keep the car from going off on a tangent. It is also proportional to the square of the speed; the faster the car is going, the more force is required to keep it on the road.

Passengers experience the opposite of these forces in a vivid way, by feeling a push against the back of the seat when the car accelerates and a push in the outward direction of the normal as the car rounds the curve.

□

□ **Example 5**   Find the tangential and normal components of acceleration of the bumblebee in Example 2, Section 15.2.

Solution: The bee's path is given by the vector function

$$\mathbf{x} = (2t^2, t^3) \qquad t \geq 0$$

Since the velocity is $\mathbf{v} = d\mathbf{x}/dt = (4t, 3t^2)$, the speed is

$$\frac{ds}{dt} = |\mathbf{v}| = \sqrt{16t^2 + 9t^4} = t(16 + 9t^2)^{1/2}$$

The (scalar) tangential component of acceleration is

$$a_T = \frac{d^2s}{dt^2} = \frac{1}{2}(16t^2 + 9t^4)^{-1/2}(32t + 36t^3) = \frac{2(8 + 9t^2)}{\sqrt{16 + 9t^2}}$$

To find the (scalar) normal component, $a_N = \kappa(ds/dt)^2$, it appears that we have to know the curvature. There is a way around this, however. Since $\mathbf{T}$ and $\mathbf{N}$ are perpendicular unit vectors, and since

$$\mathbf{a} = a_T\mathbf{T} + a_N\mathbf{N}$$

the magnitude of the acceleration satisfies

$$\boxed{|\mathbf{a}|^2 = a_T{}^2 + a_N{}^2}$$

(Draw a picture!) We have already computed a_T, and we can find $\mathbf{a}$ by differentiating the velocity:

$$\mathbf{a} = \frac{d\mathbf{v}}{dt} = (4, 6t) = 2(2, 3t)$$

Hence $|\mathbf{a}|^2 = 4(4 + 9t^2)$ and

$$a_N{}^2 = |\mathbf{a}|^2 - a_T{}^2 = 4(4 + 9t^2) - \frac{4(8 + 9t^2)^2}{16 + 9t^2}$$

$$= 4\left[\frac{(4 + 9t^2)(16 + 9t^2) - (8 + 9t^2)^2}{16 + 9t^2}\right]$$

$$= \frac{144t^2}{16 + 9t^2}$$

from which

$$a_N = \frac{12t}{\sqrt{16 + 9t^2}} \qquad \text{(no absolute value needed because } a_N \geq 0 \text{ and } t \geq 0)$$

We can even find the curvature now. For we know that $a_N = \kappa(ds/dt)^2$. Since

$$\left(\frac{ds}{dt}\right)^2 = t^2(16 + 9t^2)$$

it follows that

$$\kappa = \frac{a_N}{(ds/dt)^2} = \frac{12}{t(16 + 9t^2)^{3/2}} \qquad (t > 0) \qquad \square$$

Problem Set 15.3

Find the velocity, speed, and acceleration in each of the following motions.

1. $\mathbf{F}(t) = (2t, 3t + 1)$ 2. $\mathbf{F}(t) = (t^2, t)$
3. $\mathbf{F}(t) = (t, t^3)$
4. $\mathbf{F}(t) = (\cos t^2, \sin t^2)$, $t \geq 0$
5. $\mathbf{F}(t) = (a \cos t, b \sin t)$, $a > b > 0$
6. $\mathbf{F}(t) = (\cosh t, \sinh t)$ 7. $\mathbf{F}(t) = (e^t, e^{-t})$
8. The position of a marble at time t is given by

$$\mathbf{F}(t) = (t, \sin t) \qquad -\pi \leq t \leq \pi$$

Sketch the path, and indicate the direction in which it is traversed.

9. In Problem 8 find the maximum and minimum values of the speed, and the points where they occur.

10. In Problem 8 find where the marble experiences no force.

11. In Problem 8 find where the marble experiences a force that is entirely normal. Find the magnitude and direction of the acceleration vector at these points.

12. The graph of $\mathbf{x} = (\sin t, \sin 2t)$, $0 \leq t \leq 2\pi$, is a figure eight. (See Problem 12, Section 15.2.) Assuming that t is time, find where the moving point experiences no force, and where the force is directed from the point toward the origin.

13. In the motion $\mathbf{x} = (t, t^3)$ find the tangential component of acceleration a_T.

14. In Problem 13 find the normal component of acceleration a_N.

15. In Problem 13 find the curvature κ.

16. The position of a particle at time t is $\mathbf{F}(t) = (e^t, e^{-t})$. (See Problem 7 in this section and the last.) Show that the force is always directed from the particle away from the origin.

17. In Problem 16 show that the speed is $ds/dt = \sqrt{2} \cosh 2t$.

18. In Problem 16 show that the tangential component of acceleration is

$$a_T = \frac{2 \sinh 2t}{\sqrt{2} \cosh 2t}$$

On what part of the path is the particle slowing down? speeding up? Where does it change from one to the other?

19. In Problem 16 show that the normal component of acceleration is $a_N = \sqrt{2} \operatorname{sech} 2t$.

20. In Problem 16 find the curvature.

21. The position of a particle at time $t \geq 0$ is

$$\mathbf{F}(t) = (\cos t + t \sin t, \sin t - t \cos t)$$

Show that the speed is $ds/dt = t$.

22. In Problem 21 how far has the particle traveled at time t? (The result should check with Problem 19, Section 7.3, when $t = 2\pi$.)

23. In Problem 21 show that $a_T = 1$ and $a_N = t$.

24. In Problem 21 show that $\kappa = 1/t$.

25. A projectile is fired from the origin with initial speed v_0 and angle of elevation α. (The x axis is the ground and the y axis points upward.) Let g be the magnitude of the acceleration due to gravity and neglect air resistance.

(a) Explain why the acceleration of the projectile is $\mathbf{a} = -g\mathbf{j}$.

(b) Integrate to obtain the velocity

$$\mathbf{v} = -gt\mathbf{j} + \mathbf{v}_0 \qquad \text{where } \mathbf{v}_0 = v_0(\cos \alpha, \sin \alpha)$$

(c) Integrate again to obtain the position

$$\mathbf{x} = -\frac{1}{2}gt^2\mathbf{j} + t\mathbf{v}_0$$

(d) Explain why parametric equations of the path are

$$x = (v_0 \cos \alpha)t, \quad y = (v_0 \sin \alpha)t - \frac{1}{2}gt^2$$

and show that it is parabolic.

26. In Problem 25 show that the horizontal range (distance measured on the ground to the point of impact) is

$$R = \frac{v_0^2}{g} \sin 2\alpha$$

27. In Problem 25 show that the altitude reached by the projectile is

$$H = \frac{v_0^2}{2g} \sin^2 \alpha$$

28. A projectile is launched with initial speed 160 ft/sec and angle of elevation 30°. Taking $g = 32$, find **(a)** the time of flight; **(b)** the horizontal range; **(c)** the highest point reached; **(d)** the speed and angle at impact; and **(e)** the equation of the path in rectangular coordinates.

29. Let $v = ds/dt$ be the speed in a smooth motion $\mathbf{x} = \mathbf{F}(t)$.
  (a) Explain why $\mathbf{v} \cdot \mathbf{T} = v$ and $\mathbf{v} \cdot \mathbf{N} = 0$.
  (b) Explain why $\mathbf{a} \cdot \mathbf{T} = dv/dt$ and $\mathbf{a} \cdot \mathbf{N} = \kappa v^2$.
  (c) Show that $\mathbf{a} \cdot \mathbf{v} = v(dv/dt)$. Then explain why $\mathbf{a} \cdot \mathbf{v}$ is positive or negative depending on whether the moving point is speeding up or slowing down. What if the speed is constant?

30. If the speed of a particle is constant, the force keeping it in its path must be a direction-changing force only. Why is it geometrically reasonable to conclude that $\mathbf{a}$ and $\mathbf{v}$ are perpendicular? Prove that this conclusion is correct.

31. A particle moves in such a way that its velocity and acceleration vectors are always perpendicular. Explain why its speed must be constant.

32. Under what conditions are $\mathbf{a}$ and $\mathbf{v}$ parallel?

33. Newton did not actually state his Second Law of Motion in the form $\mathbf{F} = m\mathbf{a}$ but in the form.

$$\mathbf{F} = \frac{d}{dt}(m\mathbf{v})$$

(Force is the rate of change of *momentum,* which is mass times velocity.)
  (a) Explain why $\mathbf{F} = m\mathbf{a} + (dm/dt)\mathbf{v}$.
  (b) Why does Newton's law reduce to $\mathbf{F} = m\mathbf{a}$ when the mass is constant? (The mass is not necessarily constant! A rocket burning fuel, for example, loses mass as it travels. In relativistic mechanics the mass of an object increases with speed.)

34. Newton's First Law of Motion says that if the (net) force on a moving object is zero, the object travels in a straight line with constant speed. Assuming that the path of the object is the graph of a smooth vector function $\mathbf{x} = \mathbf{F}(t)$, prove this law as follows.
  (a) Why does zero force imply that $\mathbf{F}''(t) = \mathbf{0}$ for all t?
  (b) Explain why it follows that $\mathbf{F}'(t)$ is a constant vector, say $\mathbf{F}'(t) = \mathbf{m}$ for all t. Why is $\mathbf{m} \neq \mathbf{0}$? Why is the speed constant?
  (c) Show that $\mathbf{F}(t) = t\mathbf{m} + \mathbf{x}_0$ for all t, where $\mathbf{x}_0$ is some constant vector. Why is the path a straight line? (See Problems 23 and 24 in the last section.)

35. Prove Newton's First Law of Motion differently, as follows.
  (a) Why does zero force imply that the tangential and normal components of acceleration are both zero for all t?
  (b) Explain why $a_T = 0$ implies constant speed and why $a_N = 0$ implies zero curvature (and hence a straight line path).

36. Explain why the curvature of a smooth path is

$$\kappa = \left| \frac{d\mathbf{T}}{ds} \right|$$

(This will be our *definition* of curvature in three-dimensional space. See the next chapter.)

15.4

MORE ON CURVATURE AND MOTION

In the last section we defined the *curvature* of a smooth curve $\mathbf{x} = \mathbf{F}(t)$ to be

$$\kappa = \left| \frac{d\phi}{ds} \right|$$

where ϕ is the angle from $\mathbf{i}$ to the unit tangent vector $\mathbf{T}$ and s is distance measured along the curve. In the case of motion (when the parameter t is time) we showed that curvature is involved in the *normal component of acceleration,* namely

$$a_N = \kappa \left(\frac{ds}{dt} \right)^2$$

This enabled us to find the curvature indirectly (as in Example 5 of the last section) by computing ds/dt and a_N independently and then solving for κ.

Despite our success in doing this, it is important to be able to compute curvature directly in terms of the vector function $\mathbf{x} = \mathbf{F}(t)$. For except in simple cases the definition $\kappa = |d\phi/ds|$ is not very helpful.

Figure 1
Computing the curvature of a circle

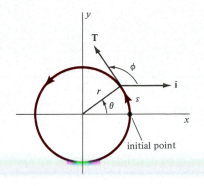
initial point

□ **Example 1**   Show that the curvature of a circle of radius r is $\kappa = 1/r$.

Solution: See Figure 1, in which we show ϕ and s at a typical point of the circle. Since $\phi = \theta + \pi/2$ and $s = r\theta$, we have

$$\phi = \frac{s}{r} + \frac{\pi}{2}$$

and hence $d\phi/ds = 1/r$. In Figure 1 ϕ is increasing and $d\phi/ds > 0$; if the direction of motion is reversed, we find $d\phi/ds = -1/r$. (Confirm!) In any case the curvature is

$$\kappa = \left|\frac{d\phi}{ds}\right| = \frac{1}{r}$$
    □

For any curve but a circle (or a straight line, when $\kappa = 0$), the computation of κ is more complicated. To derive a formula for it, we start by writing

$$\frac{d\phi}{ds} = \frac{d\phi}{dt}\frac{dt}{ds} = \frac{d\phi/dt}{ds/dt}$$

Since we know how to find ds/dt, the problem reduces to computation of $d\phi/dt$. Recall that if $\mathbf{x} = (x, y)$ is the position vector in terms of the parameter t,

$$\frac{d\mathbf{x}}{dt} = (dx/dt, dy/dt)$$

and also

$$\frac{d\mathbf{x}}{dt} = \frac{ds}{dt}\mathbf{T} = \frac{ds}{dt}(\cos\phi, \sin\phi)$$

Equating the components of $d\mathbf{x}/dt$, we have

$$\frac{dx}{dt} = \frac{ds}{dt}\cos\phi \quad \text{and} \quad \frac{dy}{dt} = \frac{ds}{dt}\sin\phi$$

By differentiating these components with respect to t, we can obtain expressions involving $d\phi/dt$. It will simplify the notation if we use primes:

$$x' = s'\cos\phi, \; y' = s'\sin\phi$$

The Product Rule yields

$$x'' = -s'\sin\phi\frac{d\phi}{dt} + s''\cos\phi, \; y'' = s'\cos\phi\frac{d\phi}{dt} + s''\sin\phi$$

To eliminate the terms involving s'', multiply the first equation by $\sin\phi$ and the second by $\cos\phi$, and subtract the first from the second:

$$y''\cos\phi - x''\sin\phi = s'\cos^2\phi\frac{d\phi}{dt} + s'\sin^2\phi\frac{d\phi}{dt} = s'\frac{d\phi}{dt}$$

Since s' is never zero (why?), we have

$$\frac{d\phi}{dt} = \frac{y''\cos\phi - x''\sin\phi}{s'}$$

and hence

$$\frac{d\phi}{ds} = \frac{d\phi/dt}{ds/dt} = \frac{y''\cos\phi - x''\sin\phi}{s'^2}$$

But $\cos\phi = x'/s'$ and $\sin\phi = y'/s'$, so

$$\frac{d\phi}{ds} = \frac{x'y'' - y'x''}{s'^3}$$

Since $s' = ds/dt = \sqrt{(dx/dt)^2 + (dy/dt)^2} = (x'^2 + y'^2)^{1/2}$, we find

$$\frac{d\phi}{ds} = \frac{x'y'' - y'x''}{(x'^2 + y'^2)^{3/2}}$$

Thus our formula for curvature (the absolute value of $d\phi/ds$) is as follows.

The curvature of the parametric path $x = f(t)$, $y = g(t)$ is

$$\kappa = \frac{|x'y'' - y'x''|}{(x'^2 + y'^2)^{3/2}}$$

where the primes mean differentiation with respect to the parameter. (Note that the parameter need not be time.)

□ **Example 2**   Find the curvature of the flight of the bumblebee in Example 2, Section 15.2.

Solution: Parametric equations of the bee's path are

$$x = 2t^2, \, y = t^3 \qquad t \geq 0$$

Computing $x' = 4t$, $y' = 3t^2$ and $x'' = 4$, $y'' = 6t$, we find

$$\kappa = \frac{|(4t)(6t) - (3t^2)(4)|}{(16t^2 + 9t^4)^{3/2}} = \frac{12t^2}{t^3(16 + 9t^2)^{3/2}}$$

$$= \frac{12}{t(16 + 9t^2)^{3/2}} \qquad (t > 0)$$

(Note that this checks with our indirect computation of κ in Example 5, Section 15.3.) At $t = 1$ (when the bee flies off on a tangent in search of a flower) the curvature is

$$\kappa = \frac{12}{125} = 0.096$$

The curvature of this path blows up at $t = 0$. Figure 4 in Section 15.2 does not indicate any good reason for this (the path is seemingly normal at the origin). We graphed only the part corresponding to $t \geq 0$, however; if t is unrestricted, the graph has a symmetric branch below the x axis and there is a cusp at the origin. In other words, the curve

$$x = 2t^2, \ y = t^3 \qquad (t \in \mathcal{R})$$

is not smooth at $t = 0$ and we should expect trouble there.   □

Sometimes it is necessary to find the curvature of a graph in nonparametric form $y = f(x)$. We leave it to you in the problem set to derive the following formula.

The curvature of the graph of $y = f(x)$ is

$$\kappa = \frac{|y''|}{(1 + y'^2)^{3/2}}$$

□ **Example 3**   Show that the maximum curvature of a parabola occurs at its vertex, and has the value $\kappa = 1/(2c)$, where c is the distance between its vertex and focus. What is the curvature at points a long way from the vertex?

Solution: Choose the coordinate axes so that an equation of the parabola is $4cy = x^2 \ (c > 0)$. Then

$$4cy' = 2x \qquad y' = \frac{x}{2c} \qquad y'' = \frac{1}{2c}$$

from which
$$\kappa = \frac{1/(2c)}{(1 + x^2/4c^2)^{3/2}}$$

The maximum value of κ is clearly $1/(2c)$ and occurs when $x = 0$ (at the vertex). On the other hand, $\kappa \to 0$ as $x \to \infty$, so the parabola is virtually straight a long way from the vertex.   □

An interesting concept in connection with curvature of a graph is the **circle of curvature.** This is the circle tangent to the graph at the point in question (with its center in the direction of **N**) and having the same curvature as the graph. Since the curvature of a circle is the reciprocal of its radius (Example 1), the radius of the circle of curvature is $r = 1/\kappa$, where κ is the

Figure 2
Circle of curvature

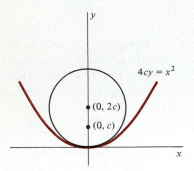

curvature of the graph. (This is called the *radius of curvature* of the graph.) See Figure 2, in which we show the parabola $4cy = x^2$ and the circle of curvature at the origin. Its radius (from Example 3) is $r = 1/\kappa = 2c$.

Because the circle is in such intimate contact with the parabola near the origin (having the same slope and curvature at $x = 0$), it is sometimes called the **osculating circle.** Feeble, to be sure, but how often do we have a chance to make jokes in calculus?

□ **Example 4**   Find the tangential and normal components of acceleration in the motion $x = 3 \cos t$, $y = 2 \sin t$ (the elliptical path shown in Figure 3 of the last section).

Solution: In Example 2, Section 15.3, we found $ds/dt = \sqrt{5 \sin^2 t + 4}$. Hence the tangential component of acceleration is

$$a_T = \frac{d^2 s}{dt^2} = \frac{5 \sin t \cos t}{\sqrt{5 \sin^2 t + 4}}$$

At this point we have a choice. We can compute a_N indirectly (as in Example 5, Section 15.3) using the formula $|\mathbf{a}|^2 = a_T^2 + a_N^2$. Or we can find the curvature and use the definition $a_N = \kappa(ds/dt)^2$. To illustrate the latter approach, we compute

$$x' = -3 \sin t, \quad y' = 2 \cos t$$

and

$$x'' = -3 \cos t, \quad y'' = -2 \sin t$$

from which

$$\kappa = \frac{|(-3 \sin t)(-2 \sin t) - (2 \cos t)(-3 \cos t)|}{(9 \sin^2 t + 4 \cos^2 t)^{3/2}} = \frac{6}{(5 \sin^2 t + 4)^{3/2}}$$

Hence

$$a_N = \frac{6}{\sqrt{5 \sin^2 t + 4}}$$

A particle traveling in this elliptical path experiences its greatest normal force (mass times acceleration) when $\sin^2 t = 0$, which corresponds to the vertices $(\pm 3, 0)$ of the ellipse. In Example 2 of Section 15.3 we found these to be the points where the speed is smallest. The formula

$$a_N = \kappa \left(\frac{ds}{dt} \right)^2$$

suggests large normal force corresponding to high speed, but in this case the large curvature at the vertices of the ellipse is the dominating factor.   □

Problem Set 15.4

1. Show that the curvature of a straight line is zero by making use of the fact that the angle ϕ from $\mathbf{i}$ to $\mathbf{T}$ is constant.

2. Every straight line has parametric equations of the form

$$x = at + x_0, \ y = bt + y_0 \quad (a \text{ and } b \text{ not both zero})$$

(See Problem 23, Section 15.2.) Use these equations to show that the curvature of a straight line is zero.

3. Show that the curvature of a nonvertical straight line is zero by making use of the equation $y = mx + b$.

4. Explain why a path with zero curvature at every point must lie in a straight line.

5. Use the parametric equations

$$x = r \cos t, \ y = r \sin t$$

of a circle of radius r and center at the origin to show that the curvature is $\kappa = 1/r$.

6. An electric train set is often laid out by joining a straight section of track to a section that is part of a circle. The motion of the train past this junction is not smooth, but involves a noticeable jerk. Why?

7. Derive the formula $\kappa = |y''|/(1 + y'^2)^{3/2}$ for the curvature of the graph of $y = f(x)$. *Hint:* Parametric equations are $x = t, \ y = f(t)$.

8. At what points of a curve $y = f(x)$ is the curvature given by $\kappa = |y''|$?

9. Show that the circle of curvature at the point $(0,1)$ of the graph of $y = \cos x$ is the unit circle $x^2 + y^2 = 1$. Draw a picture showing the graph and this circle.

10. Show that the curvature of the graph of $y = \cosh x$ is $\kappa = \operatorname{sech}^2 x$. What are the center and radius of the circle of curvature at the point $(0,1)$? Draw a picture showing the graph and this circle.

11. Show that the radius of curvature of the hyperbola $xy = 1$ at $(1,1)$ is $\sqrt{2}$. Draw a picture showing the graph and the circle of curvature at this point.

12. Find the center and radius of the circle of curvature at the point $(0,1)$ of the graph of $y = e^x$.

13. At what point of the graph of $y = e^x$ is the curvature largest?

14. At what point of the graph of $y = \ln x$ is the curvature largest? Why do the points found in this problem and the preceding problem have the same coordinates in reverse order?

15. Parametric equations of the horizontal ellipse $x^2/a^2 + y^2/b^2 = 1$ are

$$x = a \cos t, \ y = b \sin t \quad a > b > 0$$

Show that the curvature is

$$\kappa = \frac{ab}{(b^2 + c^2 \sin^2 t)^{3/2}}$$

where c is the distance from center to focus. Why should this reduce to $\kappa = 1/a$ if $a = b$? Does it?

16. In Problem 15 at what points of the ellipse is the curvature greatest? smallest? Show that the radius of curvature at these points is b^2/a and a^2/b, respectively.

17. When a wheel of radius a rolls on level ground, a point on its rim traces out the *cycloid*

$$x = a(\theta - \sin \theta), \ y = a(1 - \cos \theta)$$

(See Problem 37, Section 7.3, and Problem 13, Section 15.2.) Show that the curvature of the cycloid is $\kappa = 1/\sqrt{8ay}$.

18. In Problem 17 draw a picture showing the wheel and the cycloid when the point on the rim is at the top of the wheel.

    (a) What would you guess is the radius of curvature of the cycloid at this point?

    (b) Use Problem 17 to show that the answer to part (a) is $4a$. Did you guess correctly?

19. Show that the curvature of the hypocycloid

$$x = a \cos^3 t, \ y = a \sin^3 t \quad 0 \le t \le 2\pi$$

is

$$\kappa = \frac{2}{3a} |\csc 2t|$$

This is undefined at $t = 0, \ \pi/2, \ \pi, \ 3\pi/2, \ 2\pi$. What is the behavior of the curve at the points corresponding to these values?

20. Suppose that an object of mass m is whirled in a circle of radius r at constant speed v.

    (a) Explain why the force on the object is directed toward the center of the circle.

    (b) Show that the magnitude of the force is mv^2/r.

21. Suppose that a car on a circular track can go up to 50 mph without rolling over. How fast can it go (without rolling) on a circular track of twice the radius?

22. Suppose that a car on the parabolic track $4cy = x^2 \ (c > 0)$ will roll over if its normal component of acceleration exceeds c units. How fast can it go as it passes the vertex?

23. It is intuitively clear that if a passenger in a car experiences no sideways force, the car is traveling in a straight line. Give a mathematical reason.

24. A particle moves along the parabola $y = x^2$ with constant speed 2. Explain why $a_T = 0$.

25. In Problem 24 show that $a_N = 8/(1 + 4x^2)^{3/2}$.

26. A particle moves around the ellipse $x^2/16 + y^2/9 = 1$ with constant speed 3. What are the maximum and minimum values of the magnitude of its acceleration? *Hint:* Parametric equations are $x = 4 \cos u, y = 3 \sin u$, but u is not time.

27. The position of a particle at time $t \geq 0$ is

$$F(t) = (\cos t^2, \sin t^2)$$

(a) Show that the speed is $ds/dt = 2t$.

(b) Show that at time t the particle has traveled a distance $s = t^2$.

28. In Problem 27 show that the tangential and normal components of acceleration are $a_T = 2$ and $a_N = 4t^2$.

29. In Problem 27 the graph of F is a circle, but the parametrization is different than in Problem 5. Confirm that the curvature is still the reciprocal of the radius. (This illustrates the fact that curvature is an intrinsic property of the path, independent of the parameter.)

30. The coordinates (x, y) of a particle at time t are

$$x = a \cos^3 t, \quad y = a \sin^3 t \qquad 0 \leq t \leq \pi/2$$

Show that the speed is $ds/dt = \frac{3}{2}a \sin 2t$ and use the result to find the distance traveled by the particle. (This should check with Problem 35, Section 7.3.)

31. In Problem 30 show that the tangential component of acceleration is $a_T = 3a \cos 2t$.

32. In Problem 30 show that the normal component of acceleration is $a_N = \frac{3}{2}a \sin 2t$. *Hint:* Use Problem 19.

33. The position of a particle at time t is $F(t) = (\cosh t, \sinh t)$. Show that the force is always directed from the particle away from the origin.

34. In Problem 33 show that the speed is $ds/dt = \sqrt{1 + 2 \sinh^2 t}$. What is the minimum speed and where does it occur?

35. In Problem 33 show that the tangential component of acceleration is

$$a_T = \frac{2 \sinh t \cosh t}{\sqrt{1 + 2 \sinh^2 t}}$$

36. In Problem 33 show that the curvature is

$$\kappa = \frac{1}{(1 + 2 \sinh^2 t)^{3/2}}$$

Sketch the curve and the circle of curvature at the point where κ is largest.

37. In Problem 33 find the normal component of acceleration.

38. The coordinates of a particle at time t are

$$x = \sin t, \ y = \cos 2t, \ -\pi/2 \leq t \leq \pi/2$$

Explain why the path is the parabolic arc $y = 1 - 2x^2$, $-1 \leq x \leq 1$. Sketch the path and indicate the direction in which it is traversed.

39. In Problem 38 show that the speed is $ds/dt = \cos t \sqrt{1 + 16 \sin^2 t}$.

40. In Problem 38 find the distance traveled by the particle. *Hint:* Use the formula

$$\int \sqrt{x^2 + a^2} \, dx$$
$$= \frac{x}{2} \sqrt{x^2 + a^2} + \frac{a^2}{2} \ln (x + \sqrt{x^2 + a^2}) + C$$

41. In Problem 38 use the equation $y = 1 - 2x^2$ to find the curvature, and use the result to show that the normal component of acceleration is

$$a_N = \frac{4 \cos^2 t}{\sqrt{1 + 16 \sin^2 t}}$$

42. The position of a particle at time $t \geq 0$ is $F(t) = (t \cos t, t \sin t)$. Show that the speed is $ds/dt = \sqrt{t^2 + 1}$.

43. In Problem 42 how far has the particle traveled at time t? (Use the integration formula given in Problem 40.)

44. In Problem 42 show that

$$\kappa = \frac{t^2 + 2}{(t^2 + 1)^{3/2}}$$

What does κ approach as $t \to \infty$? Is this consistent with the fact that the path is the spiral $r = \theta, \theta \geq 0$? (See Problem 10, Section 15.2.)

45. In Problem 42 find the tangential and normal components of acceleration.

46. The position of a particle at time t is

$$F(t) = (e^t \cos t, e^t \sin t)$$

Show that the speed is $ds/dt = \sqrt{2}e^t$.

47. In Problem 46 the path is the spiral $r = e^\theta$. (See Problem 11, Section 15.2.) Explain why it is reasonable to say that at time t the particle has traveled a distance

$$s = \int_{-\infty}^{t} \sqrt{2}e^u \, du$$

and show that the result is $s = \sqrt{2}e^t$.

48. In Problem 46 show that the curvature is $\kappa = 1/s$. What is κ when the particle is near the origin? far away from the origin?

49. In Problem 46 show that the tangential and normal components of acceleration are $a_T = a_N = \sqrt{2}e^t$.

50. In Problem 46 the position vector and tangent vector make a constant angle of 45°. (See Problem 14, Section 15.2.) Explain why the position vector and acceleration vector are always perpendicular. Draw a picture showing these three vectors at a typical point of the spiral.

51. The coordinates of a particle at time t are

$$x = \frac{1}{\sqrt{1+t^2}}, y = \frac{t}{\sqrt{1+t^2}}$$

Show that the speed is $ds/dt = 1/(1+t^2)$.

52. In Problem 51 explain why, as t runs through $(-\infty,\infty)$, the particle traverses an arc joining $(0,-1)$ and $(0,1)$ without stopping or changing direction. Are the endpoints of the arc on the path? What is the speed of the particle when it is near these points? Where does it have maximum speed?

53. In Problem 51 explain why it is reasonable to find the length of the path by evaluating the improper integral

$$\int_{-\infty}^{\infty} \frac{dt}{1+t^2}$$

and show that the result is π.

54. In Problem 51 show that the tangential component of acceleration is

$$a_T = \frac{-2t}{(1+t^2)^2}$$

On what part of the path is the particle speeding up? slowing down? Where does it change from one to the other?

55. In Problem 51, without finding the curvature, show that $a_N = 1/(1+t^2)^2$. Then conclude that $\kappa = 1$.

56. In Problem 51 the path is the right-hand half of the unit circle (excluding the endpoints). (See Problem 9, Section 15.2.) Use this fact to confirm the length π and curvature 1.

15.5
PLANETARY MOTION (Optional)

Modern astronomy may be said to have begun in 1609 when Johann Kepler (1571–1630) published his first two laws of planetary motion. His first law says that *each planet travels in an elliptical orbit with the sun at one focus,* while the second says that *the radius vector from the sun to the planet sweeps out equal areas in equal times.* Our objective in this section is to derive these laws from Newton's law of gravitational attraction.

Suppose that the only force on a planet is the gravitational attraction of the sun. As we will prove in Chapter 16 (when we discuss vectors in three-dimensional space), this implies that the orbit of the planet lies in a plane containing the sun. We impose a polar coordinate system on the plane, with the sun at the origin and the perihelion of the planet on the positive x axis. (See Figure 1.) Measure the time t from the perihelion, that is, take $t = 0$ when the planet is closest to the sun.

The coordinates (r,θ) of the planet are functions of t; the position vector of the planet is

$$\mathbf{r} = r\mathbf{u}$$

where $\mathbf{u} = (\cos\theta,\sin\theta)$ is the *radial unit vector,* that is, the vector of unit length in the direction of $\mathbf{r}$.

Newton's law of gravitational attraction says that the force exerted on the planet by the sun is

$$\mathbf{F} = -\frac{km}{r^2}\mathbf{u}$$

where m is the mass of the planet and k is a positive constant. (In other words the force is directed toward the sun and its magnitude is proportional

Figure 1
Orbit of planet around the sun

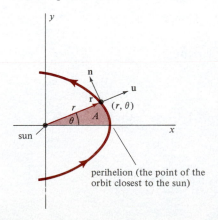

perihelion (the point of the orbit closest to the sun)

to the mass of the planet and inversely proportional to the square of the distance between the planet and the sun.)

Newton's Second Law of Motion says that

$$\mathbf{F} = m\mathbf{a}$$

where $\mathbf{a}$ is the acceleration of the planet. Hence

$$m\mathbf{a} = -\frac{km}{r^2}\mathbf{u}$$

that is,

$$\frac{d\mathbf{v}}{dt} = -\frac{k}{r^2}\mathbf{u} \tag{1}$$

where $\mathbf{v}$ is the velocity. This equation is the heart of the matter, as we will see. To explain its implications, we need an expression for $d\mathbf{v}/dt$ in terms of the perpendicular unit vectors

$$\mathbf{u} = (\cos\theta, \sin\theta)$$

and

$$\mathbf{n} = \left(\cos\left(\theta + \frac{\pi}{2}\right), \sin\left(\theta + \frac{\pi}{2}\right)\right) = (-\sin\theta, \cos\theta)$$

shown in Figure 1. Starting with

$$\mathbf{r} = r\mathbf{u}$$

we find

$$\mathbf{v} = \frac{d\mathbf{r}}{dt} = r\frac{d\mathbf{u}}{dt} + \frac{dr}{dt}\mathbf{u}$$

But

$$\frac{d\mathbf{u}}{dt} = \left(-\sin\theta\,\frac{d\theta}{dt}, \cos\theta\,\frac{d\theta}{dt}\right) = \omega(-\sin\theta, \cos\theta) = \omega\mathbf{n}$$

where $\omega = d\theta/dt$ is the angular velocity. Hence

$$\mathbf{v} = \frac{dr}{dt}\mathbf{u} + r\omega\mathbf{n} \tag{2}$$

and

$$\frac{d\mathbf{v}}{dt} = \frac{dr}{dt}\frac{d\mathbf{u}}{dt} + \frac{d^2r}{dt^2}\mathbf{u} + (r\omega)\frac{d\mathbf{n}}{dt} + \frac{d}{dt}(r\omega)\mathbf{n}$$

Since

$$\frac{d\mathbf{n}}{dt} = \left(-\cos\theta\,\frac{d\theta}{dt}, -\sin\theta\,\frac{d\theta}{dt}\right) = -\omega\mathbf{u}$$

we find

$$\frac{d\mathbf{v}}{dt} = \omega \frac{dr}{dt}\mathbf{n} + \frac{d^2r}{dt^2}\mathbf{u} - r\omega^2\mathbf{u} + \left(r\frac{d\omega}{dt} + \omega\frac{dr}{dt}\right)\mathbf{n}$$

$$= \left(\frac{d^2r}{dt^2} - r\omega^2\right)\mathbf{u} + \left(r\frac{d\omega}{dt} + 2\omega\frac{dr}{dt}\right)\mathbf{n}$$

Thus Equation (1) becomes

$$\left(\frac{d^2r}{dt^2} - r\omega^2\right)\mathbf{u} + \left(r\frac{d\omega}{dt} + 2\omega\frac{dr}{dt}\right)\mathbf{n} = -\frac{k}{r^2}\mathbf{u}$$

Equating the components in the direction of **n**, we obtain the differential equation

$$r\frac{d\omega}{dt} + 2\omega\frac{dr}{dt} = 0 \tag{3}$$

Equation (3) leads directly to Kepler's Second Law, that "equal areas are swept out in equal times." For we know from Sec. 14.2 that the area swept out by **r** in any time interval $[0,t]$ is

$$A = \int_0^{f(t)} \frac{1}{2}r^2 \, d\theta$$

where $f(t)$ is the value of θ at time t. (See Figure 1.) Kepler's Law is equivalent to the statement that A changes at a constant rate, so the question is whether

$$\frac{dA}{dt} = \text{constant, that is, } \frac{d^2A}{dt^2} = 0$$

By the Fundamental Theorem of Calculus (and the Chain Rule) we have

$$\frac{dA}{dt} = \frac{1}{2}r^2 f'(t) = \frac{1}{2}r^2\frac{d\theta}{dt} = \frac{1}{2}r^2\omega$$

and

$$\frac{d^2A}{dt^2} = \frac{1}{2}\left(r^2\frac{d\omega}{dt} + 2r\omega\frac{dr}{dt}\right) = \frac{r}{2}\left(r\frac{d\omega}{dt} + 2\omega\frac{dr}{dt}\right) = 0$$

by Equation (3).

Kepler's First Law (that the orbit is an ellipse with the sun at one focus) is harder. To derive it, we must find a way to write r in terms of θ, our objective being an equation of the form

$$r = \frac{ep}{1 + e\cos\theta}$$

For this represents a conic with focus at the origin (which must be an ellipse because the orbit is obviously neither a parabola nor a hyperbola).

We begin by returning to Equation (1),

$$\frac{d\mathbf{v}}{dt} = -\frac{k}{r^2}\mathbf{u}$$

with the object of integrating it. Since

$$\frac{d\mathbf{n}}{dt} = -\omega\mathbf{u}$$

we may write

$$\frac{d\mathbf{v}}{dt} = -\frac{k}{r^2}\left(-\frac{1}{\omega}\frac{d\mathbf{n}}{dt}\right) = \frac{k}{r^2\omega}\frac{d\mathbf{n}}{dt}$$

But

$$r^2\omega = 2\frac{dA}{dt}$$

which we just showed is constant, say $r^2\omega = a$. Hence

$$\frac{d\mathbf{v}}{dt} = \frac{k}{a}\frac{d\mathbf{n}}{dt}$$

and we may integrate to obtain

$$\mathbf{v} = \frac{k}{a}\mathbf{n} + \mathbf{C} \qquad \text{(where } \mathbf{C} \text{ is a vector constant)} \qquad (4)$$

To evaluate $\mathbf{C}$, we use the values of the vector functions $\mathbf{v}(t)$ and $\mathbf{n}(t)$ at $t = 0$. Figure 1 shows that

$$\mathbf{n}(0) = \mathbf{j}$$

Equation (2) says that

$$\mathbf{v}(t) = r'(t)\mathbf{u}(t) + r(t)\omega(t)\mathbf{n}(t)$$

Since r is a minimum at $t = 0$ (because the planet is at perihelion), we have $r'(0) = 0$, so

$$\mathbf{v}(0) = r(0)\omega(0)\mathbf{n}(0) = b\mathbf{j}, \text{ where } b = r(0)\omega(0)$$

Hence

$$\mathbf{C} = \mathbf{v}(0) - \frac{k}{a}\mathbf{n}(0) = b\mathbf{j} - \frac{k}{a}\mathbf{j} = c\mathbf{j} \qquad \left(c = b - \frac{k}{a}\right)$$

and (4) becomes

$$\mathbf{v} = \frac{k}{a}\mathbf{n} + c\mathbf{j} \qquad (5)$$

According to Equation (2) the (scalar) component of $\mathbf{v}$ in the direction of $\mathbf{n}$ is $r\omega$, whereas Equation (5) says it is

$$\mathbf{v} \cdot \mathbf{n} = \frac{k}{a}(\mathbf{n} \cdot \mathbf{n}) + c(\mathbf{j} \cdot \mathbf{n})$$

$$= \frac{k}{a} + c \cos \theta \qquad [\text{because } \mathbf{n} = (-\sin \theta, \cos \theta)]$$

Hence

$$r\omega = \frac{k}{a} + c \cos \theta$$

Multiplying by r, we have

$$r^2\omega = r\left(\frac{k}{a} + c \cos \theta\right)$$

But $r^2\omega = a$, so

$$a = r\left(\frac{k}{a} + c \cos \theta\right) = \frac{k}{a}r(1 + e \cos \theta) \qquad \left(\text{where } e = \frac{ac}{k}\right)$$

Solving for r, we find

$$r = \frac{a^2/k}{1 + e \cos \theta} = \frac{ep}{1 + e \cos \theta} \qquad \left(\text{where } p = \frac{a^2}{ek}\right)$$

This is the equation of a conic we sought; the proof of Kepler's First Law is complete.

Problem Set 15.5

Suppose that nothing is known about the orbit of a planet except that it lies in a plane containing the sun. Problems 1 through 5 are designed to help you derive Newton's inverse square law of gravitation from Kepler's laws.

1. Assume Kepler's Second Law, that the vector from the sun to the planet sweeps out area at a constant rate. Prove that the acceleration vector is radial, as follows.

   (a) Examine the argument in the text to confirm that the equations

   $$\frac{d\mathbf{v}}{dt} = \left(\frac{d^2r}{dt^2} - r\omega^2\right)\mathbf{u} + \left(r\frac{d\omega}{dt} + 2\omega\frac{dr}{dt}\right)\mathbf{n}$$

   and

   $$\frac{d^2A}{dt^2} = \frac{r}{2}\left(r\frac{d\omega}{dt} + 2\omega\frac{dr}{dt}\right)$$

   are true without any assumptions about force.

   (b) Explain why Kepler's Second Law implies that

   $$\mathbf{a} = \left(\frac{d^2r}{dt^2} - r\omega^2\right)\mathbf{u}$$

   (Thus the acceleration vector is radial regardless of the shape of the orbit, simply as a consequence of Kepler's Second Law.)

2. Assume Kepler's First Law, that the orbit is an ellipse with the sun at one focus, say

   $$r = \frac{ep}{1 + e \cos \theta} \qquad (p > 0 \text{ and } 0 < e < 1)$$

   Prove that the acceleration vector points from the planet toward the sun, with magnitude inversely proportional to the square of the distance between the planet and the sun, as follows.

   (a) Explain why Kepler's Second Law implies that $r^2\omega$ is constant, say $r^2\omega = a$.

(b) Differentiate the formula for r in terms of θ (using the fact that $r^2\omega = a$) to show that

$$\frac{dr}{dt} = \frac{a}{p}\sin\theta \quad \text{and} \quad \frac{d^2r}{dt^2} = \frac{a}{p}\omega\cos\theta$$

(c) Use part (b) (and $r^2\omega = a$) to show that

$$\frac{d^2r}{dt^2} - r\omega^2 = -\frac{k}{r^2} \quad \left(k = \frac{a^2}{ep}\right)$$

and hence (from Problem 1)

$$\mathbf{a} = -\frac{k}{r^2}\mathbf{u}$$

3. Let T be the period of the planet (the time required to complete one orbit) and let $2A$ and $2B$ be the lengths of the major and minor diameters of its orbit.

   (a) Explain why the area of the orbit is $\frac{1}{2}aT$. *Hint:* $a = r^2\omega$, which is twice the rate of change of area swept out by the position vector $\mathbf{r}$. (Why?)

   (b) The area of the orbit is also πAB. (See Problem 13, Section 6.2.) Use appropriate data from Section 14.4 to show that

$$A = \frac{ep}{1-e^2} \quad \text{and} \quad B = \sqrt{1-e^2}\,A$$

and conclude that

$$T = \left(\frac{2\pi ep}{a\sqrt{1-e^2}}\right)A$$

(c) Explain why

$$\frac{T^2}{A^3} = \frac{4\pi^2}{k}$$

where k is the constant of proportionality in Problem 2.

4. The number $k = a^2/ep$ is constant for a given planet, but appears to depend on the orbital data for that planet. Kepler's Third Law, however, says that T^2 is proportional to A^3. Why does it follow that k is the same for all planets?

5. At this point you are struck by an apple falling from a tree, and by the idea that force is mass times acceleration. Show that

$$\mathbf{F} = -\frac{km}{r^2}\mathbf{u}$$

where $\mathbf{F}$ is the force exerted by the sun on a planet of mass m and k is the same for all planets. (This is the law of gravitational attraction that we used in the text to derive Kepler's first two laws.)

True-False Quiz

1. The vectors $\mathbf{i} - 2\mathbf{j}$ and $4\mathbf{i} + 2\mathbf{j}$ are perpendicular.

2. If θ is the angle between the unit vectors $\mathbf{u}$ and $\mathbf{v}$, then $\cos\theta = \mathbf{u}\cdot\mathbf{v}$.

3. A unit vector in the same direction as $\mathbf{x} = (3,-4)$ is $\mathbf{u} = (-\frac{3}{5},\frac{4}{5})$.

4. If a is a scalar and $\mathbf{x}$ and $\mathbf{y}$ are vectors in $\mathcal{R}^2$, then $a(\mathbf{x} + \mathbf{y}) = a\mathbf{x} + a\mathbf{y}$.

5. The vector from $P_1(-1,3)$ to $P_2(1,4)$ is $\overrightarrow{P_1P_2} = (-2,-1)$.

6. If $\mathbf{x}_1\cdot\mathbf{x}_2 = 0$, then either $\mathbf{x}_1 = \mathbf{0}$ or $\mathbf{x}_2 = \mathbf{0}$.

7. If P_1 and P_2 are two points of the coordinate plane, then $\overrightarrow{P_1P_2} + \overrightarrow{P_2P_1} = \mathbf{0}$.

8. The graph of the vector function $\mathbf{F}(t) = (2\sin t, 3\cos t)$ is an ellipse.

9. The vector $(3,2)$ is parallel to the tangent to the graph of $x = t^3$, $y = t^2$ at the point $(1,1)$.

10. If $\mathbf{x} = \mathbf{F}(t)$ is differentiable, then $\frac{d}{dt}|\mathbf{x}|^2 = 2\mathbf{x}\cdot\frac{d\mathbf{x}}{dt}$.

11. If $\mathbf{T} = (\cos\phi, \sin\phi)$ is the unit tangent vector (in a smooth motion), then $d\mathbf{T}/d\phi = (-\sin\phi, \cos\phi)$ is the inward-directed unit normal vector.

12. The curvature of the path $x = t + 1$, $y = 2t$ is zero.

13. The curvature of the path

$$x = 2\cos\pi t, \quad y = 2\sin\pi t$$

is $\frac{1}{2}$.

14. If v is the speed of a particle moving along a smooth curve, then dv/dt is the magnitude of the acceleration.

15. In a smooth plane motion the tangential component of acceleration is zero at points of maximum or minimum speed.

Additional Problems

If $\mathbf{x}_1 = (2,5)$ and $\mathbf{x}_2 = (-3,1)$, find the following.

1. $3\mathbf{x}_1 - 2\mathbf{x}_2$ 2. $\mathbf{x}_1 \cdot \mathbf{x}_2$

3. The angle between $\mathbf{x}_1$ and $\mathbf{x}_2$.

4. If $\mathbf{x}_1$ has length 4 and $\mathbf{x}_2 = -\frac{1}{2}\mathbf{x}_1$, find $\mathbf{x}_1 \cdot \mathbf{x}_2$.

5. The vertices of a triangle are $A(1,4)$, $B(0,2)$, $C(3,-2)$. Find the angle at A.

6. Find all unit vectors perpendicular to $\mathbf{x} = (4,-3)$.

7. The formula $|c\mathbf{x}| = |c|\,|\mathbf{x}|$ is derived in Section 15.1 without using coordinates. Give an argument using $\mathbf{x} = (x,y)$ in coordinate form.

In each of the following, find $d\mathbf{x}/dt$ in two ways.

8. $\mathbf{x} = (u^3, u^2)$, where $u = e^{-t}$

9. $\mathbf{x} = (\cosh u, \sinh u)$, where $u = 1 - t$

10. $\mathbf{x} = t(e^t, e^{-t})$ 11. $\mathbf{x} = \ln t(t, t^2)$

In each of the following, find $f'(t)$ in two ways.

12. $f(t) = (\cos 2t, \sin 2t) \cdot (e^{2t}, e^{-2t})$

13. $f(t) = (t^2, t - 1) \cdot (e^{-t}, e^t)$

14. If $\mathbf{x} = (u^2, 1 - u)$ and u is a function of t such that $du/dt = 3$ when $u = 1$, find $d\mathbf{x}/dt$ when $u = 1$.

15. If $\mathbf{x}$ is a differentiable function of t, show that

$$\frac{d}{dt}|\mathbf{x}| = \frac{\mathbf{x}}{|\mathbf{x}|} \cdot \frac{d\mathbf{x}}{dt}$$

16. The position vector of a particle at time t is $\mathbf{x} = e^t(t, e^{-t})$. Find a vector that is tangent to the path of the particle at $t = 0$. What is an equation of the path in rectangular coordinates?

17. The position of a particle at time t is given by $x = 4\cos t$, $y = \sin t$. What is an equation of the path in rectangular coordinates? Sketch the path and indicate the direction of travel.

18. In Problem 17 show that the acceleration vector always points from the particle toward the origin.

19. If $\mathbf{v}$ is the velocity vector and $\mathbf{T}$ and $\mathbf{N}$ are the unit tangent and normal vectors defined in the text, explain why $\mathbf{v} \cdot \mathbf{T}$ is the speed and $\mathbf{v} \cdot \mathbf{N}$ is zero.

20. If $\mathbf{a}$ is the acceleration vector, show that

$$\mathbf{a} \cdot \mathbf{T} = \frac{d^2s}{dt^2} \quad \text{and} \quad \mathbf{a} \cdot \mathbf{N} = \kappa\left(\frac{ds}{dt}\right)^2$$

21. The position of a particle at time $t \geq 0$ is given by $x = 3\cos t^2$, $y = 3\sin t^2$. Show that the speed is $ds/dt = 6t$.

22. In Problem 21 find how far the particle has traveled at time t.

23. In Problem 21 find the tangential component of acceleration a_T.

24. In Problem 21 find the normal component of acceleration a_N.

25. In Problem 21 find the curvature κ.

26. The position of a particle at time t is given by $x = e^t$, $y = t$. Find an equation of the path in rectangular coordinates, sketch it, and indicate the direction of travel.

27. In Problem 26 suppose that the particle flies off on a tangent when $t = 0$ (and maintains constant speed thereafter). Where will it be when $t = 1$?

28. In Problem 26 find the tangential component of acceleration a_T.

29. In Problem 26 find the curvature κ.

30. In Problem 26 find the normal component of acceleration a_N.

31. The position of a particle at time t is $\mathbf{F}(t) = (3\cosh t, 2\sinh t)$. Find an equation of the path in rectangular coordinates, sketch it, and indicate the direction of travel.

32. In Problem 31 find the minimum speed of the particle. Where does it occur?

33. In Problem 31 find the maximum curvature of the path. Where does it occur? As time goes on, what does the curvature approach?

34. In Problem 31 show that the force on the particle is directed away from the origin.

35. In Problem 31 find the tangential component of acceleration a_T.

36. In Problem 31 find the normal component of acceleration a_N.

37. In Problem 31 on what part of the path is the particle slowing down? speeding up? Where does it change from one to the other?

38. The position of a particle at time $t \geq 0$ is given by $x = t^3 - 3t^2$, $y = t^3 + 3t^2$. How far has the particle traveled when $t = 2$?

39. In Problem 38 find the tangential component of acceleration a_T.

40. In Problem 38 find the normal component of acceleration a_N.

41. In Problem 38 find the curvature κ.

42. The position of a particle at time $t \geq 0$ is given by $x = t^3$,

$y = t^2$. Find an equation of the path in rectangular coordinates, sketch it, and indicate the direction of travel.

43. In Problem 42 find the tangential component of acceleration a_T.

44. In Problem 42 find the normal component of acceleration a_N.

45. In Problem 42 find the curvature at $t > 0$. What is its limit as $t \downarrow 0$?

46. A particle moves with constant speed 3 along a curve whose curvature at a certain point is $\frac{2}{3}$. Find the magnitude of the acceleration at that point.

47. Find the curvature of the graph of $y = 1 - x^2$ and name the center of the circle of curvature at the point $(0,1)$.

48. Find the curvature of the graph of $y = \ln \cos x$, $-\pi/2 < x < \pi/2$. At what point of the graph is the curvature largest? Draw a picture of the graph and the circle of curvature at that point.

49. Find the curvature of the graph of $y = \sqrt{1 - x^2}$ by using the general formula for curvature. Then explain how the result could have been predicted.

50. Show that the curvature of the graph of $y = \sin^{-1} x$ increases in the interval $[0,1]$. What are its maximum and minimum values? Find the center and radius of the circle of curvature at the point $(1,\pi/2)$ and draw a picture showing the graph and this circle.

16 Geometry in Space

Let no one ignorant of geometry enter here.

(Inscription over the door of Plato's Academy)

There is no branch of mathematics, however abstract, which may not some day be applied to phenomena of the real world.

NICOLAI LOBACHEVSKY (1793–1856)

As long as a branch of science offers an abundance of problems, so long is it alive.

DAVID HILBERT (1862–1943)

This chapter is like the last, but refers to a higher dimension. We will discuss coordinates in space (rectangular, cylindrical, and spherical), vector functions (curves in space), and equations of planes and surfaces (analogous to equations of lines and curves in the plane). Although this material is interesting and important in its own right, its primary purpose is to lay the groundwork for multivariable calculus, which begins in Chapter 17.

16.1
COORDINATES AND VECTORS IN SPACE

Just as single real numbers serve as coordinates in one-dimensional space, and pairs of real numbers locate points of the plane, *triples* of real numbers are used as coordinates in three-dimensional space. All we need is a third coordinate line (the z axis) perpendicular to the axes already used in the plane. Its origin is taken at the common origin of the x and y axes, and its positive direction is chosen so that the unit vectors

$$\mathbf{i} = (1, 0, 0), \mathbf{j} = (0, 1, 0), \mathbf{k} = (0, 0, 1)$$

form a "right-handed triple." Figure 1 shows a common arrangement of the axes. Right-handed orientation is most easily understood by thinking of the z axis as a threaded screw. Turning the x and y axes counterclockwise tends to drive the screw upward; had we pointed the positive z axis downward we would have a left-handed system.

Figure 1
Locating a point in space

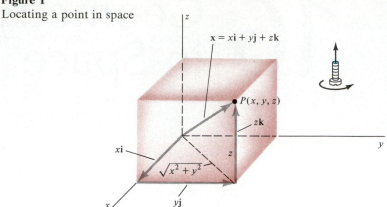

Figure 2
Another right-handed system

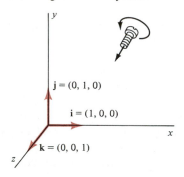

The orientation of a coordinate system is not important for plotting points. But some of our theorems about vectors require a right-handed system, so we will assume such an orientation from the start. The axes do not have to be labeled as in Figure 1, however. The system shown in Figure 2 (which is sometimes more convenient) is also right-handed. (Why?)

A triple of real numbers (x, y, z) locates a point P in space as indicated in Figure 1. The scalar multiples $x\mathbf{i}, y\mathbf{j}, z\mathbf{k}$ may be regarded as sides of a box whose vertex diagonally opposite from the origin is the point P. (We show the case where x, y, and z are all positive in Figure 1. The collection of all such points is called the **first octant.** There are seven others, but they are not ordinarily numbered.)

Conversely, every point P in space determines a unique triple of real numbers (x, y, z). The value of x, for example, is determined by passing a plane through P parallel to the yz plane (the plane of the paper in Figure 1). This plane intersects the x axis in a point whose coordinate in 1-space is x. The values of y and z are found in the same way.

Thus we are justified in calling the set

$$\mathcal{R}^3 = \{(x, y, z): x \in \mathcal{R}, y \in \mathcal{R}, z \in \mathcal{R}\}$$

three-dimensional space, identifying a point with its coordinate representation in the same way as in lower dimensions.

If addition and multiplication by a scalar are defined as in $\mathcal{R}^2$, we may also regard $\mathcal{R}^3$ as the set of *vectors*

$$\mathbf{x} = (x, y, z) = x\mathbf{i} + y\mathbf{j} + z\mathbf{k}$$

The geometric representation may be taken to be an arrow from the origin to $P(x, y, z)$ or from a point $P_1(x_1, y_1, z_1)$ to a point $P_2(x_2, y_2, z_2)$, provided that $x = x_2 - x_1$, $y = y_2 - y_1$, and $z = z_2 - z_1$. In other words,

$$\mathbf{x} = (x, y, z) \qquad \text{and} \qquad \overrightarrow{P_1P_2} = (x_2 - x_1, y_2 - y_1, z_2 - z_1)$$

are the same vector.

The *length* or *magnitude* of $\mathbf{x}$ (and the *distance* between O and P, or between P_1 and P_2) is found by the Pythagorean Theorem applied to the vertical right triangle labeled in Figure 1:

$$|\mathbf{x}|^2 = (\sqrt{x^2 + y^2})^2 + z^2 = x^2 + y^2 + z^2$$

Hence
$$|\mathbf{x}| = \sqrt{x^2 + y^2 + z^2}$$

and
$$d(P_1, P_2) = \sqrt{(x_2 - x_1)^2 + (y_2 - y_1)^2 + (z_2 - z_1)^2}$$

□ **Example 1**   Identify the graph in $\mathfrak{R}^3$ of the equation $x^2 + y^2 + z^2 = r^2$, where r is a positive constant.

Solution: A point $P(x, y, z)$ satisfies the equation if and only if

$$d(O, P) = \sqrt{x^2 + y^2 + z^2} = r$$

Hence the graph is a sphere of radius r and center O. More generally, the equation $(x - x_0)^2 + (y - y_0)^2 + (z - z_0)^2 = r^2$ represents a sphere of radius r and center (x_0, y_0, z_0).   □

Still more generally, an equation in three variables x, y, and z usually represents a *surface* of some kind. Intuitively speaking, a surface in $\mathfrak{R}^3$ (like the surface of the earth in space) is two-dimensional; this is analogous to the situation in $\mathfrak{R}^2$, where we expect an equation in x and y to represent a curve (which is one-dimensional). That raises the question of what represents a curve in $\mathfrak{R}^3$. As we will see, the answer is a vector function. An equation in x, y, and z *does not represent a curve but a surface.* We will discuss curves and surfaces in later sections of this chapter.

The **dot product** of two vectors $\mathbf{x}_1 = (x_1, y_1, z_1)$ and $\mathbf{x}_2 = (x_2, y_2, z_2)$ is defined as in $\mathfrak{R}^2$, namely

$$\mathbf{x}_1 \cdot \mathbf{x}_2 = x_1 x_2 + y_1 y_2 + z_1 z_2$$

The angle θ between (nonzero) vectors $\mathbf{x}_1$ and $\mathbf{x}_2$ is then given by

$$\cos \theta = \frac{\mathbf{x}_1 \cdot \mathbf{x}_2}{|\mathbf{x}_1| \, |\mathbf{x}_2|} \qquad 0 \leq \theta \leq \pi$$

The proof of this formula for $\cos \theta$ proceeds exactly as in Section 15.1, using the Law of Cosines. The same terminology is used as well:

Two (nonzero) vectors $\mathbf{x}_1$ and $\mathbf{x}_2$ are said to be *perpendicular* if $\theta = \pi/2$ and *parallel* if $\theta = 0$ or $\theta = \pi$. They have the *same direction* if $\theta = 0$ (that is, if one is a positive scalar multiple of the other) and *opposite directions* if

$\theta = \pi$ (when one is a negative multiple of the other). We adopt the convention that the zero vector has *any* direction, in which case we can say that $\mathbf{x}_1$ and $\mathbf{x}_2$ are perpendicular $\Leftrightarrow \mathbf{x}_1 \cdot \mathbf{x}_2 = 0$

Other laws involving the dot product are proved as before:

$$\mathbf{x}_1 \cdot \mathbf{x}_2 = \mathbf{x}_2 \cdot \mathbf{x}_1$$

$$\mathbf{x}_1 \cdot (\mathbf{x}_2 + \mathbf{x}_3) = \mathbf{x}_1 \cdot \mathbf{x}_2 + \mathbf{x}_1 \cdot \mathbf{x}_3$$

$$(c\mathbf{x}_1) \cdot \mathbf{x}_2 = c(\mathbf{x}_1 \cdot \mathbf{x}_2) = \mathbf{x}_1 \cdot (c\mathbf{x}_2)$$

$$\mathbf{x} \cdot \mathbf{x} = |\mathbf{x}|^2$$

$\dfrac{\mathbf{x}}{|\mathbf{x}|}$ is a unit vector in the direction of $\mathbf{x}$ (if $\mathbf{x} \neq \mathbf{0}$).

□ **Example 2**   Find the angles of the triangle with vertices $(1, -1, 3)$, $(2, 1, 7)$, and $(4, 2, 6)$.

Solution: Rather than drawing an accurate figure in three-dimensional space (which is often hard), we place the triangle in the plane of the paper, as shown in Figure 3. No attempt is made to orient the points properly, or to get the angles and lengths right; the purpose is merely to help us get started on the problem. (You will often find this kind of schematic diagram to be useful. It is legitimate as long as you are aware of its inaccuracies.)

To find the angle θ at P_1, we compute

$$\vec{P_1 P_2} = (2 - 1, 1 + 1, 7 - 3) = (1, 2, 4)$$
$$\vec{P_1 P_3} = (4 - 1, 2 + 1, 6 - 3) = (3, 3, 3) = 3(1, 1, 1)$$

The angle between $\vec{P_1 P_2}$ and $\vec{P_1 P_3}$ is the same as the angle between $(1, 2, 4)$ and $(1, 1, 1)$ and satisfies

$$\cos \theta = \frac{(1, 2, 4) \cdot (1, 1, 1)}{|(1, 2, 4)||(1, 1, 1)|} = \frac{1 + 2 + 4}{\sqrt{21}\sqrt{3}} = \frac{7}{3\sqrt{7}} = \frac{\sqrt{7}}{3}$$

Hence $\theta = \cos^{-1}(\sqrt{7}/3) \approx 0.49$ (or about $28°$). Since

$$\vec{P_2 P_1} = -(1, 2, 4) \quad \text{and} \quad \vec{P_2 P_3} = (4 - 2, 2 - 1, 6 - 7) = (2, 1, -1)$$

we have $\vec{P_2 P_1} \cdot \vec{P_2 P_3} = -(2 + 2 - 4) = 0$, so the angle at P_2 is $90°$. Hence the angle at P_3 is (approximately) $90° - 28° = 62°$.   □

Figure 3
Schematic diagram of triangle in space

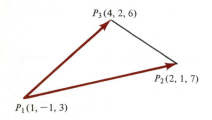

$P_3(4, 2, 6)$

$P_2(2, 1, 7)$

$P_1(1, -1, 3)$

Problem Set 16.1

Describe the following sets in $\Re^3$.

1. $\{(x, y, z) : x = 0\}$ **2.** $\{(x, y, z) : y = 0\}$

3. $\{(x, y, z) : z = 0\}$ **4.** $\{(x, y, z) : z = 1\}$

5. $\{(x, y, z) : x = 0 \text{ and } y = 0\}$

6. $\{(x, y, z) : x^2 + y^2 + z^2 = 0\}$

7. $\{(x, y, z) : x^2 + y^2 = 1 \text{ and } z = 0\}$

8. Sketch the box

$$\{(x, y, z): 1 \leq x \leq 2, 2 \leq y \leq 4, 0 \leq z \leq 3\}$$

9. Sketch the tetrahedron (solid with four triangular faces) whose vertices are $(0, 0, 0)$, $(3, 0, 0)$, $(0, 2, 0)$, $(0, 0, 5)$.

10. Sketch the hemisphere

$$\{(x, y, z): x^2 + y^2 + z^2 = 1 \text{ and } z \geq 0\}$$

11. Find the distance between the points $(3, -1, 1)$ and $(2, 0, 5)$.

12. The dimensions of a rectangular room are 9 ft, 12 ft, and 8 ft. Can a pole vaulter put his 16-ft pole in the room and shut the door?

13. Explain why the distance from (x, y, z) to the z axis is $\sqrt{x^2 + y^2}$. What is the distance from (x, y, z) to the x axis? the y axis?

14. Find an equation of the sphere with center $(2, -1, 0)$ and radius 3.

15. Find an equation of the sphere having $(1, -2, 5)$ and $(3, 4, -1)$ as endpoints of a diameter.

16. Find the center and radius of the sphere $x^2 + y^2 + z^2 + 2x - 6z + 6 = 0$.

17. Find the center and radius of the sphere $x^2 + y^2 + z^2 - 6x + 10y - 2 = 0$.

18. The sphere $x^2 + y^2 + z^2 = 9$ intersects the plane $z = 2$ in a circle. What are the center and radius of the circle?

19. Do the points $(1, 3, 0)$, $(-1, 2, 1)$, and $(5, 5, -2)$ lie in a straight line?

20. Show that if P_1, P_2, P_3 are any three points in space, then $\overrightarrow{P_1 P_2} = \overrightarrow{P_1 P_3} + \overrightarrow{P_3 P_2}$.

21. Show that

$$\mathbf{u}_1 = \tfrac{1}{3}(1, 2, 2), \mathbf{u}_2 = \tfrac{1}{3}(2, -2, 1), \mathbf{u}_3 = \tfrac{1}{3}(2, 1, -2)$$

are mutually perpendicular unit vectors.

22. Find a unit vector in the same direction as the vector $\mathbf{x} = (8, -9, 12)$.

23. Find two unit vectors perpendicular to the vectors $\mathbf{x}_1 = (1, 0, 1)$ and $\mathbf{x}_2 = (2, 2, -1)$.

24. The medians of a triangle intersect in the point that is two-thirds of the way from any vertex to the midpoint of the opposite side. Find the point of intersection of the medians of the triangle with vertices $(1, 0, 2)$, $(-1, 3, 1)$, $(3, 5, -1)$.

25. Find the angles of the triangle with vertices $(1, 1, 1)$, $(2, 3, 3)$, $(3, -1, 2)$.

26. Find the acute angle between the diagonal of a cube and an edge.

27. Show that if $\mathbf{T}$, $\mathbf{N}$, and $\mathbf{B}$ are mutually perpendicular unit vectors and if $\mathbf{x} = a\mathbf{T} + b\mathbf{N} + c\mathbf{B}$, then $a = \mathbf{x} \cdot \mathbf{T}$, $b = \mathbf{x} \cdot \mathbf{N}$, and $c = \mathbf{x} \cdot \mathbf{B}$.

28. In the plane every unit vector can be written in the form $\mathbf{u} = (\cos \alpha, \sin \alpha)$ for some α.

   (a) Show that every unit vector in space can be written in the form $\mathbf{u} = (\cos \alpha, \cos \beta, \cos \gamma)$, where α, β, γ are the angles between $\mathbf{u}$ and $\mathbf{i}$, $\mathbf{j}$, $\mathbf{k}$, respectively. Explain why $\cos^2 \alpha + \cos^2 \beta + \cos^2 \gamma = 1$.

   (b) The same thing can be done in the plane, that is, $\mathbf{u} = (\cos \alpha, \cos \beta)$, where α and β are the angles between $\mathbf{u}$ and $\mathbf{i}$, $\mathbf{j}$, respectively, and $\cos^2 \alpha + \cos^2 \beta = 1$. How does this fit in with the formula $\mathbf{u} = (\cos \alpha, \sin \alpha)$?

29. Let $P_1(x_1, y_1, z_1)$ and $P_2(x_2, y_2, z_2)$ be two points in space and let $M(x, y, z)$ be the midpoint of the segment joining them.

   (a) Draw a picture to convince yourself that $\overrightarrow{P_1 M} = \overrightarrow{M P_2}$.

   (b) Use the vector equation in part (a) to derive the midpoint formulas

$$x = \tfrac{1}{2}(x_1 + x_2), y = \tfrac{1}{2}(y_1 + y_2), \text{ and } z = \tfrac{1}{2}(z_1 + z_2)$$

30. Confirm that $\mathbf{x} \cdot \mathbf{x} = |\mathbf{x}|^2$.

31. Show that $|c\mathbf{x}| = |c||\mathbf{x}|$.

32. Explain why

$$\mathbf{x}_1 \text{ and } \mathbf{x}_2 \text{ are perpendicular} \Leftrightarrow \mathbf{x}_1 \cdot \mathbf{x}_2 = 0$$

33. Show that $\mathbf{x}$ and $c\mathbf{x}$ have the same direction if $c \geq 0$ and opposite directions if $c < 0$.

34. Let θ be the angle between the nonzero vectors $\mathbf{x}_1$ and $\mathbf{x}_2$.

   (a) If $\theta = 0$ (so that $\mathbf{x}_1$ and $\mathbf{x}_2$ have the same direction), name a positive scalar c such that $\mathbf{x}_2 = c\mathbf{x}_1$.

   (b) If $\theta = \pi$ (so that $\mathbf{x}_1$ and $\mathbf{x}_2$ have opposite directions), name a negative scalar c such that $\mathbf{x}_2 = c\mathbf{x}_1$.

35. Prove the Cauchy-Schwarz Inequality,

$$|\mathbf{x}_1 \cdot \mathbf{x}_2| \leq |\mathbf{x}_1||\mathbf{x}_2|$$

36. Use the Cauchy-Schwarz Inequality to prove the Triangle Inequality, $|\mathbf{x}_1 + \mathbf{x}_2| \leq |\mathbf{x}_1| + |\mathbf{x}_2|$.

16.2
LINES AND PLANES

Figure 1
Straight line in space

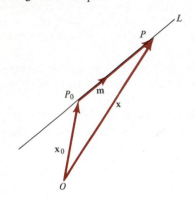

Let L be a straight line in $\mathfrak{R}^3$. To represent it analytically, we proceed as in $\mathfrak{R}^2$ (where we used a fixed point and the slope to write an equation of the line). In 3-space the direction of L is not described by a number, but by a nonzero vector parallel to L, say $\mathbf{m} = (a, b, c)$. Let $P_0(x_0, y_0, z_0)$ be a fixed point of L. In Figure 1 we show a typical point $P(x, y, z)$ of L and the position vectors

$$\mathbf{x}_0 = \overrightarrow{OP_0} = (x_0, y_0, z_0) \qquad \text{and} \qquad \mathbf{x} = \overrightarrow{OP} = (x, y, z)$$

Evidently $\overrightarrow{P_0P}$ is a scalar multiple of $\mathbf{m}$, say $\overrightarrow{P_0P} = t\mathbf{m}$, where t may be positive (as in Figure 1), negative (if P is on the other side of P_0), or zero (if P coincides with P_0). The geometric interpretation of vector addition (which is the same in $\mathfrak{R}^3$ as in $\mathfrak{R}^2$) says that

$$\overrightarrow{OP} = \overrightarrow{OP_0} + \overrightarrow{P_0P}$$

(Also see Problem 20, Section 16.1, which is independent of pictures.) Hence a vector equation of the line is $\mathbf{x} = \mathbf{x}_0 + t\mathbf{m}$, $t \in \mathfrak{R}$. Equivalent parametric equations are $x = x_0 + at$, $y = y_0 + bt$, $z = z_0 + ct$.

Conversely, any vector equation of the form $\mathbf{x} = \mathbf{x}_0 + t\mathbf{m}$ ($\mathbf{m} \neq \mathbf{0}$) represents a line in space, namely the line parallel to $\mathbf{m}$ and containing the point whose position vector is $\mathbf{x}_0$. We have proved the following.

Every straight line in $\mathfrak{R}^3$ is the graph of a vector equation
$$\mathbf{x} = \mathbf{x}_0 + t\mathbf{m} \qquad t \in \mathfrak{R}$$
where $\mathbf{x}_0$ is the position vector of a fixed point on the line and $\mathbf{m}$ is a nonzero vector parallel to the line. Conversely, every such vector equation represents a line. Equivalent parametric equations are $x = x_0 + at$, $y = y_0 + bt$, $z = z_0 + ct$, where (x_0, y_0, z_0) is the fixed point and (a, b, c) is the direction vector.

In $\mathfrak{R}^2$ we may eliminate the parameter from the equations

$$x = x_0 + at, \quad y = y_0 + bt \qquad \text{(Problem 23, Section 15.2)}$$

and obtain a linear equation $Ax + By + C = 0$ representing the line. It is important to realize that this cannot be done in $\mathfrak{R}^3$. As we pointed out in the last section, an equation in x, y, and z represents a *surface*, not a curve; in fact we show in this section that the graph of a linear equation $Ax + By + Cz + D = 0$ is a plane.

If a, b, and c are all different from zero (which is not always the case), we can eliminate the parameter from $x = x_0 + at$, $y = y_0 + bt$, $z = z_0 + ct$ by solving for t in each equation and writing

$$\frac{x - x_0}{a} = \frac{y - y_0}{b} = \frac{z - z_0}{c}$$

Note, however, that these are simultaneous equations, not a single equation in x, y, and z. While they are occasionally useful, it is generally easier to stick with the parametric representation of a straight line.

□ **Example 1**   Find parametric equations of the line through $(2, 0, -1)$ and $(3, -3, -1)$.

Solution: A vector parallel to the line is

$$\mathbf{m} = (3 - 2, -3 - 0, -1 + 1) = (1, -3, 0)$$

Taking P_0 to be the point $(2, 0, -1)$, we find the equations

$$x = 2 + 1t, \, y = 0 - 3t, \, z = -1 + 0t$$

or simply $x = 2 + t$, $y = -3t$, $z = -1$. Note that the line lies in the plane $z = -1$ (parallel to the xy plane and 1 unit below it).   □

□ **Example 2**   Show that the parametric equations

$$x = t, \, y = t - 1, \, z = 2t + 5$$

and $$x = 2t + 1, \, y = 2t, \, z = 4t + 7$$

represent the same line.

Solution: Writing the first set of equations in the form

$$x = 1t + 0, \, y = 1t - 1, \, z = 2t + 5$$

we see that they represent the line through the point $(0, -1, 5)$ parallel to the vector $\mathbf{m} = (1, 1, 2)$. The second set of equations represents the line through $(1, 0, 7)$ parallel to $(2, 2, 4)$. Since $(2, 2, 4) = 2\mathbf{m}$, the lines are at least parallel; in fact they are identical because the point $(1, 0, 7)$ satisfies the first set of equations (take $t = 1$).   □

The point of Example 2 is that the parameter appearing in different representations of the same line need not be the same. To emphasize the point, we might label one of the parameters differently, say s in place of t in the first set of equations: $x = s$, $y = s - 1$, $z = 2s + 5$. The relation between s and the parameter t in the second set is $s = 2t + 1$.

□ **Example 3**   Show that the direction vectors of the lines

$$L_1: x = 2t, \, y = t - 3, \, z = 1 - t$$
$$L_2: x = t, \, y = t + 1, \, z = 3t - 2$$

are perpendicular and determine whether the lines intersect.

Solution: Vectors parallel to L_1 and L_2 are $\mathbf{m}_1 = (2, 1, -1)$ and $\mathbf{m}_2 = (1, 1, 3)$. Since $\mathbf{m}_1 \cdot \mathbf{m}_2 = (2)(1) + (1)(1) + (-1)(3) = 0$, the vectors are perpendicular.

The second part of the question sounds strange, since we are accustomed to lines intersecting unless they are parallel. In three-dimensional space, however, we can have **skew lines,** that is, lines which are not parallel and never meet. As an example, visualize the line of intersection of one wall of a (rectangular) room and the floor, and a line running diagonally across the ceiling.

To determine whether L_1 and L_2 intersect, note that the parameters in the representations of L_1 and L_2 (although they are both labeled t) are not necessarily the same. To emphasize that fact, we will relabel the first one s. Then our representations of L_1 and L_2 are

$$L_1: x = 2s, \; y = s - 3, \; z = 1 - s$$
$$L_2: x = t, \; y = t + 1, \; z = 3t - 2$$

A point of intersection (if one exists) must be

$$(2s, s - 3, 1 - s) = (t, t + 1, 3t - 2)$$

where s and t satisfy the system of equations

$$2s = t, \; s - 3 = t + 1, \; 1 - s = 3t - 2$$

Substituting $t = 2s$ from the first equation into the second, we have $s - 3 = 2s + 1$, from which $s = -4$ (and $t = 2s = -8$). Thus the pair $(s,t) = (-4, -8)$ satisfies the first two equations. Does it also satisfy the third? We find out by substitution in $1 - s = 3t - 2$:

$$1 - (-4) \overset{?}{=} 3(-8) - 2$$
$$5 \overset{?}{=} -26$$

Since the answer is no, we conclude that the lines do not intersect. (Evidently the answer will be no in most cases! Parametric equations of two lines must be concocted to make the lines meet.)   □

□ **Example 4**   Determine whether the lines

$$L_1: x = 3t + 3, \; y = t, \; z = t + 3$$
$$L_2: x = t - 2, \; y = 2t - 5, \; z = t$$

intersect.

Solution: As in Example 3, replace t by s in the first set of parametric equations:

$$L_1: x = 3s + 3, \; y = s, \; z = s + 3$$
$$L_2: x = t - 2, \; y = 2t - 5, \; z = t$$

A point of intersection (if one exists) is

$$(3s + 3, s, s + 3) = (t - 2, 2t - 5, t)$$

where s and t satisfy the system

$$3s + 3 = t - 2, \; s = 2t - 5, \; s + 3 = t$$

Substitute $s = 2t - 5$ from the second equation into the first, obtaining

$$3(2t - 5) + 3 = t - 2$$

from which $t = 2$ (and $s = 2t - 5 = -1$). The pair $(s,t) = (-1,2)$ satisfies the first two equations *and also the third* (as you can check). Hence the lines intersect in the point $(0, -1, 2)$.

What if we had not relabeled the parameter in the equations representing L_1? A point of intersection would apparently be

$$(3t + 3, t, t + 3) = (t - 2, 2t - 5, t)$$

where t satisfies the equations

$$3t + 3 = t - 2, \quad t = 2t - 5, \quad t + 3 = t$$

This is impossible, which suggests that the lines do not intersect. A student reaching this conclusion and looking in the answer section to find the point of intersection $(0, -1, 2)$ is likely to be confused! Perhaps the best way to explain what went wrong is to think of t as time. Particles moving along L_1 and L_2 reach the point $(0, -1, 2)$ when $t = -1$ and $t = 2$, respectively. Their paths intersect but the particles do not collide.   □

Figure 2
Plane in space

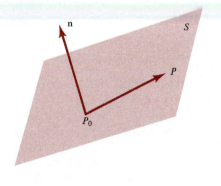

Now we turn to the problem of the analytic representation of planes in space. If S is a given plane, we may specify its orientation in space by naming a nonzero vector perpendicular to it (a **normal vector**). There are infinitely many such vectors (all parallel); let $\mathbf{n} = (a, b, c)$ be one of them, and let $P_0(x_0, y_0, z_0)$ be a fixed point of the plane. If $P(x, y, z)$ is any point in space, then (see Figure 2)

$$P \in S \Leftrightarrow \mathbf{n} \perp \overrightarrow{P_0P}$$
$$\Leftrightarrow (a, b, c) \cdot (x - x_0, y - y_0, z - z_0) = 0$$
$$\Leftrightarrow a(x - x_0) + b(y - y_0) + c(z - z_0) = 0$$

Hence S is the graph of the equation

$$a(x - x_0) + b(y - y_0) + c(z - z_0) = 0$$

or simply

$$ax + by + cz + d = 0 \quad [\text{where } d = -(ax_0 + by_0 + cz_0)]$$

Conversely, any equation of the form

$$ax + by + cz + d = 0 \quad (a, b, c \text{ not all zero}) \tag{1}$$

represents a plane. For the equation is satisfied by some triple (x_0, y_0, z_0), that is,

$$ax_0 + by_0 + cz_0 + d = 0 \tag{2}$$

(Why does such a triple exist?) Subtracting Equation (2) from Equation (1), we obtain the equation $a(x - x_0) + b(y - y_0) + c(z - z_0) = 0$ equivalent to Equation (1). This represents the plane through (x_0, y_0, z_0) perpendicular to the nonzero vector $\mathbf{n} = (a, b, c)$.

Every plane in $\mathcal{R}^3$ is the graph of a linear equation

$$ax + by + cz + d = 0 \qquad (a, b, c \text{ not all zero})$$

and every such equation represents a plane. The nonzero vector $\mathbf{n} = (a, b, c)$ is normal to the plane. If (x_0, y_0, z_0) is a point of the plane, the equation is equivalent to

$$a(x - x_0) + b(y - y_0) + c(z - z_0) = 0$$

□ **Example 5**   Find an equation of the plane perpendicular to $\mathbf{n} = (2, 1, 1)$ and containing the point $(-1, 3, 1)$.

Solution: An equation is

$$2(x + 1) + 1(y - 3) + 1(z - 1) = 0$$

or simply $2x + y + z - 2 = 0$.   □

□ **Example 6**   Solve the system of equations

$$2x - 5y + z = 4$$
$$x + 2y - 3z = 1$$

and give a geometric interpretation of the result.

Solution: Let's *start* with the geometric interpretation. The given equations represent planes with normal vectors $\mathbf{n}_1 = (2, -5, 1)$ and $\mathbf{n}_2 = (1, 2, -3)$. Since the planes are not parallel (why?), they intersect in a straight line. We expect the solutions of our system to be the points of this line.

These remarks make it plain that "solving" the system will not lead to a unique triple (x, y, z), but to an infinite set of points. In other words, there are too many unknowns (or too few equations). This suggests one way to proceed; why not write our system in the form

$$2x - 5y = 4 - z$$
$$x + 2y = 1 + 3z$$

and solve for x and y without worrying about z? The result (as you can check) is $x = \frac{13}{9} + \frac{13}{9}z, y = -\frac{2}{9} + \frac{7}{9}z$. Since z is arbitrary, say $z = t$ (where t is any real number), the solutions of our system are points (x, y, z) of the form

$$x = \tfrac{13}{9} + \tfrac{13}{9}t, \; y = -\tfrac{2}{9} + \tfrac{7}{9}t, \; z = t \qquad (t \in \mathcal{R})$$

You should recognize these as parametric equations of a line, namely the line of intersection of the given planes.

Algebra students often find systems of this kind to be mysterious. Note how the geometry of lines and planes removes the mystery!   □

□ **Example 7**   Find the point of intersection of the line

$$L: x = 3t - 1, \; y = 2t, \; z = 2 - t$$

and the plane

$$S: 3x - y + z - 5 = 0$$

and find the (acute) angle between L and S.

Figure 3
Angle between a line and a plane

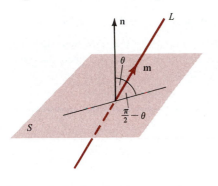

Solution: To find where L and S intersect, we need only substitute the point $(x, y, z) = (3t - 1, 2t, 2 - t)$ in the equation $3x - y + z - 5 = 0$. (Why?) This gives

$$3(3t - 1) - 2t + (2 - t) - 5 = 0$$

from which $t = 1$. Hence the point of intersection is $(2, 2, 1)$. The angle between L and S is the complement of the angle θ between $\mathbf{m}$ and $\mathbf{n}$ (Figure 3), where $\mathbf{m} = (3, 2, -1)$ is a vector parallel to L and $\mathbf{n} = (3, -1, 1)$ is a normal vector to S. Since

$$\cos \theta = \frac{\mathbf{m} \cdot \mathbf{n}}{|\mathbf{m}||\mathbf{n}|} = \frac{(3)(3) + (2)(-1) + (-1)(1)}{\sqrt{14}\sqrt{11}} = \frac{6}{\sqrt{154}} = \sin\left(\frac{\pi}{2} - \theta\right)$$

we find $\pi/2 - \theta = \sin^{-1}(6/\sqrt{154}) \approx 0.5$ (or about $29°$).   □

Problem Set 16.2

Find parametric equations of each of the following lines.

1. Containing $(1, -1, 2)$ and parallel to $\mathbf{m} = (2, 3, 2)$.
2. Containing $(0, 4, 1)$ and parallel to $\mathbf{m} = (-1, 2, 1)$.
3. Containing $(1, -1, 3)$ and $(2, 0, 5)$.
4. Containing $(0, 0, 0)$ and $(0, 0, 1)$.
5. Containing $(2, -1, 5)$ and parallel to $\mathbf{i}$.
6. Containing $(0, 2, -1)$ and parallel to the line $x = 3t$, $y = 2 - t$, $z = t + 1$.
7. Containing $(4, 0, 1)$ and parallel to the line
$$\frac{x - 1}{2} = y = \frac{z + 2}{3}$$
8. Containing $(-1, 2, 1)$ and parallel to the line
$$\frac{x + 2}{3} = \frac{y - 1}{2} = \frac{z + 1}{2}$$
9. Containing $(3, 1, 1)$ and perpendicular to the plane $3x + y + z - 5 = 0$.
10. Containing $(0, 1, 1)$ and perpendicular to the plane $x - 3y - z = 1$.
11. The line of intersection of the planes $y = 0$ and $z = 2$.

12. The line of intersection of the plane $2x - y + z - 1 = 0$ and the xy plane. What is the equation in x and y of this line regarded as a graph in $\mathfrak{R}^2$?
13. The line of intersection of the planes $x - y + z = 0$ and $3x + y - z = 1$.
14. The line of intersection of the planes $x + y - 2z = 5$ and $2x - y - z = 1$.
15. Where does the line $x = 2t$, $y = 5t - 1$, $z = 2 - t$ intersect the xy plane? What acute angle does it make with this plane?
16. Explain why the line
$$L: x = t, \; y = 2t, \; z = 3t$$
is perpendicular to the plane
$$S: x + 2y + 3z - 14 = 0$$
Where do L and S intersect?
17. The positions of two airplanes at time t are given by
$$x = 2t - 1, \; y = 1 - t, \; z = 3t$$
and $x = t + 1, \; y = 2t, \; z = 3 - 2t$

Do the paths of the airplanes intersect? Do the airplanes collide?

18. In Problem 17 find the acute angle between the paths.

19. Do the lines $x = 3t$, $y = t$, $z = t$ and $x = t - 3$, $y = 2t$, $z = t + 1$ intersect?

Find an equation of each of the following planes.

20. Containing the origin and perpendicular to **k**.

21. Containing $(2, -1, 1)$ and perpendicular to the vector $\mathbf{n} = (-1, 1, 3)$.

22. Containing $(1, -1, 3)$ and perpendicular to the vector $\mathbf{n} = (2, 1, 4)$.

23. Containing $(1, 0, 3)$ and parallel to the plane $z = 5$.

24. Containing $(2, -1, 5)$ and parallel to the plane $3x + y - 2z + 1 = 0$.

25. Containing $(5, -1, 0)$ and perpendicular to the line $x = 2t$, $y = t + 3$, $z = 3t - 1$.

26. Containing $(-1, 5, 2)$ and perpendicular to the line

$$\frac{x - 1}{2} = y + 3 = z - 5$$

27. Containing $(1, -1, 3)$, $(0, 2, 2)$, and $(1, -2, 2)$. *Hint:* Find a vector $\mathbf{n} = (a, b, c)$ normal to the plane by observing that $\mathbf{n}$ must be perpendicular to both $\overrightarrow{P_1P_2}$ and $\overrightarrow{P_1P_3}$, where P_1, P_2, P_3 are the given points.

28. Containing $(2, -1, 1)$, $(5, 0, -1)$, and $(-2, 3, 3)$.

29. Containing $(1, -1, 2)$ and the line $x = t - 1$, $y = 2t + 1$, $z = 2t$.

30. Containing the lines $x = 3t + 3$, $y = t$, $z = t + 3$ and $x = t - 2$, $y = 2t - 5$, $z = t$.

31. Tangent to the sphere $(x - 2)^2 + (y + 1)^2 + (z - 1)^2 = 9$ at the point $(3, 1, -1)$.

32. What happens when you try to find a plane containing the points $(0, 3, -1)$, $(2, 4, 2)$, and $(-2, 2, -4)$? What is the geometric explanation?

33. What can you say about the planes that have equations of the form $x - 3y + z + d = 0$, where d is arbitrary?

34. Find the point of intersection of the line

$$x = 2t, \ y = t - 5, \ z = -t$$

and the plane

$$2x + 5y - z - 5 = 0$$

and the acute angle between them.

35. Find the acute angle between the planes $2x + y - z - 5 = 0$ and $x + y + z - 1 = 0$.

36. Find an equation of the set of points equidistant from $(2, 2, -1)$ and $(4, 0, 3)$.

37. Show that the distance between the point $P_0(x_0, y_0, z_0)$ and the plane $ax + by + cz + d = 0$ is

$$\frac{|ax_0 + by_0 + cz_0 + d|}{\sqrt{a^2 + b^2 + c^2}}$$

as follows.

(a) Explain why the line through P_0 perpendicular to the plane is

$$x = x_0 + at, \ y = y_0 + bt, \ z = z_0 + ct$$

(b) Letting $k = ax_0 + by_0 + cz_0 + d$, show that the point $P(x, y, z)$ where the line in part (a) intersects the plane satisfies the equations

$$x - x_0 = \frac{-ak}{a^2 + b^2 + c^2}$$

$$y - y_0 = \frac{-bk}{a^2 + b^2 + c^2}$$

$$z - z_0 = \frac{-ck}{a^2 + b^2 + c^2}$$

(c) Finish the proof by showing that

$$d(P_0, P) = \frac{|k|}{\sqrt{a^2 + b^2 + c^2}}$$

38. Use the formula in Problem 37 to find the distance between the point $(2, 1, -1)$ and the plane $x - y + 3z - 5 = 0$.

39. Explain why the planes $x + 2y - z + 1 = 0$ and $x + 2y - z - 3 = 0$ are parallel and find the distance between them.

40. Show that the distance between the parallel planes $ax + by + cz + d_1 = 0$ and $ax + by + cz + d_2 = 0$ is

$$\frac{|d_1 - d_2|}{\sqrt{a^2 + b^2 + c^2}}$$

16.3
THE CROSS PRODUCT

There are many situations in which we need a vector perpendicular to two given vectors. An illustration is the following.

□ **Example 1**   Find an equation of the plane containing the points $(1, 0, 1)$, $(3, 1, 4)$, and $(0, 1, 2)$.

Solution: See Figure 1, in which we show vectors $\overrightarrow{P_1P_2}$ and $\overrightarrow{P_1P_3}$ lying in the plane and a vector $\mathbf{n} = (a, b, c)$ perpendicular to both of them. If we knew $\mathbf{n}$, the problem would be solved. (Why?) Hence we seek equations involving a, b, and c that can be solved to find $\mathbf{n}$. Since $\mathbf{n}$ is perpendicular to both

$$\overrightarrow{P_1P_2} = (3 - 1, 1 - 0, 4 - 1) = (2, 1, 3)$$

and
$$\overrightarrow{P_1P_3} = (0 - 1, 1 - 0, 2 - 1) = (-1, 1, 1)$$

we have

$$(2, 1, 3) \cdot (a, b, c) = 0 \quad \text{and} \quad (-1, 1, 1) \cdot (a, b, c) = 0$$

Figure 1
Plane determined by three points

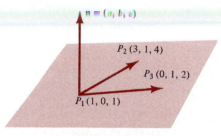

Hence a, b, and c must satisfy the system of equations

$$2a + b + 3c = 0$$
$$-a + b + c = 0$$

As in Example 6 of the last section, we write this system in the form

$$2a + b = -3c$$
$$-a + b = -c$$

and solve for a and b in terms of c. The result is $a = -\frac{2}{3}c$, $b = -\frac{5}{3}c$, as you can check. Of course, there are infinitely many vectors perpendicular to the plane we seek, but we need only *one*. Choosing $c = 3$ (any nonzero value will do), we find $\mathbf{n} = (-2, -5, 3)$. Using $(1, 0, 1)$ as the fixed point in the standard equation of the plane, we obtain

$$-2(x - 1) - 5(y - 0) + 3(z - 1) = 0 \quad \text{or} \quad 2x + 5y - 3z + 1 = 0$$

□

In this section we are interested in the general question of finding a vector $\mathbf{n} = (a, b, c)$ perpendicular to two given vectors $\mathbf{x}_1 = (x_1, y_1, z_1)$ and $\mathbf{x}_2 = (x_2, y_2, z_2)$. We approach the problem as in Example 1. Since

$$(x_1, y_1, z_1) \cdot (a, b, c) = 0 \qquad \text{and} \qquad (x_2, y_2, z_2) \cdot (a, b, c) = 0$$

we seek a, b, and c satisfying the system of equations

$$\begin{aligned} x_1 a + y_1 b + z_1 c = 0 \\ x_2 a + y_2 b + z_2 c = 0 \end{aligned} \tag{1}$$

Writing the system in the form

$$\begin{aligned} x_1 a + y_1 b = -z_1 c \\ x_2 a + y_2 b = -z_2 c \end{aligned}$$

we solve for a and b in terms of c by the usual process of elimination. Multiplying the first equation by y_2 and the second by y_1, we have

$$\begin{aligned} x_1 y_2 a + y_1 y_2 b = -y_2 z_1 c \\ x_2 y_1 a + y_1 y_2 b = -y_1 z_2 c \end{aligned}$$

Subtract the second equation from the first to eliminate b:

$$(x_1 y_2 - x_2 y_1) a = (y_1 z_2 - y_2 z_1) c$$

Assuming that $x_1 y_2 - x_2 y_1 \neq 0$, we find

$$a = \left(\frac{y_1 z_2 - y_2 z_1}{x_1 y_2 - x_2 y_1} \right) c$$

Eliminating a from the original system (1) in the same way, we find

$$b = \left(\frac{x_2 z_1 - x_1 z_2}{x_1 y_2 - x_2 y_1} \right) c$$

The reader who is familiar with determinants will recognize that these formulas may be rewritten as

$$a = \frac{\begin{vmatrix} y_1 & z_1 \\ y_2 & z_2 \end{vmatrix}}{\begin{vmatrix} x_1 & y_1 \\ x_2 & y_2 \end{vmatrix}} c, \quad b = -\frac{\begin{vmatrix} x_1 & z_1 \\ x_2 & z_2 \end{vmatrix}}{\begin{vmatrix} x_1 & y_1 \\ x_2 & y_2 \end{vmatrix}} c$$

For those not acquainted with the notation, we define the 2×2 determinant

$$\begin{vmatrix} a_1 & b_1 \\ a_2 & b_2 \end{vmatrix} = a_1 b_2 - a_2 b_1$$

(the product of the entries a_1 and b_2 on the "main diagonal" minus the product of the entries on the other diagonal).

Now suppose that we choose the convenient value

$$c = \begin{vmatrix} x_1 & y_1 \\ x_2 & y_2 \end{vmatrix} = x_1 y_2 - x_2 y_1$$

Then the formulas for a and b reduce to

$$a = \begin{vmatrix} y_1 & z_1 \\ y_2 & z_2 \end{vmatrix}, \quad b = -\begin{vmatrix} x_1 & z_1 \\ x_2 & z_2 \end{vmatrix}$$

and we have found a vector $\mathbf{n} = (a, b, c)$ with the desired property. Our analysis required that

$$c = \begin{vmatrix} x_1 & y_1 \\ x_2 & y_2 \end{vmatrix} \neq 0$$

but in fact it doesn't matter if $c = 0$. You can check that $\mathbf{n}$ is perpendicular to $\mathbf{x}_1$ and $\mathbf{x}_2$ in any case by substituting for a, b, and c in the original system (1):

$$x_1 \begin{vmatrix} y_1 & z_1 \\ y_2 & z_2 \end{vmatrix} - y_1 \begin{vmatrix} x_1 & z_1 \\ x_2 & z_2 \end{vmatrix} + z_1 \begin{vmatrix} x_1 & y_1 \\ x_2 & y_2 \end{vmatrix} \overset{?}{=} 0$$

$$x_2 \begin{vmatrix} y_1 & z_1 \\ y_2 & z_2 \end{vmatrix} - y_2 \begin{vmatrix} x_1 & z_1 \\ x_2 & z_2 \end{vmatrix} + z_2 \begin{vmatrix} x_1 & y_1 \\ x_2 & y_2 \end{vmatrix} \overset{?}{=} 0$$

that is,

$$x_1(y_1z_2 - y_2z_1) - y_1(x_1z_2 - x_2z_1) + z_1(x_1y_2 - x_2y_1) \overset{?}{=} 0$$
$$x_2(y_1z_2 - y_2z_1) - y_2(x_1z_2 - x_2z_1) + z_2(x_1y_2 - x_2y_1) \overset{?}{=} 0$$

By expanding the products and collecting terms, you can see that each left side is indeed zero.

Thus we have proved that if $\mathbf{x}_1 = (x_1, y_1, z_1)$ and $\mathbf{x}_2 = (x_2, y_2, z_2)$ are any vectors in $\mathfrak{R}^3$, a vector perpendicular to both of them is

$$\mathbf{n} = (a, b, c) = a\mathbf{i} + b\mathbf{j} + c\mathbf{k}$$

where

$$a = \begin{vmatrix} y_1 & z_1 \\ y_2 & z_2 \end{vmatrix}, \quad b = -\begin{vmatrix} x_1 & z_1 \\ x_2 & z_2 \end{vmatrix}, \quad c = \begin{vmatrix} x_1 & y_1 \\ x_2 & y_2 \end{vmatrix}$$

This vector is called the **cross product** of $\mathbf{x}_1$ and $\mathbf{x}_2$, written

$$\mathbf{n} = \mathbf{x}_1 \times \mathbf{x}_2$$

A convenient way to remember the formula for $\mathbf{n}$ is to write

$$\mathbf{x}_1 \times \mathbf{x}_2 = \begin{vmatrix} \mathbf{i} & \mathbf{j} & \mathbf{k} \\ x_1 & y_1 & z_1 \\ x_2 & y_2 & z_2 \end{vmatrix}$$

Of course this will be meaningless to students who have not been exposed to 3×3 determinants, so we hasten to give another definition.

The third-order determinant

$$\begin{vmatrix} a_1 & b_1 & c_1 \\ a_2 & b_2 & c_2 \\ a_3 & b_3 & c_3 \end{vmatrix}$$

is defined to be the number

$$a_1 \begin{vmatrix} b_2 & c_2 \\ b_3 & c_3 \end{vmatrix} - b_1 \begin{vmatrix} a_2 & c_2 \\ a_3 & c_3 \end{vmatrix} + c_1 \begin{vmatrix} a_2 & b_2 \\ a_3 & b_3 \end{vmatrix}$$

This expression may be described as the sum of the entries of the first row (a_1, b_1, c_1) multiplied by their *cofactors*

$$\begin{vmatrix} b_2 & c_2 \\ b_3 & c_3 \end{vmatrix}, \quad -\begin{vmatrix} a_2 & c_2 \\ a_3 & c_3 \end{vmatrix}, \quad \begin{vmatrix} a_2 & b_2 \\ a_3 & b_3 \end{vmatrix}$$

Each cofactor is a second-order determinant multiplied by ± 1 (in the pattern $+$, $-$, $+$); the determinant itself is obtained by striking out the row and column (of the original 3×3 determinant) in which the entry associated with the cofactor appears. This is harder to say in words than it is to do! To see what we mean, consider the determinants in order:

$$\begin{vmatrix} b_2 & c_2 \\ b_3 & c_3 \end{vmatrix} = \begin{vmatrix} \cancel{a_1} & \cancel{b_1} & \cancel{c_1} \\ \cancel{a_2} & b_2 & c_2 \\ \cancel{a_3} & b_3 & c_3 \end{vmatrix}, \quad \begin{vmatrix} a_2 & c_2 \\ a_3 & c_3 \end{vmatrix} = \begin{vmatrix} \cancel{a_1} & \cancel{b_1} & \cancel{c_1} \\ a_2 & \cancel{b_2} & c_2 \\ a_3 & \cancel{b_3} & c_3 \end{vmatrix}$$

$$\begin{vmatrix} a_2 & b_2 \\ a_3 & b_3 \end{vmatrix} = \begin{vmatrix} \cancel{a_1} & \cancel{b_1} & \cancel{c_1} \\ a_2 & b_2 & \cancel{c_2} \\ a_3 & b_3 & \cancel{c_3} \end{vmatrix}$$

There are other ways to "expand" a third-order determinant; you may have learned them in an algebra course before starting calculus. Our purpose, however, is not to discuss determinants, but to explain the formula

$$\mathbf{x}_1 \times \mathbf{x}_2 = \begin{vmatrix} \mathbf{i} & \mathbf{j} & \mathbf{k} \\ x_1 & y_1 & z_1 \\ x_2 & y_2 & z_2 \end{vmatrix}$$

This is a "pseudo-determinant," in the sense that the first row is not a triple of numbers, but the vector triple $\mathbf{i}, \mathbf{j}, \mathbf{k}$. The pattern of expansion is the same, however. Multiply each vector by its cofactor (obtained as described above) and add the products; the result is the formula for $\mathbf{x}_1 \times \mathbf{x}_2$.

□ **Example 2**   Find a vector perpendicular to both $\mathbf{x}_1 = (2, 1, 3)$ and $\mathbf{x}_2 = (-1, 1, 1)$.

Solution: This is a repetition of Example 1. Now, however, we have a mechanical way of proceeding. Simply compute the cross product

$$\mathbf{x}_1 \times \mathbf{x}_2 = \begin{vmatrix} \mathbf{i} & \mathbf{j} & \mathbf{k} \\ 2 & 1 & 3 \\ -1 & 1 & 1 \end{vmatrix} = \begin{vmatrix} 1 & 3 \\ 1 & 1 \end{vmatrix} \mathbf{i} - \begin{vmatrix} 2 & 3 \\ -1 & 1 \end{vmatrix} \mathbf{j} + \begin{vmatrix} 2 & 1 \\ -1 & 1 \end{vmatrix} \mathbf{k}$$

$$= (1 - 3)\mathbf{i} - (2 + 3)\mathbf{j} + (2 + 1)\mathbf{k} = -2\mathbf{i} - 5\mathbf{j} + 3\mathbf{k}$$

$$= (-2, -5, 3)$$

□

Figure 2
Plane determined by two lines

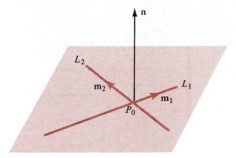

□ **Example 3**   Find an equation of the plane determined by the lines

$$L_1: x = 3t + 3, \; y = t, \; z = t + 3$$
$$L_2: x = t - 2, \; y = 2t - 5, \; z = t$$

Solution: See Figure 2, which shows two lines lying in a plane. To find an equation of the plane, we need a point P_0 in the plane and a normal vector **n**. The point may be taken to be the intersection of the lines, which we found to be $P_0(0, -1, 2)$ in Example 4 of the last section. (An easier point to choose, if we did not know this one, would be $(3, 0, 3)$, obtained by letting $t = 0$ in the parametric equations for L_1.)

The normal vector must be perpendicular to the direction vectors

$$\mathbf{m_1} = (3, 1, 1) \quad \text{and} \quad \mathbf{m_2} = (1, 2, 1)$$

of the lines, so we may take it to be

$$\mathbf{n} = \mathbf{m_1} \times \mathbf{m_2} = \begin{vmatrix} \mathbf{i} & \mathbf{j} & \mathbf{k} \\ 3 & 1 & 1 \\ 1 & 2 & 1 \end{vmatrix} = \begin{vmatrix} 1 & 1 \\ 2 & 1 \end{vmatrix} \mathbf{i} - \begin{vmatrix} 3 & 1 \\ 1 & 1 \end{vmatrix} \mathbf{j} + \begin{vmatrix} 3 & 1 \\ 1 & 2 \end{vmatrix} \mathbf{k}$$

$$= (1 - 2)\mathbf{i} - (3 - 1)\mathbf{j} + (6 - 1)\mathbf{k} = -\mathbf{i} - 2\mathbf{j} + 5\mathbf{k}$$
$$= (-1, -2, 5)$$

Hence an equation of the plane is

$$-1(x - 0) - 2(y + 1) + 5(z - 2) = 0$$

or $x + 2y - 5z + 12 = 0$. You can check that this plane contains L_1 and L_2 by substituting $(3t + 3, t, t + 3)$ and $(t - 2, 2t - 5, t)$ for (x, y, z) in its equation.   □

As you can see, cross products are useful. Because they occur in so many applications, we devote the rest of this section to a discussion of their properties.

The first thing to observe is that the cross product of two vectors is *another vector* (not a scalar, like the dot product). For that reason it is often called the *vector product* (while the dot product is sometimes referred to as the *scalar product*). The most important property of the cross product is that *it is perpendicular to both the given vectors;* that is what makes it so important in applications. Of course in some cases it may be zero, but that does not affect the property of perpendicularity (since we have agreed that the zero vector is perpendicular to every vector).

Assuming that $\mathbf{x_1}$ and $\mathbf{x_2}$ are not zero, the magnitude of the cross product is

$$|\mathbf{x_1} \times \mathbf{x_2}| = |\mathbf{x_1}| \, |\mathbf{x_2}| \sin \theta$$

where θ is the angle between $\mathbf{x_1}$ and $\mathbf{x_2}$. To see why, let

$$\mathbf{n} = \mathbf{x_1} \times \mathbf{x_2} = \begin{vmatrix} \mathbf{i} & \mathbf{j} & \mathbf{k} \\ x_1 & y_1 & z_1 \\ x_2 & y_2 & z_2 \end{vmatrix} = \begin{vmatrix} y_1 & z_1 \\ y_2 & z_2 \end{vmatrix} \mathbf{i} - \begin{vmatrix} x_1 & z_1 \\ x_2 & z_2 \end{vmatrix} \mathbf{j} + \begin{vmatrix} x_1 & y_1 \\ x_2 & y_2 \end{vmatrix} \mathbf{k}$$

Then

$$|\mathbf{n}|^2 = \begin{vmatrix} y_1 & z_1 \\ y_2 & z_2 \end{vmatrix}^2 + \begin{vmatrix} x_1 & z_1 \\ x_2 & z_2 \end{vmatrix}^2 + \begin{vmatrix} x_1 & y_1 \\ x_2 & y_2 \end{vmatrix}^2 \tag{2}$$
$$= (y_1 z_2 - y_2 z_1)^2 + (x_1 z_2 - x_2 z_1)^2 + (x_1 y_2 - x_2 y_1)^2$$

On the other hand,

$$|\mathbf{x}_1|^2|\mathbf{x}_2|^2 \sin^2 \theta = |\mathbf{x}_1|^2|\mathbf{x}_2|^2(1 - \cos^2 \theta) \tag{3}$$
$$= |\mathbf{x}_1|^2|\mathbf{x}_2|^2 - |\mathbf{x}_1|^2|\mathbf{x}_2|^2 \cos^2 \theta$$
$$= |\mathbf{x}_1|^2|\mathbf{x}_2|^2 - (\mathbf{x}_1 \cdot \mathbf{x}_2)^2 \qquad \text{(Why?)}$$
$$= (x_1{}^2 + y_1{}^2 + z_1{}^2)^2(x_2{}^2 + y_2{}^2 + z_2{}^2)^2 - (x_1 x_2 + y_1 y_2 + z_1 z_2)^2$$

Expanding the products and collecting terms in Equations (2) and (3) (this involves a fair amount of algebra, which you should fill in), we find

$$|\mathbf{n}|^2 = |\mathbf{x}_1|^2|\mathbf{x}_2|^2 \sin^2 \theta$$

from which

$$|\mathbf{n}| = |\mathbf{x}_1||\mathbf{x}_2| \sin \theta \qquad \text{(because } 0 \le \theta \le \pi)$$

If $\mathbf{x}_1$ and $\mathbf{x}_2$ are nonzero vectors in $\mathcal{R}^3$, then

$$|\mathbf{x}_1 \times \mathbf{x}_2| = |\mathbf{x}_1||\mathbf{x}_2| \sin \theta$$

where θ is the angle between them.

In the problem set we ask you to explain why

$|\mathbf{x}_1 \times \mathbf{x}_2|$ is the area of any parallelogram having $\mathbf{x}_1$ and $\mathbf{x}_2$ as adjacent sides.

The argument is based on the above formula. We may use the result to find the area of a triangle as well as that of a parallelogram, as shown in the following example.

□ **Example 4**   Find the area of the triangle with vertices $(1, -2, 2)$, $(3, 0, -1)$, and $(2, -3, 2)$.

Solution: From Figure 3 we find $\mathbf{x}_1 = \overrightarrow{P_1 P_2} = (2, 2, -3)$ and $\mathbf{x}_2 = \overrightarrow{P_1 P_3} = (1, -1, 0)$ to be adjacent sides of a parallelogram having twice the area of the triangle. Since

$$\mathbf{x}_1 \times \mathbf{x}_2 = \begin{vmatrix} \mathbf{i} & \mathbf{j} & \mathbf{k} \\ 2 & 2 & -3 \\ 1 & -1 & 0 \end{vmatrix} = -3\mathbf{i} - 3\mathbf{j} - 4\mathbf{k} = (-3, -3, -4)$$

Figure 3
Triangle in space

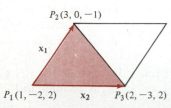

$P_2(3, 0, -1)$

$\mathbf{x}_1$

$P_1(1, -2, 2)$ $\mathbf{x}_2$ $P_3(2, -3, 2)$

we find the area of the triangle to be

$$\tfrac{1}{2}|\mathbf{x}_1 \times \mathbf{x}_2| = \tfrac{1}{2}\sqrt{34}$$

□

Since the cross product of the zero vector with any vector is $\mathbf{0}$ (Why?), we may also conclude from the formula $|\mathbf{x}_1 \times \mathbf{x}_2| = |\mathbf{x}_1||\mathbf{x}_2|\sin\theta$ that

$$\boxed{\mathbf{x}_1 \text{ and } \mathbf{x}_2 \text{ are parallel} \Leftrightarrow \mathbf{x}_1 \times \mathbf{x}_2 = \mathbf{0}}$$

It is important to note that the cross product is not commutative, that is, $\mathbf{x}_1 \times \mathbf{x}_2 \neq \mathbf{x}_2 \times \mathbf{x}_1$. Instead it is "anticommutative," a term that means

$$\boxed{\mathbf{x}_1 \times \mathbf{x}_2 = -(\mathbf{x}_2 \times \mathbf{x}_1)}$$

We leave it to you to show why, as well as to verify the failure of the associative law:

$$(\mathbf{x}_1 \times \mathbf{x}_2) \times \mathbf{x}_3 \neq \mathbf{x}_1 \times (\mathbf{x}_2 \times \mathbf{x}_3)$$

An "associative law" that works, on the other hand, is

$$\boxed{(\mathbf{x}_1 \times \mathbf{x}_2) \cdot \mathbf{x}_3 = \mathbf{x}_1 \cdot (\mathbf{x}_2 \times \mathbf{x}_3)}$$

as we ask you to show in the problem set. This is called the **scalar triple product,** and is occasionally useful in applications. One way to read the formula is to say that the cross and dot may be interchanged (with appropriate placement of parentheses to ensure that each side makes sense).

In view of the failure of the commutative and associative laws, it may be surprising to learn that the distributive laws hold, that is,

$$\boxed{\begin{aligned} \mathbf{x}_1 \times (\mathbf{x}_2 + \mathbf{x}_3) &= (\mathbf{x}_1 \times \mathbf{x}_2) + (\mathbf{x}_1 \times \mathbf{x}_3) \\ (\mathbf{x}_2 + \mathbf{x}_3) \times \mathbf{x}_1 &= (\mathbf{x}_2 \times \mathbf{x}_1) + (\mathbf{x}_3 \times \mathbf{x}_1) \end{aligned}}$$

Moreover, scalars can be moved around (as in the case of the dot product):

$$\boxed{(c\mathbf{x}_1) \times \mathbf{x}_2 = c(\mathbf{x}_1 \times \mathbf{x}_2) = \mathbf{x}_1 \times (c\mathbf{x}_2)}$$

Finally, we observe that if $\mathbf{n} = \mathbf{x}_1 \times \mathbf{x}_2 \neq \mathbf{0}$ (which means that $\mathbf{x}_1$ and $\mathbf{x}_2$ are nonzero and not parallel), the vectors $\mathbf{x}_1$, $\mathbf{x}_2$, and $\mathbf{n}$ form a right-handed triple, as shown in Figure 4. To see why, confirm that $\mathbf{i} \times \mathbf{j} = \mathbf{k}$ and $\mathbf{i} \times (-\mathbf{j}) = -\mathbf{k}$. We know that $\mathbf{i}, \mathbf{j}, \mathbf{k}$ and $\mathbf{i}, -\mathbf{j}, -\mathbf{k}$ are right-handed triples (because we have agreed that our coordinate system in $\mathfrak{R}^3$ has right-handed orientation). Visualize the plane containing $\mathbf{x}_1$ and $\mathbf{x}_2$ in Figure 4; then $\mathbf{n}$ is perpendicular to the plane. Rotate the figure until the plane coincides with the xy plane and $\mathbf{x}_1$ has the same direction as $\mathbf{i}$. If $\mathbf{x}_2$ is counterclockwise

Figure 4
The right-handed triple $\mathbf{x}_1$, $\mathbf{x}_2$, $\mathbf{x}_1 \times \mathbf{x}_2$

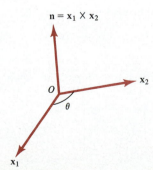

from x_1 (like j is from i), then n has the same direction as k. If x_2 is clockwise from x_1 (like $-j$ is from i), then n has the same direction as $-k$. In any case the triple x_1, x_2, n is right-handed.

> If the cross product of two vectors x_1 and x_2 is not zero, then x_1 and x_2 determine a plane. The vector $n = x_1 \times x_2$ is perpendicular to the plane and is directed so that the triple x_1, x_2, n is right-handed.

Problem Set 16.3

Let $x_1 = (2, -1, 1)$, $x_2 = (3, 0, 5)$, $x_3 = (1, 4, -2)$. Compute each of the following.

1. $x_1 \times x_2$ **2.** $x_2 \times x_1$

3. $(x_1 \times x_2) \times x_3$ **4.** $x_1 \times (x_2 \times x_3)$

5. $x_1 \times (x_2 + x_3)$ **6.** $(x_1 \times x_2) + (x_1 \times x_3)$

7. $(x_1 \times x_2) \cdot x_3$ **8.** $x_1 \cdot (x_2 \times x_3)$

Use the definition of the cross product as a determinant to confirm each of the following.

9. $i \times j = k$ **10.** $j \times k = i$

11. $k \times i = j$ **12.** $j \times i = -k$

13. $k \times j = -i$ **14.** $i \times k = -j$

15. $x \times 0 = 0$ for every x in $\mathcal{R}^3$.

16. $x \times x = 0$ for every x in $\mathcal{R}^3$.

Find an equation of each of the following planes.

17. Containing $(1, -1, 3)$, $(0, 2, 2)$, and $(1, -2, 2)$. (This is a repetition of Problem 27, Section 16.2, but now the cross product is available.)

18. Containing $(2, -1, 1)$, $(5, 0, -1)$, and $(-2, 3, 3)$. (See Problem 28, Section 16.2.)

19. Containing $(3, 5, 0)$, $(1, -2, 1)$, and $(2, 2, 3)$.

20. Containing $(1, -1, 2)$ and the line $x = t - 1$, $y = 2t + 1$, $z = 2t$. (See Problem 29, Section 16.2.)

21. Containing the origin and the line $x = t - 1$, $y = 2t$, $z = 5t + 2$.

22. Containing the lines $x = 3t + 3$, $y = t$, $z = t + 3$ and $x = t - 2$, $y = 2t - 5$, $z = t$. (See Problem 30, Section 16.2.)

23. Containing the lines $x = 2t + 5$, $y = t - 3$, $z = -t$ and $x = t - 1$, $y = 2t$, $z = 1 - t$.

24. Containing $(1, -1, 4)$ and $(2, 0, 1)$ and parallel to the z axis.

25. Containing $(1, 1, 3)$ and $(2, -4, 1)$ and perpendicular to the plane $3x - y + z - 2 = 0$.

26. Show that an equation of the plane with nonzero intercepts $(a, 0, 0)$, $(0, b, 0)$, $(0, 0, c)$ is $x/a + y/b + z/c = 1$.

27. Find parametric equations of the line of intersection of the planes $2x + y - z - 6 = 0$ and $x + 4y + 2z - 1 = 0$.

28. Find parametric equations of the line perpendicular to the lines

$$L_1: x = 2t, \; y = t + 2, \; z = 2t + 1$$
$$L_2: x = t, \; y = t - 1, \; z = t + 1$$

at their point of intersection.

Prove each of the following statements.

29. The cross product is anticommutative, that is,

$$x_1 \times x_2 = -(x_2 \times x_1)$$

30. $(cx_1) \times x_2 = c(x_1 \times x_2) = x_1 \times (cx_2)$

31. The cross product is "left-distributive," that is,

$$x_1 \times (x_2 + x_3) = (x_1 \times x_2) + (x_1 \times x_3)$$

32. The cross product is "right-distributive," that is,

$$(x_2 + x_3) \times x_1 = (x_2 \times x_1) + (x_3 \times x_1)$$

Hint: Use Problem 29.

33. $(x_1 \times x_2) \cdot x_3 = x_1 \cdot (x_2 \times x_3)$ *Hint:* The equation is equivalent to

$$\begin{vmatrix} x_3 & y_3 & z_3 \\ x_1 & y_1 & z_1 \\ x_2 & y_2 & z_2 \end{vmatrix} = \begin{vmatrix} x_1 & y_1 & z_1 \\ x_2 & y_2 & z_2 \\ x_3 & y_3 & z_3 \end{vmatrix}$$ (Why?)

Hence you need only explain why these determinants are equal.

34. Use Problem 33 to prove that

$$(\mathbf{x}_1 \times \mathbf{x}_2) \cdot \mathbf{x}_1 = 0 \qquad \text{and} \qquad (\mathbf{x}_1 \times \mathbf{x}_2) \cdot \mathbf{x}_2 = 0$$

thus confirming that $\mathbf{x}_1 \times \mathbf{x}_2$ is perpendicular to both $\mathbf{x}_1$ and $\mathbf{x}_2$.

35. In the text we derived the formula

$$|\mathbf{x}_1 \times \mathbf{x}_2| = |\mathbf{x}_1||\mathbf{x}_2| \sin \theta$$

and then concluded that

$$\mathbf{x}_1 \text{ and } \mathbf{x}_2 \text{ are parallel} \Leftrightarrow \mathbf{x}_1 \times \mathbf{x}_2 = 0$$

Explain why this conclusion follows from the formula.

36. If $\mathbf{x}_1 \times \mathbf{x}_2 \neq 0$, explain why $|\mathbf{x}_1 \times \mathbf{x}_2|$ is the area of the parallelogram having $\mathbf{x}_1$ and $\mathbf{x}_2$ as adjacent sides.

37. Use Problem 36 to find the area of the triangle with vertices $(2, -1, 5)$, $(1, 0, -1)$, and $(4, 1, 1)$.

38. Find the area of the triangle with vertices $(1, -1, 2)$, $(4, 0, 1)$, and $(2, -1, 2)$.

39. A particle moves along a path in the xy plane with velocity $\mathbf{v}$ and acceleration $\mathbf{a}$. Show that the curvature of the path is

$$\kappa = \frac{|\mathbf{v} \times \mathbf{a}|}{|\mathbf{v}|^3}$$

16.4
QUADRIC SURFACES

In Section 16.2 we proved that the graph of a linear equation $ax + by + cz + d = 0$ is a plane. The graph of a second-degree equation in x, y, and z is called a **quadric surface**. It can be shown that every plane section of such a surface is a conic section (possibly degenerate), so these surfaces have names that are reminiscent of our study of circles, parabolas, ellipses, and hyperbolas in $\mathcal{R}^2$. In this section we present some examples to give you an idea of how quadric surfaces are classified. Our purpose is not to be systematic or exhaustive, but simply to supply enough information for you to be able to recognize a quadric surface when it comes up in practice.

□ **Example 1**   The equation $x^2 + y^2 = 1$ represents the unit circle in $\mathcal{R}^2$. In three-dimensional space, however, its graph is the set of points

$$\{(x, y, z): x^2 + y^2 = 1, z \in \mathcal{R}\}$$

The intersection of this graph with any horizontal plane $z = k$ is the unit circle $x^2 + y^2 = 1$ lying in that plane; as k varies, the circle moves parallel to the z axis and generates the *circular cylinder* shown in Figure 1.

In general the term **cylinder** refers to a surface generated by *any* plane curve that is moved through space in a direction perpendicular to its plane. The curve in Example 1 is a circle, hence the name *circular cylinder*.   □

□ **Example 2**   The equation $z = 1 - x^2$ represents a parabola in $\mathcal{R}^2$ (lying in the xz plane). In three-dimensional space its graph is the *parabolic cylinder*

$$\{(x, y, z): z = 1 - x^2, y \in \mathcal{R}\}$$

generated by moving the parabola parallel to the y axis. (See Figure 2.)   □

Figure 1
The circular cylinder $x^2 + y^2 = 1$

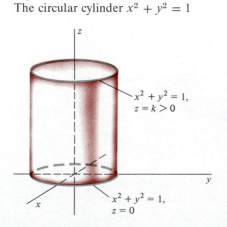

$x^2 + y^2 = 1$,
$z = k > 0$

$x^2 + y^2 = 1$,
$z = 0$

Figure 2
The parabolic cylinder $z = 1 - x^2$

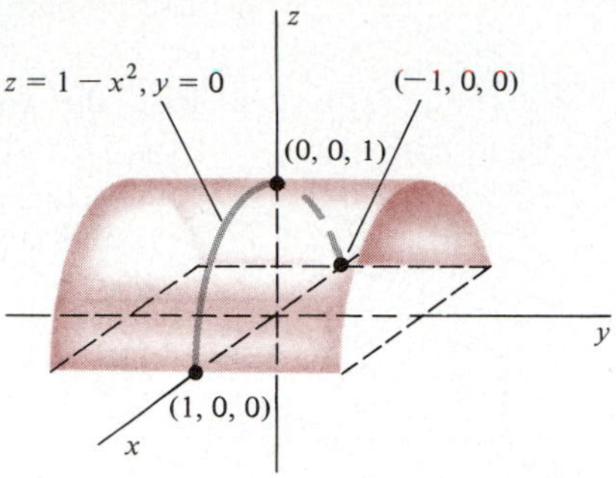

Figure 3
The cylinder $z = \sin y$

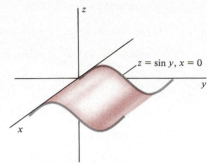

When the motion of the curve generating a cylinder is parallel to a coordinate axis, the corresponding variable will be missing from the equation of the cylinder. (Thus z is missing in Example 1, and y in Example 2.) These are the only types of cylinders you will encounter in this book. Of course not every cylinder is a quadric surface, but only the ones that are generated by second-degree curves in the plane. The cylinder $z = \sin y$, for example (Figure 3), is not a quadric surface.

☐ **Example 3**   The equation $x^2/a^2 + y^2/b^2 + z^2/c^2 = 1$ represents a *sphere* if $a = b = c$. (See Section 16.1.) In general its graph is a surface called an **ellipsoid.** See Figure 4, which illustrates a procedure that is often helpful in visualizing and sketching surfaces. The "trace" of the surface in the xy plane, for example, is obtained by putting $z = 0$ in the equation of the ellipsoid. This yields the ellipse

$$\frac{x^2}{a^2} + \frac{y^2}{b^2} = 1, \; z = 0$$

Figure 4
The ellipsoid $\dfrac{x^2}{a^2} + \dfrac{y^2}{b^2} + \dfrac{z^2}{c^2} = 1$

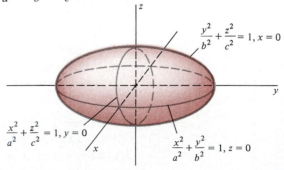

Similarly, putting $x = 0$ and $y = 0$ in the original equation gives elliptical cross sections in the other coordinate planes; these traces help us figure out what the surface looks like.   ☐

Remark   Note that in three-dimensional space we do not write the equation $x^2/a^2 + y^2/b^2 = 1$ alone and call it an ellipse. (By itself it is an elliptical cylinder, with axis parallel to the z axis.) Instead we write

$$\frac{x^2}{a^2} + \frac{y^2}{b^2} = 1, z = 0$$

which indicates that we are thinking of the intersection of the cylinder and the plane $z = 0$. This emphasizes a remark that we have made before, that a single equation in x, y, and z does not represent a curve, but a surface.

□ **Example 4**   The graph of $z = x^2/a^2 + y^2/b^2$ is called an **elliptical paraboloid** (circular if $a = b$). It opens upward from its vertex at the origin (because $z \geq 0$ for all x and y) and its horizontal cross sections are the ellipses (or circles)

$$\frac{x^2}{a^2} + \frac{y^2}{b^2} = k, z = k > 0$$

Its traces in the yz and xz planes are the parabolas

$$z = \frac{y^2}{b^2}, x = 0 \qquad \text{and} \qquad z = \frac{x^2}{a^2}, y = 0$$

respectively. (See Figure 5.)   □

Figure 5

The elliptical paraboloid $z = \dfrac{x^2}{a^2} + \dfrac{y^2}{b^2}$

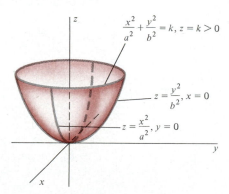

□ **Example 5**   The equation $z = y^2/b^2 - x^2/a^2$ represents a **hyperbolic paraboloid,** a saddle-shaped surface like the one shown in Figure 6. Horizontal cross sections are the hyperbolas

$$\frac{y^2}{b^2} - \frac{x^2}{a^2} = k, z = k > 0 \qquad \text{(opening in the direction of the } y \text{ axis)}$$

Figure 6

The hyperbolic paraboloid $z = \dfrac{y^2}{b^2} - \dfrac{x^2}{a^2}$

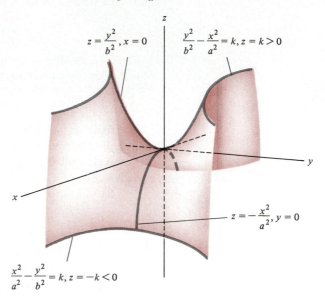

$z = \dfrac{y^2}{b^2}, x = 0$ $\dfrac{y^2}{b^2} - \dfrac{x^2}{a^2} = k, z = k > 0$

$z = -\dfrac{x^2}{a^2}, y = 0$

$\dfrac{x^2}{a^2} - \dfrac{y^2}{b^2} = k, z = -k < 0$

or

$$\frac{x^2}{a^2} - \frac{y^2}{b^2} = k, \; z = -k < 0 \qquad \text{(opening in the direction of the } x \text{ axis)}$$

Traces in the yz and xz planes are the parabolas

$$z = \frac{y^2}{b^2}, \; x = 0 \qquad \text{and} \qquad z = -\frac{x^2}{a^2}, \; y = 0$$

respectively.   □

□ **Example 6**   The graph of $x^2/a^2 + y^2/b^2 - z^2/c^2 = 1$ is a **hyperboloid of one sheet.** (See Figure 7.) Horizontal cross sections are ellipses (circles if $a = b$), the smallest being

$$\frac{x^2}{a^2} + \frac{y^2}{b^2} = 1, \; z = 0$$

in the xy plane. Traces in the yz and xz planes are the hyperbolas

$$\frac{y^2}{b^2} - \frac{z^2}{c^2} = 1, \; x = 0 \qquad \text{and} \qquad \frac{x^2}{a^2} - \frac{z^2}{c^2} = 1, \; y = 0$$

respectively.   □

Figure 7

The hyperboloid $\dfrac{x^2}{a^2} + \dfrac{y^2}{b^2} - \dfrac{z^2}{c^2} = 1$

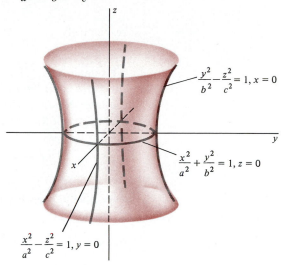

$\dfrac{y^2}{b^2} - \dfrac{z^2}{c^2} = 1, x = 0$

$\dfrac{x^2}{a^2} + \dfrac{y^2}{b^2} = 1, z = 0$

$\dfrac{x^2}{a^2} - \dfrac{z^2}{c^2} = 1, y = 0$

□ **Example 7**   The graph of $z^2/c^2 - x^2/a^2 - y^2/b^2 = 1$ is a **hyperboloid of two sheets**, as shown in Figure 8. Since

$$\frac{x^2}{a^2} + \frac{y^2}{b^2} = \frac{z^2}{c^2} - 1 \geq 0$$

values of z between c and $-c$ are not allowed; each value of z satisfying $|z| > c$ yields an ellipse in a plane parallel to the xy plane (a circle if $a = b$).

Figure 8

The hyperboloid $\dfrac{z^2}{c^2} - \dfrac{x^2}{a^2} - \dfrac{y^2}{b^2} = 1$

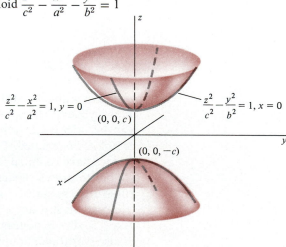

$\dfrac{z^2}{c^2} - \dfrac{x^2}{a^2} = 1, y = 0$

$(0, 0, c)$

$\dfrac{z^2}{c^2} - \dfrac{y^2}{b^2} = 1, x = 0$

$(0, 0, -c)$

Traces in the yz and xz planes are the hyperbolas

$$\frac{z^2}{c^2} - \frac{y^2}{b^2} = 1, \ x = 0 \qquad \text{and} \qquad \frac{z^2}{c^2} - \frac{x^2}{a^2} = 1, \ y = 0$$

respectively.   ☐

☐ **Example 8**   The graph of $z^2 = x^2/a^2 + y^2/b^2$ is an **elliptical cone** (circular if $a = b$). Horizontal cross sections are ellipses (or circles) except for the intersection of the cone with the plane $z = 0$, which is the origin alone (the *vertex* of the cone). Traces in the yz and xz planes are the straight lines

$$z = \pm\frac{y}{b}, \ x = 0 \qquad \text{and} \qquad z = \pm\frac{x}{a}, \ y = 0$$

respectively. (See Figure 9.)   ☐

Figure 9

The cone $z^2 = \dfrac{x^2}{a^2} + \dfrac{y^2}{b^2}$

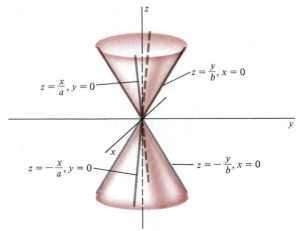

Every quadric surface is like one of the types presented in these examples (cylinder, ellipsoid, paraboloid, hyperboloid, or cone). Of course they can be oriented in other ways relative to the coordinate axes, as well as translated or rotated. The problem set contains examples of such variations, but we have confined them to surfaces that are reasonably easy to describe (if not to draw). Sketching surfaces on a two-dimensional sheet of paper often requires artistic ability that we do not expect you to have. In most applications it is not so important to be able to draw an elegant figure as it is to visualize in your imagination what the three-dimensional situation is. It is also worth mentioning that computer-generated pictures are now practical; they are often startlingly realistic.

Problem Set 16.4

Identify each of the following surfaces as one of the types listed below, and sketch it.

• Plane
• Cylinder (circular, elliptical, parabolic, hyperbolic, or not quadric)
• Ellipsoid (or sphere)
• Paraboloid (circular, elliptical, or hyperbolic)
• Hyperboloid (one or two sheets)
• Cone (circular or elliptical)

1. $z = x^2 + y^2$
2. $y = x^2 + z^2$
3. $x^2 + y^2 + (z - 1)^2 = 1$
4. $x^2 + z^2 = 4$
5. $x^2 + (y - 1)^2 = 1$
6. $z = y^2$
7. $y^2 - z^2 = 1$
8. $z = 1 - y^2$
9. $z - 1 = x^2 + y^2$
10. $z = 4 - x^2 - y^2$
11. $z = \cos y$
12. $x/2 + y/3 + z/4 = 1$
13. $y + z = 1$
14. $z^2 = x^2 + y^2$
15. $x^2 + y^2 + z^2 = 9$
16. $x^2 + y^2 - z^2 = 4$
17. $z^2 - x^2 - y^2 = 4$
18. $yz = 1$
19. $z = y^2 - x^2$
20. $z = x^2 - y^2$
21. $(z - 1)^2 = x^2 + y^2$
22. $x^2/4 + y^2/9 + z^2/25 = 1$
23. $x^2 + y^2 + z^2/4 = 1$
24. $z = e^y$
25. $z^2 = x^2 + y^2 + 1$
26. $x^2 + y^2 = z^2 + 1$
27. $z = \ln y$
28. $36x^2 + 9y^2 + 4z^2 = 36$
29. $36z^2 = 9x^2 + 4y^2$
30. $x + y = 1$
31. $z = 4 - x^2$
32. $x^2 + y^2 - z^2 = 4$

Sketch the graph of each of the following surfaces.

33. $z = \sqrt{x^2 + y^2}$
34. $y = \sqrt{x^2 + z^2}$
35. $z = \sqrt{4 - x^2 - y^2}$
36. $x^2 - y^2 = 0$
37. $(x - 1)^2 + (y - 1)^2 = 0$

Sketch each of the following.

38. The "ice cream cone" bounded by the hemisphere $x^2 + y^2 + z^2 = 2$, $z \geq 0$, and the cone $z^2 = x^2 + y^2$.
39. The cap of the sphere $x^2 + y^2 + z^2 = 25$ above the plane $z = 3$.
40. The part of the paraboloid $z = x^2 + y^2$ below the plane $z = 1$.
41. The region above the xy plane, between the planes $y = 0$ and $y = 2$, and below the parabolic cylinder $z = 1 - x^2$.
42. The part of the first octant below the plane $x + z = 1$ and inside the cylinder $x^2 + y^2 = 1$.
43. The wedge in the first octant cut from the cylinder $y^2 + z^2 = 4$ by the yz plane and the plane $y = x$.
44. The region above the xy plane bounded by the cone $z^2 = x^2 + y^2$ and the paraboloid $z = 2 - x^2 - y^2$.
45. The region above the xy plane, inside the cylinder $x^2 + y^2 = 1$, and below the paraboloid $z = 4 - x^2 - y^2$.
46. The region above the paraboloid $z = x^2 + y^2$ and inside the sphere $x^2 + y^2 + z^2 = 2$.
47. The region bounded by the hyperboloid $x^2 + y^2 = z^2 - 1$ and the plane $z = 2$.

16.5

CYLINDRICAL AND SPHERICAL COORDINATES

Polar coordinates in the plane are a natural way of simplifying the equations of many graphs, as we saw in Chapter 14. We can take advantage of them in space by imposing a polar coordinate system on the xy plane in the usual way, leaving the third space coordinate alone. In other words, we replace the triple (x, y, z) by (r, θ, z) where

$$x = r \cos \theta \quad \text{and} \quad y = r \sin \theta$$

(as in $\mathfrak{R}^2$). The numbers (r, θ, z) are called **cylindrical coordinates** of the point whose rectangular coordinates are (x, y, z).

The reason for the name is that the graph of $r = a$ (where a is a nonzero constant) is a circular cylinder (rather than a circle, as in $\mathfrak{R}^2$). You can see why by observing that it is the set of points

$$\{(r, \theta, z): r = a, \theta \in \mathfrak{R}, z \in \mathfrak{R}\}$$

Figure 1
The cylinder $r = a$ and the plane $\theta = \alpha$

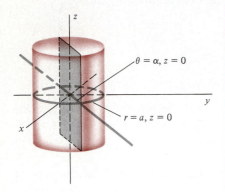

In rectangular coordinates this is the set

$$\{(x, y, z): x^2 + y^2 = a^2, z \in \mathcal{R}\}$$

which as we saw in the last section is the vertical cylinder consisting of all points at a fixed distance from the z axis. (See Figure 1, which also shows the plane $\theta = \alpha$, where α is constant.) In $\mathcal{R}^2$ the equation $\theta = \alpha$ represents a line through the origin, but its graph in 3-space is the set

$$\{(r, \theta, z): r \in \mathcal{R}, \theta = \alpha, z \in \mathcal{R}\}$$

or (in rectangular coordinates) the plane

$$\{(x, y, z): y = mx, z \in \mathcal{R}\}$$

where $m = \tan \alpha$. (If m is undefined, which happens when the terminal side of α lies in the y axis, the plane is $x = 0$.)

□ **Example 1**   Describe the graph (in cylindrical coordinates) of the equation $z = r^2$.

Solution: This is the same as the circular paraboloid $z = x^2 + y^2$ in rectangular coordinates. (See Example 4, Section 16.4.)   □

Figure 2
The hemisphere $z = \sqrt{1 - r^2}$

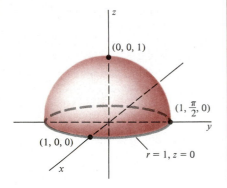

□ **Example 2**   Describe the graph (in cylindrical coordinates) of the equation $z = \sqrt{1 - r^2}$.

Solution: Square both sides to obtain $z^2 = 1 - r^2$, or (in rectangular coordinates)

$$x^2 + y^2 + z^2 = 1$$

Since $z \geq 0$ in the original equation, the graph is the upper half of the unit sphere. (See Figure 2.)   □

□ **Example 3**   The graph (in cylindrical coordinates) of the equation $z = r$ is the circular cone $z^2 = x^2 + y^2$ in rectangular coordinates. (See Example 8, Section 16.4.) Note that although we squared both sides of $z = r$ to obtain $z^2 = x^2 + y^2$, the equations have the same graph. The reason for this is that r can be negative, which means that the bottom half of the cone is included. (Why does the equation $z = -r$ also represent the whole cone?) If we restrict $r \geq 0$ (which is often done in applications of cylindrical coordinates), the equation $z = r$ represents only the top half of the cone.   □

Figure 3
Spherical coordinates of a point P

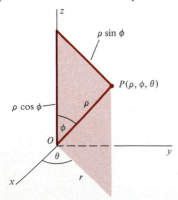

Spherical coordinates (ρ, ϕ, θ) of a point $P(x, y, z)$ in space are shown in Figure 3. The first coordinate ρ is defined to be the length of the vector $\overrightarrow{OP}$, which means that

$$\boxed{\rho^2 = x^2 + y^2 + z^2 \qquad (\rho \geq 0)}$$

The second coordinate ϕ is defined to be the angle between $\overrightarrow{OP}$ and $\mathbf{k}$, which means that

$$\cos\phi = \frac{\overrightarrow{OP}\cdot\mathbf{k}}{|\overrightarrow{OP}||\mathbf{k}|} = \frac{z}{\rho}, \quad 0 \le \phi \le \pi$$

Hence $z = \rho\cos\phi$. The third coordinate θ is the usual polar angle in the xy plane; its value is unrestricted.

One way of describing spherical coordinates is to observe that for each value of θ, the numbers ρ and ϕ are polar coordinates of P in the half-plane containing the terminal side of θ and the z axis.

We have already seen that $z = \rho\cos\phi$. To express x and y in terms of spherical coordinates, observe from Figure 3 that

$$r = \rho\sin\phi \qquad (r \ge 0)$$

(For an argument not depending on pictures, see the problem set.) Hence

$$x = r\cos\theta = \rho\sin\phi\cos\theta \qquad \text{and} \qquad y = r\sin\theta = \rho\sin\phi\sin\theta$$

Figure 4
The graph of $\phi = \alpha$

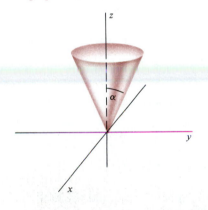

Relations Between Rectangular, Cylindrical, and Spherical Coordinates

Rectangular-cylindrical	Rectangular-spherical
$x = r\cos\theta$	$x = \rho\sin\phi\cos\theta$
$y = r\sin\theta$	$y = \rho\sin\phi\sin\theta$
$z = z$	$z = \rho\cos\phi$

The connection between cylindrical and spherical coordinates is by way of the equation $r = \rho\sin\phi$, but note that the restrictions $\rho \ge 0$ and $0 \le \phi \le \pi$ require $r \ge 0$. Hence we cannot allow r to be negative in any context where cylindrical and spherical coordinates are used together.

The reason for the term *spherical* coordinates is that the graph of $\rho = a$ (where a is a positive constant) is the sphere of radius a and center at the origin. (Why?) It is also worth noting that the graph of $\phi = \alpha$ (where $0 < \alpha < \pi, \alpha \ne \pi/2$) is a half-cone, as shown in Figure 4. (This degenerates into the positive z axis if $\alpha = 0$, the xy plane if $\alpha = \pi/2$, and the negative z axis if $\alpha = \pi$. Why?) Similarly, the graph of $\theta = \beta$ (where β is any constant) is a half-plane with the z axis as edge, as in Figure 5. (Why just a *half*-plane?) Thus the location of a point $P(\rho, \phi, \theta)$ in spherical coordinates may be described as the intersection of the sphere determined by ρ, the half-cone determined by ϕ, and the half-plane determined by θ.

Figure 5
The graph of $\theta = \beta$

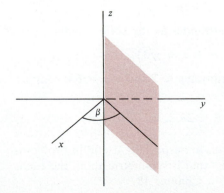

□ **Example 4**   Find rectangular and cylindrical coordinates of the point with spherical coordinates $(2, \pi/3, \pi/6)$.

Solution: Since $\rho = 2$, $\phi = \pi/3$, and $\theta = \pi/6$, we find

$$x = \rho \sin \phi \cos \theta = (2)(\sqrt{3}/2)(\sqrt{3}/2) = 3/2$$
$$y = \rho \sin \phi \sin \theta = (2)(\sqrt{3}/2)(1/2) = \sqrt{3}/2$$
$$z = \rho \cos \phi = (2)(1/2) = 1$$

Cylindrical coordinates are

$$r = \rho \sin \phi = (2)(\sqrt{3}/2) = \sqrt{3}, \quad \theta = \frac{\pi}{6}, \quad z = 1 \qquad \square$$

□ **Example 5**   What is the graph of the equation $\rho = \sec \phi$ in spherical coordinates?

Solution: The equation is equivalent to $\rho \cos \phi = 1$, so the graph is the plane $z = 1$. $\qquad\qquad\qquad\qquad\qquad\qquad\qquad\qquad\qquad\qquad\qquad\quad \square$

□ **Example 6**   Describe the graph of the equation $\rho = 2 \cos \phi$ in spherical coordinates.

Solution: Multiply each side by ρ to obtain $\rho^2 = 2\rho \cos \phi$, or (in rectangular coordinates) $x^2 + y^2 + z^2 = 2z$. This is the sphere

$$x^2 + y^2 + (z - 1)^2 = 1$$

of radius 1 and center $(0, 0, 1)$ in rectangular coordinates. $\qquad\qquad\quad \square$

□ **Example 7**   Describe the set of points

$$\{(\rho, \phi, \theta): \rho = 2 \text{ and } \phi = \pi/6\}$$

Solution: This is the intersection of the sphere $\rho = 2$ and the half-cone $\phi = \pi/6$. Hence it is the colored circle in Figure 6. The radius of the circle is

$$r = \rho \sin \theta = (2)(\tfrac{1}{2}) = 1$$

and its height above the xy plane is

$$z = \rho \cos \phi = (2)(\sqrt{3}/2) = \sqrt{3}$$

It may be described in rectangular coordinates as the set of points

$$\{(x, y, z): x^2 + y^2 = 1, z = \sqrt{3}\}$$

Note that the spherical coordinate description $\rho = 2$ and $\phi = \pi/6$ is simpler. A cylindrical coordinate description is

$$\{(r, \theta, z): r = 1, 0 \leq \theta \leq 2\pi, z = \sqrt{3}\} \qquad\qquad\qquad \square$$

Cylindrical and spherical coordinates are especially useful in problems involving integration over a region in $\mathcal{R}^3$ that is symmetric about the z axis or the origin. We shall return to them in Chapter 18.

Figure 6
The circle $\rho = 2$, $\phi = \pi/6$

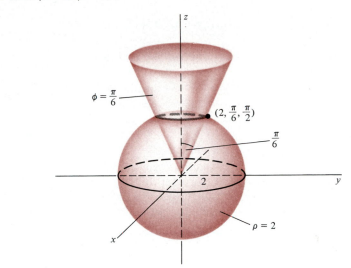

Problem Set 16.5

1. Find cylindrical and spherical coordinates of the point with rectangular coordinates $(0, 1, 1)$.

2. Repeat Problem 1 with the rectangular coordinates $(1, 0, 1)$.

3. Find rectangular and spherical coordinates of the point with cylindrical coordinates $(1, \pi, 1)$.

4. Repeat Problem 3 with the cylindrical coordinates $(1, \pi/2, -1)$.

5. Find rectangular and cylindrical coordinates of the point with spherical coordinates $(2, \pi/3, \pi/2)$.

6. Repeat Problem 5 with the spherical coordinates $(4, \pi/6, \pi)$.

Describe the graph of each of the following equations in cylindrical coordinates.

7. $r = 2$

8. $r = 0$

9. $\theta = 0$

10. $\theta = \pi/2$

11. $\theta = \pi/4$

12. $z = 1 - r^2$

13. $r^2 + z^2 = 4$

14. $r = 2 \sin \theta$

15. $z = r \sin \theta$

16. $z = \sqrt{9 - r^2}$

17. $z = 2r$

18. $z = r \cos \theta$

Describe the graph of each of the following equations in spherical coordinates.

19. $\rho = 2$

20. $\rho = 0$

21. $\phi = \pi/4$

22. $\phi = \pi/2$

23. $\phi = 3\pi/4$

24. $\rho = 4 \cos \phi$

25. $\rho = 2 \sec \phi$

26. $\rho = \csc \phi$

27. $\rho \sin^2 \phi = \cos \phi$

28. $\rho = \sin 2\phi$ *Hint:* For each fixed $\theta = \alpha$, $\rho = \sin 2\phi$ describes what polar coordinate graph in the half-plane $\theta = \alpha$?

29. Describe the set of points

$$\{(\rho, \phi, \theta): \rho = 2 \text{ and } \rho \sin^2 \phi = 3 \cos \phi\}$$

30. Describe the set of points

$$\{(\rho, \phi, \theta): 0 \le \rho \le 1, 0 \le \phi \le \pi/2, 0 \le \theta \le \pi/2\}$$

31. Describe the set of points

$$\{(\rho, \phi, \theta): 0 \le \rho \le 1, 0 \le \phi \le \pi/6, 0 \le \theta \le 2\pi\}$$

32. Describe the set of points

$$\{(\rho, \phi, \theta): \sec \phi \le \rho \le 2, 0 \le \phi \le \pi/3, 0 \le \theta \le 2\pi\}$$

33. Let P be a point with cylindrical coordinates (r, θ, z), where $r \geq 0$, and spherical coordinates (ρ, ϕ, θ). Show that $r = \rho \sin \phi$ as follows.
   (a) Explain why $|\mathbf{k} \times \overrightarrow{OP}| = \rho \sin \phi$, where O is the origin.
   (b) Show that $\mathbf{k} \times \overrightarrow{OP} = (-y, x, 0)$ and use the result to finish the proof.
34. In Chapter 18 we will have occasion to find "partial derivatives" of the rectangular coordinates $x = r \cos \theta$, $y = r \sin \theta$, $z = z$ with respect to the cylindrical coordinates, r, θ, z:

$$D_r x = \cos \theta, \; D_r y = \sin \theta, \; D_r z = 0$$
$$\text{(holding } \theta \text{ and } z \text{ fixed)}$$
$$D_\theta x = -r \sin \theta, \; D_\theta y = r \cos \theta, \; D_\theta z = 0$$
$$\text{(holding } r \text{ and } z \text{ fixed)}$$
$$D_z x = 0, \; D_z y = 0, \; D_z z = 1 \quad \text{(holding } r \text{ and } \theta \text{ fixed)}$$

The determinant

$$\begin{vmatrix} D_r x & D_\theta x & D_z x \\ D_r y & D_\theta y & D_z y \\ D_r z & D_\theta z & D_z z \end{vmatrix}$$

is called the **Jacobian** of the transformation from cylindrical to rectangular coordinates. Show that its value is r.

35. Show that the Jacobian of the transformation from spherical to rectangular coordinates is

$$\begin{vmatrix} \sin \phi \cos \theta & \rho \cos \phi \cos \theta & -\rho \sin \phi \sin \theta \\ \sin \phi \sin \theta & \rho \cos \phi \sin \theta & \rho \sin \phi \cos \theta \\ \cos \phi & -\rho \sin \phi & 0 \end{vmatrix} = \rho^2 \sin \phi$$

16.6
CURVES IN SPACE

Curves in $\mathscr{R}^3$ are studied in much the same way as in $\mathscr{R}^2$ (in fact a great deal of the terminology is word-for-word what it is in the plane). We will present most of it without comment, trusting you to look up anything you have forgotten in Section 7.3 and Chapter 15. The main thing that is different in $\mathscr{R}^3$ is that the path of a point can "twist" as well as "turn." The turning (measured by the rate of change of the unit tangent vector) is described by curvature, as in the plane. The twisting requires something new.

□ **Example 1**   The parametric equations $x = 2t$, $y = t - 1$, $z = t + 1$ represent a straight line parallel to the vector $\mathbf{m} = (2, 1, 1)$, as we saw in Section 16.2. This is a "curve" in space, but it neither turns nor twists, and is therefore not very interesting. It does suggest, however, that parametric equations are the heart of the matter. In general they involve the parameter in a nonlinear way, in which case we need heavier artillery to deal with them.   □

□ **Example 2**   The parametric equations $x = \cos t$, $y = \sin t$, $z = t$ represent a curve called a **circular helix.** Since

$$x^2 + y^2 = \cos^2 t + \sin^2 t = 1 \qquad \text{for all } t$$

the graph lies on the surface $x^2 + y^2 = 1$, which is a circular cylinder of radius 1 symmetric about the z axis. As t goes through the interval $(-\infty, \infty)$, z increases; the point (x, y, z) spirals up and around the cylinder as shown in Figure 1.   □

Figure 1
Circular helix $x = \cos t$, $y = \sin t$, $z = t$

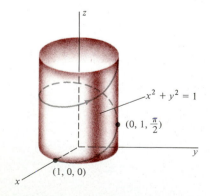

To analyze this kind of thing, we start by repeating (in the context of three-dimensional space) the ideas about curves that were developed in the plane. A **curve** in $\mathcal{R}^3$ is a vector function

$$\mathbf{F}(t) = (f(t), g(t), h(t))$$

where f, g, and h are real functions defined in some closed interval I. Its **graph** (or **track**) is the set of points (x, y, z) satisfying the parametric equations $x = f(t), y = g(t), z = h(t), t \in I$. We assume that the curve is **smooth,** which means that f', g', and h' are continuous in I and

$$\left(\frac{dx}{dt}\right)^2 + \left(\frac{dy}{dt}\right)^2 + \left(\frac{dz}{dt}\right)^2 > 0 \qquad \text{for all interior points of } I$$

The **position vector** of a given point (x, y, z) of the graph is

$$\mathbf{x} = \mathbf{F}(t) = (x, y, z)$$

(understood to start at the origin and terminate at the given point). Its **derivative** is

$$\frac{d\mathbf{x}}{dt} = \mathbf{F}'(t) = \left(\frac{dx}{dt}, \frac{dy}{dt}, \frac{dz}{dt}\right) = \lim_{z \to t} \frac{\mathbf{F}(z) - \mathbf{F}(t)}{z - t}$$

(the limit being evaluated "componentwise," as in the case of functions from I to $\mathcal{R}^2$). Since the magnitude of this vector is

$$|\mathbf{F}'(t)| = \sqrt{(dx/dt)^2 + (dy/dt)^2 + (dz/dt)^2} > 0$$

the vector is never zero (except possibly at the endpoints of I). Hence it has a definite direction. The **tangent line** at (x, y, z) is defined to be the line through (x, y, z) parallel to $\mathbf{F}'(t)$.

Remark   In $\mathcal{R}^2$ we defined "tangent line" long before considering vector functions. Hence there was something to prove when we asserted that the vector derivative is parallel to the tangent. In $\mathcal{R}^3$ the tangent is *defined* by the vector derivative.

□ **Example 3**   Find the tangent line to the graph of

$$\mathbf{F}(t) = (\cos t, \sin t, t)$$

at the point $(1, 0, 0)$ corresponding to the parameter value $t = 0$.

Solution: Since $\mathbf{F}'(t) = (-\sin t, \cos t, 1)$, a direction vector of the tangent line is $\mathbf{m} = \mathbf{F}'(0) = (0, 1, 1)$. The point of tangency is the terminal point of the position vector $\mathbf{F}(0) = (1, 0, 0)$, so the tangent line is defined by the parametric equations

$$x = 1 + 0u, \ y = 0 + 1u, \ z = 0 + 1u$$

or simply $x = 1, y = u, z = u$. This is the line $z = y$ in the plane $x = 1$. (See Figure 2.) *Note:* It is legitimate to write the line as $x = 1, y = t, z = t$. We used u to avoid confusion with the parameter t in $\mathbf{F}(t)$.   □

Figure 2
Tangent to $\mathbf{F}(t) = (\cos t,\ \sin t,\ t)$ at $(1, 0, 0)$

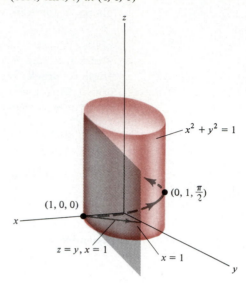

The **length** of the curve $\mathbf{x} = \mathbf{F}(t) = (x, y, z)$, $a \leq t \leq b$, is

$$s = \int_a^b \sqrt{(dx/dt)^2 + (dy/dt)^2 + (dz/dt)^2}\ dt = \int_a^b |\mathbf{F}'(t)|\ dt$$

If the upper limit is replaced by t, the length is a function of t, namely

$$s(t) = \int_a^t |\mathbf{F}'(u)|\ du$$

with derivative

$$\frac{ds}{dt} = |\mathbf{F}'(t)| = \sqrt{(dx/dt)^2 + (dy/dt)^2 + (dz/dt)^2}$$

□ **Example 4**   Find ds/dt for the curve $\mathbf{F}(t) = (\cos t, \sin t, t)$ and compute the length from $(1, 0, 0)$ to $(0, 1, \pi/2)$.

Solution: Since $\mathbf{F}'(t) = (-\sin t, \cos t, 1)$, we have

$$|\mathbf{F}'(t)|^2 = (-\sin t)^2 + \cos^2 t + 1 = 2$$

and hence $ds/dt = \sqrt{2}$. The length corresponding to the parameter interval $[0, \pi/2]$ is

$$\int_0^{\pi/2} |\mathbf{F}'(t)|\ dt = \int_0^{\pi/2} \sqrt{2}\ dt = \frac{\pi}{\sqrt{2}} \approx 2.2 \qquad \square$$

The **unit tangent vector** at a point (x, y, z) of the curve $\mathbf{x} = \mathbf{F}(t)$ is

$$\mathbf{T} = \frac{\mathbf{F}'(t)}{|\mathbf{F}'(t)|}$$

from which we have

$$\mathbf{F}'(t) = \frac{ds}{dt}\mathbf{T}$$

Hence (by the Product Rule and Chain Rule)

$$\mathbf{F}''(t) = \frac{d^2s}{dt^2}\mathbf{T} + \frac{ds}{dt}\frac{d\mathbf{T}}{dt} = \frac{d^2s}{dt^2}\mathbf{T} + \left(\frac{ds}{dt}\right)^2\frac{d\mathbf{T}}{ds}$$

Since $\mathbf{T} \cdot \mathbf{T} = |\mathbf{T}|^2 = 1$ for all t, the Product Rule for dot products yields

$$\mathbf{T} \cdot \frac{d\mathbf{T}}{ds} + \frac{d\mathbf{T}}{ds} \cdot \mathbf{T} = 0$$

from which $\mathbf{T} \cdot (d\mathbf{T}/ds) = 0$. Hence $\mathbf{T}$ and $d\mathbf{T}/ds$ are perpendicular. The **unit normal vector** (sometimes called the *principal* normal) is

$$\mathbf{N} = \frac{d\mathbf{T}/ds}{|d\mathbf{T}/ds|}$$

Remark $\mathbf{N}$ is defined only if $|d\mathbf{T}/ds| \neq 0$, a restriction we assume throughout this section. (If $d\mathbf{T}/ds$ is the zero vector for all t, the tangent vector $\mathbf{T}$ is constant and hence the path is a straight line.)

We also remark that the rules for differentiation of vector functions (used in the preceding discussion) are proved as in Section 15.2. A new one should be added, however:

$$\frac{d}{dt}\mathbf{F}(t) \times \mathbf{G}(t) = \mathbf{F}(t) \times \mathbf{G}'(t) + \mathbf{F}'(t) \times \mathbf{G}(t)$$

Note that since the cross product is not commutative, the order of the factors in each term is important.

We define **curvature** at (x, y, z) to be $\kappa = |d\mathbf{T}/ds|$. In Section 15.3 we defined it by the formula $\kappa = |d\phi/ds|$, where ϕ is the angle from $\mathbf{i}$ to $\mathbf{T}$. That would not make sense here, since $\mathbf{T}$ is a space vector whose direction cannot be specified by a single angle. The formula $\kappa = |d\mathbf{T}/ds|$ reduces to $\kappa = |d\phi/ds|$ in the plane, however. (See Problem 36, Section 15.3.) Note that since $\mathbf{T}$ has constant length 1, it can change only in direction. Hence κ is a measure of the rate at which the tangent line to the graph of $\mathbf{F}$ turns (with respect to distance measured along the graph).

It follows from the definition of curvature that $d\mathbf{T}/ds = \kappa\mathbf{N}$ and hence

$$\mathbf{F}''(t) = \frac{d^2s}{dt^2}\mathbf{T} + \kappa\left(\frac{ds}{dt}\right)^2\mathbf{N} \tag{1}$$

To find a manageable formula for curvature, we proceed differently than we did in Section 15.4 (where we did not have cross products to work with). The first step is not obvious, but note how smoothly it works! Cross each side of Equation (1) with the unit tangent vector **T**:

$$\mathbf{T} \times \mathbf{F}'' = \mathbf{T} \times (s''\mathbf{T} + \kappa s'^2 \mathbf{N})$$

$$\frac{\mathbf{F}'}{|\mathbf{F}'|} \times \mathbf{F}'' = \mathbf{T} \times (s''\mathbf{T}) + \mathbf{T} \times (\kappa s'^2 \mathbf{N})$$

$$\frac{1}{s'}(\mathbf{F}' \times \mathbf{F}'') = s''(\mathbf{T} \times \mathbf{T}) + \kappa s'^2 (\mathbf{T} \times \mathbf{N}) = \kappa s'^2 (\mathbf{T} \times \mathbf{N})$$

$$(\text{because } \mathbf{T} \times \mathbf{T} = \mathbf{0})$$

$$\mathbf{F}' \times \mathbf{F}'' = \kappa s'^3 (\mathbf{T} \times \mathbf{N}) \qquad (2)$$

Taking the magnitude of each side, we have

$$|\mathbf{F}' \times \mathbf{F}''| = \kappa s'^3 |\mathbf{T} \times \mathbf{N}|$$

But $|\mathbf{T} \times \mathbf{N}| = |\mathbf{T}||\mathbf{N}| \sin 90° = 1$ (why?), so

$$\kappa = \frac{|\mathbf{F}' \times \mathbf{F}''|}{s'^3} = \frac{|\mathbf{F}' \times \mathbf{F}''|}{|\mathbf{F}'|^3}$$

If $\mathbf{F}(t)$ is a smooth vector function, the curvature of its graph is

$$\kappa = \frac{|\mathbf{F}' \times \mathbf{F}''|}{|\mathbf{F}'|^3}$$

□ **Example 5**   Find the curvature of the graph of

$$\mathbf{F}(t) = (\cos t, \sin t, t)$$

Solution: Since

$$\mathbf{F}'(t) = (-\sin t, \cos t, 1) \qquad \text{and} \qquad \mathbf{F}''(t) = (-\cos t, -\sin t, 0)$$

we find

$$\mathbf{F}'(t) \times \mathbf{F}''(t) = \begin{vmatrix} \mathbf{i} & \mathbf{j} & \mathbf{k} \\ -\sin t & \cos t & 1 \\ -\cos t & -\sin t & 0 \end{vmatrix} = (\sin t)\mathbf{i} - (\cos t)\mathbf{j} + 1\mathbf{k}$$

$$= (\sin t, -\cos t, 1)$$

from which

$$|\mathbf{F}'(t) \times \mathbf{F}''(t)| = \sqrt{\sin^2 t + (-\cos t)^2 + 1} = \sqrt{2}$$

Moreover, $|\mathbf{F}'(t)| = \sqrt{2}$ (from Example 4). Hence

$$\kappa = \frac{|\mathbf{F}' \times \mathbf{F}''|}{|\mathbf{F}'|^3} = \frac{\sqrt{2}}{2\sqrt{2}} = \frac{1}{2}$$

Thus the circular helix has constant curvature, which means that the tangent line turns at a constant rate with respect to distance measured along the graph. If the path were in the plane, it would have to be a circle. (Why?) In fact it *is* circular (winding around the cylinder $x^2 + y^2 = 1$) except for the fact that it rises as it winds. A bug flying along this path (and unable to distinguish changes in altitude) would think that it is traveling in a circle.

$\square$

The perpendicular unit vectors $\mathbf{T}$ and $\mathbf{N}$ determine a plane containing the tangent line and (principal) normal line through (x, y, z). This is the plane in which the path "momentarily" lies, in the sense that if there were no twisting going on, the path would stay in the plane. (In $\Re^2$ it is the xy plane and the path *does* stay in the plane.) Because it contains the circle of curvature defined in Section 15.4 (the "osculating circle") it is called the *osculating plane*. In space the plane can turn; by measuring the rate of change of its direction we get an idea of how the path twists.

The direction of a plane is specified by a normal vector. Since the osculating plane is determined by $\mathbf{T}$ and $\mathbf{N}$, a natural choice of a vector normal to the plane is $\mathbf{B} = \mathbf{T} \times \mathbf{N}$. This has unit length (why?) and it is perpendicular to both $\mathbf{T}$ and $\mathbf{N}$; we call it the **unit binormal vector.** Since it can change only in direction (its length is fixed), the derivative $d\mathbf{B}/ds$ contains the information we need to measure the rate of twist. In the problem set we ask you to prove that $d\mathbf{B}/ds$ is perpendicular to $\mathbf{B}$ and $\mathbf{T}$. Since $\mathbf{N}$ is also perpendicular to $\mathbf{B}$ and $\mathbf{T}$, the vectors $d\mathbf{B}/ds$ and $\mathbf{N}$ are parallel, that is, one is a scalar multiple of the other. For reasons that are explained in "differential geometry" (the study of curves and surfaces by the methods of calculus), we express this fact in the form $d\mathbf{B}/ds = -\tau \mathbf{N}$, where τ (the Greek letter *tau*) is a scalar called the **torsion** of the curve. Since $|\tau| = |d\mathbf{B}/ds|$, the torsion is a measure of the rate at which $\mathbf{B}$ turns, just as the curvature $\kappa = |d\mathbf{T}/ds|$ is a measure of the rate at which $\mathbf{T}$ turns. The only difference in the definitions is that κ is never negative, whereas τ can be any real number.

A formula for computing τ in terms of the vector function $\mathbf{F}$ is harder to derive than the formula for κ. We have put an outline in the problem set; it is an elegant exercise in vector calculus.

If $\mathbf{F}(t)$ is a smooth vector function, the torsion of its graph is

$$\tau = \frac{(\mathbf{F}' \times \mathbf{F}'') \cdot \mathbf{F}'''}{|\mathbf{F}' \times \mathbf{F}''|^2}$$

$\square$ **Example 6**   Find the torsion of the graph of $\mathbf{F}(t) = (\cos t, \sin t, t)$.

Solution: In Example 5 we found

$$\mathbf{F}'(t) \times \mathbf{F}''(t) = (\sin t, -\cos t, 1)$$

and

$$|\mathbf{F}'(t) \times \mathbf{F}''(t)| = \sqrt{2}$$

Since $\mathbf{F}'''(t) = (\sin t, -\cos t, 0)$, we have

$$(\mathbf{F}'(t) \times \mathbf{F}''(t)) \cdot \mathbf{F}'''(t) = \sin^2 t + \cos^2 t = 1$$

and hence

$$\tau = \frac{(\mathbf{F}' \times \mathbf{F}'') \cdot \mathbf{F}'''}{|\mathbf{F}' \times \mathbf{F}''|^2} = \frac{1}{2}$$

We saw in Example 5 that $\kappa = \frac{1}{2}$, which means that the tangent line turns at a constant rate. Now we have shown that the osculating plane (determined by $\mathbf{T}$ and $\mathbf{N}$) also turns at a constant rate, that is, the helix twists uniformly as it winds around the cylinder $x^2 + y^2 = 1$. □

We now have formulas for κ and τ in terms of the original vector function $\mathbf{F}$ (and its derivatives). It is convenient to state such formulas for $\mathbf{T}$, $\mathbf{N}$, and $\mathbf{B}$ as well. We already have (by definition)

$$\boxed{\mathbf{T} = \frac{\mathbf{F}'}{|\mathbf{F}'|}}$$

and it is not hard to show that

$$\boxed{\mathbf{B} = \frac{\mathbf{F}' \times \mathbf{F}''}{|\mathbf{F}' \times \mathbf{F}''|}}$$

(See the Additional Problems at the end of the chapter.) We have also defined $\mathbf{B} = \mathbf{T} \times \mathbf{N}$, from which it follows that

$$\boxed{\mathbf{N} = \mathbf{B} \times \mathbf{T}}$$

(See the problem set.)

□ **Example 7**   Find $\mathbf{T}$, $\mathbf{N}$, and $\mathbf{B}$ for the curve $\mathbf{F}(t) = (\cos t, \sin t, t)$.

Solution: Since $\mathbf{F}'(t) = (-\sin t, \cos t, 1)$ and $|\mathbf{F}'(t)| = \sqrt{2}$, we find

$$\mathbf{T} = \frac{1}{\sqrt{2}}(-\sin t, \cos t, 1)$$

In Example 5 we found

$$\mathbf{F}'(t) \times \mathbf{F}''(t) = (\sin t, -\cos t, 1) \quad \text{and} \quad |\mathbf{F}'(t) \times \mathbf{F}''(t)| = \sqrt{2}$$

from which

$$\mathbf{B} = \frac{1}{\sqrt{2}}(\sin t, -\cos t, 1)$$

Finally,

$$\mathbf{N} = \frac{1}{2}\begin{vmatrix} \mathbf{i} & \mathbf{j} & \mathbf{k} \\ \sin t & -\cos t & 1 \\ -\sin t & \cos t & 1 \end{vmatrix} = -(\cos t, \sin t, 0) \qquad \square$$

Everything we have said so far is independent of any interpretation of the parameter t as time; we have been discussing purely geometric properties of curves in space. However, if $\mathbf{x} = \mathbf{F}(t)$ is the position of (x, y, z) at time t, we call

$$\mathbf{v} = \frac{d\mathbf{x}}{dt} = \mathbf{F}'(t)$$

the *velocity,*

$$\frac{ds}{dt} = |\mathbf{F}'(t)| = |\mathbf{v}|$$

the *speed,* and

$$\mathbf{a} = \frac{d\mathbf{v}}{dt} = \mathbf{F}''(t)$$

the *acceleration* of the moving point (just as in $\mathcal{R}^2$). Equation (1) then reads

$$\mathbf{a} = \frac{d^2s}{dt^2}\mathbf{T} + \kappa\left(\frac{ds}{dt}\right)^2\mathbf{N}$$

which is the same formula for acceleration in terms of tangential and normal components as given in $\mathcal{R}^2$. It is interesting to note that in $\mathcal{R}^3$ the acceleration has no "binormal" component, that is, the coefficient of $\mathbf{B}$ in the above formula is 0. Thus the acceleration vector (and hence the force vector) always lies in the osculating plane. At first glance this is puzzling, for if the force has only $\mathbf{T}$ and $\mathbf{N}$ components, why doesn't the moving particle stay in the plane determined by $\mathbf{T}$ and $\mathbf{N}$?

The answer is that if $\tau \neq 0$, the osculating plane is turning with time. See Figure 3, which shows the triple $\mathbf{T}$, $\mathbf{N}$, $\mathbf{B}$ (and the plane determined by $\mathbf{T}$ and $\mathbf{N}$) at a given instant. At a different time this triple (and the plane) have a new orientation. Thus the force can cause the particle to follow a twisting curve even though at every instant it has no component in the direction of $\mathbf{B}$.

Figure 3
The unit vectors $\mathbf{T}$, $\mathbf{N}$, $\mathbf{B}$

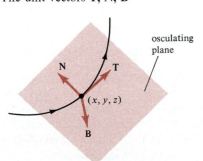

osculating plane

$\square$ **Example 8**   Suppose that the position of a particle at time $t \geq 0$ is given in spherical coordinates by the parametric equations $\rho = t$, $\phi = \pi/6$, $\theta = t$. Discuss the motion of the particle as it leaves the origin.

Figure 4
The conical helix $\rho = t$, $\phi = \pi/6$, $\theta = t$

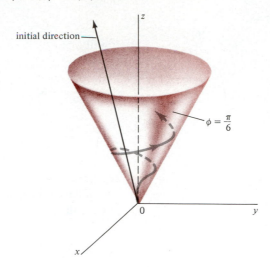

Solution: When $t = 0$ the particle is at the origin because $\rho = 0$. Since $\phi = \pi/6$ is the equation of the half-cone shown in Figure 4, the particle is always on this surface; it spirals up and around it because ρ and θ increase with t. The curve is called a **conical helix.**

To discuss the motion, observe that the position vector in this case is

$$\mathbf{x} = (x, y, z) = (\rho \sin \phi \cos \theta, \rho \sin \phi \sin \theta, \rho \cos \phi) = \tfrac{1}{2}(t \cos t, t \sin t, \sqrt{3}t)$$

Hence

$$\mathbf{x}' = \tfrac{1}{2}(\cos t - t \sin t, \sin t + t \cos t, \sqrt{3})$$
$$\mathbf{x}'' = \tfrac{1}{2}(-2 \sin t - t \cos t, 2 \cos t - t \sin t, 0)$$
$$\mathbf{x}''' = \tfrac{1}{2}(-3 \cos t + t \sin t, -3 \sin t - t \cos t, 0)$$

At $t = 0$ these vectors are

$$\mathbf{x} = (0, 0, 0) \qquad \mathbf{x}' = \tfrac{1}{2}(1, 0, \sqrt{3}) \qquad \mathbf{x}'' = \tfrac{1}{2}(0, 2, 0) \qquad \mathbf{x}''' = \tfrac{1}{2}(-3, 0, 0)$$

Since $\mathbf{x}'$ is a direction vector of the tangent line, parametric equations of the tangent line at the origin are $x = \tfrac{1}{2}u$, $y = 0$, $z = (\sqrt{3}/2)u$. This is the line $z = \sqrt{3}x$ in the xz plane (lying in the cone $\phi = \pi/6$). Hence the particle starts off as though it were going straight up the cone from the origin. Its speed as it leaves the origin is $ds/dt = |\mathbf{x}'| = 1$. It immediately veers off this course, however, because θ (which is 0 at $t = 0$) increases with t.

At $t = 0$ we have

$$\mathbf{x}' \times \mathbf{x}'' = \tfrac{1}{4}\begin{vmatrix} \mathbf{i} & \mathbf{j} & \mathbf{k} \\ 1 & 0 & \sqrt{3} \\ 0 & 2 & 0 \end{vmatrix} = \tfrac{1}{2}(-\sqrt{3}, 0, 1)$$

and $|\mathbf{x}' \times \mathbf{x}''| = 1$. Hence the curvature at $t = 0$ is

$$\kappa = \frac{|\mathbf{x}' \times \mathbf{x}''|}{|\mathbf{x}'|^3} = 1$$

The tangent line is turning at this rate as the particle leaves the origin. At $t = 0$ we also have

$$(\mathbf{x}' \times \mathbf{x}'') \cdot \mathbf{x}''' = \tfrac{1}{2}(-\sqrt{3}, 0, 1) \cdot \tfrac{1}{2}(-3, 0, 0) = \tfrac{3}{4}\sqrt{3}$$

Hence the torsion at $t = 0$ is

$$\tau = \frac{(\mathbf{x}' \times \mathbf{x}'') \cdot \mathbf{x}'''}{|\mathbf{x}' \times \mathbf{x}''|^2} = \tfrac{3}{4}\sqrt{3}$$

This is the rate at which the path is twisting as the particle leaves the origin.

The speed at any time t is $ds/dt = |\mathbf{x}'| = \tfrac{1}{2}\sqrt{4 + t^2}$, so the particle speeds up as it leaves the origin. Since

$$\frac{d^2s}{dt^2} = \frac{t}{2\sqrt{4 + t^2}} = 0 \qquad \text{at } t = 0$$

the acceleration at $t = 0$ is

$$\mathbf{a} = \frac{d^2s}{dt^2}\mathbf{T} + \kappa\left(\frac{ds}{dt}\right)^2\mathbf{N} = 0\mathbf{T} + (1)(1)^2\mathbf{N} = \mathbf{N}$$

It is also $\mathbf{x}'' = \tfrac{1}{2}(0, 2, 0) = \mathbf{j}$, so we have $\mathbf{N} = \mathbf{j}$ at $t = 0$. Thus the osculating plane at $t = 0$ is the plane containing the tangent line and the y axis. Since

$$\mathbf{T} = \tfrac{1}{2}(1, 0, \sqrt{3}) \text{ at } t = 0 \qquad \text{(why?)}$$

a normal vector to this plane is

$$\mathbf{B} = \mathbf{T} \times \mathbf{N} = \tfrac{1}{2}(-\sqrt{3}, 0, 1)$$

An equation of the plane is therefore $z = \sqrt{3}x$. $\square$

You may feel that Example 8 tells you more than you want to know (as James Thurber remarked in a review of an article about pigeons). Our purpose is to illustrate how the ideas of this chapter are all relevant to a concrete situation: rectangular and spherical coordinates; lines and planes; surfaces; the dot product, cross product, and scalar triple product of vectors; vector functions and their derivatives; speed, curvature, and torsion; tangential and normal components of acceleration; the unit tangent, normal, and binormal vectors.

Problem Set 16.6

1. The position of a bee at time t is given by $\mathbf{F}(t) = (\cos t, \sin t, t)$. At $t = \pi/2$ the bee flies off on a tangent. Find parametric equations of its line of flight and show that it is the line $z = \pi/2 - x$ in the plane $y = 1$.

2. In Problem 1 assume that the bee maintains constant speed after flying off on a tangent. How long does it take (from $t = \pi/2$) to reach a flower at $(-\pi/2, 1, \pi)$?

3. The position of a particle at time t is given by

$$\mathbf{F}(t) = (\cos t, \sin t, t)$$

Use the formula

$$\mathbf{a} = \frac{d^2s}{dt^2}\mathbf{T} + \kappa\left(\frac{ds}{dt}\right)^2\mathbf{N}$$

to show that $\mathbf{a} = \mathbf{N}$ for all t. *Hint:* In the text we showed that $ds/dt = \sqrt{2}$ and $\kappa = \frac{1}{2}$ for all t.

4. In Problem 3 explain why $\mathbf{N} = (-\cos t, -\sin t, 0)$.

5. In Problem 3 show that $\mathbf{T} = (1/\sqrt{2})(-\sin t, \cos t, 1)$.

6. In Problem 3 show that $\mathbf{B} = (1/\sqrt{2})(\sin t, -\cos t, 1)$.

7. In Problems 4–6 differentiate the formulas for $\mathbf{T}$, $\mathbf{N}$, and $\mathbf{B}$ to confirm the Frenet-Serret formulas $d\mathbf{T}/ds = \kappa\mathbf{N}$, $d\mathbf{N}/ds = \tau\mathbf{B} - \kappa\mathbf{T}$, $d\mathbf{B}/ds = -\tau\mathbf{N}$. (See Problems 54–56 for derivations of the last two; the first is proved in the text.)

8. In the motion $\mathbf{F}(t) = (\cos t, \sin t, t)$ find the equation of the osculating plane at $t = 0$.

9. The graph of $\mathbf{F}(t) = (a\cos t, a\sin t, bt)$ $(a > 0$ and $b > 0)$ is a circular helix like the one discussed in the text. Show that the tangent at $(a, 0, 0)$ is the line $z = (b/a)y$ in the plane $x = a$.

10. In Problem 9 show that ds/dt is constant.

11. In Problem 9 show that the curvature is $\kappa = a/(a^2 + b^2)$.

12. In Problem 9 show that the torsion is $\tau = b/(a^2 + b^2)$.

13. Find where the graphs of $\mathbf{F}(t) = (\cos t, \sin t, t)$ and $\mathbf{G}(t) = (\cosh t, \sinh t, t)$ intersect and show that they have identical tangent lines there.

14. Let $\mathbf{F}(t) = (\cos t, \sin t, \sin t)$. Explain why the graph of $\mathbf{F}$ is the intersection of the cylinder $x^2 + y^2 = 1$ and the plane $z = y$. Sketch the graph (which is an ellipse) and name the center and the endpoints of the major and minor diameters. What are their lengths?

15. In Problem 14 show that the curvature is

$$\kappa = \frac{\sqrt{2}}{(1 + \cos^2 t)^{3/2}}$$

16. In Problem 14, since the graph lies in a plane, its torsion should be zero. (Why?) Confirm that it is.

17. The position of a particle at time t is given by $\mathbf{F}(t) = (t, t, t^2)$. Explain why the path of the particle is a parabola in the plane $y = x$. Sketch the path and indicate the direction in which it is traversed.

18. In Problem 17 show that the curvature is

$$\kappa = \frac{1}{(2t^2 + 1)^{3/2}}$$

and explain why its maximum value occurs at the vertex of the parabola.

19. In Problem 17, since the path lies in a plane, its torsion should be zero. Confirm that it is.

20. The curve $\mathbf{F}(t) = (t, t^2, t^3)$ is called a *twisted cubic*. Explain why the graph of $\mathbf{F}$ lies in the parabolic cylinder $y = x^2$. Sketch the portion between $(0, 0, 0)$ and $(1, 1, 1)$.

21. In Problem 20 find the curvature at $t = 0$.

22. In Problem 20 find the torsion at $t = 0$.

23. Suppose that the position of a particle at time t is given by

$$\mathbf{x} = (r\cos \omega t, r\sin \omega t, h)$$

where r, ω, and h are positive constants. Explain why the path of the particle is the circle $x^2 + y^2 = r^2$ in the plane $z = h$.

24. In Problem 23 what is a geometric interpretation of $\theta = \omega t$? Show that the speed of the particle is

$$\frac{ds}{dt} = r\frac{d\theta}{dt} = r\omega$$

25. In Problem 23 the vector $\boldsymbol{\omega} = \omega\mathbf{k}$ is called the *angular velocity* of the particle. What is its magnitude? Show that the velocity of the particle is $\mathbf{v} = \boldsymbol{\omega} \times \mathbf{x}$.

26. In Problem 23, if ϕ is the angle between $\boldsymbol{\omega}$ and $\mathbf{x}$, explain why $|\mathbf{v}| = |\boldsymbol{\omega}||\mathbf{x}|\sin\phi$. Show that this checks with the formula $ds/dt = r\omega$ in Problem 24.

27. In Problem 23 use the definition $\mathbf{T} = \mathbf{v}/|\mathbf{v}|$ to show that $\mathbf{T} = (-\sin\theta, \cos\theta, 0)$.

28. In Problem 23 differentiate $\mathbf{T}$ to obtain

$$\frac{d\mathbf{T}}{ds} = \frac{1}{r}(-\cos\theta, -\sin\theta, 0)$$

and then use the definition

$$\mathbf{N} = \frac{d\mathbf{T}/ds}{|d\mathbf{T}/ds|}$$

to show that $\mathbf{N} = (-\cos\theta, -\sin\theta, 0)$.

29. In Problem 23 compute $\mathbf{B} = \mathbf{T} \times \mathbf{N}$ to show that $\mathbf{B} = \mathbf{k}$.

30. In Problem 23 use the formula $d\mathbf{T}/ds = \kappa\mathbf{N}$ to show that $\kappa = 1/r$. How could this have been predicted?

31. In Problem 23 use the formula $d\mathbf{B}/ds = -\tau\mathbf{N}$ to show that $\tau = 0$. How could this have been predicted?

32. The position of a particle at time t is given in spherical coordinates by $\rho = 2$, $\phi = \pi/3$, $\theta = t$. Explain why the path of the particle is the circle $x^2 + y^2 = 3$ in the plane $z = 1$. What is its length?

33. In Problem 32 show that the position vector of the particle is $\mathbf{x} = (x, y, z) = (\sqrt{3}\cos t, \sqrt{3}\sin t, 1)$.

34. In Problem 32 how fast is the particle traveling at time t?

35. In Problem 32 use the formula

$$\kappa = \frac{|\mathbf{x}' \times \mathbf{x}''|}{|\mathbf{x}'|^3}$$

to show that $\kappa = 1/\sqrt{3}$. How could this have been predicted?

36. In Problem 32 use the formula

$$\tau = \frac{(\mathbf{x}' \times \mathbf{x}'') \cdot \mathbf{x}'''}{|\mathbf{x}' \times \mathbf{x}''|^2}$$

to show that $\tau = 0$. How could this have been predicted?

37. The position of a particle at time $t \geq 0$ is given in cylindrical coordinates by $r = t$, $\theta = t$, $z = t$. Explain why the particle spirals up and around the upper half of the cone $z^2 = x^2 + y^2$. Where does it start?

38. In Problem 37 show that the position vector of the particle is $\mathbf{x} = (x, y, z) = (t\cos t, t\sin t, t)$.

39. In Problem 37 find parametric equations of the tangent at the origin. Show that it is the line $z = x$ in the xz plane and lies in the cone $z^2 = x^2 + y^2$.

40. In Problem 37 what is the speed of the particle as it leaves the origin?

41. In Problem 37 what is the curvature at the origin?

42. In Problem 37 what is the torsion at the origin?

43. In Problem 37 find $\mathbf{T}$, $\mathbf{N}$, and $\mathbf{B}$ at $t = 0$ and the equation of the osculating plane.

44. The position of a particle at time $t \geq 0$ is given by

$$x = e^{-t}\cos t, \quad y = e^{-t}\sin t, \quad z = e^{-t}$$

Explain why the particle spirals down the upper half of the cone $z^2 = x^2 + y^2$ toward the origin. Where does it start?

45. In Problem 44 show that the speed of the particle is

$$\frac{ds}{dt} = \sqrt{3}e^{-t}$$

How does this behave as $t \to \infty$?

46. In Problem 44 show that the distance traveled by the particle is $\sqrt{3}$. *Hint:* Evaluate an appropriate improper integral.

47. In Problem 44 show that the curvature is $\kappa = (\sqrt{2}/3)e^t$. How does this behave as $t \to \infty$?

48. In Problem 44 show that the torsion is $\tau = -\frac{1}{3}e^t$. How does this behave as $t \to \infty$?

49. Use the definition $\kappa = |d\mathbf{T}/ds|$ to explain why the curvature of a straight line is zero.

50. Use the formula

$$\kappa = \frac{|\mathbf{F}' \times \mathbf{F}''|}{|\mathbf{F}'|^3}$$

to show that the curvature of a straight line is zero. *Hint:* The line may be represented by a vector function $\mathbf{x} = \mathbf{x}_0 + t\mathbf{m}$, where $\mathbf{x}_0$ and $\mathbf{m}$ are constant vectors ($\mathbf{m} \neq \mathbf{0}$). (See Section 16.2.)

51. Suppose that the torsion of a space curve is always zero. Explain why the track lies in a plane.

52. Suppose that the graph of a smooth vector function $\mathbf{F}(t)$ lies on a sphere with center at the origin.

   (a) Give a geometric reason why $\mathbf{F}(t) \cdot \mathbf{F}'(t) = 0$ for all t.

   (b) Prove it analytically.

53. Explain why the binormal vector $\mathbf{B}$ is a unit vector perpendicular to $\mathbf{T}$ and $\mathbf{N}$.

54. Prove that $d\mathbf{B}/ds$ is perpendicular to $\mathbf{B}$ and $\mathbf{T}$, as follows.

   (a) Explain why $\mathbf{B} \cdot \mathbf{B} = 1$ for all t. Differentiate with respect to s to show that

$$\mathbf{B} \cdot \frac{d\mathbf{B}}{ds} = 0$$

   (b) Explain why $\mathbf{B} \cdot \mathbf{T} = 0$ for all t. Differentiate with respect to s (using the fact that $d\mathbf{T}/ds = \kappa\mathbf{N}$) to show that

$$\mathbf{T} \cdot \frac{d\mathbf{B}}{ds} = 0$$

55. The preceding problem shows that $d\mathbf{B}/ds$ is parallel to $\mathbf{N}$ (why?), so we can write $d\mathbf{B}/ds = -\tau\mathbf{N}$, where τ is some scalar constant. Derive the formula

$$\tau = \frac{(\mathbf{F}' \times \mathbf{F}'') \cdot \mathbf{F}'''}{|\mathbf{F}' \times \mathbf{F}''|^2}$$

as follows.

   (a) Explain why it follows from Equation (2) in the text that

$$(\mathbf{F}' \times \mathbf{F}'') \cdot \mathbf{F}''' = \kappa s'^3 (\mathbf{B} \cdot \mathbf{F}''')$$

(b) To find $\mathbf{B} \cdot \mathbf{F}'''$, explain why $\mathbf{B} \cdot \mathbf{F}'' = 0$ and differentiate with respect to t to obtain $\mathbf{B} \cdot \mathbf{F}''' = \kappa s'^3 \tau$.

(c) Substitute part (b) in part (a) and solve for τ.

56. The unit vectors $\mathbf{T}$ and $\mathbf{B}$ change at rates involving the curvature and torsion, namely $d\mathbf{T}/ds = \kappa \mathbf{N}$ and $d\mathbf{B}/ds = -\tau \mathbf{N}$. One might expect that another geometric property of the curve must be introduced to obtain the rate of change of $\mathbf{N}$. Show, however, that $d\mathbf{N}/ds = \tau \mathbf{B} - \kappa \mathbf{T}$ as follows.

(a) Explain why the definition $\mathbf{B} = \mathbf{T} \times \mathbf{N}$ implies that $\mathbf{N} = \mathbf{B} \times \mathbf{T}$.

(b) Differentiate with respect to s to obtain

$$\frac{d\mathbf{N}}{ds} = \kappa(\mathbf{B} \times \mathbf{N}) - \tau(\mathbf{N} \times \mathbf{T})$$

(c) Explain why $\mathbf{B} \times \mathbf{N} = -\mathbf{T}$ and $\mathbf{N} \times \mathbf{T} = -\mathbf{B}$ to finish the proof.

57. Derive the Product Rule for cross products,

$$\frac{d}{dt}\mathbf{F}(t) \times \mathbf{G}(t) = \mathbf{F}(t) \times \mathbf{G}'(t) + \mathbf{F}'(t) \times \mathbf{G}(t)$$

True–False Quiz

1. Each point (x, y, z) satisfying $x^2 + y^2 + z^2 - 4x > 0$ is more than 2 units from $(2, 0, 0)$.

2. If P, Q, R are three points in space, then $\overrightarrow{PQ} - \overrightarrow{RQ} = \overrightarrow{PR}$.

3. A unit vector opposite in direction to the vector $\mathbf{x} = (2, 1, -2)$ is $\mathbf{u} = \frac{1}{3}(-2, -1, 2)$.

4. A linear equation in x, y, and z represents a line in space.

5. A vector parallel to the line $x = 2t + 1$, $y = t + 2$, $z = t - 1$ is $\mathbf{m} = (1, 2, -1)$.

6. A normal vector to the plane $x - 2y + z - 1 = 0$ is $\mathbf{n} = (-1, 2, -1)$.

7. $\mathbf{i} \times \mathbf{k} = \mathbf{j}$.

8. If $\mathbf{y}$ is a scalar multiple of $\mathbf{x}$, then $\mathbf{x} \times \mathbf{y} = \mathbf{0}$.

9. If $\mathbf{x}_1$ and $\mathbf{x}_2$ are vectors in $\mathbb{R}^3$, then $\mathbf{x}_1 \times \mathbf{x}_2 = \mathbf{x}_2 \times \mathbf{x}_1$.

10. The cross product of two perpendicular unit vectors is a unit vector.

11. The set of points $\{(x, y, z): x^2 + y^2 = 1\}$ is a circle.

12. The equation $z = r$ in cylindrical coordinates represents a half-cone.

13. The equation $z = \rho$ in spherical coordinates represents the z axis.

14. The vectors $\mathbf{T}$, $\mathbf{N}$, and $\mathbf{B}$ (unit tangent, normal, and binormal, respectively) satisfy $\mathbf{N} \times \mathbf{B} = \mathbf{T}$.

15. The torsion of the curve $x = 2 \cos \pi t$, $y = 3 \sin \pi t$, $z = 1$ is zero.

Additional Problems

1. Sketch the tetrahedron whose vertices are $(0, 0, 0)$, $(4, 0, 0)$, $(0, 5, 0)$, and $(0, 0, 3)$. Then find an equation of the plane in which the front faces lies. (This is the face not lying in a coordinate plane.)

2. Find an equation of the sphere having $(3, -1, 0)$ and $(1, 1, -2)$ as endpoints of a diameter.

3. Find all unit vectors perpendicular to the vectors $\mathbf{x}_1 = (1, 3, 0)$ and $\mathbf{x}_2 = (2, -1, 2)$.

4. A triangle has vertices $A(-1, 2, 2)$, $B(0, 5, -3)$, $C(2, 4, 1)$. Find the angle at A.

Find parametric equations of each of the following lines.

5. Containing $(5, 0, 2)$ and $(1, -2, 1)$.

6. Containing $(1, 4, -4)$ and parallel to the line

$$\frac{x - 3}{5} = \frac{y + 1}{2} = \frac{z}{2}$$

7. Containing $(-1, 4, 2)$ and perpendicular to the plane $5x - y + 2z - 6 = 0$.

8. The line of intersection of the planes $2x - 3y + 5z - 2 = 0$ and $x + 2y + z + 1 = 0$.

9. Containing $(2, -5, 1)$ and parallel to the planes $x + y - z + 1 = 0$ and $3x - y + 2z - 8 = 0$.

10. Do the lines

$$L_1: x = 2t - 1, \; y = t + 1, \; z = 2t$$

and

$$L_2: x = t - 2, \; y = 3t - 5, \; z = 2t + 2$$

intersect?

11. Find the point of intersection of the lines

$$\frac{x - 1}{2} = \frac{y + 2}{3} = z - 2$$

and $x = t - 1, y = -t, z = 2t - 2$

(or conclude that they do not meet).

12. What point of the line with parametric equations $x = t - 2, y = 1 - t, z = t$ is closest to the origin?

Find an equation of each of the following planes.

13. Containing $(-2, 2, 3)$ and parallel to the plane $x - 3y + 4z - 2 = 0$.

14. Containing $(2, 0, 1)$ and perpendicular to the line $x = t - 1, y = 2t + 3, z = 2 - t$.

15. Containing $(2, -1, 1)$, $(1, -2, 2)$, and $(2, 1, 2)$.

16. Tangent to the sphere $(x - 2)^2 + y^2 + (z + 1)^2 = 14$ at the point $(4, -1, 2)$.

17. Find the point of intersection of the line $x = t - 1$, $y = 2t, z = -t$ and the plane $2x + y - 3z - 5 = 0$, and find the (acute) angle between the line and plane.

18. Find the area of the triangle with vertices $(2, 3, 5)$, $(-2, 2, 1)$, and $(4, -3, -3)$.

19. Prove that if $\mathbf{x}_2 = c\mathbf{x}_1$, then $\mathbf{x}_1 \times \mathbf{x}_2 = \mathbf{0}$.

20. Give an example disproving the associative law $(\mathbf{x}_1 \times \mathbf{x}_2) \times \mathbf{x}_3 = \mathbf{x}_1 \times (\mathbf{x}_2 \times \mathbf{x}_3)$.

Graph and name each of the following quadric surfaces.

21. $z = \frac{1}{2}y^2$

22. $z = 1 - x^2 - y^2$

23. $y^2 = x^2 + z^2$

24. $x^2 + y^2 - \frac{1}{2}z^2 = 1$

25. $4x^2 + 4y^2 + 9z^2 = 36$

26. Describe the graph of the equation $r^2 - z^2 = 1$ in cylindrical coordinates.

27. Describe the graph of the equation $\rho = 4 \cos \phi$ in spherical coordinates.

28. Let (ρ, ϕ, θ) be spherical coordinates of a point P on the surface $\rho = a$ and extend the segment OP until it intersects the surface $r = a$. Find cylindrical coordinates of the point of intersection.

29. The surfaces $\rho = 4$ (spherical coordinates) and $z = r$ (cylindrical coordinates) intersect in a certain curve. Find its length.

30. The position vectors of two objects in space are $\mathbf{F}(t) = (1, t, t^2)$ and $\mathbf{G}(t) = (t, t^2, t^3)$, where $t > 0$ is time.
   (a) When do the objects collide?
   (b) Prior to the collision, an observer at the origin looking at the objects continually experiences an eclipse of one by the other. Explain why, and figure out which one is eclipsed.

31. A hawk's position at time t is given by $\mathbf{x} =$ $(t^2, -t, 3 - t^2)$. At $t = 1$ the hawk looks straight ahead and sees a rabbit on the ground (the xy plane).
   (a) Give parametric equations of the hawk's line of sight.
   (b) Where is the rabbit?
   (c) What is the hawk's speed at $t = 1$? Assuming that it heads straight for the rabbit at this speed, how long will it take to reach the rabbit?

32. Suppose that you are standing at the top of the hill whose equation is $x^2 + y^2 + z - 2 = 0$. (The z axis points up, the y axis north, the x axis east.) If you walk northeast down the hill, what are parametric equations of your path of descent? Draw a picture showing the hill and the curve. When you reach the ground (the xy plane), how far have you walked?

33. The first-octant portion of the surfaces $y = 2x$ and $x^2 + y^2 = 5z^2$ intersect in a certain curve.
   (a) Give parametric equations of the curve.
   (b) Find the length of the part of the curve lying between the planes $z = 0$ and $z = 1$. Can you do this in two ways?

34. Find the length of the curve $x = \frac{1}{3}t^3, y = t^2, z = 2t$, $0 \le t \le 3$.

35. Find the curvature and torsion of the curve in Problem 34 at $t = 2$.

36. Find the curvature and torsion of the path $x = \cos t$, $y = 2 \sin t, z = t^2$ at $t = 0$. What is an equation of the osculating plane at this point?

37. Explain why the graph of $\mathbf{F}(t) = (\sin t, \cos t, \cos t)$ is the intersection of the cylinder $x^2 + y^2 = 1$ and the plane $z = y$. What type of curve is it?

38. In Problem 37 find the curvature.

39. In Problem 37 find the torsion. How could the result have been predicted?

40. Find the curvature of the graph of $\mathbf{F}(t) = (\cosh t, \sinh t, t)$.

41. In Problem 40 find the torsion.

42. Suppose that the graph of the smooth vector function $\mathbf{x} = \mathbf{F}(t)$ lies on the sphere $x^2 + y^2 + z^2 = 1$. Prove that $\mathbf{x}$ and $d\mathbf{x}/dt$ are perpendicular.

43. Prove that if $\mathbf{F}(t)$ is a smooth curve, the unit binormal vector is given by

$$\mathbf{B} = \frac{\mathbf{F}' \times \mathbf{F}''}{|\mathbf{F}' \times \mathbf{F}''|}$$

Hint: Use the equation $\mathbf{F}' \times \mathbf{F}'' = \kappa s'^3 (\mathbf{T} \times \mathbf{N})$ together with the equation obtained by taking the magnitude of each side. (See the argument for curvature in Section 16.6.)

17 | Functions of Several Variables

All the measurements in the world are not the equivalent of a single theorem that produces a significant advance in our greatest of sciences.

KARL FRIEDRICH GAUSS (1777–1855)
(generally ranked with Archimedes and Newton at the pinnacle of mathematical achievement)

I never realized that in America there was so much interest in tensor analysis.

ALBERT EINSTEIN
(upon seeing a great crowd come to hear him lecture at Princeton)

There is a quite peculiar duplicity in the nature of mathematics

JOHN VON NEUMANN (1903–1957)

Almost every function we have considered in previous chapters has been a function of one independent variable, either a real function $y = f(x)$ or a vector function $\mathbf{x} = \mathbf{F}(t)$. In other words, the *domain* has been restricted to a subset of $\mathcal{R}$ (usually an interval or union of intervals on the number line). The *range* has also been a subset of $\mathcal{R}$ in the case of real functions, but in the last two chapters we allowed it to be a subset of $\mathcal{R}^2$ or $\mathcal{R}^3$ (in the case of vector functions).

Despite all we have accomplished, it is unrealistic to study functions of only one variable. Applications of mathematics very often involve several independent variables; to analyze the functions that arise, we need *multivariable calculus*. This is the study of functions with domains in $\mathcal{R}^2$ or $\mathcal{R}^3$ (or even $\mathcal{R}^n$, where $n > 3$), such as

$$z = f(x,y) = x^2 + y^2 \qquad \text{or} \qquad w = f(x,y,z) = e^{xyz}$$

Note that in these examples the range is $\mathcal{R}$ again, that is, the functional outputs are real numbers.

The symbol $\mathcal{R}^n$ stands for the set of all "*n*-tuples" of real numbers, namely

$$\mathcal{R}^n = \{(x_1, x_2, \ldots, x_n): x_1 \in \mathcal{R}, \, x_2 \in \mathcal{R}, \ldots, x_n \in \mathcal{R}\}$$

When $n = 1, 2, 3$ we are dealing with the number line, the coordinate plane, or 3-space, respectively. Geometric visualization can go no higher, but there is no reason why analytic definitions cannot be given in n-space. The "distance" between $P(x_1, x_2, x_3, x_4)$ and $Q(y_1, y_2, y_3, y_4)$ in 4-space, for example, is defined to be

$$d(P,Q) = \sqrt{(y_1 - x_1)^2 + (y_2 - x_2)^2 + (y_3 - x_3)^2 + (y_4 - x_4)^2}$$

Nobody pretends that such ideas can be illustrated by pictures, but it is remarkable how well geometric intuition stands up in higher-dimensional analogies. This is important in the analysis of functions of several variables. For example, the temperature at (x, y, z) in 3-space, at time t, might be given by

$$T = f(x, y, z, t) = x^2 + y^2 + z^2 + t^2$$

No pictures can be drawn of this function. Suppose, however, that we have studied the simpler function $T = x^2 + y^2 + z^2$ and have observed (for example) that it is constant on the sphere $x^2 + y^2 + z^2 = 1$ and is most rapidly increasing in directions normal to this sphere. This suggests a similar statement about T as a space-time function. The "hypersphere" $x^2 + y^2 + z^2 + t^2 = 1$ in 4-space cannot be visualized, but maybe it plays the same role relative to the rate of change of T as does the ordinary sphere when T involves only three independent variables.

Such speculations turn out to be true in a great many cases, that is, we *discover* theorems about functions of many variables by thinking about what goes on when there are only two or three. To put it differently, the quantum leap in multivariable calculus occurs when we go from functions of one variable to functions of two or three. After that, it is relatively easy to generalize to higher dimensions.

17.1 PARTIAL DERIVATIVES

Suppose that an anthill is in the shape of the graph of

$$z = 1 - \tfrac{1}{4}x^2 - y^2 \qquad z \geq 0$$

(the portion of the elliptical paraboloid shown in Figure 1). We assume that the xy plane is ground level, that the z axis points upward, and that the x and y axes point east and north, respectively. An ant at $(0, 0, 1)$ proposing to go down the hill might head straight east, in which case its curve of descent would be the parabola $z = 1 - \tfrac{1}{4}x^2$ in the xz plane. Single-variable calculus is sufficient to describe its path, since z is a function of x alone (y being fixed at the value $y = 0$).

Similarly, an ant at $(0, 0, 1)$ heading north would descend along the curve $z = 1 - y^2$ in the yz plane. In this case z is a function of y alone (since $x = 0$).

In general, however, a descent of the hill along an arbitrary path involves altitude as a function of two variables, namely

$$z = f(x, y) = 1 - \tfrac{1}{4}x^2 - y^2 \qquad z \geq 0$$

Figure 1
Graph of $z = 1 - \frac{1}{4}x^2 - y^2$

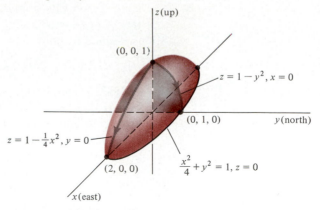

The domain of this function is a set of points in the xy plane,

$$D = \left\{(x,y): \frac{x^2}{4} + y^2 \leq 1\right\}$$

[the points inside and on the ellipse $(x^2/4) + y^2 = 1$]. The notation f: $D \to \Re$ indicates that f converts each point $(x,y) \in D$ into a real number $z = f(x,y)$, the third coordinate of a point (x,y,z) on the anthill.

We have not developed any apparatus for discussing the rate of change of a function of two variables. If one of the variables is fixed, however (as in the case of the first ant, when $z = 1 - \frac{1}{4}x^2$ and $y = 0$), we can find the rate of change of altitude as an ordinary derivative, $D_x z = -\frac{1}{2}x$ (obtained without any reference to y). To indicate that y was originally involved (but has been fixed at $y = 0$), we can write

$$D_x f(x,0) = D_x(1 - \frac{1}{4}x^2) = -\frac{1}{2}x$$

More generally, suppose that y is fixed at an arbitrary value $(-1 < y < 1)$. Then (regarding x as the variable) we write

$$D_x f(x,y) = D_x(1 - \frac{1}{4}x^2 - y^2) = -\frac{1}{2}x \qquad \text{(because } y \text{ is constant)}$$

Such a derivative, obtained by activating only one variable, is called a **partial derivative.** To distinguish between ordinary and partial derivatives, we often replace the notation $D_x f(x,y)$ by $f_1(x,y)$ (meaning the derivative of f with respect to its first variable) or $f_x(x,y)$ (meaning the same thing, if x is the first variable) or $\partial z/\partial x$ (instead of the usual dz/dx). Other symbols in common use are z_x and $\partial f/\partial x$; as in single-variable calculus, you should learn to live with all of them.

The geometrical significance of $\partial z/\partial x = -\frac{1}{2}x$ is that the curve of intersection of the surface

$$z = 1 - \frac{1}{4}x^2 - y^2$$

Figure 2
Geometrical significance of partial derivatives

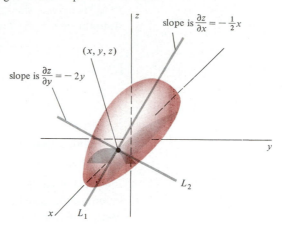

and a plane parallel to the xz plane (corresponding to the fixed value of y) has slope $-\frac{1}{2}x$ at the point (x, y, z). (See Figure 2.) Similarly, the partial derivative $\partial z/\partial y = -2y$ (obtained by holding x fixed and activating y) is the slope of the curve of intersection of the surface and the plane parallel to the yz plane corresponding to the fixed value of x ($-2 < x < 2$).

□ **Example 1**   Find the partial derivatives of the function

$$f(x, y) = x^2 - xy + y^2$$

Solution: Holding y fixed and differentiating with respect to x, we find $f_1(x, y) = 2x - y$, while if x is fixed and y is activated, we obtain $f_2(x, y) = -x + 2y$. This is simple enough if you understand it, but it is easy to let one's attention wander and make a mistake! Note that when y is fixed the term y^2 may be ignored (because it is a constant). The term $-xy$ might better be thought of as $-yx$, since x is the variable and $-y$ is the coefficient. Hence its derivative with respect to x is $-y$ (just as the derivative of, say, $-5x$ is -5).   □

□ **Example 2**   Find the partial derivatives of $f(x, y) = e^{x^2y}$.

Solution: Holding y fixed and differentiating with respect to x, we use the Chain Rule in the standard way:

$$\frac{\partial}{\partial x}(e^{x^2y}) = e^{x^2y}\frac{\partial}{\partial x}(x^2y) = e^{x^2y}(2xy) = 2xye^{x^2y}$$

Similarly, if x is fixed and y is activated, we find

$$\frac{\partial}{\partial y}(e^{x^2y}) = e^{x^2y}\frac{\partial}{\partial y}(x^2y) = x^2e^{x^2y}$$ □

Formal definitions of partial derivatives are unnecessary in most applications. They are important, however, when we have to go back to the idea of derivative as the limit of a difference quotient. Let (x, y) be a point of the domain of $f(x, y)$. Then

$$f_1(x, y) = \lim_{h \to 0} \frac{f(x + h, y) - f(x, y)}{h}$$

$$f_2(x, y) = \lim_{k \to 0} \frac{f(x, y + k) - f(x, y)}{k}$$

(provided that the limits exist).

Although these limits look more complicated than in single-variable calculus, they are in fact nothing new. In the first one, for example, *y is fixed.* Hence we are dealing with a function of x alone. To find its derivative, we compute the usual difference quotient (subtracting the functional value at x from the value at $x + h$ and dividing by h). Then we evaluate the limit as $h \to 0$.

□ **Example 3**   Use the definition of $f_1(x, y)$ as the limit of a difference quotient to find $\partial z / \partial x$ if $z = x^2 - xy + y^2$.

Solution: Fix y. Then

$$\frac{\partial z}{\partial x} = \lim_{h \to 0} \frac{[(x + h)^2 - (x + h)y + y^2] - (x^2 - xy + y^2)}{h}$$

$$= \lim_{h \to 0} \frac{2hx + h^2 - hy}{h} = \lim_{h \to 0} (2x + h - y) = 2x - y$$

This checks with the derivative found in Example 1. As you can see, it involves nothing that we have not already discussed in Chapter 2, when we evaluated derivatives as limits of difference quotients. Having developed rules of differentiation that bypass the definition, we have no reason for not using them in multivariable calculus. The beauty of a partial derivative is that every variable but one is suppressed while differentiation proceeds with respect to that one.   □

Higher-order partial derivatives are defined as in single-variable calculus, but the notation requires some comment. Suppose, for example, that we have differentiated $z = f(x, y)$ with respect to x to find $\partial z / \partial x = f_1(x, y)$. This is another function of x and y (as you can see in the preceding examples), so it also has partial derivatives. Its derivative with respect to x (keeping y fixed) is

$$\frac{\partial}{\partial x} \left(\frac{\partial z}{\partial x} \right) = \frac{\partial^2 z}{\partial x^2} = f_{11}(x, y) \qquad \text{(Read this as "f one-one," not "f eleven"!)}$$

Its derivative with respect to y (keeping x fixed) is

$$\frac{\partial}{\partial y}\left(\frac{\partial z}{\partial x}\right) = \frac{\partial^2 z}{\partial y\,\partial x} = f_{12}(x,y)$$

Other symbols for these derivatives are

$$z_{xx} = f_{xx}(x,y) \quad\text{and}\quad z_{xy} = f_{xy}(x,y)$$

Note that in the symbol $\partial^2 z/\partial y\,\partial x$ we read the order of differentiation from right to left (first x, then y), but in the symbol $f_{12}(x,y)$ or $f_{xy}(x,y)$ it is read from left to right.

Similarly, the partial derivatives of $\partial z/\partial y = f_2(x,y)$ are

$$\frac{\partial}{\partial x}\left(\frac{\partial z}{\partial y}\right) = \frac{\partial^2 z}{\partial x\,\partial y} = f_{21}(x,y) \quad\text{and}\quad \frac{\partial}{\partial y}\left(\frac{\partial z}{\partial y}\right) = \frac{\partial^2 z}{\partial y^2} = f_{22}(x,y)$$

Hence a given function of two variables has four second-order partial derivatives.

□ **Example 4**   Find the second partial derivatives of

$$f(x,y) = x^2 - xy + y^2$$

Solution: Since $f_1(x,y) = 2x - y$ and $f_2(x,y) = -x + 2y$, we find

$$f_{11}(x,y) = 2, \; f_{12}(x,y) = -1, \; f_{21}(x,y) = -1, \; f_{22}(x,y) = 2 \qquad\qquad □$$

□ **Example 5**   Find the second partial derivatives of $f(x,y) = e^{x^2 y}$.

Solution: From Example 2 we have $f_1(x,y) = 2xye^{x^2 y}$ and $f_2(x,y) = x^2 e^{x^2 y}$. Using the Product Rule, we find

$$f_{11}(x,y) = (2xy)\frac{\partial}{\partial x}(e^{x^2 y}) + e^{x^2 y}\frac{\partial}{\partial x}(2xy) = 4x^2 y^2 e^{x^2 y} + 2ye^{x^2 y}$$

$$= 2ye^{x^2 y}(2x^2 y + 1)$$

$$f_{12}(x,y) = (2xy)\frac{\partial}{\partial y}(e^{x^2 y}) + e^{x^2 y}\frac{\partial}{\partial y}(2xy) = 2x^3 ye^{x^2 y} + 2xe^{x^2 y}$$

$$= 2xe^{x^2 y}(x^2 y + 1)$$

$$f_{21}(x,y) = x^2\frac{\partial}{\partial x}(e^{x^2 y}) + e^{x^2 y}\frac{\partial}{\partial x}(x^2) = 2x^3 ye^{x^2 y} + 2xe^{x^2 y}$$

$$= 2xe^{x^2 y}(x^2 y + 1)$$

$$f_{22}(x,y) = x^2\frac{\partial}{\partial y}(e^{x^2 y}) + e^{x^2 y}\frac{\partial}{\partial y}(x^2) = x^4 e^{x^2 y} + e^{x^2 y}\cdot 0$$

$$= x^4 e^{x^2 y} \qquad\qquad □$$

Note that in both Examples 4 and 5 we have $f_{12}(x,y) = f_{21}(x,y)$. This could be an accident in the first case (after all, f_{11} and f_{22} are the same in that example, too), but it is unlikely to happen in derivatives as complicated as those in Example 5 unless there is a conspiracy. What we have stumbled across is a general theorem, that the "mixed" second partial derivatives of a function of two variables are always equal if the function is reasonably well behaved.

This statement raises questions that we have not yet mentioned. In single-variable calculus a function is "well behaved" if it is (at least) continuous. It is downright civilized if it is differentiable (and that property implies continuity, as you should recall). The existence of higher-order derivatives improves its behavior still more, especially if they all exist (for then we can work out a Taylor series for the function, and in ordinary circumstances this series represents the function).

In multivariable calculus the only one of these ideas that is reasonably simple is continuity:

The function $f(x,y)$ is said to be **continuous** at a point (a,b) of its domain if

$$\lim_{\mathbf{x} \to \mathbf{a}} f(\mathbf{x}) = f(\mathbf{a}), \text{ where } \mathbf{x} = (x,y) \text{ and } \mathbf{a} = (a,b)$$

that is, if $f(x,y)$ can be made arbitrarily close to $f(a,b)$ by keeping (x,y) sufficiently close to (a,b).

In other words, continuity of f at (a,b) means that the *number* $f(x,y)$ approaches the *number* $f(a,b)$ when the *point* (x,y) approaches the *point* (a,b). What makes this seemingly straightforward definition more complicated than in single-variable calculus is that the simultaneous behavior of two independent variables (x and y) is involved. There are infinitely many paths along which (x,y) may approach (a,b), whereas in single-variable calculus [when we talk about $f(x)$ approaching $f(a)$ as x approaches a] the variable x is confined to the real number line.

□ **Example 6**   Define the function $f \colon \mathfrak{R}^2 \to \mathfrak{R}$ by the rule

$$f(x,y) = \begin{cases} \dfrac{xy}{x^2 + y^2} & \text{if } (x,y) \neq (0,0) \\ 0 & \text{if } (x,y) = (0,0) \end{cases}$$

Is f continuous at $(0,0)$?

Solution: The question is whether $\lim_{\mathbf{x} \to \mathbf{0}} f(\mathbf{x})$ exists and is equal to $f(\mathbf{0}) = 0$, where $\mathbf{x} = (x,y)$ and $\mathbf{0} = (0,0)$. To evaluate the limit (if it exists), we must determine how $f(x,y)$ behaves when (x,y) is close to $(0,0)$. If we confine (x,y) to the x axis (in which case $y = 0$), we have

$$f(x,y) = f(x,0) = \frac{x \cdot 0}{x^2 + 0^2} = 0 \qquad (x \neq 0)$$

and the limit of $f(x,y)$ appears to be 0. If (x,y) is made to approach $(0,0)$ along the y axis (in which case $x = 0$), we have

$$f(x,y) = f(0,y) = 0 \qquad (y \neq 0)$$

and again the limit appears to be 0. But suppose we let (x,y) approach $(0,0)$ along the line $y = x$. Then

$$f(x,y) = f(x,x) = \frac{x \cdot x}{x^2 + x^2} = \frac{x^2}{2x^2} = \frac{1}{2} \qquad (x \neq 0)$$

and the limit is apparently $\frac{1}{2}$. (!) We must conclude that $\lim_{\mathbf{x}\to 0} f(\mathbf{x})$ does not exist and hence f is not continuous at $(0,0)$.

In the problem set we ask you to show that $f_1(0,0) = 0$ and $f_2(0,0) = 0$. Thus a function may have partial derivatives at a point where it is not continuous, a situation that is profoundly different than in single-variable calculus (where the existence of f' implies continuity of f). □

Thus even the "simple" notion of continuity is not so simple. We are going to ignore the difficulties by assuring you that the functions you will encounter in this book are continuous wherever they are defined (with rare exceptions that will be pointed out when they occur). It would be nice to be able to say that this is a consequence of differentiability, as it is in single-variable calculus. When you think about it, however, what does it mean for a function $f(x,y)$ to be "differentiable?" The only kind of derivative that we have mentioned so far is a *partial* derivative, one with respect to x and the other with respect to y. Is f differentiable if one of them exists but not the other? Surely not. How about if they both exist? That turns out to be closer to the mark, but even then something more is needed (for reasons that are not obvious).

As you can see, we are getting into questions that call for some discussion. On the whole, we will leave them for a course in advanced calculus, stating only what is essential for an understanding of the practical procedures we develop.

For the record, here is the theorem about equality of mixed partial derivatives that got us into this discussion:

If $f(x,y)$ has continuous mixed partial derivatives in a neighborhood of the point (a,b), then $f_{12}(a,b) = f_{21}(a,b)$.

Briefly put, f_{12} and f_{21} are the same whenever they are continuous at (and near) the point in question. Since we will be dealing almost exclusively with functions that are continuous in their domains (together with their partial derivatives), you can ordinarily forget about the theoretical distinction between f_{12} and f_{21}. Examples of functions for which $f_{12} \neq f_{21}$ are not easy to invent!

The ideas and notation of this section also apply to functions of more than two variables, although we are no longer able to interpret them geometrically. (A function of *one* variable requires $\mathcal{R}^2$ for a picture of its graph, and a function of two variables requires $\mathcal{R}^3$. Evidently the concept of "graph" must be abandoned when more than two independent variables occur, except by way of analogy.)

□ **Example 7** Find the partial derivatives of

$$w = f(x, y, z) = x \sin yz - z \ln xy$$

Solution:

$$\frac{\partial w}{\partial x} = \sin yz - z \cdot \frac{y}{xy} = \sin yz - \frac{z}{x} \qquad \text{(keeping } y \text{ and } z \text{ fixed)}$$

$$\frac{\partial w}{\partial y} = xz \cos yz - z \cdot \frac{x}{xy} = xz \cos yz - \frac{z}{y} \qquad \text{(keeping } x \text{ and } z \text{ fixed)}$$

$$\frac{\partial w}{\partial z} = xy \cos yz - \ln xy \qquad \text{(keeping } x \text{ and } y \text{ fixed)}$$

You should study these results until you are sure you understand how they are obtained. If necessary, substitute other letters for the "fixed variables" (to emphasize the distinction between constants and variables). For example, let $y = b$ and $z = c$ in the first computation:

$$f(x, b, c) = x \sin bc - c \ln bx$$

This is a function of x alone; its derivative is found by the rules of single-variable calculus:

$$f_1(x, b, c) = \sin bc - c \cdot \frac{b}{bx} = \sin bc - \frac{c}{x}$$

Restoring the original variables, we have

$$f_1(x, y, z) = \sin yz - \frac{z}{x}$$

as before. The use of b and c for y and z is a mental crutch that is unnecessary after you have the idea. But it may be useful in the beginning. □

□ **Example 8** Demonstrate the equality of f_{123} and f_{321} in Example 7.

Solution: Since $f_1(x, y, z) = \sin yz - (z/x)$, we find

$$f_{12}(x, y, z) = z \cos yz \qquad \text{(keeping } x \text{ and } z \text{ fixed)}$$

and hence

$$f_{123}(x, y, z) = -yz \sin yz + \cos yz \qquad \text{(keeping } x \text{ and } y \text{ fixed)}$$

On the other hand, $f_3(x, y, z) = xy \cos yz - \ln xy$, from which

$$f_{32}(x, y, z) = -xyz \sin yz + x \cos yz - \frac{1}{y} \quad \text{(keeping } x \text{ and } z \text{ fixed)}$$

and hence

$$f_{321}(x, y, z) = -yz \sin yz + \cos yz \quad \text{(keeping } y \text{ and } z \text{ fixed)}$$

It looks as though mixed partial derivatives of any order are equal, regardless of the number of variables and the order in which we differentiate. As in the case of functions of two variables, this is correct as long as all derivatives are continuous. □

Problem Set 17.1

The domain of each of the following functions is a subset of R^2 or R^3. Find the domain and describe it geometrically.

1. $f(x,y) = \dfrac{1}{x^2 + y^2}$ 2. $f(x,y) = \sqrt{1 - x^2 - y^2}$

3. $f(x,y) = \ln xy$ 4. $f(x,y) = y \ln x$

5. $f(x,y) = \dfrac{x + y}{x - y}$

6. $f(x, y, z) = \sqrt{9 - x^2 - y^2 - z^2}$

7. $f(x, y, z) = \dfrac{xy}{z}$

8. $f(x, y, z) = \dfrac{x}{y - z}$

Find the first partial derivatives of each of the following functions.

9. $z = x^2 + xy + y^2$ 10. $w = \sin uv$

11. $z = \dfrac{x}{y} - \dfrac{y}{x}$ 12. $f(s,t) = \dfrac{s - t}{s + t}$

13. $f(x,y) = e^{xy}$ 14. $z = e^{x^2 + y^2}$
15. $z = ye^{y/x}$ 16. $z = \ln xy$

17. $g(u,v) = e^u \cos v$ 18. $z = \ln \sin \dfrac{x}{y}$

19. $z = x^y$ 20. $w = xyz$
21. $f(x, y, z) = \sqrt{x^2 + y^2 + z^2}$
22. $F(u, v, w) = e^{uv} \sin vw$

23. $g(x, y, z) = (xy + yz + zx)^2$
24. Show that if $r = \sqrt{x^2 + y^2}$, then

$$x \frac{\partial r}{\partial x} + y \frac{\partial r}{\partial y} = r$$

25. Show that if $z = \ln \sqrt{x^2 + y^2}$, then

$$x \frac{\partial z}{\partial x} + y \frac{\partial z}{\partial y} = 1$$

26. Show that if $z = \tan^{-1}(y/x)$, then

$$x \frac{\partial z}{\partial x} + y \frac{\partial z}{\partial y} = 0$$

27. Show that if $f(x, y, z) = (x - y)(y - z)(z - x)$, then $f_1 + f_2 + f_3 = 0$.
28. The area of a triangle is $K = \frac{1}{2}ab \sin C$, where a and b are two sides and C is the included angle.
   (a) Find $\partial K/\partial a$, $\partial K/\partial b$, $\partial K/\partial C$.
   (b) Regarding b as a function of a, C, and K, show that $\partial b/\partial a = -b/a$.
29. The *Ideal Gas Law* says that the pressure, volume, and temperature of a confined gas satisfy the equation $PV = kT$, where k is a constant. Show that

$$\frac{\partial P}{\partial V} \cdot \frac{\partial V}{\partial T} \cdot \frac{\partial T}{\partial P} = -1$$

where each derivative is found by regarding the appropriate dependent variable as a function of the other two.

30. The surface of a mountain is given by $z = 6 - 3x^2 - 2y^2$. If a mountaineer at the point $(1, 1, 1)$ has a choice of going

east or north, which direction is best if the rate of descent at that point is to be greatest? (Recall our agreement at the beginning of this section that the xy plane is ground level, the z axis is pointing upward, and the x and y axes are pointing east and north, respectively.)

31. The temperature at each point of the disk

$$D = \{(x,y): x^2 + y^2 \le 25\}$$

is given by $T = x^2 + y^2$. A bug starts at $(-3,4)$ and moves east along the line $y = 4$. What partial derivative represents the rate of change of temperature with respect to distance traveled? What is the temperature when the bug starts and how is it changing? What is the temperature when the bug reaches the other side of the disk and how is it changing? What is the highest temperature encountered by the bug? the lowest?

32. In Problem 31 another bug starts at $(3,-4)$ and moves north along the line $x = 3$. Answer the same questions as in Problem 31.

33. In Problem 31 a bug starts at $(1,0)$ and (disliking change) follows a path along which the temperature is constant. What is the temperature and what is the path? In general, what are the isothermal ("equal temperature") paths?

34. In Problem 31 a bug starts at $(3,4)$ and moves across the disk along a diameter. What function of a single variable describes the temperature variation along this path? Is its derivative a partial derivative of the original temperature function? What are the highest and lowest temperatures encountered by the bug along this path?

The second-order partial differential equation

$$\frac{\partial^2 z}{\partial x^2} + \frac{\partial^2 z}{\partial y^2} = 0$$

is called *Laplace's equation*. Show that each of the following functions satisfies it. (Such a function is called *harmonic*.)

35. $z = ax + by + c$ **36.** $z = e^x \cos y$

37. $z = \tan^{-1}(y/x)$ **38.** $z = \sin x \sinh y$

39. $z = \ln(x^2 + y^2)$

40. Given the functions $u = f(x,y)$ and $v = g(x,y)$, suppose that $\partial u/\partial x = \partial v/\partial y$ and $\partial u/\partial y = -\partial v/\partial x$. (These are called the *Cauchy-Riemann equations*.) Show that f and g both satisfy Laplace's equation (assuming, of course, that they are "well behaved").

41. Show that the function $f(x,y,z) = (x^2 + y^2 + z^2)^{-1/2}$ satisfies Laplace's equation in 3-space, namely $f_{11} + f_{22} + f_{33} = 0$.

42. Suppose that the temperature of a straight wire at point x and time t is

$$T = e^{-c^2 t} \sin x \qquad \text{(where } c \text{ is a constant)}$$

Show that T satisfies the *heat-flow equation*

$$\frac{\partial T}{\partial t} = c^2 \frac{\partial^2 T}{\partial x^2}$$

43. Suppose that $f(x,y) = x^2 y - (x/y^2)$. Confirm that $f_{12} = f_{21}$.

44. In Problem 43 confirm that $f_{112} = f_{121} = f_{211}$.

45. If $f(x,y,z) = xy^2 z^3$, confirm that $f_{123} = f_{132} = f_{213} = f_{231} = f_{312} = f_{321}$.

46. Suppose that $f(x,y)$ is a function with continuous first and second partial derivatives. Explain why f cannot have the property that

$$f_1(x,y) = e^{-x} \sin y \qquad \text{and} \qquad f_2(x,y) = e^{-x} \cos y$$

47. Use the definition of $f_1(x,y)$ as the limit of a difference quotient to find $\partial z/\partial x$ if $z = x^2 \sin y$.

48. Use the definition of $f_2(x,y)$ as the limit of a difference quotient to find $\partial z/\partial y$ if $z = y^2 e^x$.

49. It is geometrically apparent that the half-cone $z = \sqrt{x^2 + y^2}$ is not "smooth" at the origin. (Draw a picture!) Hence we may reasonably expect that if $f(x,y) = \sqrt{x^2 + y^2}$, then $f_1(0,0)$ and $f_2(0,0)$ do not exist. Confirm this by using their definitions as limits of difference quotients.

50. Define the function $f : \mathcal{R}^2 \to \mathcal{R}$ by

$$f(x,y) = \begin{cases} \dfrac{xy}{x^2 + y^2} & \text{if } (x,y) \ne (0,0) \\ 0 & \text{if } (x,y) = (0,0) \end{cases}$$

Use the definitions

$$f_1(0,0) = \lim_{h \to 0} \frac{f(0 + h,0) - f(0,0)}{h}$$

and

$$f_2(0,0) = \lim_{k \to 0} \frac{f(0,0 + k) - f(0,0)}{k}$$

to confirm that $f_1(0,0) = 0$ and $f_2(0,0) = 0$. (This shows that a function may have partial derivatives at a point without being continuous there. See Example 6.)

51. If $f(x,y,z)$ is a function of three variables, which of its partial derivatives is given by

$$\lim_{k \to 0} \frac{f(x,y + k,z) - f(x,y,z)}{k}?$$

52. Given the function $f(x,y)$, identify

$$\lim_{h \to 0} \frac{f_1(x + h,y) - f_1(x,y)}{h}$$

and

$$\lim_{k \to 0} \frac{f_1(x,y + k) - f_1(x,y)}{k}$$

as second partial derivatives of *f*. Give similar definitions of its other second partial derivatives.

53. Define the function $f: \mathbb{R}^2 \to \mathbb{R}$ by

$$f(x,y) = \begin{cases} \dfrac{xy(x^2 - y^2)}{x^2 + y^2} & \text{if } (x,y) \neq (0,0) \\ 0 & \text{if } (x,y) = (0,0) \end{cases}$$

Prove that $f_{12}(0,0) \neq f_{21}(0,0)$ as follows.

(a) Use the definition

$$f_1(0,y) = \lim_{h \to 0} \frac{f(0 + h, y) - f(0, y)}{h}$$

to show that $f_1(0,y) = -y$ for all *y*.

(b) Similarly, show that $f_2(x,0) = x$ for all *x*.

(c) Use the definition

$$f_{12}(0,0) = \lim_{k \to 0} \frac{f_1(0,0 + k) - f_1(0,0)}{k}$$

to show that $f_{12}(0,0) = -1$.

(d) Similarly, show that $f_{21}(0,0) = 1$.

17.2

CHAIN RULES AND THE GRADIENT

The Chain Rule in single-variable calculus says that if $y = f(x)$ and $x = g(t)$, then

$$\frac{dy}{dt} = \frac{dy}{dx}\frac{dx}{dt}$$

A similar situation in multivariable calculus is $z = f(x,y)$, where $x = g(t)$ and $y = h(t)$. Using vector notation (which often suggests analogies with single-variable calculus), we can write $z = f(\mathbf{x})$, where $\mathbf{x} = (x,y) = (g(t),h(t)) = \mathbf{g}(t)$. A Chain Rule for this situation might be expected to read

$$\frac{dz}{dt} = \frac{dz}{d\mathbf{x}}\frac{d\mathbf{x}}{dt}$$

except for the unfortunate fact that the symbol $dz/d\mathbf{x}$ is meaningless. (We have never discussed differentiation with respect to a vector variable!)

Note that the other two derivatives in our proposed Chain Rule are familiar, and make sense in the present context. For if *z* is a function of *x* and *y*, and they are both functions of *t*, then *z* is a function of *t* and the ordinary derivative dz/dt is a reasonable thing to compute. Moreover, the vector function $\mathbf{x} = (x,y)$ has a derivative (defined in Section 15.2), namely

$$\frac{d\mathbf{x}}{dt} = \left(\frac{dx}{dt}, \frac{dy}{dt}\right)$$

□ **Example 1**   Suppose that $z = ye^{x^2}$, where $x = 2t$ and $y = 1 - t$. Then

$$z = (1 - t)e^{4t^2}$$

and (by the Product Rule)

$$\frac{dz}{dt} = 8t(1 - t)e^{4t^2} - e^{4t^2}$$

Thus z is an ordinary function of t, with an ordinary derivative. The vector function $\mathbf{g}(t)$ in this case is $\mathbf{x} = (2t, 1 - t)$ with derivative $d\mathbf{x}/dt = (2, -1)$. If any sense is to be made of the formula

$$\frac{dz}{dt} = \frac{dz}{d\mathbf{x}} \frac{d\mathbf{x}}{dt}$$

it is plain that the symbol $dz/d\mathbf{x}$ (whatever it means) must be a *vector,* and the product $(dz/d\mathbf{x})(d\mathbf{x}/dt)$ must be a *dot product* (in order to produce the scalar dz/dt).   □

Can you think of a sensible way to differentiate a function of two variables so as to produce a vector? The only derivatives of $z = f(x, y)$ that we know are *partial* derivatives $\partial z/\partial x$ and $\partial z/\partial y$. How about using them as components of a vector? More precisely, replace the symbol $dz/d\mathbf{x}$ by $(\partial z/\partial x, \partial z/\partial y)$ and use the dot product:

$$\frac{dz}{dt} = \left(\frac{\partial z}{\partial x}, \frac{\partial z}{\partial y}\right) \cdot \left(\frac{dx}{dt}, \frac{dy}{dt}\right) = \frac{\partial z}{\partial x}\frac{dx}{dt} + \frac{\partial z}{\partial y}\frac{dy}{dt}$$

Let's see how this works in Example 1. Since $\partial z/\partial x = 2xye^{x^2}$ and $\partial z/\partial y = e^{x^2}$, our formula reads

$$\frac{dz}{dt} = 2xye^{x^2}\frac{dx}{dt} + e^{x^2}\frac{dy}{dt}$$

Substituting $x = 2t$, $y = 1 - t$, $dx/dt = 2$, and $dy/dt = -1$, we have

$$\frac{dz}{dt} = 2(2t)(1 - t)e^{(2t)^2}(2) + e^{(2t)^2}(-1) = 8t(1 - t)e^{4t^2} - e^{4t^2}$$

Interestingly enough, this is what we found in Example 1 (which may be an accident, but if so, it is an exceedingly lucky one). It is more likely that we have stumbled across a valid pattern (simply by guesswork based on a vector analogy to a rule in single-variable calculus).

Mathematicians do not use the symbol $dz/d\mathbf{x}$. Instead they write

$$\nabla z = (\partial z/\partial x, \partial z/\partial y)$$

and call ∇z the **gradient** of the function $z = f(x, y)$. The Chain Rule that we are suggesting then reads

$$\frac{dz}{dt} = \nabla z \cdot \frac{d\mathbf{x}}{dt} = \left(\frac{\partial z}{\partial x}, \frac{\partial z}{\partial y}\right) \cdot \left(\frac{dx}{dt}, \frac{dy}{dt}\right) = \frac{\partial z}{\partial x}\frac{dx}{dt} + \frac{\partial z}{\partial y}\frac{dy}{dt}$$

We do not claim to have proved anything. But the formula does seem reasonable when you think about it in words. To differentiate z with respect to t (when it is given in terms of x and y), differentiate it *partially* with respect to x and y, multiplying by dx/dt and dy/dt as in the case of the ordinary Chain Rule. Since

$$\frac{\partial z}{\partial x}\frac{dx}{dt} \qquad \text{and} \qquad \frac{\partial z}{\partial y}\frac{dy}{dt}$$

involve partial derivatives, it is not unreasonable to expect that they should be added to give the "total" derivative dz/dt.

□ **Example 2**   Let's try the same idea in the case of a function of three variables, say $w = x^2 + y^2 + z^2$, where $x = \cos t$, $y = \sin t$, $z = t$. The straightforward way to find dw/dt is to substitute before differentiating:

$$w = \cos^2 t + \sin^2 t + t^2 = 1 + t^2$$

$$\frac{dw}{dt} = 2t$$

The Chain Rule in this case calls for the computation of a three-dimensional gradient, namely

$$\nabla w = (\partial w/\partial x, \partial w/\partial y, \partial w/\partial z) = (2x, 2y, 2z) = 2(x, y, z)$$

The vector function $\mathbf{g}(t)$ is $\mathbf{x} = (x, y, z) = (\cos t, \sin t, t)$ with derivative $d\mathbf{x}/dt = (-\sin t, \cos t, 1)$. The formula

$$\frac{dw}{dt} = \nabla w \cdot \frac{d\mathbf{x}}{dt}$$

reads $\dfrac{dw}{dt} = 2(x, y, z) \cdot (-\sin t, \cos t, 1) = 2(-x \sin t + y \cos t + z)$

$$= 2(-\cos t \sin t + \sin t \cos t + t) = 2t \qquad □$$

Thus the formula $\dfrac{dw}{dt} = \dfrac{\partial w}{\partial x} \dfrac{dx}{dt} + \dfrac{\partial w}{\partial y} \dfrac{dy}{dt} + \dfrac{\partial w}{\partial z} \dfrac{dz}{dt}$

works. You may feel that it is the hard way (compared to substitution followed by ordinary differentiation). On the other hand, one wonders *why* it works! We offer a proof (for the two-variable case) in an optional note at the end of this section, the results of which we state as a theorem:

The Chain Rule

Suppose that $z = f(x, y)$ is a function with continuous partial derivatives. If $x = g(t)$ and $y = h(t)$ are differentiable, then z is a differentiable function of t and

$$\frac{dz}{dt} = \frac{\partial z}{\partial x} \frac{dx}{dt} + \frac{\partial z}{\partial y} \frac{dy}{dt} = \nabla z \cdot \frac{d\mathbf{x}}{dt} \qquad \text{where } \nabla z = (\partial z/\partial x, \partial z/\partial y)$$

is the **gradient** of $z = f(x, y)$ and $d\mathbf{x}/dt = (dx/dt, dy/dt)$.

□ **Example 3**   If $w = \ln(u^2 + v^2)$, where $u = 1 - x$ and $v = 2x$, find dw/dx at $x = 0$.

Solution: The letters differ from those in the statement of the Chain Rule, but the idea is the same:

$$\frac{dw}{dx} = \frac{\partial w}{\partial u}\frac{du}{dx} + \frac{\partial w}{\partial v}\frac{dv}{dx} = \frac{2u}{u^2 + v^2}(-1) + \frac{2v}{u^2 + v^2}(2) = \frac{4v - 2u}{u^2 + v^2}$$

When $x = 0$, we find $u = 1$ and $v = 0$, so $dw/dx = -2$.   □

□ **Example 4**   The radius of the base of a right circular cylinder is 10 cm and is decreasing at the rate of 2 cm/sec. The height of the cylinder is 15 cm and is increasing at the rate of 3 cm/sec. Is the volume of the cylinder increasing or decreasing? at what rate?

Solution: The volume is $V = \pi r^2 h$, where r is the radius of the base and h is the height. Since r and h depend on the time t, the volume does, too. Its derivative is

$$\frac{dV}{dt} = \frac{\partial V}{\partial r}\frac{dr}{dt} + \frac{\partial V}{\partial h}\frac{dh}{dt} = (2\pi r h)(-2) + (\pi r^2)(3)$$

(because $dr/dt = -2$ and $dh/dt = 3$). Since $r = 10$ and $h = 15$ at the instant in question, we find

$$\frac{dV}{dt} = (2\pi \cdot 10 \cdot 15)(-2) + (\pi \cdot 100)(3) = -300\pi$$

Thus the volume is decreasing at the rate of 300π cm^3/sec.   □

□ **Example 5**   When we change from rectangular to polar coordinates in the plane, the function $z = x^2 - y^2$ becomes a function of r and θ. Find $\partial z/\partial r$ and $\partial z/\partial \theta$.

Solution: As in our earlier examples, the conservative approach is to substitute before differentiating. Since $x = r \cos \theta$ and $y = r \sin \theta$, we have

$$z = r^2 \cos^2 \theta - r^2 \sin^2 \theta = r^2 \cos 2\theta$$

Hence $\partial z/\partial r = 2r \cos 2\theta$ and $\partial z/\partial \theta = -2r^2 \sin 2\theta$.

    Suppose, however, that we choose not to substitute. To find $\partial z/\partial r$, we fix θ, which makes x and y (and therefore z) functions of r alone. The Chain Rule therefore applies as stated above, except that each ordinary derivative is now a partial derivative:

$$\frac{\partial z}{\partial r} = \frac{\partial z}{\partial x}\frac{\partial x}{\partial r} + \frac{\partial z}{\partial y}\frac{\partial y}{\partial r} = 2x \cos \theta - 2y \sin \theta$$
$$= 2(r \cos \theta) \cos \theta - 2(r \sin \theta) \sin \theta$$
$$= 2r(\cos^2 \theta - \sin^2 \theta) = 2r \cos 2\theta$$

(as before). Similarly,

$$\frac{\partial z}{\partial \theta} = \frac{\partial z}{\partial x}\frac{\partial x}{\partial \theta} + \frac{\partial z}{\partial y}\frac{\partial y}{\partial \theta} = 2x(-r\sin\theta) - 2y(r\cos\theta)$$

$$= 2(r\cos\theta)(-r\sin\theta) - 2(r\sin\theta)(r\cos\theta)$$

$$= -2r^2 \cdot 2\sin\theta\cos\theta = -2r^2\sin 2\theta$$

(also as before).   □

Example 5 illustrates a version of the Chain Rule that is worth stating separately. This time we will assume that all functions are "well behaved" and omit the technical hypotheses:

Suppose that z is a function of x and y, which are in turn functions of s and t. Then

$$\frac{\partial z}{\partial s} = \frac{\partial z}{\partial x}\frac{\partial x}{\partial s} + \frac{\partial z}{\partial y}\frac{\partial y}{\partial s} \qquad \text{and} \qquad \frac{\partial z}{\partial t} = \frac{\partial z}{\partial x}\frac{\partial x}{\partial t} + \frac{\partial z}{\partial y}\frac{\partial y}{\partial t}$$

You should not memorize the various versions of the Chain Rule. The important thing is to recognize the pattern, which applies to any number of variables. We will give one more version (deliberately changing the letters to emphasize freedom from notational restrictions). Suppose that w is a function of s, t, u, v, which are in turn functions of x, y, z. Then

$$\frac{\partial w}{\partial x} = \frac{\partial w}{\partial s}\frac{\partial s}{\partial x} + \frac{\partial w}{\partial t}\frac{\partial t}{\partial x} + \frac{\partial w}{\partial u}\frac{\partial u}{\partial x} + \frac{\partial w}{\partial v}\frac{\partial v}{\partial x}$$

$$\frac{\partial w}{\partial y} = \frac{\partial w}{\partial s}\frac{\partial s}{\partial y} + \frac{\partial w}{\partial t}\frac{\partial t}{\partial y} + \frac{\partial w}{\partial u}\frac{\partial u}{\partial y} + \frac{\partial w}{\partial v}\frac{\partial v}{\partial y}$$

$$\frac{\partial w}{\partial z} = \frac{\partial w}{\partial s}\frac{\partial s}{\partial z} + \frac{\partial w}{\partial t}\frac{\partial t}{\partial z} + \frac{\partial w}{\partial u}\frac{\partial u}{\partial z} + \frac{\partial w}{\partial v}\frac{\partial v}{\partial z}$$

It is hard to state this kind of thing in complete generality (without digressing to introduce simplifying notation from advanced mathematics). Notice, however, that each formula is a sum involving all possible partial derivatives of the given function with respect to its independent variables, and each term of the sum is like the Chain Rule in single-variable calculus.

□ **Example 6**   Suppose that $w = st + tu + uv + vs$, where $s = x^2 + y^2 + z^2$, $t = xyz$, $u = \dfrac{xy}{z}$, and $v = e^{xyz}$. Find $\partial w/\partial y$ when $x = 0$, $y = 1$, and $z = 1$.

Solution: Since w is a function of s, t, u, and v, we need the partial derivatives

$$\frac{\partial w}{\partial s} = t + v \qquad \frac{\partial w}{\partial t} = s + u \qquad \frac{\partial w}{\partial u} = t + v \qquad \frac{\partial w}{\partial v} = u + s$$

Since $\partial w/\partial y$ is wanted, we also need

$$\frac{\partial s}{\partial y} = 2y \qquad \frac{\partial t}{\partial y} = xz \qquad \frac{\partial u}{\partial y} = \frac{x}{z} \qquad \frac{\partial v}{\partial y} = xze^{xyz}$$

The version of the Chain Rule stated just before this example now yields

$$\frac{\partial w}{\partial y} = (t + v)(2y) + (s + u)(xz) + (t + v)\left(\frac{x}{z}\right) + (u + s)(xze^{xyz})$$

When $x = 0$, $y = 1$, and $z = 1$, we have $s = 2$, $t = 0$, $u = 0$, and $v = 1$. Hence

$$\frac{\partial w}{\partial y} = (1)(2) + (2)(0) + (1)(0) + (2)(0) = 2 \qquad\qquad \square$$

From what we have said in this section, the gradient of a function seems to have little more significance than as a notational shortcut in the Chain Rule:

$$\frac{dz}{dt} = \nabla z \cdot \frac{d\mathbf{x}}{dt} \qquad \text{instead of} \qquad \frac{dz}{dt} = \frac{\partial z}{\partial x}\frac{dx}{dt} + \frac{\partial z}{\partial y}\frac{dy}{dt}$$

Its true importance remains to be seen; in the next section we begin to develop the remarkable role it plays in multivariable calculus.

Optional Note (on the proof of the Chain Rule)   Let $z = f(x,y)$, where $x = g(t)$ and $y = h(t)$. Then z is a function of t, say

$$z = F(t) = f(g(t),h(t))$$

The definition of its derivative as the limit of a difference quotient is

$$\frac{dz}{dt} = F'(t) = \lim_{s \to t} \frac{F(s) - F(t)}{s - t} = \lim_{s \to t} \frac{f(g(s),h(s)) - f(g(t),h(t))}{s - t}$$

The problem is to deal with the functional difference

$$f(g(s),h(s)) - f(g(t),h(t)) = f(u,v) - f(x,y)$$

where $u = g(s)$, $v = h(s)$, $x = g(t)$, and $y = h(t)$.

Functional differences in single-variable calculus are simplified by the Mean Value Theorem:

$$f(b) - f(a) = f'(c)(b - a) \qquad \text{where } c \text{ is between } a \text{ and } b$$

We can apply this notion to $f(u,v) - f(x,y)$ by rewriting it so that only one variable changes at a time:

$$f(u,v) - f(x,y) = [f(u,v) - f(x,v)] + [f(x,v) - f(x,y)]$$

In the first bracket f may be regarded as a function of its first variable alone (the second variable being fixed at the value v). Apply the Mean Value Theorem to write

$$f(u,v) - f(x,v) = f_1(c,v)(u - x) \qquad \text{where } c \text{ is between } x \text{ and } u$$

Similarly (in the second bracket) f is a function of its second variable alone, its first variable being fixed at the value x. Hence

$$f(x,v) - f(x,y) = f_2(x,d)(v - y) \qquad \text{where } d \text{ is between } y \text{ and } v$$

This yields

$$f(u,v) - f(x,y) = f_1(c,v)(u - x) + f_2(x,d)(v - y)$$

and hence

$$\frac{f(u,v) - f(x,y)}{s - t} = f_1(c,v)\frac{g(s) - g(t)}{s - t} + f_2(x,d)\frac{h(s) - h(t)}{s - t}$$

You can already see the Chain Rule taking shape! We need only evaluate the limit as $s \to t$:

$$\frac{dz}{dt} = \lim_{s \to t} f_1(c,v) \cdot \lim_{s \to t} \frac{g(s) - g(t)}{s - t} + \lim_{s \to t} f_2(x,d) \cdot \lim_{s \to t} \frac{h(s) - h(t)}{s - t}$$

$$= \left[\lim_{s \to t} f_1(c,v) \right] g'(t) + \left[\lim_{s \to t} f_2(x,d) \right] h'(t)$$

(assuming that g and h are differentiable). When $s \to t$, we have

$$u = g(s) \to g(t) = x \qquad \text{and} \qquad v = h(s) \to h(t) = y$$

(because g and h are continuous). This forces $c \to x$ (because c is between x and u) and $d \to y$ (because d is between y and v). Hence

$$f_1(c,v) \to f_1(x,y) \text{ and } f_2(x,d) \to f_2(x,y)$$

(assuming that the partial derivatives f_1 and f_2 are continuous). Thus

$$\frac{dz}{dt} = f_1(x,y)g'(t) + f_2(x,y)h'(t) = \frac{\partial z}{\partial x}\frac{dx}{dt} + \frac{\partial z}{\partial y}\frac{dy}{dt}$$

You can see that the same proof can be given for a function of more than two variables; the only difficulty is one of notation.

Problem Set 17.2

In each of the following, find dz/dt or dw/dt in two ways, first by substituting before differentiating, then by the Chain Rule. In each case reconcile the answers.

1. $z = e^{xy}$, where $x = t^2$, $y = t^3$

2. $z = e^{x-y}$, where $x = 1 - t$, $y = 1 - t^2$

3. $z = \ln(x^2 + y^2)$, where $x = e^t$, $y = e^{-t}$

4. $z = \ln xy$, where $x = e^t$, $y = e^{-t}$

5. $z = x/(x^2 + y^2)$, where $x = \cos t$, $y = \sin t$

6. $z = x^2 - y^2$, where $x = \cosh t$, $y = \sinh t$

7. $w = x^2 + y^2 + z^2$, where $x = t$, $y = t^2$, $z = t^3$

8. $w = \ln(x^2 + y^2 + z^2)$, where $x = \cos t$, $y = \sin t$, $z = t$

9. $w = xyz$, where $x = \cos t$, $y = \sin t$, $z = t$

10. $w = e^{xyz}$, where $x = t$, $y = t^2$, $z = t^3$

11. $w = \sqrt{x^2 + y^2 + z^2}$, where $x = \cos t$, $y = \sin t$, $z = \sin t$

12. $w = xy + yz + zx$, where $x = 2t - 1$, $y = t$, $z = 1 - t$

13. Suppose that $w = xy + yz + zx$, where $x = e^t \cos t$, $y = e^t \sin t$, $z = e^t$. Find dw/dt when $t = 0$.

14. Suppose that the temperature at each point of the plane is given by $T = \sqrt{x^2 + y^2}$ and that a bug's position at time t is $(x, y) = (t^2, t^3)$. Find the rate of change of temperature experienced by the bug as it passes through the point $(4, 8)$.

15. In Problem 14 sketch the path of the bug, noting the cusp at the origin. Show that the rate of change of temperature experienced by the bug is nevertheless defined for all t, and is given by

$$\frac{dT}{dt} = \frac{t(2 + 3t^2)}{\sqrt{1 + t^2}}$$

What temperature behavior does the bug experience as it passes through the origin?

16. Suppose that the pressure, volume, and temperature of a gas are related by the law $PV = kT$, where k is a constant. At the instant when $P = 20$, $V = 80$, and $T = 300$, the pressure and volume are changing at the rates $dP/dt = 2$ and $dV/dt = -3$ (where t is time). Is the temperature increasing or decreasing at that instant? at what rate?

17. The radius of the base of a right circular cone is 30 cm and is increasing at 2 cm/sec. The altitude is 40 cm and is decreasing at 5 cm/sec. Is the volume of the cone increasing or decreasing? at what rate?

18. Two sides of a triangle are 10 cm and 15 cm, and are increasing at 3 cm/sec and 4 cm/sec, respectively, while the included angle is $\pi/3$ and is decreasing at $\frac{1}{2}$ rad/sec. Is the third side increasing or decreasing? at what rate? *Hint:* Use the Law of Cosines.

19. The Product Rule in single-variable calculus says that if $u = f(x)$ and $v = g(x)$, then

$$\frac{d}{dx}(uv) = u\frac{dv}{dx} + v\frac{du}{dx}$$

Derive this formula by using the Chain Rule, letting $z = uv$ and finding dz/dx.

20. As in the preceding problem, use the Chain Rule to derive the Quotient Rule,

$$\frac{d}{dx}\left(\frac{u}{v}\right) = \frac{v(du/dx) - u(dv/dx)}{v^2}$$

21. If $w = u^2 e^v$, where $u = x/y$ and $v = y \ln x$, find $\partial w/\partial x$ and $\partial w/\partial y$ when $x = 1$ and $y = 2$.

22. Suppose that

$$z = \frac{x - y}{x + y} \qquad \text{where } x = uvw \text{ and } y = u^2 + v^2 + w^2$$

Find $\partial z/\partial u$, $\partial z/\partial v$, $\partial z/\partial w$ when $u = 2$, $v = -1$, $w = 1$.

23. Suppose that

$$z = \ln(x^2 + y^2) \qquad \text{where } x = r \cos \theta \text{ and } y = r \sin \theta$$

Find $\partial z/\partial r$ by the Chain Rule and check by substituting before differentiating.

24. Repeat Problem 23 for $\partial z/\partial \theta$.

25. Suppose that $z = \tan^{-1}(y/x)$, where $x = r \cos \theta$ and $y = r \sin \theta$. Find $\partial z/\partial r$ by the Chain Rule and check by substituting before differentiating.

26. Repeat Problem 25 for $\partial z/\partial \theta$.

27. When rectangular coordinates are changed to polar coordinates, the function $z = f(x, y)$ becomes a function of r and θ. Show that

$$\left(\frac{\partial z}{\partial r}\right)^2 + \frac{1}{r^2}\left(\frac{\partial z}{\partial \theta}\right)^2 = \left(\frac{\partial z}{\partial x}\right)^2 + \left(\frac{\partial z}{\partial y}\right)^2$$

28. In the preceding problem show that

$$z_{rr} = z_{xx} \cos^2 \theta + 2z_{xy} \sin \theta \cos \theta + z_{yy} \sin^2 \theta$$

as follows.

(a) Use the Chain Rule to obtain

$$z_r = z_x \cos \theta + z_y \sin \theta$$

(b) Differentiate with respect to r again, keeping in mind that z_x and z_y are themselves functions of x and y (and hence of r and θ). Thus the Chain Rule is needed in each term of part (a).

29. When rectangular coordinates are changed to spherical coordinates, the function $w = x^2 + y^2 + z^2$ becomes a function of ρ, ϕ, θ by virtue of the formulas

$$x = \rho \sin \phi \cos \theta, \ y = \rho \sin \phi \sin \theta, \ z = \rho \cos \phi$$

(a) Use the Chain Rule to show that $\partial w/\partial \rho = 2\rho$, $\partial w/\partial \phi = 0$, $\partial w/\partial \theta = 0$

(b) Check by substituting before differentiating.

30. If $P = f(u, v, w)$, where $u = x - y$, $v = y - z$, $w = z - x$, show that

$$\frac{\partial P}{\partial x} + \frac{\partial P}{\partial y} + \frac{\partial P}{\partial z} = 0$$

31. Suppose that $z = f(x, y)$, where $x = u + v$ and $y = u - v$. Show that (a) $z_u + z_v = 2z_x$; (b) $z_u z_v = z_x^2 - z_y^2$; and (c) $z_{uv} = z_{xx} - z_{yy}$.

Find the gradient of each of the following functions.

32. $z = 2x - 3y$

33. $z = x^2 + y^2$

34. $z = e^{xy}$

35. $w = 5x + y - z$

36. $w = \ln(x^2 + y^2 + z^2)$ **37.** $w = \sin xyz$

38. Show that the gradient of $z = e^{x^2+y^2}$ is $\nabla z = 2z(x, y)$.

39. Suppose that $f(\mathbf{x}) = \mathbf{x} \cdot \mathbf{x}$, where $\mathbf{x} \in \mathcal{R}^2$. Show that $\nabla f(\mathbf{x}) = 2\mathbf{x}$.

Assuming that $f(x, y)$ and $g(x, y)$ have partial derivatives, derive each of the following formulas.

40. $\nabla(f + g) = \nabla f + \nabla g$ **41.** $\nabla(fg) = f(\nabla g) + g(\nabla f)$

42. $\nabla\left(\dfrac{f}{g}\right) = \dfrac{g(\nabla f) - f(\nabla g)}{g^2}$

Assuming that $z = f(x, y)$ has partial derivatives, derive each of the following formulas.

43. $\nabla(z^r) = rz^{r-1}(\nabla z)$ where r is any real number

44. $\nabla(e^z) = e^z(\nabla z)$

45. $\nabla(\sin z) = (\cos z)(\nabla z)$

17.3

TANGENT PLANES AND TANGENT LINES

When the "problem of tangents" first came up in Section 2.1, we pointed out that the tangent to a circle may be found by ordinary analytic geometry. For if (a, b) is the point of tangency, the radius from the center of the circle to (a, b) is perpendicular to the tangent. (See Figure 1.) The slope of the tangent is therefore the negative reciprocal of the slope of the radius and we have the information needed to write an equation of the tangent.

The same idea applies to a sphere. Suppose, for example, that we want an equation of the plane tangent to the sphere $x^2 + y^2 + z^2 = 9$ at the point $(1, 2, 2)$. (See Figure 2.) The "radius vector" from $(0, 0, 0)$ to $(1, 2, 2)$ is $\mathbf{n} = (1, 2, 2)$, which is normal to the plane. Since $(1, 2, 2)$ is a point of the plane, an equation is

$$1(x - 1) + 2(y - 2) + 2(z - 2) = 0 \quad \text{or} \quad x + 2y + 2z - 9 = 0$$

It may be objected (as noted in Section 2.1 when we discussed tangent lines) that a definition is needed. For we are using the term *tangent plane* without having explained what it means. As usual we are depending on your intuition; the definition will arrive shortly.

Generally speaking, a surface in 3-space is described by an equation involving x, y, and z. Such an equation may always be written in the form $F(x, y, z) = 0$, where $F: \mathcal{R}^3 \to \mathcal{R}$ is a function with real values. For example, the above sphere is represented by the equation

$$F(x, y, z) = x^2 + y^2 + z^2 - 9 = 0$$

We assume that F is well behaved, that is, both F and its first partial derivatives are continuous in a neighborhood of whatever point we are considering.

Let $P = (x, y, z)$ be a point of the surface $F(x, y, z) = 0$ at which we want to find the tangent plane (assuming for the moment that this term is understood). It is geometrically plausible to think of the plane as determined by the family of tangent lines to curves lying in the surface and passing through P. (See Figure 3.) Let C be any such curve; we propose to show that the gradient of F, namely

$$\mathbf{n} = \nabla F = (F_1, F_2, F_3)$$

Figure 1
Tangent to a circle

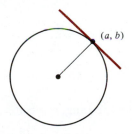

Figure 2
Tangent plane to a sphere

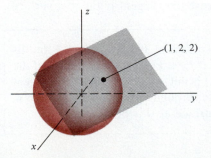

Figure 3
Tangent plane determined by tangent lines

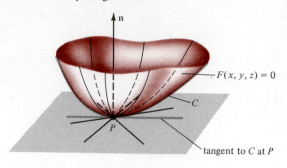

is perpendicular to the tangent line to C at P. If this is correct, then $\mathbf{n}$ is a normal vector to the tangent plane. For if C is *any* curve (through P) lying in the surface, $\mathbf{n}$ must be perpendicular to all tangent lines in the family that determines the plane, hence perpendicular to the plane itself.

We know from Section 16.6 that C may be defined by a vector function $\mathbf{x} = \mathbf{g}(t) = (x, y, z)$, and that the derivative $d\mathbf{x}/dt$ is parallel to the tangent line to C. Since each point of C is on the surface $F(x, y, z) = 0$, we have $F(\mathbf{x}) = 0$ for all values of the parameter t. Differentiate each side of this equation with respect to t:

$$\frac{d}{dt} F(\mathbf{x}) = 0 \qquad \text{for all } t$$

According to the Chain Rule in the last section,

$$\frac{d}{dt} F(\mathbf{x}) = \frac{\partial F}{\partial x}\frac{dx}{dt} + \frac{\partial F}{\partial y}\frac{dy}{dt} + \frac{\partial F}{\partial z}\frac{dz}{dt} = \nabla F \cdot \frac{d\mathbf{x}}{dt}$$

so we have
$$\nabla F \cdot \frac{d\mathbf{x}}{dt} = 0$$

This equation says that ∇F is perpendicular to the tangent vector $d\mathbf{x}/dt$, which is what we wanted to show.

Having arrived at this conclusion by plausible geometrical reasoning, we now make a definition. In other words (as is often the case in discussions of this kind), we have not proved anything; instead we have prepared the ground for defining something (in this case the tangent plane to a surface).

Let $F(x, y, z) = 0$ be an equation of a surface in 3-space, where F and its first partial derivatives are continuous. If $P_0 = (x_0, y_0, z_0)$ is a point of the surface at which the vector $\mathbf{n} = \nabla F(P_0)$ is not zero, we define the **tangent plane** to the surface at P_0 to be the plane through P_0 with normal vector $\mathbf{n}$. We also say that $\mathbf{n}$ is **normal to the surface** at this point.

We could have included an equation of the tangent plane in this definition, but that would look like something new to be learned, when in fact you already know how to write the equation. Simply compute the numbers

$$a = F_1(P_0), \quad b = F_2(P_0), \quad c = F_3(P_0)$$

Then $\mathbf{n} = (a, b, c)$ is a normal vector to the plane and an equation is

$$a(x - x_0) + b(y - y_0) + c(z - z_0) = 0$$

(See Section 16.2.) Note the role of our assumption that $\mathbf{n} \neq \mathbf{0}$. This guarantees that a, b, c are not all zero and leads to an equation that represents a definite plane.

□ **Example 1**   Find the tangent plane to the sphere $x^2 + y^2 + z^2 = 9$ at the point $(1, 2, 2)$.

Solution: Writing the equation of the sphere in the form

$$F(x, y, z) = x^2 + y^2 + z^2 - 9 = 0$$

we have $\partial F/\partial x = 2x$, $\partial F/\partial y = 2y$, $\partial F/\partial z = 2z$. Evaluating these derivatives at $(1, 2, 2)$, we find the normal vector

$$\mathbf{n} = \nabla F(1, 2, 2) = (2, 4, 4)$$

Hence an equation of the tangent plane is

$$2(x - 1) + 4(y - 2) + 4(z - 2) = 0 \qquad \text{or} \qquad x + 2y + 2z - 9 = 0$$

This checks with our earlier result when we used the radius vector $\mathbf{n} = (1, 2, 2)$ as a normal to the plane. The gradient of F at $(1, 2, 2)$ is not the same $\mathbf{n}$, but a multiple of it (as it must be if both vectors are normal to the plane).   □

□ **Example 2**   Find the tangent plane to the elliptical paraboloid $z = 1 - \frac{1}{4}x^2 - y^2$ at the point $(1, \frac{1}{2}, \frac{1}{2})$.

Solution: Rewrite the equation of the surface in the form

$$F(x, y, z) = x^2 + 4y^2 + 4z - 4 = 0$$

Then $\nabla F(x, y, z) = (2x, 8y, 4)$ and a normal vector to the tangent plane is

$$\nabla F(1, \tfrac{1}{2}, \tfrac{1}{2}) = (2, 4, 4)$$

An equation of the plane is

$$2(x - 1) + 4(y - \tfrac{1}{2}) + 4(z - \tfrac{1}{2}) = 0 \quad \text{or} \quad x + 2y + 2z - 3 = 0 \quad □$$

A common error should be noted in connection with Example 2. Surfaces in space are often represented by equations of the form $z = f(x, y)$. In Example 2

$$f(x, y) = 1 - \tfrac{1}{4}x^2 - y^2$$

It is not unusual for a student to compute the gradient of this function,

$$\nabla f(x, y) = (-\tfrac{1}{2}x, -2y) = (-\tfrac{1}{2}, -1) \text{ at } (1, \tfrac{1}{2})$$

and to try to use it as a normal vector to the tangent plane. Note, however, that the gradient of f is a vector in $\mathcal{R}^2$, not $\mathcal{R}^3$, and could not play the role of a normal vector to a plane.

Despite this remark, there is no reason why we cannot deal with surfaces in the form $z = f(x, y)$. Simply rewrite the equation in the form

$$F(x, y, z) = f(x, y) - z = 0$$

and compute

$$\frac{\partial F}{\partial x} = f_1(x, y), \ \frac{\partial F}{\partial y} = f_2(x, y), \ \frac{\partial F}{\partial z} = -1$$

If $P_0 = (x_0, y_0, z_0)$ is a point of the surface, where $z_0 = f(x_0, y_0)$, a normal vector to the tangent plane at P_0 is $\mathbf{n} = (a, b, c)$, where

$$a = f_1(x_0, y_0), \ b = f_2(x_0, y_0), \ c = -1$$

□ **Example 3**   Rework Example 2 by the above method.

Solution: Letting $f(x, y) = 1 - \tfrac{1}{4}x^2 - y^2$, we compute

$$f_1(1, \tfrac{1}{2}) = -\tfrac{1}{2} \quad \text{and} \quad f_2(1, \tfrac{1}{2}) = -1$$

A normal vector to the tangent plane is $\mathbf{n} = (-\tfrac{1}{2}, -1, -1)$ and an equation of the plane is

$$-\tfrac{1}{2}(x - 1) - 1(y - \tfrac{1}{2}) - 1(z - \tfrac{1}{2}) = 0 \quad \text{or} \quad x + 2y + 2z - 3 = 0 \quad □$$

Now we turn to some interesting results of specializing our discussion to a lower dimension. If the tangent plane to a surface $F(x, y, z) = 0$ is found by using the gradient of F as a normal vector to the plane, why not find the tangent line to a curve (in $\mathcal{R}^2$) in the same way? We investigate the implications of this idea in a series of remarks.

1. Suppose that we are back in single-variable calculus looking for the tangent line to a curve in $\mathcal{R}^2$. The equation of the curve may be written in the form $F(x, y) = 0$, where $F: \mathcal{R}^2 \to \mathcal{R}$ is a function with real values. For example, the circle $x^2 + y^2 = 1$ can be described by the equation

$$F(x, y) = x^2 + y^2 - 1 = 0$$

The curve $F(x, y) = 0$ can also be described by a vector function $\mathbf{x} = \mathbf{g}(t) = (x, y)$. Since each point $\mathbf{x} = (x, y)$ lies on the curve, we have $F(\mathbf{x}) = 0$ for all t. Differentiate with respect to t (by the Chain Rule of

Section 17.2) to find

$$\frac{d}{dt}F(\mathbf{x}) = \nabla F \cdot \frac{d\mathbf{x}}{dt} = 0 \qquad \text{for all } t$$

Hence ∇F and $d\mathbf{x}/dt$ are perpendicular. But $d\mathbf{x}/dt$ is tangent to the curve (Section 15.2), so we conclude that ∇F is normal to the tangent line at (x,y).

Thus when F is a function of three variables (as in the first part of this section), its gradient is normal to the tangent *plane* to the *surface* $F(x,y,z) = 0$. When it is a function of two variables, its gradient is normal to the tangent *line* to the *curve* $F(x,y) = 0$.

2. Let L be the tangent line to the curve $F(x,y) = 0$ at the point $P_0 = (x_0, y_0)$. According to Remark (1), the vector

$$\mathbf{n} = (a,b) \qquad \text{where } a = F_1(P_0) \text{ and } b = F_2(P_0)$$

is normal to L. Assuming that $\mathbf{n} \neq \mathbf{0}$, we conclude that if $P = (x,y)$ is any point of the plane, then

$$P \in L \Leftrightarrow \mathbf{n} \perp \overrightarrow{P_0P} \Leftrightarrow (a,b) \cdot (x - x_0, y - y_0) = 0$$
$$\Leftrightarrow a(x - x_0) + b(y - y_0) = 0$$

(See Figure 4.) Hence an equation of L is $a(x - x_0) + b(y - y_0) = 0$. (Note the resemblance to the equation of the tangent plane to a surface.)

3. When a curve in the plane is given in explicit functional form $y = f(x)$, we ordinarily characterize its tangent line at P_0 by the equation

$$y - y_0 = m(x - x_0) \qquad \text{where } m = f'(x_0)$$

This is a special case of Remark (2), as you can see by writing the equation $y = f(x)$ in the form

$$F(x,y) = f(x) - y = 0$$

Since $F_1(x,y) = f'(x)$ and $F_2(x,y) = -1$, the tangent line at P_0 is

$$a(x - x_0) + b(y - y_0) = 0 \qquad \text{where } a = f'(x_0) \text{ and } b = -1$$

This reduces to $y - y_0 = m(x - x_0)$.

4. Suppose that the relation $F(x,y) = 0$ implicitly defines y as a function of x. Then dy/dx is the slope of the tangent line at (x,y), as we know from single-variable calculus. Since $\nabla F = (F_1, F_2)$ is normal to the tangent, the vector $(F_2, -F_1)$ is parallel to the tangent. (Why?) If $F_2(x,y) \neq 0$, the slope of this vector is $-F_1/F_2$, so we have

$$\boxed{\frac{dy}{dx} = -\frac{F_1(x,y)}{F_2(x,y)}}$$

This formula gives us a new look at implicit differentiation (Section 3.4), as the following example illustrates.

Figure 4
Characterization of tangent line to a curve

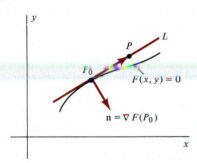

□ **Example 4** Find dy/dx from the equation $x^2 + y^2 = 1$ in two ways.

Solution: Differentiating implicitly in the given equation, we find

$$2x + 2y\frac{dy}{dx} = 0$$

from which $dy/dx = -x/y$. A second method is to write the given equation in the form

$$F(x,y) = x^2 + y^2 - 1 = 0$$

and to find $F_1(x,y) = 2x$ and $F_2(x,y) = 2y$. According to the formula in Remark (4),

$$\frac{dy}{dx} = -\frac{F_1(x,y)}{F_2(x,y)} = -\frac{2x}{2y} = -\frac{x}{y}$$

as before. □

At this point it would be worthwhile to go back to Section 3.4 and read the optional note on the Implicit Function Theorem. In that note we stated that if (x_0, y_0) is a point of the graph of $F(x,y) = 0$ in a neighborhood of which F and its partial derivatives are continuous, and if $F_2(x_0, y_0) \neq 0$, then the equation $F(x,y) = 0$ defines y as a differentiable function of x in a neighborhood of x_0. You can see from the above remarks why the condition $F_2(x_0, y_0) \neq 0$ is needed; moreover, we now have a theoretical formula for dy/dx, the derivative of our implicitly defined function. In Section 3.4 we presented this theorem to help you believe that implicitly defined functions make sense, but of course we did not expect you to understand the background. This discussion should supply it.

Problem Set 17.3

In each of the following, find an equation of the tangent plane to the given surface at the given point.

1. $x^2 + y^2 + 2z^2 = 9$ at $(1, 0, 2)$

2. $x^2 + 2y^2 + 3z^2 = 6$ at $(1, -1, 1)$

3. $2x^2 + y^2 + 2z^2 + 2xz = 7$ at $(-1, 1, 2)$

4. $xyz = 6$ at $(1, 2, 3)$

5. $xy + yz + zx = 8$ at $(2, 1, 2)$

6. $yz = 1$ at $(2, \frac{3}{2}, \frac{2}{3})$

7. $y = x^2 + z^2$ at $(0, 4, -2)$

8. $z = 5 - 2x^2 - y^2$ at $(1, -1, 2)$

9. $z = \sqrt{x^2 + y^2}$ at $(3, 4, 5)$

10. $z = \sqrt{4 - x^2 - y^2}$ at $(1, -1, \sqrt{2})$

11. $\dfrac{x^2}{4} + \dfrac{y^2}{9} + \dfrac{z^2}{25} = 1$ at $(2, 0, 0)$

12. $\dfrac{x}{2} + \dfrac{y}{3} + \dfrac{z}{4} = 1$ at $(0, \frac{3}{2}, 2)$

13. Sketch the circular paraboloid $z = 4 - x^2 - y^2$ to see why the tangent plane at its vertex should be horizontal. Confirm that it is by finding its equation.

14. Show that the tangent plane to the hyperbolic paraboloid $z = y^2/b^2 - x^2/a^2$ at the origin is the xy plane. (Then look up Figure 6 in Section 16.4 to see that this makes sense geometrically. The origin is a "saddle point" of the surface.)

15. Show that the tangent plane to the cylinder $x^2 + y^2 = 1$ at (x_0, y_0, z_0) is $x_0 x + y_0 y = 1$. This plane is parallel to the z axis; draw a picture to illustrate why.

16. Show that the tangent plane to the cone

$$z^2 = \frac{x^2}{a^2} + \frac{y^2}{b^2} \qquad \text{at } (x_0, y_0, z_0) \neq (0, 0, 0)$$

is

$$z_0 z = \frac{x_0 x}{a^2} + \frac{y_0 y}{b^2}$$

What is the geometric reason for the fact that this plane contains the origin? Why do we restrict the point of tangency to be different from the origin?

17. Show that the tangent plane to the ellipsoid $x^2/a^2 + y^2/b^2 + z^2/c^2 = 1$ at (x_0, y_0, z_0) is

$$\frac{x_0 x}{a^2} + \frac{y_0 y}{b^2} + \frac{z_0 z}{c^2} = 1$$

At what points of the ellipsoid is this plane horizontal? What does its equation reduce to in these cases?

18. Use gradients to show that at each point of intersection of the sphere $x^2 + y^2 + z^2 = 1$, the cone $z^2 = x^2 + y^2$, and the plane $y = x$, the surfaces are mutually perpendicular.

19. Explain why the line perpendicular to the surface $F(x, y, z) = 0$ at (x_0, y_0, z_0) is given by the parametric equations

$$x = x_0 + at, \ y = y_0 + bt, \ z = z_0 + ct$$

where $(a, b, c) = \nabla F(x_0, y_0, z_0) \neq \mathbf{0}$.

20. Explain why the line perpendicular to the surface $z = f(x, y)$ at (x_0, y_0, z_0) is given by the parametric equations

$$x = x_0 + at, \ y = y_0 + bt, \ z = z_0 - t$$

where $a = f_1(x_0, y_0)$ and $b = f_2(x_0, y_0)$.

21. Use Problem 19 to find parametric equations of the line perpendicular to the surface $z = 1 - \frac{1}{4}x^2 - y^2$ at $(1, \frac{1}{2}, \frac{1}{2})$.

22. Repeat Problem 21 using Problem 20.

23. A bug at the point $(1, 1, 1)$ on the egg $x^2/4 + y^2/4 + z^2/2 = 1$ flies off along the normal line. Does it reach a flower at $(5, 5, 10)$?

24. Find the distance from the point $(5, 5, 4)$ to the paraboloid $z = 4 - x^2 - y^2$. *Hint:* Find the point of the paraboloid such that the normal at the point passes through $(5, 5, 4)$.

25. Use the gradient to show that every circle of the type $x^2 + y^2 = r^2$ intersects every line of the type $ax + by = 0$ at right angles.

26. Use the gradient to show that every hyperbola of the type $x^2 - y^2 = k^2$ intersects every hyperbola of the type $xy = c$ at right angles.

In each of the following, use the formula in Remark (4) to find dy/dx from the given equation.

27. $xy - y - x = 0$ (Check with Example 1, Section 3.4.)
28. $2x^2 - 2xy + y^2 = 5$ (Check with Example 2, Section 3.4.)
29. $y - x \sin y + 1 = 0$ (Check with Example 3, Section 3.4.)
30. $x/y = 2$
31. $x - \sqrt{y} = 2$
32. $1/x - 1/y = 1$
33. $x = \dfrac{y - 1}{y + 1}$
34. $x^3 + y^3 = 1$
35. $xy - 2x + 3y = 0$
36. $x^2 - xy + y^2 = 1$
37. $2x^2 - 5xy - y^2 = 2$
38. $y = (x + y)^2$

39. Suppose that the equation $F(x, y) = 0$ defines y implicitly as a function of x, say $y = f(x)$. Then $F(x, y)$ becomes a function of x alone when $f(x)$ is substituted for y. Use the Chain Rule of Section 17.2 to explain why

$$F_1(x, y) + F_2(x, y) \frac{dy}{dx} = 0$$

and solve for dy/dx. This is another way to derive the formula for dy/dx in Remark (4).

Implicit differentiation can be used to find partial derivatives as well as ordinary derivatives. The following problems illustrate what we mean.

40. The equation $x^2 + y^2 + z^2 = 1$, $z \geq 0$, defines z implicitly as a function of x and y.
   (a) Use implicit differentiation to show that $\partial z/\partial x = -x/z$ and $\partial z/\partial y = -y/z$.
   (b) Confirm the results in part (a) by solving for z in terms of x and y and then differentiating.

41. The equation $xy + yz + zx = 0$ defines z implicitly as a function of x and y.
   (a) Use implicit differentiation to show that

$$\frac{\partial z}{\partial x} = -\frac{y + z}{x + y} \qquad \text{and} \qquad \frac{\partial z}{\partial y} = -\frac{x + z}{x + y}$$

   (b) Confirm the results in part (a) by solving for z in terms of x and y and then differentiating.

42. The equations $x = r \cos \theta$, $y = r \sin \theta$ define r and θ im-

plicitly as functions of x and y. Find the partial derivatives of these functions as follows.

(a) Differentiate implicitly in the given equations to show that

$$\cos\theta \, \frac{\partial r}{\partial x} - r\sin\theta \, \frac{\partial\theta}{\partial x} = 1$$

$$\sin\theta \, \frac{\partial r}{\partial x} + r\cos\theta \, \frac{\partial\theta}{\partial x} = 0$$

and

$$\cos\theta \, \frac{\partial r}{\partial y} - r\sin\theta \, \frac{\partial\theta}{\partial y} = 0$$

$$\sin\theta \, \frac{\partial r}{\partial y} + r\cos\theta \, \frac{\partial\theta}{\partial y} = 1$$

(b) Solve these systems to find

$$\frac{\partial r}{\partial x} = \cos\theta, \quad \frac{\partial\theta}{\partial x} = \frac{-\sin\theta}{r}$$

and

$$\frac{\partial r}{\partial y} = \sin\theta, \quad \frac{\partial\theta}{dy} = \frac{\cos\theta}{r}$$

(c) Confirm these results by differentiating implicitly in the equations $r^2 = x^2 + y^2$ and $\tan\theta = y/x$.

43. If $x = r\cos\theta$ and $y = r\sin\theta$, show that

$$\nabla\theta = \frac{1}{r}(-\sin\theta, \cos\theta)$$

Why is $\nabla\theta$ perpendicular to the terminal side of θ? What is its magnitude?

44. Find an equation of the tangent plane to the sphere $x^2 + y^2 + z^2 = 9$ at the point $(1, 2, 2)$ as follows.

(a) Explain why the tangent line to the curve of intersection of the sphere and the plane $y = 2$ at the point $(1, 2, 2)$ has slope $-\frac{1}{2}$. Show that it is described by the equations

$$z - 2 = -\tfrac{1}{2}(x - 1), \ y = 2$$

and that parametric equations of the line are

$$x = t, \ y = 2, \ z = \tfrac{5}{2} - \tfrac{1}{2}t$$

Show that a vector parallel to the line is $\mathbf{m}_1 = (2, 0, -1)$.

(b) Similarly, show that a vector parallel to the tangent line to the curve of intersection of the sphere and the plane $x = 1$ at $(1, 2, 2)$ is $\mathbf{m}_2 = (0, 1, -1)$.

(c) Now compute $\mathbf{n} = \mathbf{m}_1 \times \mathbf{m}_2$ to find a normal vector to the plane and write an equation of the plane. (The result should check with Example 1.)

45. Use the method outlined in the preceding problem to find an equation of the tangent plane to the surface $z = 1 - \frac{1}{4}x^2 - y^2$ at the point $(1, \frac{1}{2}, \frac{1}{2})$. (The result should check with Example 2.)

46. Use the method outlined in Problem 44 to show that the tangent plane to the surface $z = f(x, y)$ at (x_0, y_0, z_0) is given by

$$a(x - x_0) + b(y - y_0) - (z - z_0) = 0$$

where $a = f_1(x_0, y_0)$ and $b = f_2(x_0, y_0)$.

17.4
THE DIRECTIONAL DERIVATIVE

The partial derivatives of a function of several variables are useful, as we have seen. They are special cases of its rate of change, however, since only one of the independent variables is activated; the others are kept fixed. Differentiation with respect to x, for example, tells us the rate of change of a function per unit increase in x. We might say that we are finding its rate in the direction of the unit vector $\mathbf{i}$. Similarly, the partial derivative with respect to y is a rate in the direction of $\mathbf{j}$.

The idea of this section is to allow other directions (either in $\mathcal{R}^2$ or $\mathcal{R}^3$, depending on whether our function involves two independent variables or three). It is possible to generalize to n variables, but to visualize the situation geometrically we will keep $n \leq 3$.

Suppose that f is a function (with domain D) whose rate of change we seek at a point $P_0 \in D$ in a given direction away from this point. We specify the direction by a unit vector $\mathbf{u}$; what we want is a reasonable definition of the rate at which f changes as we move away from P_0 in the direction $\mathbf{u}$.

Figure 1
Computing the derivative at P_0 in the direction **u**

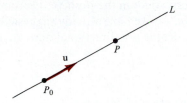

We proceed by analogy with single-variable calculus. Let L be the line through P_0 parallel to **u** and let P be a point of L near P_0. (See Figure 1.) Then $\overrightarrow{P_0P} = t\mathbf{u}$ for some nonzero scalar t. The number t is the directed distance from P_0 to P (positive or negative depending on whether P is on one side of P_0 or the other). In single-variable calculus L is the x axis, P_0 and P have coordinates x_0 and x, and $t = x - x_0$. We then compute

$$f'(x_0) = \lim_{x \to x_0} \frac{f(x) - f(x_0)}{x - x_0} = \lim_{t \to 0} \frac{f(x_0 + t) - f(x_0)}{t}$$

The same idea in higher dimensions is

$$D_{\mathbf{u}}f(P_0) = \lim_{t \to 0} \frac{f(P) - f(P_0)}{t}$$

where $P_0 = (x_0, y_0)$ and $P = (x, y)$ if the domain is in $\mathfrak{R}^2$, while $P_0 = (x_0, y_0, z_0)$ and $P = (x, y, z)$ if it is in $\mathfrak{R}^3$. Thus we have arrived at the following definition.

> Let $f: D \to \mathfrak{R}$ be a function whose domain D is a subset of $\mathfrak{R}^n$ (where $n \leq 3$). If $P_0 \in D$ and **u** is a unit vector in $\mathfrak{R}^n$, the **directional derivative** of f at P_0 in the direction **u** is
>
> $$D_{\mathbf{u}}f(P_0) = \lim_{t \to 0} \frac{f(P) - f(P_0)}{t}$$
>
> where $\overrightarrow{P_0P} = t\mathbf{u}$.

This definition omits any reference to the line L through P_0 parallel to **u**. Note, however, that the vector equation $\overrightarrow{P_0P} = t\mathbf{u}$ forces P to be a point of L. (Why?) Hence the derivative is *directional*.

□ **Example 1**   Show that the directional derivative of f at P_0 in the direction $\mathbf{u} = \mathbf{i}$ is just the partial derivative $f_1(P_0)$.

Solution: We will do this for a function of two variables; the argument is the same when $n = 3$. Writing $P_0 = (x_0, y_0)$ and $P = (x, y)$, observe that the vector equation $\overrightarrow{P_0P} = t\mathbf{u}$ reduces to

$$(x - x_0, y - y_0) = t(1, 0)$$

from which $x = x_0 + t$ and $y = y_0$. Hence

$$D_{\mathbf{i}}f(P_0) = \lim_{t \to 0} \frac{f(x_0 + t, y_0) - f(x_0, y_0)}{t}$$

This is the same as the partial derivative $f_1(x_0, y_0)$, as defined in Section 17.1. In the problem set we ask you to show that

$$D_{\mathbf{j}}f(P_0) = f_2(P_0)$$

and (in the case of a function of three variables)

$$D_{\mathbf{k}}f(P_0) = f_3(P_0)$$

Hence partial derivatives are directional derivatives in the positive direction of the coordinate axes in $\mathcal{R}^2$ or $\mathcal{R}^3$. Moreover (as we have already suggested) the ordinary derivative in single-variable calculus is a directional derivative in the positive direction of the number line.   □

□ **Example 2**   Find the directional derivative of $f(x,y) = 4 - x^2 - y^2$ at $(1,1)$ in the direction "northeast."

Solution: "Northeast" in the xy plane is specified by the unit vector

$$\mathbf{u} = \frac{1}{\sqrt{2}}(1,1)$$

Letting $P_0 = (1,1)$ and $P = (x,y)$, we see that the vector equation $\overrightarrow{P_0P} = t\mathbf{u}$ becomes

$$(x - 1, y - 1) = \frac{t}{\sqrt{2}}(1,1)$$

from which $x = 1 + (t/\sqrt{2})$ and $y = 1 + (t/\sqrt{2})$. We compute

$$f(P) - f(P_0) = f(x,y) - f(1,1) = 4 - \left(1 + \frac{t}{\sqrt{2}}\right)^2 - \left(1 + \frac{t}{\sqrt{2}}\right)^2 - 2$$
$$= -2\sqrt{2}t - t^2$$

Hence $$D_{\mathbf{u}}f(1,1) = \lim_{t \to 0} \frac{f(P) - f(P_0)}{t} = \lim_{t \to 0}(-2\sqrt{2} - t) = -2\sqrt{2}$$

Thus as we move away from the point $(1,1)$ in the northeast (NE) direction, the function $f(x,y) = 4 - x^2 - y^2$ is decreasing, its rate of change (with respect to distance measured along the NE line) being $-2\sqrt{2}$.   □

Remark   There is a nice geometrical interpretation of the result in Example 2. Imagine the surface $z = 4 - x^2 - y^2$ as a mountain and suppose that a climber is at the point $(1, 1, 2)$. (See Figure 2.) In choosing to go northeast, the climber is opting to descend the mountain. If distances are measured in meters, his altitude is changing at the instantaneous rate of $-2\sqrt{2}$ meters per meter as he leaves $(1, 1, 2)$. The term "meters per meter" means that if the rate were constant, each meter in the NE direction in the xy plane results in an altitude change of $-2\sqrt{2}$ meters. (Of course the rate is not constant, but differs from one point to another.)

Another way of saying the same thing is that $D_{\mathbf{u}}f(1,1) = -2\sqrt{2}$ is the slope of the curve of intersection of the surface $z = 4 - x^2 - y^2$ and the plane $y = x$. This is the same interpretation we gave to partial derivatives in Section 17.1, but now the plane in which the curve lies is not parallel to a coordinate plane.

Finding directional derivatives by evaluating the limit of a difference quotient is painful. As in single-variable calculus we seek a more routine procedure that will enable us to take advantage of differentiation formulas. To this end, let's look at the definition as it applies to a function $f(x, y, z)$

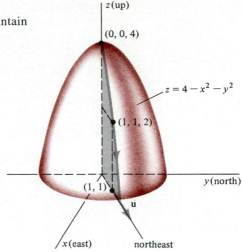

Figure 2
Descent of a mountain

whose domain is in $\mathcal{R}^3$. If $P_0 = (x_0, y_0, z_0)$ is in the domain and $\mathbf{u} = (a, b, c)$ is a given unit vector, we may represent the line L through P_0 parallel to $\mathbf{u}$ by the parametric equations

$$x = x_0 + at, \; y = y_0 + bt, \; z = z_0 + ct$$

If $P = (x, y, z)$ is a point of L (and also in the domain), we have

$$f(P) - f(P_0) = f(x_0 + at, y_0 + bt, z_0 + ct) - f(x_0, y_0, z_0)$$

This expression is a function of t alone, say $g(t) = f(P) - f(P_0)$. Since $g(0) = 0$ (why?), our definition of directional derivative may be written in the form

$$D_{\mathbf{u}} f(P_0) = \lim_{t \to 0} \frac{g(t) - g(0)}{t} = g'(0)$$

Hence we can find $D_{\mathbf{u}} f(P_0)$ by differentiating $g(t)$ and evaluating the result at $t = 0$.

Now comes the interesting part! The expression

$$w = g(t) = f(x, y, z) - f(x_0, y_0, z_0)$$

depends on t by virtue of its dependence on x, y, and z, which in turn are functions of t, namely

$$x = x_0 + at, \; y = y_0 + bt, \; z = z_0 + ct$$

This is a situation calling for the Chain Rule:

$$\frac{dw}{dt} = \frac{\partial w}{\partial x} \frac{dx}{dt} + \frac{\partial w}{\partial y} \frac{dy}{dt} + \frac{\partial w}{\partial z} \frac{dz}{dt} = \nabla w \cdot \frac{d\mathbf{x}}{dt}$$

where (as usual) $\mathbf{x} = (x, y, z)$ and

$$\frac{d\mathbf{x}}{dt} = \left(\frac{dx}{dt}, \frac{dy}{dt}, \frac{dz}{dt} \right) = (a, b, c) = \mathbf{u}$$

Since $f(x_0, y_0, z_0)$ is a constant, we have

$$\frac{\partial w}{\partial x} = f_1(x, y, z), \quad \frac{\partial w}{\partial y} = f_2(x, y, z), \quad \frac{\partial w}{\partial z} = f_3(x, y, z)$$

and hence $\nabla w = \nabla f(x, y, z)$. Thus our formula for the derivative of $g(t)$ is

$$\frac{dw}{dt} = g'(t) = \nabla f(x, y, z) \cdot \mathbf{u}$$

Evaluating this at $t = 0$ gives

$$D_{\mathbf{u}} f(P_0) = g'(0) = \nabla f(x_0, y_0, z_0) \cdot \mathbf{u} = \nabla f(P_0) \cdot \mathbf{u}$$

The same argument applies to functions of two variables (simply by suppressing the third coordinate throughout). Hence we have proved the following.

If f is a function of two or three variables (with continuous partial derivatives), the directional derivative of f at P in the direction of the unit vector $\mathbf{u}$ is

$$D_{\mathbf{u}} f(P) = \nabla f(P) \cdot \mathbf{u}$$

□ **Example 3**   Use our new formula to confirm the result in Example 2.

Solution: The problem is to find the directional derivative of

$$f(x, y) = 4 - x^2 - y^2$$

at $(1,1)$ in the direction of

$$\mathbf{u} = \frac{1}{\sqrt{2}}(1,1)$$

Since $\nabla f(x, y) = (-2x, -2y) = -2(x, y)$, we find

$$D_{\mathbf{u}} f(1,1) = \nabla f(1,1) \cdot \mathbf{u} = -2(1,1) \cdot \frac{1}{\sqrt{2}}(1,1) = -\sqrt{2}(1 + 1) = -2\sqrt{2}$$

□

□ **Example 4**   The temperature at each point of the spherical region

$$D = \{(x, y, z): x^2 + y^2 + z^2 \leq 9\}$$

is given by $T = e^{xyz}$. A bug flies across the sphere on a diameter from $(2, -1, 2)$ to $(-2, 1, -2)$. At what rate (with respect to distance) is the temperature changing as the bug leaves $(2, -1, 2)$? as it passes through the origin? as it arrives at $(-2, 1, -2)$?

Solution: The direction of flight is from $P(2, -1, 2)$ toward the origin, given by the vector $\vec{PO} = (-2, 1, -2)$. A unit vector in this direction is $\mathbf{u} = \frac{1}{3}(-2, 1, -2)$. Since

$$\nabla T = (yze^{xyz}, xze^{xyz}, xye^{xyz}) = e^{xyz}(yz, xz, xy)$$

the rate of change of T in the direction $\mathbf{u}$ (at any point) is

$$D_{\mathbf{u}}T = \nabla T \cdot \mathbf{u} = e^{xyz}(yz, xz, xy) \cdot \tfrac{1}{3}(-2, 1, -2) = \tfrac{1}{3}e^{xyz}(-2yz + xz - 2xy)$$

As the bug leaves $(2, -1, 2)$, the temperature is

$$T(2, -1, 2) = e^{-4} \approx 0.02$$

and is changing at the rate

$$D_{\mathbf{u}}T(2, -1, 2) = \tfrac{1}{3}e^{-4}(4 + 4 + 4) = 4e^{-4} \approx 0.07$$

while at the origin it is $T(0, 0, 0) = 1$ with rate of change $D_{\mathbf{u}}T(0, 0, 0) = 0$. When the bug reaches $(-2, 1, -2)$, the temperature is

$$T(-2, 1, -2) = e^4 \approx 55$$

and is changing at the rate

$$D_{\mathbf{u}}T(-2, 1, -2) = \tfrac{1}{3}e^4(4 + 4 + 4) = 4e^4 \approx 218 \qquad \square$$

Remark   Each rate found in Example 4 is *with respect to distance,* that is, we are describing the change in T per unit of distance along the line of flight (which is what a directional derivative is). This has nothing to do with time; the bug's speed does not enter into it. If we want to know the *time* rate of change experienced by the bug (as in Problem 14, Section 17.2), we need to know its position at time t. Then the rate is simply dT/dt.

The formula $D_{\mathbf{u}}f = \nabla f \cdot \mathbf{u}$ exhibits the central role played by the gradient in the computation of directional derivatives. If $\nabla f(P) \neq \mathbf{0}$, we have

$$D_{\mathbf{u}}f(P) = \nabla f(P) \cdot \mathbf{u} = |\nabla f(P)||\mathbf{u}| \cos \theta = |\nabla f(P)| \cos \theta$$

where θ is the angle between $\nabla f(P)$ and $\mathbf{u}$. Letting $M = |\nabla f(P)|$, we see that the derivative at P takes on all values between $-M$ and M. Its maximum and minimum values occur at $\theta = 0$ and $\theta = \pi$ (when $\mathbf{u}$ is in the direction of the gradient or opposite to that direction, respectively). The derivative is zero when $\theta = \pi/2$, that is, when $\mathbf{u}$ is perpendicular to the gradient.

The directional derivative of f at P has its maximum value

$$M = |\nabla f(P)|$$

when it is computed in the direction of the gradient. Hence the function increases most rapidly when we move away from P in that direction. It decreases most rapidly, at the rate

$$-M = -|\nabla f(P)|$$

when we move away from P in the opposite direction. Its rate of change is zero when we move away from P in a direction perpendicular to the gradient.

For a vivid geometrical interpretation of these statements in the case of a function of two variables, return to the "mountain" $z = f(x,y) = 4 - x^2 - y^2$ in Figure 2. A climber standing at the point (1, 1, 2) has infinitely many choices of direction. These are unit vectors in $\mathcal{R}^2$ pointing away from (1,1) in the domain of f, but we may imagine them moved upward to the plane $z = 2$, in which case they point away from (1, 1, 2). By choosing $\mathbf{u}$ the climber is deciding how his altitude z will change as he leaves the point (1, 1, 2). If he wants to climb the steepest possible route, he will choose $\mathbf{u}$ in the direction of $\nabla f(1,1) = -2(1,1)$, namely,

$$\mathbf{u} = -\frac{1}{\sqrt{2}}(1,1)$$

(This is the direction "southwest.") The rate of change of altitude in this case is $M = |\nabla f(1,1)| = 2\sqrt{2}$, a result that you can check by computing

$$D_{\mathbf{u}}f(1,1) = \nabla f(1,1) \cdot \mathbf{u} = -2(1,1) \cdot \left[-\frac{1}{\sqrt{2}}(1,1) \right] = 2\sqrt{2}$$

If he wants to descend the steepest possible route, he will choose $\mathbf{u}$ in the opposite direction (northeast), in which case the altitude changes at the rate $-M = -2\sqrt{2}$ as he leaves (1, 1, 2). (This is the choice made in Example 2.) If he wants to putter around looking for alpine flowers without changing altitude, he will choose $\mathbf{u}$ perpendicular to the gradient (northwest or southeast). In that case the derivative is zero, as you can check by taking

$$\mathbf{u} = \pm\frac{1}{\sqrt{2}}(-1,1)$$

and computing

$$D_{\mathbf{u}}f(1,1) = \nabla f(1,1) \cdot \mathbf{u} = -2(1,1) \cdot \left[\pm\frac{1}{\sqrt{2}}(-1,1) \right]$$
$$= \mp\sqrt{2}(1,1) \cdot (-1,1) = 0$$

This geometrical interpretation is the source of the term "gradient":

The grade is steepest in the direction of the gradient.

Many other equivalent interpretations may be given. A skier wishing to take the steepest route downhill will follow the negative of the gradient all the way. A mountain stream tends to do the same thing, cutting a gully that is everywhere perpendicular to the "contour lines" (which are "level curves" of constant altitude like those you have probably seen on a topographical map). Heat flows in the direction of greatest temperature drop (perpendicular to "isothermal lines," that is, curves of equal temperature). Electric charges move in the direction of greatest "potential drop" (perpendicular to "equipotential lines," that is, curves of equal electric potential).

The preceding terminology calls for a general definition.

Given a function $f(x, y)$ of two variables, the graph (in $\mathcal{R}^2$) of the equation $f(x, y) = c$ (where c is constant) is called a **level curve** of f. The gradient of f is perpendicular to it. Similarly, given a function $f(x, y, z)$ of three variables, the graph (in $\mathcal{R}^3$) of the equation $f(x, y, z) = c$ is called a **level surface** of f. The gradient of f is perpendicular to it. (See Remark 1 in Section 17.3.)

□ **Example 5**   If the bug in Example 4 is at $(2, -1, 2)$ in the domain of the temperature function $T = e^{xyz}$, what direction of flight will result in the greatest rate of temperature change as it leaves this point? What is this rate?

Solution: From Example 4 we know that

$$\nabla T = e^{xyz}(yz, xz, xy) = e^{-4}(-2, 4, -2) = 2e^{-4}(-1, 2, -1)$$

at the point $(2, -1, 2)$. For maximum rate of change of T, the bug should fly in the direction of this gradient, namely

$$\mathbf{u} = \frac{1}{\sqrt{6}}(-1, 2, -1)$$

The maximum rate is the magnitude of the gradient,

$$M = |\nabla T(2, -1, 2)| = |2e^{-4}(-1, 2, -1)| = 2e^{-4}\sqrt{6} \approx 0.09$$

If T is measured in degrees and distance (in $\mathcal{R}^3$) in meters, the units of this derivative are degrees/meter.   □

Note that the bug in Example 5 is in the *domain* of the temperature function (unlike the mountain climber of the preceding discussion, who is on the graph of the altitude function, "above" the domain). We could hardly put the bug on the graph of $T = e^{xyz}$, since that is a set of points in 4-space. Thus when we are dealing with a function of three variables, the mountain-climbing interpretation must be abandoned. However, as in the case of a temperature function of two variables, we can still say that heat flows in the direction of greatest temperature drop (perpendicular to "isothermal *surfaces*," that is, level surfaces of the function $T = e^{xyz}$). Such a surface has the form $e^{xyz} = c$, where c is constant.

The implications of this kind of language are significant in scientific applications, particularly in physics.

Problem Set 17.4

In each of the following, find the directional derivative of f at P in the given direction.

1. $f(x, y) = xy$, $P = (1, -1)$, $\mathbf{u} = \frac{1}{5}(3, 4)$.

2. $f(x, y) = xy$, $P = (1, 2)$, in the direction from P toward $(0, 3)$.

3. $f(x, y) = \sin(xy^2)$, $P = (\pi/4, 2)$, $\mathbf{u} = \mathbf{i}$.

4. $f(x,y) = \sin(xy^2)$, $P = (\pi/4, 2)$, southeast.

5. $f(x,y) = x^2 + xy + y^2$, $P = (0,1)$, in the direction of $-3\mathbf{i} + 4\mathbf{j}$.

6. $f(x,y) = \sqrt{x^2 + y^2}$, P is any point except the origin, in a direction perpendicular to $\nabla f(P)$.

7. $f(x,y) = \tan^{-1}(y/x)$, P is any point not on the y axis, in the direction of $\overrightarrow{OP}$.

8. $f(x,y) = xy(1 - x^2 - y^2)$, P is any point of the unit circle, in the direction of $\overrightarrow{PO}$.

9. $f(x,y,z) = ax + by + cz$, P is any point, in the direction of $a\mathbf{i} + b\mathbf{j} + c\mathbf{k}$.

10. $f(x,y,z) = xyz$, $P = (1, 2, 3)$, in the direction from P toward $(3, 3, 1)$.

11. $f(x,y,z) = (x - y)(y - z)(z - x)$, $P = (1, 1, 1)$, any direction.

12. $f(x,y,z) = xy + yz + zx$, P is any point of the surface $xy + yz + zx = 0$, in the direction of $\nabla f(P)$.

13. $f(x,y,z) = \ln(x^2 + y^2 + z^2)$, P is any point except the origin, in the direction of $\overrightarrow{OP}$.

14. Suppose that $z = e^{xy + x - y}$. How fast is z changing when we move away from the origin toward $(2,1)$?

15. In Problem 14 in what direction should we move away from the origin for z to change most rapidly? What is this maximum rate? In what directions is the derivative zero at the origin?

16. Find the maximum rate of change of the function $z = e^{2x} \sin y$ at the origin. In what direction does it occur? In what directions is the derivative zero at the origin?

17. Suppose that you are standing at the point $(1, 1, 3)$ on the hill $z = 5y - x^2 - y^2$. If you decide to climb in the direction of steepest ascent, what is your rate of descent (with respect to distance) as you start?

18. In Problem 17, if you decide to go northwest, will you be ascending or descending? at what rate (with respect to distance)?

19. In Problem 17, if you decide to maintain your altitude, in what directions can you go?

20. Suppose that you are climbing the hill with equation $x^2 + y^2 - 6x + z = 0$. Your route is such that as you pass through the point $(1, 1, 4)$ you are heading northeast. At what rate (with respect to distance) is your altitude changing at that point?

21. In Problem 20 your partner at $(1, 2, 1)$ wants to start upward as quickly as possible. In what direction should she take off? At what rate (with respect to distance) is her altitude changing when she starts?

22. Suppose that $z = f(x,y) = x^2y$. If $\mathbf{u} = (u,v)$ is a unit vector, write a formula for $D_{\mathbf{u}} f(1,-1)$ in terms of u and v.

23. In Problem 22 in what direction(s) should we go from $(1,-1)$ if we want the rate of change of z (with respect to distance) to be 2?

24. Frodo and Sam are resting blissfully at the point $(1,1)$ in the Field of Cormallen (the xy plane). The temperature at each point of the plane is $T = x^2 + y^2$. Gandalf awaits them at $(4, -3)$. At what rate (with respect to distance) is the temperature changing as they leave $(1,1)$ in his direction? Is it increasing or decreasing?

25. In Problem 24 Sam gets upset after reaching $(4, -3)$ and decides to cool off as fast as possible. In what direction should he go? How fast (with respect to distance) is the temperature changing as he leaves?

26. In Problem 24 Frodo is content, and decides to wander off from $(4, -3)$ along a path of constant temperature. What is an equation of his path?

27. In Problem 24 the King, naturally enough, is at the origin. He answers to no man (and only one woman), but walks away as he pleases. What is the rate of change of temperature as he leaves?

28. Suppose that the origin is the deepest point of hell and that the temperature elsewhere is

$$T = \frac{1}{x^2 + y^2 + z^2}$$

It is obvious to Dante in his position that to cool off as rapidly as possible he should head directly away from the origin. Confirm that fact by using the directional derivative. At what rate (with respect to distance) is the temperature changing as Dante flies from hell?

29. Virgil leaves hell (Problem 28) along the twisted cubic

$$x = t, \ y = t^2, \ z = t^3 \qquad \text{(where } t \text{ is time)}$$

As he passes through the point $(1, 1, 1)$, at what rate (with respect to distance) is the temperature changing?

30. In Problem 29 at what rate with respect to *time* is Virgil's temperature changing at $(1, 1, 1)$?

31. The voltage at each point in space is

$$V = xyz(9 - x^2 - y^2 - z^2)$$

Show that if P is on the sphere $x^2 + y^2 + z^2 = 9$,

$$\nabla V(P) = -2xyz(\overrightarrow{OP})$$

32. A positively charged particle in an electric field (Problem 31) always moves in the direction in which the voltage is decreasing most rapidly (the direction of greatest "potential drop"). Assuming that its path intersects the sphere $x^2 + y^2 + z^2 = 9$ at a point P in the first octant, in what direction is it moving when it reaches P? Show that the rate of change of voltage at P is $-6xyz$.

33. Let θ be the counterclockwise angle from $\mathbf{i}$ to the unit vector $\mathbf{u}$ in $\mathfrak{R}^2$. If f is a function of two variables, explain why

$$D_{\mathbf{u}}f(P) = f_1(P)\cos\theta + f_2(P)\sin\theta$$

34. Suppose that $\mathbf{u}$ is a unit vector in $\mathfrak{R}^3$ and α, β, γ are the angles between $\mathbf{u}$ and $\mathbf{i}, \mathbf{j}, \mathbf{k}$, respectively. If f is a function of three variables, explain why

$$D_{\mathbf{u}}f(P) = f_1(P)\cos\alpha + f_2(P)\cos\beta + f_3(P)\cos\gamma$$

35. Suppose that $\mathbf{u}$ and $\mathbf{v}$ are perpendicular unit vectors in the plane and f is a function of two variables. Show that

$$(D_{\mathbf{u}}f)^2 + (D_{\mathbf{v}}f)^2 = |\nabla f|^2$$

Assuming that $f(x,y)$ and $g(x,y)$ have continuous partial derivatives, derive the following formulas.

36. $D_{\mathbf{u}}(f + g) = D_{\mathbf{u}}f + D_{\mathbf{u}}g$
37. $D_{\mathbf{u}}(fg) = f(D_{\mathbf{u}}g) + g(D_{\mathbf{u}}f)$
38. $D_{\mathbf{u}}\left(\dfrac{f}{g}\right) = \dfrac{g(D_{\mathbf{u}}f) - f(D_{\mathbf{u}}g)}{g^2}$
39. Suppose that $\nabla f(P) = \mathbf{0}$.
    (a) Explain why $D_{\mathbf{u}}f(P) = 0$ for every unit vector $\mathbf{u}$.
    (b) In the text we said that if $\nabla f(P) \neq \mathbf{0}$, the maximum and minimum directional derivatives occur in the di-

rection of $\nabla f(P)$ and opposite to that direction, respectively, while the derivative is zero in directions perpendicular to $\nabla f(P)$. Explain why these statements are still true if $\nabla f(P) = \mathbf{0}$.

40. Suppose that $P_0 = (x_0, y_0)$ is a point at which the partial derivatives of $f(x,y)$ are continuous. Show that

$$D_{\mathbf{j}}f(P_0) = f_2(P_0)$$

in two ways, as follows.
    (a) Use the definition of the directional derivative as the limit of a difference quotient.
    (b) Use the formula $D_{\mathbf{u}}f(P_0) = \nabla f(P_0) \cdot \mathbf{u}$.

41. Repeat Problem 40 for a function of three variables and the formula $D_{\mathbf{k}}f(P_0) = f_3(P_0)$.

Use the definition of the directional derivative as the limit of a difference quotient to find each of the following.

42. $D_{\mathbf{u}}f(2, -1)$, where $f(x,y) = x^2 - y^2$ and $\mathbf{u} = \frac{1}{5}(4, -3)$.
43. $D_{\mathbf{u}}f(1, 0, -2)$, where $f(x,y,z) = x - y + z^2$ and $\mathbf{u} = \frac{1}{3}(1, 2, -2)$.
44. Suppose that P is a point of a level curve of $f(x,y)$. Why is $\nabla f(P)$ perpendicular to the curve?
45. Suppose that P is a point of a level surface of $f(x,y,z)$. Why is $\nabla f(P)$ perpendicular to the surface?

17.5
EXTREME VALUES

Maximum and minimum values of a function of several variables are found in much the same way as in single-variable calculus. We begin with a theoretical statement about *global* extreme values (which are the largest and smallest values of a function anywhere in its domain). Two technical terms are needed for an understanding of this theorem, namely that a set of points in $\mathfrak{R}^n$ (where n is 2 or 3) is **closed** when it includes all boundary points (if any), and **bounded** if it can be enclosed in a circle (in $\mathfrak{R}^2$) or a sphere (in $\mathfrak{R}^3$). In higher-dimensional space these geometrical terms are replaced by equivalent analytic terms. We state the theorem without digressing to explain the details; its proof is given in advanced calculus.

Extreme Value Theorem

Suppose that $f: D \to \mathfrak{R}$ is a function whose domain D is a closed bounded subset of $\mathfrak{R}^n$. If f is continuous in D, it has a maximum and a minimum, that is, there are points P_1 and P_2 in D such that

$$f(P_1) \leq f(P) \leq f(P_2) \qquad \text{for all } P \in D$$

□ **Example 1**   The domain of $f(x,y) = \sqrt{4 - x^2 - 2y^2}$ is the subset of $\mathbb{R}^2$ defined by

$$D = \{(x,y): x^2 + 2y^2 \le 4\}$$

This is the set of points inside and on the ellipse $x^2/4 + y^2/2 = 1$, so it is clearly closed and bounded. Since f is continuous in D, the Extreme Value Theorem guarantees a global maximum and minimum. In this case they are easy to find. The radical is never negative; its minimum value is zero (occurring when $x^2 + 2y^2 = 4$, that is, at all points of the boundary of D). Since $x^2 + 2y^2 \ge 0$ for all (x,y), the maximum value of f is $f(0,0) = 2$, occurring at the interior point $(0,0)$. These results are clear from Figure 1, which shows the graph of $z = f(x,y)$, namely the upper half of the ellipsoid $x^2/4 + y^2/2 + z^2/4 = 1$. (Why?) The figure shows the high point $(0,0,2)$ and low points everywhere on the boundary of D in the xy plane.   □

Figure 1
Graph of $f(x,y) = \sqrt{4 - x^2 - 2y^2}$

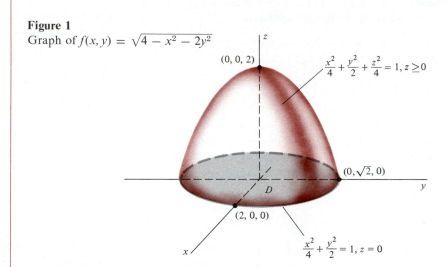

A *local* extreme value of a function $f: D \to \mathbb{R}$ is a number $f(P)$ which is the largest or smallest value of f in a neighborhood of P. Such values may occur on the boundary of D or at an interior point. In the latter case we can prove a theorem reminiscent of Theorem 1 in Section 4.3, which says that if f is a function of one variable with an extreme value at an interior point c where the derivative exists, then $f'(c) = 0$. In multivariable calculus the gradient replaces the derivative.

Suppose that P is an interior point of the domain of f, in a neighborhood of which f has continuous partial derivatives. If $f(P)$ is a local extreme value, then $\nabla f(P) = \mathbf{0}$.

There are several ways to prove this theorem. Assuming that f is a function of two variables (the same argument works in higher dimensions), we observe that if $f(x_0,y_0)$ is an extreme value of f, it is also an extreme value of the single-variable function $g(x) = f(x,y_0)$. Theorem 1 of Section 4.3 says that $g'(x_0)$ must be zero, that is, $f_1(x_0,y_0) = 0$. Similarly, if $h(y) = f(x_0,y)$, then $h'(y_0) = 0$, that is, $f_2(x_0,y_0) = 0$. Hence

$$\nabla f(x_0,y_0) = (f_1(x_0,y_0), f_2(x_0,y_0)) = (0,0)$$

A more geometrically appealing argument (in the case of a function of two variables) is to look at the surface $z = f(x,y)$. An extreme value $z_0 = f(x_0,y_0)$ is the third coordinate of a high or low point of this surface. (See Figure 2.) Since (x_0,y_0) is an interior point of the domain of f, and since continuity of f and its partial derivatives guarantees the existence of a tangent plane at (x_0, y_0, z_0), it is geometrically apparent that this plane is horizontal. Its equation is therefore $z = z_0$, that is,

$$0(x - x_0) + 0(y - y_0) - (z - z_0) = 0$$

We know from Section 17.3 that an equation of the plane is

$$a(x - x_0) + b(y - y_0) - (z - z_0) = 0$$

where $a = f_1(x_0,y_0)$ and $b = f_2(x_0,y_0)$. Hence $a = b = 0$, that is,

$$\nabla f(x_0,y_0) = (0,0)$$

Figure 2
Horizontal tangent plane at a high point of a surface

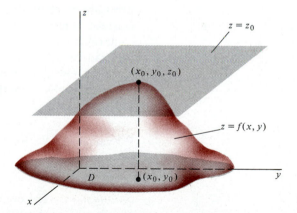

As in single-variable calculus, you should not read more into our theorem than it says. Interior points at which $\nabla f(P) = \mathbf{0}$ are points at which f *may* have an extreme value (not *must*). A mountain pass is an example of a surface with a horizontal tangent plane at a point that is neither the highest nor the lowest in its neighborhood. (See Figure 3.) It is a **saddle point**, analogous to the origin on the curve $y = x^3$ (which is an example of a function with a critical point at which there is no extreme value, as shown in Figure 4).

Figure 3
Critical point at which there is no extreme value

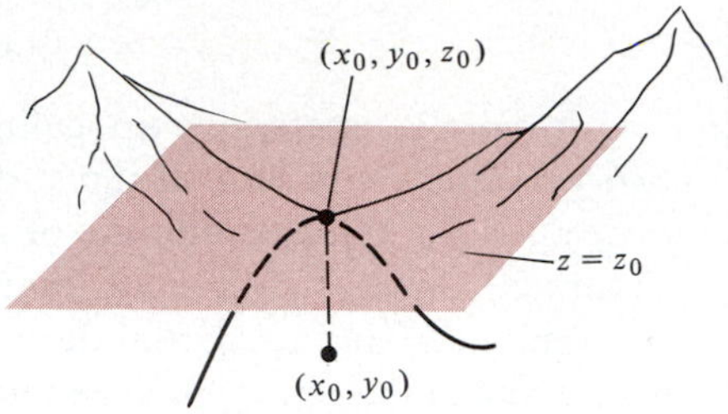

Figure 4
$f'(0) = 0$ but $f(0)$ is not an extreme value

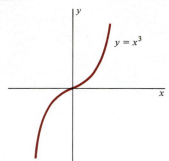

What our theorem says is that as far as interior points of the domain are concerned, the search for extreme values may be confined to points at which the gradient is zero. (If the function is not well behaved, we may have to consider points where the gradient does not exist. By analogy to "corners" and "cusps" in single-variable calculus, like the origin on the graph of $y = |x|$ or $y = x^{2/3}$, there may be extreme values at such points.) Extreme values may also occur at boundary points of the domain (as in Example 1), so the procedures we follow are pretty much the same as in single-variable calculus. Of course it is more complicated to find where a gradient is zero than it is to solve an equation $f'(x) = 0$ involving only one variable. Boundaries are more of a problem, too, since they are whole curves or surfaces (instead of just endpoints of intervals on the number line). It is also harder to develop and apply tests that distinguish between high points, low points, and saddle points. In short, although the theory nicely parallels what we already know about functions of one variable, the actual solution of extreme value problems is another matter. We will confine our discussion to examples that are manageable.

◻ **Example 2**   Find the extreme values of the function

$$f(x,y) = 2x - x^2 - y^2$$

Solution: The domain is $\mathscr{R}^2$, so there are no boundary points to worry about; all points are interior points. Since $f_1(x,y) = 2 - 2x$ and $f_2(x,y) = -2y$, the gradient of f is zero when $x = 1$ and $y = 0$. Hence if f has an extreme value, it must be $f(1,0) = 1$.

Without some way of testing the critical point $(1,0)$, we don't know whether $f(1,0) = 1$ is a maximum, minimum, or neither. In this example, however, we may complete the square in the equation $z = 2x - x^2 - y^2$ to obtain

$$z = -(x^2 - 2x + 1) - y^2 + 1 = 1 - (x - 1)^2 - y^2$$

The graph is a circular paraboloid opening downward from the high point $(1, 0, 1)$, so $f(1,0) = 1$ is a maximum. (See Figure 5.)

Figure 5
Graph of $f(x,y) = 1 - (x - 1)^2 - y^2$

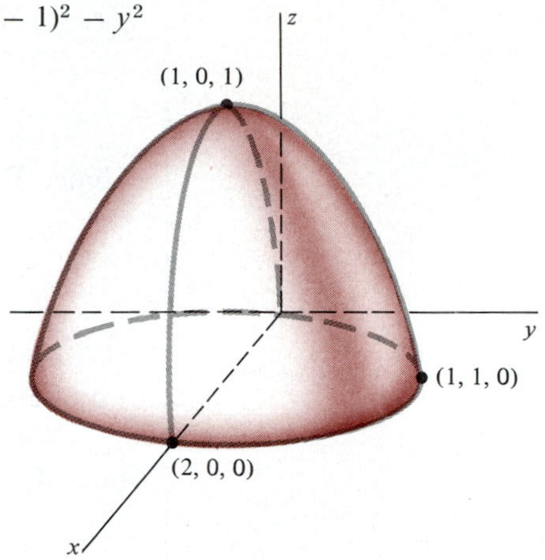

Of course this way of doing the problem does not require calculus. In general we cannot expect things to be so easy.   □

□ **Example 3**   A rectangular box with no top is to be made from 20 ft² of cardboard. What is the maximum volume of such a box?

Figure 6
Rectangular box of dimensions x, y, z

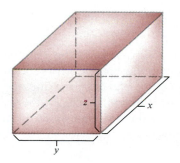

Solution: Label the dimensions of the box as in Figure 6. Its volume is $V = xyz$; the problem is to maximize V. We do this by setting the gradient of V equal to zero, but it is important not to barge into the procedure without thinking! We do *not* find the gradient of V as a function of three variables and set it equal to zero. (What would happen if we did?) For V is not a function of three independent variables. There is a *constraint* governing the variables, namely the condition that the surface area of the box is 20 ft². Since the box has no top, this condition is $xy + 2xz + 2yz = 20$, from which

$$z = \frac{20 - xy}{2(x + y)}$$

Upon substitution of the constraint into $V = xyz$, we have a function of *two* variables,

$$V = f(x,y) = \frac{xy(20 - xy)}{2(x + y)} = \frac{20xy - x^2y^2}{2(x + y)}$$

It is this function that we maximize. Using the Quotient Rule, we find

$$f_1(x,y) = \frac{2(x + y)(20y - 2xy^2) - (20xy - x^2y^2) \cdot 2}{4(x + y)^2}$$

$$= \frac{2y[2(x + y)(10 - xy) - (20x - x^2y)]}{4(x + y)^2}$$

$$= \frac{y^2(20 - x^2 - 2xy)}{2(x + y)^2}$$

Since x and y may be interchanged without affecting $f(x,y)$, we obtain $f_2(x,y)$ by symmetry:

$$f_2(x,y) = \frac{x^2(20 - y^2 - 2xy)}{2(x + y)^2}$$

Hence $\nabla f(x,y) = (0,0)$ when (x,y) satisfies the system of equations

$$20 - x^2 - 2xy = 0 \tag{1}$$

$$20 - y^2 - 2xy = 0 \tag{2}$$

(We exclude $x = 0$ and $y = 0$ because the dimensions of the box are positive numbers.)

We solve the above system by subtracting Equation (2) from Equation (1) to obtain

$$y^2 - x^2 = 0$$
$$y^2 = x^2$$
$$y = x \qquad \text{(because } x \text{ and } y \text{ are positive)}$$

Substituting this into Equation (1), we find

$$20 - x^2 - 2x^2 = 0$$
$$3x^2 = 20$$
$$x = \tfrac{2}{3}\sqrt{15}$$

Since $y = x$, the maximum volume of the box is

$$V = f(\tfrac{2}{3}\sqrt{15}, \tfrac{2}{3}\sqrt{15}) = \frac{20}{9}\sqrt{15} \approx 8.6 \text{ ft}^3$$

How do we know that this value of V is a *maximum?* In this example it is geometrically apparent that no box of minimum volume exists (because we can squeeze the volume as close to zero as we please by making one of the dimensions small). Moreover, the volume is bounded because there are only 20 ft² of cardboard available. Hence a maximum volume exists. In such circumstances the only possibility is the functional value found from the critical point.   □

□ **Example 4**   A triangular plate occupies the region D shown in Figure 7. The temperature at each point of the plate is given by $T = x^2 + xy + 2y^2 - 3x + 2y$. Find the hottest and coldest points of the plate.

Solution: The problem is to locate the global extreme values of the function T in its domain D. Since D is closed and bounded, we know that such values exist; we will find them by locating the local extreme values and picking out the largest and smallest.

There are no constraints that will reduce the number of variables, but there are boundary conditions (which we will look at shortly). First we locate the interior critical points (if any). Since

$$\nabla T = (2x + y - 3, x + 4y + 2) = \mathbf{0}$$

Figure 7
Hotplate in the region D

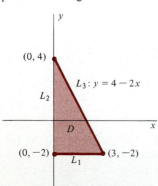

when $2x + y - 3 = 0$ and $x + 4y + 2 = 0$, we solve the system

$$2x + y = 3$$
$$x + 4y = -2$$

The solution is $(x, y) = (2, -1)$, as you can check, so the number $T(2, -1) = -4$ is a candidate for an extreme value.

The only other places an extreme value can occur is on the boundary of the plate. This consists of the line segments

$$L_1: y = -2 \qquad (0 \le x \le 3)$$
$$L_2: x = 0 \qquad (-2 \le y \le 4)$$
$$L_3: y = 4 - 2x \qquad (0 \le x \le 3)$$

The temperature at each point of L_1 is

$$T(x, -2) = x^2 - 5x + 4 \qquad (0 \le x \le 3)$$

a function of x alone whose extreme values occur at $x = \frac{5}{2}$ (where the derivative $2x - 5$ is zero) and at the endpoints $x = 0$ and $x = 3$. Hence we have three more candidates for extreme values, namely $T(\frac{5}{2}, -2) = -\frac{9}{4}$, $T(0, -2) = 4$, and $T(3, -2) = -2$.

At each point of L_2 the temperature is

$$T(0, y) = 2y^2 + 2y \qquad (-2 \le y \le 4)$$

a function of y alone whose extreme values are $T(0, -\frac{1}{2}) = -\frac{1}{2}$, $T(0, -2) = 4$ (already listed), and $T(0, 4) = 40$. (Confirm!) At each point of L_3 the temperature is

$$T(x, 4 - 2x) = x^2 + x(4 - 2x) + 2(4 - 2x)^2 - 3x + 2(4 - 2x)$$
$$= 7x^2 - 35x + 40 \qquad (0 \le x \le 3)$$

a function whose extreme values are $T(\frac{5}{2}, -1) = -\frac{15}{4}$, $T(0, 4) = 40$, and $T(3, -2) = -2$ (the last two being already listed).

Our list of possible extreme values consists of the numbers

$$T(2, -1) = -4, \; T(\tfrac{5}{2}, -2) = -\tfrac{9}{4}, \; T(0, -2) = 4, \; T(3, -2) = -2,$$
$$T(0, -\tfrac{1}{2}) = -\tfrac{1}{2}, \; T(0, 4) = 40, \; T(\tfrac{5}{2}, -1) = -\tfrac{15}{4}$$

Hence the hottest point of the plate is $(0, 4)$ (where the temperature is 40) and the coldest is $(2, -1)$ (where the temperature is -4). ◻

These examples illustrate some general procedures for solving extreme value problems. (As usual, we advise you not to treat the following list of steps like a cookbook recipe. Simply be aware of the ideas.)

- The quantity to be maximized or minimized is expressed in terms of variables (as few as possible).
- Any constraints that are present in the problem are used to reduce the number of variables to the point where they are independent.
- The partial derivatives of the resulting function are computed and set equal to zero (to find where the gradient is zero).

- The resulting system of equations is solved. (This is a purely algebraic problem, but it may be hard.)
- The critical points found in the preceding step are examined to see whether they yield maxima, minima, or neither. (So far we have given no theoretical procedures for doing this.)
- Boundary points (if any) are examined to see if they yield other extreme values. (That can be a separate problem of some complexity!)
- The extreme values are sorted out to answer the original question (which may call for a global maximum or minimum, or local extreme values, depending on the question).

The Second Derivative Test in single-variable calculus (Section 5.2) says that if $f'(c) = 0$, then

$f''(c) > 0 \Rightarrow f(c)$ is a minimum   (the graph of f is concave up)
$f''(c) < 0 \Rightarrow f(c)$ is a maximum   (the graph of f is concave down)

while if $f''(c) = 0$, the test fails. Similar tests exist for distinguishing among the extreme values of functions of several variables, but they are more complicated. We state one for functions of two variables, leaving the proof for an advanced course. (See the problem set for a geometric argument that shows why the test reads as it does.)

A Second Derivative Test

Suppose that P is an interior point of the domain of $f(x, y)$ in a neighborhood of which the second partial derivatives of f are continuous, and suppose that $\nabla f(P) = \mathbf{0}$. (In other words, P is an interior critical point.) Let

$$H(P) = \begin{vmatrix} f_{11}(P) & f_{12}(P) \\ f_{21}(P) & f_{22}(P) \end{vmatrix} = AC - B^2$$

where $A = f_{11}(P)$, $B = f_{12}(P) = f_{21}(P)$, $C = f_{22}(P)$.

1. If $H(P) > 0$, then

$$A \text{ and } C \text{ positive} \Rightarrow f(P) \text{ is a local minimum}$$
$$A \text{ and } C \text{ negative} \Rightarrow f(P) \text{ is a local maximum}$$

2. If $H(P) < 0$, the graph of f has a saddle point at P.
3. If $H(P) = 0$, the test fails.

□ **Example 5**   Find the top of the mountain defined by $z = 2x^2 + xy - y^2$ in the square domain

$$D = \{(x, y): -3 \le x \le 3, -3 \le y \le 3\}$$

Solution: Since $\nabla z = (4x + y, x - 2y) = \mathbf{0}$ when $4x + y = 0$ and $x - 2y = 0$, the only interior critical point is $(0,0)$. We compute

Figure 8
Square domain D

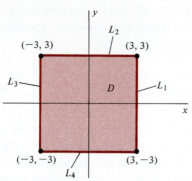

$A = z_{xx} = 4$, $B = z_{xy} = 1$, $C = z_{yy} = -2$, from which

$$H(0,0) = AC - B^2 = -8 - 1 < 0$$

Hence $(0, 0, 0)$ is a saddle point (a "pass") on the mountain.

The top must correspond to a boundary point (where the upward slopes of the mountain end abruptly at the edge of a vertical cliff). The boundary of D (Figure 8) consists of the line segments

$$L_1: x = 3 \qquad (-3 \le y \le 3)$$
$$L_2: y = 3 \qquad (-3 \le x \le 3)$$
$$L_3: x = -3 \qquad (-3 \le y \le 3)$$
$$L_4: y = -3 \qquad (-3 \le x \le 3)$$

Letting $f(x,y) = 2x^2 + xy - y^2$, we see that on L_1 the function is

$$f(3,y) = 18 + 3y - y^2 \qquad (-3 \le y \le 3)$$

the extreme values of which are

$$f(3,\tfrac{3}{2}) = \tfrac{81}{4} \qquad \text{(corresponding to the derivative } 3 - 2y = 0)$$
$$f(3,3) = 18 \qquad \text{(corresponding to the endpoint } y = 3)$$
$$f(3,-3) = 0 \qquad \text{(corresponding to the endpoint } y = -3)$$

We leave it to you to examine the function on L_2, L_3, and L_4, the remaining extreme values being

$$f(-\tfrac{3}{4},3) = -\tfrac{81}{8}$$
$$f(-3,3) = 0$$
$$f(-3,-\tfrac{3}{2}) = \tfrac{81}{4}$$
$$f(-3,-3) = 18$$
$$f(\tfrac{3}{4},-3) = -\tfrac{81}{8}$$

Looking through the list, we see that the mountain has two tops, namely

$$(3, \tfrac{3}{2}, \tfrac{81}{4}) \qquad \text{and} \qquad (-3, -\tfrac{3}{2}, \tfrac{81}{4})$$

It also has the pass $(0, 0, 0)$ discovered earlier. Note that we do not try to draw a picture of it! Except in simple cases, surfaces are not easy to sketch, and of course if we are dealing with a function of more than two variables, no graph can be drawn at all. Multivariable calculus becomes more analytic and less geometric as we get deeper into it.   □

□ **Example 6**   Use the Second Derivative Test to confirm that a local cold point of the plate in Example 4 is $(2, -1)$.

Solution: The temperature function in Example 4 is

$$T = x^2 + xy + 2y^2 - 3x + 2y$$

with partial derivatives $T_x = 2x + y - 3$ and $T_y = x + 4y + 2$. By setting these equal to zero, we found the interior critical point $(2, -1)$, but at the time we had no direct way of testing whether it yields a maximum or a minimum or neither. To apply the Second Derivative Test, we compute

$$A = T_{xx} = 2, \ B = T_{xy} = 1, \ C = T_{yy} = 4$$

from which $H(2, -1) = AC - B^2 = 8 - 1 > 0$. This tells us that we have an extreme value; since A and C are positive, it is a minimum.   □

It is important to note that the Second Derivative Test does not tell us anything about *global* extreme values. All it says in Example 6 is that $T(2, -1) = -4$ is a local minimum; there might be colder points somewhere else. We argued in Example 4 that $(2, -1)$ is in fact the coldest point of the plate, but our reasoning was based on the Extreme Value Theorem, which refers to functions with closed and bounded domains. Because we knew that a coldest point exists, and because we checked all possible extreme values (both inside and on the boundary of the domain), our conclusion was legitimate.

Problem Set 17.5

In each of the following, find all points where the gradient of the given function is zero. (You need not investigate whether these critical points yield extreme values; however, see Problems 7–12.)

1. $f(x, y) = x^2 - y^2 + 4y$
2. $f(x, y) = 3x^2 + 2y^2 - 6x$
3. $f(x, y) = x^2 + xy + y^2 - 3x$
4. $f(x, y) = x^2 - 3xy + 2y^2 + 4y$
5. $f(x, y) = x^4 + 8xy^2 + 2y^4$
6. $f(x, y) = 2x^4 - 6x^2y + y^4$

7–12. In Problems 1 through 6, use the Second Derivative Test to determine local high points, local low points, and saddle points of the graph of f. Note that the test fails at one critical point in Problems 11 and 12. Can you tell by inspection of the function itself what happens at that point?

13. Find the peak of the surface $z = 2xy - x^2 - 2y^2 + 6x - 8y + 2$.

14. Find the saddle point of the surface $z = x^2 + 2xy - 3y^2 - 8x + 14$.

15. When it rains on the surface

$$z = \frac{1}{x} + \frac{1}{y} + xy$$

a puddle will form at one point. Where?

16. The surface $z = x^3 - 3xy + y^3$ has a low point and a saddle point. Find them.

17. The surface $z = xy(1 - x^2 - y^2)$ has two peaks, five passes, and two low points. Find them.

18. Show that the hyperbolic paraboloid $z = y^2/b^2 - x^2/a^2$ (Example 5, Section 16.4) has a saddle point at the origin.

19. Find the highest point of the circular paraboloid $x^2 + y^2 - 4x + 6y + z + 10 = 0$ by expressing z as a function of x and y and finding its maximum value.

20. Do Problem 19 by completing the square in x and y and finding the vertex of the paraboloid.

21. Find the highest point of the ellipsoid $2x^2 + y^2 + 4z^2 - 4x + 4y + 2 = 0$ by completing the square in x and y to find the center. (The high point is then apparent.)

22. The equation in Problem 21 defines z as a function of x and y. Find the highest point of the ellipsoid by differentiating implicitly to find where $\nabla z = \mathbf{0}$.

23. The temperature at each point of the circular disk

$$D = \{(x, y): x^2 + y^2 \leq 1\}$$

is given by $T = x^2 + 2y^2 - x$. Find the hottest and coldest points of the disk.

24. The temperature at each point of the triangular region with vertices $(0,0)$, $(3,0)$, $(0,1)$ is given by $T = x^2 - 4xy + y^2 + 6x$. Find the hottest and coldest points of the region.

25. The temperature at each point in space is given by $T = x^4 + y^4 + z^4$. Find the hottest and coldest points of the sphere $x^2 + y^2 + z^2 = 1$.

26. In Problem 25 find the hottest and coldest points of the spherical ball $D = \{(x,y,z): x^2 + y^2 + z^2 \leq 1\}$.

27. What points of the surface $xy - z^2 - 3y + 12 = 0$ are closest to the origin?

28. Find the distance from the point $(1, -1, 3)$ to the plane $2x - 2y + z - 5 = 0$ by minimizing the square of the distance between a point (x,y,z) of the plane and the given point.

29. Do Problem 28 by using the formula in Problem 37, Section 16.2.

30. Find the distance between the skew lines

$$L_1: x = t, \ y = t, \ z = t - 1$$

and

$$L_2: x = t, \ y = -t, \ z = t$$

Hint: Use different letters for the parameters!

31. Two particles are free to move on the curves $y = x^2$ and $x - y = 1$, respectively. What are their positions when they are closest together? *Hint:* Represent their positions by (s,s^2) and $(t,t - 1)$.

32. Find the distance from the point $(3, -3, 1)$ to the paraboloid $z = x^2 + y^2$.

33. What plane through the point $(2, 1, 2)$ cuts off the least volume from the first octant? *Hint:* Write the equation of the plane in the form $x/a + y/b + z/c = 1$. Then a, b, c are its intercepts on the coordinate axes (Problem 26, Section 16.3). The tetrahedron formed by this plane and the coordinate planes has volume $V = \frac{1}{6}abc$ (a formula we will derive in the next chapter).

34. Find the dimensions of a rectangular box with no top having fixed volume and minimum surface area.

35. Show that a rectangular box of fixed volume and minimum surface area must be a cube.

36. A rectangular box with no top is to be made from 48 ft² of wood. What should be its dimensions for maximum volume?

37. The Post Office will not accept a rectangular package whose combined length and girth (perimeter of a cross section perpendicular to the length) exceeds 84 in. What are the dimensions of the box of largest volume that can be mailed?

38. Show that the product of the sines of the angles of a triangle is largest when the triangle is equilateral.

39. Suppose that when a consumer purchases x units of one product, y of another, and z of a third, the *utility* of the purchases is given by

$$u = 5x^{1/3}y^{2/3}z^{1/2}$$

If the price per unit of the products is $2, $5, and $1, respectively, and the consumer has $100 to spend, how many units of each product should be purchased to achieve maximum utility?

40. Let $f(x,y) = 1/\sqrt{1 - x^2 - y^2}$.
    (a) What is the domain of f?
    (b) Does f have a global maximum? a global minimum?
    (c) Does the Extreme Value Theorem apply? Explain.

41. Let $f(x,y) = (x - y)^2$ and $g(x,y) = -(x - y)^2$.
    (a) Explain why the global minimum of f and the global maximum of g occur at every point P of the line $y = x$.
    (b) Show that when the Second Derivative Test is applied to f and g, every point P of the line $y = x$ is a critical point and $H(P) = 0$. (This shows why the test fails when H is zero.)

42. Suppose that several measurements are made of related quantities x and y, the data being reported in the form of number pairs

$$(x_1, y_1), (x_2, y_2), \ldots, (x_n, y_n)$$

If these points appear to lie in a straight line $y = mx + b$, the vertical "deviation" of each point (x_k, y_k) from the theoretical position $(x_k, mx_k + b)$ is

$$mx_k + b - y_k \qquad k = 1, 2, \ldots, n$$

We define the *best-fitting line* (determined by our choice of m and b) to be the line which minimizes the sum of the squares of the deviations,

$$f(m,b) = \sum_{k=1}^{n} (mx_k + b - y_k)^2$$

(Hence the process of finding this line is called the **method of least squares.**)
    (a) Show that

$$f_1(m,b) = 2\left(m\sum x_k^2 + b\sum x_k - \sum x_k y_k\right)$$

and $\quad f_2(m, b) = 2\left(m\sum x_k + nb - \sum y_k\right)$

(We suppress the index of summation to save writing.)
    (b) Explain why $\nabla f(m,b) = \mathbf{0}$ when m and b satisfy the system of equations

$$\left(\sum x_k^2\right)m + \left(\sum x_k\right)b = \sum x_k y_k$$

$$\left(\sum x_k\right)m + nb = \sum y_k$$

(c) It is tedious to use the Second Derivative Test to prove that these choices of m and b minimize $f(m,b)$. Why is it geometrically apparent that they do?

43. Use the preceding problem to find the best-fitting line for the data $(-1,1)$, $(0,0)$, $(1,-2)$, $(2,-3)$. Sketch this line (and plot the given points) to see how the method works.

44. Use Problem 42 to find the best-fitting line for the data $(0,1)$, $(1,3)$, $(2,2)$, $(3,3)$, $(4,5)$. Sketch the line and plot the given points.

45. Show that if $y = mx + b$ is the best-fitting line for the data

$$(x_1,y_1), (x_2,y_2), \dots, (x_n,y_n)$$

the sum of the deviations is zero, that is,

$$\sum_{k=1}^{n} (mx_k + b - y_k) = 0$$

(Thus the positive and negative deviations cancel. That is why the deviations are *squared* before the sum is minimized.)

46. Given the points (x_1,y_1), (x_2,y_2), $\dots$, (x_n,y_n), let

$$\bar{x} = \frac{1}{n} \sum_{k=1}^{n} x_k \qquad \text{and} \qquad \bar{y} = \frac{1}{n} \sum_{k=1}^{n} y_k$$

(the *mean values* of the x_k and y_k, respectively).

(a) Confirm that $(\bar{x},\bar{y})$ in Problem 43 lies on the best-fitting line.

(b) Repeat part (a) for Problem 44.

(c) Show that $(\bar{x},\bar{y})$ always lies on the best-fitting line. (Thus the line passes through the centroid of the system of points. See Section 7.6.)

47. This counterexample to a frequently proposed "theorem" is due to Prof. David A. Smith. Let

$$f(x,y) = e^{-y^2}(2x^3 - 3x^2 + 1) + e^{-y}(2x^3 - 3x^2)$$

(a) Show that $(0,0)$ is the only critical point and $f(0,0) = 1$ is a local maximum.

(b) By analogy with single-variable calculus, we might be tempted to conclude that $f(0,0)$ is a *global* maximum. (Surely a smooth surface with only one peak and no other points where the terrain bottoms out to a local minimum cannot rise higher than the peak!) Confirm, however, that $f(2,0) > f(0,0)$, thus showing that there *is* a point of the surface higher than the peak.

48. Suppose that $H(P) > 0$ in the Second Derivative Test. Show that A and C have the same sign. (Thus it is sufficient to assume $A \neq 0$ for a local extreme value; C need not be mentioned.)

49. To see why the Second Derivative Test reads as it does, let $P = (x_0,y_0)$ be a point of the domain of $f(x,y)$ with the property that $\nabla f(P) = \mathbf{0}$, $H(P) > 0$, and $A > 0$. Use the following argument to explain why every curve of intersection of the surface $z = f(x,y)$ and a vertical plane through P has a horizontal tangent at (x_0,y_0,z_0), where $z_0 = f(x_0,y_0)$, and is concave up near this point. (Such behavior suggests a local *minimum,* as predicted by the test.)

(a) Show that $D_{\mathbf{u}} f(P) = 0$ for every unit vector $\mathbf{u}$. [Thus the vertical section determined by $\mathbf{u}$ has a horizontal tangent at (x_0,y_0,z_0).]

(b) Let $\mathbf{u} = (u,v)$ be an arbitrary unit vector. Show that

$$D_{\mathbf{u}}{}^2 f(P) = Au^2 + 2Buv + Cv^2$$

Hint: $D_{\mathbf{u}}{}^2 f$ means $D_{\mathbf{u}}(D_{\mathbf{u}} f)$.

(c) Complete the square in part (b) to show that

$$D_{\mathbf{u}}{}^2 f(P) = \frac{1}{A}[(Au + Bv)^2 + (AC - B^2)v^2]$$

and explain why it follows that $D_{\mathbf{u}}{}^2 f(P) > 0$. [Thus the vertical section determined by $\mathbf{u}$ is concave up near (x_0,y_0,z_0).]

50. Let $f(x,y) = (y - x^2)(y - 3x^2)$.

(a) Confirm that $\nabla f(0,0) = \mathbf{0}$. What does the Second Derivative Test say about the critical point $(0,0)$?

(b) Confirm that $f(0,y) > 0$ if $y \neq 0$, while $f(x,2x^2) < 0$ if $x \neq 0$. Why does it follow that $f(0,0)$ is not an extreme value? (Thus there is a saddle point at the origin.)

(c) Show that if $\mathbf{u} = (u,v)$ is any unit vector, then $D_{\mathbf{u}}{}^2 f(0,0) = 2v^2$. (Thus if $v \neq 0$, the vertical section through the origin determined by $\mathbf{u}$ is concave up near the origin.)

(d) If $v = 0$, then $\mathbf{u} = \pm\mathbf{i}$ (why?) and we are dealing with the intersection of the graph of f and the xz plane. Why is this curve also concave up? (Thus *every* vertical section through the origin is concave up near the origin.)

This problem shows that the Second Derivative Test is not really proved by the argument in Problem 49. The test is correct as stated, but some tinkering is needed to fix the proof.

17.6
LAGRANGE
MULTIPLIERS
(Optional)

Suppose that the temperature at each point of space is $T = x^4 + y^4 + z^4$ and we are interested in finding the hottest and coldest points of the sphere $x^2 + y^2 + z^2 = 1$. (See Problem 25 in the last section.) This is an example of an extreme value problem with a *constraint*. The most direct way to handle it is to write the constraint in the form $z^2 = 1 - x^2 - y^2$ and substitute into the function whose extreme values are sought:

$$T = x^4 + y^4 + (1 - x^2 - y^2)^2$$

This reduces the number of variables to the point where they are independent and enables us to finish the analysis in the usual way.

Sometimes, however, substitution of the constraint into the function is difficult or impossible to carry out explicitly. While implicit differentiation may come to the rescue, there is another method which is often the best approach, involving an elegant device called **Lagrange multipliers.**

Suppose that we are looking for the extreme values of a function $f(x, y, z)$ on a surface S defined by the equation $F(x, y, z) = 0$. In other words, we want to maximize or minimize f subject to the constraint $F(x, y, z) = 0$. If P is an interior point of S at which f has an extreme value (and if f and F are, as usual, well behaved), we claim that $\nabla f(P)$ and $\nabla F(P)$ are parallel, that is,

$$\nabla f(P) = \lambda \nabla F(P)$$

where λ (the Greek letter *lambda*) is some scalar.

To see why this is true, we proceed as in Section 17.3, where we proved that the gradient of a function at a given point of a level surface is normal to the surface. Let C be any curve through P lying in the surface S, defined by the vector function $\mathbf{x} = \mathbf{g}(t) = (x, y, z)$. Then the function of t defined by

$$f(x, y, z) = f(\mathbf{x}) = f(\mathbf{g}(t))$$

has an extreme value at P (that is, at the parameter value corresponding to P). Its derivative with respect to t must therefore be zero at this point. Since the derivative is found by the Chain Rule of Section 17.2, it follows that

$$f_1(P)\frac{dx}{dt} + f_2(P)\frac{dy}{dt} + f_3(P)\frac{dz}{dt} = 0$$

or simply

$$\nabla f(P) \cdot \frac{d\mathbf{x}}{dt} = 0$$

We know from Section 16.6 that $d\mathbf{x}/dt$ is parallel to the tangent line to C at P (which lies in the tangent plane to the surface S at P). Since C is an arbitrary curve (through P) in this surface, $\nabla f(P)$ is perpendicular to every such line and hence normal to the surface $F(x, y, z) = 0$ at P. According to the theorem already quoted from Section 17.3, so is $\nabla F(P)$. Hence $\nabla f(P)$ and $\nabla F(P)$ are parallel, as advertised.

Thus if $f(x, y, z)$ has an extreme value at a point $P(x, y, z)$ of the surface $F(x, y, z) = 0$, there is a scalar λ which (together with the coordinates of P) satisfies the equations

$$F(x, y, z) = 0 \quad \text{and} \quad \nabla f(x, y, z) = \lambda \nabla F(x, y, z)$$

This leads to the following criterion for a critical point.

Suppose that we are looking for critical points of the function $f(x, y, z)$ subject to the constraint $F(x, y, z) = 0$. If P is such a point (on the surface $F(x, y, z) = 0$ but not on its boundary), and if f and F have continuous partial derivatives in a neighborhood of P, there is a scalar λ which (together with the coordinates x, y, z of P) satisfies the system of equations

$$F(x, y, z) = 0$$
$$f_1(x, y, z) = \lambda F_1(x, y, z)$$
$$f_2(x, y, z) = \lambda F_2(x, y, z)$$
$$f_3(x, y, z) = \lambda F_3(x, y, z)$$

This scalar is called a **Lagrange multiplier.**

The same statement holds for a function $f(x, y)$ subject to the constraint $F(x, y) = 0$ (in which case "surface" becomes "curve"). We merely suppress the third variable (and the last equation of the system):

$$F(x, y) = 0$$
$$f_1(x, y) = \lambda F_1(x, y)$$
$$f_2(x, y) = \lambda F_2(x, y)$$

We begin with an example of this type (because the algebra is simpler).

□ **Example 1**   A wire in the form of the unit circle $x^2 + y^2 = 1$ is heated in such a way that its temperature at (x, y) is $T = x^2 + 2y^2 - x$. Find the hottest and coldest points of the wire.

Solution: The constraint is $F(x, y) = x^2 + y^2 - 1 = 0$ and the function whose extreme values are sought is $T = f(x, y) = x^2 + 2y^2 - x$. We therefore solve the system

$$x^2 + y^2 - 1 = 0$$
$$2x - 1 = 2\lambda x$$
$$4y = 2\lambda y$$

If $y \neq 0$, we may solve the third equaton for $\lambda = 2$. The second equation then reduces to $2x - 1 = 4x$, from which $x = -\frac{1}{2}$. The first equation yields $\frac{1}{4} + y^2 - 1 = 0$, from which $y = \pm\sqrt{3}/2$. Hence we have found the critical points $(-\frac{1}{2}, \pm\sqrt{3}/2)$, at which the temperature is $T = x^2 + 2y^2 - x = \frac{9}{4}$.
    If $y = 0$, the system to be solved reduces to

$$x^2 - 1 = 0$$
$$2x - 1 = 2\lambda x$$

and we don't bother with λ; the first equation yields $x = \pm 1$. Thus there are two more critical points $(\pm 1, 0)$, at which the temperature is $f(1, 0) = 0$ and $f(-1, 0) = 2$. The hottest points of the wire (where $T = \frac{9}{4}$) are $(-\frac{1}{2}, \pm\sqrt{3}/2)$ and the coldest (where $T = 0$) is $(1, 0)$.   □

□ **Example 2**   Find the distance from the point $(5, 5, 4)$ to the paraboloid $z = 4 - x^2 - y^2$.

Solution: The required distance is the minimum of all distances from $(5, 5, 4)$ to points (x, y, z) of the paraboloid. We will find it by minimizing the *square* of these distances, namely

$$f(x, y, z) = (x - 5)^2 + (y - 5)^2 + (z - 4)^2$$

subject to the constraint $z = 4 - x^2 - y^2$, that is,

$$F(x, y, z) = x^2 + y^2 + z - 4 = 0$$

The system to be solved is

$$x^2 + y^2 + z - 4 = 0$$
$$2(x - 5) = 2\lambda x$$
$$2(y - 5) = 2\lambda y$$
$$2(z - 4) = \lambda$$

Solving the last three equations for x, y, and $z - 4$, respectively, we find

$$x = \frac{5}{1 - \lambda}, \, y = \frac{5}{1 - \lambda}, \, z - 4 = \frac{\lambda}{2}$$

(Why is $\lambda \neq 1$?) The first equation then becomes

$$\frac{25}{(1 - \lambda)^2} + \frac{25}{(1 - \lambda)^2} + \frac{\lambda}{2} = 0$$

$$\frac{50}{(1 - \lambda)^2} + \frac{\lambda}{2} = 0$$

$$100 + \lambda(1 - \lambda)^2 = 0$$

$$\lambda^3 - 2\lambda^2 + \lambda + 100 = 0$$

It is not obvious how to solve this cubic equation, so we will simply tell you that one root is $\lambda = -4$. To find the others (if any), divide the polynomial by $\lambda + 4$ to obtain the factorization

$$(\lambda + 4)(\lambda^2 - 6\lambda + 25) = 0$$

The quadratic factor has no real roots, so we conclude that the solution of our system is $\lambda = -4$, $x = 1$, $y = 1$, $z = 2$. Hence the only critical point is $(x, y, z) = (1, 1, 2)$ and the minimum distance (squared) is $f(1, 1, 2) = 36$. The distance itself is $d = 6$, which should check with Problem 24, Section 17.3. (Also see Problem 32 in the last section, which is similar but was supposed to be done without the help of Lagrange multipliers.)   □

□ **Example 3**   Find the hottest and coldest points of the sphere $x^2 + y^2 + z^2 = 1$ if the temperature is given by $T = x^4 + y^4 + z^4$.

Solution: The constraint is $F(x, y, z) = x^2 + y^2 + z^2 - 1 = 0$ and the function whose extreme values are sought is $T = f(x, y, z) = x^4 + y^4 + z^4$.

We therefore solve the system

$$x^2 + y^2 + z^2 - 1 = 0$$
$$4x^3 = 2\lambda x$$
$$4y^3 = 2\lambda y$$
$$4z^3 = 2\lambda z$$

This is messy enough to call for a systematic consideration of cases.

1. Suppose that x, y, and z are nonzero. Then the system reduces to

$$x^2 + y^2 + z^2 = 1$$
$$x^2 = y^2 = z^2 = \frac{\lambda}{2}$$

Substituting from the second group into the first, we find $\lambda = \frac{2}{3}$. The corresponding critical points (x, y, z) satisfy $x^2 = y^2 = z^2 = \frac{1}{3}$, and the temperature at these points is $T = x^4 + y^4 + z^4 = \frac{1}{3}$.

2. Suppose that $z = 0$, in which case our system reduces to

$$x^2 + y^2 = 1$$
$$4x^3 = 2\lambda x$$
$$4y^3 = 2\lambda y$$

(a) If x and y are nonzero, the system reduces further to

$$x^2 + y^2 = 1$$
$$x^2 = y^2 = \frac{\lambda}{2}$$

from which $\lambda = 1$. The corresponding critical points satisfy $x^2 = y^2 = \frac{1}{2}$, $z = 0$, and the temperature is $T = \frac{1}{2}$.

(b) If $y = 0$ in Case (2), the system becomes

$$x^2 = 1$$
$$4x^3 = 2\lambda x$$

and we don't bother with λ. The critical points satisfy $x^2 = 1$, $y = z = 0$, and the temperature is $T = 1$.

(c) If $x = 0$ in Case (2), we find $x = z = 0$, $y^2 = 1$ and temperature $T = 1$, as in (b).

3. Suppose that $y = 0$. By symmetry with Case (2) we find

$$x^2 = z^2 = \tfrac{1}{2}, \ y = 0, \ T = \tfrac{1}{2}$$
$$x^2 = 1, \ y = z = 0, \ T = 1 \qquad \text{(already listed)}$$
$$x = y = 0, \ z^2 = 1, \ T = 1$$

4. Suppose that $x = 0$. By symmetry with Case (2) we find

$$x = 0, \ y^2 = z^2 = \tfrac{1}{2}, \ T = \tfrac{1}{2}$$
$$x = z = 0, \ y^2 = 1, \ T = 1 \qquad \text{(already listed)}$$
$$x = y = 0, \ z^2 = 1, \ T = 1 \qquad \text{(already listed)}$$

Thus the hottest points of the sphere (when $T = 1$) are

$$(\pm 1, 0, 0), \ (0, \pm 1, 0), \ (0, 0, \pm 1)$$

and the coldest points (when $T = \frac{1}{3}$) are

$$\left(\frac{1}{\sqrt{3}}, \frac{1}{\sqrt{3}}, \frac{1}{\sqrt{3}}\right), \left(\frac{1}{\sqrt{3}}, \frac{1}{\sqrt{3}}, -\frac{1}{\sqrt{3}}\right),$$

$$\left(\frac{1}{\sqrt{3}}, -\frac{1}{\sqrt{3}}, \frac{1}{\sqrt{3}}\right), \left(\frac{1}{\sqrt{3}}, -\frac{1}{\sqrt{3}}, -\frac{1}{\sqrt{3}}\right),$$

$$\left(-\frac{1}{\sqrt{3}}, \frac{1}{\sqrt{3}}, \frac{1}{\sqrt{3}}\right), \left(-\frac{1}{\sqrt{3}}, \frac{1}{\sqrt{3}}, -\frac{1}{\sqrt{3}}\right),$$

$$\left(-\frac{1}{\sqrt{3}}, -\frac{1}{\sqrt{3}}, \frac{1}{\sqrt{3}}\right), \left(-\frac{1}{\sqrt{3}}, -\frac{1}{\sqrt{3}}, -\frac{1}{\sqrt{3}}\right) \quad \square$$

□ **Example 4**   Show that the sum of the sines of the angles of a triangle is largest when the triangle is equilateral.

Solution: Let α, β, γ be the angles and $S = \sin \alpha + \sin \beta + \sin \gamma$. The problem is to maximize S subject to the constraint $\alpha + \beta + \gamma = 180°$. Hence we solve the system

$$\alpha + \beta + \gamma = 180°$$
$$\cos \alpha = \lambda$$
$$\cos \beta = \lambda$$
$$\cos \gamma = \lambda$$

The last three equations yield $\cos \alpha = \cos \beta = \cos \gamma$, from which $\alpha = \beta = \gamma$ (because each angle is between $0°$ and $180°$). Hence the triangle is equilateral at the critical point of S.   □

Note that in Example 4 (unlike the previous ones) we did not solve for the Lagrange multiplier λ, but eliminated it. Sometimes that is more convenient; the decision is made by looking at the system of equations to be solved (and the original question) and using one's judgment. Much time can be saved in some problems by making the right judgment, but there is no way to be systematic about it. It comes with experience.

The method of Lagrange multipliers can be generalized to any number of variables, and to situations involving more than one constraint. The complications are mainly algebraic (because the system to be solved gets larger); since this is not an algebra book, we will go no farther.

Problem Set 17.6

Solve each of the following problems by using Lagrange multipliers. (Many have been solved before; the point of this problem set is to provide practice with a new method.)

1. Find the critical points of $f(x, y) = xy$ on the curve $y = x^2$. Check by substitution of the constraint into the function.

2. Find the critical points of $f(x, y) = x/y$ on the curve $y = x^2 + 1$. Check by substitution of the constraint into the function.

3. Find the critical points of $f(x, y) = x + 2y$ on the curve $x^2 + y^2 = 5$. (Note that substitution of the constraint into the function is inconvenient, to say the least.)

4. Find the critical points of $f(x,y) = x^2 + y^2$ on the curve $xy - x - y = 1$.

5. Find the critical points of $f(x,y,z) = 2x - y + 3z$ on the surface $z = x^2 + y^2$. Check by substitution of the constraint into the function.

6. Find the critical points of $f(x,y,z) = x + 2y - z$ on the surface $z = 1 - x^2 - y^2$. Check by substitution of the constraint into the function.

7. Find the critical points of

$$f(x,y,z) = \ln (x^2 + y^2 + z^2)$$

on the surface $x + y - z - 3 = 0$.

8. Find the critical points of $f(x,y,z) = e^{xyz}$ on the surface $x^2 + 2y^2 + z^2 = 4$.

Each of the following extreme value problems from Section 5.5 involves a constraint. Use Lagrange multipliers to solve them.

9. Example 1   10. Example 2

11. Example 3   *Hint:* Maximize θ by maximizing the function $f(x, y) = \tan \theta = y/(1 + x)$, subject to the constraint $y^2 = x$.

12. Example 4   13. Problem 2   14. Problem 7

15. Problem 8   16. Problem 9   17. Problem 11

18. Problem 13   19. Problem 16   20. Problem 17

21. Problem 18   22. Problem 19   23. Problem 20

24. Problem 21   25. Problem 22   26. Problem 23

27. Problem 24   28. Problem 25   29. Problem 26

30. Problem 27   31. Problem 28   32. Problem 29

33. Problem 30   34. Problem 31   35. Problem 32

36. Problem 33   37. Problem 34   38. Problem 35

39. Problem 36

40. The temperature at each point of the wire $x^2 + y^2 = 1$ is $T = xy$. Find the hottest and coldest points of the wire.

41. Find the ellipse $x^2/a^2 + y^2/b^2 = 1$ containing the point $(1,2)$ and having minimum area. *Hint:* The area of the ellipse is πab.

Each of the following extreme value problems from Section 17.5 involves a constraint. Use Lagrange multipliers to solve them.

42. Example 3   43. Problem 28   44. Problem 32

45. Problem 33   46. Problem 34   47. Problem 35

48. Problem 36   49. Problem 37   50. Problem 38

51. Problem 39

52. Find the lowest temperature on the plane $x + y + z = 1$ if the temperature in space is $T = x^2 + y^2 + z^2$.

53. What point of the surface $1/x + 1/y + 1/z = 1$ is closest to the origin?

54. Show that the rectangular box of maximum surface area that can be inscribed in a sphere is a cube.

55. A rectangular box with sides parallel to the coordinate planes is inscribed in the ellipsoid $x^2/a^2 + y^2/b^2 + z^2/c^2 = 1$. Show that its largest volume is $8abc/(3 \sqrt{3})$.

56. Show that the distance from the point (x_0, y_0) to the line $ax + by + c = 0$ is

$$\frac{|ax_0 + by_0 + c|}{\sqrt{a^2 + b^2}}$$

(See Problem 56, Section 1.2.)

57. Show that the distance from the point (x_0, y_0, z_0) to the plane $ax + by + cz + d = 0$ is

$$\frac{|ax_0 + by_0 + cz_0 + d|}{\sqrt{a^2 + b^2 + c^2}}$$

(See Problem 37, Section 16.2.)

True–False Quiz

1. The domain of the function

$$f(x, y, z) = \frac{1}{\sqrt{1 - x^2 - y^2 - z^2}}$$

is the interior of a sphere.

2. If $f(x,y)$ has second partial derivatives, then $f_{12} = f_{21}$.

3. If $f(x,y) = x^2/y$, then $f_{12}(-1,2) = \frac{1}{2}$.

4. The function $z = e^x \sin y$ satisfies Laplace's equation, $\partial^2 z/\partial x^2 + \partial^2 z/\partial y^2 = 0$.

5. If $z = \sin xy$, where $x = e^t$ and $y = e^{-t}$, then $dz/dt = 0$.

6. If $f(x,y)$ and $g(x,y)$ have the same gradient, they are identical.

7. The gradient of f is perpendicular to the graph of $z = f(x,y)$.

8. The tangent plane to the surface $z = f(x,y)$ is horizontal at points (x, y, z) such that $\nabla f(x,y) = \mathbf{0}$.

9. If $f(x,y)$ is a function with continuous partial derivatives, and $\mathbf{u} = -\mathbf{i}$, then $D_{\mathbf{u}}f(x,y) = -f_1(x,y)$.

10. As we move away from the origin, the function $z = e^{2x} \sin y$ increases most rapidly in the direction $\mathbf{u} = \mathbf{j}$.

11. The function $f(x,y) = \sqrt{1 - x^2 - y^2}$ has a maximum value and a minimum value.

12. If P_0 is a point at which $\nabla f(P_0) = \mathbf{0}$, then f has an extreme value at P_0.

Additional Problems

In each of the following, find $f_1(x,y)$ and $f_2(x,y)$.

1. $f(x,y) = xye^{-x^2}$

2. $f(x,y) = \dfrac{x}{y} + \dfrac{y}{x}$

3. $f(x,y) = x\sqrt{x^2 + y^2}$

4. $f(x,y) = \tan^{-1} xy$

5. $f(x,y) = \dfrac{x}{y} \sin \dfrac{y}{x}$

In each of the following, find $f_1(x,y,z)$, $f_2(x,y,z)$, and $f_3(x,y,z)$.

6. $f(x,y,z) = \dfrac{z}{x^2 + y^2 + z^2}$

7. $f(x,y,z) = \ln xyz$

8. $f(x,y,z) = \dfrac{xy}{z}$

9. $f(x,y,z) = e^{xy} \cos yz$

10. $f(x,y,z) = \sin xy + \sin yz + \sin zx$

11. What is the domain of the function $f(x,y) = x^y$? Find $f_{12}(x,y)$ and $f_{21}(x,y)$ and see if they are equal.

12. Show that the function $z = e^y \cos x$ satisfies Laplace's equation,

$$\frac{\partial^2 z}{\partial x^2} + \frac{\partial^2 z}{\partial y^2} = 0$$

13. Give the definition of $f_1(x,y,z)$ as the limit of a difference quotient and use it to find $\partial w / \partial x$ if $w = e^{xyz}$.

14. Given the functions $f(x,y)$ and $g(x,y)$, suppose that

$$f_1(x,y) = g_2(x,y) \quad \text{and} \quad f_2(x,y) = -g_1(x,y)$$

(These are known as the *Cauchy-Riemann equations*.) Show that the level curves $f(x,y) = a$ and $g(x,y) = b$ intersect at right angles.

15. Suppose that $y = f(x)$ and $x = g(u,v)$, where f and g are well-behaved. Show that

$$\begin{vmatrix} \partial y/\partial u & \partial x/\partial u \\ \partial y/\partial v & \partial x/\partial v \end{vmatrix} = 0$$

16. Confirm Problem 15 in the case $y = \ln x$ and $x = u^2 + v^2$.

In each of the following, find dz/dt in two ways, first by substituting before differentiating, then by the Chain Rule. Reconcile the results.

17. $z = e^x \ln y$, where $x = \ln t$, $y = e^t$

18. $z = \dfrac{1}{x^2 + y^2}$, where $x = \cos t$, $y = \sin t$

19. $z = \ln \dfrac{x}{y}$, where $x = 1 - t^2$, $y = 1 + t$

20. $z = \sqrt{x^2 + y^2}$, where $x = 2t$, $y = t - 1$

21. $z = \sin xy$, where $x = e^t$, $y = e^{-t}$

22. $z = \tan^{-1} \dfrac{y}{x}$, where $x = \cos t$, $y = \sin t$

23. Suppose that $z = (x^2 + y^2)^{-1/2}$, where $x = r \cos \theta$ and $y = r \sin \theta$. Find $\partial z/\partial r$ and $\partial z/\partial \theta$. (Assume that $r > 0$.)

In each of the following, find $\partial w/\partial x$ and $\partial w/\partial y$ at the indicated point.

24. $w = ve^{u/v}$, $u = xy$, $v = x - y$, $(x,y) = (1,0)$

25. $w = \sqrt{u^2 + v^2}$, $u = x^2 y$, $v = xy^2$, $(x,y) = (1,1)$

26. $w = \sin(u + v)$, $u = 2x + y$, $v = x + 2y$, $(x,y) = (\pi/6, \pi/6)$

27. Suppose that

$$z = y/x \qquad \text{where } x = e^u \cos v \text{ and } y = e^u \sin v$$

Find $\partial z/\partial u$ and $\partial z/\partial v$ in two ways and reconcile the answers.

28. If $z = F(x,y)$ and $y = f(x)$, show that

$$\frac{d^2 z}{dx^2} = z_{xx} + 2z_{xy}\frac{dy}{dx} + z_{yy}\left(\frac{dy}{dx}\right)^2 + z_y \frac{d^2 y}{dx^2}$$

(Assume that F and f are suitably well-behaved.) Confirm this formula in the case $z = x^2 + xy$, where $y = 1/x$.

29. Find an equation of the tangent plane to the surface $z = x^2 - 2xy + y^2$ at the point $(1, 2, 1)$.

30. Find an equation of the tangent plane to the surface $x^2y + y^2z + z^2x = 6$ at the point $(2, -1, 2)$.

31. Two sides of a triangle are 15 cm and 20 cm, and are increasing at 4 cm/sec and 5 cm/sec, respectively, while the included angle is $\pi/4$ and is decreasing at $\frac{1}{10}$ rad/sec. Is the third side increasing or decreasing? at what rate?

32. Assuming that $z = f(x, y)$ has partial derivatives, show that

$$\nabla (\cos z) = (-\sin z)(\nabla z)$$

33. The equations $x = r\cos\theta$, $y = r\sin\theta$ give rectangular coordinates in terms of polar coordinates.

  (a) Find the partial derivative of each rectangular coordinate with respect to each polar coordinate.

  (b) Show that these partial derivatives satisfy the equation

$$\begin{vmatrix} \partial x/\partial r & \partial x/\partial\theta \\ \partial y/\partial r & \partial y/\partial\theta \end{vmatrix} = r$$

34. Show that the partial derivatives found in Problem 42, Section 17.3, satisfy

$$\begin{vmatrix} \partial r/\partial x & \partial r/\partial y \\ \partial\theta/\partial x & \partial\theta/\partial y \end{vmatrix} = \frac{1}{r}$$

(Question for discussion: Derivatives of inverse functions in single-variable calculus are in some sense reciprocals. What inverse relationship is involved here and what reciprocal relation occurs in Problems 33 and 34?)

35. The equations $x = r\cos\theta$, $y = r\sin\theta$, $z = z$ give rectangular coordinates in terms of cylindrical coordinates.

  (a) Show that

$$\begin{vmatrix} \partial x/\partial r & \partial x/\partial\theta & \partial x/\partial z \\ \partial y/\partial r & \partial y/\partial\theta & \partial y/\partial z \\ \partial z/\partial r & \partial z/\partial\theta & \partial z/\partial z \end{vmatrix} = r$$

  (b) If cylindrical coordinates are regarded as functions of rectangular coordinates, what 3×3 determinant comes out to be $1/r$?

36. The equations

$$x = \rho\sin\phi\cos\theta, \quad y = \rho\sin\phi\sin\theta, \quad z = \rho\cos\phi$$

give rectangular coordinates in terms of spherical coordinates.

  (a) Find the partial derivative of each rectangular coordinate with respect to each spherical coordinate.

  (b) Show that

$$\begin{vmatrix} \partial x/\partial\rho & \partial x/\partial\phi & \partial x/\partial\theta \\ \partial y/\partial\rho & \partial y/\partial\phi & \partial y/\partial\theta \\ \partial z/\partial\rho & \partial z/\partial\phi & \partial z/\partial\theta \end{vmatrix} = \rho^2\sin\phi$$

Find the directional derivative of f at the given point P in the given direction.

37. $f(x, y) = e^{xy}$, $P = (-1, 1)$, in the direction "northwest"

38. $f(x, y) = x^2/y^2$, $P = (1, 1)$, toward the point $(1, 0)$

39. $f(x, y) = \ln(x^2 + y^2)$, $P = (-1, 1)$, in the direction of the origin

40. $f(x, y, z) = \dfrac{x}{y} + \dfrac{y}{z} + \dfrac{z}{x}$, $P = (1, -1, 1)$, in the direction of $\mathbf{i} + 2\mathbf{j} - 2\mathbf{k}$

41. $f(x, y, z) = \sqrt{x^2 + y^2 + z^2}$, $P = (1, 1, 1)$, in the direction of the gradient

42. Suppose that $P_0(x_0, y_0, z_0)$ is a point at which the partial derivatives of $f(x, y, z)$ are continuous. Show that $D_{\mathbf{j}}f(P_0) = f_2(P_0)$ in two ways, as follows.

  (a) Use the definition of the directional derivative as the limit of a difference quotient.

  (b) Use the formula $D_{\mathbf{u}}f(P_0) = \nabla f(P_0) \cdot \mathbf{u}$.

43. Find the derivative of $w = \ln xyz$ at the point $(2, -1, -2)$ in the direction of the vector from the point to the origin. Is the function increasing or decreasing in that direction? In what direction does it increase most rapidly? at what rate? In what directions is its derivative zero?

44. A skier is at the point $(1, 2, 1)$ on the hill $z = 6x - x^2 - y^2$. (The z axis points up, the y axis north, the x axis east.)

  (a) What is the direction of steepest descent? What is the steepness of the hill in that direction?

  (b) If the skier heads west, will she be going up the hill or down? at what rate (with respect to distance)?

  (c) In what directions may the skier go in order to stay on the level?

45. Find all high points, low points, and saddle points of the surface $z = e^{x+y} - e^x - e^y$.

46. Repeat Problem 45 for the surface $z = x^2 + y^2 + \dfrac{2}{xy}$.

47. Find the extreme values of $f(x, y) = xy$ in the triangular region with vertices $(0, 0)$, $(1, 0)$, and $(0, 1)$.

48. Show that the function $f(x, y) = x^4 - x^2y^2 + y^4$ has only one critical point. What does the Second Derivative Test say about this point? Does f have an extreme value at this point? *Hint:* $f(x, y) = (x^2 - y^2)^2 + x^2y^2$.

49. The temperature at each point (x, y) in the first quadrant is $T = \ln xy - x - y$. Find all local hot points. Why is there no coldest point?

50. Find the points of the ellipsoid $3x^2 + 2y^2 + z^2 = 9$ that are farthest from the point $(-2, 1, 0)$.

51. Find the maximum volume of a rectangular box inscribed in the ellipsoid $x^2 + y^2 + 4z^2 = 12$.

52. A rectangular box with no top is made to hold 1 cubic meter. The bottom is made from a material that costs twice as much as the material for the sides. Find the dimensions of the cheapest box.

53. What point of the parabola $y = -x^2$ is closest to the line $x + y = 1$?

54. Find the extreme values of $f(x, y) = x^2 + xy + y^2$ on the circle $x^2 + y^2 = 2$.

55. Find the extreme values of $f(x, y, z) = x + y - z$ on the sphere $x^2 + y^2 + z^2 = 3$.

56. Find the distance between the origin and the surface $xy - z^2 = 1$ in three ways, as follows.

    (a) Express the square of the distance between the origin and a point (x, y, z) of the surface as a function of x and y, and minimize this function. *Hint:* What is the domain of z as a function of x and y? Where does the minimum occur?

    (b) Use Lagrange multipliers to do part (a) without expressing the squared distance in terms of x and y.

    (c) Find the points of the surface at which the normal line passes through the origin, and use them to find the distance. Do you see any connection between this method and part (b)?

18 | Multiple Integrals

'Obvious' is the most dangerous word in mathematics.
E. T. BELL

A mathematician is one to whom that

$$\left[referring\ to\ \int_0^\infty e^{-x^2}\,dx = \frac{\sqrt{\pi}}{2} \right]$$

is as obvious as that twice two makes four is to you.
LORD KELVIN (physicist)

Many things are not accessible to intuition at all, the value of

$$\int_0^\infty e^{-x^2}\,dx$$

for instance.
J. E. LITTLEWOOD (mathematician)

The last chapter was about differentiation of functions of several variables; this one is about integration. The integral of a function of one variable will now be called a single integral, to distinguish it from the multiple integrals of functions of several variables that we will introduce. These are defined in much the same way as single integrals, but the domain of integration is a region in the plane (or in space) instead of an interval on the number line. The theoretical ideas are similar to those in single-variable calculus, too. They are harder to discuss with precision, however, and for the most part are left for an advanced course. We will depend heavily on intuition, the main object of the chapter being the development of techniques for evaluation of multiple integrals in the context of applications. These techniques are not hard if you know how to evaluate single integrals, but of course that may require some review.

18.1

ITERATED
INTEGRALS
IN THE PLANE

Figure 1
Region between two curves

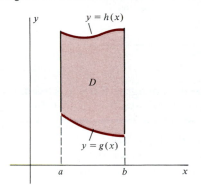

Suppose that $z = f(x,y)$ is a function of two variables defined in a region D of the xy plane. At the outset we assume that D is a region like that shown in Figure 1, namely

$$D = \{(x,y): a \leq x \leq b,\ g(x) \leq y \leq h(x)\}$$

In other words, the left and right boundaries of D are vertical lines $x = a$ and $x = b$, while the lower and upper boundaries are allowed to be curved, $y = g(x)$ and $y = h(x)$, respectively. We also assume that g and h are continuous, and that $f(x,y)$ is continuous and nonnegative in D.

The graph of $z = f(x,y)$ for $(x,y) \in D$ is a surface lying above D, a "roof" that caps a vertical cylinder whose base is D. See Figure 2, which also shows a vertical section of the cylinder by a plane parallel to the yz plane (passing through a point on the x axis between a and b). As this vertical section moves from $x = a$ to $x = b$ it "sweeps out" the volume of the cylinder under the graph of f. More precisely, if $A(x)$ is the area of the typical vertical section, we may set up the volume as the limit of Riemann sums of $A(x)$ associated with partitions of $[a,b]$. The typical subinterval (of length Δx) leads to a "slice" of the cylinder parallel to the yz plane. If x is a point of this subinterval, the volume of the slice is $\Delta V \approx A(x)\,\Delta x$. The total volume is

$$V = \sum \Delta V \approx \sum A(x)\,\Delta x$$

and the limit of this Riemann sum is

$$V = \int_a^b A(x)\,dx$$

So far this should be plain sailing (although you may have to review the ideas from single-variable calculus that are involved in what we are saying).

Figure 2
Vertical cylinder with base D and roof $z = f(x,y)$

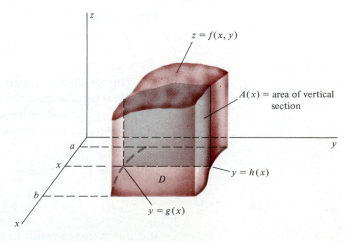

Figure 3
$A(x)$ as an area in the yz plane

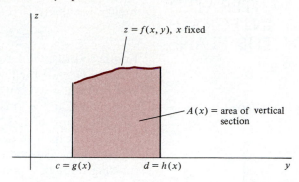

What leads to a new idea is that $A(x)$ is itself an integral. For if x is fixed, the vertical section (projected into the yz plane) looks like Figure 3. Its left and right boundaries are the lines $y = c = g(x)$ and $y = d = h(x)$, while its lower and upper boundaries are the y axis and the curve $z = f(x,y)$, respectively. This doesn't look right at first glance, but remember that x is *fixed*. Hence the equations $y = g(x)$ and $y = h(x)$ are of the form

$$y = \text{constant} \qquad \text{(a vertical line in the } yz \text{ plane)}$$

while the equation $z = f(x,y)$ is of the form

$$z = \text{function of } y \qquad \text{(a curve in the } yz \text{ plane)}$$

Accordingly, the area of the cross section is

$$A(x) = \int_c^d f(x,y)\, dy = \int_{g(x)}^{h(x)} f(x,y)\, dy$$

and our formula for volume becomes

$$V = \int_a^b \left[\int_{g(x)}^{h(x)} f(x,y)\, dy \right] dx$$

The brackets are ordinarily omitted, in which case we write

$$V = \int_a^b \int_{g(x)}^{h(x)} f(x,y)\, dy\, dx$$

This is called an **iterated** ("repeated") **integral,** the understanding being that we integrate from the inside out. In other words x is fixed while we integrate $f(x,y)$ with respect to y from $g(x)$ to $h(x)$; the result (a function of x) is integrated with respect to x from a to b.

□ **Example 1**   Evaluate the iterated integral

$$\int_0^1 \int_0^x 2\, dy\, dx$$

and interpret the result as a volume.

Figure 4

$D = \{(x, y): 0 \le x \le 1, 0 \le y \le x\}$

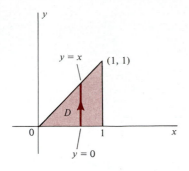

Figure 5

Vertical cylinder with base D and roof $z = 2$

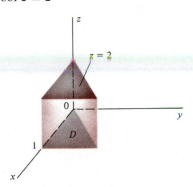

Solution: First we fix x and evaluate the inside integral:

$$\int_0^x 2\, dy = 2y \Big|_{y=0}^{y=x} = 2x$$

Then we integrate this with respect to x from 0 to 1:

$$\int_0^1 \int_0^x 2\, dy\, dx = \int_0^1 2x\, dx = x^2 \Big|_0^1 = 1$$

To interpret the result as a volume, we must figure out the domain of integration. For a fixed x between 0 and 1, y runs from $y = 0$ to $y = x$; then x runs from 0 to 1. Hence the domain is

$$D = \{(x, y): 0 \le x \le 1, 0 \le y \le x\}$$

as shown in Figure 4.

The integrand in this case is the constant function $z = f(x, y) = 2$, the graph of which is a horizontal plane two units above the xy plane. In finding

$$\int_0^1 \int_0^x 2\, dy\, dx = 1$$

we are computing the volume of a vertical cylinder with base D and roof $z = 2$, as shown in Figure 5. The result can be checked by ordinary geometry; multiply the area of the base by the height of the cylinder to obtain

$$V = (\text{area of } D) \cdot 2 = \tfrac{1}{2}(1)(1) \cdot 2 = 1 \qquad \square$$

□ **Example 2**   Evaluate the iterated integral

$$\int_0^1 \int_0^x (2 - x^2 - y^2)\, dy\, dx$$

and interpret the result as a volume.

Solution: First we fix x and work on the inside integral. Its evaluation involves nothing new except that x appears in an integration with respect to y; simply treat x as a constant:

$$\int_0^x (2 - x^2 - y^2)\, dy = \left[(2 - x^2)y - \frac{y^3}{3} \right]\Big|_{y=0}^{y=x} = (2 - x^2)x - \frac{x^3}{3}$$

$$= \frac{x}{3}[3(2 - x^2) - x^2] = \frac{x}{3}(6 - 4x^2) = \frac{2x}{3}(3 - 2x^2)$$

Now integrate with respect to x from 0 to 1:

$$\int_0^1 \int_0^x (2 - x^2 - y^2)\, dy\, dx = \frac{2}{3} \int_0^1 x(3 - 2x^2)\, dx = \frac{2}{3} \int_0^1 (3x - 2x^3)\, dx$$

$$= \frac{2}{3} \left(\frac{3x^2}{2} - \frac{x^4}{2} \right)\Big|_0^1 = \frac{2}{3}\left(\frac{3}{2} - \frac{1}{2} \right) = \frac{2}{3}$$

The domain of integration is the same as in Example 1, but the integrand is $z = f(x, y) = 2 - x^2 - y^2$. Hence we are finding the volume of

Figure 6
Vertical cylinder with base D and roof $z = 2 - x^2 - y^2$

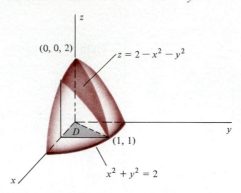

the vertical cylinder with base D and roof lying in the circular paraboloid shown in Figure 6.   □

□ **Example 3**   Evaluate the iterated integral

$$\int_0^1 \int_y^1 (2 - x^2 - y^2) \, dx \, dy$$

and interpret the result as a volume.

Solution: The variable of integration in the first integral is x, so we fix y:

$$\int_y^1 (2 - x^2 - y^2) \, dx = \left[(2 - y^2)x - \frac{x^3}{3} \right]\Big|_{x=y}^{x=1} = \frac{x}{3}[3(2 - y^2) - x^2]\Big|_{x=y}^{x=1}$$

$$= \frac{1}{3}[3(2 - y^2) - 1] - \frac{y}{3}[3(2 - y^2) - y^2]$$

$$= \frac{1}{3}(5 - 3y^2) - \frac{y}{3}(6 - 4y^2) = \frac{1}{3}(5 - 6y - 3y^2 + 4y^3)$$

Now integrate with respect to y from 0 to 1:

$$\int_0^1 \int_y^1 (2 - x^2 - y^2) \, dx \, dy = \frac{1}{3} \int_0^1 (5 - 6y - 3y^2 + 4y^3) \, dy$$

$$= \frac{1}{3}(5y - 3y^2 - y^3 + y^4)\Big|_0^1 = \frac{2}{3}$$

To figure out the domain of integration, note that for a fixed y between 0 and 1, x runs from $x = y$ to $x = 1$; then y runs from 0 to 1. Hence the domain is

$$D = \{(x,y): 0 \le y \le 1, \, y \le x \le 1\}$$

as shown in Figure 7. Although it is described differently, it is the same D as in Example 2, so of course the volume is the same.   □

Figure 7
$D = \{(x,y): 0 \le y \le 1, \, y \le x \le 1\}$

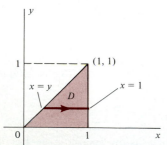

These examples show the same region D described in different ways:

$$D = \{(x,y): 0 \leq x \leq 1, 0 \leq y \leq x\} \quad \text{(Figure 4)}$$

$$D = \{(x,y): 0 \leq y \leq 1, y \leq x \leq 1\} \quad \text{(Figure 7)}$$

While some regions cannot be described both ways, it is important to be able to change from one to the other in those cases where it is possible.

□ **Example 4**   Let D be the region bounded by the upper half of the unit circle and the x axis. Describe D in two ways in rectangular coordinates and also describe it in polar coordinates.

Solution: In Figure 8 we have fixed x between -1 and 1. Then y runs from $y = 0$ (the x axis) to $y = \sqrt{1 - x^2}$ (the upper half of the unit circle $x^2 + y^2 = 1$). Releasing x to run from -1 to 1, we obtain the description

$$D = \{(x,y): -1 \leq x \leq 1, 0 \leq y \leq \sqrt{1 - x^2}\}$$

Now look at Figure 9, in which y is fixed between 0 and 1. Then x runs from $x = -\sqrt{1 - y^2}$ (the left half of the unit circle) to $x = \sqrt{1 - y^2}$ (the right half of the circle). Releasing y to run from 0 to 1, we find

$$D = \{(x,y): 0 \leq y \leq 1, -\sqrt{1 - y^2} \leq x \leq \sqrt{1 - y^2}\}$$

Figure 8
$D = \{(x,y): -1 \leq x \leq 1,$
$\qquad 0 \leq y \leq \sqrt{1 - x^2}\}$

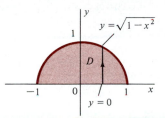

Figure 9
$D = \{(x,y): 0 \leq y \leq 1,$
$\qquad -\sqrt{1 - y^2} \leq x \leq \sqrt{1 - y^2}\}$

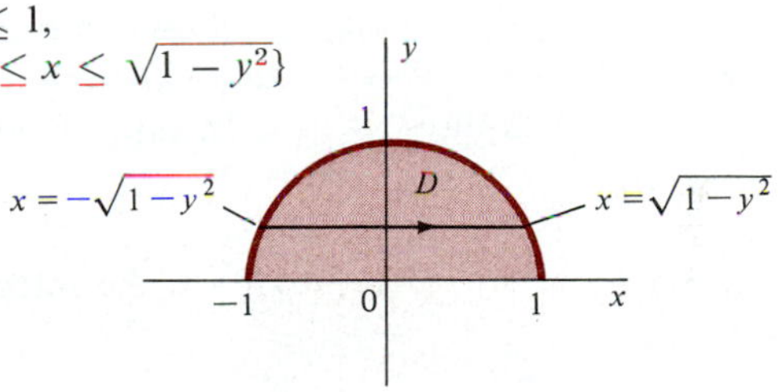

Finally, consider Figure 10, in which θ is fixed between 0 and π. After r runs from $r = 0$ (the origin) to $r = 1$ (the unit circle), θ runs from 0 to π. Hence

$$D = \{(r, \theta): 0 \leq r \leq 1, 0 \leq \theta \leq \pi\} \qquad \square$$

In Examples 2 and 3 we obtained the results

$$\int_0^1 \int_0^x (2 - x^2 - y^2) \, dy \, dx = \frac{2}{3}$$

and

$$\int_0^1 \int_y^1 (2 - x^2 - y^2) \, dx \, dy = \frac{2}{3}$$

respectively. Since the domain of integration is the same both times (Figures 4 and 7) and the integrands are the same, we may legitimately speculate that

Figure 10
$D = \{(r,\theta): 0 \leq r \leq 1, 0 \leq \theta \leq \pi\}$

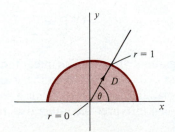

reversing the order of integration has no effect on the answer. Although we won't prove it, the speculation turns out to be true for continuous integrands in regions that can be described two ways in rectangular coordinates.

□ **Example 5**   Reverse the order of integration in the iterated integral

$$\int_0^{\pi/2} \int_0^{\sin x} \cos x \, dy \, dx$$

and evaluate it both ways.

Solution: As the integral stands, we start by fixing x:

$$\int_0^{\sin x} \cos x \, dy = (\cos x) y \Big|_{y=0}^{y=\sin x} = \sin x \cos x$$

Then we integrate with respect to x:

$$\int_0^{\pi/2} \int_0^{\sin x} \cos x \, dy \, dx = \int_0^{\pi/2} \sin x \cos x \, dx = \frac{\sin^2 x}{2} \Big|_0^{\pi/2} = \frac{1}{2}$$

To reverse the order of integration, observe that the domain is

$$D = \{(x,y): 0 \le x \le \pi/2, \, 0 \le y \le \sin x\}$$

as shown in Figure 11. A different way of describing it is to fix y between 0 and 1 and let x run from $x = \sin^{-1} y$ to $x = \pi/2$, then let y run from 0 to 1. In other words

$$D = \{(x,y): 0 \le y \le 1, \, \sin^{-1} y \le x \le \pi/2\}$$

Hence the integral in reverse order is

$$\int_0^1 \int_{\sin^{-1} y}^{\pi/2} \cos x \, dx \, dy$$

To evaluate this, start by fixing y:

$$\int_{\sin^{-1} y}^{\pi/2} \cos x \, dx = \sin x \Big|_{x=\sin^{-1} y}^{x=\pi/2} = 1 - \sin(\sin^{-1} y) = 1 - y$$

Then integrate with respect to y:

$$\int_0^1 \int_{\sin^{-1} y}^{\pi/2} \cos x \, dx \, dy = \int_0^1 (1 - y) \, dy = -\frac{1}{2}(1 - y)^2 \Big|_0^1 = \frac{1}{2} \qquad □$$

Note that in Example 5 we said nothing about volume. The integrand is $z = f(x,y) = \cos x$ (the graph of which is a cylinder parallel to the y axis generated by the curve $z = \cos x$ in the xz plane). This surface forms a roof over the domain D in the xy plane and we can interpret the integral as the

Figure 11
$D = \{(x,y): 0 \le x \le \pi/2,$
$\qquad\qquad 0 \le y \le \sin x\}$

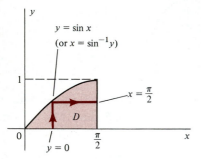

volume under this roof. But it is unnecessary to do so. Iterated integrals have an existence of their own, independent of any interpretation as volume; we used that notion simply to get the discussion off the ground. Like integrals in single-variable calculus, iterated integrals have many interpretations and applications (of which volume is only one). Hence we feel free to relax the restriction that $z = f(x,y)$ should be nonnegative (made at the beginning of this section in order to talk about volume), and to compute the iterated integral of any continuous function over any (reasonable) domain in the xy plane.

We end this section with a fairly complicated computation of volume, to give you some experience in dealing with the algebraic manipulations that often arise in this chapter.

□ **Example 6**   Find the volume of the tetrahedron cut off from the first octant by the plane $x/a + y/b + z/c = 1$ (where a, b, and c are positive).

Solution: See Figure 12, which shows the plane in the first octant as a roof over the domain D in the xy plane. The volume is found by thinking of this roof as the graph of a function $z = f(x,y)$ which is to be integrated over D. Hence we solve for z in the equation of the plane:

$$z = f(x,y) = c\left(1 - \frac{x}{a} - \frac{y}{b}\right)$$

Now think of the upper boundary of D (the line $x/a + y/b = 1$ in the xy plane) as the graph of a function of x, namely $y = b(1 - x/a)$. Then the volume is

$$V = \int_0^a \int_0^{b(1-x/a)} c\left(1 - \frac{x}{a} - \frac{y}{b}\right) dy \, dx$$

Figure 12
Tetrahedron in the first octant

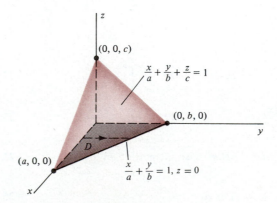

Fixing x in the first integral, we find

$$\int_0^{b(1-x/a)} c\left(1 - \frac{x}{a} - \frac{y}{b}\right) dy = c\left[\left(1 - \frac{x}{a}\right)y - \frac{y^2}{2b}\right]\Big|_{y=0}^{y=b(1-x/a)}$$

$$= \frac{cy}{2b}\left[2b\left(1 - \frac{x}{a}\right) - y\right]\Big|_{y=0}^{y=b(1-x/a)}$$

$$= \frac{c}{2b} \cdot b\left(1 - \frac{x}{a}\right)\left[2b\left(1 - \frac{x}{a}\right) - b\left(1 - \frac{x}{a}\right)\right]$$

$$= \frac{c}{2}\left(1 - \frac{x}{a}\right) \cdot b\left(1 - \frac{x}{a}\right) = \frac{bc}{2}\left(1 - \frac{x}{a}\right)^2$$

Now integrate with respect to x to obtain

$$V = \int_0^a \frac{bc}{2}\left(1 - \frac{x}{a}\right)^2 dx = \frac{bc}{2}(-a) \cdot \frac{1}{3}\left(1 - \frac{x}{a}\right)^3\Big|_0^a = \frac{abc}{6}\left(1 - \frac{x}{a}\right)^3\Big|_a^0$$

$$= \frac{1}{6}abc \qquad \square$$

Problem Set 18.1

Evaluate each of the following integrals.

1. $\int_2^6 \int_0^1 (x - y)\, dy\, dx$

2. $\int_0^3 \int_0^2 xy\, dy\, dx$

3. $\int_0^{1/2} \int_0^2 \frac{dy\, dx}{\sqrt{1 - x^2}}$

4. $\int_0^1 \int_0^2 \frac{dx\, dy}{1 + y^2}$

Evaluate each of the following integrals.

5. $\int_0^2 \int_0^{x^2} x\, dy\, dx$

6. $\int_0^1 \int_0^{\sqrt{1-y^2}} y\, dx\, dy$

7. $\int_0^2 \int_0^{4-2x} (x + y)\, dy\, dx$

8. $\int_{-1}^1 \int_0^{x^2} x^2 y\, dy\, dx$

9. $\int_0^1 \int_0^y x\sqrt{x^2 + y^2}\, dx\, dy$

10. $\int_0^1 \int_0^y \frac{dx\, dy}{\sqrt{4 - y^2}}$

11. $\int_0^1 \int_0^x \frac{dy\, dx}{1 + x^2}$

12. $\int_0^1 \int_1^{e^x} \frac{dy\, dx}{y}$

13. $\int_1^e \int_0^{\ln x} x\, dy\, dx$ *Hint:* Use integration by parts.

14. $\int_0^{\pi/4} \int_0^y \cos(x + y)\, dx\, dy$

15. $\int_0^1 \int_{-x}^x e^{x+y}\, dy\, dx$

16. $\int_1^2 \int_0^{1/x} xe^{xy}\, dy\, dx$

17. $\int_0^1 \int_0^{\sin x} \frac{x\, dy\, dx}{\sqrt{1 - y^2}}$

18. $\int_0^1 \int_0^{\tan y} \frac{y\, dx\, dy}{1 + x^2}$

19. $\int_1^2 \int_0^x \frac{dy\, dx}{x^2 + y^2}$

Reverse the order of integration and evaluate the result.

20. $\int_2^6 \int_0^1 (x - y)\, dy\, dx$ (Check with Problem 1.)

21. $\int_0^3 \int_0^2 xy\, dy\, dx$ (Check with Problem 2.)

22. $\int_0^{1/2} \int_0^2 \frac{dy\, dx}{\sqrt{1 - x^2}}$ (Check with Problem 3.)

23. $\int_0^1 \int_0^2 \frac{dx\, dy}{1 + y^2}$ (Check with Problem 4.)

24. $\int_0^2 \int_0^{x^2} x\, dy\, dx$ (Check with Problem 5.)

25. $\int_0^1 \int_0^{\sqrt{1-y^2}} y\, dx\, dy$ (Check with Problem 6.)

26. $\int_0^2 \int_0^{4-2x} (x + y) \, dy \, dx$ (Check with Problem 7.)

27. $\int_0^1 \int_1^{e^x} \frac{dy \, dx}{y}$ (Check with Problem 12.)

28. $\int_1^e \int_0^{\ln x} x \, dy \, dx$ (Check with Problem 13.)

29. $\int_0^{\pi/4} \int_0^y \cos(x + y) \, dx \, dy$ (Check with Problem 14.)

30. $\int_0^1 \int_{-x}^x e^{x+y} \, dy \, dx$ Hint: The domain should be broken into two parts. (Check with Problem 15.)

31. Reverse the order of integration in

$$\int_0^{\pi/4} \int_0^{\tan x} \sec x \, dy \, dx$$

and evaluate whichever integral seems easier.

32. Reverse the order of integration in

$$\int_0^1 \int_x^1 \cos y^2 \, dy \, dx$$

and evaluate the result. Can you evaluate the integral as it stands?

33. Reverse the order of integration in

$$\int_0^2 \int_{y/2}^1 e^{-x^2} \, dx \, dy$$

and evaluate the result. Can you evaluate the integral as it stands?

Use the volume interpretation of integral to evaluate each of the following.

34. $\int_{-1}^1 \int_0^{\sqrt{1-x^2}} 5 \, dy \, dx$

35. $\int_0^2 \int_0^{\sqrt{4-y^2}} 3 \, dx \, dy$

36. $\int_0^1 \int_0^{\sqrt{1-x^2}} \sqrt{1 - x^2 - y^2} \, dy \, dx$

37. $\int_0^1 \int_0^2 \sqrt{1 - x^2} \, dy \, dx$

38. Use integration to find the volume cut off from the first octant by the plane $2x + y + z = 4$. (Check by the formula derived in Example 6.)

39. Find the volume inside the cylinder $x^2 + y^2 = 1$, below the plane $z = y$, and above the xy plane.

40. Find the volume below the cylinder $z = 1 - x^2$, between the xz plane and the plane $x + y = 1$, and above the xy plane.

41. Find the volume bounded by the paraboloid $z = 1 - x^2 - y^2$ and the xy plane. Hint: Use a trigonometric substitution in the second integration (Section 10.3), together with the reduction formulas given at the end of Section 10.1.

42. Show that the volume of the ellipsoid $x^2/a^2 + y^2/b^2 + z^2/c^2 = 1$ is $V = \frac{4}{3}\pi abc$. Hint: One way to do this is to write

$$V = 8 \int_0^a \int_0^{b\sqrt{1-x^2/a^2}} c \sqrt{1 - x^2/a^2 - y^2/b^2} \, dy \, dx$$

While this looks messy, the trigonometric substitution

$$\frac{y}{b} = k \sin t \qquad (k = \sqrt{1 - x^2/a^2})$$

quickly reduces it to a manageable problem.

18.2

DOUBLE INTEGRALS

In this section we propose to define the *double integral* of a function $f(x, y)$ over a closed and bounded region of the plane. Recall from Section 6.2 that the (single) integral of a function $f(x)$ over an interval $[a,b]$ is found by means of partitions $P = \{x_0, x_1, \ldots, x_n\}$. If c_k is any point of the kth subinterval (of length Δx_k), then

$$\int_a^b f(x) \, dx = \lim_{\|P\| \to 0} \sum_{k=1}^n f(c_k) \, \Delta x_k$$

Figure 1
Partition of a rectangular region

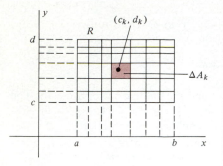

where $\|P\|$ is the norm of the partition (the length of the longest subinterval).

To generalize this definition to functions of two variables, we start by assuming that our region is a rectangle, say

$$R = \{(x,y): a \leq x \leq b, c \leq y \leq d\}$$

A *partition* of R (Figure 1) is produced by partitioning the intervals $[a,b]$ and $[c,d]$ on the x and y axes and using vertical and horizontal lines through the points of each partition to form rectangles covering R. Number the rectangles in some orderly way and let (c_k,d_k) be an arbitrary point of the kth rectangle ($k = 1, 2, \ldots, n$). If ΔA_k is the area of the kth rectangle, we call

$$\sum_{k=1}^{n} f(c_k,d_k)\, \Delta A_k$$

a *Riemann sum* of f associated with the partition P.

We define $\|P\|$, the *norm* of P, to be the length of the longest diagonal of the rectangles of P; by making $\|P\|$ small we force each rectangle to be small in both dimensions. As in single-variable calculus, continuity of f in R is enough to guarantee the existence and uniqueness of the limit of Riemann sums as the norm approaches zero. This limit is called the *double integral* of f over R, written

$$\iint_R f(x,y)\, dA = \lim_{\|P\| \to 0} \sum_{k=1}^{n} f(c_k,d_k)\, \Delta A_k$$

Extending this definition to a nonrectangular domain D is easily done (less easily defended!). Simply enclose D in a rectangle R (which can always be done if D is bounded). Define the function $\overline{f}$ by

$$\overline{f}(x,y) = \begin{cases} f(x,y) \text{ if } (x,y) \in D \\ 0 \text{ if } (x,y) \notin D \end{cases}$$

The integral of f over D is defined to be the integral of $\overline{f}$ over R, that is,

$$\iint_D f(x,y)\, dA = \iint_R \overline{f}(x,y)\, dA$$

The problem with this definition is that $\overline{f}$ is likely to be discontinuous at points of the boundary of D. Geometrically speaking, we are on the surface $z = f(x,y)$ as long as $(x,y) \in D$, then suddenly we drop down to the plane $z = 0$ when $(x,y) \notin D$. The graph of $\overline{f}$ is thus a surface together with the part of R outside D in the xy plane (with vertical cliffs at the boundary of D unless the surface happens to meet the xy plane there).

Thus we are forced to deal with discontinuities at boundary points, a subject that is more complicated than in single-variable calculus (where the only boundary points are endpoints of the interval of integration). Since this is more properly done in advanced calculus, we will merely state that it is a question of how pathological the boundary is allowed to be. In this book all boundaries will be sufficiently simple so that the discontinuities of $\bar{f}$ at the boundary of D are of no consequence. In other words, continuity of f in D is all we need to guarantee that the double integral of $\bar{f}$ over R (and hence the double integral of f over D) exists.

Now we turn to the question of how double integrals are evaluated. In the case of the rectangular region R shown in Figure 2 it is clear that the typical element of volume under the surface $z = f(x,y)$ is

$$\Delta V_k \approx f(c_k, d_k)\, \Delta A_k \qquad (k = 1, 2, \ldots, n)$$

The total volume under the surface is

$$V \approx \sum_{k=1}^{n} f(c_k, d_k)\, \Delta A_k$$

an approximation that improves as the partition is refined. The exact volume is

$$V = \lim_{\|P\| \to 0} \sum_{k=1}^{n} f(c_k, d_k)\, \Delta A_k = \int\!\!\int_{R} f(x,y)\, dA$$

We already know (from Section 18.1) that the volume is also given by

$$V = \int_a^b \int_c^d f(x,y)\, dy\, dx = \int_c^d \int_a^b f(x,y)\, dx\, dy$$

Hence *the double integral is equal to the iterated integral in either order.*

Figure 2
Typical element of volume under the surface $z = f(x,y)$

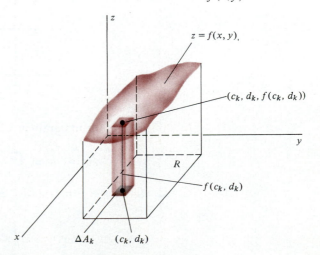

Remark   The preceding volume interpretation requires the restriction $f(x,y) \geq 0$ in R. In general we appeal to the definition

$$\iint_R f(x,y)\, dA = \lim_{\|P\| \to 0} \sum_{k=1}^{n} f(c_k, d_k)\, \Delta A_k$$

The idea (which we will not pursue) is that by running through the rectangles of P by *rows* we obtain one iterated integral; the other is found by running through the rectangles by *columns*.

If D is not rectangular, we have defined

$$\iint_D f(x,y)\, dA = \iint_R \overline{f}(x,y)\, dA$$

where R is a rectangle containing D and $\overline{f}$ is the same as f in D and zero elsewhere. Assume that D is a region of the type discussed in the last section, namely

$$D = \{(x,y) \colon a \leq x \leq b,\ g(x) \leq y \leq h(x)\}$$

where g and h are continuous. Then (see Figure 3) we have

$$\iint_D f(x,y)\, dA = \iint_R \overline{f}(x,y)\, dA = \int_a^b \int_c^d \overline{f}(x,y)\, dy\, dx$$

Figure 3
Reducing a double integral to an iterated integral

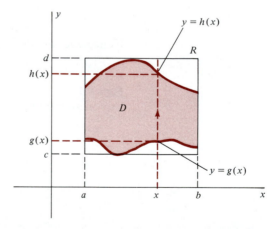

The inner integral (in which x is fixed) may be broken up by writing

$$\int_c^d \overline{f}(x,y)\, dy = \int_c^{g(x)} \overline{f}(x,y)\, dy + \int_{g(x)}^{h(x)} \overline{f}(x,y)\, dy + \int_{h(x)}^d \overline{f}(x,y)\, dy$$

Since $\overline{f} = f$ in D and $\overline{f} = 0$ outside D, this reduces to

$$\int_c^d \overline{f}(x,y)\, dy = \int_{g(x)}^{h(x)} f(x,y)\, dy$$

Figure 4

Reducing a double integral to an iterated integral

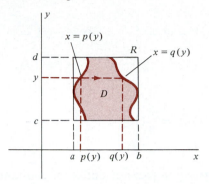

and hence

$$\iint_D f(x,y)\, dA = \int_a^b \int_{g(x)}^{h(x)} f(x,y)\, dy\, dx$$

Thus *a double integral may be evaluated by an iterated integral* if D has the right shape. Moreover (by the same argument) the double integral is an iterated integral in the other order if D can be written in the form

$$D = \{(x,y) : c \le y \le d,\ p(y) \le x \le q(y)\}$$

(See Figure 4.) That is,

$$\iint_D f(x,y)\, dA = \int_c^d \int_{p(y)}^{q(y)} f(x,y)\, dx\, dy$$

□ **Example 1**   If D is the region bounded by the upper half of the unit circle and the x axis, find

$$\iint_D (x - y)\, dA$$

Figure 5

Computing a double integral over a region

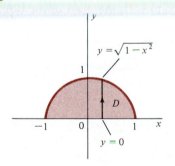

Solution: From Figure 5 we see that D may be written in the form

$$D = \{(x,y) : -1 \le x \le 1,\ 0 \le y \le \sqrt{1 - x^2}\}$$

Hence the given double integral may be replaced by an iterated integral as follows:

$$\iint_D (x - y)\, dA = \int_{-1}^1 \int_0^{\sqrt{1-x^2}} (x - y)\, dy\, dx$$

The inside integral is

$$\int_0^{\sqrt{1-x^2}} (x - y)\, dy = \left(xy - \frac{y^2}{2}\right)\Big|_{y=0}^{y=\sqrt{1-x^2}} = x\sqrt{1 - x^2} - \frac{1}{2}(1 - x^2)$$

Thus the original double integral is

$$\iint_D (x - y)\, dA = \int_{-1}^1 x\sqrt{1 - x^2}\, dx - \frac{1}{2}\int_{-1}^1 (1 - x^2)\, dx$$

$$= 0 - \int_0^1 (1 - x^2)\, dx \quad \text{(odd and even integrands)}$$

$$= \left(\frac{x^3}{3} - x\right)\Big|_0^1 = -\frac{2}{3}$$

Alternate solution: Since D may also be written in the form

$$D = \{(x,y) : 0 \le y \le 1,\ -\sqrt{1 - y^2} \le x \le \sqrt{1 - y^2}\}$$

we have

$$\iint_D (x - y)\, dA = \int_0^1 \int_{-\sqrt{1-y^2}}^{\sqrt{1-y^2}} (x - y)\, dx\, dy$$

The inside integral is

$$\int_{-\sqrt{1-y^2}}^{\sqrt{1-y^2}} (x - y) \, dx = \left(\frac{x^2}{2} - xy\right)\Bigg|_{x=-\sqrt{1-y^2}}^{x=\sqrt{1-y^2}}$$

$$= \frac{1}{2}(1 - y^2) - y\sqrt{1 - y^2} - \frac{1}{2}(1 - y^2) - y\sqrt{1 - y^2}$$

$$= -2y\sqrt{1 - y^2}$$

Hence

$$\iint_D (x - y) \, dA = \int_0^1 \sqrt{1 - y^2}(-2y) \, dy = \frac{2}{3}(1 - y^2)^{3/2}\Bigg|_0^1 = -\frac{2}{3} \quad \square$$

□ **Example 2**   Find the center of mass of a piece of tinfoil (of constant density) occupying the semicircular region

$$D = \{(x,y): x^2 + y^2 \leq a^2, \, y \geq 0\}$$

Solution: You should review Example 3 in Section 7.6 to see how we set this up in terms of single integrals. It is more natural, however, to partition D into rectangles and to calculate moments of the typical rectangle relative to the coordinate axes. (See Figure 6.) Let ΔA be the area of this rectangle. If $\delta(x,y) = k$ is the density of the tinfoil at $(x,y) \in D$, the mass of the typical rectangle is $\Delta m = k \, \Delta A$. Letting (x,y) be a point of the rectangle, we find moments of the mass relative to the coordinate axes to be

$$\Delta M_y \approx x \, \Delta m = kx \, \Delta A \qquad \text{and} \qquad \Delta M_x \approx y \, \Delta m = ky \, \Delta A$$

respectively. The total moments are

$$M_y = \sum \Delta M_y \approx \sum kx \, \Delta A \qquad \text{and} \qquad M_x = \sum \Delta M_x \approx \sum ky \, \Delta A$$

Taking limits as the partition is refined, we find

$$M_y = \iint_D kx \, dA \qquad \text{and} \qquad M_x = \iint_D ky \, dA$$

Each of these double integrals may be written as an iterated integral:

$$M_y = \int_{-a}^a \int_0^{\sqrt{a^2-x^2}} kx \, dy \, dx = k \int_{-a}^a x\sqrt{a^2 - x^2} \, dx = 0 \qquad \text{(odd integrand)}$$

$$M_x = \int_{-a}^a \int_0^{\sqrt{a^2-x^2}} ky \, dy \, dx = \frac{k}{2} \int_{-a}^a (a^2 - x^2) \, dx = k\left(a^2 x - \frac{x^3}{3}\right)\Bigg|_0^a = \frac{2}{3}ka^3$$

Since the density is constant, we may find the mass of the tinfoil by elementary geometry, namely

$$m = k(\text{area of } D) = \frac{k\pi a^2}{2}$$

Figure 6
Computing moments relative to the coordinate axes

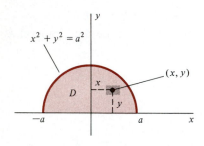

Hence the coordinates of the center of mass are

$$\bar{x} = \frac{M_y}{m} = \frac{0}{m} = 0 \quad \text{and} \quad \bar{y} = \frac{M_x}{m} = \frac{4a}{3\pi} \qquad \square$$

□ **Example 3**   Use double integration to find the area of the region bounded by the curve $y = 2 - x^2$ and the line $y = x$.

Solution: See Figure 7, in which we show a typical element of area as a small rectangle with area ΔA. The total area is $A \approx \sum \Delta A$ and the limit of this sum (as the partition is refined) is

$$A = \iint_D dA$$

Evaluating the double integral by an iterated integral, we find

$$A = \int_{-2}^{1} \int_{x}^{2-x^2} dy \, dx = \int_{-2}^{1} (2 - x^2 - x) \, dx = \left(2x - \frac{x^3}{3} - \frac{x^2}{2}\right)\Big|_{-2}^{1}$$

$$= \frac{x}{6}(12 - 2x^2 - 3x)\Big|_{-2}^{1}$$

$$= \frac{1}{6}(12 - 2 - 3) + \frac{2}{6}(12 - 8 + 6) = \frac{9}{2} \qquad \square$$

Figure 7
Finding area by a double integral

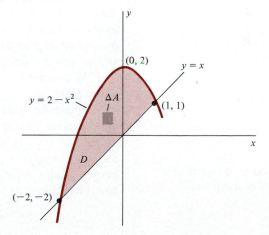

The formula

$$A = \iint_D dA$$

in Example 3 is quite general, giving the area of any (reasonable) region D. If D is covered by a thin material of density $\delta(x,y)$ at each point (x,y), we may use the same reasoning to obtain the formula

$$m = \iint_D \delta(x,y)\, dA$$

for mass. With the formulas

$$M_y = \iint_D x\, \delta(x,y)\, dA \qquad \text{and} \qquad M_x = \iint_D y\, \delta(x,y)\, dA$$

for moments relative to the coordinate axes (implicit in our discussion of Example 2), we have new versions of the formulas for center of mass given in Section 7.6. In particular (if the density is constant), the *centroid* of D is the point $(\overline{x}, \overline{y})$ whose coordinates are given by

$$\overline{x} = \frac{M_y}{m} = \frac{\displaystyle\iint_D x\, dA}{\displaystyle\iint_D dA}, \qquad \overline{y} = \frac{M_x}{m} = \frac{\displaystyle\iint_D y\, dA}{\displaystyle\iint_D dA}$$

The **moment of inertia** of a point mass m relative to an axis r units from the point is defined in physics to be $I = mr^2$. Sometimes I is called a *second moment,* in contrast to the *first moment* $M = mr$. It plays a role in rotational motion analogous to mass in linear motion. Higher moments (third, fourth, and so on) are important in statistics.

We can use double integrals to extend this idea to distributions of mass in the plane, as follows. Let D be a region in the xy plane that is covered by a thin material of density $\delta(x,y)$ at each point (x,y). To find the moment of inertia relative to the z axis, partition D into small pieces; let (x,y) be a point in the typical piece (of area ΔA and mass Δm). The distance from (x,y) to the z axis is $r = \sqrt{x^2 + y^2}$. (See Figure 8, in which the z axis is understood to be perpendicular to the plane of the paper.) Imagining the mass of the piece to be concentrated at (x,y), we find its moment of inertia to be

$$\Delta I_z \approx r^2\, \Delta m \approx (x^2 + y^2)\, \delta(x,y)\, \Delta A$$

Adding the contributions of each piece, and taking the limit as the partition is refined, we have

$$I_z = \iint_D (x^2 + y^2)\, \delta(x,y)\, dA$$

Moments of inertia about the x axis and y axis are similarly defined:

Figure 8
Finding the moment of inertia of a mass distribution in the plane

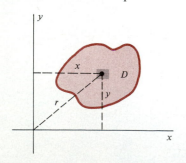

$$I_x = \iint_D y^2 \, \delta(x,y) \, dA$$

$$I_y = \iint_D x^2 \, \delta(x,y) \, dA$$

□ **Example 4**   Find the moment of inertia (relative to the y axis) of a thin material of constant density 1 covering the region in Example 3.

Solution: Since $\delta(x,y) = 1$, we find

$$I_y = \iint_D x^2 \, dA = \int_{-2}^{1} \int_{x}^{2-x^2} x^2 \, dy \, dx$$

The first integral is

$$\int_{x}^{2-x^2} x^2 \, dy = x^2 y \Big|_{y=x}^{y=2-x^2} = x^2(2 - x^2) - x^3 = 2x^2 - x^3 - x^4$$

Hence

$$I_y = \int_{-2}^{1} (2x^2 - x^3 - x^4) \, dx = \left(\frac{2x^3}{3} - \frac{x^4}{4} - \frac{x^5}{5} \right) \Big|_{-2}^{1}$$

$$= \frac{x^3}{60}(40 - 15x - 12x^2) \Big|_{-2}^{1}$$

$$= \frac{1}{60}(40 - 15 - 12) + \frac{8}{60}(40 + 30 - 48)$$

$$= \frac{63}{20}$$

□

□ **Example 5**   The density at each point of the square region

$$D = \{(x,y): \; -1 \le x \le 1, \; -1 \le y \le 1\}$$

is proportional to the square of its distance from the origin. Find the moment of inertia relative to the z axis.

Solution: We are told that $\delta(x,y) = k(x^2 + y^2)$, where k is a constant. Hence

$$I_z = \iint_D (x^2 + y^2) \cdot k(x^2 + y^2) \, dA = k \iint_D (x^2 + y^2)^2 \, dA$$

$$= k \int_{-1}^{1} \int_{-1}^{1} (x^4 + 2x^2 y^2 + y^4) \, dy \, dx$$

$$= 4k \int_{0}^{1} \int_{0}^{1} (x^4 + 2x^2 y^2 + y^4) \, dy \, dx \qquad \text{(Why?)}$$

We leave it to you to confirm that $I_z = \frac{112}{45} k$.

□

The following properties of double integrals are easy to believe on the basis of what we know about limits. We omit the proofs (which are somewhat technical and not very enlightening).

If f and g are continuous in D and k is a real number, then

$$\iint_D (f + g)\, dA = \iint_D f\, dA + \iint_D g\, dA$$

$$\iint_D kf\, dA = k \iint_D f\, dA$$

Our last theorem about double integrals is analogous to the formula

$$\int_a^b f = \int_a^c f + \int_c^b f$$

in single-variable calculus. It is geometrically obvious (but nontrivial to prove).

Suppose that f is continuous in D, where $D = D_1 \cup D_2$ and D_1 and D_2 have no interior points in common. In other words D_1 and D_2 are nonoverlapping regions whose union is D. Then

$$\iint_D f\, dA = \iint_{D_1} f\, dA + \iint_{D_2} f\, dA$$

Problem Set 18.2

Use double integration to find the area of the region bounded by the given curves.

1. $y = x^2 - 2x$ and the x axis
2. $y = x^3 - 2x$ and the x axis
3. $x = y^2 - 4y$ and the y axis
4. $y^2 = 4x$ and $y = 2x - 4$
5. $y = x^3 - 3x$ and $y = 2x^2$
6. $y = 1/x^2$, $y = x$, $y = 0$, and $x = 2$
7. $y = \sqrt{x}$, $y = 1/\sqrt{x}$, $y = 0$, and $x = 2$
8. $y^2 = x$ and $x - y = 2$
9. $y = x^3$ and its tangent line at $(1,1)$

10. $y^2 = x$, $(y - 2)^2 = x$, and the y axis
11. $y = 1/\sqrt{x + 1}$, $y = x + 1$, and $y = 2$
12. $y = \sqrt{x}$ and $y = x^3$
13. $y = x/\sqrt{9 - x^2}$, $x = \pm 2$, and $y = 0$
14. $y = x^2 - 2x$ and $y = 2x - x^2$
15. $\sqrt{x} + \sqrt{y} = 1$ and the coordinate axes
16. $|x| + |y| = 1$
17. $y = |x|$ and $x = 2 - y^2$
18. $y = x/\sqrt{4 - x^2}$ and $y = x$
19. $x^4 - 9x^2 + y^2 = 0$
20. $y^2 = x^2(3 - x)$

21. $y^2 = x + 2$, $y^3 = x$, and the x axis   *Hint: $y^3 - y^2 + 2 = (y + 1)(y^2 - 2y + 2)$.* The region is in the third quadrant.

22. $y = \cos x$, $y = 1$, $x = 0$, and $x = 2\pi$

23. $y = \sin x$, $y = \cos x$, $x = 0$, and $x = \pi/4$

24. $y = \sin 2x$ and $y = \cos x$ (between consecutive points of intersection on the x axis)

25. $y = \cos 2x$ and $y = \sin x$ (between consecutive points of intersection on opposite sides of the x axis)

Use double integration to find the area of each of the following regions.

26. Bounded by the parabola $x = 4 - y^2$ and the y axis.

27. Bounded by the parabolas $y = 4 - x^2$ and $4y = 4 - x^2$.

28. Bounded by the line $x + y = 5$ and the hyperbola $xy = 4$.

29. The smaller of the two regions bounded by the parabola $y^2 = 2x$ and the circle $x^2 + y^2 = 8$.

Use double integration to find the centroid of each of the following regions.

30. The region of Problem 26.

31. The region of Problem 27.

32. The region of Problem 28.

33. The region of Problem 29.

34. The first quadrant portion of the disk $x^2 + y^2 \leq a^2$.

35. Bounded by the parabolas $y = x^2 - 2x$ and $y = 6x - 3x^2$.

36. Bounded by the x axis and the curve $y = \sin x$, $0 \leq x \leq \pi$.

37. Bounded by the x axis and the curve $y = \cos x$, $0 \leq x \leq \pi/2$.

Each of the following regions is covered by a thin material of constant density 1. Find the moment of inertia relative to the given axis.

38. The region of Problem 26; the x axis.

39. The region of Problem 27; the y axis.

40. The region of Problem 28; the z axis.

41. The region of Problem 29; the x axis.

42. The region of Problem 36; the x axis.

43. The density at each point of a square region of side a is proportional to the distance from one side. Find the mass.

44. In Problem 43 find the center of mass.

45. In Problem 43 find the moment of inertia relative to the side from which distance is measured.

46. The density at each point of a semicircular region of radius a is proportional to the distance from the diameter. Find the mass.

47. In Problem 46 find the center of mass.

48. In Problem 46 find the moment of inertia relative to the diameter.

49. A right triangle of base b and altitude h is covered by a thin material of constant density. Show that the moment of inertia relative to the base is $\frac{1}{6}mh^2$, where m is the mass.

50. Suppose that $f(x,y) = x^2 + y^2$ and
$$D = \{(x,y): -1 \leq x \leq 1, -1 \leq y \leq 1\}$$

(a) Using a partition of D into four squares of equal area (and evaluating f at the midpoint of each square), compute a Riemann sum that approximates the double integral of f over D.

(b) With the same partition as in part (a), what is the smallest Riemann sum that can be computed? the largest?

(c) Find the value of $\iint_D f(x,y)\, dA$.

51. Repeat Problem 50 for the function $f(x,y) = xy$ and the region
$$D = \{(x,y): 0 \leq x \leq 4, 0 \leq y \leq 4\}$$

52. Suppose that $f(x,y)$ is defined in a square region D, partitioned into n squares of equal area ΔA, and let P_k be a point of the kth square.

(a) What is the average value of the numbers $f(P_1), f(P_2), \ldots, f(P_n)$?

(b) Show that the average value in part (a) can be written in the form
$$\frac{1}{A} \sum_{k=1}^{n} f(P_k)\, \Delta A, \qquad \text{where } A \text{ is the area of } D$$

(c) What is the limit of the expression in part (b) as the norm of the partition approaches zero?

53. The preceding problem suggests that in general we should define the *average value* of a function $f(x,y)$ in a region D (not necessarily square) to be
$$\frac{1}{A} \iint_D f(x,y)\, dA \qquad \text{where } A \text{ is the area of } D$$

Explain how this definition is analogous to the formula
$$\frac{1}{b - a} \int_a^b f(x)\, dx$$
for the average value of a function of one variable in the interval $[a,b]$. (See Section 6.3.)

54. Show that the average value of x in the region D is the coordinate $\bar{x}$ of the centroid of D. What is the average value of y in D?

55. Find the average value of xy in the region

$$D = \{(x,y): x^2 + y^2 \leq 1, x \geq 0, y \geq 0\}$$

56. Show that the average value of e^{x+y} in the square region

$$D = \{(x,y): -1 \leq x \leq 1, -1 \leq y \leq 1\}$$

is $\sinh^2 1$.

57. Suppose that $f(x,y)$ has continuous second partial derivatives in the rectangular region

$$D = \{(x,y): a \leq x \leq b, c \leq y \leq d\}$$

Show that

$$\iint_D f_{12}(x,y) \, dA = f(P_1) - f(P_2) + f(P_3) - f(P_4)$$

where P_1, P_2, P_3, P_4 are the vertices of D numbered counterclockwise from (a,c).

18.3

DOUBLE INTEGRALS IN POLAR COORDINATES

Many double integrals that are hard to evaluate in terms of rectangular coordinates can be simplified by converting to polar coordinates. Suppose, for example, that we want to find the volume of the region bounded by the paraboloid $z = 1 - x^2 - y^2$ and the xy plane. This volume is four times its portion in the first octant (Figure 1), so we can write

$$V = 4 \iint_D (1 - x^2 - y^2) \, dA = 4 \int_0^1 \int_0^{\sqrt{1-x^2}} (1 - x^2 - y^2) \, dy \, dx$$

If you solved Problem 41 in Section 18.1, you know that evaluation of the iterated integral is not trivial. The situation seems tailor-made for polar coordinates, since the integrand becomes $1 - r^2$ upon substitution of $x = r \cos\theta$ and $y = r \sin\theta$, and the region

$$D = \{(x,y): 0 \leq x \leq 1, 0 \leq y \leq \sqrt{1 - x^2}\}$$

is more easily described in polar coordinates by

$$D = \{(r,\theta): 0 \leq r \leq 1, 0 \leq \theta \leq \pi/2\}$$

It is not unreasonable to expect that an iterated integral for V in terms of polar coordinates will read

$$V = 4 \int_0^{\pi/2} \int_0^1 (1 - r^2)(?) \, dr \, d\theta$$

The question mark is inserted because there is some doubt about the conversion of the element of area $\Delta A = \Delta x \, \Delta y$ (in rectangular coordinates) to polar coordinates. One's first guess might be that $\Delta A = \Delta r \, \Delta\theta$ but that is incorrect, as we will see. Before discussing why, we observe that the limits on the iterated integral are found exactly as in the case of rectangular coordinates. The variable of integration in the first integral is r, so we fix θ between 0 and $\pi/2$ and imagine r running from 0 to 1. (See Figure 2.) This radial line segment sweeps out D when θ runs from 0 to $\pi/2$.

Figure 1
First octant portion of the region below a paraboloid

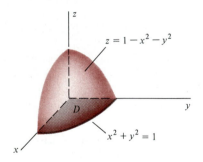

Figure 2
For fixed θ, r runs from 0 to 1

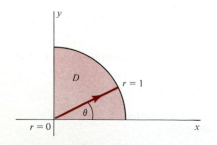

Figure 3
Polar coordinate partition of a region D

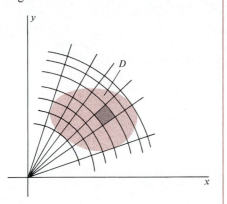

Figure 4
"Polar rectangle" with area $\Delta A \approx r \, \Delta r \, \Delta \theta$

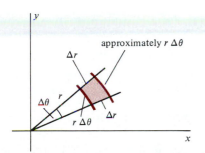

What about the element of area? Figure 3 shows a partition of a region D by radial lines (fixed θ) and circles (fixed r), while Figure 4 shows the typical "polar rectangle." As indicated in the figure, the sides of this "rectangle" are Δr and $r \, \Delta \theta$ (the latter coming from the formula for arc length, $s = r\theta$, on a circle of radius r and central angle θ). The inner and outer arcs are not the same, of course (only the inner one being exactly $r \, \Delta \theta$), but if the partition is "fine," the error in calling the outer one $r \, \Delta \theta$ is negligible. If the polar rectangle were a true rectangle, its area would be the product of these sides; as it is, the approximation $\Delta A \approx r \, \Delta r \, \Delta \theta$ is reasonable (and improves as the partition is refined).

It therefore appears that the question mark in our formula for V should be filled in by r, that is, dA becomes $r \, dr \, d\theta$ when a double integral in rectangular coordinates is transformed by changing to polar coordinates:

$$\iint_D f(x, y) \, dA = \iint_D f(r \cos \theta, r \sin \theta) r \, dr \, d\theta$$

Note that in the geometrical discussion leading to this result, we assume that $r > 0$. It is important to avoid using polar coordinates in such a way as to allow r to be negative (unless the factor r in $dA = r \, dr \, d\theta$ is replaced by $|r|$).

Remark   In more formal treatments of the subject, the transformation

$$T: \Re^2 \rightarrow \Re^2 \text{ defined by } T(r, \theta) = (x, y) = (r \cos \theta, r \sin \theta)$$

is regarded as a "mapping" from the $r\theta$ plane to the xy plane, carrying a region partitioned in terms of r and θ into a different region partitioned in terms of x and y. In that case the same symbol D should not be used on both sides of the equation. We mention it because of what you may encounter in other books; our intuitive approach fudges such details.

It is also worth mentioning that in general a transformation from $\Re^2$ to $\Re^2$ multiplies dA by $|J|$, where J is the *Jacobian* of the transformation. (We have mentioned Jacobians in Problems 34 and 35, Sec. 16.5, and in Additional Problems 33–36 at the end of Chapter 17.) In the polar coordinate transformation we have

$$J = \begin{vmatrix} \partial x/\partial r & \partial x/\partial \theta \\ \partial y/\partial r & \partial y/\partial \theta \end{vmatrix} = \begin{vmatrix} \cos \theta & -r \sin \theta \\ \sin \theta & r \cos \theta \end{vmatrix} = r \cos^2 \theta + r \sin^2 \theta = r$$

Hence the correct factor multiplying dA is $|r|$, as noted above.

□ **Example 1**   Find the volume of the region bounded by the paraboloid $z = 1 - x^2 - y^2$ and the xy plane.

Solution: If we are going to change to polar coordinates, there is little point in restricting our attention to the volume in the first octant and multiplying by 4 (as at the beginning of this section). It is easier to let D be the whole unit disk in the xy plane, that is,

$$D = \{(x, y): x^2 + y^2 \leq 1\}$$

Then the volume is

$$V = \iint_D (1 - x^2 - y^2)\, dA$$

Since

$$D = \{(r,\theta): 0 \le r \le 1, 0 \le \theta \le 2\pi\}$$

in polar coordinates, we have

$$V = \int_0^{2\pi} \int_0^1 (1 - r^2) r\, dr\, d\theta$$

The inside integral is

$$\int_0^1 (r - r^3)\, dr = \left(\frac{r^2}{2} - \frac{r^4}{4}\right)\Big|_0^1 = \frac{1}{4}$$

and hence

$$V = \int_0^{2\pi} \frac{1}{4}\, d\theta = \frac{\pi}{2}$$

Compare the simplicity of this calculation with Problem 41, Section 18.1. □

Figure 5
Polar coordinate region

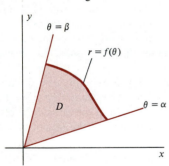

□ **Example 2**   Use double integration to derive the formula

$$A = \frac{1}{2} \int_\alpha^\beta f(\theta)^2\, d\theta$$

for the area of the region D shown in Figure 5. (See Section 14.2.)

Solution: As noted in Section 18.2, the area is $A = \iint_D dA$. Evaluating this double integral by an iterated integral in polar coordinates, we write

$$A = \int_\alpha^\beta \int_0^{f(\theta)} r\, dr\, d\theta$$

The inside integral is

$$\int_0^{f(\theta)} r\, dr = \frac{r^2}{2}\Big|_0^{f(\theta)} = \frac{1}{2} f(\theta)^2$$

and hence

$$A = \frac{1}{2} \int_\alpha^\beta f(\theta)^2\, d\theta$$ □

□ **Example 3**   Find the centroid of the semicircular region

$$D = \{(x,y): x^2 + y^2 \le a^2, y \ge 0\}$$

(See Figure 6, Section 18.2.)

Solution: The area of D (by geometry) is $A = \frac{1}{2}\pi a^2$. Since $\bar{x} = 0$ (by symmetry), we need only calculate

$$M_x = \iint_D y\, dA = \int_0^\pi \int_0^a r \sin\theta \cdot r\, dr\, d\theta = \frac{a^3}{3} \int_0^\pi \sin\theta\, d\theta = \frac{2a^3}{3}$$

Hence $\bar{y} = M_x/A = 4a/3\pi$. (Compare this calculation with Example 2, Section 18.2.)   □

□ **Example 4**   The density at each point of a circular disk of radius a is proportional to the distance from the center of the disk. Find the moment of inertia relative to the line through the center perpendicular to the disk.

Solution: Place the disk D with its center at the origin of the xy plane. We are told that the density at (x,y) is $\delta(x,y) = kr$, where r is the first polar coordinate of (x,y) and k is a constant. Hence

$$I_z = \iint_D r^2 \cdot kr\, dA = k \int_0^{2\pi} \int_0^a r^3 \cdot r\, dr\, d\theta = \frac{ka^5}{5} \int_0^{2\pi} d\theta = \frac{2\pi ka^5}{5}$$

While this answer is adequate, it is useful in physics to express the moment of inertia in terms of the mass. Since

$$m = \iint_D kr\, dA = k \int_0^{2\pi} \int_0^a r \cdot r\, dr\, d\theta = \frac{2\pi ka^3}{3}$$

we find $I_z = \frac{3}{5}ma^2$.   □

□ **Example 5**   Find the centroid of the region D inside the cardioid

$$r = a(1 + \cos\theta)$$

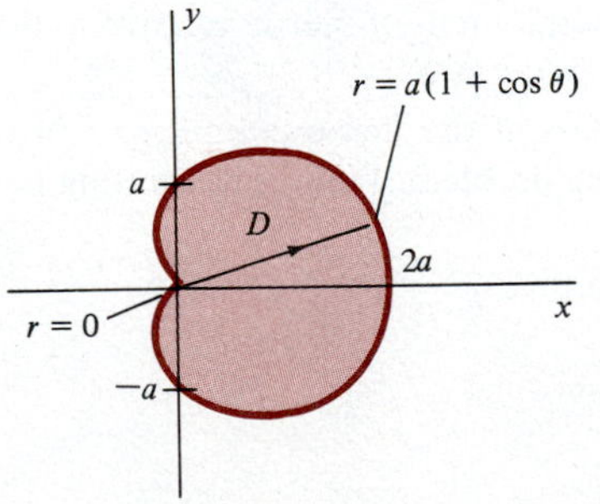

Solution: See Figure 6, from which it is clear that $\bar{y} = 0$. To find $\bar{x}$, we compute

$$M_y = \iint_D x\, dA = \int_0^{2\pi} \int_0^{a(1+\cos\theta)} r\cos\theta \cdot r\, dr\, d\theta$$

$$= \frac{a^3}{3} \int_0^{2\pi} (1 + \cos\theta)^3 \cos\theta\, d\theta$$

$$= \frac{a^3}{3} \int_0^{2\pi} (\cos\theta + 3\cos^2\theta + 3\cos^3\theta + \cos^4\theta)\, d\theta$$

Using the reduction formula

$$\int \cos^n x\, dx = \frac{\cos^{n-1} x \sin x}{n} + \frac{n-1}{n} \int \cos^{n-2} x\, dx$$

we find $M_y = \frac{5}{4}\pi a^3$. The area of D is

$$A = \iint_D dA = \int_0^{2\pi} \int_0^{a(1+\cos\theta)} r\, dr\, d\theta = \frac{a^2}{2} \int_0^{2\pi} (1 + \cos\theta)^2\, d\theta = \frac{3\pi a^2}{2}$$

Hence $\bar{x} = M_y/A = \frac{5}{6}a$.   □

Figure 6
Region inside cardioid
$$r = a(1 + \cos\theta)$$

Problem Set 18.3

Use double integration in polar coordinates to find the area of the region bounded by the given polar curves.

1. $r = 2(1 + \cos \theta)$ **2.** $r = 3(1 + \sin \theta)$

3. $r = a(1 - \cos \theta)$

4. $r = 2 \sin \theta$ (Check by geometry.)

5. $r = 2a \cos \theta$ (Check by geometry.)

6. $r = \sin 2\theta$ **7.** $r = \cos 2\theta$

8. $r = \sin 3\theta$ **9.** $r^2 = \cos 2\theta$

10. $r^2 \cos^4 \theta - 2 \cos^2 \theta + 1 = 0$

11. The inside loop of $r = 1 - 2 \cos \theta$

12. The inside loop of $r = 1 + 2 \cos \theta$

13. $r = 2/(1 - \cos \theta)$ and $\theta = \pi/2$

Use double integration in polar coordinates to solve the following problems.

14. Find the area inside the circle $x^2 + y^2 = a^2$.

15. Find the area inside the circle $r = 2a \cos \theta$.

16. Find the area inside the circle $x^2 + y^2 = 4$ and to the right of the line $x = 1$.

17. Find the area inside the cardioid $r = a(1 + \sin \theta)$.

18. Find the area inside both of the curves $r = 2a \cos \theta$ and $r = 2a \sin \theta$.

19. Find the area inside both of the curves $r = 2 \sin \theta$ and $r = 2(1 - \sin \theta)$.

20. Find the area inside both of the curves $r = 2(1 + \cos \theta)$ and $r = 3$.

21. Find the area of the smaller of the two regions bounded by the curves $r = 4 \cos \theta$ and $r = 3 \sec \theta$.

22. Find the area outside the circle $r = 2 \sin \theta$ and inside the cardioid $r = 1 + \sin \theta$.

23. Find the area outside the cardioid $r = 2(1 + \cos \theta)$ and inside the circle $r = 6 \cos \theta$.

Use double integration in polar coordinates to find the volume of each of the following regions.

24. Bounded by the paraboloid $z = 4 - x^2 - y^2$ and the xy plane.

25. Inside the cylinder $x^2 + y^2 = 4$, below the paraboloid $z = 9 - x^2 - y^2$, and above the xy plane.

26. Inside a sphere of radius a.

27. Inside the ellipsoid $x^2 + y^2 + z^2/4 = 1$.

28. Bounded by the hyperboloid $z^2 - x^2 - y^2 = 1$ and the plane $z = 2$.

29. Inside the cylinder $x^2 + y^2 = a^2$, below the plane $z = y$, and above the xy plane.

30. Inside the cylinder $r = 2a \sin \theta$ and between the planes $z = 0$ and $z = h$.

31. The region removed from a bowling ball of radius $\sqrt{5}$ by drilling a hole of radius 1 through the center.

32. The "ice cream cone" (above the xy plane) bounded by the sphere $x^2 + y^2 + z^2 = 1$ and the cone $z^2 = x^2 + y^2$.

33. Above the xy plane, bounded by the cylinder $r = 2 \sin \theta$ and the sphere $x^2 + y^2 + z^2 = 4$.

34. Inside the cone $a^2(z - h)^2 = h^2(x^2 + y^2)$ and between the planes $z = 0$ and $z = h$. *Hint:* Be careful about the sign of the square root involved in solving for z.

35. The density at each point of the region inside the circle $r = 2 \cos \theta$ is proportional to the distance from the origin. Find the mass.

36. The density at each point of a semicircular disk of radius a is proportional to the distance from the midpoint of the diameter. Find the mass.

37. In Problem 36 find the center of mass.

38. In Problem 36 find the moment of inertia relative to the diameter (in terms of the mass).

39. In Problem 36 find the moment of inertia relative to a line through the midpoint of the diameter perpendicular to the disk (in terms of the mass).

40. The density at each point of the region between the circles $r = a$ and $r = b$ $(0 < a < b)$ is inversely proportional to the distance from the origin. Find the mass.

41. In Problem 40 show that the mass remains finite as $a \to 0$ (even though the density is unbounded in a neighborhood of the origin).

42. In Problem 40 find the moment of inertia relative to the z axis. Does it remain finite as $a \to 0$?

43. Find the center of mass of the first quadrant part of the region in the preceding problem. What is its limiting position as $a \to 0$?

44. Show that if D is the disk $x^2 + y^2 \le a^2$, then

$$\iint_D e^{-(x^2+y^2)} \, dA = \pi(1 - e^{-a^2})$$

(This is a remarkable result, because the single-variable problem $\int_0^a e^{-x^2} \, dx$ cannot be computed in terms of elementary functions.)

45. Let D be the square region $0 \le x \le a$, $0 \le y \le a$.

(a) Confirm that the region in the first quadrant bounded by the coordinate axes and the circle $x^2 + y^2 = a^2$ is inside D, while the similar region bounded by the axes and $x^2 + y^2 = 2a^2$ contains D. Use that fact (together with Problem 44) to explain why

$$\frac{\pi}{4}(1 - e^{-a^2}) < \iint_D e^{-(x^2+y^2)}\, dA < \frac{\pi}{4}(1 - e^{-2a^2})$$

(b) It can be shown (perhaps you can show it) that

$$\int_0^a \int_0^a e^{-(x^2+y^2)}\, dx\, dy = \left(\int_0^a e^{-x^2}\, dx\right)\left(\int_0^a e^{-y^2}\, dy\right)$$

Use that fact together with part (a) to deduce that

$$\int_0^\infty e^{-x^2}\, dx = \frac{\sqrt{\pi}}{2}$$

(This result is important in probability and statistics. As noted in Problem 44, it is remarkable that we can compute it at all. Also see the quotations concerning it at the beginning of this chapter.)

18.4
SURFACE AREA

In this section we develop a formula for calculating the area of a surface $z = f(x, y)$ lying above a domain D in the xy plane. As an example, suppose that we want the surface area of the sphere $x^2 + y^2 + z^2 = a^2$. The upper hemisphere is represented by the function

$$z = f(x, y) = \sqrt{a^2 - x^2 - y^2}$$

and lies above the disk

$$D = \{(x, y): x^2 + y^2 \leq a^2\}$$

in the xy plane, as shown in Figure 1. We have often used the formula $S = 4\pi a^2$ for the surface area of a sphere of radius a, so in this case we know the answer. The question is *how* do we know it, and how do we find surface area in less simple cases?

We proceed as usual by partitioning the domain D into small rectangles. Let ΔA be the area of the typical rectangle and let ΔS be the area of the portion of the surface $z = f(x, y)$ lying above this rectangle. Choose a point (x, y) in the rectangle and let $P = (x, y, z)$ be the point of the surface above it. The crux of our argument is that ΔS is closely approximated by the area ΔT of the piece of the tangent plane at P lying above the typical rectangle in D. (See Figure 2.)

Figure 1
Hemisphere above the disk D

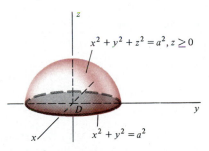

Figure 2
Approximating surface area by the tangent plane

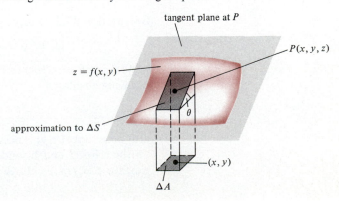

Let θ be the (acute) angle between the tangent plane and the xy plane, as shown in the figure. Then

$$\Delta A = (\cos \theta)\, \Delta T \qquad \text{(Why?)}$$

Hence our approximation for the area of the typical piece of surface is

$$\Delta S \approx \Delta T = (\sec \theta)\, \Delta A$$

Summing over the partition of D, we have

$$S = \sum \Delta S \approx \sum (\sec \theta)\, \Delta A$$

and (taking the limit as the partition is refined) $S = \iint_D \sec \theta\, dA$.

□ **Example 1**   Find the area of the first octant portion of the plane $2x + 2y + z - 2 = 0$.

Solution: The surface whose area we seek is shown in Figure 3, together with the region D in the xy plane below it. Since θ is constant, we have

$$S = (\sec \theta) \iint_D dA = (\sec \theta)(\text{area of } D) = \frac{1}{2} \sec \theta$$

Hence the problem reduces to finding $\sec \theta$. The angle between the given plane and the xy plane is the same as the angle between their normal vectors, namely $\mathbf{n} = (2, 2, 1)$ and $\mathbf{k} = (0, 0, 1)$. We compute

$$\cos \theta = \frac{\mathbf{n} \cdot \mathbf{k}}{|\mathbf{n}||\mathbf{k}|} = \frac{1}{3}$$

and find $S = \frac{1}{2}\sec \theta = \frac{3}{2}$.   □

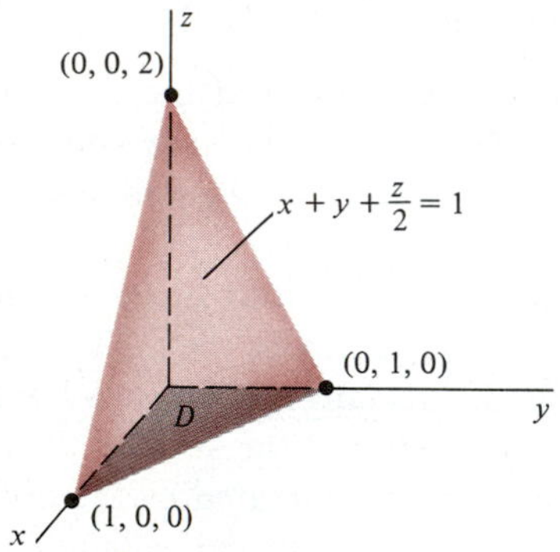

In general θ is not constant, because the surface $z = f(x, y)$ is curved unless it is a plane. The tangent plane at $P(x, y, z)$, however, has normal vector

$$\mathbf{n} = (-a, -b, 1) \qquad \text{where } a = f_1(x, y) \text{ and } b = f_2(x, y)$$

(See Section 17.3.) Since

$$|\mathbf{n}| = \sqrt{a^2 + b^2 + 1},\ |\mathbf{k}| = 1, \text{ and } \mathbf{n} \cdot \mathbf{k} = 1$$

we find $\sec \theta = \sqrt{(\partial z/\partial x)^2 + (\partial z/\partial y)^2 + 1}$. Thus we have arrived at the following result.

Figure 3
First octant portion of the plane $2x + 2y + z - 2 = 0$

Suppose that $f(x, y)$ is nonnegative and has continuous partial derivatives in the (closed and bounded) region D in the xy plane. The area of the surface $z = f(x, y)$ above D is

$$S = \iint_D \sqrt{(\partial z/\partial x)^2 + (\partial z/\partial y)^2 + 1}\ dA$$

The formula also applies if $f(x,y) \leq 0$ in D, that is, if the surface is below D. (Why?)

It sounds as though we have proved a theorem. In fact, however, surface area needs defining; for our purposes this formula supplies the definition. The discussion preceding it was deliberately kept free of the paraphernalia of precision (subscripts, indices of summation, and so on), because we were trying to motivate a definition, not prove a theorem. More advanced treatments of the subject start with a general definition of surface area that does not depend on the functional form $z = f(x,y)$. Our formula is then a special case that requires proof.

□ **Example 2**   Show that the surface area of a sphere of radius a is $S = 4\pi a^2$.

Solution: As at the beginning of this section, we consider the hemisphere $z = \sqrt{a^2 - x^2 - y^2}$ lying above the disk D shown in Figure 1. Since $\partial z/\partial x = -x/z$ and $\partial z/\partial y = -y/z$, we have

$$\left(\frac{\partial z}{\partial x}\right)^2 + \left(\frac{\partial z}{\partial y}\right)^2 + 1 = \frac{x^2}{z^2} + \frac{y^2}{z^2} + 1 = \frac{a^2}{z^2} \quad \text{(because } x^2 + y^2 + z^2 = a^2\text{)}$$

Hence the area of the sphere (twice the area of the hemisphere) is

$$S = 2\iint_D \frac{a}{z}\ dA = 2a \int_0^{2\pi} \int_0^a \frac{r}{\sqrt{a^2 - r^2}}\ dr\ d\theta \quad \text{(polar coordinates)}$$

The inside integral is improper (why?); we leave it to you to show that it nevertheless exists and has the value

$$\int_0^a \frac{r\ dr}{\sqrt{a^2 - r^2}} = a$$

Hence $$S = 2a^2 \int_0^{2\pi} d\theta = 4\pi a^2 \qquad \qquad \square$$

□ **Example 3**   Find the area of the surface cut from the cone $z = 1 - r$ by the cylinder $r = \sin\theta$.

Solution: The first octant portion of the surface is shown in Figure 4; the other half is behind the yz plane. Since $z = 1 - \sqrt{x^2 + y^2}$, we have $\partial z/\partial x = -x/r$ and $\partial z/\partial y = -y/r$, from which

$$\left(\frac{\partial z}{\partial x}\right)^2 + \left(\frac{\partial z}{\partial y}\right)^2 + 1 = \frac{x^2}{r^2} + \frac{y^2}{r^2} + 1 = 2$$

Figure 4
Surface cut from a cone by a cylinder

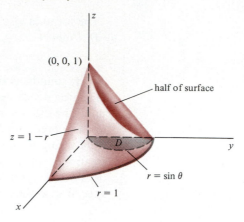

Taking D to be the first quadrant region inside the circle $r = \sin\theta$, we find

$$S = 2 \iint_D \sqrt{2}\, dA = 2\sqrt{2} \int_0^{\pi/2} \int_0^{\sin\theta} r\, dr\, d\theta = \sqrt{2} \int_0^{\pi/2} \sin^2\theta\, d\theta = \frac{\pi\sqrt{2}}{4}$$

$\square$

Figure 5
Curve to be rotated about x axis

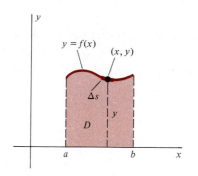

We end this section by returning to a formula for the area of a surface of revolution (derived in Section 7.4). Suppose that $f(x) \geq 0$ in the interval $[a,b]$, as shown in Figure 5. If the curve $y = f(x)$, $a \leq x \leq b$, is rotated about the x axis, it generates a surface (of a solid of revolution). For example, the semicircle $y = \sqrt{a^2 - x^2}$ generates the sphere $x^2 + y^2 + z^2 = a^2$ when it is rotated about the x axis. The simplest way to find the area of the surface is by the methods of single-variable calculus. Let Δs be the length of a small piece of the curve and let (x,y) be a point in this piece (as shown in Figure 5). When the piece rotates about the x axis it generates a strip of the surface of revolution, the area of the strip being

$$\Delta S \approx 2\pi y\, \Delta s \qquad \text{(Why?)}$$

Add up the strips and take the limit as the pieces get smaller to obtain

$$S = 2\pi \int_{x=a}^{x=b} y\, ds$$

Since $ds = \sqrt{1 + y'^2}\, dx$ (Section 7.3), we find

$$S = 2\pi \int_a^b y\sqrt{1 + y'^2}\, dx$$

It is interesting to see how our general formula in this section applies to the same problem. An equation of the surface of revolution is $y^2 + z^2 = f(x)^2$, because each point (x, y, z) of the surface is the same distance from the x axis as the point $(x, f(x), 0)$ in the xy plane. (Why?) To find $\partial z/\partial x$ and

$\partial z / \partial y$, differentiate implicitly:

$$2z \frac{\partial z}{\partial x} = 2f(x)f'(x) \qquad \text{and} \qquad 2y + 2z \frac{\partial z}{\partial y} = 0$$

Hence $\partial z / \partial x = f(x)f'(x)/z$ and $\partial z / \partial y = -y/z$, and we find

$$\left(\frac{\partial z}{\partial x}\right)^2 + \left(\frac{\partial z}{\partial y}\right)^2 + 1 = \frac{f(x)^2}{z^2}[1 + f'(x)^2] \qquad \text{(Confirm!)}$$

The part of the surface above the xy plane ($z \geq 0$) is half the total; half of this part lies above the region D shown in Figure 5. Hence

$$S = 4 \iint_D \frac{f(x)}{z} \sqrt{1 + f'(x)^2}\, dA = 4 \int_a^b \int_0^{f(x)} \frac{f(x)\sqrt{1 + f'(x)^2}}{\sqrt{f(x)^2 - y^2}}\, dy\, dx$$

To evaluate the inner integral (in which x is fixed), we write

$$\int_0^{f(x)} \frac{dy}{\sqrt{f(x)^2 - y^2}} = \sin^{-1} \frac{y}{f(x)}\Bigg|_{y=0}^{y=f(x)} = \frac{\pi}{2}$$

(The integral is improper. We bypassed formal evaluation of the limit of a proper integral by going directly to the inverse sine.) Hence

$$\boxed{S = 2\pi \int_a^b f(x)\sqrt{1 + f'(x)^2}\, dx}$$

□ **Example 4**   Find the surface area of a sphere of radius a.

Solution: The sphere is generated by rotating the semicircle $x^2 + y^2 = a^2$, $y \geq 0$, about the x axis. Differentiating implicitly to find y', we have $2x + 2yy' = 0$ or $y' = -x/y$. Hence

$$y\sqrt{1 + y'^2} = y\sqrt{1 + x^2/y^2} = \sqrt{x^2 + y^2} = a$$

and $$S = 2\pi \int_{-a}^a a\, dx = 4\pi a^2$$

As you can see, this is easier than the double integration in Example 2. Its chief advantage, however, is that we need not work directly with the surface if it is generated by rotation; the equation $y = f(x)$ is sufficient.

□

Problem Set 18.4

Find the area of each of the following surfaces by double integration.

1. Cut from the plane $z = 2$ by the cylinder $x^2 + y^2 = 1$.

2. Cut from the plane $z = 3$ by the cylinder $y = 1 - x^2$.

3. Cut from the plane $z = y$ by the cylinder $x^2 + y^2 = 9$.

4. Cut from the plane $2x + 3y + z - 6 = 0$ by the cylinder $r = \cos \theta$.

5. Cut from the plane $x + y + 2z - 2 = 0$ by the cylinder $x^2 + y^2 = 1$.

6. The part of the plane $x/2 + y + z/3 = 1$ in the first octant.

7. The part of the plane $4x - 2y + z = 4$ above the region in the xy plane bounded by the lines $y = x$, $y = 0$, and $x = 1$.

8. The part of the paraboloid $z = 4 - x^2 - y^2$ above the xy plane.

9. The part of the paraboloid $z = x^2 + y^2$ below the plane $z = 1$.

10. Cut from the paraboloid $z = 2 - x^2 - y^2$ by the cylinder $x^2 + y^2 = 1$.

11. The part of the parabolic cylinder $z = 1 - y^2$ above the xy plane and between the planes $x = 0$ and $x = 2$.

12. The part of the parabolic cylinder $z = x^2$ above the square $0 \le x \le 1$, $0 \le y \le 1$ in the xy plane.

13. The part of the cylinder $x^2 + z^2 = 4$ above the xy plane and between the planes $y = 0$ and $y = 2$.

14. Cut from the cone $z = 1 - r$ by the cylinder $r = \cos \theta$.

15. The cap of the sphere $x^2 + y^2 + z^2 = 25$ above the plane $z = 3$.

16. Cut from the hemisphere $x^2 + y^2 + z^2 = 2$, $z \ge 0$, by the cone $z^2 = x^2 + y^2$.

17. Cut from the hemisphere $x^2 + y^2 + z^2 = 4$, $z \ge 0$, by the cylinder $r = 2 \cos \theta$.

18. Cut from the hemisphere $x^2 + y^2 + z^2 = 1$, $z \ge 0$, by the cylinder $r = \sin \theta$.

19. The part of the sphere $x^2 + y^2 + z^2 - 2z = 0$ inside the paraboloid $z = x^2 + y^2$.

20. The lateral surface area of a right circular cone of base radius a and altitude h is given in geometry by the formula $S = \pi a s$, where $s = \sqrt{a^2 + h^2}$ is the "slant height". Derive this formula by double integration. *Hint:* One way to place the cone in a coordinate system is to use the equation $az = h(a - r)$. (Confirm!)

21. Show that the portion of the surface $z = x^2 + y^2$ lying above the region D in the xy plane has the same area as the portion of the surface $z = 2xy$ lying above D. (Assume that D is in the first quadrant.)

22. Show that the area of the surface cut from the (nonverti-cal) plane $ax + by + cz + d = 0$ by the vertical cylinder with base D in the xy plane is

$$S = \frac{\sqrt{a^2 + b^2 + c^2}}{|c|} \times (\text{area of } D)$$

Why doesn't the constant term d enter into this formula?

23. Show that the area of the first octant portion of the plane $x/a + y/b + z/c = 1$ (where a, b, c are positive) is

$$S = \tfrac{1}{2} \sqrt{a^2 b^2 + b^2 c^2 + c^2 a^2}$$

Use the formula for the area of a surface of revolution to find the area of each of the following surfaces.

24. The right circular cylinder of radius a and altitude h.

25. The right circular cone of base radius a and altitude h.

26. Generated by rotating the curve $y^2 = x$, $0 \le x \le 2$, about the x axis.

27. Generated by rotating the curve $y = \cosh x$, $0 \le x \le 1$, about the x axis.

28. Generated by rotating the curve $y = x^2$, $0 \le x \le 1$, about the y axis. *Hint:* The correct formula for revolution about the y axis is

$$S = 2\pi \int x \, ds \quad (\text{Why?})$$

29. Generated by rotating the curve $y = \ln x$, $1 \le x \le 2$, about the y axis.

30. Generated by rotating the hypocycloid $x = a \cos^3 t$, $y = a \sin^3 t$ about the x axis. *Hint:* Return to the formula

$$S = 2\pi \int y \, ds$$

and express everything in terms of the parameter.

31. Generated by rotating one arch of the cycloid

$$x = a(t - \sin t), \; y = a(1 - \cos t)$$

about the x axis.

18.5
TRIPLE INTEGRALS IN RECTANGULAR AND CYLINDRICAL COORDINATES

One of the applications of calculus to physics is the computation of center of mass and moment of inertia of a distribution of mass. You may have noticed that we have confined our discussion of the subject (in Section 7.6 and so far in this chapter) to distributions along a straight line (or curve) or in a region of the plane. The reason for this rather unrealistic restriction (most objects are not "wires" or "membranes") is that integration in three dimensions has not been available.

Now we are going to discuss a three-dimensional problem leading to a *triple integral*. Suppose that the density at each point of the solid rectangular domain

$$D = \{(x, y, z): 0 \le x \le a, 0 \le y \le b, 0 \le z \le c\}$$

is given by the function $\delta(x, y, z) = x^2 + y^2 + z^2$. To find the mass in D, we proceed much as before, the only difference being that we are dealing with a function of three variables whose domain is in $\mathcal{R}^3$. Partition D into small boxes, the typical box having dimensions Δx, Δy, Δz and volume $\Delta V = \Delta x\, \Delta y\, \Delta z$. If (x, y, z) is a point in this box, the mass of the box is

$$\Delta m \approx \delta(x, y, z)\, \Delta V$$

an approximation that improves as the boxes become smaller. The total mass in D is $m = \sum \Delta m \approx \sum \delta(x, y, z)\, \Delta V$, a Riemann sum whose limit (as the partition is refined) can be expected to exist because δ is continuous. We call this limit the **triple integral** of $\delta(x, y, z)$ over D, written

$$m = \iiint_D \delta(x, y, z)\, dV$$

It is reasonable to expect (by analogy with double integrals) that this limit can be evaluated by the *iterated integral*

$$m = \int_0^a \int_0^b \int_0^c \delta(x, y, z)\, dz\, dy\, dx$$

the understanding being that x and y are fixed while we integrate with respect to z, and that the resulting function of x and y is integrated in the usual way. Look at Figure 1 to see what we mean. Fixed values of x and y correspond to planes parallel to the yz and xz planes, respectively; these planes intersect in a vertical line along which z varies from 0 to c. Since $\delta(x, y, z) = x^2 + y^2 + z^2$, the inner integral is

$$\int_0^c (x^2 + y^2 + z^2)\, dz = \left[(x^2 + y^2)z + \frac{z^3}{3}\right]\Bigg|_{z=0}^{z=c} = (x^2 + y^2)c + \frac{c^3}{3}$$

$$= \frac{c}{3}[3(x^2 + y^2) + c^2]$$

This reduces the problem to the twice-iterated integral

$$m = \frac{c}{3}\int_0^a \int_0^b (3x^2 + 3y^2 + c^2)\, dy\, dx$$

$$= \frac{bc}{3}\int_0^a (3x^2 + b^2 + c^2)\, dx = \frac{abc}{3}(a^2 + b^2 + c^2)$$

As you can see, the ideas involved in this calculation are not all that new. Because the details of the theory are unenlightening (in view of their similarity to our discussion of double integrals in Section 18.2), we are going to omit them and proceed by geometric intuition.

Figure 1
When x and y are fixed, z runs from 0 to c

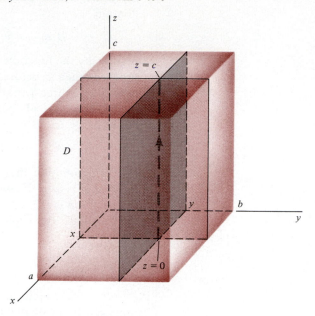

The center of mass of a three-dimensional mass distribution (in a region D) is defined in terms of moments relative to the coordinate *planes* (instead of *axes,* as in $\Re^2$). Partition D into small boxes and let ΔV be the volume of the typical box. Then (as before)

$$\Delta m \approx \delta(x, y, z) \, \Delta V$$

where $\delta(x, y, z)$ is the density at a point (x, y, z) in the box. The (directed) distance from the yz plane to (x, y, z) is x, so the moment of Δm relative to this plane is

$$\Delta M_{yz} \approx x \, \Delta m \approx x \, \delta(x, y, z) \, \Delta V$$

Add up the contributions of all the boxes and take the limit as the partition is refined. The result is

$$M_{yz} = \iiint_D x \, \delta(x, y, z) \, dV$$

Moments relative to the other coordinate planes are similarly defined, namely

$$M_{xz} = \iiint_D y \, \delta(x, y, z) \, dV \qquad \text{and} \qquad M_{xy} = \iiint_D z \, \delta(x, y, z) \, dV$$

The **center of mass** is the point $(\bar{x}, \bar{y}, \bar{z})$ satisfying

$$\bar{x} = M_{yz}/m, \; \bar{y} = M_{xz}/m, \; \bar{z} = M_{xy}/m$$

where
$$m = \iiint_D \delta(x, y, z)\, dV$$

is the total mass. If the density is constant, we call this point the **centroid** of D, as before. (The centroid is a purely geometric property of D, since the density cancels out when it is constant and its value is irrelevant.)

□ **Example 1**   Find the centroid of the region D cut from the first octant by the plane $x + y + z = 1$.

Solution: The region is shown in Figure 2. We take the density to be $\delta(x, y, z) = 1$ (since it cancels anyway). To compute

$$M_{yz} = \iiint_D x\, dV$$

as an iterated integral, fix (x, y) and look at Figure 2 to see that z runs from the xy plane ($z = 0$) to the plane $z = 1 - x - y$. Once this variation is accounted for by our first integration, the problem reduces to a double integration over the two-dimensional region

$$R = \{(x, y): 0 \leq x \leq 1,\, 0 \leq y \leq 1 - x\}$$

(the projection of D onto the xy plane). Hence

$$M_{yz} = \iint_R \left[\int_0^{1-x-y} x\, dz \right] dA = \int_0^1 \int_0^{1-x} \int_0^{1-x-y} x\, dz\, dy\, dx$$

Figure 2
The tetrahedron D

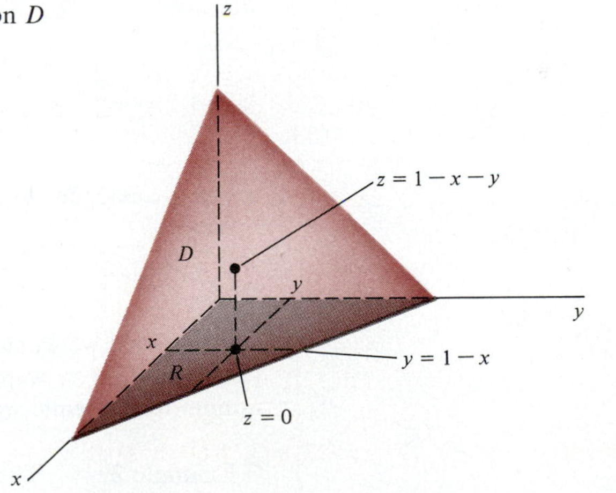

The first integral (in which z is the variable) is

$$\int_0^{1-x-y} x\, dz = xz \Big|_{z=0}^{z=1-x-y} = x(1 - x - y)$$

The second integral is

$$\int_0^{1-x} x(1 - x - y)\, dy = x\left[(1-x)y - \frac{y^2}{2}\right]\Bigg|_{y=0}^{y=1-x}$$

$$= \frac{xy}{2}[2(1-x) - y]\Bigg|_{y=0}^{y=1-x}$$

$$= \frac{x(1-x)}{2}[2(1-x) - (1-x)] = \frac{1}{2}x(1-x)^2$$

Hence

$$M_{yz} = \frac{1}{2}\int_0^1 x(1-x)^2\, dx = \frac{1}{24}$$

The volume of D is

$$V = \iiint_D dV \qquad \text{(Why?)}$$

$$= \int_0^1 \int_0^{1-x} \int_0^{1-x-y} dz\, dy\, dx = \int_0^1 \int_0^{1-x} (1 - x - y)\, dy\, dx$$

$$= \frac{1}{2}\int_0^1 (1-x)^2\, dx = \frac{1}{6}$$

(We could have put this down immediately by using the result of Example 6, Section 18.1.) Since $m = V$ (because the density is 1), we have $\bar{x} = M_{yz}/m = \frac{1}{4}$. It is clear from the symmetry of D that $\bar{x} = \bar{y} = \bar{z}$, so the centroid is $(\frac{1}{4}, \frac{1}{4}, \frac{1}{4})$. ☐

Remark   We found the volume of D in Example 1 by using the formula

$$\boxed{V = \iiint_D dV}$$

which is analogous to the area formula $A = \iint_D dA$ in Section 18.2. Note that it reduces to

$$V = \int_0^1 \int_0^{1-x} (1 - x - y)\, dy\, dx$$

which is what we would put down for the volume under the plane $z = 1 - x - y$ (and over R) if we were using a double integral to find it. But the triple integral is a more fundamental way to express volume.

☐ **Example 2**   The conical region shown in Figure 3 is filled with water of constant density $\delta(x, y, z) = 1$. Find the moment of inertia relative to the z axis.

Solution: In Section 18.2 we defined the moment of inertia of a point mass m relative to an axis r units from the point to be $I = mr^2$. To apply that idea to a three-dimensional mass distribution (in a region D), partition D

Figure 3
Cone filled with water

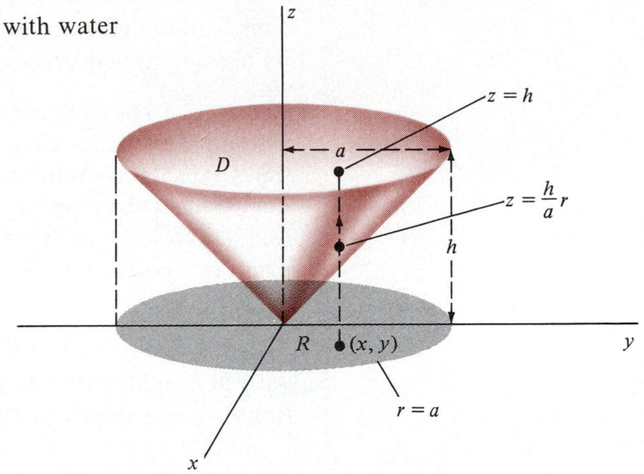

into small boxes. Let ΔV be the volume of the typical box and imagine its mass

$$\Delta m \approx \delta(x, y, z)\, \Delta V$$

to be concentrated at a point (x, y, z) in the box. The moment of inertia of this mass relative to the z axis is

$$\Delta I_z \approx (x^2 + y^2)\, \Delta m \approx (x^2 + y^2)\delta(x, y, z)\, \Delta V$$

and hence the total moment is

$$I_z = \iiint_D (x^2 + y^2)\delta(x, y, z)\, dV$$

In the present example (constant density 1) we have

$$I_z = \iiint_D (x^2 + y^2)\, dV$$

Fixing $(x, y) \in R$, we see from Figure 3 that z runs from the conical surface

$$az = hr \qquad (r = \sqrt{x^2 + y^2})$$

to the plane $z = h$. Hence

$$I_z = \iint_R \left[\int_{hr/a}^{h} (x^2 + y^2)\, dz \right] dA$$

The inside integral (in which x and y are fixed) is

$$\int_{hr/a}^{h} r^2\, dz = r^2 \left(h - \frac{hr}{a} \right) = \frac{hr^2}{a}(a - r) \qquad \text{(because } x^2 + y^2 = r^2\text{)}$$

Thus
$$I_z = \frac{h}{a} \iint_R r^2(a - r)\, dA$$

$$= \frac{h}{a} \int_0^{2\pi} \int_0^a r^2(a - r)r\, dr\, d\theta = \frac{ha^4}{20} \int_0^{2\pi} d\theta = \frac{\pi h a^4}{10}$$

Since the density is 1, we have $m = V = \frac{1}{3}\pi a^2 h$ and hence $I_z = \frac{3}{10}ma^2$. This formula (for the moment of inertia of a solid conical region relative to its axis) is useful in physics. □

Remark The equation $az = hr$ of the cone in Example 2 is in cylindrical coordinates (Section 16.5). Since these are merely polar coordinates in $\mathcal{R}^2$ together with the rectangular coordinate z, we are not doing anything new. (It is not until we finish integrating with respect to z that polar coordinates are used.) Hence we omit any formal discussion of "triple integrals in cylindrical coordinates." Spherical coordinates are another matter, to be taken up in the next section.

□ **Example 3** A bowl in the form of a hemisphere of radius a is filled with glop that settles in such a way that its density at each point is proportional to the depth at that point. Find the mass of the glop.

Figure 4
Bowl full of glop

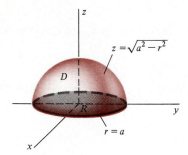

Solution: Turn the bowl upside down, as shown in Figure 4 (trusting that the glop is like jello and won't make a mess). Its surface is given by $x^2 + y^2 + z^2 = a^2$, $z \geq 0$, and the density at (x, y, z) is $\delta(x, y, z) = kz$ (where k is a constant). The mass is

$$m = \iiint_D \delta(x, y, z)\, dV = \iint_R \left[\int_0^{\sqrt{a^2-r^2}} kz\, dz \right] dA = \frac{k}{2} \iint_R (a^2 - r^2)\, dA$$

$$= \frac{k}{2} \int_0^{2\pi} \int_0^a (a^2 - r^2) r\, dr\, d\theta = \frac{ka^4}{8} \int_0^{2\pi} d\theta = \frac{k\pi a^4}{4} \qquad □$$

Problem Set 18.5

Use triple integration to find the volume of each of the following regions.

1. In the first octant, bounded by the planes $x = 2$, $y = 3$, and $z = y$.

2. In the first octant, bounded by the cylinder $z = 1 - x^2$ and the plane $y = 2$.

3. In the first octant, below the cylinder $z = y^2$ and bounded by the planes $x = 2$ and $y = 3$.

4. Above the xy plane, inside the cylinder $x^2 + y^2 = 4$, and below the plane $z = y$.

5. Cut from the first octant by the cylinder $x^2 + y^2 = 1$ and the plane $x + z = 1$.

6. The wedge in the first octant cut from the cylinder $y^2 + z^2 = 4$ by the yz plane and the plane $y = x$.

7. Bounded by the paraboloid $z = 4 - x^2 - y^2$ and the xy plane.

8. Bounded by the paraboloids $z = x^2 + y^2$ and $z = 2 - x^2 - y^2$.

9. Inside the cone $az = hr$ and between the planes $z = 0$ and $z = h$.

10. Cut from the first octant by the plane $x/a + y/b + z/c = 1$ (where a, b, and c are positive).

Find the centroid of each of the following regions.

11.–20. The regions in Problems 1–10.

21. Cut from the first octant by the ellipsoid $x^2/a^2 + y^2/b^2 + z^2/c^2 = 1$. *Hint:* The volume of the ellipsoid is $\frac{4}{3}\pi abc$ (from Problem 42, Section 18.1).

22. Bounded by the paraboloid $z = x^2 + y^2$ and the plane $z = 4$.

23. The hemispherical region $x^2 + y^2 + z^2 \leq a^2$, $z \geq 0$.

24. Above the xy plane, bounded by the cone $z^2 = x^2 + y^2$ and the paraboloid $z = 2 - x^2 - y^2$.

25. Above the xy plane, below the paraboloid $z = x^2 + y^2$, and inside the cylinder $r = 2 \sin \theta$.

In each of the following problems assume that the density at each point is 1.

26. Find the moment of inertia (relative to the side of length a) of a solid rectangular box of dimensions a, b, c.

27. Find the moment of inertia (relative to the z axis) of the region bounded by the coordinate planes, the plane $z = 2$, and the plane $x + y = 1$.

28. Show that the moment of inertia (relative to its axis) of a solid right circular cylinder of radius a (and given altitude) is $I = \frac{1}{2}ma^2$, where m is the mass.

29. Show that the moment of inertia (relative to a diameter) of a solid spherical ball of radius a is $I = \frac{2}{5}ma^2$, where m is the mass.

30. Find the moment of inertia (relative to the z axis) of the region inside the cylinder $x^2 + y^2 = 1$, above the xy plane, and below the paraboloid $z = 4 - x^2 - y^2$.

31. The density at each point of the solid rectangular region

$$D = \{(x,y,z): 0 \le x \le 3, 0 \le y \le 4, 0 \le z \le 5\}$$

is $\delta(x,y,z) = xyz$. Find the mass.

32. In Problem 31 find the center of mass.

33. In Problem 31 find the moment of inertia relative to the z axis.

34. A cylindrical container of radius a and altitude h is filled with glop whose density is proportional to the depth. Find the mass.

35. In Problem 34 find the center of mass.

36. In Problem 34 find the moment of inertia relative to the axis of the container.

37. The density at each point of a solid right circular cone of base radius a and altitude h is proportional to the distance from the base. Find the mass.

38. In Problem 37 find the center of mass.

39. In Problem 37 find the moment of inertia relative to the axis of the cone.

40. Suppose that $f(x,y,z)$ is defined in a cubical box D, partitioned into n cubes of equal volume ΔV, and let P_k be a point of the kth cube.
    (a) What is the average value of the numbers $f(P_1), f(P_2),$ $\ldots, f(P_n)$?
    (b) Show that the average value in part (a) can be written in the form

$$\frac{1}{V} \sum_{k=1}^{n} f(P_k)\,\Delta V \qquad \text{where } V \text{ is the volume of } D$$

    (c) What is the limit of the expression in part (b) as the partition is refined?

41. The preceding problem suggests that in general we should define the *average value* of a function $f(x,y,z)$ in a region D (not necessarily cubical) to be

$$\frac{1}{V} \iiint_D f(x,y,z)\,dV \quad \text{where } V \text{ is the volume of } D$$

Show that the average value of x in the region D is the coordinate $\bar{x}$ of the centroid of D. What are the average values of y and z in D?

42. Find the average value of xyz in the region

$$D = \{(x,y,z): x^2 + y^2 + z^2 \le 1, x \ge 0, y \ge 0, z \ge 0\}$$

43. Show that the average value of e^{x+y+z} in the cubical region

$$D = \{(x,y,z): -1 \le x \le 1, -1 \le y \le 1, -1 \le z \le 1\}$$

is $\sinh^3 1$. In view of this result and Problem 56, Section 18.2, what do you suppose is the average value of e^x in the interval $[-1,1]$?

44. Show that the average distance of a point on the sphere $x^2 + y^2 + z^2 = a^2$ from the point $(a,0,0)$ is $\frac{4}{3}a$. *Hint:* This reduces to a problem in single-variable calculus!

18.6
TRIPLE INTEGRALS IN SPHERICAL COORDINATES

Just as polar coordinates simplify the evaluation of many double integrals, spherical coordinates (Section 16.5) are useful in triple integration. The reason for this is that applications often involve spheres, cones, and other regions in which there is symmetry that can be exploited by spherical coordinates.

Our first task is to discover how the volume element $\Delta V = \Delta x\,\Delta y\,\Delta z$ is changed when an integral

$$\iiint_D f(x, y, z)\,dV$$

is transformed by the equations

$$x = \rho \sin \phi \cos \theta, \; y = \rho \sin \phi \sin \theta, \; z = \rho \cos \phi$$

There are two ways to do this, one geometric (based on Figure 1) and the other analytic, based on the theorem about Jacobians that we mentioned in the Remark in Sec. 18.3. You can take your choice; neither approach is trivial. Figure 1 shows a typical "box" in a partition of a three-dimensional region by spheres (fixed ρ), half-cones (fixed ϕ), and half-planes (fixed θ). (Review Section 16.5 if you don't know what we're talking about!) The dimensions of the box in the direction of increasing ρ, ϕ, θ, respectively, are $\Delta\rho, \; \rho\,\Delta\phi, \; \rho \sin \phi\,\Delta\theta$. Rather than explaining this at length (which won't help), we suggest that you study Figure 1 until you understand it. It may help to look back to Section 18.3, where we explained why the area element $\Delta A = \Delta x\,\Delta y$ becomes $\Delta A \approx r\,\Delta r\,\Delta\theta$ in a polar coordinate transformation. (The ideas are similar.)

Figure 1
Spherical coordinate "box"

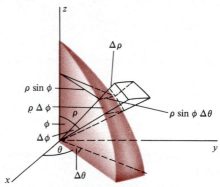

Assuming that the dimensions of the box are understood, it is easy to finish; simply treat the box as though it has straight sides, obtaining the approximation

$$\Delta V \approx \rho^2 \sin \phi\,\Delta\rho\,\Delta\phi\,\Delta\theta$$

This can be expected to improve as the partition is refined, so we conclude that dV becomes $\rho^2 \sin \phi\,d\rho\,d\phi\,d\theta$ when a triple integral in rectangular coordinates is transformed by changing to spherical coordinates:

$$\iiint_D f(x, y, z)\,dV$$
$$= \iiint_D f(\rho \sin \phi \cos \theta, \; \rho \sin \phi \sin \theta, \; \rho \cos \phi)\rho^2 \sin \phi\,d\rho\,d\phi\,d\theta$$

The other approach to this result is the theorem that a transformation from $\mathcal{R}^3$ to $\mathcal{R}^3$ multiplies dV by $|J|$, where J is the Jacobian. In the present case we find $J = \rho^2 \sin \phi$ (Problem 35, Section 16.5, and Additional Problem 36, Chapter 17). Since $0 \le \phi \le \pi$, the Jacobian is never negative, so $|J| = J$ and our geometrical approach is confirmed.

□ **Example 1**   Find the volume of a sphere of radius a.

Solution: Place the sphere in a coordinate system so that it encloses the region

$$D = \{(x, y, z): x^2 + y^2 + z^2 \le a^2\}$$

In rectangular coordinates the calculation of

$$V = \iiint_D dV = 8 \int_0^a \int_0^{\sqrt{a^2-x^2}} \int_0^{\sqrt{a^2-x^2-y^2}} dz \, dy \, dx$$

(based on the first octant portion of D) is gruesome; we will not finish it. In cylindrical coordinates the calculation becomes

$$V = 2 \int_0^{2\pi} \int_0^a \int_0^{\sqrt{a^2-r^2}} r \, dz \, dr \, d\theta = 2 \int_0^{2\pi} \int_0^a r \sqrt{a^2 - r^2} \, dr \, d\theta$$

which is not hard. (See Problem 26, Section 18.3.) In spherical coordinates, however, we have

$$V = \int_0^{2\pi} \int_0^{\pi} \int_0^a \rho^2 \sin \phi \, d\rho \, d\phi \, d\theta$$

(an expression that does not even require a picture if you are clear in your mind about the definition of these coordinates). The inside integral is

$$\int_0^a \rho^2 \sin \phi \, d\rho = \frac{\rho^3}{3} \sin \phi \Big|_{\rho=0}^{\rho=a} = \frac{a^3}{3} \sin \phi$$

The second integral is

$$\frac{a^3}{3} \int_0^{\pi} \sin \phi \, d\phi = \frac{a^3}{3} \cos \phi \Big|_{\pi}^0 = \frac{2}{3} a^3$$

and hence

$$V = \frac{2}{3} a^3 \int_0^{2\pi} d\theta = \frac{4}{3} \pi a^3$$

Elegant!   □

□ **Example 2**   A right circular cone of base radius a and altitude h is filled with water of constant density 1. Find the moment of inertia relative to its axis.

Solution: This is the same problem we solved in Example 2 of the last section. Place the cone in a coordinate system as shown in Figure 2, letting D be the region inside it. The moment we seek is

$$I_z = \iiint_D (x^2 + y^2)\, dV$$

Since $r = \rho \sin \phi$ (see Figure 1, and also Section 16.5), the integrand becomes

$$\rho^2 \sin^2 \phi \cdot \rho^2 \sin \phi \, d\rho \, d\phi \, d\theta = \rho^4 \sin^3 \phi \, d\rho \, d\phi \, d\theta$$

upon changing to spherical coordinates. In the first integration ϕ and θ are fixed; you can see from Figure 2 that ρ runs from 0 to its value on the plane $z = h$. Since $z = \rho \cos \phi$, the plane is $\rho = h \sec \phi$ in spherical coordinates. Releasing ϕ for the second integration (but keeping θ fixed), we see that ϕ runs from 0 to its value on the cone, namely $\tan^{-1}(a/h)$. Finally, θ runs from 0 to 2π. Hence

$$I_z = \int_0^{2\pi} \int_0^{\tan^{-1}(a/h)} \int_0^{h \sec \phi} \rho^4 \sin^3 \phi \, d\rho \, d\phi \, d\theta$$

The first integral is

$$\int_0^{h \sec \phi} \rho^4 \sin^3 \phi \, d\rho = \frac{h^5}{5} \sec^5 \phi \sin^3 \phi = \frac{h^5}{5} \tan^3 \phi \sec^2 \phi$$

The second integral is

$$\frac{h^5}{5} \int_0^{\tan^{-1}(a/h)} \tan^3 \phi \sec^2 \phi \, d\phi = \frac{h^5}{5} \cdot \frac{\tan^4 \phi}{4} \Big|_0^{\tan^{-1}(a/h)}$$

$$= \frac{h^5}{20} \tan^4 \left(\tan^{-1} \frac{a}{h} \right) = \frac{h^5}{20} \cdot \frac{a^4}{h^4} = \frac{ha^4}{20}$$

Figure 2
Cone filled with water

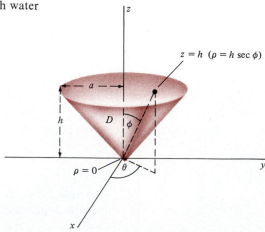

Hence (as in Example 2 of the last section)

$$I_z = \frac{ha^4}{20} \int_0^{2\pi} d\theta = \frac{\pi ha^4}{10} = \frac{3}{10} ma^2$$

where m is the mass.   □

□ **Example 3**   A bowl in the form of a hemisphere of radius a is filled with glop whose density at each point is proportional to the depth at that point. Find the mass, the center of mass, and the moment of inertia relative to the vertical line of symmetry of the glop.

Figure 3
Bowl full of glop

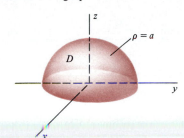

Solution: We found the mass in Example 3 of the last section (by turning the bowl upside down, as in Figure 3). The equation of its surface is $x^2 + y^2 + z^2 = a^2$, $z \ge 0$, and the density at (x, y, z) is $\delta(x, y, z) = kz$ (where k is a constant). In spherical coordinates we have

$$m = \iiint_D \delta(x, y, z) \, dV = \int_0^{2\pi} \int_0^{\pi/2} \int_0^a k\rho \cos\phi \cdot \rho^2 \sin\phi \, d\rho \, d\phi \, d\theta$$

$$= \frac{ka^4}{4} \int_0^{2\pi} \int_0^{\pi/2} \sin\phi \cos\phi \, d\phi \, d\theta = \frac{ka^4}{8} \int_0^{2\pi} d\theta = \frac{k\pi a^4}{4}$$

(a computation which, unlike the one in the last section, can be followed without filling in any steps).

The center of mass is clearly on the z axis, so $\bar{x} = \bar{y} = 0$. To find $\bar{z}$, compute

$$M_{xy} = \iiint_D z \, \delta(x, y, z) \, dV$$

$$= \int_0^{2\pi} \int_0^{\pi/2} \int_0^a \rho \cos\phi \cdot k\rho \cos\phi \cdot \rho^2 \sin\phi \, d\rho \, d\phi \, d\theta$$

$$= \frac{ka^5}{5} \int_0^{2\pi} \int_0^{\pi/2} \cos^2\phi \sin\phi \, d\phi \, d\theta = \frac{ka^5}{15} \int_0^{2\pi} d\theta = \frac{2k\pi a^5}{15}$$

Hence $\bar{z} = M_{xy}/m = \frac{8}{15}a$.

The moment of inertia relative to the z axis is

$$I_z = \iiint_D (x^2 + y^2) \, \delta(x, y, z) \, dV$$

$$= \int_0^{2\pi} \int_0^{\pi/2} \int_0^a \rho^2 \sin^2\phi \cdot k\rho \cos\phi \cdot \rho^2 \sin\phi \, d\rho \, d\phi \, d\theta$$

$$= \frac{ka^6}{6} \int_0^{2\pi} \int_0^{\pi/2} \sin^3\phi \cos\phi \, d\phi \, d\theta = \frac{ka^6}{24} \int_0^{2\pi} d\theta = \frac{k\pi a^6}{12} = \frac{1}{3} ma^2 \quad □$$

Problem Set 18.6

Use integration in spherical coordinates to do the following problems (some of which have been done before in other ways).

1. Find the volume of the hemispherical region $x^2 + y^2 + z^2 \leq a^2$, $z \geq 0$.

2. Find the volume of the first octant region inside the sphere $x^2 + y^2 + z^2 = a^2$.

3. Find the centroid of the region in Problem 1.

4. Find the centroid of the region in Problem 2.

5. Find the moment of inertia (relative to the z axis) of the region in Problem 1. (Assume that the density is 1.)

6. Find the moment of inertia (relative to the z axis) of the region in Problem 2. (Assume that the density is 1.)

7. Show that the moment of inertia (relative to a diameter) of a solid spherical ball of radius a (and constant density 1) is $I = \frac{2}{5}ma^2$, where m is the mass.

8. Find the volume of the "ice cream cone" above the xy plane, inside the cone $x^2 + y^2 = 3z^2$, and capped by the sphere $x^2 + y^2 + z^2 = 1$.

9. Find the volume cut from the cone $\phi = \pi/4$ by the sphere $\rho = 2a \cos \phi$.

10. Find the centroid of the region in Problem 8.

11. Find the centroid of the region in Problem 9.

12. Find the moment of inertia (relative to the z axis) of the region in Problem 8. (Assume that the density is 1.)

13. Find the moment of inertia (relative to the z axis) of the region in Problem 9. (Assume that the density is 1.)

14. Show that a right circular cone of base radius a and altitude h has volume $V = \frac{1}{3}\pi a^2 h$.

15. Find the volume of the region above the paraboloid $z = x^2 + y^2$ and inside the sphere $x^2 + y^2 + z^2 = 2$. *Hint:* Break the region into two parts.

16. Show that the centroid of the region inside the cone of Problem 14 is at a distance $\frac{3}{4}h$ from the vertex.

17. Find the centroid of the region in Problem 15.

18. Find the moment of inertia (relative to the axis) of the conical region in Problem 14. (Assume that the density is 1.)

19. Find the moment of inertia (relative to the z axis) of the region in Problem 15. (Assume that the density is 1.)

20. Find the volume of the region inside both the spheres $\rho = a$ and $\rho = 2a \cos \phi$.

21. Find the centroid of the region in Problem 20.

22. Find the moment of inertia (relative to the z axis) of the region in Problem 20. (Assume that the density is 1.)

23. The density at each point of a solid spherical ball of radius a is proportional to the distance from the center. Find the mass.

24. In Problem 23 find the moment of inertia relative to a diameter (in terms of the mass).

25. The density at each point of the ball $x^2 + y^2 + z^2 \leq a^2$ is proportional to the distance from the xy plane. Find the mass.

26. In Problem 25 find the center of mass.

27. In Problem 25 find the moment of inertia relative to the z axis (in terms of the mass).

28. The density at each point of the first octant region inside the sphere $x^2 + y^2 + z^2 = 1$ is given by

$$\delta(x,y,z) = \frac{1}{x^2 + y^2 + z^2 + 1}$$

Find the mass.

29. In Problem 28 find the center of mass.

30. In Problem 28 find the moment of inertia relative to the z axis.

31. The density at each point (other than the center) of a solid spherical ball of radius a is inversely proportional to the square of the distance from the center. Show that the mass is finite even though the density is unbounded near the center.

32. The density at each point inside the sphere $\rho = 2a \cos \phi$ is proportional to the distance from the origin. Find the mass.

33. In Problem 32 find the center of mass.

34. In Problem 32 find the moment of inertia relative to the z axis (in terms of the mass).

35. Let D be the hemispherical shell of inner radius a and outer radius b,

$$D = \{(x,y,z): a^2 \leq x^2 + y^2 + z^2 \leq b^2, z \geq 0\}$$

(a) Show that the centroid of D has coordinates

$$\bar{x} = \bar{y} = 0, \bar{z} = \frac{3(b^4 - a^4)}{8(b^3 - a^3)}$$

(b) What is the limiting position of the centroid as $a \to 0$? (This is the centroid of the solid hemisphere of radius b, and should check with Problem 3.)

(c) What is the limiting position of the centroid as $b \to a$? (This is the centroid of the *surface* $x^2 + y^2 + z^2 = a^2$, $z \ge 0$. You might be interested in setting it up directly in terms of surface area as defined in Section 18.4. The computation turns out to be trivial.)

36. Show that the average distance of a point in the solid spherical ball $x^2 + y^2 + z^2 \le a^2$ from the center is $\frac{3}{4}a$.

37. Let D be the solid spherical ball $x^2 + y^2 + z^2 \le a^2$.

(a) Evaluate $\iiint_D x^2 \, dV$.

(b) Explain why $\iiint_D x^2 \, dV = \frac{1}{3} \iiint_D (x^2 + y^2 + z^2) \, dV$ and evaluate the expression on the right to do part (a) with less effort.

True–False Quiz

1. $\displaystyle \int_0^1 \int_0^x x \, dy \, dx = \int_0^1 \int_0^y x \, dx \, dy$.

2. If $D = \{(x, y): x^2 + y^2 \le 1, x \ge 0, y \ge 0\}$, then

$$\iint_D 2 \, dA = 2\pi$$

3. If D is the unit disk $x^2 + y^2 \le 1$, then

$$\iint_D \sqrt{1 - x^2 - y^2} \, dA = \frac{2\pi}{3}$$

4. If the density of a region D is constant, its value is irrelevant in the computation of the center of mass of D.

5. The area of the first octant portion of the plane $x + y + z = 1$ is $\sqrt{3}/2$.

6. If D is the ball $x^2 + y^2 + z^2 \le 9$, then

$$\iiint_D dV = 36\pi$$

7. If D is the unit ball $x^2 + y^2 + z^2 \le 1$, then

$$\iiint_D z \, dV = 0$$

8. The moment of inertia (relative to the x axis) of a three-dimensional region D with density $\delta(x, y, z)$ at each point is

$$I_x = \iiint_D y^2 \, \delta(x, y, z) \, dV$$

9. The moment of inertia (relative to the z axis) of a three-dimensional region D with constant density 1 is

$$I_z = \iiint_D \rho^4 \sin^3 \phi \, d\rho \, d\phi \, d\theta$$

in spherical coordinates.

10. The moment (relative to the xy plane) of the region in Problem 9 is

$$M_{xy} = \iiint_D rz \, dz \, dr \, d\theta$$

in cylindrical coordinates.

Additional Problems

In each of the following, reverse the order of integration in the iterated integral and evaluate it both ways.

1. $\displaystyle \int_0^1 \int_0^{1-y} (x + y) \, dx \, dy$

2. $\displaystyle \int_0^{\pi/2} \int_0^{\cos x} \sin x \, dy \, dx$

3. $\displaystyle \int_0^2 \int_0^x \frac{dy \, dx}{1 + x^2}$

4. Evaluate

$$\int_0^4 \int_{\sqrt{y}}^2 dx \, dy$$

as it stands and also with the order of integration reversed. What does the answer represent geometrically?

5. Evaluate

$$\int_0^1 \int_0^{e^x} dy \, dx$$

as it stands and also with the order of integration reversed.

6. It is difficult to evaluate

$$\int_0^\infty \frac{1}{x}(e^{-x} - e^{-3x}) \, dx$$

directly. Evaluate it indirectly as follows.

(a) Show that as far as the lower limit is concerned, the integral (despite appearances) is proper.

(b) Show that for each $x > 0$,

$$\frac{1}{x}(e^{-x} - e^{-3x}) = \int_1^3 e^{-xy} \, dy$$

thus converting the given integral into the iterated integral

$$\int_0^\infty \int_1^3 e^{-xy} \, dy \, dx$$

(c) Reverse the order of integration and evaluate the result.

7. If D is the unit disk $x^2 + y^2 \leq 1$, find

$$\iint_D \sqrt{x^2 + y^2} \, dA$$

8. Evaluate

$$\int_0^{\pi/4} \int_0^{\sec \theta} r \, dr \, d\theta$$

by carrying out the integration and also by elementary geometry.

Find the volume of each of the following regions.

9. Cut from the first octant by the cylinder $z = 4 - y^2$ and the plane $y = x$.

10. Bounded by the paraboloid $z = 9 - x^2 - y^2$ and the xy plane.

11. With semicircular base

$$D = \{(x,y): -1 \leq x \leq 1, 0 \leq y \leq \sqrt{1 - x^2}\}$$

capped by the plane $z = y$.

12. Cut from the first octant by the cylinder $x^2 + 4y^2 = 4$ and the plane $y + z = 1$.

13. Inside the ellipsoid $x^2 + y^2 + z^2/9 = 1$.

14. Suppose that we want to find the volume of the tetrahedron cut off from the first octant by the plane $2x + 3y + z = 6$.

(a) One way to do this is to slice the region by planes parallel to the yz plane. If $A(x)$ is the area of the cross section x units from the yz plane, the volume is given by an integral of the form

$$\int_a^b A(x) \, dx$$

Draw a picture showing the typical cross section and deduce from your picture a formula for $A(x)$ in terms of x. Then evaluate the integral to find the volume.

(b) Another way to find the volume is to evaluate an appropriate double integral. Write this as an iterated integral (in the order of integration $dy \, dx$), but do not evaluate it.

(c) The volume may also be found by means of a triple integral. Write this as an iterated integral (in the order of integration $dz \, dy \, dx$), but do not evaluate it.

(d) Evaluate the first stage of the triply iterated integral in part (c) to show that it reduces to the doubly iterated integral in part (b).

(e) Evaluate the first stage of the doubly iterated integral in part (b) to show that it reduces to the single integral in part (a).

15. Find the volume of the region above the xy plane bounded by the surfaces $x^2 + y^2 = 1$ and $z = x$, as follows.

(a) Use double integration in rectangular coordinates.

(b) Use double integration in polar coordinates (or triple integration in cylindrical coordinates).

(c) Use triple integration in spherical coordinates.

Use double integration to find the centroid of each of the following regions.

16. Bounded by the line $x + y = 2$ and the coordinate axes.

17. Bounded by the curve $y = x^2 - 3x$ and the x axis.

18. Inside the circle $x^2 + y^2 = 16$ and to the right of the line $x = 2$.

19. Bounded by the curve $y = 2 - x^2$ and the line $y = x$.

20. Bounded by the graph of $y = x/\sqrt{4 - x^2}$, the lines $x = 0$ and $x = 1$, and the x axis.

21. The density at each point of

$$D = \{(x,y): 0 \leq x \leq \pi, 0 \leq y \leq \sin x\}$$

is $\delta(x,y) = 2y$. Find the center of mass of D.

22. In Problem 21 find the moment of inertia of D relative to the x axis.

23. The density at each point of the region inside the graph of $r = \sin \theta$ is proportional to the distance from the origin. Find the center of mass.

24. In Problem 23 find the moment of inertia relative to the y axis.

Use double integration to find the area of each of the following surfaces.

25. Cut from the plane $z = 1$ by the cylinder $r = 2 \sin \theta$. Check by geometry.

26. Cut from the first octant by the plane $2x - 3y + 5z - 30 = 0$. Check by vector methods. (See Example 4, Section 16.3.)

27. The triangle with vertices $(1,3,4)$, $(3,3,2)$, $(3,1,5)$. Check by vector methods.

28. The part of the paraboloid $\frac{1}{2}z = x^2 + y^2$ below the plane $z = 2$.

29. The cap of the sphere $x^2 + y^2 + z^2 = 16$ above the plane $z = 2$.

30. Cut from the surface $z = xy$ by the cylinder $x^2 + y^2 = 1$.

31. The part of the hyperboloid $z^2 = x^2 + y^2 + 1$ between the planes $z = 1$ and $z = 2$.

32. Use integration to find the area of the surface generated by rotating the curve $r = 2a \cos \theta$ about the x axis. Check by geometry. *Hint:* The differential of arc length in polar coordinates (Section 14.2) is $ds = \sqrt{r^2 + (dr/d\theta)^2} \, d\theta$.

Use triple integration to find the volume of each of the following regions.

33. Above the paraboloid $\frac{1}{2}z = x^2 + y^2$ and below the plane $z = 2$.

34. Bounded by the paraboloid $z = x^2 + y^2$ and the sphere $x^2 + y^2 + z^2 = 2$.

35. The "ice cream cone" cut from the sphere $x^2 + y^2 + z^2 = 16$ by the cone $z^2 = 3(x^2 + y^2)$.

36. The solid that is left after a hole of radius 1 is bored through the ball $x^2 + y^2 + z^2 \leq 4$.

37. Bounded by the cylinder $y = x^2$, the plane $y + z = 4$, and the xy plane.

Find the centroid of each of the following regions.

38. The first-octant part of the ball $x^2 + y^2 + z^2 \leq a^2$.

39. Between the xy plane and the paraboloid $z = 9 - x^2 - y^2$.

40. The solid in Problem 33.

41. The solid in Problem 34.

42. The solid in Problem 35.

43. The solid in Problem 36.

44. The solid in Problem 37.

Find the moment of inertia relative to the z axis of each of the following regions. Assume that the density is $\delta(x, y) = 1$.

45. The solid in Problem 33.

46. The solid in Problem 34.

47. The solid in Problem 35.

48. The solid in Problem 36.

49. The solid in Problem 37.

50. The density at each point of the region above the xy plane, inside the cone $z^2 = x^2 + y^2$, and below the sphere $x^2 + y^2 + z^2 = 1$ is inversely proportional to the distance from the origin. Find **(a)** the mass; **(b)** the center of mass; and **(c)** the moment of inertia relative to the z axis.

19 | Vector Fields

There is something fascinating about science. One gets such wholesale returns of conjecture out of such a trifling investment of fact.
MARK TWAIN

We see it as Columbus saw America from the shores of Spain. Its movements have been felt, trembling along the far-reaching line of our analysis, with a certainty hardly inferior to that of ocular demonstration.
SIR JOHN HERSCHEL (prophesying the discovery of Neptune)

Gravitation influences both our bodies and our minds . . . it was gravity which taught us to think of three-dimensional space.
D'ARCY THOMPSON (1860–1949)

This chapter is a combination (and culmination) of multidimensional differentiation and integration that from a mathematical point of view might be called *The Fundamental Theorem of Calculus Revisited.* The main results, however (the theorems of Green, Gauss, and Stokes), are associated with mathematical physics as much as with mathematics itself, both in their origin and in their applications. From that point of view (in which gravitational and electromagnetic fields, together with their potential energy functions, are the important ideas) we might call the subject *Force Fields and Potential Theory.* Another possibility is *Line and Surface Integrals,* since these generalizations of ordinary integration provide the focus of our discussion.

What's in a name? A great deal, actually! We are trying to suggest that the subject matter of this chapter is rich in possible interpretations and applications. By calling it *Vector Fields* we have adopted the typical compromise of the mathematician who is interested in calculus for its own sake, but who is at the same time eager to point out that it arose historically, and is alive today, because of its analytical power in connection with real problems. Since this is not a physics book, we will do no more than suggest what these problems are. They supplied the original impetus, however, for most of the mathematics you will encounter in this chapter.

19.1
LINE INTEGRALS IN THE PLANE

The notation $x = x(t)$, $y = y(t)$ is common in mathematics, but we have not used it before. It has the same meaning as, say, $x = f(t)$, $y = g(t)$, but we conserve letters by writing $x(t)$ instead of $f(t)$ and $y(t)$ instead of $g(t)$.

Suppose that a particle moves in the plane along the graph of a smooth curve

$$C: x = x(t), \; y = y(t) \qquad \text{where } t \text{ is time}$$

and suppose that at each point (x, y) it is subject to a force that is a vector function of position, say

$$\mathbf{F(x)} = (F(x, y), G(x, y)) \qquad \text{where } \mathbf{x} = (x, y)$$

What is the work done by the force (on the particle) during the time interval $[t_1, t_2]$?

Before answering this question, we remark that the function $\mathbf{F(x)}$ is called a **force field,** terminology intended to evoke our everyday experience. We live in a gravitational force field, which we may imagine to be represented by an "earthward-directed" vector at each point of space (the length of the vector being inversely proportional to the square of the distance from the center of the earth). See Figure 1, which shows a few such vectors. More generally, a **vector field** (in the plane) is any function $\mathbf{F}: D \to \Re^2$ which associates a vector $\mathbf{F(x)}$ with each point $\mathbf{x} \in D$ (a subset of $\Re^2$). For example, the velocity at each point of a smoothly flowing river is a vector in the direction of the flow at that point and whose length is the speed of the water passing that point. (See Figure 2.) There is nothing new about this idea except the terminology; a vector field is just a function. Unlike most of the functions we have considered before, however, both its domain and range are multidimensional.

Similar terminology for a function with numerical ("scalar") values is a **scalar field.** The function $f(x, y) = x^2 + y^2$, for example, is a scalar field

Figure 1
Gravitational force field of the earth

Figure 2
Vector field representing velocity at points of a river

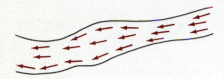

whose domain is the plane. When we discussed the gradient of such a function, $\nabla f(x,y) = (2x,2y)$, that is, $\nabla f(\mathbf{x}) = 2\mathbf{x}$, we were associating a vector field with a scalar field, but at the time we did not refer to it that way.

In linear motion the answer to our question about work is

$$W = \int_a^b F(x)\, dx$$

where $F(x)$ is the force at point x (in the direction of motion) and a and b are the initial and terminal positions, respectively. (See Section 7.5.) In that case we do not need vector notation, because the path lies in the x axis and only one direction is allowed (from a to b).

By analogy to the linear case, we might expect the work done in motion along the curve C to be

$$W = \int_{\mathbf{a}}^{\mathbf{b}} \mathbf{F}(\mathbf{x}) \cdot d\mathbf{x}$$

where $\mathbf{a} = \mathbf{x}(t_1)$ and $\mathbf{b} = \mathbf{x}(t_2)$ are the initial and terminal position vectors of the moving point. (The dot product is used to generalize ordinary multiplication to the context of vector force and position.) We can make sense of the integrand by recalling from Section 15.2 that

$$\frac{d\mathbf{x}}{dt} = \left(\frac{dx}{dt}, \frac{dy}{dt}\right)$$

Hence

$$\mathbf{F}(\mathbf{x}) \cdot d\mathbf{x} = \left[\mathbf{F}(\mathbf{x}) \cdot \frac{d\mathbf{x}}{dt}\right] dt = \left[F(x,y)\frac{dx}{dt} + G(x,y)\frac{dy}{dt}\right] dt$$

and the integral for work becomes

$$W = \int_{t_1}^{t_2} \left[F(x,y)\frac{dx}{dt} + G(x,y)\frac{dy}{dt}\right] dt$$

where $x = x(t)$ and $y = y(t)$. This is an ordinary integral of a function of one variable and offers no theoretical difficulty. The original formula

$$W = \int_{\mathbf{a}}^{\mathbf{b}} \mathbf{F}(\mathbf{x}) \cdot d\mathbf{x}$$

however, must be understood in the context of the curve C. Unlike the linear case (where the path is confined to the x axis between a and b), there are infinitely many paths from $\mathbf{a}$ to $\mathbf{b}$. In general we expect different answers for different paths. For that reason the integral for work might be called a "path integral," the idea being that the calculation involves not only the endpoints but also the path. The actual terminology used by mathematicians and physicists is *line integral* (a misleading term, since the path need not be straight). More precisely, the definition and notation are as follows.

Let C be a smooth curve $x = x(t), y = y(t), t_1 \le t \le t_2$, and suppose that

$$\mathbf{F}(\mathbf{x}) = (F(x,y), G(x,y))$$

is a continuous vector field defined in a plane region containing the graph of C. The **line integral** of $\mathbf{F}$ along C is

$$\int_C \mathbf{F}(\mathbf{x}) \cdot d\mathbf{x} = \int_{t_1}^{t_2} \left[\mathbf{F}(\mathbf{x}) \cdot \frac{d\mathbf{x}}{dt} \right] dt$$

where $\mathbf{x} = (x(t), y(t))$. Alternate notation is

$$\int_C F(x,y)\, dx + G(x,y)\, dy$$

obtained from the vector differential $d\mathbf{x} = (dx, dy)$ and the dot product

$$\mathbf{F}(\mathbf{x}) \cdot d\mathbf{x} = F(x,y)\, dx + G(x,y)\, dy$$

□ **Example 1**   Evaluate the line integral of the vector field

$$\mathbf{F}(x,y) = (xy, 2x - y)$$

along the parabola $y = x^2$ from $(0,0)$ to $(1,1)$.

Solution: We may "parametrize" the path in any way we like. (It is not hard to show that the line integral does not depend on what parametric equations are used, provided that the "orientation" of the path from initial point to terminal point is kept the same.) The simplest choice is

$$C: x = t, \ y = t^2 \qquad 0 \le t \le 1$$

Letting $\mathbf{x} = (x,y) = (t, t^2)$, we find

$$\mathbf{F}(\mathbf{x}) \cdot \frac{d\mathbf{x}}{dt} = (xy, 2x - y) \cdot (1, 2t) = xy + (2x - y)(2t)$$

$$= t^3 + (2t - t^2)(2t) = 4t^2 - t^3$$

Hence

$$\int_C \mathbf{F}(\mathbf{x}) \cdot d\mathbf{x} = \int_0^1 (4t^2 - t^3)\, dt = \frac{13}{12}$$

To illustrate the fact that the integral is independent of the parameter, suppose that we represent the path by

$$C^*: x = \cos s, \ y = \cos^2 s$$

where s runs from $\pi/2$ to 0 in order for the path to start at $(0,0)$ and end at $(1,1)$. Then

$$\mathbf{F}(\mathbf{x}) \cdot \frac{d\mathbf{x}}{ds} = (xy, 2x - y) \cdot (-\sin s, -2 \sin s \cos s)$$

$$= \sin s \cos^3 s - 4 \sin s \cos^2 s$$

and
$$\int_{C^*} \mathbf{F(x)} \cdot d\mathbf{x} = \int_{\pi/2}^{0} (\sin s \cos^3 s - 4 \sin s \cos^2 s) \, ds$$
$$= \left(-\frac{1}{4} \cos^4 s + \frac{4}{3} \cos^3 s \right) \Big|_{\pi/2}^{0} = \frac{13}{12}$$

as before.   □

Remark   We said in Example 1 that changing the parameter does not affect the integral, provided that orientation is preserved. However, what about a change that *reverses* the orientation? As an illustration, consider the path

$$C^*: x = \cos s, \ y = \cos^2 s$$

where s runs from 0 to $\pi/2$. This has the same "track" as the curve C in Example 1, but it is not the same curve because the initial and terminal points are reversed. You can see that the effect on the integral is to multiply it by -1. In general, whenever C^* is C with its orientation reversed, we write $C^* = -C$ and obtain

$$\boxed{\int_{-C} \mathbf{F(x)} \cdot d\mathbf{x} = -\int_{C} \mathbf{F(x)} \cdot d\mathbf{x}}$$

□ **Example 2**   Find the line integral of $\mathbf{F}(x,y) = (xy, 2x - y)$ along the straight line $y = x$ from (0,0) to (1,1).

Solution: This is the same vector field as in Example 1, but the path is different. Letting $\mathbf{x} = (x,y)$ and regarding x as the parameter, we have

$$\mathbf{F(x)} \cdot d\mathbf{x} = (xy, 2x - y) \cdot (dx, dy) = xy \, dx + (2x - y) \, dy$$
$$= (x^2 + x) \, dx \qquad (\text{because } y = x \text{ and } dy = dx)$$

Hence
$$\int_{C} \mathbf{F(x)} \cdot d\mathbf{x} = \int_{0}^{1} (x^2 + x) \, dx = \frac{5}{6}$$

If $\mathbf{F(x)}$ is a force field, the work done by $\mathbf{F}$ in moving the particle from (0,0) to (1,1) is different in Examples 1 and 2. This illustrates the fact that in general the line integral of a vector field along a curve depends on the curve as well as on the field.   □

In many applications the curve has "corners" at which it is not smooth. Assuming that the number of corners is finite, we define the line integral along C to be the sum of the integrals along the smooth pieces. Such a curve is called **piecewise smooth.**

□ **Example 3**   Find the line integral of $\mathbf{F}(x,y) = (4x + y, x + 2y)$ around the square shown in Figure 3.

Solution: The path may be broken into the smooth curves

$$C_1: y = 0 \qquad (x \text{ running from 0 to 1})$$
$$C_2: x = 1 \qquad (y \text{ running from 0 to 1})$$

Figure 3
The curve C in Example 3

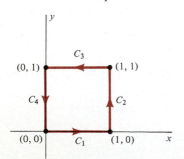

$$C_3: y = 1 \qquad (x \text{ running from 1 to 0})$$
$$C_4: x = 0 \qquad (y \text{ running from 1 to 0})$$

Letting $\mathbf{x} = (x, y)$, we have

$$\mathbf{F}(\mathbf{x}) \cdot d\mathbf{x} = (4x + y, x + 2y) \cdot (dx, dy) = (4x + y)\, dx + (x + 2y)\, dy$$

Since $dy = 0$ on C_1 and C_3, while $dx = 0$ on C_2 and C_4, we find

$$\int_C \mathbf{F}(\mathbf{x}) \cdot d\mathbf{x} = \int_C (4x + y)\, dx + (x + 2y)\, dy$$

$$= \int_0^1 4x\, dx + \int_0^1 (1 + 2y)\, dy + \int_1^0 (4x + 1)\, dx + \int_1^0 2y\, dy$$

$$= \int_0^1 dy + \int_1^0 dx = 0 \qquad\qquad \square$$

The answer zero in Example 3 is no accident; it could have been predicted. Not merely, however, because the curve is closed (although that is part of the reason). In general the line integral of a vector field around a closed path is not zero. What makes the answer zero in Example 3 is that the function

$$\mathbf{F}(x, y) = (4x + y, x + 2y)$$

is the *gradient of a scalar field,* namely the function

$$f(x, y) = 2x^2 + xy + y^2$$

As you can see,

$$\nabla f(x, y) = (4x + y, x + 2y) = \mathbf{F}(x, y)$$

To grasp the significance of this, recall that in multivariable calculus the gradient plays the role of the derivative. We know from the Fundamental Theorem of Calculus that if $f'(x) = F(x)$, then

$$\int_a^b F(x)\, dx = f(x) \Big|_a^b = f(b) - f(a)$$

a result that depends only on the endpoints a and b. The analogous result here is that if $\nabla f(\mathbf{x}) = \mathbf{F}(\mathbf{x})$, then

$$\int_C \mathbf{F}(\mathbf{x}) \cdot d\mathbf{x} = f(\mathbf{b}) - f(\mathbf{a})$$

where $\mathbf{a}$ and $\mathbf{b}$ are the initial and terminal points of C. More precisely, we can prove the following theorem.

Fundamental Theorem of Calculus for Line Integrals

Suppose that the function $\mathbf{F}(\mathbf{x})$ is a gradient field, that is,

$$\mathbf{F}(\mathbf{x}) = \nabla f(\mathbf{x})$$

for some scalar field $f(\mathbf{x})$ (in a region D of the plane). If $\mathbf{a}$ and $\mathbf{b}$ are any points of D, then

$$\int_C \mathbf{F}(\mathbf{x}) \cdot d\mathbf{x} = f(\mathbf{b}) - f(\mathbf{a})$$

for any piecewise smooth curve C that runs from $\mathbf{a}$ to $\mathbf{b}$ in D.

The proof is enlightening. Suppose first that the curve is smooth,

$$C: \mathbf{x} = (x(t), y(t)) \qquad t_1 \le t \le t_2$$

where $\mathbf{x}(t_1) = \mathbf{a}$ and $\mathbf{x}(t_2) = \mathbf{b}$. Then

$$\int_C \mathbf{F}(\mathbf{x}) \cdot d\mathbf{x} = \int_C \nabla f(\mathbf{x}) \cdot d\mathbf{x} = \int_{t_1}^{t_2} \left[\nabla f(\mathbf{x}) \cdot \frac{d\mathbf{x}}{dt} \right] dt$$

$$= \int_{t_1}^{t_2} \left[\frac{d}{dt} f(\mathbf{x}) \right] dt \qquad \text{(the Chain Rule of Section 17.2)}$$

$$= f(\mathbf{x}) \Big|_{t=t_1}^{t=t_2} \qquad \text{(the Fundamental Theorem of Calculus)}$$

$$= f(\mathbf{x}(t_2)) - f(\mathbf{x}(t_1)) = f(\mathbf{b}) - f(\mathbf{a})$$

If the path is piecewise smooth, the same argument applies to the individual pieces; the functional values at the junction points cancel out. (Why?)

□ **Example 4**   Use the Fundamental Theorem for Line Integrals to predict the result of Example 3.

Solution: As we have already noted, the function

$$\mathbf{F}(x,y) = (4x + y, x + 2y)$$

is a gradient field, that is,

$$\mathbf{F}(\mathbf{x}) = \nabla f(\mathbf{x}) \qquad \text{where } f(x,y) = 2x^2 + xy + y^2$$

Hence its line integral around the square in Figure 3 is

$$\int_C \nabla f(\mathbf{x}) \cdot d\mathbf{x} = f(0,0) - f(0,0) = 0 \qquad \qquad \square$$

Now we return to the motivating idea of this section (the definition of work as the line integral of force). We are in good position to collect a remarkable bonus in physics! Suppose that a particle moves along the curve

$$C: \mathbf{x} = (x(t), y(t)), \ t_1 \leq t \leq t_2 \qquad \text{(where } t \text{ is time)}$$

in the force field $\mathbf{F(x)}$. The work done by $\mathbf{F}$ during the time interval $[t_1, t_2]$ is

$$W = \int_C \mathbf{F(x)} \cdot d\mathbf{x} = \int_{t_1}^{t_2} \left[\mathbf{F(x)} \cdot \frac{d\mathbf{x}}{dt} \right] dt$$

According to Newton's Second Law, $\mathbf{F(x)} = m \, (d\mathbf{v}/dt)$, where m is the mass of the particle and $\mathbf{v} = d\mathbf{x}/dt$ is its velocity at time t. If $v = |\mathbf{v}|$ is the speed, then

$$W = \int_{t_1}^{t_2} \left(m \frac{d\mathbf{v}}{dt} \cdot \mathbf{v} \right) dt = \frac{1}{2} m \int_{t_1}^{t_2} \left[\frac{d}{dt} (\mathbf{v} \cdot \mathbf{v}) \right] dt \qquad \text{(why?)}$$

$$= \frac{1}{2} m \int_{t_1}^{t_2} \left[\frac{d}{dt} (v^2) \right] dt = \frac{1}{2} m v^2 \Big|_{t=t_1}^{t=t_2} = \frac{1}{2} m v_2^2 - \frac{1}{2} m v_1^2$$

where v_1 and v_2 are the initial and terminal speeds, respectively. Physicists define the **kinetic energy** of a moving object to be $\frac{1}{2} m v^2$, so we have proved that *the work done is the change in the kinetic energy.* (See Section 7.5 for a similar discussion in the case of linear motion.)

Another definition from physics is that if $\mathbf{F}(x, y)$ is a force field, its **potential energy** function is a scalar field $V(x, y)$ with the property

$$\nabla V(x, y) = -\mathbf{F}(x, y)$$

(provided that such a function exists). From our point of view, V is just the negative of what we have been calling f; it exists if and only if $\mathbf{F}$ is a gradient field. When it does, we have

$$\int_C \mathbf{F(x)} \cdot d\mathbf{x} = -\int_C \nabla V(\mathbf{x}) \cdot d\mathbf{x} = -[V(\mathbf{b}) - V(\mathbf{a})] = V(\mathbf{a}) - V(\mathbf{b})$$

where C is any curve that runs from $\mathbf{a}$ to $\mathbf{b}$.

We may combine this result with

$$W = \int_C \mathbf{F(x)} \cdot d\mathbf{x} = \frac{1}{2} m v_2^2 - \frac{1}{2} m v_1^2$$

to obtain $V(\mathbf{a}) - V(\mathbf{b}) = \frac{1}{2} m v_2^2 - \frac{1}{2} m v_1^2$, that is,

$$V(\mathbf{a}) + \tfrac{1}{2} m v_1^2 = V(\mathbf{b}) + \tfrac{1}{2} m v_2^2$$

This says that the sum of the potential energy and the kinetic energy is the same at the initial point as at the terminal point. Since these points were arbitrary, *the sum is constant.* In other words, the total energy of the particle does not change during the motion. All that happens is that the force does work to convert potential energy into kinetic energy (or vice versa). This is the *Law of Conservation of Energy,* a fundamental result in physics.

For this reason a force field that is a gradient (and thus has a potential energy function) is said to be **conservative.** In such a field, motion obeys the energy conservation law. The earth's gravitational field is conservative; in fact so is any field satisfying the inverse-square law. Gravity, electricity, and magnetism all behave this way, which is why these results are so profound in physics.

Problem Set 19.1

Use the definition to find the line integral of $F(x,y) = (y,x)$ along each of the following paths.

1. The parabola $y = x^2$ from $(-1,1)$ to $(1,1)$.

2. The parabola $y = x^2$ from $(1,1)$ to $(-1,1)$.

3. The line $y = 1$ from $(-1,1)$ to $(1,1)$.

4. Counterclockwise around the circle $x^2 + y^2 = 1$.

5. Counterclockwise around the triangle with vertices $(0,0)$, $(1,0)$, $(0,1)$.

6. Confirm that $F(x,y) = (y,x)$ is the gradient of $f(x,y) = xy$. Then use the Fundamental Theorem for Line Integrals to evaluate

$$\int_{(-1,1)}^{(1,1)} F(x) \cdot dx$$

7. Why do we get the same result in Problems 1 and 3?

8. Why do we get zero in Problems 4 and 5?

Use the definition to find

$$\int_C (x + y)\, dx + (x - y)\, dy$$

where C is defined as follows.

9. The line $y = x$ from $(0,0)$ to $(1,1)$.

10. The parabola $y = x^2$ from $(0,0)$ to $(1,1)$.

11. $x = t^2$, $y = t^3$, $0 \leq t \leq 1$.

12. $x = \cos t$, $y = \sin t$, $0 \leq t \leq 2\pi$

13. $x = 1 - t$, $y = 1 - t$, $0 \leq t \leq 1$

14. The rectangle with vertices $(0,0)$, $(a,0)$, (a,b), $(0,b)$, where a and b are positive and the direction is counterclockwise.

15. Confirm that $F(x,y) = (x + y, x - y)$ is the gradient of $f(x,y) = \frac{1}{2}x^2 + xy - \frac{1}{2}y^2$. Then use the Fundamental Theorem for Line Integrals to evaluate

$$\int_{(0,0)}^{(1,1)} F(x) \cdot dx$$

16. Why do we get the same result in Problems 9, 10, and 11?

17. Why do we get zero in Problems 12 and 14?

18. Why are the answers to Problems 9 and 13 opposite in sign?

In Problems 19 through 22 use the definition to find

$$\int_C (x^2 + y^2)\, dx + 2xy\, dy$$

where C is defined as follows.

19. The semicircle $x^2 + y^2 = 1$, $y \geq 0$, oriented counterclockwise.

20. $x = \sin t$, $y = \cos t$, $-\pi/2 \leq t \leq \pi/2$.

21. The parabola $y = 1 - x^2$ from $(1,0)$ to $(-1,0)$.

22. The closed path that bounds the semicircular region $x^2 + y^2 \leq 1$, $y \geq 0$.

23. Confirm that $\nabla(\frac{1}{3}x^3 + xy^2) = (x^2 + y^2, 2xy)$ and use the Fundamental Theorem for Line Integrals to do Problem 19.

24. Use the Fundamental Theorem to do Problem 20.

25. Use the Fundamental Theorem to do Problem 21.

26. Use the Fundamental Theorem to do Problem 22.

27. Use the definition to find

$$\int_C e^x \sin y\, dx + e^x \cos y\, dy$$

where C is the square with vertices $(0,0)$, $(1,0)$, $(1,1)$, $(0,1)$, oriented counterclockwise. Can you name a scalar field whose gradient is

$$F(x,y) = (e^x \sin y, e^x \cos y)?$$

28. Evaluate the line integral of

$$F(x,y) = (\cos y, \cos x)$$

around the triangle with vertices $(0,0)$, $(\pi,0)$, (π,π) in the counterclockwise direction. Can you name a scalar field whose gradient is $F(x,y)$?

29. Evaluate $\int_C y \cos x \, dx + x \sin y \, dy$, where C is the square with vertices $(0,0)$, $(\pi,0)$, (π,π), $(0,\pi)$, oriented counterclockwise. Can you name a scalar field whose gradient is

$$F(x,y) = (y \cos x, x \sin y)?$$

30. Evaluate $\int_C x^3 \, dy - y^3 \, dx$, where C is the triangle with vertices $(0,0)$, $(1,0)$, $(0,1)$, oriented counterclockwise. Can you name a scalar field whose gradient is

$$F(x,y) = (-y^3, x^3)?$$

31. Evaluate the line integral of

$$F(x,y) = \frac{1}{x^2 + y^2}(1,1)$$

along the upper half of the circle $x^2 + y^2 = a^2$ from $(a,0)$ to $(-a,0)$.

32. Evaluate $\int_C dx/y + dy/x$, where C is the hyperbola $xy = 2$ from $(1,2)$ to $(2,1)$.

33. Given the vector $\mathbf{k}$, confirm that $\nabla(\mathbf{k} \cdot \mathbf{x}) = \mathbf{k}$ for all $\mathbf{x}$ and use the result to show that

$$\int_C \mathbf{k} \cdot d\mathbf{x} = \mathbf{k} \cdot (\mathbf{b} - \mathbf{a})$$

where C is any path from $\mathbf{a}$ to $\mathbf{b}$.

34. Confirm that $\nabla(\mathbf{x} \cdot \mathbf{x}) = 2\mathbf{x}$ for all $\mathbf{x}$ and use the result to show that

$$\int_C 2\mathbf{x} \cdot d\mathbf{x} = \mathbf{a} \cdot \mathbf{a}$$

where C is any path from $\mathbf{0}$ to $\mathbf{a}$.

35. Prove that if $F(\mathbf{x})$ is a gradient field, then

$$\int_C F(\mathbf{x}) \cdot d\mathbf{x}$$

is independent of the path, that is, the integral depends only on the initial and terminal points $\mathbf{a}$ and $\mathbf{b}$, not on C.

36. Prove that if $F(\mathbf{x})$ is a gradient field, then

$$\int_C F(\mathbf{x}) \cdot d\mathbf{x} = 0$$

for every closed curve C.

37. Suppose that the path of a particle through a force field is always perpendicular to the lines of force. Explain why no work is done by the force.

38. Suppose that the force $F(\mathbf{x})$ is directed toward the origin and has magnitude inversely proportional to the square of the distance from the origin. Show that $F(\mathbf{x}) = (-k/r^3)\mathbf{x}$, where $r = |\mathbf{x}|$ and k is a positive constant.

39. In Problem 38 let C be any circle centered at the origin. Use Problem 37 to explain why the line integral of F around C is zero.

40. Confirm Problem 39 by evaluating the line integral directly.

41. In Problem 38 confirm that $F(\mathbf{x}) = \nabla(k/r)$. Why is this another proof of Problem 39?

42. Suppose that an object of mass m moves in the xy plane subject to the constant gravitational force $F = -mg\mathbf{j}$. (The x axis is ground level and the y axis points up.) Show that the work done by gravity when the object goes from (x_1, y_1) to (x_2, y_2) is

$$W = mg(y_1 - y_2)$$

regardless of the path.

43. Let $V(x,y)$ be a potential energy function of the force field $F(\mathbf{x})$, that is,

$$\nabla V(x,y) = -F(x,y)$$

If $\mathbf{a}$ is a point at which the potential energy is zero, show that the work done against F in moving from $\mathbf{a}$ to $\mathbf{x}$ is $V(\mathbf{x})$. (In other words, the potential energy at a point is the work done in getting to that point from a point of zero potential.)

44. Let $T(\mathbf{x})$ be the unit tangent vector (in the direction of motion) at each point of the curve

$$C: \mathbf{x} = (x(t), y(t)) \qquad t_1 \leq t \leq t_2$$

Show that the length of C is given by $\int_C T(\mathbf{x}) \cdot d\mathbf{x}$.

45. Confirm that

$$\nabla\left(\tan^{-1}\frac{x}{y}\right) = F(x,y) = \left(\frac{y}{x^2 + y^2}, \frac{-x}{x^2 + y^2}\right)$$

a result that appears to prove that the line integral of $F(\mathbf{x})$ around a closed path is zero. Show, however, that the integral of F around the unit circle is not zero. What is the explanation?

19.2
GRADIENT FIELDS

The Fundamental Theorem of Calculus for Line Integrals, proved in Section 19.1, says that if the vector field $\mathbf{F}$ is the gradient of a scalar field f, then

$$\int_C \mathbf{F}(\mathbf{x}) \cdot d\mathbf{x} = f(\mathbf{b}) - f(\mathbf{a})$$

where C is a piecewise smooth curve running from $\mathbf{a}$ to $\mathbf{b}$.

An implication of the Fundamental Theorem is that *the line integral of a gradient field is independent of the path.* For the expression

$$\int_C \nabla f(\mathbf{x}) \cdot d\mathbf{x} = f(\mathbf{b}) - f(\mathbf{a})$$

depends only on the initial and terminal points $\mathbf{a}$ and $\mathbf{b}$, its value being the same for every path C from $\mathbf{a}$ to $\mathbf{b}$.

□ **Example 1**　　Prove that the vector field $\mathbf{F}(x,y) = (xy, 2x - y)$ is not a gradient field.

Solution:　　In Example 1 of Section 19.1 we computed the line integral of $\mathbf{F}$ from $(0,0)$ to $(1,1)$ along the parabola $y = x^2$ and got $\frac{13}{12}$. In Example 2 we changed the path to the line $y = x$ and got $\frac{5}{6}$. If $\mathbf{F}$ were a gradient field, these results would be the same. Since they are not, there is no scalar field $f(x,y)$ whose gradient is $\mathbf{F}(x,y)$.　　　　　　　□

Here is one of the places where you should note the enormous distinction between single-variable and multivariable calculus. Every (continuous) function of one variable has an antiderivative, that is, given $F(x)$ we can always find a function $f(x)$ such that $f'(x) = F(x)$. The formula

$$\int_a^b F(x) \, dx = f(b) - f(a)$$

is therefore universal. "Independence of path" is an empty concept on the number line, because there is only one way to go from a to b. When we change from $\mathcal{R}$ to $\mathcal{R}^2$ the possibilities increase by a quantum leap. Not every vector function is the gradient of another function; not every line integral is independent of the path. That makes life in higher dimensions altogether richer!

□ **Example 2**　　Find the line integral of the vector field

$$\mathbf{F}(x,y) = (y \cos xy, x \cos xy)$$

along the curve $y = 1 - x^2$, $0 \le x \le 1$.

Solution:　　Letting $\mathbf{x} = (x,y)$, we have

$$\mathbf{F}(\mathbf{x}) \cdot d\mathbf{x} = y \cos xy \, dx + x \cos xy \, dy$$

Since $y = 1 - x^2$ and $dy = -2x \, dx$, this becomes

$$\mathbf{F}(\mathbf{x}) \cdot d\mathbf{x} = (1 - x^2) \cos(x - x^3) \, dx - 2x^2 \cos(x - x^3) \, dx$$
$$= (1 - 3x^2) \cos(x - x^3) \, dx$$

and hence

$$\int_C \mathbf{F}(\mathbf{x}) \cdot d\mathbf{x} = \int_0^1 (1 - 3x^2) \cos(x - x^3) \, dx$$

$$= \int_0^0 \cos u \, du = 0 \qquad [u = x - x^3, \, du = (1 - 3x^2) \, dx]$$

However, this is the hard way. Had we noticed that $\mathbf{F}(x, y)$ is the gradient of $f(x, y) = \sin xy$ (confirm!), we could have integrated over any path from $(0,1)$ to $(1,0)$, as shown in Figure 1. Take the easy route consisting of

$$C_1: x = 0 \qquad (y \text{ running from 1 to 0})$$
$$C_2: y = 0 \qquad (x \text{ running from 0 to 1})$$

Since $x = 0$ and $dx = 0$ on C_1, while $y = 0$ and $dy = 0$ on C_2, we have

$$\int_C y \cos xy \, dx + x \cos xy \, dy = \int_1^0 0 \, dy + \int_0^1 0 \, dx = 0$$

This is still not the easiest way! We need only use the Fundamental Theorem:

$$\int_C \mathbf{F}(\mathbf{x}) \cdot d\mathbf{x} = \sin xy \Big|_{(0,1)}^{(1,0)} = \sin 0 - \sin 0 = 0 \qquad \square$$

Another implication of the Fundamental Theorem is that *the line integral of a gradient field around a closed path is zero*. For if the initial and terminal points $\mathbf{a}$ and $\mathbf{b}$ are equal, we have

$$\int_C \nabla f(\mathbf{x}) \cdot d\mathbf{x} = f(\mathbf{a}) - f(\mathbf{a}) = 0$$

□ **Example 3**   If C is the unit circle (oriented counterclockwise), find

$$\int_C (2x + y) \, dx + (x + 2y) \, dy$$

Solution: The function $\mathbf{F}(x, y) = (2x + y, x + 2y)$ is the gradient of $f(x, y) = x^2 + xy + y^2$, as you can check. The integral is therefore zero.

□

So far we have shown that

Gradient Field $\Rightarrow$ Integral Around Closed Path Is Zero

$$\Downarrow$$

Integral Is Independent of Path

Figure 1
Alternate paths from $(0,1)$ to $(1,0)$

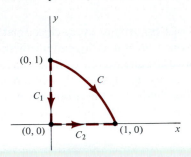

Figure 2

Closed path broken into two paths between **a** and **b**

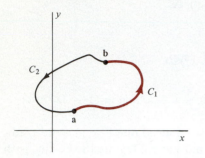

Actually these three concepts are equivalent (that is, each implies the other two). Suppose, for example, that the line integral of the vector field $\mathbf{F}(\mathbf{x})$ is independent of the path. Then its integral around a closed path must be zero. To see why, look at Figure 2, in which we have chosen two points **a** and **b** on the closed path C and have broken C into the paths C_1 from **a** to **b** and C_2 from **b** to **a**. Since C_1 and $-C_2$ are both paths from **a** to **b**, independence of path implies that

$$\int_{C_1} \mathbf{F}(\mathbf{x}) \cdot d\mathbf{x} = \int_{-C_2} \mathbf{F}(\mathbf{x}) \cdot d\mathbf{x} = -\int_{C_2} \mathbf{F}(\mathbf{x}) \cdot d\mathbf{x}$$

Hence

$$\int_C \mathbf{F}(\mathbf{x}) \cdot d\mathbf{x} = \int_{C_1} \mathbf{F}(\mathbf{x}) \cdot d\mathbf{x} + \int_{C_2} \mathbf{F}(\mathbf{x}) \cdot d\mathbf{x} = 0$$

Conversely, suppose that the integral of $\mathbf{F}(\mathbf{x})$ around every closed path is zero. Then the integral along a path from **a** to **b** is independent of the path, as you can see by reversing the above argument.

To finish the argument for equivalence, we must show that independence of path implies that $\mathbf{F}(\mathbf{x})$ is a gradient field. (See the problem set for an outline of the proof.)

Let $\mathbf{F}$ be a continuous vector field. Then each of the following conditions is equivalent to the other two.

1. $\mathbf{F}$ is a gradient field.
2. The line integral of $\mathbf{F}$ around every (piecewise smooth) closed curve is zero.
3. The line integral of $\mathbf{F}$ from one point to another is independent of the path.

By now the reader, although satisfied with the theory, may feel that a critical question is being overlooked. Given a vector field $\mathbf{F}(x,y)$, how do we discover a function $f(x,y)$ whose gradient is $\mathbf{F}(x,y)$? In other words, how do we find an *antigradient* of $\mathbf{F}$? We have been putting them down without explanation, as in Example 2, where we said that $f(x,y) = \sin xy$ is an antigradient of

$$\mathbf{F}(x,y) = (y \cos xy, x \cos xy)$$

What do we do if it is not obvious what f is, or whether it exists?

The answer is deceptively simple. For if

$$\mathbf{F}(\mathbf{x}) = (F(\mathbf{x}), G(\mathbf{x}))$$

is a gradient, say $\mathbf{F}(\mathbf{x}) = \nabla f(\mathbf{x})$, where f is some scalar field, then

$$F(\mathbf{x}) = f_1(\mathbf{x}) \qquad \text{and} \qquad G(\mathbf{x}) = f_2(\mathbf{x})$$

Simply differentiate these functions in such a way as to take advantage of the equality of mixed partial derivatives:

$$F_2(\mathbf{x}) = f_{12}(\mathbf{x}) = f_{21}(\mathbf{x}) = G_1(\mathbf{x})$$

(We assume that F and G have continuous partial derivatives, so that f_{12} and f_{21} are the same.)

Test for a Gradient Field

Suppose that $\mathbf{F}(\mathbf{x}) = (F(x,y), G(x,y))$ is a gradient field with continuous partial derivatives. Then

$$\frac{\partial G}{\partial x} = \frac{\partial F}{\partial y}$$

This is a *necessary* condition for a gradient field, that is, any field that fails the test is not a gradient. Whether it is also sufficient (does $\partial G/\partial x = \partial F/\partial y$ imply that $\mathbf{F}$ is a gradient?) is a fascinating question that leads to the central theorem of this chapter. First, however, let's see how our test works.

☐ **Example 4**   Show that the function $\mathbf{F}(x,y) = (xy, 2x - y)$ is not a gradient field.

Solution: We have already done this in Example 1, but the argument depended on the calculation of line integrals along different paths (Examples 1 and 2, Section 19.1). It is easier to use our test:

$$\frac{\partial}{\partial x}(2x - y) = 2 \qquad \text{and} \qquad \frac{\partial}{\partial y}(xy) = x$$

Since these are different, $\mathbf{F}$ cannot be a gradient.   ☐

☐ **Example 5**   Find a scalar field whose gradient is

$$\mathbf{F}(x,y) = (4x + y, x + 2y)$$

Solution: Following Example 3 in the last section we gave such a field, namely $f(x,y) = 2x^2 + xy + y^2$, but we did not explain where it came from. Perhaps you could have named it yourself, by inspection, but in general that is hard to do (just as antidifferentiation is usually not obvious in single-variable calculus). Let's act ignorant and see how such an "anti-gradient" can be discovered.
First we test $\mathbf{F}$ to see if it even has a chance to be a gradient. Since

$$\frac{\partial}{\partial x}(x + 2y) = 1 \qquad \text{and} \qquad \frac{\partial}{\partial y}(4x + y) = 1$$

the necessary condition is satisfied. That doesn't prove that **F** is a gradient (at least not yet, because we haven't discussed whether the condition is sufficient). But **F** *may* be a gradient; hence we have some reason to search for an antigradient.

The obvious thing to do (by analogy with single-variable calculus) is to integrate the components of **F**. Since they are functions of two variables, it will have to be "partial integration," that is, we will hold one variable fixed and integrate with respect to the other. As it turns out, we need only integrate one of the components; the other will fall into place (or else it will not) depending on whether an antigradient exists.

We begin by integrating $F(x,y) = 4x + y$ with respect to x (holding y fixed). This produces a *candidate* for an antigradient:

$$f(x,y) = \int (4x + y)\, dx = 2x^2 + xy + C(y)$$

where $C(y)$ is a "constant" as far as x is concerned, that is, a function of y alone. This function has the correct partial derivative with respect to x:

$$f_1(x,y) = 4x + y = F(x,y)$$

(Our integration guarantees that much.) Does it also have the correct partial derivative with respect to y? That is, does

$$f_2(x,y) = x + 2y = G(x,y)?$$

To find out, we differentiate our candidate with respect to y:

$$f_2(x,y) = x + C'(y) \overset{?}{=} x + 2y$$

This looks promising because the terms involving x drop out, leaving us with a differential equation $C'(y) = 2y$ that involves y alone. The general solution is $C(y) = y^2 + k$ (where k is an arbitrary constant). Hence our candidate for an antigradient takes the explicit form

$$f(x,y) = 2x^2 + xy + y^2 + k$$

Both of its partial derivatives are correct,

$$f_1(x,y) = 4x + y = F(x,y) \qquad \text{and} \qquad f_2(x,y) = x + 2y = G(x,y)$$

so we are done: $\nabla f(x,y) = \mathbf{F}(x,y)$.

Note that we could have started this problem by integrating $G(x,y) = x + 2y$ with respect to y:

$$f(x,y) = \int (x + 2y)\, dy = xy + y^2 + C(x)$$

where $C(x)$ is a function of x alone. The additional condition to be satisfied is then $f_1(x,y) = F(x,y)$, that is, $y + C'(x) = 4x + y$. Hence

$$C'(x) = 4x, \quad C(x) = 2x^2 + k$$

and we have found $f(x,y) = 2x^2 + xy + y^2 + k$ as before.   □

□ **Example 6**   Find out what goes wrong if we try to find a scalar field whose gradient is $\mathbf{F}(x,y) = (xy, 2x - y)$.

Solution: We know from Example 4 that $\mathbf{F}$ is not a gradient, so something will go wrong. It is interesting to see where it does. Integrate the first component of $\mathbf{F}$ with respect to x:

$$f(x,y) = \int xy \, dx = \frac{1}{2}x^2y + C(y)$$

Then

$$f_2(x,y) = \frac{1}{2}x^2 + C'(y) \overset{?}{=} 2x - y$$

The terms involving x do not cancel and we cannot solve for $C(y)$. Try it the other way, integrating the second component of $\mathbf{F}$ with respect to y:

$$f(x,y) = \int (2x - y) \, dy = 2xy - \frac{1}{2}y^2 + C(x)$$

Then

$$f_1(x,y) = 2y + C'(x) \overset{?}{=} xy$$

This time the terms involving y do not cancel and we cannot solve for $C(x)$. Either way it is hopeless.   □

We still haven't settled the question of whether the condition $\partial G/\partial x = \partial F/\partial y$ is sufficient for the vector function $\mathbf{F} = (F,G)$ to be a gradient. It will be the first order of business in the next section.

Problem Set 19.2

Which of the following vector fields fails to satisfy the necessary condition $\partial G/\partial x = \partial F/\partial y$ for a gradient field?

1. $\mathbf{F}(x,y) = (y,x)$

2. $\mathbf{F}(x,y) = (x + y, x - y)$

3. $\mathbf{F}(x,y) = (x^2 + y^2, 2xy)$

4. $\mathbf{F}(x,y) = (y \cos xy, x \cos xy)$

5. $\mathbf{F}(x,y) = (e^x \sin y, e^x \cos y)$

6. $\mathbf{F}(x,y) = (\cos y, \cos x)$

7. $\mathbf{F}(x,y) = (y \cos x, x \sin y)$

8. $\mathbf{F}(x,y) = (-y^3, x^3)$

9. $\mathbf{F}(x,y) = \frac{1}{x^2 + y^2}(1,1)$

10. $\mathbf{F}(x,y) = \left(\frac{1}{y}, \frac{1}{x}\right)$

11. $\mathbf{F}(x,y) = \left(\frac{x}{x^2 + y^2}, \frac{y}{x^2 + y^2}\right)$

12. $\mathbf{F}(x,y) = \left(\frac{y}{x^2 + y^2}, \frac{-x}{x^2 + y^2}\right)$

13. $\mathbf{F}(x,y) = (ye^{xy} + 2x, xe^{xy} - 2y)$

In each of the following, use "partial integration" (Examples 5 and 6) to find an antigradient of the given vector field, or determine that no antigradient exists.

14. $\mathbf{F}(x,y) = (y,x)$

15. $\mathbf{F}(x,y) = (x + y, x - y)$

16. $\mathbf{F}(x,y) = (x^2 + y^2, 2xy)$

17. $\mathbf{F}(x,y) = (y \cos xy, x \cos xy)$

18. $\mathbf{F}(x,y) = (e^x \sin y, e^x \cos y)$

19. $\mathbf{F}(x,y) = (\cos y, \cos x)$

20. $\mathbf{F}(x,y) = (y \cos x, x \sin y)$

21. $\mathbf{F}(x,y) = (-y^3, x^3)$

22. $\mathbf{F}(x,y) = \frac{1}{x^2 + y^2}(1,1)$

23. $\mathbf{F}(x,y) = \left(\frac{1}{y}, \frac{1}{x}\right)$

24. $\mathbf{F}(x,y) = \left(\frac{x}{x^2 + y^2}, \frac{y}{x^2 + y^2}\right)$

25. $F(x, y) = \left(\dfrac{y}{x^2 + y^2}, \dfrac{-x}{x^2 + y^2} \right)$

26. $F(x, y) = (ye^{xy} + 2x, xe^{xy} - 2y)$

27. Assume that the earth's gravitational field is conservative. (See Section 19.1.) Why does this imply that the work done against gravity in lifting an object is the same regardless of the route chosen?

28. In Problem 27 explain why the work done by gravity in pulling a satellite around one orbit is zero. Neglecting friction, explain why the satellite will remain in orbit indefinitely.

29. Show that

$$\int_C 2xy\, dx + x^2\, dy = x_2{}^2 y_2 - x_1{}^2 y_1$$

for any smooth curve C from (x_1, y_1) to (x_2, y_2).

30. Show that

$$\int_C x\, dx + y\, dy = 0$$

for every (smooth) closed curve C.

31. Let C be the unit circle $x = \cos t$, $y = \sin t$, $0 \leq t \leq 2\pi$. Find

$$\int_C (x + y)\, dx + xy\, dy$$

What does the result imply about the vector field $F(x, y) = (x + y, xy)$? How could we have arrived at the same conclusion with less work?

32. Let C be any smooth curve from $(1, -1)$ to $(1, 1)$. Explain why

$$\int_C y\, dx + x\, dy = 2$$

33. Let C be any (smooth) closed curve. Explain why

$$\int_C ye^{xy}\, dx + xe^{xy}\, dy = 0$$

34. In the text we proved that if the line integral of the vector field $F(x)$ is independent of the path, its integral around a closed path must be zero. Reverse the argument to prove the converse.

35. Suppose that $F(x) = (F(x), G(x))$ is a continuous vector field whose line integral is independent of the path. Show that F is the gradient of the function

$$f(x) = \int_a^x F(u) \cdot du$$

(where a is some fixed point of the domain of F) as follows.

(a) By the definition of partial derivative,

$$f_1(x, y) = \lim_{h \to 0} \frac{f(x + h, y) - f(x, y)}{h}$$

Explain why the numerator of the difference quotient is

$$f(x + h, y) - f(x, y) = \int_{(x, y)}^{(x+h, y)} F(u) \cdot du$$

(b) To find the integral in part (a), take the straight path from (x, y) to $(x + h, y)$. Explain why this yields

$$f(x + h, y) - f(x, y) = \int_x^{x+h} F(u, y)\, du = F(c, y)h$$

where c is between x and $x + h$. *Hint:* Use the Theorem of the Mean for Integrals (Section 6.3).

(c) Divide by h and let $h \to 0$ to conclude that $f_1(x, y) = F(x, y)$. (The same argument shows that $f_2 = G$, so $\nabla f = F$.)

19.3

GREEN'S THEOREM

In the last section we proved that if $F = (F, G)$ is a vector field, the condition $\partial G / \partial x = \partial F / \partial y$ is *necessary* for F to be a gradient field. We did not settle the question of whether the condition is also *sufficient,* that is, whether it implies that F is a gradient. Another way of posing the question is to ask whether the condition $\partial G / \partial x = \partial F / \partial y$ implies that the line integral of F from one point to another is independent of the path. (See Section 19.2.)

Given two points in the domain of F (see Figure 1), suppose that we integrate F along the paths C_1 and C_2 shown in the figure. If the integral is independent of the path (we don't know whether it is or not, but let's assume

Figure 1

Alternate routes from (a,c) to (b,d)

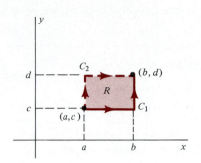

it for the moment), we have

$$\int_{C_1} \mathbf{F}(\mathbf{x}) \cdot d\mathbf{x} = \int_{C_2} \mathbf{F}(\mathbf{x}) \cdot d\mathbf{x}$$

$$\int_{C_1} F(x,y)\,dx + G(x,y)\,dy = \int_{C_2} F(x,y)\,dx + G(x,y)\,dy$$

$$\int_a^b F(x,c)\,dx + \int_c^d G(b,y)\,dy = \int_c^d G(a,y)\,dy + \int_a^b F(x,d)\,dx$$

$$\int_c^d [G(b,y) - G(a,y)]\,dy = \int_a^b [F(x,d) - F(x,c)]\,dx$$

Now comes the clever part! By the Fundamental Theorem of Calculus we can write

$$G(b,y) - G(a,y) = G(x,y)\,\Big|_{x=a}^{x=b} = \int_a^b \frac{\partial}{\partial x} G(x,y)\,dx$$

$$F(x,d) - F(x,c) = F(x,y)\,\Big|_{y=c}^{y=d} = \int_c^d \frac{\partial}{\partial y} F(x,y)\,dy$$

Hence our equation takes the form

$$\int_c^d \int_a^b G_1(x,y)\,dx\,dy = \int_a^b \int_c^d F_2(x,y)\,dy\,dx$$

These are iterated integrals that can be written as double integrals over the region R in Figure 1:

$$\iint_R G_1(x,y)\,dA = \iint_R F_2(x,y)\,dA$$

Suddenly the light should dawn. The condition $\partial G/\partial x = \partial F/\partial y$ implies that these double integrals are equal. By reversing the steps, we conclude that

$$\int_{C_1} \mathbf{F}(\mathbf{x}) \cdot d\mathbf{x} = \int_{C_2} \mathbf{F}(\mathbf{x}) \cdot d\mathbf{x}$$

It is tempting to infer that $\mathbf{F}$ is a gradient. We have not yet proved independence of path, however, but only that the integrals of $\mathbf{F}$ along the curves C_1 and C_2 in Figure 1 are the same. Let's postpone any conclusion about our sufficient condition and look at something else that we *have* proved.

The above argument shows that

$$\int_{C_1} \mathbf{F} \cdot d\mathbf{x} - \int_{C_2} \mathbf{F} \cdot d\mathbf{x} = \iint_R G_1\,dA - \iint_R F_2\,dA$$

in any case (regardless of whether $\mathbf{F}$ is a gradient, whether $\partial G/\partial x = \partial F/\partial y$, or whether the integrals along C_1 and C_2 are the same). Let C be the closed path that runs counterclockwise around the rectangle R in Figure 1 (consisting of the pieces C_1 and $-C_2$). Then our equation reads

$$\int_C \mathbf{F} \cdot d\mathbf{x} = \iint_R (G_1 - F_2)\,dA$$

a formula known as **Green's Theorem** (in honor of the English physicist George Green, 1793–1841).

Before generalizing this theorem to nonrectangular regions, we need a definition.

Any region D that can be written in the two ways

$$D = \{(x,y): a \leq x \leq b,\ g(x) \leq y \leq h(x)\}$$
$$D = \{(x,y): c \leq y \leq d,\ p(y) \leq x \leq q(y)\}$$

is called a *standard* region. (An example is shown in Figure 2. This is the type of region needed for iteration of a double integral in either order. See Section 18.2.)

Now we are ready to give an argument for Green's Theorem in standard regions that is similar to the argument already given for rectangles. We start with double integrals this time (a mysterious place to start if we did not have Green's Theorem in hand for rectangular regions). Let C be the boundary of D oriented counterclockwise, where D is a standard region labeled as in Figure 2. Break C into the curves

$$C_1: y = g(x) \qquad (x \text{ running from } a \text{ to } b)$$
$$C_2: y = h(x) \qquad (x \text{ running from } b \text{ to } a)$$

as shown in Figure 2. Then

$$-\iint_D F_2(x,y)\, dA = -\int_a^b \int_{g(x)}^{h(x)} F_2(x,y)\, dy\, dx$$

$$= -\int_a^b [F(x,h(x)) - F(x,g(x))]\, dx \qquad \text{(why?)}$$

$$= \int_a^b F(x,g(x))\, dx + \int_b^a F(x,h(x))\, dx$$

Figure 2
"Standard" region D with boundary C

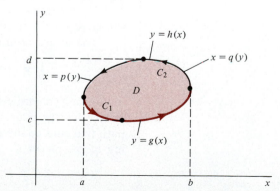

$$= \int_{C_1} F(x,y)\, dx + \int_{C_2} F(x,y)\, dx$$

$$= \int_{C} F(x,y)\, dx$$

Similarly, by breaking C into the curves $x = p(y)$ (y running from d to c) and $x = q(y)$ (y running from c to d), you can show that

$$\iint_{D} G_1(x,y)\, dA = \int_{C} G(x,y)\, dy$$

It follows that

$$\iint_{D} (G_1 - F_2)\, dA = \int_{C} F(x,y)\, dx + G(x,y)\, dy = \int_{C} \mathbf{F}(\mathbf{x}) \cdot d\mathbf{x}$$

Green's Theorem

Suppose that $\mathbf{F}(\mathbf{x}) = (F(x,y), G(x,y))$ is a vector field with continuous partial derivatives in a "standard" region D. If C is the boundary of D (oriented counterclockwise), then

$$\int_{C} \mathbf{F}(\mathbf{x}) \cdot d\mathbf{x} = \iint_{D} \left(\frac{\partial G}{\partial x} - \frac{\partial F}{\partial y} \right) dA$$

□ **Example 1**   Evaluate the line integral of $\mathbf{F}(x,y) = (x - y, x + y)$ around the unit circle

$$C: x = \cos t, \ y = \sin t \qquad 0 \leq t \leq 2\pi$$

Solution: The direct approach is to write

$$\int_{C} \mathbf{F}(\mathbf{x}) \cdot d\mathbf{x} = \int_{C} (x - y)\, dx + (x + y)\, dy$$

$$= \int_{0}^{2\pi} [(\cos t - \sin t)(-\sin t) + (\cos t + \sin t)(\cos t)]\, dt$$

$$= \int_{0}^{2\pi} dt = 2\pi$$

While this is not hard, Green's Theorem makes it easier. Let D be the unit disk enclosed by C. Since

$$\frac{\partial}{\partial x}(x + y) = 1 \qquad \text{and} \qquad \frac{\partial}{\partial y}(x - y) = -1$$

we have

$$\int_{C} \mathbf{F}(\mathbf{x}) \cdot d\mathbf{x} = \iint_{D} [1 - (-1)]\, dA = 2 \iint_{D} dA = 2(\text{area of } D) = 2\pi \quad \square$$

□ **Example 2**   Let D be any standard region with boundary C (oriented counterclockwise). Show that the area of D is

$$A = \frac{1}{2} \int_C x \, dy - y \, dx$$

Solution: Let $\mathbf{F}(x,y) = (-y,x)$. Then

$$\int_C \mathbf{F}(\mathbf{x}) \cdot d\mathbf{x} = \int_C x \, dy - y \, dx$$

$$= \iint_D \left[\frac{\partial}{\partial x}(x) - \frac{\partial}{\partial y}(-y) \right] dA = \iint_D 2 \, dA = 2A$$

from which the desired formula follows.   □

□ **Example 3**   Use the formula in Example 2 to find the area of the ellipse $x^2/a^2 + y^2/b^2 = 1$.

Solution: Parametrize the ellipse by writing it as the curve

$$C\colon x = a \cos t, \, y = b \sin t \qquad 0 \leq t \leq 2\pi$$

According to Example 2, its area is

$$A = \frac{1}{2} \int_C x \, dy - y \, dx = \frac{1}{2} \int_0^{2\pi} [a \cos t \, (b \cos t) - b \sin t \, (-a \sin t)] \, dt$$

$$= \frac{1}{2} \int_0^{2\pi} ab \, dt = \pi ab \qquad \qquad □$$

To generalize Green's Theorem further, we need another definition.

A curve $\mathbf{x}(t) = (x(t), y(t))$, $t_1 \leq t \leq t_2$, is said to be *simple* if it does not intersect itself, except that its initial and terminal points may coincide, $\mathbf{x}(t_1) = \mathbf{x}(t_2)$. In that case it is called a *simple closed* curve. (A circle is simple, a figure eight is not.)

Figure 3
Region with one hole

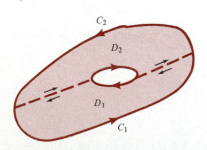

By piecing together a finite number of standard regions, we can extend Green's Theorem to quite complicated domains, including regions with holes. See Figure 3, which shows a region D with one hole; the boundary C consists of two simple closed curves, one inside the other. To see how Green's Theorem applies, we insert two slits, thus dividing D into regions D_1 and D_2 with boundaries C_1 and C_2. Integrating counterclockwise around C_1 accounts for the lower part of the outside curve, the lower part of the inside curve, and the slits oriented from right to left. Integrating around C_2 accounts for the upper parts of the original boundary and the slits oriented from left to right. Addition of these integrals cancels the contribution of the slits, the result being

$$\int_C \mathbf{F} \cdot d\mathbf{x} = \int_{C_1} \mathbf{F} \cdot d\mathbf{x} + \int_{C_2} \mathbf{F} \cdot d\mathbf{x}$$

$$= \iint_{D_1} (G_1 - F_2)\, dA + \iint_{D_2} (G_1 - F_2)\, dA = \iint_D (G_1 - F_2)\, dA$$

Thus Green's Theorem works if we agree to orient C as indicated by the arrows in the figure. This looks like clockwise orientation on the inside curve, but in fact it is not. If you are moving along the inside curve, your motion relative to nearby points of the region is counterclockwise. Perhaps it is less confusing to say that each curve must be traversed in such a way as to keep the region on your left. (This is called "positive" orientation.)

□ **Example 4**   Suppose that

$$\mathbf{F}(x,y) = \left(\frac{-y}{x^2 + y^2}, \frac{x}{x^2 + y^2} \right)$$

and let C be any simple closed curve around the origin (positively oriented). Show that

$$\int_C \mathbf{F}(\mathbf{x}) \cdot d\mathbf{x} = 2\pi$$

Solution: This seems hopeless at first glance, since we are not told anything about C except that the origin is inside. Moreover (even if we knew its parametric equations), Green's Theorem does not apply to the region enclosed by C because $\mathbf{F}(0,0)$ is undefined. We get around both objections by drawing a circle of radius a centered at the origin and inside C. (See Figure 4.) Parametric equations of the circle (oriented as indicated in the figure) are

$$C^*\colon x = a \cos t,\ y = a \sin t \qquad 0 \le t \le 2\pi$$

Figure 4
Circle inside curve around the origin

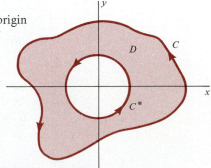

Applying Green's Theorem to the region D (whose boundary consists of C and $-C^*$), we have

$$\int_C \mathbf{F} \cdot d\mathbf{x} - \int_{C^*} \mathbf{F} \cdot d\mathbf{x} = \iint_D \left[\frac{\partial}{\partial x}\left(\frac{x}{x^2 + y^2} \right) - \frac{\partial}{\partial y}\left(\frac{-y}{x^2 + y^2} \right) \right] dA$$

$$= \iint_D \left[\frac{y^2 - x^2}{(x^2 + y^2)^2} + \frac{x^2 - y^2}{(x^2 + y^2)^2} \right] dA = 0$$

Hence

$$\int_C \mathbf{F} \cdot d\mathbf{x} = \int_{C^*} \mathbf{F} \cdot d\mathbf{x}$$

Thus we need not evaluate the integral of $\mathbf{F}$ around C; integrating around C^* will give the same result. Since

$$\mathbf{F}(\mathbf{x}) \cdot d\mathbf{x} = \frac{-y\,dx}{x^2 + y^2} + \frac{x\,dy}{x^2 + y^2} = \frac{1}{a^2}(-y\,dx + x\,dy)$$

we find

$$\int_{C^*} \mathbf{F} \cdot d\mathbf{x} = \frac{1}{a^2} \int_{C^*} x\,dy - y\,dx = \frac{2}{a^2}(\text{area inside } C^*) \qquad \text{(Example 2)}$$

$$= \frac{2}{a^2}(\pi a^2) = 2\pi$$

□

We have left a loose end dangling. What happened to our attempt to discover whether the condition $\partial G/\partial x = \partial F/\partial y$ is sufficient to make

$$\mathbf{F}(\mathbf{x}) = (F(x,y), G(x,y))$$

a gradient field? Green's Theorem supplies the answer. For if C is any closed path in the domain of $\mathbf{F}$, and if the region D enclosed by C is also in the domain, we have

$$\int_C \mathbf{F}(\mathbf{x}) \cdot d\mathbf{x} = \iint_D \left(\frac{\partial G}{\partial x} - \frac{\partial F}{\partial y} \right) dA$$

If $\partial G/\partial x = \partial F/\partial y$, the double integral (and hence the line integral) is zero, which proves that $\mathbf{F}$ is a gradient field. Notice, however, that the region enclosed by C must be in the domain of $\mathbf{F}$. To make sure of that, we restrict the domain to be *simply connected,* meaning that the region enclosed by every simple closed curve in the domain is also in the domain. (In other words, the domain has no holes.)

> Let $\mathbf{F}(\mathbf{x}) = (F(x,y), G(x,y))$ be a vector field with continuous partial derivatives in a simply connected region with a piecewise smooth boundary. If $\partial G/\partial x = \partial F/\partial y$, then $\mathbf{F}$ is a gradient field.

Thus the necessary condition given at the beginning of this section is also sufficient (provided that the domain of $\mathbf{F}$ is as described).

Problem Set 19.3

Use Green's Theorem to evaluate each of the following line integrals. (In each case C is understood to be oriented counterclockwise.)

1. $\int_C (2x - y)\, dx + (x + y)\, dy$, where C is the square with vertices $(1,1)$, $(-1,1)$, $(-1,-1)$, $(1,-1)$.

2. $\int_C y^2\, dx + x^2\, dy$, where C is the boundary of $D = \{(x,y):$ $x^2 + y^2 \le 1, y \ge 0\}$.

3. $\int_C dx + x\, dy$, where C is the unit circle $x^2 + y^2 = 1$.

4. $\displaystyle\int_C \frac{x\, dx + y\, dy}{(x^2 + y^2)^{3/2}}$, where C is the graph of $|x| + |y| = 1$.

5. $\int_C 2x^2 y\, dx + (x^3 - y)\, dy$, where C is the boundary of the part of the first quadrant cut off by the line $x + y = 1$.

6. $\int_C xy\, dx + x\, dy$, where C is the square with vertices $(0,0)$, $(1,0)$, $(1,1)$, $(0,1)$.

7. $\int_C xy\, dx + y\, dy$, where C is the boundary of $D = \{(x,y):$ $x^2 + y^2 \le 1, x \ge 0, y \ge 0\}$.

8. $\int_C x^3\, dy - y^3\, dx$, where C is the unit circle $x^2 + y^2 = 1$.

9. $\int_C e^u \sin y\, dx + e^u \cos y\, dy$, where C is the square with vertices $(0,0)$, $(1,0)$, $(1,1)$, $(0,1)$.

10. $\int_C \cos y\, dx + \cos x\, dy$, where C is the triangle with vertices $(0,0)$, $(\pi,0)$, (π,π).

11. $\int_C y \cos x\, dx + x \sin y\, dy$, where C is the square with vertices $(0,0)$, $(\pi,0)$, (π,π), $(0,\pi)$.

12. $\int_C (e^{x^2} + y^2)\, dx + (x + \sqrt{y^3 + 1})\, dy$, where C is the triangle with vertices $(0,0)$, $(1,0)$, $(1,1)$. (Contemplate the problem of evaluating this integral directly!)

13. $\int_C \sin x \cos y\, dx + \cos x \sin y\, dy$, where C is any circle.

14. $\int_C (e^x - y)\, dx + (x - e^y)\, dy$, where C is any circle of radius 1 (not necessarily centered at the origin).

15. $\int_C x\, dy/(x^2 + y^2) - y\, dx/(x^2 + y^2)$, where C is any circle with the origin outside.

16. Suppose that f and g are functions of one variable that are everywhere differentiable. Explain why

$$\int_C f(x)\, dx + g(y)\, dy = 0$$

where C is any circle.

17. Suppose that $f(x,y)$ is a scalar field in a standard region D with boundary C (positively oriented). Show that

$$\int_C f_1\, dy - f_2\, dx = \iint_D (f_{11} + f_{22})\, dA$$

18. Let D be a standard region with boundary C (positively oriented). Show that the area of D is $A = -\int_C y\, dx$.

19. In Problem 18 show that the area of D is also

$$A = \int_C x\, dy$$

Why does it follow that

$$A = \frac{1}{2}\int_C x\, dy - y\, dx?$$

(See Example 2.)

Use one of the formulas in Problems 18 and 19 to find the area of the given region.

20. Bounded by $y = x^2$ and $y^2 = x$.

21. Bounded by the x axis and the curve $y = \sin x$, $0 \le x \le \pi$.

22. Inside the hypocycloid $x = a \cos^3 t$, $y = a \sin^3 t$.

23. Inside the loop of the curve $y^2 = x^2(3 - x)$.

24. Bounded by the graph of $\sqrt{x} + \sqrt{y} = 1$ and the coordinate axes.

25. Bounded by the graph of $y = 2 - |x|$ and the x axis.

26. Bounded by the curves $y = x^3$, $y = (x - 2)^2$, and the x axis.

27. Let D be a standard region with boundary C (positively oriented). Show that the centroid of D is given by

$$\bar{x} = \frac{1}{2A}\int_C x^2\, dy, \quad \bar{y} = -\frac{1}{2A}\int_C y^2\, dx$$

Use Problem 27 to find the centroid of each of the following regions.

28. A semicircular region of radius a.

29. Bounded by the x axis and the curve $y = \sin x$, $0 \le x \le \pi$.

30. Bounded by the graph of $y = \ln x$ and the lines $x = 1$ and $x = e$.

31. Bounded by the curve $y = 2x - x^2$ and the x axis.

32. In Problem 27 show that if the density at each point of D is 1, the moment of inertia relative to the z axis is

$$I_z = \frac{1}{3}\int_C x^3\, dy - y^3\, dx$$

Use Problem 32 to find the moment of inertia (relative to the z axis) of each of the following regions. Assume constant density 1.

33. The disk $x^2 + y^2 \leq a^2$.

34. Bounded by the curve $y = 2x - x^2$ and the x axis.

35. Bounded by the x axis and the curve $y = \sin x$, $0 \leq x \leq \pi$.

36. Inside the hypocycloid $x = a \cos^3 t$, $y = a \sin^3 t$.

37. Suppose that a force $\mathbf{F}(\mathbf{x})$ is directed toward the origin and has magnitude proportional to the distance from the origin. Show that $\mathbf{F}$ is conservative and find a potential energy function.

38. Suppose that a force $\mathbf{F}(\mathbf{x})$ is directed away from the origin and has magnitude proportional to the square of the distance from the origin. Show that $\mathbf{F}$ is conservative and find a potential energy function.

39. In Example 4 we showed that if C is any simple closed curve around the origin (positively oriented), then

$$\int_C \mathbf{F}(\mathbf{x}) \cdot d\mathbf{x} = 2\pi$$

where $\mathbf{F}(x, y) = \left(\dfrac{-y}{x^2 + y^2}, \dfrac{x}{x^2 + y^2} \right)$

(a) Confirm that

$$\frac{\partial}{\partial x} \left(\frac{x}{x^2 + y^2} \right) = \frac{\partial}{\partial y} \left(\frac{-y}{x^2 + y^2} \right)$$

In view of this, shouldn't $\mathbf{F}$ be a gradient field and doesn't it follow that its integral around C is zero? Explain.

(b) If your explanation in part (a) included the statement that $\mathbf{F}$ is not a gradient field, how do you account for the fact that $\nabla f = \mathbf{F}$, where $f(x, y) = -\tan^{-1}(x/y)$?

19.4
GREEN'S THEOREM REVISITED

Suppose that a stream of some kind (like air or water) is flowing through a region in the plane. If its density is the scalar function δ and its velocity is the vector function $\mathbf{v}$, the vector field $\mathbf{F} = \delta \mathbf{v}$ is called the *flux* of the fluid. As we will see, it is involved in the question of the rate at which the fluid is entering or leaving the region.

Let D be the region (assumed to be standard) and suppose that its boundary is the smooth curve C (oriented counterclockwise). Let Δs be the length of a small section of C and let P be a point of this section. (See Figure 1.) The tangential component of $\mathbf{F}$ at P is of no interest to us if we are talking about fluid entering or leaving the region, because a flow along the tangent simply goes past the region. The normal component, however, is a measure of the flow *across the boundary*. More precisely, let $\mathbf{N}$ be the outward unit normal at P. Then $\mathbf{F} \cdot \mathbf{N}$ is the (scalar) normal component that we are talking about; fluid is leaving or entering the region at P depending on whether $\mathbf{F} \cdot \mathbf{N}$ is positive or negative.

The amount of fluid (per unit of time) passing through the section of C containing P is $\Delta m \approx (\mathbf{F} \cdot \mathbf{N}) \Delta s$. (Remember that $\mathbf{F} = \delta \mathbf{v}$, which has units of mass per unit length per unit time.) Adding the contributions of all the sections, we find that the total amount of fluid passing through the curve C (per unit of time) is

$$m = \Sigma \, \Delta m \approx \Sigma (\mathbf{F} \cdot \mathbf{N}) \, \Delta s$$

Figure 1
Tangential and normal components of flux

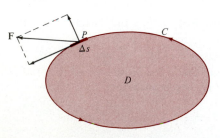

This suggests that in the limit (as the partition of C is refined and the sections get smaller) we have

$$m = \int_C (\mathbf{F} \cdot \mathbf{N}) \, ds$$

The trouble with this result is that we have not defined the line integral of a scalar field. That is easily remedied, however. Let $g(x,y)$ be a continuous scalar field and suppose that the curve

$$C: \mathbf{x}(t) = (x(t), y(t)), \; t_1 \leq t \leq t_2$$

is smooth. By partitioning the parameter interval $[t_1, t_2]$, divide C into small sections and let Δs be the length of the typical section. If (x, y) is a point of this piece, compute $g(x, y) \, \Delta s$, then sum over the partition to obtain the Riemann sum $\Sigma \, g(x, y) \, \Delta s$. The limit of this as the partition is refined is the line integral we are defining:

Let $g(x, y)$ be a continuous scalar field and suppose that $C: \mathbf{x}(t) = (x(t), y(t)), \; t_1 \leq t \leq t_2$ is a smooth curve. The *line integral* of $g(x, y)$ along C is defined by

$$\int_C g(x, y) \, ds = \int_{t_1}^{t_2} \left[g(x, y) \frac{ds}{dt} \right] dt$$

where $x = x(t)$, $y = y(t)$, and s is distance measured along C.

□ **Example 1**   Let $g(x, y) = -2(x^2 + y^2)$ and suppose that C is the unit circle $x = \cos t$, $y = \sin t$, $0 \leq t \leq 2\pi$. Find $\int_C g(x, y) \, ds$.

Solution: The problem is to compute

$$\int_0^{2\pi} \left[g(x, y) \frac{ds}{dt} \right] dt, \text{ where } x = \cos t \text{ and } y = \sin t$$

Since

$$g(x, y) = -2(x^2 + y^2) = -2(\cos^2 t + \sin^2 t) = -2$$

and

$$\frac{ds}{dt} = \sqrt{(dx/dt)^2 + (dy/dt)^2} = \sqrt{(-\sin t)^2 + (\cos t)^2} = 1$$

we find

$$\int_C g(x, y) \, ds = \int_0^{2\pi} (-2) \, dt = -4\pi$$

□

Returning to the fluid flow problem, note that we have arrived at the following solution:

If **F** is the flux of a fluid crossing the boundary C of a region D, the rate at which the fluid leaves the region (in units of mass per unit time) is

$$\int_C (\mathbf{F} \cdot \mathbf{N})\, ds$$

where **N** is the outward unit normal to C.

We can compute this integral as it stands, as in Example 1. It is usually easier, however, to apply Green's Theorem. For that purpose we need to do some rewriting. Recall from Section 15.3 that if $\mathbf{x}(t) = (x(t), y(t))$ is a vector function defining C, the unit tangent vector in the positive direction satisfies

$$\frac{d\mathbf{x}}{dt} = \frac{ds}{dt}\mathbf{T}$$

From this it follows that

$$\mathbf{T} = \frac{(dx/dt, dy/dt)}{ds/dt} = (dx/ds, dy/ds)$$

The outward unit normal to C is 90° clockwise from **T** (as shown in Figure 2), so it is given by

$$\mathbf{N} = (dy/ds, -dx/ds)$$

(This is not the same **N** as in Section 15.3. There we had to be more picky: sometimes **N** was clockwise from **T,** sometimes counterclockwise, chosen so as to make **N** inward, not outward.)

If $\mathbf{F} = (F, G)$, then

$$\mathbf{F} \cdot \mathbf{N} = (F, G) \cdot (dy/ds, -dx/ds) = F\frac{dy}{ds} - G\frac{dx}{ds}$$

and

$$\int_C (\mathbf{F} \cdot \mathbf{N})\, ds = \int_C F\, dy - G\, dx$$

Apply Green's Theorem to the line integral on the right, to obtain

$$\int_C (\mathbf{F} \cdot \mathbf{N})\, ds = \iint_D \left(\frac{\partial F}{\partial x} + \frac{\partial G}{\partial y}\right) dA$$

□ **Example 2**   Suppose that the flux of a fluid is given by $\mathbf{F}(x, y) = -2(x, y)$. Compute in two ways the rate at which the fluid is crossing the

Figure 2
Outward unit normal vector

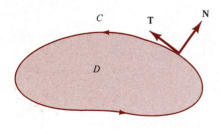

Figure 3

Flux of a fluid flowing toward the origin (not drawn to scale)

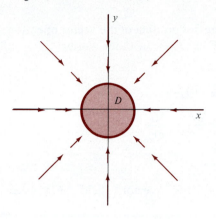

Figure 4

Two arrows representing $\mathbf{N} = (x, y)$

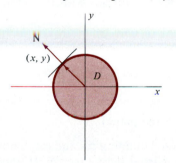

unit circle

$$C: x = \cos t, \; y = \sin t, \; 0 \le t \le 2\pi$$

Solution: First we observe that $\mathbf{F} = -2\mathbf{x}$ is directed toward the origin. (See Figure 3.) Since it has the same direction as the velocity vector (because $\mathbf{F} = \delta\mathbf{v}$, where δ is density), the fluid must be flowing toward the origin. It will therefore be *entering* the region D inside the unit circle and we should expect the answer to be negative.

We begin by computing

$$\int_C (\mathbf{F} \cdot \mathbf{N}) \, ds$$

where $\mathbf{N}$ is the outward unit normal to C. In Figure 4 we show the position vector (starting at the origin) of a typical point (x, y) on the unit circle. This vector has unit length and is perpendicular to the tangent at (x, y). The arrow marked $\mathbf{N}$ has the same length and direction; hence $\mathbf{N} = (x, y)$.

Since $\mathbf{F} = -2(x, y)$, we find

$$\mathbf{F} \cdot \mathbf{N} = -2(x, y) \cdot (x, y) = -2(x^2 + y^2)$$

The line integral of this function around C has already been computed in Example 1, the result is

$$\int_C (\mathbf{F} \cdot \mathbf{N}) \, ds = -4\pi$$

To confirm this result, we compute

$$\iint_D \left(\frac{\partial F}{\partial x} + \frac{\partial G}{\partial y} \right) dA$$

where $F(x, y) = -2x$ and $G(x, y) = -2y$ are the components of $\mathbf{F}(x, y) = -2(x, y)$. Since $\partial F / \partial x = -2$ and $\partial G / \partial y = -2$, we find

$$\iint_D \left(\frac{\partial F}{\partial x} + \frac{\partial G}{\partial y} \right) dA = \iint_D (-4) \, dA = -4(\text{area of } D) = -4\pi \quad \square$$

Remark   As we have already observed, the negative result indicates that fluid is entering the region D, which is what to expect from a flux directed toward the origin. Physicists refer to this situation as a *sink* in the region (as though there were a "black hole" at the origin into which mass is disappearing). If the rate comes out positive (so that fluid is leaving the region), we say there is a *source* in the region. When the rate is zero the sources and sinks cancel, the net flow of fluid through the boundary being zero.

The equation

$$\int_C (\mathbf{F} \cdot \mathbf{N}) \, ds = \iint_D \left(\frac{\partial F}{\partial x} + \frac{\partial G}{\partial y} \right) dA$$

is usually abbreviated by defining the *divergence* of $\mathbf{F} = (F,G)$ to be

$$\text{div } \mathbf{F} = \frac{\partial F}{\partial x} + \frac{\partial G}{\partial y}$$

It is common practice to regard this as the dot product of the vector operator $\mathbf{\nabla} = (\partial/\partial x, \partial/\partial y)$ and the vector field $\mathbf{F} = (F,G)$. In other words,

---

The *divergence* of the vector field $\mathbf{F} = (F,G)$ is

$$\text{div } \mathbf{F} = \mathbf{\nabla} \cdot \mathbf{F} = \frac{\partial F}{\partial x} + \frac{\partial G}{\partial y}$$

□ **Example 3**   The divergence of the vector field $\mathbf{F}(x,y) = (xy, 2x - y)$ is

$$\mathbf{\nabla} \cdot \mathbf{F} = \frac{\partial}{\partial x}(xy) + \frac{\partial}{\partial y}(2x - y) = y - 1$$

Notice that $\mathbf{\nabla} \cdot \mathbf{F}$ is a *scalar;* it should not be confused with the similar-looking symbol $\mathbf{\nabla} f$. The latter is the gradient of the scalar function f and is a *vector*.   □

Now we are ready for a new version of Green's Theorem:

Divergence Theorem (in the plane)

Let C be a (positively oriented) closed curve bounding a standard region D. If $\mathbf{F}$ is a vector field with continuous partial derivatives in D, then

$$\int_C (\mathbf{F} \cdot \mathbf{N}) \, ds = \int\!\!\int_D (\mathbf{\nabla} \cdot \mathbf{F}) \, dA$$

where $\mathbf{N}$ is the outward unit normal to C.

The Divergence Theorem was derived from Green's Theorem to simplify the computation of the line integral of the normal component of flux. The theorem is true, however, for any vector field (with continuous partial derivatives); there is no need to interpret it physically.

□ **Example 4**   Let $\mathbf{F}(x,y) = (xy, 2x - y)$ and suppose that D is the region bounded by the line $x + y = 1$ and the coordinate axes. If C is the (positively oriented) boundary of D, confirm the Divergence Theorem.

Figure 5
Confirming the Divergence Theorem

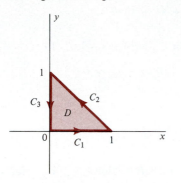

Solution: As shown in Figure 5, we may break C into the smooth paths

$$C_1: y = 0 \quad (x \text{ running from } 0 \text{ to } 1)$$
$$C_2: y = 1 - x \quad (x \text{ running from } 1 \text{ to } 0)$$
$$C_3: x = 0 \quad (y \text{ running from } 1 \text{ to } 0)$$

The outward unit normal at $(x,0)$ on C_1 is $\mathbf{N} = (0,-1)$, so

$$\mathbf{F} \cdot \mathbf{N} = (xy, 2x - y) \cdot (0,-1) = y - 2x = -2x$$

(because $y = 0$ on C_1). A parametrization of C_1 is

$$x = t, \, y = 0, \, 0 \le t \le 1$$

Hence

$$\int_{C_1} (\mathbf{F} \cdot \mathbf{N}) \, ds = \int_0^1 \left[(-2t) \frac{ds}{dt} \right] dt$$

Since

$$\left(\frac{ds}{dt} \right)^2 = \left(\frac{dx}{dt} \right)^2 + \left(\frac{dy}{dt} \right)^2 = 1$$

we find

$$\int_{C_1} (\mathbf{F} \cdot \mathbf{N}) \, ds = \int_0^1 (-2t) \, dt = -1$$

The outward unit normal at (x,y) on C_2 may be found by inspection of Figure 5. It is more educational, however, to use the methods of Section 17.3. The gradient of the function $x + y - 1$, namely $(1,1)$, is perpendicular to the level curve $x + y - 1 = 0$. Since this vector is outward, we find $\mathbf{N} = \dfrac{1}{\sqrt{2}}(1,1)$. Then

$$\mathbf{F} \cdot \mathbf{N} = (xy, 2x - y) \cdot \frac{1}{\sqrt{2}}(1,1) = \frac{1}{\sqrt{2}}(xy + 2x - y)$$

Since $y = 1 - x$ on C_2, we find

$$\mathbf{F} \cdot \mathbf{N} = \frac{1}{\sqrt{2}}(-x^2 + 4x - 1)$$

A parametrization of C_2 is

$$x = t, \, y = 1 - t \quad (t \text{ running from } 1 \text{ to } 0)$$

This is a poor choice, however, because (as you can check) $ds/dt = -\sqrt{2}$, the minus sign reflecting the fact that s decreases with increasing t. It is better to parametrize so that s increases with the parameter:

$$x = 1 - t, \, y = t, \, 0 \le t \le 1$$

Now we find $ds/dt = \sqrt{2}$ and hence

$$\int_{C_2} (\mathbf{F} \cdot \mathbf{N}) \, ds = \int_0^1 (-t^2 + 4t - 1) \, dt = \frac{2}{3}$$

The outward unit normal at $(0, y)$ on C_3 is $\mathbf{N} = (-1, 0)$, so

$$\mathbf{F} \cdot \mathbf{N} = (xy, 2x - y) \cdot (-1, 0) = -xy = 0$$

(because $x = 0$ on C_3). Hence

$$\int_{C_3} (\mathbf{F} \cdot \mathbf{N}) \, ds = 0$$

The final answer is the sum of the line integrals along C_1, C_2, and C_3, namely

$$\int_C (\mathbf{F} \cdot \mathbf{N}) \, ds = -1 + \frac{2}{3} + 0 = -\frac{1}{3}$$

To confirm the Divergence Theorem, we compute

$$\iint_D (\mathbf{\nabla} \cdot \mathbf{F}) \, dA = \int_0^1 \int_0^{1-x} (y - 1) \, dy \, dx = \int_0^1 \frac{1}{2}(x^2 - 1) \, dx = -\frac{1}{3} \quad \square$$

We end this section with another version of Green's Theorem, and another application. Our original definition of the line integral of the vector field $\mathbf{F}$ along the curve

$$C \colon \mathbf{x}(t) = (x(t), y(t)), \ t_1 \le t \le t_2$$

was

$$\int_C \mathbf{F} \cdot d\mathbf{x} = \int_{t_1}^{t_2} \left(\mathbf{F} \cdot \frac{d\mathbf{x}}{dt} \right) dt$$

Since

$$\frac{d\mathbf{x}}{dt} = \frac{ds}{dt} \mathbf{T}$$

we have

$$\int_C \mathbf{F} \cdot d\mathbf{x} = \int_{t_1}^{t_2} \left(\mathbf{F} \cdot \frac{ds}{dt} \mathbf{T} \right) dt = \int_{t_1}^{t_2} \left[(\mathbf{F} \cdot \mathbf{T}) \frac{ds}{dt} \right] dt = \int_C (\mathbf{F} \cdot \mathbf{T}) \, ds$$

(by definition of the line integral of a scalar field).

In other words, our original definition of the line integral of a vector field may be regarded as a special case of our new definition of the line integral of a scalar field, the scalar in this case being the tangential component of the vector field, $\mathbf{F} \cdot \mathbf{T}$.

If C is a simple closed curve and D is the (standard) region inside it, we can apply Green's Theorem to the vector field $\mathbf{F} = (F, G)$ by writing

$$\int_C (\mathbf{F} \cdot \mathbf{T})\, ds = \int\int_D \left(\frac{\partial G}{\partial x} - \frac{\partial F}{\partial y} \right) dA$$

The Divergence Theorem, on the other hand, says that

$$\int_C (\mathbf{F} \cdot \mathbf{N})\, ds = \int\int_D \left(\frac{\partial F}{\partial x} + \frac{\partial G}{\partial y} \right) dA$$

You can see the similarity between these two versions of Green's Theorem; it is also important to notice the differences.

If $\mathbf{F}$ is interpreted as a force field, the integral of its tangential component, $\mathbf{F} \cdot \mathbf{T}$, is the work done by $\mathbf{F}$ around C. If $\mathbf{F}$ is the flux of a fluid instead, the integral of its normal component, $\mathbf{F} \cdot \mathbf{N}$, is the rate at which the fluid crosses C. We may also interpret $\mathbf{F}$ as flux in the first case. Remembering that $\mathbf{F} \cdot \mathbf{T}$ is a measure of fluid flow along C (instead of across it), we call the integral of $\mathbf{F} \cdot \mathbf{T}$ the *circulation* around C. There is a *whirlpool* inside C if the integral is nonzero.

□ **Example 5**   Would you expect a whirlpool in the fluid flow whose flux is $\mathbf{F}(x, y) = -2(x, y)$? Consider the region inside the unit circle.

Solution: This is the vector field we considered in Example 2. It is fairly apparent from a picture of the flux (Figure 3) that the flow is entirely radial, that is, there is no tendency for the fluid to rotate. Hence we expect the integral of $\mathbf{F} \cdot \mathbf{T}$ around the unit circle C to be zero. To confirm this expectation, we will compute

$$\int_C (\mathbf{F} \cdot \mathbf{T})\, ds = \int\int_D \left(\frac{\partial G}{\partial x} - \frac{\partial F}{\partial y} \right) dA$$

where D is the region inside C and $F(x, y) = -2x$ and $G(x, y) = -2y$. Since $\partial G / \partial x = 0$ and $\partial F / \partial y = 0$, the computation doesn't amount to much; the integral is clearly 0.   □

□ **Example 6**   Determine the circulation of $\mathbf{F}(x, y) = (xy, 2x - y)$ around the region D bounded by the line $x + y = 1$ and the coordinate axes.

Solution: In Example 4 we found the integral of $\mathbf{F} \cdot \mathbf{N}$ around C (the boundary of D) to be $-\frac{1}{3}$. If $\mathbf{F}$ is interpreted as a flux, then fluid must be entering D, that is, there is a sink in the region.

To find the circulation, we compute

$$\int_C (\mathbf{F} \cdot \mathbf{T})\, ds = \int\int_D \left(\frac{\partial G}{\partial x} - \frac{\partial F}{\partial y} \right) dA$$

where $F(x,y) = xy$ and $G(x,y) = 2x - y$. Since $\partial G/\partial x = 2$ and $\partial F/\partial y = x$, we find

$$\int_C (\mathbf{F} \cdot \mathbf{T})\, ds = \iint_D (2 - x)\, dA = \int_0^1 \int_0^{1-x} (2 - x)\, dy\, dx$$

$$= \int_0^1 (2 - x)(1 - x)\, dx = \frac{5}{6}$$

The positive answer tells us that there is a whirlpool in D and that it is rotating counterclockwise. (Why?)   □

From a mathematical point of view none of these interpretations matters (although of course they are important in applications). Each of the versions of Green's Theorem is correct for any vector field. When you read other books on the subject you will find numerous variations on these themes, and many different approaches. (Flux and circulation, for example, are not always defined precisely as we have given them.) This lack of unanimity concerning what to emphasize, and what names and symbols to use, is an indication of the richness of the subject. In physics, particularly, you will find these ideas popping up repeatedly (in mechanics, fluid flow, heat, electricity, and magnetism). The mathematics is virtually the same each time, but the interpretation is different.

Problem Set 19.4

1. Suppose that the flux of a fluid is $\mathbf{F}(x,y) = (x,y)$. Sketch the field $\mathbf{F}$ and identify any sources or sinks.

2. In Problem 1 would you expect the rate at which the fluid passes the boundary of the square region $-1 \le x \le 1$, $-1 \le y \le 1$ to be positive, negative, or zero? Check by computing an appropriate line integral.

3. Check your answer in Problem 2 by using the Divergence Theorem.

4. In Problem 2 would you expect any whirlpools in the region? Check by computing an appropriate line integral.

5. Check your answer in Problem 4 by using Green's Theorem.

6. Suppose that the flux of a fluid is $\mathbf{F}(x,y) = -(x,y)$. Sketch the field $\mathbf{F}$ and identify any sources or sinks.

7. In Problem 6 would you expect the rate at which the fluid passes the boundary of the disk $x^2 + y^2 \le 1$ to be positive, negative, or zero? Check by computing an appropriate line integral.

8. Check your answer in Problem 7 by using the Divergence Theorem.

9. In Problem 7 would you expect any whirlpools in the region? Check by computing an appropriate line integral.

10. Check your answer in Problem 9 by using Green's Theorem.

11. Suppose that the flux of a fluid is $\mathbf{F}(x,y) = (0,y)$. Sketch the field $\mathbf{F}$ and identify any sources or sinks.

12. In Problem 11 would you expect the rate at which the fluid passes the boundary of the disk $x^2 + y^2 \le 1$ to be positive, negative, or zero? Check by computing an appropriate line integral.

13. Check your answer in Problem 12 by using the Divergence Theorem.

14. In Problem 12 would you expect any whirlpools in the region? Check by computing an appropriate line integral.

15. Check your answer in Problem 14 by using Green's Theorem.

16. Suppose that the flux of a fluid is $F(x,y) = (x^2,0)$. Sketch the field F and identify any sources or sinks.

17. In Problem 16 would you expect the rate at which the fluid passes the boundary of the square region $-1 \leq x \leq 1$, $-1 \leq y \leq 1$ to be positive, negative, or zero? Check by computing an appropriate line integral.

18. Check your answer in Problem 17 by using the Divergence Theorem.

19. In Problem 17 would you expect any whirlpools in the region? Check by computing an appropriate line integral.

20. Check your answer in Problem 19 by using Green's Theorem.

21. Repeat Problem 17 for the rectangular region $0 \leq x \leq 1$, $-1 \leq y \leq 1$.

22. Check your answer in Problem 21 by using the Divergence Theorem.

23. Repeat Problem 17 for the rectangular region $-1 \leq x \leq 0$, $-1 \leq y \leq 1$.

24. Check your answer in Problem 23 by using the Divergence Theorem.

25. Suppose that the flux of a fluid is $F(x,y) = (x - y, x + y)$. By evaluating an appropriate line integral, find the rate at which the fluid passes the boundary of the unit disk $x^2 + y^2 \leq 1$. Is there a source, sink, or neither in this region?

26. Check Problem 25 by using the Divergence Theorem.

27. In Problem 25 evaluate an appropriate line integral to find the circulation around the disk $x^2 + y^2 \leq 1$.

28. Check Problem 27 by using Green's Theorem.

29. Suppose that $F(x) = x/r^2$, where $r = |x|$. Show that the divergence of F is zero.

30. In Problem 29 show that if C is the unit circle, then

$$\int_C (F \cdot N)\, ds = 2\pi$$

Why doesn't this result (in view of Problem 29) contradict the Divergence Theorem?

31. Show that if the flux of a fluid is constant, the rate at which the fluid passes through a closed curve is zero. What is the physical interpretation?

32. Suppose that the vector field $F(x,y)$ is the gradient of the scalar field $f(x,y)$. Show that div $F = f_{11} + f_{22}$.

33. Suppose that f and F are scalar and vector fields, respectively. Derive the product rule

$$\nabla \cdot (fF) = f(\nabla \cdot F) + (\nabla f) \cdot F$$

34. Verify the formula in Problem 33 in the case $f(x,y) = \sin xy$ and $F(x,y) = (y,x)$.

35. Verify the formula in Problem 33 in the case $f(x,y) = x/y$ and $F(x,y) = (xy, 2x - y)$.

36. Explain why the length of the curve C is

$$s = \int_C ds$$

37. Use the formula in Problem 36 to find the length of the first-quadrant arc of the hypocycloid $x = a\cos^3 t$, $y = a\sin^3 t$.

38. Suppose that the density at each point of the curve C is $\delta(x,y)$. By considering the mass Δm of a typical section of C (of length Δs), explain the formula

$$m = \int_C \delta(x,y)\, ds$$

39. The density at each point of the unit circle is proportional to the square of the distance from the x axis. Find the mass.

40. Explain why the centroid of a curve C is the point $(\overline{x}, \overline{y})$ satisfying

$$\overline{x} = \frac{1}{s}\int_C x\, ds, \quad \overline{y} = \frac{1}{s}\int_C y\, ds$$

where s is the length of C.

41. Use Problem 40 to find the centroid of the semicircle

$$x = a\cos t, \quad y = a\sin t, \quad 0 \leq t \leq \pi$$

42. Use Problem 40 to find the centroid of the first-quadrant arc of the hypocycloid $x = a\cos^3 t$, $y = a\sin^3 t$. Then check with Example 5, Section 7.6. (The discussion of mass distributions along a curve in Section 7.6 really involves line integrals, but we did not say so at the time.)

We begin this section almost word-for-word the way we began the last one, but now we will be talking about a vector field in three-dimensional space, say

$$\mathbf{F(x)} = (F(x,y,z), G(x,y,z), H(x,y,z))$$

Our object is to generalize the Divergence Theorem.

Suppose that a stream of some kind (like air or water) is flowing through a region in space. If its density is the scalar function δ and its velocity is the vector function $\mathbf{v}$, the vector function $\mathbf{F} = \delta\mathbf{v}$ is called the *flux* of the fluid. As we will see, it is involved in the question of the rate at which the fluid is entering or leaving the region.

Let R be the region and suppose that its boundary is a smooth surface σ. (For example, R might be the spherical region $x^2 + y^2 + z^2 \leq a^2$ and σ the sphere $x^2 + y^2 + z^2 = a^2$.) Let ΔS be the area of a small patch of surface and let P be a point of this patch. (See Figure 1.) The tangential component of $\mathbf{F}$ at P is of no interest to us if we are talking about fluid entering or leaving the region, because a flow in the tangent plane simply goes past the region. The normal component, however, is a measure of the flow *through the surface*. More precisely, let $\mathbf{N}$ be the outward unit normal vector at P. Then $\mathbf{F} \cdot \mathbf{N}$ is the (scalar) normal component that we are talking about; fluid is leaving or entering the region at P depending on whether $\mathbf{F} \cdot \mathbf{N}$ is positive or negative.

The amount of fluid passing through the patch containing P (per unit of time) is $\Delta m \approx (\mathbf{F} \cdot \mathbf{N}) \, \Delta S$. Adding the contributions of all the patches, we find that the total amount of fluid passing through the surface (per unit of time) is

$$m = \sum \Delta m \approx \sum (\mathbf{F} \cdot \mathbf{N}) \, \Delta S$$

This suggests that in the limit (as the partition of the surface is refined and the patches get smaller) we have

$$m = \iint_\sigma (\mathbf{F} \cdot \mathbf{N}) \, dS$$

Figure 1
Tangential and normal components of flux

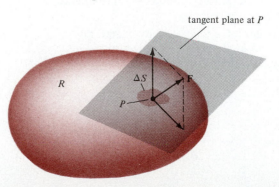

The trouble with this result is that we have never defined an integral over a surface (except in the special case of a flat domain in the plane). In Section 18.4, however, we developed a formula for the area of a surface. You should review that discussion unless you remember it clearly. Given a surface $z = f(x,y)$ lying above a domain D in the xy plane (or below it), we partition D into small rectangles. Letting ΔS be the area of the portion of the surface above the typical rectangle of area ΔA, we obtain the approximation

$$\Delta S \approx (\sec \theta)\, \Delta A$$

where θ is the (acute) angle between the tangent plane and the xy plane. (See Figure 2 in Section 18.4.) This leads to the formula

$$S = \iint_D \sec \theta\, dA = \iint_D \sqrt{(\partial z/\partial x)^2 + (\partial z/\partial y)^2 + 1}\; dA$$

We did not mention it at the time, but this analysis involves the concept of surface integral. If σ is the surface in question, the area formula might be written

$$S = \iint_\sigma dS$$

in much the same way that we write

$$A = \iint_D dA$$

for the area of a flat region in the plane.

The integrand in this formula for S is 1. To define the surface integral of a general function $g(x,y,z)$, we proceed as so often before. Let (x,y) be a point of the typical subrectangle in D, and (x,y,z) the point of the surface above it. Form the product $g(x,y,z)\,\Delta S$ and add the products corresponding to all the subrectangles of D. The limit of the result (as the partition is refined) is written

$$\iint_\sigma g(x,y,z)\, dS = \lim \sum g(x,y,z)\,\Delta S$$

and is called the *surface integral* of $g(x,y,z)$ over the surface σ. Replacing dS by $\sec \theta\, dA$ and z by $f(x,y)$, we arrive at the formula

$$\iint_\sigma g(x,y,z)\, dS = \iint_D g(x,y,f(x,y))\sec \theta\, dA$$

$$= \iint_D g(x,y,f(x,y)) \sqrt{(\partial z/\partial x)^2 + (\partial z/\partial y)^2 + 1}\; dA$$

Despite its complicated appearance, this is an ordinary double integral of a function of two variables.

□ **Example 1**   Suppose that the flux of a fluid at each point in space is given by the vector field

$$\mathbf{F}(x, y, z) = -k(x, y, z) \qquad \text{where } k \text{ is a positive constant}$$

(Note that since $\mathbf{F} = \delta \mathbf{v}$ is directed toward the origin and has the same direction as the velocity vector, the fluid is flowing toward the origin.) Find the rate at which the fluid is passing through the surface $x^2 + y^2 + z^2 = a^2$.

Solution: Consider only the top hemisphere, namely the surface

$$\sigma: z = \sqrt{a^2 - x^2 - y^2}$$

The rate at which the fluid flows through σ is given by the surface integral

$$\iint_\sigma (\mathbf{F} \cdot \mathbf{N}) \, dS \qquad \text{where } \mathbf{N} \text{ is the outward unit normal}$$

Letting $\mathbf{x} = (x, y, z)$, we have $\mathbf{N} = \mathbf{x}/|\mathbf{x}| = \mathbf{x}/a$ (why is this *outward?*) and hence

$$\mathbf{F} \cdot \mathbf{N} = (-k\mathbf{x}) \cdot (\mathbf{x}/a) = -ka$$

To evaluate the integral of this function over σ, we compute $\partial z/\partial x = -x/z$ and $\partial z/\partial y = -y/z$, from which

$$\left(\frac{\partial z}{\partial x}\right)^2 + \left(\frac{\partial z}{\partial y}\right)^2 + 1 = \frac{x^2}{z^2} + \frac{y^2}{z^2} + 1 = \frac{a^2}{z^2} \qquad \text{(because } x^2 + y^2 + z^2 = a^2\text{)}$$

Hence $$\iint_\sigma (\mathbf{F} \cdot \mathbf{N}) \, dS = \iint_D (-ka)\frac{a}{\sqrt{a^2 - x^2 - y^2}} \, dA$$

where D is the projection of the hemisphere in the xy plane, namely

$$D = \{(x, y): x^2 + y^2 \leq a^2\}$$

Using polar coordinates, we find

$$\iint_\sigma (\mathbf{F} \cdot \mathbf{N}) \, dS = -ka^2 \int_0^{2\pi} \int_0^a \frac{r \, dr \, d\theta}{\sqrt{a^2 - r^2}} = -ka^3 \int_0^{2\pi} d\theta = -2k\pi a^3$$

(The inner integral is improper; we did not show the steps involved in evaluating it.) The rate of flow through the entire sphere is $-4k\pi a^3$. (Why?)

□

Remark   As in the last section, the negative result indicates that fluid is entering the region and that there is a sink there. When the rate is positive, fluid is leaving the region and there is a source.

Given a vector field in space, say

$$\mathbf{F}(\mathbf{x}) = (F(x, y, z), G(x, y, z), H(x, y, z))$$

we define the **divergence** of $\mathbf{F}$ to be the scalar field

$$\text{div } \mathbf{F} = \frac{\partial F}{\partial x} + \frac{\partial G}{\partial y} + \frac{\partial H}{\partial z}$$

Regarding this as the dot product of the vector operator

$$\nabla = (\partial/\partial x, \, \partial/\partial y, \, \partial/\partial z)$$

and the field $\mathbf{F} = (F,G,H)$, we write

$$\text{div } \mathbf{F} = \nabla \cdot \mathbf{F} = \frac{\partial F}{\partial x} + \frac{\partial G}{\partial y} + \frac{\partial H}{\partial z}$$

The Divergence Theorem in the plane (Section 19.4) has the form

$$\int_C (\mathbf{F} \cdot \mathbf{N}) \, ds = \int\int_D (\nabla \cdot \mathbf{F}) \, dA$$

We may reasonably expect (by analogy) that the line integral on the left becomes a surface integral in space and the double integral on the right becomes a triple integral. Before stating this formally, however, we need two more definitions.

- A "standard" region in space (like a standard region in the plane, defined in Section 19.3) is a closed and bounded region in which triple integrals may be iterated in any order. A spherical ball, for example, is such a region.
- We have defined surface integrals only for surfaces $z = f(x, y)$ lying above (or below) a domain in the xy plane. To extend this definition to a closed surface σ bounding a standard region, we agree that the integral over σ is the sum of the integrals over its upper and lower parts (like the upper and lower hemispheres in Example 1).

The Divergence Theorem (The Theorem of Gauss)

Let σ be a closed surface bounding a "standard" region R in space. If $\mathbf{F}$ is a vector field with continuous partial derivatives in R, then

$$\int\int_\sigma (\mathbf{F} \cdot \mathbf{N}) \, dS = \int\int\int_R (\nabla \cdot \mathbf{F}) \, dV$$

where $\mathbf{N}$ is the outward unit normal to σ.

The proof of this theorem is tricky, so we have put it in an optional note at the end of this section in order to get on with the discussion. First, some examples of the power of the Divergence Theorem.

□ **Example 2**   Use the Divergence Theorem to confirm the result of Example 1.

Solution: The divergence of the vector field $\mathbf{F}(x, y, z) = -k(x, y, z)$ is

$$\nabla \cdot \mathbf{F} = -k(\partial/\partial x, \, \partial/\partial y, \, \partial/\partial z) \cdot (x, y, z) = -k\left(\frac{\partial x}{\partial x} + \frac{\partial y}{\partial y} + \frac{\partial z}{\partial z}\right) = -3k$$

The closed surface σ in this case is the sphere $x^2 + y^2 + z^2 = a^2$ (not the same σ as in Example 1) and the region R is the solid spherical ball $x^2 + y^2 + z^2 \le a^2$. Hence

$$\iint_\sigma (\mathbf{F} \cdot \mathbf{N}) \, dS = \iiint_R (\nabla \cdot \mathbf{F}) \, dV = -3k \iiint_R dV = -3k(\text{volume of } R)$$

$$= -3k(\tfrac{4}{3}\pi a^3) = -4k\pi a^3$$

This is considerably less work than our direct evaluation of the surface integral in Example 1 (and that was a simple integral). □

□ **Example 3**   Suppose that the flux of a fluid is given by

$$\mathbf{F}(x, y, z) = (x + y, y + z, z + x)$$

Find the rate at which the fluid is passing through the tetrahedron formed by the coordinate planes and the plane $x + y + z = 1$.

Solution: The problem is to find

$$\iint_\sigma (\mathbf{F} \cdot \mathbf{N}) \, dS$$

where $\mathbf{N}$ is the outward unit normal to the tetrahedron σ. First, we will compute the integral directly. Let $\sigma_1, \sigma_2, \sigma_3, \sigma_4$ be the surfaces indicated in Figure 2. The outward unit normal to σ_1 is $\mathbf{N} = -\mathbf{k}$. (Why?) Hence

$$\mathbf{F} \cdot \mathbf{N} = (x + y, y + z, z + x) \cdot (0, 0, -1) = -(z + x) = -x$$

(because $z = 0$ on σ_1). Since the equation of σ_1 is $z = f(x, y) = 0$, we have $\partial z / \partial x = \partial z / \partial y = 0$ and hence

$$\left(\frac{\partial z}{\partial x}\right)^2 + \left(\frac{\partial z}{\partial y}\right)^2 + 1 = 1$$

Figure 2
The tetrahedron σ

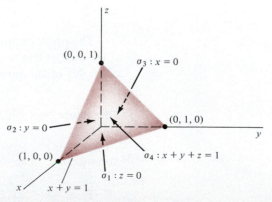

(More directly, $\sec^2 \theta = 1$ because the angle between σ_1 and the xy plane is $\theta = 0$.) Therefore

$$\iint_{\sigma_1} (\mathbf{F} \cdot \mathbf{N})\, dS = \iint_D (-x)(1)\, dA$$

where D is the projection of σ_1 on the xy plane, namely σ_1 itself:

$$D = \{(x,y): 0 \le x \le 1, 0 \le y \le 1 - x\}$$

We find

$$\iint_{\sigma_1} (\mathbf{F} \cdot \mathbf{N})\, dS = -\int_0^1 \int_0^{1-x} x\, dy\, dx = \int_0^1 x(x-1)\, dx = -\frac{1}{6}$$

The symmetry of the situation (in both the function $\mathbf{F}$ and the surface σ) shows that

$$\iint_{\sigma_2} (\mathbf{F} \cdot \mathbf{N})\, dS = \iint_{\sigma_3} (\mathbf{F} \cdot \mathbf{N})\, dS = -\frac{1}{6}$$

so we move on to σ_4. The outward unit normal to this surface (the plane $x + y + z = 1$) is

$$\mathbf{N} = \frac{1}{\sqrt{3}}(1, 1, 1)$$

$$\mathbf{F} \cdot \mathbf{N} = \frac{1}{\sqrt{3}}[(x + y) + (y + z) + (z + x)] = \frac{2}{\sqrt{3}}(x + y + z) = \frac{2}{\sqrt{3}}$$

(because $x + y + z = 1$ on σ_4). Writing the equation of σ_4 in the form $z = 1 - x - y$, we find $\partial z/\partial x = \partial z/\partial y = -1$, from which

$$\left(\frac{\partial z}{\partial x}\right)^2 + \left(\frac{\partial z}{\partial y}\right)^2 + 1 = 3$$

Hence

$$\iint_{\sigma_4} (\mathbf{F} \cdot \mathbf{N})\, dS = \iint_D \left(\frac{2}{\sqrt{3}}\right)(\sqrt{3})\, dA = 2(\text{area of } D) = 1$$

The sum of these four integrals is the answer,

$$\iint_{\sigma} (\mathbf{F} \cdot \mathbf{N})\, dS = -\frac{1}{6} - \frac{1}{6} - \frac{1}{6} + 1 = \frac{1}{2}$$

The Divergence Theorem makes the problem much easier. The divergence of $\mathbf{F}$ is

$$\nabla \cdot \mathbf{F} = (\partial/\partial x, \partial/\partial y, \partial/\partial z) \cdot (x + y, y + z, z + x) = 1 + 1 + 1 = 3$$

Letting R be the region bounded by the tetrahedron σ, we have

$$\iint_{\sigma} (\mathbf{F} \cdot \mathbf{N})\, dS = \iiint_R (\nabla \cdot \mathbf{F})\, dV = 3(\text{volume of } R)$$

$$= 3\left(\frac{1}{6}\right) = \frac{1}{2} \qquad \text{(Example 6, Section 18.1)}$$

The positive answer indicates that fluid is leaving the region R, so there is a source in R.   □

□ **Example 4**   Suppose that the flux of a fluid is given by

$$\mathbf{F}(x, y, z) = (z, -x, y)$$

Find the rate at which the fluid is passing through the cylindrical surface formed by $x^2 + y^2 = 1$, $z = 0$, and $z = 1$.

Solution: Again the problem is to evaluate $\iint_\sigma (\mathbf{F} \cdot \mathbf{N}) \, dS$. Let $\sigma_1, \sigma_2, \sigma_3$ be the surfaces shown in Figure 3. The outward unit normal to σ_1 is $\mathbf{N} = -\mathbf{k}$, so on σ_1 we have

$$\mathbf{F} \cdot \mathbf{N} = (z, -x, y) \cdot (0, 0, -1) = -y$$

As in Example 3 we find $\sec^2 \theta = 1$, so

$$\iint_{\sigma_1} (\mathbf{F} \cdot \mathbf{N}) \, dS = \iint_D (-y)(1) \, dA = -\int_0^{2\pi} \int_0^1 (r \sin \theta) r \, dr \, d\theta$$

$$= -\frac{1}{3} \int_0^{2\pi} \sin \theta \, d\theta = 0$$

Similarly, the integral over σ_2 is zero. The outward unit normal to σ_3 at (x, y, z) is

$$\mathbf{N} = \frac{1}{\sqrt{x^2 + y^2}}(x, y, 0) = (x, y, 0)$$

(because $x^2 + y^2 = 1$ on σ_3). Hence

$$\mathbf{F} \cdot \mathbf{N} = (z, -x, y) \cdot (x, y, 0) = xz - xy$$

At this point we have to modify our theory to apply to a surface that is not of the form $z = f(x, y)$ over a region in the xy plane. Thinking of the y axis as "up," we regard the cylinder $x^2 + y^2 = 1$ as the union of the surfaces $y = \sqrt{1 - x^2}$ and $y = -\sqrt{1 - x^2}$ "above" and "below" the rectangular

Figure 3
The cylindrical surface σ

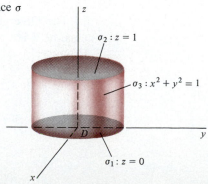

Figure 4
"Upper" half of the cylinder
$x^2 + y^2 = 1$

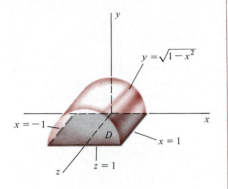

region D shown in Figure 4. Hence we express everything in terms of the independent variables x and z:

$$\mathbf{F} \cdot \mathbf{N} = xz - x\sqrt{1 - x^2} \qquad \text{(on the upper surface)}$$

Since the equation of this surface as a function of x and z is $y = \sqrt{1 - x^2}$, we find $\partial y / \partial x = -x/y$ and $\partial y / \partial z = 0$, from which

$$\left(\frac{\partial y}{\partial x}\right)^2 + \left(\frac{\partial y}{\partial z}\right)^2 + 1 = \frac{x^2}{y^2} + 1 = \frac{1}{1 - x^2}$$

(because $x^2 + y^2 = 1$). The integral of $\mathbf{F} \cdot \mathbf{N}$ over this surface is

$$\iint_D (xz - x\sqrt{1 - x^2})\left(\frac{1}{\sqrt{1 - x^2}}\right) dA = \int_{-1}^1 \int_0^1 \frac{xz\, dz\, dx}{\sqrt{1 - x^2}} - \int_{-1}^1 \int_0^1 x\, dz\, dx$$

$$= \frac{1}{2} \int_{-1}^1 \frac{x\, dx}{\sqrt{1 - x^2}} - \int_{-1}^1 x\, dx$$

$$= 0 - 0 = 0 \qquad \text{(the integrands are odd)}$$

Similarly, the integral of $\mathbf{F} \cdot \mathbf{N}$ over the lower surface $y = -\sqrt{1 - x^2}$ is zero. The answer to the question is

$$\iint_\sigma (\mathbf{F} \cdot \mathbf{N})\, dS = 0$$

The application of the Divergence Theorem is trivial in this example. Since

$$\nabla \cdot \mathbf{F} = (\partial/\partial x, \partial/\partial y, \partial/\partial z) \cdot (z, -x, y) = 0 + 0 + 0 = 0$$

we find

$$\iint_\sigma (\mathbf{F} \cdot \mathbf{N})\, dS = \iiint_R (\nabla \cdot \mathbf{F})\, dV = 0$$

where R is the region bounded by σ. (!)   □

Optional Note (*on the proof of the Divergence Theorem*)   We prove the Divergence Theorem by looking at the expansions of $\mathbf{F} \cdot \mathbf{N}$ and $\nabla \cdot \mathbf{F}$. The unit vector $\mathbf{N}$ can be written in the form

$$\mathbf{N} = (\cos \alpha, \cos \beta, \cos \gamma)$$

where α, β, γ are the angles between $\mathbf{N}$ and $\mathbf{i}, \mathbf{j}, \mathbf{k}$, respectively. (See Problem 28, Section 16.1.) If $\mathbf{F} = (F, G, H)$, we have

$$\mathbf{F} \cdot \mathbf{N} = F \cos \alpha + G \cos \beta + H \cos \gamma$$

Similarly,

$$\nabla \cdot \mathbf{F} = \frac{\partial F}{\partial x} + \frac{\partial G}{\partial y} + \frac{\partial H}{\partial z}$$

We propose to prove that

$$\iint_\sigma F \cos \alpha\, dS = \iiint_R \frac{\partial F}{\partial x}\, dV$$

$$\iint_\sigma G \cos \beta\, dS = \iiint_R \frac{\partial G}{\partial y}\, dV \qquad \iint_\sigma H \cos \gamma\, dS = \iiint_R \frac{\partial H}{\partial z}\, dV$$

from which the theorem will follow by addition.

Consider the third one. Since R is a standard region, we can break up its boundary σ into a lower surface σ_1: $z = g(x,y)$, an upper surface σ_2: $z = h(x,y)$, and a lateral surface σ_3 (which may not be present). See Figure 5. By the definition of surface integral, we have

$$\iint_{\sigma_1} H(x, y, z) \cos \gamma \, dS = \iint_D H(x, y, g(x,y)) \cos \gamma \sec \theta \, dA$$

where θ is the (acute) angle between the tangent plane to σ_1 and the xy plane. This is the same as the (acute) angle between a normal vector to σ_1 and the vector $\mathbf{k}$, whereas γ is the angle (not necessarily acute) between the *outward* normal $\mathbf{N}$ and $\mathbf{k}$. Since σ_1 is the lower part of σ, the angle γ is in fact obtuse (why?), so $\gamma = \pi - \theta$ and

$$\cos \gamma \sec \theta = -\cos \theta \sec \theta = -1$$

Hence

$$\iint_{\sigma_1} H(x, y, z) \cos \gamma \, dS = -\iint_D H(x, y, g(x,y)) \, dA$$

Similarly,

$$\iint_{\sigma_2} H(x, y, z) \cos \gamma \, dS = \iint_D H(x, y, h(x,y)) \, dA \qquad \text{(because } \gamma = \theta \text{ on } \sigma_2 \text{)}$$

Figure 5
Proof of the Divergence Theorem

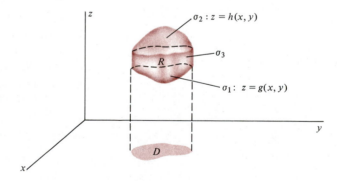

The integral of a function over the cylindrical surface σ_3 must be treated differently, because the surface is not of the form $z = f(x,y)$ over a region in the xy plane. (See Example 4 for an illustration of what we mean.) In this case, however, we are integrating only the term $H \cos \gamma$. Since $\gamma = \pi/2$ on σ_3, the integrand is zero and σ_3 does not contribute to the result. Hence

$$\iint_\sigma H \cos \gamma \, dS = \iint_{\sigma_1} H \cos \gamma \, dS + \iint_{\sigma_2} H \cos \gamma \, dS$$

$$= \iint_D [H(x, y, h(x,y)) - H(x, y, g(x,y))] \, dA$$

$$= \iint_D \left[\int_{g(x,y)}^{h(x,y)} \frac{\partial H}{\partial z} \, dz \right] dA = \iiint_R \frac{\partial H}{\partial z} \, dV$$

The integrals of $F \cos \alpha$ and $G \cos \beta$ over σ are similarly treated; the result is the Divergence Theorem.

Problem Set 19.5

In each of the following, evaluate the surface integral of the given function on the given surface.

1. $g(x,y,z) = x$ on the first-octant part of the sphere $x^2 + y^2 + z^2 = 1$.

2. $g(x,y,z) = z$ on the first-octant part of the sphere $x^2 + y^2 + z^2 = 1$.

3. $g(x,y,z) = x^2 + y^2$ on the hemisphere $x^2 + y^2 + z^2 = 1$, $z \geq 0$.

4. $g(x,y,z) = x + y + z$ on the sphere $x^2 + y^2 + z^2 = 1$.

5. $g(x,y,z) = (2x^2 + 2y^2 + z)/\sqrt{4x^2 + 4y^2 + 1}$ on the part of the paraboloid $z = 4 - x^2 - y^2$ above the xy plane.

6.–10. The *average value* of a function $g(x,y,z)$ on the surface σ is

$$\frac{1}{S} \iint_\sigma g(x,y,z) \, dS$$

where S is the area of σ. In Problems 1 through 5, find the average value of the given function on the given surface.

In each of the following, evaluate an appropriate surface integral to find the rate at which a fluid with flux $\mathbf{F}$ passes through the given surface.

11. $\mathbf{F}(x,y,z) = (x,y,z)$ through the sphere $x^2 + y^2 + z^2 = 1$.

12. $\mathbf{F}(x,y,z) = (0,0,z)$ through the closed surface formed by the hemisphere $z = \sqrt{1 - x^2 - y^2}$ and the xy plane.

13. $\mathbf{F}(x,y,z) = (x,y,0)$ through the tetrahedron formed by the coordinate planes and the plane $x + y + z = 1$.

14. $\mathbf{F}(x,y,z) = (x,y,z)$ through the cube formed by the coordinate planes and the planes $x = 1$, $y = 1$, $z = 1$.

15. $\mathbf{F}(x,y,z) = (1,1,z)$ through the sphere $x^2 + y^2 + z^2 = 1$.

16. $\mathbf{F}(x,y,z) = (x,y,z)$ through the closed surface formed by the paraboloid $z = 1 - x^2 - y^2$ and the xy plane.

17. $\mathbf{F}(x,y,z) = (x + y, y + z, z + x)$ through the sphere $x^2 + y^2 + z^2 = 1$.

18. $\mathbf{F}(x,y,z) = (x - y, y - z, z - x)$ through the tetrahedron formed by the coordinate planes and the plane $2x + 2y + z = 2$.

19. $\mathbf{F}(x,y,z) = (x,y,z)$ through the closed surface formed by the cylinder $x^2 + y^2 = 1$ and the planes $z = 0$ and $z = 1$.

20.–28. In Problems 11 through 19, use the Divergence Theorem to find the rate at which a fluid with flux $\mathbf{F}$ passes through the given surface.

In each of the following, use the Divergence Theorem to find the rate at which a fluid with flux $\mathbf{F}$ passes through the given surface.

29. $\mathbf{F}(x,y,z) = (x^3, y^3, z^3)$ through the sphere $x^2 + y^2 + z^2 = a^2$.

30. $\mathbf{F}(x,y,z) = (y - z, z - x, x - y)$ through the ellipsoid $x^2/a^2 + y^2/b^2 + z^2/c^2 = 1$.

31. $\mathbf{F}(x,y,z) = (x^2 + e^{yz}, y^2 + e^{zx}, z^2 + e^{xy})$ through the cube formed by the coordinate planes and the planes $x = 1$, $y = 1$, $z = 1$.

32. $\mathbf{F}(x,y,z) = xyz(1, 1, 1)$ through the cube formed by the planes $x = \pm 1$, $y = \pm 1$, $z = \pm 1$.

33. $\mathbf{F}(x,y,z) = (z,y,x)$ through the tetrahedron formed by the coordinate planes and the plane $x/2 + y/3 + z/5 = 1$.

34. Show that if the flux of a fluid is constant, the rate at which the fluid passes through a closed surface is zero. What is the physical interpretation?

35. Suppose that $\mathbf{F}(x) = x/\rho^3$, where $\rho = |x|$.
    (a) Show that the divergence of $\mathbf{F}$ is zero.
    (b) If σ is the unit sphere $x^2 + y^2 + z^2 = 1$, show that

    $$\iint_\sigma (\mathbf{F} \cdot \mathbf{N}) \, dS = 4\pi$$

    where $\mathbf{N}$ is the outward unit normal to σ.
    (c) Why don't these results contradict the Divergence Theorem?

36. Suppose that the vector field $\mathbf{F}(x,y,z)$ is the gradient of the scalar field $f(x,y,z)$. Show that div $\mathbf{F} = f_{11} + f_{22} + f_{33}$.

37. Explain why the area of the surface $\sigma: z = f(x,y)$ is given by

    $$\iint_\sigma dS$$

38. Suppose that the density at each point of the surface σ is $\delta(x,y,z)$. By considering the mass Δm of a typical patch of surface (of area ΔS), explain the formula

    $$m = \iint_\sigma \delta(x,y,z) \, dS$$

39. The density at each point of the sphere $x^2 + y^2 + z^2 = a^2$ is proportional to the square of the distance from the xy plane. Find the mass.

40. Explain why the centroid of a surface σ is the point $(\bar{x}, \bar{y}, \bar{z})$ satisfying

    $$\bar{x} = \frac{1}{S} \iint_\sigma x \, dS, \quad \bar{y} = \frac{1}{S} \iint_\sigma y \, dS, \quad \bar{z} = \frac{1}{S} \iint_\sigma z \, dS$$

    where S is in the area of σ.

41. Use Problem 40 to find the centroid of the hemisphere $x^2 + y^2 + z^2 = a^2$, $z \geq 0$.

42. Use Problem 40 to find the centroid of the portion of the cone $a^2 z^2 = h^2(x^2 + y^2)$ between the planes $z = 0$ and $z = h$.

19.6
LINE INTEGRALS IN SPACE AND STOKES' THEOREM

The Divergence Theorem in the plane (Section 19.4) is a means of evaluating a line integral around a closed curve by a double integral over the region bounded by the curve:

$$\int_C (\mathbf{F} \cdot \mathbf{N}) \, ds = \iint_D (\nabla \cdot \mathbf{F}) \, dA$$

In the last section we derived a natural extension of that idea to space, namely the evaluation of a surface integral over a closed surface by a triple integral over the region bounded by the surface:

$$\iint_\sigma (\mathbf{F} \cdot \mathbf{N}) \, dS = \iiint_R (\nabla \cdot \mathbf{F}) \, dV \qquad \text{(Divergence Theorem)}$$

Generalizing from line integrals to surface integrals is not the only "natural extension" we might seek, however. Why not consider line integrals along curves in space? More precisely, suppose that

$$\mathbf{F}(\mathbf{x}) = (F(x, y, z), G(x, y, z), H(x, y, z))$$

is a continuous vector field in space and

$$C: \mathbf{x} = (x(t), y(t), z(t)) \qquad t_1 \leq t \leq t_2$$

is a smooth curve in space whose graph lies in the domain of $\mathbf{F}$. The line integral of $\mathbf{F}$ along C is defined by

$$\int_C \mathbf{F}(\mathbf{x}) \cdot d\mathbf{x} = \int_{t_1}^{t_2} \left[\mathbf{F}(\mathbf{x}) \cdot \frac{d\mathbf{x}}{dt} \right] dt$$

exactly as in Section 19.1. Alternate notation is

$$\int_C \mathbf{F}(\mathbf{x}) \cdot d\mathbf{x} = \int_C F(x, y, z) \, dx + G(x, y, z) \, dy + H(x, y, z) \, dz$$

obtained from the vector differential $d\mathbf{x} = (dx, dy, dz)$ and the dot product

$$\mathbf{F}(\mathbf{x}) \cdot d\mathbf{x} = F(x, y, z) \, dx + G(x, y, z) \, dy + H(x, y, z) \, dz$$

□ **Example 1**   Find the line integral of the vector field $\mathbf{F}(x, y, z) = (yz, zx, xy)$ along the circular helix

$$C: x = \cos t, \, y = \sin t, \, z = t \qquad 0 \leq t \leq 2\pi$$

Solution: Since

$$\begin{aligned} \mathbf{F}(\mathbf{x}) \cdot d\mathbf{x} &= [yz(-\sin t) + zx(\cos t) + xy(1)] \, dt \\ &= (-t \sin^2 t + t \cos^2 t + \sin t \cos t) \, dt \\ &= (t \cos 2t + \tfrac{1}{2} \sin 2t) \, dt \end{aligned}$$

we find $\displaystyle \int_C \mathbf{F}(\mathbf{x}) \cdot d\mathbf{x} = \int_0^{2\pi} t \cos 2t \, dt + \frac{1}{2} \int_0^{2\pi} \sin 2t \, dt = 0$ □

As you can see by inspection, the vector field $\mathbf{F}(x, y, z) = (yz, zx, xy)$ is the gradient of the scalar field $f(x, y, z) = xyz$. Whenever that happens, we can write (as in Section 19.1)

$$\int_C \mathbf{F}(\mathbf{x}) \cdot d\mathbf{x} = \int_C \nabla f(\mathbf{x}) \cdot d\mathbf{x} = \int_{t_1}^{t_2} \left[\nabla f(\mathbf{x}) \cdot \frac{d\mathbf{x}}{dt} \right] dt = \int_{t_1}^{t_2} \left[\frac{d}{dt} f(\mathbf{x}) \right] dt$$

$$= f(\mathbf{x}) \Big|_{t=t_1}^{t=t_2} = f(\mathbf{b}) - f(\mathbf{a})$$

where $\mathbf{a} = \mathbf{x}(t_1)$ and $\mathbf{b} = \mathbf{x}(t_2)$ are the initial and terminal points of C. This is the **Fundamental Theorem of Calculus for Line Integrals.**

Thus Example 1 could have been done by writing

$$\int_C \mathbf{F}(\mathbf{x}) \cdot d\mathbf{x} = \int_C \nabla(xyz) \cdot d\mathbf{x} = xyz \Big|_{t=0}^{t=2\pi} = 0 - 0 = 0$$

since $\mathbf{a} = (1, 0, 0)$ and $\mathbf{b} = (1, 0, 2\pi)$ both give zero in the function $f(x, y, z) = xyz$.

Again as in Section 19.1, we can restate the Fundamental Theorem by saying that the line integral of a gradient field around a closed path is zero, or (equivalently) the line integral from one point to another is independent of the path. The converse is also true (by an argument virtually identical to the one suggested in Problem 35, Section 19.2). Hence (as before)

Gradient Field $\Leftrightarrow$ Integral Around Closed Path Is Zero
$\Leftrightarrow$ Integral Is Independent of Path

To develop a test for gradient fields (as in Section 19.2), we suppose that the vector field

$$\mathbf{F}(\mathbf{x}) = (F(\mathbf{x}), G(\mathbf{x}), H(\mathbf{x}))$$

is the gradient of the scalar field $f(x, y, z)$, that is, $\mathbf{F}(\mathbf{x}) = \nabla f(\mathbf{x})$. Then

$$F(\mathbf{x}) = f_1(\mathbf{x}), \ G(\mathbf{x}) = f_2(\mathbf{x}), \ H(\mathbf{x}) = f_3(\mathbf{x})$$

from which (differentiating to take advantage of the equality of mixed partial derivatives) we have

$$F_2(\mathbf{x}) = f_{12}(\mathbf{x}) = f_{21}(\mathbf{x}) = G_1(\mathbf{x})$$
$$G_3(\mathbf{x}) = f_{23}(\mathbf{x}) = f_{32}(\mathbf{x}) = H_2(\mathbf{x})$$
$$H_1(\mathbf{x}) = f_{31}(\mathbf{x}) = f_{13}(\mathbf{x}) = F_3(\mathbf{x})$$

If the region in which these *necessary* conditions hold is simply connected, they are also *sufficient*:

Let $\mathbf{F}(\mathbf{x}) = (F(x, y, z), G(x, y, z), H(x, y, z))$ be a vector field with continuous partial derivatives in a simply connected region of space. Then $\mathbf{F}$ is a gradient field if and only if

$$\frac{\partial G}{\partial x} = \frac{\partial F}{\partial y}, \ \frac{\partial H}{\partial y} = \frac{\partial G}{\partial z}, \ \frac{\partial F}{\partial z} = \frac{\partial H}{\partial x}$$

Note that the first of these conditions, $\partial G/\partial x = \partial F/\partial y$, is the same as for a vector field in the plane. In fact, if we suppress the third variable z and the third component $H(x,y,z)$, we are back to the theory of Section 19.2.

□ **Example 2**   Find an antigradient of the vector field

$$\mathbf{F}(x, y, z) = (3x^2 + 2xy + z^2, 3y^2 + 2yz + x^2, 3z^2 + 2zx + y^2)$$

(or determine that none exists).

Solution: First we compute

$$\frac{\partial}{\partial x}(3y^2 + 2yz + x^2) = 2x = \frac{\partial}{\partial y}(3x^2 + 2xy + z^2)$$

$$\frac{\partial}{\partial y}(3z^2 + 2zx + y^2) = 2y = \frac{\partial}{\partial z}(3y^2 + 2yz + x^2)$$

$$\frac{\partial}{\partial z}(3x^2 + 2xy + z^2) = 2z = \frac{\partial}{\partial x}(3z^2 + 2zx + y^2)$$

This shows that an antigradient $f(x, y, z)$ exists. To find it, use "partial integration" as in Section 19.2. We begin by integrating

$$F(x, y, z) = 3x^2 + 2xy + z^2$$

with respect to x, keeping y and z fixed:

$$f(x, y, z) = \int (3x^2 + 2xy + z^2)\, dx = x^3 + x^2y + xz^2 + C(y,z)$$

where $C(y,z)$ is a "constant" relative to x, that is, a function of y and z alone. Differentiate f with respect to y and see if it fits $G(x, y, z) = 3y^2 + 2yz + x^2$:

$$f_2(x, y, z) = x^2 + \frac{\partial C}{\partial y} \overset{?}{=} 3y^2 + 2yz + x^2$$

As expected, the terms involving x drop out, leaving $\partial C/\partial y = 3y^2 + 2yz$. Now integrate this with respect to y to obtain

$$C(y,z) = \int (3y^2 + 2yz)\, dy = y^3 + y^2z + D(z)$$

where $D(z)$ is a function of z alone. This yields

$$f(x, y, z) = x^3 + x^2y + xz^2 + y^3 + y^2z + D(z)$$

which when differentiated with respect to z should fit $H(x, y, z) = 3z^2 + 2zx + y^2$:

$$f_3(x, y, z) = 2xz + y^2 + D'(z) \overset{?}{=} 3z^2 + 2zx + y^2$$

The terms involving x and y drop out, leaving $D'(z) = 3z^2$. Hence $D(z) = z^3 + k$ and our antigradient is

$$f(x, y, z) = x^3 + x^2y + xz^2 + y^3 + y^2z + z^3 + k$$

The expected cancellation of the right variables at the right time doesn't work when the vector field is not a gradient.   □

With these generalizations from Sections 19.1 and 19.2 in place, we are ready for our three-dimensional version of Green's Theorem. To see what to expect, recall that if

$$\mathbf{F}(\mathbf{x}) = (F(x,y), G(x,y))$$

is a vector field in the plane, then

$$\int_C \mathbf{F}(\mathbf{x}) \cdot d\mathbf{x} = \iint_D \left(\frac{\partial G}{\partial x} - \frac{\partial F}{\partial y} \right) dA$$

where D is a region with boundary curve C. The integrand is zero precisely when $\partial G/\partial x = \partial F/\partial y$, which is the condition that $\mathbf{F}$ should be a gradient. This condition in the case of a three-dimensional field

$$\mathbf{F}(\mathbf{x}) = (F(x, y, z), G(x, y, z), H(x, y, z))$$

is

$$\frac{\partial G}{\partial x} = \frac{\partial F}{\partial y}, \quad \frac{\partial H}{\partial y} = \frac{\partial G}{\partial z}, \quad \frac{\partial F}{\partial z} = \frac{\partial H}{\partial x}$$

as we have seen. Therefore we may expect Green's Theorem in space to read

$$\int_C \mathbf{F}(\mathbf{x}) \cdot d\mathbf{x} = \text{surface integral involving } \frac{\partial G}{\partial x} - \frac{\partial F}{\partial y}, \quad \frac{\partial H}{\partial y} - \frac{\partial G}{\partial z}, \quad \frac{\partial F}{\partial z} - \frac{\partial H}{\partial x}$$

Remark   Why *surface* integral instead of *triple* integral? One might expect the double integral in Green's Theorem to become a triple integral in the spatial generalization (as in the Divergence Theorem of the last section). However, a double integral can also be regarded as a surface integral (the surface being a flat region D in the plane and the integrand involving only two variables for that reason). If we imagine this flat surface pushed out into space (like a bubble blown from a child's ring immersed in soapy water), we obtain a surface σ that is still "bounded" in some sense by the curve C. See Figure 1, which shows the hemisphere $x^2 + y^2 + z^2 = a^2$, $z \geq 0$, and its "boundary curve" $x^2 + y^2 = a^2$ in the xy plane. Obviously the hemisphere is not inside the circle; it is nevertheless constrained by the circle in size and shape (although of course so are infinitely many other surfaces we might form by blowing bubbles from a circular ring).

In general we cannot require C to remain in the plane (otherwise our line integral would be confined to two-dimensional vector fields). So what we are saying is that a closed *space* curve (like a child's bubble ring bent out of shape) is the "boundary" of a surface (like a bubble blown from the deformed ring).

Figure 1
Hemisphere "bounded" by a circle

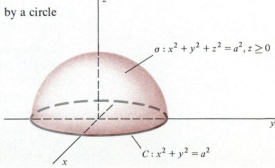

It might occur to a clever mind that the differences involved in our surface integral come from the cross product

$$\nabla \times \mathbf{F} = \begin{vmatrix} \mathbf{i} & \mathbf{j} & \mathbf{k} \\ \partial/\partial x & \partial/\partial y & \partial/\partial z \\ F & G & H \end{vmatrix} \qquad (!!)$$

In any case we ordinary mortals can check it out:

$$\nabla \times \mathbf{F} = \left(\frac{\partial H}{\partial y} - \frac{\partial G}{\partial z}, \frac{\partial F}{\partial z} - \frac{\partial H}{\partial x}, \frac{\partial G}{\partial x} - \frac{\partial F}{\partial y} \right)$$

This vector function is called the **curl** of **F,** written

$$\boxed{\operatorname{curl} \mathbf{F} = \nabla \times \mathbf{F}}$$

Before jumping to a conclusion, let's return to Green's Theorem for a two-dimensional field $\mathbf{F(x)} = (F(x,y), G(x,y))$, namely

$$\int_C \mathbf{F(x)} \cdot d\mathbf{x} = \iint_D \left(\frac{\partial G}{\partial x} - \frac{\partial F}{\partial y} \right) dA$$

and see if we can write it in terms of curl **F**. Since cross products are defined only for space vectors, we must embed the field in space. That is, take $\mathbf{x} = (x, y, 0)$ instead of (x, y) and write

$$\mathbf{F(x)} = (F(x, y, 0), G(x, y, 0), 0)$$

Then

$$\operatorname{curl} \mathbf{F} = \nabla \times \mathbf{F} = \begin{vmatrix} \mathbf{i} & \mathbf{j} & \mathbf{k} \\ \partial/\partial x & \partial/\partial y & \partial/\partial z \\ F & G & 0 \end{vmatrix} = \left(-\frac{\partial G}{\partial z}, \frac{\partial F}{\partial z}, \frac{\partial G}{\partial x} - \frac{\partial F}{\partial y} \right)$$

To obtain the integrand we want, dot this vector with $\mathbf{k} = (0, 0, 1)$:

$$(\nabla \times \mathbf{F}) \cdot \mathbf{k} = \frac{\partial G}{\partial x} - \frac{\partial F}{\partial y}$$

Thus Green's Theorem may be put in the form

$$\int_C \mathbf{F} \cdot d\mathbf{x} = \iint_D (\nabla \times \mathbf{F}) \cdot \mathbf{k} \, dA$$

The expression $(\nabla \times \mathbf{F}) \cdot \mathbf{k}$ is the (scalar) component of $\nabla \times \mathbf{F}$ *normal to the surface D* (since D lies in the xy plane). Moreover, the direction of the normal ($\mathbf{k}$ instead of $-\mathbf{k}$) is the direction of a right-handed screw induced by the counterclockwise orientation of C (the same right-handed orientation that the x, y, z coordinate system has). This observation fits the last piece into the puzzle. We may now predict (although we are far from a proof) that Green's Theorem in space reads

$$\int_C \mathbf{F} \cdot d\mathbf{x} = \iint_\sigma (\nabla \times \mathbf{F}) \cdot \mathbf{N} \, dS$$

where $\mathbf{N}$ is the outward unit normal to the surface σ and C is "positively" oriented by the right-hand rule.

Remark   "Counterclockwise" is inadequate for the orientation of a space curve. If you have ever seen a *Möbius strip* (a twisted ribbon with only one side), you will appreciate the difficulty of orientation in space. The simple closed curve bounding this strip can be traced in a definite direction, but the surface cannot be kept on our left as we go around its boundary. (See Figure 2.) Such "nonorientable" surfaces require some rethinking of intuitive notions of curves and surfaces in space. We avoid the question here by assuming that our surface is part of the boundary of a standard region. If $\mathbf{N}$ is the outward unit normal, C may be oriented by the right-hand rule.

Figure 2
Möbius strip

Stokes' Theorem

Suppose that $\mathbf{F}(\mathbf{x})$ is a vector field with continuous partial derivatives in a standard region of space. Let σ be a part of the surface of this region that is bounded by a closed curve C, positively oriented relative to the outward unit normal $\mathbf{N}$. Then

$$\int_C \mathbf{F} \cdot d\mathbf{x} = \iint_\sigma (\nabla \times \mathbf{F}) \cdot \mathbf{N} \, dS$$

The proof of this theorem, interestingly enough, is based on Green's Theorem (even though it is a generalization to space). Since it contributes nothing further to an understanding of these matters, we omit it.

□ **Example 3**   Let $\mathbf{F}(x, y, z) = (xy, yz, zx)$ and let C be the boundary of the first-octant part of the plane $x + y + z = 1$. See Figure 3, which shows C broken into three positively oriented curves

$$
\begin{array}{lll}
C_1: x + y = 1 & (y = 1 - x,\ x \text{ running from 1 to 0}) \\
C_2: y + z = 1 & (z = 1 - y,\ y \text{ running from 1 to 0}) \\
C_3: z + x = 1 & (x = 1 - z,\ z \text{ running from 1 to 0})
\end{array}
$$

Figure 3
Boundary curve of first-octant portion of the surface $x + y + z = 1$

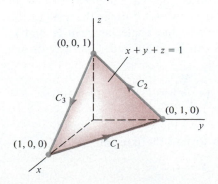

The line integral of $\mathbf{F}$ around C is

$$\int_C \mathbf{F} \cdot d\mathbf{x} = \int_{C_1} \mathbf{F} \cdot d\mathbf{x} + \int_{C_2} \mathbf{F} \cdot d\mathbf{x} + \int_{C_3} \mathbf{F} \cdot d\mathbf{x}$$

The first of these is

$$\int_{C_1} xy \, dx + yz \, dy + zx \, dz = \int_{C_1} xy \, dx \qquad \text{(because } z = 0 \text{ on } C_1)$$

$$= \int_1^0 x(1 - x) \, dx = -\frac{1}{6}$$

The other two are also $-\frac{1}{6}$ (because of the symmetry of $\mathbf{F}$ and C), so we find

$$\int_C \mathbf{F} \cdot d\mathbf{x} = -\frac{1}{2}$$

To compute the integral by Stokes' Theorem, observe that the "outward" unit normal (the one pointing toward the reader) is

$$\mathbf{N} = \frac{1}{\sqrt{3}}(1, 1, 1)$$

Rather than computing $\nabla \times \mathbf{F}$ and then dotting it with $\mathbf{N}$, we use the *scalar triple product*

$$(\nabla \times \mathbf{F}) \cdot \mathbf{N} = \frac{1}{\sqrt{3}} \begin{vmatrix} 1 & 1 & 1 \\ \partial/\partial x & \partial/\partial y & \partial/\partial z \\ xy & yz & zx \end{vmatrix} \qquad \text{(Section 16.3)}$$

$$= \frac{1}{\sqrt{3}}[(0 - y) - (z - 0) + (0 - x)]$$

$$= -\frac{1}{\sqrt{3}}(x + y + z) = -\frac{1}{\sqrt{3}}$$

(because $x + y + z = 1$ on the surface σ). Hence

$$\int_C \mathbf{F} \cdot d\mathbf{x} = \iint_\sigma (\nabla \times \mathbf{F}) \cdot \mathbf{N} \, dS = -\frac{1}{\sqrt{3}} \iint_\sigma dS = -\frac{1}{\sqrt{3}}(\text{area of } \sigma)$$

The area of σ can be found directly by observing that the graph of C is an equilateral triangle of side length $\sqrt{2}$. It is also $\sec \theta$ times the area of the projection of σ on the xy plane, where θ is the angle between $\mathbf{N}$ and $\mathbf{k}$. (Why?) Or the surface integral $\iint_\sigma dS$ can be evaluated. One way or another, we find

$$\int_C \mathbf{F} \cdot d\mathbf{x} = -\frac{1}{\sqrt{3}}\left(\frac{\sqrt{3}}{2}\right) = -\frac{1}{2} \qquad \square$$

□ **Example 4**   Use Stokes' Theorem to find the line integral of

$$\mathbf{F}(x, y, z) = (x - y, y - z, z - x)$$

around the curve of intersection of the sphere $x^2 + y^2 + z^2 = 2$ and the paraboloid $z = x^2 + y^2$.

Solution: This curve is the circle $x^2 + y^2 = 1$ in the plane $z = 1$ (as you can discover by solving the equations of the sphere and paraboloid simultaneously). To use Stokes' Theorem, we must choose a surface σ having C as boundary. Of the infinitely many possibilities, the simplest is the disk

$$\sigma = \{(x, y, z): x^2 + y^2 \leq 1, z = 1\}$$

(the plane region inside C). Then $\mathbf{N} = \mathbf{k}$ and we have

$$(\mathbf{\nabla} \times \mathbf{F}) \cdot \mathbf{N} = \begin{vmatrix} 0 & 0 & 1 \\ \partial/\partial x & \partial/\partial y & \partial/\partial z \\ x - y & y - z & z - x \end{vmatrix} = \frac{\partial}{\partial x}(y - z) - \frac{\partial}{\partial y}(x - y) = 1$$

from which $\displaystyle \int_C \mathbf{F} \cdot d\mathbf{x} = \iint_\sigma dS = \text{area of } \sigma = \pi$ □

We end this section by observing that

$$\int_C \mathbf{F} \cdot d\mathbf{x} = \int_C (\mathbf{F} \cdot \mathbf{T}) \, ds$$

where $\mathbf{T}$ is the unit tangent vector in the direction of motion along C and s is distance measured along C. Hence Stokes' Theorem can be written in the form

$$\int_C (\mathbf{F} \cdot \mathbf{T}) \, ds = \iint_\sigma (\mathbf{\nabla} \times \mathbf{F}) \cdot \mathbf{N} \, dS$$

As in the two-dimensional case, the left side is called the **circulation** around C (if $\mathbf{F}$ is interpreted as the flux of a fluid). Since it is the integral of the tangential component of $\mathbf{F}$, it measures whirlpools (being zero if there are none). The right side is the integral of the normal component of curl $\mathbf{F}$. The source of the term "curl" is that we are measuring the tendency of the fluid to rotate. Some books refer to (curl $\mathbf{F}$) $\cdot \mathbf{N}$ as the *rotation* of $\mathbf{F}$ about $\mathbf{N}$. Note that the vector equation curl $\mathbf{F} = \mathbf{0}$ is equivalent to the three conditions for a gradient field. (Why?) Such fields (called *conservative* when $\mathbf{F}$ is interpreted as a force) are often said to be **irrotational** when $\mathbf{F}$ is interpreted as a flux.

Problem Set 19.6

Test each of the following vector fields to determine whether it is a gradient. When it is, find an antigradient by "partial integration."

1. $\mathbf{F}(x, y, z) = (y + z, x + y, z + x)$

2. $\mathbf{F}(x, y, z) = (z, y, x)$

3. $\mathbf{F}(x, y, z) = 2(x + y + z)(1, 1, 1)$

4. $\mathbf{F}(x, y, z) = (x - y, y - z, z - x)$

5. $\mathbf{F}(x, y, z) = (yz + y + z, xz + x + z, xy + x + y)$

6. $\mathbf{F}(x, y, z) = (xy, yz, zx)$

7. Find the line integral of $\mathbf{F}(x, y, z) = (z, y, x)$ along the curve

$$C: x = e^t \cos t, \ y = e^t \sin t, \ z = e^t \quad 0 \le t \le \pi$$

   Hint: $\mathbf{F}$ is a gradient!

8. Find the line integral of $\mathbf{F}(x, y, z) = (z, x, y)$ around the curve of intersection of the hemisphere $x^2 + y^2 + z^2 = a^2$, $z \ge 0$, and the xy plane (positively oriented relative to the outward unit normal to the hemisphere).

9. In Problem 8 verify Stokes' Theorem by computing an appropriate surface integral over the hemisphere.

10. Find the line integral of $\mathbf{F}(x, y, z) = (x - y, y - z, z - x)$ around the triangular curve of intersection of the plane $x + y + z = 1$ and the coordinate planes (positively oriented relative to the outward unit normal to the plane).

11. In Problem 10 verify Stokes' Theorem by computing an appropriate surface integral over the plane.

12. Find the line integral of $\mathbf{F}(x, y, z) = (y, z, x)$ around the curve of intersection of the paraboloid $z = 4 - x^2 - y^2$ and the plane $z = 1$ (positively oriented relative to the outward unit normal to the paraboloid).

13. In Problem 12 verify Stokes' Theorem by computing an appropriate surface integral over the paraboloid.

14. Find the line integral of $\mathbf{F}(x, y, z) = (x + y, y + z, z + x)$ around the curve of intersection of the half-cone $z = \sqrt{x^2 + y^2}$ and the plane $z = 1$ (positively oriented relative to the outward unit normal to the cone).

15. In Problem 14 verify Stokes' Theorem by computing an appropriate surface integral over the cone. (If your answers to Problems 14 and 15 have opposite signs, check the orientation in Problem 14!)

16. Find the line integral of $\mathbf{F}(x, y, z) = (x - y, y - z, z - x)$ around the ellipse $x^2/a^2 + y^2/b^2 = 1$, $z = 0$ (oriented counterclockwise).

17. In Problem 16 verify Stokes' Theorem by computing an appropriate surface integral.

18. Find the line integral of $\mathbf{F}(x, y, z) = (x - y, y - z, z - x)$ around the curve of intersection of the hemisphere $z = \sqrt{2 - x^2 - y^2}$ and the paraboloid $z = x^2 + y^2$ (positively oriented relative to the outward unit normal to the hemisphere). Why should the answer check with Example 4?

19. In Problem 18 verify Stokes' Theorem by computing an appropriate surface integral over the upper part of the hemisphere bounded by the curve of intersection.

20. In Problem 18 reverse the orientation of the curve of intersection and verify Stokes' Theorem by computing an appropriate surface integral over the part of the paraboloid bounded by the curve.

21. Let σ be the sphere $x^2 + y^2 + z^2 = a^2$, with outward unit normal $\mathbf{N}$. Show that if $\mathbf{F}$ is any vector field (with continuous partial derivatives), then

$$\iint_\sigma (\mathbf{\nabla} \times \mathbf{F}) \cdot \mathbf{N} \, dS = 0$$

   Hint: Apply Stokes' Theorem to the upper and lower hemispheres.

22. Prove that no vector field can have curl equal to $\mathbf{x}$.

23. Suppose that $\mathbf{F} = (F, G, H)$ is a vector field with continuous partial derivatives in a simply connected region of space. Explain why Stokes' Theorem implies that the conditions

$$\frac{\partial G}{\partial x} = \frac{\partial F}{\partial y}, \ \frac{\partial H}{\partial y} = \frac{\partial G}{\partial z}, \ \frac{\partial F}{\partial z} = \frac{\partial H}{\partial x}$$

   are sufficient for $\mathbf{F}$ to be a gradient field.

24. If $\mathbf{F}$ is a vector field with continuous second partial derivatives, show that $\text{div}(\text{curl} \ \mathbf{F}) = 0$.

25. If f is a scalar field with continuous second partial derivatives, show that $\text{curl}(\mathbf{\nabla} f) = \mathbf{0}$.

True–False Quiz

1. If $\mathbf{F}(\mathbf{x})$ is a continuous vector field in a region D, then $\int_C \mathbf{F}(\mathbf{x}) \cdot d\mathbf{x} = 0$ for every (smooth) closed curve lying in D.

2. The line integral of a conservative force field is independent of the path.

3. If C is the unit circle (oriented counterclockwise), then

$$\int_C (e^{-x^2} + y) \, dx + (2x + \sqrt{y^3 + 1}) \, dy = \pi$$

4. The divergence of the three-dimensional vector field $\mathbf{F}(\mathbf{x}) = \mathbf{x}$ is 3.

5. If $\mathbf{F}(x, y, z)$ is the flux of a fluid and $\text{div} \ \mathbf{F} = 0$ at each point of a region R, there are no sources or sinks in R.

6. If σ is a sphere with outward unit normal $\mathbf{N}$, and $\mathbf{F}$ is a constant vector field, then

$$\iint_\sigma (\mathbf{F} \cdot \mathbf{N}) \, dS = 0$$

7. $F(x, y, z) = (xz - y, yx - z, zy - x)$ is a gradient field.

8. If $x = (x, y, z)$ and a is a constant vector, then $\text{curl}(a \times x) = 2a$.

9. The curl of a gradient field is zero.

10. If T is the unit tangent vector to the smooth curve C (pointing in the positive direction) and F is a vector field with continuous partial derivatives in a region containing C, then

$$\int_C F(x) \cdot dx = \int_C (F \cdot T)\, ds$$

Additional Problems

Use the definition of line integral to find

$$\int_C (3x^2 + y)\, dx + x\, dy$$

where C is defined as follows.

1. The semicircle $x^2 + y^2 = 1$, $y \geq 0$, oriented counterclockwise.

2. The parabola $y = x^2 - 1$ from $(-1,0)$ to $(1,0)$.

3. The triangle with vertices $(0,0)$, $(1,0)$, $(0,1)$, oriented counterclockwise.

4. Find a scalar field whose gradient is $F(x, y) = (3x^2 + y, x)$.

5. Why do the answers to Problems 1 and 2 have opposite signs?

6. Why do we get zero in Problem 3?

7. Use the Fundamental Theorem for Line Integrals to do Problems 1, 2, and 3.

8. Use Green's Theorem to do Problem 3.

Use the definition to find the line integral of $F(x, y) = (x - y, 2y - x)$ along each of the following paths.

9. The line $x + y = 1$ from $(0,1)$ to $(1,0)$.

10. The parabola $y = 1 - x^2$ from $(0,1)$ to $(1,0)$.

11. The curve $x = \cos t$, $y = \sin t$, $0 \leq t \leq \pi/2$.

12. Counterclockwise around the rectangle $-2 \leq x \leq 2$, $-1 \leq y \leq 1$.

13. Counterclockwise around the ellipse $x = 2 \cos t$, $y = 3 \sin t$, $0 \leq t \leq 2\pi$.

14. Find an antigradient of $F(x, y) = (x - y, 2y - x)$ by using partial integration.

15. Why do the answers to Problems 9 and 11 have opposite signs?

16. Why do Problems 9 and 10 have the same answer?

17. Why do we get zero in Problems 12 and 13?

18. Use the Fundamental Theorem for Line Integrals to do Problems 9 through 13.

19. Use Green's Theorem to do Problems 12 and 13.

20. Use Green's Theorem to evaluate

$$\int_C xy\, dx + x^2\, dy$$

where C is the unit circle oriented counterclockwise. Does the answer imply that $F(x, y) = (xy, x^2)$ is a gradient field? Explain.

21. Let C be the (positively oriented) curve $|x| + |y| = 1$. Evaluate

$$\int_C (x^2 - y^2)\, dx + 2xy\, dy$$

22. Does the result in Problem 21 tell you whether $F(x, y) = (x^2 - y^2, 2xy)$ is a gradient field? Explain.

23. Evaluate the line integral of $F(x, y) = (x - y, x + y)$ around the unit circle oriented counterclockwise.

24. Use Green's Theorem to do Problem 23. Can the integral be computed by expressing F as a gradient and using the Fundamental Theorem for Line Integrals?

25. Suppose that $F(x, y) = (x, y)$ and let C be any closed curve (with the usual properties). Explain why

$$\int_C F \cdot dx = 0$$

Is F a gradient field?

26. Determine whether

$$F(x, y) = (\sin y - y \sin x, x \cos y + \cos x)$$

is a gradient field. If it is, find an antigradient.

27. Let C be the boundary (oriented counterclockwise) of the region between the parabola $y = x(x - 2)$ and the x axis. Evaluate

$$\int_C xy^2\, dx + \sin y\, dy$$

28. Find an antigradient of

$$F(x, y) = \frac{(x, y)}{x^2 + y^2}$$

29. In Problem 28 use the definition of line integral to compute

$$\int_C \mathbf{F(x)} \cdot d\mathbf{x}$$

where C is the unit circle (positively oriented).

30. In Problem 29 can Green's Theorem be used to confirm the answer? Explain.

31. In Problem 28 find

$$\int_C (\mathbf{F} \cdot \mathbf{N}) \, ds$$

where C is the unit circle (positively oriented) and $\mathbf{N}$ is the outward unit normal to C.

32. In Problem 28 find

$$\iint_D (\mathbf{\nabla} \cdot \mathbf{F}) \, dA$$

where D is the region inside the unit circle. Does the Divergence Theorem apply?

33. In Problem 28 embed $\mathbf{F}$ in space by writing

$$\mathbf{F}(x,y,z) = \frac{(x,y,0)}{x^2 + y^2}$$

Find curl $\mathbf{F}$.

34. Evaluate

$$\int_C \tan^{-1} y \, dx + \frac{x}{1 + y^2} \, dy$$

where C is the unit circle oriented counterclockwise.

35. Show that

$$\mathbf{F}(x,y) = \left(\tan^{-1} y, \frac{x}{1 + y^2}\right)$$

is a gradient field and name a function $f(x,y)$ whose gradient is $\mathbf{F}(x,y)$.

36. Let D be a standard region with boundary C, oriented counterclockwise. Show that if the density at each point of D is 1, the moment of inertia relative to the x axis is

$$I_x = \int_C xy^2 \, dy$$

37. In Problem 36 show that the moment of inertia relative to the y axis is

$$I_y = -\int_C x^2 y \, dx$$

38. In Problem 36 show that the moment of inertia relative to the z axis is

$$I_z = \int_C xy^2 \, dy - x^2 y \, dx$$

39. Suppose that $\mathbf{F} = (F,G)$ is a gradient field. Show that

$$\int_{(a,b)}^{(c,d)} \mathbf{F(x)} \cdot d\mathbf{x} = \int_a^c F(x,b) \, dx + \int_b^d G(c,y) \, dy$$

In each of the following, find the rate at which a fluid with flux $\mathbf{F}$ passes through the given surface by evaluating an appropriate surface integral.

40. $\mathbf{F}(x,y,z) = (x,y,z)$ through the surface formed by the hemisphere $z = \sqrt{1 - x^2 - y^2}$ and the xy plane.

41. $\mathbf{F}(x,y,z) = (x,0,0)$ through the sphere $x^2 + y^2 + z^2 = 1$.

42. $\mathbf{F}(x,y,z) = (x + z, y + x, z + y)$ through the tetrahedron formed by the coordinate planes and the plane $x + y + z = 1$.

43. $\mathbf{F}(x,y,z) = (0,0,z)$ through the box formed by the coordinate planes and the planes $x = 1$, $y = 1$, $z = 1$.

44. $\mathbf{F}(x,y,z) = (z,x,y)$ through the surface formed by the paraboloid $z = 1 - x^2 - y^2$ and the xy plane.

45.–49. In Problems 40 through 44, use the Divergence Theorem to find the rate at which a fluid with flux $\mathbf{F}$ passes through the given surface.

50. Evaluate

$$\iint_\sigma (\mathbf{F} \cdot \mathbf{N}) \, dS$$

where $\mathbf{F}(x,y,z) = (1/x, 1/y, 1/z)$ and $\mathbf{N}$ is the outward unit normal to the sphere $\sigma: x^2 + y^2 + z^2 = 1$. Why can't the Divergence Theorem be used to answer the question?

51. Suppose that $\sigma: z = f(x,y)$ is a surface above the region D in the xy plane. If $\mathbf{F} = (F,G,H)$ is a vector field, show that

$$\iint_\sigma (\mathbf{F} \cdot \mathbf{N}) \, dS = \pm \iint_D \left(-F \frac{\partial z}{\partial x} - G \frac{\partial z}{\partial y} + H\right) dA$$

where $\mathbf{N}$ is the outward unit normal to σ.

52. Let σ be a closed surface bounding the region R. Explain why the volume of R is

$$V = \frac{1}{3} \iint_\sigma (\mathbf{x} \cdot \mathbf{N}) \, dS$$

where $\mathbf{x} = (x,y,z)$ and $\mathbf{N}$ is the outward unit normal to σ.

53. Show that the divergence of the vector field $\mathbf{F}(\mathbf{x}) = \mathbf{x}/|\mathbf{x}|$ is div $\mathbf{F} = 2/|\mathbf{x}|$.

54. Use Problem 53 and the Divergence Theorem to evaluate

$$\iiint_R \frac{dV}{\sqrt{x^2 + y^2 + z^2}}$$

where R is the spherical ball $x^2 + y^2 + z^2 \le a^2$. Check by evaluating the integral directly.

55. Find an antigradient of

$$\mathbf{F}(x,y,z) = \frac{(x,y,z)}{x^2 + y^2 + z^2}$$

56. In Problem 55 use the definition of line integral to compute

$$\int_C \mathbf{F}(\mathbf{x}) \cdot d\mathbf{x}$$

where C is the (positively oriented) unit circle in the xy plane.

57. In Problem 56 use the Fundamental Theorem for Line Integrals to confirm your answer.

58. In Problem 55 find

$$\iint_\sigma (\nabla \times \mathbf{F}) \cdot \mathbf{N} \, dS$$

where σ is the hemisphere $z = \sqrt{1 - x^2 - y^2}$ and $\mathbf{N}$ is its outward unit normal.

59. If $\mathbf{F}(\mathbf{x}) = \mathbf{x}/\rho^3$, where $\rho = |\mathbf{x}|$, show that div $\mathbf{F} = 0$.

60. In Problem 59 show that curl $\mathbf{F} = \mathbf{0}$.

61. If $\mathbf{F}(x,y,z)$ is a constant vector field, show that

$$\int_C \mathbf{F}(\mathbf{x}) \cdot d\mathbf{x} = 0$$

where C is a closed curve with the usual properties.

62. Use the definition of line integral to find

$$\int_C (x - yz) \, dx + (y - zx) \, dy + (z - xy) \, dz$$

where C is the curve $x = \cos t$, $y = \sin t$, $z = t$, $0 \le t \le 2\pi$.

63. In Problem 62 use the Fundamental Theorem for Line Integrals to check your answer.

Use the definition of line integral to compute

$$\int_C \mathbf{F} \cdot d\mathbf{x}$$

where $\mathbf{F}$ and C are as follows.

64. $\mathbf{F}(x,y,z) = (x,y,z)$ around the unit circle in the xy plane (oriented counterclockwise).

65. $\mathbf{F}(x,y,z) = (z,x,y)$ around the curve of intersection of the sphere $x^2 + y^2 + z^2 = 8$ and the cone $z^2 = x^2 + y^2$ (positively oriented relative to $\mathbf{k}$).

66. $\mathbf{F}(x,y,z) = (xz,yx,zy)$ around the boundary of the first-octant part of the plane $x + y + z = 1$ (positively oriented relative to the outward unit normal).

67.–69. Use Stokes' Theorem to compute the given line integral in Problems 64 through 66.

70. Use Stokes' Theorem to find

$$\int_C (x - z) \, dx + (y - x) \, dy + (z - y) \, dz$$

where C is the (positively oriented) triangle cut from the plane $2x + y + 2z = 2$ by the coordinate planes.

71. Use Stokes' Theorem to show that if σ is a sphere with outward unit normal $\mathbf{N}$, and $\mathbf{F}$ is a vector field, then

$$\iint_\sigma (\text{curl } \mathbf{F}) \cdot \mathbf{N} \, dS = 0$$

72. Verify Stokes' Theorem in the case where $\mathbf{F}$ is a gradient field.

20 Differential Equations

In the future, as in the past, the great ideas must be simplifying ideas.
ANDRÉ WEIL

There is no permanent place in the world for ugly mathematics.
G. H. HARDY

Everything of importance has been said before by somebody who did not discover it.
ALFRED NORTH WHITEHEAD

A natural sequel to a calculus course is the study of differential equations as a subject in its own right. It is a vast field, with numerous important applications and many unanswered questions that are of current interest to research mathematicians and scientists. This chapter is an introduction covering some of the simpler types of equations that can be solved by methods accessible to a calculus student.

20.1 ORDINARY DIFFERENTIAL EQUATIONS

In Section 7.7 (which you should probably review at this point) we said that a *differential equation* is an equation involving an independent variable, an unknown function of that variable, and one or more derivatives of the unknown function. This is really a definition of an *ordinary* differential equation (as opposed to one in which there are partial derivatives). *Partial* differential equations, like Laplace's equation $\partial^2 z/\partial x^2 + \partial^2 z/\partial y^2 = 0$ mentioned in Problem Set 17.1, have occurred from time to time in this book, but only in the context of known solutions. Finding unknown solutions is a subject for another course.

For purposes of general discussion, we usually write ordinary differential equations in a form in which the highest derivative is expressed in terms of the other variables that may occur. Thus (for example) the general *first-order* equation (in which the first derivative of the unknown function is the only one to occur) has the form $y' = f(x,y)$, which expresses y' in terms of x and y.

A *solution* of such an equation is a function $y = \phi(x)$ which reduces the equation to an identity (in some interval I) when it is substituted: $\phi'(x) = f(x,\phi(x))$ for all $x \in I$.

□ **Example 1**   The equation $y' = 3x - (y/x)$ is a first-order differential equation in which

$$f(x,y) = 3x - \frac{y}{x}$$

A solution is $\phi(x) = x^2 - (3/x)$, as you can check by computing

$$\phi'(x) = 2x + \frac{3}{x^2}$$

and substituting in the equation:

$$2x + \frac{3}{x^2} \overset{?}{=} 3x - \frac{1}{x}\left(x^2 - \frac{3}{x}\right)$$

$$2x + \frac{3}{x^2} \overset{?}{=} 3x - x + \frac{3}{x^2}$$

$$2x + \frac{3}{x^2} = 2x + \frac{3}{x^2}$$

This identity holds for all $x \neq 0$; an *interval* in which it holds is $I = (0,\infty)$ or $I = (-\infty,0)$. Note that $\phi(x)$ is not a solution in any interval containing 0.

□

It is a comparatively easy matter to check whether a proposed solution actually works. The problem is to *discover* solutions. Even the simple differential equation in Example 1 is a tough nut to crack. You might ask yourself whether you know any method that will solve it. *Separation of variables* (which we discussed in Section 7.7 and have used throughout the book) is of no help in this equation, yet it is easy to solve when the method is explained. That is what this chapter is about: methods of solution that apply to certain classes of differential equations.

In this section we are not primarily interested in *how* to solve differential equations. Our purpose is to explain a general theorem that states conditions under which a (first-order) equation is guaranteed to have a solution. This theorem is not easy to prove, so we are going to discuss it with the limited objective of stating it correctly.

An example from earlier in the book will help focus the problem. Newton's Law of Cooling (Section 8.6) says that when a hot object is brought into a room at constant temperature T_0, its temperature T decreases at a rate proportional to the difference between T and T_0. In other words, $dT/dt = -k(T - T_0)$, where k is a positive constant and t is time. The independent variable in this case is t and the unknown function is T; the question is whether we can express T in terms of t.

Suppose, for example, that the constant room temperature is $T_0 = 20°C$ and $k = 2$. Then our differential equation is $dT/dt = -2(T - 20)$, and the problem is to find a function $T = \phi(t)$ satisfying it. Separation of variables (which we presume you remember) yields

$$\frac{dT}{T - 20} = -2dt \qquad (T \neq 20 \text{ because the object is hotter than } 20°)$$

$$\int \frac{dT}{T - 20} = -2 \int dt$$

$$\ln(T - 20) = -2t + C \qquad (\text{no absolute value needed because } T > 20)$$

$$T - 20 = e^{-2t+C} = Ae^{-2t} \qquad (A = e^C)$$

$$T = \phi(t) = 20 + Ae^{-2t}$$

This is a solution in the interval $(-\infty, \infty)$, which we cut down to $[0, \infty)$ if the cooling process started at $t = 0$ (not for any mathematical reason, but because values of $t < 0$ are physically meaningless).

Thus a solution certainly exists. In fact (since A is an arbitrary positive constant) there are infinitely many solutions, as indicated in Figure 1. Of course only one of these curves fits the cooling process of a given object. We don't know which one unless we are told something more about the object. An *initial condition* would help, say the information that $T = 100°C$ when $t = 0$. This provides the point $(0, 100)$ in Figure 1, which singles out the curve that describes the cooling process actually taking place. Putting this into the function $T = 20 + Ae^{-2t}$, we have $100 = 20 + A$, or $A = 80$. Hence the *unique* solution of the problem is

$$\phi(t) = 20 + 80e^{-2t}$$

Figure 1
Solutions of $dT/dt = -2(T - 20)$ in the interval $[0, \infty)$

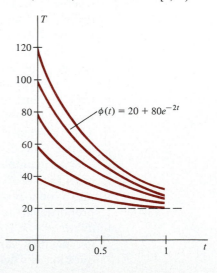

Such a problem, consisting of a differential equation together with an initial condition (which need not have anything to do with time) is called an **initial value problem.**

This example illustrates something general about first-order differential equations. If $y' = f(x,y)$ is such an equation, and if f is sufficiently well behaved to guarantee a solution $y = \phi(x)$, an initial condition (x_0, y_0) pinpoints the unique solution satisfying $\phi(x_0) = y_0$ (that is, the curve passing through the given point).

It is difficult to explain what "well behaved" means in the preceding statement. Mathematicians have discovered that what is needed is an open region (in $\mathfrak{R}^2$) in which $f(x,y)$ and its partial derivative $f_2(x,y)$ are continuous. Why these conditions are important is not obvious; we leave the explanation for a course in differential equations.

The theorem we have in mind reads as follows.

Existence and Uniqueness Theorem

Suppose that $f(x,y)$ and its partial derivative $f_2(x,y)$ are continuous in the open region D in the xy plane. Given any point $(x_0, y_0) \in D$, there is an interval I containing x_0 (on the x axis), and exactly one function $\phi(x)$, differentiable in I, such that

$$\phi'(x) = f(x, \phi(x)) \qquad \text{for all } x \in I \qquad \text{and} \qquad \phi(x_0) = y_0$$

In other words, the differential equation $y' = f(x,y)$ has a unique solution $y = \phi(x)$ (in the interval I) satisfying the initial condition $y = y_0$ when $x = x_0$.

□ **Example 2**   The differential equation $y' = \sqrt{y - 1}$ is of the form $y' = f(x,y)$ with $f(x,y) = \sqrt{y - 1}$. The domain of f is

$$\{(x,y): y \geq 1\}$$

which is the half-plane consisting of the horizontal line $y = 1$ and all points above it. Notice that this domain is not open, that is, boundary points (on the line $y = 1$) are included. Hence it is not a region of the type required in the Existence and Uniqueness Theorem. Since

$$f_2(x,y) = \frac{\partial}{\partial y} \sqrt{y - 1} = \frac{1}{2\sqrt{y - 1}}$$

however, the functions f and f_2 are continuous in $D = \{(x,y): y > 1\}$, which *is* open. Given a point in D, say $(1,2)$, our theorem guarantees several things (all tied together):

1. A solution $y = \phi(x)$ of the differential equation $y' = \sqrt{y - 1}$, meaning that (in some interval) $\phi'(x) = \sqrt{\phi(x) - 1}$.

2. Satisfaction of the initial condition $\phi(1) = 2$ by this solution, that is, the graph of $y = \phi(x)$ passes through the point $(1,2)$.
3. Uniqueness of ϕ, that is, if $y = \psi(x)$ is "another" solution satisfying $\psi(1) = 2$, then $\psi = \phi$.
4. An interval I on the x axis (containing the point $x = 1$) in which these statements hold.

To find ϕ, separate the variables:

$$\frac{dy}{dx} = \sqrt{y - 1}$$

$$\frac{dy}{\sqrt{y - 1}} = dx$$

$$\int (y - 1)^{-1/2}\, dy = \int dx$$

$$2\sqrt{y - 1} = x + C$$

Putting the initial condition $(1,2)$ into this equation, we have $2 = 1 + C$, or $C = 1$. Hence

$$2\sqrt{y - 1} = x + 1$$

$$4(y - 1) = (x + 1)^2$$

$$y = \phi(x) = \tfrac{1}{4}(x + 1)^2 + 1$$

At this point an interesting question arises. Since D is the half-plane $y > 1$ (which involves no restriction on x) and since $\phi(x)$ is defined for all x, it is reasonable to expect that the interval I is $(-\infty,\infty)$. A check reveals that this is not the case, however. For if we put

$$y = \tfrac{1}{4}(x + 1)^2 + 1 \qquad \text{and} \qquad y' = \tfrac{1}{2}(x + 1)$$

into the original equation $y' = \sqrt{y - 1}$, we have

$$\tfrac{1}{2}(x + 1) \overset{?}{=} \sqrt{\tfrac{1}{4}(x + 1)^2 + 1 - 1} = \tfrac{1}{2}|x + 1|$$

This holds only for $x + 1 \geq 0$, that is, in the interval $I = [-1,\infty)$.

Thus the interval I guaranteed by the Existence and Uniqueness Theorem is not (in general) predictable. Nothing in either the original differential equation $y' = \sqrt{y - 1}$ or the solution $\phi(x) = \tfrac{1}{4}(x + 1)^2 + 1$ suggests that the interval of validity is $I = [-1,\infty)$. We discovered it only by checking the solution, which is awkward. Worse yet, the interval is not "stable," that is, it changes with the initial condition. The solution of $y' = \sqrt{y - 1}$ satisfying $\phi(1) = 5$, for example, is

$$y = \phi(x) = \tfrac{1}{4}(x + 3)^2 + 1$$

which turns out to be valid in the interval $I = [-3,\infty)$.

In general the solutions of this equation are the half-parabolas

$$\phi(x) = \tfrac{1}{4}(x - a)^2 + 1 \qquad \text{valid in } I = [a,\infty)$$

Figure 2
Solutions of $y' = \sqrt{y - 1}$

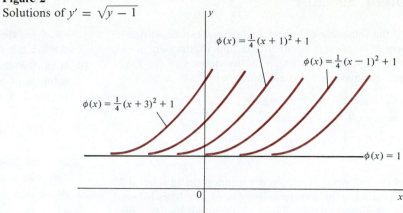

together with the special (constant) solution

$$\phi(x) = 1 \qquad \text{valid in } I = (-\infty, \infty)$$

(See Figure 2.)   □

Remark   The solution $\phi(x) = 1$ got lost when we separated the variables. For if $y = 1$, we cannot go from

$$\frac{dy}{dx} = \sqrt{y - 1} \qquad \text{to} \qquad \frac{dy}{\sqrt{y - 1}} = dx$$

The special solution $\phi(x) = 1$ may be noted at the beginning, before division.

Students sometimes misunderstand the nature of the restriction $y \neq 1$ in justifying division by $\sqrt{y - 1}$. It is not just a question of keeping y away from 1 at the initial point, but *throughout some interval* in which the equation

$$\frac{dy}{\sqrt{y - 1}} = dx$$

can be integrated. Fortunately we have a property of continuity to help us here. If the initial condition is (for example) $\phi(1) = 2$, the value $y = 2$ at $x = 1$ guarantees that $\phi(x)$ is *near* 2 (and hence different from 1) for all values of x in some neighborhood of $x = 1$ (because ϕ, being differentiable, is continuous). As you can see, the solution

$$\phi(x) = \tfrac{1}{4}(x + 1)^2 + 1$$

is in fact different from 1 for all $x \neq -1$, so the "neighborhood" of $x = 1$ in which our separation of variables is legitimate turns out to be $(-1, \infty)$. We did not know this when we started, however, which is an illustration of the importance of theory. Continuity guarantees a neighborhood of $x = 1$ in which $y \neq 1$, so we can divide by $\sqrt{y - 1}$ with confidence that the results will hold in some interval containing the initial value.

The usefulness of the Existence and Uniqueness Theorem is not that it helps us solve differential equations, but that it gives us some idea of what to expect. It is also important in the development of some of the ideas to come.

Problem Set 20.1

Each of the following differential equations can be written in the form $y' = f(x,y)$. Name f and find its domain in $\mathcal{R}^2$. Name an open region in which the hypotheses of the Existence and Uniqueness Theorem hold.

1. $xy' + y = \sqrt{x}$ **2.** $y' = 1 + y^2$

3. $y' + 3xy\sqrt{1 - x^2} = 0$ **4.** $xy' = \sqrt{1 - y^2}$

5. $y' - 2xy = 1$ **6.** $y' = |y - 2|$

7. $y' = (y - 1)^{3/2}$ **8.** $y' = \ln(1 - x^2 - y^2)$

Each of the following functions is a solution (in an interval I) of the equation in the corresponding one of Problems 1–7, and satisfies the given initial condition. Confirm this by direct substitution in the equation and name the largest interval (containing the initial value of x) in which the solution is valid.

9. Problem 1: $\phi(x) = \dfrac{2}{3}\sqrt{x} + \dfrac{1}{3x}$, $\phi(1) = 1$

10. Problem 2: $\phi(x) = \tan x$, $\phi(0) = 0$

11. Problem 3: $\phi(x) = e^{(1-x^2)^{3/2}}$, $\phi(-1) = 1$

12. Problem 4: $\phi(x) = -\cos(\ln x)$, $\phi(1) = -1$

13. Problem 5: $\phi(x) = e^{x^2} + e^{x^2}\displaystyle\int_0^x e^{-t^2}\, dt$, $\phi(0) = 1$

14. Problem 6: $\phi(x) = 2$, $\phi(1) = 2$

15. Problem 7: $\phi(x) = 1 + \dfrac{4}{(x - 3)^2}$, $\phi(1) = 2$

16. In which of Problems 9 through 15 does the Existence and Uniqueness Theorem guarantee that the given solution is the only one satisfying the given initial condition?

17. Confirm that the constant function $\phi(x) = -1$ is a solution of the differential equation

$$xy' = \sqrt{1 - y^2}$$

Both this function and the solution

$$\phi(x) = -\cos(\ln x)$$

in Problem 12 satisfy the initial condition $\phi(1) = -1$. Why doesn't this contradict the Uniqueness Theorem?

18. Confirm that the functions

$$\phi(x) = \tfrac{1}{4}x^2 + 1 \quad \text{and} \quad \psi(x) = 1$$

are solutions of the differential equation

$$y' = \sqrt{y - 1}$$

satisfying the same initial condition $y = 1$ at $x = 0$. Why doesn't this contradict the Uniqueness Theorem?

19. It can be shown that the function $\phi(x) = 2$ is the only solution of $y' = |y - 2|$ satisfying the initial condition $\phi(1) = 2$. Yet $(1,2) \notin D$, where D is the region in which

$$f(x,y) = |y - 2| \quad \text{and} \quad f_2(x,y) = \frac{|y - 2|}{y - 2}$$

are continuous. Why doesn't this contradict the Uniqueness Theorem?

20. Prove that no constant function can be a solution of

$$y' = \ln(1 - x^2 - y^2)$$

Hint: Can you find an interval (other than a single point) in which such a solution is valid?

In each of the following, find the solution satisfying the given initial condition.

21. $y' = 2\sqrt{y}$, $\phi(1) = 4$

22. $y' = 2xy$, $\phi(0) = 1$

23. $y' = (y - 2)\cos x$, $\phi(0) = 1$

24. $y' = (y - 2)\cos x$, $\phi(0) = 2$

25. $y' = e^{-y}$, $\phi(0) = 0$

26. $y' = 1 + y^2$, $\phi(0) = 0$

27. $y' = 1 - y^2$, $\phi(0) = 1$

28. $xy' = \sqrt{1 - y^2}$, $\phi(\tfrac{1}{2}) = 0$

29. $y' = (y - 1)^{3/2}$, $\phi(1) = 2$

30. $y' = \dfrac{2x}{y^2 - 1}$, $\phi(0) = 2$ (Leave the solution in the form of an equation implicitly defining y as a function of x.)

20.2

EXACT DIFFERENTIAL EQUATIONS

The general first-order differential equation is $y' = f(x,y)$, where $f(x,y)$ is a function whose form dictates the method of attack. One form of particular importance is

$$y' = -\frac{F(x,y)}{G(x,y)}$$

because then the equation can be written

$$F(x,y)\,dx + G(x,y)\,dy = 0$$

Of course *every* equation $y' = f(x,y)$ can be written this way by naming $F(x,y) = f(x,y)$ and $G(x,y) = -1$. The importance of the form (as we will see) is in the special case when F and G are components of a gradient.

The expression $F(x,y)\,dx + G(x,y)\,dy$ should remind you of the line integral of the vector field $\mathbf{F}(\mathbf{x}) = (F(x,y),G(x,y))$ along a curve C:

$$\int_C \mathbf{F}(\mathbf{x}) \cdot d\mathbf{x} = \int_C F(x,y)\,dx + G(x,y)\,dy \qquad \text{(Section 19.1)}$$

The equation $F(x,y)\,dx + G(x,y)\,dy = 0$ can be solved immediately if $\mathbf{F}$ is a gradient field. For then there is a scalar function $f(x,y)$ [not the same f as in the form $y' = f(x,y)$!] such that

$$\mathbf{F}(x,y) = \nabla f(x,y)$$

that is, $F(x,y) = f_1(x,y)$ and $G(x,y) = f_2(x,y)$, and our equation becomes

$$f_1(x,y)\,dx + f_2(x,y)\,dy = 0$$

If t is the parameter in the curve $C\colon \mathbf{x} = (x(t),y(t))$, we can write this in the form

$$f_1(x,y)\,\frac{dx}{dt} + f_2(x,y)\,\frac{dy}{dt} = 0$$

$$\frac{d}{dt}\,f(x,y) = 0 \qquad \text{(the Chain Rule in Section 17.2)}$$

and integrate to obtain $f(x,y) = k$.

Note that the curve C is of no significance in the final result; we used it only to justify the integration. Moreover, the line integral is independent of the path (why?), so the choice of C is immaterial.

Such a differential equation (in which F and G are components of a gradient) is called *exact;* its solutions are implicitly defined (y in terms of x) by the equation $f(x,y) = k$.

From Section 19.2 we have a criterion for exactness, namely the condition $\partial G/\partial x = \partial F/\partial y$ (which in a simply connected region of the plane is both necessary and sufficient for $\mathbf{F}$ to be a gradient). Hence we can recognize exact differential equations by a simple test, and solve them by "partial integration" to find an antigradient of $\mathbf{F}$.

A differential equation

$$F(x,y)\, dx + G(x,y)\, dy = 0$$

is said to be *exact* if $\partial G/\partial x = \partial F/\partial y$ in some (simply connected) region of the plane. Its solutions $y = \phi(x)$ are implicitly defined by $f(x,y) = k$, where f is a function satisfying

$$\nabla f(x,y) = (F(x,y), G(x,y))$$

□ **Example 1**   Solve the differential equation

$$\frac{dy}{dx} = -\frac{4x + y}{x + 2y}$$

Solution: Writing this in the form $(4x + y)\, dx + (x + 2y)\, dy = 0$, we observe that

$$\frac{\partial}{\partial x}(x + 2y) = 1 = \frac{\partial}{\partial y}(4x + y)$$

Hence the equation is exact. We find a function $f(x,y)$ whose gradient is $(4x + y, x + 2y)$ by "partial integration":

$$f(x,y) = \int (4x + y)\, dx = 2x^2 + xy + C(y)$$

$$f_2(x,y) = x + C'(y) = x + 2y$$
$$C'(y) = 2y$$
$$C(y) = y^2 \qquad \text{(plus an arbitrary constant we need not insert here)}$$
$$f(x,y) = 2x^2 + xy + y^2$$

(See Example 5, Section 19.2, for a fuller explanation of this method of integration.) The general solution of

$$(4x + y)\, dx + (x + 2y)\, dy = 0$$

is therefore $2x^2 + xy + y^2 = k$, where k is an arbitrary constant.   □

□ **Example 2**   Explain why the differential equation

$$y' = \frac{xy}{y - 2x}$$

cannot be solved as an exact equation.

Solution: Writing the equation in the form $xy\, dx + (2x - y)\, dy = 0$, we observe that

$$\frac{\partial}{\partial x}(2x - y) = 2 \qquad \text{and} \qquad \frac{\partial}{\partial y}(xy) = x$$

Hence the equation is not exact; there is no function $f(x, y)$ whose gradient is $(xy, 2x - y)$. (See Examples 4 and 6, Section 19.2.)   □

□ **Example 3**   Find the solution of $dy/dx = -y/x$ satisfying $\phi(1) = 3$.

Solution: The elementary approach is to separate the variables, writing

$$\frac{dy}{y} = -\frac{dx}{x}$$

$$\int \frac{dy}{y} = -\int \frac{dx}{x}$$

$$\ln y = -\ln x + C \qquad \text{(no absolute values needed because } x > 0 \text{ and } y > 0 \text{ near (1,3))}$$

$$\ln xy = C$$

Since $y = 3$ when $x = 1$, we find $C = \ln 3$, so

$$\ln xy = \ln 3$$

$$xy = 3$$

$$y = \phi(x) = \frac{3}{x}$$

This solution is valid in the interval $I = (0, \infty)$, as you can check. It is also valid in $(-\infty, 0)$, but that is of no interest here because the initial value $x = 1$ is in I.

The method of this section suggests writing the equation in the form

$$y\, dx + x\, dy = 0$$

Since

$$\frac{\partial}{\partial x}(x) = 1 = \frac{\partial}{\partial y}(y)$$

the equation is exact. We find a function $f(x, y)$ whose gradient is (y, x) by writing

$$f(x, y) = \int y\, dx = xy + C(y)$$

$$f_2(x, y) = x + C'(y) = x$$

$$C'(y) = 0$$

$$C(y) = 0 \qquad \text{(or any constant)}$$

$$f(x, y) = xy$$

Hence the general solution is $xy = k$. The initial condition $\phi(1) = 3$ yields $k = 3$, so (as before) we find $xy = 3$ and $y = \phi(x) = 3/x$.   □

Example 3 suggests that a "separable" differential equation is a special case of an exact equation. To see that this is correct, recall from Section 7.7

that a differential equation is called *separable* if it has the form $y' = g(x)h(y)$, where $g(x)$ is a function of x alone and $h(y)$ is a function of y alone. For then we can separate the variables by writing

$$\frac{dy}{h(y)} = g(x)\,dx$$

and integrate to find the solutions. However, the equation is also exact, as you can see by writing it in the form $g(x)\,dx - dy/h(y) = 0$ and applying our criterion for exactness.

Problem Set 20.2

Find the general solution of each of the following differential equations.

1. $x\,dx + y\,dy = 0$ **2.** $y\,dx - x\,dy = 0$

3. $(x + y)\,dx + (x - y)\,dy = 0$

4. $(x^2 + y^2)\,dx + 2xy\,dy = 0$

5. $(ye^{xy} + 2x)\,dx + (xe^{xy} - 2y)\,dy = 0$

6. $(ye^x + e^y)\,dx + (xe^y + e^x)\,dy = 0$

7. $e^x \sin y\,dx + e^x \cos y\,dy = 0$

8. $y \sin x\,dx + (\cos y - \cos x)\,dy = 0$

Solve the following initial value problems.

9. $(3x^2 - y)\,dx + (2y - x)\,dy = 0$, $y = 1$ when $x = 0$.

10. $2xy\,dx + (x^2 - y^2)\,dy = 0$, $y = 2$ when $x = 1$.

11. $(2xy + 1)\,dx + (x^2 - 1)\,dy = 0$, $y = 1$ when $x = 2$.

12. $e^x\,dx - e^{-y}\,dy = 0$, $y = 0$ when $x = 0$.

13. $(y \cos x - 2x)\,dx + (\sin x + \cos y)\,dy = 0$, $y = 0$ when $x = 0$.

14. $\sin x \cos y\,dx + \cos x \sin y\,dy = 0$, $y = \pi$ when $x = 0$.

15. $\left(2x + \dfrac{x}{x^2 + y^2}\right)dx + \left(2y + \dfrac{y}{x^2 + y^2}\right)dy = 0$, $y = 1$ when $x = 0$. Without merely checking that it works, can

you show how to put the solution in the form $y = \sqrt{1 - x^2}$?

16. The equation $y = x + Ce^{-x}$ (where C is an arbitrary constant) describes a family of curves. Find their **orthogonal trajectories** (the curves that intersect them at right angles) as follows.

(a) Show that the orthogonal trajectories satisfy the differential equation

$$\frac{dy}{dx} = \frac{1}{y - x - 1}$$

Hint: See Example 6, Section 7.7.

(b) Confirm that the equation in part (a) is not exact, but that multiplication of each side by e^y makes it exact.

(c) Solve the exact equation obtained in part (b).

17. Suppose that the equation

$$F(x,y)\,dx + G(x,y)\,dy = 0$$

is exact. Show that the solution whose graph passes through the point (x_0, y_0) satisfies

$$\int_{x_0}^{x} F(t,y)\,dt + \int_{y_0}^{y} G(x_0,t)\,dt = 0$$

20.3
LINEAR
FIRST-ORDER
EQUATIONS

The first-order differential equation $y' = f(x,y)$ is said to be *linear* if $f(x,y)$ involves y only to the first degree, that is,

$$f(x,y) = a(x)y + b(x)$$

where $a(x)$ and $b(x)$ are functions of x alone. For example, the equation

$$y' = x^2 y + \sin x$$

is linear; the "coefficients" are $a(x) = x^2$ and $b(x) = \sin x$.

Linear differential equations (not only first-order but arbitrary-order) are important for two reasons. First, they are easier to analyze, as you will see. Nonlinear equations have not been studied systematically except in special cases, and are still the subject of intensive research. Second, linear equations often serve as good mathematical models in nonlinear situations. The motion of a pendulum, for example, may be represented by a nonlinear (second-order) differential equation whose solutions are hard to find. By replacing the equation with a linear one whose solutions are reasonable approximations to the ones wanted, we simplify the problem considerably.

If we assume that $a(x)$ and $b(x)$ are continuous in some (open) interval I, the functions

$$f(x, y) = a(x)y + b(x) \quad \text{and} \quad f_2(x, y) = a(x)$$

are continuous in the open region

$$D = \{(x, y): x \in I\}$$

This is an infinite vertical strip in the xy plane (half or all of the plane if I is half or all of the x axis). According to the Existence and Uniqueness Theorem of Section 20.1, we may impose an arbitrary initial condition $(x_0, y_0) \in D$, where $x_0 \in I$ and $y_0 \in \mathcal{R}$, and are guaranteed a unique solution of $y' = a(x)y + b(x)$ satisfying $\phi(x_0) = y_0$ (in some interval containing x_0). As we pointed out in Section 20.1, this interval is in general neither predictable nor stable. In the linear case, however, we will prove that it always coincides with I, regardless of the initial condition; thus it is both predictable and stable.

As an example of what we mean, consider the linear equation

$$y' = \frac{y}{x} + \sqrt{x}$$

the coefficients of which are

$$a(x) = \frac{1}{x} \quad \text{and} \quad b(x) = \sqrt{x}$$

Since these are continuous in $I = (0, \infty)$, we may choose any point (x_0, y_0) with $x_0 > 0$ and expect a unique solution satisfying $\phi(x_0) = y_0$. The general solution turns out to be $y = \phi(x) = 2x^{3/2} + Cx$, where C is a constant to be evaluated by using the initial condition. We will soon explain how this solution is obtained, but our point now is that it satisfies the differential equation for all $x \in I$, regardless of the initial condition. We assert that this always happens in the linear case; the interval of validity can be named ahead of time as the interval I in which the coefficients are continuous.

To prove this assertion (and more importantly, to obtain some practical results), we are going to solve the linear first-order equation completely and explicitly. It is a rare type of differential equation for which this can be done; in the rest of the chapter we will have to be content with less general results.

The standard form in which linear first-order equations are usually written is $y' + p(x)y = q(x)$, where $p(x)$ and $q(x)$ are continuous in an interval I. Let (x_0, y_0) be any point with $x_0 \in I$; the problem is to find the solution $y = \phi(x)$ satisfying $\phi(x_0) = y_0$. We begin by taking $q(x) = 0$ (the so-called *homogeneous* case), so our equation reads $y' + p(x)y = 0$. The reason we start here is that we know how to solve the equation because it is separable. Assuming that $y_0 \neq 0$ (which implies $y \neq 0$ in a neighborhood of x_0), we write

$$\frac{dy}{y} = -p(x)\, dx$$

$$\int_{y_0}^{y} \frac{du}{u} = -\int_{x_0}^{x} p(t)\, dt$$

The integral on the right side is a function of x, which we denote by

$$P(x) = \int_{x_0}^{x} p(t)\, dt$$

Then our equation becomes

$$\int_{y_0}^{y} \frac{du}{u} = -P(x)$$

$$\ln |u| \Big|_{y_0}^{y} = -P(x)$$

$$\ln \left| \frac{y}{y_0} \right| = -P(x)$$

Since $y_0 \neq 0$, y and y_0 have the same sign in a neighborhood of x_0. (Why?) Hence we may drop the absolute value to obtain

$$\ln \frac{y}{y_0} = -P(x)$$

$$\frac{y}{y_0} = e^{-P(x)}$$

$$y = \phi(x) = y_0 e^{-P(x)}$$

This formula was derived by assuming that $y_0 \neq 0$, but it is correct even when $y_0 = 0$, since the solution of $y' + p(x)y = 0$ satisfying $\phi(x_0) = 0$ is the constant function $y = \phi(x) = 0$. (Confirm!) Moreover, although our argument referred to an unknown neighborhood of x_0 in which y and y_0 have the same sign, the solution $\phi(x) = y_0 e^{-P(x)}$ is valid in the original interval I. (We leave this for you to check in the problem set.) Hence the homogeneous case is solved.

The nonhomogeneous case cannot be solved by separating the variables, but now we have a clue for attacking it differently. The clue is not obvious, but a clever rewriting of what we have done will reveal it. Observe that the formula $y = y_0 e^{-P(x)}$ is equivalent to $e^{P(x)} y = y_0$, from which

$$\frac{d}{dx}[e^{P(x)} y] = 0$$

The Product Rule yields $e^{P(x)}y' + ye^{P(x)}P'(x) = 0$, and since

$$P'(x) = \frac{d}{dx}\int_{x_0}^{x} p(t)\,dt = p(x)$$

we have $e^{P(x)}[y' + p(x)y] = 0$.

The fascinating thing about this is that it is equivalent to the original equation, $y' + p(x)y = 0$, because $e^{P(x)}$ is never zero. Working backwards, we see that we could have solved the homogeneous case by *multiplying by* $e^{P(x)}$. For this changes the problem to the equation

$$\frac{d}{dx}[e^{P(x)}y] = 0$$

which is easily integrated. We call $e^{P(x)}$ an **integrating factor.**

Of course there is no point in using an integrating factor to solve the homogeneous case, since we have already solved it by separating the variables. However, why not tackle the nonhomogeneous case by this method?

To see how it goes, recall that we defined

$$P(x) = \int_{x_0}^{x} p(t)\,dt$$

Multiplying each side of $y' + p(x)y = q(x)$ by $e^{P(x)}$, we obtain

$$\frac{d}{dx}[e^{P(x)}y] = e^{P(x)}q(x)$$

and integration (from x_0 to x) yields

$$e^{P(x)}y - e^{P(x_0)}y_0 = \int_{x_0}^{x} e^{P(t)}q(t)\,dt$$

Since $P(x_0) = 0$ (why?), we have

$$e^{P(x)}y - y_0 = \int_{x_0}^{x} e^{P(t)}q(t)\,dt$$

$$e^{P(x)}y = y_0 + \int_{x_0}^{x} e^{P(t)}q(t)\,dt$$

$$y = \phi(x) = y_0 e^{-P(x)} + e^{-P(x)}\int_{x_0}^{x} e^{P(t)}q(t)\,dt$$

As in the homogeneous case, this may be checked out as a solution valid in the interval I, so we are done. The formula even includes the homogeneous case, since it reduces to $\phi(x) = y_0 e^{-P(x)}$ when $q(x) = 0$. Hence it serves as a complete answer to the initial value problem

$$y' + p(x)y = q(x) \qquad \phi(x_0) = y_0$$

Note that we have not put any of these results in neon lights, even though the argument has yielded an explicit formula for the solutions of an arbitrary linear equation. The reason for this is that the formula is a monstrosity. Keep in mind that $P(x)$ is itself an integral. Our formula involves it as an exponent, and the exponential expression (multiplied by q) is integrated again. There is no point in learning such a complicated result; in practice it is better to remember the device by which it was obtained.

An integrating factor of the linear first-order equation

$$y' + p(x)y = q(x)$$

is $e^{P(x)}$, where $P(x) = \int p(x)\,dx$.

Note that in this statement we have dropped the limits of integration in the formula for $P(x)$. In the solution of concrete problems it is easier to use a simple antiderivative.

□ **Example 1**   Find the general solution of $y' = (y/x) + \sqrt{x}$.

Solution: We have already asserted that the answer is

$$y = \phi(x) = 2x^{3/2} + Cx \quad \text{in } I = (0, \infty)$$

Now we are in a position to explain why. Rewrite the equation in the form $y' - (y/x) = \sqrt{x}$ and note that the coefficients $p(x) = -1/x$ and $q(x) = \sqrt{x}$ are continuous in I. Compute

$$P(x) = \int p(x)\,dx = -\int \frac{dx}{x} = -\ln x \qquad \text{(no absolute value needed because } x \in I)$$

and form the integrating factor

$$e^{P(x)} = e^{-\ln x} = e^{\ln x^{-1}} = x^{-1}$$

This is the key to the problem. For if we multiply each side of $y' - (y/x) = \sqrt{x}$ by x^{-1}, we are guaranteed to have a left side that we can integrate:

$$x^{-1}y' - x^{-2}y = x^{-1/2}$$

$$\frac{d}{dx}(x^{-1}y) = x^{-1/2}$$

$$x^{-1}y = \int x^{-1/2}\,dx = 2x^{1/2} + C$$

$$y = 2x^{3/2} + Cx \qquad \text{(for all } x > 0) \qquad\qquad □$$

□ **Example 2**   Find the solution of $y' + 2xy = x$ satisfying $\phi(0) = 2$.

Solution: Since $p(x) = 2x$, we compute $\int 2x\,dx = x^2$ and form the integrating factor e^{x^2}. Multiplication of the given equation by e^{x^2} yields

$$\frac{d}{dx}(e^{x^2}y) = xe^{x^2}$$

$$e^{x^2}y = \int xe^{x^2}\,dx = \tfrac{1}{2}e^{x^2} + C$$

Since $y = 2$ when $x = 0$, we find $C = \tfrac{3}{2}$ and hence

$$y = \phi(x) = \tfrac{1}{2} + \tfrac{3}{2}e^{-x^2}$$

The original coefficients $p(x) = 2x$ and $q(x) = x$ are continuous every-where, so our solution is valid for all x. (No need to check that it is; our theoretical results guarantee it.)   □

□ **Example 3**   Find the solution of $y' + y\cos x = x^2$ satisfying $\phi(0) = -1$.

Solution: This time an integrating factor is $e^{\sin x}$ (why?) and we write the equivalent equation

$$\frac{d}{dx}(e^{\sin x}y) = x^2 e^{\sin x}$$

Since the right side cannot be integrated in terms of elementary functions, we revert to our original procedure of integration between limits (from the initial value 0 to x):

$$e^{\sin x}y - e^{\sin 0}(-1) = \int_0^x t^2 e^{\sin t}\,dt$$

$$y = \phi(x) = e^{-\sin x}\left(\int_0^x t^2 e^{\sin t}\,dt - 1\right) \qquad \text{(valid for all } x\text{)}$$

While this is not a very attractive answer, it is explicit. The point is that we can solve *any* linear first-order equation (with continuous coefficients). Such an achievement is to be cherished; the subject of differential equations is usually less satisfying.   □

□ **Example 4**   A boat weighing 160 lb is pushed in the direction of motion by a constant force of 20 lb exerted by its motor. The water resists the motion with a backward force whose magnitude is always half the speed. If the boat starts from rest, find its speed t sec later. How fast can it go?

Solution: Since weight is the force of gravity, Newton's Second Law tells us that the mass of the boat satisfies $mg = 160$ (where g is the accelera-tion due to gravity). Taking $g = 32$, we find $m = 5$. Again using Newton's Law, we find the force on the boat (in the direction of motion) to be

$$5\frac{dv}{dt} = 20 - \frac{1}{2}v \qquad \text{(where } v \text{ is the speed at time } t\text{)}$$

This is a linear (also separable) first-order differential equation which we write in the standard form

$$\frac{dv}{dt} + \frac{1}{10}v = 4$$

An integrating factor is $e^{t/10}$, so we solve the equivalent equation

$$\frac{d}{dt}(e^{t/10}v) = 4e^{t/10}$$

$$e^{t/10}\,v = 40e^{t/10} + C$$

The initial condition $v = 0$ at $t = 0$ yields $C = -40$, so the speed of the boat at time t is

$$v = 40(1 - e^{-t/10})$$

As time goes on, the boat approaches a speed of 40 ft/sec.   □

□ **Example 5**   A tank contains 100 liters of brine (salt water) holding 10 kilograms of dissolved salt. Water containing 0.02 kg/liter of salt enters the tank at the rate of 2 liters/min and the brine (kept uniformly mixed by stirring) flows out at the same rate. How much salt is in the tank after one hour?

Solution: Let Q be the number of kilograms of salt in the tank t minutes after the process starts. The problem is to find $Q = \phi(t)$ as a function of t and then evaluate $\phi(60)$. It seems reasonable to write

$$\frac{dQ}{dt} = \text{(rate at which salt enters)} - \text{(rate at which salt leaves)}$$

Since the concentration of the entering salt solution is 0.02 kg/liter, and the solution enters at the rate of 2 liters/min, the rate at which salt enters is 0.04 kg/min. The concentration of the brine in the tank at time t is $Q/100$ kg/liter. Since the brine leaves at the rate of 2 liters/min, the rate at which salt leaves is $Q/50$ kg/min (at time t). Thus at every instant we have

$$\frac{dQ}{dt} = 0.04 - \frac{Q}{50} \qquad \text{with } Q = 10 \text{ at } t = 0$$

This is a linear (also separable) first-order initial value problem whose solution is

$$Q = \phi(t) = 2 + 8e^{-t/50} \qquad \text{(Confirm!)}$$

The number of kilograms of salt in the tank at the end of one hour is

$$\phi(60) = 2 + 8e^{-1.2} \approx 4.4$$

Note that as time goes on, Q approaches 2, that is, the concentration of the 100-liter solution is eventually the same (approximately) as the concentration of the entering solution (which is what we should expect).   □

Problem Set 20.3

Solve the following initial value problems, in each case specifying the largest interval of validity.

1. $y' = x + 2xy$, $\phi(0) = 1$.
2. $y' = (x^3 - 2y)/x$, $\phi(1) = 0$.

3. $y' - 3y = e^{2x}$, $\phi(2) = -1$.

4. $y' + y \tan x = \sec x$, $\phi(\pi/4) = 0$.

5. $y' - (2x + 1/x)y = x$, $\phi(1) = 3$.

6. $y' + y = \sin x$, $\phi(0) = 1$.

7. $y' - 2y/x = \sqrt{y}$, $\phi(1) = 4$. *Hint:* The substitution $z = \sqrt{y}$ changes this to a linear equation in z. (This is a special case of Bernoulli's equation $y' + p(x)y = q(x)y^n$. The substitution $z = y^{1-n}$ was discovered by Leibniz in 1696.)

Find the general solution of each of the following differential equations.

8. $y' + y = e^{-x}$

9. $xy' - 2y = x^2$ in the interval $(0, \infty)$

10. $y' + xy/(1 + x^2) = x$

11. $y' + \dfrac{y}{x} = \cos x$ in the interval $(0, \infty)$

12. $y' + y \cos x = \cos x$

13. $y' + y \cot x = \csc x$ in the interval $(0, \pi)$

14. The differential equation $y' = 2 - y$ is both separable and linear. Find its general solution by treating it both ways. Which is easier?

15. The graph of a function satisfying the differential equation $xy' + y = 2x$ crosses the y axis. What is its equation?

16. The tangent at each point (x, y) of a certain graph has slope $y - x$. If the curve passes through the origin, what is its equation?

17. A fish swimming at the rate of 2 m/sec stops swimming and glides horizontally through the water. If the magnitude of the water resistance is 0.2 times the speed, how far does the fish glide? (Take its mass to be 1.)

18. A dog pulling an 80-lb sled on level ice at 10 ft/sec stops pulling and the sled coasts on. Frictional resistance and air resistance (both opposite to the direction of motion) are 0.04 times the weight and $\frac{1}{5}$ of the speed, respectively.

(a) Show that the speed of the sled t sec after the dog stops pulling is

$$v = 26e^{-0.08t} - 16$$

(b) How far will the sled coast?

19. Solve Example 5 if fresh water enters the tank instead of a salt solution. How much salt is in the tank after a long time has passed?

20. Suppose that the outlet valve on the tank in Example 5 is turned down so that the brine flows out at only 1 liter/min. Then the tank fills; assuming that it has 500-liter capacity, how many kilograms of salt does it contain when it is full?

21. Suppose that the coefficient $p(x)$ in the linear homogeneous equation $y' + p(x)y = 0$ is continuous in the interval I, and let (x_0, y_0) be any point with $x_0 \in I$. Show by direct substitution in the equation that the function

$$\phi(x) = y_0 e^{-P(x)} \qquad \text{where } P(x) = \int_{x_0}^{x} p(t) \, dt$$

is a solution that is valid in I, and that $\phi(x_0) = y_0$.

22. Suppose that the coefficients $p(x)$ and $q(x)$ in the linear equation $y' + p(x)y = q(x)$ are continuous in the interval I, and let (x_0, y_0) be any point with $x_0 \in I$. Show by direct substitution in the equation that the function

$$\phi(x) = y_0 e^{-P(x)} + e^{-P(x)} \int_{x_0}^{x} e^{P(t)} q(t) \, dt$$

where $P(x) = \int_{x_0}^{x} p(t) \, dt$, is a solution that is valid in I, and that $\phi(x_0) = y_0$.

20.4
SECOND-ORDER
EQUATIONS

Since acceleration is the second derivative of position, it is not surprising that a great many differential equations arising in physics are second-order. Such equations involve not only an independent variable, an unknown function of that variable, and the derivative of the function, but also the second derivative.

As in the first-order case, the general form of a second-order equation is given by writing y'' explicitly in terms of the other variables, that is, $y'' = f(x, y, y')$, where f is a function whose domain is in $\mathcal{R}^3$. A solution of such an equation is a function $y = \phi(x)$, twice-differentiable in some interval I, such that

$$\phi''(x) = f(x, \phi(x), \phi'(x)) \qquad \text{for all } x \in I$$

An example is the equation

$$xy'' - (x + 1)y' + y = x^2$$

in which f is defined by

$$f(x, y, y') = x - \frac{y}{x} + \left(1 + \frac{1}{x}\right)y'$$

The domain of f is $D = \{(x, y, z): x \neq 0\}$, which is 3-space with the yz plane deleted. A solution in $I = (0, \infty)$ is

$$y = \phi(x) = -(x^2 + 2x + 2)$$

as is easily confirmed. What is not so easy is to explain where it came from (or what the other solutions are). You may also note that it satisfies the original equation $xy'' - (x + 1)y' + y = x^2$ for all x, not just in $I = (0, \infty)$. There are theoretical reasons, however, for writing y'' explicitly in terms of x, y, and y' (which requires division by x). In this form the solution is valid only in intervals not containing 0.

The question of initial conditions for the solution of a second-order equation is more complicated than in the first-order case. To see why, consider the equation

$$\frac{d^2s}{dt^2} = -32$$

which describes the motion of a body falling with constant acceleration. Ordinary antidifferentiation yields

$$\frac{ds}{dt} = -32t + C_1 \qquad \text{(where } C_1 \text{ is an arbitrary constant)}$$

and hence

$$s = \phi(t) = -16t^2 + C_1 t + C_2 \qquad \text{(where } C_2 \text{ is another constant)}$$

To get a unique solution we must impose two requirements, which opens up possibilities that are not present in the first-order case. For example, we might be told the *initial* velocity and the *terminal* position of the body, say $ds/dt = -5$ when $t = 0$, and $s = 0$ when $t = 10$. In terms of ϕ, these conditions are $\phi'(0) = -5$ and $\phi(10) = 0$, which lead to the solution

$$s = \phi(t) = -16t^2 - 5t + 1650$$

The interval I in this case may be taken to be $[0,10]$. Since the requirements $\phi(10) = 0$ and $\phi'(0) = -5$ are imposed at the endpoints of I, we refer to them as **boundary conditions.** Although in this case they lead to a unique solution, we do not call them *initial conditions;* that terminology is reserved for the situation in which ϕ and ϕ' are specified *at the same point.* Thus we might have been told the position as well as the velocity at $t = 0$, say $\phi(0) = 1650$ and $\phi'(0) = -5$. The solution would be the same.

While these two situations may appear to be equivalent, in general they are not. Boundary value problems (as they are called) differ significantly from initial value problems, as the following example shows.

□ **Example 1**   In the next section we will prove that every solution of the second-order equation $y'' + y = 0$ has the form

$$y = \phi(x) = c_1 \cos x + c_2 \sin x$$

(a) Find all solutions satisfying the initial conditions $\phi(0) = 2$ and $\phi'(0) = 5$.

(b) Find all solutions satisfying the boundary conditions $\phi(0) = 2$ and $\phi'(\pi/2) = -2$.

Solution: Putting the initial condition $\phi(0) = 2$ into the general solution

$$\phi(x) = c_1 \cos x + c_2 \sin x$$

we have $2 = c_1 \cos 0 + c_2 \sin 0$, from which $c_1 = 2$. Since

$$\phi'(x) = -c_1 \sin x + c_2 \cos x$$

the initial condition $\phi'(0) = 5$ yields

$$5 = -c_1 \sin 0 + c_2 \cos 0$$

and hence $c_2 = 5$. Thus the answer to part (a) is the unique solution

$$\phi(x) = 2 \cos x + 5 \sin x$$

In part (b) we are not given the values of ϕ and ϕ' at the same point, but at $x = 0$ and $x = \pi/2$. The first condition yields $c_1 = 2$ as before. But the condition $\phi'(\pi/2) = -2$, substituted into

$$\phi'(x) = -c_1 \sin x + c_2 \cos x$$

yields
$$-2 = -c_1 \sin \frac{\pi}{2} + c_2 \cos \frac{\pi}{2}$$

from which $c_1 = 2$ again! In other words, neither condition tells us anything about c_2, which therefore remains an arbitrary constant. There are infinitely many solutions

$$\phi(x) = 2 \cos x + c_2 \sin x$$

satisfying the boundary conditions given in part (b).   □

This debacle should throw some additional light on the Existence and Uniqueness Theorem (for first-order equations) in Section 20.1, and should help you realize that such theorems must be stated carefully in order to work. In the second-order case our theorem reads as follows. (Again we have to leave its explanation and proof for an advanced course.)

Existence and Uniqueness Theorem

Suppose that $f(x, y, z)$ and its partial derivatives $f_2(x, y, z)$ and $f_3(x, y, z)$ are continuous in the open region D in space. Given any point $(x_0, y_0, z_0) \in D$, there is an interval I containing x_0 (on the x axis), and exactly one function $\phi(x)$, twice-differentiable in I, such that

$$\phi''(x) = f(x, \phi(x), \phi'(x)) \qquad \text{for all } x \in I$$

with $\phi(x_0) = y_0$ and $\phi'(x_0) = z_0$. In other words, the differential equation $y'' = f(x, y, y')$ has a unique solution $y = \phi(x)$ (in the interval I) satisfying the initial conditions $y = y_0$ and $y' = z_0$ when $x = x_0$.

This long-winded statement is much the same as the corresponding theorem for first-order equations in Section 20.1. The difference is that the dimension of the domain D has increased by one, and there are two initial conditions instead of one. The same sort of theorem can be proved for higher-order equations, but then D is in n-space with $n > 3$ and we can no longer visualize it geometrically. That is why we are not drawing pictures to illustrate these theorems. The proof even in the first-order case is more analytic then geometric; pictures are not much help.

You probably noticed that we did not do much with the Existence and Uniqueness Theorem in Section 20.1, nor are we going to dwell on this one. The reason we have stated it is that it is a powerful tool for deriving some important facts about *linear* second-order equations (in the next section). In the first-order case we did not need it, because we were able to solve the linear case explicitly. The second-order linear equation is more difficult.

In this section we confine our attention to two special types of second-order equations (not necessarily linear) that can be reduced to first-order equations by a substitution.

Type I.   Suppose that y does not appear explicitly in the second-order equation $y'' = f(x, y, y')$, so that it has the form $y'' = f(x, y')$. The substitution $v = y'$ reduces the equation to $v' = f(x, v)$, which is first-order in v. If we can solve this, we are in business.

□ **Example 2**   Solve the initial value problem

$$y'' = x + y', \; \phi(0) = 1, \; \phi'(0) = 2$$

Solution: Noting that y is missing, we let $v = y'$. Since $v' = y''$, the equation becomes $v' = x + v$, which is not only first-order, but linear, with initial condition $v = 2$ when $x = 0$. (Why?) Writing it in the standard form $v' - v = x$, we use the integrating factor e^{-x} to change it to

$$\frac{d}{dx}(e^{-x}v) = xe^{-x}$$

Integration by parts yields $e^{-x}v = -xe^{-x} - e^{-x} + C_1$ and the initial condition $v = 2$ when $x = 0$ gives $C_1 = 3$. Hence

$$v = 3e^x - x - 1$$

Returning to the original substitution $v = y'$, we have

$$y' = 3e^x - x - 1$$

and another integration gives

$$y = 3e^x - \tfrac{1}{2}x^2 - x + C_2$$

The initial condition $y = 1$ when $x = 0$ yields $C_2 = -2$, so our final answer is

$$y = \phi(x) = 3e^x - \tfrac{1}{2}x^2 - x - 2 \qquad \square$$

Type II. Suppose that x does not appear explicitly in the second-order equation $y'' = f(x, y, y')$, so that it has the form $y'' = f(y, y')$. The same substitution as in Type I equations is used, namely $v = y'$, but this time an additional device (due to Newton) is needed. The reduced equation is $v' = f(y, v)$, which looks like a first-order equation in v. However, the position of y as the first variable in $f(y, v)$ corresponds to the independent variable in Section 20.1. Since the left side is not dv/dy but $v' = dv/dx$, we need the Chain Rule:

$$v' = \frac{dv}{dx} = \frac{dv}{dy}\frac{dy}{dx} = v\frac{dv}{dy} \qquad \left(\text{because } v = \frac{dy}{dx}\right)$$

Our equation now reads $v\,(dv/dy) = f(y, v)$ and can be treated as first-order in v as a function of y.

Remark   Our use of the Chain Rule in Type II equations requires that $y' \neq 0$ (which guarantees that y as a function of x can be inverted to x as a function of y). In other words, the change from x to y as independent variable needs this condition. This fits in with our earlier remarks about expressing the highest derivative in terms of the other variables. If $v = y' \neq 0$, the equation $v\,(dv/dy) = f(y, v)$ can be solved for dv/dy in terms of y and v.

□ **Example 3**   Solve the initial value problem

$$y'' = -y, \ \phi(1) = 0, \ \phi'(1) = -2$$

Solution: In Example 1 we stated that the general solution of $y'' + y = 0$ is

$$y = c_1 \cos x + c_2 \sin x$$

That was based on advance information from the next section, however; for the present let's ignore it and see how far we can get by treating the equation as Type II (x missing). Letting $v = y'$, we reduce our equation to the form $v' = -y$, that is, $dv/dx = -y$. Recognizing that x as the independent variable is not convenient (whereas y would be) we use the device $v' = v\,(dv/dy)$ explained above. This changes the equation to

$$v \frac{dv}{dy} = -y \qquad \text{or} \qquad v\,dv = -y\,dy$$

which can be integrated to yield

$$\tfrac{1}{2}v^2 = -\tfrac{1}{2}y^2 + C_1$$

The initial condition $v = -2$ when $y = 0$ gives $C_1 = 2$, so we have $v^2 = 4 - y^2$. Since v is negative near the initial point, we extract the negative square root:

$$v = -\sqrt{4 - y^2}$$

$$\frac{dy}{dx} = -\sqrt{4 - y^2}$$

$$\frac{dy}{\sqrt{4 - y^2}} = -dx$$

$$\sin^{-1} \frac{y}{2} = -x + C_2$$

Since $y = 0$ when $x = 1$, we find $C_2 = 1$ and hence

$$\sin^{-1} \frac{y}{2} = 1 - x$$

$$\frac{y}{2} = \sin(1 - x)$$

$$y = \phi(x) = 2 \sin(1 - x)$$

Note that although heavy restrictions were imposed as we worked toward the answer ($y' \neq 0$ to ensure invertibility and $v < 0$ to extract the correct square root), the solution is valid for all x, as you can check by substitution in the original equation. This is not uncommon in differential equations. We have to be careful while we solve them, but what counts in the end is whether (and where) the solution works.   □

□ **Example 4**   A space vehicle is fired straight up from the surface of the earth. What must be its initial velocity (the "escape velocity") if it is to leave the earth's gravitational field? (Assume that it has no power of its own.)

Solution: If x is the distance between the vehicle and the center of the earth, the gravitational attraction exerted by the earth is $F = -km/x^2$, where m is the mass of the vehicle and k is a positive constant. By Newton's Second Law we can write this in the form

$$m \frac{dv}{dt} = -\frac{km}{x^2} \qquad \text{or} \qquad \frac{dv}{dt} = -\frac{k}{x^2}$$

where v is the velocity of the vehicle at time t. This is really a Type II second-order equation

$$x'' = \frac{d^2x}{dt^2} = -\frac{k}{x^2}$$

with x as the dependent variable and t independent. We have already made the substitution $v = x' = dx/dt$.

Since t is inconvenient as the independent variable (whereas x would be convenient), we write

$$\frac{dv}{dt} = \frac{dv}{dx}\frac{dx}{dt} = v\frac{dv}{dx}$$

This changes the equation to

$$v\frac{dv}{dx} = -\frac{k}{x^2}$$

$$v\, dv = -\frac{k\, dx}{x^2}$$

$$\int v\, dv = -k\int \frac{dx}{x^2}$$

$$\frac{1}{2}v^2 = \frac{k}{x} + C$$

Let v_0 be the (unknown) initial velocity in miles per second. Then the initial conditions are $v = v_0$ and $x = 3960$ at $t = 0$. (The radius of the earth is about 3960 miles.) This yields

$$\frac{1}{2}v_0^2 = \frac{k}{3960} + C$$

from which

$$C = \frac{1}{2}v_0^2 - \frac{k}{3960}$$

and hence

$$v^2 = v_0^2 + 2k\left(\frac{1}{x} - \frac{1}{3960}\right)$$

In order for the vehicle to escape from the earth's gravitational field, we cannot allow $v = 0$, that is, v must remain positive. Hence we have to satisfy the inequality

$$v_0^2 > 2k\left(\frac{1}{3960} - \frac{1}{x}\right) \qquad \text{for all } x$$

When x is large, this may be replaced by $v_0^2 > k/1980$ with negligible error. If we knew the value of k, the problem would be solved.

Return to the original force equation, $F = -(km/x^2)$. At the surface of the earth this reads $-mg = -km/(3960)^2$, where g is the acceleration due to gravity at ground level. Taking $g = 32$ ft/sec^2, we find

$$k = (3960)^2 g = (3960)^2\left(\frac{32}{5280}\right)$$

and hence our inequality becomes

$$v_0{}^2 > \frac{(3960)^2(32)}{(1980)(5280)} = 48$$

$$v_0 > 4\sqrt{3} \approx 6.93$$

Thus the escape velocity is about 7 miles per second (neglecting air resistance).   □

Problem Set 20.4

1. Name a solution of $y'' + (x^2 - 1)y' + y = 0$ whose graph is tangent to the x axis at the origin. Why is this the only such solution?

Solve the following initial value problems.

2. $y'' - y' = 0$, $\phi(0) = 2$, $\phi'(0) = -1$.
3. $y'' = 1 + y'^2$, $\phi(1) = 2$, $\phi'(1) = 0$.
4. $xy'' + y' = x \sin x$, $\phi(\pi/2) = -1$, $\phi'(\pi/2) = 0$.
5. $y'' = y$, $\phi(0) = -1$, $\phi'(0) = 1$.
6. $y'' = 6y^2$, $\phi(0) = 1$, $\phi'(0) = 2$.
7. $y'' + 2yy' = 0$, $\phi(0) = 2$, $\phi'(0) = -3$.

Find all solutions of each of the following differential equations.

8. $y'' = \cos x$ **9.** $y'' - 2y' = 0$
10. $y'' - y' = e^x$ **11.** $y'' - y' = \sin x$
12. $y'' + 2xy' = 2x$ **13.** $yy'' = 2y'^2$
14. An object weighing 16 lb is dropped at the surface of a

lake and begins to sink. If its buoyancy is 8 lb and the magnitude of the water resistance is twice the square of the speed, find the speed in terms of the time.

15. An object 16 ft off the ground is released at the top of a 32-ft frictionless inclined plane. Just as it starts to slide, a pebble is dislodged from the top of the plane and falls to the ground with negligible air resistance. Show that although the objects do not reach the ground at the same time, they attain the same speed. (*Note:* This can be done in two ways, one involving differential equations, the other involving the Law of Conservation of Energy as described in Section 19.1.)

16. A chain 60 ft long weighing 1 lb/ft hangs over a small frictionless pulley with 30 ft on each side and a 4-lb weight attached to each end. One of the weights falls off. Find the velocity of the remaining weight in terms of its distance from the pulley. How fast is this weight falling when the chain comes off the pulley? *Hint:* The entire system (weighing 64 lb) is in motion. Use Newton's Law to find the force and set it equal to the net force due to the weight on each side of the pulley.

20.5

LINEAR SECOND-ORDER EQUATIONS

A second-order differential equation of the form

$$a(x)y'' + b(x)y' + c(x)y = g(x)$$

is said to be *linear*. We assume that the functions a, b, c, and g are continuous in an (open) interval I and that $a(x)$ is nowhere zero in I. This last condition ensures that we can divide by $a(x)$ to solve for y'' in terms of x, y, and y', as required by the Existence and Uniqueness Theorem of the last section. The theorem then applies in the region

$$D = \{(x, y, z): x \in I\}$$

which is an infinite slab in 3-space parallel to the yz plane (half or all of 3-space if I is half or all of the x axis). Hence we may impose an arbitrary initial condition $(x_0, y_0, z_0) \in D$, where $x_0 \in I$, $y_0 \in \mathcal{R}$, and $z_0 \in \mathcal{R}$, and are guaranteed a unique solution of $ay'' + by' + cy = g$ satisfying $\phi(x_0) = y_0$

and $\phi'(x_0) = z_0$ (in some interval containing x_0). As in the first-order linear case, this interval turns out to be I itself, regardless of the initial conditions, so it is both predictable and stable. Unlike the first-order case, however, we are unable to prove this by solving the equation. The proof is based on other considerations and is left for an advanced course.

A **homogeneous** linear equation is of the form

$$a(x)y'' + b(x)y' + c(x)y = 0$$

that is, the function $g(x)$ in the general linear equation is zero. This section is devoted to the theory of such equations, which requires a sudden dose of algebra that you may be unprepared to take. We apologize for that, with the hope that the hard parts will interest you in further study along these lines.

First we define **linear independence** of two functions ϕ_1 and ϕ_2 to mean that neither is a scalar multiple of the other (in a given interval I). The functions $\phi_1(x) = \cos x$ and $\phi_2(x) = 2 \cos x$, for example, are linearly *dependent,* since the second is simply two times the first. But $\phi_1(x) = \cos x$ and $\phi_2(x) = \sin x$ are independent, there being no scalar k such that $\phi_2 = k\phi_1$ or $\phi_1 = k\phi_2$. (Why?) Another way of stating linear independence (which can be generalized to more than two functions) is that no scalars c_1 and c_2 exist (other than $c_1 = c_2 = 0$) such that $c_1\phi_1 + c_2\phi_2 = 0$. We will use this form of the definition in the following discussion.

Throughout this section it is understood that we are talking about a given homogeneous linear equation

$$a(x)y'' + b(x)y' + c(x)y = 0$$

where a, b, and c are continuous in an interval I and $a(x)$ is nowhere zero in I. For the sake of brevity, we denote the left side of this equation by the "operator"

$$L(y) = ay'' + by' + cy$$

If ϕ is a solution of the equation $L(y) = 0$, we will write $L(\phi) = 0$, meaning that $a\phi'' + b\phi' + c\phi = 0$.

□ **Theorem 1** L is a *linear* operator, that is,

$$L(\phi_1 + \phi_2) = L(\phi_1) + L(\phi_2)$$

and (for every scalar k)

$$L(k\phi) = kL(\phi)$$

where ϕ, ϕ_1, ϕ_2 are twice-differentiable in I.

Proof: This is just a fancy way of saying that differentiation has these properties (which we already know). For if ϕ_1 and ϕ_2 are functions that are twice-differentiable in I, we have

$$\begin{aligned}
L(\phi_1 + \phi_2) &= a(\phi_1 + \phi_2)'' + b(\phi_1 + \phi_2)' + c(\phi_1 + \phi_2) \\
&= a(\phi_1'' + \phi_2'') + b(\phi_1' + \phi_2') + c(\phi_1 + \phi_2) \\
&= (a\phi_1'' + b\phi_1' + c\phi_1) + (a\phi_2'' + b\phi_2' + c\phi_2) = L(\phi_1) + L(\phi_2)
\end{aligned}$$

The second part of the theorem is proved in the same way.   □

□ **Corollary**   Sums and scalar multiples of solutions of $L(y) = 0$ are also solutions. More generally, any *linear combination* $\phi = c_1\phi_1 + c_2\phi_2$ of two solutions ϕ_1 and ϕ_2 is itself a solution.   □

□ **Example 1**   It is easy to check that $\phi_1(x) = \cos x$ and $\phi_2(x) = \sin x$ are solutions of the differential equation $y'' + y = 0$. The corollary to Theorem 1 says that any function of the form

$$\phi(x) = c_1 \cos x + c_2 \sin x$$

is also a solution. This can be checked directly (by substituting ϕ in the equation), but a more elegant (and general) observation is to use the linearity of the operator $L(y) = y'' + y$:

$$L(\phi) = L(c_1\phi_1 + c_2\phi_2) = L(c_1\phi_1) + L(c_2\phi_2) = c_1 L(\phi_1) + c_2 L(\phi_2)$$
$$= c_1(0) + c_2(0) = 0$$

(because ϕ_1 and ϕ_2 are solutions of $y'' + y = 0$). Hence ϕ is also a solution.

□

The point of the corollary is that once two solutions of a homogeneous equation are known, a whole class of solutions is generated by forming linear combinations of these two. It is a remarkable fact that this class contains *all* the solutions of the equation if its generators are linearly independent. Thus (for example) every solution of $y'' + y = 0$ is of the form

$$\phi(x) = c_1 \cos x + c_2 \sin x$$

(because $\cos x$ and $\sin x$ are independent solutions). This fact is far from obvious, however. Moreover, how do we find the crucial generators? If you did not know any solutions of $y'' + y = 0$, you might guess that $\cos x$ and $\sin x$ will work. In general, however, it is difficult to get started. We will return to that question later; our objective now is to prove that independent solutions generate the whole class of solutions.

To develop the proof, we need two theorems from algebra. You may have encountered these in an earlier course (and we came close to proving them in Section 16.3 when we used determinants to discuss the cross product of vectors). One of them is known as *Cramer's Rule* for solving the system of equations

$$ax + by = e$$
$$cx + dy = f$$

What this rule says is that if the *determinant of coefficients* is not zero, then the system has a unique solution (x, y) given by

$$x = \frac{\begin{vmatrix} e & b \\ f & d \end{vmatrix}}{\begin{vmatrix} a & b \\ c & d \end{vmatrix}}, \quad y = \frac{\begin{vmatrix} a & e \\ c & f \end{vmatrix}}{\begin{vmatrix} a & b \\ c & d \end{vmatrix}}$$

When e and f are zero, the system reduces to

$$ax + by = 0$$
$$cx + dy = 0$$

which always has the "trivial" solution $(x, y) = (0, 0)$. Cramer's Rule says that this is the only solution if the determinant of coefficients is not zero. (Why?) If the determinant is zero, however, another theorem from algebra says that there are nontrivial solutions $(x, y) \neq (0, 0)$. An example is the system

$$2x + 3y = 0$$
$$6x + 9y = 0$$

whose determinant of coefficients is

$$\begin{vmatrix} 2 & 3 \\ 6 & 9 \end{vmatrix} = 18 - 18 = 0$$

This system has the trivial solution $(0, 0)$ and also infinitely many others of the form $(3t, -2t)$ where $t \in R$. The reason for this is that both equations represent the same straight line. Every point of the line (including the origin) is a solution of the system.

Now we are ready to proceed. Suppose that we have found two solutions ϕ_1 and ϕ_2 of the differential equation $L(y) = 0$ and are interested in whether they are linearly independent. The definition of independence involves the question of whether there are scalars c_1 and c_2 (other than $c_1 = c_2 = 0$) such that

$$c_1\phi_1 + c_2\phi_2 = 0$$

If there are (that is, if ϕ_1 and ϕ_2 are dependent), it follows by differentiation that

$$c_1\phi_1' + c_2\phi_2' = 0$$

Evaluating these functions at an arbitrary point $x_0 \in I$, we have

$$c_1\phi_1(x_0) + c_2\phi_2(x_0) = 0$$
$$c_1\phi_1'(x_0) + c_2\phi_2'(x_0) = 0$$

which may be regarded as a system of equations in the "unknowns" c_1 and c_2. Its determinant of coefficients is the number

$$\begin{vmatrix} \phi_1(x_0) & \phi_2(x_0) \\ \phi_1'(x_0) & \phi_2'(x_0) \end{vmatrix}$$

and we know that the only way the system can have a nontrivial solution $(c_1, c_2) \neq (0, 0)$ is for this determinant to be zero. Define the *Wronskian* of ϕ_1 and ϕ_2 to be the function

$$W(x) = \begin{vmatrix} \phi_1(x) & \phi_2(x) \\ \phi_1'(x) & \phi_2'(x) \end{vmatrix}$$

Then we have proved that

Linear dependence of ϕ_1 and $\phi_2 \Rightarrow W(x) = 0$ for all $x \in I$

An equivalent statement is that if there is any point $x \in I$ for which $W(x) \neq 0$, then ϕ_1 and ϕ_2 are linearly independent. (Why?) This is a practical test for independence (particularly when it is generalized to more than two functions).

□ **Theorem 2**   Suppose that ϕ_1 and ϕ_2 are differentiable in the interval I. If their Wronskian is nonzero somewhere in I, they are linearly independent.

□

□ **Example 2**   Show that $\phi_1(x) = \cos x$ and $\phi_2(x) = \sin x$ are linearly independent.

Solution: The Wronskian is

$$W(x) = \begin{vmatrix} \cos x & \sin x \\ -\sin x & \cos x \end{vmatrix} = \cos^2 x + \sin^2 x = 1 \neq 0$$

Interestingly enough, the Wronskian of $\cos x$ and $\sin x$ isn't just nonzero *somewhere* (as called for by Theorem 2), but *everywhere*. The reason for this is not obvious; the following theorem will explain it.   □

□ **Theorem 3**   Suppose that ϕ_1 and ϕ_2 are linearly independent solutions of the homogeneous linear equation $L(y) = 0$ (in the interval I). Then their Wronskian is nowhere zero in I.

Proof: Suppose the contrary. Then there is a point $x_0 \in I$ for which $W(x_0) = 0$. As we have already observed, this implies that the system

$$c_1\phi_1(x_0) + c_2\phi_2(x_0) = 0$$
$$c_1\phi_1'(x_0) + c_2\phi_2'(x_0) = 0$$

has a nontrivial solution $(c_1, c_2) \neq (0,0)$. Choose such a solution and define the function $\phi = c_1\phi_1 + c_2\phi_2$. Since ϕ_1 and ϕ_2 are solutions of $L(y) = 0$, so is ϕ (by the corollary to Theorem 1). Moreover,

$$\phi(x_0) = c_1\phi_1(x_0) + c_2\phi_2(x_0) = 0$$
$$\phi'(x_0) = c_1\phi_1'(x_0) + c_2\phi_2'(x_0) = 0$$

(because c_1 and c_2 satisfy the above system). Hence ϕ satisfies the initial conditions $\phi(x_0) = 0$ and $\phi'(x_0) = 0$.

Now comes the deep part! According to the Uniqueness Theorem, there is only one solution of $L(y) = 0$ satisfying given initial conditions. It is obvious that the zero function $\psi(x) = 0$ is a solution satisfying the initial conditions $\psi(x_0) = 0$ and $\psi'(x_0) = 0$. Since ϕ is, too, ϕ *must be the zero function.* Hence

$$c_1\phi_1 + c_2\phi_2 = 0 \qquad \text{where } c_1 \text{ and } c_2 \text{ are not both zero}$$

But that's impossible! For if it were true, ϕ_1 and ϕ_2 would be linearly dependent, whereas by hypothesis they are independent. Our opening supposition (that the theorem is false) is incorrect; the theorem is true.   □

□ **Theorem 4**   If ϕ_1 and ϕ_2 are linearly independent solutions of $L(y) = 0$, then every solution is a linear combination of ϕ_1 and ϕ_2.

Proof: Let ϕ be any solution of $L(y) = 0$. The problem is to name scalars c_1 and c_2 such that $\phi = c_1\phi_1 + c_2\phi_2$. Choose an arbitrary point $x_0 \in I$ and note that such scalars (if they exist) must satisfy the system of equations

$$c_1\phi_1(x_0) + c_2\phi_2(x_0) = a$$
$$c_1\phi_1'(x_0) + c_2\phi_2'(x_0) = b$$

where $a = \phi(x_0)$ and $b = \phi'(x_0)$. The determinant of coefficients is $W(x_0)$; since ϕ_1 and ϕ_2 are linearly independent solutions of $L(y) = 0$, we know by Theorem 3 that $W(x_0) \neq 0$ (because W is nowhere zero in I). Cramer's Rule then guarantees that the system has a unique solution (c_1, c_2).

We have not yet asserted that c_1 and c_2 exist such that $\phi = c_1\phi_1 + c_2\phi_2$. The above argument shows, however, that for a given choice of $x_0 \in I$ we can name (unique) scalars c_1 and c_2 such that

$$c_1\phi_1(x_0) + c_2\phi_2(x_0) = a = \phi(x_0)$$

(the first equation of the above system). It is easy to jump to the conclusion that we are done. The trouble is that c_1 and c_2 may change with the choice of x_0 (in which case they would not be constants).

Again here comes the deep part! Choose $x_0 \in I$ and name c_1 and c_2 as described above. The function $\psi = c_1\phi_1 + c_2\phi_2$ is then definite and we know that it agrees with ϕ at x_0, that is,

$$\psi(x_0) = c_1\phi_1(x_0) + c_2\phi_2(x_0) = a = \phi(x_0)$$
$$\psi'(x_0) = c_1\phi_1'(x_0) + c_2\phi_2'(x_0) = b = \phi'(x_0)$$

Since ϕ_1 and ϕ_2 are solutions of $L(y) = 0$, so is ψ. Moreover, it satisfies the same initial conditions as ϕ, namely $\psi(x_0) = a$ and $\psi'(x_0) = b$. By the Uniqueness Theorem, $\phi = \psi$, that is, $\phi = c_1\phi_1 + c_2\phi_2$. *Now* we are done.   □

Where do we stand? The corollary to Theorem 1 says that any two solutions of $L(y) = 0$ generate a whole class of solutions. Theorem 4 says that if the two solutions are independent, this class includes all solutions. We are therefore in a position to find all solutions of a homogeneous linear equation *if we can find two independent ones.* Our next theorem does not tell us how to do this, but guarantees that it can be done.

□ **Theorem 5**   Every homogeneous linear equation $L(y) = 0$ has two linearly independent solutions.

Proof: In Theorems 3 and 4 we used the Uniqueness Theorem; now we use the Existence Theorem. Choose a point $x_0 \in I$. According to the theorem, a solution ϕ_1 of $L(y) = 0$ exists satisfying the initial conditions $\phi_1(x_0) = 1$ and $\phi_1'(x_0) = 0$. (We have taken $y_0 = 1$ and $z_0 = 0$ in the theorem; see Section 20.4.) By the same reasoning (taking $y_0 = 0$ and $z_0 = 1$) we know that a solution ϕ_2 exists satisfying $\phi_2(x_0) = 0$ and $\phi_2'(x_0) = 1$. We claim that ϕ_1 and ϕ_2 are linearly independent. The reason is that

$$W(x_0) = \begin{vmatrix} \phi_1(x_0) & \phi_2(x_0) \\ \phi_1'(x_0) & \phi_2'(x_0) \end{vmatrix} = \begin{vmatrix} 1 & 0 \\ 0 & 1 \end{vmatrix} = 1 \neq 0$$

which implies independence by Theorem 2.   □

□ **Example 3**   Solve the initial value problem

$$y'' - y = 0, \; \phi(0) = 3, \; \phi'(0) = 1$$

Solution: Since $\cos x$ and $\sin x$ are solutions of $y'' + y = 0$, we may guess that $\cosh x$ and $\sinh x$ are solutions of $y'' - y = 0$. It is easy to confirm the guess; then we need only check that our solutions are independent. Compute the Wronskian

$$W(x) = \begin{vmatrix} \cosh x & \sinh x \\ \sinh x & \cosh x \end{vmatrix} = \cosh^2 x - \sinh^2 x = 1 \neq 0$$

(Note that it is zero *nowhere,* as predicted by Theorem 3.) This guarantees independence, so the general solution is

$$\phi(x) = c_1 \cosh x + c_2 \sinh x$$

Since $\phi'(x) = c_1 \sinh x + c_2 \cosh x$, we find from the initial conditions that

$$3 = c_1 \cosh 0 + c_2 \sinh 0$$
$$1 = c_1 \sinh 0 + c_2 \cosh 0$$

Hence $c_1 = 3$ and $c_2 = 1$, which yields the answer

$$\phi(x) = 3 \cosh x + \sinh x$$ □

You can see how beautifully the theory works. The only problem (it is a tough one) is to name independent functions ϕ_1 and ϕ_2 satisfying $L(y) = 0$. In the next section we will show you how to do this when the coefficients in

$$L(y) = a(x)y'' + b(x)y' + c(x)y = 0$$

are constants, but there is no systematic procedure that applies to all cases.

Problem Set 20.5

1. Prove the second part of Theorem 1, $L(k\phi) = kL(\phi)$, where k is a scalar and ϕ is twice-differentiable in I.

2. Confirm that $x^2 - 2$ and $\sin x + x^2 - 2$ are solutions of $y'' + y = x^2$, but their sum is not. Why doesn't this contradict the corollary to Theorem 1?

3. If e^x and e^{-x} are solutions of $L(y) = 0$, why does it follow that $\cosh x$ and $\sinh x$ are, too?

4. Confirm that e^x and e^{-x} are solutions of $y'' - y = 0$ and show that they are linearly independent. What is the general solution? What is the solution satisfying $\phi(0) = 3$ and $\phi'(0) = 1$? (This should check with Example 3.)

5. Solve the initial value problem

$$y'' + y = 0, \ \phi(1) = 0, \ \phi'(1) = -2$$

by using the fact that $\cos x$ and $\sin x$ are linearly independent solutions of $y'' + y = 0$. Then show that your answer is the same as $\phi(x) = 2 \sin (1 - x)$ found in Example 3, Section 20.4.

6. The functions ϕ_1 and ϕ_2 are linearly dependent when one is a scalar multiple of the other. Compute their Wronskian in this case, showing it to be zero for all $x \in I$. Why is this a proof of Theorem 2?

7. Let $\phi_1(x) = e^{3x} + e^{-2x}$ and $\phi_2(x) = e^{3x} - e^{-2x}$.

   (a) Confirm that ϕ_1 and ϕ_2 are solutions of $y'' - y' - 6y = 0$ in $I = (-\infty, \infty)$.

   (b) Confirm that $W(x) = 10e^x$ and note that (as predicted by Theorem 3) the Wronskian is nowhere zero.

   (c) The computation in part (b) is messy. If we are merely trying to confirm the linear independence of ϕ_1 and ϕ_2, why is it sufficient to evaluate them at $x = 0$ and compute $W(0)$?

Use the Wronskian to prove linear independence of each of the following pairs of functions in the given interval. According to Theorem 3, which pairs cannot be solutions of a homogeneous linear equation $L(y) = 0$ in I?

8. $\phi_1(x) = 1, \ \phi_2(x) = e^x, \ I = (-\infty, \infty)$.

9. $\phi_1(x) = \sin x, \ \phi_2(x) = \tan x, \ I = (-\pi/2, \pi/2)$.

10. $\phi_1(x) = \cos x, \ \phi_2(x) = \tan x, \ I = (-\pi/2, \pi/2)$.

11. $\phi_1(x) = e^{r_1 x}, \ \phi_2(x) = e^{r_2 x}, \ I = (-\infty, \infty)$, where r_1 and r_2 are distinct constants.

12. $\phi_1(x) = e^{rx}, \ \phi_2(x) = xe^{rx}, \ I = (-\infty, \infty)$, where r is a constant.

13. $\phi_1(x) = e^{px} \cos qx, \quad \phi_2(x) = e^{px} \sin qx, \quad I = (-\infty, \infty)$, where p and $q \neq 0$ are constants.

14. Confirm that the Wronskian of $\phi_1(x) = x$ and $\phi_2(x) = x^2$ is $W(x) = x^2$. Then answer the following questions.

   (a) Explain why ϕ_1 and ϕ_2 cannot be solutions of a homogeneous linear equation $L(y) = 0$ in an interval I containing 0.

   (b) Confirm that ϕ_1 and ϕ_2 are solutions of $x^2 y'' - 2xy' + 2y = 0$ in $I = (-\infty, \infty)$.

   (c) Parts (a) and (b) look contradictory. What is the explanation? *Hint:* Remember our restrictions on L, stated before Theorem 1.

15. Prove that the derivative of the Wronskian of two functions ϕ_1 and ϕ_2 is

$$W' = \begin{vmatrix} \phi_1 & \phi_2 \\ \phi_1'' & \phi_2'' \end{vmatrix}$$

16. Suppose that ϕ_1 and ϕ_2 are solutions of $L(y) = ay'' + by' + cy = 0$.

   (a) Use Problem 15 to show that their Wronskian satisfies $W' = -(b/a)W$. *Hint:* Since $L(\phi_1) = 0$,

$$\phi_1'' = -\frac{b}{a}\phi_1' - \frac{c}{a}\phi_1$$

and similarly for ϕ_2''.

   (b) Part (a) shows that the Wronskian is a solution of the first-order equation $W' + (b/a)W = 0$. Use Section 20.3 to explain why

$$W(x) = W(x_0)e^{-P(x)}$$

where x_0 is an arbitrarily selected point of I and

$$P(x) = \int_{x_0}^x \frac{b(t)}{a(t)} dt$$

(This expression for the Wronskian is known as **Abel's Formula,** discovered by the Norwegian mathematician Niels Abel, 1802–1829. He was one of the great figures of his time, despite his brief life.)

   (c) Use Abel's Formula to explain why the Wronskian of ϕ_1 and ϕ_2 is either nowhere zero in I or everywhere zero in I.

17. Taking $x_0 = 1$, confirm Abel's Formula for the Wronskian of $\phi_1(x) = x$ and $\phi_2(x) = x^2$ as solutions of $x^2 y'' - 2xy' + 2y = 0$ in $I = (0, \infty)$.

Suppose that one solution ϕ_1 of $L(y) = 0$ is known. D'Alembert (1717–1783) devised a method for finding a second solution of the form $\phi_2 = v\phi_1$, where v is an unknown function to be found by substitution in $L(y) = 0$. The following problems illustrate how it works.

18. Confirm that $\phi_1(x) = e^x$ is a solution of $L(y) = y'' - 2y' + y = 0$ and answer the following questions.

  (a) If $y = \phi_2(x) = ve^x$ is a second solution, show that substitution in $L(y) = 0$ yields $e^x v'' = 0$.

  (b) Solve for v (omitting arbitrary constants because only one answer is needed) to obtain $v = x$. Then confirm that $\phi_2(x) = xe^x$ is indeed a solution of $L(y) = 0$.

  (c) Confirm that ϕ_1 and ϕ_2 are linearly independent and give the general solution of $L(y) = 0$.

19. Confirm that $\phi_1(x) = x$ is a solution of

$$L(y) = (x^2 + 1)y'' - 2xy' + 2y = 0$$

and answer the following questions.

  (a) If $y = \phi_2(x) = vx$ is a second solution, show that substitution in $L(y) = 0$ yields $x(x^2 + 1)v'' + 2v' = 0$.

  (b) Let $w = v'$ in part (a) and solve the resulting equation in $I = (0, \infty)$ to obtain $w = 1 + (1/x^2)$ as one solution. (Decomposition of a rational function is involved.)

  (c) Integrate to obtain $v = x - (1/x)$ and hence $\phi_2(x) = x^2 - 1$.

  (d) Check that ϕ_1 and ϕ_2 are linearly independent and give the general solution of $L(y) = 0$. Where is it valid?

Use D'Alembert's method to find the general solution of each of the following equations, using the solution given.

20. $xy'' - (x + 1)y' + y = 0$, $\phi_1(x) = e^x$

21. $x^2y'' + xy' - 4y = 0$, $\phi_1(x) = x^2$

22. $y'' - y' \tan x + 2y = 0$, $\phi_1(x) = \sin x$, in $I = (-\pi/2, \pi/2)$

23. Prove Cramer's Rule by solving the system

$$ax + by = e$$
$$cx + dy = f$$

under the assumption that $ad - bc \neq 0$.

24. Show that the system

$$ax + by = 0$$
$$cx + dy = 0$$

has nontrivial solutions $(x, y) \neq (0, 0)$ if $ad - bc = 0$.

25. Suppose that $p(x)$ is continuous in the interval I and let $L(y) = y' + py$.

  (a) Show that L is a linear operator (as in Theorem 1).

  (b) Why does it follow that sums and scalar multiples of solutions of the homogeneous linear equation $y' + p(x)y = 0$ are also solutions?

  (c) Choosing $x_0 \in I$, explain why there is a solution of $L(y) = 0$ satisfying $\phi(x_0) = 1$. Then show that every solution is a scalar multiple of this one. (See Section 20.3.)

This problem shows that the theory of homogeneous linear equations developed in this section has a simpler version in the first-order case. Instead of generating all solutions as linear combinations of *two* independent ones, we use scalar multiples of *one*. Generalizing in the other direction, we may reasonably expect that the solutions of an nth-order homogeneous linear equation are linear combinations of n independent ones.

20.6
CONSTANT COEFFICIENTS

In general the solutions of the homogeneous linear equation

$$L(y) = a(x)y'' + b(x)y' + c(x)y = 0$$

are hard to find. There is one case that turns out to be easy, however; its analysis has been known for more than 200 years. (Euler published it in 1743.) Suppose that the coefficients in $L(y) = 0$ are *constants*. Since e^x is a function that is not changed by differentiation, a good candidate for a solution is $y = e^{rx}$, where r is a constant. When we substitute this in $ay'' + by' + cy = 0$ we find

$$e^{rx}(ar^2 + br + c) = 0$$

as you can check. Hence e^{rx} is a solution if and only if r is a root of the *characteristic equation*

$$ar^2 + br + c = 0$$

□ **Example 1**   Find the general solution of $y'' - 5y' - 6y = 0$.

Solution: Since $a = 1$, $b = -5$, $c = -6$, the characteristic equation is $r^2 - 5r - 6 = 0$, the roots of which are $r_1 = 6$ and $r_2 = -1$. Corresponding solutions of the differential equation are e^{6x} and e^{-x}, which are linearly independent by Problem 11 in the last section. Hence the general solution is

$$\phi(x) = c_1 e^{6x} + c_2 e^{-x}$$

□

This seems too good to be true! Suddenly we have encountered a class of second-order equations that can be solved systematically, simply by finding the roots of a quadratic equation. There are complications, however. The equation $ar^2 + br + c = 0$ has roots

$$r_1 = \frac{-b + \sqrt{b^2 - 4ac}}{2a} \quad \text{and} \quad r_2 = \frac{-b - \sqrt{b^2 - 4ac}}{2a}$$

and these are distinct real numbers (as in Example 1) only if $b^2 - 4ac > 0$. If $b^2 - 4ac = 0$, they are identical and if $b^2 - 4ac < 0$, they are imaginary. Hence we have some work to do.

□ **Example 2**   Find the general solution of $y'' - 4y' + 4y = 0$.

Solution: This time the characteristic equation is $r^2 - 4r + 4 = 0$, with double root $r = r_1 = r_2 = 2$. We can name *one* solution, namely $\phi_1(x) = e^{2x}$, but how do we find another? The answer is to be found in D'Alembert's method described in Problems 18–22 of the last section. A second solution is of the form $\phi_2 = v\phi_1$, where v is an unknown function to be found by substitution in the given differential equation:

$$y = ve^{2x}$$
$$y' = 2ve^{2x} + v'e^{2x}$$
$$y'' = 4ve^{2x} + 4v'e^{2x} + v''e^{2x}$$
$$y'' - 4y' + 4y = (4ve^{2x} + 4v'e^{2x} + v''e^{2x}) - 4(2ve^{2x} + v'e^{2x}) + 4ve^{2x}$$
$$= e^{2x}v'' = 0$$
$$v'' = 0$$
$$v' = 1 \quad \text{(Only one solution is needed, so we omit the arbitrary constant.)}$$
$$v = x \quad \text{(Ditto.)}$$

Hence a second solution is $\phi_2(x) = xe^{2x}$. Linear independence of ϕ_1 and ϕ_2 is easy to confirm (see Problem 12 in the last section), so the general solution is

$$\phi(x) = c_1 e^{2x} + c_2 x e^{2x}$$

□

The result in Example 2 can be generalized to any problem where the characteristic equation has a double root r. One solution is $\phi_1(x) = e^{rx}$ and a second is $\phi_2(x) = xe^{rx}$. Since ϕ_1 and ϕ_2 are independent, the general solution is

$$\phi(x) = c_1 e^{rx} + c_2 x e^{rx}$$

(We ask you to confirm these statements in the problem set.)

◻ **Example 3**   Find the general solution of $y'' - 2y' + 5y = 0$.

Solution: The characteristic equation is $r^2 - 2r + 5 = 0$, with conjugate imaginary roots $r_1 = 1 + 2i$ and $r_2 = 1 - 2i$. (The *conjugate* of a complex number $a + bi$, where a and b are real, is $a - bi$.) If we ignore any difficulties with imaginary exponents, our theory seems to say that

$$e^{r_1 x} = e^{(1+2i)x} \qquad \text{and} \qquad e^{r_2 x} = e^{(1-2i)x}$$

are two solutions of the differential equation. Theorem 5 of Section 20.5, however, guarantees two *real-valued* solutions, so we are not yet done.

To make sense of the above complex-valued functions, we digress to introduce **Euler's Formula,** which says that

$$e^{ix} = \cos x + i \sin x$$

(See Problem 40, Section 8.6, Problem 55, Section 9.3, and Problem 40, Section 13.2. You may be interested in looking at Problems 34–40 in Section 8.6, where a preview of this section was provided.) This formula is a definition of e^{ix}, so it needs no defense. Nevertheless there are good reasons for it, the simplest being the power series manipulation

$$e^{ix} = 1 + ix + \frac{(ix)^2}{2!} + \frac{(ix)^3}{3!} + \cdots$$

$$= \left(1 - \frac{x^2}{2!} + \frac{x^4}{4!} - \cdots\right) + i\left(x - \frac{x^3}{3!} + \frac{x^5}{5!} - \cdots\right)$$

$$= \cos x + i \sin x$$

(See Problem 40, Section 13.2.)

The more general exponential symbol e^{u+vi} is defined by

$$e^{u+vi} = e^u(\cos v + i \sin v)$$

Again there are good reasons for this. If $u = 0$, it reduces to Euler's Formula, and in general it preserves the usual exponential rule $e^a e^b = e^{a+b}$ (Why?)

In view of these agreements we can return to Example 3 and write

$$e^{(1+2i)x} = e^{x+2xi} = e^x(\cos 2x + i \sin 2x) = e^x \cos 2x + ie^x \sin 2x$$

The interesting thing about this expression is that its component parts,

$$\phi_1(x) = e^x \cos 2x \qquad \text{and} \qquad \phi_2(x) = e^x \sin 2x$$

are themselves solutions of the original differential equation. The reason is a theorem to the effect that if the complex-valued function

$$f(x) = g(x) + ih(x)$$

is a solution of $L(y) = 0$, so are its real-valued parts, $g(x)$ and $h(x)$. (See Problem 40, Section 8.6.) Since $e^x \cos 2x$ and $e^x \sin 2x$ are linearly independent (Problem 13, Section 20.5), we have discovered the general solution

$$\phi(x) = c_1 e^x \cos 2x + c_2 e^x \sin 2x \qquad \square$$

This is a remarkable achievement. Neither the original problem nor the final answer involve complex numbers, but the problem was solved by detouring from real-variable calculus into the headier realm of complex analysis. If you go on in mathematics, you will see that this is commonplace; a great deal of insight concerning real-variable theory can be had by learning something about complex variables.

We summarize this section as follows. Suppose that the coefficients in the homogeneous linear equation $L(y) = ay'' + by' + cy = 0$ are constants ($a \neq 0$).

1. If $b^2 - 4ac > 0$, the roots r_1 and r_2 of the characteristic equation $ar^2 + br + c = 0$ are distinct real numbers. Then

$$\phi_1(x) = e^{r_1 x} \qquad \text{and} \qquad \phi_2(x) = e^{r_2 x}$$

are linearly independent solutions of $L(y) = 0$.

2. If $b^2 - 4ac = 0$, the characteristic equation has a double root r. Then

$$\phi_1(x) = e^{rx} \qquad \text{and} \qquad \phi_2(x) = xe^{rx}$$

are linearly independent solutions of $L(y) = 0$.

3. If $b^2 - 4ac < 0$, the roots of the characteristic equation are conjugate imaginary numbers. If $p + qi$ is one of them (where p and q are real, $q \neq 0$), then

$$\phi_1(x) = e^{px} \cos qx \qquad \text{and} \qquad \phi_2(x) = e^{px} \sin qx$$

are linearly independent solutions of $L(y) = 0$.

In any case the general solution of $L(y) = 0$ is

$$\phi(x) = c_1 \phi_1(x) + c_2 \phi_2(x)$$

□ **Example 4**   The *Euler equation* is $ax^2 y'' + bxy' + cy = 0$, where a, b, and c are constants ($a \neq 0$). Show that in $I = (0, \infty)$ the substitution $t = \ln x$ reduces this equation to one with constant coefficients, namely

$$a\frac{d^2 y}{dt^2} + (b - a)\frac{dy}{dt} + cy = 0$$

Solution: If $t = \ln x$, then $dt/dx = 1/x$ and hence

$$y' = \frac{dy}{dx} = \frac{dy}{dt}\frac{dt}{dx} = \frac{1}{x}\frac{dy}{dt}$$

$$y'' = \frac{d}{dx}\left(\frac{dy}{dx}\right) = \frac{d}{dx}\left(\frac{1}{x}\frac{dy}{dt}\right) = \frac{1}{x}\frac{d}{dx}\left(\frac{dy}{dt}\right) - \frac{1}{x^2}\frac{dy}{dt}$$

$$= \frac{1}{x}\frac{d}{dt}\left(\frac{dy}{dt}\right)\frac{dt}{dx} - \frac{1}{x^2}\frac{dy}{dt} = \frac{1}{x^2}\frac{d^2y}{dt^2} - \frac{1}{x^2}\frac{dy}{dt}$$

The equation $ax^2y'' + bxy' + cy = 0$ becomes

$$a\frac{d^2y}{dt^2} - a\frac{dy}{dt} + b\frac{dy}{dt} + cy = a\frac{d^2y}{dt^2} + (b-a)\frac{dy}{dt} + cy = 0$$

As an example of how this result may be used, consider the equation

$$x^2y'' + xy' + 4y = 0$$

in which $a = 1$, $b = 1$, $c = 4$. The substitution $t = \ln x$ changes this to

$$\frac{d^2y}{dt^2} + 4y = 0$$

the characteristic equation of which is $r^2 + 4 = 0$. One of the imaginary roots is $2i$, which yields the linearly independent solutions

$$\cos 2t = \cos(2\ln x) = \cos(\ln x^2)$$

and

$$\sin 2t = \sin(2\ln x) = \sin(\ln x^2)$$

Hence the general solution is

$$\phi(x) = c_1 \cos(\ln x^2) + c_2 \sin(\ln x^2) \qquad \square$$

□ **Example 5**   Suppose that a pendulum bob is suspended from a rigid support by a weightless inflexible wire of length L. Find a differential equation representing its motion if it is pulled aside from its equilibrium position and then released.

Solution: In Figure 1 we have shown the bob at a typical point P of its circular path. Its position may be designated by naming θ $(-\theta_0 \leq \theta \leq \theta_0)$, so we will try to say something about θ as a function of the time t, subject to the initial conditions $\theta = \theta_0$ and $d\theta/dt = 0$ at $t = 0$. (The reason that $d\theta/dt = 0$ at $t = 0$ is that the bob is released, not pushed.)

The forces on the bob (neglecting friction and air resistance) are its weight (the downward pointing vector in the figure) and the tension in the wire (shown as a pull on the bob by the wire). Since the tension has no tangential component, the magnitude of the tangential force on the bob is

$$|\mathbf{F}_T| = |mg \sin \theta| = \pm mg \sin \theta$$

the sign depending on whether $\theta \geq 0$ (as in Figure 1) or $\theta < 0$ (when P is on the other side of O).

Now recall from Section 15.3 that the tangential component of acceleration is $(d^2s/dt^2)\mathbf{T}$, where $\mathbf{T}$ is the unit tangent vector in the direction of

Figure 1
Motion of a pendulum bob

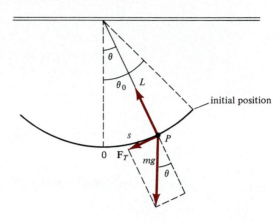

increasing s (counterclockwise in Figure 1). Newton's Second Law implies that the tangential components of force and acceleration are related by

$$\mathbf{F}_T = m\frac{d^2s}{dt^2}\mathbf{T}$$

Since d^2s/dt^2 is positive or negative depending on whether $\mathbf{F}_T$ and $\mathbf{T}$ have the same or opposite directions, we find

$$|\mathbf{F}_T| = \left|m\frac{d^2s}{dt^2}\right| = \mp m\frac{d^2s}{dt^2}$$

the sign depending on whether $\theta \geq 0$ (when $\mathbf{F}_T$ and $\mathbf{T}$ have opposite directions) or $\theta < 0$ (when they have the same direction). Hence in any case we have

$$m\frac{d^2s}{dt^2} = -mg\sin\theta$$

Since $s = L\theta$, we find

$$\frac{ds}{dt} = L\frac{d\theta}{dt} \qquad \text{and} \qquad \frac{d^2s}{dt^2} = L\frac{d^2\theta}{dt^2}$$

so our equation becomes

$$\frac{d^2\theta}{dt^2} = -\frac{g}{L}\sin\theta$$

This differential equation, together with the given initial conditions, represents the motion of the pendulum bob. Note that it is nonlinear (because the dependent variable θ appears in the function $\sin\theta$ rather than in a linear expression). For small values of θ, however, the approximation

$\sin \theta \approx \theta$ is a good one. In that case we may represent the motion by the linear equation

$$\frac{d^2\theta}{dt^2} + \frac{g}{L}\theta = 0$$

which is easy to solve by the methods of this section. Suppose, for example, that we take $\theta_0 = 5°$, $g = 32$ ft/sec^2, and $L = 2$ ft. Then the equation reads

$$\frac{d^2\theta}{dt^2} + 16\theta = 0$$

and its general solution is

$$\theta = c_1 \cos 4t + c_2 \sin 4t$$

(Confirm!) Since

$$\frac{d\theta}{dt} = -4c_1 \sin 4t + 4c_2 \cos 4t$$

the initial conditions $\theta = \pi/36$ and $d\theta/dt = 0$ at $t = 0$ yield

$$\frac{\pi}{36} = c_1 \cos 0 + c_2 \sin 0 \qquad \text{and} \qquad 0 = -4c_1 \sin 0 + 4c_2 \cos 0$$

from which $c_1 = \pi/36$ and $c_2 = 0$. The solution satisfying these conditions is

$$\theta = \frac{\pi}{36} \cos 4t$$

a result which enables us to find (for example) the *period* of the pendulum (the time required for the bob to return to its original position $\theta = \pi/36$). The first time this occurs is when $4t = 2\pi$, so the period is $\pi/2$ sec.   □

Problem Set 20.6

1. Show that e^{rx} is a solution of $ay'' + by' + cy = 0$ if and only if r is a root of $ar^2 + br + c = 0$.

Find the general solution of each of the following differential equations.

2. $y'' - 4y = 0$

3. $y'' - y' - 6y = 0$

4. $y'' - 2y' = 0$ (See Problem 9, Section 20.4.)

5. $y'' - 2y' + y = 0$ (See Problem 18, Section 20.5.)

6. $4y'' - 4y' + y = 0$

7. $y'' + 9y = 0$

8. $y'' - 4y' + 5y = 0$

9. $y'' - 6y' + 25y = 0$

10. $x^2y'' + xy' - 4y = 0$ (See Problem 21, Section 20.5.)

11. Suppose that $k \neq 0$. Show that the general solution of $y'' + k^2y = 0$ is $c_1 \cos kx + c_2 \sin kx$. Then show that the general solution of $y'' - k^2y = 0$ can be written in the form $c_1 \cosh kx + c_2 \sinh kx$,

12. Suppose that the pendulum bob in Example 5 starts at $\theta = 3°$ with an angular velocity $d\theta/dt = -0.5$ radians/sec. If $L = 8$, find θ as a function of t.

13. Suppose that r is a double root of the characteristic equation of $ay'' + by' + cy = 0$. Then $\phi_1(x) = e^{rx}$ is one solution.

  (a) Show by direct substitution that $\phi_2(x) = xe^{rx}$ is another solution. *Hint:* Since r is a root of the characteristic equation, $ar^2 + br + c = 0$. Since it is a double root, $2ar + b = 0$. (Why?)

  (b) Show how this second solution can be *discovered* (not merely checked) by substituting $y = ve^{rx}$ in the equation to find the unknown function v.

14. An interesting alternative to Problem 13 is the following.

(a) Show that for every $t \in \mathcal{R}$,

$$a(e^{tx})'' + b(e^{tx})' + c(e^{tx}) = ae^{tx}(t - r)^2$$

where the primes mean differentiation with respect to x. *Hint:* Since r is a double root of the characteristic equation, $at^2 + bt + c = a(t - r)^2$. (Why?)

(b) Define the function f by

$$f(x,t) = a(e^{tx})'' + b(e^{tx})' + c(e^{tx})$$

Show that

$$f_2(x,t) = a(xe^{tx})'' + b(xe^{tx})' + c(xe^{tx})$$

Hint: Reverse the order of differentiation with respect to x and t.

(c) Conclude from parts (a) and (b) that

$$a(xe^{tx})'' + b(xe^{tx})' + c(xe^{tx})$$
$$= 2ae^{tx}(t - r) + axe^{tx}(t - r)^2$$

Why does it follow that xe^{rx} is a solution of $ay'' + by' + cy = 0$?

15. Prove that if $z = a + bi$ and $w = c + di$ are complex numbers (where a, b, c, and d are real), then $e^z e^w = e^{z+w}$.

16. The derivative of a complex-valued function $f(x) = g(x) + ih(x)$, where g and h are real, is defined to be $f'(x) = g'(x) + ih'(x)$.

(a) Use this definition to prove that $D_x(e^{ix}) = ie^{ix}$ as in real-variable calculus.

(b) More generally, show that $D_x(e^{rx}) = re^{rx}$, where $r = p + qi$ (p and q being real).

20.7 THE NONHOMOGENEOUS CASE

When g is not the zero function, the linear equation

$$L(y) = a(x)y'' + b(x)y' + c(x)y = g(x)$$

is nonhomogeneous. Suppose that we have already found two independent solutions ϕ_1 and ϕ_2 of the homogeneous equation $L(y) = 0$, and by some means have also found *one* solution ϕ_p of the nonhomogeneous equation. Then we can find *all* solutions of the nonhomogeneous equation, as the following argument shows.

Let ϕ be any solution of $L(y) = g$. Then $L(\phi) = g$. Since $L(\phi_p) = g$, we have

$$L(\phi - \phi_p) = L(\phi) - L(\phi_p) \qquad \text{(because of the linearity of } L)$$
$$= g - g = 0$$

This means that $\phi - \phi_p$ is a solution of the homogeneous equation $L(y) = 0$. We know that every such solution is a linear combination of ϕ_1 and ϕ_2,

$$\phi - \phi_p = c_1\phi_1 + c_2\phi_2$$

so we have $\phi = c_1\phi_1 + c_2\phi_2 + \phi_p$.

The general solution of the nonhomogeneous linear equation

$$L(y) = a(x)y'' + b(x)y' + c(x)y = g(x)$$

is $\phi(x) = c_1\phi_1(x) + c_2\phi_2(x) + \phi_p(x)$, where ϕ_1 and ϕ_2 are linearly independent solutions of the homogeneous equation $L(y) = 0$ and ϕ_p is a particular solution of $L(y) = g$.

□ **Example 1**   One solution of $y'' - y = x$ is $\phi_p(x) = -x$, as you can check. What is the general solution? What is the solution satisfying $\phi(0) = 2$ and $\phi'(0) = 1$?

Solution: Linearly independent solutions of the homogeneous equation $y'' - y = 0$ are e^x and e^{-x}. (Why?) Hence the general solution of $y'' - y = x$ is

$$\phi(x) = c_1 e^x + c_2 e^{-x} - x$$

Since $\phi'(x) = c_1 e^x - c_2 e^{-x} - 1$, the given initial conditions yield $2 = c_1 + c_2$ and $1 = c_1 - c_2 - 1$, from which $c_1 = 2$ and $c_2 = 0$. The desired solution is

$$\phi(x) = 2e^x - x$$

□

The problem (assuming that we can solve the homogeneous equation) is to find ϕ_p, a particular solution of the nonhomogeneous equation. There are two approaches to this question, each with its virtues and defects. The first, known as the method of **variation of parameters,** is more general, applying whenever two independent solutions of $L(y) = 0$ are known (regardless of whether the coefficients are constant). But it is often gruesome to carry out. The second, the method of **undetermined coefficients** (or the *annihilator* method), is usually easier when it works, but it applies only in special cases.

Variation of Parameters

Suppose that two independent solutions ϕ_1 and ϕ_2 of the homogeneous linear equation

$$L(y) = a(x)y'' + b(x)y' + c(x)y = 0$$

are known. Then of course the function $\phi = v_1\phi_1 + v_2\phi_2$ is also a solution if v_1 and v_2 are constants. The idea (due to Lagrange) is to try a solution of the *nonhomogeneous* equation $L(y) = g$ in this form, but with v_1 and v_2 as unknown functions to be determined. Let

$$y = \phi_p(x) = v_1\phi_1(x) + v_2\phi_2(x)$$

be this trial solution. In order to determine v_1 and v_2, two conditions must be imposed. One is the fact that ϕ_p is supposed to be a solution of the nonhomogeneous equation, that is, $L(\phi_p) = g$. The other may be imposed more or less arbitrarily; Lagrange's idea was to treat the variables v_1 and v_2 as "pseudo-constants."

To see what we mean, note that

$$y' = v_1\phi_1' + v_1'\phi_1 + v_2\phi_2' + v_2'\phi_2$$

If v_1 and v_2 were true constants, the terms $v_1'\phi_1$ and $v_2'\phi_2$ would not appear.

The condition we impose is that their *sum* drops out, that is,

$$v_1'\phi_1 + v_2'\phi_2 = 0 \qquad (1)$$

This reduces the formula for y' to $y' = v_1\phi_1' + v_2\phi_2'$, from which

$$y'' = v_1\phi_1'' + v_1'\phi_1' + v_2\phi_2'' + v_2'\phi_2'$$

The beauty of this is that no second derivatives of the unknown functions v_1 and v_2 appear. When we substitute in

$$L(y) = ay'' + by' + cy = g$$

we will have only a first-order problem to solve:

$$L(y) = a(v_1\phi_1'' + v_1'\phi_1' + v_2\phi_2'' + v_2'\phi_2') + b(v_1\phi_1' + v_2\phi_2') + c(v_1\phi_1 + v_2\phi_2)$$
$$= v_1(a\phi_1'' + b\phi_1' + c\phi_1) + v_2(a\phi_2'' + b\phi_2' + c\phi_2) + a(v_1'\phi_1' + v_2'\phi_2') = g$$

Since ϕ_1 and ϕ_2 are solutions of $L(y) = 0$, the first two parentheses are zero, so the condition that our trial solution satisfies $L(y) = g$ reduces to

$$a(v_1'\phi_1' + v_2'\phi_2') = g$$

or
$$v_1'\phi_1' + v_2'\phi_2' = g/a \qquad (2)$$

Equations (1) and (2) constitute a system in the unknowns v_1' and v_2', namely

$$\phi_1 v_1' + \phi_2 v_2' = 0$$
$$\phi_1' v_1' + \phi_2' v_2' = g/a$$

A remarkable aspect of this sytem is that its determinant of coefficients is the Wronskian of ϕ_1 and ϕ_2,

$$W = \begin{vmatrix} \phi_1 & \phi_2 \\ \phi_1' & \phi_2' \end{vmatrix} \quad (!)$$

Since ϕ_1 and ϕ_2 are linearly independent solutions of $L(y) = 0$, the Wronskian is nowhere zero (Theorem 3, Section 20.5), so we know that the system can be solved by Cramer's Rule. In other words we can find v_1' and v_2'. Integration will yield v_1 and v_2, so the method works.

☐ **Example 2**   Find a particular solution of the nonhomogeneous linear equation $y'' + y = x$.

Solution: Linearly independent solutions of the homogeneous equation $y'' + y = 0$ are $\cos x$ and $\sin x$, so our trial solution has the form

$$\phi_p(x) = v_1 \cos x + v_2 \sin x$$

where v_1 and v_2 are unknown functions satisfying the system

$$(\cos x)\, v_1' + (\sin x)\, v_2' = 0$$
$$(-\sin x)\, v_1' + (\cos x)\, v_2' = x$$

The determinant of coefficients is the Wronskian

$$W(x) = \begin{vmatrix} \cos x & \sin x \\ -\sin x & \cos x \end{vmatrix} = \cos^2 x + \sin^2 x = 1$$

Cramer's Rule (Section 20.5) yields

$$v_1' = \frac{\begin{vmatrix} 0 & \sin x \\ x & \cos x \end{vmatrix}}{W(x)} = -x \sin x, \quad v_2' = \frac{\begin{vmatrix} \cos x & 0 \\ -\sin x & x \end{vmatrix}}{W(x)} = x \cos x$$

Integration by parts gives

$$v_1 = x \cos x - \sin x, \ v_2 = x \sin x + \cos x$$

(We omit arbitrary constants because only one solution is needed.) Hence our trial solution is

$$\phi_p(x) = v_1 \cos x + v_2 \sin x$$
$$= (x \cos x - \sin x) \cos x + (x \sin x + \cos x) \sin x = x$$

This is a case of a mountain laboring and bringing forth a mouse, but we are not trying to be efficient; the idea is to illustrate Lagrange's method. The general solution of $y'' + y = x$, by the way, is now known:

$$\phi(x) = c_1 \cos x + c_2 \sin x + x \qquad \square$$

□ **Example 3**   Find a particular solution of $y'' - y = \tan x$ in the interval $I = (-\pi/2, \pi/2)$.

Solution: Linearly independent solutions of $y'' - y = 0$ are e^x and e^{-x}, so we try a solution of $y'' - y = \tan x$ in the form

$$\phi_p(x) = v_1 e^x + v_2 e^{-x}$$

where v_1 and v_2 satisfy

$$e^x v_1' + e^{-x} v_2' = 0$$
$$e^x v_1' - e^{-x} v_2' = \tan x$$

The determinant of coefficients is

$$\begin{vmatrix} e^x & e^{-x} \\ e^x & -e^{-x} \end{vmatrix} = -2$$

so Cramer's Rule yields

$$v_1' = -\frac{1}{2} \begin{vmatrix} 0 & e^{-x} \\ \tan x & -e^{-x} \end{vmatrix} = \frac{1}{2} e^{-x} \tan x$$
$$v_2' = -\frac{1}{2} \begin{vmatrix} e^x & 0 \\ e^x & \tan x \end{vmatrix} = -\frac{1}{2} e^x \tan x$$

Antiderivatives are hard to find in this case; instead we write

$$v_1 = \frac{1}{2} \int_0^x e^{-t} \tan t \, dt, \qquad v_2 = -\frac{1}{2} \int_0^x e^t \tan t \, dt$$

Our trial solution is

$$\phi_p(x) = \frac{1}{2} e^x \int_0^x e^{-t} \tan t \, dt - \frac{1}{2} e^{-x} \int_0^x e^t \tan t \, dt$$

which is not very good-looking, but nevertheless it works. □

As you can see, Lagrange's method of variation of parameters may be messy. Its virtue is that it is infallible.

Undetermined Coefficients

Unlike the method of variation of parameters, this one is restricted to the case of constant coefficients. Moreover, the function g on the right side of $L(y) = g$ must be of a special form (fortunately one that occurs frequently in applications). Namely, g should be a solution of some homogeneous linear equation with constant coefficients, say $M(y) = 0$ (where M is a linear operator like L, but not necessarily second-order).

The implications of this statement about g may not be clear, because we have not discussed higher-order equations. What it amounts to is that g should appear among a "shopping list" of functions that arise when such equations are solved. Suppose, for example, that $M(y) = 0$ is the third-order equation $y''' + y' = 0$. This is of the homogeneous linear form

$$ay''' + by'' + cy' + dy = 0 \qquad \text{with } a = 1, \, b = 0, \, c = 1, \, d = 0$$

its characteristic equation being $r^3 + r = 0$ or $r(r^2 + 1) = 0$. Just as in the second-order case, the roots 0 and i lead to the independent solutions 1, $\cos x$, $\sin x$. More generally, given the nth-order homogeneous linear equation $M(y) = 0$ (with constant coefficients), we expect its characteristic equation to have n roots (real numbers r, pure imaginary numbers qi, complex numbers $p + qi$—any one of which may be a multiple root). Corresponding solutions of the differential equation are of the form

$$e^{rx}, \ \cos qx, \ \sin qx, \ e^{px} \cos qx, \ e^{px} \sin qx$$

any one of which must be multiplied by $x, x^2, x^3, \ldots$ if it corresponds to a multiple root. The equation $y''' - 3y'' + 3y' - y = 0$, for example, has characteristic equation

$$r^3 - 3r^2 + 3r - 1 = (r - 1)^3 = 0$$

with triple root $r = 1$. The corresponding solutions are e^x, xe^x, $x^2 e^x$.

To put it concisely, we are assuming that $g(x)$ is of the form

$$x^m e^{rx}, \ x^m e^{px} \cos qx, \ x^m e^{px} \sin qx$$

where m is a nonnegative integer ($m = 0, 1, 2, \ldots$) and p and q are real (or a linear combination of such functions). This is what it means to say that g is a solution of $M(y) = 0$, where M is an nth-order linear operator with constant coefficients.

To see why this assumption about g is helpful, consider what it tells us about the solution ϕ_p of $L(y) = g$ that we are trying to find. Since $L(\phi_p) = g$ and $M(g) = 0$, we have

$$M[L(\phi_p)] = M(g) = 0$$

which means that ϕ_p is itself a solution of a homogeneous linear equation with constant coefficients, namely $M[L(y)] = 0$. Thus ϕ_p must be a linear combination of the independent solutions of this equation, and they are functions of the type described above. The problem is to discover the right linear combination; by substituting one with "undetermined coefficients" into $L(y) = g$, we find out.

You can see where this method gets its name. It is also called the "annihilator" method because the linear operator $M(y)$ *annihilates* the nonhomogeneous term g in $L(y) = g$, that is, $M(g) = 0$.

□ **Example 4**   Find a solution of the nonhomogeneous linear equation $y'' - y' - 6y = 2 \sin x$.

Solution: We begin by figuring out an annihilator of $g(x) = 2 \sin x$. The function $\sin x$ arises from a characteristic polynomial with root i (and its conjugate $-i$), that is, a polynomial with factors

$$(r - i)(r + i) = r^2 + 1$$

The corresponding linear operator is $M(y) = y'' + y$, which "annihilates" $\sin x$ because $M(\sin x) = 0$. According to the above discussion, every solution ϕ_p of

$$L(y) = y'' - y' - 6y = 2 \sin x$$

is also a solution of

$$M[L(y)] = M(y'' - y' - 6y) = (y'' - y' - 6y)'' + (y'' - y' - 6y)$$
$$= y^{(4)} - y''' - 5y'' - y' - 6y = 0$$

The characteristic polynomial of this equation is

$$r^4 - r^3 - 5r^2 - r - 6 = (r^2 + 1)(r^2 - r - 6)$$

the product of the characteristic polynomials of M and L. This observation makes it unnecessary to go through the above computation of $M[L(y)]$. Instead we simply form the product

$$(r^2 + 1)(r^2 - r - 6) = (r^2 + 1)(r - 3)(r + 2)$$

and list its roots i, $-i$, 3, -2. The corresponding solutions of $M[L(y)] = 0$ are $\cos x$, $\sin x$, e^{3x}, e^{-2x} and hence ϕ_p must be a linear combination of these. The general form is

$$\phi_p(x) = c_1 \cos x + c_2 \sin x + c_3 e^{3x} + c_4 e^{-2x}$$

but the last two terms may be dropped because e^{3x} and e^{-2x} are solutions of $L(y) = 0$. To see why, observe that substitution of ϕ_p in $L(y) = g$ involves the computation

$$L(c_1 \cos x + c_2 \sin x + c_3 e^{3x} + c_4 e^{-2x})$$
$$= c_1 L(\cos x) + c_2 L (\sin x) + c_3 L(e^{3x}) + c_4 L(e^{-2x})$$

Since $L(e^{3x}) = 0$ and $L(e^{-2x}) = 0$, the last two terms drop out.

Thus the linear combination whose coefficients we will try to determine is

$$\phi_p(x) = c_1 \cos x + c_2 \sin x$$

Substituting

$$y = c_1 \cos x + c_2 \sin x$$
$$y' = -c_1 \sin x + c_2 \cos x$$
$$y'' = -c_1 \cos x - c_2 \sin x$$

in the equation $L(y) = y'' - y' - 6y = 2 \sin x$, we have

$$(-c_1 \cos x - c_2 \sin x) - (-c_1 \sin x + c_2 \cos x) - 6(c_1 \cos x + c_2 \sin x)$$
$$= 2 \sin x$$

from which

$$-(7c_1 + c_2) \cos x + (c_1 - 7c_2) \sin x = 0 \cos x + 2 \sin x$$

Since $\cos x$ and $\sin x$ are independent, we may equate corresponding coefficients to find

$$-(7c_1 + c_2) = 0$$
$$c_1 - 7c_2 = 2$$

The solution of this system yields the "undetermined coefficients" $c_1 = \frac{1}{25}$ and $c_2 = -\frac{7}{25}$, so the solution we sought is

$$\phi_p(x) = \tfrac{1}{25} \cos x - \tfrac{7}{25} \sin x$$

Looking back over the discussion of Example 4, you can see that all we really did was to annihilate the right side of

$$L(y) = y'' - y' - 6y = 2 \sin x$$

by $M(y) = y'' + y$, which led us to try a solution of the form

$$\phi_p(x) = c_1 \cos x + c_2 \sin x$$

The rest is window dressing. There are occasional complications (see Example 7), but basically the method is simple (and ingenious).   □

□ **Example 5**   Find a solution of $y'' - y' - 6y = x^2$.

Solution: The function $g(x) = x^2$ arises from a characteristic polynomial with triple root 0, that is, a polynomial with factors

$$(r - 0)(r - 0)(r - 0) = r^3$$

This root produces solutions

$$e^{0x} = 1, \ xe^{0x} = x, \ x^2 e^{0x} = x^2$$

of the corresponding homogeneous equation $M(y) = y''' = 0$. (Note that M annihilates x^2. Why?) Hence we form the product of the characteristic polynomials of M and L, namely

$$r^3(r^2 - r - 6) = r^3(r - 3)(r + 2)$$

Its roots are 0, 0, 0, 3, −2, so ϕ_p must be of the form

$$\phi_p(x) = c_1 + c_2 x + c_3 x^2 + c_4 e^{3x} + c_5 e^{-2x}$$

Again we may drop the last two terms, so our trial solution is

$$\phi_p(x) = c_1 + c_2 x + c_3 x^2$$

Substituting
$$y = c_1 + c_2 x + c_3 x^2$$
$$y' = c_2 + 2c_3 x$$
$$y'' = 2c_3$$

in $L(y) = y'' - y' - 6y = x^2$, we find

$$(-6c_1 - c_2 + 2c_3) + (-6c_2 - 2c_3)x + (-6c_3)x^2 = 0 + 0x + x^2$$

from which
$$-6c_1 - c_2 + 2c_3 = 0$$
$$-6c_2 - 2c_3 = 0$$
$$-6c_3 = 1$$

(because the functions 1, x, x^2 are linearly independent). The solution of this system is $c_1 = -\frac{7}{108}$, $c_2 = \frac{1}{18}$, $c_3 = -\frac{1}{6}$, so

$$\phi_p(x) = -\tfrac{7}{108} + \tfrac{1}{18}x - \tfrac{1}{6}x^2 \qquad\qquad □$$

□ **Example 6**   Find a solution of $y'' - y' - 6y = xe^{2x}$.

Solution: An annihilator of $g(x) = xe^{2x}$ is the linear operator M whose characteristic polynomial has the double root 2. (Why?) The polynomial is $(r - 2)^2$, so we form the product

$$(r - 2)^2(r^2 - r - 6) = (r - 2)^2(r - 3)(r + 2)$$

with roots 2, 2, 3, −2 and corresponding solutions e^{2x}, xe^{2x}, e^{3x}, e^{-2x}. Ignoring the last two, we try

$$\phi_p(x) = c_1 e^{2x} + c_2 xe^{2x}$$

which when substituted in $L(y) = g$ gives

$$(-4c_1 + 3c_2)e^{2x} + (-4c_2)xe^{2x} = 0e^{2x} + xe^{2x}$$

Equating coefficients of the independent functions e^{2x} and xe^{2x}, we have

$$-4c_1 + 3c_2 = 0$$
$$-4c_2 = 1$$

from which $c_1 = -\frac{3}{16}$ and $c_2 = -\frac{1}{4}$. Hence

$$\phi_p(x) = -\tfrac{3}{16}e^{2x} - \tfrac{1}{4}xe^{2x} \qquad \square$$

These examples may leave the impression that forming the product of the characteristic polynomials of M and L is a waste of time, since we always drop the solutions of $L(y) = 0$. Why not work with the annihilator alone? Our last example shows why it is important (in general) to follow the procedure suggested.

□ **Example 7**   Find a solution of $y'' - y' - 6y = e^{3x}$.

Solution: This time an annihilator of $g(x) = e^{3x}$ is L itself, since e^{3x} is a solution of $L(y) = 0$. A simpler one, however, is the first-order linear operator $M(y) = y' - 3y$, because e^{3x} arises from the simple root 3 of the characteristic equation $r - 3 = 0$. The characteristic polynomial of $M[L(y)] = 0$ is

$$(r - 3)(r^2 - r - 6) = (r - 3)^2(r + 2)$$

with roots $3, 3, -2$, so our trial solution is

$$\phi_p(x) = c_1 e^{3x} + c_2 x e^{3x} + c_3 e^{-2x}$$

The terms that can be dropped are the first and last. (Why?) Substituting $\phi_p(x) = c_2 x e^{3x}$ into $L(y) = g$, we find

$$\phi_p(x) = \tfrac{1}{5}x e^{3x}$$

as you can check. $\qquad \square$

These examples should give you the idea. The essential steps are as follows.

1. Find an annihilator of $g(x)$, the right side of the equation $L(y) = g$ to be solved. This can be done if g is a function of the type described in our opening remarks.
2. Form the product of the characteristic polynomials of M and L, where M is the annihilator found in the first step. List the roots of this product, and the corresponding solutions of the homogeneous equation $M[L(y)] = 0$.
3. Form a linear combination of the solutions listed in the preceding step, dropping whichever terms are solutions of $L(y) = 0$. The result is the trial solution ϕ_p of $L(y) = g$.
4. Substitute ϕ_p in $L(y) = g$ and find its coefficients from the result.

Problem Set 20.7

Use the method of variation of parameters to find a particular solution of each of the following differential equations.

1. $y'' - y' = e^x$ **2.** $y'' - y = xe^x$

3. $y'' + y = x^2$

4. $y'' + y = \sec x$ in $I = (-\pi/2, \pi/2)$

5. $y'' + 4y = \sin x$

6. $y'' - 2y' + y = e^x$

7. $y'' - 3y' + 2y = 2$

8. $x^2 y'' + xy' - 4y = x^3$ in $I = (0, \infty)$ (Independent solutions of the homogeneous equation are x^2 and x^{-2}.)

9. $x^2 y'' - 2xy' + 2y = x^2$ in $I = (0, \infty)$ (Independent solutions of the homogeneous equation are x and x^2.)

10. $xy'' - (x + 1)y' + y = x^2$ (Independent solutions of the homogeneous equation are e^x and $x + 1$.)

Use the method of undetermined coefficients to find a particular solution of each of the following differential equations.

11. $y'' - y' = e^x$ **12.** $y'' - y = xe^x$

13. $y'' + y = x^2$ **14.** $y'' + 4y = \sin x$

15. $y'' + y = \sin x$ **16.** $y'' - 2y' + y = x^2$

17. $y'' - 2y' + y = e^x$ **18.** $y'' - 3y' + 2y = 2$

19. $y'' - y' - 6y = x \cos x$

Solve each of the following initial value problems.

20. $y'' - y' = e^x$, $\phi(0) = 1$, $\phi'(0) = -2$

21. $y'' + y = \sec x$, $\phi(0) = 2$, $\phi'(0) = -1$

22. $y'' - 2y' + y = e^x$, $\phi(0) = 3$, $\phi'(0) = 0$

23. $x^2 y'' + xy' - 4y = x^3$, $\phi(1) = 0$, $\phi'(1) = 1$
(See Problem 8.)

24. $x^2 y'' - 2xy' + 2y = x^2$, $\phi(1) = 2$, $\phi'(1) = -1$
(See Problem 9.)

25. Why can't the annihilator method be used to solve $y'' - y = \tan x$? Can it be used to solve $y'' - y = \cosh x$?

26. Show that the general solution of the linear equation $L(y) = g_1 + g_2$ is the sum of a particular solution of $L(y) = g_1$, a particular solution of $L(y) = g_2$, and the general solution of $L(y) = 0$.

27. Use Problem 26 to solve $y'' + y = x + e^x$ by the annihilator method.

28. Let $P(r)$ be the characteristic polynomial of the linear operator $L(y)$, and suppose that k is not one of its roots. Prove that

$$\phi_p(x) = \frac{ce^{kx}}{P(k)}$$

is a solution of $L(y) = ce^{kx}$.

29. Show that every solution of $y'' + k^2 y = \cos kx$ is unbounded as $x \to \infty$. (This always happens if the nonhomogeneous term is a sine or cosine which is a solution of the homogeneous equation. In physics it corresponds to an applied frequency in "resonance" with a natural frequency of the system.)

20.8

VIBRATIONS (Optional)

In Section 7.5 we mentioned *Hooke's Law* for an elastic spring, which says that when a spring is stretched or compressed (within certain limits) it exerts a restoring force proportional to the displacement. If the displacement is x (as shown in Figure 1), Hooke's Law may be written in the form

$$\textit{Restoring force} = -kx, \qquad \text{where } k \text{ is a positive constant}$$

Newton's Second Law (force equals mass times acceleration) reduces this to the homogeneous linear equation

$$m \frac{d^2 x}{dt^2} + kx = 0$$

where m is the mass and t is time.

The characteristic equation of this differential equation is

$$mr^2 + k = 0, \text{ with roots } r = \pm \omega i, \qquad \text{where } \omega = \sqrt{\frac{k}{m}}$$

Figure 1(a)
Coiled spring in natural position ($x = 0$)

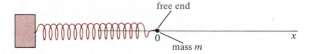

Figure 1(b)
Coiled spring in stretched position ($x > 0$)

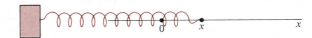

Figure 1(c)
Coiled spring in compressed position ($x < 0$)

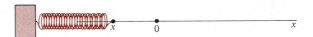

Hence the general solution is

$$x = a \cos \omega t + b \sin \omega t = A \cos (\omega t - \phi) \tag{1}$$

where $A = \sqrt{a^2 + b^2}$ and ϕ satisfies

$$\cos \phi = \frac{a}{A} \text{ and } \sin \phi = \frac{b}{A}$$

The motion described by (1) is called *simple harmonic motion,* with *amplitude A, period $T = 2\pi/\omega$,* and *phase angle ϕ.* It represents the simplest kind of vibration (or oscillation), with fixed amplitude and frequency (the reciprocal of the period T), and indefinite duration.

Simple harmonic motion is unrealistic because it leaves out of account all forces on the mass except that exerted by the spring. Internal friction and fluid resistance (air or liquid depending on the medium surrounding the spring), together with "damping" like that supplied by the shock absorbers on a car, may be combined into a single *damping force* that we take to be proportional to the velocity (and opposed to it). In other words we assume that

$$Damping \ force = -c \frac{dx}{dt}, \qquad \text{where } c \text{ is a positive constant}$$

The total force on the mass m is then

$$Restoring \ force + Damping \ force = -kx - c \frac{dx}{dt}$$

and Newton's Law leads to the equation

$$m \frac{d^2x}{dt^2} + c \frac{dx}{dt} + kx = 0 \tag{2}$$

The characteristic equation is $mr^2 + cr + k = 0$, with roots

$$r_1 = \frac{-c + \sqrt{c^2 - 4km}}{2m} \quad \text{and} \quad r_2 = \frac{-c - \sqrt{c^2 - 4km}}{2m}$$

We shall discuss the motion in three cases.

(i) (*Overdamping*) Suppose that $c^2 - 4km > 0$. Then r_1 and r_2 are distinct real numbers and the general solution of (2) is

$$x = c_1 e^{r_1 t} + c_2 e^{r_2 t}$$

Since r_1 and r_2 are negative (why?), x approaches 0 as $t \to \infty$. Thus in this case the motion is not oscillatory and the mass returns to its equilibrium position rather rapidly (unless r_1 and r_2 are nearly zero). We say that the motion is "overdamped" because the damping force is so large in comparison to the restoring force of the spring that it prevents vibration.

(ii) (*Critical damping*) Suppose that $c^2 - 4km = 0$. Then $r_1 = r_2 = -c/2m$ and the general solution of (2) is

$$x = c_1 e^{rt} + c_2 t e^{rt}, \quad \text{where } r = -\frac{c}{2m} < 0$$

Again there is no oscillation and $x \to 0$ as $t \to \infty$. (Why?) We say that the motion is "critically damped" because if the damping force is slightly decreased, we are into the next case (in which vibrations occur).

(iii) (*Underdamping*) Suppose that $c^2 - 4km < 0$. Then the roots of the characteristic equation are imaginary,

$$r_1 = \alpha + \omega i \quad \text{and} \quad r_2 = \alpha - \omega i,$$

$$\text{where } \alpha = -\frac{c}{2m} \quad \text{and} \quad \omega = \sqrt{\frac{k}{m} - \frac{c^2}{4m^2}}$$

The general solution of (2) is

$$\begin{aligned} x &= a e^{\alpha t} \cos \omega t + b e^{\alpha t} \sin \omega t = e^{\alpha t}(a \cos \omega t + b \sin \omega t) \\ &= A e^{\alpha t} \cos (\omega t - \phi) \end{aligned}$$

where A and ϕ are defined as in the case of simple harmonic motion.

Note that if $c = 0$ (so that there is no damping force), we have $\alpha = 0$ and $\omega = \sqrt{k/m}$. The general solution is then $x = A \cos (\omega t - \phi)$ and we are back to simple harmonic motion. In general, however, $c > 0$ and $\alpha < 0$, which means that $x \to 0$ as $t \to \infty$. Thus the mass eventually returns to its equilibrium position as in cases (i) and (ii). The difference is that oscillation occurs indefinitely because of the term $\cos (\omega t - \phi)$ in the general solution. Hence (iii) is sometimes called the case of "damped vibration."

The above analysis is based on the *homogeneous* equation

$$m \frac{d^2x}{dt^2} + c \frac{dx}{dt} + kx = 0$$

Suppose, however, that the mass is subject to an additional external force (such as a push each time it reaches the equilibrium position). Denoting this force by $g(t)$, we have

Restoring force + Damping force + External force $= -kx - c \dfrac{dx}{dt} + g(t)$

and hence our differential equation is *nonhomogeneous*,

$$m \frac{d^2x}{dt^2} + c \frac{dx}{dt} + kx = g(t)$$

The analysis of this equation is more complicated (depending on the function g), and is more properly left for a course in differential equations. It is interesting, however, involving such things as "beats" (as in the amplitude modulation used by AM radio stations) and "resonance" (mentioned in Problem 29 of the last section).

It is also worth mentioning that if Q is the charge at time t in an electric circuit with inductance L, resistance R, and capacitance C, Kirchhoff's Law (from physics) leads to the differential equation

$$L \frac{d^2Q}{dt^2} + R \frac{dQ}{dt} + \frac{Q}{C} = E(t)$$

where $E(t)$ is the applied electromotive force. This is mathematically indistinguishable from the above equation for the motion of a spring, which means that analysis of one applies equally well to the other. Note the striking analogy of the two sets of quantities (electrical and mechanical):

Charge Q	Position x
Inductance L	Mass m
Resistance R	Damping constant c
Inverse capacitance $1/C$	Spring constant k
Electromotive force $E(t)$	External force $g(t)$

Perhaps it is appropriate to end this book with such a remarkable example of the unifying power of mathematics.

Problem Set 20.8

1. A 4-lb force is known to stretch a certain spring 2 ft from its natural length. If the spring is stretched 3 ft from its natural length and released, find the displacement of its free end t sec later. What are the amplitude, period, and phase angle in this motion? (Take $g = 32$; then $m = \frac{1}{8}$.)

2. If the spring in Problem 1 is stretched 3 ft from its natural length and given a push that starts it back toward the equilibrium position at 4 ft/sec, find the displacement of its free end t sec later. What are the amplitude, period, and phase angle?

3. A spring with mass $m = 1$ kg and damping constant $c = 4$ is stretched 1 meter by a force of 13 newtons. If it is compressed $\frac{1}{2}$ meter from its equilibrium position and released, what is the displacement of its free end t sec later? What is the period of its damped vibration?

4. A spring with mass $m = 1$ kg and damping constant $c = 5$ is stretched 1 meter by a force of 6 newtons. If it is set in motion from this position with an initial velocity of 2 meters per sec (away from the equilibrium position), what is the displacement of its free end t sec later? Does the free end ever pass through its equilibrium position?

True–False Quiz

1. The equation $2xy\,dx - (x^2 + y^2)\,dy = 0$ is exact.

2. Separable first-order equations can be written in a form that is exact.

3. The equation $y' + 2xy = e^{-x^2}$ is linear.

4. An integrating factor of the linear equation $xy' + y = 2$ is e^x.

5. The interval of validity of solutions of a linear differential equation may change if the initial conditions are changed.

6. If ϕ is a solution of $y' + p(x)y = 0$, so is $c\phi$, where c is any constant.

7. If ϕ_1 and ϕ_2 are solutions of $y'' + y = x^2$, so is $\phi_1 + \phi_2$.

8. If r is a double root of the characteristic equation associated with $ay'' + by' + cy = 0$, then e^{rx} and xe^{rx} are linearly independent solutions of the differential equation.

9. If ϕ and ϕ_p are solutions of the linear second-order equation $L(y) = g$, then $\phi - \phi_p$ is a solution of $L(y) = 0$.

10. The method of annihilators can be used to find a solution of $y'' + y = \sinh x$.

Additional Problems

1. Confirm that $\phi(x) = 2$ and $\psi(x) = 2\cos x$ are both solutions of $y' = \sqrt{4 - y^2}$ satisfying the initial condition $y = 2$ when $x = 0$. Why doesn't this contradict the Uniqueness Theorem for first-order differential equations?

2. The temperature T of a hot object t minutes after being placed in a room at $25°C$ satisfies the differential equation $dT/dt = -3(T - 25)$. If $T = 30$ when $t = 10$, find T as a function of t. What does T approach as time goes on?

Solve the following initial value problems.

3. $y' = \sqrt{4 - y}$, $\phi(2) = 0$ **4.** $y' = \sqrt{4 - y}$, $\phi(2) = 4$

5. $y' = e^y$, $\phi(1) = 0$ **6.** $y' = 4 + y^2$, $\phi(0) = 0$

7. $y' = \sqrt{1 - y^2}$, $\phi(0) = \frac{1}{2}$ **8.** $y' = 3x^2y$, $\phi(0) = 0$

9. $y' = 2x(1 - y^2)$, $\phi(0) = 2$

10. Find the general solution of $y' = 1 - y$ by treating the equation as separable.

11. Do Problem 10 by treating the equation as linear.

Solve the following initial value problems.

12. $y' = \dfrac{x^2 - y^2}{2xy}$, $\phi(2) = 1$

13. $y' = -\dfrac{3x^2 + 2y}{2x + 4y^3}$, $\phi(0) = 2$

14. $y' = \dfrac{y - 3}{y^2 - x}$, $\phi(1) = 0$

15. $y' - 2xy = x$, $\phi(1) = 0$

16. $y' - y = \cos x$, $\phi(0) = 2$

17. $y' + \dfrac{2xy}{1 + x^2} = \dfrac{1}{1 + x^2}$, $\phi(1) = 1$

18. $y' + y \cot x = 2 \cos x$, $\phi(\pi/2) = 1$

19. $y' - y \tan x = \sec x$, $\phi(0) = 0$

20. $y' - y/x = x$, $\phi(1) = 0$

Find the general solution of each of the following differential equations.

21. $y' - y = e^x$

22. $y' + y \sin x = \sin x$

23. $y' - 2y/x = x^2 \ln x$

24. $y' + 3x^2y = x^2$

25. $xy' + y = x^3$

26. Solve the initial value problem $y'' = y'$, $\phi(1) = 0$, $\phi'(1) = 1$.

27. Find the solution of $y'' = 3y^2$ satisfying $\phi(1) = 2$ and $\phi'(1) = -4$. In what interval is it valid?

28. Find all solutions of the differential equation $y'' + y' = e^{-x}$.

29. If $L(y) = 0$ is a linear second-order differential equation, and ϕ_1 and ϕ_2 are solutions, show that $\phi_1 + \phi_2$ is also a solution.

30. If $L(y) = 0$ is a linear second-order differential equation, and ϕ is a solution, show that $k\phi$ is also a solution (where k is an arbitrary constant).

In each of the following, determine whether the given functions are linearly independent in the given interval.

31. $\tan x$ and $\cot x$ in $I = (0, \pi/2)$

32. 1 and x in $I = (-\infty, \infty)$

33. e^x and xe^x in $I = (-\infty, \infty)$

34. e^x and $2e^x$ in $I = (-\infty, \infty)$

35. $\cosh x$ and $e^x + e^{-x}$ in $I = (-\infty, \infty)$

36. Are the functions x^2 and $x^2 - 1$ linearly independent in the interval $(-\infty, \infty)$? in the interval $(0, \infty)$?

37. Find the Wronskian of $\phi_1(x) = x$ and $\phi_2(x) = e^x$. Then explain why it is impossible to name a linear, homogeneous second-order differential equation satisfied by ϕ_1 and ϕ_2 in the interval $(0, \infty)$.

Find the general solution of each of the following differential equations.

38. $y'' + 2y' - 15y = 0$

39. $y'' - 4y' = 0$

40. $y'' - 8y' + 25y = 0$

41. $y'' + 6y' + 9y = 0$

42. $y'' - 2y' + 10y = 0$

43. $2y'' - 3y' + y = 0$

44. $y'' - 6y' + 10y = 0$

Solve each of the following initial value problems.

45. $y'' + 2y' = 0$, $\phi(0) = 2$, $\phi'(0) = 0$.

46. $y'' + 2y' + y = 0$, $\phi(0) = 0$, $\phi'(0) = 1$.

47. $y'' + 4y = 0$, $\phi(0) = 1$, $\phi'(0) = 1$.

Find one solution of each of the following differential equations. Then give the general solution.

48. $y'' + y = x^2$

49. $y'' - y = \cos x$

50. $y'' + y' - 6y = e^x$

51. $y'' - 4y = \sinh x$

52. $y'' + y = x^2 - 1$

53. In Problem 52 what solution passes through the point $(0,2)$ with slope 3?

54. Two linearly independent solutions of the homogeneous equation associated with $x^2y'' - 3xy' + 3y = x^2$ are x and x^3. Use that fact, together with the method of variation of parameters, to find a particular solution of the given equation.

55. In Problem 54 what solution satisfies the initial conditions $\phi(1) = 0$ and $\phi'(1) = 1$?

56. In Problem 54 find all solutions satisfying $\phi(0) = 0$ and $\phi'(0) = 0$. Why are there infinitely many such solutions? Why doesn't this contradict the Uniqueness Theorem?

57. In Problem 54 find all solutions satisfying $\phi(0) = 1$ and $\phi'(0) = 0$. Why doesn't the result contradict the Existence Theorem?

Answers to Selected Odd-Numbered Problems
(and all True-False Quizzes)

Section 1.1

1. $(-\infty, \frac{2}{5})$ **3.** $(-\infty, 1)$ **5.** all $x \neq 3$ **7.** $(1, \infty)$
11. $\pi - 3$ **13.** 9   **15.** $\sqrt[3]{3} - \sqrt{2}$ **17.** 0   **19.** 2
21. $\frac{1}{3}$ **23.** $3\sqrt{2}/4$ **27.** $[-2, \frac{8}{3}]$ **29.** union of
$(-\infty, -\frac{1}{5})$ and $(1, \infty)$ **31.** $[-2, 7]$ **33.** union of
$(-\infty, -\frac{9}{5}]$ and $[-\frac{7}{5}, \infty)$ **35.** $\{0, -8\}$ **39.** $(-2, 2)$
41. union of $(-\infty, -4)$ and $(4, \infty)$ **43.** $[-2, 8]$
45. $(-2, 4)$ **47.** union of $(-\infty, -1)$ and $(1, \infty)$
49. empty set; $\{3\}$ **53.** irrational

Section 1.2

1. $\sqrt{13}; (\frac{1}{2}, 4); \frac{2}{3}$ **3.** $\sqrt{5}; (\frac{1}{2}\sqrt{2}, \frac{1}{2}\sqrt{3}); -\frac{1}{2}\sqrt{6}$
5. $4; (2, 1);$ undefined   **9.** yes   **11.** 3   **13.** $\frac{2}{3}$
15. $-a/b$ **17.** $y = x$ **19.** $x + 2y - 1 = 0$
21. $y = 3x + 1$ **23.** $2x - y - 2 = 0$ **25.** $y = x + 1$
27. $2x - y - 6 = 0$ **29.** $y = -2x + 5$ **31.** $x = 2$
33. $y = 1$ **35.** $y = -3x + 4$ **37.** $2x - y + 2 = 0$
39. $3x - 4y - 25 = 0$ **41.** $y = \sqrt{3}x - 1$ and $y = -\sqrt{3}x - 1$
45. $45°$ **53.** $0 \cdot \infty = -1$, which is nonsense
55. $7/\sqrt{13}$ **57.** $1/\sqrt{2}$ **59.** $\sqrt{5}$ **61.** 0   **63.** 4

Section 1.3

1. $25; 5$ **3.** $\frac{25}{4}; \frac{5}{2}$ **5.** $36; 6$ **7.** circle with center
$(1, -2)$ and radius $2; (x - 1)^2 + (y + 2)^2 = 4$
9. circle with center $(0, 3)$ and radius $\sqrt{2};$
   $x^2 + (y - 3)^2 = 2$
11. circle with center $(0, 0)$ and radius 3   **13.** circle
with center $(-\frac{3}{2}, 0)$ and radius $\frac{1}{2}$ **15.** circle with center
$(-2, 1)$ and radius $1/\sqrt{3}$ **17.** straight line $y = x$

19. left half of circle with center $(0, 0)$ and radius 1
21. all points of the coordinate plane outside or on the
circle with center $(0, 0)$ and radius 1   **23.** $(x - 1)^2 +$
$(y + 1)^2 = 8$ **25.** $x^2 + (y - 4)^2 = 25$ or $(x - 1)^2 +$
$(y + 3)^2 = 25$ **27.** $(x - 1)^2 + (y - 4)^2 = 10$
29. $(0, 0)$ and $(4, 2)$
33. $(0, 0); (\pm 2, 0); (0, \pm 3)$

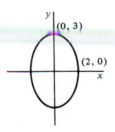

35. $(-1, 2); (-5, 2)$ and $(3, 2); (-1, 0)$ and $(-1, 4)$

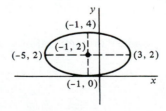

37. $(3, 1); (2, 1)$ and $(4, 1); (3, -2)$ and $(3, 4)$

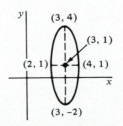

A-1

41. $(0, 0); y = 0$

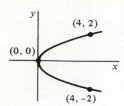

43. $(0, 1); x = 0$

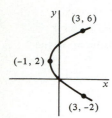

59.

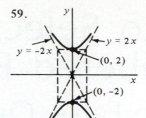

61.

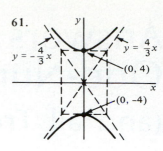

45. $(-1, 2); y = 2$

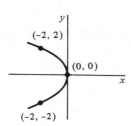

47. $(0, 0); y = 0$

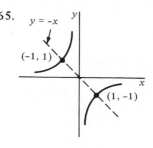

63.

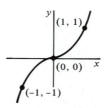

65.

67. the single point $(1, -3)$ **69.** the coordinate axes

49. $4(y + 1) = (x - 2)^2$

51. upper half of parabola $y^2 = x$

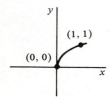

53. right half of parabola $y = x^2$ together with left half of parabola $y = -x^2$

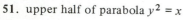

55. **(a)** $y = x - x^2/500$ **(b)** 500 meters   **(c)** $t = 5$; $y = 125$

Section 1.4

1. $\mathbb{R}; -8; -1; 0; 1; 8$

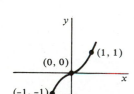

3. all $x \neq 0; 1; \frac{1}{2}; 2; -1; -\frac{1}{2}; -2$

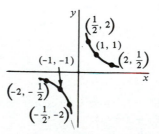

5. **(a)** $x^2 + 2hx + h^2$ **(b)** $2x + h$

7. **(a)** $x^2 + 2hx + h^2 - 1$ **(b)** $2x + h$

9. **(a)** $\{0, 1, 2, 3, 4\}; \{0, 2, 4, 6, 8\}$ **(b)** $2x$

11. (a) $\{-2, -1, 0, 1, 2\}$; $\{\frac{1}{4}, \frac{1}{2}, 1, 2, 4\}$ **(b)** 2^x
(d) about 1.4   **(e)** $\sqrt{2}$ **(f)** about 8.8   **13. (a)** 1;
4; 1; 5   **(b)** $\{1, 2, 3, 4, \ldots\}$ **15.** 1; 1; 2; 3; 5; 8; 13
17. (a) $\$1.25$; $\$1.25$; $\$2.00$; $\$2.00$; $\$2.75$; $\$2.75$
(b) $(0, 5]$; $\{\$1.25, \$2.00, \$2.75, \$3.50, \$4.25\}$
(c)

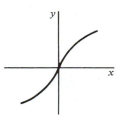

19. $\mathbb{R}$; $[-1, \infty)$ **21.** $\mathbb{R}$; $[0, \infty)$ **23.** $[-2, 2]$; $[0, 2]$
25. all $x \neq 3$ **27.** all $x \neq \pm 3$ **29.** $(0, 2]$ **31.** $\mathbb{R}$

33. $f(x) = 1$ **35.** $f(x) = \sqrt{x}$

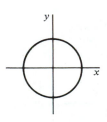

37. $f(x) = \sqrt[3]{x}$ **39.** not a function

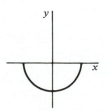

41. $f(x) = -\sqrt{4 - x^2}$ **43.** $f(x) = 4/x$

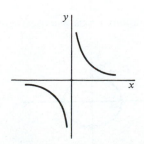

45. $f(x) = 1 - |x|$ **47.** not a function

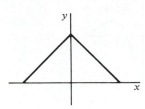   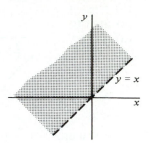

49. the line $x - 2y - 2 = 0$ **51.** the circle $x^2 + y^2 = 1$
53. $y = \frac{5}{3}|x|$, with domain $-3 \leq x \leq 3$ and range
$-4 \leq y \leq 4$

Section 1.5

1.   **3.**

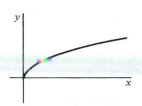   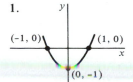

5.   **7.**

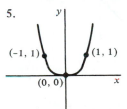

9.   **11. (a)** all x
(b) about the y axis
(c) If $x \neq 0$, $x^2 + 1 > 1$,
so $1/(x^2 + 1) < 1$.
(d) $y \to 0$

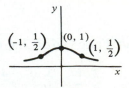

13. (a) $x \neq 0$
(b) $y \to -\infty$
(c) at $(1,0)$
(d) $y \to 0$

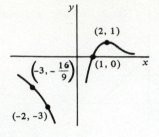

15. (a) $x \leq 0$ or $x > 1$ **(b)** When $x \downarrow 1, y \to \infty$.
(c) When $x \to \pm\infty, y \to 1$. **(d)** $(0,0)$ **(e)** the upper half; vertical

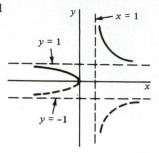

17. the line $y = x + 2$ with a hole at $(2, 4)$
19.

21.

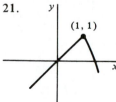

23.

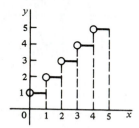

27. even **29.** odd **31.** even **33.** odd **35.** even
37. neither **39.** $f(x) = 0$ **41.** all are even

True-False Quiz (Chapter 1)

1. F **2.** F **3.** T **4.** F **5.** T **6.** T **7.** F
8. F **9.** T **10.** T **11.** F **12.** F **13.** T

14. T **15.** T **16.** F **17.** T **18.** F **19.** T
20. F **21.** F **22.** F **23.** T **24.** F **25.** T
26. T **27.** T **28.** T **29.** T **30.** F **31.** F
32. F **33.** T **34.** T

Additional Problems (Chapter 1)

1. $-\dfrac{4}{3} < x < 2$ **3.** $x > \sqrt{5}$ or $x < -\sqrt{5}$

5. 0–1000 miles **7.** x might be negative. The correct solution is $0 < x < 2$. **9.** $x > 1$ or $x < -1$
11. $x \geq 8$ or $x < -2$ **13.** $0 < x < 3$ or $x < -2$

15. $x < 0$ **17.** $-\dfrac{1}{2} < x < 2$ or $x < -1$ **19.** $1 <$ $x < 2$ or $-3 < x < -2$ **21.** $2\sqrt{3}$ **23.** $2x + 5y - 11 = 0$ **25.** $3x - 4y + 25 = 0$
29. about x axis, y axis, $y = x$, origin

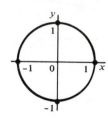

31. about y axis

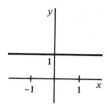

33. about x axis, y axis, origin

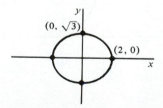

35. about $y = x$, origin

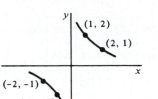

39. $(2, 0); \sqrt{2}$

41. $(2, -1); y = -1$

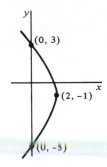

43. (a) $t^2 + 1$ (b) $(0, 0)$

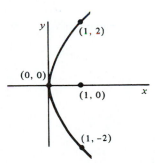

45. domain $= \mathcal{R}$; range $= [-1, \infty)$

47. domain $= \mathcal{R}$; range $= [0, \infty)$

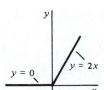

49. $2x + h$

37. about origin

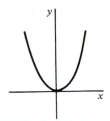

Section 2.1

1. (a) $Q(x) = x + 2, x \neq 2$ (b) 4   (c) $4x - y - 4 = 0$

3. $4; 4x - y - 2 = 0$ 5. $0; y = 1$; graph is a parabola with high point $(0, 1)$ 7. $3; 3x - y - 2 = 0$

9. $\frac{1}{2}; x - 2y + 1 = 0$ 11. 2   13. 12   15. $-\frac{1}{4}$

17. $\frac{1}{4}$ 19. (a) $Q(x) = 2(x + x_0), x \neq x_0$ (b) $4x_0$

(c) $4x$; yes   21. $-2x_0; -2x$ 23. $3x_0^2; 3x^2$

25. $\dfrac{1}{2\sqrt{x_0}}; \dfrac{1}{2\sqrt{x}}; x > 0$ 27. $2t_0; 2t$ 29. $3t_0^2; 3t^2$

31. $-1/t_0^2; -1/t^2$ 33. $\dfrac{1}{2\sqrt{t_0}}; \dfrac{1}{2\sqrt{t}}$

35. rising when $x > 0$, falling when $x < 0$, flat at $x = 0$

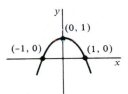

37. rising when $x < 0$, falling when $x > 0$, flat at $x = 0$

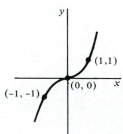

39. rising when $x \neq 0$, never falling, flat at $x = 0$

41. rising when $x > 0$, never falling, never flat

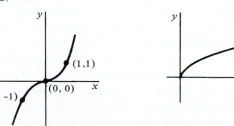

43. positive direction when $t > 0$, negative direction when $t < 0$, stops when $t = 0$

45. positive direction when $t \neq 0$, never negative direction, stops when $t = 0$
47. never positive direction, negative direction when $t > 0$, never stops
49. positive direction when $t > 0$, never negative direction, never stops
53. (a) 96 ft   **(c)** when $t = \frac{1}{2}$; 100 ft   **(d)** when $t = 1$; $v = -16$, $|v| = 16$ **(e)** when $t = 3$; $v = -80$; $|v| = 80$
55. $m(x) = 0$ when $x = \pm 1$ and changes sign at each of these points.

57.

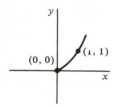

61. rising when $x > 0$; flat at $x = 0$

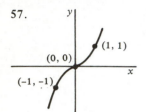

67. (a) $0; 1; 0; -1; 0$ **(c)** $v = \cos t$ **(d)** $1; 0; -1; 0; 1$

Section 2.2

1. 0; graph is a parabola with low point $(0, -4)$ **3.** 2
5. -12 **7.** 80   **9.** $\sqrt{2}$ **11.** -1 **13.** 0; make it flatten out at $(0, 0)$ **15.** undefined; with vertical tangent at $(0, 0)$ **17.** undefined; graph is upper half of parabola opening left from $(1, 0)$ **19.** -0.7699; -0.7520; -0.7502 **21.** 52.6717; 62.7359; 63.8722
23. 0.9983; 0.999983; 0.99999983
25. $Q(0.1) = 0.7177$, $Q(-0.1) = 0.6697$; $Q(0.01) = 0.6955$, $Q(-0.01) = 0.6908$; $Q(0.001) = 0.693387$, $Q(-0.001) = 0.692907$ **27.** $2x + 1$ **29.** $6x^2$ **31.** $4t^3 - 5$
33. $4x^3 + 2x$ **35.** $-2/t^3$ **37.** $5/(x + 5)^2$ **39.** $\frac{5}{2}t^{3/2}$
41. $\dfrac{2}{3\sqrt[3]{x}}$ **43.** graph is a horizontal line
47. (a) upper half of unit circle   **(b)** 0; undefined

Section 2.3

1.

Degrees	0°	30°	45°	60°	90°	120°	135°	150°	180°
Radians	0	$\pi/6$	$\pi/4$	$\pi/3$	$\pi/2$	$2\pi/3$	$3\pi/4$	$5\pi/6$	π

210°	225°	240°	270°	300°	315°	330°	360°
$7\pi/6$	$5\pi/4$	$4\pi/3$	$3\pi/2$	$5\pi/3$	$7\pi/4$	$11\pi/6$	2π

3. (a) $100°$ **(b)** $585°$ **(c)** $1080°$ **(d)** $70°$
(e) $5°$ **(f)** $160°$

11.

t	$\sin t$	$\cos t$	$\tan t$	$\cot t$	$\sec t$	$\csc t$
0	0	1	0		1	
$\pm\pi/2$	± 1	0		0		± 1
$\pm\pi$	0	-1	0		-1	
$\pm 3\pi/2$	∓ 1	0		0		∓ 1
$\pm 2\pi$	0	1	0		1	

15. $\dfrac{90°}{\pi} \approx 28.6°$ **17.** $3\pi \approx 9.4$ (square units)   **59.** 3
61. 0   **63.** 1

Section 2.4

1. $M = (2 - \epsilon/3, 2 + \epsilon/3)$
3. $M = (2 - \epsilon/3, 2 + \epsilon/3)$ with 3 excluded
5. $M = (1 - \epsilon/5, 1 + \epsilon/5)$ **7.** $M = \left(\dfrac{3}{2} - \dfrac{\epsilon}{2}, \dfrac{3}{2} + \dfrac{\epsilon}{2}\right)$ with $\dfrac{3}{2}$ excluded   **9.** $M = (4 - 4\epsilon + \epsilon^2, 4 + 4\epsilon + \epsilon^2)$
11. $M = \left(\dfrac{1}{1 + \epsilon}, \dfrac{1}{1 - \epsilon}\right)$ **13.** $M = (5, 5 + \epsilon^2)$
15. $M = (\sqrt{1 - \epsilon^2}, 1)$ **17.** $\delta = \epsilon/3$ **19.** $\delta = \sqrt{\epsilon}$
21. $\delta = \epsilon^2$ **23.** -4 **25.** 8   **27.** 2   **29.** 1
31. 1   **35.** 1   **37.** 4   **39.** does not exist   **41.** 1
43. 1   **45.** -1 **51.** $-1; 1$; does not exist
53. none exists

Section 2.5

1. 10   **3.** 6   **5.** 1   **7.** 0   **9.** 8   **11.** 1
13. –3   **15.** 3   **17.** 0   **19.** 0   **21.** 0   **23.** 0
25. 1   **27.** 1   **29.** 2   **33.** 3   **35.** 0   **37.** 1

True-False Quiz (Chapter 2)

1. T   **2.** T   **3.** F   **4.** T   **5.** T   **6.** F   **7.** T
8. T   **9.** T   **10.** T   **11.** T   **12.** T   **13.** T
14. F   **15.** F   **16.** T   **17.** T   **18.** T   **19.** F
20. F   **21.** F   **22.** T   **23.** T   **24.** F   **25.** F
26. F   **27.** T   **28.** F   **29.** T   **30.** F   **31.** F

Additional Problems (Chapter 2)

1. $x + y + 1 = 0$ **3.** $x + y - 3 = 0$ **5.** $x = 2$
7. $y = 0$ **9.** $x + y - \pi/2 = 0$ **11.** 2   **13.** –1
15. 2   **17.** $2x + 1$ **19.** $3x^2 + 2x$ **21.** $-5/(x - 5)^2$
23. $\frac{4}{3}x^{1/3}$
25. rising when $-1 < x < 1$, falling when $x > 1$ or $x < -1$

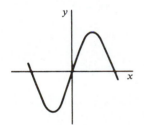

27.

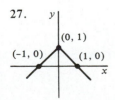

29. (b) $t = 3, h = 144$ **(c)** $t = 6, v = -96$
31. $\frac{1}{2}$; negative direction   **33.** 0   **35.** 0   **37.** 0
39. $-\cos t$ **41.** $-\sin t$ **43.** $-\cos t$ **47.** $(4, 12)$
49. $1; -1;$ no   **51.** 3

Section 3.1

1. $2x - 2$ **3.** $6x^2 + 1$ **5.** $7x^6 - 10x^4 + 12x^2 - 5$
7. $2 \cos x - 3 \sin x$ **9.** $-6/x^3$ **11.** $3x^2 - 3/x^4$
13. $\dfrac{1}{3\sqrt[3]{x^2}}$ **15.** $\dfrac{x - 3}{2x\sqrt{x}}$ **17.** 0   **19.** $\frac{1}{2}$ **21.** $\dfrac{5}{2\sqrt{x}}$
23. $2x - y - 1 = 0$ **25.** $x + 2y - 7 = 0$

29.

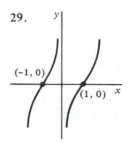

35. $y = x^4 + 1$ **37. (a)** $s = 5; |v| = 3$ **(b)** negative
(c) $t = 4; s = 1$ **(d)** $t = 9;$ positive   **(e)** object
continues in positive direction, speeding up

Section 3.2

1. $\sec^2 x - 3 \sec x \tan x$ **3.** $-3 \sin x - 2 \csc^2 x$
5. $-x^3 \sin x + 3x^2 \cos x$ **7.** $\sec^3 x + \sec x \tan^2 x$
9. $\frac{5}{2}x^{3/2} - 2x$ **11.** $6x^5 - 4x^3 + 2x$ **13.** $\dfrac{9x^2 + 1}{4x^{3/4}}$
15. $\dfrac{-x^2 + 2x + 2}{(x^2 + x + 1)^2}$ **17.** $\dfrac{1 - 3x}{2\sqrt{x}(3x + 1)^2}$
19. $\dfrac{1 + 16x^{15/4} - 15x^4}{4x^{3/4}(x^4 + 1)^2}$ **21.** $-x^{3/2} \sin x + \frac{3}{2}x^{1/2} \cos x$
23. $\dfrac{\sec x (x \tan x - 2)}{x^3}$ **25.** $\dfrac{\cos x + \sin x \sec x}{\tan x - \sec x}$
27. $\dfrac{2 \cos x + 5x \sin x}{5x^{3/5} \cos^2 x}$ **29.** $-x^{1/3} \sin x + \frac{1}{3}x^{-2/3} \cos x -$
$x^{-1/3} \cos x + \frac{1}{3}x^{-4/3} \sin x$ **31.** $\csc \theta (\cot \theta - \csc \theta)$
33. $x \sec x (x \tan x + 2)$ **35.** $\dfrac{x \cos x - \sin x}{x^2}$
37. $-2/(t - 2)^2$ **39.** $4x/(x^2 + 1)^2$
41. $-\dfrac{3x^2 + 1}{2\sqrt{x}(x^2 - 1)^2}$ **43.** $\dfrac{5x^2 + 1}{3x^{4/3}}$ **45.** $-1/\pi^{3/2}$

47. $\dfrac{x(2 - 3x^2)}{\sqrt{1 - x^2}}$ **49.** $\dfrac{x(2 - x^2)}{(1 - x^2)^{3/2}}$ **53.** $\frac{1}{2}$

63.

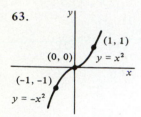

Section 3.3

1. $-3(1 - x)^2$ **3.** $2 \sec x \tan^2 x$ **5.** $2x^3/\sqrt{x^4 + 2}$

7. $-2x/(x^2 - 2)^2$ **9.** $2\pi \cos (2\pi x)$

11. $-6 \sin (2x) \cos^2 (2x)$ **13.** $\dfrac{2(2x - 1)}{|2x - 1|}$

15. $\dfrac{\sin x \cos x}{|\sin x|}$ **17.** $\sin x (2 \cos^2 x - \sin^2 x)$

19. $\dfrac{3(1 - x^2)}{(x^2 + 3)^3}$ **21.** $\dfrac{2x(\tan x - x \sec^2 x)}{\tan^3 x}$

23. $x^2(14x - 3)(2x - 1)^3$ **25.** $9/(9 - x^2)^{3/2}$

27. $3\left(x + \dfrac{2}{x}\right)^2 \left(1 - \dfrac{2}{x^2}\right)$ **29.** $\dfrac{28(2x - 1)^3}{(x + 3)^5}$

31. $v = -\pi; |v| = \pi$ **33. (a)** $x \le 3$ **(b)** $(0, 0)$ and $(3, 0)$ **(c)** $y' > 0$ when $x < 2; y' = 0$ when $x = 2$; $y' < 0$ when $2 < x < 3$ **(d)** $(2, 2)$ **(e)** y' decreases without bound as $x \uparrow 3$

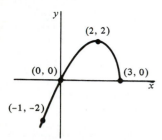

35.

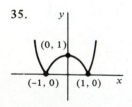

43. 1; up the curve   **45.** 1 radian/sec

Section 3.4

1. $-\frac{1}{2}$ **3.** $-y/x$ **5.** $1/(2y)$ **7.** $-x^2/y^2$ **9.** $2\sqrt{y}$

11. $-x/y$ **13.** $\dfrac{y - 2x}{2y - x}$ **15.** y^2/x^2 **17.** $\dfrac{2(x + y)}{1 - 2x - 2y}$

19. $\dfrac{\cos x - \cos (x + y)}{\cos (x + y) - \cos y}$ **21.** $12x + 5y + 8 = 0$

23. $2x + 5y - 12 = 0$ **25.** $x + y - 2 = 0$ and $y = -2$

29. $y = \pm\dfrac{1}{\sqrt{2}}(x + 2)$ **33.** $x + y + 2 = 0$

35. $5x - 12y - 9 = 0$

True-False Quiz (Chapter 3)

1. F   **2.** F   **3.** T   **4.** T   **5.** T   **6.** T   **7.** T
8. F   **9.** T   **10.** T   **11.** F   **12.** T   **13.** F
14. F   **15.** F   **16.** F   **17.** T   **18.** F

Additional Problems (Chapter 3)

1. $1 - \dfrac{1}{x^2}$ **3.** $x^{1/2} - x^{-1/3}$ **5.** $x^3 \cos x + 3x^2 \sin x$

7. $6 \sin^2 2x \cos 2x$ **9.** $x(x^2 + 1)(7x^3 + 3x - 4)$

11. $\dfrac{3(x - 1)}{2\sqrt{x}}$ **13.** $\dfrac{x^3}{(x^2 + 1)^{3/2}}$ **15.** $\dfrac{x^2 - 4x + 6}{(x - 2)^2}$

17. $\dfrac{1 - x^2}{2\sqrt{x}(x^2 + 1)^{3/2}}$ **19.** $\dfrac{1}{x} \sin \dfrac{1}{x} + \cos \dfrac{1}{x}$

21. $-\dfrac{\sin x \cos x}{|\cos x|}$ **23.** $\dfrac{x(18 - x^2)}{(9 - x^2)^{3/2}}$ **29.** π^2

31.

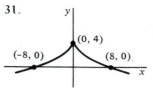

33. (a) $-1 < x < 1$ **(b)** $x = \pm 1$ **(c)** rising when $0 < x < 1$; falling when $-1 < x < 0$

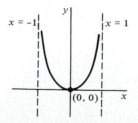

37. $\sin 2x = 2 \sin x \cos x$ 39. $-\sqrt{y}/\sqrt{x}$

41. $-\dfrac{2x + 3y}{3x + 2y}$ 43. $\dfrac{y}{2y - x}$ 45. $-\frac{1}{8}(y - 4)^2$

47. $\dfrac{1 + y^2}{\sqrt{1 + y^2} - xy}$ 49. $x = -2$

Section 4.1

1. all x 3. $x \neq 0$ 5. all x 7. $x \neq \pm 1$
9. $x \neq 0$ or 2   11. $x \geq 1$ 13. $x \leq 0$ or $x \geq 1$
15. $x > 0$ 17. all x 19. $x \neq \pi/2$ (and $0 \leq x \leq 2\pi$)
21. $x \neq \pm\pi/4, \pm 3\pi/4, \pm 5\pi/4, \ldots$ 23. $x \neq 3$
25. $x \neq 1$ 27. all x 29. $x \neq 0$ 31. all x
33. $x \neq -1$ 35. $x \neq 0$ 37. $x \neq \pm 1, \pm 2, \pm 3, \ldots$
39. $0 \leq x < a$ 41. (a) $0 \leq x \leq 4$ (and $x \neq 2$)
(b) $-\$1000$ (c) $\frac{1}{2}$ kg   (d) more than $\frac{1}{2}$ kg

Section 4.2

1. $[-1, 3]$ 3. $[0, 4]$ 5. $[1, 3]$ 7. $[0, 1]$
9. $[-1, -1]$ 11. $[0, 5]$ 13. $[0, 2]$ 15. $[1, 1]$
17. $[0, 4]$ 21. min -1; no max; bounded
23. min 0; no max; unbounded
25. min 1; max 3; no, it misses 2

Section 4.3

1. $c = 1$ 3. $c = -1$ 5. $c = \frac{25}{4}$ 7. $c = \pi/2$
11. $c = 0$ 13. (a) 75 mph   15. (a) 1   (b) 0.56
25. max 1; min 0   27. $f(x) = \sin x$; $a = 0$; $\sin 0.2 \approx 0.2$
29. $f(x) = \sqrt[3]{x}$; $a = 8$; $\sqrt[3]{8.5} \approx 2.042$ 33. $c = \pi/4$

Section 4.4

1. decreasing in $(-\infty, \infty)$

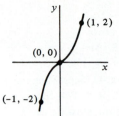

3. increasing in $(-\infty, 0]$,
   decreasing in $[0, \infty)$

---

5. increasing in $(-\infty, \infty)$

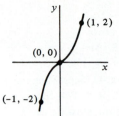

7. increasing in $(-\infty, 0]$

9. increasing in $[0, \infty)$

11. increasing in $(-\infty, 0)$,
    decreasing in $(0, \infty)$

13. increasing in $[0, \pi/2]$ and in $[3\pi/2, 2\pi]$, decreasing
in $[\pi/2, 3\pi/2]$

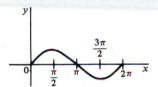

15. decreasing in $(0, \pi)$

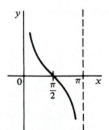

17. $3x^2 + C$ 19. $\frac{3}{2}x^2 + 2x + C$ 21. $\frac{2}{3}x^{3/2} + C$
23. $-\cos x + C$ 25. $\frac{1}{2}\sin 2x + C$
27. $3 \sin x + \cos x + C$ 29. $x^3 - \dfrac{1}{x} + C$
31. $y = x^4 - x - 9$ 33. $y = (x - 1)^2 - 1$
35. $f(x) = 1 - \cot x$ 37. (a) $v = -32t + 64$; $h = -16t^2 + 64t + 80$ (b) $t = 2$; $h = 144$ (c) $t = 5$; $|v| = 96$
41. 100 ft   43. yes, when $t = 1$ 45. $\frac{1}{2}$

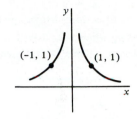

Section 4.5

1. 0   **3.** $12x$ **5.** $-\frac{1}{4}x^{-3/2}$ **7.** $-2/x^3$
9. $2\sec^2 x \tan x$ **11.** $-x\sin x + 2\cos x$
13. 4; v is increasing in $(-\infty, \infty)$ **15.** $6t - 6$; v is
increasing in $(1, \infty)$ and decreasing in $(-\infty, 1)$
17. $-3\cos t$; v is increasing in $(\pi/2, 3\pi/4)$, decreasing in
$(0, \pi/2)$ and $(3\pi/2, 2\pi)$ **19.** $v = 1/2$, $a = -1/4$; the
object is moving in the positive direction, slowing down

23. $\dfrac{n!}{(2-x)^{n+1}}$ **25.** $\dfrac{1 \cdot 3 \cdot 5 \cdots (2n-1)}{2^n(1-x)^{n+1/2}}$

27. $2^n \sin\left(2x + \tfrac{1}{2}n\pi\right)$ **29. (a)** left; right
(b) The direction of turning changes.
35. rising when $x < -1$ or $x > 1$, falling when $-1 < x < 1$;
concave up when $x > 0$, down when $x < 0$; turning points
$(-1, 4)$, $(1, 0)$; inflection point $(0, 2)$

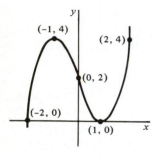

37. rising when $-2 < x < 0$ or $x > 2$, falling when $x < -2$
or $0 < x < 2$; concave up when $x < -2\sqrt{3}/3$ or $x >$
$2\sqrt{3}/3$, down when $-2\sqrt{3}/3 < x < 2\sqrt{3}/3$; turning points
$(\pm 2, 0)$, $(0, 16)$; inflection points $(\pm 2\sqrt{3}/3, 64/9)$.

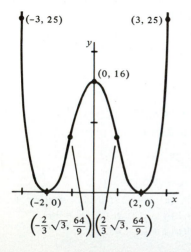

39. rising when $x < -1$ or $x > 1$, falling when $-1 < x <$
0 or $0 < x < 1$; concave up when $x > 0$, down when
$x < 0$; turning points $(1, 2)$, $(-1, -2)$; no inflection point

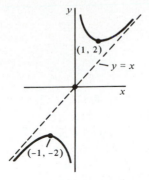

41. always rising and concave down; no turning point or
inflection point

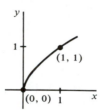

43. rising when $-2 < x < 0$, falling when $0 < x < 2$;
always concave down; turning point $(0, 2)$; no inflection
point

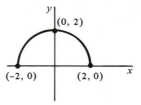

45. always rising; concave up when $0 < x < \pi/2$, down
when $-\pi/2 < x < 0$; no turning point; inflection point
$(0, 0)$

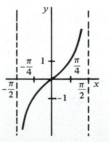

Section 4.6

1. (a) $L(x) = 9 + 6(x - 3)$ (b) $9.72; 0.1\%$ error
3. $L(x) = 1 + 3(x - 1); 1.18; 0.9\%$ error   5. $L(x) = 3 + \frac{1}{6}(x - 9); 3.0\overline{3}; 0.006\%$ error   7. $L(x) = \frac{1}{5} - \frac{1}{25}(x - 5);$
$0.1956; 0.05\%$ error   9. $L(x) = \frac{1}{2} - \frac{\sqrt{3}}{2}\left(x - \frac{\pi}{3}\right);$
$0.5409; 0.1\%$ error   11. $f(x) = x^2; x_0 = 3; 9.156;$
0.007% error   13. $f(x) = x^{3/2}; x_0 = 4; 8.3; 0.02\%$ error
15. $f(x) = \tan x; x_0 = 0; 0.092; 0.3\%$ error
17. $f(x) = \cos x; x_0 = \pi/6; 0.8778; 0.003\%$ error
19. $f(x) = \sqrt[3]{x}; x_0 = 8; 1.99; 0.003\%$ error
21. error $\leq 0.02; (3.12)^2 \approx 9.7$ 23. error $\leq 0.02;$
$(1.06)^3 \approx 1.2$ 25. error $\leq 0.0002; \sqrt{9.2} \approx 3.033$
27. error $\leq 0.0001; \frac{1}{5.11} \approx 0.196$ 29. error $\leq 0.0007;$
$\cos 1 \approx 0.54$ 31. one   33. two   35. five

Section 4.7

1. $3\, dx$ 3. $-\sin t\, dt$ 5. $\frac{dt}{(1 - t)^2}$ 7. $2x\, dx$
9. $\sec t \tan t\, dt$ 11. $-dx/x^2$ 13. $dy = 12\, dx$
15. $dy = dx$ 17. $dy = -dx$ 19. (a) $\Delta A = 2x\, \Delta x + (\Delta x)^2$ (b) $\Delta A \approx 2x\, \Delta x;$ error $= (\Delta x)^2$, the area of the small square in Figure 2   21. (a) Each of the six faces (of area x^2) contributes $x^2 t$ to the volume of paint. (This neglects the paint built up at the edges.)   25. 4%
29. p increases by 0.3 39. $\frac{dy}{dx} = \frac{3x^2 - 1}{3y^2}; \frac{dx}{dy} = \frac{3y^2}{3x^2 - 1}$
41. $\frac{dy}{dx} = \frac{1 + 2y}{3 - 2x - 2y}; \frac{dx}{dy} = \frac{3 - 2x - 2y}{1 + 2y}$
43. $\frac{dy}{dx} = \frac{y \cos (xy) - 2}{1 - x \cos (xy)}; \frac{dx}{dy} = \frac{1 - x \cos (xy)}{y \cos (xy) - 2}$
47. (a) $dx = 2\, dt, dy = 2t\, dt, dy/dx = t$
(b) $y = \frac{1}{4}(x + 1)^2, dy/dx = \frac{1}{2}(x + 1) = t$
49. (b) $-\cot t$ (c) $-\csc^3 t$

True-False Quiz (Chapter 4)

1. F   2. F   3. F   4. T   5. T   6. F   7. T
8. F   9. F   10. T   11. T   12. F   13. T
14. F   15. T   16. F   17. F   18. F   19. T

20. T   21. F   22. F   23. T   24. T   25. T
26. T   27. T   28. F   29. T   30. F   31. T
32. T   33. T   34. T   35. F

Additional Problems (Chapter 4)

1. all x 3. all x 5. $x \geq 2$ 7. all $x \neq 0, \pm\pi/2,$
$\pm 3\pi/2, \ldots$ 9. $x \neq 0$ 11. all x 13. all $x \neq \pm 1,$
$\pm 2, \pm 3, \ldots$ 15. $f(1)$ cannot be defined to make f continuous at $1; f(2)$ should be defined to be 1
19. $[0, 1]$ 21. $[0, 1]$ 23. when f is continuous and increasing
25. yes; no; yes

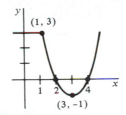

27. f is continuous in $[1, 2]$; min 0 at $x = 1, 2$
29. tangent is continuous in $[0, 1.57]$ and 100 is between $\tan 0 = 0$ and $\tan 1.57 \approx 1256$
31. between 0.4 and 0.5   33. $\pi/2$; Mean Value Theorem
35. max 1; min 0; yes at $x = 0$, no at $x = 1$; doesn't apply
37. $y = 2x - \sin x + 1$ 39. within one unit
43. $t < 0$: going west, slowing down; $t = 0$: momentarily stopped (at $s = 0$); $0 < t < 2$: going west, speeding up; $t = 2$: changing from speeding up to slowing down (at $s = -16$); $2 < t < 3$: going west, slowing down; $t = 3$: momentarily stopped (at $s = -27$); $t > 3$: going east, speeding up   45. $n!(3 - x)^{-(n + 1)}$
47. (a) $x = 2$ and $y = x + 2$ (b) rising when $x < 0$ or $x > 4$; falling when $0 < x < 2$ or $2 < x < 4$ (c) up when $x > 2$, down when $x < 2$

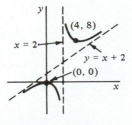

49. 9.975 51. $dy = 2\, dx$ 53. $6x^2 (\Delta x)^2 + 4x(\Delta x)^3 + (\Delta x)^4$ 55. about 1%

Section 5.1

1. $\dfrac{3}{10\pi}$ cm/sec   3. $-1/\sqrt{2}$ cm/sec   5. 18π cm³/min

7. 250 ft³/min   9. 7.2π ft³/hr   11. decreasing at $2\pi/3$ cm³/hr   13. $\sqrt{10}$ ft/sec   15. (a) -40 ft/sec
(b) 0 ft/sec   (c) -25 ft/sec   17. $5\pi/3$ meters/sec
19. 12.5 knots   21. $\frac{20}{7}$ cubic units per second;
increasing   23. (a) $600\pi \sec^2 \theta$ km/hr
(b) 600π km/hr   (c) $\sec \theta \to \infty$ as $\theta \uparrow \pi/2$
25. (a) $19.70   (b) $20.00

Section 5.2

1. critical point 0; maximum 1 (global)

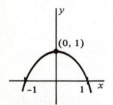

3. no critical points; no extreme values

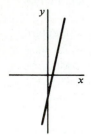

5. critical points 0, 1; maximum 0 (local), minimum -1 (local)

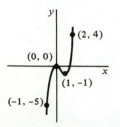

7. critical point 1; no extreme values

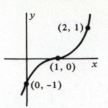

9. critical points 0, ± 1; maximum 1 (local), minimum 0 (global)

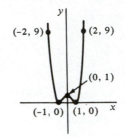

11. critical point 0; maximum 2 (global)

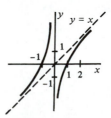

13. no critical points; no extreme values

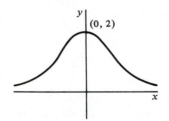

15. critical point 0; minimum 0 (global)

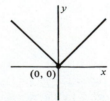

17. critical point 1; minimum –1 (global)

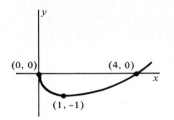

19. critical point 0; maximum 1 (global)

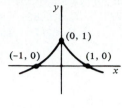

5. (a) about the origin

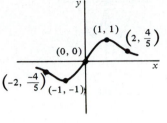

7. (a) about the origin **(d)** $(\pm 2, 0)$

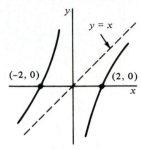

21. critical points $\pi/4$, $5\pi/4$; maximum $\sqrt{2}$ (global), minimum $-\sqrt{2}$ (global)

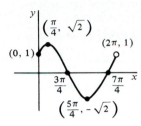

9. (a) none **(c)** falling when $x < 0$ or $0 < x < 1$, rising when $x > 1$; minimum 3 **(d)** $-\sqrt[3]{2} \approx 1.26$

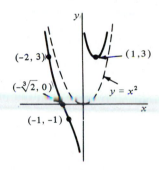

25. $f'(0)$ is undefined; nothing **27. (a)** $f(0)$ is a maximum **(b)** nothing

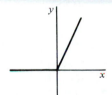

Section 5.3

1.

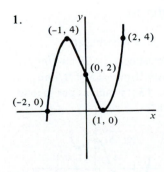

3.

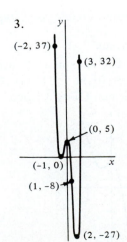

11.

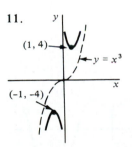

13.

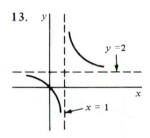

15.

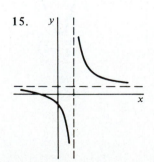

17.

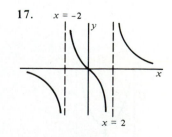

19. (a) $(0, 0), (4, 0)$

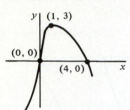

21.

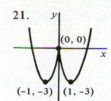

23.

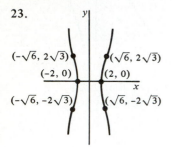

25. (b) about the x axis; $x = 2$

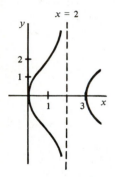

27.

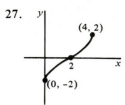

29.

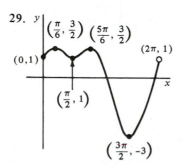

31.

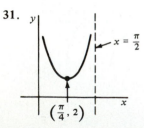

Section 5.4

1. Each application of the formula yields r.   **3.** 3.1623
5. 4.1231   **7.** 6.1885   **9.** 3.0541   **11.** 2.7321;
$1 + \sqrt{3}$ **13.** 1.3247   **15.** 1.3301   **17.** 0.7391
19. 0.6367

Section 5.5

1. 36   **3.** 1000   **5.** 1750; $182.50   **7.** 120 yd
9. 7.5×10 (ft)   **11.** 12 ft^2 **13.** length 28″,
girth 56″   **15.** 5:00 P.M. (when both ships have reached
the harbor)   **17. (a)** 8   **(b)** no; $6 < P \le 8$
23. arbitrarily large; min 4   **25.** $\frac{1}{3}P \times \frac{1}{6}P$, where P is
the perimeter   **31.** $(\frac{1}{2}, 1/\sqrt{2})$ **33. (a)** $2x^3 + x - 1 = 0$
(b) $(0.59, 0.35)$ **37. (a)** on land for $100 - 350/\sqrt{51} \approx$
51 meters, then across the river   **(b)** on land for 50
meters, then across the river   **39.** $\frac{8}{3}$ miles
43. $c\sqrt[3]{a}/(\sqrt[3]{a} + \sqrt[3]{b})$ units from the source with strength a
45. (b) no   **(c)** yes   **47. (b)** $m \approx 1.72; y \approx 8.6$
49. 50 mph   **51.** $x = 71$

True-False Quiz (Chapter 5)

1. F   **2.** T   **3.** T   **4.** F   **5.** T   **6.** T   **7.** F
8. F   **9.** F   **10.** T   **11.** T   **12.** T   **13.** T
14. F

Additional Problems (Chapter 5)

1. $\frac{1}{150}$ ft/min   **3.** 34 knots   **5.** increasing at
$\frac{3}{2}(36 + 11\sqrt{3})$ knots   **7.** decreasing at $7\pi/30$ cm^3/hr
9. $81\pi/50$ ft^3/hr   **11. (a)** 300π ft/min; 150π ft/min
(b) 0 ft/min; $150\pi\sqrt{3}$ ft/min   **13.** $\frac{2}{3}$ radians per second
15.

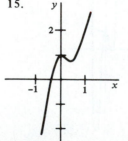

17.

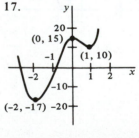

19.

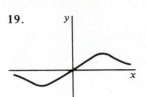

21.

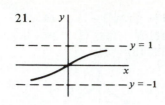

23.

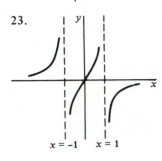

$x = -1$ $x = 1$

25.

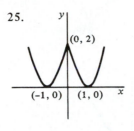

(0, 2)

(-1, 0)   (1, 0)

27.

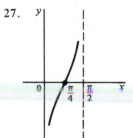

29.

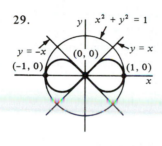

$x^2 + y^2 = 1$

$y = -x$ (0, 0) $y = x$

(-1, 0)   (1, 0)

31.

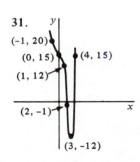

(-1, 20)
(0, 15)   (4, 15)
(1, 12)
(2, -1)
(3, -12)

37. All dimensions should be the same (a cube).
39. $\frac{2450}{27}$ cm^3 **43.** $(a^{2/3} + b^{2/3})^{3/2}$ **47.** 1000 units;
$30   **49.** 1772 units   **51.** half of it

Section 6.1

1. 25   **3.** 30   **5.** $\sum_{k=1}^{n} k^2$ **7.** $\sum_{k=1}^{n} f(x_k)\Delta x$

9. (a) 1   **(b)** 6   **(c)** 10   **(d)** The average is exact.
(e) 8   **11. (a)** 1   **(b)** 1   **(c)** 9   **(d)** The average

differs from $A = 4$ by 25%.   **(e)** 3.5   **13.** $\frac{17}{16}; \frac{137}{16}$
15. (a) $\pi/3$ **(b)** $\pi/3$ **(c)** $5\pi/6$ **(d)** The average

differs from $A = 2$ by about 8.4%.   **17.** $\frac{\pi}{6}(1 + \sqrt{3})$;

$\frac{\pi}{6}(3 + \sqrt{3})$ **21. (d)** 4   **23. (c)** 68

Section 6.2

1. 6   **3.** $\frac{1}{2}$ **5.** 1   **7.** 4   **9.** $25\pi/4$ **11.** $2 - \pi/4$
15. 8   **17.** $9\pi/4$ **19.** exists   **21.** does not exist
23. does not exist   **25.** exists

Section 6.3

1. 4   **3.** 0   **5.** -2   **7.** -6   **9.** 3   **11.** -1
13. $\frac{33}{2}$ **15.** $4 - 3\pi$ **17.** $27\pi/2$ **19.** 5; any point
of I **21.** 2; $c = 1$ **23.** $\frac{5}{6}$; $c = \pm\frac{5}{6}$ **25.** $\pi/4$;
$c = \pm\frac{1}{4}\sqrt{16 - \pi^2}$

Section 6.4

1. 8   **3.** 4   **5.** 2   **7.** 4   **9.** 68   **11.** 2   **13.** $\frac{3}{5}$
15. $\frac{3}{8}$ **17.** 1   **19.** $\frac{1}{2}$ **21.** $\frac{1}{2}(2 + \sqrt{3})$ **23.** 1
25. $\frac{1}{4}(4 - \pi)$ **27.** 36   **29.** $\frac{4}{5}$ **31.** 4   **33.** 7
35. $\frac{3}{2}$ **37.** $2/\pi$ **43.** $\sin(x^2)$ **45.** $\sqrt{t^2 + 1}$
47. $-\tan x$ **49.** $\sec^4 x$ **51.** $\cos x \,|\cos x| - \sqrt{1 - x^2}$
53. $f(x)$

Section 6.5

1. $\frac{1}{10}(2x - 1)^5 + C$ **3.** $\frac{1}{3}(1 + x^2)^{3/2} + C$
5. $\frac{2}{3}\sqrt{3x - 1} + C$ **7.** $-\frac{1}{5}\cos^5 \theta + C$ **9.** $2 \tan \frac{1}{2}t + C$
11. $\frac{1}{4}(2 - \sqrt{3})$ **13.** $3 - \sqrt{5}$ **15.** $\frac{8}{3}$ **17.** $\pi/6$
19. $\frac{1}{3}(16 - 9\sqrt{3})$ **21.** $\frac{1}{15}(1 + x^3)^5 + C$
23. $-\frac{1}{2}\cos t^2 + C$ **25.** $\frac{1}{2}\tan 2t + C$
27. $2\sqrt{x^2 - x} + C$ **29.** $3(\sqrt{2} - 1)$ **31.** $\frac{9}{28}$ **33.** 0
35. $-\frac{1}{2}\cot^2 t$ or $-\frac{1}{2}\csc^2 t$ (plus an arbitrary constant in
each case). Since $\csc^2 t - \cot^2 t = 1$, these differ only by a
constant.   **39.** 1   **41.** $\frac{662}{63} \approx 10.5$ km   **43.** $\frac{1}{2}$

True-False Quiz (Chapter 6)

1. T   2. T   3. T   4. T   5. T   6. T   7. T
8. T   9. T   10. T   11. T   12. F   13. T
14. T   15. T   16. T   17. T   18. T   19. T
20. T   21. T   22. T

Additional Problems (Chapter 6)

1. $\frac{19}{20}; \frac{77}{60}$; average $= \frac{67}{60} = 1.11\overline{6}$ (about 1.6% error)

3. 1.09   5. 16   9. $D = 1$ 11. $\frac{2}{3}$ 13. 40

15. –4   17. $25\pi/4$ 19. π 21. 12   23. 0

25. $\frac{1}{4}(4 - \pi)$ 27. $\frac{3}{16}$ 29. $4 - \sqrt{7}$ 31. $2(\sqrt{3} - 1)$

33. $\frac{1}{3}(x^2 + 2)\sqrt{x^2 - 1}$ 35. undefined   37. 10

39. $2/\pi$; twice   45. $\sec^2 x$ 47. $1/\pi$ 49. $\frac{5}{6}$

51. $1 - 1/b; 1$

Section 7.1

1. $\frac{4}{3}$ 3. $\frac{4}{3}$ 5. $\frac{9}{2}$ 7. $\frac{3}{2} - 2\int_1^2 \frac{dx}{x} \approx 0.11$

(The integral, which cannot be evaluated by antidifferentiation at present, can be approximated by a Riemann sum.)   9. $\frac{16}{3}$ 11. $\frac{4}{3}$ 13. $\frac{4}{3}$ 15. $\frac{9}{2}$

17. $\frac{3}{2} - 2\int_1^2 \frac{dy}{y} \approx 0.11$ (as in Problem 7)   19. $\frac{16}{3}$

21. $\frac{4}{3}$ 23. $\frac{32}{3}$ 25. $\frac{71}{6}$ 27. $\frac{2}{3}(3\sqrt{2} - 2)$ 29. $\frac{27}{4}$

31. 1   33. $2(3 - \sqrt{5})$ 35. $\frac{1}{6}$ 37. $\frac{13}{6}$ 39. 36

41. $\frac{17}{12}$ 43. $\sqrt{2} - 1$ 45. $\frac{3}{4}\sqrt{3}$

Section 7.2

1. $16\pi/3$ 3. $32\pi/3$ 5. 8π 7. $128\pi/5$

9. $40\pi/3$ 11. 8π 13. $16\pi/15$ 15. π 17. $\frac{1}{2}\pi^2$

19. $\pi/2$ 21. $28\pi/3$ 23. $256\pi/15$ 25. $26\pi/3$

27. $\frac{4}{3}\pi a^2 b$ 29. $5\pi/6$ 31. $324\pi/5$ 33. $32\pi/5$

35. $4\pi/15$ 37. (a) $2\pi^2 a^2 b$ 39. $3500\pi/3$ cm³

41. $\frac{16}{3}$

Section 7.3

1. 3   3. $\sqrt{13}$ 5. $\sqrt{2}$ 7. $4\sqrt{2}$ 9. 3π

11. $\frac{1}{3}(2\sqrt{2} - 1)$ 13. $\frac{1}{27}(13\sqrt{13} - 8)$ 15. 12

17. $\frac{8}{3}(2\sqrt{2} - 1)$ 19. $2\pi^2$ 21. $\frac{2}{27}(10\sqrt{10} - 1)$

23. $\frac{14}{3}$ 25. $\frac{33}{16}$ 29. $\frac{1}{27}(80\sqrt{10} - 13\sqrt{13})$ 31. $\frac{387}{20}$

35. (b)   (d) 6a

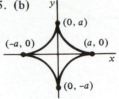

41. (a) $y = x^2, x \geq 0$ (b) large; $\sqrt{5}/2$

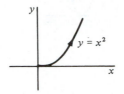

43. (a) $x^2/a^2 + y^2/b^2 = 1$; the track is an ellipse
(traversed counterclockwise)   (b) $\sqrt{a^2 \sin^2 t + b^2 \cos^2 t}$
(c) maximum a, at $(0, \pm b)$; minimum b at $(\pm a, 0)$

(d) $\int_0^{2\pi} \sqrt{a^2 \sin^2 t + b^2 \cos^2 t}\, dt$

Section 7.4

1. 4π 3. $2\pi\sqrt{5}$ 5. $\frac{\pi}{6}(5\sqrt{5} - 1)$

7. $\frac{\pi}{27}(10\sqrt{10} - 1)$ 9. 36π 11. $12\pi a^2/5$

13. $4\pi\sqrt{2}$ 15. $\frac{\pi}{9}(2\sqrt{2} - 1)$ 17. $56\pi/3$

19. $\frac{128\pi}{1215}(125\sqrt{10} + 1)$ 21. $1179\pi/256$

23. $\frac{8\pi}{3}(2\sqrt{2} - 1)$ 25. $\frac{\pi}{27}(13\sqrt{13} - 8)$ 27. 2π

29. $64\pi/3$ 31. $\frac{2\pi}{3}(5\sqrt{5} - 1)$ 33. 4π

Section 7.5

3. 0.03 newton-meters   5. $\frac{384}{5}$ dyne-cm

7. $\frac{84}{5}$ dyne-cm   9. $\frac{175}{2}$ in-lb   11. 30 in-lb

13. 11×10^5 ft-lb   15. $\frac{2}{81} \times 10^4$ mile-tons

17. $\frac{5650}{27}$ mile-tons   19. 5 dyne-cm

21. (a) $30 - \dfrac{300}{r}$ dyne-cm   (b) 30 dyne-cm

23. $39,375\pi$ ft-lb   **25.** 2,296,875 ft-lb
27. $9765.625\pi \approx 30,680$ ft-lb   **29.** 3750π ft-lb
33. 250,000/3 newton-meters

Section 7.6

1. 7.2 ft from the 120-lb person   **3.** $\bar{x} = \frac{5}{2}$
5. $\frac{377}{14} \approx 27$ miles from city A   **9.** $(1, -25)$
11. $(\frac{1}{3}, \frac{1}{3})$ **13.** $\left(\frac{4a}{3\pi}, \frac{4a}{3\pi}\right)$ **15.** $(\frac{4}{5}, \frac{2}{7})$ **17.** $(\frac{1}{2}, \frac{2}{5})$
19. $(0, \pi/8)$ **21.** $(0, \frac{1}{3})$ **23.** $3a/8$ **25.** $(\frac{5}{6}, \frac{5}{16})$
27. $\frac{3}{5}$ **37.** $(\pi, 4a/3)$ **39.** $\frac{2}{35}(5 + 3\sqrt{2})$

Section 7.7

13. $y = \frac{1}{27}(x + 3)^3$ **15.** $y = \frac{1}{4}(x + 1)^2 + 1$
17. $y = \sec x$ **23.** $y = \frac{1}{1 - x}$ **25.** $y = \tan x - 1$
27. (b) $\phi(x) = \frac{1}{4}(x + 1)^2 + 1$ (c) $\phi(x) = \frac{1}{4}x^2 + 1$
(d) $(0, 0)$ is not in the domain of $f(x, y) = \sqrt{y - 1}$,
(e) the domain of f **29.** (c) $y = x - 1/x$ **31.** 400 ft
33. 4 sec; 104 ft/sec   **35.** yes   **37.** 25.6 ft
39. (b) $x = \cos at$ (c) $x = \sin at - \cos at$ **41.** 0.8 ft
43. $(x - 2)^2 + y^2 = 9$

True-False Quiz (Chapter 7)

1. T   **2.** F   **3.** T   **4.** T   **5.** F   **6.** T   **7.** T
8. T   **9.** T   **10.** T   **11.** T   **12.** T   **13.** T
14. F   **15.** F   **16.** F   **17.** T

Additional Problems (Chapter 7)

1. $\frac{5}{12}$ **3.** $\frac{17}{12}$ **5.** $\frac{5}{6}$ **7.** $\frac{125}{6}$ **9.** 1
11. $\frac{1}{2}(2 - \sqrt{2})$ **13.** $4\pi/3$ **15.** $\pi/4$ **17.** $28\pi/3$
19. $2\pi/3$ **21.** $4\pi/3$ **23.** $\pi^2/4$ **25.** 144
27. $\pi^2/2$ **29.** $\frac{2}{27}(13\sqrt{13} - 8)$ **31.** $2\sqrt{5}$
33. $5\sqrt{5} - 2\sqrt{2}$ **35.** $\frac{5}{3}$ **37.** 4π
39. $\frac{2\pi}{3}(5\sqrt{5} - 1)$ **41.** 2π **43.** 0.18 newton-meters
45. 17,500 ft-lb   **47.** 4950π ft-lb   **49.** mass $= 3$;
center of mass is at midpoint   **51.** $(0, \frac{2}{5})$ **53.** $(0, \frac{2}{7})$
55. $(0, \frac{4}{7})$ **57.** $y = \frac{1}{4}(x^2 + 2)^2$ **59.** $y = \frac{-1}{x^2 + C}$; $y = 0$

Section 8.1

1. (a) $\ln 2 = \int_1^2 \frac{dt}{t}$ (b) 0.691   **3.** 2.89
5. 0.41   **7.** 1.61   **9.** $\ln x + \ln y + 3 \ln z$
11. $\ln x + 2 \ln y - \ln z$ **13.** $\ln x + \frac{1}{2} \ln y + \frac{1}{2} \ln z$
17.   **19.**

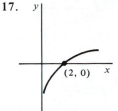

21. $\frac{2}{2x - 5}$ **23.** $-\tan x$ **25.** $\frac{2}{x} \ln x$ **27.** $\frac{1 - \ln x}{x^2}$
29. $1/x$ **31.** $3/x$ **33.** $\frac{1}{x} + \cot x$ **35.** $\frac{x}{x^2 + 1}$
37. $\frac{-2x}{x^2 + 1}$ **39.** (a) $f(x) = 0$ (c) the x axis with
holes at $0, \pm\pi/2, \pm\pi, \pm3\pi/2, \ldots$ **43.** $\frac{y(1 - 2x)}{x(y - 1)}$
47. (a) $x > 0$ (b) $y \to \infty$ (c) minimum $f(1) = 1$
(d) always concave up

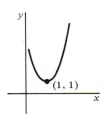

51. $f(x) = \ln x - \ln a$

Section 8.2

1. $f^{-1}(x) = \frac{1}{2}x$ **3.** $f^{-1}(x) = x^3$

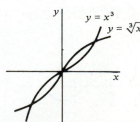

5. no inverse

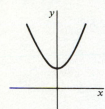

7. $f^{-1}(x) = \sqrt{x-1}$

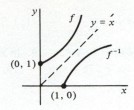

43. exists

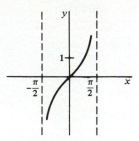

53. $dy/dx = -x/y; dx/dy = -y/x$

9. $f^{-1}(x) = \sqrt[3]{x-1}$

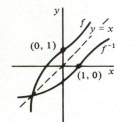

11. $f^{-1}(x) = 1/x$

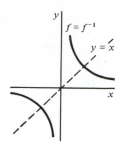

Section 8.3

1. The original number is returned.   **3.** The domains are $(0, \infty)$ and $(-\infty, \infty)$. When $x > 0, f_1(x) = f_2(x)$.
5. same as Problem 3   **7.** same as Problem 3

9. $f^{-1}(x) = -\ln x$ **11.** $f^{-1}(x) = \dfrac{1}{2}\ln\dfrac{x+1}{x-1}$

13. $f^{-1}(x) = \frac{1}{2}e^x$ **15.** $f^{-1}(x) = 1 - e^{x/2}$ **17.** $-e^{-x}$

19. $-2xe^{-x^2}$ **21.** $-e^{\cos x}\sin x$ **23.** $(2x+1)e^{2x}$

25. $e^{-x}(\cos x - \sin x)$ **27.** $\frac{1}{2}(e^x + e^{-x})$

29. $\dfrac{4}{(e^x + e^{-x})^2}$ **31.** e

13. $f^{-1}(x) = \dfrac{x}{x-1}$

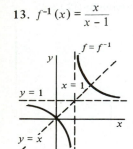

27. $\frac{1}{2}$ **29.** $3x^2$ **31.** $\dfrac{1}{2\sqrt{x-1}}$ **33.** $\dfrac{1}{3(x-1)^{2/3}}$

35. $-1/x^2$ **37.** $-1/(x-1)^2$

39. does not exist

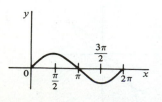

41. exists

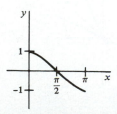

35.

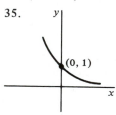

37.

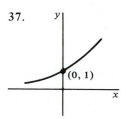

39.

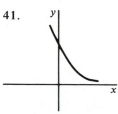

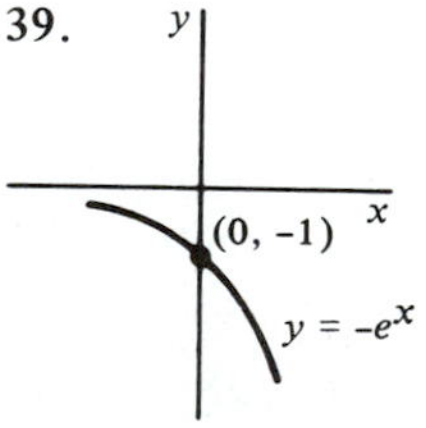

41.

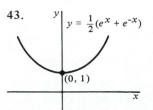

43.

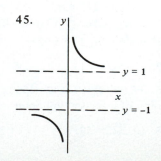

45.

47.

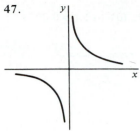

49.

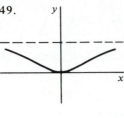

21. $x^{\ln x}\left(\dfrac{2\ln x}{x}\right)$ **23.** 0; meaningless

25. $x^2\sqrt{1-x}\left[\dfrac{2}{x}-\dfrac{1}{2(1-x)}\right]$

27. $\dfrac{x(x^2-1)^3}{\sqrt{x^2+1}}\left(\dfrac{1}{x}+\dfrac{6x}{x^2-1}-\dfrac{x}{x^2+1}\right)$ **29.** $\dfrac{1}{x\ln 2}$

31. $\dfrac{1+\ln x}{\ln 3}$ **37.** $2.7182804\cdots$ and $2.7182831\cdots$

(both correct to five places)

51.

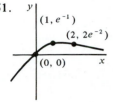

53.

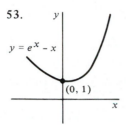

Section 8.5

1. $\ln|x+3|+C$ **3.** $\frac{1}{2}\ln|2x+1|+C$
5. $\frac{1}{2}\ln(x^2+1)+C$ **7.** $\frac{1}{3}\ln|x^3-1|+C$
9. $\frac{1}{2}e^{2x}+C$ **11.** $\frac{1}{2}e^{x^2}+C$ **13.** $-e^{\cos t}+C$
15. $\ln|\ln x|+C$ **17.** $\frac{1}{2}\ln|1+\sin 2t|+C$
19. $\ln|\tan t+2|+C$ **21.** $-2\ln|4-\sec(t/2)|+C$
23. $\frac{1}{2}\ln|\sec 2t|+C$ **25.** $\ln|\sec t(\sec t+\tan t)|+C$
27. $\frac{1}{2}\ln|\csc 2t-\cot 2t|+C$ **29.** $\ln 3$ **31.** $\frac{1}{2}\ln\frac{5}{4}$
33. $\frac{1}{3}\ln\frac{7}{8}$ **35.** 1 **37.** $2(e-1)$ **39.** $\frac{1}{3}(e-1)$
41. $\ln 2$ **43.** $\ln(1+\sqrt{2})$ **45.** $\frac{1}{2}\ln 2$ **47.** $\frac{1}{2}\ln 3$
49. $e-1$ **51.** $\frac{3}{2}-2\ln 2$ **53.** $\pi\ln 4$
55. $\pi(1-1/e)$ **57.** $\ln(2+\sqrt{3})$

55.

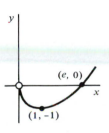

57.

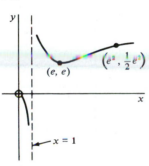

Section 8.6

1. $P=50e^{2t}$ **3.** $T=30+60e^{-2t}$; 30 **5.** $m=10(2^{-t})$
7. **(a)** $P=20,000(2.4)^{t/3}$ **(b)** $154,237$ **(c)** about
13.4 hr   **9.** about 64.3 lb; about 6.7 months
11. in about 49 min   **13.** **(a)** $\$4558$ **(b)** $V=1000+$
$49,000(\frac{49}{29})^{-t/2}$; approaches $\$1000$ **15.** **(a)** $\$2012.20$
(b) $\$2032.79$ **(c)** $\$2047.10$ **(d)** $\$2054.19$
(e) $\$2054.43$ **19.** $m=m_0(2^{-t/1600})$
21. about $15,053$ yr   **25.** $x=e^{-1/2}$ **27.** $I=I_0e^{-Rt/L}$
35. $y=e^x,\ y=e^{-x}$ **37.** $y=1,\ y=e^x$

Section 8.4

1. $31.544\cdots$ **3.** $4.7288\cdots$
5.

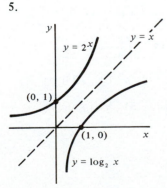

7.

9. $3^x\ln 3$ **11.** $2^{x^2+1}(x\ln 2)$ **13.** ex^{e-1}

15. $\pi(\sin x)^{\pi-1}\cos x$ **19.** $x^{\sin x}\left[\dfrac{\sin x}{x}+(\cos x)(\ln x)\right]$

True-False Quiz (Chapter 8)

1. T   **2.** T   **3.** F   **4.** T   **5.** T   **6.** T   **7.** T
8. T   **9.** T   **10.** F   **11.** F   **12.** T   **13.** T
14. F   **15.** F   **16.** F   **17.** T   **18.** T   **19.** T
20. T

Additional Problems (Chapter 8)

1. (a) $\ln 5 = \int_1^5 \frac{dt}{t}$ (b) 1.575 (2% error)   3. $\frac{x}{x^2-1}$

7. $x = \pi/4 + n\pi$, where n is an integer   9. $x = 1$

11. $(\frac{1}{2}e^{-1}, \frac{1}{2}e^{-1})$

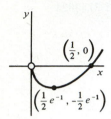

13. (a) $y > 0$ when $x > 0$; $y < 0$ when $x < 0$
(b) rising when $x < 1$, falling when $x > 1$ (c) concave
up when $x > 2$, down when $x < 2$ (d) It appears from
the graph that $y \to 0$ when $x \to \infty$. (This will be confirmed
in Problem 32, Section 11.2.) When $x \to -\infty$, $y \to -\infty$
(no guesswork needed).

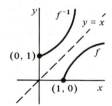

15. $f^{-1}(x) = x^2 + 1$

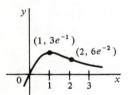

17. $f^{-1}(x) = \frac{3}{x-1}$ 19. $f^{-1}(x) = 1 - \ln \frac{x}{2}$

21. $f^{-1}(x) = e^{2(1-x)} - 2$ 23. $\log_2 5 = \frac{\ln 5}{\ln 2}; \log_2 5 =$

$\frac{\log 5}{\log 2}$ 29. $-\frac{1}{2}e^{-x^2} + C$ 31. $\sec t + C$

33. $\frac{1}{2}\ln(e^{2x} + 1) + C$ 35. $y = \ln(x^2 - 3)$ 37. $\frac{1}{2}\ln 2$

39. $\ln 7$ 41. $\frac{1}{2}(e^2 - e^{-2})$ 43. $\ln(1 + \sqrt{2})$

45. $\frac{7}{3} - \ln 2$ 47. $\pi(1 - e^{-1})$ 49. $y = 1$

51. $Q = 10(2^{-t})$; about 3.3 sec   53. $1391.71

55. (d) $\tan^{-1} x = \int_0^x \frac{dt}{1+t^2}$ 57. (d) $\ln 1.1 = 0.0953$

(correct to four places)   59. (b) error $\leq \frac{(x-1)^{n+1}}{n+1}$

Section 9.1

1. $\pi/6$ 3. $\pi/3$ 5. 0   7. undefined   9. $5\pi/6$
11. $-\pi/6$ 13. 2   15. $\frac{1}{3}$ 17. $-\frac{16}{65}$

19. $1/\sqrt{4-x^2}$ 21. $\frac{2}{x\sqrt{x^4-1}}$ 23. $\frac{1}{1+(1-x)^2}$

25. $\frac{1-x}{\sqrt{1-x^2}}$ 27. $\sin^{-1} x$ 29. $\frac{1}{x^2+1}$ 31. -1

33. $\frac{-\sin x}{|\sin x|}$ 35. $\frac{1+x}{1+x^2}$ 37. $\frac{dy}{dx} = \frac{x+y}{x-y}$

45.

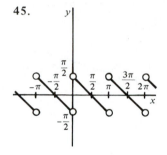

Section 9.2

1. $\sin^{-1} \frac{x}{5} + C$ 3. $\frac{1}{2}\sin^{-1} \frac{x^2}{3} + C$ 5. $\frac{1}{12}\tan^{-1} \frac{4x}{3} + C$

7. $\sec^{-1} 3x + C$ 9. $\frac{1}{2}\tan^{-1} \frac{x-1}{2} + C$ 11. $\pi/3$

13. $\pi/6$ 15. $\frac{1}{2}\ln 2 + \frac{\pi}{4}$ 17. $\frac{1}{2}(\frac{\pi}{3} - \cos^{-1} \frac{2}{3})$

19. $\pi/8$ 21. $\frac{2}{3}(3\sqrt{3} - \pi)$ 23. $\pi^2/12$ 25. πr

27. $x = \cos at$ 29. $\frac{9}{20}$ rad/sec   31. $\frac{16}{5}$ ft/sec
41. $\pi/2$

Section 9.3

1. right-hand branch of the hyperbola $x^2 - y^2 = 1$; upward
7. $\sinh t = -3/\sqrt{7}$, $\cosh t = 4/\sqrt{7}$, $\coth t = -4/3$, $\text{sech } t =$
$\sqrt{7}/4$, $\text{csch } t = -\sqrt{7}/3$ 27. $2\cosh 2x$

29. $2x \operatorname{sech}^2 x^2$ **31.** $\cosh t - \sinh t$ **33.** $x \cosh x + \sinh x$ **35.** $\tanh x$ **37.** $e^{-x}(\cosh x - \sinh x)$
39. $x = -\frac{1}{2}\ln 3$ **41.** $\frac{1}{4}$ **43.** $\tanh 1$ **45.** $\ln(\cosh 1)$
49. $\pi(1 - \tanh 1)$ **51.** $\frac{\pi}{2}(\sinh 2 + 2)$

Section 9.4

1. 0   **3.** 0   **5.** $\frac{1}{2}\ln 3$ **7.** undefined
9. $x = \pm\frac{1}{2}\ln(3 + 2\sqrt{2})$ **13.** $f^{-1}(x) = 2\tanh^{-1}\frac{1}{2}x$
15. upper half   **17.** $\dfrac{6}{\sqrt{4x^2 + 1}}$ **19.** $\dfrac{2}{4 - x^2}$
21. $\sinh^{-1} x$ **23.** $|\sec x|$ **25.** $\frac{1}{6}\ln 5$ **27.** $-\frac{1}{10}\ln 4$
29. $\ln(2 + \sqrt{5})$ **31.** $\ln(7 + 4\sqrt{3}) - \ln(2 + \sqrt{3})$
33. $\frac{1}{3}(\cosh^{-1} 3 - \cosh^{-1}\frac{3}{2})$ **35. (a)** a

Section 9.5

1. (c) 1.18   **(d)** 13.23 ft   **5.** 14.1 lb; 22.9 lb

True-False Quiz (Chapter 9)

1. F   **2.** T   **3.** F   **4.** T   **5.** F   **6.** T   **7.** F
8. F   **9.** F   **10.** F   **11.** T   **12.** T   **13.** T
14. F   **15.** F   **16.** F   **17.** T   **18.** F   **19.** T
20. T   **21.** F   **22.** F   **23.** T   **24.** F   **25.** T

Additional Problems (Chapter 9)

1. $-\pi/4$ **3.** $\pi/4$ **5.** 0   **7.** undefined   **9.** 1
11. 0   **15.** $-1/\sqrt{2x - x^2}$ **17.** $\dfrac{1}{x\sqrt{4x^2 - 1}}$
19. $\dfrac{-2x}{1 + (x^2 - 1)^2}$ **21.** $\dfrac{x}{\sqrt{1 - x^2}} + \sin^{-1} x$ **23.** $\tan^{-1} x$
25. $\dfrac{\operatorname{sech}^2\sqrt{x}}{2\sqrt{x}}$ **27.** $-\tanh x$ **29.** $2/\sqrt{4x^2 - 1}$
31. $\pi/6$ **33.** $\frac{1}{2}\sin^{-1} x^2$ **35.** $\frac{1}{6}\ln\left|\dfrac{x}{x + 6}\right| + C$
37. $2(\cosh 1 - 1)$ **41.** $y = \sin(x - 1)$ **43.** $\pi/6$
45. $\frac{\pi}{6}\ln 5$ **47.** $f^{-1}(x) = (\sinh^{-1}\frac{1}{2}x)^2$

Section 10.1

1. $\frac{1}{2}xe^{2x} - \frac{1}{4}e^{2x} + C$ **3.** $-x\cos x + \sin x + C$
5. $\frac{1}{5}x(x + 1)^5 - \frac{1}{30}(x + 1)^6 + C$ **7.** $(1 - x)\cot x + \ln|\sin x| + C$ **9.** $\frac{2}{3}x(x - 1)^{3/2} - \frac{4}{15}(x - 1)^{5/2} + C$
11. $\frac{1}{2}x^2 \ln x - \frac{1}{4}x^2 + C$ **13.** $x^2 e^x - 2xe^x + 2e^x + C$
15. $\frac{1}{2}e^x(\sin x - \cos x) + C$ **17.** $\frac{1}{2}x\sin(\ln x) - \frac{1}{2}x\cos(\ln x) + C$ **19.** $x\cos^{-1} x - \sqrt{1 - x^2} + C$
21. $x\sinh^{-1} x - \sqrt{x^2 + 1} + C$ **23.** $-\frac{3}{8}\sin x\cos 3x + \frac{1}{8}\cos x\sin 3x + C$ **25.** $\frac{1}{2}\sec x\tan x + \frac{1}{2}\ln|\sec x + \tan x| + C$ **27.** $1 - 2e^{-1}$ **29.** $\frac{1}{10}$
31. $\frac{8}{3}\ln 2 - \frac{7}{9}$ **33.** $\frac{1}{2}(e^{-\pi} + 1)$ **35.** $\frac{1}{2}\sqrt{2} - \frac{1}{2}\ln(\sqrt{2} - 1)$ **37.** $\frac{1}{2}\tanh^{-1}\frac{1}{2} + \frac{1}{2}\ln\frac{3}{4}$
39. $\frac{1}{4}e^{2x} - \frac{1}{2}x + C$ **41.** 1   **43.** 2π
45. $\bar{x} = \pi/2 - 1, \bar{y} = \pi/8$ **47.** $-e^{-x}(x^2 + 2x + 2) + C$
49. $e^x[(1 - x)^3 + 3(1 - x)^2 + 6(1 - x) + 6] + C$
51. $(1 - x^2)\cos x + 2x\sin x + C$ **53.** $(2x - 1)\cos x + (x^2 - x - 1)\sin x + C$ **55.** $\frac{1}{5}x^3(x - 1)^5 - \frac{1}{10}x^2(x - 1)^6 + \frac{1}{35}x(x - 1)^7 - \frac{1}{280}(x - 1)^8 + C$ **61.** $5\pi/32$ **63.** $\frac{16}{15}$

Section 10.2

1. $\frac{2}{35}$ **3.** $\frac{3}{4}(\sin x)^{4/3} - \frac{3}{10}(\sin x)^{10/3} + C$
5. $\frac{1}{4}\sin^4 x - \frac{1}{6}\sin^6 x + C$ **7.** $\frac{1}{7}\cos^7 x - \frac{1}{5}\cos^5 x + C$
9. $3\pi/8$ **11.** $\frac{2}{3}$ **13.** $\frac{1}{4}\sec^4 x - \sec^2 x + \ln|\sec x| + C$
15. $\tan x - \frac{3}{2}x + \frac{1}{4}\sin 2x + C$ **17.** $\frac{1}{2}\sin^2 t + C$
19. $\frac{1}{6}$ **21.** $\ln|\csc 2t - \cot 2t| + C$ **23.** $\frac{1}{3}\sec^3 x + C$
25. $\frac{6}{35}(33\sqrt{3} - 2)$ **27.** $\frac{1}{2}\tan^2 x + \ln|\cos x| + C$
29. $\frac{28}{15}$ **31.** $\frac{2}{7}(\sec x)^{7/2} - \frac{2}{3}(\sec x)^{3/2} + C$
33. $1 - \pi/4$ **35.** $-\frac{1}{5}\cot^5 t - \frac{1}{3}\cot^3 t + C$ **37.** $\frac{4}{3}$
39. $\frac{7}{24}$

Section 10.3

1. $-\dfrac{\sqrt{4 - x^2}}{4x} + C$ **3.** $\ln|\sqrt{1 + x^2} + x| - \dfrac{\sqrt{1 + x^2}}{x} + C$
5. $\frac{1}{2}x\sqrt{x^2 - 1} + \frac{1}{2}\ln|x + \sqrt{x^2 - 1}| + C$

7. $x/\sqrt{1-x^2} + C$ 9. $\dfrac{x}{4\sqrt{4+9x^2}} + C$

11. $-\dfrac{\sqrt{1-4x^2}}{x} - 2\sin^{-1} 2x + C$ 13. $\frac{1}{2} x\sqrt{9+x^2} -$

$\frac{9}{2}\ln |\sqrt{9+x^2} + x| + C$ 15. $\dfrac{(x^2-9)^{3/2}}{27x^3} + C$

17. $\frac{1}{2}x^2 \sin^{-1} x - \frac{1}{4}\sin^{-1} x + \frac{1}{4}x\sqrt{1-x^2} + C$

19. $1/\sqrt{3} - \pi/3$ 21. $\frac{1}{4}\ln\dfrac{\sqrt{6}}{2}$ 23. $\sqrt{3}/3 - \pi/6$

25. $\frac{1}{24}(40\sqrt{5} - 17\sqrt{17})$ 27. $24\sqrt{3} - 24\ln(2 + \sqrt{3})$

29. (a) $\frac{1}{2}\ln(x^2 + 9) + C$ (b) $\ln\dfrac{\sqrt{x^2+9}}{3} + C$

(c) The answers in (a) and (b) differ by the constant ln 3.
31. $9(2 - \sqrt{2})$ 33. $\sqrt{5} + \frac{1}{2}\ln(2 + \sqrt{5})$

35. $y = \ln(x + \sqrt{x^2-1}) - \dfrac{\sqrt{x^2-1}}{x}, x \geq 1$

49. $\frac{1}{12}(3\sqrt{2} - \sqrt{10})$ 51. $\tan(x/2) + C$

53. $\frac{1}{2}\ln |\tan(x/2)| - \frac{1}{4}\tan^2(x/2) + C$

Section 10.4

1. $\frac{1}{4}\ln\left|\dfrac{x-1}{x+3}\right| + C$ 3. $\ln\left|\dfrac{x^2-1}{x}\right| + C$

5. $\dfrac{1}{x} + \ln\left|\dfrac{x-1}{x}\right| + C$ 7. $\frac{1}{8}\ln\left(\dfrac{x^2}{x^2+4}\right) + C$

9. $-\dfrac{1}{9x} - \dfrac{1}{27}\tan^{-1}\dfrac{x}{3} + C$ 11. $\frac{1}{2}\ln(x^2+1) +$

$\dfrac{1}{2(x^2+1)} + C$ 13. $\dfrac{1}{24}\ln\dfrac{(x-2)^2}{x^2+2x+4} -$

$\dfrac{\sqrt{3}}{12}\tan^{-1}\left(\dfrac{x+1}{\sqrt{3}}\right) + C$ 15. $-\frac{1}{3}\ln 2$ 17. $\frac{1}{2} + \ln 2$

19. $\frac{1}{2}\ln\frac{8}{5}$ 21. $\dfrac{1}{16} - \dfrac{1}{8}\tan^{-1} 2 + \dfrac{\pi}{32}$

23. $\frac{1}{3}\ln 2 + \dfrac{\pi\sqrt{3}}{9}$ 27. $\dfrac{1}{10}\ln\dfrac{x}{10-x} + C$

33. (a) $x \to 10$ (b) in the year 2090   (c) in the year 2040   35. (a) $x = x_0 e^{kt}$ (c) when $t =$

$\dfrac{1}{k}\ln[1 + k/(ax_0)]$ 37. $-\ln |1 - \tan(x/2)| + C$

Section 10.5

1. $\frac{1}{2}\ln\left(\dfrac{7+4\sqrt{3}}{3}\right)$ 3. $2\sin\sqrt{x} - 2\sqrt{x}\cos\sqrt{x} + C$

5. $\ln 2 - \frac{1}{2}$ 7. $\frac{1}{2}\ln\left|\dfrac{x-2}{x}\right| + C$ 9. $\pi/32$ 11. 1

13. $x + \frac{1}{2}\ln\left|\dfrac{x-1}{x+1}\right| - \tan^{-1} x + C$ 15. $\dfrac{2\pi}{3} - \dfrac{\sqrt{3}}{2}$

17. $e^x(x^3 - 3x^2 + 6x - 5) + C$ 19. $\frac{2}{15}(1 + \sqrt{2})$

21. $\frac{1}{5}\tan^5 t + \frac{2}{3}\tan^3 t + \tan t + C$ 23. $\frac{1}{3}x^3 + x^2 +$

$4x + 16\ln |x-2| + C$ 25. $\frac{5}{2}\ln |x^2 - 1| - 4\ln |x| + C$

27. $\ln |e^t - 1| + C$ 29. $\pi^2/32$ 31. $3\sin^{-1}\dfrac{x-3}{3} -$

$\sqrt{6x - x^2} + C$ 33. $-\dfrac{\sqrt{4-x^2}}{2x^2} - \dfrac{1}{4}\ln\left|\dfrac{2-\sqrt{4-x^2}}{x}\right| + C$

35. $\frac{1}{4}\ln\left(\dfrac{2+\sin t}{2-\sin t}\right) + C$ 37. $\frac{1}{5}e^x(2\sin 2x + \cos 2x) + C$

39. $-\frac{1}{8}e^{-2x}(4x^3 + 6x^2 + 6x + 3) + C$

Section 10.6

1. 0.697024 (true value: ln 2; 0.6% error)
3. 1.722257 (true value: $e - 1$; 0.2% error)
5. 1.954097 (true value: 2; 2.3% error)
7. 0.782794 (true value: $\pi/4$; 0.3% error)   9. 0.748927
(true value: $\pi/4$; 4.6% error)   11. 1.096795
13. 0.693254 (true value: ln 2; 0.02% error)
15. 1.718289 (true value: $e - 1$; 0.0004% error)
17. 2.000863 (true value: 2; 0.04% error)
19. 0.785392 (true value: $\pi/4$; 0.0008% error)
21. 0.770899 (true value: $\pi/4$; 1.8% error)
23. 1.089413   25. $E_n < 0.02$ (round off answer to Problem 1 to 0.7)   27. $E_n < 0.007$ (round off answer to Problem 3 to 1.7)   29. $E_n < 0.08$ (round off answer to Problem 5 to 2)   31. $E_n < 0.0006$ (round off answer to Problem 13 to 0.69)   33. $E_n < 0.00002$ (round off answer to Problem 15 to 1.7183)
35. $E_n < 0.002$ (round off answer to Problem 17 to 2.00)
37. 3.131176 (accurate to one place)   39. 50 in^2
41. 13.7 meters   43. 35.25   45. 3.83   47. 14

Additional Problems (Chapter 10)

1. $x - 2\ln |x+2| + C$ 3. $2\tan x + 2\sec x - x + C$
5. $-\sqrt{x^2+1}/x + C$ 7. $2\ln |\csc t - \cot t| - \ln |\sin t| + C$

9. $2e^{\sqrt{x}}(\sqrt{x}-1)+C$ 11. $\ln 2 - 2 + \pi/2$
13. $-e^{1/x}+C$ 15. $x \sinh x - \cosh x + C$
17. $\frac{1}{5}\sec^5 x - \frac{2}{3}\sec^3 x + \sec x + C$ 19. $\frac{5}{2}\ln|x-5| -$
$\frac{3}{2}\ln|x-3|+C$ 21. $\ln(3/2)$ 23. $\frac{1}{2}(e^{-\pi/2}+1)$
25. $\pi/12$ 27. $\frac{1}{3}\ln|3x+\sqrt{9x^2-16}|+C$
29. $\frac{2}{3}(2\pi - 3\sqrt{3})$ 31. $2\ln(4/3)$
33. $\frac{1}{2}x^2\sin^{-1}x^2 + \frac{1}{2}\sqrt{1-x^4}+C$ 35. $\frac{1}{6}\ln(5/2)$
37. $\frac{5}{2}t + \frac{1}{4}\sin 2t + \tan t + C$ 39. $\ln|x+\sqrt{x^2-4}|+C$
41. $\frac{19}{324}$ 43. $x\tan x - \ln|\sec x|+C$ 45. $\frac{5}{3}(3\sqrt{3}-\pi)$
47. $\frac{1}{4}x^2 + \frac{5}{4}x + \frac{21}{8}\ln|2x-5|+C$ 49. $\sqrt{4+x^2}+C$
51. $\frac{1}{3}$ 53. $\frac{1}{2}\sin^{-1}\frac{x^2}{2}+C$ 55. $\sqrt{2}/18$
57. $\frac{2}{3}\ln|x+1| + \frac{1}{6}\ln(x^2-x+1) +$
$\frac{1}{\sqrt{3}}\tan^{-1}\left(\frac{2x-1}{\sqrt{3}}\right)+C$ 59. $\pi/(3\sqrt{3})$
61. $-\frac{1}{x} - \tan^{-1}x + C$ 63. $\frac{1}{2}\ln(2+\sqrt{3})$
65. $2\sqrt{x^2-5x+4}+C$ 67. $-e^{-x}(x^3+3x^2+6x+6)+C$
69. $\frac{1}{2}(e^\pi+1)$ 71. $7\sinh 1 - 9\cosh 1 + 6$
73. $\frac{1}{4}x^2\sqrt{x^4-16} - 4\ln(x^2+\sqrt{x^4-16})+C$
75. $\ln 2 - \frac{1}{2}$ 77. $4\ln 2 - \frac{15}{16}$ 79. $\frac{1}{30}$
81. $\sec x - \tan x + x + C$ 83. $\frac{44}{3}$ 85. $-\cot\frac{t}{2}+C$
87. $\tan\frac{t}{2}+C$ 89. $\frac{16}{35}$ 91. $105\pi/384$
93. $\frac{\pi}{4} - \frac{1}{2}\ln 2$ 95. (a) $\frac{1}{2}I_0 V_0$ (b) $\frac{1}{2}I_0 V_0 \cos\alpha$
97. 0.646459   99. 6.194562   101. 1.149717

Section 11.1

1. 1   3. $-\infty$ 5. $-\infty$ 7. $-\infty$ 9. $-\infty$ 11. 0
13. ∞ 15. 0   17. ∞ 19. ∞ 27. $0; 1; \infty$

Section 11.2

1. 2   3. 0   5. 0   7. 1   9. 0   11. 1   13. 1
15. $\frac{1}{2}$ 17. 6   19. 0   21. $-\infty$ 23. 0   25. 0
27. $1/e$ 29. 1   31. $y \to \infty$ when $x \to \infty$; $y \to 0$
when $x \to -\infty$ 33. $y \to \infty$ when $x \downarrow 0$; $y \to -\infty$ when
$x \uparrow 0$; $y \to \infty$ when $x \to \infty$; $y \to 0$ when $x \to -\infty$

35. $y \to 0$ when $x \downarrow 0$; $y \to \infty$ when $x \to \infty$ 37. $y \to -\infty$
when $x \downarrow 0$; $y \to 0$ when $x \to \infty$ 39. 1   41. (b) No;
the derivative of sine depends on it.   45. When $x = 100$,
$x - \sinh x \approx -1.3 \times 10^{43}$.   49. (a) P and Q both
approach $(1, 0)$.   (b) B approaches $(0, -1)$.

Section 11.3

1. $2\sqrt{2}$ 3. diverges   5. diverges   7. diverges
9. $\pi/2$ 11. diverges   13. diverges   15. 0
17. diverges   19. π 21. diverges   23. diverges
25. $\ln(2+\sqrt{3})$ 27. $9\pi/4$ 29. diverges   31. $\frac{1}{4}$
33. no   35. $20/\sqrt{3}$ 37. $x = e$

Section 11.4

1. $\pi/4$ 3. 1   5. 0   7. diverges   9. π 11. $\frac{1}{2}$
13. $\frac{1}{2}\ln 3$ 15. diverges   17. diverges   19. π
21. diverges   27. yes; 2   29. yes; $\pi/2$ 39. $1/x^2$
41. $\frac{x}{x^2+1}$ $(x > 0)$

Section 11.5

1. If $f(x)$ is a polynomial of degree n, the nth-order
Taylor polynomial associated with f at $a = 0$ is $f(x)$.

5. Since $e^{0.2} < e < 3$, $|R_4(0.2)| < \frac{3(0.2)^5}{5!} = 0.8 \times 10^{-5} <$

0.00001. Hence the computation of $e^{0.2}$ by the formula
in Problem 3 (with $n = 4$) is correct to at least four places,
$e^{0.2} \approx 1.2214$.   17. $\sin 0.2 \approx 0.1987$
23. 0.980066578 (correct to the places shown)

29. $\sqrt{x} \approx 1 + \frac{1}{2}(x-1) - \frac{1}{8}(x-1)^2$; $\sqrt{0.98} \approx 0.98995$

33. $\ln 1.1 \approx 0.09531$ 45. (a) 0.461272
(b) 0.461286

True-False Quiz (Chapter 11)

1. T   2. T   3. F   4. F   5. T   6. F   7. F
8. F   9. T   10. T   11. T   12. T

Additional Problems (Chapter 11)

1. 0   **3.** 1   **5.** ∞ **7.** ∞ **9.** 0   **11.** ∞ **13.** ∞

15. ∞ **17.** $\frac{1}{2}\ln B$ **19.** $1/B$ **21.** $\dfrac{2}{1+B}$ **23.** ∞

25. $-\infty$ **27.** 0   **29.** $\frac{1}{2}$ **31.** 1   **33.** 1   **35.** 0
37. $\pi/2$ **39.** 1   **41.** $-\infty$ **43.** 1   **45.** 2   **47.** 2
49. diverges   **51.** $\frac{1}{2}$ **53.** 4   **55.** $\frac{1}{3}\sec^{-1}\frac{5}{3}$
57. diverges   **59.** diverges   **61.** π **63.** 0.826
(true value: 0.826446 $\cdots$; 0.05% error)
65. $0.201\overline{3}$ (0.03% error)   **67.** 1.0954375
(0.0007% error)

Section 12.1

1. 0   **3.** 3   **5.** no limit   **7.** 1   **9.** 0
11. no limit   **13.** no limit   **15.** 1
17. (a) $s_n = 1000(0.9)^n$ (b) 0

Section 12.2

1. diverges   **3.** diverges   **5.** $\frac{1}{2}$ **7.** 1   **9.** $\frac{3}{4}$

11. 1   **13.** 4   **15.** 3   **17.** diverges   **19.** $\dfrac{\pi}{\pi-3}$

21. $\cot^2 x$ **23.** diverges   **25.** diverges

29. $1 < x < 3; \dfrac{1}{3-x}$ **31.** $\dfrac{1}{1+x} = 1 - x + x^2 - x^3 +$

$\cdots$ $(-1 < x < 1)$ **33.** $\dfrac{x}{1-x} = x + x^2 + x^3 +$

$\cdots$ $(-1 < x < 1)$ **37.** $\dfrac{a(1+r)}{1-r}$ **41.** $4 billion

43. $\frac{5}{11}$ (Going first is a losing proposition about 55% of the time.)

Section 12.3

1. converges   **3.** converges   **5.** diverges
7. converges   **9.** diverges   **11.** diverges
13. converges   **15.** converges   **17.** diverges
19. diverges   **21.** converges   **23.** converges
25. 1.07

Section 12.4

1. diverges   **3.** converges   **5.** diverges   **7.** diverges
9. converges   **11.** converges   **13.** converges
15. diverges   **17.** diverges   **19.** diverges
21. converges   **23.** converges   **25.** converges
27. converges

Section 12.5

1. absolutely convergent   **3.** divergent
5. conditionally convergent   **7.** absolutely convergent
9. absolutely convergent   **11.** absolutely convergent
13. conditionally convergent   **15.** no   **17.** converges
19. converges   **21.** converges   **23.** diverges
25. converges   **27.** diverges   **33.** 0.84147

Section 12.6

1. converges   **3.** converges   **5.** converges
7. converges   **9.** converges   **11.** converges
13. converges   **15.** converges   **17.** converges
19. diverges   **21.** converges absolutely for all x
23. converges absolutely for all x **25.** converges
absolutely for $-1 < x < 1$, diverges otherwise
27. converges absolutely for $-1 < x < 1$, diverges
otherwise   **29.** converges absolutely for $-1 < x < 1$,
conditionally at $x = -1$, diverges otherwise
31. converges absolutely for $1 \le x \le 3$, diverges otherwise
33. converges absolutely for $-\frac{1}{2} < x < \frac{1}{2}$, diverges
otherwise   **35.** converges absolutely for all x
37. converges absolutely for $-\sqrt{2} < x < \sqrt{2}$ ($x \ne 0$),
diverges otherwise   **39.** converges absolutely at $x = 0$,
diverges otherwise   **43.** no

True-False Quiz (Chapter 12)

1. F   **2.** F   **3.** F   **4.** F   **5.** F   **6.** F   **7.** F
8. F   **9.** F   **10.** F   **11.** F   **12.** F   **13.** F
14. F   **15.** F

Additional Problems (Chapter 12)

1. 0   3. no limit   5. 2   7. 0   9. diverges
13. $\frac{1}{4}$ 15. $\frac{5}{3}$; undefined   17. $f(x) = 1 + x^3 + x^6 +$
$x^9 + \cdots$ $(-1 < x < 1)$ 19. 30 ft   21. 1.20
23. The first three tests all show convergence; the Ratio
Test does not apply.   25. 0.54030   27. conditionally
convergent   29. converges   31. converges
33. diverges   35. converges   37. converges
39. diverges   41. converges absolutely for $-1 \leq x \leq 1$,
diverges otherwise   43. converges absolutely for $1 <$
$x < 3$, conditionally at $x = 1$, diverges otherwise
45. converges absolutely for $-2 \leq x \leq 4$, diverges
otherwise

Section 13.1

1. all x 3. all x 5. $-2 < x < 2$ 7. $0 \leq x < 2$
9. $2 \leq x \leq 0$ 11. 0.36788   13. $f(x) = \dfrac{2}{2 + x}$

15. $f(x) = \dfrac{1}{3 - x}$ $(1 < x < 3)$ 17. yes; no   19. no

21. $f'(x) = \displaystyle\sum_{n=1}^{\infty} \dfrac{(-1)^n x^{n-1}}{(n-1)!}$, all x

23. $f'(x) = \displaystyle\sum_{n=1}^{\infty} \dfrac{(x-1)^{n-1}}{n^{n-1}}$, all x

25. $f'(x) = \displaystyle\sum_{n=1}^{\infty} \dfrac{(-1)^n n x^{n-1}}{2^n}$, $-2 < x < 2$

27. $f'(x) = \displaystyle\sum_{n=1}^{\infty} (x-1)^{n-1}$, $0 < x < 2$

29. $f'(x) = \displaystyle\sum_{n=1}^{\infty} (-1)^{n-1} (x+1)^{2n-2}$, $-2 < x < 0$

31. $\displaystyle\sum_{n=1}^{\infty} \dfrac{x^n}{n(n-1)!}$ 33. $\displaystyle\sum_{n=1}^{\infty} \dfrac{(x-1)^{n+1}}{n+1}$

Section 13.2

1. $1 + x + \dfrac{x^2}{2!} + \dfrac{x^3}{3!} + \cdots$ 3. $1 + \dfrac{x^2}{2!} + \dfrac{x^4}{4!} + \dfrac{x^6}{6!} + \cdots$

5. $1 - \dfrac{x^2}{2!} + \dfrac{x^4}{4!} - \dfrac{x^6}{6!} + \cdots$ 7. $1 + x + x^2 + x^3 + \cdots$

9. $1 + 2x + 3x^2 + 4x^3 + \cdots$ 11. $1 + \dfrac{1}{2}x - \dfrac{1}{2\cdot4}x^2 +$

$\dfrac{1\cdot3}{2\cdot4\cdot6}x^3 - \dfrac{1\cdot3\cdot5}{2\cdot4\cdot6\cdot8}x^4 + \cdots$

13. $e\left[1 + (x-1) + \dfrac{(x-1)^2}{2!} + \dfrac{(x-1)^3}{3!} + \cdots\right]$

15. $-1 + \dfrac{(x-\pi)^2}{2!} - \dfrac{(x-\pi)^4}{4!} + \dfrac{(x-\pi)^6}{6!} - \cdots$

17. $(x-1) - \dfrac{1}{2}(x-1)^2 + \dfrac{1}{3}(x-1)^3 - \dfrac{1}{4}(x-1)^4 + \cdots$

19. $x + \dfrac{1}{3}x^3 + \dfrac{2}{15}x^5 + \cdots$ 21. $1 + x - \dfrac{1}{3}x^3 - \dfrac{1}{6}x^4 + \cdots$

23. 0 and 24   39. $1 + x(\ln 2) + \dfrac{x^2}{2!}(\ln 2)^2 +$

$\dfrac{x^3}{3!}(\ln 2)^3 + \cdots$

Section 13.3

1. $1 + x^2 + x^4 + x^6 + \cdots, -1 < x < 1$
7. $1 + (x-1)^2 + (x-1)^4 + (x-1)^6 + \cdots$

9. $\dfrac{1}{2}\left(\dfrac{x^2}{2!} - \dfrac{x^4}{4!} + \dfrac{x^2}{6!} - \cdots\right)$ 11. $\dfrac{1}{2}\left(2 - \dfrac{x^2}{2!} + \dfrac{x^4}{4!} - \cdots\right)$

15. $\sin c\left(1 - \dfrac{x^2}{2!} + \dfrac{x^4}{4!} - \cdots\right) + \cos c\left(x - \dfrac{x^3}{3!} + \dfrac{x^5}{5!} - \cdots\right)$

19. $1 - x + \dfrac{1}{3}x^3 - \dfrac{1}{6}x^4 + \cdots$ 27. $1 + \dfrac{1}{2}x^2 + \dfrac{5}{24}x^4 +$
$\dfrac{61}{720}x^6 + \cdots$ 29. 0.75   31. 0.4940   33. 0.6931
35. $\dfrac{1}{120}$

Section 13.4

1. $1 - x + x^2 - x^3 + \cdots$ 3. $1 + \dfrac{1}{2}x - \dfrac{1}{2\cdot4}x^2 +$
$\dfrac{1\cdot3}{2\cdot4\cdot6}x^3 - \dfrac{1\cdot3\cdot5}{2\cdot4\cdot6\cdot8}x^4 + \cdots$ 5. $1 - \dfrac{1}{2}x^2 +$
$\dfrac{1\cdot3}{2\cdot4}x^4 - \dfrac{1\cdot3\cdot5}{2\cdot4\cdot6}x^6 + \cdots$ 7. $1 + 4x + 6x^2 + 4x^3 + x^4$

9. $2\left(1 + \frac{1}{8}x^2 - \frac{1}{8^2 \cdot 2!}x^4 + \frac{1 \cdot 3}{8^3 \cdot 3!}x^6 - \frac{1 \cdot 3 \cdot 5}{8^4 \cdot 4!}x^8 + \cdots\right)$, $-2 < x < 2$ **13.** 2.04 **17.** 0.49204

19. $x - \frac{1}{2} \cdot \frac{x^3}{3} + \frac{1 \cdot 3}{2 \cdot 4} \cdot \frac{x^5}{5} - \frac{1 \cdot 3 \cdot 5}{2 \cdot 4 \cdot 6} \cdot \frac{x^7}{7} + \cdots$

21. 3.1623 **25.** 2.0805

37. $1 + x + \frac{1}{2}x^2 + \cdots$ **41.** $1 + \frac{1}{4}x^2 - \frac{3}{4^2(2!)}x^4 + \frac{3 \cdot 7}{4^3(3!)}x^6 - \frac{3 \cdot 7 \cdot 11}{4^4(4!)}x^8 + \cdots$ **45.** 0.5236

47. (a) $1, 2, 2, 2, \ldots$ (b) $\phi(x) = 1 + 2\left(x + \frac{x^2}{2!} + \frac{x^3}{3!} + \cdots\right) = 2e^x - 1$ **49.** $y = x - \frac{x^2}{2!} + \frac{x^3}{3!} - \cdots = 1 - e^{-x}$ **51.** $y = 1 - x + \frac{x^2}{2!} - \frac{x^3}{3!} + \cdots = e^{-x}$

Section 13.5

1. $y = 1 + x^2 + \frac{x^4}{2!} + \frac{x^6}{3!} + \cdots$ **3.** $y = x + \frac{x^3}{3!} + \frac{x^5}{7!} + \frac{x^7}{7!} + \cdots$ **5.** $y = x + \frac{2^2}{4!}x^4 + \frac{2^3 \cdot 5}{7!}x^7 + \frac{2^4 \cdot 5 \cdot 8}{10!}x^{10} + \cdots$ **7.** $y = x + \frac{2 \cdot 1}{3!}x^3 + \frac{2^2 \cdot 1 \cdot 3}{5!}x^5 + \frac{2^3 \cdot 1 \cdot 3 \cdot 5}{7!}x^7 + \cdots$ **9.** $y = 2e^x - x - 1$

11. $y = c_0 + (c_0 + 1)x + (c_0 + 2)\frac{x^2}{2!} + (c_0 + 3)\frac{x^3}{3!} + \cdots = c_0 e^x + xe^x$ **13.** $y = 0$

True-False Quiz (Chapter 13)

1. F **2.** T **3.** T **4.** F **5.** T **6.** T **7.** F **8.** T **9.** F **10.** T

Additional Problems (Chapter 13)

1. $-\frac{1}{2} \le x < \frac{1}{2}$ **3.** $-1 < x < 1$ **5.** $-3 \le x \le 1$

7. $0 < x \le 2$ **9.** $-(x - \pi) + \frac{(x - \pi)^3}{3!} - \frac{(x - \pi)^5}{5!} + \frac{(x - \pi)^7}{7!} - \cdots$ **11.** $\frac{1}{2}\left[2! - 3!(x - 1) + \frac{4!}{2!}(x - 1)^2 - \frac{5!}{3!}(x - 1)^3 + \cdots\right]$ **13.** $-\frac{1}{2}x^2 - \frac{1}{12}x^4$ **15.** 24

17. $t - t^3 + t^5 - t^7 + \cdots$ **23.** $1 - 2x + 3x^2 - 4x^3 + \cdots$ **25.** $x + x^2 + \frac{5}{6}x^3 + \frac{5}{6}x^4 + \cdots$ **27.** $x + x^2 + \frac{1}{3}x^3 + \cdots$ **29.** 0.382 **31.** 0.7635 **33.** 1.65 **35.** 5.2

Section 14.1

1. $(-3, 0)$ **3.** $(1, 0)$ **5.** $(-\sqrt{2}, \sqrt{2})$ **7.** $(2\sqrt{3}, -2)$ **9.** $(1, \sqrt{3})$ **11.** $(3, 0)$ **13.** $(\sqrt{2}, \pi/4)$ **15.** $(2, -\pi/6)$ **17.** $(1, -\pi/2)$ **19.** circle $x^2 + y^2 = 4$ **21.** straight line $y = 2x$ **23.** straight line $x + y = 1$ **25.** circle $(x + 3)^2 + y^2 = 9$ **27.** hyperbola $xy = 1$ **29.** parabola $y^2 = 1 + 2x$

31.

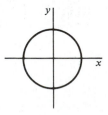

33.

35.

37.

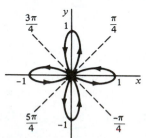

39.

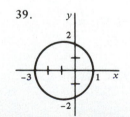

41.

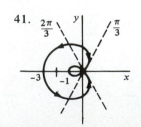

43.

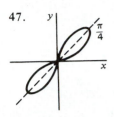

45.

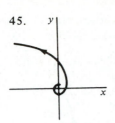

47.

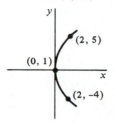

49. origin and $(\sqrt{2}, \pi/4)$ **51.** $(0, \pi/2)$, $(\sqrt{3}/2, \pi/6)$, $(-\sqrt{3}/2, 5\pi/6)$ **53.** infinitely many; one

Section 14.2

1. $\frac{4}{3}\pi^3$ **3.** $\frac{1}{2}\ln(1 + 2\pi)$ **5.** 6π **7.** π **9.** $\pi/4$

11. $\frac{4}{3}$ **13.** $\frac{1}{2}a^2(\pi - 2)$ **15.** $7\pi - \frac{9}{2}\sqrt{3}$

17. $4\pi/3 - \sqrt{3}$ **19.** 4π **21.** 6π **23.** 2
25. $\sqrt{2}(e^{2\pi} - 1)$ **27.** 8 **29.** 24 **31.** $\sqrt{2}$

Section 14.3

1. vertex $(0, 4)$, axis $x = 0$, focus $(0, \frac{15}{4})$, directrix $y = \frac{17}{4}$

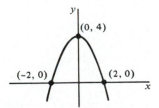

3. vertex $(1, 0)$, axis $y = 0$, focus $(0, 0)$, directrix $x = 2$

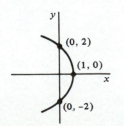

5. vertex $(1, -1)$, axis $x = 1$, focus $(1, -\frac{3}{4})$, directrix $y = -\frac{5}{4}$

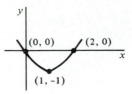

7. vertex $(0, 1)$, axis $y = 1$, focus $(2, 1)$, directrix $x = -2$

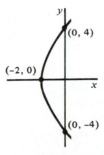

9. vertex $(-\frac{3}{2}, 1)$, axis $x = -\frac{3}{2}$, focus $(-\frac{3}{2}, \frac{5}{2})$, directrix $y = -\frac{1}{2}$

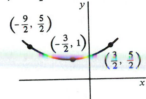

11. $4(y - 1) = (x - 1)^2$ **13.** $-2(y - 1) = x^2$
15. $16x = y^2$ **17.** $-3(x - \frac{1}{3}) = (y - 1)^2$ **21.** $y = x + 2$
23. vertex $(-2, 0)$, axis $y = 0$, focus $(0, 0)$, directrix $x = -4$

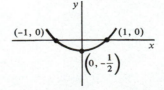

25. vertex $(0, -\frac{1}{2})$, axis $x = 0$, focus $(0, 0)$, directrix $y = -1$

31. $90°$

Section 14.4

1. center $(0, 0)$, endpoints $(\pm 3, 0)$ and $(0, \pm 2)$, foci $(\pm\sqrt{5}, 0)$, directrices $x = \pm 9/\sqrt{5}$, $e = \sqrt{5}/3$

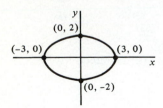

3. center and endpoints shown in the figure below, foci $(2 \pm 2\sqrt{3}, 1)$, directrices $x = 2 \pm 8/\sqrt{3}$, $e = \frac{1}{2}\sqrt{3}$

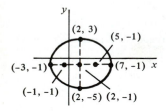

5. center, endpoints, and foci shown in the figure below, directrices $x = 2 \pm \dfrac{25}{3}$, $e = \dfrac{3}{5}$

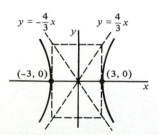

7. $y^2/16 + x^2/12 = 1$ **9.** $(x - 2)^2/4 + y^2/3 = 1$
11. $(y - 2)^2/5 + (x + 1)^2/4 = 1$ or $(y - 2)^2/20 + (x + 1)^2/4 = 1$
13. center $(0, 0)$, vertices $(\pm 3, 0)$, foci $(\pm 5, 0)$, directrices $x = \pm\frac{9}{5}$, $e = \frac{5}{3}$

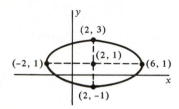

15. center and vertices shown in the figure below, foci $(3 \pm \frac{5}{3}, 0)$, directrices $x = 3 \pm \frac{3}{5}$, $e = \frac{5}{3}$

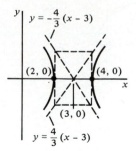

17. center $(\frac{1}{2}, \frac{3}{2})$, vertices $(\frac{1}{2}, \frac{3}{2} \pm \sqrt{3})$, foci $(\frac{1}{2}, \frac{3}{2} \pm \sqrt{11})$, directrices $y = \frac{3}{2} \pm 3/\sqrt{11}$, $e = \frac{1}{3}\sqrt{33}$

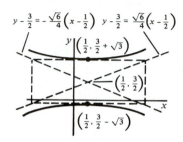

19. $y^2/4 - x^2/16 = 1$ **21.** $(x - 3)^2/4 - y^2/12 = 1$
23. right branch of the hyperbola $x^2/4 - (y - 2)^2/9 = 1$, traversed upward   **25.** $\sqrt{2}$ **27.** ellipse
29. center $(3, 0)$, foci $(0, 0)$ and $(6, 0)$, directrices $x = -\dfrac{16}{3}$ and $x = \dfrac{34}{3}$, $e = \dfrac{3}{5}$

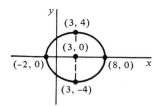

31. center $(0, 2)$, foci $(0, 0)$ and $(0, 4)$, directrices $y = \dfrac{3}{2}$ and $y = \dfrac{5}{2}$, $e = 2$

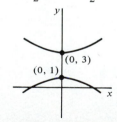

39. $y^2 = 4(c - a)(x - a)$, where $c = \sqrt{a^2 + b^2}$
41. $y^2 = 4(c - a)(x - a)$, where $c = \sqrt{a^2 - b^2}$
43. about 91.35 and 94.45 million miles, respectively; average 92.9 million miles

True-False Quiz (Chapter 14)

1. T   **2.** T   **3.** T   **4.** T   **5.** T   **6.** F   **7.** F
8. T   **9.** T   **10.** T   **11.** T   **12.** F   **13.** F
14. F   **15.** F

Additional Problems (Chapter 14)

1. $(2, \pm\pi/2)$

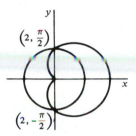

3. origin and $(3\sqrt{3}, \pm\pi/6)$

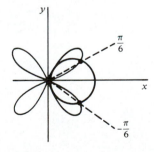

7. 9   **9.** π **11.** $\pi/4$ **13.** $4\sqrt{3}$ **15.** $\sqrt{5}/2$
17. $8a$
19. vertex $(3, 3)$, axis $x = 3$, focus $(3, \frac{9}{4})$, directrix $y = \frac{15}{4}$

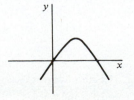

21. center and endpoints shown in the figure below, foci $(-2, 3 \pm \sqrt{7})$, directrices $y = 3 \pm 16/\sqrt{7}$, eccentricity $\sqrt{7}/4$

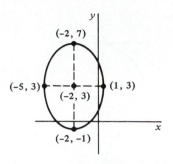

23. center $(2, 0)$, endpoints $(6, 0)$ and $(-2, 0)$, foci $(7, 0)$ and $(-3, 0)$, directrices $x = \frac{26}{5}$ and $x = -\frac{6}{5}$, asymptotes $y = \pm\frac{3}{4}(x - 2)$, eccentricity $\frac{5}{4}$

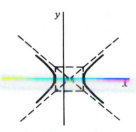

29. $16x^2 - 4xy + 19y^2 - 48x - 24y + 36 = 0$
39. parabola $2\bar{y}^2 - \sqrt{2}\bar{x} + \sqrt{2}\bar{y} + 5 = 0$
41. ellipse $(3 - \sqrt{2})\bar{x}^2 + (3 + \sqrt{2})\bar{y}^2 = 2$
43. degenerate parabola; union of vertical lines $x = 3$ and $x = -2$ **45.** degenerate parabola; union of lines $x - y = \pm 1$

Section 15.1

1. 0   **3.** π **5.** -3 **7.** $(3, -9)$ **9.** $(0, 0)$
11. $1/\sqrt{10}$ **13.** $(5, -2)$ **15.** $-\mathbf{j}$ **17.** $\frac{5}{13}\mathbf{i} - \frac{12}{13}\mathbf{j}$
19. $\mathbf{u}_1 \cdot \mathbf{u}_2 = 0$ and $|\mathbf{u}_1| = |\mathbf{u}_2| = 1$
21. the line $ax + by = 0$; $\mathbf{a}$ is perpendicular to the line
23. $45°$, $\cos^{-1}(2/\sqrt{5}) \approx 26.6°$, $\cos^{-1}(-1/\sqrt{10}) \approx 108.4°$
35. If $\mathbf{x}_1$ or $\mathbf{x}_2$ is $\mathbf{0}$, then $\mathbf{x}_1 \cdot \mathbf{x}_2 = 0$. But $\mathbf{x}_1 \cdot \mathbf{x}_2 = 0$ does not imply that $\mathbf{x}_1$ or $\mathbf{x}_2$ is $\mathbf{0}$. An example is $\mathbf{x}_1 = \mathbf{i}$ and $\mathbf{x}_2 = \mathbf{j}$.   **39.** Neither product makes sense. The left side, for example, is the dot product of the *number* $\mathbf{x}_1 \cdot \mathbf{x}_2$ and the *vector* $\mathbf{x}_3$, which is not defined.

Section 15.2

1.
 $y = \frac{3}{2}x + 1$

3.
 $y = x^3$

5.
 $\frac{x^2}{a^2} + \frac{y^2}{b^2} = 1$, $(0, b)$, $(a, 0)$

7.
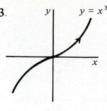 $xy = 1$

15. $(1, 3t^2)$ **17.** $e^t(\cos t - \sin t, \cos t + \sin t)$

19. $2t(-\sin t^2, \cos t^2)$ **21.** $4t^2 \cos t + (2t - t^3) \sin t$

Section 15.3

1. $\mathbf{v} = (2, 3), |\mathbf{v}| = \sqrt{13}, \mathbf{a} = (0, 0)$ **3.** $\mathbf{v} = (1, 3t^2), |\mathbf{v}| = \sqrt{1 + 9t^4}, \mathbf{a} = (0, 6t)$ **5.** $\mathbf{v} = (-a \sin t, b \cos t), |\mathbf{v}| = \sqrt{a^2 \sin^2 t + b^2 \cos^2 t}, \mathbf{a} = -F(t)$ **7.** $\mathbf{v} = (e^t, -e^{-t}), |\mathbf{v}| = \sqrt{e^{2t} + e^{-2t}}, \mathbf{a} = F(t)$ **9.** maximum $\sqrt{2}$, at $t = 0$ or $t = \pm\pi$; minimum 1 at $t = \pm\pi/2$ **11.** At $t = 0, \pm\pi$ the force is zero (which is "entirely normal" but doesn't mean much). At $t = \pm\pi/2$ the force is nonzero and entirely normal. The magnitude of $\mathbf{a}$ is 1 at these points; its direction is perpendicular to the path.

13. $a_T = \dfrac{18t^3}{\sqrt{1 + 9t^4}}$ **15.** $\kappa = \dfrac{|6t|}{(1 + 9t^4)^{3/2}}$

Section 15.4

9.

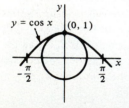 $y = \cos x$, $(0, 1)$, $-\frac{\pi}{2}$, $\frac{\pi}{2}$

11.
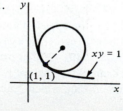 $xy = 1$, $(1, 1)$

13. $(-\frac{1}{2}\ln 2, 1/\sqrt{2})$ **19.** The curve has a cusp at each point mentioned.   **21.** $50\sqrt{2} \approx 71$ mph

37. $a_N = \dfrac{1}{\sqrt{1 + 2 \sinh^2 t}}$ **41.** $\kappa = \dfrac{4}{(1 + 16 \sin^2 t)^{3/2}}$

43. $\frac{1}{2}t\sqrt{t^2 + 1} + \frac{1}{2}\ln (t + \sqrt{t^2 + 1})$

45. $a_T = \dfrac{t}{\sqrt{t^2 + 1}}, a_N = \dfrac{t^2 + 2}{\sqrt{t^2 + 1}}$

True-False Quiz (Chapter 15)

1. T   **2.** T   **3.** F   **4.** T   **5.** F   **6.** F   **7.** T   **8.** T   **9.** T   **10.** T   **11.** F   **12.** T   **13.** T   **14.** F   **15.** T

Additional Problems (Chapter 15)

1. $(12, 13)$ **3.** $\pi - \cos^{-1}(1/\sqrt{290}) \approx 93°$ **5.** $45°$ **9.** $-(\sinh(1 - t), \cosh(1 - t))$ **11.** $(\ln t + 1, 2t \ln t + t)$ **13.** $te^t + (2t - t^2)e^{-t}$

17. $\dfrac{x^2}{16} + y^2 = 1$

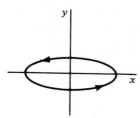

23. 6   **25.** $\frac{1}{3}$ **27.** at $(2, 1)$ **29.** $\dfrac{e^t}{(e^{2t} + 1)^{3/2}}$

31. $\dfrac{x^2}{9} - \dfrac{y^2}{4} = 1$

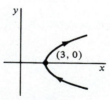 $(3, 0)$

33. $\frac{3}{4}$ at $(3, 0); \kappa \to 0$ as $t \to \infty$ **35.** $\dfrac{13 \sinh t \cosh t}{\sqrt{13 \sinh^2 t + 4}}$

37. slowing down before reaching $(3, 0)$, speeding up

thereafter   39. $\dfrac{12(t^2 + 2)}{\sqrt{2(t^2 + 4)}}$ 41. $\dfrac{\sqrt{2}}{3t(t^2 + 4)^{3/2}}$

43. $\dfrac{18t^2 + 4}{\sqrt{9t^2 + 4}}$ 45. $\dfrac{6}{t(9t^2 + 4)^{3/2}}; \kappa \to \infty$ as $t \downarrow 0$

47. $\dfrac{2}{(1 + 4x^2)^{3/2}}; (0, \tfrac{1}{2})$ 49. 1; graph is the upper half

of a circle of radius 1

Section 16.1

1. yz plane   3. xy plane   5. z axis
7. unit circle in xy plane

9.
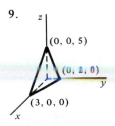

11. $3\sqrt{2}$ 15. $(x - 2)^2 + (y - 1)^2 + (z - 2)^2 = 19$
17. center $(3, -5, 0)$, radius 6   19. yes
23. $\dfrac{1}{\sqrt{17}} (2, -3, -2)$ and $\dfrac{1}{\sqrt{17}}(-2, 3, 2)$
25. $90°$ at $(1, 1, 1)$, $45°$ at $(2, 3, 3)$, $45°$ at $(3, -1, 2)$

Section 16.2

1. $x = 2t + 1, y = 3t - 1, z = 2t + 2$ 3. $x = t + 1,$
$y = t - 1, z = 2t + 3$ 5. $x = t + 2, y = -1, z = 5$
7. $x = 2t + 4, y = t, z = 3t + 1$ 9. $x = 3t + 3, y = t + 1,$
$z = t + 1$ 11. $x = t, y = 0, z = 2$ 13. $x = \tfrac{1}{4}, y = t + \tfrac{1}{4},$
$z = t$ 15. $(4, 9, 0); \pi/2 - \cos^{-1}(1/\sqrt{30}) \approx 57.7°$
17. Yes, at $(1, 0, 3)$, but the airplanes reach this point at
$t = 1$ and $t = 0$.   19. no   21. $-x + y + 3z = 0$
23. $z = 3$ 25. $2x + y + 3z - 9 = 0$
27. $-4x - y + z = 0$ 29. $4x + y - 3z + 3 = 0$
31. $x + 2y - 2z - 7 = 0$ 33. They are parallel.
35. $\cos^{-1}(\sqrt{2}/3) \approx 61.9°$ 39. $\tfrac{2}{3}\sqrt{6}$

Section 16.3

1. $(-5, -7, 3)$ 3. $(2, -7, -13)$ 5. $(-7, -2, 12)$
7. -39 17. $-4x - y + z = 0$
19. $-18x + 5y - z + 29 = 0$ 21. $4x - 7y + 2z = 0$
23. $x + y + 3z - 2 = 0$ 25. $x + y - 2z + 4 = 0$
27. $x = \tfrac{23}{7} + \tfrac{6}{7}t, y = -\tfrac{4}{7} - \tfrac{5}{7}t, z = t$ 37. $2\sqrt{21}$

Section 16.4

1. circular paraboloid   3. sphere

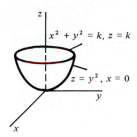

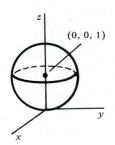

5. circular cylinder   7. hyperbolic cylinder

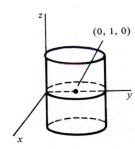

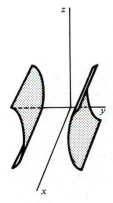

9. circular paraboloid   11. cylinder (not quadric)

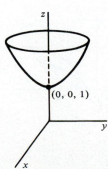

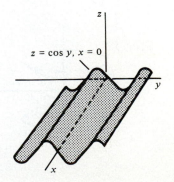

13. plane

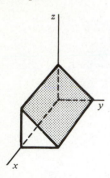

15. sphere

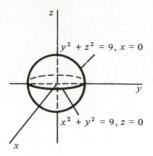

$y^2 + z^2 = 9, x = 0$

$x^2 + y^2 = 9, z = 0$

29. elliptical cone

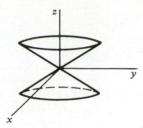

31. parabolic cylinder

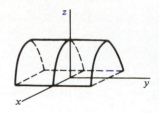

17. hyperboloid of two sheets

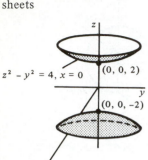

$z^2 - y^2 = 4, x = 0$

$(0, 0, 2)$

$(0, 0, -2)$

19. hyperbolic paraboloid

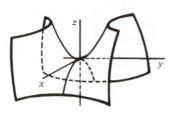

33.

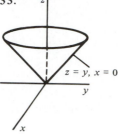

$z = y, x = 0$

35.

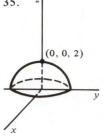

$(0, 0, 2)$

37.

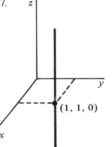

$(1, 1, 0)$

39.

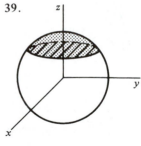

21. circular cone

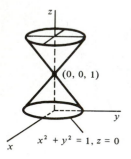

$(0, 0, 1)$

$x^2 + y^2 = 1, z = 0$

23. ellipsoid

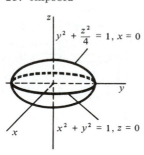

$y^2 + \dfrac{z^2}{4} = 1, x = 0$

$x^2 + y^2 = 1, z = 0$

41.

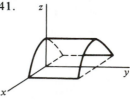

43.

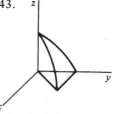

25. hyperboloid of two sheets

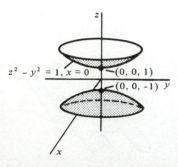

$z^2 - y^2 = 1, x = 0$

$(0, 0, 1)$

$(0, 0, -1)$

27. cylinder (not quadric)

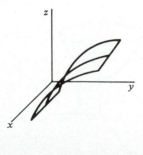

45.

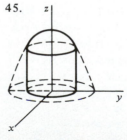

47.

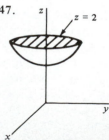

$z = 2$

Section 16.5

1. cylindrical: $(1, \pi/2, 1)$; spherical: $(\sqrt{2}, \pi/4, \pi/2)$
3. rectangular: $(-1, 0, 1)$; spherical: $(\sqrt{2}, \pi/4, \pi)$
5. rectangular: $(0, \sqrt{3}, 1)$; cylindrical: $(\sqrt{3}, \pi/2, 1)$
7. circular cylinder generated by moving circle $r = 2$
(in xy plane) parallel to z axis   9. xz plane
11. plane $y = x$ 13. sphere of center $(0, 0, 0)$ and
radius 2   15. plane $z = y$ 17. circular cone $z^2 = 4(x^2 + y^2)$ 19. sphere of center $(0, 0, 0)$ and radius 2
21. upper half of $45°$ cone with vertex $(0, 0, 0)$ and axis
the z axis   23. lower half of $45°$ cone with vertex
$(0, 0, 0)$ and axis the z axis   25. plane $z = 2$
27. circular paraboloid opening upward from vertex
$(0, 0, 0)$ 29. circle in plane $z = 1$ with center $(0, 0, 1)$
and radius $\sqrt{3}$ (intersection of sphere $x^2 + y^2 + z^2 = 4$
and paraboloid $x^2 + y^2 = 3z$)   31. "ice cream cone"
above xy plane, inside $30°$ cone opening upward from
vertex at origin, and capped by unit sphere

Section 16.6

1. $x = -u, y = 1, z = \pi/2 + u$ 13. $(1, 0, 0)$ 21. 2
25. ω 39. $x = u, y = 0, z = u$ 41. 1

43. $\mathbf{T} = \dfrac{1}{\sqrt{2}}(1, 0, 1), \mathbf{N} = (0, 1, 0), \mathbf{B} = \dfrac{1}{\sqrt{2}}(-1, 0, 1); z = x$

True-False Quiz (Chapter 16)

1. T   2. T   3. T   4. F   5. F   6. T   7. F
8. T   9. F   10. T   11. F   12. F   13. F
14. T   15. T

Additional Problems (Chapter 16)

1. $\dfrac{x}{4} + \dfrac{y}{5} + \dfrac{z}{3} = 1$

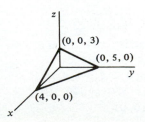

3. $\mathbf{u_1} = \dfrac{1}{\sqrt{89}}(-6, 2, 7), \mathbf{u_2} = \dfrac{1}{\sqrt{89}}(6, -2, -7)$

5. $x = 4t + 5, y = 2t, z = t + 2$ 7. $x = 5t - 1, y = 4 - t,$
$z = 2t + 2$ 9. $x = t + 2, y = -5t - 5, z = 1 - 4t$
11. $(1, -2, 2)$ 13. $x - 3y + 4z - 4 = 0$
15. $3x - y + 2z - 9 = 0$ 17. $(0, 2, -1); \pi/2 -$
$\cos^{-1}(\sqrt{21}/6) \approx 49.8°$

21. parabolic cylinder   23. circular cone

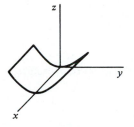   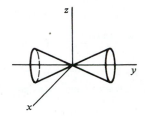

25. ellipsoid

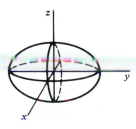

27. sphere of center $(0, 0, 2)$ and radius 2   29. $4\pi\sqrt{2}$
31. (a) $x = 1 + 2u, y = -1 - u, z = 2 - 2u$
(b) at $(3, -2, 0)$ (c) 3; 1 unit of time
33. (a) $x = t, y = 2t, z = t, t \geq 0$ (b) $\sqrt{6}$
35. $\kappa = \frac{1}{18}, \tau = -\frac{1}{18}$ 37. ellipse   39. 0
41. $\frac{1}{2} \text{sech}^2 t$

Section 17.1

1. all of $\mathcal{R}^2$ except the origin   3. first and third
quadrants of the plane   5. all of $\mathcal{R}^2$ except the line
$y = x$ 7. all of $\mathcal{R}^3$ except the xy plane   9. $z_x = 2x +$
$y, z_y = x + 2y$ 11. $z_x = \dfrac{1}{y} + \dfrac{y}{x^2}, z_y = -\dfrac{x}{y^2} - \dfrac{1}{x}$
13. ye^{xy}, xe^{xy} 15. $-y^2 e^{y/x}/x^2, e^{y/x}(y/x + 1)$
17. $e^u \cos v, -e^u \sin v$ 19. $yx^{y-1}, x^y \ln x$
21. $x/\rho, y/\rho, z/\rho$, where $\rho = \sqrt{x^2 + y^2 + z^2}$
23. $2(y + z)(xy + yz + zx), 2(x + z)(xy + yz + zx),$
$2(x + y)(xy + yz + zx)$ 31. $\partial T/\partial x = 2x; T = 25$ with

rate -6; at the other side $T = 25$ with rate 6; highest and lowest values 25 and 16   **33.** $T = 1$ along circle $x^2 + y^2 = 1$; isothermal paths are circles $x^2 + y^2 = a^2$, $0 < a \leq 5$ **51.** $f_2(x, y, z)$

Section 17.2

1. $5t^4 e^{t^5}$ **3.** $2 \tanh 2t$ **5.** $-\sin t$
7. $2t + 4t^3 + 6t^5$ **9.** $t \cos 2t + \frac{1}{2} \sin 2t$

11. $\dfrac{\sin t \cos t}{\sqrt{1 + \sin^2 t}}$ **13.** 4

15. As the bug passes through the origin, the temperature falls to 0 and then rises again.

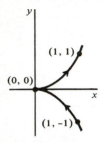

17. increasing at 100π cm^3/sec   **21.** $\partial w/\partial x = 1$, $\partial w/\partial y = -1/4$ **23.** $2/r$ **25.** 0   **33.** $(2x, 2y)$
35. $(5, 1, -1)$ **37.** $\cos xyz\ (yz, xz, xy)$

Section 17.3

1. $x + 4z - 9 = 0$ **3.** $y + 3z - 7 = 0$ **5.** $3x + 4y + 3z - 16 = 0$ **7.** $y + 4z + 4 = 0$ **9.** $3x + 4y - 5z = 0$
11. $x = 2$ **13.** $z = 4$ **17.** at $(0, 0, \pm c)$; $z = \pm c$
21. $x = 1 + \frac{1}{2} t, y = \frac{1}{2} + t, z = \frac{1}{2} + t$ **23.** no

27. $\dfrac{y - 1}{1 - x}$ **29.** $\dfrac{\sin y}{1 - x \cos y}$ **31.** $2\sqrt{y}$ **33.** $\dfrac{y + 1}{1 - x}$

35. $\dfrac{2 - y}{x + 3}$ **37.** $\dfrac{4x - 5y}{5x + 2y}$

Section 17.4

1. $\frac{1}{5}$ **3.** -4 **5.** 1   **7.** 0   **9.** $\sqrt{a^2 + b^2 + c^2}$
11. 0   **13.** $2/\sqrt{x^2 + y^2 + z^2}$ **15.** southeast; $\sqrt{2}$;
northeast or southwest   **17.** $\sqrt{13}$ **19.** in the

directions of the unit vectors $\pm \dfrac{1}{\sqrt{13}}(3, 2)$

21. southeast; $4\sqrt{2}$ **23.** in the directions of the unit vectors $(-1, 0)$ and $(-\frac{3}{5}, \frac{4}{5})$ **25.** in the direction of the unit vector $(-\frac{4}{5}, \frac{3}{5})$; -10 **27.** 0   **29.** $-\frac{2}{21}\sqrt{14}$
(degrees per unit of distance)   **43.** $\frac{7}{3}$

Section 17.5

1. $(0, 2)$ **3.** $(2, -1)$ **5.** $(0, 0), (-2, \pm 2)$
7. saddle point $(0, 2, 4)$ **9.** local low point $(2, -1, -3)$
11. local low points $(-2, \pm 2, -16)$; saddle point $(0, 0, 0)$
(by inspection)   **13.** $(2, -1, 12)$ **15.** $(1, 1, 3)$
17. peaks $(\frac{1}{2}, \frac{1}{2}, \frac{1}{8}), (-\frac{1}{2}, -\frac{1}{2}, \frac{1}{8})$; passes $(0, 0, 0)$,
$(\pm 1, 0, 0), (0, \pm 1, 0)$; low points $(\frac{1}{2}, -\frac{1}{2}, -\frac{1}{8}), (-\frac{1}{2}, \frac{1}{2}, -\frac{1}{8})$
19. $(2, -3, 3)$ **21.** $(1, -2, 1)$ **23.** hottest:
$(-1/2, \pm\sqrt{3}/2)$; coldest: $(1/2, 0)$ **25.** hottest: $(\pm 1, 0, 0)$,
$(0, \pm 1, 0), (0, 0, \pm 1)$; coldest: $(1/\sqrt{3}, 1/\sqrt{3}, \pm 1/\sqrt{3})$,
$(1/\sqrt{3}, -1/\sqrt{3}, \pm 1/\sqrt{3}), (-1/\sqrt{3}, 1/\sqrt{3}, \pm 1/\sqrt{3})$,
$(-1/\sqrt{3}, -1/\sqrt{3}, \pm 1/\sqrt{3})$ **27.** $(-1, 2, \pm 2)$ **29.** $\frac{2}{3}$
31. $(\frac{1}{2}, \frac{1}{4})$ and $(\frac{7}{8}, -\frac{1}{8})$ **33.** $x/6 + y/3 + z/6 = 1$
37. $14 \times 14 \times 28$ **39.** Maximum u occurs when $x = \frac{100}{9}, y = \frac{80}{9}, z = \frac{100}{3}$. Since x, y, z must be integers, the consumer should buy 11 of the first product, 9 of the second, 33 of the third.
43. $y = -\frac{7}{5}x - \frac{3}{10}$

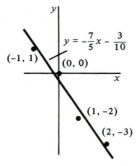

Section 17.6

1. $(0, 0)$ **3.** $(1, 2), (-1, -2)$ **5.** $(-\frac{1}{3}, \frac{1}{6}, \frac{5}{36})$
7. $(1, 1, -1)$ **9.** 1250 square meters   **11.** from $0°$ to $\tan^{-1} \frac{1}{2} \approx 27°$ **13.** 50, 50   **15.** 3750 ft^2
17. 12 ft^2 **19. (a)** 4   **(b)** no; $0 < A \leq 4$
27. shortest segment has endpoints $(5, 0)$ and $(0, 10)$
33. $(2/\sqrt{5}, 1/\sqrt{5})$ **35.** $(0, 0); (0, 0)$

39. (a) $\pi L/(4 + \pi)$, where L is the length of the wire
(b) All the wire should be used for the circle.
41. $x^2/2 + y^2/8 = 1$ **43.** $\frac{2}{3}$ **45.** $x/6 + y/3 + z/6 = 1$
49. $14 \times 14 \times 28$ **51.** Maximum u occurs when $x =$
$\frac{100}{9}, y = \frac{80}{9}, z = \frac{100}{3}$. Since x, y, z must be integers, the
consumer should buy 11 of the first product, 9 of the
second, 33 of the third.   **53.** $(3, 3, 3)$

True-False Quiz (Chapter 17)

1. T   **2.** F   **3.** T   **4.** T   **5.** T   **6.** F   **7.** F
8. T   **9.** T   **10.** T   **11.** T   **12.** F

Additional Problems (Chapter 17)

1. $y(1 - 2x^2)e^{-x^2}; xe^{-x^2}$ **3.** $\dfrac{2x^2 + y^2}{\sqrt{x^2 + y^2}}; \dfrac{xy}{\sqrt{x^2 + y^2}}$

5. $-\dfrac{1}{x}\cos\dfrac{y}{x} + \dfrac{1}{y}\sin\dfrac{y}{x}; \dfrac{1}{y}\cos\dfrac{y}{x} - \dfrac{x}{y^2}\sin\dfrac{y}{x}$

7. $1/x; 1/y; 1/z$ **9.** $ye^{xy}\cos yz; xe^{xy}\cos yz -$
$ze^{xy}\sin yz; -ye^{xy}\sin yz$
11. $\{(x, y): x > 0\}; f_{12}(x, y) = f_{21}(x, y) =$

$x^{y-1}(1 + y \ln x)$ **13.** $f_1(x, y, z) =$

$\lim\limits_{h\to 0}\dfrac{f(x + h, y, z) - f(x, y, z)}{h} = yze^{xyz}$

17. $2t$ **19.** $\dfrac{1}{t - 1}$

21. 0   **23.** $-1/r^2; 0$ **25.** $3/\sqrt{2}; 3/\sqrt{2}$
27. $0; \sec^2 v$ **29.** $2x - 2y + z + 1 = 0$

31. increasing, at $\dfrac{64 - 37\sqrt{2}}{2\sqrt{25 - 12\sqrt{2}}} \approx 2$ cm/sec

33. (a) $\partial x/\partial r = \cos\theta, \partial x/\partial\theta = -r\sin\theta, \partial y/\partial r = \sin\theta,$
$\partial y/\partial\theta = r\cos\theta$ **37.** $-\sqrt{2}/e$ **39.** $-\sqrt{2}$ **41.** 1
43. -1; decreasing; increases most rapidly in the direction
of $\nabla w = (\frac{1}{2}, -1, -\frac{1}{2})$, the rate being $\frac{1}{2}\sqrt{6}$; derivative is
zero in any direction given by a vector of the form $\mathbf{v} =$
$(x - 2, y + 1, z + 2)$, where $x - 2y - z - 6 = 0$
45. saddle point $(0, 0, -1)$ **47.** maximum $\frac{1}{4}$,
minimum 0   **49.** local hot point $(1, 1)$; no coldest point
because $T \to -\infty$ when $(x, y) \to (0, 0)$ **51.** 32
53. $(\frac{1}{2}, -\frac{1}{4})$ **55.** maximum 3, minimum -3

Section 18.1

1. 14   **3.** $\pi/3$ **5.** 4   **7.** 8   **9.** $\frac{1}{12}(2\sqrt{2} - 1)$
11. $\frac{1}{2}\ln 2$ **13.** $\frac{1}{4}(e^2 + 1)$ **15.** $\frac{1}{2}(e^2 - 3)$ **17.** $\frac{1}{3}$

19. $\dfrac{\pi}{4}\ln 2$ **21.** $\displaystyle\int_0^2 \int_0^3 xy \, dx \, dy = 9$

23. $\displaystyle\int_0^2 \int_0^1 \dfrac{dy \, dx}{1 + y^2} = \dfrac{\pi}{2}$ **25.** $\displaystyle\int_0^1 \int_0^{\sqrt{1-x^2}} y \, dy \, dx = \frac{1}{3}$

27. $\displaystyle\int_1^e \int_{\ln y}^1 \dfrac{dx \, dy}{y} = \dfrac{1}{2}$

29. $\displaystyle\int_0^{\pi/4} \int_x^{\pi/4} \cos(x + y)dy \, dx = \dfrac{1}{2}(\sqrt{2} - 1)$

31. $\displaystyle\int_0^1 \int_{\tan^{-1} y}^{\pi/4} \sec x \, dx \, dy = \sqrt{2} - 1$

33. $\displaystyle\int_0^1 \int_0^{2x} e^{-x^2} \, dy \, dx = 1 - 1/e$

35. 3π **37.** $\pi/2$ **39.** $\frac{2}{3}$ **41.** $\pi/2$

Section 18.2

1. $\frac{4}{3}$ **3.** $\frac{32}{3}$ **5.** $\frac{71}{6}$ **7.** $2\sqrt{2} - \frac{1}{3}$ **9.** $\frac{27}{4}$ **11.** 1
13. $2(3 - \sqrt{5})$ **15.** $\frac{1}{6}$ **17.** $\frac{13}{6}$ **19.** 36   **21.** $\frac{17}{12}$
23. $\sqrt{2} - 1$ **25.** $3\sqrt{3}/4$ **27.** 8   **29.** $\frac{2}{3}(2 + 3\pi)$

31. $(0, 2)$ **33.** $\bar{x} = \dfrac{88}{5(2 + 3\pi)}, \bar{y} = 0$ **35.** $(1, \frac{4}{5})$

37. $\bar{x} = \pi/2 - 1, \bar{y} = \pi/8$ **39.** $\frac{32}{5}$ **41.** $4\pi - \frac{32}{5}$
43. $\frac{1}{2}ka^3$, where k is the constant in the formula for

density   **45.** $\frac{1}{4}ka^5$ **47.** on the perpendicular

bisector of the diameter of the region, $\frac{3}{16}\pi a$ units from
the diameter   **51. (a)** 64   **(b)** smallest 16; largest 144

(c) 64   **55.** $\dfrac{1}{2\pi}$

Section 18.3

1. 6π **3.** $\frac{3}{2}\pi a^2$ **5.** πa^2 **7.** $\pi/2$ **9.** 1

11. $\pi - \dfrac{3\sqrt{3}}{2}$ **13.** $\frac{8}{3}$ **15.** πa^2 **17.** $\frac{3}{2}\pi a^2$

19. $\dfrac{7\pi}{6} - 2\sqrt{3}$ **21.** $\dfrac{4\pi}{3} - \sqrt{3}$ **23.** 4π **25.** 28π

27. $8\pi/3$ **29.** $\frac{2}{3}a^3$ **31.** $\frac{4\pi}{3}(5\sqrt{5}-8)$

33. $\frac{8}{9}(3\pi-4)$ **35.** $32k/9$, where k is the constant in the formula for density   **37.** on the perpendicular bisector of the diameter, $\frac{3a}{2\pi}$ units above it   **39.** $\frac{3}{5}ma^2$

43. $\left(\dfrac{a+b}{\pi},\dfrac{a+b}{\pi}\right)\to\left(\dfrac{b}{\pi},\dfrac{b}{\pi}\right)$ when $a\to 0$

Section 18.4

1. π **3.** $9\pi\sqrt{2}$ **5.** $\frac{1}{2}\pi\sqrt{6}$ **7.** $\frac{1}{2}\sqrt{21}$

9. $\frac{\pi}{6}(5\sqrt{5}-1)$ **11.** $2\sqrt{5}+\ln(2+\sqrt{5})$ **13.** 4π

15. 20π **17.** $4\pi-8$ **19.** 2π **25.** $\pi a\sqrt{a^2+h^2}$

27. $\frac{\pi}{2}(2+\sinh 2)$ **29.** $\pi\left(2\sqrt{5}-\sqrt{2}+\ln\dfrac{2+\sqrt{5}}{1+\sqrt{2}}\right)$

31. $\frac{64}{3}\pi a^2$

Section 18.5

1. 9   **3.** 18   **5.** $\frac{\pi}{4}-\frac{1}{3}$ **7.** 8π **9.** $\frac{1}{3}\pi a^2 h$

11. $(1,2,1)$ **13.** $(1,\frac{9}{4},\frac{27}{10})$ **15.** $\bar{x}=\dfrac{16-3\pi}{4(3\pi-4)}$, $\bar{y}=\dfrac{5}{2(3\pi-4)}$ **17.** $(0,0,\frac{4}{3})$ **19.** $(0,0,\frac{3}{4}h)$

21. $\bar{x}=\frac{3}{64}a$, $\bar{y}=\frac{3}{64}b$, $\bar{z}=\frac{3}{64}c$ **23.** $(0,0,\frac{3}{8}a)$

25. $(0,\frac{4}{3},\frac{10}{9})$ **27.** $\frac{1}{3}$ **31.** 450   **33.** 5625

35. on the axis of the cylinder, $\frac{2}{3}h$ units from the base

37. $\frac{5}{12}k\pi a^2 h^2$, where k is the constant in the formula for density   **39.** $\frac{1}{60}k\pi a^4 h^2$

Section 18.6

1. $\frac{2}{3}\pi a^3$ **3.** $(0,0,\frac{3}{8}a)$ **5.** $\frac{4}{15}\pi a^5$ **9.** πa^3

11. $(0,0,\frac{7}{6}a)$ **13.** $\frac{11}{60}a^5$ **15.** $\frac{\pi}{6}(8\sqrt{2}-7)$

17. $\bar{x}=0$, $\bar{y}=0$, $\bar{z}=\dfrac{7}{2(8\sqrt{2}-7)}$ **19.** $\frac{\pi}{15}(16\sqrt{2}-19)$

21. $\pi a^4/48$ **23.** $k\pi a^4$, where k is the constant in the formula for density   **25.** $\frac{1}{2}k\pi a^4$, where k is the constant in the formula for density   **27.** $\frac{1}{3}ma^2$ **29.** $\bar{x}=\bar{y}=$
$\bar{z}=\dfrac{1-\ln 2}{4-\pi}$ **33.** $(0,0,\frac{8}{7}a)$ **35.** (b) $(0,0,\frac{3}{8}b)$
(c) $(0,0,\frac{1}{2}a)$ **37.** $\frac{4}{15}\pi a^5$

True-False Quiz (Chapter 18)

1. F   **2.** F   **3.** T   **4.** T   **5.** T   **6.** T   **7.** T
8. F   **9.** T   **10.** T

Additional Problems (Chapter 18)

1. $\displaystyle\int_0^1\int_0^{1-x}(x+y)\,dy\,dx=\frac{1}{3}$ **3.** $\displaystyle\int_0^2\int_y^2\frac{dx\,dy}{1+x^2}=$
$\frac{1}{2}\ln 5$ **5.** $e-1$ **7.** $2\pi/3$ **9.** 4   **11.** $\frac{2}{3}$ **13.** 4π

15. $\frac{2}{3}$ **17.** $(\frac{3}{2},-\frac{9}{10})$ **19.** $(-\frac{1}{2},\frac{2}{5})$ **21.** $\bar{x}=\frac{\pi}{2}$,
$\bar{y}=\dfrac{16}{9\pi}$ **23.** $(0,\frac{3}{5})$ **25.** π **27.** $\sqrt{17}$ **29.** 16π

31. $\pi\left[-1+2\sqrt{7}+\dfrac{1}{\sqrt{2}}\ln(4+2\sqrt{2}-\sqrt{7}-\sqrt{14})\right]$

33. π **35.** $\dfrac{64\pi}{3}(2-\sqrt{3})$ **37.** $\frac{64}{15}$ **39.** $(0,0,3)$

41. $\bar{x}=0$, $\bar{y}=0$, $\bar{z}=\frac{7}{158}(7+8\sqrt{2})$ **43.** $(0,0,0)$

45. $\pi/3$ **47.** $\dfrac{256\pi}{15}(16-9\sqrt{3})$ **49.** $\dfrac{23,552}{315}$

Section 19.1

1. 2   **3.** 2   **5.** 0   **9.** 1   **11.** 1   **13.** -1
15. 1   **19.** $-\frac{2}{3}$ **21.** $-\frac{2}{3}$ **27.** 0; yes, $f(x,y)=$
$e^x\sin y$ **29.** 2π; no   **31.** $-2/a$

Section 19.2

1. satisfies   **3.** satisfies   **5.** satisfies   **7.** fails to satisfy   **9.** fails to satisfy   **11.** satisfies

13. satisfies   15. $f(x, y) = \frac{1}{2}x^2 + xy - \frac{1}{2}y^2 + k$
17. $f(x, y) = \sin xy + k$ 19. no antigradient
21. no antigradient   23. no antigradient
25. $f(x, y) = \tan^{-1}(x/y) + k$ 31. $-\pi$; **F** is not a gradient;
use the necessary condition

Section 19.3

1. 8   3. π 5. $\frac{1}{12}$ 7. $-\frac{1}{3}$ 9. 0   11. 2π
13. 0   15. 0   21. 2   23. $\frac{24}{5}\sqrt{3}$ 25. 4
29. $(\pi/2, \pi/8)$ 31. $(1, \frac{2}{5})$ 33. $\frac{1}{2}\pi a^4$ 35. $\pi^2 - \frac{32}{9}$
37. $\frac{1}{2}k(x^2 + y^2)$

Section 19.4

1. source at the origin

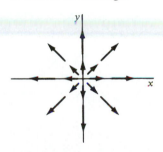

3. 8   5. no; 0   7. negative; -2π 9. no; 0
11. source at each point of the x axis

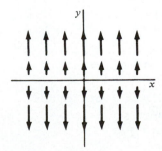

13. π 15. no; 0   17. zero; 0   19. no; 0
21. positive; 2   23. negative; -2 25. 2π; source
27. 2π 37. $\frac{3}{2}a$ 39. $k\pi$, where k is the constant in
the formula for density   41. $(0, 2/\pi)$

Section 19.5

1. $\pi/4$ 3. $4\pi/3$ 5. 24π 7. $\frac{1}{2}$ 9. 0   11. 4π
13. $\frac{1}{3}$ 15. $4\pi/3$ 17. 4π 19. 3π 21. $2\pi/3$
23. 3   25. $3\pi/2$ 27. 1   29. $\frac{12}{5}\pi a^5$ 31. 3
33. 5   39. $\frac{4}{3}k\pi a^4$, where k is the constant in the
formula for density   41. $(0, 0, \frac{1}{2}a)$

Section 19.6

1. $xy + xz + \frac{1}{2}y^2 + \frac{1}{2}z^2$ 3. $x^2 + y^2 + z^2 +$
$2(xy + yz + xz)$ 5. $xyz + xy + yz + xz$ 7. $-(e^{2\pi} + 1)$
9. πa^2 11. $\frac{3}{2}$ 13. -3π 15. π 17. $-\pi ab$
19. π

True-False Quiz (Chapter 19)

1. F   2. T   3. T   4. T   5. F   6. T   7. F
8. T   9. T   10. T

Additional Problems (Chapter 19)

1. -2 3. 0   9. $-\frac{1}{2}$ 11. $\frac{1}{2}$ 13. 0   21. 0
23. 2π 27. $\frac{16}{15}$ 29. 0   31. 0   33. 0
35. $f(x, y) = x \tan^{-1} y$ 41. $4\pi/3$ 43. 1   45. 2π
47. $\frac{1}{2}$ 49. 0   55. $f(x, y, z) = \frac{1}{2}\ln(x^2 + y^2 + z^2)$
57. 0   63. $2\pi^2$ 65. 4π 67. 0   69. $\frac{1}{2}$

Section 20.1

1. $f(x, y) = \frac{\sqrt{x - y}}{x}$; $D = \{(x, y): x > 0\}$ 3. $f(x, y) =$
$-3xy\sqrt{1 - x^2}$, with domain $\{(x, y): -1 \leq x \leq 1\}$; $D =$
$\{(x, y): -1 < x < 1\}$ 5. $f(x, y) = 2xy + 1$; $D = \mathbb{R}^2$
7. $f(x, y) = (y - 1)^{3/2}$, with domain $\{(x, y): y \geq 1\}$; $D =$
$\{(x, y): y > 1\}$ 9. $(0, \infty)$ 11. $[-1, 1]$
13. $(-\infty, \infty)$ 15. $(-\infty, 3)$ 21. $y = (x + 1)^2$, valid in
$[-1, \infty)$ 23. $y = 2 - e^{\sin x}$, valid in $(-\infty, \infty)$

25. $y = \ln(x + 1)$, valid in $(-1, \infty)$ 27. $y = 1$, valid in $(-\infty, \infty)$ 29. $y = \dfrac{4}{(x-3)^2} + 1$, valid in $(-\infty, 3)$

11. $y = c_1 + c_2 e^x + \frac{1}{2}(\cos x - \sin x)$

13. $y = \dfrac{1}{ax + b}$ and $y = 0$

Section 20.2

1. $x^2 + y^2 = a^2$ 3. $x^2 + 2xy - y^2 = k$
5. $e^{xy} + x^2 - y^2 = k$ 7. $e^x \sin y = k$
9. $x^3 - xy + y^2 = 1$ 11. $y = \dfrac{5 - x}{x^2 - 1}$ 13. $y \sin x - x^2 + \sin y = 0$ 15. $x^2 + \frac{1}{2}\ln(x^2 + y^2) + y^2 = 1$. The

original equation can be written $x\left(2 + \dfrac{1}{x^2 + y^2}\right) dx + y\left(2 + \dfrac{1}{x^2 + y^2}\right) dy = 0$, or $x\, dx + y\, dy = 0$. The solution

of this satisfying $y = 1$ when $x = 0$ is implicitly defined by $x^2 + y^2 = 1$, or explicitly by $y = \sqrt{1 - x^2}$ (since $y > 0$ near $x = 0$).

Section 20.3

1. $\phi(x) = \frac{1}{2}(3e^{x^2} - 1)$, all x 3. $\phi(x) = e^{3x - 2} - e^{3x - 6} - e^{2x}$, all x 5. $\phi(x) = xe^{x^2} \displaystyle\int_1^x e^{-t^2}\, dt + 3xe^{x^2 - 1}$, $x > 0$

7. $\phi(x) = \frac{1}{4}x^2 (\ln x + 4)^2$, $x > 0$ 9. $y = Cx^2 + x^2 \ln x$

11. $y = \dfrac{C}{x} + \sin x + \dfrac{\cos x}{x}$ 13. $y = C \csc x + x \csc x$

15. $y = x$ 17. approaches 10 meters as time goes on
19. $10e^{-1.2} \approx 3$ kg; virtually none

Section 20.4

1. $\phi(x) = 0$, unique because the initial conditions $\phi(0) = 0$ and $\phi'(0) = 0$ represent a point $(0, 0, 0)$ in the region $D = \mathbb{R}^3$ in which the Existence and Uniqueness Theorem applies   3. $\phi(x) = \ln \sec(x - 1) + 2$, $1 - \pi/2 < x < 1 + \pi/2$ 5. $\phi(x) = -e^{-x}$, all x 7. $\phi(x) = \dfrac{3e^{2x} + 1}{3e^{2x} - 1}$, $x > -\frac{1}{2}\ln 3$ 9. $y = c_1 + c_2 e^{2x}$

Section 20.5

3. $\cosh x = \frac{1}{2}e^x + \frac{1}{2}e^{-x}$ and $\sinh x = \frac{1}{2}e^x - \frac{1}{2}e^{-x}$ are linear combinations of e^x and e^{-x}.   5. $\phi(x) = 2 \sin 1 \cos x - 2 \cos 1 \sin x = 2 \sin(1 - x)$
7. (c) $W(0) \neq 0 \Rightarrow$ linear independence by Theorem 2
9. $W(\pi/4) \neq 0$. But $W(0) = 0$, so ϕ_1 and ϕ_2 cannot be solutions of any equation $L(y) = 0$ in I.
11. $W(0) = r_2 - r_1 \neq 0$ 13. $W(0) = q \neq 0$
17. $W(x) = x^2$ and $W(1) = 1$. Since $P(x) = -2 \ln x$, Abel's Formula reads $W(x) = 1e^{2\ln x} = x^2$.
19. (d) $\phi(x) = c_1 x + c_2(x^2 - 1)$ for all x
21. $\phi(x) = c_1 x^2 + c_2 x^{-2}$ in $(0, \infty)$ or $(-\infty, 0)$

Section 20.6

3. $y = c_1 e^{3x} + c_2 e^{-2x}$ 5. $y = c_1 e^x + c_2 xe^x$
7. $y = c_1 \cos 3x + c_2 \sin 3x$ 9. $y = c_1 e^{3x} \cos 4x + c_2 e^{3x} \sin 4x$

Section 20.7

1. $\phi_p(x) = xe^x$. If you find $\phi_p(x) = -e^x + xe^x$, you are not wrong, but the term $-e^x$ may be dropped. (Why?)
3. $\phi_p(x) = x^2 - 2$ 5. $\phi_p(x) = \frac{1}{3}\sin x$ 7. $\phi_p(x) = 1$
9. $\phi_p(x) = x^2 \ln x$ 11. $\phi_p(x) = xe^x$
13. $\phi_p(x) = x^2 - 2$ 15. $\phi_p(x) = -\frac{1}{2}x \cos x$
17. $\phi_p(x) = \frac{1}{2}x^2 e^x$
19. $\phi_p(x) = \frac{1}{250}(2 \cos x + 11 \sin x - 35x \cos x - 5x \sin x)$
21. $\phi(x) = 2 \cos x - \sin x - (\cos x) \ln(\sec x) + x \sin x$, $-\pi/2 < x < \pi/2$
23. $\phi(x) = -\frac{1}{5}x^{-2} + \frac{1}{5}x^3$ 25. $\tan x$ is not a solution of any homogeneous linear equation; yes
27. $\phi(x) = c_1 \cos x + c_2 \sin x + x + \frac{1}{2}e^x$

Section 20.8

1. $x = 3 \cos 4t$; $A = 3$, $T = \pi/2$, $\phi = 0$
3. $x = -\frac{1}{6}e^{-2t}(3 \cos 3t + 2 \sin 3t)$; $T = 2\pi/3$

True-False Quiz (Chapter 20)

1. F   2. T   3. T   4. F   5. F   6. T   7. F
8. T   9. T   10. T

Additional Problems (Chapter 20)

3. $y = 4 - \frac{1}{4}(x - 6)^2$ 5. $y = -\ln(2 - x)$

7. $y = \sin(x + \pi/6)$ 9. $y = \dfrac{3e^{2x^2} + 1}{3e^{2x^2} - 1}$

11. $y = Ce^{-x} + 1$ 13. $x^3 + 2xy + y^4 = 16$

15. $y = \frac{1}{2}(e^{x^2 - 1} - 1)$ 17. $y = \dfrac{x + 1}{x^2 + 1}$

19. $y = x \sec x$ 21. $y = Ce^x + xe^x$ 23. $y = Cx^2 + x^3 \ln x - x^3$ 25. $y = Cx^{-1} + \frac{1}{4}x^3$ 27. $y = 2x^{-2}$, valid in $(0, \infty)$ 31. yes   33. yes   35. no
37. $(x - 1)e^x$ 39. $y = c_1 + c_2 e^{4x}$ 41. $y = c_1 e^{-3x} + c_2 xe^{-3x}$ 43. $y = c_1 e^x + c_2 e^{x/2}$ 45. $y = 2$
47. $y = \cos 2x + \frac{1}{2} \sin 2x$ 49. $-\frac{1}{2}\cos x; y = c_1 e^x + c_2 e^{-x} - \frac{1}{2}\cos x$ 51. $-\frac{1}{3}\sinh x; y = c_1 e^{2x} + c_2 e^{-2x} - \frac{1}{3}\sinh x$ 53. $y = 5 \cos x + 3 \sin x + x^2 - 3$
55. $y = x^3 - x^2$ 57. No such solutions exist.

Appendix Tables

Square Roots and Cube Roots

n	$\sqrt{n}$	$\sqrt[3]{n}$	n	$\sqrt{n}$	$\sqrt[3]{n}$	n	$\sqrt{n}$	$\sqrt[3]{n}$
1	1.00000	1.00000	35	5.91608	3.27107	69	8.30662	4.10157
2	1.41421	1.25992	36	6.00000	3.30193	70	8.36660	4.12129
3	1.73205	1.44225	37	6.08276	3.33222	71	8.42615	4.14082
4	2.00000	1.58740	38	6.16441	3.36198	72	8.48528	4.16017
5	2.23607	1.70998	39	6.24500	3.39121	73	8.54400	4.17934
6	2.44949	1.81712	40	6.32456	3.41995	74	8.60233	4.19834
7	2.64575	1.91293	41	6.40312	3.44822	75	8.66025	4.21716
8	2.82843	2.00000	42	6.48074	3.47603	76	8.71780	4.23582
9	3.00000	2.08008	43	6.55744	3.50340	77	8.77496	4.25432
10	3.16228	2.15443	44	6.63325	3.53035	78	8.83176	4.27266
11	3.31662	2.22398	45	6.70820	3.55689	79	8.88819	4.29084
12	3.46410	2.28943	46	6.78233	3.58305	80	8.94427	4.30887
13	3.60555	2.35133	47	6.85565	3.60883	81	9.00000	4.32675
14	3.74166	2.41014	48	6.92820	3.63424	82	9.05539	4.34448
15	3.87298	2.46621	49	7.00000	3.65931	83	9.11043	4.36207
16	4.00000	2.51984	50	7.07107	3.68403	84	9.16515	4.37952
17	4.12311	2.57128	51	7.14143	3.70843	85	9.21954	4.39683
18	4.24264	2.62074	52	7.21110	3.73251	86	9.27362	4.41400
19	4.35890	2.66840	53	7.28011	3.75629	87	9.32738	4.43105
20	4.47214	2.71442	54	7.34847	3.77976	88	9.38083	4.44796
21	4.58258	2.75892	55	7.41620	3.80295	89	9.43398	4.46475
22	4.69042	2.80204	56	7.48331	3.82586	90	9.48683	4.48140
23	4.79583	2.84387	57	7.54983	3.84850	91	9.53939	4.49794
24	4.89898	2.88450	58	7.61577	3.87088	92	9.59166	4.51436
25	5.00000	2.92402	59	7.68115	3.89300	93	9.64365	4.53065
26	5.09902	2.96250	60	7.74597	3.91487	94	9.69536	4.54684
27	5.19615	3.00000	61	7.81025	3.93650	95	9.74679	4.56290
28	5.29150	3.03659	62	7.87401	3.95789	96	9.79796	4.57886
29	5.38516	3.07232	63	7.93725	3.97906	97	9.84886	4.59470
30	5.47723	3.10723	64	8.00000	4.00000	98	9.89949	4.61044
31	5.56776	3.14138	65	8.06226	4.02073	99	9.94987	4.62607
32	5.65685	3.17480	66	8.12404	4.04124	100	10.0000	4.64159
33	5.74456	3.20753	67	8.18535	4.06155	101	10.0499	4.65701
34	5.83095	3.23961	68	8.24621	4.08166	102	10.0995	4.67233

Square Roots and Cube Roots (*continued*)

n	$\sqrt{n}$	$\sqrt[3]{n}$	n	$\sqrt{n}$	$\sqrt[3]{n}$	n	$\sqrt{n}$	$\sqrt[3]{n}$
103	10.1489	4.68755	136	11.6619	5.14256	169	13.0000	5.52877
104	10.1980	4.70267	137	11.7047	5.15514	170	13.0384	5.53966
105	10.2470	4.71769	138	11.7473	5.16765	171	13.0767	5.55050
106	10.2956	4.73262	139	11.7898	5.18010	172	13.1149	5.56130
107	10.3441	4.74746	140	11.8322	5.19249	173	13.1529	5.57205
108	10.3923	4.76220	141	11.8743	5.20483	174	13.1909	5.58277
109	10.4403	4.77686	142	11.9164	5.21710	175	13.2288	5.59344
110	10.4881	4.79142	143	11.9583	5.22932	176	13.2665	5.60408
111	10.5357	4.80590	144	12.0000	5.24148	177	13.3041	5.61467
112	10.5830	4.82028	145	12.0416	5.25359	178	13.3417	5.62523
113	10.6301	4.83459	146	12.0830	5.26564	179	13.3791	5.63574
114	10.6771	4.84881	147	12.1244	5.27763	180	13.4164	5.64622
115	10.7238	4.86294	148	12.1655	5.28957	181	13.4536	5.65665
116	10.7703	4.87700	149	12.2066	5.30146	182	13.4907	5.66705
117	10.8167	4.89097	150	12.2474	5.31329	183	13.5277	5.67741
118	10.8628	4.90487	151	12.2882	5.32507	184	13.5647	5.68773
119	10.9087	4.91868	152	12.3288	5.33680	185	13.6015	5.69802
120	10.9545	4.93242	153	12.3693	5.34848	186	13.6382	5.70827
121	11.0000	4.94609	154	12.4097	5.36011	187	13.6748	5.71848
122	11.0454	4.95968	155	12.4499	5.37169	188	13.7113	5.72865
123	11.0905	4.97319	156	12.4900	5.38321	189	13.7477	5.73879
124	11.1355	4.98663	157	12.5300	5.39469	190	13.7840	5.74890
125	11.1803	5.00000	158	12.5698	5.40612	191	13.8203	5.75897
126	11.2250	5.01330	159	12.6095	5.41750	192	13.8564	5.76900
127	11.2694	5.02653	160	12.6491	5.42884	193	13.8924	5.77900
128	11.3137	5.03968	161	12.6886	5.44012	194	13.9284	5.78896
129	11.3578	5.05277	162	12.7279	5.45136	195	13.9642	5.79889
130	11.4018	5.06580	163	12.7671	5.46256	196	14.0000	5.80879
131	11.4455	5.07875	164	12.8062	5.47370	197	14.0357	5.81865
132	11.4891	5.09164	165	12.8452	5.48481	198	14.0712	5.82848
133	11.5326	5.10447	166	12.8841	5.49586	199	14.1067	5.83827
134	11.5758	5.11723	167	12.9228	5.50688	200	14.1421	5.84804
135	11.6190	5.12993	168	12.9615	5.51785			

Trigonometric Functions of Numerical Input

t	$\sin t$	$\cos t$	$\tan t$
0.00	0.0000	1.0000	0.0000
0.01	0.0100	1.0000	0.0100
0.02	0.0200	0.9998	0.0200
0.03	0.0300	0.9996	0.0300
0.04	0.0400	0.9992	0.0400
0.05	0.0500	0.9988	0.0500
0.06	0.0600	0.9982	0.0601
0.07	0.0699	0.9976	0.0701
0.08	0.0799	0.9968	0.0802
0.09	0.0899	0.9960	0.0902
0.10	0.0998	0.9950	0.1003
0.11	0.1098	0.9940	0.1104
0.12	0.1197	0.9928	0.1206
0.13	0.1296	0.9916	0.1307
0.14	0.1395	0.9902	0.1409
0.15	0.1494	0.9888	0.1511
0.16	0.1593	0.9872	0.1614
0.17	0.1692	0.9856	0.1717
0.18	0.1790	0.9838	0.1820
0.19	0.1889	0.9820	0.1923
0.20	0.1987	0.9801	0.2027
0.21	0.2085	0.9780	0.2131
0.22	0.2182	0.9759	0.2236
0.23	0.2280	0.9737	0.2341
0.24	0.2377	0.9713	0.2447
0.25	0.2474	0.9689	0.2553
0.26	0.2571	0.9664	0.2660
0.27	0.2667	0.9638	0.2768
0.28	0.2764	0.9611	0.2876
0.29	0.2860	0.9582	0.2984
0.30	0.2955	0.9553	0.3093
0.31	0.3051	0.9523	0.3203
0.32	0.3146	0.9492	0.3314
0.33	0.3240	0.9460	0.3425
0.34	0.3335	0.9428	0.3537
0.35	0.3429	0.9394	0.3650
0.36	0.3523	0.9359	0.3764
0.37	0.3616	0.9323	0.3879
0.38	0.3709	0.9287	0.3994
0.39	0.3802	0.9249	0.4111
0.40	0.3894	0.9211	0.4228
0.41	0.3986	0.9171	0.4346
0.42	0.4078	0.9131	0.4466
0.43	0.4169	0.9090	0.4586
0.44	0.4259	0.9048	0.4708
0.45	0.4350	0.9004	0.4831
0.46	0.4439	0.8961	0.4954
0.47	0.4529	0.8916	0.5080
0.48	0.4618	0.8870	0.5206
0.49	0.4706	0.8823	0.5334
0.50	0.4794	0.8776	0.5463
0.51	0.4882	0.8727	0.5594
0.52	0.4969	0.8678	0.5726

t	$\sin t$	$\cos t$	$\tan t$
0.53	0.5055	0.8628	0.5859
0.54	0.5141	0.8577	0.5994
0.55	0.5227	0.8525	0.6131
0.56	0.5312	0.8473	0.6269
0.57	0.5396	0.8419	0.6410
0.58	0.5480	0.8365	0.6552
0.59	0.5564	0.8309	0.6696
0.60	0.5646	0.8253	0.6841
0.61	0.5729	0.8196	0.6989
0.62	0.5810	0.8139	0.7139
0.63	0.5891	0.8080	0.7291
0.64	0.5972	0.8021	0.7445
0.65	0.6052	0.7961	0.7602
0.66	0.6131	0.7900	0.7761
0.67	0.6210	0.7838	0.7923
0.68	0.6288	0.7776	0.8087
0.69	0.6365	0.7712	0.8253
0.70	0.6442	0.7648	0.8423
0.71	0.6518	0.7584	0.8595
0.72	0.6594	0.7518	0.8771
0.73	0.6669	0.7452	0.8949
0.74	0.6743	0.7385	0.9131
0.75	0.6816	0.7317	0.9316
0.76	0.6889	0.7248	0.9505
0.77	0.6961	0.7179	0.9697
0.78	0.7033	0.7109	0.9893
0.79	0.7104	0.7038	1.0092
0.80	0.7174	0.6967	1.0296
0.81	0.7243	0.6895	1.0505
0.82	0.7311	0.6822	1.0717
0.83	0.7379	0.6749	1.0934
0.84	0.7446	0.6675	1.1156
0.85	0.7513	0.6600	1.1383
0.86	0.7578	0.6524	1.1616
0.87	0.7643	0.6448	1.1853
0.88	0.7707	0.6372	1.2097
0.89	0.7771	0.6294	1.2346
0.90	0.7833	0.6216	1.2602
0.91	0.7895	0.6137	1.2864
0.92	0.7956	0.6058	1.3133
0.93	0.8016	0.5978	1.3409
0.94	0.8076	0.5898	1.3692
0.95	0.8134	0.5817	1.3984
0.96	0.8192	0.5735	1.4284
0.97	0.8249	0.5653	1.4592
0.98	0.8305	0.5570	1.4910
0.99	0.8360	0.5487	1.5237
1.00	0.8415	0.5403	1.5574
1.01	0.8468	0.5319	1.5922
1.02	0.8521	0.5234	1.6281
1.03	0.8573	0.5148	1.6652
1.04	0.8624	0.5062	1.7036
1.05	0.8674	0.4976	1.7433

t	$\sin t$	$\cos t$	$\tan t$
1.06	0.8724	0.4889	1.7844
1.07	0.8772	0.4801	1.8270
1.08	0.8820	0.4713	1.8712
1.09	0.8866	0.4625	1.9171
1.10	0.8912	0.4536	1.9648
1.11	0.8957	0.4447	2.0143
1.12	0.9001	0.4357	2.0660
1.13	0.9044	0.4267	2.1198
1.14	0.9086	0.4176	2.1759
1.15	0.9128	0.4085	2.2345
1.16	0.9168	0.3993	2.2958
1.17	0.9208	0.3902	2.3600
1.18	0.9246	0.3809	2.4273
1.19	0.9284	0.3717	2.4979
1.20	0.9320	0.3624	2.5722
1.21	0.9356	0.3530	2.6503
1.22	0.9391	0.3436	2.7328
1.23	0.9425	0.3342	2.8198
1.24	0.9458	0.3248	2.9119
1.25	0.9490	0.3153	3.0096
1.26	0.9521	0.3058	3.1133
1.27	0.9551	0.2963	3.2236
1.28	0.9580	0.2867	3.3413
1.29	0.9608	0.2771	3.4672
1.30	0.9636	0.2675	3.6021
1.31	0.9662	0.2579	3.7471
1.32	0.9687	0.2482	3.9033
1.33	0.9711	0.2385	4.0723
1.34	0.9735	0.2288	4.2556
1.35	0.9757	0.2190	4.4552
1.36	0.9779	0.2092	4.6734
1.37	0.9799	0.1994	4.9131
1.38	0.9819	0.1896	5.1774
1.39	0.9837	0.1798	5.4707
1.40	0.9854	0.1700	5.7979
1.41	0.9871	0.1601	6.1654
1.42	0.9887	0.1502	6.5811
1.43	0.9901	0.1403	7.0555
1.44	0.9915	0.1304	7.6018
1.45	0.9927	0.1205	8.2381
1.46	0.9939	0.1106	8.9886
1.47	0.9949	0.1006	9.8874
1.48	0.9959	0.0907	10.9834
1.49	0.9967	0.0807	12.3499
1.50	0.9975	0.0707	14.1014
1.51	0.9982	0.0608	16.4281
1.52	0.9987	0.0508	19.6695
1.53	0.9992	0.0408	24.4984
1.54	0.9995	0.0308	32.4611
1.55	0.9998	0.0208	48.0785
1.56	0.9999	0.0108	92.6205
1.57	1.0000	0.0008	1255.77
$\pi/2$	1.0000	0.0000	

Natural Logarithms

x	$\ln x$	x	$\ln x$	x	$\ln x$	x	$\ln x$
0.01	-4.60517	0.45	-0.79851	0.89	-0.11653	1.34	0.29267
0.02	-3.91202	0.46	-0.77653			1.35	0.30010
0.03	-3.50656	0.47	-0.75502	0.90	-0.10536	1.36	0.30748
0.04	-3.21888	0.48	-0.73397	0.91	-0.09431	1.37	0.31481
0.05	-2.99573	0.49	-0.71335	0.92	-0.08338	1.38	0.32208
0.06	-2.81341			0.93	-0.07257	1.39	0.32930
0.07	-2.65926	0.50	-0.69315	0.94	-0.06188		
0.08	-2.52573	0.51	-0.67334	0.95	-0.05129	1.40	0.33647
0.09	-2.40795	0.52	-0.65393	0.96	-0.04082	1.41	0.34359
		0.53	-0.63488	0.97	-0.03046	1.42	0.35066
0.10	-2.30259	0.54	-0.61619	0.98	-0.02020	1.43	0.35767
0.11	-2.20727	0.55	-0.59784	0.99	-0.01005	1.44	0.36464
0.12	-2.12026	0.56	-0.57982			1.45	0.37156
0.13	-2.04022	0.57	-0.56212	1.00	0.00000	1.46	0.37844
0.14	-1.96611	0.58	-0.54473	1.01	0.00995	1.47	0.38526
0.15	-1.89712	0.59	-0.52763	1.02	0.01980	1.48	0.39204
0.16	-1.83258			1.03	0.02956	1.49	0.39878
0.17	-1.77196	0.60	-0.51083	1.04	0.03922		
0.18	-1.71480	0.61	-0.49430	1.05	0.04879	1.50	0.40547
0.19	-1.66073	0.62	-0.47804	1.06	0.05827	1.51	0.41211
		0.63	-0.46204	1.07	0.06766	1.52	0.41871
0.20	-1.60944	0.64	-0.44629	1.08	0.07696	1.53	0.42527
0.21	-1.56065	0.65	-0.43078	1.09	0.08618	1.54	0.43178
0.22	-1.51413	0.66	-0.41552			1.55	0.43825
0.23	-1.46968	0.67	-0.40048	1.10	0.09531	1.56	0.44469
0.24	-1.42712	0.68	-0.38566	1.11	0.10436	1.57	0.45108
0.25	-1.38629	0.69	-0.37106	1.12	0.11333	1.58	0.45742
0.26	-1.34707			1.13	0.12222	1.59	0.46373
0.27	-1.30933	0.70	-0.35667	1.14	0.13103	1.60	0.47000
0.28	-1.27297	0.71	-0.34249	1.15	0.13976	1.61	0.47623
0.29	-1.23787	0.72	-0.32850	1.16	0.14842	1.62	0.48243
		0.73	-0.31471	1.17	0.15700	1.63	0.48858
0.30	-1.20397	0.74	-0.30111	1.18	0.16551	1.64	0.49470
0.31	-1.17118	0.75	-0.28768	1.19	0.17395	1.65	0.50078
0.32	-1.13943	0.76	-0.27444	1.20	0.18232	1.66	0.50682
0.33	-1.10866	0.77	-0.26136	1.21	0.19062	1.67	0.51282
0.34	-1.07881	0.78	-0.24846	1.22	0.19885	1.68	0.51879
0.35	-1.04982	0.79	-0.23572	1.23	0.20701	1.69	0.52473
0.36	-1.02165			1.24	0.21511		
0.37	-0.99425	0.80	-0.22314	1.25	0.22314	1.70	0.53063
0.38	-0.96758	0.81	-0.21072	1.26	0.23111	1.71	0.53649
0.39	-0.94161	0.82	-0.19845	1.27	0.23902	1.72	0.54232
		0.83	-0.18633	1.28	0.24686	1.73	0.54812
0.40	-0.91629	0.84	-0.17435	1.29	0.25464	1.74	0.55389
0.41	-0.89160	0.85	-0.16252			1.75	0.55962
0.42	-0.86750	0.86	-0.15082	1.30	0.26236	1.76	0.56531
0.43	-0.84397	0.87	-0.13926	1.31	0.27003	1.77	0.57098
0.44	-0.82098	0.88	-0.12783	1.32	0.27763	1.78	0.57661
				1.33	0.28518	1.79	0.58222

Natural Logarithms (*continued*)

x	ln x	x	ln x	x	ln x	x	ln x
1.80	0.58779	4.50	1.50408	9.00	2.19722	45.0	3.80666
1.81	0.59333	4.60	1.52606	9.10	2.20827	46.0	3.82864
1.82	0.59884	4.70	1.54756	9.20	2.21920	47.0	3.85015
1.83	0.60432	4.80	1.56862	9.30	2.23001	48.0	3.87120
1.84	0.60977	4.90	1.58924	9.40	2.24071	49.0	3.89182
1.85	0.61519			9.50	2.25129		
1.86	0.62058	5.00	1.60944	9.60	2.26176	50.0	3.91202
1.87	0.62594	5.10	1.62924	9.70	2.27213	51.0	3.93183
1.88	0.63127	5.20	1.64866	9.80	2.28238	52.0	3.95124
1.89	0.63658	5.30	1.66771	9.90	2.29253	53.0	3.97029
		5.40	1.68640			54.0	3.98898
1.90	0.64185	5.50	1.70475	10.0	2.30259	55.0	4.00733
1.91	0.64710	5.60	1.72277	11.0	2.39790	56.0	4.02535
1.92	0.65233	5.70	1.74047	12.0	2.48491	57.0	4.04305
1.93	0.65752	5.80	1.75786	13.0	2.56495	58.0	4.06044
1.94	0.66269	5.90	1.77495	14.0	2.63906	59.0	4.07754
1.95	0.66783			15.0	2.70805		
1.96	0.67294	6.00	1.79176	16.0	2.77259	60.0	4.09434
1.97	0.67803	6.10	1.80829	17.0	2.83321	61.0	4.11087
1.98	0.68310	6.20	1.82455	18.0	2.89037	62.0	4.12713
1.99	0.68813	6.30	1.84055	19.0	2.94444	63.0	4.14313
		6.40	1.85630			64.0	4.15888
2.00	0.69315	6.50	1.87180	20.0	2.99573	65.0	4.17439
2.10	0.74194	6.60	1.88707	21.0	3.04452	66.0	4.18965
2.20	0.78846	6.70	1.90211	22.0	3.09104	67.0	4.20469
2.30	0.83291	6.80	1.91692	23.0	3.13549	68.0	4.21951
2.40	0.87547	6.90	1.93152	24.0	3.17805	69.0	4.23411
2.50	0.91629			25.0	3.21888		
2.60	0.95551	7.00	1.94591	26.0	3.25810	70.0	4.24850
2.70	0.99325	7.10	1.96009	27.0	3.29584	71.0	4.26268
2.80	1.02962	7.20	1.97408	28.0	3.33220	72.0	4.27667
2.90	1.06471	7.30	1.98787	29.0	3.36730	73.0	4.29046
		7.40	2.00148			74.0	4.30407
3.00	1.09861	7.50	2.01490	30.0	3.40120	75.0	4.31749
3.10	1.13140	7.60	2.02815	31.0	3.43399	76.0	4.33073
3.20	1.16315	7.70	2.04122	32.0	3.46574	77.0	4.34381
3.30	1.19392	7.80	2.05412	33.0	3.49651	78.0	4.35671
3.40	1.22378	7.90	2.06686	34.0	3.52636	79.0	4.36945
3.50	1.25276			35.0	3.55535		
3.60	1.28093	8.00	2.07944	36.0	3.58352	80.0	4.38203
3.70	1.30833	8.10	2.09186	37.0	3.61092	81.0	4.39445
3.80	1.33500	8.20	2.10413	38.0	3.63759	82.0	4.40672
3.90	1.36098	8.30	2.11626	39.0	3.66356	83.0	4.41884
		8.40	2.12823			84.0	4.43082
4.00	1.38629	8.50	2.14007	40.0	3.68888	85.0	4.44265
4.10	1.41099	8.60	2.15176	41.0	3.71357	86.0	4.45435
4.20	1.43508	8.70	2.16332	42.0	3.73767	87.0	4.46591
4.30	1.45862	8.80	2.17475	43.0	3.76120	88.0	4.47734
4.40	1.48160	8.90	2.18605	44.0	3.78419	89.0	4.48864

Natural Logarithms (*continued*)

x	ln x	x	ln x	x	ln x	x	ln x
90.0	4.49981	300.	5.70378	600.	6.39693	900.	6.80239
91.0	4.51086	310.	5.73657	610.	6.41346	910.	6.81344
92.0	4.52179	320.	5.76832	620.	6.42972	920.	6.82437
93.0	4.53260	330.	5.79909	630.	6.44572	930.	6.83518
94.0	4.54329	340.	5.82895	640.	6.46147	940.	6.84588
95.0	4.55388	350.	5.85793	650.	6.47697	950.	6.85646
96.0	4.56435	360.	5.88610	660.	6.49224	960.	6.86693
97.0	4.57471	370.	5.91350	670.	6.50728	970.	6.87730
98.0	4.58497	380.	5.94017	680.	6.52209	980.	6.88755
99.0	4.59512	390.	5.96615	690.	6.53669	990.	6.89770
100.	4.60517	400.	5.99146	700.	6.55108	1000.	6.90776
110.	4.70048	410.	6.01616	710.	6.56526		
120.	4.78749	420.	6.04025	720.	6.57925		
130.	4.86753	430.	6.06379	730.	6.59304		
140.	4.94164	440.	6.08677	740.	6.60665		
150.	5.01064	450.	6.10925	750.	6.62007		
160.	5.07517	460.	6.13123	760.	6.63332		
170.	5.13580	470.	6.15273	770.	6.64639		
180.	5.19296	480.	6.17379	780.	6.65929		
190.	5.24702	490.	6.19441	790.	6.67203		
200.	5.29832	500.	6.21461	800.	6.68461		
210.	5.34711	510.	6.23441	810.	6.69703		
220.	5.39363	520.	6.25383	820.	6.70930		
230.	5.43808	530.	6.27288	830.	6.72143		
240.	5.48064	540.	6.29157	840.	6.73340		
250.	5.52146	550.	6.30992	850.	6.74524		
260.	5.56068	560.	6.32794	860.	6.75693		
270.	5.59842	570.	6.34564	870.	6.76849		
280.	5.63479	580.	6.36303	880.	6.77992		
290.	5.66988	590.	6.38012	890.	6.79122		

Exponential and Hyperbolic Functions

x	e^x	e^{-x}	sinh x	cosh x	tanh x
0.00	1.0000	1.00000	0.0000	1.0000	.00000
0.01	1.0101	0.99005	0.0100	1.0001	.01000
0.02	1.0202	.98020	0.0200	1.0002	.02000
0.03	1.0305	.97045	0.0300	1.0005	.02999
0.04	1.0408	.96079	0.0400	1.0008	.03998
0.05	1.0513	.95123	0.0500	1.0013	.04996
0.06	1.0618	.94176	0.0600	1.0018	.05993
0.07	1.0725	.93239	0.0701	1.0025	.06989
0.08	1.0833	.92312	0.0801	1.0032	.07983
0.09	1.0942	.91393	0.0901	1.0041	.08976
0.10	1.1052	.90484	0.1002	1.0050	.09967
0.11	1.1163	.89583	0.1102	1.0061	.10956
0.12	1.1275	.88692	0.1203	1.0072	.11943
0.13	1.1388	.87810	0.1304	1.0085	.12927
0.14	1.1503	.86936	0.1405	1.0098	.13909
0.15	1.1618	.86071	0.1506	1.0113	.14889
0.16	1.1735	.85214	0.1607	1.0128	.15865
0.17	1.1853	.84366	0.1708	1.0145	.16838
0.18	1.1972	.83527	0.1810	1.0162	.17808
0.19	1.2092	.82696	0.1911	1.0181	.18775
0.20	1.2214	.81873	0.2013	1.0201	.19738
0.21	1.2337	.81058	0.2115	1.0221	.20697
0.22	1.2461	.80252	0.2218	1.0243	.21652
0.23	1.2586	.79453	0.2320	1.0266	.22603
0.24	1.2712	.78663	0.2423	1.0289	.23550
0.25	1.2840	.77880	0.2526	1.0314	.24492
0.26	1.2969	.77105	0.2629	1.0340	.25430
0.27	1.3100	.76338	0.2733	1.0367	.26362
0.28	1.3231	.75578	0.2837	1.0395	.27291
0.29	1.3364	.74826	0.2941	1.0423	.28213
0.30	1.3499	.74082	0.3045	1.0453	.29131
0.31	1.3634	.73345	0.3150	1.0484	.30044
0.32	1.3771	.72615	0.3255	1.0516	.30951
0.33	1.3910	.71892	0.3360	1.0549	.31852
0.34	1.4049	.71177	0.3466	1.0584	.32748
0.35	1.4191	.70469	0.3572	1.0619	.33638
0.36	1.4333	.69768	0.3678	1.0655	.34521
0.37	1.4477	.69073	0.3785	1.0692	.35399
0.38	1.4623	.68386	0.3892	1.0731	.36271
0.39	1.4770	.67706	0.4000	1.0770	.37136
0.40	1.4918	.67032	0.4108	1.0811	.37995
0.41	1.5068	.66365	0.4216	1.0852	.38847
0.42	1.5220	.65705	0.4325	1.0895	.39693
0.43	1.5373	.65051	0.4434	1.0939	.40532
0.44	1.5527	.64404	0.4543	1.0984	.41364
0.45	1.5683	.63763	0.4653	1.1030	.42190
0.46	1.5841	.63128	0.4764	1.1077	.43008
0.47	1.6000	.62500	0.4875	1.1125	.43820
0.48	1.6161	.61878	0.4986	1.1174	.44624
0.49	1.6323	.61263	0.5098	1.1225	.45422
0.50	1.6487	.60653	0.5211	1.1276	.46212
0.51	1.6653	.60050	0.5324	1.1329	.46995
0.52	1.6820	.59452	0.5438	1.1383	.47770
0.53	1.6989	.58860	0.5552	1.1438	.48538
0.54	1.7160	.58275	0.5666	1.1494	.49299
0.55	1.7333	.57695	0.5782	1.1551	.50052
0.56	1.7507	.57121	0.5897	1.1609	.50798
0.57	1.7683	.56553	0.6014	1.1669	.51536
0.58	1.7860	.55990	0.6131	1.1730	.52267
0.59	1.8040	.55433	0.6248	1.1792	.52990
0.60	1.8221	.54881	0.6367	1.1855	.53705
0.61	1.8404	.54335	0.6485	1.1919	.54413
0.62	1.8589	.53794	0.6605	1.1984	.55113
0.63	1.8776	.53259	0.6725	1.2051	.55805
0.64	1.8965	.52729	0.6846	1.2119	.56490
0.65	1.9155	.52205	0.6967	1.2188	.57167
0.66	1.9348	.51685	0.7090	1.2258	.57836
0.67	1.9542	.51171	0.7213	1.2330	.58498
0.68	1.9739	.50662	0.7336	1.2402	.59152
0.69	1.9937	.50158	0.7461	1.2476	.59798
0.70	2.0138	.49659	0.7586	1.2552	.60437
0.71	2.0340	.49164	0.7712	1.2628	.61068
0.72	2.0544	.48675	0.7838	1.2706	.61691
0.73	2.0751	.48191	0.7966	1.2785	.62307
0.74	2.0959	.47711	0.8094	1.2865	.62915
0.75	2.1170	.47237	0.8223	1.2947	.63515
0.76	2.1383	.46767	0.8353	1.3030	.64108
0.77	2.1598	.46301	0.8484	1.3114	.64693
0.78	2.1815	.45841	0.8615	1.3199	.65271
0.79	2.2034	.45384	0.8748	1.3286	.65841

Exponential and Hyperbolic Functions (*continued*)

x	e^x	e^{-x}	sinh x	cosh x	tanh x	x	e^x	e^{-x}	sinh x	cosh x	tanh x
0.80	2.2255	.44933	0.8881	1.3374	.66404	3.00	20.086	.04979	10.018	10.068	.99505
0.81	2.2479	.44486	0.9015	1.3464	.66959	3.10	22.198	.04505	11.076	11.122	.99595
0.82	2.2705	.44043	0.9150	1.3555	.67507	3.20	24.533	.04076	12.246	12.287	.99668
0.83	2.2933	.43605	0.9286	1.3647	.68048	3.30	27.113	.03688	13.538	13.575	.99728
0.84	2.3164	.43171	0.9423	1.3740	.68581	3.40	29.964	.03337	14.965	14.999	.99777
0.85	2.3396	.42741	0.9561	1.3835	.69107	3.50	33.115	.03020	16.543	16.573	.99818
0.86	2.3632	.42316	0.9700	1.3932	.69626	3.60	36.598	.02732	18.285	18.313	.99851
0.87	2.3869	.41895	0.9840	1.4029	.70137	3.70	40.447	.02472	20.211	20.236	.99878
0.88	2.4109	.41478	0.9981	1.4128	.70642	3.80	44.701	.02237	22.339	22.362	.99900
0.89	2.4351	.41066	1.0122	1.4229	.71139	3.90	49.402	.02024	24.691	24.711	.99918
0.90	2.4596	.40657	1.0265	1.4331	.71630	4.00	54.598	.01832	27.290	27.308	.99933
0.91	2.4843	.40252	1.0409	1.4434	.72113	4.10	60.340	.01657	30.162	30.178	.99945
0.92	2.5093	.39852	1.0554	1.4539	.72590	4.20	66.686	.01500	33.336	33.351	.99955
0.93	2.5345	.39455	1.0700	1.4645	.73059	4.30	73.700	.01357	36.843	36.857	.99963
0.94	2.5600	.39063	1.0847	1.4753	.73522	4.40	81.451	.01228	40.719	40.732	.99970
0.95	2.5857	.38674	1.0995	1.4862	.73978	4.50	90.017	.01111	45.003	45.014	.99975
0.96	2.6117	.38289	1.1144	1.4973	.74428	4.60	99.484	.01005	49.737	49.747	.99980
0.97	2.6379	.37908	1.1294	1.5085	.74870	4.70	109.95	.00910	54.969	54.978	.99983
0.98	2.6645	.37531	1.1446	1.5199	.75307	4.80	121.51	.00823	60.751	60.759	.99986
0.99	2.6912	.37158	1.1598	1.5314	.75736	4.90	134.29	.00745	67.141	67.149	.99989
1.00	2.7183	.36788	1.1752	1.5431	.76159	5.00	148.41	.00674	74.203	74.210	.99991
1.10	3.0042	.33287	1.3356	1.6685	.80050	5.10	164.02	.00610	82.008	82.014	.99993
1.20	3.3201	.30119	1.5095	1.8107	.83365	5.20	181.27	.00552	90.633	90.639	.99994
1.30	3.6693	.27253	1.6984	1.9709	.86172	5.30	200.34	.00499	100.17	100.17	.99995
1.40	4.0552	.24660	1.9043	2.1509	.88535	5.40	221.41	.00452	110.70	110.71	.99996
1.50	4.4817	.22313	2.1293	2.3524	.90515	5.50	244.69	.00409	122.34	122.35	.99997
1.60	4.9530	.20190	2.3756	2.5775	.92167	5.60	270.43	.00370	135.21	135.22	.99997
1.70	5.4739	.18268	2.6456	2.8283	.93541	5.70	298.87	.00335	149.43	149.44	.99998
1.80	6.0496	.16530	2.9422	3.1075	.94681	5.80	330.30	.00303	165.15	165.15	.99998
1.90	6.6859	.14957	3.2682	3.4177	.95624	5.90	365.04	.00274	182.52	182.52	.99998
2.00	7.3891	.13534	3.6269	3.7622	.96403	6.00	403.43	.00248	201.71	201.72	.99999
2.10	8.1662	.12246	4.0219	4.1443	.97045	6.25	518.01	.00193	259.01	259.01	.99999
2.20	9.0250	.11080	4.4571	4.5679	.97574	6.50	665.14	.00150	332.57	332.57	1.0000
2.30	9.9742	.10026	4.9370	5.0372	.98010	6.75	854.06	.00117	427.03	427.03	1.0000
2.40	11.023	.09072	5.4662	5.5569	.98367	7.00	1096.6	.00091	548.32	548.32	1.0000
2.50	12.182	.08208	6.0502	6.1323	.98661	7.50	1808.0	.00055	904.02	904.02	1.0000
2.60	13.464	.07427	6.6947	6.7690	.98903	8.00	2981.0	.00034	1490.5	1490.5	1.0000
2.70	14.880	.06721	7.4063	7.4735	.99101	8.50	4914.8	.00020	2457.4	2457.4	1.0000
2.80	16.445	.06081	8.1919	8.2527	.99263	9.00	8103.1	.00012	4051.5	4051.5	1.0000
2.90	18.174	.05502	9.0596	9.1146	.99396	9.50	13360.	.00007	6679.9	6679.9	1.0000
						10.00	22026.	.00005	11013.	11013.	1.0000

Index

Abel, Niels (1802–1829), 893
Abel's Formula, 893
Absolute value, 6
Absolute value function, 47
   derivative of, 115
   graph of, 47
Absolutely convergent series, 535
   rearrangement of, 539
Acceleration, 153, 638, 697
   due to gravity, 153, 175, 179, 312
   tangential and normal components
     of, 641, 697
Algebraic function, 335
Alternating series, 534
Alternating Series Test, 536
Analytic geometry, 11
Angle, 62–63
   between a line and plane, 669
   measure of, 63
   as rotation, 62–63
   between vectors, 626, 661
Angular velocity, 700
Annihilator, 906
Anticommutative law of the cross
    product, 677
Antiderivative, 151, 256
   existence of, 257
Antidifferentiation, 151, 255, 260
Antigradient, 819, 852
Arc length
   of a curve, 294, 296, 692
   on a circle, 71
Archimedes (c. 287–212 B.C.), 704
Area
   of a circle, 237
   of a circular sector, 71
   between curves, 278
   as a double integral, 775
   of an ellipse, 244
   in the plane, 226–27
   in polar coordinates, 598
   of a surface, 786–87
   of a surface of revolution, 300, 789

of an unbounded region, 486–87,
   490
Asymptote, 24, 38, 196
   of a hyperbola, 28, 203, 614
Average value of a function, 250, 253,
   779, 797
Average velocity, 48
Axiom of Completeness, 511
Axis
   of a conic, 608
   coordinate, 11
   of a parabola, 604
   rotation of, 615, 619

Bacon, Roger (c. 1214–c. 1294), 276
Base
   change of, 367
   of the natural exponential function,
    354
   of the natural logarithm, 365
Base point
   in linear approximation, 163, 494
   of a power series, 553
   in Taylor's Formula, 496
Bell, E. T. (1883–1960), 97, 760
Berkeley, George (1685–1753), 127, 169
Bernoulli, Jakob (1654–1705), 405
Bernoulli, Johann (1667–1748), 226,
   362
Bernoulli's equation, 879
Best-fitting line, 749
Best value of several measurements,
   222
Binomial series, 578
Binomial Theorem, 575
Binormal vector, 695
Boundary conditions in a differential
    equation, 331, 880
Boundary points of a region, 739
Bounded
   function, 139
   region, 739
   sequence, 510, 511

Bowditch, Nathaniel (1773–1838), 180
Boyle's Law, 175, 377
Burke, Edmund (1729–1797), 335

Cantor, Georg (1845–1918), 506
Carbon dating, 374
Cardioid, 595
   area of, 598–99
   length of, 601
Catenary, 405, 416
Cauchy-Riemann equations, 714
Cauchy-Schwarz Inequality, 629, 663
Cauchy's Mean Value Theorem, 148
Cayley, Arthur (1821–1895), 551
Center
   of a conic, 608
   of gravity, 313
   of mass, 312, 314, 316, 319, 776,
    792
Centroid, 318, 319, 776, 793, 803, 839
Chain Rule, 112, 114
   in multivariable calculus, 717, 719
   for vector functions, 634
Change of variable in an integral, 265
Characteristic equation, 894
Circle, 21
   arc length on, 71
   area of, 237
   area of sector of, 71
   curvature of, 162, 645
   of curvature, 647
   osculating, 648
   in polar coordinates, 593–94
Circular functions, 402 (*see*
   Trigonometric functions)
Circulation in a vector field, 837, 857
Closed
   curve, 811, 817–18, 826
   interval, 6
   region, 739
Coefficients of a power series, 553,
   562
Cofactor in a determinant, 674

Cofunction identities in trigonometry, 67

Common logarithm, 336

Comparison Test for series, 528–29

Completing the square, 21

Complex analysis, 897

Complex-valued function, 381, 896

Composite function, 89

Composite Function Theorem, 90, 471

Compound interest, 377

Concavity, 157, 200

Conditionally convergent series, 535
  rearrangement of, 539, 541

Cone, 684
  sections of, 25–26, 603

Conic, 608
  axis of, 608
  center of, 608
  characterized by discriminant, 619
  directrix of, 608
  eccentricity of, 608
  focus of, 608
  latus rectum of, 616
  in polar coordinates, 614
  section, 25–26, 603
  vertex of, 608

Conservation of energy, 813

Conservative force field, 814

Constant function, 36

Constraint, 211, 743, 751–52

Continuity
  of a curve, 292
  of a differentiable function, 105
  of a function of one variable, 87, 127, 130
  of a function of two variables, 710
  Fundamental Principle of, 137
  of a power series, 557
  of a vector function, 629

Contour line, 736

Contraction property of sine and cosine, 73

Convergence
  absolute, 535
  conditional, 535
  of an improper integral, 483
  of an infinite series, 513
  necessary condition for, 516
  radius of, 555

Coordinate(s)
  axis, 11

cylindrical, 685

line, 2

plane, 11

polar, 591

rectangular, 11

spherical, 686

Corner of a graph, 47

Courant, Richard (1888–1972), 385

Cramer's Rule, 888

Critical point, 188, 742

Cross product of vectors, 673
  anticommutative law of, 677

Curl of a vector field, 854

Current, 181, 380, 481

Curvature
  circle of, 647
  of a circle, 162, 645
  of a plane curve, 162, 181, 640, 646, 647
  radius of, 648
  of a space curve, 693–94

Curve
  area under, 227
  centroid of, 319, 839
  closed, 811, 817–18, 826
  continuous, 292
  curvature of, 162, 181, 640, 646, 647, 693–94
  direction of, 46
  length of, 294, 296, 692
  level, 737
  orientation of, 810, 827, 855
  piecewise smooth, 810
  in the plane, 292
  rectification of, 293
  second-degree, 25, 615
  simple, 826
  smooth, 134, 142, 292, 629, 691
  in space, 691
  tangent to, 45, 691
  trace of, 292
  track of, 292, 691
  torsion of, 695

Cusp of a graph, 100

Cycloid, 298–99, 634–35

Cylinder, 679
  circular, 679
  elliptical, 681
  parabolic, 679

Cylindrical coordinates, 685
  in triple integration, 796

D'Alembert, Jean (1717–1783), 893

D'Alembert's method, 895

Damped sine wave, 82

Damping force, 911

Decimal
  infinite, 3, 518
  nonperiodic, 3
  nonterminating, 518
  periodic, 3, 520

Decrease without bound, 24

Decreasing function, 148

Degree measure of an angle, 63

De Moivre, Abraham (1667–1754), 408

De Morgan, Augustus (1806–1871), 335, 422

Density, 314, 315, 319, 791, 792

Dependent variable, 29, 30

Depreciation, 379

Derivative(s)
  of absolute value function, 115
  algebra of, 100
  of a composite function, 112, 113, 114
  defined, 54
  directional, 731
  of exponential functions, 356, 361
  first, 156
  higher-order, 156
  of hyperbolic functions, 404
  of an implicitly defined function, 120, 727
  of an integral, 257
  of an inverse function, 349
  of inverse hyperbolic functions, 409–13
  of inverse trigonometric functions, 390–94
  of irrational powers, 361–62
  of logarithmic functions, 337, 365
  one-sided, 54, 56–57
  partial, 122, 706, 708
  of a power series, 556
  of a product, 104
  of a quotient, 106
  as rate of change, 181
  of rational powers, 97
  second, 155
  of a sum, 100
  symbols for, 58
  of trigonometric functions, 107
  of a vector function, 630–31, 691

Descartes, René (1596–1650), 11, 25
Determinant, 672, 673–74
  of coefficients, 888
Deviation, 749
Difference quotient, 53, 54
Differentiable function, 54
Differential, 170
  as an approximation to the increment, 172
  of arc length, 297
  used in integration, 266, 280–81
Differential equation, 154, 322, 862
  annihilator method, 902, 906
  boundary conditions, 331, 880
  exact, 870
  Existence and Uniqueness
    Theorem, 329, 865, 882
  first-order, 323, 862
  general solution, 324, 329
  homogeneous, 874, 887
  initial conditions, 322, 331, 864, 880
  integrating factor, 875, 876
  linear, 872, 886
  nonhomogeneous, 874, 901
  order, 323
  ordinary, 862
  partial, 862
  second-order, 323, 879
  separable, 329, 872
  separation of variables, 325
  series solutions of, 581
  solution, 322, 329, 863
  undetermined coefficients, 905
  variation of parameters, 902
Differentiation, 59
  implicit, 120, 727
  as inverse of integration, 259–60
  logarithmic, 362
  of vector functions, 630–31, 691
Direction
  of a curve, 46
  of motion, 50
  of a vector, 626, 661–62
Directional derivative, 731
Directrix
  of a conic, 608
  of a parabola, 603
Discriminant, 619
Displacement, 48
Distance
  on the coordinate line, 7

  in the coordinate plane, 12
  along a curve, 294, 296, 692
  between a point and a line, 21, 756
  between a point and a plane, 670, 756
  between a point and a surface, 753
  in polar coordinates, 597
  in space, 661
  traveled by a bouncing ball, 511–12
  traveled by a falling body, 52, 155, 323
  traveled in one direction, 233
Divergence
  of an improper integral, 483
  of an infinite series, 513
  test for, 516
Divergence of a vector field, 834, 842
Divergence Theorem, 843
  in the plane, 834
Domain
  of a function, 29, 30
  of a power series, 555
Domination of one function by another, 478, 544
Dot product of vectors, 625, 661
Double integral, 770
  for area, 775
  evaluated by an iterated integral, 773
  in polar coordinates, 781
  as volume under a surface, 771
Dyson, Freeman, 551

e, 354
  decimal representation of, 3, 355
  as an infinite series, 545
  as a limit, 364
Eccentricity of a conic, 608
Einstein, Albert (1879–1955), 180, 580, 591, 620, 704
Elementary functions of analysis, 335
Ellipse, 24, 608
  alternate definition of, 618
  analyzed by calculus, 160
  area of, 244
  center of, 608
  directrices of, 608, 610–11
  eccentricity of, 608, 609
  foci of, 608, 610
  major and minor diameters of, 612
  optical property of, 618

  in polar coordinates, 614
  standard forms of, 610
  tangent to, 124, 617
  vertices of, 608
Ellipsoid, 680
  volume of, 769
Elliptical cone, 684
Elliptical paraboloid, 681
Energy
  conservation of, 813
  equivalence to mass, 580
  kinetic, 309, 813
  potential, 813
Equipotential line, 736
Escape velocity, 884
Euler, Leonhard (1707–1783), 506, 551, 894
Euler equation, 897
Euler's constant, 527
Euler's formula, 380, 408, 570, 896
Even function, 41, 66
  integral of, 270
Exact differential equation, 870
Existence and Uniqueness Theorem
  for differential equations, 329, 865, 882
Exponential function
  derivative of, 356, 361
  general, 361
  natural, 353
Exponential growth, 450
Exponents
  imaginary, 380, 408, 570, 896
  irrational, 337, 355, 359
  laws of, 362
Extent of a graph, 195
Extreme value
  global, 138, 187, 739
  local, 138, 187, 740
Extreme Value Theorem, 138, 739

Factorial, 157, 496
  approximated by Stirling's Formula, 359
  defined by gamma function, 493–94
Fermat's Principle of Optics, 221
Fibonacci sequence, 35
First derivative, 156
First Derivative Test, 190
First octant, 660

First-order differential equation, 323, 862
  exact, 870
  linear, 872
  separable, 329, 872
Fluid flow, 830, 832, 840
Flux of a fluid, 830, 832, 840
Focus
  of a conic, 608
  of a parabola, 603
Force, 181
  exerted by a spring, 306
  of gravity, 307
  inverse-square law of, 1, 307, 488
  moment of, 311
  in motion of a car, 641–42
  as a vector, 621–22
  work done by, 305, 307, 808, 813
Force field, 807
Fourier, Jean-Baptiste-Joseph (1768–1830), 591
Fourier series, 526
Four-leaf rose, 595–96
Frenet-Serret formulas, 700
Function(s), 29, 30
  absolute value, 47
  age, 41
  algebraic, 335
  average value of, 250, 253, 779, 797
  bounded, 139
  circular, 402
  complex-valued, 381, 896
  composite, 89
  constant, 36
  continuous (see Continuity)
  decreasing, 148
  derivative of (see Derivative)
  differentiable, 54
  differential of, 170
  domain of, 29, 30, 704, 706
  elementary, 335
  even, 41, 66
  exponential, 353, 361
  extreme value of, 138, 187, 739, 740
  gamma, 493
  graph of, 32, 292, 691, 705
  greatest integer, 35
  harmonic, 714
  hyperbolic, 403
  identity, 37
  implicitly defined, 118

  increasing, 148
  integrable, 240, 482
  integral of (see Integral)
  inverse, 345
  inverse hyperbolic, 408
  inverse trigonometric, 387, 390, 392
  linear, 37
  linear combination of, 101, 888
  linearly independent, 887
  logarithmic, 155, 260, 337, 365
  monotonic, 148, 199
  nonalgebraic, 335, 385
  odd, 41, 66
  of one variable, 31, 36
  period of, 66
  postage, 40–41
  power, 37
  range of, 29, 30
  rational, 38
  real, 31, 36
  represented by a power series, 562
  root, 37
  of several variables, 704
  as a subset of $\mathcal{R}^2$, 33
  sum, difference, product, quotient of, 41, 86
  transcendental, 335, 385
  trigonometric, 62
  vector, 629, 691
Fundamental Theorem of Calculus, 245, 257
  for line integrals, 812, 851

Gabriel's Horn, 370
Galileo (1564–1642), 127, 322, 379–80
Galton, Francis (1822–1911), 1
Gamma function, 493
Gauss, Karl Friedrich (1777–1855), 704
Gauss' Theorem, 843
General solution of a differential equation, 324, 329
Geometric progression, 512
Geometric series, 517–18
Global extreme value, 138, 187, 739
Goethe, Johann Wolfgang von (1749–1832), 422
Gradient, 716, 736
  normal to a surface or curve, 727

Gradient field, 811–12
  necessary and sufficient conditions for, 819, 828, 851
Graph
  concavity of, 157, 200
  extent of, 195
  of a function of two variables, 705
  of a real function, 32
  of a relation, 33
  slope of, 12, 46, 53
  of a vector function, 292, 691
Gravity
  acceleration due to, 153, 175, 179, 312, 323
  center of, 313
  force of, 307, 651
Greatest integer function, 35
Greatest lower bound, 511
Green, George (1793–1841), 824
Green's Theorem, 825, 834, 837
  in space, 853–55
Gregory's series, 559

Half-life of a radioactive substance, 373–74
Halley, Edmund (1656–1742), 1
Hanging cable, 405, 416
Hardy, G. H. (1877–1947), 465, 862
Harmonic function, 714
Harmonic series, 519, 523
Heat-flow equation, 714
Helix, 690, 698
Herschel, John (1792–1871), 806
Hertz, Heinrich (1857–1894), 97
Higher-order derivative, 156
Hilbert, David (1862–1943), 506, 659
Homogeneous differential equation, 874, 887
Hooke's Law, 135, 306, 323, 399, 910
Hyperbola, 24, 608
  alternate definition of, 618
  asymptotes of, 28, 203, 613–14
  center of, 608
  directrices of, 608, 610–11
  eccentricity of, 608, 609
  foci of, 608, 610
  optical property of, 618
  in polar coordinates, 614
  standard forms of, 610–11
  tangent to, 124, 617

transverse diameter of, 613
vertices of, 608
Hyperbolic functions, 403
Hyperbolic paraboloid, 681–82
Hyperbolic substitution in an integral, 441
Hyperboloid, 682–83
Hypocycloid, 124, 162, 176, 319

Ideal Gas Law, 713
Identity function, 37
Implication, 4
Implicit differentiation, 120, 727, 729
Implicit Function Theorem, 122, 728
Improper integral, 452, 483, 485, 489
Inclination of a straight line, 20
Increase without bound, 24
Increasing function, 148
Increment, 168
   approximated by differential, 172
Independence of path in a vector
      field, 816
Independent functions, 887
Independent variable, 29, 30
Indeterminate forms, 473–74, 476–79, 481
Inequalities, 4–5
Infinite decimal, 3
Infinite limit, 469
   "existence" of, 470
   properties of, 473
Infinite series, 506, 513 (see Series)
Infinity, 6
Inflection point, 157
Inhibited growth, 442
Initial conditions, 322, 331, 864, 880
Initial value problem, 323, 865
Instantaneous velocity, 49
Integers, 2
Integrable function, 240, 482
   necessary and sufficient conditions
      for, 242–44
Integral
   derivative of, 257
   double, 770
   existence and uniqueness of, 233, 242
   improper, 452, 483, 485, 489
   iterated, 762, 791
   line, 809, 831, 850
   of a power series, 558

proper, 483
of a real function, 240
surface, 841
triple, 791
Integral Test for series, 523
Integrand, 241
Integrating factor, 875, 876
Integration, 240, 770, 791
   by parts, 423
   by substitution, 267
   by trigonometric substitution, 433
Intercept form of equation of a line, 20
Interior point
   of an interval, 10, 54
   of a region, 740
Intermediate Value Theorem, 139
Interval, 5
   closed, 6
   consisting of one point, 8
   of convergence of a power series, 555
   half-open, 6
   interior point of, 10, 54
   open, 6
Inverse function, 345
Inverse hyperbolic functions, 408
Inverse relation, 347
Inverse-Square Law, 1, 307, 488
Inverse trigonometric functions, 387, 390, 392
Irrational exponent, 337, 355, 359
Irrational numbers, 2
   as nonperiodic decimals, 3
Irreducible quadratic factor, 444
Irrotational vector field, 857
Isothermal
   line, 736
   path, 36
   surface, 737
Iterated integral, 762, 791

Jacobi, K. G. J. (1804–1851), 620
Jacobian, 690, 781
Jeans, James (1877–1946), 276

Kelvin, Lord (1824–1907), 760
Kepler, Johann (1571–1630), 651
Kepler's Laws, 612, 651, 656
Kinetic energy, 309, 813
Kirchhoff's Law, 913
Kronecker, Leopold (1823–1891), 1

Lagrange, Joseph Louis (1736–1813), 902
Lagrange form of the remainder, 498, 564
Lagrange multipliers, 752
Laplace, Pierre Simon (1749–1827), 97, 180
Laplace Transform, 492, 494
Laplace's equation, 714
Latus rectum
   of a conic, 616
   of a parabola, 607
Law of Cosines, 72–73
Law of Mass Action, 450
Least squares, 222, 749
Least upper bound, 139, 511
Left-handed limit, 81
Left neighborhood, 10, 75
Leibniz, Gottfried Wilhelm von
      (1646–1716), 44, 58, 114, 169, 226, 256, 879
Leibniz' formula, 559
Length
   of a curve, 294, 296, 692
   in polar coordinates, 600
   of a vector, 624, 661
Less than, 4
Level curve, 737
Level surface, 737
L'Hôpital's Rule, 474, 475–77, 482
Libby, Willard (1908–1980), 374
Limit(s)
   algebra of, 86, 470
   alternate definition of, 79
   of a composite function, 90, 471
   "existence" of, 470
   finite, 469
   of a function of two variables, 710
   infinite, 469
   left-handed, 81
   neighborhood definition of, 78
   one-sided, 54, 56–7, 77, 81
   right-handed, 81
   of a sequence, 507
   of a vector function, 630
Limit Comparison Test for series, 529
Line(s)
   coordinate, 2
   intercept form of, 20
   parallel, 14
   parametric equations of, 664

Line(s) (*Continued*)
  perpendicular, 14
  point-slope form of, 17
  skew, 665
  slope of, 13
  slope-intercept form of, 18
  in space, 664
  tangent, 44, 691
Line integral
  of a scalar field, 831
  of a vector field, 809, 850
Linear approximation, 163
  error in, 164
Linear combination of functions, 101, 888
Linear differential equation, 872, 886
  homogeneous, 874, 887
  nonhomogeneous, 874, 901
Linear equation, 19, 664, 668
Linear function, 37
Linear independence of functions, 887
Linear operator, 102, 887
Linearity
  of differentiation, 102
  of integration, 247–48
  of the operator $L(y) = ay'' + by' + cy$, 887
Littlewood, J. E., (b. 1885), 760
Lobachevsky, Nicolai (1793–1856), 659
Local extreme value, 138, 187, 740
Logarithm(s)
  baseless, 337, 365
  common, 336
  general, 337, 365
  laws of, 336, 338–39
  natural, 155, 260, 337
Logarithmic differentiation, 362–63
Lord of the Rings, The, 738
Lower sum, 228, 232

Machin, John (d. 1751), 561
Maclaurin, Colin (1698–1746), 563
Maclaurin series, 563
Magnitude of a vector, 624, 661
Marginal
  cost, 181, 187
  profit, 181
  revenue, 181, 187, 222
Mass, 312, 314, 315, 319, 776, 791
  center of, 312, 314, 316, 319, 776, 792

equivalence to energy, 580
  as a measure of inertia, 312
  in relativity theory, 580
Mass distribution
  along a curve, 319
  homogeneous, 318
  in the plane, 315, 776
  in space, 792
  along a straight line, 314
Mathematical induction, 110
Mathematical model, 210, 376
Maximum value, 138, 187, 739
Mean value, 750
Mean Value Theorem, 142, 145
  Cauchy's, 148
  for Integrals, 248
Mechanic's Rule, 205
Median of a triangle, 21
Method of least squares, 222, 749
Midpoint
  of an interval on the coordinate line, 7
  of a line segment in the plane, 12
Minimum value, 138, 187, 739
Minuit, Peter (1580–1638), 379
Mixed partial derivatives, 710
  equality of, 711, 713
Möbius strip, 855
Moment of a mass distribution
  along a curve, 319
  in the plane, 316, 776
  in space, 792
  along a straight line, 314
Moment of force, 311
Moment of inertia, 776–77, 795
Momentum, 181, 644
Monotonic function, 148, 199
Multivariable calculus, 704

Natural exponential function, 353
  derivative of, 356
  integral of, 369
Natural logarithm, 155, 260, 337
  base of, 365
  derivative of, 337
  graph of, 341
  inverse of, 348
  range of, 341
Natural numbers, 2
Neighborhood, 10, 54, 75
  left, 10, 75

punctured, 75
  right, 10, 75
Newton, Isaac (1642–1727), 1, 44, 176, 226, 256, 399, 576, 704, 883
Newton's First Law of Motion, 644
Newton's Law of Cooling, 375
Newton's Law of Gravitational Attraction, 307, 651
Newton's Method, 205
  convergence of, 208
Newton's Second Law of Motion, 161, 181, 309, 312, 323, 644, 652, 910
Nonalgebraic function, 335, 385
Nonhomogeneous differential equation, 874, 901
Nonperiodic decimal, 3
Nonstandard analysis, 169
Nonterminating decimal, 518
Norm of a partition, 230, 770
Normal component of acceleration, 641, 697
Normal vector
  to a curve, 693
  to a plane, 667
  to a surface, 724
n-space, 704

Octant of space, 660
Odd function, 41, 66
  integral of, 270
Ohm's Law, 380
One-sided derivative, 54, 56–57
One-sided limit, 54, 56–57, 77, 81
Open interval, 6
Optical property
  of the ellipse, 618
  of the hyperbola, 618
  of the parabola, 605
Order of a differential equation, 323
Order properties of the real numbers, 4–5
Ordered pair, 11
Ordinary differential equation, 862
Orientation of a curve, 810, 827, 855
Origin
  of the coordinate line, 2
  of the coordinate plane, 11
  in three-dimensional space, 659
Orthogonal trajectory, 327, 810

Osculating
  circle, 648
  plane, 695

Pappus' Theorem, 321
Parabola
  axis of, 604
  directrix of, 603
  eccentricity of, 608
  focus of, 603
  latus rectum of, 607
  optical property of, 605
  in polar coordinates, 606
  standard forms of, 604
  tangent to, 124, 607
  vertex of, 603
Parabolic Rule, 460
Paraboloid, 681
Paradox, 376–77
Parallel,
  lines, 14
  planes, 670
  vectors, 626, 661, 677
Parallelogram Law of vector addition, 622
Parameter, 291, 292
Parametric equations, 291, 292
Partial derivative, 122, 706, 708
  as directional derivative, 731
  higher-order, 708–709
Partial sum of a series, 513
Partition, 230, 770, 791
  norm of, 230, 770
  refinement of, 236
Peirce, Benjamin (1809–1880), 465
Pendulum, motion of, 898
Perihelion, 651
Period of a function, 66
Periodic decimal, 3
  as rational number, 520
Perpendicular
  lines, 14
  vectors, 626, 661
Pi (π), 2
  computation of, 561
  decimal representation of, 3
  as an infinite product, 429
  as an infinite series, 559
Piecewise smooth curve, 810
Plane
  coordinate, 11

in space, 668
  tangent, 724
Planetary motion, 651
Plato (428–348 B.C.), 620, 659
Point of inflection, 157
Point-slope form of equation of a line, 17
Polar coordinates, 541–92
  arc length in, 600
  area in, 598
  circles in, 593, 594
  conics in, 614
  in double integration, 781
Polynomial(s), 38, 551
  approximation by, 494
  factorization of, 444
  Taylor, 495, 496, 563
Pope, Alexander (1688–1744), 44
Population growth, 161, 372, 442, 450
Position vector, 623, 691
Positive series, 521
Potential energy, 813
Power, 181
Power function, 37
Power Rule, 97
  for irrational exponents, 361–62
Power series, 551, 553
  base point of, 553
  binomial, 578
  coefficients of, 553, 562
  continuity of, 557
  differentiation of, 556
  domain of, 555
  for elementary functions, 565, 568, 573, 579
  integration of, 558
  Maclaurin, 563
  multiplication of, 571
  rearrangement of, 570
  Taylor, 563
Present value, 491
  of future profits, 491
Prime number, 35
Prime Number Theorem, 383
Principia Mathematica, 1
Principle of Mathematical Induction, 110
Product Rule, 104
  for vector functions, 633, 693
p-series, 523–24
Punctured neighborhood, 75
Pythagorean Theorem, 11

Quadric surface, 679
Quotient Rule, 106

$\mathscr{R}$, 2
$\mathscr{R}^2$, 11
$\mathscr{R}^3$, 660
Radial unit vector, 651
Radian measure of an angle, 63
Radioactive decay, 373–74
Radius of convergence of a power series, 555
Radius of curvature, 648
Range of a function, 29, 30
Rate of change, 181
  of arc length, 297
  time, 182
Ratio Test for series, 542
Rational function, 38
  decomposition of, 443–48
  proper, 443
Rational numbers, 2
  as periodic decimals, 3, 520
Real function, 31, 36
Real numbers, 2
  addition and multiplication of, 4
  as infinite decimals, 3
  order of, 4–5
  as points of a coordinate line, 3
Rectangular coordinates, 11, 659–60
Rectification of a curve, 293–94
Recursion formula, 205, 583
Refinement of a partition, 236
Region
  boundary point of, 739
  closed and bounded, 739
  under a curve, 227
  interior point of, 740
  simply connected, 828
  standard, 824, 843
  unbounded, 486, 487, 489–91
Related rates, 181
Relation, 33
  inverse of, 347
Relative error, 163
Relativity, theory of, 580
Remainder in Taylor's Formula, 498, 564
Representation problem in Taylor series, 564
Riemann sum, 239, 278, 770
Right-handed limit, 81

Right-handed triple of vectors, 659, 677–78
Right neighborhood, 10, 75
Rolle's Theorem, 143, 145
Root function, 37
Root Test for series, 546
Rotation of axes, 615–16, 619
Russell, Bertrand (1872–1953), 127, 385

Saddle point of a surface, 741
Scalar
   field, 807
   function, 632
   product of vectors, 675
   as synonym for number, 632
   triple product of vectors, 677
Second-degree curve, 25, 615
Second-degree equation, 25, 615, 679
Second derivative, 155
Second Derivative Test, 192, 746
Second-order differential equation, 323, 879
   linear, 886
Sensitivity to a drug, 181
Separable differential equation, 329, 872
Separation of variables, 325
Sequence, 507
   bounded monotonic, 510, 511
   limit of, 507, 510
   of partial sums of a series, 513, 514
   terms of, 507
Series
   absolutely convergent, 535
   alternating, 534
   Alternating Series Test for, 536
   approximation to, 524–25
   binomial, 578
   Comparison Test for, 528
   conditionally convergent, 535
   convergent, 513
   divergent, 513
   Fourier, 526
   geometric, 517–18
   Gregory's, 559
   harmonic, 519, 523
   Integral Test for, 523
   interval of convergence of, 555
   Limit Comparison Test for, 529
   Maclaurin, 563

.p-, 523–24
   partial sum of, 513
   positive, 521
   power, 551, 553
   radius of convergence of, 555
   Ratio Test for, 542
   rearrangement of, 539, 541
   Root Test for, 546
   sum of, 513
   Taylor, 563
   test for divergence of, 516
Series solutions of differential equations, 581
Set notation, 6
Shaw, George Bernard (1856–1950), 335
Simple curve, 826
Simple harmonic motion, 911
   amplitude of, 911
   frequency of, 911
   period of, 911
   phase angle of, 911
Simply connected region, 828
Simpson's Rule, 460
Sink in a vector field, 833
Skew lines, 665
Slope
   of a curve, 46, 53
   of a horizontal line, 14
   of parallel lines, 14
   of perpendicular lines, 14
   of a straight line, 13
   of a tangent, 46
   of a vertical line, 14
Slope-intercept form of equation of a line, 18
Smith, David A., 442, 750
Smooth curve, 134, 142, 292, 629, 691
Snell's Law of Refraction, 221
Solid of revolution, 283
Solution of a differential equation, 322, 329, 863
Source in a vector field, 833
Speed, 48, 50, 297, 637, 697
Sphere, 661, 680
Spherical coordinates, 686
   in triple integration, 798
Squeeze Play Theorem, 91, 471
Standard region, 824, 843
Stirling's Formula, 359
Stokes' Theorem, 855, 857

Substitution in an integral, 267
Sum of an infinite series, 513
Summation notation, 231
Surface, 661
   area of, 300, 786–87, 849
   centroid of, 849
   integral, 841
   level, 737
   quadric, 679
   of revolution, 299, 789
   saddle point of, 741
   volume under, 761–62, 771
Surface area, 300, 786–87, 849
   of a cone, 304, 790
   of a sphere, 301, 787
Surface integral, 841
Symmetry, 22, 23, 195
   of a mass distribution, 318

Tangent
   to a curve, 45, 691
   line, 45, 691
   plane, 724
   vector, 636, 693
   vertical, 47
Tangential component of acceleration, 641, 697
Taylor, Brook (1685–1731), 495, 559
Taylor polynomial, 495, 496, 563
Taylor series, 563
Taylor's Formula, 498
Telescoping sum, 238, 245
Tension, 417
Theorem of the Mean for Integrals, 248
Thompson, D'Arcy (1860–1949), 806
Three-dimensional space, 660
Thurber, James (1894–1961), 699
Time rate of change, 182
Torsion of a space curve, 695
Torus, 289
Trace
   of a curve, 292
   of a surface, 680
Track of a curve, 292, 691
Transcendental function, 335, 385
Transitive Law, 4
Transverse diameter of a hyperbola, 613
Trapezoidal Rule, 457

Derivatives

1. Product Rule: $\dfrac{d}{dx}(uv) = u\dfrac{dv}{dx} + v\dfrac{du}{dx}$

2. Quotient Rule: $\dfrac{d}{dx}\left(\dfrac{u}{v}\right) = \dfrac{v\,(du/dx) - u\,(dv/dx)}{v^2}$

3. Power Rule: $\dfrac{d}{dx}(x^r) = rx^{r-1}$

4. Chain Rule: $\dfrac{dy}{dx} = \dfrac{dy}{du}\dfrac{du}{dx}$

5. $D_x|x| = \dfrac{|x|}{x}$

6. $D_x \sin x = \cos x \qquad D_x \cos x = -\sin x$

   $D_x \tan x = \sec^2 x \qquad D_x \cot x = -\csc^2 x$

   $D_x \sec x = \sec x \tan x \qquad D_x \csc x = -\csc x \cot x$

7. $D_x \sinh x = \cosh x \qquad D_x \coth x = -\operatorname{csch}^2 x$

   $D_x \cosh x = \sinh x \qquad D_x \operatorname{sech} x = -\operatorname{sech} x \tanh x$

   $D_x \tanh x = \operatorname{sech}^2 x \qquad D_x \operatorname{csch} x = -\operatorname{csch} x \coth x$

8. $D_x \ln x = \dfrac{1}{x} \qquad D_x \log_a x = \dfrac{1}{x \ln a}$

9. $D_x e^x = e^x \qquad D_x a^x = a^x \ln a$

10. $D_x \sin^{-1} x = \dfrac{1}{\sqrt{1-x^2}} \qquad D_x \cos^{-1} x = \dfrac{-1}{\sqrt{1-x^2}}$

    $D_x \tan^{-1} x = \dfrac{1}{1+x^2} \qquad D_x \cot^{-1} x = \dfrac{-1}{1+x^2}$

    $D_x \sec^{-1} x = \dfrac{1}{x\sqrt{x^2-1}} \qquad D_x \csc^{-1} x = \dfrac{-1}{x\sqrt{x^2-1}}$

11. $D_x \sinh^{-1} x = \dfrac{1}{\sqrt{x^2+1}}$

    $D_x \cosh^{-1} x = \dfrac{1}{\sqrt{x^2-1}} \quad (x>1)$

    $D_x \tanh^{-1} x = \dfrac{1}{1-x^2} \quad (|x|<1)$

    $D_x \coth^{-1} x = \dfrac{1}{1-x^2} \quad (|x|>1)$

Date Due

~01 '93			
MAR 4 '94			
MAY 2 '9~			
NOV 1 '9~			
NOV 29 '9~			
APR 28 '96			
DE~ 5 '96			
APR 21 '97			
MAR 28 '05			
10/3/05 ILL			
APR 3 0 2010			

BRODART, INC.　　　Cat. No. 23 233　　　Printed in U.S.A.